36.00

APPLIED MATHEMATICS

FOR BUSINESS AND ECONOMICS, LIFE SCIENCES,
AND SOCIAL SCIENCES   SECOND EDITION

RAYMOND A. BARNETT
Merritt College

MICHAEL R. ZIEGLER
Marquette University

CHARLES J. BURKE
City College of San Francisco

DELLEN PUBLISHING COMPANY
San Francisco, California

COLLIER MACMILLAN PUBLISHERS
London

divisions of Macmillan, Inc.

© Copyright 1986 by Dellen Publishing Company, a division of Macmillan, Inc.

Printed in the United States of America

Permissions: Dellen Publishing Company
              400 Pacific Avenue
              San Francisco, California 94133
Orders: Macmillan Publishing Company
         Front and Brown Streets
         Riverside, New Jersey 08075

Collier Macmillan Canada, Inc.

Library of Congress Cataloging-in-Publication Data

Barnett, Raymond A.
   Applied mathematics for business and economics, life sciences, and social sciences.

   Includes index.
   1. Mathematics—1961–   . I. Burke, Charles J.
II. Ziegler, Michael R.   III. Title.
QA39.2.B366   1986   510   85-13322
ISBN 0-02-305590-1

Printing: 4 5 6 7 8   Year: 7 8 9 0

ISBN 0-02-305590-1

Contents

PART II   Finite Mathematics

CHAPTER 9   Probability   477

PART III   Calculus

CHAPTER 10   The Derivative   549

CHAPTER 11   Additional Derivative Topics   629

Chapter Dependencies

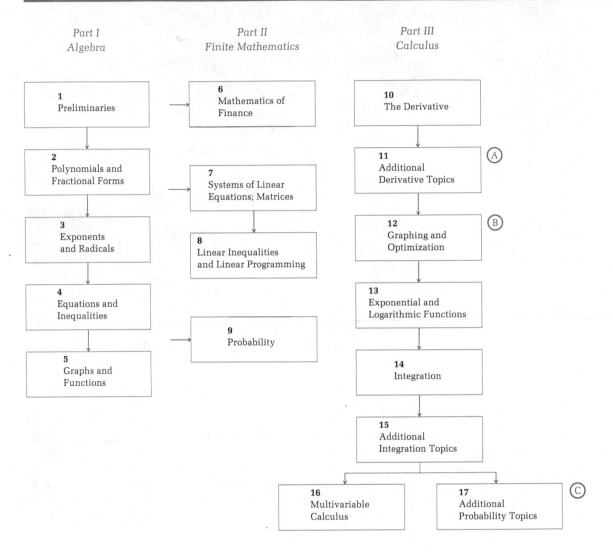

Part I
Algebra

Part II
Finite Mathematics

Part III
Calculus

1 Preliminaries

6 Mathematics of Finance

10 The Derivative

2 Polynomials and Fractional Forms

7 Systems of Linear Equations; Matrices

11 Additional Derivative Topics — Ⓐ

3 Exponents and Radicals

8 Linear Inequalities and Linear Programming

12 Graphing and Optimization — Ⓑ

4 Equations and Inequalities

9 Probability

13 Exponential and Logarithmic Functions

5 Graphs and Functions

14 Integration

15 Additional Integration Topics

16 Multivariable Calculus

17 Additional Probability Topics — Ⓒ

Ⓐ: Section 11-2 can be omitted without loss of continuity.

Ⓑ: Section 12-4 can be omitted without loss of continuity.

Ⓒ: Chapter 9 is also a prerequisite for Chapter 17.

Note: The instructor's manual has a detailed discussion of chapter and section interdependencies to aid instructors and departments in designing a course for their own particular needs.

Preface

Many colleges and universities now offer mathematics courses that emphasize topics that are most useful to students in business and economics, life sciences, and social sciences. Because of this trend, the authors have reviewed course outlines and college catalogs from a large number of colleges and universities, and on the basis of this survey, selected the topics, applications, and emphasis found in this text.

The material in this book is suitable for mathematics courses that include topics from algebra, finite mathematics, and calculus. Part I provides a substantial review of the fundamentals of algebra, which may be treated in a systematic way or may be referred to as needed. In addition, certain key prerequisite topics are reviewed immediately before their use (see, for example, Section 8-1 and Section 13-1), while others are discussed in Appendix A. Part II contains ample material for a finite mathematics course covering the topics that have become standard in this area: mathematics of finance, linear systems, matrices, linear programming, and probability. Part III consists of a thorough presentation of calculus for functions of one variable, including the exponential and logarithmic functions, followed by an introduction to multivariable calculus and some additional probability topics that involve calculus concepts. The choice and organization of topics in all three parts make the book readily adaptable to a variety of courses. (See the diagram on the facing page for chapter dependencies.)

■ Major Changes from the First Edition

The second edition of *Applied Mathematics for Business and Economics, Life Sciences, and Social Sciences* reflects the experiences and recommendations of a large number of the users of the first edition. Additional examples and exercises have been included in many sections to increase student support and to give students a better understanding of the material. In particular, a concentrated effort has been made to increase the student's ability to visualize mathematical relationships and to deal with graphs.

The major changes in Part II are in the chapter on linear programming. This chapter has been extensively rewritten; it now contains an expanded development of the basic simplex method and new sections on the dual and big M methods. Increased attention has been paid to the development of the simplex method and its relationship to the geometric method, which should make the simplex method seem much less mysterious to the student. The material on probability has been expanded and rearranged. Chapter 9 contains a new section on union, intersection, and complement of events, and the section on random variables has been moved to the new chapter on additional probability concepts (Chapter 17).

In Part III, the most noticeable change from the first edition is the reorganization of the calculus material. The material on graphing has been expanded and rewritten and now occupies most of Chapter 12. The exponential and logarithmic functions are introduced at an earlier point so that Chapters 10–13 now deal exclusively with differential calculus and Chapters 14 and 15 deal with integral calculus. There are new sections on asymptotes, elasticity of demand, use of integral tables, continuous random variables, and binomial, uniform, beta, and normal probability distributions.

■ General Comments

Part I of this book presents the algebraic concepts used in Parts II and III. If a minimal review is deemed desirable, then Chapter 4 could be covered before beginning Part II and Chapter 5 before beginning Part III.

Part II deals with three areas that are independent of each other (see the diagram on page viii). The mathematics of finance is presented in Chapter 6. Standard angle notation is used for the compound interest factor and the present value factor. All the required exercises can be solved using either the tables in the back of the book or a hand calculator. Some optional problems have been included that require the use of a calculator.

Chapters 7 and 8 cover topics from linear algebra and linear programming. Elementary row operations are used for solving systems of equations, inverting matrices, and solving linear programming problems. The material on linear programming is organized so as to provide the instructor with maximum flexibility. Those who want a good intuitive introduction to the subject can cover only the material up to the dual method. On the other hand, those who wish to emphasize the development of computational skills can also cover the dual method or the big M method (or both). Finally, those who wish to concentrate on problem solving (setting up problems) and applications can cover the applications in Section 8-6 or 8-7 (or both) and omit the computational methods entirely. In order to facilitate these approaches, the answer section contains an appropriate model for each applied problem, as well as the numerical solution. Section 8-7 also contains optional applications which lead to linear programming problems that are too complex to solve by hand. These applications provide a natural place to introduce the use of a computer program to solve linear programming problems. Such a program is available to institutions adopting this book at no charge from the publisher.

Chapter 9 covers counting techniques and the basic concepts of probability. More advanced topics are covered in Chapter 17.

In Part III, Chapters 10–13 present differential calculus for functions of one variable, including the exponential and logarithmic functions. Trigonometric functions are not discussed in this book. Limits and continuity are presented in an intuitive fashion, utilizing numerical approximations and one-sided limits. All the rules of differentiation are covered in Chapter 10. Various applications of differentiation are then presented in Chapters 11 and 12, with a strong emphasis on graphing concepts. Finally, the exponential and logarithmic functions are covered in Chapter 13.

Chapters 14 and 15 deal with integral calculus. In Chapter 14, differential equations and exponential growth and decay are included as an application of antidifferentiation. The definite integral is intuitively introduced in terms of an area function and then is later formally defined as the limit of a sum. Techniques of integration and improper integrals are covered in Chapter 15. Since the integral table used in Section 15-3 contains formulas for a variety of rational functions, the method of partial fractions is not included among the techniques of integration.

Chapter 16 introduces multivariable calculus, including partial derivatives, differentials, optimization, Lagrange multipliers, least squares, and double integrals. If desired, this chapter can be covered immediately after Chapter 14. (See the diagram on page viii.)

Finally, Chapter 17 presents some additional probability topics, most of which involve applications of calculus. Chapter 9 is also a prerequisite for this chapter. (See the diagram on page viii.)

■ Important Features

Emphasis

Emphasis is on computational skills, ideas, and problem solving rather than on mathematical theory. Most derivations and proofs are omitted except where their inclusion adds significant insight into a particular concept. General concepts and results are usually presented only after particular cases have been discussed.

Examples and Matched Problems

This book contains over 460 completely worked out examples. Each example is followed by a similar problem for the student to work while reading the material. The answers to these matched problems are included at the end of each section for easy reference.

Exercise Sets

This book contains over 5,000 exercises. Each exercise set is designed so that an average or below-average student will experience success and a very capable student will be challenged. They are mostly divided into A (routine, easy mechanics), B (more difficult mechanics), and C (difficult mechanics and some theory) levels.

Applications

Enough applications are included in this book to convince even the most skeptical student that mathematics is really useful. The majority of the applications are included at the end of exercise sets and are generally divided into business and economics, life science, and social science groupings. An instructor with students from all three disciplines can let them choose applications from their own field of interest; if most students are from one of the three areas, then special emphasis can be placed there. Most of the applications are simplified versions of actual real-world problems taken from professional journals and books. No specialized experience is required to solve any of the applications included in this book.

■ Student and Instructor Aids

Student Aids

Dotted **"think boxes"** are used to enclose steps that are usually performed mentally (see Section 1-2).

Examples and developments are often **annotated** to help students through critical stages (see Section 1-4).

A **second color** is used to indicate key steps (see Section 1-4).

Boldface type is used to introduce new terms and highlight important comments.

Answers to odd-numbered problems are included in the back of the book.

Chapter review sections include a review of all important terms and symbols, a comprehensive review exercise set, and a practice test. Answers to all review exercises and practice test problems are included in the back of the book.

A **solutions manual** is available at a nominal cost through a book store. The manual includes detailed solutions to all odd-numbered problems, all review exercises, and all practice test problems.

A **computer applications supplement** by Carolyn L. Meitler and Michael R. Ziegler is available at a nominal cost through a book store. The supplement contains examples, computer program listings, and exercises that demonstrate the use of a computer to solve a variety of problems in finite mathematics and calculus. No previous computing experience is necessary to use this supplement.

Instructor Aids

A **test battery** designed by Carolyn L. Meitler can be obtained from the publisher without charge. The test battery contains six different tests with varying degrees of difficulty for each chapter. The format is $8\frac{1}{2} \times 11$ inches for ease of reproduction.

An **instructor's manual** can be obtained from the publisher without charge. The instructor's manual contains some remarks on selection of topics and answers to the even-numbered problems, which are not included in the text.

A **solutions manual** (see Student Aids) is available to instructors without charge from the publisher.

A **computer applications supplement** by Carolyn L. Meitler and Michael R. Ziegler (see Student Aids) is available to instructors without charge from the publisher. The programs in this supplement are also available on diskettes for APPLE II® and IBM® PC computers.* The publisher will supply one of these diskettes without charge to institutions using this book.

■ Acknowledgments

In addition to the authors, many others are involved in the successful publication of a book. We wish to thank personally: Susan Boren, University of Tennessee at Martin; Gary Brown, Washington State University; David Cochener, Angelo State University; Henry Decell, University of Houston; Gary Etgen, University of Houston; Sandra Gossum, University of Tennessee at Martin; Freida Holly, Metropolitan State College; David Johnson, University of Kentucky; Stanley Lukawecki, Clemson University; Lyle Mauland, University of North Dakota at Grand Forks; Carolyn L. Meitler, Marquette University; R. A. Moreland, Texas Technological University; Marian Paysinger, University of Texas at Arlington; John Plachy, Metropolitan State College; Walter Roth, University of North Carolina at Charlotte; Wesley Sanders, Sam Houston State College; Arthur Sparks, University of Tennessee at Martin; Martha Stewart, University of North Carolina at Charlotte; James Strain, Midwestern State College; Michael Vose, Austin Community College; Dennis Weiss, Indianapolis, Indiana; Scott Wright, Loyola Marymount University; Donald Zalewski, Northern Michigan University; Dennis Zill, Loyola Marymount University.

We also wish to thank:

Janet Bollow for another outstanding book design.

John Williams for a strong and effective cover design.

John Drooyan for the many sensitive and beautiful photographs seen throughout the book.

Phillip Bender, Gary Etgen, Robert Mullins, Mary Utzerath, and Caroline Woods for carefully checking all examples and problems (a tedious but extremely important job).

* APPLE II is a registered trademark of Apple Computer, Incorporated. IBM is a registered trademark of the International Business Machine Corporation.

Phyllis Niklas for her ability to guide the book smoothly through all production details.

Don Dellen, the publisher, who continues to provide all the support services and encouragement an author could hope for.

Producing this new edition with the help of all these extremely competent people has been a most satisfying experience.

R. A. Barnett
C. J. Burke
M. R. Ziegler

ALGEBRA

I

Preliminaries

CHAPTER 1   Contents

This chapter introduces the basic tools of algebra that will be used in subsequent chapters. It includes a discussion of sets, properties of real numbers, operations on signed numbers, evaluating algebraic expressions, and an introduction to exponents.

1-1   Sets

- Set Properties and Set Notation
- Set Operations
- Application

In this section we will review a few key ideas from set theory. Set concepts and notation not only help us talk about certain mathematical ideas with greater clarity and precision, but are indispensable to a clear understanding of probability.

■ Set Properties and Set Notation

We can think of a **set** as any collection of objects specified in such a way that we can tell whether any given object is or is not in the collection. Capital letters, such as A, B, and C, are often used to designate particular sets. Each object in a set is called a **member** or **element** of the set. Symbolically,

$a \in A$	means	"a is an element of set A"
$a \notin A$	means	"a is not an element of set A"

A set without any elements is called the **empty** or **null set.** For example, the set of all people over 10 feet tall is an empty set. Symbolically,

$\emptyset$ represents "the empty or null set"

A set is usually described either by listing all its elements between braces { } or by enclosing a rule within braces that determines the elements of the set. Thus, if $P(x)$ is a statement about x, then

$S = \{x|P(x)\}$ means "S is the set of all x such that $P(x)$ is true"

Recall that the vertical bar in the symbolic form is read "such that." The following example illustrates the rule and listing methods of representing sets.

Example 1

Rule	Listing

$$\{x|x \text{ is a weekend day}\} = \{\text{Saturday, Sunday}\}$$

$$\{x|x^2 = 4\} = \{-2, 2\}$$

$$\{x|x \text{ is an odd positive counting number}\} = \{1, 3, 5, \ldots\}$$

The three dots . . . in the last set in Example 1 indicate that the pattern established by the first three entries continues indefinitely. The first two sets in Example 1 are **finite sets** (we intuitively know that the elements can be counted); the last set is an **infinite set** (we intuitively know that there is no end in counting the elements). When listing the elements in a set, we do not list an element more than once.

Problem 1   Let G be the set of all numbers such that $x^2 = 9$.*

(A) Denote G by the rule method.
(B) Denote G by the listing method.
(C) Indicate whether the following are true or false: $3 \in G$, $9 \notin G$.

If each element of a set A is also an element of set B, we say that A is a **subset** of B. For example, the set of all women students in a class is a subset of the whole class. Note that the definition allows a set to be a subset of itself. If set A and set B have exactly the same elements, then the two sets are said to be **equal.** Symbolically,

* Answers to matched problems are found near the end of each section just before the exercise set.

$A \subset B$	means	"A is a subset of B"
$A = B$	means	"A and B have exactly the same elements"
$A \not\subset B$	means	"A is not a subset of B"
$A \neq B$	means	"A and B do not have exactly the same elements"

It can be proved that $\varnothing$ **is a subset of every set.**

Example 2 If $A = \{-3, -1, 1, 3\}$, $B = \{3, -3, 1, -1\}$, and $C = \{-3, -2, -1, 0, 1, 2, 3\}$, then each of the following statements is true:

$$
\begin{array}{lll}
A = B & A \subset C & A \subset B \\
C \neq A & C \not\subset A & B \subset A \\
\varnothing \subset A & \varnothing \subset C & \varnothing \not\subset A
\end{array}
$$

Problem 2 Given $A = \{0, 2, 4, 6\}$, $B = \{0, 1, 2, 3, 4, 5, 6\}$, and $C = \{2, 6, 0, 4\}$, indicate whether the following relationships are true (T) or false (F):

(A) $A \subset B$ (B) $A \subset C$ (C) $A = C$
(D) $C \subset B$ (E) $B \not\subset A$ (F) $\varnothing \subset B$

Example 3 List all the subsets of the set $\{a, b, c\}$.

Solution $\{a, b, c\}, \{a, b\}, \{a, c\}, \{b, c\}, \{a\}, \{b\}, \{c\}, \varnothing$

Problem 3 List all the subsets of the set $\{1, 2\}$.

■ Set Operations

The **union** of sets A and B, denoted by $A \cup B$, is the set of all elements formed by combining all the elements of A and all the elements of B into one set. Symbolically,

Union

$$A \cup B = \{x | x \in A \quad \textbf{or} \quad x \in B\}$$

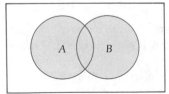

Figure 1 $A \cup B$ is the shaded region.

Here we use the word or in the way it is always used in mathematics; that is, x may be an element of set A or set B or both.

Venn diagrams are useful in visualizing set relationships. The union of two sets can be illustrated as shown in Figure 1. Note that

$$A \subset A \cup B \quad \text{and} \quad B \subset A \cup B$$

The **intersection** of sets A and B, denoted by $A \cap B$, is the set of elements in set A that are also in set B. Symbolically,

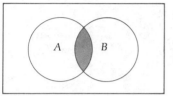

Figure 2 $A \cap B$ is the shaded region.

Intersection

$$A \cap B = \{x | x \in A \ \textbf{and} \ x \in B\}$$

This relationship is easily visualized in the Venn diagram shown in Figure 2. Note that

$$A \cap B \subset A \quad \text{and} \quad A \cap B \subset B$$

If $A \cap B = \varnothing$, then the sets A and B are said to be **disjoint;** this is illustrated in Figure 3.

The set of all elements under consideration is called the **universal set** U. Once the universal set is determined for a particular discussion, all other sets in that discussion must be subsets of U.

We now define one more operation on sets, called the *complement*. The **complement** of A (relative to U), denoted by A', is the set of elements in U that are not in A (see Fig. 4). Symbolically,

---

Complement

$$A' = \{x \in U | x \notin A\}$$

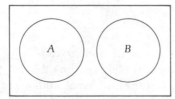

Figure 3 $A \cap B = \varnothing$

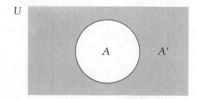

Figure 4 The complement of A is A'.

Example 4 If $A = \{3, 6, 9\}$, $B = \{3, 4, 5, 6, 7\}$, $C = \{4, 5, 7\}$, and $U = \{1, 2, 3, 4, 5, 6, 7, 8, 9\}$, then

$A \cup B = \{3, 4, 5, 6, 7, 9\}$

$A \cap B = \{3, 6\}$

$A \cap C = \varnothing$ $\quad$ A and C are disjoint

$B' = \{1, 2, 8, 9\}$

Problem 4 $\quad$ If $R = \{1, 2, 3, 4\}$, $S = \{1, 3, 5, 7\}$, $T = \{2, 4\}$, and $U = \{1, 2, 3, 4, 5, 6, 7, 8, 9\}$, find:

(A) $\;R \cup S$ $\qquad$ (B) $\;R \cap S$ $\qquad$ (C) $\;S \cap T$ $\qquad$ (D) $\;S'$

■ Application

Example 5 $\quad$ From a survey of 100 college students, a marketing research company found that 75 students owned stereos, 45 owned cars, and 35 owned cars and stereos.

(A) $\;$ How many students owned either a car or a stereo?
(B) $\;$ How many students did not own either a car or a stereo?

Solutions $\quad$ Venn diagrams are very useful for this type of problem. If we let

$U =$ Set of students in sample (100)

$S =$ Set of students who own stereos (75)

$C =$ Set of students who own cars (45)

$S \cap C =$ Set of students who own cars and stereos (35)

then

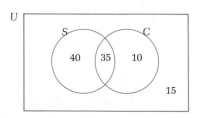

Place the number in the intersection first, then work outward:

$40 = \;75 - 35$

$10 = \;45 - 35$

$15 = 100 - (40 + 35 + 10)$

(A) $\;$ The number of students who own either a car or a stereo is the number of students in the set $S \cup C$. You might be tempted to say that this is just the number of students in S plus the number of students in C, $75 + 45 = 120$, but this sum is larger than the sample we started with! What is wrong? We have actually counted the number in the intersection (35) twice. The correct answer, as seen in the Venn diagram, is

$40 + 35 + 10 = 85$

(B)   The number of students who do not own either a car or a stereo is the number of students in the set $(S \cup C)'$; that is, 15.

Problem 5   Referring to Example 5:

(A)   How many students owned a car but not a stereo?
(B)   How many students did not own both a car and a stereo?

Note in Example 5 and Problem 5 that the word *and* is associated with intersection and the word *or* is associated with union.

Answers to   1.   (A) $\{x|x^2 = 9\}$ (B) $\{-3, 3\}$ (C)   True, True
Matched Problems
2.   All are true

3. $\{1, 2\}, \{1\}, \{2\}, \varnothing$

4.   (A) $\{1, 2, 3, 4, 5, 7\}$ (B) $\{1, 3\}$ (C) $\varnothing$ (D) $\{2, 4, 6, 8, 9\}$

5.   (A)   10 [the number in $S' \cap C$]   (B)   65 [the number in $(S \cap C)'$]

Exercise 1-1

A   *Indicate true (T) or false (F).*

1. $4 \in \{2, 3, 4\}$
2. $6 \notin \{2, 3, 4\}$
3. $\{2, 3\} \subset \{2, 3, 4\}$
4. $\{3, 2, 4\} = \{2, 3, 4\}$
5. $\{3, 2, 4\} \subset \{2, 3, 4\}$
6. $\{3, 2, 4\} \in \{2, 3, 4\}$
7. $\varnothing \subset \{2, 3, 4\}$
8. $\varnothing = \{0\}$

In Problems 9–14 write the resulting set using the listing method.

9. $\{1, 3, 5\} \cup \{2, 3, 4\}$
10. $\{3, 4, 6, 7\} \cup \{3, 4, 5\}$
11. $\{1, 3, 4\} \cap \{2, 3, 4\}$
12. $\{3, 4, 6, 7\} \cap \{3, 4, 5\}$
13. $\{1, 5, 9\} \cap \{3, 4, 6, 8\}$
14. $\{6, 8, 9, 11\} \cap \{3, 4, 5, 7\}$

B   *In Problems 15–20 write the resulting set using the listing method.*

15. $\{x|x - 2 = 0\}$
16. $\{x|x + 7 = 0\}$
17. $\{x|x^2 = 49\}$
18. $\{x|x^2 = 100\}$
19. $\{x|x$ is an odd number between 1 and 9, inclusive$\}$
20. $\{x|x$ is a month starting with M$\}$
21.   For $U = \{1, 2, 3, 4, 5\}$ and $A = \{2, 3, 4\}$, find A'.
22.   For $U = \{7, 8, 9, 10, 11\}$ and $A = \{7, 11\}$, find A'.

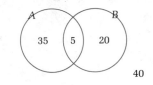

U

A   B

35   5   20

40

Problems *23–34 refer to the Venn diagram in the margin. How many elements are in each of the indicated sets?*

23. A **24.** U **25.** A' **26.** B'

27. $A \cup B$ **28.** $A \cap B$ **29.** $A' \cap B$ **30.** $A \cap B'$

31. $(A \cap B)'$ **32.** $(A \cup B)'$ **33.** $A' \cap B'$ **34.** U'

35. If $R = \{1, 2, 3, 4\}$ and $T = \{2, 4, 6\}$, find:

(A) $\{x | x \in R$ or $x \in T\}$ (B) $R \cup T$

36. If $R = \{1, 3, 4\}$ and $T = \{2, 4, 6\}$, find:

(A) $\{x | x \in R$ and $x \in T\}$ (B) $R \cap T$

37. For $P = \{1, 2, 3, 4\}$, $Q = \{2, 4, 6\}$, and $R = \{3, 4, 5, 6\}$, find $P \cup (Q \cap R)$.

38. For P, Q, and R in Problem 37, find $P \cap (Q \cup R)$.

C   *Venn diagrams may be of help in Problems 39–44.*

39. If $A \cup B = B$, can we always conclude that $A \subset B$?

40. If $A \cap B = B$, can we always conclude that $B \subset A$?

41. If A and B are arbitrary sets, can we always conclude that $A \cap B \subset B$?

42. If $A \cap B = \varnothing$, can we always conclude that $B = \varnothing$?

43. If $A \subset B$ and $x \in A$, can we always conclude that $x \in B$?

44. If $A \subset B$ and $x \in B$, can we always conclude that $x \in A$?

45. How many subsets does each of the following sets have? Also, try to discover a formula in terms of n for a set with n elements.

(A) $\{a\}$ (B) $\{a, b\}$ (C) $\{a, b, c\}$

46. How do the sets $\varnothing$, $\{\varnothing\}$, and $\{0\}$ differ from each other?

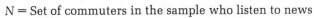

Applications

Business & Economics

Problems *47–58 refer to the following survey: A marketing survey of 1,000 car commuters found that 600 listen to the news, 500 listen to music, and 300 listen to both. Let*

$N =$ Set of commuters in the sample who listen to news

$M =$ Set of commuters in the sample who listen to music

Following the procedures in Example 5, find the number of commuters in each set described below.

47. $N \cup M$ **48.** $N \cap M$ **49.** $(N \cup M)'$

50. $(N \cap M)'$ **51.** $N' \cap M$ **52.** $N \cap M'$

53. Set of commuters who listen to either news or music

54. Set of commuters who listen to both news and music

55. Set of commuters who do not listen to either news or music

56. Set of commuters who do not listen to both news and music
57. Set of commuters who listen to music but not news
58. Set of commuters who listen to news but not music
59. The management of a company, a president and three vice-presidents, denoted by the set $\{P, V_1, V_2, V_3\}$, wish to select a committee of two people from among themselves. How many ways can this committee be formed; that is, how many two-person subsets can be formed from a set of four people?
60. The management of the company in Problem 59 decides for or against certain measures as follows: The president has two votes and each vice-president has one vote. Three favorable votes are needed to pass a measure. List all minimal winning coalitions; that is, list all subsets of $\{P, V_1, V_2, V_3\}$ that represent exactly three votes.

Life Sciences | *Blood types.* When receiving a blood transfusion, a recipient must have all the antigens of the donor. A person may have one or more of the three antigens A, B, and Rh, or none at all. Eight blood types are possible, as indicated in the following Venn diagram, where U is the set of all people under consideration:

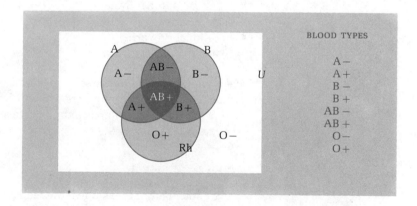

An A— person has A antigens but no B or Rh; an O+ person has Rh but neither A nor B; an AB— person has A and B antigens but no Rh; and so on.

Using the Venn diagram, indicate which of the eight blood types are included in each set.

61. $A \cap Rh$
62. $A \cap B$
63. $A \cup Rh$
64. $A \cup B$
65. $(A \cup B)'$
66. $(A \cup B \cup Rh)'$
67. $A' \cap B$
68. $Rh' \cap A$

Social Sciences | *Group structures.* R. D. Luce and A. D. Perry, in a study on group structure (*Psychometrika*, 1949, 14: 95–116), used the idea of sets to formally define the notion of a clique within a group. Let G be the set of all persons in the group and let $C \subset G$. Then C is a clique provided that:

1. C contains at least three elements.
2.   For every $a, b \in C$, $a \mathrel{R} b$ and $b \mathrel{R} a$.
3.   For every $a \notin C$, there is at least one $b \in C$ such that $a \mathrel{\not R} b$ or $b \mathrel{\not R} a$ or both.

[*Note:*   Interpret "$a \mathrel{R} b$" to mean "a relates to b," "a likes b" "a is as wealthy as b," and so on. Of course, "$a \mathrel{\not R} b$" means "a does not relate to b," and so on.]

69.   Translate statement 2 into ordinary English.
70.   Translate statement 3 into ordinary English.

1-2   Real Numbers and the Rules of Algebra

- The Real Number System
- Basic Properties
- Additional Properties

In algebra we are interested in manipulating symbols in order to simplify algebraic expressions and to solve algebraic equations. Since many of these symbols represent real numbers, we will briefly review the real number system and some of its important properties. These properties provide the basic rules for much of the manipulation of symbols in algebra.

The Real Number System

Table 1 and Figure 5 break down the real number system into its important subsets and show how these sets are related to each other. Note that the set of natural numbers is a subset of the set of integers, the set of integers is a subset of the set of rational numbers, and the set of rational numbers is a subset of the set of real numbers.

It is interesting to note that rational numbers have repeating decimal representations, while irrational numbers have infinite nonrepeating decimal representations. For example, the decimal representations for the rational numbers 3, $\frac{5}{6}$, and $\frac{8}{11}$ are

$$3 = 3.000. . . \qquad \tfrac{5}{6} = 0.8333. . . \qquad \tfrac{8}{11} = 0.727272. . .$$

while those for the irrational numbers $\sqrt{5}$ and π are

$$\sqrt{5} = 2.23606797. . . \qquad \pi = 3.14159265. . .$$

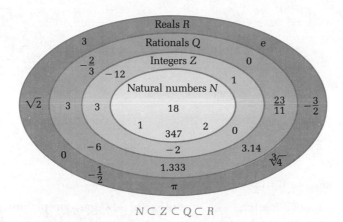

$$N \subset Z \subset Q \subset R$$

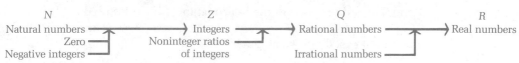

Figure 5 The set of real numbers

Table 1 The Set of Real Numbers

Symbol	Number System	Description	Examples
N	Natural numbers	Counting numbers (also called positive integers)	1, 2, 3, . . .
Z	Integers	Set of natural numbers, their negatives, and zero	. . . , -2, -1, 0, 1, 2, . . .
Q	Rationals	Any number that can be represented as a/b where a and b are integers and $b \neq 0$	-4; $\frac{-3}{5}$; 0; 1; $\frac{2}{3}$; 3.67
R	Reals	Set of all rational and irrational numbers (irrational numbers are all real numbers that are not rational)	-4; $\frac{-3}{5}$; 0; 1; $\frac{2}{3}$; 3.67; $\sqrt{2}$; π; $\sqrt[3]{5}$

A one-to-one correspondence exists between the set of real numbers and the points on a line; that is, each real number corresponds to exactly one point, and each point to exactly one real number. A line with a real number associated with each point, as in Figure 6, is called a **real number line,** or simply, a **real line.**

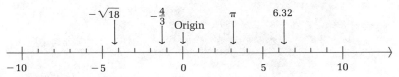

Figure 6 A real number line

■ Basic Properties

In applying the basic operations of addition, subtraction, multiplication, and division to real numbers, we find that there are many basic rules that always hold true. These rules are often referred to as **axioms** or **properties,** and they provide us with the basic manipulative rules used in algebra.

The **equality symbol,** =, is used to join two expressions if they both represent exactly the same thing. Thus,

$$a = b$$

means a and b represent the same thing—that is, a **is equal to** b. The expression

$$a \neq b$$

means a **is not equal to** b. Thus, we write

$$7 = 2 + 5$$

since 7 and $2 + 5$ represent the same number. But we write

$$7 \neq 3 + 5$$

since 7 and $3 + 5$ do not represent the same number.

Three important properties of the equality symbol, =, must hold any time it is used. They may be stated as follows:

1. *Symmetry property.*   If $a = b$, then $b = a$. We may reverse the left and right members of an equation any time we wish—a useful process when solving certain types of equations. For example, if

    $$A = P + Prt \quad \text{then} \quad P + Prt = A$$

2. *Transitive property.*   If $a = b$ and $b = c$, then $a = c$. This property is used extensively whenever algebra is used. For example, if

    $$5x + 2x = (5 + 2)x \quad \text{and} \quad (5 + 2)x = 7x \quad \text{then} \quad 5x + 2x = 7x$$

3. *Substitution property.*   If $a = b$, then either may replace the other in any statement without changing the truth or falsity of the statement. The substitution property is also used extensively whenever algebra is

used. For example, if $A = P + I$ and $I = Prt$, then I in the first formula may be replaced by Prt from the second formula to obtain

$$A = P + Prt$$

One of the first rules learned in arithmetic is that the order in which we add or multiply two numbers does not affect the result. For example, we have

$$7 + 5 = 5 + 7 \quad \text{and} \quad 5 \cdot 7 = 7 \cdot 5$$

These examples illustrate the **commutative properties** for addition and multiplication of real numbers. In general,

---

Commutative Properties

If a and b represent real numbers, then

$$a + b = b + a \qquad ab = ba$$

On the other hand, the order in which we subtract or divide numbers *does* affect the result. For instance,

$$12 - 4 \neq 4 - 12 \quad \text{and} \quad 12 \div 4 \neq 4 \div 12$$

Thus, subtraction and division are not commutative.

Example 6   If x and y represent real numbers, then, using the commutative properties, we have:

(A) $7 + x = x + 7$ (B) $y5 = 5y$
(C) $4y + 7x = 7x + 4y$ (D) $5 + yx = 5 + xy$

Problem 6   Using the commutative properties, determine the expression that would replace each question mark.

(A) $y + 8 = 8 + ?$ (B) $7(y + x) = 7(? + y)$
(C) $2x + y3 = 2x + 3?$ (D) $(x + y)8 = ?(x + y)$

If we were to compute

$$3 + 2 + 4 \quad \text{and} \quad 3 \cdot 2 \cdot 4$$

we would not hesitate to give the answers 9 and 24, respectively. However, we might hesitate to say whether the first two numbers or the last two numbers are to be added or multiplied first. The fact is that it does not matter. We can express this fact by writing

$$(3 + 2) + 4 = 3 + (2 + 4) \quad \text{and} \quad (3 \cdot 2) \cdot 4 = 3 \cdot (2 \cdot 4)$$

These examples illustrate the **associative properties** for addition and multiplication of real numbers. In general,

---

Associative Properties

If a, b, and c represent real numbers, then

$$(a + b) + c = a + (b + c) \qquad (ab)c = a(bc)$$

Subtraction and division, on the other hand, are not associative; for example,

$$(5 - 4) - 1 \neq 5 - (4 - 1) \qquad \text{and} \qquad (12 \div 6) \div 2 \neq 12 \div (6 \div 2)$$

Example 7   If x, y, and z represent real numbers, then, using the associative properties, we have:

(A) $(x + 8) + 7 = x + (8 + 7)$ (B) $9(2z) = (9 \cdot 2)z$
(C) $y + (y + 9) = (y + y) + 9$ (D) $(4x)x = 4(xx)$

Problem 7   Using the associative properties, determine the expression that would replace each question mark.

(A) $(z + 3) + 8 = z + (?)$ (B) $5(7x) = (?)x$
(C) $(4 + y) + y = 4 + (?)$ (D) $(5z)z = 5(?)$

Remark

The commutative and associative properties for addition allow us to rearrange the order of addition in any way we please and to insert or remove parentheses as desired. The commutative and associative properties for multiplication allow us to do the same for any product. However, we cannot apply these properties to subtraction or division.

The commutative and associative properties enable us to simplify many algebraic expressions.

Example 8   Simplify the following expressions mentally:

(A) $(5m + 3) + (3n + 9)$ $\begin{aligned} &= 5m + 3 + 3n + 9 \\ &= 5m + 3n + 3 + 9 \end{aligned}$ *   Remove parentheses.
Rearrange terms.

$= 5m + 3n + 12$ Add.

* The dashed boxes shown in color are used throughout the book to indicate steps that are usually done mentally. Think of them as "think boxes."

(B) $(7x)(5y)$ $\begin{aligned} &= 7 \cdot x \cdot 5 \cdot y \\ &= 7 \cdot 5 \cdot x \cdot y \end{aligned}$ Remove parentheses.
  Reaarrange factors.

$= 35xy$ Multiply.

Problem 8   Simplify the following expressions mentally:

(A) $(8x + 5) + (3y + 7)$ (B) $(4u)(7v)$

The next property is used extensively in algebra and involves both the operations of addition and multiplication. Let us compare the following two calculations:

$$\begin{array}{c|c} 7(3+6) & 7 \cdot 3 + 7 \cdot 6 \\ = 7 \cdot 9 & = 21 + 42 \\ = 63 & = 63 \end{array}$$

We see that

$$7(3+6) = 7 \cdot 3 + 7 \cdot 6$$

This result holds in general, and we have the **distributive property:**

Distributive Property

If a, b, and c represent real numbers, then

$$a(b+c) = ab + ac \qquad (b+c)a = ba + ca$$

We say that the factor a *distributes* over the sum $(b+c)$.

Example 9   Multiply using the distributive property.

(A) $5(x + y) = 5x + 5y$

(B) $3(2m + 9)$ $= 3 \cdot 2m + 3 \cdot 9$ $= 6m + 27$

(C) $x(y + z) = xy + xz$

(D) $(3 + 5)x = 3x + 5x$

Problem 9   Multiply using the distributive property.

(A) $7(u + v)$ (B) $9(4x + 3)$ (C) $c(m + n)$ (D) $(7 + 2)y$

The distributive property may be written in either of the following forms (using the symmetry property of equality):

$$ab + ac = a(b + c) \qquad \text{or} \qquad ba + ca = (b + c)a$$

In both cases, the sum of two terms has been converted into a product of two factors.

Example 10   Using the distributive property, write each of the following as a product:

(A) $3m + 3n = 3(m + n)$

(B) $15y + 35 \;\boxed{= 5 \cdot 3y + 5 \cdot 7}\; = 5(3y + 7)$

(C) $20x + 16y \;\boxed{= 4 \cdot 5x + 4 \cdot 4y}\; = 4(5x + 4y)$

(D) $5x + 2x = (5 + 2)x = 7x$

Problem 10   Using the distributive property, write each of the following as a product:

(A) $7x + 7y$ (B) $10x + 16$ (C) $21m + 14n$ (D) $4y + 7y$

■ Additional Properties

To complete our survey of the properties of real numbers, we list some additional properties:

Additional Properties

If a represents a real number, then:

1.   The number zero, 0, satisfies the properties

$$a + 0 = 0 + a = a \qquad a \cdot 0 = 0 \cdot a = 0$$

2.   The number one, 1, satisfies the property

$$1 \cdot a = a \cdot 1 = a$$

3.   There is exactly one number, $-a$, called the **additive inverse** or **negative of** a, that satisfies

$$a + (-a) = (-a) + a = 0$$

4.   If $a \neq 0$, there is exactly one number, $\dfrac{1}{a}$, called the **multiplicative inverse** or **reciprocal of** a, that satisfies

$$a \cdot \frac{1}{a} = \frac{1}{a} \cdot a = 1$$

Example 11   (A) $0 + m = m$ (B) $0x = 0$ (C) $1xy = xy$
(D) $(-p) + p = 0$ (E) $2 \cdot \frac{1}{2} = 1$

Problem 11   Simplify each of the following:

(A) $x + 0$ (B) $b \cdot 0$ (C) $1uv$ (D) $t + (-t)$ (E) $\frac{1}{5} \cdot 5$

Example 12   (A)   The negative of 7 is -7.

                (B)   The negative of -3 is $-(-3) = 3$.

                (C)   The reciprocal of $\dfrac{2}{3}$ is $\dfrac{1}{\frac{2}{3}} = \dfrac{3}{2}$.

                (D)   The reciprocal of $-\dfrac{5}{4}$ is $\dfrac{1}{-\frac{5}{4}} = -\dfrac{4}{5}$.

Problem 12   Find each of the following:

                (A)   The negative of -13 (B)   The negative of 8

                (C)   The reciprocal of $\frac{3}{11}$ (D)   The reciprocal of $-\frac{9}{5}$

Answers to
Matched Problems

6.   (A) y (B) x (C) y (D)   8
7.   (A) $3 + 8$ (B) $5 \cdot 7$ (C) $y + y$ (D) zz
8.   (A) $8x + 3y + 12$ (B) $28uv$
9.   (A) $7u + 7v$ (B) $36x + 27$ (C) $cm + cn$ (D) $7y + 2y$
10.   (A) $7(x + y)$ (B) $2(5x + 8)$ (C) $7(3m + 2n)$
     (D) $(4 + 7)y = 11y$
11.   (A) x (B)   0   (C) uv (D)   0   (E)   1
12.   (A)   13   (B) -8 (C) $\frac{11}{3}$ (D) $-\frac{5}{9}$

Exercise 1-2

A   *State whether each statement is true (T) or false (F).*

     1. $\frac{3}{7}$ is a rational number   2. -10 is an integer

     3. $-\frac{2}{5}$ is an integer   4. -8 is a natural number

     5.   9 is a rational number   6. $\frac{1}{3}$ is an integer

Each statement in Problems 7–14 is justified by the commutative or associative property of real numbers. Indicate which.

     7. $5 + x = x + 5$ 8. $sr = rs$

     9. $(8x)y = 8(xy)$ 10. $(x + 7) + 9 = x + (7 + 9)$

    11. $x8 = 8x$ 12. $6 + 2m = 2m + 6$

    13. $u + (u + 5) = (u + u) + 5$ 14. $(8p)p = 8(pp)$

In Problems 15–20 use the associative and commutative properties to simplify each expression mentally.

    15. $5(7x)$ 16. $(9u)4$

    17. $(8 + t) + 5$ 18. $13 + (x + 5)$

    19. $(3x + 7) + (2y + 9)$ 20. $(5a + 7) + (8b + 3)$

Give the negative of each.

21. -5 **22.** 13 **23.** $\frac{15}{2}$ **24.** $-\frac{9}{11}$

Give the reciprocal of each.

25. 8 **26.** 13 **27.** $-\frac{3}{7}$ **28.** $-\frac{9}{4}$

B   *Each statement in Problems 29–36 is justified by the commutative or associative property of real numbers. Indicate which.*

29. $8 + (z + 2) = (z + 2) + 8$ **30.** $(5y)(7x) = (7x)(5y)$
31. $(2u + 3v) + 4v = 2u + (3v + 4v)$ **32.** $(4y)(y + 5) = 4[y(y + 5)]$
33. $(2y + z) + 3y = (z + 2y) + 3y$ **34.** $(5x)(y8) = (5x)(8y)$
35. $(x + y) + (y + z) = x + [y + (y + z)]$
36. $x + [y + (y + z)] = x + [(y + y) + z]$

In Problems 37–42 use the commutative and associative properties to simplify each expression mentally.

37. $p + (q + 5) + (r + 10)$ **38.** $(6 + m) + (7 + n) + (2 + p)$
39. $(5x)(8y)$ **40.** $(6u)(9v)$
41. $(2m)(5n)(6p)$ **42.** $(3x)(8y)(2z)$

Multiply using the distributive property.

43. $9(2x + 3)$ **44.** $8(5 + 3x)$
45. $7(3u + 4v)$ **46.** $6(5x + 3y)$
47. $a(m + n)$ **48.** $(u + v)w$

Using the distributive property, write each expression as a product.

49. $15u + 25v$ **50.** $14x + 21y$
51. $10m + 5$ **52.** $18x + 9$
53. $32x + 24y$ **54.** $48u + 36v$
55. $ah + ak$ **56.** $rt + st$

C   *Give the reciprocal of each.*

57. $4\frac{1}{3}$ **58.** $5\frac{1}{6}$ **59.** $-3\frac{3}{4}$ **60.** $-7\frac{2}{5}$

State whether each statement is true (T) or false (F) for all real numbers. If false, give examples for which the statement is false, using numbers in place of the letters.

61. $a - b = b - a$ **62.** $a + b = b + a$
63. $ab = ba$ **64.** $a \div b = b \div a$
65. $(a - b) - c = a - (b - c)$ **66.** $(a + b) + c = a + (b + c)$
67. $a(bc) = (ab)c$ **68.** $(a \div b) \div c = a \div (b \div c)$

1-3   Inequality Statements and Line Graphs

- Inequalities and the Real Number Line
- Inequalities and Line Graphs
- Interval Notation

When we wish to indicate that two quantities are equal, we use the equality symbol, $=$. However, it is often necessary to compare two quantities that are not equal. Although we could use the symbol $\neq$, we might be more interested in indicating which of the two quantities is smaller or which is larger. Our objective in this section is to introduce the inequality symbols that will allow us to make such comparisons. We will also describe how inequality forms can be interpreted using number lines.

■ Inequalities and the Real Number Line

If a and b are real numbers, then there are four **inequality symbols** that may be used to compare a and b:

Inequality	Interpretation
$a < b$	a is less than b
$a > b$	a is greater than b
$a \leq b$	a is less than or equal to b
$a \geq b$	a is greater than or equal to b

It should be clear that

$$7 < 10 \quad \text{and} \quad 10{,}000 > 10$$

It may not be quite as obvious that

$$-15 < -5 \qquad 0 > -10 \qquad -40{,}000 < 1$$

Using a real number line, it is fairly easy to determine an inequality relationship between two numbers. We have $\boldsymbol{a < b}$ if a is to the left of b, and $\boldsymbol{c > d}$ if c is to the right of d, as shown in Figure 7.*

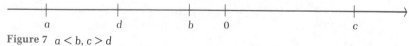

Figure 7 $a < b, c > d$

* The inequality symbols $<$ and $>$ are defined more formally in Section 4-2.

Example 13 Referring to Figure 7, we have:

(A) $a < c$ a is to the left of c
(B) $0 > d$ 0 is to the right of d
(C) $c > b$ c is to the right of b
(D) $b < 0$ b is to the left of 0

Problem 13 Referring to Figure 7, replace each question mark with either $<$ or $>$.

(A) $c\,?\,a$ (B) $a\,?\,0$ (C) $b\,?\,a$ (D) $0\,?\,c$

Example 14 (A) $3 < 5$ 3 is to the left of 5 on a number line
(B) $-6 < -2$ -6 is to the left of -2 on a number line
(C) $0 > -10$ 0 is to the right of -10 on a number line
(D) $-5 > -25$ -5 is to the right of -25 on a number line

Problem 14 Replace each question mark with $<$ or $>$.

(A) $8\,?\,2$ (B) $-20\,?\,0$ (C) $-3\,?\,-30$ (D) $0\,?\,-15$

■ Inequalities and Line Graphs

Let us now consider the inequality

$x \geq -4$

The **solution set** for this inequality is the set of all real numbers which when substituted for the letter x make the statement true. This set includes the number -4 and all real numbers to the right of -4 on a real number line. We can **graph the inequality** on a real number line by placing a solid dot at -4 and drawing a heavy line to the right of -4, as shown in Figure 8.

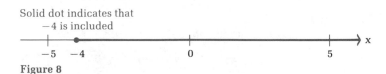

Solid dot indicates that
  -4 is included

Figure 8

Example 15 Graph $x < 4$ on a real number line.

Solution The solution set is the set of all real numbers to the left of 4 on a real number line. The graph is obtained by placing an open dot at 4 and then drawing a heavy line to the left of 4.

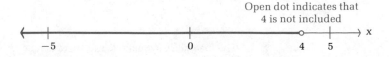

Open dot indicates that
4 is not included

Problem 15   Graph each inequality on a real number line.

(A) $x > 2$ (B) $x \leq -2$

The inequality statement

$$-3 \leq x < 2$$

is called a **double inequality,** since two inequality symbols occur. This means that both

$$-3 \leq x \quad \text{and} \quad x < 2$$

The solution set for this inequality is the set of all real numbers between -3 and 2, including -3, but excluding 2. The graph of this inequality is obtained by placing a solid dot at -3 and an open dot at 2 on a real number line. Then, a heavy line is drawn from -3 to 2, as shown in Figure 9.

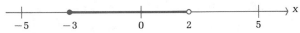

Figure 9

Example 16   Graph $-3 < x < 4$ on a real number line.

Solution   The solution set is the set of all real numbers between -3 and 4, excluding -3 and 4. Thus, the graph is

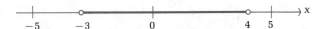

Problem 16   Graph each double inequality on a real number line.

(A) $-4 < x \leq 0$ (B) $-2 \leq x \leq 5$

▪ Interval Notation

Another useful notation for representing inequalities such as those described above is **interval notation.** For example, the inequality $-3 \leq x < 2$ may be represented by $[-3, 2)$. Notice that the square bracket on the left, [, corresponds to $\leq$, and the round parenthesis on the right,), corresponds to $<$. Variations of the interval notation and the equivalent inequality notation are given in Table 2. The symbols ∞ and $-\infty$, called **infinity** and **minus infinity,** are used for convenience in the interval notation when the corresponding line graph extends indefinitely to the right or indefinitely to the left. The symbols ∞ and $-\infty$ do not represent real numbers; hence, they are never enclosed with square brackets.

Table 2

Interval Notation	Inequality Notation	Line Graph
$[a, b]$	$a \leq x \leq b$	
$[a, b)$	$a \leq x < b$	
$(a, b]$	$a < x \leq b$	
(a, b)	$a < x < b$	
$(-\infty, a]$	$x \leq a$	
$(-\infty, a)$	$x < a$	
$[b, \infty)$	$x \geq b$	
(b, ∞)	$x > b$	

Example 17 Graph each interval.

(A) $[0, 6)$ (B) $(-4, 8)$ (C) $(-\infty, -3]$ (D) $(-2, \infty)$

Solutions (A) $[0, 6)$ is equivalent to $0 \leq x < 6$. Thus, the graph is

(B) $(-4, 8)$ is equivalent to $-4 < x < 8$. Thus,

(C) $(-\infty, -3]$ is equivalent to $x \leq -3$. Thus,

(D) $(-2, \infty)$ is equivalent to $x > -2$. Thus,

Problem 17 Graph each interval.

(A) $(-3, 5]$ (B) $[-5, 7]$ (C) $(-\infty, 3)$ (D) $[2, \infty)$

The following examples illustrate situations where more than one inequality statement is involved.

Example 18

Graph the values of x that satisfy both the inequalities

$$-3 < x < 2 \quad \text{and} \quad 0 \leqslant x \leqslant 5$$

That is, graph $(-3, 2) \cap [0, 5]$, the intersection of two intervals.

Solution

Graphing one inequality above the other, we have:

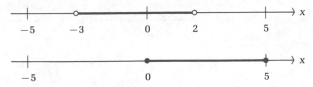

For x to satisfy both inequalities, we must have $0 \leqslant x < 2$. Graphically, the intersection of the two graphs is shown as

Problem 18

Graph the values of x that satisfy both the inequalities

$$-5 \leqslant x < 0 \quad \text{and} \quad -3 < x \leqslant 4$$

That is, graph $[-5, 0) \cap (-3, 4]$, the intersection of two intervals.

Example 19

Graph the values of x that satisfy either

$$x < -2 \quad \text{or} \quad x > 3$$

That is, graph $(-\infty, -2) \cup (3, \infty)$, the union of two intervals.

Solution

Since x must be less than -2 or greater than 3, we obtain

Notice that this line graph consists of two distinct parts.

Problem 19

Graph the values of x that satisfy either

$$x \leqslant -4 \quad \text{or} \quad x > -1$$

That is, graph $(-\infty, -4] \cup (-1, \infty)$, the union of two intervals.

Answers to Matched Problems

13. (A) $>$ (B) $<$ (C) $>$ (D) $<$
14. (A) $>$ (B) $<$ (C) $>$ (D) $>$

15. (A)   (B)

16. (A)

                 $-5\ -4 \qquad\quad 0 \qquad\qquad 5$

    (B)

                 $-5 \qquad -2\ 0 \qquad\quad 5$

17. (A)   (B)

        $-3 \qquad 5$ $-5 \qquad\quad 7$

    (C)   (D)

           3 2

18.   19.

    $-3 \qquad 0$ $-4 \qquad -1$

Exercise 1-3

A *Write each statement using an inequality symbol.*

1. -5 is greater than -30 **2.** -18 is less than -9
3. x is greater than or equal to -6 **4.** x is less than or equal to 7

Write each inequality using ordinary language.

5. $8 > -8$ **6.** $-15 < -5$ **7.** $x \geqslant 8$ **8.** $x \leqslant -9$

Replace each question mark with $<$ or $>$.

9. $6 \ ? \ 11$ **10.** $15 \ ? \ 4$ **11.** $-5 \ ? \ -8$ **12.** $-13 \ ? \ -10$
13. $-3 \ ? \ 9$ **14.** $5 \ ? \ -8$ **15.** $-10 \ ? \ 0$ **16.** $0 \ ? \ -10{,}000$

Problems 17–22 refer to the number line below. Replace each question mark with $<$ or $>$.

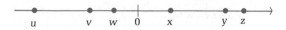

17. $v \ ? \ y$ **18.** $z \ ? \ w$ **19.** $x \ ? \ u$
20. $u \ ? \ v$ **21.** $x \ ? \ 0$ **22.** $v \ ? \ 0$

B *Represent each inequality using interval notation and as a graph on a real number line.*

23. $x \leqslant 5$ **24.** $x \leqslant -3$ **25.** $x > -5$
26. $x > 2$ **27.** $-2 < x < 3$ **28.** $4 < x < 7$
29. $-5 \leqslant x \leqslant -1$ **30.** $-3 \leqslant x \leqslant 8$ **31.** $2 < x \leqslant 8$
32. $-5 \leqslant x < 2$ **33.** $-7 \leqslant x < -2$ **34.** $-8 < x \leqslant 5$

Represent each interval as an inequality and as a graph on a real number line.

35. $(5, \infty)$	**36.** $(-7, \infty)$	**37.** $(-\infty, 4]$	**38.** $(-\infty, -5]$
39. $[-2, 5]$	**40.** $[-5, -1]$	**41.** $(-7, -2)$	**42.** $(3, 8)$
43. $[-2, 2)$	**44.** $(-3, 3]$	**45.** $(2, 10]$	**46.** $[-5, -2)$

Represent each line graph as an inequality and using interval notation.

47. ◄————●————→ x
　　　　　　8

48. ————●————→ x
　　　　　　　−5

49. ————○————→ x
　　　　　−6

50. ◄————○————→ x
　　　　　　4

51. ————●———●————→ x
　　　　　−3　9

52. ————○———○————→ x
　　　　　−12　−6

53. ————○———●————→ x
　　　　　−5　15

54. ————●———○————→ x
　　　　　−10　10

Represent each line graph using inequality notation.

55. ◄————●———○————→ x
　　　　　−3　5

56. ◄————○———●————→ x
　　　　　　4　10

57. ◄————●———●————→ x
　　　　　−5　0

58. ◄————○———○————→ x
　　　　　　−4　4

C　*Represent each pair of inequalities as a single double inequality and graph.*

59. $x \leq 5$ and $x \geq -5$　　　　**60.** $x < 10$ and $x > 1$

61. $x > -2$ and $x \leq 5$　　　　**62.** $x \geq -4$ and $x < 4$

63. $-2 < x < 4$ and $0 < x < 8$

64. $3 \leq x \leq 10$ and $5 \leq x \leq 15$

Graph on a real number line.

65. $x \geq 7$ or $x \leq -7$　　　　**66.** $x > 5$ or $x < 0$

67. $x \leq 20$ or $x > 50$　　　　**68.** $x \geq 100$ or $x < 20$

69. $-10 \leq x \leq 5$ or $-5 \leq x \leq 10$

70. $-20 < x < 0$ or $-5 < x < 5$

Applications

Business & Economics
71. *Salary.* A job announcement indicates a starting salary, S, from \$14,000 to \$18,500, depending on qualifications and experience. Represent the salary range using an inequality.

72. *Sales.* The best salespeople in a given company generally have weekly sales, S, from \$15,000 to \$25,000. Express this range in terms of an inequality.

Life Sciences

Social Sciences

73. *Temperature control.* One food item is to be stored at temperatures ranging from 40°F to 70°F, and another item is to be stored at temperatures ranging from 50°F to 80°F. Use an inequality to represent the allowable temperature, *T*, at which both items can be stored.

74. *Temperature control.* One pharmaceutical drug is to be stored at a temperature ranging from 2°C to 10°C, and another drug is to be stored at 5°C to 15°C. Use an inequality to represent the allowable temperature, *T*, at which both drugs can be stored.

75. *Population.* A forest ranger counted forty foxes on an island. Thus, she concluded that the number of foxes on the island was at least forty. Represent the number of foxes, *n*, using an inequality.

76. *Climate.* One day last week the humidity, *h*, in San Jose ranged from 30% to 55%. Represent the humidity using an inequality and decimal forms for the percentages.

77. *Anthropology.* In the study of human genetic groupings, anthropologists use a ratio called the **cephalic index,** *C.* This is the ratio of the width of the head to its length (looking down from the top) expressed as a percent. A long-headed person is one with *C* less than 75, an intermediate person is one with *C* from 75 to 80, and a round-headed person is one with *C* larger than 80. Represent these classifications in terms of inequality statements.

78. *Psychology.* The IQ of a group of 12-year-olds ranges from 70 to 120. Represent the IQ range using an inequality.

1-4   Basic Operations on Signed Numbers

- Signed Numbers and Absolute Value
- Basic Operations
- Evaluating Algebraic Expressions

In this section we will review the addition, subtraction, multiplication, and division of signed numbers. Then we will describe how to evaluate algebraic expressions for various real numbers. Before we discuss the basic operations on signed numbers, we need to describe the meaning of "the negative of a number" and the "absolute value of a number."

Signed Numbers and Absolute Value

One of the properties of real numbers discussed in Section 1-2 states that for each real number a there is exactly one real number $-a$ that satisfies

$$a + (-a) = (-a) + a = 0$$

The number $-a$ is called the **negative of, opposite of,** or **additive inverse of** the number a. The negative of a number can be obtained simply by changing its sign. Recall that if a sign does not appear in front of a number, then a $+$ sign is assumed. Thus, 5 and $+5$ represent the same number. The number 0 is unique, because it is its own negative; thus, $-0 = +0 = 0$. Note that for any real number a, we always have

$$-(-a) = a$$

Example 20

(A) The negative of 13 is -13.
(B) The negative of $+9$ is $-(+9) = -9$.
(C) The negative of -10 is $-(-10) = 10$.
(D) The negative of $-(-6)$ is $-[-(-6)] = -[+6] = -6$.

Problem 20

Give the negative of each number.

(A) 29   (B) $+47$ (C) -35 (D) $-[-(-8)]$

You should keep in mind that the minus sign, $-$, is used in different ways. It is used to indicate subtraction, as in $5 - 2 = 3$; to represent the negative of a number, as in $-(+9)$; and as part of a symbol representing a number, as in -7.

If a represents a real number, the *absolute value* of a is denoted by $|a|$ and is defined as follows:

Absolute Value

$$|a| = \begin{cases} a & \text{if } a \text{ is positive} \\ 0 & \text{if } a \text{ is } 0 \\ -a & \text{if } a \text{ is negative} \end{cases}$$

Note that when a is negative, $-a$ is positive. Thus, we see that **the absolute value of a number is never negative.** In fact, if $a \neq 0$, we see that $|a|$ is always a positive number. The absolute value of a number is often described as the distance of the number from 0 on a real number line, as illustrated in Figure 10.

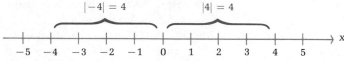

Figure 10

Example 21

(A) $|25| = 25$ (B) $|-31| = 31$ (C) $|0| = 0$

(D) $-|14| = -(14) = -14$ (E) $-|-36| = -(+36) = -36$

Problem 21   Give the value of each expression.

(A) $|72|$ (B) $|-84|$ (C) $|-0|$ (D) $-|29|$ (E) $-|-13|$

The following example involves the familiar operations of addition of two positive numbers, subtraction of one positive number from a larger positive number, and the use of absolute values.

Example 22   (A) $|-9|+|-5| \boxed{=9+5} = 14$

(B) $-(|-13|+|-6|) \boxed{=-(13+6)=-(19)} = -19$

(C) $|+8|-|-3| \boxed{=8-3} = 5$

(D) $-(|-11|-|+5|) \boxed{=-(11-5)=-(6)} = -6$

Problem 22   Evaluate each expression.

(A) $|-6|+|-11|$ (B) $-(|-3|+|-8|)$
(C) $|-15|-|+7|$ (D) $-(|-19|-|-8|)$

■ Basic Operations

Recall from Section 1-2 that if a denotes a real number, then the number 0 satisfies

$$a + 0 = 0 + a = a$$

For example, we have

$$29 + 0 = 29 \qquad (-13) + 0 = -13 \qquad 0 + (-8) = -8$$

The addition of nonzero signed numbers is described as follows:

The Addition of Two Signed Numbers

1. When both numbers are positive, we add them as in ordinary arithmetic.
2. When both numbers are negative, we add their absolute values and attach a minus sign to the result.
3. When the numbers have opposite signs, we take their absolute values, subtract the smaller absolute value from the larger absolute value, and then attach the same sign to this result as that of the number with the larger absolute value.

Example 23   (A) $13 + 9 = 22$

(B) $(-15) + (-8)$ $\boxed{= -(|-15| + |-8|) = -(15 + 8)}$ $= -23$

(C) $(-29) + 15$ $\boxed{= -(|-29| - |15|) = -(29 - 15)}$ $= -14$

(D) $10 + (-5)$ $\boxed{= +(|10| - |-5|) = +(10 - 5)}$ $= 5$

Problem 23   Evaluate each expression.

(A) $19 + 13$ (B) $(-11) + (-23)$
(C) $(-50) + 29$ (D) $18 + (-13)$

To add three or more numbers with mixed signs, we first rearrange the numbers so that the positive numbers are grouped together and the negative numbers are grouped together (this is justified by the commutative and associative properties). Next, we add the positives together and add the negatives together. This leaves us with the sum of two numbers with opposite signs. These two numbers are then added according to rule 3 above.

Example 24 $(-5) + 7 + (-9) + 4 + (-3)$ $\boxed{= [(-5) + (-9) + (-3)] + (7 + 4)}$

$$= (-17) + 11$$

$$\boxed{= -(17 - 11)}$$

$$= -6$$

Problem 24   Evaluate each expression.

(A) $11 + (-8) + 5 + (-3) + 9$ (B) $(-15) + (-9) + 4 + (-6) + 26$

The subtraction of signed numbers is easily accomplished by means of the following definition, which gives the meaning of subtraction in terms of addition:

Subtraction of Two Signed Numbers

If a and b represent real numbers, then

$$a - b = a + (-b)$$

Thus, to subtract a number b from a number a, we change the sign of b and add this to a; that is, we add the negative of b to a. For example,

$$\text{Change the sign of } -12$$

$$(-8) - (-12) = (-8) + (+12) = 4$$

Change subtraction
to addition

With a little practice, this procedure can be accomplished mentally.

Example 25 (A) $13 - (-8) \boxed{= 13 + 8} = 21$

(B) $(-5) - 15 \boxed{= (-5) + (-15)} = -20$

(C) $(-16) - (-9) \boxed{= (-16) + 9} = -7$

(D) $0 - (-12) \boxed{= 0 + 12} = 12$

Problem 25 Evaluate each expression.

(A) $6 - (-9)$ (B) $(-7) - 11$ (C) $(-21) - (-19)$ (D) $0 - 17$

When three or more numbers are combined using both addition and subtraction, we can perform the desired computation by first converting all subtractions to additions, and then adding the resulting numbers. For example,

$$(-3) + 8 - (-5) - 6 \boxed{\begin{aligned} &= (-3) + 8 + 5 + (-6) \\ &= (-9) + 13 \end{aligned}}$$

$$= 4$$

Example 26 (A) $5 - 9 - 6 + 11 \boxed{\begin{aligned} &= 5 + (-9) + (-6) + 11 \\ &= 16 + (-15) \end{aligned}}$

$$= 1$$

(B) $-15 - (-16) + 8 - 11 \boxed{\begin{aligned} &= (-15) + 16 + 8 + (-11) \\ &= (-26) + 24 \end{aligned}}$

$$= -2$$

Problem 26 Evaluate each expression.

(A) $11 - 6 - 13 + 18$ (B) $-16 - (-9) + 11 - 17$

In Section 1-2 we mentioned the fact that the number 0 satisfies

$$a \cdot 0 = 0 \cdot a = 0$$

for any real number a. For example,

$$5 \cdot 0 = 0 \qquad 0 \cdot (-9) = 0$$

We have the following rules for multiplying two nonzero signed numbers:

Multiplication of Two Signed Numbers

1. If two numbers have the same sign, their product is obtained by multiplying their absolute values as in ordinary arithmetic.
2. If two numbers have opposite signs, their product is obtained by multiplying their absolute values and then attaching a minus sign to this result.

Notice that the product of two numbers is positive if both numbers are positive or if both numbers are negative. The product is negative if one number is positive and the other is negative.

Example 27

(A) $8 \cdot 7 = 56$

(B) $(-5)(-9) \boxed{= +(5 \cdot 9)} = 45$
  $\llcorner$ Since -5 and -9 have the same sign

(C) $(-7) \cdot 6 \boxed{= -(7 \cdot 6) = -(42)} = -42$
  $\llcorner$ Since -7 and 6 have opposite signs

(D) $4 \cdot (-9) \boxed{= -(4 \cdot 9) = -(36)} = -36$
  $\llcorner$ Since 4 and -9 have opposite signs

Problem 27

Evaluate each expression.

(A) $11 \cdot 5$ (B) $(-8)(-3)$ (C) $(-12) \cdot 4$ (D) $5 \cdot (-13)$

Some important properties of products involving minus signs are listed in the box.

Properties of Multiplication with Minus Signs

If a and b represent real numbers, then:

1. $(-1)a = -a$
2. $(-a)b = a(-b) = (-1)ab = -(ab) = -ab$
3. $(-a)(-b) = ab$

The division of two numbers may be symbolized in several ways. If a and b represent real numbers with $b \neq 0$, then

$$a/b \qquad \frac{a}{b} \qquad a \div b \qquad b\overline{)a}$$

all indicate that the number a is divided by the number b. The number obtained by dividing one number (the **dividend**) by another (the **divisor**) is called the **quotient**. Notice that **division by 0 is not defined.** If b represents a nonzero real number, we have

$$\frac{0}{b} = 0 \qquad \frac{b}{0} \text{ is undefined} \qquad \frac{0}{0} \text{ is undefined}$$

We have the following rules for dividing two nonzero signed numbers:

Division of Two Signed Numbers

1. If two nonzero numbers have the same sign, their quotient is obtained by dividing their corresponding absolute values as in ordinary arithmetic.
2. If two nonzero numbers have opposite signs, their quotient is obtained by dividing their corresponding absolute values and then attaching a minus sign to this result.

Notice that the quotient of two numbers is positive if both numbers are positive or if both numbers are negative. The quotient is negative if one number is negative and the other is positive.

Example 28

(A) $\dfrac{28}{7} = 4$

(B) $\dfrac{-36}{-3} = +\left(\dfrac{36}{3}\right) = 12$

Since -36 and -3 have the same sign

(C) $\dfrac{-48}{12} = -\left(\dfrac{48}{12}\right) = -(4) = -4$

Since -48 and 12 have opposite signs

(D) $\dfrac{42}{-6} = -\left(\dfrac{42}{6}\right) = -(7) = -7$

Since 42 and -6 have opposite signs

(E) $\dfrac{0}{-113} = 0$

Problem 28 Find each quotient.

(A) $\dfrac{36}{4}$ (B) $\dfrac{-48}{-6}$ (C) $\dfrac{-72}{9}$ (D) $\dfrac{100}{-20}$ (E) $\dfrac{0}{-57}$

(F) $\dfrac{-11}{0}$

The following are some important properties of quotients involving minus signs:

Properties of Division with Minus Signs

If a and b represent real numbers with $b \neq 0$, then

1. $\dfrac{-a}{-b} = -\dfrac{-a}{b} = -\dfrac{a}{-b} = \dfrac{a}{b}$

2. $\dfrac{-a}{b} = \dfrac{a}{-b} = -\dfrac{-a}{-b} = -\dfrac{a}{b}$

We will now consider expressions where various combinations of addition, subtraction, multiplication, and division occur. In evaluating such expressions, we will use the following convention:

Order of Operations

Unless indicated otherwise, multiplication and division are performed before addition and subtraction.

Example 29 (A) $(-3)(-5) - (-4)(2) = 15 - (-8)$ Multiplication is
performed before
$= 15 + 8 = 23$ subtraction.

(B) $\dfrac{-15}{3} - \dfrac{36}{-4} = -5 - (-9)$ Division is performed
before subtraction.

$= -5 + 9 = 4$

(C) $\dfrac{(-12) - (-4)}{-2} = \dfrac{-12 + 4}{-2} = \dfrac{-8}{-2} = 4$ Simplify numerator first.

(D) $\dfrac{(-5)(-4) - (-3)(2)}{(-4) - (-17)} = \dfrac{20 - (-6)}{-4 + 17}$ Simplify numerator and
denominator first.

$= \dfrac{20 + 6}{13}$

$= \dfrac{26}{13} = 2$

Problem 29   Evaluate each expression.

(A) $(-7)4 - (-5)(-3) + 0(-8)$ (B) $\dfrac{24}{-3} - \dfrac{32}{-8} + \dfrac{-16}{-4}$

(C) $\dfrac{(-18) - (-4)}{2 - 9}$ (D) $\dfrac{13(-3) - (-2)(-5)}{12 + (-19)}$

■ Evaluating Algebraic Expressions

We have already seen many expressions involving real numbers and expressions where letters have been used to represent real numbers. For example,

$$a + b \qquad \frac{a}{b} \qquad (a + b) + c \qquad 7x + 3y \qquad \frac{(-2)3 + 4(-2)}{-7}$$

Expressions such as these are called **algebraic expressions.** The letters are used as placeholders for real numbers and are called **variables.** Specific numerals that appear in an expression are called **constants.** For example, in

$$2x + 3y$$

x and y are variables, and 2 and 3 are constants. When two or more algebraic expressions are joined by addition or subtraction, the individual expressions are called **terms.** For example,

$$\underbrace{3x}\ +\ \underbrace{2y}\ -\ \underbrace{3z}$$
$$\text{Terms}$$

When two or more algebraic expressions are joined by multiplication, the individual expressions are called **factors.** For example

$$\underbrace{ab}\underbrace{(c + d)}$$
$$\text{Factors}$$

When we replace the variables in an algebraic expression with numerals and perform the indicated operations, we say that the expression is being **evaluated** for the given numbers. For example, we may evaluate

$$2x + 3y$$

using $x = -5$ and $y = 7$. The result is

$$2(-5) + 3 \cdot 7 = -10 + 21 = 11$$

In many algebraic expressions, **symbols of grouping** are used to indicate which operations are to be performed before others. For example, consider the two expressions

$$(12 - 6) \div 2 \qquad 12 - (6 \div 2)$$

Both expressions involve the same operations; however, the order in which the operations are performed is different and affects the result. Besides

parentheses, ( ), we also use **brackets,** [ ], and **braces,** { }, as symbols of grouping. When evaluating expressions involving various operations and symbols of grouping, we follow the conventions listed below.

Order of Operations

1. Simplify expressions inside the innermost symbols of grouping first, then proceed to work outward to the next set of innermost symbols of grouping, and so on.
2. Multiplication and division are performed before addition and subtraction unless symbols of grouping indicate otherwise.

Example 30 Evaluate each expression.

(A) $15 - 3(-4) = 15 - (-12)$ Multiply 3 and -4 first.
$$= 15 + 12 = 27$$

(B) $-3 - 2[-4 - (-6)] = -3 - 2[-4 + 6]$ Work on innermost
$$= -3 - 2[2]$$ grouping symbols first.
$$= -3 - 4 = -7$$

(C) $(-3 - 2)[-4 - (-8)] = (-5)[-4 + 8]$
$$= (-5)[4] = -20$$

(D) $4\{-30 + 2[8 - 2(-5 - 4)]\}$ Work from innermost grouping
$$= 4\{-30 + 2[8 - 2(-9)]\}$$ symbols outward.
$$= 4\{-30 + 2[8 - (-18)]\}$$
$$= 4\{-30 + 2[8 + 18]\}$$
$$= 4\{-30 + 2[26]\}$$
$$= 4\{-30 + 52\}$$
$$= 4\{22\} = 88$$

(E) $(-15)(-3) - 2[(-2)3 - (-3)(-4)] = 45 - 2[-6 - 12]$
$$= 45 - 2[-18]$$
$$= 45 - [-36]$$
$$= 45 + 36 = 81$$

(F) $\dfrac{-6 - 3[(-2)3 - 4(-2)]}{2[3 - (-4)] - 2} = \dfrac{-6 - 3[-6 - (-8)]}{2[3 + 4] - 2}$ Simplify numerator and denominator first.

$$= \dfrac{-6 - 3[-6 + 8]}{2[7] - 2}$$

$$= \dfrac{-6 - 3[2]}{14 - 2}$$

$$= \dfrac{-6 - 6}{12}$$

$$= \dfrac{-12}{12} = -1$$

Problem 30 Evaluate each expression.

(A) $-29 - 3(-11)$ (B) $12 - 2[5 - (-9)]$

(C) $(12 - 2)[5 - (-9)]$ (D) $3\{15 - 5[11 - 3(-5 + 9)]\}$

(E) $8(-10) - 3[(-5)2 - (-4)(-5)]$ (F) $\dfrac{-2[(-3)4 - 5(-4)] - 8}{5 + 3[-8 - (-5)]}$

We now turn our attention to the evaluation of algebraic expressions for various replacements of variables by constants.

Example 31 Evaluate each expression using $x = -2$, $y = 3$, and $z = -36$.

(A) $4x + 3y$ (B) $z - xy$ (C) $\dfrac{z}{xy} + xy - \dfrac{0}{z}$ (D) $\dfrac{z + 6x}{2y - 5x} - \dfrac{z}{y}$

Solutions For $x = -2$, $y = 3$, and $z = -36$, we have:

(A) $4x + 3y = 4(-2) + 3 \cdot 3 = -8 + 9 = 1$

(B) $z - xy = -36 - (-2) \cdot 3 = -36 - (-6)$
$$= -36 + 6 = -30$$

(C) $\dfrac{z}{xy} + xy - \dfrac{0}{z} = \dfrac{-36}{(-2) \cdot 3} + (-2) \cdot 3 - \dfrac{0}{-36}$

$$= \dfrac{-36}{-6} + (-6) - 0$$

$$= 6 - 6 = 0$$

(D) $\dfrac{z + 6x}{2y - 5x} - \dfrac{z}{y} = \dfrac{(-36) + 6(-2)}{2 \cdot 3 - 5(-2)} - \dfrac{-36}{3}$

$$= \dfrac{-36 + (-12)}{6 - (-10)} - (-12)$$

$$= \dfrac{-36 - 12}{6 + 10} + 12$$

$$= \dfrac{-48}{16} + 12 = -3 + 12 = 9$$

Problem 31 Evaluate each expression using $u = -48$, $v = -6$, and $w = 4$.

(A) $3v - 5w$ (B) $wv - u$

(C) $\dfrac{0}{uv} - vw + \dfrac{u}{wv}$ (D) $\dfrac{9w - 4v}{u + 7w} + \dfrac{u}{4w}$

Answers to Matched Problems

20. (A) -29 (B) -47 (C) $+35$ or 35 (D) 8

21. (A) 72 (B) 84 (C) 0 (D) -29
 (E) -13

22. (A) 17 (B) -11 (C) 8 (D) -11

23. (A) 32 (B) -34 (C) -21 (D) 5

24. (A) 14   (B) 0
25. (A) 15   (B) -18 (C) -2 (D) -17
26. (A) 10   (B) -13
27. (A) 55   (B) 24   (C) -48 (D) -65
28. (A) 9   (B) 8   (C) -8 (D) -5
    (E) 0   (F) Undefined
29. (A) -43 (B) 0   (C) 2   (D) 7
30. (A) 4   (B) -16 (C) 140   (D) 60
    (E) 10   (F) 6
31. (A) -38 (B) 24   (C) 26   (D) -6

Exercise 1-4

A   *Evaluate each expression.*

1. $-(+20)$
2. $-(-10)$
3. $-[-(-7)]$
4. $-[-(+10)]$
5. $|-9|$
6. $|+9|$
7. $-|-7|$
8. $-|-13|$
9. $-|-(-5)|$
10. $-|-(+9)|$
11. $|-8|+|-12|$
12. $|-8|-|-12|$
13. $-(|-18|-|+8|)$
14. $-(|-7|+|-9|)$
15. $9+(-30)$
16. $(-12)+(-9)$
17. $7-11$
18. $(-6)-8$
19. $(-9)(-7)$
20. $(-5)\cdot 4$
21. $\dfrac{120}{-30}$
22. $\dfrac{-60}{-15}$
23. $\dfrac{0}{-5}$
24. $\dfrac{-5}{0}$
25. $8\div 0$
26. $0\cdot(-8)$
27. $(-6)(2)(-4)$
28. $(-6)(0)8$
29. $(-10)-(-6)(-2)$
30. $(-5)(-2)+(-9)$
31. $\dfrac{-36}{-4}-9$
32. $8-\dfrac{-24}{8}$
33. $(-15)(-3)-(-9)(-5)$
34. $(-8)(4)-(-16)(-2)$
35. $\dfrac{-12}{-4}-\dfrac{-6}{-2}$
36. $\dfrac{-15}{3}-\dfrac{8}{-2}$
37. $13-9-12+4$
38. $-6+15-20+3$
39. $6-3[5-(-2)]$
40. $7-4[(-8)-(-3)]$

B   *In Problems 41–44 select the appropriate word to make the statement true.*

41. The negative of a number is (*always, sometimes, never*) a negative number.
42. The absolute value of a negative number is (*always, sometimes, never*) a positive number.

43. The absolute value of a number is (always, sometimes, never) a positive number.

44. The negative of a negative number is (always, sometimes, never) a positive number.

Evaluate each expression.

45. $\dfrac{(-2) \cdot 14 + (-4) \cdot 2}{15 - (-2)(-3)}$

46. $[7 - (-3)] - [(-5) - (-2)]$

47. $[(-8) - (-2)] + [(-10) - 15]$

48. $[(-5) - (-10)][(-8) + (-2)]$

49. $[6 + (-3)][(-9) - (-5)]$

50. $\dfrac{(-12) - (-8)}{(-5) - (-7)}$

51. $\dfrac{18 - (-12)}{6 - (-4)}$

52. $\dfrac{(-6)(-4) - (-3) \cdot 2}{(-19) - (-4)}$

53. $-3 + 5\{(-6) - 2[(-5) + (-6)]\}$

54. $10 - 3\{5 - 2[3 - (4 - 5)]\}$

Evaluate using $t = 2$, $u = 0$, $v = -3$, and $w = -48$.

55. $v - (t - w)$

56. $w - (v - t)$

57. $\dfrac{u}{w}$

58. $\dfrac{v}{u}$

59. $w - uv$

60. $\dfrac{u}{v} - tw$

61. $uw - \dfrac{tuv}{w}$

62. $wvu - \dfrac{w}{t}$

63. $uvw - \dfrac{w}{v} + \dfrac{w}{t}$

64. $\dfrac{w}{tv} + \dfrac{w}{t} + \dfrac{w}{v}$

C   *Evaluate each expression.*

65. $\dfrac{10 - 2[(-3) \cdot 3 - (-5)(-2)]}{2[3 - (-3)] - (-4)}$

66. $\dfrac{-2[2 \cdot (-12) - (-2)(-6)] - 2}{7 - 6[(-2) - (-90)]}$

Evaluate using $t = 2$, $u = 0$, $v = -3$, and $w = -48$.

67. $\dfrac{w}{8v} - \dfrac{w - 6v}{t - v}$

68. $\dfrac{v - t}{v + t} + \dfrac{w}{6t}$

Applications

Business & Economics

69. *Stock prices.* At the close of trading on Friday a stock was worth $33 per share. On Monday it fell $7, on Tuesday it was down another $6, on Wednesday it rose $5, and on Thursday it rose another $3. Using addition of signed numbers, determine the closing value of the stock on Thursday.

70. *Inventory.* At the start of business on Monday a local gas station had an inventory of 9,500 gallons of gasoline. On Monday 3,500 gallons were sold and on Tuesday 2,600 gallons were sold. On Wednesday a delivery of 15,000 gallons was received and 2,900 gallons were sold. On Thursday 3,200 gallons were sold. Using addition of signed numbers, determine the amount of gasoline on hand at the start of business on Friday.

Life Sciences

71. *Laboratory management.* In a supply house for laboratory animals there were 2,400 animals at the beginning of the week. During the week, 350 new animals were born, 105 died, 840 were sold, and a shipment of 750 new animals was received. Using addition of signed numbers, determine the number of animals on hand at the beginning of the following week.

72. *Diet.* In an experiment a laboratory rat weighing 273.6 grams gained 4.2 grams on Monday, another 1.8 grams on Tuesday, lost 2.1 grams on Wednesday, another 3.4 grams on Thursday, and gained 2.8 grams on Friday. Use addition of signed numbers to find the weight of the rat by the end of Friday.

Social Sciences

73. *Politics.* In an attempt to have a piece of controversial legislation passed, a lobbyist organization tries to convince members of the House of Representatives to vote in favor of a particular bill. Early Monday morning, it is estimated that there are 175 votes in favor of the bill. On Monday 5 votes are gained but 7 are lost. On Tuesday 10 more votes are gained with 4 lost. On Wednesday 13 more votes are gained with a loss of 5. Use addition of signed numbers to find the estimated number of votes in favor of the bill as of Wednesday.

1-5 Positive Integer Exponents

- Definition of a^n
- Properties of Exponents
- Common Errors
- Summary

In this section we will introduce the use of *exponents*, which will enable us to write many products more simply. We will also introduce several important properties of exponents that will permit us to manipulate and simplify algebraic expressions involving exponents.

■ Definition of a^n

Consider the product

$$a \cdot a \cdot a \cdot a \cdot a \cdot a$$

where a represents a real number. Notice that the same factor a occurs six times. Another way to express this product is to write a^6. Thus,

$$a^6 = a \cdot a \cdot a \cdot a \cdot a \cdot a$$

In the expression a^6, the 6 is called the **exponent** or **power** and a is called the **base.** We can generalize this as follows:

Definition of a^n

If a represents a real number and n represents a positive integer, then a^n is defined by

Exponent ─────┐
              ↓
$$a^n = a \cdot a \cdot a \cdot \cdots \cdot a \qquad n \text{ factors of } a$$
              ↑
            Base

We have

$a^1 = a$
$a^2 = a \cdot a$ Often read "a squared"
$a^3 = a \cdot a \cdot a$ Often read "a cubed"
$a^4 = a \cdot a \cdot a \cdot a$

and so on

Example 32

(A) $5^3 = 5 \cdot 5 \cdot 5 = 125$
(B) $(-4)^4 = (-4)(-4)(-4)(-4) = 256$
(C) $(\frac{1}{2})^5 = (\frac{1}{2})(\frac{1}{2})(\frac{1}{2})(\frac{1}{2})(\frac{1}{2}) = \frac{1}{32}$
(D) $8^1 = 8$

Problem 32

Find the value of each expression.

(A) 2^6 (B) $(-3)^3$ (C) $(\frac{3}{2})^3$ (D) 23^1

Example 33

Write each expression using exponents.

(A) $xxxxyyzzzzz = x^4y^2z^5$
(B) $aaabccccd = a^3bc^4d$ The exponent 1 is understood when no exponent occurs.

Problem 33

Write each expression using exponents.

(A) $uuuvvwwwwww$ (B) $pqqqqrsssss$

■ Properties of Exponents

We will now illustrate and state five properties of exponents. In this discussion, a and b will represent real numbers and m and n will denote positive integers.

To begin, let us consider the product a^5a^3. We have

$$a^5a^3 = \underbrace{(a \cdot a \cdot a \cdot a \cdot a)}_{5 \text{ factors}}\underbrace{(a \cdot a \cdot a)}_{3 \text{ factors}} = \underbrace{a \cdot a \cdot a \cdot a \cdot a \cdot a \cdot a \cdot a}_{5 + 3 = 8 \text{ factors}} = a^{5+3} = a^8$$

Thus, $a^5a^3 = a^8$. This illustrates property 1:

Property 1

$$a^m a^n = a^{m+n}$$

Next, let us consider $(a^2)^4$. We have

$$(a^2)^4 = \underbrace{a^2 \cdot a^2 \cdot a^2 \cdot a^2}_{4 \text{ factors of } a^2} = \underbrace{(a \cdot a)(a \cdot a)(a \cdot a)(a \cdot a)}_{4 \text{ groups of 2 factors}}$$

$$= \underbrace{a \cdot a \cdot a \cdot a \cdot a \cdot a \cdot a \cdot a}_{4 \cdot 2 = 8 \text{ factors}} = a^{4 \cdot 2} = a^8$$

Thus, $(a^2)^4 = a^8$. This illustrates property 2:

Property 2

$$(a^n)^m = a^{mn}$$

Consider $(ab)^4$. We have

$$(ab)^4 = \underbrace{(ab)(ab)(ab)(ab)}_{4 \text{ factors of } ab} = a \cdot b \cdot a \cdot b \cdot a \cdot b \cdot a \cdot b$$

$$= \underbrace{(a \cdot a \cdot a \cdot a)}_{\substack{4 \text{ factors} \\ \text{of } a}}\underbrace{(b \cdot b \cdot b \cdot b)}_{\substack{4 \text{ factors} \\ \text{of } b}} = a^4b^4$$

Thus, $(ab)^4 = a^4b^4$. This illustrates property 3:

Property 3

$$(ab)^m = a^m b^m$$

Now, let us consider $\left(\dfrac{a}{b}\right)^5$. We have, assuming $b \neq 0$,

$$\left(\frac{a}{b}\right)^5 = \underbrace{\left(\frac{a}{b}\right)\left(\frac{a}{b}\right)\left(\frac{a}{b}\right)\left(\frac{a}{b}\right)\left(\frac{a}{b}\right)}_{5 \text{ factors of } \frac{a}{b}} = \frac{\overbrace{a \cdot a \cdot a \cdot a \cdot a}^{5 \text{ factors of } a}}{\underbrace{b \cdot b \cdot b \cdot b \cdot b}_{5 \text{ factors of } b}} = \frac{a^5}{b^5}$$

Thus, $\left(\dfrac{a}{b}\right)^5 = \dfrac{a^5}{b^5}$. This illustrates property 4:

Property 4

$$\left(\frac{a}{b}\right)^m = \frac{a^m}{b^m} \qquad b \neq 0$$

The next property involves three possible situations. For $a \neq 0$, we have:

(A) $\quad \dfrac{a^8}{a^5} = \dfrac{a \cdot a \cdot a \cdot a \cdot a \cdot a \cdot a \cdot a}{a \cdot a \cdot a \cdot a \cdot a}$

$$= \frac{\overset{1}{\cancel{(a \cdot a \cdot a \cdot a \cdot a)}}(a \cdot a \cdot a)}{\underset{1}{\cancel{(a \cdot a \cdot a \cdot a \cdot a)}}} = \frac{a \cdot a \cdot a}{1} = a^{8-5} = a^3$$

Thus, $\dfrac{a^8}{a^5} = a^{8-5} = a^3$.

(B) $\quad \dfrac{a^4}{a^4} = \dfrac{\overset{1}{\cancel{a \cdot a \cdot a \cdot a}}}{\underset{1}{\cancel{a \cdot a \cdot a \cdot a}}} = 1$

Thus, $\dfrac{a^4}{a^4} = 1$.

(C) $\dfrac{a^3}{a^7} = \dfrac{a \cdot a \cdot a}{a \cdot a \cdot a \cdot a \cdot a \cdot a \cdot a}$

$$= \dfrac{\overset{1}{(\cancel{a \cdot a \cdot a})}}{\underset{1}{(\cancel{a \cdot a \cdot a})}(a \cdot a \cdot a \cdot a)} = \dfrac{1}{a \cdot a \cdot a \cdot a} = \dfrac{1}{a^{7-3}} = \dfrac{1}{a^4}$$

Thus, $\dfrac{a^3}{a^7} = \dfrac{1}{a^{7-3}} = \dfrac{1}{a^4}$.

These three examples illustrate property 5:

Property 5

If $a \neq 0$, then

$$\dfrac{a^m}{a^n} = \begin{cases} a^{m-n} & \text{if } m \text{ is larger than } n \\ 1 & \text{if } m = n \\ \dfrac{1}{a^{n-m}} & \text{if } n \text{ is larger than } m \end{cases}$$

■ Common Errors

Many errors in algebra occur because the properties of exponents are applied incorrectly, particularly to sums and differences. We now list several pairs of expressions that are not, in general, equal. Common errors occur when these pairs of expressions are assumed to be equal. Compare these with the properties of exponents discussed above.

Expressions That Are Not Equal

$(a + b)^m$	is *not* equal to	$a^m + b^m$
$(a - b)^m$	is *not* equal to	$a^m - b^m$
$a^m + a^n$	is *not* equal to	a^{m+n}
$a^m a^n$	is *not* equal to	a^{mn}
$(a^m)^n$	is *not* equal to	a^{m+n}
$(2b)^3$	is *not* equal to	$2b^3$ or $6b^3$

For example, to compute $(2b)^3$ correctly, we have

$(2b)^3 = 2^3 b^3 = 8b^3$

Another common error occurs in computing an expression such as -5^2.

$$-5^2 \quad \textbf{is not equal to} \quad (-5)^2$$

The expression -5^2 means $-(5 \cdot 5) = -25$. On the other hand, $(-5)^2$ means $(-5)(-5) = 25$.

▪ Summary

For easy reference, we summarize the exponent properties discussed above.

Properties of Exponents

If a and b represent real numbers and m and n denote positive integers, then:

1. $a^m a^n = a^{m+n}$
2. $(a^n)^m = a^{mn}$
3. $(ab)^m = a^m b^m$
4. $\left(\dfrac{a}{b}\right)^m = \dfrac{a^m}{b^m} \qquad b \neq 0$
5. $\dfrac{a^m}{a^n} = \begin{cases} a^{m-n} & \text{if } m \text{ is larger than } n \\ 1 & \text{if } m = n \\ \dfrac{1}{a^{n-m}} & \text{if } n \text{ is larger than } m \end{cases} \qquad a \neq 0$

Example 34

(A) $u^8 u^{12} \boxed{= u^{8+12}} = u^{20}$

(B) $(y^3)^7 \boxed{= y^{7 \cdot 3}} = y^{21}$

(C) $(xy)^8 = x^8 y^8$

(D) $\left(\dfrac{r}{s}\right)^5 = \dfrac{r^5}{s^5}$

(E) $\dfrac{t^{15}}{t^7} \boxed{= t^{15-7}} = t^8$

(F) $\dfrac{c^5}{c^5} = 1$

(G) $\dfrac{y^5}{y^{11}} \boxed{= \dfrac{1}{y^{11-5}}} = \dfrac{1}{y^6}$

Problem 34 Simplify each expression using the properties of exponents.

(A) $x^9 x^{13}$

(B) $(z^5)^6$

(C) $(bc)^{10}$

(D) $\left(\dfrac{u}{v}\right)^6$

(E) $\dfrac{y^{17}}{y^{11}}$

(F) $\dfrac{z^7}{z^7}$

(G) $\dfrac{x^6}{x^{13}}$

Example 35 (A) $(u^3v^4)^3 \boxed{= (u^3)^3(v^4)^3} = u^9v^{12}$ (B) $\left(\dfrac{x^5}{y^3}\right)^4 \boxed{= \dfrac{(x^5)^4}{(y^3)^4}} = \dfrac{x^{20}}{y^{12}}$

(C) $(-x^2y)^2(2xy^3)^3 \boxed{= [(-1)^2x^4y^2](2^3x^3y^9)} = [x^4y^2](8x^3y^9)$

$\boxed{= 8x^4x^3y^2y^9} = 8x^7y^{11}$

(D) $\dfrac{35x^5y^7}{25x^8y^3} \boxed{= \left(\dfrac{35}{25}\right)\left(\dfrac{x^5}{x^8}\right)\left(\dfrac{y^7}{y^3}\right) = \left(\dfrac{7}{5}\right)\left(\dfrac{1}{x^3}\right)\left(\dfrac{y^4}{1}\right)} = \dfrac{7y^4}{5x^3}$

(E) $\dfrac{(4x^2y)^2}{(2xy^2)^4} \boxed{= \dfrac{4^2(x^2)^2y^2}{2^4x^4(y^2)^4}} = \dfrac{16x^4y^2}{16x^4y^8}$

$\boxed{= \left(\dfrac{16}{16}\right)\left(\dfrac{x^4}{x^4}\right)\left(\dfrac{y^2}{y^8}\right) = 1 \cdot 1 \cdot \dfrac{1}{y^6}} = \dfrac{1}{y^6}$

Problem 35 Simplify each expression using the properties of exponents.

(A) $(x^5y^3)^8$ (B) $\left(\dfrac{u^5}{v^7}\right)^3$ (C) $(-u^3v)^3(3u^2v^3)^2$ (D) $\dfrac{28x^8y^3}{35x^4y^9}$

(E) $\dfrac{(3x^3y^2)^3}{(6x^4y^3)^2}$

Answers to Matched Problems
32. (A) 64   (B) -27 (C) $\frac{27}{8}$ (D) 23
33. (A) $u^3v^2w^6$ (B) pq^4rs^5

34. (A) x^{22} (B) z^{30} (C) $b^{10}c^{10}$ (D) $\dfrac{u^6}{v^6}$

  (E) y^6 (F) 1   (G) $\dfrac{1}{x^7}$

35. (A) $x^{40}y^{24}$ (B) $\dfrac{u^{15}}{v^{21}}$ (C) $-9u^{13}v^9$ (D) $\dfrac{4x^4}{5y^6}$

  (E) $\dfrac{3x}{4}$

Exercise 1-5

A *Replace each question mark with an appropriate expression.*

1. $m^6m = m^?$ 2. $ww^5 = w^?$ 3. $y^{12} = y^7y^?$
4. $n^{11} = n^?n^8$ 5. $(x^6)^3 = x^?$ 6. $(p^4)^2 = ?$
7. $z^{12} = (z^?)^4$ 8. $(ab)^8 = ?$ 9. $w^{16} = (w^2)^?$
10. $(mn)^8 = m^8n^?$ 11. $x^6y^6 = (xy)^?$ 12. $t^7v^7 = (tv)^?$

13. $\left(\dfrac{b}{c}\right)^5 = ?$ 14. $\left(\dfrac{y}{z}\right)^7 = \dfrac{y^7}{z^?}$ 15. $\dfrac{t^4}{u^4} = \left(\dfrac{t}{u}\right)^?$

16. $\dfrac{x^5}{y^5} = \left(\dfrac{x}{y}\right)^?$ 17. $\dfrac{m^6}{m^2} = m^?$ 18. $\dfrac{p^{10}}{p^5} = p^?$

19. $\dfrac{q^4}{q^7} = \dfrac{1}{q^?}$ 20. $\dfrac{b^6}{b^{12}} = \dfrac{1}{b^?}$ 21. $y^5 = \dfrac{y^?}{y^4}$

22. $a^6 = \dfrac{a^8}{a^?}$ 23. $\dfrac{1}{b^4} = \dfrac{b^?}{b^5}$ 24. $\dfrac{1}{w^3} = \dfrac{w^6}{w^?}$

Simplify each expression using the properties of exponents.

25. $(7y^2)(4y^3)$ 26. $(5x^4)(4x)$ 27. $\dfrac{18w^4}{9w^2}$ 28. $\dfrac{36x^6}{12x^3}$

29. $\dfrac{6m^5}{15m^7}$ 30. $\dfrac{\cdot 8a^4}{24a^7}$ 31. $(uv)^9$ 32. $(xy)^4$

33. $\left(\dfrac{p}{q}\right)^7$ 34. $\left(\dfrac{w}{z}\right)^3$ 35. $(m^8)^3$ 36. $(p^4)^8$

B   *Simplify each expression using the properties of exponents.*

37. $(5x^2)(2x^3)(7x^4)$ 38. $(3z^5)(5z)(5z^4)$ 39. $(x^2y^3)^4$
40. $(p^3q)^5$ 41. $(2x^2)^3$ 42. $(3y^4)^2$

43. $\left(\dfrac{a^4}{b^2}\right)^3$ 44. $\left(\dfrac{w^2}{x^5}\right)^4$ 45. $\dfrac{27x^3y^7}{18x^5y^2}$

46. $\dfrac{35u^8v^3}{25u^3v^7}$ 47. $(3u^2v^3w)^3$ 48. $(2a^4b^3c^2)^4$

49. $(-4a^3b^2)^2$ 50. $(-3t^4u^2)^3$ 51. $4(x^3y)^4$

52. $2(u^5v)^6$ 53. $\left(\dfrac{m^4n^2}{mn^5}\right)^2$ 54. $\left(\dfrac{a^7b^9}{a^8b^2}\right)^3$

C   *Simplify each expression using the properties of exponents.*

55. $\dfrac{(4x^2y^3)^2}{(2xy^2)^4}$ 56. $\dfrac{(3u^4v^2)^4}{(6u^2v^5)^2}$ 57. $\dfrac{-y^2}{(-y)^2}$ 58. $\dfrac{-x^3}{(-x)^3}$

59. $(2x^2y)^2(3y^2z)^2(x^3z)$ 60. $(3a^2b^4)^2(b^4c^2)^5(2ac^2)^3$
61. $(-2u^2v)^2(-3u^2v^4)^3(u^3v)^4$ 62. $(m^4n^3)(-4m^2n^5)^3(-2mn^4)^2$

63. $\dfrac{(2t^2u^3)^3(3u^4v^2)^2}{(6t^2u^4v^3)^2}$ 64. $\dfrac{(3x^2y)^3(3y^4z)^2}{(9xy^8z^3)^2}$

1-6   Chapter Review

| Important Terms and Symbols | 1-1   *Sets.* element of, member of, null set, empty set, finite set, infinite set, subset, equal sets, union, intersection, disjoint, universal set, complement, $\varnothing$, $A \subset B$, $A \cup B$, $A \cap B$, U, A' |

1-2 *Real numbers and the rules of algebra.* natural number, integer, rational number, irrational number, real number, real number line, equality symbol, symmetry property, transitive property, substitution property, commutative properties, associative properties, distributive property, addition and multiplication with the number 0, multiplication by the number 1, additive inverse, negative, multiplicative inverse, reciprocal

1-3 *Inequality statements and line graphs.* inequality symbols, less than, greater than, less than or equal to, greater than or equal to, solution set, line graph, double inequality, interval notation, the infinity symbol, $a < b$, $a > b$, $a \leq b$, $a \geq b$, $a \leq x \leq b$, $a \leq x < b$, $a < x \leq b$, $a < x < b$, $[a, b]$, $[a, b)$, $(a, b]$, (a, b), $(-\infty, a]$, $(-\infty, a)$, $[b, \infty)$, (b, ∞)

1-4 *Basic operations on signed numbers.* negative, opposite, additive inverse, absolute value, addition, subtraction, multiplication, division, quotient, combined operations, order of operations, algebraic expressions, variable, constant, term, factor, symbols of grouping, evaluation of algebraic expressions, $|a|$

1-5 *Positive integer exponents.* exponent, power, base, properties of exponents, a^n

Exercise 1-6   Chapter Review

Work through all the problems in this chapter review and check your answers in the back of the book. (Answers to all review problems are there.) Where weaknesses show up, review appropriate sections in the text. When you are satisfied that you know the material, take the practice test following this review.

A

1. Indicate whether each statement is true (T) or false (F).

(A) $7 \in \{4, 6, 8\}$ (B) $\{8\} \subset \{4, 6, 8\}$
(C) $\emptyset \in \{4, 6, 8\}$ (D) $\emptyset \in \{4, 6, 8\}$

2. Indicate whether each statement is true (T) or false (F).

(A) -9 is a rational number   (B) -5 is a natural number
(C) 5.237 is a real number

Each statement is justified by either the associative property or the commutative property of real numbers. Indicate which.

3. $y8 = 8y$ 4. $(9a)b = 9(ab)$
5. $(z + 6) + 3 = z + (6 + 3)$ 6. $7 + y = y + 7$

In Problems 7–10 use the associative and commutative properties to simplify each expression mentally.

7. $4(5y)$

8. $(w6)7$

9. $(y + 11) + 7$

10. $(2x + 3) + 8$

11.   Give the negative of:
    (A) -16 (B) 9

12.   Give the reciprocal of:
    (A) $\frac{2}{7}$ (B) -18

Write each statement using an inequality symbol.

13.   x is greater than or equal to 3

14. -13 is less than -5

Write in ordinary language.

15. $20 > 7$

16. $x \leq -2$

Replace each question mark with $<$ or $>$.

17. $10 \; ? \; -1$

18. $-100 \; ? \; 10$

Evaluate each expression.

19. $-(-|-3|)$

20. $|-5| - |-10|$

21. $-[5 - 2(7 - 10)]$

22. $(-5)(2) - (-3)$

23. $(-6)(0)(4)(-3)$

24. $\dfrac{-6}{0}$

25. $-7 + \dfrac{-8}{-2}$

26. $\dfrac{0}{-4} - (-3)(8)$

Replace each question mark with an appropriate expression.

27. $u^9 = u^4 u^?$

28. $u^4 v^4 = (uv)^?$

29. $w^7 = \dfrac{w^?}{w^4}$

30. $\dfrac{1}{a^3} = \dfrac{a^7}{a^?}$

In Problems 31–36 simplify each expression using the properties of exponents.

31. $(6m^3)(7m^4)$

32. $(2x^2y)^3$

33. $\left(\dfrac{a^6}{b^2}\right)^3$

34. $\dfrac{8a^4}{32a^2}$

35. $\dfrac{48n^3}{16n^5}$

36. $\left(\dfrac{2x^2}{3y^3}\right)^3$

B   37.   If $A = \{1, 2, 3\}$ and $B = \{2, 3, 4\}$, find:

    (A) $A \cup B$ (B) $\{x | x \in A \;$ and $\; x \in B\}$

Each statement is justified by either the associative or commutative property of real numbers. Indicate which.

38. $(7a)(b + 9) = 7[a(b + 9)]$

39. $4(y6) = 4(6y)$

40. $3v + 8u = 8u + 3v$
41. $(2x + 3y) + (2y + 5z) = 2x + [3y + (2y + 5z)]$

In Problems 42–45 use the distributive property to multiply.

42. $2(3x + 5y)$ 43. $5(4a + b)$ 44. $(h + k)m$ 45. $p(q + r)$

Use the distributive property to write each expression as a product.

46. $9r + 9s$ 47. $15x + 5$ 48. $24a + 16b$ 49. $km + kn$

Represent each inequality using interval notation and as a graph on a number line.

50. $x < -8$ 51. $x \geqslant 2$ 52. $-5 \leqslant x < 5$
53. $8 < x < 15$

Represent each interval as an inequality and as a graph on a number line.

54. $(-\infty, 5]$ 55. $(-3, \infty)$ 56. $[-4, 3]$ 57. $(5, 15)$

Represent each line graph as an inequality.

58. ![number line graph with open circle at 8, arrow pointing left] $\to x$
     8

59. ![number line graph with open circle at -6, solid circle at 6] $\to x$
     $-6 \quad 6$

Evaluate each expression.

60. $(-9) \cdot 4 - (-6)(-6)$ 61. $[(-8) - (-5)] + (9 - 3)$

62. $[(-8) + (-5)][(-9) - (-7)]$ 63. $\dfrac{(-24) - (-3)}{1 - (-3)(2)}$

64. $8 - 2\{7 - 4[3 - (5 - 4)]\}$

Evaluate using $w = 10$, $x = 100$, $y = 0$, and $z = -5$.

65. $w + z$ 66. $x + wz$ 67. $x - (w - z)$

68. $xyz = \dfrac{x}{w}$ 69. $\dfrac{y}{z} - \dfrac{x}{z}$ 70. $3w - 5z$

Simplify each expression in Problems 71–74 using the properties of exponents.

71. $(3x^2y^4)(2x^5y)$ 72. $\dfrac{4x^5y}{6x^3y^3}$ 73. $(-2x^3yz^2)^3$ 74. $\left(\dfrac{u^3v^5}{u^6v^2}\right)^4$

75. In a freshman class of 100 students, 70 are taking English, 45 are taking math, and 25 are taking both English and math.

   (A) How many students are taking either English or math?
   (B) How many students are taking English and not math?

76. Given the Venn diagram shown, with the number of elements indi-

cated in each part, how many elements are in each of the following sets?

(A) $M \cup N$ (B) $M \cap N$ (C) $(M \cup N)'$ (D) $M \cap N'$

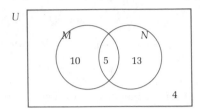

C   77. Give the reciprocal of: (A) $\frac{3}{5}$ (B) $-7\frac{2}{3}$

Represent each pair of inequalities as a single double inequality and graph.

78. $x < 3$ and $x > -5$ 79. $-5 \leqslant x \leqslant 1$ and $-1 \leqslant x \leqslant 5$

Use a real number line to represent the values of x that satisfy the inequalities in Problems 80 and 81.

80. $x < -5$ or $x > 3$ 81. $-8 \leqslant x < 0$ or $-5 < x \leqslant 5$

82. Evaluate: $\dfrac{17 - 3[(-5)(-3) - (-4)(6)]}{2[5 - (-6)] + 3}$

83. Evaluate using $w = 10$, $x = 100$, $y = 0$, and $z = -5$:
$$\frac{x - 5w}{4z - 5} - \frac{6z + y}{w}$$

Simplify each expression using the properties of exponents.

84. $\dfrac{(2x^2y^5)^3}{(4x^3y^2)^2}$ 85. $(-2m^2n)^3(3mn^4)^2(mn)^4$

Practice Test: Chapter 1

1. If $U = \{2, 4, 5, 6, 8\}$, $M = \{2, 4, 5\}$, and $N = \{5, 6\}$, find:

   (A) $M \cup N$ (B) $M \cap N$ (C) $(M \cup N)'$ (D) $M \cap N'$

2. Indicate true (T) or false (F) for U, M, and N in Problem 1:

   (A) $N \subset M$ (B) $\emptyset \subset U$ (C) $6 \notin M$ (D) $5 \in N$

3. Each of the following statements is justified by the associative or commutative property of real numbers. Indicate which one.

   (A) $z + (9 + z) = z + (z + 9)$ (B) $(w + 5) + 7 = w + (5 + 7)$
   (C) $(9y)(7x) = (7x)(9y)$ (D) $(5a)(a + b) = 5[a(a + b)]$

4. Use the distributive property to write each expression as a product.

   (A) $36m + 27n$ (B) $18x + 6$

5. Simplify the following expressions mentally:

   (A) $(5x)(3y)(4z)$ (B) $(a + 6) + (b + 3) + (c + 5)$

6. Evaluate each expression.

   (A) $(-5) \cdot 3 - (-3)(-6)$ (B) $\dfrac{18}{-3} - \dfrac{24}{-4}$

   (C) $\dfrac{(5)(-5) - (-3)(-5)}{14 - (-2)(3)}$

7. Evaluate using $a = 50$, $b = -2$, and $c = 5$.

   (A) $a - bc$ (B) $\dfrac{a - 6c}{2bc}$

8. Represent each inequality using interval notation and as a graph on a real number line.

   (A) $x \leq -5$ (B) $-3 \leq x < 5$

9. Represent each interval as an inequality and as a graph on a real number line.

   (A) $(3, \infty)$ (B) $(-2, 4]$

10. Graph on a real number line:

    (A) $x > 5$ or $x < 2$ (B) $x \geq -2$ and $x \leq 3$

11. Simplify each expression using the properties of exponents.

    (A) $\dfrac{45x^3y^7}{25x^6y^2}$ (B) $\dfrac{(2u^2v^3)^2}{(uv^2)^4}$

12. A survey company sampled 1,000 students at a university. Out of the sample, it was found that 500 smoked cigarettes, 820 drank alcoholic beverages, and 470 did both.

    (A) How many smoked or drank?
    (B) How many drank, but did not smoke?

Polynomials and Fractional Forms

CHAPTER 2   Contents

Polynomials and fractional expressions are found throughout mathematics and its applications. Consequently, it is of great importance to have a complete understanding of these expressions and the many mathematical operations that can be performed with them. Developing this understanding is our goal in this chapter.

2-1   Basic Operations on Polynomials

- Polynomials
- Simplifying Polynomials
- Addition and Subtraction of Polynomials
- Multiplication of Polynomials
- Multiplying Binomials

In this section we will consider polynomial forms and the basic operations of addition, subtraction, and multiplication.

■ Polynomials

In Chapter 1 we discussed the real number system and worked with simple algebraic expressions such as $2x^4$, $5x^2y^3$, and $-7z^3$, which involve only the operation of multiplication. Numbers and expressions such as these are called **monomials.** *Polynomials* are formed by combining monomials using the operations of addition and subtraction. The individual monomials that make up a polynomial are called **terms.** For convenience, monomials are considered to be single-term polynomials. A polynomial with two terms is called a **binomial,** and a polynomial with three terms is a **trinomial.** Several types of polynomials are listed below, along with some nonpolynomial expressions for comparison.

Polynomials

$x^3 - 3x^2 + 5x - 1$	$5x^2$	$3m^2 - 5n^2$
4 terms	1 term	2 terms
	Monomial	Binomial

$$7 \qquad \tfrac{1}{2}x^2 - \tfrac{2}{3}xy + \tfrac{1}{4}y^2 \qquad x$$

1 term	3 terms	1 term
Monomial	Trinomial	Monomial

Nonpolynomials

$$\frac{1}{x} + x \qquad 1 + \sqrt{x} \qquad 5x^{-3} + 2x^2 - 5 \qquad \sqrt{x^2 - 2x + 1} \qquad \frac{2x - 1}{3x + 1}$$

As you can see from the examples above:

Polynomials

Polynomials (in one or two variables) are constructed by adding or subtracting numbers and monomials of the form ax^n or $bx^p y^q$, where a and b denote real numbers, x and y denote variables, and the exponents n, p, and q are positive integers.

A monomial of the form ax^n, with $a \neq 0$, is said to have **degree** n. A monomial of the form $bx^p y^q$, with $b \neq 0$, is said to have **degree** $p + q$. In general, the **degree of a nonzero monomial** with one or more variables is defined to be the sum of the exponents of the variables. If a monomial consists of only a nonzero number, its degree is said to be 0. The number 0 itself is not assigned a degree—that is, the degree of 0 is undefined. The **degree of a polynomial** is defined to be the same as that of the nonzero term having the highest degree.

Example 1
(A) $4x^6$ has degree 6   Sixth-degree monomial
(B)   9 has degree 0   Zero-degree monomial
(C) $5x^2 yz^4$ has degree 7   (Not 6. Why?)   Seventh-degree monomial
(D) $6x^3 + 4x^2 - 7x + 9$ has degree 3   Third-degree polynomial
(Note that, from left to right, the terms have degree 3, 2, 1, and 0. The highest degree of any term is 3.)
(E) $5x^2 + 7xy - 3y^2 + 8x - 5y + 8$ is a second-degree polynomial
(Note that the first three terms have degree 2, the next two terms have degree 1, and the last term has degree 0. The highest degree of any term is 2.)

Problem 1   Give the degree of each polynomial.

(A) $3y^4$ (B) x (C) $6x^3 y^4 z$
(D) $8t^4 + 3t^3 - 5t + 9$ (E) $3u^3 - 5u^2 v^3 + 6v^3 - 9u^2 v^2$

The number at the front of each term, including the sign preceding the term, is called the **numerical coefficient,** or simply, the **coefficient** of the term. If a number does not appear, or only a + sign appears, the coefficient is understood to be a 1. If only a $-$ sign appears at the front of a term, the coefficient is understood to be a -1. Given the polynomial

$$5x^4 - x^3 - 6x^2 + x + 8 = 5x^4 + (-x^3) + (-6x^2) + x + 8$$

the coefficient of the first term is 5, the coefficient of the second term is -1, that of the third term is -6, and that of the fourth term is 1.

■ Simplifying Polynomials

Two terms are called **like terms** if they have the same variables with exactly the same exponents. For example, $3x^2y$ and $5x^2y$ are like terms, but $4x^2y$ and $2xy^2$ are not. **Two or more like terms can be combined using addition and subtraction to form a single term.** This is justified by the distributive property, as illustrated in the next example.

Example 2 (A) $5x + 9x \boxed{= (5 + 9)x} = 14x$ Recall that the steps in the dashed boxes are usually done mentally.

(B) $3t - 7t \boxed{= (3 - 7)t} = -4t$

(C) $8xy^2 - 5xy^2 + 7xy^2 \boxed{= (8 - 5 + 7)xy^2} = 10xy^2$

Problem 2 Combine like terms.

(A) $13t + 5t$ (B) $5y - 11y$ (C) $5x^2y^3 + 7x^2y^3 - 10x^2y^3$

From Example 2, you can see that we can combine like terms by simply adding their numerical coefficients.

When a polynomial contains many terms of different types, the commutative and associative properties allow us to rearrange the terms so that like terms are grouped together. The polynomial can then be **simplified** by combining like terms.

Example 3 (A) $3x + 5y + 7x - 9y \boxed{= 3x + 7x + 5y - 9y}$

$$= 10x - 4y$$

(B) $6y^2 - 4y + 7 - 5y^2 + 8y - 10 \boxed{= 6y^2 - 5y^2 - 4y + 8y + 7 - 10}$

$$= y^2 + 4y - 3$$

(C) $9x^2 - 5xy + 3x^2 - 5y^2 + 7xy \boxed{= 9x^2 + 3x^2 - 5xy + 7xy - 5y^2}$

$$= 12x^2 + 2xy - 5y^2$$

Problem 3 Simplify each polynomial by combining like terms.

(A) $7u - 9v - 5u + 3v$ (B) $4t^2 - 8t + 5 - 7t^2 + 3t - 9$
(C) $3x^2y - 5xy^2 + 7xy - xy^2 + 2x^2y$

Using the distributive property, we can multiply any number (or a monomial) times a polynomial, as illustrated in the next example.

Example 4 Multiply.

(A) $5(3t^2 + 4t - 5) = 15t^2 + 20t - 25$

(B) $2x(3x^2 - 2xy + y^2) = 6x^3 - 4x^2y + 2xy^2$

(C) $-3(6x - 3y + 4z) = -18x + 9y - 12z$

(D) $-(x^2 - 3xy + y^2 \boxed{= (-1)(x^2 - 3xy + y^2)}$

$= -x^2 + 3xy - y^2$

Problem 4 Multiply.

(A) $8(5m^3 - 3m - 6)$ (B) $3x^2(2x^2 + 3x - 5)$

(C) $-4(2m - 3n + 4p)$ (D) $-(u^2 + 6uv - 5v^2)$

In Example 4 we were able to remove the parentheses simply by multiplying the expression inside the grouping symbols by the number (or monomial) appearing in front of the symbols of grouping. This process can be easily extended to more complicated situations. For example, the parentheses in the expression

$3(4x - 5y) + 2(2x + 3y)$

can be removed by multiplying $4x - 5y$ by 3 and multiplying $2x + 3y$ by 2 to obtain

$12x - 15y + 4x + 6y$

We can now simplify this expression by combining like terms to obtain

$16x - 9y$

As another example, to simplify

$2(3m + 4n) - 4(2m - 5n)$

we multiply $3m + 4n$ by 2 and $2m - 5n$ by -4 to obtain

$6m + 8n - 8m + 20n$

Then, combining like terms, we have

$-2m + 28n$

This process is described as **removing symbols of grouping and combining like terms.** It can be applied to much more general expressions, as illustrated in Example 5. But before looking at more general examples, we point out a very simple rule:

Rule for Removing Symbols of Grouping

To remove a pair of parentheses (or other symbols of grouping), multiply each term within the parentheses by the number (or monomial) in front of the parentheses.

Example 5

(A) $2(x^2 - 3) - (x^2 + 2x - 3)$ $\boxed{= 2(x^2 - 3) - 1(x^2 + 2x - 3)}$

$= 2x^2 - 6 - x^2 - 2x + 3$

$= x^2 - 2x - 3$

(B) $(2x - 3y) - [x - 2(3x - y)]$

$= (2x - 3y) - [x - 6x + 2y]$ Work from the inside out.

$= 2x - 3y - x + 6x - 2y$

$= 7x - 5y$

(C) $3x - \{5 - 3[x - x(3 - x)]\}$

$= 3x - \{5 - 3[x - 3x + x^2]\}$ Work from the inside out.

$= 3x - \{5 - 3x + 9x - 3x^2\}$

$= 3x - 5 + 3x - 9x + 3x^2$

$= 3x^2 - 3x - 5$

Problem 5   Remove symbols of grouping and simplify.

(A) $3(x^2 + 2x) - (x^2 - 2x + 1)$ (B) $(4x + 2y) - [3x - 5(x - 3y)]$

(C) $2m - \{7 - 2[m - m(4 + m)]\}$

■ Addition and Subtraction of Polynomials

Addition or subtraction of polynomials can be represented by expressions involving symbols of grouping. The desired operation is performed by removing the symbols of grouping and combining like terms. For example, to add the polynomials

$x^2 - 3x + 5$ and $3x^2 + 4x - 6$

we would first write

$(x^2 - 3x + 5) + (3x^2 + 4x - 6)$

Removing the parentheses, we have

$x^2 - 3x + 5 + 3x^2 + 4x - 6$

Then, combining like terms, we obtain the desired sum:

$4x^2 + x - 1$

This procedure is often referred to as the **horizontal method** of adding polynomials. Another procedure, called the **vertical method,** is often preferred when many polynomials are to be added. The vertical method consists of writing one polynomial above the other so that like terms line up. The like terms are then combined vertically to obtain the desired sum. This is illustrated below:

$$
\begin{array}{r}
x^2 - 3x + 5 \\
\underline{3x^2 + 4x - 6} \\
4x^2 + x - 1
\end{array}
$$

To add, write one polynomial above the other so that like terms line up; then combine like terms by adding the coefficients.

Example 6   Add $2x^3 + 3x - 5$, $-x^2 + 5x - 6$, and $3x^3 - 5x^2 + 9$.

Solution   Adding horizontally, we have

$$
\begin{aligned}
&(2x^3 + 3x - 5) + (-x^2 + 5x - 6) + (3x^3 - 5x^2 + 9) \\
&= 2x^3 + 3x - 5 - x^2 + 5x - 6 + 3x^3 - 5x^2 + 9 \\
&= 5x^3 - 6x^2 + 8x - 2
\end{aligned}
$$

Clear parentheses and combine like terms.

Adding vertically, we have

$$
\begin{array}{r}
2x^3 + 3x - 5 \\
- x^2 + 5x - 6 \\
\underline{3x^3 - 5x^2 + 9} \\
5x^3 - 6x^2 + 8x - 2
\end{array}
$$

Leave space where necessary so that like terms line up.

Problem 6   Add $5t^2 - t + 8$, $-3t^3 + 7t - 6$, and $t^3 - 3t^2 - 4$, using both the horizontal and vertical methods.

The subtraction of polynomials can be handled by either of the two methods used for addition.

Example 7   Subtract $3x^2 - 5x + 8$ from $-5x^2 + 3x - 4$.

Solution   Note that the first polynomial is to be subtracted from the second polynomial. Subtracting horizontally, we have

$$
\begin{aligned}
&(-5x^2 + 3x - 4) - (3x^2 - 5x + 8) \\
&= -5x^2 + 3x - 4 - 3x^2 + 5x - 8 \\
&= -8x^2 + 8x - 12
\end{aligned}
$$

Clear parentheses and note sign changes.

Subtracting vertically, we have

$$
\begin{array}{r}
-5x^2 + 3x - 4 \\
- + - \\
\underline{3x^2 - 5x + 8} \\
-8x^2 + 8x - 12
\end{array}
$$

⟵ Change the signs; then combine like terms by adding coefficients.

Problem 7   Subtract $5m^2 - 8mn - 4n^2$ from $9m^2 + 3mn - 7n^2$, using both the horizontal and vertical methods.

■ Multiplication of Polynomials

In Section 1-5 we learned to multiply monomials by applying the rules for exponents. In this section we have multiplied a monomial times a polynomial using the distributive property. Example 8 reviews both processes.

Example 8

(A) $(3m^2n^4)(5mn^3) = 15m^3n^7$

(B) $(-2x^3z^4)(3x^2y^3z) = -6x^5y^3z^5$

(C) $4uv(-u^2 + 3uv + 5v^2) = -4u^3v + 12u^2v^2 + 20uv^3$

(D) $-2t^4(t^3 - 3t^2 + t + 4) = -2t^7 + 6t^6 - 2t^5 - 8t^4$

Problem 8 Multiply.

(A) $(8a^3b^2)(3ab^4)$ (B) $(-5w^3y^3)(-4x^2y^4)$

(C) $3mn^2(4m^2 - mn - 2n^2)$ (D) $-4x^2(-x^3 + 2x^2 - 5x + 2)$

In order to multiply two polynomials we need to utilize the distributive property once again. For example, to find the product of two binomials, say $3x + 5y$ and $2x - 3y$, we have

$(3x + 5y)(2x - 3y) = 3x(2x - 3y) + 5y(2x - 3y)$ — Apply the distributive property.

$= 6x^2 - 9xy + 10xy - 15y^2$ — Apply the distributive property again.

$= 6x^2 + xy - 15y^2$ — Combine like terms.

Note that the product of these two first-degree polynomials is a second-degree polynomial.

Although the distributive property can be used to find the product of polynomials of any length, we can simplify the procedure by noticing that **the product of two polynomials can be obtained by multiplying each term in one polynomial by each term in the other polynomial.** The result is then simplified by combining like terms. For example, we have

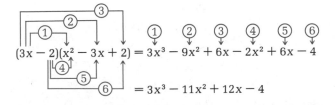

$(3x - 2)(x^2 - 3x + 2) = 3x^3 - 9x^2 + 6x - 2x^2 + 6x - 4$

$= 3x^3 - 11x^2 + 12x - 4$

Note that the product of a first-degree polynomial and a second-degree polynomial is a third-degree polynomial.

Multiplying polynomials as shown above is again referred to as the **horizontal method.** The **vertical method** (which many people prefer) is illustrated next.

$$\begin{array}{l} x^2 - 3x + 2 \\ \underline{3x - 2} \\ 3x^3 - 9x^2 + 6x \end{array}$$ Write one polynomial above the other.

Multiply each term in the top polynomial by $3x$.

$$\underline{-2x^2 + 6x - 4}$$ Multiply each term in the top polynomial by -2 and line up like terms.

$$3x^3 - 11x^2 + 12x - 4$$ Combine like terms by adding coefficients.

Example 9 Multiply: $(2x - 3y)(x^2 - 5xy + 3y^2)$

Solution By the horizontal method, we have

$$(2x - 3y)(x^2 - 5xy + 3y^2) = 2x^3 - 10x^2y + 6xy^2 - 3x^2y + 15xy^2 - 9y^3$$
$$= 2x^3 - 13x^2y + 21xy^2 - 9y^3$$

By the vertical method, we have

$$\begin{array}{l} x^2 - 5xy + 3y^2 \\ \underline{2x - 3y} \\ 2x^3 - 10x^2y + 6xy^2 \\ \underline{-3x^2y + 15xy^2 - 9y^3} \\ 2x^3 - 13x^2y + 21xy^2 - 9y^3 \end{array}$$

Problem 9 Multiply $(4m - 5n)(2m^2 - 3mn - 2n^2)$, using both the horizontal and vertical methods.

▪ Multiplying Binomials

The product of two binomials such as

$$(3x + 5y)(2x - 3y)$$

where the corresponding terms in each binomial are like terms, can be found mentally by following the pattern illustrated below:

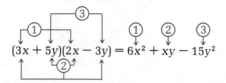

Notice that the steps labeled ① and ③ are obtained by multiplying the first terms and the last terms in each binomial. Step ② involves two products—often called the **inner product** and the **outer product**—which give like terms that can be combined mentally. The ability to multiply binomials mentally will be very important and valuable for future work in this book, so this process should be practiced until it can be done with ease.

Example 10   (A) $(3x - 4)(5x - 2) = 15x^2 - 26x + 8$

(B) $(a + 7b)(3a + 5b) = 3a^2 + 26ab + 35b^2$

Problem 10   Multiply mentally:

(A) $(x - 3)(2x + 4)$ (B) $(3m - 4n)(2m - 5n)$

(C) $(2u + 3v)(5u + 2v)$ (D) $(6a + 5b)(2a + 3b)$

Certain products occur frequently enough to deserve a special note.

Special Products

1. $(a + b)(a - b) = a^2 - b^2$
2. $(a + b)^2 = (a + b)(a + b) = a^2 + 2ab + b^2$
3. $(a - b)^2 = (a - b)(a - b) = a^2 - 2ab + b^2$

Note: In equations 2 and 3, the middle terms on the right are twice the product ab.

Example 11   (A) $(2x - 3)(2x + 3) \boxed{= (2x)^2 - 3^2} = 4x^2 - 9$

(B) $(m + 5n)(m - 5n) \boxed{= m^2 - (5n)^2} = m^2 - 25n^2$

(C) $(2x + 3y)^2 \boxed{= (2x)^2 + 2(2x)(3y) + (3y)^2}$

$= 4x^2 + 12xy + 9y^2$

(D) $(5u - 3v)^2 \boxed{= (5u)^2 - 2(5u)(3v) + (3v)^2}$

$= 25u^2 - 30uv + 9v^2$

Problem 11   Find each product mentally.

(A) $(4w - 5)(4w + 5)$ (B) $(3a + 4b)(3a - 4b)$

(C) $(7u + 3v)^2$ (D) $(2x - 7y)^2$

Answers to Matched Problems

1. (A) 4   (B) 1   (C) 8   (D) 4   (E) 5
2. (A) $18t$ (B) $-6y$ (C) $2x^2y^3$
3. (A) $2u - 6v$ (B) $-3t^2 - 5t - 4$ (C) $5x^2y + 7xy - 6xy^2$
4. (A) $40m^3 - 24m - 48$ (B) $6x^4 + 9x^3 - 15x^2$
   (C) $-8m + 12n - 16p$ (D) $-u^2 - 6uv + 5v^2$
5. (A) $2x^2 + 8x - 1$ (B) $6x - 13y$ (C) $-2m^2 - 4m - 7$
6. $-2t^3 + 2t^2 + 6t - 2$ 7. $4m^2 + 11mn - 3n^2$
8. (A) $24a^4b^6$ (B) $20w^3x^2y^7$
   (C) $12m^3n^2 - 3m^2n^3 - 6mn^4$ (D) $4x^5 - 8x^4 + 20x^3 - 8x^2$
9. $8m^3 - 22m^2n + 7mn^2 + 10n^3$
10. (A) $2x^2 - 2x - 12$ (B) $6m^2 - 23mn + 20n^2$
    (C) $10u^2 + 19uv + 6v^2$ (D) $12a^2 + 28ab + 15b^2$
11. (A) $16w^2 - 25$ (B) $9a^2 - 16b^2$
    (C) $49u^2 + 42uv + 9v^2$ (D) $4x^2 - 28xy + 49y^2$

Exercise 2-1

A   For the polynomial $3x^5 - 2x^4 + x^3 - x^2 + 8x + 3$ indicate:

1. The coefficient of the third term
2. The coefficient of the fourth term
3. The degree of the fifth term
4. The degree of the first term

Perform the indicated operations and simplify.

5. $(4u - 3v) + (7u + 2v)$
6. $(3a - 5b) + 3(2a + b)$
7. $(3m - 7n) - (5m - 2n)$
8. $6(2m - 1) - 4(m - 3)$
9. $(5a^2)(-6a^3)$
10. $(-7x^5)(5x^3)$
11. $(3m^2n)(5mn^4)$
12. $(-3u^4v^2)(-7uv^4)$
13. $2x(5x^2 - 3x + 2)$
14. $-5w(3w^2 - 5w - 6)$
15. $2a - b - 5(3a - 4b)$
16. $3u + v - 6(3u - 2v)$

Add.

17. $3a + 7$ and $9a - 12$
18. $11x - 5$ and $4x - 9$
19. $5x^2 - 3x + 9$, $-4x + 7$, and $8x^2 - 9$
20. $6x^2 - 5x$, $3x - 8$, and $2x^2 - 4x + 7$

Subtract.

21. $7x - 9$ from $12x - 8$
22. $5x - 11$ from $4x + 6$
23. $2t^2 - 3t + 1$ from $5t^2 - 6t - 9$
24. $7x^2 + 3x - 5$ from $3x^2 - 6x + 4$

B   *Perform the indicated operations and simplify.*

25. $5z^2 - 8z - 3 - 9z^2 + 10z - 6$
26. $5x^2 - 3xy + 8y^2 - 7x^2 + 9xy - 5y^2$
27. $(3x - 4y) - (5x + 2y) + (6x + 7y)$
28. $(9a - 7b) + (3a + 5b) - (4a - 8b)$
29. $x - 3(2x - y) + 3y$ 30. $z - 4(2x - z) + 5x$
31. $3x - 2[5x - (3x - 5)]$ 32. $5w - 3[2w - (5w - 3)]$
33. $-3(4m - 2n) + 2(2m + 3n) - 2(5m - 4n)$
34. $5(r - 3s) - 7(2r - 5s) + 3(5r + 4s)$
35. $a(a - 4b) - b(3a - 2b)$ 36. $3m(2m + 5n) - 2n(3m - 2n)$
37. $5u^2v(3u^2 - 2uv + 4v^2)$ 38. $-7a^2b^3(a^2 + 3ab - 5b^2)$
39. $(x - y)(x^2 + xy + y^2)$ 40. $(a + b)(a^2 - ab + b^2)$
41. $(3x + 2y)(4x^2 - xy - 3y^2)$ 42. $(2u - 5v)(u^2 + 3uv - 2v^2)$
43. $(2x^2 + 2x - 1)(3x^2 - 2x + 1)$
44. $(3x^2 - xy + y^2)(x^2 + 2xy - 3y^2)$

Add.

45. $3x^4 - 5x + 7$, $2x^3 + 3x^2 - 9$,   and $6x^3 - 5x^2 + 3$
46. $2x^3 - 10x - 12$, $5x^2 + 7x - 3$,   and $9x^3 - 6x^2 + 3x + 5$

Subtract.

47. $2x^3 - 3x + 4$ from $8x^3 + 5x^2 - 9$
48. $5x^3 + 7x^2 - 9$ from $2x^3 - 7x + 4$

Multiply mentally.

49. $(y - 5)(y + 7)$ 50. $(z + 8)(z + 3)$ 51. $(5x - 1)(2x + 1)$
52. $(6x - 1)(7x - 1)$ 53. $(4r - 3)(5r - 2)$ 54. $(8a - 3)(4a + 5)$
55. $(2x - 1)(x - 3)$ 56. $(5w - 2)(2w + 3)$
57. $(3a - 2b)(3a + 2b)$ 58. $(6x + 5y)(6x - 5y)$
59. $(2x + 3)^2$ 60. $(4y - 3)^2$
61. $(5x - 4y)^2$ 62. $(7u + 2v)^2$

C   *Perform the indicated operations and simplify.*

63. $2(2x - 5y) - 3[2(3x - y) - (x - 5y)]$
64. $5[2(3t - 1) - (5t + 6)] - 2(4t - 3)$
65. $3x - 5\{2x - 4[3 - (5 - 4x)]\} - 2(3x - 1)$
66. $5t^2 - 3\{4t - t[5 - 2(3 - 2t)]\} - 2t(t - 2)$
67. $2x - 3\{x + 2[x - (x + 5)] + 1\}$
68. $u - \{u - [u - (u - 1)]\}$
69. $(2x + 3)(x - 5) - (3x - 1)^2$
70. $(2x + 5)^2 - (2x - 1)(x + 5)$
71. $2\{(x - 3)(x^2 - 2x + 1) - x[3 - x(x - 2)]\}$
72. $-3x\{x[x - x(2 - x)] - (x + 2)(x^2 - 3)\}$
73. $(x + y)^3$ 74. $(x - y)^3$ 75. $(x - y)^4$ 76. $(x + y)^4$

2-2   Factoring Polynomials

- Factoring Out Common Factors
- Factoring by Grouping
- Factoring Second-Degree Trinomials
- The ac Test
- Special Factoring Formulas

To **factor a polynomial** we write it as a product of two or more "simpler" expressions called **factors.** Factoring is often described as the opposite, or reverse, of multiplying polynomials, but it usually requires a little more skill and ingenuity. In this section we will consider factoring as it applies to several types of polynomials. Remember that the ability to factor polynomials is an acquired skill, and it is learned only through lots of practice.

■ Factoring Out Common Factors

If we write the distributive property in the form

$$ab + ac = a(b + c)$$

we can interpret the right side as the **factored form** of the left side. We say that the factor a, which is common to both terms on the left, has been **factored out.** This process, called **factoring out common factors,** may be applied to many polynomials, as illustrated in the next example.

Example 12   Factor out all factors common to each term.

(A) $25x - 35y \;\boxed{= 5 \cdot 5x - 5 \cdot 7y} = 5(5x - 7y)$

(B) $x^2 - x \;\boxed{= x \cdot x - x \cdot 1} = x(x - 1)$

(C) $6x^2 + 2xy - 4x \;\boxed{= 2x \cdot 3x + 2x \cdot y - 2x \cdot 2} = 2x(3x + y - 2)$

(D) $2x^3y - 8x^2y^2 - 6xy^3 \;\boxed{= 2xy \cdot x^2 - 2xy \cdot 4xy - 2xy \cdot 3y^2}$

$$= 2xy(x^2 - 4xy - 3y^2)$$

Problem 12   Factor out all factors common to each term.

(A) $12a - 30b + 24c$ (B) $3m^2 + m$
(C) $8m^2 - 4mn + 6m$ (D) $3u^3v - 6u^2v^2 - 3uv^3$

Let us now consider

$$3x(x - 5) - 4(x - 5)$$

Here, we see that $(x - 5)$ is a common factor of both terms; hence, it may be factored out. Thus,

$$3x(x - 5) - 4(x - 5) = (x - 5)(3x - 4)$$

Example 13   Factor out all factors common to each term.

(A) $2x(3x - 5) - (3x - 5)\; \boxed{= 2x(3x - 5) - 1(3x - 5)}$

$\qquad\qquad\qquad\qquad\quad = (3x - 5)(2x - 1)$

(B) $3x(2x - y) + 5y(2x - y) = (2x - y)(3x + 5y)$

Problem 13   Factor out all factors common to each term.

(A) $3m(4m - 1) + (4m - 1)$ (B) $2x(3x - 2) - 7(3x - 2)$

■ Factoring by Grouping

Many polynomials can be factored by grouping terms in such a way that the result is similar to that in Example 13. This procedure will be particularly useful later in this section where we will discuss a general method for factoring second-degree polynomials. The method of **factoring by grouping** is illustrated below.

$6x^2 - 3x + 8x - 4 = (6x^2 - 3x) + (8x - 4)$ Group the first two terms and the last two terms.

$\qquad\qquad\qquad = 3x(2x - 1) + 4(2x - 1)$ Factor all common factors from each group.

$\qquad\qquad\qquad = (2x - 1)(3x + 4)$ Factor out $(2x - 1)$.

Now study the following example carefully and observe any differences from the example given above:

$4x^2 - 4x - 3x + 3 = (4x^2 - 4x) - (3x - 3)$ Group the first two terms and the last two terms; factor -1 from the last two terms.*

Note signs

$\qquad\qquad\qquad = 4x(x - 1) - 3(x - 1)$ Factor common factors from each group.

$\qquad\qquad\qquad = (x - 1)(4x - 3)$ Factor out $(x - 1)$.

Example 14   Factor by grouping.

(A) $2x^2 + 10x + 3x + 15 = (2x^2 + 10x) + (3x + 15)$

$\qquad\qquad\qquad\qquad\quad = 2x(x + 5) + 3(x + 5)$

$\qquad\qquad\qquad\qquad\quad = (x + 5)(2x + 3)$

---

* Note that when parentheses are inserted and are preceded by a minus sign, then the sign of each term within the parentheses must be changed. This is the same as factoring -1 from all the terms within the parentheses.

Note signs

(B) $6x^2 - 3x - 4x + 2 = (6x^2 - 3x) - (4x - 2)$
$$= 3x(2x - 1) - 2(2x - 1)$$
$$= (2x - 1)(3x - 2)$$

Note signs

(C) $4x^2 + 6xy - 2xy - 3y^2 = (4x^2 + 6xy) - (2xy + 3y^2)$
$$= 2x(2x + 3y) - y(2x + 3y)$$
$$= (2x + 3y)(2x - y)$$

Problem 14 Factor by grouping.

(A) $3x^2 - 9x + 2x - 6$ (B) $10m^2 - 2m - 15m + 3$
(C) $9u^2 + 6uv - 3uv - 2v^2$

■ Factoring Second-Degree Trinomials

We have discussed how to multiply binomials where the product is a trinomial. For example,

$$(x + 5)(x - 3) = x^2 + 2x - 15$$
$$(2x - y)(x + 3y) = 2x^2 + 5xy - 3y^2$$

Our objective now is to start with a trinomial with integer coefficients and to write it as a product of two binomials with integer coefficients. For example, we would like to determine integers a, b, c, and d so that

$$3x^2 - 11x + 6 = (ax + b)(cx + d)$$

With practice, determining the integers a, b, c, and d (if they exist) will become almost automatic for problems like this and others that are not too complicated. We should point out that not every trinomial can be factored in this way. For example,

$$x^2 + x + 1$$

is such a trinomial.

Before we attempt to factor the trinomial $3x^2 - 11x + 6$, let us consider a trinomial with factors that are more easily determined. Consider

$$x^2 + 10x + 21$$

If this trinomial can be factored into a product of binomials with integer coefficients, we must have

x must be the first term
of each binomial. Why?

$$x^2 + 10x + 21 = (x + \quad)(x + \quad) \qquad \text{Both signs must be positive. Why?}$$

?   ?

We must determine factors of 21 which when placed in the blank spaces give us two binomials whose product is the trinomial on the left. The factors of 21 to be considered are:

21
1 · 21
3 · 7

We do not have to consider the products 7 · 3 and 21 · 1 here. Why?

If we insert each pair of factors into the blank spaces (mentally) and multiply the binomials (mentally), we can determine whether one of the pairs of factors produces the desired middle term 10x. Doing this, we find that

$$x^2 + 10x + 21 = (x + 3)(x + 7)$$

Of course, it would be just as correct to write the product on the right as $(x + 7)(x + 3)$, since, by the commutative property, both products are equal.

Now, let us consider the trinomial

$$3x^2 - 11x + 6$$

If this is factorable using integers, then we must have

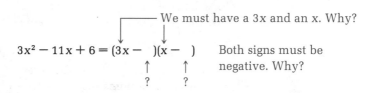

The possible pairs of integers to be tested in the blank spaces are the factors of 6:

6
1 · 6
2 · 3
3 · 2
6 · 1

Notice that we must consider pairs of integers in both orders here. Why?

Inserting each pair of integers in the blank spaces (mentally) and multiplying the binomials on the right side (mentally), we find that

$$3x^2 - 11x + 6 = (3x - 2)(x - 3)$$

As another example, let us try to factor

$$x^2 - x + 3$$

If this is factorable using integers, we must have

$$x^2 - x + 3 = (x - \quad)(x - \quad)$$
$$\qquad\qquad\qquad \uparrow \qquad \uparrow$$
$$\qquad\qquad\qquad ? \qquad ?$$

Testing the factors of 3, we need only consider $1 \cdot 3$. Since this pair of factors does not produce binomials whose product is the trinomial on the left, we conclude that $x^2 - x + 3$ cannot be factored using integers.

In applying the above procedure, we utilize the following rule:

Rule for Factoring Trinomials

First write down as much information as possible about the factored form of the trinomial—whatever can be determined immediately by inspection. Then try to determine the remaining parts of the product (if possible).

Example 15 Factor, if possible, using integer coefficients.

(A) $3x^2 + xy - 2y^2$ (B) $2x^2 + 3x + 4$ (C) $4x^2 + 5xy - 6y^2$

Signs must be opposite
$$\qquad\qquad\qquad\qquad \downarrow \qquad\quad \downarrow$$

Solutions   (A) $3x^2 + xy - 2y^2 = (3x \quad y)(x \quad y)$
$$\qquad\qquad\qquad\qquad\qquad\quad \uparrow \qquad\quad \uparrow$$
$$\qquad\qquad\qquad\qquad\qquad\quad ? \qquad\quad ?$$

We need to test the factors of 2:

2
$1 \cdot 2$
$2 \cdot 1$

Now, we must test each pair of factors using a $+$ and $-$, and then a $-$ and $+$ with each combination. We find that

$$3x^2 + xy - 2y^2 = (3x - 2y)(x + y)$$

(B) $2x^2 + 3x + 4 = (2x + ?)(x + ?)$

4
$1 \cdot 4$
$2 \cdot 2$
$4 \cdot 1$

Testing each pair of factors, we find that no pair will produce the desired middle term $3x$. Thus, $2x^2 + 3x + 4$ cannot be factored using integers.

Signs must be opposite

(C) $4x^2 + 5xy - 6y^2 = (?x \quad ?y)(?x \quad ?y)$

To determine the coefficients of the x and y terms in each binomial (if possible), we need to consider the factors of 4 and 6:

4
$1 \cdot 4$
$2 \cdot 2$
$4 \cdot 1$

6
$1 \cdot 6$
$2 \cdot 3$
$3 \cdot 2$
$6 \cdot 1$

Testing each pair of factors of 4 as x coefficients and each pair of factors of 6 as y coefficients, where each pair of factors of 6 is tested with a $+$ and $-$, and a $-$ and $+$ combination, we eventually find that

$$4x^2 + 5xy - 6y^2 = (x + 2y)(4x - 3y)$$

Problem 15 Factor, if possible, using integer coefficients.

(A) $5x^2 + 9xy - 2y^2$ (B) $3x^2 - x + 4$ (C) $6x^2 + 7xy - 10y^2$

■ The ac Test

In factoring $4x^2 + 5xy - 6y^2$ in Example 15C, we had to consider twenty-four different combinations of coefficients and sign arrangements. Although, with practice, many trinomials can be easily factored using this procedure, it becomes more tedious as the number of possible combinations that must be considered increases. And, in fact, it may turn out that no combination will work! It would be nice to know ahead of time whether a given trinomial with integer coefficients is factorable using integers. Fortunately, there is a test, called the **ac test,** that will tell us whether a given trinomial with integer coefficients is factorable using integers, and if it is factorable, the ac test also will provide a very efficient process for obtaining the factored form.

We will now consider the general problem of factoring trinomials of the form

$$ax^2 + bx + c \qquad \text{and} \qquad ax^2 + bxy + cy^2 \tag{1}$$

where a, b, and c denote integers. If the product ac has integer factors that add up to b, that is, if there are integers p and q such that

$$ac = pq \qquad \text{and} \qquad p + q = b \tag{2}$$

then it can be shown that both polynomials (1) can be factored into binomial factors with integer coefficients. If no integers p and q exist so that conditions (2) are satisfied, then polynomials (1) will not have binomial factors with integer coefficients. Once we find p and q, if they exist, our work is almost finished. We can then write polynomials (1) in the form

$$ax^2 + px + qx + c \quad \text{and} \quad ax^2 + pxy + qxy + cy^2 \tag{3}$$

and factoring can be completed in a couple of steps by the method of grouping. An example will help clarify the process.

Consider the trinomial

$$x^2 - 4x - 12$$

Comparing this with the standard forms in (1), we see that

$$a = 1 \qquad b = -4 \qquad c = -12$$

We want to find factors p and q of

$$ac = 1 \cdot (-12) = -12$$

that add up to b; that is, such that $p + q = -4$. The factors of $ac = -12$ to be considered are:

pq
$(-1) \cdot 12$
$1 \cdot (-12)$
$(-2) \cdot 6$
$2 \cdot (-6)$
$(-3) \cdot 4$
$3 \cdot (-4)$

Checking this list, we see that $p = 2$ and $q = -6$ satisfy

$$2 \cdot (-6) = -12 \qquad pq = ac$$

and

$$2 + (-6) = -4 \qquad p + q = b$$

Thus, $x^2 - 4x - 12$ has binomial factors with integer coefficients. Now, we write $x^2 - 4x - 12$ in the form shown in equations (3) and proceed as in factoring by grouping:

$$
\begin{aligned}
x^2 - 4x - 12 &= x^2 \overset{p}{+ 2x} \overset{q}{- 6x} - 12 \\
&= (x^2 + 2x) - (6x + 12) \\
&= x(x + 2) - 6(x + 2) \qquad \text{$(x + 2)$ is a common factor and} \\
&= (x + 2)(x - 6) \qquad\qquad \text{can be factored out.}
\end{aligned}
$$

Note that if we let $p = -6$ and $q = 2$ instead of $p = 2$ and $q = -6$, we obtain the same result:

$$
\begin{aligned}
x^2 - 4x - 12 &= x^2 \overset{p}{-\ 6x} + \overset{q}{2x} - 12 \\
&= (x^2 - 6x) + (2x - 12) \\
&= x(x - 6) + 2(x - 6) \\
&= (x - 6)(x + 2)
\end{aligned}
$$

This procedure is summarized in the box for easy reference.

The *ac* Test

Given the trinomials

$$ax^2 + bx + c \qquad \text{and} \qquad ax^2 + bxy + cy^2 \qquad\qquad (1)$$

if there exist integers p and q satisfying

$$pq = ac \qquad \text{and} \qquad p + q = b \qquad\qquad (2)$$

then trinomials (1) can be written in the form

$$ax^2 + px + qx + c \qquad \text{and} \qquad ax^2 + pxy + qxy + cy^2 \qquad\qquad (3)$$

and factored by the method of grouping.

Example 16   Using the *ac* test, factor (if possible) using integer coefficients.

(A) $x^2 - 10xy - 24y^2$ (B) $x^2 - 3x + 4$ (C) $6x^2 + 5xy - 4y^2$

Solutions   (A) $x^2 - 10xy - 24y^2$
We have $a = 1$, $b = -10$, and $c = -24$. Thus,

$$ac = 1 \cdot (-24) = -24$$

By checking the factors of -24, we find that $p = -12$ and $q = 2$ satisfy

$$pq = (-12) \cdot 2 = -24 = ac$$

and

$$p + q = (-12) + 2 = -10 = b$$

Thus,

$$
\begin{aligned}
x^2 - 10xy - 24y^2 &= x^2 - 12xy + 2xy - 24y^2 \\
&= x(x - 12y) + 2y(x - 12y) \\
&= (x - 12y)(x + 2y)
\end{aligned}
$$

(B) $x^2 - 3x + 4$

We have $a = 1$, $b = -3$, and $c = 4$. Thus,

$$ac = 1 \cdot 4 = 4$$

The factors of 4 are:

pq
$1 \cdot 4$
$(-1) \cdot (-4)$
$2 \cdot 2$
$(-2) \cdot (-2)$

Since no pair of these factors add up to $b = -3$, the trinomial $x^2 - 3x + 4$ cannot be factored using integer coefficients.

(C) $6x^2 + 5xy - 4y^2$

We have $a = 6$, $b = 5$, and $c = -4$. Thus,

$$ac = 6 \cdot (-4) = -24$$

We find that $p = -3$ and $q = 8$ satisfy

$$pq = (-3) \cdot 8 = -24 = ac$$

and

$$p + q = (-3) + 8 = 5 = b$$

Thus,

$$\begin{aligned}
6x^2 + 5xy - 4y^2 &= 6x^2 - 3xy + 8xy - 4y^2 \\
&= 3x(2x - y) + 4y(2x - y) \\
&= (2x - y)(3x + 4y)
\end{aligned}$$

Problem 16   Using the ac test, factor (if possible) using integer coefficients.

(A) $x^2 + 3x - 18$ (B) $x^2 - 2xy + 12y^2$ (C) $8x^2 + 6xy - 9y^2$

■ Special Factoring Formulas

From the special products listed in Section 2-1, we have the corresponding factored forms:

1. $a^2 - b^2 = (a + b)(a - b)$
2. $a^2 + 2ab + b^2 = (a + b)^2$
3. $a^2 - 2ab + b^2 = (a - b)^2$

The first formula is referred to as the **difference of two squares** and occurs frequently in algebra. The next two formulas are referred to as **perfect squares.** It should be noted that the sum of two squares,

$$a^2 + b^2$$

cannot be factored using integer coefficients unless a and b have common integer factors. Try it to see why.

Example 17   Factor, if possible, using integer coefficients.

(A) $x^2 - 25 = (x + 5)(x - 5)$
(B) $9m^2 - 49n^2 = (3m + 7n)(3m - 7n)$
(C) $9x^2 + 30xy + 25y^2 = (3x + 5y)^2$
(D) $4u^2 - 12uv + 9v^2 = (2u - 3v)^2$
(E) $4m^2 + n^2$ does not factor

Problem 17   Factor, if possible, using integer coefficients.

(A) $r^2 - t^2$ (B) $x^2 + 4y^2$ (C) $16m^2 - 24mn + 9n^2$
(D) $4x^2 + 20xy + 25y^2$ (E) $16t^2 - 81u^2$

We will now consider some examples that combine the techniques discussed above for factoring polynomials. In general, we first factor out factors that are common to all terms. Then we apply the techniques outlined above.

Example 18   Factor as far as possible using integer coefficients.

(A) $4x^3 - 14x^2 + 6x$ (B) $18x^3 - 8x$ (C) $x^4 - 81$
(D) $3xy^3 - 15xy^2 - 6xy$

Solutions   (A) $4x^3 - 14x^2 + 6x = 2x(2x^2 - 7x + 3)$ Factor out the common factor $2x$ first.

We can now apply the ac test to

$$2x^2 - 7x + 3$$

We have $a = 2$, $b = -7$, and $c = 3$. Thus,

$$ac = 2 \cdot 3 = 6$$

Then, $p = -1$ and $q = -6$ satsify

$$pq = (-1) \cdot (-6) = 6 = ac$$

and

$$p + q = (-1) + (-6) = -7 = b$$

Thus,

$$2x^2 - 7x + 3 = 2x^2 - x - 6x + 3$$
$$= x(2x - 1) - 3(2x - 1)$$
$$= (2x - 1)(x - 3)$$

Therefore, we have

$$4x^3 - 14x^2 + 6x = 2x(2x - 1)(x - 3)$$

(B) $18x^3 - 8x = 2x(9x^2 - 4)$
$$= 2x(3x + 2)(3x - 2)$$

(C) $x^4 - 81 = (x^2 + 9)(x^2 - 9)$
$$= (x^2 + 9)(x + 3)(x - 3)$$

(D) $3xy^3 - 15xy^2 - 6xy = 3xy(y^2 - 5y - 2)$ Cannot be factored further using integer coefficients

Problem 18 Factor as far as possible using integer coefficients.

(A) $8x^3y + 20x^2y^2 - 12xy^3$ (B) $3x^3 - 48x$
(C) $4x^3y - 12x^2y^2 + 9xy^3$ (D) $2x^2y + 14xy - 8y$

Answers to Matched Problems

12. (A) $6(2a - 5b + 4c)$ (B) $m(3m + 1)$
    (C) $2m(4m - 2n + 3)$ (D) $3uv(u^2 - 2uv - v^2)$
13. (A) $(4m - 1)(3m + 1)$ (B) $(3x - 2)(2x - 7)$
14. (A) $(x - 3)(3x + 2)$ (B) $(5m - 1)(2m - 3)$
    (C) $(3u + 2v)(3u - v)$
15. (A) $(5x - y)(x + 2y)$ (B) Cannot be factored using integers.
    (C) $(x + 2y)(6x - 5y)$
16. (A) $(x + 6)(x - 3)$ (B) Cannot be factored using integer
    (C) $(4x - 3y)(2x + 3y)$ coefficients.
17. (A) $(r + t)(r - t)$ (B) Does not factor.
    (C) $(4m - 3n)^2$ (D) $(2x + 5y)^2$
    (E) $(4t + 9u)(4t - 9u)$
18. (A) $4xy(2x - y)(x + 3y)$ (B) $3x(x + 4)(x - 4)$
    (C) $xy(2x - 3y)^2$ (D) $2y(x^2 + 7x - 4)$

Exercise 2-2

A *Factor out all factors common to each term.*

1. $6a^2 - 8a$ 2. $21m^2 + 14m$
3. $7u^3v^2 - 14u^2v^3$ 4. $22x^3y + 11xy^3$

5. $x(x + 2) + 5(x + 2)$

6. $y(y - 4) - 3(y - 4)$

7. $3a(a - 5) - 2(a - 5)$

8. $2w(w + 6) + 3(w + 6)$

9. $5y(4y - 3) - (4y - 3)$

10. $3z(2z - 1) + (2z - 1)$

11. $6x^3 - 9x^2 + 15x$

12. $8y^4 + 12y^3 - 16y^2$

13. $8u^3v + 6u^2v - 14uv$

14. $25a^4b - 30a^3b + 10a^2b$

B   *Replace each question mark with an expression so that both sides are equal.*

15. $5x^2 - 5x + 3x - 3 = (5x^2 - 5x) + (?)$

16. $6x^2 + 3x + 4x + 2 = (6x^2 + 3x) + (?)$

17. $2x^2 - 8x - x + 4 = (2x^2 - 8x) - (?)$

18. $3t^2 - 9t - t + 3 = (3t^2 - 9t) - (?)$

Factor out all factors common to the terms within the parentheses and complete the factoring if possible.

19. $(5x^2 - 5x) + (3x - 3)$

20. $(6x^2 + 3x) + (4x + 2)$

21. $(2x^2 - 8x) - (x - 4)$

22. $(3t^2 - 9t) - (t - 3)$

Factor by grouping.

23. $5x^2 - 5x + 3x - 3$

24. $6x^2 + 3x + 4x + 2$

25. $2x^2 - 8x - x + 4$

26. $3t^2 - 9t - t + 3$

27. $4x^2 - 2xy - 6xy + 3y^2$

28. $6a^2 - 2ab + 3ab - b^2$

29. $2u^2 - 3uv - 4uv + 6v^2$

30. $4r^2 - rs - 12rs + 3s^2$

Factor, if possible, using integer coefficients.

31. $x^2 + 5xy + 6y^2$

32. $x^2 - 7xy + 10y^2$

33. $u^2 - 7u + 12$

34. $w^2 - 9w + 14$

35. $x^2 + 5x + 5$

36. $y^2 - 3y + 4$

37. $u^2 + 3uv - 10v^2$

38. $r^2 - 3rs - 4s^2$

39. $x^2 + xy + y^2$

40. $u^2 + 3uv + 4v^2$

41. $x^2 - 9$

42. $y^2 - 25$

43. $a^2 + 4ab + 4b^2$

44. $x^2 - 6xy + 9y^2$

45. $x^2 - xy - 20y^2$

46. $x^2 - 3xy - 18y^2$

47. $2x^2 - 7x + 6$

48. $3x^2 - 5x - 2$

49. $6x^2 - 13xy + 6y^2$

50. $2x^2 - 7xy + 3y^2$

51. $2x^2 + x + 1$

52. $3x^2 + 5x + 3$

53. $25u^2 + 5u - 6$

54. $16v^2 + 8v - 15$

55. $4u^2 + 4uv - 3v^2$

56. $3a^2 + ab - 4b^2$

57. $2u^2 - 3uv + 4v^2$

58. $3r^2 + 4rs + 5s^2$

59. $25w^2 - 4$

60. $9u^2 - 16v^2$

61. $25u^2 - 30uv + 9v^2$

62. $9m^2 + 12mn + 4n^2$

Factor as far as possible using integer coefficients.

63. $x^3 - 9x$

64. $y^4 - 25y^2$

65. $10w^3 - 100w^2 + 250w$

66. $24u^3 + 24u^2 + 6u$

67. $6x^3 + 6x^2 - 72x$

68. $7y^3 + 21y^2 - 70y$

C   *Factor as far as possible using integer coefficients.*

69. $5u^3v + 5u^2v^2 - 30uv^3$ **70.** $16a^3b + 40a^2b^2 - 24ab^3$

71. $3m^4n + 6m^3n^2 - 9m^2n^3$ **72.** $24x^3y^3 - 4x^2y^4 - 4xy^5$

Use the following factoring formulas for the sum and difference of two cubes to factor the binomials given in Problems 73–76:

$$A^3 + B^3 = (A + B)(A^2 - AB + B^2)$$
$$A^3 - B^3 = (A - B)(A^2 + AB + B^2)$$

73. $x^3 - 8$ **74.** $x^3 + 1$ **75.** $x^3 + 27$ **76.** $8y^3 - 1$

2-3   Multiplying and Dividing Fractions

- The Fundamental Principle of Fractions
- Reducing to Lowest Terms
- Raising to Higher Terms
- Multiplication
- Division

In this section we will apply our knowledge about polynomials to algebraic fractions formed using quotients of polynomials. We will discuss the fundamental principle of fractions and its consequences, and then describe the basic operations of multiplication and division.

■ The Fundamental Principle of Fractions

A fractional form that has polynomials as both numerator and denominator is called a **rational expression.** For example,

$$\frac{1}{x} \qquad \frac{-7}{z + 3} \qquad \frac{u - 2}{u^2 - 3u + 5} \qquad \frac{x^2 - 5xy + y^2}{6x^2y^3}$$

are all rational expressions. As long as the denominator is never 0, each rational expression defines a real number whenever real numbers are substituted for the variables. Because of this, all the properties of real numbers apply to these expressions.

In arithmetic we find that the value of a fraction is unchanged if we multiply (or divide) both the numerator and the denominator by the same nonzero number. For instance,

$$\frac{3}{4} = \frac{3 \cdot 5}{4 \cdot 5} = \frac{15}{20} \quad \text{and} \quad \frac{28}{35} = \frac{28 \div 7}{35 \div 7} = \frac{4}{5}$$

The first example illustrates **raising a fraction to higher terms,** where we have multiplied both the numerator and the denominator by 5. The second example illustrates **reducing a fraction to lower terms,** where we have divided the numerator and the denominator by 7. The second example is often described by saying that *the common factor 7 is canceled* from both the numerator and the denominator. This may be represented in the following ways:

$$\frac{28}{35} = \frac{\overset{1}{4 \cdot \cancel{7}}}{\underset{1}{5 \cdot \cancel{7}}} = \frac{4}{5} \quad \text{or} \quad \frac{28}{35} = \frac{\overset{4}{\cancel{28}}}{\underset{5}{\cancel{35}}} = \frac{4}{5}$$

The above examples illustrate the *fundamental principle of fractions,* which we shall now generalize to rational expressions.

Fundamental Principle of Fractions

If P, Q, and K represent polynomials with Q and K not equal to 0, then

$$\frac{PK}{QK} = \frac{P}{Q}$$

In words, the fundamental principle states that we may multiply the numerator and the denominator of a rational form by a nonzero polynomial, or divide the numerator and the denominator by a nonzero polynomial, and the new rational form will be equivalent to the original.* This principle is the basis of all canceling used to reduce fractional forms to lower terms (canceling common factors from the numerator and the denominator). And, when it is read from right to left, this principle is used to raise rational forms to higher terms (multiplying the numerator and the denominator by the same nonzero polynomial form). The latter operation is fundamental to the process of addition and subtraction of rational forms.

■ Reducing to Lowest Terms

A rational form is said to be in lowest terms if the numerator and the denominator do not have any common factors other than 1. To *reduce a*

---

* Two rational forms are equivalent if they represent the same number for all replacements of the variable(s) by real numbers, as long as we avoid division and multiplication by 0.

rational form to lowest terms means to cancel all common factors from the numerator and the denominator. We have the following rule:

> **Reducing Rational Expressions**
>
> To reduce a rational expression to lowest terms, factor the numerator and the denominator as far as possible; then cancel all factors common to both the numerator and the denominator.

Example 19   Reduce to lowest terms.

(A) $\dfrac{25x^3y^8}{35x^5y^6} = \dfrac{\overset{5}{\cancel{25}} \cdot \overset{1}{\cancel{x^3}} \cdot \overset{y^2}{\cancel{y^8}}}{\underset{7}{\cancel{35}} \cdot \underset{x^2}{\cancel{x^5}} \cdot \underset{1}{\cancel{y^6}}}$ Cancel common factors from the numerator and the denominator.

$= \dfrac{5y^2}{7x^2}$

or

$\dfrac{25x^3y^8}{35x^5y^6} = \dfrac{\overset{1}{\cancel{(5x^3y^6)}}(5y^2)}{\underset{1}{\cancel{(5x^3y^6)}}(7x^2)} = \dfrac{5y^2}{7x^2}$

(B) $\dfrac{8m^2 - 6m}{2m} = \dfrac{2m(4m - 3)}{\underset{1}{\cancel{2m}}}$ Factor the numerator and cancel common factors.

$= 4m - 3$

(C) $\dfrac{4u^2 - 9v^2}{4u^2 + 12uv + 9v^2} = \dfrac{\overset{1}{\cancel{(2u + 3v)}}(2u - 3v)}{\underset{1}{\cancel{(2u + 3v)}}(2u + 3v)}$ Factor the numerator and the denominator and cancel common factors.

$= \dfrac{2u - 3v}{2u + 3v}$

(D) $\dfrac{2x^3 + 2x^2 - 24x}{4x^3 - 12x^2} = \dfrac{2x(x^2 + x - 12)}{4x^2(x - 3)}$ Factor the numerator and the denominator as far as possible and cancel common factors.

$= \dfrac{\overset{1}{\cancel{2x}}(x + 4)\overset{1}{\cancel{(x - 3)}}}{\underset{2x}{\cancel{4x^2}}\underset{1}{\cancel{(x - 3)}}}$

$= \dfrac{x + 4}{2x}$

Problem 19   Reduce to lowest terms.

(A) $\dfrac{24m^7n^5}{60m^3n^8}$ (B) $\dfrac{10y}{25y^2 + 15y}$

(C) $\dfrac{9x^2 - 6xy + y^2}{9x^2 - y^2}$ (D) $\dfrac{16x^3 - 8x^2}{8x^3 + 20x^2 - 12x}$

▪ Raising to Higher Terms

To *raise a rational expression to higher terms* means to multiply the numerator and the denominator by the same nonzero polynomial expression. This process will be very important when we study addition and subtraction of rational forms in Section 2-4. The following example illustrates how the fundamental principle of fractions is used to raise fractions to higher terms.

Example 20

(A) $\dfrac{5}{3x} = \dfrac{(5xy)(5)}{(5xy)(3x)} = \dfrac{25xy}{15x^2y}$

(B) $\dfrac{2x}{x - 3} = \dfrac{(x + 3)(2x)}{(x + 3)(x - 3)} = \dfrac{2x^2 + 6x}{x^2 - 9}$

(C) $\dfrac{m - n}{m + n} = \dfrac{(m - 2n)(m - n)}{(m - 2n)(m + n)} = \dfrac{m^2 - 3mn + 2n^2}{m^2 - mn - 2n^2}$

(D) $\dfrac{2y}{x(x + 2)} = \dfrac{2y}{x(x + 2)} \cdot \dfrac{2x(y - 2)}{2x(y - 2)} = \dfrac{4xy(y - 2)}{2x^2(x + 2)(y - 2)}$

Problem 20   Raise to higher terms by finding the expression that should replace the question mark.

(A) $\dfrac{2m}{3} = \dfrac{?}{30m^2n}$ (B) $\dfrac{3x}{x + 5} = \dfrac{?}{(x + 5)(x - 5)}$

(C) $\dfrac{u + v}{u - v} = \dfrac{?}{u^2 - 3uv + 2v^2}$ (D) $\dfrac{3x}{4(x - 3)} = \dfrac{?}{8x(x - 3)(x + 2)}$

▪ Multiplication

The rule for multiplying rational expressions is the same as that for multiplying rational numbers:

Multiplication

If P, Q, R, and S represent polynomials with Q and S not equal to 0, then

$$\frac{P}{Q} \cdot \frac{R}{S} = \frac{P \cdot R}{Q \cdot S}$$

To illustrate this, we have

$$\frac{3x^3y^5}{5u^2v^6} \cdot \frac{10u^5v^3}{9xy^7} = \frac{(3x^3y^5) \cdot (10u^5v^3)}{(5u^2v^6) \cdot (9xy^7)}$$

Multiply numerators and denominators.

$$= \frac{30u^5v^3x^3y^5}{45u^2v^6xy^7}$$

Reduce to lowest terms.

$$= \frac{2u^3x^2}{3v^3y^2}$$

This process can be simplified by first canceling any factors that are common to the numerators and the denominators:

$$\frac{3x^3y^5}{5u^2v^6} \cdot \frac{10u^5v^3}{9xy^7} = \frac{\overset{1}{\cancel{3}} \cdot \overset{x^2}{\cancel{x^3}} \cdot \overset{1}{\cancel{y^5}}}{\underset{1}{\cancel{5}} \cdot \underset{1}{\cancel{u^2}} \cdot \underset{v^3}{\cancel{v^6}}} \cdot \frac{\overset{2}{\cancel{10}} \cdot \overset{u^3}{\cancel{u^5}} \cdot \overset{1}{\cancel{v^3}}}{\underset{3}{\cancel{9}} \cdot \underset{1}{\cancel{x}} \cdot \underset{y^2}{\cancel{y^7}}}$$

$$= \frac{2u^3x^2}{3v^3y^2}$$

As you can see, canceling common factors before multiplying two rational expressions can greatly simplify the process. We can state this as a rule:

Multiplying Rational Expressions

To multiply two (or more) rational expressions, factor all numerators and denominators completely, cancel factors that are common to the numerators and the denominators, and then multiply the resulting expressions. (The answer should automatically be reduced to lowest terms.)

Example 21 (A) $\dfrac{6m - 9n}{4m - 2n} \cdot \dfrac{6m + 24n}{18m - 27n}$

Factor numerators and denominators, and cancel common factors.

$$= \frac{\overset{1}{\cancel{3}}(2m - 3n)}{\underset{1}{\cancel{2}}(2m - n)} \cdot \frac{\overset{1}{\cancel{6}}(m + 4n)}{\underset{\cancel{3}}{\cancel{9}}(2m - 3n)}$$

Write the answer.

$$= \frac{m + 4n}{2m - n}$$

(B) $(x^2 - 4) \cdot \dfrac{2x - 3}{x + 2}$ Factor where possible and cancel common factors.

$$= \dfrac{\overset{1}{\cancel{(x + 2)}}(x - 2)}{1} \cdot \dfrac{(2x - 3)}{\underset{1}{\cancel{(x + 2)}}}$$ Write the answer.

$$= (x - 2)(2x - 3) \quad \text{or} \quad 2x^2 - 7x + 6$$

(C) $\dfrac{6x^2y}{x^2y + xy^2} \cdot \dfrac{x^2 + 2xy + y^2}{3x^2 - 3xy}$ Factor all numerators and denominators, and cancel common factors.

$$= \dfrac{6x^2y}{xy(x + y)} \cdot \dfrac{(x + y)(x + y)}{3x(x - y)}$$

$$= \dfrac{2(x + y)}{x - y}$$

Problem 21   Multiply and reduce to lowest terms.

(A) $\dfrac{2u + 6v}{40u - 10v} \cdot \dfrac{20u - 5v}{8u - 20v}$ (B) $\dfrac{x + 5}{x^2 - 9} \cdot (x + 3)$

(C) $\dfrac{2a^2b + 4ab^2}{a^2 + 4ab + 4b^2} \cdot \dfrac{a^2 - 4b^2}{6a^2b^3}$

▪ Division

The process of dividing one rational form by another is accomplished by converting the given division problem into an equivalent multiplication problem:

Division

If P, Q, R, and S represent polynomials with Q, R, and S not equal to 0, then

$$\dfrac{P}{Q} \div \dfrac{R}{S} \qquad = \qquad \dfrac{P}{Q} \cdot \dfrac{S}{R}$$

Invert divisor and multiply

Example 22   (A) $\dfrac{14a^3b^2}{9c^2d^4} \div \dfrac{7ab^4}{27c^5d^2} = \dfrac{14a^3b^2}{9c^2d^4} \cdot \dfrac{27c^5d^2}{7ab^4} = \dfrac{6a^2c^3}{b^2d^2}$

(B) $\quad (x+4) \div \dfrac{2x^2-32}{6xy} = \dfrac{x+4}{1} \cdot \dfrac{6xy}{2(x+4)(x-4)} = \dfrac{3xy}{x-4}$

(C) $\quad \dfrac{x^2-9y^2}{x^2-6xy+9y^2} \div \dfrac{2x^2+6xy}{6x^2y} = \dfrac{(x-3y)(x+3y)}{(x-3y)(x-3y)} \cdot \dfrac{6x^2y}{2x(x+3y)}$

$$= \dfrac{3xy}{x-3y}$$

Problem 22 Divide and reduce to lowest terms.

(A) $\quad \dfrac{25x^5y^2}{9w^6z} \div \dfrac{15xy^6}{27w^3z^4}$ 　　　　(B) $\quad \dfrac{2x^2-8}{4x} \div (x+2)$

(C) $\quad \dfrac{x^2-4x+4}{4x^2y-8xy} \div \dfrac{x^2+x-6}{6x^2+18x}$

Answers to Matched Problems

19. (A) $\dfrac{2m^4}{5n^3}$ 　(B) $\dfrac{2}{5y+3}$ 　(C) $\dfrac{3x-y}{3x+y}$ 　(D) $\dfrac{2x}{x+3}$

20. (A) $(2m)(10m^2n)$ or $20m^3n$ 　　(B) $(3x)(x-5)$ or $3x^2-15x$
    (C) $(u+v)(u-2v)$ or $u^2-uv-2v^2$ 　(D) $6x^2(x+2)$ or $6x^3+12x^2$

21. (A) $\dfrac{u+3v}{4(2u-5v)}$ 　(B) $\dfrac{x+5}{x-3}$ 　(C) $\dfrac{a-2b}{3ab^2}$

22. (A) $\dfrac{5x^4z^3}{w^3y^4}$ 　(B) $\dfrac{x-2}{2x}$ 　(C) $\dfrac{3}{2y}$

Exercise 2-3

A 　*Reduce to lowest terms.*

1. $\dfrac{18x^5y^3}{24x^4y^6}$ 　　2. $\dfrac{16u^4v^3}{12u^6v}$ 　　3. $\dfrac{x^3(x+3)^2}{x^5(x+3)}$ 　　4. $\dfrac{z^5(z-3)}{z^2(z-3)^2}$

5. $\dfrac{35u^3(2u-1)^2}{42u^7(2u-1)}$ 　6. $\dfrac{18t^4(3t-2)}{12t^2(3t-2)^2}$ 　7. $\dfrac{10x}{5x^2-15x}$ 　8. $\dfrac{36m^2-27m}{18m}$

Raise to higher terms by finding the expression that should replace the question mark.

9. $\dfrac{7x}{5} = \dfrac{?}{25x^2y}$ 　10. $\dfrac{2}{3y} = \dfrac{?}{18x^2y^2}$ 　11. $\dfrac{5x}{y} = \dfrac{15x^3y^2}{?}$ 　12. $\dfrac{4u}{7v} = \dfrac{20u^3v}{?}$

Perform the indicated operations and reduce to lowest terms.

13. $\dfrac{2u^2}{5v^2} \cdot \dfrac{20v}{8u}$ 　　　14. $\dfrac{6r}{20s} \cdot \dfrac{15s^2}{9r^2}$ 　　　15. $\dfrac{7uv}{8w^2} \div \dfrac{21v^2}{24uw}$

16. $\dfrac{36a^2}{15bc} \div \dfrac{6a}{10c^2}$

17. $5x \div \dfrac{3y}{5x}$

18. $\dfrac{4x}{7y} \div 7y$

19. $\dfrac{25x^3y}{12xz^2} \cdot \dfrac{16yz}{20xy^2}$

20. $\dfrac{28a^4b}{8ac^2} \cdot \dfrac{6c^3d}{21a^2d^2}$

21. $\dfrac{6w^3}{27x^2} \div \dfrac{4w^4y^2}{18x^3y}$

22. $\dfrac{12u^2v}{8u^4w} \div \dfrac{9v^2w^2}{24uw^5}$

23. $\dfrac{18a^4}{12b^3} \div \dfrac{12a^2}{-15b}$

24. $\dfrac{-18u^4}{21v^2} \div \dfrac{27u^6}{28v^3}$

B   Reduce to lowest terms.

25. $\dfrac{4x^2 + 4x + 1}{4x^2 + 2x}$

26. $\dfrac{15x^2 - 5x}{9x^2 - 6x + 1}$

27. $\dfrac{x^2 - 3x}{x^3 - 9x}$

28. $\dfrac{y^3 - 25y}{y^2 + 5y}$

29. $\dfrac{u^2v - uv^2}{u^2 - uv}$

30. $\dfrac{5x^2 + 20x}{x^2 + 2x - 8}$

31. $\dfrac{x^2 - 25y^2}{x^2 - 10xy + 25y^2}$

32. $\dfrac{u^2 - 8uv + 16v^2}{u^2 - 16v^2}$

33. $\dfrac{x^2 - 3x + 4xy - 12y}{3x^3 + 12x^2y}$

34. $\dfrac{r^2 + rs - 5r - 5s}{4r^2 + 8rs + 4s^2}$

35. $\dfrac{6x^4 - 24x^3 + 18x^2}{4x^4 - 12x^3}$

36. $\dfrac{15u^3 - 60u^2}{10u^3 - 30u^2 - 40u}$

Raise to higher terms by finding the expression that should replace the question mark.

37. $\dfrac{3x}{x - 4} = \dfrac{?}{x^2 - 7x + 12}$

38. $\dfrac{5}{x + 2} = \dfrac{5x - 15}{?}$

39. $\dfrac{x - 2y}{x + 2y} = \dfrac{x^2 - 4y^2}{?}$

40. $\dfrac{2u + v}{u - v} = \dfrac{?}{u^2 + 2uv - 3v^2}$

41. $\dfrac{3z}{x(x - 2)} = \dfrac{?}{3x^2(x - 2)^2}$

42. $\dfrac{5w}{2y(y + 3)} = \dfrac{?}{4y^3(y + 3)(y - 2)}$

Perform the indicated operations and reduce to lowest terms.

43. $\dfrac{4a - 12}{24b^2} \cdot \dfrac{12b^3}{6a - 18}$

44. $\dfrac{25x^3}{5x - 5y} \cdot \dfrac{12x - 12y}{15x^4}$

45. $\dfrac{2x - 4}{5z^4} \div \dfrac{15x - 30}{25z^6}$

46. $\dfrac{7m^2}{6m - 18} \div \dfrac{21m^3}{8m - 24}$

47. $\dfrac{x^2 - 6x + 8}{6y^4} \cdot \dfrac{18y^2}{x - 2}$

48. $\dfrac{8u^3}{v^2 + 2v - 15} \cdot \dfrac{v + 5}{4u}$

49. $\dfrac{x^2 - 7x + 10}{5x^2 - 10x} \cdot \dfrac{20x^2}{x^2 - 10x + 25}$

50. $\dfrac{6a^2 + 18a}{a^2 - 9} \cdot \dfrac{a^2 - 6a + 9}{18a^4}$

51. $\dfrac{x^2 + 8x + 16}{x^2 - 16} \div \dfrac{4x + 16}{x^2 - 8x + 16}$

52. $\dfrac{8u - 16}{u^2 + 3u - 10} \div \dfrac{u^2 + 6u + 9}{u^2 + 8u + 15}$

53. $\dfrac{a+b}{a^2-ab} \cdot \dfrac{a^2-2ab+b^2}{a^2-b^2}$

54. $\dfrac{x^2+5xy+6y^2}{x^2-xy-12y^2} \cdot \dfrac{x^2-6xy+8y^2}{x^3+2x^2y}$

55. $\dfrac{y^2+y-6}{y^2-7y+10} \cdot \dfrac{y^2-3y-10}{y^2+7y+12}$

56. $\dfrac{n^2-2n-8}{n^2+5n+6} \cdot \dfrac{n^2-7n+10}{n^2-9n+20}$

57. $\dfrac{z^2-12z+36}{6z^2-36z} \div (z^2-7z+6)$

58. $(x^2-9x+20) \div \dfrac{x^2-25}{3x+15}$

59. $\dfrac{4x^2-4xy+y^2}{4x^2-y^2} \div \dfrac{9x^2-y^2}{6x^2+5xy+y^2}$

60. $\dfrac{4u^2-12uv+9v^2}{2u^2-5uv+3v^2} \div \dfrac{4u^2-8uv+3v^2}{2u^2+uv-v^2}$

61. $\dfrac{-25b^2}{27a^3c^6} \cdot \dfrac{-35a^5c^2}{-15b^3} \cdot \dfrac{18c^3}{12ab^2}$

62. $\dfrac{-14x^6z^5}{60y^4} \cdot \dfrac{12y^2}{-21x^5z^7} \cdot \dfrac{-20x^2y^3}{-8z^3}$

63. $\left(\dfrac{3x}{y^5} \div \dfrac{6x^2}{y^2}\right) \cdot \dfrac{4y^2}{x}$

64. $\dfrac{3x}{y^5} \div \left(\dfrac{y}{6x^2} \cdot \dfrac{x}{4y^2}\right)$

C *Reduce to lowest terms.*

65. $\dfrac{4m^4n-28m^3n^2+48m^2n^3}{6m^2n^2+12mn^3-90n^4}$

66. $\dfrac{8a^4+8a^3b-48a^2b^2}{6a^3b+42a^2b^2+72ab^3}$

67. $\dfrac{x^2-3x-xy+3y}{2x^2-6x+xy-3y}$

68. $\dfrac{a^2-2ab+2a-4b}{a^2-2ab-3a+6b}$

Perform the indicated operations and reduce to lowest terms.

69. $\dfrac{x^2-2x+xy-2y}{x^2-3x-xy+3y} \cdot \dfrac{x^2-3x+2xy-6y}{x^2+2xy-2x-4y}$

70. $\dfrac{a^2-2ab+2a-4b}{a^2+2a-3ab-6b} \div \dfrac{a^2+4a-5ab-20b}{a^2-3ab+4a-12b}$

71. $\dfrac{x^2-xy}{xy+y^2} \div \left(\dfrac{x^2-y^2}{x^2+2xy+y^2} \div \dfrac{x^2-2xy+y^2}{x^2y+xy^2}\right)$

72. $\left(\dfrac{x^2-xy}{xy+y^2} \div \dfrac{x^2-y^2}{x^2+2xy+y^2}\right) \div \dfrac{x^2-2xy+y^2}{x^2y+xy^2}$

2-4 Adding and Subtracting Fractions

- Common Denominators
- Least Common Denominator (LCD)
- Addition and Subtraction

In Section 2-3 we discussed the multiplication and division of rational forms. Now, we will discuss the addition and subtraction of rational expressions. Just as multiplication and division are based on the corresponding properties of real numbers, the addition and subtraction of rational forms will be based on the corresponding properties of real numbers.

■ Common Denominators

When two rational forms have exactly the same denominator, we say that they have a **common denominator.** The addition or subtraction of two rational expressions with a common denominator is performed according to the rules given in the box.

Addition and Subtraction (Common Denominators)

If P, Q, and D represent polynomials with D not equal to 0, then

$$\frac{P}{D} + \frac{Q}{D} = \frac{P+Q}{D} \tag{1}$$

$$\frac{P}{D} - \frac{Q}{D} = \frac{P-Q}{D} \tag{2}$$

Thus, when common denominators are present, the addition or subtraction of rational forms is obtained simply by adding or subtracting the numerators, and then placing this result over the common denominator. The resulting fractional form should then be reduced to lowest terms whenever possible.

Example 23

(A) $\dfrac{2x}{6xyz} + \dfrac{3y}{6xyz} - \dfrac{6z}{6xyz} = \dfrac{2x + 3y - 6z}{6xyz}$

(B) $\dfrac{2x}{x+4} + \dfrac{x+5}{x+4} = \dfrac{2x + (x+5)}{x+4}$

$$= \frac{2x + x + 5}{x + 4}$$

$$= \frac{3x + 5}{x + 4}$$

(C) $\dfrac{7x+6}{2x-1} - \dfrac{3x+8}{2x-1} = \dfrac{(7x+6)-(3x+8)}{2x-1}$

$= \dfrac{7x+6-3x-8}{2x-1}$

$= \dfrac{4x-2}{2x-1}$ Factor the numerator and reduce.

$= \dfrac{2(2x-1)}{2x-1} = 2$

(D) $\dfrac{4x+5}{3x-2} - \dfrac{7x+3}{3x-2} = \dfrac{(4x+5)-(7x+3)}{3x-2}$

$= \dfrac{4x+5-7x-3}{3x-2}$

$= \dfrac{-3x+2}{3x-2}$ Factor -1 from the numerator and reduce.

$= \dfrac{(-1)(3x-2)}{3x-2} = -1$

Problem 23 Combine into a single fraction and reduce to lowest terms whenever possible.

(A) $\dfrac{5u}{20uvw} - \dfrac{2v}{20uvw} + \dfrac{20w}{20uvw}$ (B) $\dfrac{5x-2}{2x+1} + \dfrac{6}{2x+1}$

(C) $\dfrac{3x-2}{4x-3} - \dfrac{4-5x}{4x-3}$ (D) $\dfrac{m+9}{2m-5} - \dfrac{7m-6}{2m-5}$

■ Least Common Denominator (LCD)

In order to add or subtract two rational forms that do not have common denominators, we must first express each rational form in terms of a common denominator. Although any common denominator would do, the process of addition and subtraction is greatly simplified if we use the *least common denominator*. The **least common denominator (LCD)** of two or more rational expressions is the polynomial of lowest degree that is exactly divisible by each of the denominators in the original expressions. The following procedure may be used to determine the LCD:

How to Determine the Least Common Denominator (LCD)

1. Factor each denominator completely, including the numerical coefficients.
2. Form the LCD by selecting as factors each different factor that occurs in the denominators, and include each such factor to the highest power that it occurs in any one denominator.

Once we have determined the LCD of two or more rational expressions, we use the fundamental principle of fractions in the form

$$\frac{P}{Q} = \frac{P \cdot K}{Q \cdot K} \qquad Q, K \neq 0$$

to express each fraction in terms of the LCD.

Example 24   Find the LCD for each group of rational expressions, and express each fraction in terms of the LCD.

(A) $\dfrac{x}{6y^2}, \quad \dfrac{5}{12xy}, \quad \dfrac{z}{9y^3}$

To determine the LCD, we factor each denominator completely:

$$6y^2 = 2 \cdot 3y^2 \qquad 12xy = 2^2 \cdot 3xy \qquad 9y^3 = 3^2y^3$$

The LCD must contain the factors 2 (twice), 3 (twice), x (once), and y (three times). Thus, the LCD is

$$\text{LCD} = 2^2 \cdot 3^2xy^3 = 36xy^3$$

We now use the fundamental principle of fractions to express each fraction in terms of the least common denominator $36xy^3$:

$$\frac{x}{6y^2} = \frac{x \cdot 6xy}{6y^2 \cdot 6xy} = \frac{6x^2y}{36xy^3} \qquad \text{Multiply numerator and denominator by } 6xy.$$

$$\frac{5}{12xy} = \frac{5 \cdot 3y^2}{12xy \cdot 3y^2} = \frac{15y^2}{36xy^3} \qquad \text{Multiply numerator and denominator by } 3y^2.$$

$$\frac{z}{9y^3} = \frac{z \cdot 4x}{9y^3 \cdot 4x} = \frac{4xz}{36xy^3} \qquad \text{Multiply numerator and denominator by } 4x.$$

(B) $\dfrac{x-3}{x^2-1}, \quad \dfrac{x+5}{x^2+2x+1}$

Factoring each denominator completely, we have

$$x^2 - 1 = (x-1)(x+1) \qquad x^2 + 2x + 1 = (x+1)^2$$

The LCD must contain the factors $x - 1$ (once) and $x + 1$ (twice). Thus,

$$\text{LCD} = (x-1)(x+1)^2$$

Using the fundamental principle of fractions, we have

$$\frac{x-3}{x^2-1} = \frac{(x-3) \cdot (x+1)}{(x-1)(x+1) \cdot (x+1)} = \frac{x^2-2x-3}{(x-1)(x+1)^2}$$

$$\frac{x+5}{x^2+2x+1} = \frac{(x+5) \cdot (x-1)}{(x+1)^2 \cdot (x-1)} = \frac{x^2+4x-5}{(x-1)(x+1)^2}$$

Notice that we have left the LCD in factored form.

(C) $\dfrac{2}{3m^2 + 12m + 12}, \quad \dfrac{1}{4m^3 - 16m}$

Factoring each denominator completely, we have

$$3m^2 + 12m + 12 = 3(m^2 + 4m + 4) = 3(m + 2)^2$$

$$4m^3 - 16m = 4m(m^2 - 4) = 2^2 m(m - 2)(m + 2)$$

Thus, the LCD is

$$\text{LCD} = 2^2 \cdot 3m(m - 2)(m + 2)^2$$
$$= 12m(m - 2)(m + 2)^2$$

Except for the numerical coefficient, it is common practice to leave the LCD in factored form.

Using the fundamental principle of fractions, we have

$$\frac{2}{3m^2 + 12m + 12} = \frac{2 \cdot [4m(m - 2)]}{3(m + 2)^2 \cdot [4m(m - 2)]}$$

$$= \frac{8m(m - 2)}{12m(m - 2)(m + 2)^2}$$

$$= \frac{8m^2 - 16m}{12m(m - 2)(m + 2)^2}$$

$$\frac{1}{4m^3 - 16m} = \frac{1 \cdot [3(m + 2)]}{4m(m - 2)(m + 2) \cdot [3(m + 2)]}$$

$$= \frac{3m + 6}{12m(m - 2)(m + 2)^2}$$

Problem 24 Find the LCD for each group of rational expressions, and express each fraction in terms of the LCD.

(A) $\dfrac{3v}{4u^3}, \quad \dfrac{5}{6uv}, \quad \dfrac{u}{8v^2}$

(B) $\dfrac{m + 3}{m^2 - 4m + 4}, \quad \dfrac{m - 5}{m^2 - 4}$

(C) $\dfrac{1}{3x^2 - 18x + 27}, \quad \dfrac{3}{4x^3 - 36x}$

■ Addition and Subtraction

We will now illustrate the procedure for adding and subtracting rational expressions with several examples and problems. In most cases, we will not include all the steps for finding the LCD or expressing the given fractions in terms of the LCD.

Example 25 $\dfrac{x}{6y^2} - \dfrac{5}{12xy} + \dfrac{z}{9y^3} = \dfrac{x \cdot 6xy}{6y^2 \cdot 6xy} - \dfrac{5 \cdot 3y^2}{12xy \cdot 3y^2} + \dfrac{z \cdot 4x}{9y^3 \cdot 4x}$ $\text{LCD} = 36xy^3$

$$= \frac{6x^2y}{36xy^3} - \frac{15y^2}{36xy^3} + \frac{4xz}{36xy^3}$$

$$= \frac{6x^2y - 15y^2 + 4xz}{36xy^3}$$

Problem 25   Combine into a single fraction and reduce to lowest terms:

$$\frac{3v}{4u^3} + \frac{5}{6uv} - \frac{u}{8v^2}$$

Example 26 $\dfrac{4}{x-2} + \dfrac{3}{x+5}$ $\quad$ LCD $= (x-2)(x+5)$

$$= \frac{4 \cdot (x+5)}{(x-2) \cdot (x+5)} + \frac{3 \cdot (x-2)}{(x+5) \cdot (x-2)}$$

$$= \frac{(4x+20) + (3x-6)}{(x-2)(x+5)}$$

$$= \frac{4x+20+3x-6}{(x-2)(x+5)}$$

$$= \frac{7x+14}{(x-2)(x+5)}$$

Problem 26   Combine into a single fraction and reduce to lowest terms:

$$\frac{5}{x+3} - \frac{2}{x-4}$$

Example 27 $\dfrac{x+2}{x^2-5x-6} - \dfrac{x-3}{x^2+5x+4}$

$$= \frac{x+2}{(x+1)(x-6)} - \frac{x-3}{(x+1)(x+4)} \quad \text{LCD} = (x+1)(x+4)(x-6)$$

$$= \frac{(x+2) \cdot (x+4)}{(x+1)(x-6) \cdot (x+4)} - \frac{(x-3) \cdot (x-6)}{(x+1)(x+4) \cdot (x-6)}$$

$$= \frac{(x^2+6x+8) - (x^2-9x+18)}{(x+1)(x+4)(x-6)}$$

$$= \frac{x^2+6x+8-x^2+9x-18}{(x+1)(x+4)(x-6)}$$

$$= \frac{15x-10}{(x+1)(x+4)(x-6)}$$

Problem 27   Combine into a single fraction and reduce to lowest terms:

$$\frac{x+4}{x^2-5x+6} + \frac{x+2}{x^2+x-12}$$

Example 28

$$\frac{x-1}{5x^2-125}+\frac{2}{3x^2-30x+75}$$

$$=\frac{x-1}{5(x+5)(x-5)}+\frac{2}{3(x-5)^2} \qquad \text{LCD}=15(x+5)(x-5)^2$$

$$=\frac{(x-1)\cdot[3(x-5)]}{5(x+5)(x-5)\cdot[3(x-5)]}+\frac{2\cdot[5(x+5)]}{3(x-5)^2\cdot[5(x+5)]}$$

$$=\frac{3(x-1)(x-5)}{15(x+5)(x-5)^2}+\frac{10(x+5)}{15(x+5)(x-5)^2}$$

$$=\frac{3(x^2-6x+5)+10(x+5)}{15(x+5)(x-5)^2}$$

$$=\frac{3x^2-18x+15+10x+50}{15(x+5)(x-5)^2}=\frac{3x^2-8x+65}{15(x+5)(x-5)^2}$$

Problem 28 Combine into a single fraction and reduce to lowest terms:

$$\frac{x}{4x^2-36}-\frac{x-1}{3x^2+18x+27}$$

Example 29

$$2x-1-\frac{2}{x+1} \left| \begin{array}{l} =\dfrac{2x-1}{1}-\dfrac{2}{x+1} \end{array} \right| \qquad \text{LCD}=x+1$$

$$=\frac{(2x-1)\cdot(x+1)}{1\cdot(x+1)}-\frac{2}{x+1}$$

$$=\frac{2x^2+x-1-2}{x+1}=\frac{2x^2+x-3}{x+1}$$

Problem 29 Combine into a single fraction and reduce to lowest terms: $\dfrac{x-3}{2x+1}-x$

Answers to Matched Problems

23. (A) $\dfrac{5u-2v+20w}{20uvw}$ (B) $\dfrac{5x+4}{2x+1}$ (C) 2 (D) -3

24. (A) LCD $=24u^3v^2$; $\dfrac{18v^3}{24u^3v^2},\ \dfrac{20u^2v}{24u^3v^2},\ \dfrac{3u^4}{24u^3v^2}$

(B) LCD $=(m+2)(m-2)^2$; $\dfrac{m^2+5m+6}{(m+2)(m-2)^2},\ \dfrac{m^2-7m+10}{(m+2)(m-2)^2}$

(C) LCD $=12x(x+3)(x-3)^2$; $\dfrac{4x^2+12x}{12x(x+3)(x-3)^2},\ \dfrac{9x-27}{12x(x+3)(x-3)^2}$

25. $\dfrac{18v^3+20u^2v-3u^4}{24u^3v^2}$ 26. $\dfrac{3x-26}{(x+3)(x-4)}$

27. $\dfrac{2x^2+8x+12}{(x-2)(x-3)(x+4)}$ 28. $\dfrac{-x^2+25x-12}{12(x-3)(x+3)^2}$ 29. $\dfrac{-2x^2-3}{2x+1}$

Exercise 2-4

A   *Combine into a single fraction and reduce to lowest terms.*

1. $\dfrac{3m}{10pq} + \dfrac{2m}{10pq}$

2. $\dfrac{5y}{7x^2} - \dfrac{8y}{7x^2}$

3. $\dfrac{3x - 1}{4y} - \dfrac{2x - 3}{4y}$

4. $\dfrac{7}{3m} - \dfrac{5 - 3m}{3m}$

5. $\dfrac{5x + 6}{2x + 5} - \dfrac{x - 4}{2x + 5}$

6. $\dfrac{11u - 6}{4u - 3} - \dfrac{3 - u}{4u - 3}$

7. $\dfrac{z}{z^2 - 25} - \dfrac{5}{z^2 - 25}$

8. $\dfrac{u}{u^2 - 16} + \dfrac{4}{u^2 - 16}$

Find the LCD for each group of rational expressions.

9. $\dfrac{2a}{3c}, \dfrac{b}{4d}$

10. $\dfrac{m}{6p}, \dfrac{2n}{15q}$

11. $\dfrac{3}{2x}, \dfrac{5}{3x^2}, \dfrac{2}{9}$

12. $\dfrac{4}{3y}, \dfrac{5}{9y^2}, \dfrac{3}{4}$

13. $\dfrac{4}{x - 1}, \dfrac{3}{x + 1}$

14. $\dfrac{6}{z + 3}, \dfrac{2}{z - 2}$

15. $\dfrac{5}{3y}, \dfrac{2}{y + 2}$

16. $\dfrac{4}{u - 5}, \dfrac{3}{5u}$

Combine into a single fraction and reduce to lowest terms.

17. $\dfrac{u}{3v} - \dfrac{2w}{9v}$

18. $\dfrac{3x}{5y} + \dfrac{z}{2y}$

19. $7 - \dfrac{2}{y}$

20. $\dfrac{5}{x} + 3$

21. $\dfrac{2a}{3c} - \dfrac{b}{4d}$

22. $\dfrac{m}{6p} + \dfrac{2n}{15q}$

23. $\dfrac{5}{9y^2} + \dfrac{4}{3y} - \dfrac{3}{4}$

24. $\dfrac{2}{9} - \dfrac{3}{2x} + \dfrac{5}{3x^2}$

25. $\dfrac{6}{z + 3} + \dfrac{2}{z - 2}$

26. $\dfrac{4}{x - 1} - \dfrac{3}{x + 1}$

27. $\dfrac{4}{u - 5} - \dfrac{3}{5u}$

28. $\dfrac{5}{3y} + \dfrac{2}{y + 2}$

B   *Find the LCD for each group of rational expressions.*

29. $\dfrac{7}{8m^2n}, \dfrac{5}{6mn^3}, \dfrac{2}{3mn}$

30. $\dfrac{5}{4r^4s}, \dfrac{1}{6rs^3}, \dfrac{2}{r^2s^2}$

31. $\dfrac{4}{3n - 6}, \dfrac{2}{n^2 - 3n + 2}$

32. $\dfrac{2}{x^2 - 5x + 6}, \dfrac{3}{5x - 15}$

33. $\dfrac{3}{x^2 - 9}, \dfrac{2}{x^2 - 6x + 9}$

34. $\dfrac{5}{x^2 - 4x + 4}, \dfrac{3}{x^2 - 4}$

35. $\dfrac{1}{2m^2 - 4m}, \dfrac{2}{m^2 - 4m + 4}, \dfrac{5}{4m^2}$

36. $\dfrac{1}{n^2 - n - 2}, \ \dfrac{2}{3n^2 + 3n}, \ \dfrac{4}{7n^2}$

Combine into a single fraction and reduce to lowest terms.

37. $\dfrac{8}{3uv} - \dfrac{5}{9v^3} + \dfrac{3}{4u^2}$

38. $\dfrac{3}{5x^3} + \dfrac{5}{6xy} - \dfrac{3}{10y^2}$

39. $\dfrac{5}{4r^2s} + \dfrac{1}{6rs^3} + \dfrac{2}{r^2s^2}$

40. $\dfrac{7}{8m^2n} - \dfrac{5}{6mn^3} - \dfrac{2}{3mn}$

41. $\dfrac{2}{x^2 - 5x + 6} - \dfrac{3}{5x - 15}$

42. $\dfrac{4}{3n - 6} + \dfrac{2}{n^2 - 3n + 2}$

43. $\dfrac{5}{x^2 - 4x + 4} + \dfrac{3}{x^2 - 4}$

44. $\dfrac{3}{x^2 - 9} - \dfrac{2}{x^2 - 6x + 9}$

45. $\dfrac{1}{2m^2 - 4m} + \dfrac{2}{m^2 - 4m + 4} - \dfrac{5}{4m^2}$

46. $\dfrac{1}{n^2 - n - 2} - \dfrac{2}{3n^2 + 3n} + \dfrac{4}{7n^2}$

47. $3x - 2 - \dfrac{x}{x + 3}$

48. $\dfrac{4x}{x - 5} + 2x - 1$

49. $\dfrac{3}{x - 4} - \dfrac{1}{x + 4} - \dfrac{2x}{x^2 - 16}$

50. $\dfrac{3x}{x - y} - \dfrac{4y}{x + y} + \dfrac{3xy}{x^2 - y^2}$

51. $\dfrac{1}{2x - 1} - \dfrac{2}{2x + 1} + \dfrac{x}{4x^2 - 1}$

52. $\dfrac{3x}{6x^2 + 5x + 1} - \dfrac{1}{3x + 1} + \dfrac{2}{2x + 1}$

C *Combine into a single fraction and reduce to lowest terms.*

53. $\dfrac{1}{x^2 - y^2} + \dfrac{1}{x^2 + 2xy + y^2} + \dfrac{1}{x^2 - 2xy + y^2}$

54. $\dfrac{1}{2u^2 + 2u} + \dfrac{1}{6u^2} - \dfrac{1}{4u^2 + 8u + 4}$

55. $\dfrac{1}{x - 3} + \dfrac{1}{x + 4} + \dfrac{1}{x - 2}$

56. $\dfrac{1}{x + 2} - \dfrac{1}{x - 1} + \dfrac{1}{x + 3}$

57. $\dfrac{1}{x^2 - 5x + 6} - \dfrac{2}{x^2 - x - 6} + \dfrac{1}{x^2 - 4}$

58. $\dfrac{2}{x^2 - 9} + \dfrac{1}{x^2 + x - 12} - \dfrac{1}{x^2 + 7x + 12}$

2-5 Chapter Review

Important Terms and Symbols

2-1 *Basic operations on polynomials.* monomial, binomial, trinomial, polynomial, term, degree of a monomial, degree of a polynomial, coefficient, like terms, symbols of grouping, combining like terms, addition, subtraction, horizontal method, vertical method, multiplication, mental multiplication, inner product, outer product, special products, $(a + b)(a - b) = a^2 - b^2$, $(a + b)^2 = a^2 + 2ab + b^2$, $(a - b)^2 = a^2 - 2ab + b^2$

2-2 *Factoring polynomials.* factor, factored form, common factors, factoring by grouping, factorable trinomials, ac test, difference of two squares, perfect squares, combined factoring

2-3 *Multiplying and dividing fractions.* rational expressions, raising to higher terms, reducing to lower terms, fundamental principle of fractions, lowest terms, multiplication, canceling common factors, division, $\dfrac{PK}{QK} = \dfrac{P}{Q}$, $Q, K \neq 0$

2-4 *Adding and subtracting fractions.* common denominators, addition and subtraction with common denominators, least common denominator (LCD)

Exercise 2-5 Chapter Review

Work through all the problems in this chapter review and check your answers in the back of the book. (Answers to all review problems are there.) Where weaknesses show up, review appropriate sections in the text. When you are satisfied that you know the material, take the practice test following this review.

A

1. Add: $4x^2 - 5x + 3$, $4x - 5$, $3x^2 - 2$
2. Subtract: $7x^2 - 3x + 6$ from $4x^2 - 3x - 5$

Perform the indicated operations and simplify.

3. $(5r + 3s) - (2r - s)$
4. $2(u - 5) - 3(2u - 7)$
5. $(3xy^5)(-5x^2y^3)$
6. $2t(t^2 - 2t + 3)$

Factor out all factors common to each term.

7. $14u^3v - 7uv^2$

8. $12x^4 - 24x^3 + 6x^2$

9. $7u^4v - 21u^3v^2 + 35u^2v^3$

10. $2x(3x - 5) + 3(3x - 5)$

11. $5a(2a + 3) - (2a + 3)$

Reduce to lowest terms.

12. $\dfrac{24a^3b^7}{16a^5b^2}$

13. $\dfrac{15y^2 - 10y}{20y}$

Perform the indicated operations and reduce to lowest terms.

14. $\dfrac{9a^3b}{15ac^2} \cdot \dfrac{35b^2c^4}{21a^2b^6}$

15. $\dfrac{16u^3v^4}{9vw^3} \div \dfrac{24u^2v^3}{15uw^2}$

16. $\dfrac{5}{6} - \dfrac{7}{9z} + \dfrac{3}{2z^2}$

17. $\dfrac{5}{x - 3} - \dfrac{3}{4x}$

B   *Perform the indicated operations and simplify.*

18. $4a(2a + 3b) - 3b(a - 4b)$

19. $3r^2s(4r^2 - 3rs + 7s^2)$

20. $2x(3x^2 - 2x + 1) - 3(3x^2 - 2x + 1)$

21. $(2y - 3)(y^2 - 3y + 1)$

22. $(x^2 - 3x + 5)(2x^2 + x - 2)$

23. $5t - 4[2t - 3(2t - 2)]$

24. $4\{2(a - 2b) - [3(a - 4b) - 4(2a - 3b)]\}$

25. $3(2u - 4v) - 2[3(u - 3v) - 4(3u - 2v)]$

In Problems 26–31 multiply mentally.

26. $(5x - 1)(x - 3)$

27. $(2u + 3v)(4u + v)$

28. $(3r - 5s)(r + 3s)$

29. $(7x - 3y)(7x + 3y)$

30. $(3m + 4n)^2$

31. $(2a - b)^2$

32. Add: $3x^3 + 4x^2 - 3$, $\ 2x^2 - 4x + 2$, $\ 2x^3 - 5x - 9$

33. Subtract: $5x^3 - 4x + 5$ from $3x^3 + 7x^2 - 8$

Factor by grouping.

34. $5z^2 - 10z + 2z - 4$

35. $6w^2 + 4w - 9w - 6$

36. $6u^2 - 3uv - 2uv + v^2$

37. $2a^2 - 4ab + ab - 2b^2$

Factor as far as possible using integer coefficients.

38. $x^2 + x - 12$

39. $x^2 - 3x + 4$

40. $a^2 + 2ab - 24b^2$

41. $4x^2 - 16x + 15$

42. $5a^2 + 34ab - 7b^2$

43. $25x^2 - 10xy + y^2$

44. $w^4 - 81w^2$

45. $u^4 - 10u^3 + 25u^2$

46. $5y^3 + 5y^2 - 60y$

47. $7m^3n + 14m^2n^2 - 56mn^3$

Reduce to lowest terms.

48. $\dfrac{z^2 - 4z}{z^2 - 16}$

49. $\dfrac{x^2 + 2x - 8}{4x^2 - 8x}$

50. $\dfrac{16a^2 - b^2}{16a^2 + 8ab + b^2}$

51. $\dfrac{8y^4 - 8y^3 - 48y^2}{6y^4 - 18y^3}$

Perform the indicated operations and reduce to lowest terms.

52. $\dfrac{5a + 10}{30a^2} \cdot \dfrac{12a^3}{8a + 16}$

53. $\dfrac{3x - 15}{18x^3} \div \dfrac{2x - 10}{24x}$

54. $\dfrac{x^2 + 6x + 9}{x^2 + 8x + 15} \cdot \dfrac{x^2 + 3x - 10}{7x - 14}$

55. $\dfrac{u^2 - 8u + 16}{5u + 20} \div \dfrac{u^2 - 16}{u^2 + 8u + 16}$

56. $\dfrac{y - 1}{y + 1} - \dfrac{y - 2}{y - 3}$

57. $\dfrac{2}{z^2 - 2z - 8} + \dfrac{4}{5z + 10}$

58. $\dfrac{3}{u^2 + 2uv + v^2} - \dfrac{1}{u^2 - v^2}$

59. $3x - 1 - \dfrac{4x}{2x - 3}$

60. $(3x - 1)(x + 2) - (2x - 3)^2$

61. $(x - 3)^2 - (x + 2)(x - 3)$

C   *Factor as far as possible using integer coefficients.*

62. $3m^4n + 6m^3n^2 - 9m^2n^3$

63. $16x^4 - 72x^2y^2 + 81y^4$

Reduce to lowest terms.

64. $\dfrac{5x^4 + 10x^3y - 40x^2y^2}{10x^3 - 70x^2y + 100xy^2}$

65. $\dfrac{x^2 - 2x + xy - 2y}{x^2 + 3x + xy + 3y}$

Perform the indicated operations and reduce to lowest terms.

66. $\dfrac{a^2 - 4ab + 4b^2}{a^2 - 4b^2} \cdot \dfrac{a^2 - ab - 6b^2}{a^2 - 7ab + 12b^2}$

67. $\dfrac{u^2 - 2uv - 3u + 6v}{u^2 - 2uv + 2u - 4v} \div \dfrac{u^2 + u - 12}{u^2 - 3u - 10}$

68. $\dfrac{1}{x + 2} - \dfrac{1}{x + 1} + \dfrac{1}{x - 3}$

69. $\dfrac{3}{x^2 - 9} + \dfrac{4}{x^2 - 5x + 6} - \dfrac{2}{x^2 + x - 6}$

Practice Test: Chapter 2

1. Add: $3x^2 - 4x - 5$, $4x + 8$, $5x^2 + 2x$, $4x^3 - 3x^2 + 2x - 4$
2. Subtract: $5x^2 - 3xy - 2y^2$ from the product $(2x - 3y)(3x + y)$
3. Multiply: $(2u - 3v)(3u^2 - 2uv + v^2)$

Factor as far as possible using integer coefficients.

4. $4x^2 + 13xy - 12y^2$

5. $10m^4 - 25m^3 - 15m^2$

6. $6a^2 - 2ab - 9a + 3b$

Reduce to lowest terms.

7. $\dfrac{5x^2 + 20x}{5x^2 + 5x - 60}$

8. $\dfrac{4u^2 - 9v^2}{4u^2 - 12uv + 9v^2}$

Perform the indicated operations and simplify.

9. $2\{(3x - 4y) - 2[2(x - 3y) - (3x - 2y)]\}$

10. $\dfrac{5ab + 15b}{30b^4} \cdot \dfrac{24b^2}{6a + 18}$

11. $\dfrac{x^2 - 8x + 15}{3x^2 - 9x} \div \dfrac{x^2 - x - 20}{27x^3}$

12. $2x - 3y - \dfrac{6xy}{x + 2y}$

13. $\dfrac{3}{m^3 + 3m^2 - 10m} - \dfrac{2}{4m^3 - 8m^2}$

Exponents and Radicals

3

CHAPTER 3	Contents

In Section 1-5 we introduced the use of positive integer exponents. In this chapter we will extend this concept to negative integer exponents, zero exponents, and rational (fractional) exponents. As an application of integer exponents, we will discuss *scientific notation*, which is a useful way of representing certain real numbers. Through fractional exponents we will introduce the concept of *radical*, and then we will discuss how to manipulate and simplify radical expressions.

3-1   Integer Exponents

- Review of Positive Integer Exponents
- Zero Exponents
- Negative Integer Exponents
- Common Errors
- Applications

In this section we will extend the concept of exponent to include all integer values. This will be done in such a way that the rules of exponents for positive integer exponents (Section 1-5) will still hold. We will begin by summarizing key results from Section 1-5.

■ Review of Positive Integer Exponents

Recall that if a represents a real number and n is a positive integer, then

$$a^n = a \cdot a \cdot a \cdot \cdots \cdot a \qquad n \text{ factors of } a$$

Also, recall the five properties of exponents listed below.

Properties of Exponents

If a and b represent real numbers and m and n denote positive integers, then:

1. $a^m a^n = a^{m+n}$
2. $(a^n)^m = a^{mn}$
3. $(ab)^m = a^m b^m$
4. $\left(\dfrac{a}{b}\right)^m = \dfrac{a^m}{b^m} \qquad b \neq 0$
5. $\dfrac{a^m}{a^n} = \begin{cases} a^{m-n} & \text{if } m \text{ is larger than } n \\ 1 & \text{is } m = n \\ \dfrac{1}{a^{n-m}} & \text{if } n \text{ is larger than } m \end{cases} \qquad a \neq 0$

We now wish to give meaning to expressions such as

$$5^0 \qquad 8^{-3} \qquad a^0 \qquad b^{-5}$$

in such a way that all the properties of exponents will continue to hold.

■ Zero Exponents

For a real number a, $a \neq 0$, what meaning should be assigned to a^0? If the properties of positive integer exponents are to hold for all integer exponents, then, according to the first property, we must have

$$a^4 \cdot a^0 = a^{4+0} = a^4$$

This is valid only if $a^0 = 1$. It turns out that defining a^0, $a \neq 0$, to be equal to 1 is compatible with all the properties of exponents. For example, we would have

$$\frac{a^4}{a^4} = a^{4-4} = a^0 = 1$$

Note that we must have $a \neq 0$ for a^0 to be defined; thus, it is meaningless to write 0^0.

Zero Exponents

$$a^0 = 1 \qquad a \neq 0$$
0^0 is not defined

Whenever we write x^0, it will be understood that $x \neq 0$, even if this fact is not explicitly stated. Thus, we can replace x^0 with 1 wherever it occurs.

Example 1

(A) $(297)^0 = 1$

(B) $(-73)^0 = 1$

(C) $(\frac{5}{8})^0 = 1$

(D) $(42 \cdot 73 \cdot 109 \cdot 506)^0 = 1$

(E) $z^0 = 1, \quad z \neq 0$

(F) $(x^4 y^7)^0 = 1, \quad x \neq 0, \quad y \neq 0$

Problem 1

Give the value of each expression.

(A) 0^0

(B) $(-749)^0$

(C) $(\frac{43}{17})^0$

(D) $(243 + 597 + 842)^0$

(E) $u^0, \quad u \neq 0$

(F) $(m^5 n^5)^0, \quad m \neq 0, \quad n \neq 0$

■ Negative Integer Exponents

Now let us turn our attention to negative integer exponents and consider a^{-4}. Again, if the meaning of this expression is to be compatible with the properties of exponents, we must have

$$a^{-4} \cdot a^4 = a^{-4+4} = a^0 = 1$$

This is true if a^{-4} is the reciprocal of a^4; that is, if

$$a^{-4} = \frac{1}{a^4}$$

Notice that again we must have $a \neq 0$ for this expression to be defined. The meaning of negative integer exponents can be stated as follows:

Negative Integer Exponents

If $a \neq 0$ and n is a positive integer, then

$$a^{-n} = \frac{1}{a^n}$$

We also have

$$\frac{1}{a^{-n}} = a^n$$

Whenever we write an expression such as x^{-5}, it will be understood that $x \neq 0$.

Example 2

(A) $2^{-6} \boxed{= \frac{1}{2^6}} = \frac{1}{64}$

(B) $(-4)^{-3} \boxed{= \frac{1}{(-4)^3} = \frac{1}{-64}} = -\frac{1}{64}$

(C) $\left(\dfrac{2}{5}\right)^{-2} \boxed{= \dfrac{1}{\left(\frac{2}{5}\right)^2} = \dfrac{1}{\frac{4}{25}}} = \dfrac{25}{4}$ (D) $\dfrac{1}{5^{-3}} \boxed{= 5^3} = 125$

Problem 2   Give the value of each expression.

(A) 3^{-4} (B) $(-2)^{-5}$ (C) $\left(\frac{4}{3}\right)^{-3}$ (D) $\dfrac{1}{2^{-7}}$

Example 3   (A) $x^{-5} = \dfrac{1}{x^5}$ (B) $\dfrac{1}{y^{-7}} = y^7$

(C) $\dfrac{y^{-6}}{x^{-5}} \boxed{= \dfrac{y^{-6}}{1} \cdot \dfrac{1}{x^{-5}} = \dfrac{1}{y^6} \cdot \dfrac{x^5}{1}} = \dfrac{x^5}{y^6}$ (D) $10^{-4} = \dfrac{1}{10^4}$

Problem 3   Write using positive integer exponents:

(A) z^{-7} (B) $\dfrac{1}{x^{-4}}$ (C) $\dfrac{v^{-8}}{u^{-3}}$ (D) 10^{-3}

With zero and negative integer exponents defined as above, all the properties stated for positive integer exponents hold for all integer exponents. Since we no longer need to be concerned about the relative size of m and n in property 5, this property can be more simply stated as

5. $\dfrac{a^m}{a^n} = a^{m-n} = \dfrac{1}{a^{n-m}}$ $a \neq 0$

Thus, we are free to use either a^{m-n} or $1/a^{n-m}$ in place of a^m/a^n.

Example 4   Simplify and express answers using positive exponents only.

(A) $\dfrac{x^5}{x^9}$ (B) $\dfrac{y^{-5}}{y^{-9}}$ (C) $\dfrac{z^{-5}}{z^9}$ (D) $\dfrac{10^5}{10^{-3}}$

Solutions   (A) $\dfrac{x^5}{x^9} \boxed{= x^{5-9}} = x^{-4} = \dfrac{1}{x^4}$ or $\dfrac{x^5}{x^9} \boxed{= \dfrac{1}{x^{9-5}}} = \dfrac{1}{x^4}$

(B) $\dfrac{y^{-5}}{y^{-9}} \boxed{= y^{-5-(-9)}} = y^4$ or $\dfrac{y^{-5}}{y^{-9}} \boxed{= \dfrac{1}{y^{-9-(-5)}}} = \dfrac{1}{y^{-4}} = y^4$

(C) $\dfrac{z^{-5}}{z^9} \boxed{= z^{-5-9}} = z^{-14} = \dfrac{1}{z^{14}}$ or $\dfrac{z^{-5}}{z^9} \boxed{= \dfrac{1}{z^{9-(-5)}}} = \dfrac{1}{z^{14}}$

(D) $\dfrac{10^5}{10^{-3}}\boxed{=10^{5-(-3)}}=10^8$ or $\dfrac{10^5}{10^{-3}}\boxed{=\dfrac{1}{10^{-3-5}}}=\dfrac{1}{10^{-8}}=10^8$

Problem 4   Simplify and express answers using positive exponents only.

(A) $\dfrac{u^4}{u^7}$ (B) $\dfrac{x^{-7}}{x^{-3}}$ (C) $\dfrac{m^3}{m^{-5}}$

(D) $\dfrac{y^{-5}}{y^4}$ (E) $\dfrac{10^4}{10^{-5}}$ (F) $\dfrac{10^{-6}}{10^{-9}}$

Example 5   Simplify and express answers using positive exponents only.

(A) $x^{-5}x^3\boxed{=x^{-5+3}}=x^{-2}=\dfrac{1}{x^2}$

(B) $(a^3b^{-4})^{-3}\boxed{=(a^3)^{-3}(b^{-4})^{-3}}=a^{-9}b^{12}=\dfrac{b^{12}}{a^9}$

(C) $\left(\dfrac{y^{-3}}{y^{-2}}\right)^{-4}\boxed{=\dfrac{(y^{-3})^{-4}}{(y^{-2})^{-4}}}=\dfrac{y^{12}}{y^8}=y^4$

(D) $\dfrac{42x^{-5}y^{-4}}{36x^{-9}y^{-2}}\boxed{=\dfrac{42}{36}\cdot\dfrac{x^{-5}}{x^{-9}}\cdot\dfrac{y^{-4}}{y^{-2}}}=\dfrac{7}{6}\cdot\dfrac{x^4}{1}\cdot\dfrac{1}{y^2}=\dfrac{7x^4}{6y^2}$

(E) $\left(\dfrac{u^{-3}u^3}{v^{-6}}\right)^{-2}\boxed{=\left(\dfrac{u^0}{v^{-6}}\right)^{-2}}=\left(\dfrac{1}{v^{-6}}\right)^{-2}\boxed{=\dfrac{1^{-2}}{v^{12}}}=\dfrac{1}{v^{12}}$

(F) $\dfrac{10^{-5}\cdot10^3}{10^{-4}\cdot10^6}\boxed{=\dfrac{10^{-5+3}}{10^{-4+6}}}=\dfrac{10^{-2}}{10^2}=\dfrac{1}{10^4}$

Problem 5   Simplify and express answers using positive exponents only.

(A) y^7y^{-5} (B) $(x^{-5}y^3)^{-4}$ (C) $\left(\dfrac{z^{-4}}{z^{-7}}\right)^{-3}$

(D) $\dfrac{21x^{-8}y^4}{28x^{-3}y^{-3}}$ (E) $\left(\dfrac{m^{-3}}{n^{-2}n^2}\right)^{-4}$ (F) $\dfrac{10^4\cdot10^{-6}}{10^{-8}\cdot10^3}$

■ Common Errors

As stated in Section 1-5, many errors occur in algebra because the properties of exponents are applied incorrectly. It should be emphasized once
again that the properties of exponents apply to products and quotients and
not to sums and differences.

Expressions That Are Not Equal

$a^{-2} + a^2$	is not equal to	a^0
$\dfrac{a^{-2} + b^2}{c}$	is not equal to	$\dfrac{b^2}{a^2 c}$
$(a^{-1} + b^{-1})^2$	is not equal to	$a^{-2} + b^{-2}$
$a^{-2} + b^{-2}$	is not equal to	$\dfrac{1}{a^2 + b^2}$
$\dfrac{a^{-1}}{a^{-2} + a^{-3}}$	is not equal to	$\dfrac{a^2 + a^3}{a}$

The correct procedures for simplifying the expressions in the box are illustrated in Example 6.

Example 6 Simplify and express answers using positive exponents only.

(A) $a^{-2} + a^2 = \dfrac{1}{a^2} + a^2 = \dfrac{1 + a^4}{a^2}$

(B) $\dfrac{a^{-2} + b^2}{c} = \dfrac{\dfrac{1}{a^2} + b^2}{c} = \dfrac{\dfrac{1 + a^2 b^2}{a^2}}{c} = \dfrac{1 + a^2 b^2}{a^2} \cdot \dfrac{1}{c} = \dfrac{1 + a^2 b^2}{a^2 c}$

(C) $(a^{-1} + b^{-1})^2 = \left(\dfrac{1}{a} + \dfrac{1}{b}\right)^2 = \left(\dfrac{b + a}{ab}\right)^2 = \dfrac{(b + a)^2}{a^2 b^2}$ or $\dfrac{(a + b)^2}{a^2 b^2}$

(D) $a^{-2} + b^{-2} = \dfrac{1}{a^2} + \dfrac{1}{b^2} = \dfrac{b^2 + a^2}{a^2 b^2}$ or $\dfrac{a^2 + b^2}{a^2 b^2}$

(E) $\dfrac{a^{-1}}{a^{-2} + a^{-3}} = \dfrac{\dfrac{1}{a}}{\dfrac{1}{a^2} + \dfrac{1}{a^3}} = \dfrac{\dfrac{1}{a}}{\dfrac{a + 1}{a^3}} = \dfrac{1}{a} \cdot \dfrac{a^3}{a + 1} = \dfrac{a^2}{a + 1}$

(F) $\dfrac{2^{-3}}{4^{-1} + 2^{-4}} = \dfrac{\dfrac{1}{2^3}}{\dfrac{1}{4} + \dfrac{1}{2^4}} = \dfrac{\dfrac{1}{8}}{\dfrac{1}{4} + \dfrac{1}{16}} = \dfrac{\dfrac{1}{8}}{\dfrac{5}{16}} = \dfrac{1}{8} \cdot \dfrac{16}{5} = \dfrac{2}{5}$

Problem 6 Simplify and express answers using positive exponents only.

(A) $a^2 - a^{-3}$ (B) $\dfrac{x^{-1} + y}{z}$ (C) $(x^{-1} - y^{-1})^2$

(D) $x^{-2} - y^{-2}$ (E) $\dfrac{x^{-1} + x^{-2}}{x^{-3}}$ (F) $\dfrac{3^{-2}}{3^{-3} + 9^{-1}}$

■

■ Applications

Exponents are often used in expressions that represent *growth* (where a given quantity increases) and *decay* (where a given quantity decreases). The following two examples illustrate these uses.

Example 7
Annual Sales

The new manager of a company with annual sales of $1 million has projected that she will double the annual sales, S, each year for the next 6 years. Thus,

$$S = \$1,000,000(2^t) \qquad 0 \leqslant t \leqslant 6$$

where t is an integer that denotes the number of years.* Determine the annual sales during the fourth year.

Solution

For $t = 4$, we have

$$S = \$1,000,000(2^t) = 1,000,000(2^4) = 1,000,000(16) = \$16,000,000$$

Problem 7

In Example 7, determine the expected annual sales during the sixth year.

Example 8
Public Health

The number of rodents in a community is estimated to be 6,561. With proper control measures, a pest control company expects to reduce the rodent population by one-third each month for the next 8 months. Thus, the expected rodent population, P, at the end of t months is given by

$$P = 6,561(3^{-t}) \qquad 0 \leqslant t \leqslant 8$$

Determine the expected rodent population at the end of 3 months.

Solution

For $t = 3$, we have

$$P = 6,561(3^{-t}) = 6,561(3^{-3}) = 6,561 \cdot \frac{1}{3^3}$$

$$= 6,561 \cdot \frac{1}{27} = 243 \text{ rodents}$$

Problem 8

In Example 8, determine the expected rodent population at the end of:

(A) 6 months   (B) 8 months

Answers to
Matched Problems

1. (A) Not defined   (B)–(F) All equal to 1

2. (A) $\frac{1}{81}$ (B) $-\frac{1}{32}$ (C) $\frac{27}{64}$ (D) 128

3. (A) $\dfrac{1}{z^7}$ (B) x^4 (C) $\dfrac{u^3}{v^8}$ (D) $\dfrac{1}{10^3}$

---

* The symbol 2^t is only defined here for t an integer. Later (Sections 3-3 and 13-1), we will extend its meaning to include rational and irrational exponents.

4. (A) $\dfrac{1}{u^3}$ (B) $\dfrac{1}{x^4}$ (C) m^8 (D) $\dfrac{1}{y^9}$ (E) 10^9 (F) 10^3

5. (A) y^2 (B) $\dfrac{x^{20}}{y^{12}}$ (C) $\dfrac{1}{z^9}$ (D) $\dfrac{3y^7}{4x^5}$ (E) m^{12} (F) 10^3

6. (A) $\dfrac{a^5-1}{a^3}$ (B) $\dfrac{1+xy}{xz}$ (C) $\dfrac{(y-x)^2}{x^2y^2}$ (D) $\dfrac{y^2-x^2}{x^2y^2}$

(E) $x(x+1)$ or x^2+x (F) $\frac{3}{4}$

7. \$64,000,000 8. (A) 9 (B) 1

Exercise 3-1

A *Give the value of each expression.*

1. 31^0
2. $(-53)^0$
3. 2^{-3}
4. 4^{-2}

5. $\dfrac{1}{3^{-3}}$
6. $\dfrac{1}{5^{-4}}$
7. $\left(\dfrac{3}{2}\right)^{-3}$
8. $\left(\dfrac{4}{5}\right)^{-2}$

Simplify and express answers using positive exponents only.

9. u^{-9}
10. y^{-5}
11. $\dfrac{1}{z^{-3}}$
12. $\dfrac{1}{w^{-6}}$

13. $10^8 \cdot 10^{-3}$
14. $10^{-7} \cdot 10^3$
15. $w^{-5}w^9$
16. $m^{-6}m^2$

17. $c^{-4}c^4$
18. v^7v^{-7}
19. $\dfrac{10^4}{10^{-5}}$
20. $\dfrac{10^{-4}}{10^{-8}}$

21. $\dfrac{w^{-9}}{w^{-5}}$
22. $\dfrac{n^4}{n^{-6}}$
23. $\dfrac{10^{-3}}{10^5}$
24. $\dfrac{10^{-5}}{10^2}$

25. $(x^{-4})^{-2}$
26. $(z^{-6})^{-3}$
27. $(w^{-4})^3$
28. $(a^4)^{-5}$

29. $(3^{-4} \cdot 2^5)^{-1}$
30. $(4^{-1} \cdot 5^2)^{-2}$
31. $(u^6v^{-3})^{-2}$
32. $(x^{-2}y^3)^{-3}$

B *Simplify and express answers using positive exponents only.*

33. $(29+32+57)^0$
34. $(x^4y^7)^0$
35. $\dfrac{10^{-4} \cdot 10^8}{10^{-15} \cdot 10^{-3}}$

36. $\dfrac{10^{-7} \cdot 10^{-10}}{10^{-9} \cdot 10^5}$
37. $\left(\dfrac{a^3}{a^{-1}}\right)^4$
38. $\left(\dfrac{m^{-1}}{m^2}\right)^2$

39. $(5m^2n^{-3})^{-2}$
40. $(2r^{-3}s^2)^4$
41. $\dfrac{1}{(6u^2v^{-1})^{-2}}$

42. $\dfrac{1}{(3y^{-2}z^3)^{-3}}$
43. $\dfrac{15x^{-3}y^{-4}}{35x^{-5}y^{-3}}$
44. $\dfrac{42u^{-4}v^{-9}}{36u^{-7}v^{-5}}$

45. $\dfrac{28m^{-4}n^2}{35m^5n^{-3}}$
46. $\dfrac{18r^5s^{-3}}{12r^{-6}s^2}$
47. $\left(\dfrac{y^{-4}}{y^{-5}}\right)^{-3}$

48. $\left(\dfrac{z^{-7}}{z^{-3}}\right)^{-4}$

49. $\left(\dfrac{x^{-3}y^{-3}}{x^{-5}y^4}\right)^3$

50. $\left(\dfrac{u^4v^{-4}}{u^{-3}v^{-2}}\right)^2$

51. $\left(\dfrac{8m^{-1}n^3}{4m^3n^{-2}}\right)^{-2}$

52. $\left(\dfrac{9r^3s^{-6}}{27r^{-2}s^{-5}}\right)^{-3}$

53. $(2x^{-3}y^4)^{-3}(x^2y^{-4})^{-2}$

54. $(3u^{-4}v^3)^{-3}(uv^{-3})^{-4}$

55. $x^{-3}+x^3$

56. y^3+y^{-4}

57. $(r^2-s^2)^{-1}$

58. $(m-3)^{-2}$

59. $\dfrac{3^{-2}}{3^{-1}+3^{-3}}$

60. $\dfrac{2^{-5}}{2^{-3}+2^{-4}}$

61. $\dfrac{x^{-1}+y^{-1}}{x+y}$

62. $\dfrac{y-x}{x^{-1}-y^{-1}}$

C Simplify and express answers using positive exponents only.

63. $\left[\left(\dfrac{x^2y^{-5}z}{x^{-2}y^{-2}z^3}\right)^{-3}\right]^2$

64. $\left[\left(\dfrac{r^{-5}s^4t^{-3}}{r^{-3}st^2}\right)^{-2}\right]^3$

65. $\left(\dfrac{x^{-5}y^0}{x^{-3}}\right)^3\left(\dfrac{3^2x^0y^{-7}}{27y^{-4}}\right)^{-2}$

66. $\left(\dfrac{8a^0b^4}{2^2b^{-3}}\right)^{-2}\left(\dfrac{a^{-3}}{a^{-7}b^0}\right)^3$

67. $(a^{-2}+b^{-2})^{-1}$

68. $(x^{-2}-y^{-2})^{-2}$

69. $(10^{-4}+10^{-3})^{-1}$

70. $(10^{-2}-10^{-3})^{-1}$

71. $\dfrac{2^{-1}+2^{-2}}{4^{-1}-4^{-2}}$

72. $\dfrac{3^{-2}-3^{-3}}{9^{-1}+9^{-2}}$

Applications

Business & Economics

73. *Sales growth.* This year the sales of a growing company will be $256,000. For the next 8 years the yearly sales, S, are expected to be

$$S = \$256{,}000(\tfrac{3}{2})^t \qquad 0 \le t \le 8$$

where t is an integer that denotes the number of years. Determine the expected sales in 1, 5, and 8 years.

74. *Sales decline.* A company plans to produce a particular model of electronic cash register for the next 5 years. This year 24,300 of the cash registers will be sold, but due to competition, the sales are expected to fall over the next 5 years according to

$$S = 24{,}300(\tfrac{2}{3})^{-t} \qquad 0 \le t \le 5$$

where t is an integer denoting the number of years. Determine the expected sales in 1, 3, and 5 years.

Life Sciences

75. *Endangered species.* The world population of an endangered species of whale is estimated to be 4,096. With appropriate control measures regulating their commercial use, it is expected that the population, P, will grow according to

$$P = 4{,}096(\tfrac{5}{4})^t \qquad 0 \le t \le 6$$

where t is an integer denoting the number of years. Determine the expected population in 1, 3, and 6 years.

76. *Population control.* The present prairie dog population in Lubbock, Texas, is estimated to be 72,900. If control measures are not taken, it is expected that the prairie dog population will grow according to

$$P = 72{,}900(\tfrac{5}{3})^t \qquad 0 \leqslant t \leqslant 6$$

where t is an integer denoting the number of years. Determine the estimated population in 1, 3, and 6 years, assuming that control measures are not taken.

Social Sciences

77. *Politics.* The number of votes cast in a local election in a city was 312,500. Because of political corruption and voter apathy, it is projected that the number of votes cast in yearly local elections for the next 5 years will be

$$V = 312{,}500(\tfrac{5}{4})^{-t} \qquad 0 \leqslant t \leqslant 5$$

where t is an integer denoting the number of years. Determine the expected number of votes cast in local elections in 1, 3, and 5 years.

3-2 Scientific Notation

- Powers of 10
- Scientific Notation
- Scientific Notation and Calculators

Scientific notation is frequently associated with science and engineering where the use of very, very small numbers and very, very large numbers is common. For example, the mass of a 5 carat diamond is 1 gram. In comparison, the mass of the smallest atom, the hydrogen atom, is approximately

  0.000 000 000 000 000 000 000 001 67 gram

whereas, the mass of the earth is approximately

  5,980,000,000,000,000,000,000,000,000 grams

In scientific notation, these two numbers are written more compactly as

  1.67×10^{-24} gram   and 5.98×10^{27} grams

The usefulness of scientific notation, however, is by no means restricted to just science and engineering. For instance, we deal with large numbers daily when we speak of contracts measured in billions of dollars and a gross national product measured in trillions of dollars.

■ Powers of 10

Scientific notation is based on the use of numbers in decimal form and *powers of 10*. Recall that if a decimal point does not appear in the decimal form of a number, it is assumed to be to the right of the last digit on the right. From Section 3-1, we have

$$10^0 = 1$$
$$10^1 = 10 \qquad\qquad 10^{-1} = \tfrac{1}{10} = 0.1$$
$$10^2 = 100 \qquad\qquad 10^{-2} = \tfrac{1}{100} = 0.01$$
$$10^3 = 1{,}000 \qquad\qquad 10^{-3} = \tfrac{1}{1{,}000} = 0.001$$
$$10^4 = 10{,}000 \qquad\qquad 10^{-4} = \tfrac{1}{10{,}000} = 0.0001$$

You should be familiar with the fact that multiplying a decimal number by 10^1 (=10), 10^2 (=100), 10^3 (=1,000), etc., has the effect of moving the decimal point 1, 2, 3, etc., places to the right. For example,

$$52.932 \times 10^2 = 52.932$$

Move the decimal point two places to the right

$$= 5{,}293.2$$

$$453 \times 10^6 = 453.000\ 000$$

Move the decimal point six places to the right; insert 0's as necessary

$$= 453{,}000{,}000$$

Similarly, multiplying a decimal number by 10^{-1} (=0.1), 10^{-2} (=0.01), 10^{-3} (=0.001), etc., has the effect of moving the decimal point 1, 2, 3, etc., places to the left. For example,

$$9.21 \times 10^{-5} = 00009.21$$

Move the decimal point five places to the left; insert 0's as necessary

$$= 0.000\ 092\ 1$$

■ Scientific Notation

We say that a number is in *scientific notation* when it is written as a (decimal) number between 1 and 10 (including 1, but not 10) times an integer power of 10. More formally, we have the following:

> **Scientific Notation**
>
> A number is in **scientific notation** if it is in the form
>
> $a \times 10^{\pm n}$
>
> where $1 \leq a < 10$ and n is 0 or a positive integer. (The $\pm$ sign indicates that a $+$ or $-$ may precede n.) To convert scientific notation to standard decimal form, we move the decimal point n places to the right if n is preceded by a $+$, or n places to the left if n is preceded by a $-$.

Example 9   Convert to standard decimal form.

(A) 7.96×10^6 $= 7.960\ 000$ $= 7,960,000$

(B) 3.47×10^{-5} $= 00003.47$ $= 0.000\ 034\ 7$

(C) 5×10^0 $= 5 \times 1$ $= 5$

Problem 9   Convert to standard decimal form.

(A) 3.92×10^8 (B) 1.68×10^{-3} (C) 6×10^0 (D) 1×10^3
(E) 1×10^{-5}

The easiest way to convert a number to scientific notation is to think: "What scientific form will produce the original number?"

Example 10   Convert to scientific notation.

(A) $43,500,000$ $= 43,500,000 \times 10^?$

> We would have to move the decimal point seven places to the right to obtain the original number; thus, 7 should replace the ?

$= 4.35 \times 10^7$

(B) $0.000\ 039\ 7$ $= 0.000\ 039\ 7 \times 10^?$

> We would have to move the decimal point five places to the left to obtain the original number; thus, -5 should replace the ?

$= 3.97 \times 10^{-5}$

Problem 10   Convert to scientific notation.

(A)   1,243,000,000   (B)   0.000 000 527

Example 11   Convert to scientific notation.

(A) 653.4×10^4 $\begin{aligned}&= (6.534 \times 10^2) \times 10^4 \\ &= 6.534 \times [(10^2)(10^4)]\end{aligned}$

$\qquad\qquad\qquad = 6.534 \times 10^6$

(B) 325×10^{-5} $\begin{aligned}&= (3.25 \times 10^2) \times 10^{-5} \\ &= 3.25 \times [(10^2)(10^{-5})]\end{aligned}$

$\qquad\qquad\qquad = 3.25 \times 10^{-3}$

Problem 11   Convert to scientific notation.

(A) $0.005\ 2 \times 10^{-6}$ (B) $0.000\ 823 \times 10^8$

Many complicated arithmetic problems can be evaluated more easily by using scientific notation.

Example 12   Evaluate using scientific notation.

(A) $(9,100,000)(0.000\ 05)\ = (9.1 \times 10^6)(5 \times 10^{-5})$

$\qquad\qquad\qquad\qquad = [(9.1)(5)] \times [(10^6)(10^{-5})]$

$\qquad\qquad\qquad\qquad = 45.5 \times 10^1 = 4.55 \times 10^2$

(B) $\dfrac{0.000\ 32}{6,400} = \dfrac{3.2 \times 10^{-4}}{6.4 \times 10^3}$

$\qquad\qquad\quad = \dfrac{3.2}{6.4} \times \dfrac{10^{-4}}{10^3}$

$\qquad\qquad\quad = 0.5 \times 10^{-7} = 5 \times 10^{-8}$

(C) $\dfrac{(0.000\ 000\ 000\ 024)(35,000)}{(150,000,000)(0.000\ 08)} = \dfrac{(2.4 \times 10^{-11})(3.5 \times 10^4)}{(1.5 \times 10^8)(8 \times 10^{-5})}$

$\qquad\qquad\qquad\qquad\qquad\qquad = \dfrac{(2.4)(3.5)}{(1.5)(8)} \times \dfrac{(10^{-11})(10^4)}{(10^8)(10^{-5})}$

$\qquad\qquad\qquad\qquad\qquad\qquad = 0.7 \times 10^{-10} = 7 \times 10^{-11}$

Problem 12   Evaluate using scientific notation.

(A) $(0.000\ 000\ 000\ 001\ 4)(60,000,000)$ (B) $\dfrac{8,400,000}{0.000\ 21}$

(C) $\dfrac{(450,000,000)(0.000\ 001\ 8)}{(0.000\ 09)(3,600,000,000,000)}$

■ Scientific Notation and Calculators

Most business and scientific hand calculators represent very large or very small numbers in scientific notation. For example, to enter the numbers

0.000 000 000 000 000 000 000 000 001 67   Mass of a hydrogen atom in grams

5,980,000,000,000,000,000,000,000,000   Mass of the earth in grams

in a hand calculator, we would first have to convert each number to scientific notation:

1.67×10^{-24}

5.98×10^{27}

Then we would enter these forms of the numbers according to the instructions for the given calculator. These numbers would appear in the hand calculator display as follows:

1.67   −24

5.98   27

Furthermore, if a calculation involving numbers in standard decimal form results in a number that exceeds the capacity of the display window, the result will automatically be displayed in scientific notation. Try multiplying 52,630 by 2,893,000 or dividing 3,401,000 by 0.000 000 73 in a hand calculator to see what happens.

Answers to Matched Problems

9. (A) 392,000,000   (B) 0.001 68   (C) 6   (D) 1,000
    (E) 0.000 01
10. (A) 1.243×10^{9} (B) 5.27×10^{-7}
11. (A) 5.2×10^{-9} (B) 8.23×10^{4}
12. (A) 8.4×10^{-5} (B) 4×10^{10} (C) 2.5×10^{-6}

Exercise 3-2

A   *Convert to scientific notation.*

1. 76   2. 40   3. 86,000
4. 130,000   5. 0.094   6. 0.009
7. 0.000 000 29   8. 0.000 03   9. 52,900,000,000
10. 863,000,000   11. 0.000 000 000 068 4
12. 0.000 000 002 79

Convert to standard decimal form.

13. 3.7×10^3	14. 6×10^2	15. 8×10^1
16. 5.4×10^5	17. 8×10^{-4}	18. 9.6×10^{-3}
19. 8.2×10^{-2}	20. 2×10^{-6}	21. 2.8×10^9
22. 5.1×10^{13}	23. 6.4×10^{-13}	24. 4.6×10^{-8}

B   *Write each number in scientific notation.*

25. The approximate distance from the sun to the earth is 92,900,000 miles.
26. The approximate land area of the earth is 57,500,000 square miles.
27. The yearly egg production in the United States is approximately 70,000,000,000.
28. The population of the United States according to the 1980 census is approximately 227,000,000.
29. It takes light approximately 0.000 000 000 849 second to travel 1 inch.
30. The radius of a hydrogen atom is approximately 0.000 000 002 09 inch.

Write each number in standard decimal form.

31. The approximate distance from the sun to the planet Pluto is 3.67×10^9 miles.
32. The approximate ocean area of the earth is 1.39×10^8 square miles.
33. The gross national product of the United States in 1980 was approximately \$$2.63 \times 10^{12}$.
34. It is estimated that the world demand for oil will soon exceed 2.5×10^{10} barrels per year (1 barrel = 42 gallons).
35. The approximate mass of an electron is 9.1×10^{-28} gram.
36. The mass of one water molecule is approximately 3×10^{-23} gram.

The following numbers are not quite in scientific notation. Adjust them so that they are.

37. 294×10^5	38. 52.4×10^7
39. 0.003×10^5	40. 0.02×10^6
41. 800×10^{-5}	42. 29.7×10^{-3}
43. $0.027\ 9 \times 10^{-3}$	44. $0.000\ 451 \times 10^{-2}$

Use scientific notation to evaluate each expression. Give the answer in both scientific notation and standard decimal form.

45. $(5,600,000)(0.000\ 03)$	46. $(0.000\ 000\ 023)(80,000)$
47. $\dfrac{0.000\ 005\ 6}{2,800}$	48. $\dfrac{72,000}{0.000\ 36}$

49. $\dfrac{(1,200,000)(0.000\ 003)}{0.000\ 06}$

50. $\dfrac{(360,000)(0.000\ 002\ 5)}{0.000\ 45}$

51. $\dfrac{(240,000)(0.000\ 001\ 5)}{(0.000\ 8)(7,500,000)}$

52. $\dfrac{(0.000\ 000\ 082)(230,000)}{(46,000,000)(0.001\ 64)}$

Applications

Business & Economics

Use scientific notation to evaluate the answer to each problem. Give the answer in both scientific notation and standard decimal form.

53. *Taxes.* In 1978 individuals in the United States paid about $182,000,000,000 in income tax. If the estimated population then was 221,000,000, what was the average amount of tax paid per person?

54. *Gross national product.* If the gross national product in the United States in 1980 was about $2,630,000,000,000 and the population was about 227,000,000 people, estimate the gross national product per person.

55. *Industry.* If it takes 0.006 barrel of oil to produce 1 kilowatt-hour of electricity, how many barrels of oil would be required to produce 15,000,000 kilowatt-hours of electricity?

56. *Industry.* An oil refinery has 12 storage tanks. If each tank can store 250,000 gallons and there are 42 gallons per barrel, approximately how many barrels of oil can be stored at the refinery?

Life Sciences

57. *Pollution.* If the water in a lake contains 15,000 bacteria per cubic foot, approximately how many bacteria would be contained within 1 foot of the surface over an area 1,000 feet by 1,000 feet?

58. *Chemistry.* If one molecule of water has a mass of 3×10^{-23} gram, approximately how many molecules are in 1 gram of water?

Social Sciences

59. *Education costs.* In the United States during the 1975–1976 school year, approximately $1,500 was spent per elementary and secondary school child. If the total enrollment was approximately 42,000,000 students, what was the total amount spent?

3-3 Rational Exponents

- Roots of Real Numbers
- Rational Exponents
- Applications

We will now extend the concept of exponent to include fractional exponents. As with integer exponents, we will again require that the meaning of fractional exponents be compatible with the five properties of exponents discussed in Sections 1-5 and 3-1.

■ Roots of Real Numbers

Let us begin with a question. What real numbers squared give 25? That is, what values of x make the following statement true?

$x^2 = 25$

With a little thought it should be clear that we can have $x = 5$ or $x = -5$, since

$5^2 = 5 \cdot 5 = 25$ and $(-5)^2 = (-5)(-5) = 25$

We say that 5 and -5 are *square roots* of 25. Thus, 25 has two real square roots, which differ in sign.

Does 0 have a square root? That is, are there any values of x for which the following statement is true?

$x^2 = 0$

Clearly, $x = 0$ is the only possibility. Thus, 0 has one square root, namely, 0.

Are there real values of x for which the following statement is true?

$x^2 = -25$

Recall that the square of a real number cannot be negative. Thus, -25 has no real square roots.*

Square Roots

If a is a real number, then a real number x is called a **square root** of a if

$x^2 = a$

There will be two real square roots (with opposite signs) if a is positive, one if $a = 0$, and none if a is negative.

Besides square roots, we can define other types of roots. For example, what are the values of x for which the following statement is true?

$x^3 = 125$

---

* There is a larger system of numbers, called the *complex numbers,* in which -25 does have square roots. but since complex numbers are not to be considered in this text, we will always assume that we are considering only real numbers and that *root* means *real root.*

Clearly, $x = 5$ is the only possibility, and 5 is called a *cube root* of 125. Similarly, -5 is a cube root of -125, since $(-5)^3 = -125$.

We may generalize the above discussion as follows:

---

*n*th Roots

Let a denote a real number and let n denote a positive integer. Then a real number x is called an **nth root** of a if

$$x^n = a$$

In general, a real number a will have two, one, or no nth roots, depending on the sign of a and whether n is even or odd. We have listed all possible situations in Table 1.

Table 1 nth Roots of a Number a

	a Positive	*a* Negative
n Even	Two nth roots differing in sign	No nth roots
n Odd	One nth root	One nth root

Of course, 0 is the nth root of 0 for every positive integer n, since $0^n = 0$.

Example 13 (A) 10 and -10 are square roots of 100, since

$$10^2 = 100 \quad \text{and} \quad (-10)^2 = 100$$

(B) 2 and -2 are 6th roots of 64, since

$$2^6 = 64 \quad \text{and} \quad (-2)^6 = 64$$

(C) -3 is a 5th root of -243, since

$$(-3)^5 = -243$$

(D) -64 has no square root, 4th root, or nth root if n is even

Problem 13 Give the indicated roots.

(A) The 4th roots of 16   (B) The cube roots of 8
(C) The cube roots of -8 (D) The 8th roots of 0
(E) The 4th roots of $-10{,}000$

■ Rational Exponents

When a number has two nth roots — one positive and one negative — the positive root is called the *principal nth root*. If a number has only one nth root, it is automatically considered to be the principal nth root. Fractional exponents may be used to denote the principal nth root of a number as defined in the box.

Principal nth root

Let a represent a real number and let n denote a positive integer. Then the expression

$$a^{1/n}$$

is defined to be the **principal nth root of a,** if one exists. If a has two nth roots, $-a^{1/n}$ denotes the negative nth root. If a is negative and n is even, a has no real nth root.

Example 14

(A) $64^{1/6} = 2$ (B) $(-125)^{1/3} = -5$
(C) $-9^{1/2} = -3$ (D) $(-9)^{1/2}$ is not a real number
(E) $0^{1/7} = 0$ (F) $\left(\frac{16}{81}\right)^{1/4} = \frac{2}{3}$

Problem 14   Find each of the following:

(A) $625^{1/4}$ (B) $(-64)^{1/3}$ (C) $-49^{1/2}$
(D) $(-49)^{1/2}$ (E) $0^{1/5}$ (F) $\left(-\frac{27}{64}\right)^{1/3}$

In our examples so far, we have chosen a and n so that $a^{1/n}$ is an integer, a fraction, or is not a real number. We will now turn our attention to expressions such as $5^{1/2}$ or $6^{1/5}$. For property 2 of exponents (Section 3-1) to hold, we must have

$$(5^{1/2})^2 = 5^{(1/2)2} = 5^1 = 5$$

Thus, $5^{1/2}$ denotes a square root of 5, and the other square root of 5 is $-5^{1/2}$. Similarly, since

$$(6^{1/5})^5 = 6^{(1/5)5} = 6^1 = 6$$

$6^{1/5}$ denotes the 5th root of 6. In general, if $a^{1/n}$ is a real number (that is, if a is not negative when n is even), we have

---

$$(a^{1/n})^n = a^{n/n} = a^1 = a$$

Numbers such as $5^{1/2}$ and $6^{1/5}$ are *irrational numbers*; hence, they cannot be expressed as integers or fractions. They can be approximated to any desired accuracy using decimal representations. For example,

$$5^{1/2} \approx 2.2361 \qquad \text{and} \qquad 6^{1/5} \approx 1.4310$$

where the symbol $\approx$ indicates "is approximately equal to."

We now wish to define the meaning of an expression such as $8^{2/3}$. Since we want to preserve the five properties of exponents listed in Section 3-1, according to property 2, we must have

$$8^{2/3} = (8^{1/3})^2 = 2^2 = 4$$

Thus, $8^{2/3}$ should represent the square of the cube root of 8. This is generalized in the definition given in the box.

Rational Exponents

Let a represent a real number, and let m and n denote positive integers. If $a^{1/n}$ is defined, then we define

$$a^{m/n} = (a^{1/n})^m \qquad \text{and} \qquad a^{-m/n} = \frac{1}{a^{m/n}} \qquad a \neq 0$$

Recall that if a is negative and n is even, then $a^{1/n}$ is not a real number.

As long as we avoid even roots of negative numbers, the five properties of exponents listed in Section 3-1 continue to hold for rational exponents. Although we will not prove these properties here, we will illustrate their use below. As a consequence of the properties of exponents, whenever $a^{1/n}$ is a real number, we can express $a^{m/n}$ by

$$a^{m/n} = (a^{1/n})^m \qquad \text{or} \qquad a^{m/n} = (a^m)^{1/n}$$

In computations, the first form is usually preferred over the second.

Example 15 (A) $16^{3/4} \boxed{= (16^{1/4})^3} = 2^3 = 8 \qquad \text{or} \qquad 16^{3/4} = (16^3)^{1/4} = (4{,}096)^{1/4} = 8$

(B) $(-27)^{4/3} \boxed{= [(-27)^{1/3}]^4} = (-3)^4 = 81$

(C) $8^{-2/3} = \dfrac{1}{8^{2/3}} \boxed{= \dfrac{1}{(8^{1/3})^2}} = \dfrac{1}{2^2} = \dfrac{1}{4}$

(D) $(6x^{2/5})(3x^{1/2}) = 18x^{(2/5)+(1/2)}$
$$= 18x^{(4/10)+(5/10)} = 18x^{9/10}$$

(E) $(3x^{3/4}y^{-1/4})^4 \boxed{= 3^4(x^{3/4})^4(y^{-1/4})^4} = 81x^3y^{-1} \qquad \text{or} \qquad \dfrac{81x^3}{y}$

(F) $\left(\dfrac{9x^{1/6}}{x^{1/2}}\right)^{1/2} = \dfrac{9^{1/2}x^{1/12}}{x^{1/4}} = 3x^{(1/12)-(1/4)}$

$\qquad\qquad\qquad = 3x^{(1/12)-(3/12)} = 3x^{-2/12} = 3x^{-1/6}$ or $\dfrac{3}{x^{1/6}}$

(G) $(3x^{1/2} + y^{1/2})(x^{1/2} + 2y^{1/2}) = 3x + 7x^{1/2}y^{1/2} + 2y$

Problem 15   Simplify and express each answer using positive exponents.

(A) $8^{5/3}$ $\qquad\qquad$ (B) $(-8)^{7/3}$ $\qquad\qquad$ (C) $9^{-3/2}$

(D) $(5a^{1/4})(3a^{3/8})$ $\qquad$ (E) $(2u^{-1/3}v^{5/6})^6$ $\qquad$ (F) $\left(\dfrac{8x^{1/4}}{x^{1/2}}\right)^{1/3}$

(G) $(a^{1/2} - 3b^{1/2})(2a^{1/2} + b^{1/2})$

We have mentioned that difficulties arise in using fractional exponents unless we avoid even roots of negative numbers. To illustrate this, consider the following:

$$-1 = (-1)^1 = (-1)^{2/2} = [(-1)^2]^{1/2} = 1^{1/2} = 1$$

But $-1 \neq 1$! The problem here is that $(-1)^{2/2}$ involves the even root of a negative number, which is not a real number. Thus, the string of equalities above is not valid. This difficulty can be avoided by requiring that all fractional exponents be reduced to lowest terms.

■ Applications

In Section 3-1 we illustrated how growth and decay can be represented using integer exponents. Examples 16 and 17 illustrate how fractional exponents may be used in similar situations.

Example 16
Sales Estimates

A tire manufacturer is about to introduce a new type of tire. It is estimated that the number of units, N, the firm will be able to sell in one outlet each month during the first year after the tire's introduction on the market is given by

$$N = 100(8^{t/6}) \qquad 1 \leq t \leq 12$$

where t is a positive integer denoting the number of months. Determine the estimated number of units to be sold during the fourth month the tire is on the market.

Solution   For $t = 4$, we have

$$N = 100(8^{t/6}) = 100(8^{4/6}) = 100(8^{2/3})$$

$$\overline{\left|= 100(8^{1/3})^2\right.}$$
$$\left.= 100(2)^2\right|$$

$$= 100 \cdot 4 = 400$$

Thus, 400 units are expected to be sold during the fourth month.

Problem 16   In Example 16, determine the estimated number of units to be sold during the:

(A) Eighth month   (B) Tenth month

Example 17
Endangered Species   The population of an endangered species of bird is estimated to be 729 at present. Unless protective steps are taken, it is estimated that the population, P, at the end of t years will be

$$P = 729(9^{-t/6}) \qquad 0 \leqslant t \leqslant 15$$

Determine the estimated population at the end of 3 years.

Solution   For $t = 3$, we have

$$P = 729(9^{-t/6}) = 729(9^{-3/6}) = 729(9^{-1/2})$$

$$= 729 \cdot \frac{1}{9^{1/2}} = 729 \cdot \frac{1}{3} = 243 \text{ birds}$$

Problem 17   In Example 17, determine the estimated population at the end of:

(A) 9 years   (B) 15 years

Answers to
Matched Problems

13. (A) 2 and -2 (B) 2   (C) -2 (D) 0   (E) None

14. (A) 5   (B) -4 (C) -7 (D) Not a real number
      (E) 0   (F) $-\frac{3}{4}$

15. (A) 32   (B) -128 (C) $\frac{1}{27}$ (D) $15a^{5/8}$

      (E) $\dfrac{64v^5}{u^2}$ (F) $\dfrac{2}{x^{1/12}}$ (G) $2a - 5a^{1/2}b^{1/2} - 3b$

16. (A) 1,600   (B) 3,200   17. (A) 27   (B) 3

Exercise 3-3

A   *Give the indicated roots.*

1. The square roots of 100   2. The cube roots of 64
3. The 6th roots of 1,000,000   4. The cube roots of -125
5. The 4th roots of -16 6. The 5th roots of 0

Find each of the following:

7. $49^{1/2}$ 8. $64^{1/2}$ 9. $(-81)^{1/2}$ 10. $(-25)^{1/2}$
11. $-81^{1/2}$ 12. $-16^{1/2}$ 13. $27^{1/3}$ 14. $64^{1/3}$
15. $(-64)^{1/3}$ 16. $(-27)^{1/3}$ 17. $9^{3/2}$ 18. $16^{3/2}$
19. $(-8)^{2/3}$ 20. $(-27)^{2/3}$

Simplify and express each answer using positive exponents.

21. $x^{1/6}x^{5/6}$ **22.** $z^{3/4}z^{1/4}$ **23.** $\dfrac{u^{4/5}}{u^{2/5}}$ **24.** $\dfrac{v^{1/3}}{v^{2/3}}$

25. $(a^8)^{1/2}$ **26.** $(b^{1/3})^6$ **27.** $(r^4s^{12})^{1/4}$ **28.** $(a^6b^{15})^{1/3}$

29. $\left(\dfrac{u^9}{v^{12}}\right)^{1/3}$ **30.** $\left(\dfrac{x^{20}}{y^{15}}\right)^{1/5}$ **31.** $(x^{1/3}y^{1/2})^{12}$ **32.** $(a^{1/2}b^{1/4})^8$

B Find each of the following:

33. $\left(\frac{9}{16}\right)^{3/2}$ **34.** $\left(\frac{4}{9}\right)^{3/2}$ **35.** $49^{-1/2}$ **36.** $100^{-1/2}$
37. $16^{-3/2}$ **38.** $9^{-3/2}$ **39.** $(-27)^{-2/3}$ **40.** $(-8)^{-5/3}$

Simplify and express each answer using positive exponents.

41. $(6x^{4/9})(5x^{1/3})$ **42.** $(4r^{3/5})(3r^{3/10})$ **43.** $(2a^{-2/3}b^{1/6})^6$
44. $(3x^{3/4}y^{-5/2})^4$ **45.** $a^{2/5}a^{-3/10}$ **46.** $x^{-2/3}x^{4/5}$

47. $\dfrac{m^{2/3}}{m^{-5/6}}$ **48.** $\dfrac{a^{-5/12}}{a^{1/3}}$ **49.** $\left(\dfrac{x^{-5/6}}{y^{-2/3}}\right)^{12}$

50. $\left(\dfrac{m^{-15}}{n^{20}}\right)^{-1/5}$ **51.** $(16u^8v^{-12})^{1/4}$ **52.** $(27a^{-9}b^{15})^{1/3}$

53. $(32u^{10}v^{-15})^{-1/5}$ **54.** $(625r^{-16}s^{12})^{-1/4}$ **55.** $\left(\dfrac{64x^{-9/10}}{x^{3/2}}\right)^{1/6}$

56. $\left(\dfrac{27y^{-5/6}}{y^{2/3}}\right)^{1/3}$ **57.** $\left(\dfrac{64a^{-4}b^{-2}}{27a^5b^{-5}}\right)^{1/3}$ **58.** $\left(\dfrac{81r^7s^{-9}}{16r^{-5}s^{-1}}\right)^{1/4}$

Perform the indicated operations and express each answer using positive exponents.

59. $5x^{2/5}(3x^{3/5} - 4x^3)$ **60.** $6u^{2/3}(2u^2 - u^{4/3})$
61. $(m^{1/2} - n^{1/2})(m^{1/2} + n^{1/2})$ **62.** $(2x^{1/2} + y^{1/2})(x^{1/2} - 3y^{1/2})$
63. $(u^{1/2} + v^{1/2})^2$ **64.** $(a^{1/2} - b^{1/2})^2$

C Perform the indicated operations and express each answer using positive exponents.

65. $(a^{1/2} - b^{-1/2})(a^{-1/2} + b^{1/2})$ **66.** $(x^{-1/2} + y^{1/2})(x^{1/2} + y^{-1/2})$
67. $(x^{1/2} + y^{-1/2})(x^{1/2} - y^{-1/2})$ **68.** $(a^{-1/2} + b^{-1/2})^2$
69. $(a^{2/3} + a^{-1/3})(a^{1/3} + a^{-2/3})$ **70.** $(x^{1/4} - 2x^{-3/4})(x^{3/4} + 3x^{-1/4})$

Applications

Business & Economics

71. *Revenue estimates.* The manufacturer of a new line of computers expects revenue to be $409,600 this month. The projected revenue, R, for the next 12 months is given by

$$R = \$409{,}600\left(\tfrac{25}{16}\right)^{t/4} \qquad 0 \leqslant t \leqslant 12$$

where t is an integer denoting the number of months. Determine the expected revenue in 2, 6, and 12 months.

72. *Sales estimates.* This month, a pharmaceutical company expects to produce and sell 156,250 units of a drug it has developed. Because of a controversial report alleging serious side effects caused by the drug, the company expects the number of units, N, sold over the next 12 months to decline according to

$$N = 156{,}250 \left(\tfrac{25}{9} \right)^{-t/4} \qquad 0 \leqslant t \leqslant 12$$

where t is an integer denoting the number of months. Determine the expected number of units to be sold in 2, 6, and 12 months.

Life Sciences

73. *Bacteria growth.* A bacteria culture with 10 bacteria is expected to increase in population, P, according to

$$P = 10(32^{t/10}) \qquad 0 \leqslant t \leqslant 20$$

where t is an integer denoting the number of hours. Determine the expected bacteria population in 2, 12, and 20 hours.

74. *Endangered species.* The population of an endangered bird is estimated to be 2,187. If protective measures are not taken, it is expected that the population, P, will decrease in coming years according to

$$P = 2{,}187(81^{-t/8}) \qquad 0 \leqslant t \leqslant 6$$

where t is an integer denoting the number of years. If protective measures are not taken, determine the expected population in 2, 4, and 6 years.

Social Sciences

75. *Communication.* In a poll it was found that only 1,280,000 people are aware of a particular piece of legislation being proposed in Congress. A political awareness group plans to increase the number of people, N, informed about the legislation according to

$$N = 1{,}280{,}000 \left(\tfrac{9}{4} \right)^{t/4} \qquad 0 \leqslant t \leqslant 12$$

where t is an integer denoting the number of months. How many people should be informed at the end of 2, 6, and 12 months?

3-4 Radicals

- Radicals
- Properties of Radicals
- Simplest Radical Form
- More about $\sqrt[n]{x^n}$

In Section 3-3 we introduced rational exponents by using the symbol $a^{1/n}$. We will now introduce *radical notation* using rational exponents. As you will see, radicals can be represented in terms of rational exponents and vice versa. Because of this, we will find that the properties of exponents give us corresponding properties for radicals.

■ Radicals

Recall that in Section 3-3 we defined $a^{1/n}$ to be the principal nth root of a. We can now state the following definition:

Radical Notation

If a represents a real number and n is a positive integer greater than 1, then

$$\sqrt[n]{a} = a^{1/n}$$

Thus, $\sqrt[n]{a}$ is another way to represent the principal nth root of a. The symbol $\sqrt{}$ is called a **radical,** n is the **index,** and a is the **radicand.** When $n = 2$, we use $\sqrt{a}$ to represent the positive square root of a, $a > 0$. (Remember, the index number n can never be negative or 0.)

Using this definition and the following relationships, we can easily convert rational exponent notation to radical form and vice versa:

$$a^{m/n} = (a^m)^{1/n} = \sqrt[n]{a^m} \qquad \text{This form is usually preferred.}$$

$$a^{m/n} = (a^{1/n})^m = (\sqrt[n]{a})^m$$

As before, we must require that a is not negative when n is even. In order to avoid any difficulties, we shall stipulate that **all variables represent positive real numbers unless stated otherwise.**

Example 18   Convert from rational exponent form to radical notation.

(A) $7^{1/2} = \sqrt{7}$

(B) $x^{3/5} = \sqrt[5]{x^3}$ or $(\sqrt[5]{x})^3$ In parts B–E, the first

(C) $-y^{4/7} = -\sqrt[7]{y^4}$ or $-(\sqrt[7]{y})^4$ form is generally preferred.

(D) $(2a^3b^2)^{3/4} = \sqrt[4]{(2a^3b^2)^3}$ or $(\sqrt[4]{2a^3b^2})^3$

(E) $z^{-2/5} = \dfrac{1}{z^{2/5}} = \dfrac{1}{\sqrt[5]{z^2}}$ or $\dfrac{1}{(\sqrt[5]{z})^2}$

(F) $(a^3 + b^3)^{1/3} = \sqrt[3]{a^3 + b^3}$ This is not equal to $a + b$.

Problem 18 Convert from rational exponent form to radical notation.

(A) $5^{1/3}$ (B) $y^{4/3}$ (C) $-x^{2/5}$
(D) $(5u^2v)^{3/2}$ (E) $w^{-5/6}$ (F) $(x^2 + y^2)^{1/2}$

Example 19 Convert from radical notation to rational exponent form.

(A) $\sqrt[5]{u} = u^{1/5}$ (B) $\sqrt[4]{a^3} = a^{3/4}$
(C) $-\sqrt[3]{x^2} = -x^{2/3}$ (D) $\sqrt[5]{(4x^2y)^3} = (4x^2y)^{3/5}$
(E) $\sqrt[4]{a^4 - b^4} = (a^4 - b^4)^{1/4}$ This is not equal to $a - b$.

Problem 19 Convert from radical notation to rational exponent form.

(A) $\sqrt[3]{a}$ (B) $\sqrt[3]{u^2}$ (C) $-\sqrt[4]{y^3}$ (D) $\sqrt[5]{(2u^2v^3)^2}$
(E) $\sqrt[3]{u^3 - v^3}$

■ Properties of Radicals

The properties of radicals listed in the box follow directly from the properties of exponents.

Properties of Radicals

If k, n, and m are natural numbers greater than or equal to 2, and if x and y are positive real numbers, then:

1. $\sqrt[n]{x^n} = x$ $\sqrt[3]{x^3} = x$

2. $\sqrt[n]{xy} = \sqrt[n]{x}\,\sqrt[n]{y}$ $\sqrt[5]{xy} = \sqrt[5]{x}\,\sqrt[5]{y}$

3. $\sqrt[n]{\dfrac{x}{y}} = \dfrac{\sqrt[n]{x}}{\sqrt[n]{y}}$ $\sqrt[4]{\dfrac{x}{y}} = \dfrac{\sqrt[4]{x}}{\sqrt[4]{y}}$

4. $\sqrt[kn]{x^{km}} = \sqrt[n]{x^m}$ $\sqrt[12]{x^8} = \sqrt[4\cdot 3]{x^{4\cdot 2}} = \sqrt[3]{x^2}$

5. $\sqrt[m]{\sqrt[n]{x}} = \sqrt[mn]{x}$ $\sqrt[4]{\sqrt[3]{x}} = \sqrt[12]{x}$

To illustrate how these properties can be derived using the properties of exponents, we will prove properties 2, 4, and 5.

Proof of 2 $\sqrt[n]{xy} = (xy)^{1/n} = x^{1/n}y^{1/n} = \sqrt[n]{x}\,\sqrt[n]{y}$

Proof of 4 $\sqrt[kn]{x^{km}} = (x^{km})^{1/kn} = x^{km/kn} = x^{m/n} = (x^m)^{1/n} = \sqrt[n]{x^m}$

Proof of 5 $\sqrt[m]{\sqrt[n]{x}} = (\sqrt[n]{x})^{1/m} = (x^{1/n})^{1/m} = x^{1/mn} = \sqrt[mn]{x}$

Using the properties of radicals we can simplify many expressions involving radicals, as illustrated below. (Remember that all variables represent positive real numbers.)

Example 20

(A) $\sqrt[3]{(5x^2y)^3} = 5x^2y$

(B) $\sqrt{3}\sqrt{15} = \sqrt{45} = \sqrt{9 \cdot 5}\ \boxed{= \sqrt{9}\sqrt{5}}\ = 3\sqrt{5}$

(C) $\sqrt[4]{\dfrac{a}{16}} = \dfrac{\sqrt[4]{a}}{\sqrt[4]{16}} = \dfrac{\sqrt[4]{a}}{2}$ or $\dfrac{1}{2} \cdot \sqrt[4]{a}$

(D) $\sqrt[9]{x^6} = \sqrt[3 \cdot 3]{x^{3 \cdot 2}} = \sqrt[3]{x^2}$

(E) $\sqrt[5]{\sqrt[3]{y}} = \sqrt[5 \cdot 3]{y} = \sqrt[15]{y}$

Problem 20

Simplify each expression

(A) $\sqrt[6]{(4xy^5)^6}$ (B) $\sqrt{5}\sqrt{15}$ (C) $\sqrt[3]{\dfrac{w}{27}}$ (D) $\sqrt[15]{y^{10}}$ (E) $\sqrt{\sqrt[4]{x}}$

▪ Simplest Radical Form

The properties of radicals allow us to convert expressions containing radicals into many equivalent forms. For example,

$$\sqrt{8} = \sqrt{4 \cdot 2} = \sqrt{4}\sqrt{2} = 2\sqrt{2} \qquad \sqrt{\dfrac{3}{2}} = \sqrt{\dfrac{3 \cdot 2}{2 \cdot 2}} = \sqrt{\dfrac{6}{4}} = \dfrac{\sqrt{6}}{\sqrt{4}} = \dfrac{\sqrt{6}}{2}$$

Radicals or expressions containing radicals are said to be in *simplest radical form* if the four conditions listed in the box are satisfied.

Simplest Radical Form

1. A radicand contains no factor to a power greater than or equal to the index of the radical.

   $\sqrt[3]{x^5}$ violates this condition.

2. The power of the radicand and the index of the radical have no common factor other than 1.

   $\sqrt[6]{x^4}$ violates this condition.

3. No radical appears in a denominator.

   $y/\sqrt[3]{x}$ violates this condition.

4. No fraction appears within a radical.

   $\sqrt[4]{\dfrac{3}{5}}$ violates this condition.

In calculations it may be desirable to use radical forms other than the simplest radical form. The choice will depend on the circumstances.

Example 21 Write each expression in simplest radical form.

(A) $\sqrt{175}$ (B) $\sqrt{27x^3}$ (C) $\sqrt[15]{a^6}$ (D) $\dfrac{6x}{\sqrt{2}}$ (E) $\sqrt{\dfrac{x}{2}}$

Solutions

(A) $\sqrt{175}$ is not in simplest radical form according to condition 1, since $175 = 5^2 \cdot 7$. We have

$$\sqrt{175} = \sqrt{5^2 \cdot 7} \;\boxed{= \sqrt{5^2}\,\sqrt{7}}\; = 5\sqrt{7}$$

or

$$\sqrt{175} = \sqrt{25 \cdot 7} \;\boxed{= \sqrt{25}\,\sqrt{7}}\; = 5\sqrt{7}$$

(B) $\sqrt{27x^3}$ is not in simplest radical form according to condition 1. Note that $27x^3 = 3^3x^3$. We have

$$\sqrt{27x^3} = \sqrt{3^3x^3} = \sqrt{(3^2x^2)(3x)}$$
$$\boxed{\begin{aligned} &= \sqrt{(3x)^2(3x)} \\ &= \sqrt{(3x)^2}\,\sqrt{3x} \end{aligned}}$$
$$= 3x\sqrt{3x}$$

or

$$\sqrt{27x^3} = \sqrt{(9x^2)(3x)} \;\boxed{= \sqrt{9x^2}\,\sqrt{3x}}\; = 3x\sqrt{3x}$$

(C) $\sqrt[15]{a^6}$ is not in simplest radical form according to condition 2, since 3 is a factor of both the power of a^6 and the index of the radical, 15. We have

$$\sqrt[15]{a^6} = \sqrt[3 \cdot 5]{a^{3 \cdot 2}} = \sqrt[5]{a^2}$$

(D) $\dfrac{6x}{\sqrt{2}}$ violates condition 3, since the denominator contains $\sqrt{2}$. We have

$$\frac{6x}{\sqrt{2}} = \frac{6x}{\sqrt{2}} \cdot \frac{\sqrt{2}}{\sqrt{2}} = \frac{6x\sqrt{2}}{2} = 3x\sqrt{2}$$

(E) $\sqrt{\dfrac{x}{2}}$ violates condition 4, since a fraction occurs within a radical. We have

$$\sqrt{\frac{x}{2}} \;\boxed{= \sqrt{\frac{x \cdot 2}{2 \cdot 2}} = \sqrt{\frac{2x}{2^2}} = \frac{\sqrt{2x}}{\sqrt{2^2}}}\; = \frac{\sqrt{2x}}{2}$$

Or, we could write

$$\sqrt{\frac{x}{2}} = \frac{\sqrt{x}}{\sqrt{2}} \;\boxed{= \frac{\sqrt{x}}{\sqrt{2}} \cdot \frac{\sqrt{2}}{\sqrt{2}}}\; = \frac{\sqrt{2x}}{2}$$

Problem 21   Write each expression in simplest radical form.

(A) $\sqrt{45}$ (B) $\sqrt{8y^3}$ (C) $\sqrt[18]{x^{12}}$ (D) $\dfrac{6z}{\sqrt{3}}$ (E) $\sqrt{\dfrac{x}{3}}$

In Example 21, we dealt primarily with square roots. Simplifying radical expressions where the index n is greater than 2 is accomplished in much the same manner. For example, we can simplify

$$\sqrt[3]{54x^6y^{10}z^2}$$

as follows:

$$\sqrt[3]{54x^6y^{10}z^2} = \sqrt[3]{(27x^6y^9)(2yz^2)}$$
$$\boxed{= \sqrt[3]{27x^6y^9}\,\sqrt[3]{2yz^2}}$$
$$= 3x^2y^3\sqrt[3]{2yz^2}$$

Write the radicand as a product so that the cube root of the first factor can be expressed without radicals and the cube root of the second factor will be in simplest radical form.

To simplify the radical expression

$$\frac{6xy}{\sqrt[3]{2x^2y}}$$

we would proceed as follows:

$$\frac{6xy}{\sqrt[3]{2x^2y}} = \frac{6xy}{\sqrt[3]{2x^2y}} \cdot \frac{\sqrt[3]{2^2xy^2}}{\sqrt[3]{2^2xy^2}}$$
$$\boxed{= \frac{6xy\sqrt[3]{2^2xy^2}}{\sqrt[3]{2^3x^3y^3}}}$$
$$= \frac{6xy\sqrt[3]{4xy^2}}{2xy}$$
$$= 3\sqrt[3]{4xy^2}$$

Notice that in order to obtain the radical $\sqrt[3]{2^3x^3y^3}$ in the denominator (which simplifies to $2xy$), we must multiply the numerator and the denominator by $\sqrt[3]{2^2xy^2}$.

Removing radicals from a denominator, as illustrated above, is commonly referred to as *rationalizing the denominator*. Keep this in mind—we will discuss this process again in Section 3-5.

Example 22   Write each expression in simplest radical form.

(A) $\sqrt[3]{375} = \sqrt[3]{125 \cdot 3} = \sqrt[3]{5^3 \cdot 3}\;\boxed{= \sqrt[3]{5^3}\,\sqrt[3]{3}}\;= 5\sqrt[3]{3}$

(B) $\sqrt{24x^5y^2z^7} = \sqrt{(4x^4y^2z^6)(6xz)}\;\boxed{= \sqrt{4x^4y^2z^6}\,\sqrt{6xz}}\;= 2x^2yz^3\sqrt{6xz}$

(C) $\sqrt[5]{64x^8y^2} = \sqrt[5]{(32x^5)(2x^3y^2)}\;\boxed{= \sqrt[5]{32x^5}\,\sqrt[5]{2x^3y^2}}\;= 2x\sqrt[5]{2x^3y^2}$

(D) $\dfrac{12x^2y\sqrt{5}}{\sqrt{6xy}} = \dfrac{12x^2y\sqrt{5}}{\sqrt{6xy}} \cdot \dfrac{\sqrt{6xy}}{\sqrt{6xy}} = \dfrac{12x^2y\sqrt{30xy}}{6xy} = 2x\sqrt{30xy}$

(E) $\dfrac{6xy^2}{\sqrt[3]{9x^2y}} = \dfrac{6xy^2}{\sqrt[3]{3^2x^2y}} \cdot \dfrac{\sqrt[3]{3xy^2}}{\sqrt[3]{3xy^2}} = \dfrac{6xy^2\sqrt[3]{3xy^2}}{\sqrt[3]{3^3x^3y^3}} = \dfrac{6xy^2\sqrt[3]{3xy^2}}{3xy} = 2y\sqrt[3]{3xy^2}$

(F) $\sqrt[3]{\dfrac{3}{4x}} = \sqrt[3]{\dfrac{3}{2^2x} \cdot \dfrac{2x^2}{2x^2}} = \sqrt[3]{\dfrac{6x^2}{2^3x^3}} = \dfrac{\sqrt[3]{6x^2}}{\sqrt[3]{2^3x^3}} = \dfrac{\sqrt[3]{6x^2}}{2x}$

Problem 22 Write each expression in simplest radical form.

(A) $\sqrt[3]{128}$ (B) $\sqrt{75a^3b^4c^5}$ (C) $\sqrt[4]{64x^9y^3}$ (D) $\dfrac{9uv^2\sqrt{7}}{\sqrt{3uv}}$

(E) $\dfrac{4a^2b}{\sqrt[3]{4ab^2}}$ (F) $\sqrt[3]{\dfrac{4x}{3y^2}}$

■ More about $\sqrt[n]{x^n}$

So far, in our discussion we have restricted all variables to positive real numbers in order to avoid any difficulties. If we removed this restriction, we would no longer be able to say in general that

$$\sqrt[n]{x^n} = x$$

Of course, this continues to hold if $x = 0$; however, if x is negative, it may not be true. To see this, let us consider $\sqrt{x^2}$. For $x = 5$, we have

$$\sqrt{5^2} = 5 = x$$

For $x = -5$, we have

$$\sqrt{(-5)^2} = \sqrt{25} = \sqrt{5^2} = 5 \neq x \qquad \text{Remember, } x = -5.$$

Thus, it is not true that

$$\sqrt{x^2} = x$$

for all real numbers.

We have the following important result:

If x represents a real number, then

$$\sqrt{x^2} = |x|$$

Recall from Chapter 1 that $|x|$ denotes the absolute value of x and is defined by

$$|x| = \begin{cases} x & \text{if } x \text{ is positive} \\ 0 & \text{if } x = 0 \\ -x & \text{if } x \text{ is negative} \end{cases}$$

Note that $-x$ *is positive when x is negative.* Thus, we have

$$\sqrt{5^2} = |5| = 5 \quad \text{and} \quad \sqrt{(-5)^2} = |-5| = 5$$

which is consistent with our previous calculations.

Now, consider $\sqrt[3]{x^3}$. For $x = 5$, we have

$$\sqrt[3]{5^3} = 5 = x$$

For $x = -5$, we have

$$\sqrt[3]{(-5)^3} = \sqrt[3]{-125} = -5 = x$$

In general, we have the following:

If x represents a real number, then

$$\sqrt[3]{x^3} = x$$

When variables represent real numbers, we must be very careful in simplifying an expression such as

$$\sqrt[3]{x^3} + \sqrt{x^2}$$

It is a common error to set this expression equal to $2x$. According to the discussion above, we have

$$\sqrt[3]{x^3} + \sqrt{x^2} = x + |x|$$

When x is positive or 0, we have

$$x + |x| = x + x = 2x \qquad |x| = x \text{ if } x \geqslant 0$$

But when x is negative, we have

$$x + |x| = x + (-x) = 0 \qquad |x| = -x \text{ if } x < 0$$

Thus,

$$\sqrt[3]{x^3} + \sqrt{x^2} = \begin{cases} 2x & \text{if } x \geqslant 0 \\ 0 & \text{if } x < 0 \end{cases}$$

Example 23 Simplify: $3\sqrt[3]{x^3} - 2\sqrt{x^2}$

Solution We have

$$3\sqrt[3]{x^3} - 2\sqrt{x^2} = 3x - 2|x|$$

For $x \geqslant 0$, we have $|x| = x$ and

$$3x - 2|x| = 3x - 2x = x$$

For $x < 0$, we have $|x| = -x$ and

$$3x - 2|x| = 3x - 2(-x) = 3x + 2x = 5x$$

Thus,

$$3\sqrt[3]{x^3} - 2\sqrt{x^2} = \begin{cases} x & \text{if } x \geqslant 0 \\ 5x & \text{if } x < 0 \end{cases}$$

Problem 23 Simplify: $2\sqrt[3]{x^3} - 3\sqrt{x^2}$

The above discussion generalizes to expressions involving $\sqrt[n]{x^n}$:

If x represents a real number, then

$$\sqrt[n]{x^n} = |x| \qquad \text{if } n \text{ is even}$$
$$\sqrt[n]{x^n} = x \qquad \text{if } n \text{ is odd}$$

For example,

$$\sqrt[100]{(-4)^{100}} = |-4| = 4 \qquad \text{and} \qquad \sqrt[99]{(-4)^{99}} = -4$$

Answers to
Matched Problems

18. (A) $\sqrt[3]{5}$ (B) $\sqrt[3]{y^4}$ or $(\sqrt[3]{y})^4$ (C) $-\sqrt[5]{x^2}$ or $-(\sqrt[5]{x})^2$

(D) $\sqrt{(5u^2v)^3}$ or $(\sqrt{5u^2v})^3$ (E) $\dfrac{1}{\sqrt[6]{w^5}}$ or $\dfrac{1}{(\sqrt[6]{w})^5}$

(F) $\sqrt{x^2 + y^2}$ (*not* $x + y$)

19. (A) $a^{1/3}$ (B) $u^{2/3}$ (C) $-y^{3/4}$ (D) $(2u^2v^3)^{2/5}$
(E) $(u^3 - v^3)^{1/3}$

20. (A) $4xy^5$ (B) $5\sqrt{3}$ (C) $\dfrac{\sqrt[3]{w}}{3}$ or $\dfrac{1}{3} \cdot \sqrt[3]{w}$ (D) $\sqrt[3]{y^2}$

(E) $\sqrt[8]{x}$

21. (A) $3\sqrt{5}$ (B) $2y\sqrt{2y}$ (C) $\sqrt[3]{x^2}$ (D) $2z\sqrt{3}$ (E) $\dfrac{\sqrt{3x}}{3}$

22. (A) $4\sqrt[3]{2}$ (B) $5ab^2c^2\sqrt{3ac}$ (C) $2x^2\sqrt[4]{4xy^3}$ (D) $3v\sqrt{21uv}$

(E) $2a\sqrt[3]{2a^2b}$ (F) $\dfrac{\sqrt[3]{36xy}}{3y}$

23. $2\sqrt[3]{x^3} - 3\sqrt{x^2} = \begin{cases} -x & \text{if } x \geqslant 0 \\ 5x & \text{if } x < 0 \end{cases}$

Exercise 3-4

Unless stated otherwise, all variables represent positive real numbers.

A *Convert from rational exponent form to radical notation. (Do not simplify.)*

1. $15^{1/2}$	2. $33^{1/3}$	3. $x^{3/7}$	4. $y^{4/5}$
5. $8m^{2/3}$	6. $15w^{3/5}$	7. $(5y)^{4/5}$	8. $(7z)^{5/8}$
9. $(6a^3b^2)^{2/3}$	10. $(25u^4v^2)^{3/7}$	11. $(x-y)^{1/2}$	12. $(m^2+n^2)^{1/2}$

Convert from radical notation to rational exponent form. (Do not simplify.)

13. $\sqrt[7]{x}$	14. $\sqrt[5]{w}$	15. $\sqrt[5]{z^3}$	16. $\sqrt[6]{a^5}$
17. $\sqrt[3]{(3x^2y^3)^2}$	18. $\sqrt[5]{(7u^4v)^4}$	19. $\sqrt{a^2+b^2}$	20. $\sqrt{x^2+4}$

Write each expression in simplest radical form.

21. $\sqrt{49z^2}$	22. $\sqrt{81a^2}$	23. $\sqrt{144u^4v^6}$	24. $\sqrt{36a^8b^{10}}$
25. $\sqrt{27}$	26. $\sqrt{32}$	27. $\sqrt{50x^3}$	28. $\sqrt{125y^5}$
29. $\sqrt{48u^4v^7}$	30. $\sqrt{150m^7n^6}$	31. $\dfrac{2}{\sqrt{a}}$	32. $\dfrac{5}{\sqrt{w}}$
33. $\sqrt{\dfrac{2x}{3y}}$	34. $\sqrt{\dfrac{5u}{3v}}$	35. $\dfrac{24ab^3}{\sqrt{3ab}}$	36. $\dfrac{6m^2n}{\sqrt{3mn}}$
37. $\sqrt[6]{x^4}$	38. $\sqrt[8]{y^6}$	39. $\sqrt[4]{\sqrt{x^3}}$	40. $\sqrt[3]{\sqrt[5]{y^4}}$

B *Convert from rational exponent form to radical notation. (Do not simplify.)*

41. $y^{-4/7}$	42. $x^{-3/5}$	43. $(3r^2s^3)^{-3/4}$	44. $(7u^4v)^{-5/6}$
45. $x^{1/2}+x^{1/3}$	46. $y^{1/3}-y^{-1/4}$	47. $a^{-1/2}+b^{-1/2}$	48. $(u^4+v^4)^{-2/3}$

Convert from radical notation to rational exponent form. (Do not simplify.)

49. $-3\sqrt[4]{3x^3y}$	50. $-5a\sqrt[3]{2a^2b}$	51. $\sqrt[4]{(u^2-v^2)^3}$
52. $\sqrt[3]{(a^3-b^3)^2}$	53. $\dfrac{5}{\sqrt[5]{y^2}}$	54. $\dfrac{3}{\sqrt[4]{z^3}}$
55. $\dfrac{x}{\sqrt{y}}-\dfrac{y}{\sqrt{x}}$	56. $\dfrac{2}{\sqrt[3]{y^2}}+\dfrac{5}{\sqrt[4]{y^3}}$	

Write each expression in simplest radical form.

57. $\sqrt[3]{8u^6v^9}$	58. $\sqrt[4]{81a^{12}b^8}$	59. $\sqrt[3]{24m^7n^9}$
60. $\sqrt[4]{32u^{12}v^7}$	61. $\sqrt[5]{64x^{10}y^{17}z^9}$	62. $\sqrt[6]{128a^{16}b^{12}c^7}$
63. $\sqrt[3]{16a^2b^4}\,\sqrt[3]{4a^3b^2}$	64. $\sqrt[5]{4m^7n^4}\,\sqrt[5]{16m^3n^4}$	65. $\sqrt[3]{\dfrac{8a^9}{27b^3}}$

66. $\sqrt[4]{\dfrac{16b^4}{625c^8}}$ **67.** $\dfrac{1}{\sqrt[3]{3x^2}}$ **68.** $\dfrac{1}{\sqrt[3]{4y}}$

69. $\dfrac{15x^2y}{\sqrt[3]{25xy^2}}$ **70.** $\dfrac{21ab^3}{\sqrt[3]{3a^2b}}$

C   *Write each expression in simplest radical form.*

71. $\sqrt{4a^2 + 4b^2}$ **72.** $\sqrt{4m^2 - 16}$ **73.** $\dfrac{x - y}{\sqrt{x^2 - y^2}}$

74. $\dfrac{a + b}{\sqrt{a^2 - b^2}}$ **75.** $\dfrac{a - b}{\sqrt[3]{a - b}}$ **76.** $\dfrac{u - v}{\sqrt[3]{(u - v)^2}}$

Simplify each expression for: (A) $x \geqslant 0$ (B) $x < 0$

77. $3\sqrt{x^2} + 4\sqrt[3]{x^3}$ **78.** $5\sqrt[3]{x^3} - 2\sqrt{x^2}$ **79.** $4\sqrt[3]{x^3} - 5\sqrt[4]{x^4}$

80. $5\sqrt[4]{x^4} + 7\sqrt[5]{x^5}$

3-5   Basic Operations on Radicals

- Addition and Subtraction
- Multiplication
- Quotients — Rationalizing Denominators

We will now discuss basic operations with expressions containing radical forms. For simplicity, we will again make the restriction that *all variables represent positive real numbers.*

■ Addition and Subtraction

Just as we were able to simplify many polynomial expressions by combining like terms, we can simplify expressions containing radicals by combining terms that have the same radical forms. This is again justified by the distributive law, as illustrated in Example 24.

Example 24   Simplify by combining as many terms as possible.

(A) $3\sqrt{5} + 7\sqrt{5} \boxed{= (3 + 7)\sqrt{5}} = 10\sqrt{5}$

(B) $8\sqrt[3]{2x^2} - 12\sqrt[3]{2x^2} \boxed{= (8 - 12)\sqrt[3]{2x^2}} = -4\sqrt[3]{2x^2}$

(C) $5\sqrt{ab} - 2\sqrt[3]{ab^2} - 3\sqrt{ab} - 6\sqrt[3]{ab^2}$

$\boxed{= 5\sqrt{ab} - 3\sqrt{ab} - 2\sqrt[3]{ab^2} - 6\sqrt[3]{ab^2}}$

$= 2\sqrt{ab} - 8\sqrt[3]{ab^2}$

Problem 24   Simplify by combining as many terms as possible.

(A) $7\sqrt{3} - 13\sqrt{3}$ (B) $5\sqrt[4]{2x^3y} + 6\sqrt[4]{2x^3y}$

(C) $4\sqrt[3]{3u^2v} - 8\sqrt{5uv} - 2\sqrt[3]{3u^2v} + 6\sqrt{5uv}$

Some expressions containing radical forms can be simplified by first writing the expressions in simplest radical form.

Example 25   Simplify by writing in simplest radical form and combining terms whenever possible.

(A) $5\sqrt{27} - 2\sqrt{75} = 5\sqrt{9 \cdot 3} - 2\sqrt{25 \cdot 3}$ Express each radical in simplest radical form.

$\qquad\qquad\qquad = 5 \cdot 3\sqrt{3} - 2 \cdot 5\sqrt{3}$

$\qquad\qquad\qquad = 15\sqrt{3} - 10\sqrt{3} = 5\sqrt{3}$

(B) $4\sqrt{32} + \dfrac{8}{\sqrt{2}} = 4\sqrt{16 \cdot 2} + \dfrac{8}{\sqrt{2}} \cdot \dfrac{\sqrt{2}}{\sqrt{2}}$ Write in simplest radical form.

$\qquad\qquad\qquad = 4 \cdot 4\sqrt{2} + \dfrac{8\sqrt{2}}{2}$

$\qquad\qquad\qquad = 16\sqrt{2} + 4\sqrt{2} = 20\sqrt{2}$

(C) $5\sqrt[3]{2x^4} - x\sqrt[3]{16x} = 5\sqrt[3]{x^3 \cdot 2x} - x\sqrt[3]{8 \cdot 2x}$ Write in simplest radical form.

$\qquad\qquad\qquad = 5x\sqrt[3]{2x} - 2x\sqrt[3]{2x}$

$\qquad\qquad\qquad = 3x\sqrt[3]{2x}$

(D) $\sqrt[3]{\dfrac{1}{9}} + \sqrt[3]{81} = \sqrt[3]{\dfrac{1}{3^2} \cdot \dfrac{3}{3}} + \sqrt[3]{27 \cdot 3}$

$\qquad\qquad\qquad = \sqrt[3]{\dfrac{3}{3^3}} + 3\sqrt[3]{3}$

$\qquad\qquad\qquad = \dfrac{1}{3}\sqrt[3]{3} + 3\sqrt[3]{3} = \dfrac{10}{3}\sqrt[3]{3}$ or $\dfrac{10\sqrt[3]{3}}{3}$

Problem 25   Simplify by writing in simplest radical form and combining terms whenever possible.

(A) $6\sqrt{125} - 3\sqrt{45}$ (B) $\dfrac{12}{\sqrt{3}} - 2\sqrt{27}$ (C) $7\sqrt[4]{2y^5} + 2y\sqrt[4]{32y}$

(D) $\sqrt[3]{625} - 10\sqrt[3]{\frac{1}{25}}$

■ Multiplication

Many expressions involving radicals can be multiplied in the same manner in which we multiplied polynomials. The distributive property justifies this procedure. Example 26 illustrates several different products involving radicals.

Example 26 Multiply and simplify whenever possible.

(A) $\sqrt{5}(\sqrt{15}-4)\;\boxed{=\sqrt{5}\cdot\sqrt{15}-\sqrt{5}\cdot 4}$

$$=\sqrt{75}-4\sqrt{5}$$
$$=\sqrt{25\cdot 3}-4\sqrt{5}$$
$$=5\sqrt{3}-4\sqrt{5}$$

(B) $(\sqrt{3}+4)(\sqrt{3}-5)\;\boxed{=\sqrt{3}\cdot\sqrt{3}+4\sqrt{3}-5\sqrt{3}-20}$

$$=3-\sqrt{3}-20=-17-\sqrt{3}$$

(C) $(3\sqrt{5}-2\sqrt{3})(\sqrt{5}+\sqrt{3})\;\boxed{\begin{aligned}&=3\sqrt{5}\cdot\sqrt{5}-2\sqrt{15}+3\sqrt{15}-2\sqrt{3}\cdot\sqrt{3}\\&=3\cdot 5+\sqrt{15}-2\cdot 3\end{aligned}}$

$$=15+\sqrt{15}-6=9+\sqrt{15}$$

(D) $(\sqrt{a}-5)(\sqrt{a}+3)\;\boxed{=\sqrt{a}\cdot\sqrt{a}-5\sqrt{a}+3\sqrt{a}-15}$

$$=a-2\sqrt{a}-15$$

(E) $(\sqrt[3]{x^2}+\sqrt[3]{y})(\sqrt[3]{x}-\sqrt[3]{y^2})\;\boxed{=\sqrt[3]{x^3}+\sqrt[3]{xy}-\sqrt[3]{x^2y^2}-\sqrt[3]{y^3}}$

$$=x+\sqrt[3]{xy}-\sqrt[3]{x^2y^2}-y$$

Problem 26 Multiply and simplify whenever possible.

(A) $\sqrt{3}(\sqrt{15}-7)$ (B) $(\sqrt{7}-2)(\sqrt{7}+4)$

(C) $(2\sqrt{3}-3\sqrt{2})(2\sqrt{3}+3\sqrt{2})$ (D) $(\sqrt{x}+6)(\sqrt{x}-3)$

(E) $(\sqrt[3]{a}-\sqrt[3]{b^2})(\sqrt[3]{a^2}+\sqrt[3]{b})$

Example 27 Evaluate x^2-6x+7 using $x=3-\sqrt{2}$.

Solution For $x=3-\sqrt{2}$, we have

$$x^2-6x+7=(3-\sqrt{2})^2-6(3-\sqrt{2})+7$$
$$=9-6\sqrt{2}+2-18+6\sqrt{2}+7$$
$$=0$$

Problem 27 Evaluate x^2-6x+7 using $x=3+\sqrt{2}$.

■ Quotients — Rationalizing Denominators

In Section 3-4 we found that an expression such as $\sqrt{5}/\sqrt{3}$ can be reduced to simplest radical form by multiplying the numerator and the denominator by $\sqrt{3}$. Thus, we obtain

$$\frac{\sqrt{5}}{\sqrt{3}}\;\boxed{=\frac{\sqrt{5}}{\sqrt{3}}\cdot\frac{\sqrt{3}}{\sqrt{3}}=\frac{\sqrt{15}}{\sqrt{3^2}}}=\frac{\sqrt{15}}{3}$$

The process by which a denominator is cleared of radicals is called **rationalizing the denominator.**

It is natural to ask if we can rationalize the denominator of an expression such as

$$\frac{4}{\sqrt{5} - \sqrt{3}}$$

That is, can we write this expression in a form where no radical appears in the denominator? The answer is yes. However, before we illustrate the procedure, it will be useful to recall the special product.

$$(a - b)(a + b) = a^2 - b^2$$

For example,

$$(\sqrt{5} - \sqrt{3})(\sqrt{5} + \sqrt{3}) \boxed{= (\sqrt{5})^2 - (\sqrt{3})^2} = 5 - 3 = 2$$

This suggests that we can rationalize the denominator of

$$\frac{4}{\sqrt{5} - \sqrt{3}}$$

by multiplying the numerator and the denominator by $\sqrt{5} + \sqrt{3}$ (which is obtained by changing the middle sign of the denominator). Thus,

$$\frac{4}{\sqrt{5} - \sqrt{3}} = \frac{4}{\sqrt{5} - \sqrt{3}} \cdot \frac{\sqrt{5} + \sqrt{3}}{\sqrt{5} + \sqrt{3}} \qquad \text{Multiply numerator and denominator by } \sqrt{5} + \sqrt{3}.$$

$$= \frac{4\sqrt{5} + 4\sqrt{3}}{5 - 3}$$

$$= \frac{4\sqrt{5} + 4\sqrt{3}}{2}$$

$$= \frac{2(2\sqrt{5} + 2\sqrt{3})}{2} \qquad \text{Simplify by factoring 2 from the numerator and canceling.}$$

$$= 2\sqrt{5} + 2\sqrt{3}$$

Example 28 further illustrates this procedure for rationalizing denominators.

Example 28   Rationalize the denominator and simplify whenever possible.

(A) $\dfrac{\sqrt{3}}{3 + \sqrt{6}} = \dfrac{\sqrt{3}}{3 + \sqrt{6}} \cdot \dfrac{3 - \sqrt{6}}{3 - \sqrt{6}}$ Multiply numerator and denominator by $3 - \sqrt{6}$, obtained by changing the middle sign of $3 + \sqrt{6}$.

$$= \frac{3\sqrt{3} - \sqrt{18}}{9 - 6}$$

$$= \frac{3\sqrt{3} - 3\sqrt{2}}{3} \qquad \text{Reduce.}$$

$$= \frac{3(\sqrt{3} - \sqrt{2})}{3}$$

$$= \sqrt{3} - \sqrt{2}$$

(B) $\dfrac{\sqrt{5}}{2\sqrt{5}+3\sqrt{3}} = \dfrac{\sqrt{5}}{2\sqrt{5}+3\sqrt{3}} \cdot \dfrac{2\sqrt{5}-3\sqrt{3}}{2\sqrt{5}-3\sqrt{3}}$ Multiply numerator and denominator by $2\sqrt{5}-3\sqrt{3}$.

$\qquad\qquad = \dfrac{10-3\sqrt{15}}{20-27}$

$\qquad\qquad = \dfrac{10-3\sqrt{15}}{-7}$

$\qquad\qquad = -\dfrac{10-3\sqrt{15}}{7} \quad\text{or}\quad \dfrac{-10+3\sqrt{15}}{7}$

(C) $\dfrac{\sqrt{a}+\sqrt{b}}{\sqrt{a}-\sqrt{b}} = \dfrac{\sqrt{a}+\sqrt{b}}{\sqrt{a}-\sqrt{b}} \cdot \dfrac{\sqrt{a}+\sqrt{b}}{\sqrt{a}+\sqrt{b}}$ Multiply numerator and denominator by $\sqrt{a}+\sqrt{b}$.

$\qquad\qquad = \dfrac{a+2\sqrt{ab}+b}{a-b}$

Problem 28   Rationalize the denominator and simplify whenever possible.

(A) $\dfrac{\sqrt{2}}{\sqrt{10}-2}$ (B) $\dfrac{\sqrt{3}}{4\sqrt{3}+3\sqrt{2}}$ (C) $\dfrac{\sqrt{x}-\sqrt{y}}{\sqrt{x}+\sqrt{y}}$

Answers to Matched Problems

24. (A) $-6\sqrt{3}$ (B) $11\sqrt[4]{2x^3y}$ (C) $2\sqrt[3]{3u^2v}-2\sqrt{5uv}$

25. (A) $21\sqrt{5}$ (B) $-2\sqrt{3}$ (C) $11y\sqrt[4]{2y}$ (D) $3\sqrt[3]{5}$

26. (A) $3\sqrt{5}-7\sqrt{3}$ (B) $-1+2\sqrt{7}$ (C) -6
    (D) $x+3\sqrt{x}-18$ (E) $a-\sqrt[3]{a^2b^2}+\sqrt[3]{ab}-b$

27. 0   28. (A) $\dfrac{\sqrt{5}+\sqrt{2}}{3}$ (B) $\dfrac{4-\sqrt{6}}{10}$ (C) $\dfrac{x-2\sqrt{xy}+y}{x-y}$

Exercise 3-5

In the following problems all variables represent positive real numbers.

A   Simplify by writing in simplest radical form and combining terms whenever possible.

1. $3\sqrt{x}-8\sqrt{x}$
2. $5\sqrt{7}-3\sqrt{7}$
3. $\sqrt{7}+3\sqrt{3}-5\sqrt{7}$
4. $\sqrt{m}-3\sqrt{n}+5\sqrt{m}$
5. $\sqrt{12}-\sqrt{3}$
6. $\sqrt{50}+\sqrt{2}$
7. $\sqrt{18}+2\sqrt{2}$
8. $\sqrt{27}-\sqrt{3}$

Multiply and simplify whenever possible.

9. $\sqrt{5}(\sqrt{5}-3)$
10. $\sqrt{11}(\sqrt{11}-2)$
11. $\sqrt{u}(\sqrt{u}+3)$
12. $\sqrt{a}(\sqrt{a}-5)$
13. $\sqrt{x}(7-\sqrt{x})$
14. $\sqrt{z}(5+\sqrt{z})$
15. $\sqrt{5}(2\sqrt{15}-3\sqrt{2})$
16. $\sqrt{7}(3\sqrt{3}-4\sqrt{14})$
17. $(\sqrt{5}-3)(\sqrt{5}+3)$
18. $(\sqrt{3}+4)(\sqrt{3}-4)$

19. $(\sqrt{2} + 5)^2$

20. $(\sqrt{5} - 3)^2$

21. $(\sqrt{w} - 2)(\sqrt{w} + 2)$

22. $(\sqrt{a} + 5)(\sqrt{a} - 5)$

Rationalize each denominator and simplify.

23. $\dfrac{1}{\sqrt{5} - 2}$

24. $\dfrac{3}{\sqrt{7} - 1}$

25. $\dfrac{5}{4 + \sqrt{6}}$

26. $\dfrac{2}{3 - \sqrt{3}}$

27. $\dfrac{\sqrt{3}}{\sqrt{6} + 2}$

28. $\dfrac{\sqrt{2}}{\sqrt{10} - 2}$

29. $\dfrac{\sqrt{z}}{\sqrt{z} - 5}$

30. $\dfrac{\sqrt{a}}{\sqrt{a} + 3}$

B *Simplify by writing in simplest radical form and combining terms whenever possible.*

31. $\sqrt{50x} - \sqrt{8x}$

32. $\sqrt{27uv} + 2\sqrt{12uv}$

33. $2\sqrt{8} + 3\sqrt{18} - \sqrt{32}$

34. $2\sqrt{27} - \sqrt{48} + 3\sqrt{75}$

35. $2\sqrt{8x^3} + 2x\sqrt{18x}$

36. $3y\sqrt{75y} - 4\sqrt{27y^3}$

37. $\sqrt[3]{24z^3} - 3\sqrt[3]{3z^3} + z\sqrt[3]{81}$

38. $x\sqrt[4]{162} - \sqrt[4]{2x^4} + 3\sqrt[4]{32x^4}$

39. $\sqrt{\tfrac{1}{3}} + \sqrt{27}$

40. $\sqrt{8} - \sqrt{\tfrac{1}{2}}$

41. $\sqrt[3]{40} - \sqrt[3]{\tfrac{1}{25}}$

42. $\sqrt[4]{32} + \dfrac{1}{\sqrt[4]{8}}$

43. $\sqrt{\dfrac{2mn}{3}} + \sqrt{54mn}$

44. $\sqrt{50xy} - \sqrt{\dfrac{xy}{2}}$

In Problems 45–56 multiply and simplify the product whenever possible.

45. $(\sqrt{x} - 3)(\sqrt{x} + 5)$

46. $(\sqrt{z} + 4)(\sqrt{z} - 7)$

47. $(2\sqrt{3} - 1)(3\sqrt{3} + 4)$

48. $(3\sqrt{5} - 5)(2\sqrt{5} - 3)$

49. $(3\sqrt{a} - 5)(2\sqrt{a} + 3)$

50. $(5\sqrt{v} + 2)(3\sqrt{v} + 5)$

51. $(5\sqrt{2} - 3\sqrt{3})(3\sqrt{2} + 4\sqrt{3})$

52. $(4\sqrt{3} - 3\sqrt{5})(2\sqrt{3} - 2\sqrt{5})$

53. $(\sqrt{m} - \sqrt{n})(\sqrt{m} + \sqrt{n})$

54. $(2\sqrt{x} - 3\sqrt{y})(2\sqrt{x} + 3\sqrt{y})$

55. $(\sqrt[3]{m^2} - \sqrt[3]{n})(\sqrt[3]{m} + \sqrt[3]{n^2})$

56. $(\sqrt[3]{a} - 2\sqrt[3]{b^2})(\sqrt[3]{a^2} - 3\sqrt[3]{b})$

57. Evaluate $x^2 - 10x + 4$ using $x = 5 + \sqrt{21}$.

58. Evaluate $x^2 - 10x + 4$ using $x = 5 - \sqrt{21}$.

Rationalize each denominator and simplify.

59. $\dfrac{\sqrt{5} - 3}{\sqrt{5} + 3}$

60. $\dfrac{\sqrt{3} + 4}{\sqrt{3} - 4}$

61. $\dfrac{\sqrt{3} - \sqrt{2}}{\sqrt{3} + \sqrt{2}}$

62. $\dfrac{\sqrt{2} + \sqrt{5}}{\sqrt{2} - \sqrt{5}}$

63. $\dfrac{\sqrt{x} + 2}{\sqrt{x} - 2}$

64. $\dfrac{\sqrt{y} - 5}{\sqrt{y} + 5}$

65. $\dfrac{2\sqrt{3} - \sqrt{2}}{3\sqrt{3} + \sqrt{2}}$

66. $\dfrac{2\sqrt{2} + 3\sqrt{3}}{4\sqrt{2} - 2\sqrt{3}}$

C *Simplify by writing in simplest radical form and combining terms whenever possible.*

67. $\sqrt{\tfrac{3}{2}} + \sqrt{\tfrac{2}{3}} - \sqrt{\tfrac{1}{6}}$

68. $\sqrt{\tfrac{3}{5}} - \sqrt{\tfrac{5}{3}} + \sqrt{\tfrac{1}{15}}$

69. $\frac{1}{4}x\sqrt{2xy^3} - \frac{1}{6}y\sqrt{2x^3y} + \frac{1}{3}xy\sqrt{18xy}$

70. $a\sqrt[3]{81ab^3} - b\sqrt[3]{24a^4} + 3\sqrt[3]{3a^4b^3}$

Multiply and simplify the product whenever possible.

71. $(\sqrt[5]{x^2} - 3\sqrt[5]{y^3})(\sqrt[5]{x^3} + 2\sqrt[5]{y^2})$ 72. $(2\sqrt[4]{x^3} + \sqrt[4]{y})(\sqrt[4]{x} - 4\sqrt[4]{y^3})$

Rationalize each denominator and simplify.

73. $\dfrac{5\sqrt{u} - 3\sqrt{v}}{4\sqrt{u} + 3\sqrt{v}}$ 74. $\dfrac{2\sqrt{x} + 3\sqrt{y}}{5\sqrt{x} + 2\sqrt{y}}$

3-6 Chapter Review

Important Terms and Symbols

3-1 *Integer exponents.* positive integer exponents, zero exponents, negative integer exponents, growth, decay, a^n, a^0, a^{-n}

3-2 *Scientific notation.* scientific notation, powers of 10, $a \times 10^n$, $a \times 10^{-n}$

3-3 *Rational exponents.* root, square root, cube root, nth root, principal nth root, irrational numbers, rational exponents, $a^{1/n}$, $\approx$, $a^{m/n}$, $a^{-m/n}$

3-4 *Radicals.* radical notation, radical, index, radicand, equivalent rational exponent form, properties of radicals, simplest radical form, rationalizing denominators, $\sqrt{a}$, $\sqrt[n]{a}$, $\sqrt[n]{a^m}$, $(\sqrt[n]{a})^m$

3-5 *Basic operations on radicals.* simplifying radical expressions, addition, subtraction, multiplication, rationalizing denominators

Exercise 3-6 Chapter Review

Work through all the problems in this chapter review and check your answers in the back of the book. (Answers to all review problems are there.) Where weaknesses show up, review appropriate sections in the text. When you are satisfied that you know the material, take the practice test following this review.

A   *Give the value of each expression.*

1. $\left(\dfrac{5}{8}\right)^0$ 2. $\left(\dfrac{6}{5}\right)^{-2}$ 3. $\dfrac{1}{5^{-3}}$ 4. $-8^{-1/3}$

5. $(-49)^{3/2}$ 6. $(-27)^{4/3}$

In Problems 7–14 simplify each expression and give each answer using positive exponents.

7. $(3a^3b^2)^0$ 8. $m^{-8}m^5$ 9. $\dfrac{r^5}{r^{-3}}$ 10. $\dfrac{u^{-7}}{u^{-5}}$

11. $(x^{-3}y^2)^{-2}$ 12. $\dfrac{a^{3/7}}{a^{4/7}}$ 13. $\left(\dfrac{x^{14}}{y^{21}}\right)^{1/7}$ 14. $(u^{1/3}v^{2/5})^{15}$

15. Convert to scientific notation:
    (A) 53,000,000,000   (B) 0.000 004 9
16. Convert to standard decimal form:
    (A) 3.8×10^7 (B) 5.7×10^{-5}
17. Convert from rational exponent form to radical notation:
    (A) $(7z)^{5/6}$ (B) $4w^{3/4}$
18. Convert from radical notation to rational exponent form:
    (A) $\sqrt[5]{(2x^2y)^3}$ (B) $\sqrt{m^2 - n^2}$

Write each expression in simplest radical form. (All variables represent positive real numbers.)

19. $\sqrt{100x^2y^6}$ 20. $\sqrt{72x^5}$ 21. $\sqrt{32x^5y^8}$ 22. $\sqrt{\dfrac{2x}{7y}}$

23. $\dfrac{\sqrt{3u}}{\sqrt{5v}}$ 24. $\dfrac{28a^3b}{\sqrt{7ab}}$ 25. $\sqrt[10]{x^4}$ 26. $\sqrt{\sqrt[5]{y^3}}$

Perform the indicated operations and write each answer in simplest radical form.

27. $3\sqrt{5} - 2\sqrt{3} - 6\sqrt{5}$ 28. $\sqrt{27} - 2\sqrt{3}$ 29. $\sqrt{7}(\sqrt{5} - 3)$

30. $\sqrt{3}(\sqrt{6} - \sqrt{2})$ 31. $(\sqrt{5} + 2)^2$ 32. $\dfrac{6}{\sqrt{11} - 3}$

33. $\dfrac{\sqrt{3}}{\sqrt{6} + 2}$

B   Give the value of each expression.

34. $16^{-3/4}$ 35. $32^{1/2} \cdot 32^{-3/10}$ 36. $(64^{-2/9})^3$

Simplify and express each answer using positive exponents.

37. $(7a^{-4}b^3)^{-2}$ 38. $\dfrac{1}{(3u^2v^{-4})^{-3}}$ 39. $\dfrac{10^{-6} \cdot 10^{-4}}{10^{-7} \cdot 10^3}$

40. $\dfrac{25x^{-5}y^2}{35x^{-2}y^{-3}}$ 41. $\left(\dfrac{16m^{-3}n^{-2}}{8mn^{-5}}\right)^{-3}$ 42. $(5u^{-2}v^3)^{-3}(u^3v^{-1})^{-2}$

43. $v^4 - v^{-3}$ 44. $\dfrac{b^{-1} - a^{-1}}{a - b}$ 45. $(2u^{-3/4}v^{5/8})^8$

46. $(27m^6n^{-9})^{1/3}$ **47.** $t^{-1/4}t^{1/3}$ **48.** $\dfrac{a^{-5/6}}{a^{-4/5}}$

49. $\left(\dfrac{64x^{-3/4}}{x^{7/12}}\right)^{1/6}$ **50.** $(8u^{-12}v^9)^{-1/3}$

In Problems 51 and 52 multiply and express each answer using positive exponents.

51. $3x^{3/4}(5x^{1/4} - 2x^{-3/4})$ **52.** $(3x^{1/2} - y^{1/2})(x^{1/2} - 3y^{1/2})$

53. Convert to correct scientific notation:

     (A)   524,000,000   (B)   0.000 583
     (C) 832×10^6 (D) 529×10^{-5}

54. Evaluate the expression below using scientific notation, and give the answer in both scientific notation and standard decimal form.

$$\frac{0.000\ 020\ 8}{260(0.000\ 04)}$$

55. Convert from rational exponent form to radical notation:

     (A) $-5y^{2/3}$ (B) $a^{1/3} - a^{-1/3}$

56. Convert from radical notation to rational exponent form:

     (A) $-6x\sqrt[4]{(2xy^2)^3}$ (B) $\dfrac{3}{\sqrt[6]{w^5}}$

Write each expression in simplest radical form.

57. $\sqrt[3]{125x^9y^6}$ **58.** $\sqrt[3]{32x^5y^{12}}$ **59.** $\sqrt[4]{8x^3y^2}\ \sqrt[4]{4xy^3}$

60. $\sqrt[3]{\dfrac{125u^{12}}{27v^9}}$ **61.** $\dfrac{1}{\sqrt[3]{2x^2}}$ **62.** $\dfrac{15m^2n}{\sqrt[3]{5m^2n}}$

Perform the indicated operations in Problems 63–73 and write each answer in simplest radical form.

63. $5\sqrt{20} - 2\sqrt{80} + 3\sqrt{45}$ **64.** $3x\sqrt{27x} - 2\sqrt{3x^3}$

65. $\sqrt{\tfrac{2}{7}} + \sqrt{\tfrac{7}{2}}$ **66.** $\sqrt{8uv} - \sqrt{\dfrac{uv}{2}}$

67. $2z\sqrt[3]{16z} + 3\sqrt[3]{2z^4}$ **68.** $(2\sqrt{5} - 3\sqrt{2})(3\sqrt{5} + 4\sqrt{2})$

69. $(5\sqrt{x} - \sqrt{y})(2\sqrt{x} + 3\sqrt{y})$ **70.** $(\sqrt[3]{x} - \sqrt[3]{y^2})(\sqrt[3]{x^2} + \sqrt[3]{y})$

71. $\dfrac{\sqrt{7} - 3}{\sqrt{7} + 2}$ **72.** $\dfrac{2\sqrt{x}}{3\sqrt{x} - \sqrt{y}}$

73. $\dfrac{2\sqrt{2} + \sqrt{3}}{3\sqrt{2} + 2\sqrt{3}}$

74. Evaluate $x^2 - 4x - 9$ using $x = 2 - \sqrt{13}$.

C *Simplify and express each answer using positive exponents.*

75. $(u^{-1} + v^{-1})^{-1}$

76. $(w^{1/5} + 2w^{-4/5})(w^{-1/5} - 3w^{4/5})$

77. $\dfrac{3^{-2} + 3^{-3}}{3^{-3} - 3^{-4}}$

Write each expression in Problems 78–82 in simplest radical form. (All variables represent positive real numbers.)

78. $\sqrt{16x^2 + 4}$

79. $\dfrac{2x + 1}{\sqrt{4x^2 - 1}}$

80. $2n\sqrt[4]{3m^5n^2} - m\sqrt[4]{3mn^6} + mn\sqrt[4]{243mn^2}$

81. $\sqrt{\tfrac{7}{5}} - \sqrt{\tfrac{5}{7}} + \sqrt{35}$

82. $\dfrac{2\sqrt{a} - 3\sqrt{b}}{3\sqrt{a} - 2\sqrt{b}}$

83. Simplify $5\sqrt[4]{x^4} - 3\sqrt[3]{x^3}$:

   (A) For $x \geq 0$ (B) For $x < 0$

Practice Test: Chapter 3

In the following problems all variables represent positive real numbers.

In Problems 1–3 simplify each expression and give each answer using positive exponents.

1. $\left(\dfrac{48x^{-3}y^{-2}}{40x^{-5}y}\right)^{-2}$

2. $\left(\dfrac{27a^5b^{-6}}{a^{-4}}\right)^{1/3}$

3. $\dfrac{p^{-1} + q^{-1}}{(p + q)^{-1}}$

4. Evaluate the expression below using scientific notation, and give the answer in scientific notation.

   $$\frac{(0.000\ 037\ 5)(80,000)}{(40,000,000)(0.025)}$$

5. Convert each expression as indicated:

   (A) $-5y\sqrt[7]{(3x^2y)^3}$;   to rational exponent form
   (B) $3(x - y)^{-2/3}$;   to radical form

Write each expression in simplest radical form.

6. $\sqrt[3]{250x^5y^{12}z^{16}}$

7. $\sqrt{\dfrac{3m^2}{5n}}$

8. $\dfrac{25ab^2}{\sqrt[3]{25ab^2}}$

In Problems 9–11 simplify each expression and give each answer in simplest radical form.

9. $7a\sqrt[3]{40a^2} - 3\sqrt[3]{5a^5}$

10. $\sqrt[3]{16} - \dfrac{8}{\sqrt[3]{4}}$

11. $\dfrac{\sqrt{3} - 3\sqrt{2}}{2\sqrt{3} - \sqrt{2}}$

12. Multiply and express the answer using positive exponents only:

$$(x^{1/4} + 3x^{-3/4})(x^{-1/4} - 4x^{3/4})$$

CHAPTER 4	Contents

4-1 Linear Equations

- Linear Equations
- Solving Linear Equations
- Equations Involving Fractions with Constant Denominators
- Equations Involving Fractions with Variables in the Denominators
- Applications

An **equation** is a mathematical statement obtained when two algebraic expressions are set equal to one another. The expressions in an equation may consist of a single number or involve one or more variables. The following are examples of equations:

$$3x - 5 = 5(2x + 1) + 4 \qquad 2x + 3y = 12 \qquad 2x^2 + 9x - 5 = 0$$

In this section we will only consider a special class of equations called *linear equations in one variable*.

Linear Equations

An equation in one variable that involves only first-degree and zero-degree terms is called a **linear** or **first-degree equation in one variable.** For example,

$$2(3x - 5) + 6 = 3x + 5 \qquad \text{and} \qquad 3 - 2(x + 3) = \frac{x}{3} - 5$$

are linear equations. A **solution** or **root** of an equation is a number which when substituted for the variable (wherever it occurs) gives the same numerical value on both sides of the equality sign. The set of all solutions of an equation is called the **solution set.** To **solve** an equation means to determine the solution set.

Before we describe the process used to solve an equation, we need to define the meaning of "equivalent equations." Two equations are said to be **equivalent** if they have exactly the same solution set. For example, the equations

$$x + 5 = 15 \qquad \text{and} \qquad x = 10$$

are equivalent since both have 10, and only 10, as a solution. The process of solving an equation will involve creating a sequence of equivalent equations resulting in a final equation, such as $x = 10$, in which the solution is obvious. The properties of equality listed in the box will serve as the basis for solving linear equations.

Properties of Equality

Let a, b, and c denote real numbers.

1. *Addition property:*

   If $a = b$, then $a + c = b + c$.   The same quantity can be added to both sides of an equation.

2. *Subtraction property:*

   If $a = b$, then $a - c = b - c$.   The same quantity can be subtracted from both sides of an equation.

3. *Multiplication property:*

   If $a = b$, then $ac = bc$.   Both sides of an equation can be multipled by the same quantity.

4. *Division property:*

   If $a = b$ and $c \neq 0$, then $\dfrac{a}{c} = \dfrac{b}{c}$.   Both sides of an equation can be divided by the same nonzero quantity.

Whenever we apply the properties of equality with $c \neq 0$, we obtain equivalent equations.

■ Solving Linear Equations

The following examples illustrate how the properties of equality are used to solve equations.

Example 1   Solve $3x - 9 = 7x + 3$ and check.

Solution   By applying the properties of equality we will isolate the variable x on the left side of the equation to obtain a simple equation of the form $x = $ Some number.

$$3x - 9 = 7x + 3$$

$$\boxed{3x - 9 + 9 = 7x + 3 + 9}$$ Addition property

$$3x = 7x + 12$$

$$\boxed{3x - 7x = 7x + 12 - 7x}$$ Subtraction property

$$-4x = 12$$

$$\boxed{\dfrac{-4x}{-4} = \dfrac{12}{-4}}$$ Division property

$$x = -3$$

Check   To check this solution, we substitute $x = -3$ back into the original equation to see if the equality holds:

$$3(-3) - 9 \overset{?}{=} 7(-3) + 3$$
$$-9 - 9 \overset{?}{=} -21 + 3$$
$$-18 \overset{\checkmark}{=} -18$$

Problem 1   Solve $2x - 8 = 5x + 4$ and check.

When symbols of grouping are present on one or both sides of an equation, we first remove them from each expression and combine like terms. Then we can proceed as in Example 1.

Example 2   Solve: $8x - 3(x - 4) = 2(x - 6) + 6$

Solution $8x - 3(x - 4) = 2(x - 6) + 6$ Remove parentheses.

$8x - 3x + 12 = 2x - 12 + 6$ Combine like terms.

$5x + 12 = 2x - 6$ Solve as before.

$5x = 2x - 18$

$3x = -8$

$x = -6$

The check is left to the reader.

Problem 2   Solve $3x - 2(2x - 5) = 2(x + 3) - 8$ and check.

■ Equations Involving Fractions with Constant Denominators

When an equation involves fractions, such as

$$\frac{x}{3} - \frac{x}{5} = 4$$

we can simplify the equation by "clearing" the denominators. This is done by multiplying both sides of the equation by the LCD of the fractions present. For the above equation the LCD is 15. Multiplying both sides of the equation by 15, we obtain

$$15 \cdot \left(\frac{x}{3} - \frac{x}{5} \right) = 15 \cdot 4$$

$$\overset{5}{\cancel{15}} \cdot \frac{x}{\underset{1}{\cancel{3}}} - \overset{3}{\cancel{15}} \cdot \frac{x}{\underset{1}{\cancel{5}}} = 60$$

$$5x - 3x = 60$$
$$2x = 60$$
$$x = 30$$

Note: Do not confuse operations on equations with operations on algebraic expressions that are not equations. For example, in the algebraic expression

$$\frac{x}{3} - \frac{x}{5} + 4$$

which looks very much like the above equation, we cannot multiply "both sides" by the LCD 15 to clear the fractions, since this expression does *not* have two sides! In this case, we combine the three fractions into a single fractional form:

$$\frac{x}{3} - \frac{x}{5} + \frac{4}{1} = \frac{5x - 3x + 60}{15} = \frac{2x + 60}{15}$$

which still has a denominator—namely 15 (the LCD).

Example 3 Solve: $\dfrac{x+2}{2} - \dfrac{x}{3} = 5$

Solution The LCD is 6. Multiplying both sides of the equation by 6, we obtain

$$6 \cdot \left(\frac{x+2}{2} - \frac{x}{3} \right) = 6 \cdot 5$$

$$\overset{3}{\cancel{6}} \cdot \left(\frac{x+2}{\underset{1}{\cancel{2}}} \right) - \overset{2}{\cancel{6}} \cdot \frac{x}{\underset{1}{\cancel{3}}} = 30$$

$$3(x + 2) - 2x = 30$$
$$3x + 6 - 2x = 30$$
$$x + 6 = 30$$
$$x = 24$$

Problem 3 Solve: $\dfrac{x+1}{3} - \dfrac{x}{4} = \dfrac{1}{2}$

Equations involving decimal fractions can sometimes be transformed into a form free of decimals by multiplying both sides by a power of 10. Consider the following example.

Example 4   Solve: $0.4(x - 30) - 0.15x = 8$

Solution   To clear all decimals, we can multiply both sides of the equation by 100:

$$0.4(x - 30) - 0.15x = 8 \qquad \text{Multiply both sides by}$$
$$100 \cdot [0.4(x - 30)] - 100 \cdot (0.15x) = 100 \cdot 8 \qquad 100.$$
$$40(x - 30) - 15x = 800 \qquad \text{Solve as before}$$
$$40x - 1{,}200 - 15x = 800$$
$$25x = 2{,}000$$
$$x = 80$$

Problem 4   Solve: $0.25x + 0.4(x - 30) = 27$

In the above examples we have always obtained a single or unique solution. This is what we normally expect when solving an equation. However, two other situations may occur. The first is when the given equation is an *identity*. A linear equation in one variable is called an *identity* if the solution set is the set of all real numbers. For example, if we replace x by any real number in the equation

$$3x + 7 = 3(x + 2) + 1$$

we will obtain the same value on both sides. Thus, this equation is an identity. If we attempt to solve this equation by the usual methods, we obtain the following result:

$$3x + 7 = 3(x + 2) + 1$$
$$3x + 7 = 3x + 6 + 1$$
$$3x + 7 = 3x + 7$$
$$3x = 3x$$
$$0 = 0$$

Clearly, the last equation is true. What this tells us about the original equation can be generalized as follows:

Identities

If a linear equation in one variable can be reduced to $0 = 0$ using the properties of equality, then the given equation is an **identity** and the solution set is the set of all real numbers.

The second situation occurs when an equation has no solution. For example, there is no real number x for which

$$x + 5 = x + 10$$

Solving this in the usual manner, we would obtain

$$x + 5 = x + 10$$
$$x = x + 5$$
$$0 = 5$$

Obviously, this last equation is not valid. In general, we have the following:

Equations with No Solution

If a linear equation in one variable can be reduced to $0 = b$ using the properties of equality, where $b \neq 0$, then the given equation has no solution.

Example 5

Solve: $5(3x - 5) + 4 = 4(5x + 7) - 5x$

Solution

$$5(3x - 5) + 4 = 4(5x + 7) - 5x$$
$$15x - 25 + 4 = 20x + 28 - 5x$$
$$15x - 21 = 15x + 28$$
$$15x = 15x + 49$$
$$0 = 49 \quad \text{Impossible!}$$

The given equation has no solution.

Problem 5

Solve: $3 - 4(2 - 3x) = 6(2x - 1) + 1$

■ Equations Involving Fractions with Variables in the Denominators

Many equations involving fractional forms with variables in the denominators can be transformed into linear equations by clearing the denominators. For example, consider the equation

$$\frac{9}{x + 2} + 5 = \frac{1}{2}$$

We can clear the denominators by multiplying both sides of this equation by the LCD, which is $2(x + 2)$. Before doing this, note that $x = -2$ cannot be a solution, since this would create a 0 in the denominator. Thus, we must include the condition $x \neq -2$, as indicated below. Solving this equation, we have

$$\frac{9}{x+2} + 5 = \frac{1}{2} \qquad x \neq -2$$

-2 cannot be a solution.

$$2(\overset{1}{\cancel{x+2}}) \cdot \left(\frac{9}{\cancel{x+2}}\right) + 2(x+2) \cdot 5 = \overset{1}{\cancel{2}}(x+2) \cdot \frac{1}{\cancel{2}}$$

Multiply both sides by the LCD $= 2(x+2)$ and simplify.

$$18 + 10x + 20 = x + 2$$
$$10x + 38 = x + 2$$
$$9x = -36$$
$$x = -4$$

Solve the linear equation as before.

Example 6 Solve each equation.

(A) $\dfrac{8}{x-1} - \dfrac{1}{x} = \dfrac{3}{x}$ (B) $5 + \dfrac{3x}{x-3} = \dfrac{9}{x-3}$

Solutions (A)

$$\frac{8}{x-1} - \frac{1}{x} = \frac{3}{x} \qquad x \neq 0, 1$$

0 and 1 cannot be solutions.

$$x(x-1) \cdot \left(\frac{8}{x-1}\right) - x(x-1) \cdot \frac{1}{x} = x(x-1) \cdot \frac{3}{x}$$

Multiply both sides by the LCD $= x(x-1)$ and simplify.

$$8x - (x-1) = 3(x-1)$$
$$8x - x + 1 = 3x - 3$$
$$7x + 1 = 3x - 3$$
$$4x = -4$$
$$x = -1$$

(B)

$$5 + \frac{3x}{x-3} = \frac{9}{x-3} \qquad x \neq 3$$

3 cannot be a solution.

$$(x-3) \cdot 5 + (x-3) \cdot \left(\frac{3x}{x-3}\right) = (x-3) \cdot \left(\frac{9}{x-3}\right)$$

Multiply both sides by the LCD $= x - 3$ and simplify.

$$5x - 15 + 3x = 9$$
$$8x - 15 = 9$$
$$8x = 24$$
$$x = 3$$
$$\text{No solution}$$

3 cannot be a solution and must be rejected.

Problem 6 Solve each equation

(A) $\dfrac{9}{x+1} - \dfrac{2}{x} = \dfrac{4}{x}$ (B) $\dfrac{8}{x+2} = 7 - \dfrac{4x}{x+2}$

▪ Applications

The methods discussed in this section can be utilized to solve a large variety of practical problems.

Example 7
Break-Even Analysis

It costs a record company $6,000 to prepare a record album for production. This includes recording costs, album design costs, etc., which represent a one-time **fixed cost.** Manufacturing, marketing, and royalty costs—all **variable costs**—are $2.50 per album. If the album is sold to record shops for $4 each, how many albums must be produced and sold for the company to break even?

Solution

Let

x = Number of records sold

C = Cost for producing x records

R = Revenue (return) on the sale of records

The company breaks even when $R = C$, where

C = Fixed costs + Variable costs   and $R = \$4x$

   = $6,000 + 2.50x$

To find the value of x for which $R = C$, we solve

$$4x = 6,000 + 2.50x$$
$$1.50x = 6,000$$
$$x = 4,000 \text{ records}$$

Check

For $x = 4,000$,

$C = 6,000 + 2.50x$ and $R = 4x$

   = $6,000 + 2.50(4,000)$ = $4(4,000)$

   = $6,000 + 10,000$ = $16,000$

   = $16,000$

Thus, the company must produce and sell 4,000 records to break even. Sales over 4,000 will result in a profit, and sales under 4,000 will result in a loss.

Problem 7

What is the break-even point in Example 7 if fixed costs are $9,000 and variable costs are $2.80 per record?

Since the variety of practical problems that can be solved using linear equations is very extensive, it is difficult to describe a single procedure that will work for all problems. Some guidelines are listed in the box at the top of the next page.

> **Guidelines for Solving Word Problems**
>
> 1. Read the problem very carefully — several times if necessary.
> 2. Write down important facts and relationships.
> 3. Identify the unknown quantities in terms of a single letter — if possible.
> 4. Write an equation relating the unknown quantities based on the facts in the problem.
> 5. Solve the equation.
> 6. Write down all desired values asked for in the original problem.
> 7. Check the solution(s).

Example 8
Investment

A retired couple has $60,000 invested, part at 10% per year and the remainder at 16% per year. At the end of 1 year the total income received from both investments is $8,100. Find the amount invested at each rate.

Solution   Let

$$x = \text{Amount invested at 10\%}$$

$$\$60,000 - x = \text{Amount invested at 16\%}$$

The total income received is the sum of the income from the 10% investment and the income from the 16% investment. Translating this into an equation and solving the equation, we have

$$0.10x + 0.16(60,000 - x) = 8,100 \qquad \text{Multiply both sides by 100.}$$

$$10x + 16(60,000 - x) = 810,000 \qquad \text{Solve for } x.$$

$$10x + 960,000 - 16x = 810,000$$

$$-6x = -150,000$$

$$x = \$25,000 \text{ invested at 10\%}$$

$$60,000 - x = \$35,000 \text{ invested at 16\%}$$

Check   10% of $25,000 = $2,500

16% of $\underline{\$35,000} = \underline{\$5,600}$

$\qquad$ $60,000 $\qquad$ $8,100

Problem 8   Solve Example 8 if the total income from both investments is $6,900.

Example 9
Blending — Food Processing

A candy company has 1,200 pounds of chocolate mix that contains 20% cocoa butter. How many pounds of pure cocoa butter must be added to the mix to obtain a final mix that is 25% cocoa butter?

Solution   Let

$$x = \text{Amount of pure cocoa butter added}$$

The amount of cocoa butter in the original mixture plus the amount of cocoa butter added must equal the amount of cocoa butter in the final mixture. That is,

$$\left(\begin{array}{c}\text{Amount of}\\ \text{cocoa butter}\\ \text{in original mix}\end{array}\right) + \left(\begin{array}{c}\text{Amount of}\\ \text{cocoa butter}\\ \text{added}\end{array}\right) = \left(\begin{array}{c}\text{Amount of}\\ \text{cocoa butter}\\ \text{in final mix}\end{array}\right)$$

$$0.20(1{,}200) \quad + \quad x \quad = 0.25(1{,}200 + x)$$

$$20(1{,}200) + 100x = 25(1{,}200 + x)$$

$$24{,}000 + 100x = 30{,}000 + 25x$$

$$75x = 6{,}000$$

$$x = 80 \text{ pounds of cocoa butter}$$
$$\text{must be added}$$

Problem 9   Solve Example 9 if the original mixture contains 18% cocoa butter.

Answers to   1. $x = -4$ 2. $x = 4$ 3. $x = 2$ 4. $x = 60$
Matched Problems   5. The equation is an identity. Every real number is a solution.
6. (A) $x = 2$ (B) No solution   7. 7,500 records
8. \$45,000 at 10%; \$15,000 at 16%
9. 112 pounds of cocoa butter must be added

Exercise 4-1

Solve each equation.

A   1. $13x + 5 = 44$ 2. $5x - 7 = 13$
3. $7y + 10 = 10$ 4. $3x - 18 = -18$
5. $23 - 18b = 15 - 14b$ 6. $94 - 6c = 100 - 3c$
7. $2x + 5 = 2x - 7$ 8. $3y - 5 = 3y + 9$
9. $7(w - 5) = 5(w + 10)$ 10. $5(u + 3) = 3(u + 3)$
11. $15x - 35 = -5(7 - 3x)$ 12. $2(3x - 5) = 6x - 10$
13. $\frac{3}{4}x = -\frac{9}{2}$ 14. $-\frac{2}{7}x = \frac{8}{21}$

15. $\frac{z}{2} + \frac{z}{8} = 15$ 16. $\frac{w}{3} + \frac{w}{9} = 12$

17. $\frac{x - 4}{3} - \frac{x}{4} = 5$ 18. $\frac{x + 2}{5} + 2 = \frac{x}{3}$

19. $0.9x = 63$ 20. $0.3y = 120$
21. $0.7x + 0.8x = 45$ 22. $0.11x - 0.07x = 36$

23. $\dfrac{5}{x} - \dfrac{1}{2} = \dfrac{3}{x}$

24. $\dfrac{3}{x} - \dfrac{1}{5} = \dfrac{8}{x}$

25. $\dfrac{2}{3n} - \dfrac{1}{9} = \dfrac{4}{9} - \dfrac{1}{n}$

26. $\dfrac{2}{z} - \dfrac{1}{4} = \dfrac{1}{2z} + \dfrac{1}{8}$

B

27. $8z - (4z - 5) = 0$

28. $2(3x - 5) - 6 = 0$

29. $7x - 3(2x + 1) = -2(x + 3)$

30. $9y - 5(y - 10) = 100 - 3(y - 2)$

31. $8(z + 1) - 3(2z - 7) = 3(4z - 20) + 25$

32. $2(5x - 3) - 3(2x + 4) = 5(3x - 6) - 21$

33. $3(3x - 5) + 7(x + 4) = 3(7x - 2) - 5(x + 3)$

34. $5(7z - 4) - 4(z - 1) = 7(3 + 2z) + 17z$

35. $4 - 6(1 - 2z) = 3(9z - 4) - 5(3z - 2)$

36. $2(3 - 4u) - 5(2 - u) = 3(5 - u) - 19$

37. $(x + 3)(x - 4) = (x - 2)(x + 5)$

38. $(x - 2)(x - 5) = (x + 2)(x + 3)$

39. $\dfrac{3x + 4}{3} - \dfrac{x - 2}{5} = \dfrac{2 - x}{15} - 1$

40. $\dfrac{2x - 3}{9} - \dfrac{x + 5}{6} = \dfrac{3 - x}{2} - 1$

41. $0.12x + 0.08(40{,}000 - x) = 4{,}200$

42. $0.15x + 0.1(16{,}000 - x) = 1{,}800$

43. $(0.2)(200) + x = 0.5(200 + x)$

44. $(0.3)(60) + x = 0.6(60 + x)$

45. $\dfrac{9}{x - 3} - \dfrac{5}{2} = 2$

46. $\dfrac{2}{3} - \dfrac{10}{x + 1} = 4$

47. $\dfrac{6x}{x - 3} - 5 = \dfrac{18}{x - 3}$

48. $8 - \dfrac{3x}{x + 4} = \dfrac{12}{x + 4}$

49. $\dfrac{9}{x + 2} - \dfrac{8}{x} = \dfrac{7}{x}$

50. $\dfrac{12}{x} - \dfrac{1}{x - 4} = \dfrac{3}{x - 4}$

51. $\dfrac{m - 1}{4m + 4} = \dfrac{1}{8} - \dfrac{m - 3}{6m + 6}$

52. $\dfrac{1}{4} - \dfrac{x - 3}{3x - 6} = \dfrac{x - 6}{2x - 4}$

C

53. $1 - 2\{2 - 1[1 - 2(y + 1)]\} = 2(y + 5) - 3$

54. $5 - 2\{6 + [2x - (x - 4)]\} = 2[(x + 2) - 3] - 1$

55. $\dfrac{x^2 + 6}{x^2 - 9} = \dfrac{x}{x + 3}$

56. $\dfrac{5}{x - 4} = \dfrac{x + 4}{x^2 - 8x + 16}$

57. $\dfrac{3 - x}{14} - \dfrac{x}{5} = \dfrac{1}{2} - \dfrac{x + 2}{5}$

58. $\dfrac{2 - x}{10} - \dfrac{x}{8} = \dfrac{1}{15} - \dfrac{5 + x}{40}$

◼ Applications

Business & Economics

59. *Investment.* If $10,000 is invested, part at 11% and the rest at 18%, how much should be invested at each rate to receive $1,345 income at the end of the year?

60. *Investment.* If $40,000 is invested, part at 12% and the rest at 9%, how

much is invested at each rate if the income from both investments totals $4,050?

61. *Resale.* An art dealer paid $15,000 for two paintings. She sold the first painting for a 15% profit and the second for an 11% profit. If her total profit was $1,890, determine how much she paid for each painting.

62. *Resale.* A used car dealer paid $7,200 for two cars. He resold the first for a profit of 15% and the second for a loss of 3%. If his total profit was $540, find how much he paid for each car.

63. *Investment.* You have $12,000 to invest. If part is invested at 10% and the rest at 15%, how much should be invested at each rate to yield 12% on the total amount invested?

64. *Investment.* An investor has $20,000 to invest. If part is invested at 8% and the rest at 12%, how much should be invested at each rate to yield 11% on the total amount invested?

65. *Break-even analysis.* A small candy manufacturer has fixed costs of $1,200 per week. The variable cost to produce one pound of candy is $1.80. If the candy is then sold for $3.30 per pound, determine how many pounds must be produced and sold each week to:
(A) Break even   (B) Earn a profit of $900 per week

66. *Break-even analysis.* A record manufacturer has determined that its weekly cost equation is $C = 300 + 1.5x$, where x is the number of records produced and sold each week. If records are sold for $4.50 each, how many records must be produced and sold each week for the manufacturer to break even?

Life Sciences

67. *Pollution control.* A fuel oil distributor has 120,000 gallons of fuel with a 0.9% sulfur content. How many gallons of fuel oil with a 0.3% sulfur content must be purchased and mixed with the 120,000 gallons to obtain fuel oil with a 0.8% sulfur content?

68. *Ecology.* One day during the winter the temperature in the Antarctic reached a high of $-67°F$. What was the temperature in Celsius degrees? [*Note:* $°F = \frac{9}{5}°C + 32$]

69. *Wildlife management.* A naturalist for a fish and game department estimated the total number of rainbow trout in a certain lake using the popular capture–mark–recapture technique. He netted, marked, and released 200 rainbow trout. A week later, allowing for thorough mixing, he again netted 200 trout and found eight marked ones among them. Assuming that the proportion of marked fish in the second sample was the same as the proportion of all marked fish in the total population, estimate the number of rainbow trout in the lake.

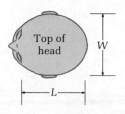

Top of head

W

L

70. *Anthropology.* In their study of genetic groupings, anthropologists use a ratio called the *cephalic index*. This is the ratio of the width of the head to its length (looking down from above) expressed as a percentage. Symbolically,

$$C = \frac{100W}{L}$$

where C is the cephalic index, W is the width, and L is the length. If an Indian tribe in Baja California, Mexico, had an average cephalic index of 66 and the average width of the Indians' heads was 6.6 inches, what was the average length of their heads?

Social Sciences

71. *Psychology.* The intelligence quotient (IQ) is found by dividing the mental age (MA), as indicated on standard tests, by the chronological age (CA) and multiplying by 100. For example, if a child has a mental age of 12 and a chronological age of 8, the calculated IQ is 150. If a 9-year-old girl has an IQ of 140, compute her mental age.

72. *City planning.* A city has just incorporated additional land so that the area of the city is now 450 square miles. At present, only 7% of this area consists of parks and recreational areas. The voters of the city have demanded that 15% of the total area of the city should consist of parks and recreational areas. How much land must be developed for parks and recreational areas to meet this demand?

4-2   Linear Inequalities

■ Properties of Inequalities
■ Solving Inequalities
■ Applications

Just as we solved linear equations in Section 4-1, we will now solve linear inequalities where one algebraic expression is greater than or less than another expression.

■ Properties of Inequalities

There are four inequality symbols: $<$, $>$, $\leq$, and $\geq$. Their meanings are reviewed in the box.

Inequality Symbols

Let a and b denote real numbers.

1. $a < b$ means a is less than b
2. $a > b$ means a is greater than b
3. $a \leq b$ means a is less than or equal to b
4. $a \geq b$ means a is greater than or equal to b

Formally, we say that $a < b$ or $b > a$ if there exists a positive real number p such that $a + p = b$. Intuitively, if we add a positive real number to *any* real number, we would expect to make it larger. That is essentially what the formal definition of $<$ and $>$ states. As we have seen in Chapter 1, $a < b$ can be interpreted geometrically by saying that a is to the left of b on a real number line. Similarly, $a > b$ means that a is to the right of b on a real number line.

An algebraic inequality is a mathematical statement where two expressions are joined using one of the inequality symbols. For example,

$$2(2x + 3) < 6(x - 2) + 10 \qquad \text{and} \qquad \frac{x - 2}{15} > \frac{x}{3} - \frac{1}{5}$$

are inequalities. An inequality is called a **linear inequality** if it contains only first-degree and zero-degree terms. The basic properties that will be used to solve inequalities are very similar to those used to solve linear equations, except for a very important difference, which is noted in the box.

Properties of Inequalities

Let a, b, and c denote real numbers.

1. *Addition property:*

   If $a < b$, then $a + c < b + c$.

2. *Subtraction property:*

   If $a < b$, then $a - c < b - c$.

3. *Multiplication properties:*

   If $a < b$ and c is positive, then $ac < bc$.
   If $a < b$ and c is negative, then $ac > bc$.*

4. *Division properties:*

   If $a < b$ and c is positive, then $\dfrac{a}{c} < \dfrac{b}{c}$.

   If $a < b$ and c is negative, then $\dfrac{a}{c} > \dfrac{b}{c}$.*

   * Note that the direction of the inequality symbol reverses.

Similar properties hold if we start with $a > b$, $a \leq b$, or $a \geq b$. From these properties, we see that the operations performed on inequalities are very much the same as those performed on equations except that **the direction of the inequality symbol is reversed whenever both sides of the inequal-**

ity are multiplied or divided by the same negative number. Otherwise, the direction of the inequality symbol does not change. To illustrate this, let us consider the inequality

$$-10 < -5$$

If we multiply both sides by the positive number 2, we obtain another true inequality,

$$-20 < -10$$

On the other hand, if we multiply both sides of $-10 < -5$ by the negative number -2, we must reverse the inequality symbol. Thus, we obtain

$$20 > 10$$

which is certainly true. If the inequality symbol had not been reversed, we would have $20 < 10$, which is clearly false!

▪ Solving Inequalities

A **solution** of an inequality is any number which, when substituted for the variable, results in a true statement. The set of all solutions is called the **solution set.** To **solve** an inequality means to determine its solution set. The solution set for a linear inequality in one variable can, in general, be represented by an interval.

Example 10   Solve and graph: $2(2x + 3) < 6(x - 2) + 10$

Solution

$2(2x + 3) < 6(x - 2) + 10$	Clear parentheses.
$4x + 6 < 6x - 12 + 10$	Combine like terms.
$4x + 6 < 6x - 2$	Apply properties of inequalities.
$-2x + 6 < -2$	
$-2x < -8$	
$x > 4$ or $(4, \infty)$	Note that since we divided both sides by -2, the direction of the inequality symbol has changed.

Problem 10   Solve and graph: $3(x - 1) \leqslant 5(x + 2) - 5$

Example 11   Solve and graph: $\dfrac{x - 2}{15} \geqslant \dfrac{x}{3} - \dfrac{1}{5}$

Solution $\dfrac{x - 2}{15} \geqslant \dfrac{x}{3} - \dfrac{1}{5}$ Multiply both sides by the LCD = 15.

$$\underset{1}{\overset{1}{\cancel{15}}} \cdot \left(\frac{x-2}{\cancel{15}}\right) \geqslant \underset{1}{\overset{5}{\cancel{15}}} \cdot \left(\frac{x}{\cancel{3}}\right) - \underset{1}{\overset{3}{\cancel{15}}} \cdot \frac{1}{\cancel{5}}$$

$$x - 2 \geqslant 5x - 3$$

$$-4x - 2 \geqslant -3$$

$$-4x \geqslant -1$$

$$x \leqslant \tfrac{1}{4} \quad \text{or} \quad (-\infty, \tfrac{1}{4}]$$

Note the reversal of the inequality symbol here.

Problem 11 Solve and graph: $\dfrac{x}{4} - \dfrac{1}{3} > \dfrac{x-3}{6}$

In Chapter 1 we encountered double inequalities such as

$$-3 < x < 5 \quad \text{and} \quad -5 \leqslant x < 3$$

which may be represented by intervals. We will now solve **double inequalities** such as

$$-3 < 2x + 3 \leqslant 9 \quad \text{and} \quad -5 \leqslant \frac{7-2x}{3} < 3$$

where the middle expression is a linear expression involving x. The rule here is that whatever arithmetical operation we perform on one expression we must perform the same operation on the other two expressions. And, again, if we multiply or divide by a negative number, the direction of both inequality symbols must be reversed. Our objective is to isolate the unknown in the middle using the properties of inequalities.

Example 12 Solve and graph:

(A) $-3 < 2x + 3 \leqslant 9$ (B) $-5 \leqslant \dfrac{7.-2x}{3} < 3$

Solutions (A)

$$-3 < 2x + 3 \leqslant 9$$

$$\boxed{-3 - 3 < 2x + 3 - 3 \leqslant 9 - 3}$$

Subtract 3 from each expression.

$$-6 < 2x \leqslant 6$$

Divide each expression by 2.

$$\boxed{\frac{-6}{2} < \frac{2x}{2} \leqslant \frac{6}{2}}$$

$$-3 < x \leqslant 3 \quad \text{or} \quad (-3, 3]$$

(B) $\qquad -5 \leqslant \dfrac{7-2x}{3} < 3$ Multiply each expression by 3.

$$3 \cdot (-5) \leqslant \overset{1}{\cancel{3}} \cdot \left(\dfrac{7-2x}{\underset{1}{\cancel{3}}}\right) < 3 \cdot 3$$

$\qquad -15 \leqslant 7 - 2x < 9$ Subtract 7 from each expression.

$$-15 - 7 \leqslant 7 - 2x - 7 < 9 - 7$$

$\qquad -22 \leqslant -2x < 2$ Divide each expression by -2 and reverse the direction of both inequality symbols.

$$\dfrac{-22}{-2} \geqslant \dfrac{-2x}{-2} > \dfrac{2}{-2}$$

$\qquad 11 \geqslant x > -1$

or $\quad -1 < x \leqslant 11 \quad$ or $\quad (-1, 11]$

Problem 12   Solve and graph:

(A) $\;-8 \leqslant 3x - 5 < 7$ (B) $\;-3 < \dfrac{1-4x}{5} \leqslant 9$

■ Applications

Example 13
Temperature Conversion

The temperature in a desert town ranged from a low at night of 77°F to a high of 122°F during the day. Find the corresponding temperature range in Celsius degrees and graph. [*Note:* °F $= \frac{9}{5}$°C $+ 32$]

Solution   We must solve

$\qquad 77 \leqslant \frac{9}{5}C + 32 \leqslant 122 \qquad 77 \leqslant F \leqslant 122$

Thus,

$\qquad 45 \leqslant \frac{9}{5}C \leqslant 90$

$\qquad 25 \leqslant C \leqslant 50$

Thus, the temperature ranged from 25°C to 50°C.

Problem 13   Solve Example 13 if the temperature ranged from -4°F to 23°F.

Example 14
Profit

A manufacturer of hand calculators can sell all units produced at $30 per calculator. If the fixed costs are $10,000 per week and it costs $20 to produce each calculator, determine how many units must be produced and sold each week for the company to make a profit.

Solution   Let

$x =$ Number of units produced

$R =$ Revenue on the sale of x units

$C =$ Cost to produce x units

Then,

$R = \$30x$ and $C = \$10,000 + \$20x$

A profit will result if revenue exceeds costs; that is, if

$R > C$

$30x > 10,000 + 20x$

$10x > 10,000$

$x > 1,000$

Thus, the company must produce and sell more than 1,000 calculators per week to make a profit.

Problem 14   Solve Example 14 if the fixed costs are $12,000 per week and the calculators can be manufactured at a cost of $15 each.

Example 15
Laboratory Management

The manager of a drug testing laboratory wants to decide whether to continue buying 6-week-old mice at 90¢ each or breed his own for testing purposes. It is estimated that breeding mice will increase overhead costs by $630 per week and food costs by 20¢ for each mouse that is bred. How many mice would be needed for testing each week to justify a decision to breed the mice?

Solution   Let x represent the number of mice needed each week. Then the cost to purchase the mice is $0.90x. The cost to breed the mice is $630 + $0.20x. We want to determine the values of x for which

Cost of purchasing $>$ Cost of breeding

That is, for which

$0.90x > 630 + 0.20x$

Solving this inequality, we find

$0.90x > 630 + 0.20x$ Multiply both sides by 10 to clear decimals.

$9x > 6,300 + 2x$

$7x > 6,300$

$x > 900$

Thus, if the number of mice needed is greater than 900 per week, a decision to breed them would be justified.

Problem 15   In Example 15 determine how many mice would be needed for testing each week to justify breeding, if the price for purchasing a mouse increases to $1.10.

Answers to Matched Problems

10. $x \geqslant -4$ or $[-4, \infty)$;

11. $x > -2$ or $(-2, \infty)$;

12. (A) $-1 \leqslant x < 4$ or $[-1, 4)$;

   (B) $-11 \leqslant x < 4$ or $[-11, 4)$;

13. $-20 \leqslant C \leqslant -5$;

14. More than 800 calculators per week   15. More than 700

Exercise 4-2

A   *Solve each inequality.*

1. $5x - 3 \geqslant 12$

2. $4x + 5 \leqslant 21$

3. $-7x \leqslant 14$

4. $-3x \geqslant 15$

5. $\dfrac{x}{4} < -5$

6. $\dfrac{x}{6} > -2$

7. $\dfrac{x}{-7} \geqslant -3$

8. $\dfrac{x}{-3} \leqslant -6$

9. $-4x \geqslant -2x + 10$

10. $-7x \leqslant -4x - 9$

11. $-5 \leqslant x - 8 \leqslant 3$

12. $3 < x + 5 \leqslant 8$

13. $-16 \leqslant 4x < 28$

14. $-10 < 5x < 25$

15. $2 \leqslant \dfrac{x}{3} < 5$

16. $-5 < \dfrac{x}{4} \leqslant 3$

17. $-9 < -3x < 12$

18. $-8 \leqslant -4x \leqslant 20$

19. $-4 \leqslant -\dfrac{x}{3} < 5$

20. $-6 < -\dfrac{x}{5} \leqslant 1$

21. $-3 < 3 - x \leqslant 8$

22. $2 \leqslant 7 - x < 9$

B   *Solve and graph each inequality.*

23. $3 + 2x \geqslant 5(x - 3)$

24. $7 - 3(n - 4) > 1$

25. $3(y - 7) - 2(y - 5) \geqslant 2(y - 1)$

26. $4(2w - 1) \leq 2(3w + 4) - (2 - 3w)$
27. $-4 < 3x + 5 \leq 17$
28. $-13 \leq 4x - 5 < 23$
29. $-5 \leq 7 - 4x \leq 15$
30. $1 < 5 - 2x < 13$
31. $\dfrac{3x - 5}{-2} > 10$
32. $\dfrac{4x + 2}{-3} < 6$
33. $\dfrac{u + 1}{2} - 1 < \dfrac{u + 3}{6}$
34. $\dfrac{1}{3} - \dfrac{2 - m}{5} > \dfrac{m - 5}{15}$
35. $0.06x + 0.09(200 - x) \geq 13.5$
36. $0.12x - 0.05(400 - x) \leq 31$
37. $-1 < \dfrac{3x + 4}{2} \leq 8$
38. $-2 \leq \dfrac{4x + 6}{3} < 6$
39. $55 \leq \frac{5}{9}(F - 32) \leq 85$
40. $-35 \leq \frac{5}{9}(F - 32) \leq 10$
41. $122 \leq \frac{9}{5}C + 32 \leq 203$
42. $-22 \leq \frac{9}{5}C + 32 \leq 95$

C *Solve and graph each inequality.*

43. $\dfrac{2x - 3}{9} - \dfrac{3 - x}{6} \geq \dfrac{3 - x}{2} - 1$
44. $\dfrac{3x + 4}{3} - \dfrac{2 - x}{15} \leq \dfrac{x - 2}{15} - 1$
45. $-1 \leq 5 - \frac{3}{4}x \leq 11$
46. $-9 < \frac{2}{5}x - 7 < -3$
47. $\dfrac{2 - x}{10} - \dfrac{1}{15} < \dfrac{x}{8} - \dfrac{5 + x}{40}$
48. $\dfrac{3 - x}{14} - \dfrac{1}{2} > \dfrac{x}{3} - \dfrac{x + 2}{5}$

■ Applications

Business & Economics

49. *Investment.* A woman has $20,000 to invest. She is considering two investments, one paying 9% per year (low risk) and the other paying 17% per year (higher risk). How much should she invest in the higher-risk investment if she wishes to earn at least $2,400 total interest per year?

50. *Investment.* If $12,000 is invested at 9% per year, what additional amount must be invested at 14% so that the total of the two investments earns the equivalent of 11% or more?

51. *Profit.* The manufacturer of a small telescope can sell all telescopes produced at a price of $80. The fixed costs are $4,500 per week and the variable costs to produce each telescope are $60. Determine how many telescopes must be produced and sold each week to:
    (A) Earn a profit   (B) Earn a profit of at least $1,000

52. *Decision analysis.* An electronics firm is considering manufacturing a particular component it uses in its products. The price paid for the component is $12. To produce the component would result in additional fixed costs of $12,000 per month and variable costs of $6 for each component produced. How many components would be required each month to justify a decision to manufacture them?

Life Sciences

53. *Nutrition.* A nutritionist is studying the protein needs of mice. She has 24 pounds of food mix that is 10% protein. In order to obtain a food mix with at least 20% protein, how much pure protein must be added to the available mix?

54. *Temperature control.* In an experiment on rats, a researcher wishes to maintain an environmental temperature that ranges from a low of 68°F at night to a high of 95°F during the day. Determine the equivalent temperature range in Celsius degrees.   [*Note:*   °F $= \frac{9}{5}$°C $+ 32$]

Social Sciences

55. *Psychology.* The intelligence quotient (IQ) is defined by

$$IQ = \frac{100 \cdot MA}{CA}$$

where MA denotes mental age and CA denotes chronological age. A psychologist is studying 12-year-olds who have an IQ range of 75 to 150. What is the corresponding range of mental age?   [*Note:* $75 \leqslant$ IQ $\leqslant 150$]

56. *Public safety.* A nuclear reactor has produced 1,000 cubic feet of gas that is 60% radioactive (by volume). It has been decided to dispose of the gas by releasing it into the atmosphere after first mixing it with enough air so that the radioactive gas is at most 2% by volume. How much air must be mixed with the 1,000 cubic feet of gas before it is released?

4-3   Quadratic Equations

- Solution by Square Root
- Solution by Factoring
- Solution by Completing the Square
- Solution by the Quadratic Formula
- Equations with Fractional Forms
- Applications

In Section 4-1 we solved linear, or first-degree, equations in one variable. When a unique solution exists, a linear equation can be transformed into the equivalent form

$$ax + b = 0 \qquad a \neq 0$$

which can be solved for x to obtain $x = -b/a$. We will now consider another special class of equations called *quadratic*, or *second-degree*, *equations*. We will discuss several methods of solution that are particularly well-suited for certain forms of these equations.

> **Quadratic Equations**
>
> A **quadratic** or **second-degree equation** in one variable is an equation that can be written in the equivalent form
>
> $$ax^2 + bx + c = 0 \qquad a \neq 0 \qquad \text{Standard form}$$
>
> where a, b, and c represent constants and x is the variable.

The equations

$$5x^2 - 3x + 7 = 0 \qquad \text{and} \qquad 18 = 32t^2 - 12t$$

are both quadratic equations. The first is in standard form, while the second can be transformed into standard form.

We will restrict our attention to real solutions of a quadratic equation. In working with quadratic equations we will find that three situations may occur. There may be two real solutions, one real solution, or no real solutions. To **solve** a quadratic equation means to determine all real solutions or to conclude that no real solution exists.

■ Solution by Square Root

When a quadratic equation in standard form has no bx term, that is, if $b = 0$, the equation takes the form

$$ax^2 + c = 0 \qquad a \neq 0$$

This equation can be solved by the **square root method** illustrated in the next example.

Example 16   Solve by the square root method.

(A) $x^2 - 25$ (B) $x^2 - 7 = 0$ (C) $2x^2 - 10 = 0$
(D) $3x^2 + 27 = 0$ (E) $(x + \frac{1}{3})^2 = \frac{5}{9}$

Solutions   (A) $x^2 - 25 = 0$ Isolate x^2 on the left.
       $x^2 = 25$ What real numbers squared give 25?
       $x = \pm 5$ ± 5 is short for 5 and -5.

Thus, $x^2 - 25 = 0$ has two real solutions: 5 and -5.

(B) $x^2 - 7 = 0$ Isolate x^2.
       $x^2 = 7$ What real numbers squared give 7?
       $x = \pm\sqrt{7}$ $\pm\sqrt{7}$ is short for $\sqrt{7}$ and $-\sqrt{7}$.

Thus, $x^2 - 7 = 0$ has two real solutions: $\sqrt{7}$ and $-\sqrt{7}$.

(C) $2x^2 - 10 = 0$ Isolate x^2 on the left.

$$2x^2 = 10$$
$$x^2 = 5$$ What real numbers squared give 5?
$$x = \pm\sqrt{5}$$

Thus, $2x^2 - 10 = 0$ has two real solutions: $\sqrt{5}$ and $-\sqrt{5}$.

(D) $3x^2 + 27 = 0$ Isolate x^2.

$$3x^2 = -27$$
$$x^2 = -9$$ What real numbers squared give -9? None!

No real solution

Thus, $3x^2 + 27 = 0$ has no real solution.

(E) $\left(x + \dfrac{1}{3}\right)^2 = \dfrac{5}{9}$ First solve for $x + \frac{1}{3}$.

$$x + \frac{1}{3} = \pm\sqrt{\frac{5}{9}} = \pm\frac{\sqrt{5}}{3}$$ Now solve for x.

$$x = -\frac{1}{3} \pm \frac{\sqrt{5}}{3}$$

$$= \frac{-1 \pm \sqrt{5}}{3}$$

Thus, we have two real solutions: $\dfrac{-1 + \sqrt{5}}{3}$ and $\dfrac{-1 - \sqrt{5}}{3}$.

Problem 16   Solve by the square root method.

(A) $x^2 - 81 = 0$ (B) $x^2 - 6 = 0$ (C) $3x^2 - 12 = 0$
(D) $2x^2 + 32 = 0$ (E) $(x - \frac{2}{5})^2 = \frac{13}{25}$

▪ Solution by Factoring

When the left side of a quadratic equation in standard form can be factored, the equation can be solved quickly. The method called **solving by factoring** is based on the following important property of multiplication:

Property of Multiplication

If a and b represent real numbers, then $ab = 0$ if and only if $a = 0$ or $b = 0$ (or both).

For example, we have $(2x - 1)(x + 3) = 0$ if and only if

$$2x - 1 = 0 \quad \text{or} \quad x + 3 = 0$$

Solving the last two equations (mentally), we obtain $x = \frac{1}{2}$ or $x = -3$. Example 17 illustrates how factoring can be utilized to solve quadratic equations.

Example 17   Solve by factoring, if possible.

(A) $x^2 - 5x - 14 = 0$ (B) $2x^2 + 7x - 15 = 0$
(C) $x^2 + x - 1 = 0$ (D) $6x^2 - 30x - 36 = 0$
(E) $5x^2 = 3x$

Solutions   (A) $x^2 - 5x - 14 = 0$ Factor the left side.
     $(x + 2)(x - 7) = 0$ The product on the left is 0 if and
    $x + 2 = 0$ or $x - 7 = 0$ only if $x + 2 = 0$ or $x - 7 = 0$.
      $x = -2$ or $x = 7$

  (B) $2x^2 + 7x - 15 = 0$ Factor the left side.
    $(2x - 3)(x + 5) = 0$ $(2x - 3)(x + 5) = 0$ if and only if
    $2x - 3 = 0$ or $x + 5 = 0$ $2x - 3 = 0$ or $x + 5 = 0$.
      $x = \frac{3}{2}$ or $x = -5$

  (C) $x^2 + x - 1 = 0$:   The left side of this equation cannot be factored using integers. We will see how to solve this equation by other methods that will be discussed later.

  (D) $6x^2 - 30x - 36 = 0$ First divide both sides of the equation by 6.
     $x^2 - 5x - 6 = 0$ Now factor the left side.
    $(x + 1)(x - 6) = 0$
   $x + 1 = 0$ or $x - 6 = 0$
     $x = -1$ or $x = 6$

  (E) $5x^2 = 3x$:   Before we solve this, we should note that it would be incorrect to divide both sides by x. Why? The equation $5x = 3$ is not equivalent to the original equation. That is, the two equations do not have the same solution set. A correct method for solving this equation is as follows:

      $5x^2 = 3x$ Put in standard form.
     $5x^2 - 3x = 0$ Factor the left side.
     $x(5x - 3) = 0$
    $x = 0$ or $5x - 3 = 0$
           $x = \frac{3}{5}$

Thus, $5x^2 = 3x$ has two solutions, $x = 0$ and $x = \frac{3}{5}$. The equation $5x = 3$ has only one solution, $x = \frac{3}{5}$.

Problem 17 Solve by factoring, if possible.

(A) $x^2 - 5x - 24 = 0$ (B) $6x^2 + 7x - 3 = 0$
(C) $2x^2 - 5x - 4 = 0$ (D) $3x^2 - 18x + 24 = 0$
(E) $4u^2 = 9u$

Solving quadratic equations using the square root method or factoring is convenient when these methods apply. However, there are many simple-looking quadratic equations, such as $x^2 + x - 1 = 0$, which require other techniques. We now turn to these techniques.

■ Solution by Completing the Square

The method of *completing the square* can be applied to any quadratic equation, and, in fact, allows us to derive a formula for finding the solutions of a general quadratic equation. The procedure involves converting a standard quadratic equation

$$ax^2 + bx + c = 0 \qquad a \neq 0$$

into the form

$$(x + k)^2 = h$$

where k and h are constants. Notice that the left side of this equation is a **perfect square.** Once we have this form we can apply the method of square roots to obtain

$$x + k = \pm \sqrt{h}$$

Thus,

$$x = -k \pm \sqrt{h}$$

The question is: How do we convert a standard quadratic equation into a form where the left side is a perfect square? Before we tackle this problem, we will first discuss how to create a perfect square given an expression of the form

$$x^2 + mx \qquad \text{or} \qquad x^2 - mx$$

where the coefficient of x^2 is 1.

The question we need to answer is: What can be added to these expressions so that they become perfect squares? Fortunately, there is a simple mechanical rule for finding what we need. We take one-half of the coefficient of x ($m/2$ or $-m/2$), square this to obtain $m^2/4$, and then add this to either expression. We then have the following:

$$x^2 + mx + \frac{m^2}{4} = \left(x + \frac{m}{2}\right)^2 \qquad \text{and} \qquad x^2 - mx + \frac{m^2}{4} = \left(x - \frac{m}{2}\right)^2$$

This can be easily verified by squaring the right side of both equations. We now state the above as a general rule.

Completing the Square

An expression of the form

$$x^2 + bx$$

will become a perfect square if we add to it the square of one-half the coefficient of x. Thus, we add

$$\frac{b^2}{4} = \left(\frac{b}{2}\right)^2$$

We then have

$$x^2 + bx + \frac{b^2}{4} = \left(x + \frac{b}{2}\right)^2$$

Example 18   Complete the square and write as a perfect square.

(A) $x^2 + 10x$ (B) $x^2 - 5x$ (C) $x^2 - \frac{4}{3}x$

Solutions   (A) To complete the square of $x^2 + 10x$, we need to add $(\frac{10}{2})^2 = 5^2 = 25$ to it. We then have

$$x^2 + 10x + 25 = (x + 5)^2$$

(B) To complete the square of $x^2 - 5x$, we need to add $(-\frac{5}{2})^2 = \frac{25}{4}$ to it. We then have

$$x^2 - 5x + \tfrac{25}{4} = (x - \tfrac{5}{2})^2$$

(C) To complete the square of $x^2 - \frac{4}{3}x$, we first divide $-\frac{4}{3}$ by 2 to obtain $-\frac{2}{3}$. Adding $(-\frac{2}{3})^2 = \frac{4}{9}$ to the expression, we obtain

$$x^2 - \tfrac{4}{3}x + \tfrac{4}{9} = (x - \tfrac{2}{3})^2$$

Problem 18   Complete the square and write as a perfect square.

(A) $x^2 + 12x$ (B) $x^2 - 11x$ (C) $x^2 - \frac{8}{5}x$

The process of solving a quadratic equation by the method of completing the square is illustrated in the following examples.

Example 19   Solve $x^2 + 8x - 2 = 0$ by completing the square.

Solution

$x^2 + 8x - 2 = 0$	Add 2 to both sides to isolate $x^2 + 8x$ on the left.
$x^2 + 8x = 2$	Add 16 to both sides to make the left side a perfect square.
$x^2 + 8x + 16 = 2 + 16$	Write the left side as a perfect square.
$(x + 4)^2 = 18$	Solve by the square root method.
$x + 4 = \pm\sqrt{18} = \pm 3\sqrt{2}$	Simplify the radical.
$x = -4 \pm 3\sqrt{2}$	

Problem 19   Solve $x^2 - 10x + 5 = 0$ by completing the square.

Example 20   Solve $2x^2 + 14x + 5 = 0$ by completing the square.

Solution

$2x^2 + 14x + 5 = 0$	First divide both sides by 2 so that x^2 will have 1 as a coefficient.
$x^2 + 7x + \frac{5}{2} = 0$	Subtract $\frac{5}{2}$ from both sides to isolate $x^2 + 7x$ on the left.
$x^2 + 7x = -\frac{5}{2}$	Add $\left(\frac{7}{2}\right)^2 = \frac{49}{4}$ to both sides to make the left side a perfect square.
$x^2 + 7x + \frac{49}{4} = -\frac{5}{2} + \frac{49}{4}$	Simplify the right side.
$= -\frac{10}{4} + \frac{49}{4}$	
$= \frac{39}{4}$	Write the left side as a perfect square.
$\left(x + \frac{7}{2}\right)^2 = \frac{39}{4}$	Solve by the square root method.

$$x + \frac{7}{2} = \pm\sqrt{\frac{39}{4}} = \pm\frac{\sqrt{39}}{2}$$

$$x = -\frac{7}{2} \pm \frac{\sqrt{39}}{2} = \frac{-7 \pm \sqrt{39}}{2}$$

Problem 20   Solve $2x^2 - 6x + 1 = 0$ by completing the square.

Example 21   Solve $x^2 + 2x + 7 = 0$ by completing the square.

Solution

$$x^2 + 2x + 7 = 0$$
$$x^2 + 2x = -7$$
$$x^2 + 2x + 1 = -7 + 1$$
$$(x + 1)^2 = -6$$
No real solution

There is no real value of x for which the last equation is true.

Problem 21   Solve $x^2 - 4x + 9 = 0$ by completing the square.

■ Solution by the Quadratic Formula

The process of completing the square can be used to establish the following result:

The Quadratic Formula

The solutions, if any, of a quadratic equation

$$ax^2 + bx + c = 0 \qquad a \neq 0$$

are given by

$$x = \frac{-b \pm \sqrt{b^2 - 4ac}}{2a}$$

If $b^2 - 4ac > 0$, there are two real solutions.

If $b^2 - 4ac = 0$, there is one real solution.

If $b^2 - 4ac < 0$, there is no real solution.

Let us derive the quadratic formula:

$$ax^2 + bx + c = 0 \qquad \text{Divide both sides by } a.$$

$$x^2 + \frac{b}{a}x + \frac{c}{a} = 0 \qquad \text{Subtract } \frac{c}{a} \text{ from both sides.}$$

$$x^2 + \frac{b}{a}x = -\frac{c}{a} \qquad \text{Add } \left(\frac{b}{2a}\right)^2 = \frac{b^2}{4a^2} \text{ to both sides.}$$

$$x^2 + \frac{b}{a}x + \frac{b^2}{4a^2} = \frac{b^2}{4a^2} - \frac{c}{a} \qquad \text{Write the left side as a perfect square and the right side as a single fraction.}$$

$$\left(x + \frac{b}{2a}\right)^2 = \frac{b^2 - 4ac}{4a^2} \qquad \text{Solve by the square root method.}$$

$$x + \frac{b}{2a} = \pm\sqrt{\frac{b^2 - 4ac}{4a^2}}$$

$$= \pm\frac{\sqrt{b^2 - 4ac}}{2a}$$

$$x = -\frac{b}{2a} \pm \frac{\sqrt{b^2 - 4ac}}{2a} \qquad \text{Write the right side as a single fraction.}$$

$$x = \frac{-b \pm \sqrt{b^2 - 4ac}}{2a}$$

We will now apply the quadratic formula to solve some quadratic equations.

Example 22   Solve, using the quadratic formula.

(A) $x^2 - 2x - 1 = 0$ (B) $3x^2 - x - 2 = 0$ (C) $x^2 + x + 1 = 0$

Solutions   (A) $x^2 - 2x - 1 = 0$

Here, $a = 1$, $b = -2$, and $c = -1$. Thus,

$$x = \frac{-b \pm \sqrt{b^2 - 4ac}}{2a}$$

$$= \frac{-(-2) \pm \sqrt{(-2)^2 - 4(1)(-1)}}{2(1)} \qquad \text{Be careful of sign errors here.}$$

$$= \frac{2 \pm \sqrt{8}}{2}$$

$$= \frac{2 \pm 2\sqrt{2}}{2} \qquad \text{Be careful in reducing this.}$$

$$= \frac{\overset{1}{\cancel{2}}(1 \pm \sqrt{2})}{\underset{1}{\cancel{2}}} = 1 \pm \sqrt{2}$$

(B) $3x^2 - x - 2 = 0$

Here, $a = 3$, $b = -1$, and $c = -2$. Thus,

$$x = \frac{-(-1) \pm \sqrt{(-1)^2 - 4(3)(-2)}}{2(3)}$$

$$= \frac{1 \pm \sqrt{25}}{6} = \frac{1 \pm 5}{6}$$

Thus,

$$x = \frac{1 + 5}{6} = \frac{6}{6} = 1 \qquad \text{or} \qquad x = \frac{1 - 5}{6} = \frac{-4}{6} = -\frac{2}{3}$$

(C) $x^2 + x + 1 = 0$

Here, $a = 1$, $b = 1$, and $c = 1$. Thus,

$$x = \frac{-1 \pm \sqrt{(1)^2 - 4(1)(1)}}{2(1)} = \frac{-1 \pm \sqrt{-3}}{2}$$

Since the number under the radical is negative, the equation has no real solution.

Problem 22   Solve, using the quadratic formula.

(A) $x^2 + 3x - 5 = 0$ (B) $8x^2 - 10x - 3 = 0$ (C) $x^2 + 2x + 6 = 0$

■ Equations with Fractional Forms

The following example illustrates how certain equations with fractional forms can be transformed into standard quadratic form and solved by one of the methods of this section.

Example 23   Solve: $2x + 5 = \dfrac{9}{x + 1}$

Solution

$$2x + 5 = \dfrac{9}{x + 1}$$ Note that $x \neq -1$.

$$(x + 1) \cdot \dfrac{(2x + 5)}{1} = (x + 1) \cdot \dfrac{9}{x + 1}$$ Multiply both sides by the LCD $= x + 1$ and simplify.

$$2x^2 + 7x + 5 = 9$$ Put in standard form.

$$2x^2 + 7x - 4 = 0$$ Factor the left side.

$$(2x - 1)(x + 4) = 0$$

$$2x - 1 = 0 \quad \text{or} \quad x + 4 = 0$$

$$x = \tfrac{1}{2} \quad \text{or} \quad x = -4$$

Problem 23   Solve: $2x - 3 = \dfrac{10}{x - 2}$

■ Applications

A large variety of practical problems lead to quadratic equations that can then be solved using the methods developed in this section. The following examples illustrate this.

Example 24
Break-Even Analysis

The manufacturer of a stereo system finds that the number of units, x, ordered per pay (demand) is given by

$$x = 125 - \dfrac{p}{4}$$

where p is the price per unit. The total cost, C, per day to manufacture x units is given by

$$C = 11{,}200 + 60x$$

How many units must be produced and sold each day to break even; that is, so that the revenue, R, is equal to the cost, C?

Solution   The demand equation

$$x = 125 - \dfrac{p}{4}$$

can be written in the form

$$p = 500 - 4x$$

The revenue R is given by

$$R = (\text{Number of units sold}) \times (\text{Price per unit})$$
$$= xp$$
$$= x(500 - 4x)$$
$$= 500x - 4x^2$$

We want to find the value(s) of x such that $R = C$; that is, such that

$$500x - 4x^2 = 11{,}200 + 60x$$

Solving this equation, we have

$$
\begin{aligned}
500x - 4x^2 &= 11{,}200 + 60x \qquad && \text{Put in standard form.} \\
-4x^2 + 440x - 11{,}200 &= 0 && \text{Divide both sides by } -4. \\
x^2 - 110x + 2{,}800 &= 0 && \text{Solve for } x. \\
(x - 40)(x - 70) &= 0 \\
x - 40 = 0 \quad \text{or} \quad x - 70 &= 0 \\
x = 40 \quad \text{or} \quad x &= 70
\end{aligned}
$$

Thus, if 40 or 70 units are produced and sold each day, the revenue will equal the cost.

Problem 24   Solve Example 24 using the demand equation

$$x = 120 - \frac{p}{5}$$

and the cost equation

$$C = 12{,}000 + 100x$$

Example 25
Revenue

The manager of a movie theater finds that she will sell all 800 tickets to the Friday evening movie if the charge for admission is $4 per person. For each $1 increase in admission price she expects 100 fewer people to buy tickets. How much should she charge for admission so that the revenue, R, from ticket sales is $3,500?

Solution   Let

$$x = \text{Number of \$1 increases}$$

Then we have

$$
\begin{aligned}
4 + x &= \text{Price per ticket} \\
800 - 100x &= \text{Number of tickets sold}
\end{aligned}
$$

Thus, the revenue on ticket sales is

$$
\begin{aligned}
R &= (\text{Number of tickets sold}) \times (\text{Price per ticket}) \\
&= (800 - 100x)(4 + x) \\
&= 3{,}200 + 400x - 100x^2
\end{aligned}
$$

To find the value(s) of x for which $R = 3{,}500$, we have

$$
\begin{aligned}
3{,}200 + 400x - 100x^2 &= 3{,}500 \qquad && \text{Put in standard form.} \\
-100x^2 + 400x - 300 &= 0 && \text{Divide by } -100. \\
x^2 - 4x + 3 &= 0 && \text{Solve for } x. \\
(x - 1)(x - 3) &= 0 \\
x - 1 = 0 \quad \text{or} \quad x - 3 &= 0 \\
x = 1 \quad \text{or} \quad x &= 3
\end{aligned}
$$

Since the price per ticket is $4 + x$, the manager should charge $5 or $7 per ticket to obtain a revenue of $3,500.

Problem 25 Solve Example 25 if the revenue is to be $3,600.

Answers to Matched Problems

16. (A) ± 9 (B) $\pm\sqrt{6}$ (C) ± 2 (D) No real solution

(E) $\dfrac{2 \pm \sqrt{13}}{5}$

17. (A) $-3, 8$ (B) $-\frac{3}{2}, \frac{1}{3}$ (C) Cannot be factored using integers
(D) $2, 4$ (E) $0, \frac{9}{4}$

18. (A) $x^2 + 12x + 36 = (x + 6)^2$ (B) $x^2 - 11x + \frac{121}{4} = (x - \frac{11}{2})^2$
(C) $x^2 - \frac{8}{5}x + \frac{16}{25} = (x - \frac{4}{5})^2$

19. $x = 5 \pm \sqrt{20} = 5 \pm 2\sqrt{5}$ 20. $x = \dfrac{3 \pm \sqrt{7}}{2}$ 21. No real solution

22. $\dfrac{-3 \pm \sqrt{29}}{2}$ (B) $\frac{3}{2}, -\frac{1}{4}$ (C) No real solution

23. $-\frac{1}{2}, 4$ 24. 40 or 60 units 25. $6 per ticket

Exercise 4-3

In the following problems, by "solve" we mean to find all real solutions. If an equation has no real solutions, say so.

A Solve by the square root method.

1. $x^2 - 36 = 0$ 2. $x^2 - 81 = 0$ 3. $x^2 + 16 = 0$
4. $x^2 + 9 = 0$ 5. $y^2 - 8 = 0$ 6. $z^2 - 27 = 0$
7. $9x^2 - 16 = 0$ 8. $36y^2 - 25 = 0$

Solve by factoring, if possible.

9. $x^2 - 7x = 0$ 10. $w^2 + 5w = 0$
11. $x^2 - 2x - 8 = 0$ 12. $x^2 + 2x - 15 = 0$
13. $x^2 - 5x - 12 = 0$ 14. $x^2 + 4x - 10 = 0$
15. $5u^2 = 4u$ 16. $8t^2 = -3t$
17. $2x^2 - x - 3 = 0$ 18. $5x^2 + 9x - 2 = 0$

Solve by completing the square.

19. $x^2 + 6x - 3 = 0$ 20. $x^2 + 8x + 6 = 0$
21. $2x^2 - 6x - 3 = 0$ 22. $2x^2 + 8x + 3 = 0$

Solve using the quadratic formula.

23. $x^2 + 10x - 8 = 0$ 24. $x^2 - 6x + 2 = 0$
25. $3y^2 - 5y - 4 = 0$ 26. $5x^2 + 3x - 4 = 0$

B Solve using the most efficient method.

27. $x^2 = -7x - 12$ 28. $z^2 - 4z = 5$ 29. $4x^2 = 5$

30. $16x^2 = 7$ **31.** $16x^2 - 13 = 0$ **32.** $9y^2 - 7 = 0$

33. $(x - 5)^2 = 36$ **34.** $(z + 3)^2 = 49$ **35.** $6x^2 - 13x + 6 = 0$

36. $8x^2 - 6x - 9 = 0$ **37.** $(x - \frac{1}{4})^2 = \frac{9}{16}$ **38.** $(x + \frac{1}{5})^2 = \frac{4}{25}$

39. $2m^2 = m + 3$ **40.** $9x^2 + 8 = 18x$ **41.** $(y - \frac{1}{3})^2 = \frac{7}{9}$

42. $(C + \frac{1}{6})^2 = \frac{13}{36}$ **43.** $x^2 + x - 1 = 0$ **44.** $x^2 - 3x + 1 = 0$

45. $16u^2 + 9 = 0$ **46.** $25a^2 + 4 = 0$ **47.** $\dfrac{8}{w} = \dfrac{w}{2}$

48. $\dfrac{x^2}{4} = 3 - x$ **49.** $u^2 = 5u + 3$ **50.** $v^2 + 7v = 3$

51. $1 + \dfrac{2}{x} = \dfrac{35}{x^2}$ **52.** $\dfrac{8}{y} - \dfrac{3}{y^2} = 4$ **53.** $2x^2 + 5x - 4 = 0$

54. $3x^2 - 2x - 7 = 0$ **55.** $\dfrac{6}{x - 3} - 2 = \dfrac{5}{x}$ **56.** $\dfrac{13}{x + 2} + \dfrac{3}{2} = \dfrac{20}{x}$

57. $2w^2 = 5w - 9$ **58.** $3t^2 + 3t + 7 = 0$ **59.** $7z^2 + 12z + 3 = 0$

60. $5u^2 - 10u + 4 = 0$

C   *Solve by factoring (a and b denote constants).*

61. $x^2 - 5ax + 6a^2 = 0$

62. $a^2x^2 - 2abx^2 - 8b^2 = 0, \ a \neq 0$

Solve by the square root method (a, b, c, m, and n denote constants).

63. $\left(x + \dfrac{b}{2a}\right)^2 = \dfrac{b^2 - 4ac}{4a^2}, \quad a \neq 0$

64. $\left(x + \dfrac{m}{2}\right)^2 = \dfrac{m^2 - 4n}{4}$

Solve using the quadratic formula (m and n denote constants).

65. $x^2 + 2mx + n = 0$ **66.** $x^2 + mx + n = 0$

Applications

Business & Economics   **67.** *Revenue.* A company manufactures and sells x television sets per month. If the revenue is given by

$$R = 200x - \dfrac{x^2}{30} \qquad 0 \leqslant x \leqslant 6{,}000$$

determine the number of sets that must be manufactured and sold each month for the revenue to be $300,000.

68. *Profit.* If the profit, P, on the manufacture and sales of x televisions per month (see Problem 67) is given by

$$P = -\dfrac{x^2}{30} + 140x - 72{,}000 \qquad 0 \leqslant x \leqslant 6{,}000$$

determine the number of sets manufactured and sold if the profit is $75,000.

69. *Break-even analysis.* The manufacturer of a short-wave radio finds that the number of units, x, ordered per week (demand) is given by

$$x = 250 - \frac{p}{4}$$

where p is the price per unit. The total cost to manufacture x units per week is $C = 30,000 + 200x$. How many radios must be produced and sold each week for the revenue, R, to be equal to the cost?   [Recall that $R = xp$.]

70. *Break-even analysis.* The demand, x, per week for a high-quality camera is given by

$$x = 350 - \frac{p}{2}$$

where p is the price per camera. The cost to produce x cameras per week is given by $C = 19,200 + 300x$. Determine the number of cameras that must be produced and sold each week to break even (for which $R = C$).   [Recall that $R = xp$.]

71. *Profit.* In Problem 69 find the profit, P, in terms of x and determine the number of radios that must be manufactured and sold each week to earn a profit of $3,600.   [Recall that $P = R - C$.]

72. *Profit.* In Problem 70 find the profit, P, in terms of x and determine the number of cameras that must be produced and sold each week to earn a profit of $600.   [Recall that $P = R - C$.]

73. *Rental income.* During the tourist season, a San Francisco hotel finds that it will rent all 120 rooms each night at a rate of $50 per night. For each $1 increase in rate, two fewer rooms will be rented. Find the rate at which the total rental income per night is $6,050.

74. *Rental income.* A car rental agency rents all 100 cars per day at a rate of $20 per day. For each $2 increase in rate, five fewer cars are rented. At what rate will the total rental income be $2,240 per day?

75. *Agriculture.* A commercial cherry grower estimates that if thirty trees are planted per acre, each tree will produce an average of 50 pounds of cherries per season. For each additional tree planted per acre, the average yield per tree will be reduced by 1 pound. How many trees should be planted per acre to obtain 1,600 pounds of cherries per acre?

76. *Small business.* Two barbers estimate that they will have an average of forty customers per day if they charge $8 for a haircut. For each additional $1 that they charge, they figure they will lose four customers. What higher rate will produce the same revenue they receive at $8 per haircut?

Life Sciences

77. *Drug concentration.* The concentration, C, in milligrams per cubic centimeter of a particular drug in a patient's bloodstream is given by

$$C = \frac{0.16t}{t^2 + 4t + 4}$$

where t is the number of hours after the drug is taken. At what time will the drug concentration be 0.02?

78. *Bacteria control.* A recreational swimming area is treated periodically to control the growth of harmful bacteria. Suppose that after treatment the bacteria count, C, per cubic centimeter is given by

$$C = 30t^2 - 240t + 500 \qquad 0 \leqslant t \leqslant 4$$

where t is the number of days after treatment. After how many days will the bacteria count be 140?

Social Sciences

79. *Fund raising.* A local chapter of a political organization estimates that it will distribute all 750 tickets to a fund-raising dinner if a \$50 donation is requested per ticket. For each additional \$5 in donation requested, it is estimated that 25 fewer tickets will be distributed. What donation per ticket should be requested in order to raise \$50,000?

4-4 Nonlinear Inequalities

- Polynomial Inequalities
- Rational Inequalities
- Application

■ Polynomial Inequalities

In Section 4-2 we solved first-degree (linear) inequalities such as

$$3x - 7 \geqslant 5(x - 2) + 3$$

But how do we solve second-degree (quadratic) inequalities such as

$$x^2 - 12 < x$$

If, after collecting all nonzero terms on the left, we find that we are able to factor the left side in terms of first-degree factors, then we will be able to solve the inequality, as shown in the following case:

$$x^2 - 12 < x \qquad \text{Move all nonzero terms to the left side.}$$
$$x^2 - x - 12 < 0 \qquad \text{Factor left side.}$$
$$(x + 3)(x - 4) < 0$$

We are looking for values of x that will make the left side less than 0 — that is, negative. What must the signs of (x + 3) and (x − 4) be so that their product is negative? They must have opposite signs.

Let us see if we can determine where each of the factors is positive, negative, and 0. The point at which either factor is 0 is called a **critical point.** We will see why in a moment.

Sign analysis for (x + 3)

Critical point	*(x + 3) is positive when*	*(x + 3) is negative when*
x + 3 = 0	x + 3 > 0	x + 3 < 0
x = −3	x > −3	x < −3

It is useful to summarize these results on a real number line:

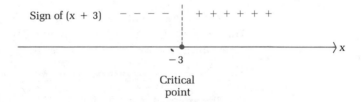

Thus, (x + 3) is negative for values of x to the left of −3 and positive for values of x to the right of −3.

Sign analysis for (x − 4)

Critical point	*(x − 4) is positive when*	*(x − 4) is negative when*
x − 4 = 0	x − 4 > 0	x − 4 < 0
x = 4	x > 4	x < 4

Geometrically, we have the results shown here:

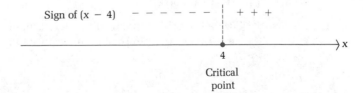

Thus, (x − 4) is negative for values of x to the left of 4 and positive for values of x to the right of 4.

Combining these results relative to a single real number line leads to a simple solution to the original problem:

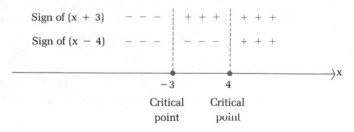

Now we see that the factors have opposite signs (and thus, their product is negative) for x between -3 and 4. We can now write the solution for the inequality $x^2 - 12 < x$:

$-3 < x < 4$ Inequality notation

$(-3, 4)$ Interval notation

$$\xrightarrow{\quad\overset{\circ}{-3}\qquad\overset{\circ}{4}\quad} x$$

Proceeding as in the example, one can easily prove Theorem 1, which is the basis for the sign-analysis method of solving second- and higher-degree inequalities, as well as other types of inequalities.

Theorem 1

> The value of x at which $(ax + b)$ is 0 is called the **critical point** for $ax + b$. To the left of this critical point on a real number line, $(ax + b)$ has one sign and to the right of this critical point, $(ax + b)$ has the opposite sign $(a \neq 0)$.

Example 26   Solve and graph: $3x^2 + 10x \geqslant 8$

Solution

$3x^2 + 10x \geqslant 8$ Move all nonzero terms to the left side.

$3x^2 + 10x - 8 \geqslant 0$ Factor the left side (if possible).

$(3x - 2)(x + 4) \geqslant 0$ Find critical points.

Critical points: $-4, \frac{2}{3}$

Locate the critical points on a real number line and determine the sign of each linear factor to the left and right of its critical point.

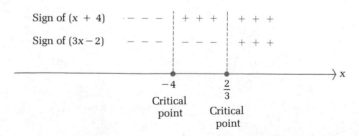

Note that the equality portion of the inequality statement is satisfied at the critical points. The inequality portion is satisfied when the product of the factors is positive—that is, when the factors have the same sign. From the figure above we see that this happens to the left of -4 or to the right of $\frac{2}{3}$. We can now write the solution:

$x \leqslant -4$ or $x \geqslant \frac{2}{3}$ Inequality notation

$(-\infty, -4] \cup [\frac{2}{3}, \infty)$ Interval notation

Problem 26 Solve and graph: $2x^2 \geqslant 3x + 9$

Example 27 Solve and graph: $x^3 - 4 \leqslant x - 4x^2$

Solution

$$x^3 - 4 \leqslant x - 4x^2$$ Move all nonzero terms to the left side.

$$x^3 + 4x^2 - x - 4 \leqslant 0$$ Factor left side by grouping first two terms and last two terms.

$$(x^3 + 4x^2) - (x + 4) \leqslant 0$$

$$x^2(x + 4) - (x + 4) \leqslant 0$$ $(x + 4)$ is a common factor.

$$(x^2 - 1)(x + 4) \leqslant 0$$

$$(x - 1)(x + 1)(x + 4) \leqslant 0$$ Find critical points.

Critical points: $-4, -1, 1$

Equality holds at all of the critical points. The inequality holds when the left side is less than 0—that is, when the left side is negative. The left side is negative when $(x - 1)$, $(x + 1)$, and $(x + 4)$ are all negative or when one is negative and two are positive. We chart the sign of each factor on a real number line:

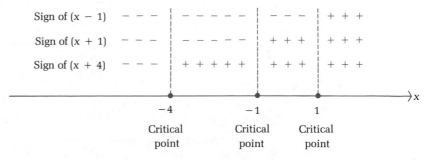

The solution is easily determined from the figure:

$x \leqslant -4$ or $-1 \leqslant x \leqslant 1$ Inequality notation

$(-\infty, -4] \cup [-1, 1]$ Interval notation

Problem 27   Solve and graph: $x^3 + 12 > 3x^2 + 4x$

Remark: The key to solving polynomial inequalities is factoring. At this point we are able to factor only a few very special types of polynomials. In a more advanced treatment of the subject, procedures can be developed that enable one to factor a fairly large class of polynomials.

■ Rational Inequalities

The sign-analysis technique described previously for solving polynomial inequalities can also be used to solve inequalities involving rational forms, such as

$$\frac{x-3}{x+5} > 0 \qquad \text{and} \qquad \frac{x^2 + 5x - 6}{5 - x} \leq 1$$

Example 28   Solve and graph: $\dfrac{x^2 - x + 1}{2 - x} \geq 1$

Solution   We might be tempted to start by multiplying both sides by $(2 - x)$, as we would do if the inequality were an equation. However, since we do not know whether $(2 - x)$ is positive or negative, we do not know if the sense of the inequality is to be changed.

We proceed instead as follows:

$$\frac{x^2 - x + 1}{2 - x} \geq 1 \qquad \text{Move all nonzero terms to the left side.}$$

$$\frac{x^2 - x + 1}{2 - x} - 1 \geq 0 \qquad \text{Combine left side into a single fraction.}$$

$$\frac{x^2 - x + 1 - (2 - x)}{2 - x} \geq 0$$

$$\frac{x^2 - 1}{2 - x} \geq 0 \qquad \text{Factor numerator.}$$

$$\frac{(x - 1)(x + 1)}{2 - x} \geq 0$$

Critical points: $-1, 1, 2$

Equality holds when $x = \pm 1$.

The left side is not defined when $x = 2$.

The inequality holds when $(x - 1)$, $(x + 1)$, and $(2 - x)$ are all positive or two are negative and one is positive. We chart the sign of each on a real number line:

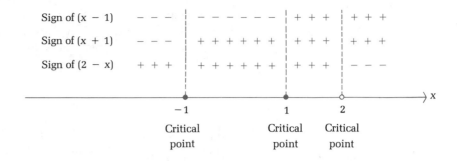

Note the sign pattern for $(2 - x)$: It is positive to the left of its critical point and negative to the right. From the figure it is easy to write the solution:

$x \leq -1$ or $1 \leq x < 2$ Inequality notation

$(-\infty, -1] \cup [1, 2)$ Interval notation

Problem 28   Solve and graph: $\dfrac{3}{2 - x} \leq \dfrac{1}{x + 4}$

■ Application

Example 29
Break-Even Analysis

The marketing research and financial departments of a company estimate that at a price of p dollars per unit, the weekly cost C and revenue R (in thousands of dollars) will be given by the equations

$C = 14 - p$ Cost equation

$R = 8p - p^2$ Revenue equation

Find prices for which the company:

(A)   Has a profit   (B)   Has a loss   (C)   Breaks even

Solutions

(A)   A profit will result if revenue is greater than cost—that is, if

$$R > C$$
$$8p - p^2 > 14 - p$$

$-p^2 + 9p - 14 > 0$ Multiply both sides by -1 (inequality
$p^2 - 9p + 14 < 0$ sign reverses).

$(p - 2)(p - 7) < 0$ Factor left side.

Critical points: 2, 7

The inequality is satisfied if $(p - 2)$ and $(p - 7)$ have opposite signs.

Sign of $(p - 2)$ $--\;|\;+ + + + + \;|\; + + + + +$

Sign of $(p - 7)$ $--\;|\;- - - - - \;|\; + + + + +$

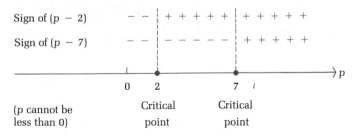

| 0   2   7 |

(p cannot be   Critical   Critical
less than 0)   point   point

Thus, a profit will result for $2 < p < 7.

(B)   A loss will result if revenue is less than cost — that is, if

$$R < C$$
$$8p - p^2 < 14 - p$$
$$-p^2 + 9p - 14 < 0$$
$$p^2 - 9p + 14 > 0$$
$$(p - 2)(p - 7) > 0$$

The inequality is satisfied if both factors have the same sign. Referring to the sign chart in part A, we see that this happens for $p < 2$ or $p > 7$. But, since p cannot be negative, a loss will result for $0 \leqslant p < 2 or $p > 7.

(C)   The company will break even if revenue equals cost — that is, if

$$R = C$$
$$8p - p^2 = 14 - p$$
$$p^2 - 9p + 14 = 0$$
$$(p - 2)(p - 7) = 0$$

The company will break even for $p = 2 or $p = 7.

Problem 29   Repeat Example 29 for $C = 8 - p$ Cost equation
$R = 5p - p^2$ Revenue equation

Answers to   26. $x \leqslant -\frac{3}{2}$ or $x \geqslant 3$ 27. $-2 < x < 2$ or $x > 3$
Matched Problems $(-\infty, -\frac{3}{2}] \cup [3, \infty)$ $(-2, 2) \cup (3, \infty)$

28. $-4 < x \leqslant -\frac{5}{2}$ or $x > 2$
$(-4, -\frac{5}{2}] \cup (2, \infty)$

29. (A) $2 < p < \$4$ (B) $\$0 \leqslant p < \$2$ or $p > \$4$
     (C) $p = \$2$ or $p = \$4$

Exercise 4-4

Solve and graph. Express answers in both inequality and interval notation.

A
1. $x^2 - x - 12 < 0$
2. $x^2 - 2x - 8 < 0$
3. $x^2 - x - 12 \geqslant 0$
4. $x^2 - 2x - 8 \geqslant 0$
5. $x^2 < 10 - 3x$
6. $x^2 + x < 12$
7. $x^2 + 21 > 10x$
8. $x^2 + 7x + 10 > 0$
9. $x^2 \leqslant 8x$
10. $x^2 + 6x \geqslant 0$
11. $x^2 + 5x \leqslant 0$
12. $x^2 \leqslant 4x$
13. $x^2 > 4$
14. $x^2 \leqslant 9$

B
15. $x^2 + 9 \geqslant 6x$
16. $x^2 + 4 \geqslant 4x$
17. $x^3 + 5 \geqslant 5x^2 + x$
18. $x^3 + x^2 < 9x + 9$
19. $x^3 + 75 < 3x^2 + 25x$
20. $x^3 + 4x^2 \geqslant 4x + 16$

21. $\dfrac{x - 2}{x + 4} \leqslant 0$
22. $\dfrac{x + 3}{x - 1} \geqslant 0$

23. $\dfrac{x^2 + 5x}{x - 3} \geqslant 0$
24. $\dfrac{x - 4}{x^2 + 2x} \leqslant 0$

25. $\dfrac{x + 4}{1 - x} \leqslant 0$
26. $\dfrac{3 - x}{x + 5} \leqslant 0$

27. $\dfrac{1}{x} < 4$
28. $\dfrac{5}{x} > 3$

29. $\dfrac{2x}{x + 3} \geqslant 1$
30. $\dfrac{2}{x - 3} \leqslant -2$

31. $\dfrac{3x + 1}{x + 4} \leqslant 1$
32. $\dfrac{5x - 8}{x - 5} \geqslant 2$

33. $\dfrac{2}{x + 1} \geqslant \dfrac{1}{x - 2}$
34. $\dfrac{3}{x - 3} \leqslant \dfrac{2}{x + 2}$

C
35. $x^2 + 1 < 2x$
36. $x^2 + 25 < 10x$
37. $x^3 + 5x > 4x^2 + 20$
38. $x^3 + 3x^2 + x + 3 < 0$
39. $4x^4 + 4 \leqslant 17x^2$
40. $x^4 + 36 \geqslant 13x^2$

Applications

Business & Economics
41. *Break-even analysis.* Repeat Example 29 for

$$C = 28 - 2p \quad \text{and} \quad R = 9p - p^2$$

42. *Break-even analysis.* Repeat Example 29 for

$$C = 27 - 2p \quad \text{and} \quad R = 10p - p^2$$

43. *Pricing.* A publisher estimates that she can sell 20,000 copies of a book per year at a price of $20 each. For each $1 increase in price, she expects to sell 500 fewer books. For what range of prices, p, will the total income on the sale of the books be at least $437,500 per year?

44. *Rental.* A Reno hotel will rent all 200 rooms each night during the summer at a rate of $30 per night. For each $1 increase in room rate, five fewer rooms will be rented each night. Find the range of rates, r, for which the total revenue from rentals will be at least $6,080 per night.

Life Sciences

45. *Drug concentration.* The concentration, C, in milligrams per cubic centimeter of a particular drug in a patient's bloodstream is given by

$$C = \frac{0.32t}{t^2 + 6t + 9}$$

where t is the number of hours after the drug is administered. During what period of time will the concentration be less than 0.02?

Social Sciences

46. *Fund raising.* A local chapter of a political organization estimates that it will distribute all 500 tickets to a fund-raising dinner if a $50 donation is requested per ticket. For each additional $5 in donation requested, it is estimated that 25 fewer tickets will be distributed. What donation, d, per ticket should be requested in order to raise at least $28,000?

4-5   Literal Equations

So far, we have worked mostly with equations having only one variable or unknown. In applications of mathematics we must often work with equations involving two or more variables. Such equations are referred to as **literal equations** or **formulas,** since they use letters to express relationships among two or more quantities. For example,

Simple interest = Principal · Rate · Time

This can be more easily expressed by using the equation

$$I = Prt$$

where I = Simple interest, P = Principal, r = Rate, and t = Time.

In working with literal equations we are often interested in **solving for one variable in terms of the others.** It is important to note that in solving for a particular variable, we must isolate it on the left side of the equal sign. It cannot appear on both sides; if it does, we have not solved for it.

In solving literal equations for a particular variable, we will often make use of the symmetric property of equality. Recall that this property states:

If $a = b$,   then $b = a$,

Using the symmetric property, we can often reverse a formula to get a letter we are solving for on a desired side.

Example 30 Solve $I = Prt$ for t.

Solution

$I = Prt$ Reverse the equation to get t on the left.

$Prt = I$ Divide both sides by Pr to isolate t.

$$\frac{Prt}{Pr} = \frac{I}{Pr}$$

$$t = \frac{I}{Pr}$$

Problem 30 Solve $I = Prt$ for r.

Example 31 Solve $A = P + Prt$ for t.

Solution

$A = P + Prt$ Reverse the equation to get t on the left.

$P + Prt = A$ Subtract P from both sides.

$Prt = A - P$ Divide both sides by Pr.

$$\frac{Prt}{Pr} = \frac{A - P}{Pr}$$

$$t = \frac{A - P}{Pr}$$

Problem 31 Solve $A = P + Prt$ for r.

Example 32 Solve $A = P + Prt$ for P.

Solution

$A = P + Prt$ Reverse the equation.

$P + Prt = A$ Factor P from the left side.

$P(1 + rt) = A$ Divide both sides by $1 + rt$.

$$\frac{P(1 + rt)}{1 + rt} = \frac{A}{1 + rt}$$

$$P = \frac{A}{1 + rt}$$

Problem 32   Solve $W = K - Kat$ for K.

Example 33   Solve $\dfrac{1}{f} = \dfrac{1}{a} + \dfrac{1}{b}$ for a.

Solution

$$\frac{1}{f} = \frac{1}{a} + \frac{1}{b}$$ Multiply both sides by fab to clear denominators.

$$ab = fb + fa$$ Subtract fa from both sides so a will not appear on the right side of the equation.

$$ab - fa = fb$$ Factor a from the left side.
$$a(b - f) = fb$$ Divide both sides by $b - f$.

$$a = \frac{fb}{b - f}$$

Problem 33   Solve $\dfrac{1}{f} = \dfrac{1}{a} + \dfrac{1}{b}$ for f.

Example 34   In a manufacturing process the total cost, C, to produce x items per day is given by

$$C = mx + b$$

where m represents the variable cost to produce one item and b represents the fixed cost. Solve this equation for x.

Solution

$$C = mx + b$$
$$mx + b = C$$
$$mx = C - b$$

$$x = \frac{C - b}{m}$$

Problem 34   Solve $C = mx + b$ for m.

Answers to Matched Problems

30. $r = \dfrac{I}{Pt}$ 31. $r = \dfrac{A - P}{Pt}$ 32. $K = \dfrac{W}{1 - at}$

33. $f = \dfrac{ab}{a + b}$ 34. $m = \dfrac{C - b}{x}$

Exercise 4-5

A   *Solve for the indicated variable.*

1. $d = rt$,   for r
3. $C = 2\pi r$,   for r

2. $Q = rt$,   for t
4. $I = Prt$,   for t

5. $C = \pi d$, for d

6. $E = mc^2$, for m

7. $V = abc$, for b

8. $A = lw$, for w

9. $ax + b = 0$, for x

10. $cx - d = 0$, for x

11. $V = \frac{1}{3}Ab$, for A

12. $\frac{V}{A} = h$, for A

13. $m = \frac{b}{a}$, for a

14. $I = \frac{E}{R}$, for R

15. $P = 2l + 2w$, for l

16. $y = 3x - 5$, for x

B   *Solve for the indicated variable.*

17. $y = mx + b$, for x

18. $y = cx - d$, for c

19. $3x - 5y + 15 = 0$, for y

20. $2x - 3y = 12$, for y

21. $Ax + By + C = 0$, for y

22. $Ax + By = 0$, for y

23. $C = \frac{100W}{L}$, for W

24. $IQ = \frac{100 \cdot MA}{CA}$, for CA

25. $A = \frac{a+b}{2}h$, for h

26. $V = \pi r^2 h$, for h

27. $\frac{a}{b} = \frac{c}{d}$, for d

28. $V = \frac{1}{3}\pi r^2 h$, for h

29. $ax + b = cx + d$, for x

30. $A = \frac{a+b}{2}h$, for a

31. $\frac{PV}{T} = k$, for P

32. $\frac{PV}{T} = k$, for T

33. $C = \frac{5}{9}(F - 32)$, for F

34. $F = \frac{9}{5}C + 32$, for C

35. $\frac{1}{R} = \frac{1}{R_1} + \frac{1}{R^2}$, for R

36. $\frac{1}{f} = \frac{1}{a} + \frac{1}{b}$, for b

37. $\frac{P_1 V_1}{T_1} = \frac{P_2 V_2}{T_2}$, for P_2

38. $\frac{P_1 V_1}{T_1} = \frac{P_2 V_2}{T_2}$, for T_1

39. $a_n = a_1 + (n - 1)d$, for d

40. $a_n = a_1 + (n - 1)d$, for n

C   *Solve for the indicated variable. Where square roots are needed, use only the positive square root.*

41. $A = \pi r^2$, for r

42. $V = \frac{1}{3}\pi r^2 h$, for r

43. $A = P(1 + i)^2$, for i

44. $V = \frac{4}{3}\pi r^3$, for r

45. $y = \frac{4x + 3}{2x - 1}$, for x

46. $x = \frac{3y - 2}{2y - 5}$, for y

Solve for x using the quadratic formula.

47. $x^2 + mx + n = 0$

48. $x^2 - 2rx - s = 0$

Applications

49. *Simple discount.* If a borrower signs a simple discount note, the lender deducts the interest at the start from the maturity value of the note and the borrower will receive the difference P, called the proceeds. In terms of a formula,

$$P = M - Mdt$$

Solve this formula for M, the maturity value.

50. *Compound interest.* The formula $A = P(1 + i)^n$ represents the amount A in an account after n periods of compounding at $100i$ percent interest per period, assuming a present value of P in the account at the beginning. Solve the formula for P to form a present value formula.

Life Sciences

51. *Ocean pressure.* The pressure P in pounds per square inch at d feet below sea level is given by

$$P = 15\left(\frac{d}{33} + 1\right)$$

Solve this formula for d.

Social Sciences

52. *Anthropology.* In their study of genetic groupings, anthropologists use a ratio called the *cephalic index.* This is the ratio of the width of the head to its length (looking down from above) expressed as a percentage. Symbolically,

$$C = \frac{100W}{L}$$

where C is the cephalic index, W is the width, and L is the length. Solve this formula for L.

4-6 Chapter Review

Important Terms
and Symbols

4-1 *Linear equations.* equation, first-degree equation, linear equation, solution, root, solution set, solving an equation, equivalent equations, properties of equality, identities, equations with no solution, fixed costs, variable costs, revenue, break even, word problems

4-2 *Linear inequalities.* inequality symbols, first-degree inequality, linear inequality, properties of inequalities, solution, solution set, solving an inequality, double inequalities, $<, >, \leq, \geq$

4-3 *Quadratic equations.* second-degree equation, quadratic equation, standard form, solving quadratic equations, square root method, solution by factoring, perfect squares, completing the square, quadratic formula, fractional forms, demand equation, $ax^2 + bx + c = 0$, $a \neq 0$,

$$x = \frac{-b \pm \sqrt{b^2 - 4ac}}{2a}$$

4-4   *Nonlinear inequalities.* polynomial inequalities, sign analysis, critical point, rational inequalities

4-5   *Literal equations.* literal equation, formula, solving for a particular variable

Exercise 4-6   Chapter Review

Work through all the problems in this chapter review and check your answers in the back of the book. (Answers to all review problems are there.) Where weaknesses show up, review appropriate sections in the text. When you are satisfied that you know the material, take the practice test following this review.

A   *Solve.*

1. $6(2x - 1) - 3 = 2(3x - 7)$

2. $-2(5x - 3) = 6 - 10x$

3. $4(3y - 8) = -3(5 - 4y)$

4. $\dfrac{m}{6} - \dfrac{m}{7} = \dfrac{1}{3}$

5. $0.16x - 0.07x = 72$

6. $\dfrac{x}{7} - 3 = \dfrac{x}{14}$

7. $\dfrac{3}{8} - \dfrac{9}{2z} = \dfrac{1}{2} - \dfrac{6}{z}$

Solve each inequality and graph.

8. $-9x > 4(5 - x)$

9. $-7 \leqslant x - 9 < 3$

10. $-5 < 3x + 1 \leqslant 10$

11. $-2 \leqslant 8 - x \leqslant 13$

Solve by the square root method.

12. $16x^2 - 49 = 0$

13. $x^2 + 25 = 0$

Solve by factoring.

14. $t^2 + 9t = 0$

15. $x^2 - 2x - 35 = 0$

16. $x^2 - 10x + 25 = 0$

17. $10w^2 = -7w$

18. $6y^2 + y - 1 = 0$

Solve by completing the square.

19. $x^2 - 4x - 8 = 0$

20. $2x^2 - 8x - 3 = 0$

Solve using the quadratic formula.

21. $x^2 - 4x + 2 = 0$ **22.** $3x^2 = 6x + 5$

In Problems 23–26 solve each inequality and graph.

23. $(x - 2)(x + 3) < 0$ **24.** $x^2 - x - 12 > 0$
25. $x(x + 3) \geqslant 0$ **26.** $x^2 \leqslant 5x + 14$

27. Solve $pv = k$ for p.   **28.** Solve $I^2 = \dfrac{W}{R}$ for R.

B   Solve.

29. $2x^2 - 7x + 10 = (2x - 1)(x - 4)$

30. $\dfrac{u + 3}{6} - \dfrac{2u - 7}{9} = 1 - \dfrac{5 - u}{2}$

31. $0.15x + 0.12(20{,}000 - x) = 2{,}850$
32. $5(3x - 5) - 7(x - 3) = 3(4x - 8) - 4(x - 5)$

33. $8 - \dfrac{15}{u + 5} = \dfrac{3u}{u + 5}$

34. $\dfrac{16}{x} - \dfrac{5}{x - 2} = \dfrac{7}{x - 2}$

Solve each inequality and graph.

35. $11 - 6(2m - 5) > 5$ **36.** $4 \leqslant 3x + 13 < 25$

37. $-9 < 3 - 4x < 11$ **38.** $-3 \leqslant \dfrac{4x + 3}{3} \leqslant 5$

39. $\dfrac{5x - 2}{-3} < -6$ **40.** $\dfrac{1}{3} - \dfrac{z - 11}{15} < \dfrac{8 - z}{5}$

41. $0.08x + 0.12(3{,}000 - x) \geqslant 280$ **42.** $-22 \leqslant \tfrac{9}{5}C + 32 \leqslant 86$

Solve by the most efficient method.

43. $(x + \tfrac{1}{4})^2 = \tfrac{7}{16}$ **44.** $20y = 3 - 7y^2$ **45.** $\dfrac{18}{z} = 2z$

46. $x^2 + 5x = 1$ **47.** $\dfrac{x}{2}(x - 2) = 4$ **48.** $z^2 = 7z + 4$

49. $(z - 3)^2 = -25$ **50.** $5y^2 + 4y - 3 = 0$

In Problems 51–54 solve each inequality and graph.

51. $x^2 \geqslant 49$ **52.** $2x^2 - 11x + 5 < 0$

53. $\dfrac{x + 2}{x - 5} < 0$ **54.** $\dfrac{x}{(x - 3)(x + 3)} > 0$

55. Solve $5x - 3y + 30 = 0$ for y.   **56.** Solve $s = \tfrac{1}{2}at^2$ for a.

C *Solve.*

57. $\dfrac{x-3}{15} - \dfrac{x-2}{8} = \dfrac{7(x-7)}{60} - \dfrac{x-5}{6}$

58. $-49 \leqslant \frac{9}{5}C + 32 \leqslant 149$

59. $(x + \frac{7}{3})^2 = \frac{5}{3}$

60. $5 + \dfrac{2-4x}{x-1} < 0$

61. $s = \frac{1}{2}at^2, \quad \text{for } t > 0$

62. $x^2 - ax - 12a^2 = 0, \quad \text{for } x$

Applications

Business & Economics

63. *Investment.* If $60,000 is invested, part at 15% and the rest at 9%, how much should be invested at each rate so that the total investment earns the equivalent of 10%?

64. *Investment.* In Problem 63, how much should be invested at 15% if the total investment is to earn the equivalent of at least 13%?

65. *Profit.* The demand and cost equations for the manufacture and sales of a commodity each week are

$$x = 200 - \frac{p}{3} \qquad C = 12{,}000 + 180x$$

Determine the number of items, x, that must be produced and sold to earn a profit.

66. *Profit.* In Problem 65, determine the number of items, x, that must be produced and sold if the profit is:
(A) Equal to $1,500   (B) At least $2,400

67. *Rentals.* A television rental service will rent all 100 television sets at $16 per month. For each $1 increase in rental, four fewer sets will be rented. What is the range of rates, r, that may be charged to obtain a gross income of at least $1,656 per month?

Life Sciences

68. *Pollution.* The concentration, C, of bacteria per cubic centimeter in a lake treated with an antibacterial agent is given by

$$C = 900 + 720t - 180t^2 \qquad 0 \leqslant t \leqslant 5$$

where t is the number of days after treatment. In how many days will the concentration be 1,440?

Social Sciences

69. *Psychology.* If

$$IQ = \frac{100 \cdot MA}{CA}$$

what would be the mental age (MA) range for a group of 10-year-olds who have an IQ range of 120 to 160?

Practice Test: Chapter 4

Solve.

1. $0.20x + 0.15(4,000 - x) = 675$

2. $\dfrac{2x - 5}{3} - (4 - x) = \dfrac{x}{2} - \dfrac{40 - 5x}{6}$

In Problems 3 and 4 solve each inequality and graph.

3. $\dfrac{x}{18} - \dfrac{1}{2} < \dfrac{x}{9} + \dfrac{1}{6}$ 4. $-8 < 8 - 2x \leqslant 12$

5.   Solve by factoring: $3x^2 = 7x - 2$
6.   Solve using the quadratic formula and simplify if possible:

$3x^2 = 2x + 4$

7.   Solve by completing the square: $2x^2 - 10x + 5 = 0$

In Problems 8 and 9 solve each inequality and graph.

8. $x^2 \geqslant x + 20$ 9. $\dfrac{x - 5}{x + 3} \leqslant 0$

10.   Solve $B = Ap - Bq$ for B.
11.   If \$30,000 is invested, part at 10% and the rest at 15%, how much should be invested at each rate if the total investment is to earn the equivalent of 12%?
12.   The weekly demand and cost equations for the production and sale of a particular item are

$$x = 200 - \frac{p}{3} \qquad C = 9,000 + 150x$$

How many items, x, should be produced and sold each week to earn a profit of at least \$7,200?

Graphs and Functions

CHAPTER 5	Contents

5-1 Cartesian Coordinate System and Straight Lines

- Cartesian Coordinate System
- Graphing Linear Equations in Two Variables
- Slope
- Equations of Lines—Special Forms
- Application

■ Cartesian Coordinate System

Recall that a **Cartesian (rectangular) coordinate system** in a plane is formed by taking two mutually perpendicular real number lines intersecting at their origins **(coordinate axes),** one horizontal and one vertical, and then assigning unique **ordered pairs** of numbers **(coordinates)** to each point P in the plane (Fig. 1). The first coordinate **(abscissa)** is the distance of P from the vertical axis, and the second coordinate **(ordinate)** is the distance of P from the horizontal axis. In Figure 1, the coordinates of point P are (a, b). By reversing the process, each ordered pair of real numbers can be associated with a unique point in the plane. The coordinate axes divide the plane into four parts **(quadrants),** numbered I to IV in a counterclockwise direction.

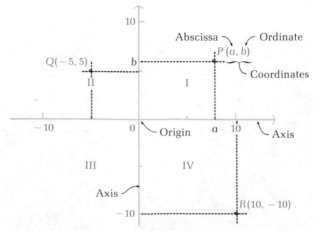

Figure 1 The Cartesian coordinate system

■ Graphing Linear Equations in Two Variables

A linear equation in two variables is an equation that can be written in the form

$Ax + By = C$ Standard form

with A and B not both zero. For example,

$2x - 3y = 5$ $x = 7$ $y = \frac{1}{2}x - 3$ $y = -3$

can all be considered linear equations in two variables. The first is in standard form, while the other three can be written in standard form as follows:

Standard form

$x = 7$ $x + 0y = 7$

$y = \frac{1}{2}x - 3$ $-\frac{1}{2}x + y = -3$ or $x - 2y = 6$

$y = -3$ $0x + y = -3$

A **solution** of an equation in two variables is an ordered pair of real numbers that satisfy the equation. For example, $(0, -3)$ is a solution of $3x - 4y = 12$. The **solution set** of an equation in two variables is the set of all solutions of the equation. When we say that we **graph an equation** in two variables, we mean that we graph its solution set on a rectangular coordinate system.

We state the following important theorem without proof:

Theorem 1

> **Graph of a Linear Equation in Two Variables**
>
> The graph of any equation of the form
>
> $Ax + By = C$ Standard form   (1)
>
> where A, B, and C are constants (A and B not both zero), is a straight line. Every straight line in a Cartesian coordinate system is the graph of an equation of this type.

Also, the graph of any equation of the form

$y = mx + b$ (2)

where m and b are constants, is a straight line. Form (2) is simply a special case of (1) for $B \neq 0$. To graph either (1) or (2), we plot any two points of their solution set and use a straightedge to draw the line through these two points. The points where the line crosses the axes—called the **intercepts**—are often the easiest to find when dealing with form (1). To find the **y intercept,** we let $x = 0$ and solve for y; to find the **x intercept,** we let $y = 0$ and solve for x. It is sometimes wise to find a third point as a check.

Example 1   (A)   The graph of $3x - 4y = 12$ is

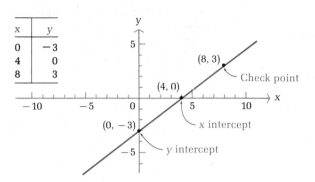

x	y
0	−3
4	0
8	3

(B)   The graph of $y = 2x - 1$ is

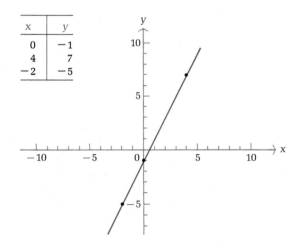

x	y
0	−1
4	7
−2	−5

Problem 1   Graph: (A) $4x - 3y = 12$ (B) $y = \dfrac{x}{2} + 2$

▪ Slope

It is very useful to have a numerical measure of the "steepness" of a line. The concept of *slope* is widely used for this purpose. The **slope** of a line through the two points (x_1, y_1) and (x_2, y_2) is given by the following formula:

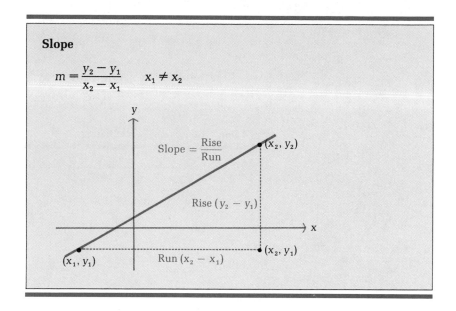

Slope

$$m = \frac{y_2 - y_1}{x_2 - x_1} \qquad x_1 \neq x_2$$

The slope of a vertical line is not defined. (Why? See Example 2B.)

Example 2   Find the slope of the line through each pair of points:

(A) $(-2, 5), (4, -7)$ (B) $(-3, -1), (-3, 5)$

Solutions   (A)   Let $(x_1, y_1) = (-2, 5)$ and $(x_2, y_2) = (4, -7)$. Then

$$m = \frac{y_2 - y_1}{x_2 - x_1} = \frac{-7 - 5}{4 - (-2)} = \frac{-12}{6} = -2$$

Note that we also could have let $(x_1, y_1) = (4, -7)$ and $(x_2, y_2) = (-2, 5)$, since this simply reverses the sign in both the numerator and the denominator and the slope does not change:

$$m = \frac{5 - (-7)}{-2 - 4} = \frac{12}{-6} = -2$$

(B)   Let $(x_1, y_1) = (-3, -1)$ and $(x_2, y_2) = (-3, 5)$. Then

$$m = \frac{y_2 - y_1}{x_2 - x_1} = \frac{5 - (-1)}{-3 - (-3)} = \frac{6}{0} \qquad \text{Not defined!}$$

Notice that $x_1 = x_2$. This is always true for a vertical line, since the abscissa (first coordinate) of every point on a vertical line is the same. Thus, the slope of a vertical line is not defined (that is, the slope does not exist).

Problem 2   Find the slope of the line through each pair of points:

(A) $(3, -6), (-2, 4)$ (B) $(-7, 5), (3, 5)$

In general, the slope of a line may be positive, negative, zero, or not defined. Each of these cases is interpreted geometrically in Table 1.

Table 1 Going from Left to Right

Line	Slope	Example
Rising	Positive	
Falling	Negative	
Horizontal	Zero	
Vertical	Not defined	

■ Equations of Lines—Special Forms

The constants m and b in the equation

$$y = mx + b \tag{3}$$

have special geometric significance.

If we let $x = 0$, then $y = b$, and we observe that the graph of (3) crosses the y axis at $(0, b)$. The constant b is the y *intercept*. For example, the y intercept of the graph of $y = -4x - 1$ is -1.

To determine the geometric significance of m, we proceed as follows: If $y = mx + b$, then by setting $x = 0$ and $x = 1$, we conclude that $(0, b)$ and $(1, m + b)$ lie on its graph (a line). Hence, the slope of this graph (line) is given by:

$$\text{Slope} = \frac{y_2 - y_1}{x_2 - x_1} = \frac{(m + b) - b}{1 - 0} = m$$

Thus, m is the slope of the line given by $y = mx + b$.

Slope–Intercept Form

The equation

$$y = mx + b \qquad \begin{array}{l} m = \text{Slope} \\ b = y \text{ intercept} \end{array} \qquad (4)$$

is called the **slope–intercept form** of an equation of a line.

Example 3   (A)   Find the slope and y intercept, and graph $y = -\frac{2}{3}x - 3$.

        (B)   Write the equation of the line with slope $\frac{2}{3}$ and y intercept -2.

Solutions   (A)   Slope $= m = -\frac{2}{3}$ (B) $m = \frac{2}{3}$ and $b = -2$;

             y intercept $= b = -3$ thus, $y = \frac{2}{3}x - 2$

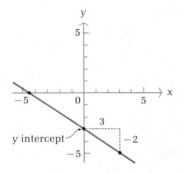

Problem 3   Write the equation of the line with slope $\frac{1}{2}$ and y intercept -1. Graph.

    Suppose a line has slope m and passes through a fixed point (x_1, y_1). If the variable point (x, y) is any other point on the line (Fig. 2), then

$$\frac{y - y_1}{x - x_1} = m$$

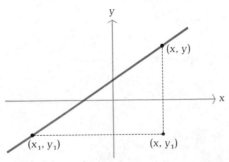

Figure 2

that is,

$$y - y_1 = m(x - x_1)$$

We now observe that (x_1, y_1) also satisfies this equation and conclude that this is an equation of a line with slope m that passes through (x_1, y_1).

Point–Slope Form

An equation of a line with slope m that passes through (x_1, y_1) is

$$y - y_1 = m(x - x_1) \qquad\qquad (5)$$

which is called the **point–slope form** of an equation of a line.

The point–slope form is extremely useful, since it enables us to find an equation for a line if we know its slope and the coordinates of a point on the line or if we know the coordinates of two points on the line.

Example 4 (A) Find an equation for the line that has slope $\frac{1}{2}$ and passes through $(-4, 3)$. Write the final answer in the form $Ax + By = C$.

(B) Write an equation for the line that passes through the two points $(-3, 2)$ and $(-4, 5)$. Write the resulting equation in the form $y = mx + b$.

Solutions (A) $y - y_1 = m(x - x_1)$
Let $m = \frac{1}{2}$ and $(x_1, y_1) = (-4, 3)$. Then

$$y - 3 = \tfrac{1}{2}[x - (-4)]$$
$$y - 3 = \tfrac{1}{2}(x + 4) \qquad\qquad \text{Multiply by 2}$$
$$2y - 6 = x + 4$$
$$-x + 2y = 10 \quad \text{or} \quad x - 2y = -10$$

(B) First, find the slope of the line by using the slope formula:

$$m = \frac{y_2 - y_1}{x_2 - x_1} = \frac{5 - 2}{-4 - (-3)} = \frac{3}{-1} = -3$$

Now use

$$y - y_1 = m(x - x_1)$$

with $m = -3$ and $(x_1, y_1) = (-3, 2)$:

$$y - 2 = -3[x - (-3)]$$
$$y - 2 = -3(x + 3)$$
$$y - 2 = -3x - 9$$
$$y = -3x - 7$$

Problem 4   (A)   Find an equation for the line that has slope $\frac{2}{3}$ and passes through $(6, -2)$. Write the resulting equation in the form $Ax + By = C, A > 0$.

   (B)   Find an equation for the line that passes through $(2, -3)$ and $(4, 3)$. Write the resulting equation in the form $y = mx + b$.

The simplest equations of a line are those for horizontal and vertical lines. A **horizontal line** has slope 0; thus its equation is of the form

$$y = 0x + c \qquad \text{Slope} = 0, \quad y \text{ intercept } c$$

or simply

$$y = c$$

Figure 3 illustrates the graphs of $y = 3$ and $y = -2$.

If a line is vertical, then its slope is not defined. All x values (abscissas) of points on a vertical line are equal, while y can take on any value (Fig. 4).

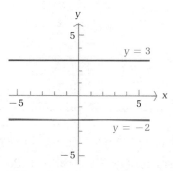

Figure 3

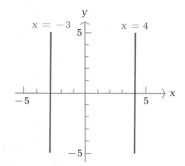

Figure 4

Thus, a **vertical line** has an equation of the form

$$x + 0y = c \qquad x \text{ intercept } c$$

or simply

$$x = c$$

Figure 4 illustrates the graphs of $x = -3$ and $x = 4$.

Equations of Horizontal and Vertical Lines

Horizontal line with y intercept c: $\qquad y = c$

Vertical line with x intercept c: $\qquad x = c$

Example 5   The equation of a horizontal line through $(-2, 3)$ is $y = 3$, and the equation of a vertical line through the same point is $x = -2$.

Problem 5   Find the equations of the horizontal and vertical lines through $(4, -5)$.

We state without proof the following important theorem regarding slope and parallel lines.

Theorem 2

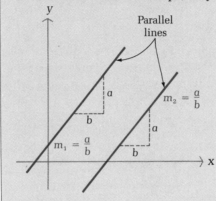

Slope and Parallel Lines

If two nonvertical lines are **parallel,** then they have the same slope; if two lines have the same slope, they are parallel.

Example 6   Find the slope of $y = mx + 7$ so that it is parallel to $3x - 2y = 4$.

Solution   To find the slope of $3x - 2y = 4$, write the equation in the form $y = mx + b$ and identify m:

$$3x - 2y = 4$$
$$-2y = -3x + 4$$
$$y = \tfrac{3}{2}x - 2$$

Thus, $m = \tfrac{3}{2}$, and

$$y = \tfrac{3}{2}x + 7$$

is parallel to $3x - 2y = 4$ (since they have the same slope).

Problem 6   Find the slope of $y = mx - 3$ so that it is parallel to $2x + 3y = 6$.

■ Application

We will now see how equations of lines occur in certain applications.

Example 7
Cost Equation
The management of a company that manufactures roller skates has fixed costs (costs at zero output) of $300 per day and total costs of $4,300 per day at an output of 100 pairs of skates per day. Assume that cost C is linearly related to output x.

(A)   Find the slope of the line joining the points associated with outputs of 0 and 100; that is, the line passing through (0, 300) and (100, 4,300).

(B)   Find an equation of the line relating output to cost. Write the final answer in the form $C = mx + b$.

(C)   Graph the cost equation from part B for $0 \leqslant x \leqslant 200$.

Solutions   (A) $m = \dfrac{y_2 - y_1}{x_2 - x_1} = \dfrac{4{,}300 - 300}{100 - 0} = \dfrac{4{,}000}{100} = 40$

(B)   We must find an equation of the line that passes through (0, 300) with slope 40. We use the slope–intercept form:

$$C = mx + b$$
$$C = 40x + 300$$

(C)

x	C
0	300
100	4,300
200	8,300

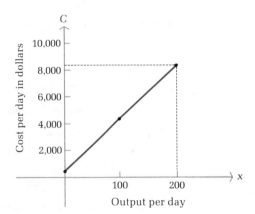

Problem 7   Answer parts A and B in Example 7 for fixed costs of $250 per day and total costs of $3,450 per day at an output of 80 pairs of skates per day.

Answers to Matched Problems

1.   (A)   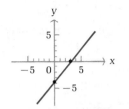   (B)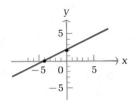

2.   (A) -2 (B)   0   (Zero is a number—it exists! It is the slope of a horizontal line.)

3. $y = \frac{1}{2}x - 1$

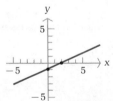

4.   (A) $2x - 3y = 18$
     (B) $y = 3x - 9$

5. $y = -5, x = 4$ 6. $m = -\frac{2}{3}$
7. (A) $m = 40$ (B) $C = 40x + 250$

Exercise 5-1

A *Graph in a rectangular coordinate system.*

1. $y = 2x - 3$

2. $y = \dfrac{x}{2} + 1$

3. $2x + 3y = 12$

4. $8x - 3y = 24$

Find the slope and y intercept of the graph of each equation.

5. $y = 2x - 3$

6. $y = \dfrac{x}{2} + 1$

7. $y = -\frac{2}{3}x + 2$

8. $y = \frac{3}{4}x - 2$

Write an equation of the line with the indicated slope and y intercept.

9. Slope $= -2$
   y intercept $= 4$

10. Slope $= -\frac{2}{3}$
    y intercept $= -2$

11. Slope $= -\frac{3}{5}$
    y intercept $= 3$

12. Slope $= 1$
    y intercept $= -2$

B *Graph in a rectangular coordinate system.*

13. $y = -\frac{2}{3}x - 2$

14. $y = -\frac{3}{2}x + 1$

15. $3x - 2y = 10$

16. $5x - 6y = 15$

17. $x = 3$ and $y = -2$

18. $x = -3$ and $y = 2$

Find the slope of the graph of each equation. (First write the equation in the form $y = mx + b$.)

19. $3x + y = 5$

20. $2x - y = -3$

21. $2x + 3y = 12$

22. $3x - 2y = 10$

Write an equation of the line through each indicated point with the indicated slope. Transform the equation into the form $y = mx + b$.

23. $m = -3$, $(4, -1)$

24. $m = -2$, $(-3, 2)$

25. $m = \frac{2}{3}$, $(-6, -5)$

26. $m = \frac{1}{2}$, $(-4, 3)$

Find the slope of the line that passes through the given points.

27. $(1, 3)$ and $(7, 5)$

28. $(2, 1)$ and $(10, 5)$

29. $(-5, -2)$ and $(5, -4)$

30. $(3, 7)$ and $(-6, 4)$

Write an equation of the line through each indicated pair of points. Write the final answer in the form Ax + By = C, A > 0.

31. (1, 3) and (7, 5)

32. (2, 1) and (10, 5)

33. (−5, −2) and (5, −4)

34. (3, 7) and (−6, 4)

Write equations of the vertical and horizontal lines through each point.

35. (3, −5)   36. (−2, 7)   37. (−1, −3)   38. (6, −4)

Find an equation of the line, given the information in each problem. Write the final answer in the form y = mx + b.

39. Line passes through (−2, 5) with slope $-\frac{1}{2}$.

40. Line passes through (3, −1) with slope $-\frac{2}{3}$.

41. Line passes through (−2, 2) parallel to $y = -\frac{1}{2}x + 5$.

42. Line passes through (−4, −3) parallel to $y = 2x - 3$.

43. Line passes through (−2, −1) parallel to $x - 2y = 4$.

44. Line passes through (−3, 2) parallel to $2x + 3y = -6$.

C

45. Graph $y = mx - 2$ for $m = 2$, $m = \frac{1}{2}$, $m = 0$, $m = -\frac{1}{2}$, and $m = -2$, all on the same coordinate system.

46. Graph $y = -\frac{1}{2}x + b$ for $b = -4$, $b = 0$, and $b = 4$, all on the same coordinate system.

Write an equation of the line through the indicated points. Be careful!

47. (2, 7) and (2, −3)

48. (−2, 3) and (−2, −1)

49. (2, 3) and (−5, 3)

50. (−3, −3) and (0, −3)

Applications

Business & Economics

51. *Simple interest.* If $P (the principal) is invested at an interest rate of r, then the amount A that is due after t years is given by

$$A = Prt + P$$

If $100 is invested at 6% (r = 0.06), then A = 6t + 100, t ≥ 0.

(A) What will $100 amount to after 5 years? After 20 years?

(B) Graph the equation for 0 ≤ t ≤ 20.

(C) What is the slope of the graph? (The slope indicates the increase in the amount A for each additional year of investment.)

52. *Cost equation.* The management of a company manufacturing surfboards has fixed costs (zero output) of $200 per day and total costs of $1,400 per day at a daily output of twenty boards.

(A) Assuming the total cost per day (C) is linearly related to the total output per day (x), write an equation relating these two quantities. [*Hint:* Find an equation of the line that passes through (0, 200) and (20, 1,400).]

(B)   What are the total costs for an output of twelve boards per day?

(C)   Graph the equation for $0 \leqslant x \leqslant 20$.

[*Note:* The slope of the line found in part A is the increase in total cost for each additional unit produced and is called the *marginal cost.* More will be said about the concept of marginal cost later.]

53. *Demand equation.* A manufacturing company is interested in introducing a new power mower. Its market research department gave the management the demand-price forecast listed in the table.

Price	Estimated Demand
$ 70	7,800
$120	4,800
$160	2,400
$200	0

(A)   Plot these points, letting *d* represent the number of mowers people are willing to buy (demand) at a price of $*p* each.

(B)   Note that the points in part A lie along a straight line. Find an equation of that line.

[*Note:* The slope of the line found in part B indicates the decrease in demand for each $1 increase in price.]

54. *Depreciation.* Office equipment was purchased for $20,000 and is assumed to have a scrap value of $2,000 after 10 years. If its value is depreciated linearly (for tax purposes) from $20,000 to $2,000:

(A)   Find the linear equation that relates value (*V*) in dollars to time (*t*) in years.

(B)   What would be the value of the equipment after 6 years?

(C)   Graph the equation for $0 \leqslant t \leqslant 10$.

[*Note:* The slope found in part A indicates the decrease in value per year.]

Life Sciences   55. *Nutrition.* In a nutrition experiment, a biologist wants to prepare a special diet for the experimental animals. Two food mixes, *A* and *B*, are available. If mix *A* contains 20% protein and mix *B* contains 10% protein, what combination of each mix will provide exactly 20 grams of protein? Let *x* be the amount of *A* used and let *y* be the amount of *B* used. Then write a linear equation relating *x*, *y*, and 20. Graph this equation for $x \geqslant 0$ and $y \geqslant 0$.

56. *Ecology.* As one descends into the ocean, pressure increases linearly. The pressure is 15 pounds per square inch on the surface and 30 pounds per square inch 33 feet below the surface.

(A)   If p is the pressure in pounds and d is the depth below the surface in feet, write an equation that expresses p in terms of d. [*Hint:* Find an equation of the line that passes through (0, 15) and (33, 30).]

(B)   What is the pressure at 12,540 feet (the average depth of the ocean)?

(C)   Graph the equation for $0 \leqslant d \leqslant 12{,}540$.

[*Note:*   The slope found in part A indicates the change in pressure for each additional foot of depth.)

Social Sciences   **57.**   *Psychology.* In an experiment on motivation, J. S. Brown trained a group of rats to run down a narrow passage in a cage to obtain food in a goal box. Using a harness, he then connected the rats to an overhead wire that was attached to a spring scale. A rat was placed at different distances d (in centimeters) from the goal box, and the pull p (in grams) of the rat toward the food was measured. Brown found that the relationship between these two variables was very close to being linear and could be approximated by the equation

$$p = -\tfrac{1}{5}d + 70 \qquad 30 \leqslant d \leqslant 175$$

(See J. S. Brown, *Journal of Comparative and Physiological Psychology*, 1948, 41:450–465.)

(A)   What was the pull when $d = 30$? When $d = 175$?

(B)   Graph the equation.

(C)   What is the slope of the line?

5-2   Relations and Functions

■ Introduction
■ Relations and Functions
■ Relations Specified by Equations
■ Function Notation
■ Application

■ Introduction

The relation–function concept is one of the most important concepts in mathematics. The idea of correspondence plays a central role in its formulation. You have already had experiences with correspondences in everyday life. For example:

To each person there corresponds an annual income.

To each item in a supermarket there corresponds a price.

To each day there corresponds a maximum temperature.

For the manufacture of x items there corresponds a cost.

For the sale of x items there corresponds a revenue.

To each square there corresponds an area.

To each number there corresponds its cube.

One of the most important aspects of any science (managerial, life, social, physical, etc.) is the establishment of correspondences among various types of phenomena. Once a correspondence is known, predictions can be made. A cost analyst would like to predict costs for various levels of output in a manufacturing process; a medical researcher would like to know the correspondence between heart disease and obesity; a psychologist would like to predict the level of performance after a subject has repeated a task a given number of times; and so on.

■ Relations and Functions

What do all of the examples of relations above have in common? Each deals with the matching of elements from one set, called the *domain* of the relation, with the elements in a second set, called the *range* of the relation. Consider the following three relations involving the cube, square, and square root. (The choice of small domains enables us to introduce two important concepts in a relatively simple setting. Shortly, we will consider relations with infinite domains.)

Relation 1		Relation 2		Relation 3	
Domain (Number)	**Range (Cube)**	**Domain (Number)**	**Range (Square)**	**Domain (Number)**	**Range (Square Root)**

Relation 1: $0 \to 0$, $1 \to 1$, $2 \to 8$

Relation 2: $-2, -1, 0, 1, 2$ mapping to $4, 1, 0$

Relation 3: $0 \to 0$; $1 \to 1, -1$; $4 \to 2, -2$; $9 \to 3, -3$

The first two relations are examples of functions, but the third is not. These two very important terms, *relation* and *function*, are now defined.

Definition of Relation and Function: Rule Form

A **relation** is a rule (process or method) that produces a correspondence between one set of elements, called the **domain,** and a second set of elements, called the **range,** such that to each element in the domain there corresponds *one or more* elements in the range. A **function** is a relation with the added restriction that to each domain element there corresponds *one and only one* range element. (All functions are relations, but some relations are not functions.)

In the cube, square, and square root examples above, we see that all three are relations according to the definition.* Relations 1 and 2 are also functions, since to each domain value there corresponds exactly one range value (for example, the square of -2 is 4 and no other number). On the other hand, relation 3 is not a function, since to at least one domain value there corresponds more than one range value (for example, to the domain value 9 there corresponds -3 and 3, both square roots of 9).

Since in a relation (or function) elements in the range are paired with elements in the domain by some rule or process, this correspondence (pairing) can be illustrated by using ordered pairs of elements where the first component represents a domain element and the second component a corresponding range element. Thus, we can write relations 1 through 3 as follows:

Relation $1 = \{(0, 0), (1, 1), (2, 8)\}$

Relation $2 = \{(-2, 4), (-1, 1), (0, 0), (1, 1), (2, 4)\}$

Relation $3 = \{(0, 0), (1, 1), (1, -1), (4, 2), (4, -2), (9, 3), (9, -3)\}$

This suggests an alternate but equivalent way of defining relations and functions that provides additional insight into these concepts.

Definition of Relation and Function: Set Form

A **relation** is *any* set of ordered pairs of elements, and a **function** is a relation with the added restriction that no two distinct ordered pairs can have the same first component. The set of first components in a relation (or function) is called the **domain** of the relation, and the set of second components is called the **range.**

---

* We have used the word *relation* earlier as a word from our ordinary language. After the formal definition, the word *relation* becomes part of our technical mathematical vocabulary. From now on when we use the term *relation* in a mathematical context, it will have the meaning specified above.

According to this definition, we see (as before) that relation 3 above is not a function, since there exist two distinct ordered pairs [(1, 1) and (1, −1), for example] that have the same first component (more than one range element is associated with a given domain element).

The rule form of the definition of a relation and function suggests a formula or a "machine" operating on domain values to produce range values — a dynamic process. On the other hand, the set definition of these concepts is closely related to graphs in a Cartesian coordinate system — a static form. Each approach has its advantages in certain situations.

Two of the main objectives of this section are to expose you to the more common ways of specifying relations and functions (including special notation) and to provide you with experience in determining whether a given relation is a function.

As a consequence of the above definitions, we find that a relation (or function) can be specified in many different ways: by an equation, by a table, by a set of ordered pairs of elements, and by a graph, to name a few of the more common ways (see Table 2). All that matters is that we are given a set of elements called the domain and a rule (method or process) of obtaining corresponding range values for each domain value. Incidentally, the **graph of a relation** specified by an equation in two variables is the graph of the set of all ordered pairs of real numbers that satisfies the equation.

Which relation in Table 2 is not a function? The relation specified by the

Table 2 Common Ways of Specifying Relations and Functions

Method	Illustration	Example
Equation	$y = x^2 + x$ $x \in R$*	If $x = 2$, then $y = 6$.
Table		If $p = 4$, then $C = 18$.

p	C
2	14
4	18
6	22

Set of ordered pairs of elements	$\{(2, 14), (4, 18), (6, 22)\}$	6 corresponds to 22.
Graph		If $x = 4$, $y = \pm 2$.

graph is not a function, since a domain value can correspond to more than one range value. (What does $x = 4$ correspond to?)

It is very easy to determine from its graph whether a relation is a function.

Vertical Line Test for a Function

A relation is a function if each vertical line in the coordinate system passes through *at most* one point on the graph of the relation. (If a vertical line passes through two or more points on the graph of a relation, then the relation is not a function.)

■ Relations Specified by Equations

Frequently, domains and ranges of relations and functions are sets of numbers, and the rules associating range values with domain values are equations in two variables. Consider the equation

$$y = x^2 - x \qquad x \in R$$

For each **input** x we obtain one **output** y. For example,

If $x = 3$, then $y = 3^2 - 3 = 6$.

If $x = -\frac{1}{2}$, then $y = (-\frac{1}{2})^2 - (-\frac{1}{2}) = \frac{1}{4} + \frac{1}{2} = \frac{3}{4}$.

The input values are domain values and the output values are range values. The equation (a rule) assigns each domain value x a range value y. The variable x is called an *independent variable* (since values are "independently" assigned to x from the domain), and y is called a *dependent variable* (since the value of y "depends" on the value assigned to x). In general, any variable used as a placeholder for domain values is called an **independent variable;** any variable that is used as a placeholder for range values is called a **dependent variable.**

Unless stated to the contrary, we shall adhere to the following convention regarding domains and ranges for relations and functions specified by equations.

Agreement on Domains and Ranges

If a relation or function is specified by an equation and the domain is not indicated; then we shall assume that the domain is the set of all real number replacements of the independent variable (inputs) that produce real values for the dependent variable (outputs). The range is the set of all outputs corresponding to input values.

Most equations in two variables specify relations, but when does an equation specify a function?

Equations and Functions

In an equation in two variables, if there corresponds exactly one value of the dependent variable (output) to each value of the independent variable (input), then the equation specifies a function. If there is more than one output for at least one input, then the equation does not specify a function.

Example 8   (A)   Is the relation specified by the equation $y^2 = x + 1$ a function, given x is the independent variable?

(B)   What is the domain of the relation?

Solutions   (A)   The relation is not a function since, for example, if $x = 3$, then $y = \pm 2$.

(B)   The domain of the relation (since it is not explicitly given) is the set of all real x that produces real y. Solving for y in terms of x, we obtain

$$y = \pm\sqrt{x + 1}$$

For y to be real, $x + 1$ must be greater than or equal to 0. That is,

$$x + 1 \geqslant 0$$
$$x \geqslant -1$$

Thus,

Domain: $x \geqslant -1$ or $[-1, \infty)$

Problem 8   (A)   Is the relation specified by the equation $x^2 + y^2 = 25$ a function, given x is the independent variable?

(B)   What is the domain of the relation?

■ Function Notation

We have just seen that a function involves two sets of elements, a domain and a range, and a rule of correspondence that enables one to assign to each element in the domain exactly one element in the range. We use different letters to denote names for numbers; in essentially the same way, we will now use different letters to denote names for functions. For example, f and g may be used to name the two functions

f: $y = 2x + 1$

g: $y = x^2 + 2x - 3$

If x represents an element in the domain of a function f, then we will often use the symbol

$f(x)$

in place of y to designate the number in the range of the function f to which x is paired (Fig. 5).

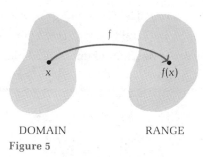

DOMAIN   RANGE

Figure 5

It is important not to think of $f(x)$ as the product of f and x. The symbol $f(x)$ is read "f of x" or "the value of f at x." The variable x is an independent variable; both y and $f(x)$ are dependent variables.

This function notation is extremely important, and its use should be mastered as quickly as possible. For example, in place of the more formal representation of the functions f and g above, we can now write

$$f(x) = 2x + 1 \quad \text{and} \quad g(x) = x^2 + 2x - 3$$

The function symbols $f(x)$ and g(x) have certain advantages over the variable y in certain situations. For example, if we write $f(3)$ and g(5), then each symbol indicates in a concise way that these are range values of particular functions associated with particular domain values. Let us find $f(3)$ and g(5).

To find $f(3)$, we replace x by 3 wherever x occurs in

$f(x) = 2x + 1$

and evaluate the right side:

$$f(3) = 2 \cdot 3 + 1$$
$$= 6 + 1$$
$$= 7$$

Thus

$f(3) = 7$ The function f assigns the range value 7 to the domain value 3; the ordered pair (3, 7) belongs to f

To find g(5), we replace x by 5 wherever x occurs in

$g(x) = x^2 + 2x - 3$

and evaluate the right side:

$$g(5) = 5^2 + 2 \cdot 5 - 3$$
$$= 25 + 10 - 3$$
$$= 32$$

Thus,

$g(5) = 32$ The function g assigns the range value 32 to the domain value 5; the ordered pair (5, 32) belongs to g

It is very important to understand and remember the definition of $f(x)$:

The $f(x)$ Symbol

For any element x in the domain of the function f, the symbol $f(x)$ represents the element in the range of f corresponding to x in the domain of f. If x is an input value, then $f(x)$ is the corresponding output value; or, symbolically, $f: x \rightarrow f(x)$. The ordered pair $(x, f(x))$ belongs to the function f.

Figure 6, which illustrates a "function machine," may give you additional insight into the nature of functions and the symbol $f(x)$. We can think of a function machine as a device that produces exactly one output (range) value for each input (domain) value on the basis of a set of instructions such as those found in an equation, graph, or table. (If more than one

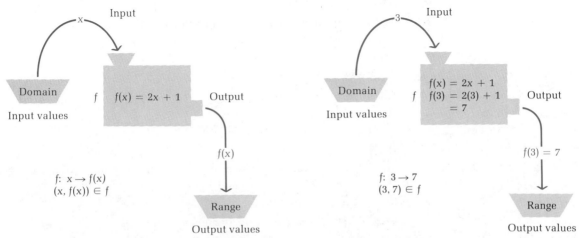

Figure 6 Function machine—exactly one output for each input

output value was produced for an input value, then the machine would be a "relation machine" instead of a function machine.)

For the function $f(x) = 2x + 1$, the machine takes each domain value (input), multiplies it by 2, then adds 1 to the result to produce the range value (output). Different rules inside the machine result in different functions.

Example 9 If

$$f(x) = \frac{12}{x - 2} \qquad g(x) = 1 - x^2 \qquad h(x) = \sqrt{x - 1}$$

then:

(A) $f(6) = \dfrac{12}{6 - 2} = \dfrac{12}{4} = 3$

(B) $g(-2) = 1 - (-2)^2 = 1 - 4 = -3$

(C) $f(0) + g(1) - h(10) = \dfrac{12}{0 - 2} + (1 - 1^2) - \sqrt{10 - 1}$

$$= \frac{12}{-2} + 0 - \sqrt{9}$$

$$= -6 - 3 = -9$$

Problem 9 Use the functions f, g, and h in Example 9 to find:

(A) $f(-2)$ (B) $g(-1)$ (C) $f(3)/h(5)$

Example 10 Find the domains of f, g, and h in Example 9.

Domain of f $12/(x - 2)$ represents a real number for all replacements of x by real numbers except for $x = 2$ (division by 0 is not defined). Thus, the domain of f is the set of all real numbers except 2. We would often indicate this by writing

$$f(x) = \frac{12}{x - 2} \qquad x \neq 2$$

Domain of g The domain is all real numbers R, since $1 - x^2$ represents a real number for all replacements of x by real numbers.

Domain of h The domain is $[1, \infty)$, since $\sqrt{x - 1}$ represents a real number for all real x such that $x - 1$ is not negative; that is, such that

$$x - 1 \geq 0$$

$$x \geq 1$$

Problem 10 Find the domains of F, G, and H defined by:

$$F(x) = x^2 - 3x + 1 \qquad G(x) = \frac{5}{x + 3} \qquad H(x) = \sqrt{2 - x}$$

Example 11   For $f(x) = 2x - 3$, find:

(A) $f(a)$ (B) $f(a + h)$ (C) $\dfrac{f(a + h) - f(a)}{h}$

Solutions   (A) $f(a) = 2a - 3$
(B) $f(a + h) = 2(a + h) - 3 = 2a + 2h - 3$

(C) $\dfrac{f(a + h) - f(a)}{h} = \dfrac{[2(a + h) - 3] - (2a - 3)}{h}$

$$= \frac{2a + 2h - 3 - 2a + 3}{h} = \frac{2h}{h} = 2$$

Problem 11   Repeat Example 11 for $f(x) = 3x - 2$.

■ Application

Example 12   A rectangular feeding pen for cattle is to be made with 100 meters of
Construction   fencing.

(A) If x represents the width of the pen, express its area $A(x)$ in terms of x.
(B) What is the domain of the function A (determined by the physical restrictions)?

Solutions   (A) Draw a figure and label the sides:

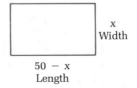

x
Width

50 − x
Length

Perimeter = 100
Half the perimeter = 50
If x = Width, then
50 − x = Length

$A(x) = (\text{Width})(\text{Length}) = x(50 - x)$ Area depends on width x

(B) To have a pen, x must be positive, but x must also be less than 50 (or the length will be zero or negative). Thus,

Domain: $0 < x < 50$ Inequality notation

$(0, 50)$ Interval notation

Problem 12   Work Example 12 with the added assumption that a large barn is to be used as one side of the pen.

Answers to   8. (A) No   (B) Domain $= [-5, 5]$
Matched Problems   9. (A) -3 (B) 0   (C) 6
10. Domain of F: R
Domain of G: All R except -3

Domain of H: $x \leqslant 2$ Inequality notation
                $(-\infty, 2]$ Interval notation

11. (A) $3a - 2$ (B) $3a + 3h - 2$ (C) 3
12. (A) $A(x) = x(100 - 2x)$
    (B) Domain: $0 < x < 50$ Inequality notation
                 $(0,50)$ Interval notation

Exercise 5-2

A   *Indicate whether each relation is a function.*

1. | **Domain** | **Range** |
    |---|---|
    | 3 ——→ 0 | |
    | 5 ——→ 1 | |
    | 7 ——→ 2 | |

2. | **Domain** | **Range** |
    |---|---|
    | −1 ——→ 5 | |
    | −2 ——→ 7 | |
    | −3 ——→ 9 | |

3. | **Domain** | **Range** |
    |---|---|
    | 3 —→ 5 | |
    | → 6 | |
    | 4 —→ 7 | |
    | 5 ——→ 8 | |

4. | **Domain** | **Range** |
    |---|---|
    | 8 ——→ 0 | |
    | 9 —→ 1 | |
    | → 2 | |
    | 10 ——→ 3 | |

5. | **Domain** | **Range** |
    |---|---|
    | 3 —→ | |
    | 6 —→ 5 | |
    | 9 —→ | |
    | 12 —→ 6 | |

6. | **Domain** | **Range** |
    |---|---|
    | −2 | |
    | −1 —→ | |
    | 0 —→ 6 | |
    | 1 —→ | |

The relations in Problems 7–12 are specified by graphs. Indicate whether each relation is a function.

7.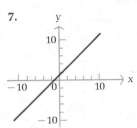

8.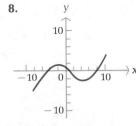

9.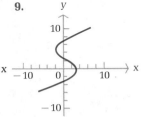

10.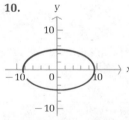

11.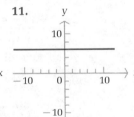

12.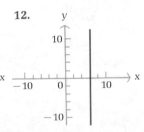

The equations in Problems 13–24 specify relations. Which equations specify functions? For each equation that does not specify a function, find a value of x that corresponds to more than one value of y (x is independent and y is dependent).

13. $y = 3x - 1$ 14. $y = \dfrac{x}{2} - 1$ 15. $y = x^2 - 3x + 1$

16. $y = x^3$ 17. $y^2 = x$ 18. $x^2 + y^2 = 25$

19. $x = y^2 - y$ 20. $x = (y - 1)(y + 2)$ 21. $y = x^4 - 3x^2$

22. $2x - 3y = 5$ 23. $y = \dfrac{x + 1}{x - 1}$ 24. $y = \dfrac{x^2}{1 - x}$

If $f(x) = 3x - 2$ and $g(x) = x - x^2$, find each of the following:

25. $f(2)$ 26. $f(1)$ 27. $f(-1)$
28. $f(-2)$ 29. $g(3)$ 30. $g(1)$
31. $f(0)$ 32. $f(\frac{1}{3})$ 33. $g(-3)$
34. $g(-2)$ 35. $f(1) + g(2)$ 36. $g(1) + f(2)$
37. $g(2) - f(2)$ 38. $f(3) - g(3)$ 39. $g(3) \cdot f(0)$
40. $g(0) \cdot f(-2)$ 41. $g(-2)/f(-2)$ 42. $g(-3)/f(2)$

B   *State the domain and range for each relation and indicate whether the relation is a function.*

43. $F = \{(1, 1), (2, 1), (3, 2) (3, 3)\}$
44. $f = \{(2, 4), (4, 2), (2, 0), (4, -2)\}$
45. $G = \{(-1, -2), (0, -1), (1, 0), (2, 1), (3, 2), (4, 1)\}$
46. $g = \{(-2, 0), (0, 2), (2, 0)\}$

47. $y = 6 - 2x$, $x \in \{0, 1, 2, 3\}$ 48. $y = \dfrac{x}{2} - 4$, $x \in \{0, 1, 2, 3, 4\}$

49. $y^2 = x$, $x \in \{0, 1, 4\}$ 50. $y = x^2$, $x \in \{-2, 0, 2\}$

If $f(x) = 2x + 1$, $g(x) = x^2 - x$, and $k(x) = \sqrt{x}$, find each of the following:

51. $f(3) + g(-2)$ 52. $g(-1) - f(1)$ 53. $k(9) - g(-2)$
54. $g(-2) - k(4)$ 55. $f[k(4)]$ 56. $k[f(4)]$
57. $k[g(2)]$ 58. $g[k(9)]$ 59. $g(e)$
60. $f(a)$ 61. $k(u)$ 62. $g(t)$
63. $g(2 + h)$ 64. $f(2 + h)$ 65. $f(a + h)$

66. $g(a + h)$ 67. $\dfrac{f(2 + h) - f(2)}{h}$ 68. $\dfrac{f(a + h) - f(a)}{h}$

69. $\dfrac{g(2 + h) - g(2)}{h}$ 70. $\dfrac{g(a + h) - g(a)}{h}$

Find the domain of each function in Problems 71–76.

71. $f(x) = \sqrt{x}$ 72. $f(x) = 1/\sqrt{x}$

73. $f(x) = \dfrac{x - 3}{(x - 5)(x + 3)}$ 74. $f(x) = \dfrac{x + 1}{x - 2}$

75. $f(x) = \sqrt{x - 1}$ 76. $f(x) = \sqrt{x + 1}$

C *Find the domain of each function in Problems 77–78.*

77. $f(x) = \dfrac{1}{x^2 - x - 6}$

78. $f(x) = \sqrt{x^2 - 1}$

79. If

$$f(x) = \begin{cases} x^2 & \text{when} \quad x < 1 \\ 2x & \text{when} \quad x \geq 1 \end{cases}$$

find: (A) $f(-1)$ (B) $f(0)$ (C) $f(1)$ (D) $f(3)$

80. If

$$f(x) = \begin{cases} -x & \text{when} \quad x \leq 0 \\ x & \text{when} \quad x > 0 \end{cases}$$

find: (A) $f(-3)$ (B) $f(-1)$ (C) $f(0)$ (D) $f(5)$

■

Applications

Each of the statements in Problems 81–88 can be described by a function. Write an equation that specifies each function.

Business & Economics

81. *Cost function.* The cost $C(x)$ of x records at \$4 per record. (The cost depends on the number of records purchased.)

82. *Cost function.* The cost $C(x)$ of manufacturing x pairs of skis if fixed costs are \$400 per day and the variable costs are \$70 per pair of skis manufactured. (The cost per day depends on the number of skis manufactured per day.)

83. *Packaging.* A candy box is to be made out of a piece of cardboard that measures 8 by 12 inches. Equal-sized squares x inches on a side will be cut out of each corner, and then the ends and sides will be folded up to form a rectangular box.

(A) Express the volume of the box $V(x)$ in terms of x.

(B) What is the domain of the function V (determined by the physical restrictions)?

(C) Complete the table:

x	V(x)
1	
2	
3	

Notice how the volume changes with different choices of x

84. *Packaging.* A parcel delivery service will only deliver packages with length plus girth (distance around) not exceeding 108 inches. A rectangular shipping box with square ends, x inches on a side, is to be used.

(A) If the full 108 inches is to be used, express the volume of the box $V(x)$ in terms of x.

(B) What is the domain of the function V (determined by the physical restrictions)?

(C) Complete the table:

x	$V(x)$
5	
10	
15	
20	
25	

Notice how the volume changes with different choices of x

Life Sciences

85. *Temperature conversion.* The temperature in degrees Celsius $C(F)$ can be found from the temperature in degrees Fahrenheit F by subtracting 32 from the Fahrenheit temperature and multiplying the difference by $\frac{5}{9}$.

86. *Ecology.* The pressure $P(d)$ in the ocean in pounds per square inch depends on the depth d. To find the pressure, divide the depth by 33, add 1 to the quotient, and multiply the result by 15.

Social Sciences

87. *Psychology.* For all 12-year-old children, IQ depends on the mental age as determined by certain standardized tests. To find an IQ, divide a mental age (MA) by 12 and multiply the quotient by 100.

88. *Politics.* The percentage of seats y won by a given party in a two-party election depends on the percentage of the two-party votes x received by the given party. The percentage of seats y can be approximated for $0.4 \leqslant x \leqslant 0.6$ by multiplying x by 2.5 and subtracting 0.7 from the product.

5-3   Graphing Functions

- Graphing Polynomial Functions
- Graphing Other Functions
- Application: Market Research

In this section we will take a look at some basic techniques of graphing relations and functions specified by equations in two variables. This discussion will be continued in Chapter 12, where calculus techniques will be used to answer questions about graphs that are either difficult or not possible to answer now.

Graphing Polynomial Functions

We already know how to graph **first-degree (linear) polynomial functions** —that is, functions specified by equations of the form

$$f(x) = mx + b$$

This is equivalent to graphing the equation

$$y = mx + b \qquad \text{Slope} = m, \quad y \text{ intercept} = b$$

which we studied in detail in Section 5-1.

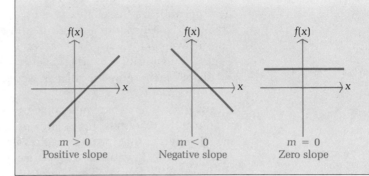

Graph of $f(x) = mx + b$

The graph of a linear function f is a nonvertical straight line with slope m and y intercept b.

$m > 0$
Positive slope

$m < 0$
Negative slope

$m = 0$
Zero slope

Example 13 Graph the linear function defined by

$$f(x) = -\frac{x}{2} + 3$$

and indicate its slope and y intercept.

Solution

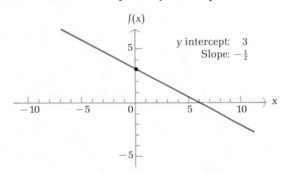

y intercept: 3
Slope: $-\frac{1}{2}$

Problem 13 Graph the linear function defined by

$$f(x) = \frac{x}{3} + 1$$

and indicate its slope and y intercept.

Now let us turn to the graphing of second- and higher-degree polynomial functions.

Example 14 Sketch a graph of the second-degree polynomial (quadratic) function defined by

$$f(x) = -x^2 + 3x + 4$$

Solution   We proceed by point-by-point plotting. The process can be speeded up by writing $f(x)$ in a "nested factored form," as follows:

$$f(x) = -x^2 + 3x + 4 \qquad \text{Factor the first two terms.}$$
$$= (-x + 3)x + 4$$

The reason for the use of the phrase "nested factored form" will become more apparent as the degree of the polynomial function increases (see Example 15). This form is well-suited to mental calculations and is even more convenient for use with a hand calculator when x is a decimal fraction. When using a hand calculator, store the chosen value of x in the calculator's memory and recall it as necessary as the calculation progresses from left to right.

To sketch a graph of the function f, we evaluate $f(x)$ for various values of x and plot the corresponding ordered pairs $(x, f(x))$. When we have plotted enough points so that the total graph is apparent, we join these points with a smooth curve. If we are in doubt in a certain region, we add more points.

Proceeding as indicated, we construct the table and graph shown. The graph is called a **parabola.**

x	$f(x)$
−3	−14
−2	− 6
−1	0
0	4
1	6
1.5	6.25
2	6
3	4
4	0
5	− 6
6	−14

Notice that it was not clear what happened between $x = 1$ and $x = 2$, so we added $x = 1.5$.

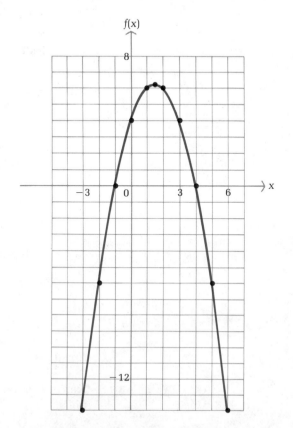

It appears that $f(x)$ has a maximum value of 6.25 at $x = 1.5$. We will say more about maximum and minimum values of functions and how they are found in Chapter 12. For now, we proceed somewhat informally, relying on our intuitive notions of these concepts.

In general:

Graph of $f(x) = ax^2 + bx + c$, $\quad a \neq 0$

The **graph of a quadratic function** f is a parabola that has its **axis** (line of symmetry) parallel to the vertical axis. It opens upward if $a > 0$ and downward if $a < 0$. The intersection point of the axis and parabola is called the **vertex.**

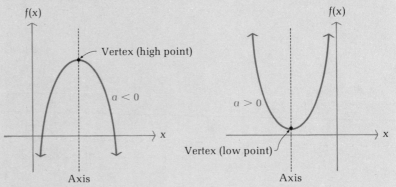

Note: If we fold the graph along the axis of the parabola, the right side will match the left side.

Problem 14   Sketch a graph of $f(x) = x^2 - 3x - 10$. Sketch in the axis, label the vertex, and estimate the maximum or minimum value of $f(x)$ from the graph.

Example 15   Graph $P(x) = x^3 + 3x^2 - x - 3$, $\quad -4 \leqslant x \leqslant 2$.

Solution   We first write $P(x)$ in a nested factored form as follows:

$$P(x) = x^3 + 3x^2 - x - 3$$

Factor the first two terms and repeat until you cannot go any further.

$$= (x + 3)x^2 - x - 3$$

$$= [(x + 3)x - 1]x - 3$$

Proceeding mentally or with a calculator, we obtain

$$P(-4) = [((-4) + 3)(-4) - 1](-4) - 3 = -15$$
$$P(-3) = [((-3) + 3)(-3) - 1](-3) - 3 = 0$$

and so on

We then construct a table of ordered pairs of numbers belonging to the function P, plot these points, and join them with a smooth curve. It is important to plot enough points so that it is clear what happens between the points when they are joined by a smooth curve.

x	−4	−3	−2	−1	0	1	2	−2.5	−1.5	−0.5	0.5
P(x)	−15	0	3	0	−3	0	15	2.6	1.9	−1.9	−2.6

Additional points
to clarify graph

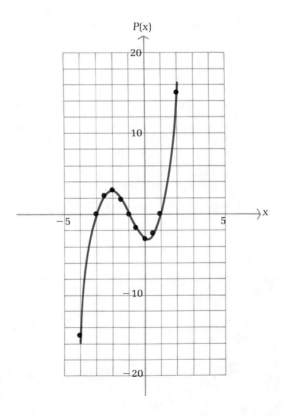

Problem 15   Graph $P(x) = x^3 - 4x^2 - 4x + 16$, $-3 \leqslant x \leqslant 5$, using the nested factoring method.

Note: Nested factorings are shown below for polynomials with missing terms:

$$P(x) = x^3 - 2x^2 - 5 \qquad\qquad Q(x) = 2x^3 - 4x + 3$$
$$= (x - 2)x^2 - 5 \qquad\qquad\quad = (2x^2 - 4)x + 3$$
$$= [(x - 2)x]x - 5$$

For simple polynomial functions, such as

$$f(x) = x^2 \qquad g(x) = x^3 - 1 \qquad h(x) = 2x^4 + 3$$

we can evaluate directly without using nested factoring.

■ Graphing Other Functions

Graphs of functions often display properties of symmetry. In particular, a graph is **symmetric with respect to the vertical axis** if $(-a, b)$ is on the graph whenever (a, b) is on the graph. A graph is **symmetric with respect to the origin** if $(-a, -b)$ is on the graph whenever (a, b) is on the graph. A function whose graph is symmetric with respect to the vertical axis is called an **even function;** a function whose graph is symmetric with respect to the origin is called an **odd function.** Convenient tests for even and odd functions follow from these definitions, as summarized in the box.

Even and Odd Functions

If $f(-x) = f(x)$, then f is an **even function.**
If $f(-x) = -f(x)$, then f is an **odd function.**

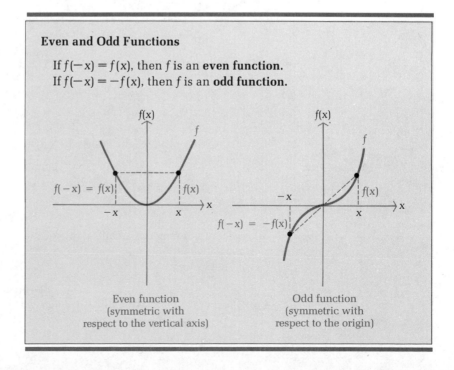

Even function
(symmetric with
respect to the vertical axis)

Odd function
(symmetric with
respect to the origin)

Of course, many functions are neither even nor odd. Why are we interested in knowing whether a function is even or odd? If we want to graph a function specified by an equation, then the even–odd test given in the box provides a useful aid for graphing. If the function is even, then its graph is symmetric with respect to the vertical axis. To graph the function we need to make a careful sketch only to the right of the vertical axis, then reflect the result across the vertical axis to obtain the whole sketch — the point-by-point plotting would be cut in half! Similarly, for odd functions, we reflect any part of a graph that we have sketched through the origin to obtain additional parts of the graph. In addition, there are certain problems and developments in calculus and more advanced mathematics that can be simplified if one recognizes the presence of either an even or an odd function.

Example 16   Without graphing, determine whether the functions f, g, and h are even, odd, or neither.

(A) $f(x) = |x|^*$ (B) $g(x) = x^3 + 1$ (C) $h(x) = \sqrt[3]{x}$

Solutions   (A) $f(-x) = |-x| = |x| = f(x)$; therefore, f is even.

(B) $g(x) = x^3 + 1$
$g(-x) = (-x)^3 + 1 = -x^3 + 1$ $g(-x) \neq g(x)$
$-g(x) = -(x^3 + 1) = -x^3 - 1$ $g(-x) \neq -g(x)$

Therefore, g is neither even nor odd.

(C) $h(-x) = \sqrt[3]{-x} = -\sqrt[3]{x} = -h(x)$; therefore, h is odd.

Problem 16   Without graphing, determine whether the functions F, G, and H are even, odd, or neither.

(A) $F(x) = x^3 + x$ (B) $G(x) = x^2 + 1$ (C) $H(x) = 2x + 4$

The following is a small sample of the many different kinds of function graphs we will encounter in this text.

---

* Recall that the **absolute value of x,** denoted by $|x|$, is defined by

$$|x| = \begin{cases} x & \text{if } x > 0 \\ 0 & \text{if } x = 0 \\ -x & \text{if } x < 0 \end{cases}$$

1. $f(x) = |x|$

    f is even, because

    $f(-x) = |-x| = |x| = f(x)$

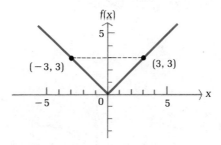

Graph to the right of the vertical axis first, then reflect across the vertical axis.

2. $g(x) = \sqrt{x-1}$

    g is neither even nor odd.

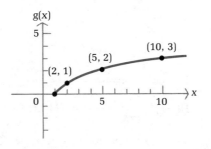

Note that x cannot be less than 1. We use point-by-point plotting.

3. $h(x) = x^{2/3}$

    h is even, because

    $h(-x) = (-x)^{2/3} = x^{2/3} = h(x)$

x	0	1	2	3	4	5
$h(x)$	0	1	1.6	2.1	2.5	2.9

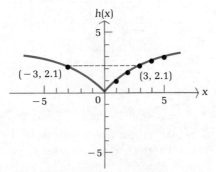

Graph to the right of the vertical axis first, then reflect across the vertical axis. Point-by-point plotting is accomplished with the aid of a calculator.

4. $G(x) = x^{1/3}$

    G is odd, because

    $G(-x) = (-x)^{1/3} = -x^{1/3} = -G(x)$

x	0	1	2	3	4	5
G(x)	0	1	1.26	1.44	1.59	1.71

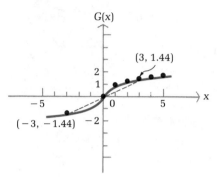

Sketch the portion in the first quadrant, then reflect across the origin. Point-by-point plotting is accomplished with the aid of a calculator.

5. $H(x) = \dfrac{1}{x} \qquad x \neq 0$

    H is odd, because

    $H(-x) = \dfrac{1}{-x} = -\dfrac{1}{x} = -H(x)$

x	5	4	3	2	1	$\frac{1}{2}$	$\frac{1}{3}$	$\frac{1}{4}$	$\frac{1}{5}$
H(x)	$\frac{1}{5}$	$\frac{1}{4}$	$\frac{1}{3}$	$\frac{1}{2}$	1	2	3	4	5

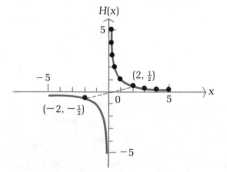

Sketch the portion in the first quadrant, then reflect across the origin.

■ Application: Market Research

The market research department of a company recommended to management that the company manufacture and market a promising new product.

After extensive surveys, the research department backed up the recommendation with the **demand equation**

$$x = f(p) = 6{,}000 - 30p \tag{1}$$

where x is the number of units that retailers are likely to buy per month at $\$p$ per unit. Notice that as the price goes up, the number of units goes down. From the financial department, the following **cost equation** was obtained:

$$C = g(x) = 72{,}000 + 60x \tag{2}$$

where $\$72{,}000$ is the fixed cost (tooling and overhead) and $\$60$ is the variable cost per unit (materials, labor, marketing, transportation, storage, etc.). The **revenue equation** (the amount of money, R, received by the company for selling x units at $\$p$ per unit) is

$$R = xp \tag{3}$$

And, finally, the **profit equation** is

$$P = R - C \tag{4}$$

where P is profit, R is revenue, and C is cost.

We notice that the cost equation (2) expresses C as a function of x and the demand equation (1) expresses x as a function of p. Substituting (1) into (2), we obtain cost C as a linear function of price p:

$$\begin{aligned} C &= 72{,}000 + 60(6{,}000 - 30p) \qquad \text{Linear function} \\ &= 432{,}000 - 1{,}800p \end{aligned} \tag{5}$$

Similarly, substituting (1) into (3), we obtain revenue R as a quadratic function of price p:

$$\begin{aligned} R &= (6{,}000 - 30p)p \qquad \text{Quadratic function} \\ &= 6{,}000p - 30p^2 \end{aligned} \tag{6}$$

When we graph equations (5) and (6) in the same coordinate system, we obtain Figure 7 (page 236). Notice how much information is contained in this graph. Let us compute the **break-even points;** that is, the prices at which cost equals revenue (the points of intersection of the two graphs).

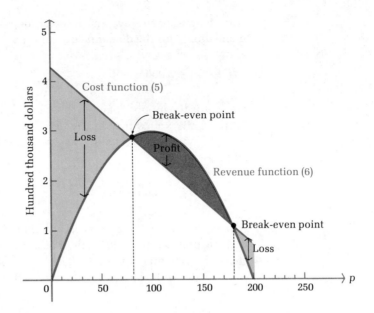

Figure 7

Find p so that

$$C = R$$

$$432,000 - 1,800p = 6,000p - 30p^2$$

$$30p^2 - 7,800p + 432,000 = 0$$

$$p^2 - 260p + 14,400 = 0$$

$$p = \frac{260 \pm \sqrt{260^2 - 4(14,400)}}{2}$$

$$= \frac{260 \pm 100}{2} = \$80, \quad \$180$$

[Recall that the solutions to the quadratic equation $ax^2 + bx + c = 0$, $a \neq 0$, are given by the quadratic formula $x = (-b \pm \sqrt{b^2 - 4ac})/2a$.]

Thus, at a price of $80 or $180 per unit, the company will break even. Between these two prices it is predicted that the company will make a profit.

Another important question (which we will consider in Chapter 12) is: At what price will the company make the maximum profit?

Answers to Matched Problems

13.

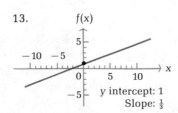

y intercept: 1
Slope: $\frac{1}{3}$

14.

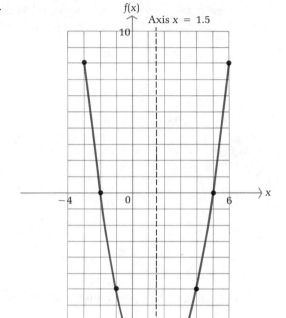

Axis x = 1.5

Vertex
(1.5, −12.25)

Min f(x) = f(1.5) = −12.25

15.

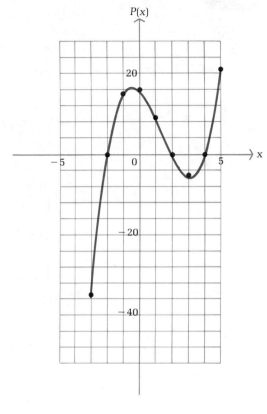

16.   (A) Odd   (B) Even   (C) Neither

Exercise 5-3

A   *Graph each linear function and indicate its slope and y intercept.*

1. $h(x) = -2x + 4$

2. $f(x) = -\dfrac{x}{2} + 3$

3. $g(x) = -\dfrac{2}{3}x + 4$

4. $f(x) = 3$

Graph each quadratic function. From the graph, estimate the coordinates of the vertex, the maximum or minimum value of $f(x)$, and the equation of the axis. Sketch in the axis.

5. $h(x) = x^2 - 2x - 3$

6. $f(u) = u^2 - 2u + 4$

7. $h(x) = -x^2 + 4x + 2$

8. $g(x) = -x^2 - 6x - 4$

9. $g(t) = t^2 + 4$

10. $F(s) = s^2 - 4$

B

11. $f(x) = 6x - x^2$

12. $G(x) = 16x - 2x^2$

13. $f(x) = -\dfrac{1}{2}x^2 + 4x - 4$

14. $f(x) = 2x^2 - 12x + 14$

15. $h(x) = -x^2 - 5x + 2$

16. $g(t) = t^2 - 5t + 2$

Graph each polynomial function using nested factoring.

17. $P(x) = x^3 - 5x^2 + 2x + 8, \quad -2 \leqslant x \leqslant 5$

18. $P(x) = x^3 + 4x^2 - x - 4, \quad -5 \leqslant x \leqslant 2$

19. $P(x) = x^3 + 2x^2 - 5x - 6, \quad -4 \leqslant x \leqslant 3$

20. $P(x) = x^3 - 2x^2 - 5x + 6, \quad -3 \leqslant x \leqslant 4$

Problems 21–22 refer to functions f, g, and p in the graphs:

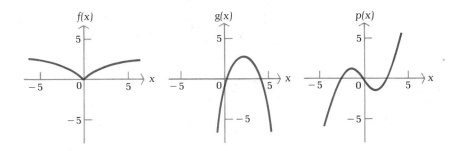

21. (A) Which functions are even?

(B) Which functions are odd?

(C) Which functions are neither even nor odd?

22. (A) Which functions are symmetric with respect to the vertical axis?

(B) Which functions are symmetric with respect to the origin?

23. Given

$$f(x) = \sqrt[3]{x} \qquad g(x) = \frac{x^2}{x^2 - 1} \qquad h(x) = x + 1$$

(A) Which functions are odd?

(B) Which functions are even?

(C) Which functions are neither even nor odd?

24. Given

$$f(x) = 3x \qquad g(x) = 2x - 1 \qquad h(x) = 3 - 2x^2$$

(A)   Which functions are odd?

(B)   Which functions are even?

(C)   Which functions are neither even nor odd?

Indicate if the function is odd, even, or neither, and sketch a graph.

25. $f(x) = -|x|$

26. $g(x) = |x| - 1$

27. $h(x) = \sqrt{x} + 1$

28. $F(x) = \sqrt{x+1}$

29. $H(x) = -x^{2/3}$

30. $G(x) = x^{2/3} + 1$

31. $M(x) = -2x^{1/3}$

32. $f(x) = x^{1/3} + 1$

33. $H(x) = \dfrac{1}{x+1}$

34. $F(x) = \dfrac{-1}{x}$

C   *Graph each polynomial function using nested factoring and a calculator.*

35. $P(x) = x^4 - 2x^2 + 16x - 15$

36. $P(x) = x^4 - 2x^3 - 2x^2 + 8x - 8$

Applications

Business & Economics

37.   *Cost equation.* The cost equation (in dollars) for a particular company that produces stereos is found to be

$$C = g(n) = 96,000 + 80n$$

where $96,000 represents fixed costs (tooling and overhead) and $80 is the variable cost per unit (material, labor, etc.). Graph this function for $0 \leq n \leq 1,000$.

38.   *Demand equation.* After extensive surveys, the research department of a stereo manufacturing company produced the demand equation

$$n = f(p) = 8,000 - 40p \qquad 100 \leq p \leq 200$$

where *n* is the number of units that retailers are likely to purchase per week at a price of $*p* per unit. Graph the function for the indicated domain.

39.   *Construction.* A rectangular pen is to be made with 100 feet of fence wire.

(A)   If x represents the width of the pen, express its area $A(x)$ in terms of x.

(B)   Considering the physical limitations, what is the domain of the function A?

(C)   Graph the function for this domain.

(D)   Determine the dimensions of the rectangle that will maximize the area enclosed by the pen (estimate from the graph).

40.   *Construction.* Work Problem 39 with the added assumption that an existing property fence will be used for one side of the pen. (*Hint:* Let x equal the width — see the figure.)

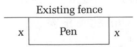

Existing fence

| x | Pen | x |

41. *Packaging.* A candy box is to be made out of a rectangular piece of cardboard that measures 8 by 12 inches. Squares of equal size (x by x inches) will be cut out of each corner, and then the ends and sides will be folded up to form a rectangular box.

   (A) Write the volume of the box V(x) in terms of x.
   (B) Considering the physical limitations, what is the domain of the function V?
   (C) Graph the function for this domain.
   (D) From the graph estimate the size square (to the nearest half inch) that must be cut from each corner to yield a box with the maximum volume. What is the maximum volume?

42. *Packaging.* A parcel delivery service will deliver a package only if the length plus girth (distance around the package) does not exceed 108 inches. A packaging company wants to design a box with a square base (x by x inches) that will have a maximum volume and will meet the delivery service's restriction.

   (A) Write the volume V(x) of the box in terms of x.
   (B) Considering the physical limitation imposed by the delivery service, what is the domain of the function V?
   (C) Graph the function for this domain.
   (D) From the graph estimate the dimensions of the box (to the nearest inch) with the maximum volume. What is the maximum volume?

43. *Market research.* Suppose that in the market research application in this section the demand equation (1) is changed to

$$x = 9,000 - 30p$$

and the cost equation (2) is changed to

$$C = 90,000 + 30x$$

   (A) Express cost C as a linear function of price p.
   (B) Express revenue R as a quadratic function of price p.
   (C) Graph the cost and revenue functions found in parts A and B in the same coordinate system, and identify the regions of profit and loss.
   (D) Find the break-even points; that is, find the prices to the nearest dollar at which $R = C$. (A hand calculator might prove useful here.)

Life Sciences

44. *Medicine.* The velocity v of blood, in centimeters per second, at x centimeters from the center of a given artery (see the figure) is given by

$$v = f(x) = 1.28 - 20{,}000x^2 \qquad 0 \leqslant x \leqslant 8 \times 10^{-3}$$

Graph this quadratic function for the indicated values of x.

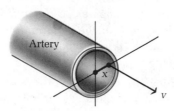

45. *Air pollution.* On an average summer day in a large city, the pollution index at 8:00 AM is 20 parts per million and it increases linearly by 15 parts per million each hour until 3:00 PM. Let $P(x)$ be the amount of pollutants in the air x hours after 8:00 AM.

(A) Express $P(x)$ as a linear function of x.
(B) What is the air pollution index at 1:00 PM?
(C) Graph the function P for $0 \leqslant x \leqslant 7$.
(D) What is the slope of the graph? (The slope is the amount of increase in pollution for each additional hour of time.)

Social Sciences

46. *Psychology—sensory perception.* One of the oldest studies in psychology concerns the following question: Given a certain level of stimulation (light, sound, weight lifting, electric shock, and so on), how much should the stimulation be increased for a person to notice the difference? In the middle of the nineteenth century, E. H. Weber (a German physiologist) formulated a law that still carries his name: If Δs is the change in stimulus that will just be noticeable at a stimulus level s, then the ratio of Δs to s is a constant:

$$\frac{\Delta s}{s} = k$$

Hence, the amount of change that will be noticed is a linear function of the stimulus level, and we note that the greater the stimulus, the more it takes to notice a difference. In an experiment on weight lifting, the constant k for a given individual was found to be $1/30$.

(A) Find Δs (the difference that is just noticeable) at the 30-pound level; at the 90-pound level.
(B) Graph $\Delta s = s/30$ for $0 \leqslant s \leqslant 120$.
(C) What is the slope of the graph?

5-4 Chapter Review

Important Terms
and Symbols

5-1 *Cartesian coordinate system and straight lines.* rectangular coordinate system, Cartesian coordinate system, coordinate axes, ordered pair, coordinates, abscissa, ordinate, quadrant, solution of an equation in two variables, solution set, graph of an equation, x intercept, y intercept, slope, slope–intercept form of the equation of a line, point–slope form of the equation of a line, horizontal line, vertical line, parallel lines, slope $= (y_2 - y_1)/(x_2 - x_1)$, $y = mx + b$, $y - y_1 = m(x - x_1)$, $y = c$, $x = c$

5-2 *Relations and functions.* relation, function, domain, range, input, output, independent variable, dependent variable, function notation, $f(x)$

5-3 *Graphing functions.* graphing first-degree (linear) polynomial functions, graphing second-degree (quadratic) polynomial functions, axis, vertex, graphing higher-degree polynomial functions, nested factoring, graphing other functions, symmetric with respect to the vertical axis, symmetric with respect to the origin, even function, odd function, demand equation, cost equation, revenue equation, profit equation, break-even points

Exercise 5-4 Chapter Review

Work through all the problems in this chapter review and check your answers in the back of the book. (Answers to all review problems are there.) Where weaknesses show up, review appropriate sections in the text. When you are satisfied that you know the material, take the practice test following this review.

A

1. Graph $y = \dfrac{x}{2} - 2$ in a rectangular coordinate system. Indicate the slope and the y intercept.

2. Write an equation of the line that passes through (4, 3) with slope $\frac{1}{2}$. Write the final answer in the form $y = mx + b$.

3. Graph $x - y = 2$ in a rectangular coordinate system. Indicate the slope.

4. For $f(x) = 2x - 1$ and $g(x) = x^2 - 2x$, find $f(-2) + g(-1)$.

Graph each quadratic polynomial. From the graph, estimate the coordinates of the vertex, the maximum or minimum value of $f(x)$, and the equation of the axis. Sketch in the axis.

5. $f(x) = x^2 - 8x + 14$ 6. $F(x) = 4 - x^2$

B

7. Find an equation of the line that passes through $(-2, 3)$ and $(6, -1)$. Write the answer in the form $Ax + By = C, A > 0$. What is the slope of the line?

8. Graph $3x - y = 9$ in a rectangular coordinate system. What is the slope of the graph?

9. Write the equations of the vertical line and the horizontal line that pass through $(-5, 2)$. Graph both equations on the same coordinate system.

10. Write an equation of the line that passes through the points $(4, -3)$ and $(4, 5)$.

11. Which of the following equations specify functions (x is an independent variable)?

   (A) $2x + y = 6$ (B) $y^2 = x + 1$

12. Evaluate for $x = -3$:

   (A) $f(x) = \sqrt{x^2 - 2x + 1}$ (B) $g(x) = 2x^{-2}$

13. Find the domain of the function f specified by each equation.

   (A) $f(x) = \dfrac{5}{x - 3}$ (B) $f(x) = \sqrt{x - 1}$

14. For $f(x) = 2x - 1$, find: $\dfrac{f(3 + h) - f(3)}{h}$

Graph each quadratic polynomial. From the graph, estimate the coordinates of the vertex, the maximum or minimum value of the function, and the equation of the axis. Sketch in the axis.

15. $f(x) = x^2 - 7x + 10$ 16. $g(t) = -t^2 + 3t + 4$

In Problems 17–19 graph each function. If the graph is a straight line, indicate the slope. Indicate whether each function is even, odd, or neither.

17. $f(x) = -\dfrac{2}{3}x + 4$ 18. $g(x) = \dfrac{|x|}{x}, \quad x \neq 0$ 19. $h(x) = x^{4/3}$

C

20. Write $P(x) = x^3 - 2x^2 - 5x + 6$ in a nested factored form and graph.

21. Write $P(x) = x^4 - 2x^3 - 8x - 1$ in a nested factored form and graph.

Applications

Business & Economics

22. **Pricing.** A sporting goods store sells a tennis racket that cost $30 for $48 and a pair of jogging shoes that cost $20 for $32.

   (A) If the markup policy of the store for items that cost over $10 is assumed to be linear and is reflected in the pricing of these two items, write an equation that relates retail price R to cost C.

   (B) What would be the retail price of a pair of skis that cost $105?

23. *Advertising.* Using past records, a company estimates that it will sell $N(x)$ units of a product after spending $x thousand on advertising, as given by

$$N(x) = 60x - x^2 \qquad 5 \leqslant x \leqslant 30$$

Graph this quadratic function for the restricted domain.

24. *Construction.* A rectangular storage yard is to be built with an existing building as one side. If 600 feet of fencing are available, what should be the dimensions of the yard in order to maximize the area? Set up an area function, graph it for an appropriate set of values, and estimate the maximum area from the graph.

Practice Test: Chapter 5

1. Graph $3x + 6y = 18$ in a rectangular coordinate system. Indicate the slope, x intercept, and y intercept.

2. Find an equation of the line that passes through $(-2, 5)$ and $(2, -1)$. Write the answer in the form $y = mx + b$.

3. Write an equation of the line that passes through $(2, -3)$ and is parallel to $2x - 4y = 5$. Write the final answer in the standard form $Ax + By = C, A > 0$.

4. Graph (in the same coordinate system) the vertical and horizontal lines that pass through $(2, -3)$. Indicate the equation of each line.

5. For $f(x) = 2x - x^2$ and $g(x) = 1/(x - 2)$, find $f(-2) - g(3)$.

6. Find the domains of the functions f and g in Problem 5.

7. Which of the following relations are functions (x is an independent variable)?

   (A) f: $2x + y = 3$ (B) g: $y^2 = x + 1$

In Problems 8 and 9 graph each function. If the graph is a straight line, indicate the slope. Indicate whether each function is even, odd, or neither.

8. $g(x) = \frac{3}{2}x - 3$ 　　　　　　　9. $f(x) = -|x| + 2$

10. Graph $f(x) = x^2 - 6x + 5$. Show the axis and vertex, and estimate the maximum or minimum value of $f(x)$ from the graph.

11. Write $f(x) = x^3 - 3x^2 - x + 3$ in a nested factored form and graph for $-2 \leqslant x \leqslant 4$.

12. An electronic computer was purchased by a company for $20,000 and is assumed to have a salvage value of $2,000 after 10 years (for tax purposes). If its value is depreciated linearly from $20,000 to $2,000:

(A) Find the linear equation that relates value V in dollars to time t in years.

(B) What would be the value of the computer after 6 years?

13. The weekly revenue R (in thousands of dollars) for items selling at $\$p$ each is estimated to be

$$R(p) = -2p^2 + 12p \qquad 0 \leqslant p \leqslant 6$$

Graph this quadratic function for the restricted domain and estimate the price that produces the maximum revenue. What is the maximum revenue?

14. A Wyoming rancher has 20 miles of fencing to fence in a rectangular piece of grazing land along a river.

(A) If no fence is required along the river and x is the width of the rectangle (at right angles to the river), express the area $A(x)$ of the rectangle in terms of x.

(B) What is the domain of the function A (due to physical restrictions)?

(C) Complete the table:

x	$A(x)$
2	
4	
5	
6	
8	

FINITE MATHEMATICS

Mathematics of Finance

CHAPTER 6	Contents

This chapter is provided primarily for those who are interested in business and managerial sciences. If you are not interested in this field, the chapter may be omitted without loss of continuity.

An inexpensive hand calculator that has the operations $+$, $-$, $\times$, and $\div$ will take most of the drudgery out of the calculations—even when tables are used. Table V in the back of the book can be used to solve most of the problems on compound interest, annuities, amortization, and so on. However, students who have financial calculators or scientific calculators will be able to work all the problems without tables. Some problems have been included that require such calculators. If desired, these problems may be omitted without loss of continuity.

If you desire and time permits, you may wish to cover arithmetic and geometric progressions, discussed in Appendixes A-1 and A-2, respectively, before beginning this chapter. Though not necessary for the chapter, these topics will provide additional insight into some of the topics covered.

6-1 Simple Interest and Simple Discount

- Simple Interest
- Simple Discount

Simple interest and simple discount are generally used only on short-term notes—often of duration less than one year. The concept of simple interest, however, forms the basis of much of the rest of the material developed in this chapter, for which time periods may be much longer than a year.

■ Simple Interest

If you deposit a sum of money P in a savings account or if you borrow a sum of money P from a lending agent, then P is referred to as the **principal.** When money is borrowed—whether it is a savings institution borrowing from you when you deposit money in your account or you borrowing from a lending agent—a fee is charged for the money borrowed. This fee is rent paid for the use of another's money, just as rent is paid for the use of another's house. The fee is called **interest.** It is usually computed as a percentage (called the **interest rate**) of the principal over a given period of time. The interest rate, unless otherwise stated, is an annual rate. Simple interest is given by the following formula:

Simple Interest

$$I = Prt \qquad (1)$$

where

$P =$ Principal

$r =$ Annual simple interest rate

$t =$ Time in years

For example, the interest on a loan of $100 at 12% for 9 months would be

$I = Prt$

$\quad = (100)(0.12)(0.75)$ Convert 12% to a decimal (0.12)

$\quad = \$9$ and 9 months to years ($\frac{9}{12} = 0.75$)

At the end of 9 months, the borrower would repay the principal ($100) plus the interest ($9), or a total of $109.

In general, if a principal P is borrowed at a rate r, then after t years the borrower will owe the lender an amount A that will include the principal P (the **face value** of the note) plus the interest I (the rent paid for the use of the money). Since P is the amount that is borrowed now and A is the amount that must be paid back in the future, P is often referred to as the **present value** and A as the **future value**. The formula relating A and P is as follows:

Amount — Simple Interest

$$A = P + Prt$$
$$ = P(1 + rt) \qquad (2)$$

where

$P =$ Principal, or present value

$r =$ Annual simple interest rate

$t =$ Time in years

$A =$ Amount, or future value

Given any three of the four variables A, P, r, and t in (2), we should be able to solve for the fourth. The following examples illustrate several types of common problems that can be solved by using formula (2).

Example 1   Find the total amount due on a loan of $800 at 18% simple interest at the end of 4 months.

Solution   To find the amount A (future value) due in 4 months, we use formula (2) with $P = 800$, $r = 0.18$, and $t = \frac{4}{12} = \frac{1}{3}$ year. Thus,

$$A = P(1 + rt)$$
$$= 800[1 + 0.18(\tfrac{1}{3})]$$
$$= 800(1.06)$$
$$= \$848$$

Problem 1   Find the total amount due on a loan of $500 at 12% simple interest at the end of 30 months.

Example 2   If you want to earn 10% simple interest on your investments, how much (to the nearest cent) should you pay for a note that will be worth $5,000 in 9 months?

Solution   We again use formula (2), but now we are interested in finding the principal P (present value), given $A = \$5,000$, $r = 0.1$, and $t = \frac{9}{12} = 0.75$ year. Thus,

$$A = P(1 + rt)$$
$$5,000 = P[1 + 0.1(0.75)] \qquad \text{Replace } A, r, \text{ and } t \text{ with the}$$
$$5,000 = (1.075)P \qquad\qquad\quad \text{given values and solve for } P$$
$$P = \$4,651.16$$

Problem 2   Repeat Example 2 with a time period of 6 months.

Example 3   If you must pay $960 for a note that will be worth $1,000 in 6 months, what annual simple interest rate will you earn? (Compute the answer to two decimal places.)

Solution   Again we use formula (2), but this time we are interested in finding r, given $P = \$960$, $A = \$1,000$, and $t = \frac{6}{12} = 0.5$ year. Thus,

$$A = P(1 + rt)$$
$$1,000 = 960[1 + r(0.5)]$$
$$1,000 = 960 + 960r(0.5)$$
$$40 = 480r$$
$$r = \frac{40}{480} \approx 0.0833 \quad \text{or} \quad 8.33\%$$

Problem 3   Repeat Example 3 assuming you have paid $952 for the note.

■ Simple Discount

If a borrower signs a **simple interest note,** he or she will receive the face value of the note (principal) and repay the face value plus interest at the end of the time period. On the other hand, if a borrower signs a **simple discount note,** the lender deducts the **discount** at the start from the face value of the note and the borrower will receive the remainder, called the **proceeds.** At the end of the time period, the borrower will pay the lender the face value (amount before the discount was deducted), called the **maturity value** of the note.

As with simple interest transactions, simple discount transactions have special formulas for their computation:

Simple Discount

$$D = Mdt \qquad\qquad (3)$$

$$P = M - D \qquad\qquad (4)$$

$$= M - Mdt = M(1 - dt)$$

where

$$D = \text{Simple discount}$$

$$M = \text{Maturity value}$$

$$d = \text{Discount rate}$$

$$t = \text{Time in years}$$

$$P = \text{Proceeds}$$

If you sign a simple discount note for M at a discount rate of d for t years, then it will cost you D (in advance) and you will receive P to use for t years. At the end of t years, you will have to pay the lender M to clear the note.

Example 4

Suppose you sign a discount note for $1,000 at 18% discount for 10 months. Find:

(A) The maturity value (amount that must be repaid at the end of 10 months)

(B) The simple discount (the cost for using the money you receive)

(C) The proceeds (actual amount received)

Solutions

(A) $M = \$1,000$ Maturity value (face value)

(B) $D = Mdt$ Simple discount

$\qquad = (1,000)(0.18)(\frac{10}{12})$

$\qquad = \$150$ Rent paid to use the proceeds (see part C) for 10 months

(C)   Proceeds = Maturity value − Simple discount

$$P = M - D$$
$$= \$1{,}000 - \$150$$
$$= \$850$$

Thus, after signing this $1,000, 18% discount note, you will receive $850, and after 10 months you will have to pay the lender $1,000 to clear the debt.

Problem 4   Repeat Example 4 for an $800, 12% discount note for 15 months.

Example 5   Suppose you need $1,000 for 9 months. If a finance company offers you a 12% discount note, compute:

(A)   The maturity value (amount you must repay at the end of 9 months to receive $1,000 now)
(B)   The simple discount (your cost for the loan)

Solutions   (A)   Use formula (4) in the form $P = M(1 - dt)$:

$$P = M(1 - dt)$$
$$1{,}000 = M[1 - (0.12)(0.75)]$$
$$1{,}000 = 0.91M$$
$$M = \$1{,}098.90 \quad \text{Amount to be repaid}$$

(B)   Use formula (4) in the form $P = M - D$, or

$$D = M - P$$
$$= \$1{,}098.90 - \$1{,}000$$
$$= \$98.90 \quad \text{Cost of using \$1,000 for 9 months}$$

Problem 5   Repeat Example 5 assuming you would like to receive and use $2,000 for a period of 6 months.

Example 6   If you sign a 6 month, $5,000 note discounted at 20%, what simple interest rate are you paying on the proceeds?

Solution   First, we must determine how much money you actually received (the proceeds) and how much that money cost you (the discount). Proceeding as before,

$$M = \$5{,}000 \qquad \text{Maturity value}$$
$$D = Mdt$$
$$= \$5{,}000(0.2)(\tfrac{6}{12})$$
$$= \$500 \qquad \text{Discount}$$
$$P = M - D$$
$$= \$5{,}000 - \$500$$
$$= \$4{,}500 \qquad \text{Proceeds}$$

Since you received $4,500 and must pay back $5,000, you paid $500 for using $4,500 for 6 months. Viewing this now as a simple interest problem and using formula (1) in the form $r = I/Pt$, we calculate the simple interest rate as follows:

$$r = \frac{I}{Pt} \qquad \text{Simple interest rate}$$

$$= \frac{500}{4,500(\frac{6}{12})}$$

$$\approx 0.2222 \quad \text{or} \quad 22.22\%$$

In other words, a 6 month, 20% simple discount note costs as much as a 6 month, 22.22% simple interest note.

Problem 6   Repeat Example 6 if the note is for 12 months.

Example 7   Suppose you decide to buy a 1.5 year, 8% simple-interest-bearing note with a face value of $3,000 by discounting it at 12% 3 months before it is due. How much should you pay for the note?

Solution   We first compute the future value (amount due in 1.5 years) for the simple-interest-bearing note. The result of this computation produces the maturity value M of the discount transaction.

Part I.   Find the future value of the simple-interest-bearing note:

$$A = P(1 + rt)$$

$$= 3,000[1 + (0.08)(1.5)]$$

$$= \$3,360$$

Part II.   Find the proceeds (the amount you will pay the holder of the simple-interest-bearing note):

$$P = M(1 - dt)$$

$$= 3,360[1 - (0.12)(\tfrac{3}{12})]$$

$$= \$3,259.20$$

So, 3 months after you buy the note for $3,259.20, you will receive $3,360 from the original borrower.

Problem 7   You own a 1 year, 10% simple-interest-bearing note with a face value of $10,000. Suppose that 6 months before the due date you need some money for another investment and decide to sell the note to another investor at 12% discount. How much will you receive for the note?

Answers to   1. $650   2. $4,761.90   3. 10.08%
Matched Problems   4. (A) $M = \$800$ (B) $D = \$120$ (C) $P = \$680$
                        5. (A) $M = \$2,127.66$ (B) $D = \$127.66$
                        6. 25%   7. $10,340

Exercise 6-1

A   Using formula (1) for simple interest and formula (3) for simple discount, find each of the indicated quantities.

1. $P = \$500$, $r = 8\%$, $t = 6$ months, $I = ?$
2. $P = \$900$, $r = 10\%$, $t = 9$ months, $I = ?$
3. $M = \$1,200$, $d = 8\%$, $t = 5$ months, $D = ?$
4. $M = \$3,600$, $d = 12\%$, $t = 10$ months, $D = ?$
5. $I = \$80$, $P = \$500$, $t = 2$ years, $r = ?$
6. $I = \$40$, $P = \$400$, $t = 4$ years, $r = ?$
7. $D = \$360$, $M = \$7,200$, $t = 6$ months, $d = ?$
8. $D = \$405$, $M = \$6,000$, $t = 9$ months, $d = ?$

B   Use formula (2) in an appropriate form to find the indicated quantities.

9. $P = \$100$, $r = 8\%$, $t = 18$ months, $A = ?$
10. $P = \$6,000$, $r = 6\%$, $t = 8$ months, $A = ?$
11. $A = \$1,000$, $r = 10\%$, $t = 15$ months, $P = ?$
12. $A = \$8,000$, $r = 12\%$, $t = 7$ months, $P = ?$

Use formula (4) in an appropriate form to find the indicated quantities.

13. $M = \$8,000$, $d = 10\%$, $t = 15$ months, $P = ?$
14. $M = \$2,400$, $d = 15\%$, $t = 9$ months, $P = ?$
15. $P = \$2,200$, $d = 12\%$, $t = 10$ months, $M = ?$
16. $P = \$5,000$, $d = 16\%$, $t = 1.2$ years, $M = ?$

C   Solve each formula for the indicated variable.

17. $I = Prt$,   for r
18. $I = Prt$,   for P
19. $D = Mdt$,   for M
20. $D = Mdt$,   for d
21. $A = P + Prt$,   for P
22. $P = M - Mdt$,   for M

■   Applications

Business & Economics

23. If you borrow $500 at 18% simple interest, how much must you repay at the end of 8 months?
24. If you borrow $1,000 at 12% simple interest, how much must you repay at the end of 9 months?
25. What is the future value of $10,000 invested at 15% simple interest for 4 months?
26. What is the future value of $500 invested at 12% simple interest for 7 months?
27. If you sign an 8 month, $500 note discounted at 18%, how much will you receive, how much will it cost you, and how much must you pay back at the end of the 8 months?
28. If you sign a 10 month, $2,000 note discounted at 14%, how much will

you receive, how much will it cost you, and how much must you pay back at the end of 10 months?

29. If you want to earn 18% simple interest on your investment, how much should you pay for a note that will be worth $3,000 in 8 months?

30. How much should you pay for a note worth $1,000 in 8 months if you want to earn 12% simple interest on your investment?

31. What is the present value of $10,000 invested at 15% simple interest for 4 months?

32. What is the present value of $500 invested at 12% simple interest for 7 months?

33. If you pay $450 for a note that will pay $500 in 6 months, what simple interest rate will you earn?

34. If you pay $920 for a note that will pay $1,000 in 9 months, what simple interest rate will you earn?

35. Suppose you need $2,400 for 15 months. If a bank offers you a 16% discount note, compute the maturity value and the simple discount.

36. What will be the maturity value of an 18% discounted note that pays you $1,200 for 9 months? What is the simple discount?

37. If you sign an 8 month, $1,500 note discounted at 14%, what simple interest rate are you paying on the proceeds?

38. If you sign a 10 month, $6,000 note discounted at 16%, what simple interest rate are you paying on the proceeds?

39. Suppose you decide to buy a 12 month, 10% simple-interest-bearing note with a face value of $5,000 by discounting it at 16% 6 months before it is due. How much should you pay for the note?

40. You own a 14 month, 12% simple-interest-bearing note with a face value of $6,000. You need some money for another investment 9 months before the due date, and you decide to sell the note to another investor at 12% discount. How much will you receive for the note?

41. If you sign a simple discount note for $M at a discount rate of d for t years, show that the simple interest rate you pay on the proceeds is

$$r = \frac{d}{1 - dt}$$

6-2 Compound Interest

- Compound Interest
- Effective Rate
- Doubling Time

■ Compound Interest

If at the end of a payment period the interest due is reinvested at the same rate, then the interest as well as the original principal will earn interest during the next payment period. Interest paid on interest reinvested is called **compound interest.**

For example, suppose you deposit $1,000 in a bank that pays 8% compounded quarterly. How much will the bank owe you at the end of a year? *Compounding quarterly* means that earned interest is paid to your account at the end of each 3 month period and that interest as well as the principal earns interest for the next quarter. Using the simple interest formula (2) from the previous section, we compute the amount in the account at the end of the first quarter after interest has been paid:

$$A = P(1 + rt)$$
$$= 1,000[1 + 0.8(\tfrac{1}{4})]$$
$$= 1,000(1.02) = \$1,020$$

Now, $1,020 is your new principal for the second quarter. At the end of the second quarter, after interest is paid, the account will have

$$A = \$1,020[1 + 0.08(\tfrac{1}{4})]$$
$$= \$1,020(1.02) = \$1,040.40$$

Similarly, at the end of the third quarter, you will have

$$A = \$1,040.40[1 + 0.8(\tfrac{1}{4})]$$
$$= \$1,040.40(1.02) = \$1,061.21$$

Finally, at the end of the fourth quarter, the account will have

$$A = \$1,061.21[1 + 0.08(\tfrac{1}{4})]$$
$$= \$1,061.21(1.02) = \$1,082.43$$

How does this compound amount compare with simple interest? The amount with simple interest would be

$$A = P(1 + rt)$$
$$= \$1,000[1 + 0.08(1)]$$
$$= \$1,000(1.08) = \$1,080$$

We see that compounding quarterly yields $2.43 more than simple interest would provide.

Let us look over the above calculations for compound interest to see if we can uncover a pattern that might lead to a general formula for computing compound interest for arbitrary cases.

$A = 1,000(1.02)$	End of first quarter
$A = [1,000(1.02)](1.02) = 1,000(1.02)^2$	End of second quarter
$A = [1,000(1.02)^2](1.02) = 1,000(1.02)^3$	End of third quarter
$A = [1,000(1.02)^3](1.02) = 1,000(1.02)^4$	End of fourth quarter

It appears that at the end of n quarters, we would have

$$A = 1,000(1.02)^n \qquad \text{End of nth quarter}$$

or

$$A = 1,000[1 + 0.08(\tfrac{1}{4})]^n$$
$$= 1,000\left[1 + \frac{0.08}{4}\right]^n$$

where $0.08/4 = 0.02$ is the interest rate per quarter. Since interest rates are generally quoted as annual rates, the **rate per compound period** is found by dividing the annual rate by the number of compounding periods per year.

The compound interest formula given in the box is the result of the above discussion. Its general proof requires a technique called *mathematical induction*, which is beyond the scope of this book.

Amount—Compound Interest

$$A = P(1 + i)^n \qquad (1)$$

where

$$i = \frac{r}{m}$$

and

r = Annual (quoted) rate
m = Number of compounding periods per year
n = Total number of compounding periods
i = Rate per compounding period
P = Principal (present value)
A = Amount (future value) at end of n periods

Several examples will illustrate different uses of formula (1). If any three of the four variables in (1) are given, we can solve for the fourth. The power form $(1 + i)^n$ in formula (1) can be evaluated for various values of i and n by using Table V in the back of the book or a financial or scientific calculator.

Example 8 If $1,000 is invested at 8% compounded

(A) annually   (B) semiannually   (C) quarterly

what is the amount after 5 years? Write answers to the nearest cent.

Solutions

(A) Compounding annually means that there is one interest payment period per year. Thus, $n = 5$ and $i = r = 0.08$.

$A = P(1 + i)^n$
$= 1,000(1 + 0.08)^5$ Use Table V or a calculator
$= 1,000(1.469\ 328)$
$= \$1,469.33$ Interest earned $= A - P = \$469.33$

(B) Compounding semiannually means that there are two interest payment periods per year. Thus, the number of payment periods in 5 years is $n = 2(5) = 10$ and the interest rate per period is

$$i = \frac{r}{m} = \frac{0.08}{2} = 0.04$$

So,

$A = P(1 + i)^n$
$= 1,000(1 + 0.04)^{10}$ Use Table V or a calculator
$= 1,000(1.480\ 244)$
$= \$1,480.24$ Interest earned $= A - P = \$480.24$

(C) Compounding quarterly means that there are four interest payments per year. Thus, $n = 4(5) = 20$ and $i = 0.08/4 = 0.02$. So,

$A = P(1 + i)^n$
$= 1,000(1 + 0.02)^{20}$ Use Table V or a calculator
$= 1,000(1.485\ 947)$
$= \$1,485.95$ Interest earned $= A - P = \$485.95$

Problem 8   Repeat Example 8 with an annual interest rate of 6% over an 8 year period.

Notice the rather significant increase in interest earned in going from annual compounding to quarterly compounding. One might wonder what happens if we compound daily, or every minute, or every second, and so on. Figure 1 shows the relative effect of increasing the number of com-

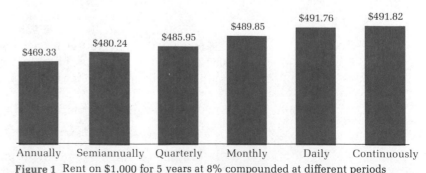

Figure 1 Rent on $1,000 for 5 years at 8% compounded at different periods

pounding periods in a year. A limit is reached at compounding *continuously*, which is not a great deal larger than that obtained through monthly compounding. Continuous compounding is discussed in the study of calculus. Compare the results in Figure 1 with simple interest earned over the same time period:

$$I = Prt = 1,000(0.08)5 = \$400$$

Another use of the compound interest formula is in determining how much you should invest now to have a given amount at a future date.

Example 9   How much should you invest now at 10% compounded quarterly to have $8,000 to buy a car in 5 years?

Solution   We are given a future value $A = \$8,000$ for a compound interest investment, and we need to find the present value (principal P) given $i = 0.10/4 = 0.025$ and $n = 4(5) = 20$.

$$A = P(1 + i)^n$$
$$8,000 = P(1 + 0.025)^{20}$$
$$P = \frac{8,000}{(1 + 0.025)^{20}} \qquad \text{Use Table V or a calculator}$$
$$= \frac{8,000}{1.638\ 616} = \$4,882.17$$

Problem 9   How much should new parents invest now at 8% compounded semiannually to have $16,000 toward their child's college education in 17 years?

■ **Effective Rate**

When interest is compounded more than once a year, the stated annual rate is called a **nominal rate**; the simple interest rate that would produce the same interest in 1 year is called the **effective rate** (also called the **annual yield** or **true interest rate**). Effective rates are used to compare various types of investments.

Example 10   An investment company pays 8% compounded quarterly. What is the effective rate? That is, what simple interest rate would produce the same interest in 1 year? (Compute the answer to three decimal places.)

Solution   We first find the total compound interest for 1 year for an arbitrary principal, say, $P = 1$. Then we find the equivalent simple interest rate that will produce the same amount of interest in 1 year using the simple interest formula:

$$A = P(1 + i)^n \qquad i = 0.08/4 = 0.02, n = 4, P = 1$$
$$= 1(1 + 0.02)^4 \qquad \text{Use Table V or a calculator}$$
$$= 1.082\ 432$$

Compound interest $= A - P = 1.082\ 432 - 1 = 0.082\ 432$

Now we use the simple interest formula $I = Prt$ with $I = 0.082\ 432, P = 1$, and $t = 1$ to find r, the effective interest rate:

$$0.082\ 432 = (1)r(1)$$
$$r = 0.082\ 432 \quad \text{or} \quad 8.243\%$$

This shows that money invested at 8.243% simple interest earns the same amount of interest in one year as money invested at 8% compounded quarterly. Thus, the effective rate of 8% compounded quarterly is 8.243%.

Problem 10   What is the effective rate of money invested at 6% compounded quarterly?

Example 11   An investor has an opportunity to purchase two different bonds. Bond A pays 15% compounded monthly, and bond B pays 15.2% compounded semiannually. Which is the better investment, assuming all else is equal?

Solution   Nominal rates with different compounding periods cannot be compared directly. We must first find the effective rate of each nominal rate and then compare the effective rates to determine which investment will yield the larger return.

Effective Rate for Bond A
$$A = P(1 + i)^n \qquad i = 0.15/12 = 0.0125, n = 12$$
$$= 1(1 + 0.0125)^{12} \qquad \text{Use Table V or a calculator}$$
$$= 1.160\ 755$$

Compound interest $= A - P = 1.160\ 775 - 1 = 0.160\ 775$

$$I = Prt \qquad\qquad\qquad I = 0.160\ 775, P = 1, t = 1$$
$$0.160\ 775 = (1)r(1)$$
$$r = 0.160\ 775 \quad \text{or} \quad 16.0775\% \qquad \text{Effective rate for bond } A$$

Effective Rate for Bond B
$$A = P(1 + i)^n \qquad i = 0.152/2 = 0.076, n = 2$$
$$= 1(1 + 0.076)^2 \qquad \text{Use a calculator}$$
$$= 1.157\ 78$$

Compound interest $= A - P = 1.157\ 78 - 1 = 0.157\ 78$

$$I = Prt \qquad\qquad\qquad I = 0.157\ 78, P = 1, t = 1$$
$$0.157\ 78 = (1)r(1)$$
$$r = 0.157\ 78 \quad \text{or} \quad 15.778\% \qquad \text{Effective rate for bond } B$$

Since the effective rate for bond A is greater than the effective rate for bond B, bond A is the preferred investment.

Problem 11   Repeat Example 11 if bond A pays 9% compounded monthly and bond B pays 9.2% compounded semiannually.

■ Doubling Time

Investments are also compared by computing their **doubling times** — the time it takes an investment to double in value. Example 12 illustrates two methods for making this calculation.

Example 12   How long will it take money to double if it is invested at 6% compounded quarterly?

Solution   The problem is to find n in $A = P(1 + i)^n$ with $A = 2P$ and $i = 0.06/4 = 0.015$.

$$2P = P(1 + 0.015)^n \qquad \text{Divide both sides by } P$$
$$2 = (1 + 0.015)^n \qquad \text{Now solve for } n$$

Method I.   Use Table V. Look down the $(1 + i)^n$ column on the page that has $i = 0.015(1\frac{1}{2}\%)$. Find the value in this column that is closest to and greater than 2 and take the n value that corresponds to it. In this case, $n = 47$ quarters, or 11 years and 9 months.

Method II.   Use logarithms:

$$2 = 1.015^n$$
$$\log_{10} 2 = \log_{10} 1.015^n$$
$$\log_{10} 2 = n \log_{10} 1.015$$
$$n = \frac{\log_{10} 2}{\log_{10} 1.015}$$

Notice how logarithmic properties are needed to solve this problem (A review of exponential and logarithmic functions can be found in Sections 13-1 and 13-2.)

$$= \frac{0.3010}{0.0065} = 46.31 \approx 47 \text{ quarters} \quad \text{or}$$
$$11 \text{ years and 9 months}$$

[*Note:* 46.31 is rounded up to 47 to guarantee doubling since interest is paid at the *end of* each quarter.]

Problem 12   How long will it take money to double if it is invested at 8% compounded semiannually? (Round to next highest half year.)

Answers to   8.   (A) $1,593.85   (B) $1,604.71   (C) $1,610.32
Matched Problems   9. $4,216.83   10.   6.1364%
          11.   Bond B (effective rate of bond A is 9.38% and of bond B is 9.412%)
          12. $8.84 \approx 9$ years

Exercise 6-2

Use the compound interest formula (1) and Table V or a calculator (or both) to find each of the indicated values (to the nearest cent).

A   1. $P = \$100$, $i = 0.01$, $n = 12$, $A = ?$
2. $P = \$1,000$, $i = 0.015$, $n = 20$, $A = ?$
3. $P = \$800$, $i = 0.06$, $n = 25$, $A = ?$
4. $P = \$10,000$, $i = 0.08$, $n = 30$, $A = ?$
5. $P = \$2,000$, $i = 0.005$, $n = 80$, $A = ?$
6. $P = \$5,000$, $i = 0.025$, $n = 75$, $A = ?$

B   7. $A = \$10,000$, $i = 0.03$, $n = 48$, $P = ?$
8. $A = \$1,000$, $i = 0.015$, $n = 60$, $P = ?$
9. $A = \$18,000$, $i = 0.01$, $n = 90$, $P = ?$
10. $A = \$50,000$, $i = 0.005$, $n = 70$, $P = ?$
11. $A = 2P$, $i = 0.06$, $n = ?$
12. $A = 2P$, $i = 0.05$, $n = ?$

C   13. $A = 3P$, $i = 0.02$, $n = ?$
14. $A = 4P$, $i = 0.06$, $n = ?$

Applications

Business & Economics

Solve using Table V or a calculator (or both). Find values to two decimal places.

15.   If $100 is invested at 6% compounded

(A)   annually   (B)   quarterly   (C)   monthly

what is the amount after 4 years? How much interest is earned?

16.   If $2,000 is invested at 7% compounded

(A)   annually   (B)   semiannually   (C)   quarterly

what is the amount after 5 years? How much interest is earned?

17.   If $5,000 is invested at 18% compounded monthly, what is the amount after

(A)   2 years?   (B)   4 years?

18.   If $20,000 is invested at 6% compounded monthly, what is the amount after

(A)   5 years?   (B)   8 years?

19. What is the future value of $1,000 invested at 8% compounded quarterly for

    (A)   10 years?   (B)   20 years?

20. What is the future value of $500 invested at 12% compounded semiannually for

    (A)   4 years?   (B)   8 years?

21. If an investment company pays 8% compounded semiannually, how much should you deposit now to have $10,000

    (A)   5 years from now?   (B)   10 years from now?

22. If an investment company pays 10% compounded quarterly, how much should you deposit now to have $6,000

    (A)   3 years from now?   (B)   6 years from now?

23. What is the present value of $5,000 invested at 15% compounded monthly for

    (A)   3 years?   (B)   6 years?

24. What is the present value of $200 invested at 12% compounded monthly for

    (A)   1 year?   (B)   2 years?

25. A business machine will have to be replaced in 5 years at an estimated cost of $10,000. How much should be invested now at 8% compounded quarterly to meet this obligation?

26. If for the past 5 years a company had an average annual increase of 8% and if sales this year are $1,000,000, what was the dollar volume in sales 5 years ago?

27. What is the effective rate of interest for money invested at

    (A)   10% compounded quarterly?
    (B)   12% compounded monthly?

28. What is the effective rate of interest for money invested at

    (A)   6% compounded monthly?
    (B)   14% compounded semiannually?

29. How long will it take money to double if it is invested at

    (A)   10% compounded quarterly?
    (B)   12% compounded quarterly?

30. How long (to the nearest year) will money take to double if it is invested at
    (A) 14% compounded semiannually?
    (B) 10% compounded semiannually?

The following problems require the use of a financial or scientific calculator. In problems that involve daily compounding, assume that there are always 365 days in a year.

31. If $5,000 is invested at 10% compounded daily, what is the amount after
    (A) 5 years?   (B) 10 years?

32. How much should be invested now at
    (A) 5.25%   (B) 8%
    compounded daily to have $10,000 in 5 years?

33. If $100 is invested at 12.6% compounded
    (A) annually   (B) semiannually   (C) quarterly
    (D) monthly   (E) daily
    what is the amount after 30 years?

34. If $100 is invested at 14.5% compounded
    (A) annually   (B) semiannually   (C) quarterly
    (D) monthly   (E) daily
    what is the amount after 25 years?

35. A savings and loan company offers rates of 10% compounded daily. What is the effective rate?

36. What is the effective rate of 7.75% compounded daily?

37. An investor bought stock at $100 a share. Five years later the stock sold for $150 a share. If interest was compounded annually, what annual rate of interest did this investment earn?

38. A family paid $10,000 in cash for a house. Twenty years later they sold the house for $80,000. If interest was compounded monthly, what annual rate of interest did the original $10,000 investment earn?

39. How long will it take money to double if it is invested at
    (A) 8%   (B) 10%   (C) 12%
    compounded daily?

6-3 Future Value of an Annuity; Sinking Funds

- Future Value of an Annuity
- Sinking Funds

■ Future Value of an Annuity

An **annuity** is any sequence of equal periodic payments. If payments are made at the end of each time interval, then the annuity is called an **ordinary annuity.** We will only consider ordinary annuities in this book. The amount, or **future value,** of an annuity is the sum of all payments plus all interest earned.

Suppose you decide to deposit $100 every 6 months into an account that pays 6% compounded semiannually. If you make six deposits, one at the end of each interest payment period, over 3 years, how much money will be in the account after the last deposit is made? To solve this problem, let us look at it in terms of a time line. Using the compound amount formula $A = P(1 + i)^n$, we can find the value of each deposit after it has earned compound interest up through the sixth deposit, as shown in Figure 2.

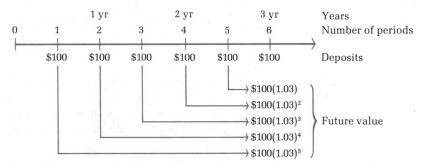

Figure 2

We could, of course, evaluate each of the future values in Figure 2 using Table V or a calculator and then add the results to find the amount in the account at the time of the sixth deposit—a tedious project at best. Instead, we take another approach that leads directly to a formula that will produce the same result in a few steps (even when the number of deposits is very large). We start by writing the total amount in the account after the sixth deposit in the form

$$S = 100 + 100(1.03) + 100(1.03)^2 + 100(1.03)^3 + 100(1.03)^4$$
$$+ 100(1.03)^5 \tag{1}$$

We would like a simple way to sum these terms. Let us multiply each side of (1) by 1.03 to obtain

$$1.03S = 100(1.03) + 100(1.03)^2 + 100(1.03)^3 + 100(1.03)^4$$
$$+ 100(1.03)^5 + 100(1.03)^6 \qquad (2)$$

Subtracting (1) from (2), left side from left side and right side from right side, we obtain

$$1.03S - S = 100(1.03)^6 - 100 \qquad \text{Notice how many terms drop out}$$
$$0.03S = 100[(1.03)^6 - 1]$$
$$S = 100\frac{(1 + 0.03)^6 - 1}{0.03} \qquad \text{We write S in this form to observe} \qquad (3)$$
$$\text{a general pattern}$$

In general, if R is the periodic deposit, i the rate per period, and n the number of periods, then the future value is given by

$$S = R + R(1 + i) + R(1 + i)^2 + \cdots + R(1 + i)^{n-1} \qquad \text{Note how this}$$
$$\text{compares to (1)}$$

and proceeding as in the above example, we obtain the general formula for the future value of an ordinary annuity:*

$$S = R\frac{(1 + i)^n - 1}{i} \qquad \text{Note how this compares to (3)} \qquad (4)$$

It is common practice to use the symbol

$$s_{\overline{n}|i} = \frac{(1 + i)^n - 1}{i}$$

for the fractional part of (4). The symbol $s_{\overline{n}|i}$, read "s angle n at i," is evaluated in Table V for various values of n and i. A financial or scientific calculator can also be used to calculate S. The advantage of a calculator, of course, is that it can handle many more situations than any table, no matter how large the table.

Returning to the example above, we now use Table V to complete the problem:

$$S = 100\frac{(1.03)^6 - 1}{0.03}$$
$$= 100s_{\overline{6}|0.03} \qquad \text{Use Table V with } i = 0.03 \text{ and } n = 6$$
$$= 100(6.468\ 410)$$
$$= \$646.84$$

[*Note:* Using a scientific calculator, we would evaluate $(1.03)^6$ first and then complete the problem using the arithmetic operations indicated. A financial calculator is even more convenient (see the instruction manual for your particular financial calculator if you have one).]

* This formula can also be obtained by using the formula in Appendix A-2 for the sum of the first n terms in a geometric progression.

We summarize the above results in the following box for convenient reference:

Future Value of an Ordinary Annuity

$$S = R\frac{(1 + i)^n - 1}{i} = Rs_{\overline{n}|i} \qquad (5)$$

where

R = Periodic payment

i = Rate per period

n = Number of payments (periods)

S = Amount or future value

(Payments are made at the end of each period.)

Example 13 What is the value of an annuity at the end of 5 years if $100 per month is deposited into an account earning 9% compounded monthly? How much of this value is interest?

Solution To find the value of the annuity, use formula (5) with $R = \$100$, $i = 0.09/12 = 0.0075$, and $n = 12(5) = 60$.

$$S = R\frac{(1 + i)^n - 1}{i} \quad \text{or} \quad Rs_{\overline{n}|i}$$

$$= 100\frac{(1.0075)^{60} - 1}{0.0075} \quad \text{or} \quad 100s_{\overline{60}|0.0075} \qquad \text{Use Table V or a calculator}$$

$$= 100(75.424\ 137)$$

$$= \$7,542.41$$

To find the interest, subtract the total amount deposited in the annuity from the value of the annuity:

$$\text{Deposits} = 60(100)$$
$$= \$6,000$$
$$\text{Interest} = \text{Value} - \text{Deposits}$$
$$= 7,542.41 - 6,000$$
$$= \$1,542.41$$

Problem 13 What is the value of an annuity at the end of 10 years if $1,000 is deposited every 6 months into an account earning 8% compounded semiannually? How much of this value is interest?

■ Sinking Funds

The formula for the future value of an ordinary annuity has another important application. Suppose the parents of a newborn child decide that on each of the child's birthdays up to the seventeenth year, they will deposit $R in an account that pays 6% compounded annually. The money is to be used for college expenses. What should the annual deposit R be in order for the amount in the account to be $16,000 after the seventeenth deposit?

We are given S, i, and n in formula (5), and our problem is to find R. Thus,

$$S = R\frac{(1 + i)^n - 1}{i} \quad \text{or} \quad Rs_{\overline{n}|i}$$

$$16,000 = R\frac{(1.06)^{17} - 1}{0.06} \quad \text{or} \quad Rs_{\overline{17}|0.06} \qquad \text{Solve for } R$$

$$R = \frac{0.06(16,000)}{(1.06)^{17} - 1} \quad \text{or} \quad \frac{16,000}{s_{\overline{17}|0.06}} \qquad \text{Use Table V or a calculator}$$

$$= 16,000(0.035\ 445)$$

$$= \$567.12 \text{ per year}$$

An annuity of seventeen $567.12 annual deposits at 6% compounded annually will amount to approximately $16,000 in 17 years.

This is one of many examples of a similar type that are referred to as *sinking fund problems*. In general, any account that is established for accumulating funds to meet future obligations or debts is called a **sinking fund**. If the payments are to be made in the form of an ordinary annuity, then we have only to solve for R in formula (5) to find the periodic payment into the fund. Doing this, we obtain the general formula:

Sinking Fund Payment

$$R = S\frac{i}{(1 + i)^n - 1} = \frac{S}{s_{\overline{n}|i}} \tag{6}$$

where

R = Sinking fund payment

S = Value of annuity after n payments

n = Number of payments (periods)

i = Rate per period

(Payments are made at the end of each period.)

Example 14 A company estimates that it will have to replace a piece of equipment at a cost of $10,000 in 5 years. To have this money available in 5 years, a sinking fund is established by making fixed monthly payments into an account paying 6% compounded monthly. How much should each payment be?

Solution We use formula (6) with $S = \$10{,}000$, $i = 0.06/12 = 0.005$, and $n = 5(12) = 60$:

$$R = S\frac{i}{(1+i)^n - 1} \quad \text{or} \quad \frac{S}{s_{\overline{n}|i}}$$

$$R = (10{,}000)\frac{0.005}{(1.005)^{60} - 1} \quad \text{or} \quad \frac{10{,}000}{s_{\overline{60}|0.005}} \qquad \text{Use Table V or a calculator}$$

$$R = 10{,}000(0.014\ 333)$$

$$= \$143.33 \text{ per month}$$

Problem 14 A bond issue is approved for building a marina in a city. The city is required to make regular payments every 6 months into a sinking fund paying 6% compounded semiannually. At the end of 10 years, the bond obligation will be retired at a cost of $5,000,000. What should each payment be?

Answers to Matched Problems
13. Value: $29,778.08; interest: $9,778.08
14. $186,078.54 every 6 months for 10 years

Exercise 6-3

Use formula (5) or (6) and Table V or a calculator (or both) to solve each problem. (Answers may vary slightly depending on whether you use a calculator or Table V.)

A
1. $S = ?$, $n = 20$, $i = 0.03$, $R = \$500$
2. $S = ?$, $n = 25$, $i = 0.04$, $R = \$100$
3. $S = ?$, $n = 40$, $i = 0.02$, $R = \$1{,}000$
4. $S = ?$, $n = 30$, $i = 0.01$, $R = \$50$

B
5. $S = \$3{,}000$, $n = 20$, $i = 0.02$, $R = ?$
6. $S = \$8{,}000$, $n = 30$, $i = 0.03$, $R = ?$
7. $S = \$5{,}000$, $n = 15$, $i = 0.01$, $R = ?$
8. $S = \$2{,}500$, $n = 10$, $i = 0.08$, $R = ?$

C
9. $S = \$4{,}000$, $i = 0.02$, $R = 200$, $n = ?$
10. $S = \$8{,}000$, $i = 0.04$, $R = 500$, $n = ?$

Applications

Business & Economics

11. What is the value of an ordinary annuity at the end of 10 years if $500 per quarter is deposited into an account earning 8% compounded quarterly? How much of this value is interest?

12. What is the value of an ordinary annuity at the end of 20 years if $1,000 per year is deposited into an account earning 7% compounded annually? How much of this value is interest?

13. In order to accumulate enough money for a down payment on a house, a couple deposits $300 per month into an account paying 6% compounded monthly. If payments are made at the end of each period, how much money will be in the account in 5 years?

14. A self-employed person has a Keogh retirement plan. (This type of plan is free of taxes until money is withdrawn.) If deposits of $7,500 are made each year into an account paying 8% compounded annually, how much will be in the account after 20 years?

15. In 5 years a couple would like to have $25,000 for a down payment on a house. What fixed amount should be deposited each month into an account paying 9% compounded monthly?

16. A person wishes to have $200,000 in an account for retirement 15 years from now. How much should be deposited quarterly in an account paying 8% compounded quarterly?

17. A company estimates it will need $100,000 in 8 years to replace a computer. If it establishes a sinking fund by making fixed monthly payments into an account paying 12% compounded monthly, how much should each payment be?

18. Parents have set up a sinking fund in order to have $30,000 in 15 years for their children's college education. How much should be paid semiannually into an account paying 10% compounded semiannually?

19. Beginning in January, a person plans to deposit $100 at the end of each month into an account earning 9% compounded monthly. Each year taxes must be paid on the interest earned during that year. Find the interest earned during each year for the first three years.

20. If $500 is deposited each quarter into an account paying 12% compounded quarterly for 3 years, find the interest earned during each of the 3 years.

Use a financial or scientific calculator to solve each of the following problems:

21. What is the value of an ordinary annuity at the end of 20 years if $50 is invested each month into an account paying 8.25% compounded monthly? How much of this value is interest?

22. What is the value of an ordinary annuity at the end of 25 years if $200

is invested each quarter into an account paying 7.75% compounded quarterly? How much of this value is interest?

23. A company establishes a sinking fund to replace machinery at an estimated cost of $1,500,000 in 5 years. How much should be invested each quarter into an account paying 9.15% compounded quarterly?

24. You wish to have $3,000 in 2 years for a down payment on a car. How much should you deposit each month into an account paying 8% compounded monthly?

25. If you establish a sinking fund with payments of $200 per month into an account paying 10% compounded monthly, how long (to the nearest year) will it take for the account to have a value of $150,000?

26. You can afford monthly deposits of only $150 into an account that pays 8.5% compounded monthly. How long (to the nearest month) will it be until you have $7,000 to buy a car?

6-4   Present Value of an Annuity; Amortization

- Present Value of an Annuity
- Amortization
- Amortization Schedules

■ Present Value of an Annuity

How much should you deposit in an account paying 6% compounded semiannually in order to be able to withdraw $1,000 every 6 months for the next 3 years? (After the last payment is made, no money is to be left in the account.)

Actually, we are interested in finding the **present value** of each $1,000 that is paid out during the 3 years. We can do this by solving for P in the compound interest formula

$$A = P(1 + i)^n$$

$$P = \frac{A}{(1 + i)^n} = A(1 + i)^{-n}$$

The rate per period is $i = 0.06/2 = 0.03$. The present value P of the first payment is $1{,}000(1.03)^{-1}$, the second payment is $1{,}000(1.03)^{-2}$, and so on. Figure 3 (page 274) shows this in terms of a time line.

We could evaluate each of the present values in Figure 3 using a calculator or Table V and add the results to find the total present values of all the payments (which will be the amount that is needed now to buy the annuity). Since this is generally a tedious process, particularly when the number of payments is large, we will use the same device we used in the

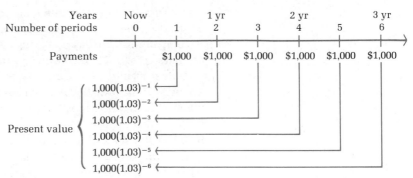

Figure 3

last section to produce a formula that will accomplish the same result in a couple of steps. We start by writing the sum of the present values in the form

$$P = 1,000(1.03)^{-1} + 1,000(1.03)^{-2} + \cdots + 1,000(1.03)^{-6} \tag{1}$$

Multiplying both sides of (1) by (1.03), we obtain

$$1.03P = 1,000 + 1,000(1.03)^{-1} + \cdots + 1,000(1.03)^{-5} \tag{2}$$

Now subtract (1) from (2):

$$1.03P - P = 1,000 - 1,000(1.03)^{-6} \qquad \text{Notice how many terms drop out}$$

$$0.03P = 1,000[1 - (1 + 0.03)^{-6}]$$

$$P = 1,000\frac{1 - (1 + 0.03)^{-6}}{0.03} \qquad \begin{matrix}\text{We write } P \text{ in this form} \\ \text{to observe a general pattern}\end{matrix} \tag{3}$$

In general, if R is the periodic payment, i the rate per period, and n the number of periods, then the present value of all payments is given by

$$P = R(1 + i)^{-1} + R(1 + i)^{-2} + \cdots + R(1 + i)^{-n} \qquad \begin{matrix}\text{Note how this} \\ \text{compares to (1)}\end{matrix}$$

Proceeding as in the above example, we obtain the general formula for the present value of an ordinary annuity:*

$$P = R\frac{1 - (1 + i)^{-n}}{i} \qquad \text{Note how this compares to (3)} \tag{4}$$

It is common practice to use the symbol

$$a_{\overline{n}|i} = \frac{1 - (1 + i)^{-n}}{i}$$

for the fractional part of (4). The symbol $a_{\overline{n}|i}$, read "a angle n at i," is evaluated for various values of n and i in Table V. The present value P can also be evaluated using a financial or scientific calculator. As we said

---

* This formula can also be obtained by using the formula in Appendix A-2 for the sum of the first n terms in a geometric progression.

before, a calculator can handle far more situations than any table, no matter how large the table.

Returning to the example above, we now use Table V to complete the problem:

$$P = 1,000 \frac{1 - (1.03)^{-6}}{0.03}$$

$$= 1,000 a_{\overline{6}|0.03} \qquad \text{Use Table V with } i = 0.03 \text{ and } n = 6$$

$$= 1,000(5.417\ 191)$$

$$= \$5,417.19$$

[*Note:*   Using a scientific calculator, we would evaluate $(1.03)^{-6}$ first and then complete the calculation using the arithmetic operations indicated. A financial calculator performs the task with even fewer steps (read the instruction manual for your particular calculator if you have one.)]

We summarize these results in the box for convenient reference:

Present Value of an Ordinary Annuity

$$P = R \frac{1 - (1 + i)^{-n}}{i} = R a_{\overline{n}|i} \qquad (5)$$

where

$R =$ Periodic payment

$i =$ Rate per period

$n =$ Number of periods

$P =$ Present value of all payments

(Payments are made at the end of each period.)

Example 15   What is the present value of an annuity that pays $200 per month for 5 years if money is worth 6% compounded monthly?

Solution   To solve this problem, use formula (5) with $R = \$200$, $i = 0.06/12 = 0.005$, and $n = 12(5) = 60$:

$$P = R \frac{1 - (1 + i)^{-n}}{i} \quad \text{or} \quad R a_{\overline{n}|i}$$

$$= 200 \frac{1 - (1.005)^{-60}}{0.005} \quad \text{or} \quad 200 a_{\overline{60}|0.005} \qquad \text{Use Table V or a calculator}$$

$$= 200(51.725\ 561)$$

$$= \$10,345.11$$

Problem 15   How much should you deposit in an account paying 8% compounded quarterly in order to receive quarterly payments of $1,000 for the next 4 years?

■ Amortization

The present value formula for an ordinary annuity (5) has another important use. Suppose you borrow $5,000 from a bank to buy a car and agree to repay the loan in 36 equal monthly payments, including all interest due. If the bank charges 1% per month on the unpaid balance (12% per year compounded monthly), how much should each payment be to retire the total debt including interest in 36 months?

Actually, the bank has bought an annuity from you. The question is, If the bank pays you $5,000 (present value) for an annuity paying them $R per month for 36 months at 12% interest compounded monthly, what are the monthly payments R? (Note that the value of the annuity at the end of 36 months is zero.) To find R, we have only to use formula (5) with $P = \$5,000$, $i = 0.01$, and $n = 36$:

$$P = R\frac{1 - (1 + i)^{-n}}{i} \quad \text{or} \quad Ra_{\overline{n}|i}$$

$$5{,}000 = R\frac{1 - (1.01)^{-36}}{0.01} \quad \text{or} \quad Ra_{\overline{36}|0.01} \qquad \text{Use Table V or a calculator;}$$
$$\text{then solve for } R$$

$$R = 5{,}000(0.033\ 214)$$

$$= \$166.07 \text{ per month}$$

At $166.07 per month, the car will be yours after 36 months. That is, you have *amortized* the debt in 36 equal monthly payments. (*Mort* means "death"; you have "killed" the loan in 36 months.) In general, **amortizing a debt** means that the debt is retired in a given length of time by equal periodic payments that include compound interest. We are usually interested in computing the equal periodic payment. Solving the present value formula (5) for R in terms of the other variables, we obtain the following amortization formula:

Amortization Formula

$$R = P\frac{i}{1 - (1 + i)^{-n}} = P\frac{1}{a_{\overline{n}|i}} \tag{6}$$

where

$P =$ Amount of loan (present value)

$i =$ Rate per period

$n =$ Number of payments (periods)

$R =$ Periodic payment

(Payments are made at the end of each period.)

Example 16

Assume that you buy a television set for $800 and agree to pay for it in 18 equal monthly payments at $1\frac{1}{2}$% interest per month on the unpaid balance.

(A) How much are your payments?
(B) How much interest will you pay?

Solutions

(A) Use formula (6) with $P = \$800$, $i = 0.015$, and $n = 18$:

$$R = P\frac{i}{1 - (1 + i)^{-n}} \quad \text{or} \quad P\frac{1}{a_{\overline{n}|i}}$$

$$= 800\frac{0.015}{1 - (1.015)^{-18}} \quad \text{or} \quad 800\frac{1}{a_{\overline{18}|0.015}} \qquad \begin{array}{l}\text{Use Table V or a}\\\text{calculator}\end{array}$$

$$= 800(0.063\ 806)$$

$$= \$51.04 \text{ per month}$$

(B) Total interest paid = Amount of all payments − Initial loan

$$= 18(\$51.04) - \$800$$

$$= \$118.72$$

Problem 16

If you sell your car to someone for $2,400 and agree to finance it at 1% per month on the unpaid balance, how much should you receive each month to amortize the loan in 24 months? How much interest will you receive?

▪ Amortization Schedules

What happens if you are amortizing a debt with equal periodic payments and at some point decide to pay off the remainder of the debt in one lump sum payment? This occurs each time a home with an outstanding mortgage is sold. In order to understand what happens in this situation, we must take a closer look at the amortization process. We begin with an example that is simple enough to allow us to examine the effect each payment has on the debt.

Example 17

If you borrow $500 that you agree to repay in 6 equal monthly payments at 1% interest per month on the unpaid balance, how much of each monthly payment is used for interest and how much is used to reduce the unpaid balance?

Solution

First, we compute the required monthly payment using formula (6) with $P = \$500$, $i = 0.01$, and $n = 6$:

$$R = P\frac{i}{1 - (1 + i)^{-n}} \quad \text{or} \quad P\frac{1}{a_{\overline{n}|i}}$$

$$= 500\frac{0.01}{1 - (1.01)^{-6}} \quad \text{or} \quad 500\frac{1}{a_{\overline{6}|0.01}} \qquad \text{Use Table V or a calculator}$$

$$= 500(0.172\ 548)$$

$$= \$86.27 \text{ per month}$$

At the end of the first month, the interest due is

$500(0.01) = $5.00

The amortization payment is divided into two parts, payment of the interest due and reduction of the unpaid balance (repayment of principal):

Monthly payment		Interest due		Unpaid balance reduction
$86.27	=	$5.00	+	$81.27

The unpaid balance for the next month is

Previous unpaid balance		Unpaid balance reduction		New unpaid balance
$500.00	−	$81.27	=	$418.73

At the end of the second month, the interest due on the unpaid balance of $418.73 is

$418.73(0.01) = $4.19

The monthly payment is divided into

$86.27 = $4.19 + $82.08

and the unpaid balance for the next month is

$418.73 − $82.08 = $366.65

This process continues until all payments have been made and the unpaid balance is reduced to zero. The calculations for each month are listed in Table 1, which is referred to as an **amortization schedule.**

Table 1 Amortization Schedule

Payment Number	Payment	Interest	Unpaid Balance Reduction	Unpaid Balance
0				$500.00
1	$ 86.27	$ 5.00	$ 81.27	418.73
2	86.27	4.19	82.08	336.65
3	86.27	3.37	82.90	253.75
4	86.27	2.54	83.73	170.02
5	86.27	1.70	84.57	85.45
6	86.30	0.85	85.45	0.00
Total	$517.65	$17.65	$500.00	

Notice that the last payment had to be increased by $0.03 in order to reduce the unpaid balance to zero. This small discrepancy is due to round-off errors that occur in the computations. In almost all cases, the last payment must be adjusted slightly in order to obtain a final unpaid balance of exactly zero.

Problem 17 Construct the amortization schedule for a $1,000 debt that is to be amortized in 6 equal monthly payments at 1.25% interest per month on the unpaid balance.

Example 18 When a family bought their home, they borrowed $25,000 at 9% compounded monthly, which was to be amortized over 30 years in equal monthly payments. Twenty years later they decided to sell the house and pay off the loan in one lump sum. Find the monthly payment and the unpaid balance after making monthly payments for 20 years.

Solution Using formula (6) with $P = \$25,000$, $i = 0.09/12 = 0.0075$ and $n = 30(12) = 360$, the monthly payment is

$$R = P\frac{i}{1 - (1 + i)^{-n}}$$

$$= 25{,}000\frac{0.0075}{1 - (1.0075)^{-360}} \qquad \text{Use a calculator}$$

$$= \$201.16 \text{ per month}$$

How can we find the outstanding balance after 20 years or $20(12) = 240$ monthly payments? One way to proceed would be to construct an amortization schedule, but this would require a table with 240 lines. Fortunately, there is an easier way. The unpaid balance after 240 payments is the amount of a loan that can be paid off with the remaining 120 payments of $201.16. Since the bank views a loan as an annuity that they bought from you, **the unpaid balance of a loan with n remaining payments is the present value of that annuity and can be computed by using formula (5).** Substituting $R = \$201.16$, $i = 0.0075$, and $n = 120$ in (5), the unpaid balance after 240 payments have been made is

$$P = R\frac{1 - (1 + i)^{-n}}{i}$$

$$= \$201.16\frac{1 - (1.0075)^{-120}}{0.0075} \qquad \text{Use a calculator}$$

$$= \$15{,}879.91$$

Problem 18 In Example 18, what was the unpaid balance after making payments for 5 years?

The answer to Example 18 may seem a surprisingly large amount to owe after having made payments for 20 years, but long-term amortizations start out with very small reductions in the unpaid balance. For example, the interest due at the end of the very first period of the loan in Example 18 was

25,000(0.0075) = 187.50

The first monthly payment was divided into

Monthly payment	Interest due	Unpaid balance reduction
$201.16	= $187.50 +	$13.66

Thus, only $13.66 was applied to the unpaid balance.

Answers to Matched Problems

15. $13,577.71   16. R = $112.98 per month; total interest = $311.52

17.

Payment Number	Payment	Interest	Unpaid Balance Reduction	Unpaid Balance
0				$1,000.00
1	$ 174.03	$12.50	$ 161.53	838.47
2	174.03	10.48	163.55	674.92
3	174.03	8.44	165.59	509.33
4	174.03	6.37	167.66	341.67
5	174.03	4.27	169.76	171.91
6	174.06	2.15	171.91	0.00
Total	$1,044.21	$44.21	$1,000.00	

18. $23,970.55

Exercise 6-4

Use formula (5) or (6) and Table V or a calculator (or both) to solve each problem. (Answers may vary slightly depending on whether you use a calculator or Table V.)

A

1. $P = ?$, $n = 30$, $i = 0.04$, $R = \$200$
2. $P = ?$, $n = 40$, $i = 0.01$, $R = \$400$
3. $P = ?$, $n = 25$, $i = 0.025$, $R = \$250$
4. $P = ?$, $n = 60$, $i = 0.0075$, $R = \$500$

B

5. $P = \$6,000$, $n = 36$, $i = 0.01$, $R = ?$
6. $P = \$1,200$, $n = 40$, $i = 0.025$, $R = ?$
7. $P = \$40,000$, $n = 96$, $i = 0.0075$, $R = ?$
8. $P = \$14,000$, $n = 72$, $i = 0.005$, $R = ?$
9. $P = \$5,000$, $i = 0.01$, $R = \$200$, $n = ?$
10. $P = \$20,000$, $i = 0.0175$, $R = \$500$, $n = ?$

Applications

Business & Economics

11. A relative wills you an annuity paying $4,000 per quarter for the next 10 years. If money is worth 8% compounded quarterly, what is the present value of this annuity?

12. How much should you deposit in an account paying 12% compounded monthly in order to receive $1,000 per month for the next 2 years?

13. Parents of a college student wish to set up an annuity that will pay $350 per month to the student for 4 years. How much should they deposit now at 9% interest compounded monthly to establish this annuity? How much will the student receive in the 4 years?

14. A person pays $120 per month for 48 months for a car, making no down payment. If the loan costs 1.5% interest per month on the unpaid balance, what was the original cost of the car? How much total interest will be paid?

15. (A) If you buy a stereo set for $600 and agree to pay for it in 18 equal installments at 1% interest per month on the unpaid balance, how much are your monthly payments? How much interest will you pay?

   (B) Repeat part A for 1.5% interest per month on the unpaid balance.

16. (A) A company buys a large copy machine for $12,000 and finances it at 12% interest compounded monthly. If the loan is to be amortized in 6 years in equal monthly payments, how much is each payment? How much interest will be paid?

   (B) Repeat part A with 18% interest compounded monthly.

17. A sailboat costs $16,000. You pay 25% down and amortize the rest with equal monthly payments over a 6 year period. If you must pay 1.5% interest per month on the unpaid balance (18% compounded monthly), what is your monthly payment? How much interest will you pay over the 6 years?

18. A law firm buys a computerized word-processing system costing $10,000. If it pays 20% down and amortizes the rest with equal monthly payments over 5 years at 9% compounded monthly, what will be the monthly payment? How much interest will the firm pay?

19. Construct the amortization schedule for a $5,000 debt that is to be amortized in 8 equal quarterly payments at 4.5% interest per quarter on the unpaid balance.

20. Construct the amortization schedule for a $10,000 debt that is to be amortized in 6 equal quarterly payments at 3.5% interest per quarter on the unpaid balance.

21. A person borrows $6,000 at 12% compounded monthly, which is to be amortized over 3 years in equal monthly payments. For tax purposes, he needs to know the amount of interest paid during each year of the loan. Find the interest paid during the first year, the second year, and the third year of the loan. [*Hint:* Find the unpaid balance after 12 payments and after 24 payments.]

22. A person establishes an annuity for retirement by depositing $50,000 into an account that pays 9% compounded monthly. Equal monthly withdrawals will be made each month for 5 years, at which time the account will have a zero balance. Each year taxes must be paid on the interest earned by the account during that year. How much interest was earned during the first year? [*Hint:* The amount in the account at the end of the first year is the present value of a 4 year annuity.]

Use a financial or scientific calculator to solve each of the following problems.

23. Some friends tell you that they paid $25,000 down on a new house and are to pay $525 per month for 30 years. If interest is 9.8% compounded monthly, what was the selling price of the house? How much interest will they pay in 30 years?

24. A family is thinking about buying a new house costing $120,000. They must pay 20% down, and the rest is to be amortized over 30 years in equal monthly payments. If money costs 9.6% compounded monthly, what will their monthly payment be? How much total interest will be paid over the 30 years?

25. A student receives a federally backed student loan of $6,000 at 3.5% interest compounded monthly. After finishing college in 2 years, the student must amortize the loan in the next 4 years by making equal monthly payments. What will the payments be and what total interest will the student pay? [*Hint:* This is a two-part problem. First find the amount of the debt at the end of the first 2 years; then amortize this amount over the next 4 years.]

26. A person establishes a sinking fund for retirement by contributing $7,500 per year at the end of each year for 20 years. For the next 20 years, equal yearly payments are withdrawn, at the end of which time the account will have a zero balance. If money is worth 9% compounded annually, what yearly payments will the person receive for the last 20 years?

27. A family has a $75,000, 30 year mortgage at 13.2% compounded monthly. Find the monthly payment. Also find the unpaid balance after

   (A) 10 years   (B) 20 years   (C) 25 years

28. A family has a $50,000, 20 year mortgage of 10.8% compounded monthly. Find the monthly payment. Also find the unpaid balance after

   (A) 5 years   (B) 10 years   (C) 15 years

29. A family has a $30,000, 20 year mortgage at 15% compounded monthly.

   (A) Find the monthly payment and the total interest paid.
   (B) Suppose the family decides to add an extra $100 to its mortgage payment each month starting with the very first payment. How long will it take the family to pay off the mortgage? How much interest will the family save?

30. At the time they retire, a couple has $200,000 in an account that pays 8.4% compounded monthly.

   (A) If they decide to withdraw equal monthly payments for 10 years, at the end of which time the account will have a zero balance, how much should they withdraw each month?
   (B) If they decide to withdraw $3,000 a month until the balance in the account is zero, how many withdrawals can they make?

6-5 Chapter Review

Important Terms and Symbols

6-1 *Simple interest and simple discount.* principal, interest, interest rate, simple interest, face value, present value, future value, simple interest note, simple discount note, discount, proceeds, maturity value, $I = Prt$, $A = P(1 + rt)$, $D = Mdt$, $P = M - D$, $P = M(1 - dt)$

6-2 *Compound interest.* compound interest, rate per compound period, nominal rate, effective rate (or annual yield), doubling time, $A = P(1 + i)^n$, $i = r/m$

6-3 *Future value of an annuity; sinking funds.* annuity, ordinary annuity, future value, sinking fund,

$$S = R\frac{(1 + i)^n - 1}{i} = Rs_{\overline{n}|i} \qquad \text{(future value)}$$

$$R = S\frac{i}{(1 + i)^n - 1} = \frac{S}{s_{\overline{n}|i}} \qquad \text{(sinking fund)}$$

6-4 *Present value of an annuity; amortization.* present value, amortizing a debt, amortization schedule,

$$P = R\frac{1 - (1 + i)^{-n}}{i} = Ra_{\overline{n}|i} \qquad \text{(present value)}$$

$$R = P\frac{i}{1 - (1 + i)^{-n}} = P\frac{1}{a_{\overline{n}|i}} \qquad \text{(amortization)}$$

Exercise 6-5   Chapter Review

Work through all the problems in this chapter review and check your answers in the back of the book. (Answers to all review problems are there.) Where weaknesses show up, review appropriate sections in the text. When you are satisfied that you know the material, take the practice test following this review.

Solve each problem using Table V or a calculator (or both).

A   Find the indicated quantity, given $A = P(1 + rt)$.

1. $A = ?$, $P = \$100$, $r = 9\%$, $t = 6$ months
2. $A = \$808$, $P = ?$, $r = 12\%$, $t = 1$ month
3. $A = \$212$, $P = \$200$, $r = 8\%$, $t = ?$
4. $A = \$4{,}120$, $P = \$4{,}000$, $r = ?$, $t = 6$ months

Find the indicated quantity, given $P = M(1 - dt)$.

5. $P = ?$, $M = \$5{,}000$, $d = 18\%$, $t = 10$ months
6. $P = \$4{,}000$, $M = ?$, $d = 15\%$, $t = 8$ months
7. $M = \$6{,}000$, $P = \$5{,}100$, $d = ?$, $t = 15$ months
8. $M = \$1{,}200$, $P = \$1{,}080$, $d = 10\%$, $t = ?$

B   Find the indicated quantity, given $A = P(1 + i)^n$ and $P = A/(1 + i)^n$.

9. $A = ?$, $P = \$1{,}200$, $i = 0.005$, $n = 30$
10. $A = \$5{,}000$, $P = ?$, $i = 0.0075$, $n = 60$

Find the indicated quantity, given

$$S = R\frac{(1 + i)^n - 1}{i} = Rs_{\overline{n}|i} \quad \text{and} \quad R = S\frac{i}{(1 + i)^n - 1} = \frac{S}{s_{\overline{n}|i}}$$

11. $S = ?$, $R = \$1{,}000$, $i = 0.005$, $n = 60$
12. $S = \$8{,}000$, $R = ?$, $i = 0.015$, $n = 48$

Find the indicated quantity, given

$$P = R\frac{1 - (1 + i)^{-n}}{i} = Ra_{\overline{n}|i} \quad \text{and} \quad R = P\frac{i}{1 - (1 + i)^{-n}} = P\frac{1}{a_{\overline{n}|i}}$$

13. $P = ?$, $R = \$2{,}500$, $i = 0.02$, $n = 16$
14. $P = \$8{,}000$, $R = ?$, $i = 0.0075$, $n = 60$

C   Use Table V or a calculator (or both) to solve for n to the nearest integer.

15. $2{,}500 = 1{,}000(1.06)^n$ 16. $5{,}000 = 100\dfrac{(1.01)^n - 1}{0.01} = 100s_{\overline{n}|0.01}$

Applications

Business & Economics

17. If you borrow $3,000 at 14% simple interest for 10 months, how much will you owe in 10 months? How much interest will you pay?

18. If you borrow $3,000 at 14% discount for 10 months, how much will you receive? How much will you owe when the debt comes due? How much will the loan cost you?

19. How much should you deposit in an account paying 8% compounded annually to have $20,000 in 20 years?

20. If $5,000 is invested at 10% compounded quarterly, what is the amount after 6 years?

21. What is the value of an annuity in 8 years if $100 per month is deposited into an account earning 6% compounded monthly?

22. Suppose you buy a stereo system costing $900. If you pay 25% down and amortize the rest in 24 monthly payments at 1.5% interest per month on the unpaid balance, how much is each payment and how much total interest will you pay?

23. How much should you pay for an annuity that pays $1,000 per quarter for 10 years if money is worth 8% compounded quarterly?

24. A company decides to establish a sinking fund to replace a piece of equipment in 6 years at an estimated cost of $50,000. To accomplish this, they decide to make fixed monthly payments into an account that pays 9% compounded monthly. How much should each payment be?

25. A savings and loan company pays 9% compounded monthly. What is the effective rate?

26. You hold a $5,000, 9 month note at 10% simple interest. You decide to sell it to another investor at 12% discount 5 months before it is due. How much will you receive for the note? How much will the other investor receive in 5 months?

27. How long (to the nearest month) will it take money to double if it is invested at 12% compounded monthly?

28. Construct the amortization schedule for a $1,000 debt that is to be amortized in 4 equal quarterly payments at 2.5% interest per quarter on the unpaid balance.

29. A car dealer offers to sell you a car for $500 down and $200 a month for 36 months. As required by law, he informs you that the effective rate of interest is 16%.

(A) What nominal rate of interest are you paying?
(B) What is the original cost of the car?

30. A business borrows $80,000 at 15% interest compounded monthly for 8 years.

(A) What is the monthly payment?

(B) What is the unpaid balance at the end of the first year?

(C) How much interest was paid during the first year?

31. An individual wants to establish an annuity for retirement purposes. He wants to make quarterly deposits for 20 years so that he can then make quarterly withdrawals of $5,000 for 10 years. The annuity earns 12% interest compounded quarterly.

(A) How much will have to be in the account at the time he retires?

(B) How much should be deposited each quarter for 20 years in order to accumulate the required amount?

(C) What is the total amount of interest earned during the 30 year period?

32. In order to save enough money for the down payment on a home, a young couple deposits $200 each month into an account that pays 9% interest compounded monthly. If they want $10,000 for a down payment, how many deposits will they have to make?

Practice Test: Chapter 6

Solve each problem using Table V or a calculator (or both).

1. How much should you deposit initially in an account paying 10% compounded semiannually in order to have $25,000 in 10 years?

2. A company decides to establish a sinking fund to replace a piece of equipment in 6 years at an estimated cost of $15,000. If they decide to make quarterly payments into an account paying 10% compounded every 3 months, how much should each payment be?

3. What is the value of an annuity in 5 years if $200 per month is deposited into an account paying 9% interest compounded monthly?

4. You decide to purchase a car costing $8,000 by paying 20% down and amortizing the rest in 4 years at 1.5% per month interest on the unpaid balance by making equal monthly payments. How much is each payment and what is your total interest?

5. How much should you pay for an annuity that pays $3,000 per quarter for 20 years if money is worth 8% compounded quarterly?

6. You hold a $10,000, 10 month note at 9% simple interest. If you sell it to another investor at 10% discount 3 months before it is due, how much will you receive? How much will the other investor receive in 3 months?

7. A savings and loan company pays 8% compounded quarterly. What is the effective rate?

8. How long will it take money to double if it is invested at 8% compounded quarterly?

9. Each quarter a couple deposits $500 into an account that pays 8% interest compounded quarterly. How long will it take them to save $10,000?

10. Two years ago you borrowed $10,000 at 12% interest compounded monthly which was to be amortized over 5 years. Now you have acquired some additional funds and decide that you want to pay off this loan. What is the unpaid balance after making payments for 2 years?

Systems of Linear Equations; Matrices

CHAPTER 7	Contents

In this chapter we will first review how systems of equations are solved by using techniques learned in elementary algebra. These techniques are suitable for systems involving two or three variables, but they are not suitable for systems involving larger numbers of variables. After this review, we will introduce techniques that are more suitable for solving systems with larger numbers of variables. These new techniques form the basis for computer solutions of large-scale systems.

7-1 Review: Systems of Linear Equations

- Systems in Two Variables
- Applications
- Systems in Three Variables
- Applications

Systems in Two Variables

To establish basic concepts, consider the following simple example: If two adult tickets and one child ticket cost $8, and if one adult ticket and three child tickets cost $9, what is the price of each?

Let x = Price of adult ticket

y = Price of child ticket

Then $2x + y = 8$

$x + 3y = 9$

We now have a system of two linear equations and two unknowns. To solve this system, we find all ordered pairs of real numbers that satisfy both equations at the same time. In general, we are interested in solving linear systems of the type

$$ax + by = h$$
$$cx + dy = k$$

where a, b, c, d, h, and k are real constants. A pair of numbers $x = x_0$ and $y = y_0$ [also written as an ordered pair (x_0, y_0)] is a **solution** of this system if each equation is satisfied by the pair. The set of all such ordered pairs of numbers is called the **solution set** for the system. To **solve** a system is to find its solution set. We will consider three methods of solving such systems, each having certain advantages in certain situations.

Solution by Graphing   To solve the ticket problem above by graphing, we graph both equations in the same coordinate system. The coordinates of any points that the graphs have in common must be solutions to the system, since they must satisfy both equations.

Example 1   Solve the ticket problem by graphing:

$$2x + \ y = 8$$
$$x + 3y = 9$$

Solution

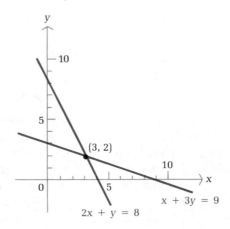

$$x = \$3 \qquad \text{Adult ticket}$$
$$y = \$2 \qquad \text{Child ticket}$$

Check

$2x + y = 8$	$x + 3y = 9$
$2(3) + 2 \overset{?}{=} 8$	$3 + 3(2) \overset{?}{=} 9$
$8 \overset{\checkmark}{=} 8$	$9 \overset{\checkmark}{=} 9$

Problem 1   Solve by graphing and check:

$$2x - \ y = -3$$
$$x + 2y = -4$$

It is clear that the above example (and problem) has exactly one solution, since the lines have exactly one point of intersection. In general, lines in a rectangular coordinate system are related to each other in one of the three ways illustrated in the next example.

Example 2 Solve each of the following systems by graphing:

(A) $x - 2y = 2$ (B) $x + 2y = -4$ (C) $2x + 4y = 8$
     $x + y = 5$ $2x + 4y = 8$ $x + 2y = 4$

Solutions

(A)

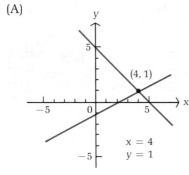

Intersection at one point only — exactly one solution

(B)

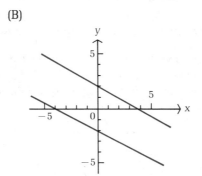

Lines are parallel (each has slope $-\frac{1}{2}$) — no solutions

(C)

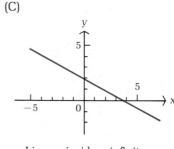

Lines coincide — infinite number of solutions

Problem 2 Solve each of the following systems by graphing:

(A) $x + y = 4$ (B) $6x - 3y = 9$ (C) $2x - y = 4$
     $2x - y = 2$ $2x - y = 3$ $6x - 3y = -18$

By geometrically interpreting a system of two linear equations in two unknowns, we gain useful information about solutions to the system. Since two lines in a coordinate system must intersect at exactly one point, be parallel, or coincide, we conclude that the system has (1) exactly one solution, (2) no solution, or (3) infinitely many solutions. In addition, graphs of problems frequently reveal relationships that might otherwise be hidden. Generally, however, graphic methods only give us rough approximations of solutions. The methods of substitution and elimination by addition yield results to any decimal accuracy desired — assuming that solutions exist.

Solution by Substitution

Choose one of two equations in a system and solve for one variable in terms of the other. (Make a choice that avoids fractions, if possible.) Then substitute the result into the other equation and solve the resulting linear equation in one variable. Now substitute this result back into either of the original equations to find the second variable. An example should make the process clear.

Example 3   Solve by substitution:

$$5x + y = 4$$
$$2x - 3y = 5$$

Solution   Solve either equation for one variable in terms of the other; then substitute into the remaining equation. In this problem we can avoid fractions by choosing the first equation and solving for y in terms of x.

$$5x + y = 4$$ Solve first equation for y in terms of x

$$y = \underline{4 - 5x}$$ Substitute into second equation

$$2x - 3y = 5$$ Second equation

$$2x - 3(4 - 5x) = 5$$ Solve for x

$$2x - 12 + 15x = 5$$

$$17x = 17$$

$$x = 1$$

Now, replace x with 1 in $y = 4 - 5x$ to find y:

$$y = 4 - 5x$$
$$y = 4 - 5(1)$$
$$y = -1$$

Check

$5x + y = 4$	$2x - 3y = 5$
$5(1) + (-1) \overset{?}{=} 4$	$2(1) - 3(-1) \overset{?}{=} 5$
$4 \overset{\checkmark}{=} 4$	$5 \overset{\checkmark}{=} 5$

Problem 3   Solve by substitution:

$$3x + 2y = -2$$
$$2x - y = -6$$

Solution by Elimination by Addition   Now we turn to **elimination by addition.** This is probably the most important method of solution, since it is readily generalized to higher-order systems. The method involves replacing systems of equations with simpler *equivalent systems* (by performing appropriate operations) until we obtain a system with an obvious solution. **Equivalent systems** of equations are, as you would expect, systems that have exactly the same solution set. Theorem 1 lists the operations that produce equivalent systems.

Theorem 1

A system of linear equations is transformed into an equivalent system if:

(A) Two equations are interchanged.
(B) An equation is multiplied by a nonzero constant.
(C) A constant multiple of another equation is added to a given equation.

Parts B and C of Theorem 1 will be of most use to us now; part A becomes useful when we generalize the theorem for larger systems. The use of the theorem is best illustrated by examples.

Example 4   Solve the following system using elimination by addition:

$$3x - 2y = 8$$
$$2x + 5y = -1$$

Solution   We use the theorem to eliminate one of the variables, thus obtaining a system with an obvious solution:

$$
\begin{array}{rl}
3x - 2y = 8 \\
2x + 5y = -1 \\
15x - 10y = 40 \\
\underline{4x + 10y = -2} \\
19x = 38 \\
x = 2
\end{array}
$$

If we multiply the top equation by 5 and the bottom by 2 and then add, we can eliminate y

Now substitute $x = 2$ back into either of the original equations, say the second equation, and solve for y ($x = 2$ paired with either of the two original equations produces an equivalent system):

$$2(2) + 5y = -1$$
$$5y = -5$$
$$y = -1$$

Check $\begin{aligned} 3x - 2y &= 8 \\ 3(2) - 2(-1) &\overset{?}{=} 8 \\ 8 &\overset{\checkmark}{=} 8 \end{aligned}$ $\begin{aligned} 2x + 5y &= -1 \\ 2(2) + 5(-1) &\overset{?}{=} -1 \\ -1 &\overset{\checkmark}{=} -1 \end{aligned}$

Problem 4   Solve the system:

$$5x - 2y = 12$$
$$2x + 3y = 1$$

Let us see what happens in the elimination process when a system has either no solution or infinitely many solutions. Consider the following system:

$$2x + 6y = -3$$
$$x + 3y = 2$$

Multiplying the second equation by -2 and adding, we obtain

$$2x + 6y = -3$$
$$\underline{-2x - 6y = -4}$$
$$0 = -7$$

We have obtained a contradiction. The assumption that the original system has solutions must be false (otherwise we have proved that $0 = -7$). Thus, the system has no solutions. The graphs of the equations are parallel. Systems with no solutions are said to be **inconsistent.**

Now consider the system

$$x - \tfrac{1}{2}y = 4$$
$$-2x + y = -8$$

If we multiply the top equation by 2 and add the result to the bottom equation, we obtain

$$2x - y = 8$$
$$\underline{-2x + y = -8}$$
$$0 = 0$$

Obtaining $0 = 0$ by addition implies that the equations are equivalent; that is, their graphs coincide. Hence, the two equations have the same solution set, and the system has infinitely many solutions. If $x = k$, then using either equation, we obtain $y = 2k - 8$; that is, $(k, 2k - 8)$ is a solution for any real number k. Such a system is said to be **dependent.** The variable k is called a **parameter;** replacing it with any real number produces a particular solution to the system.

■ Applications

Many real-world problems are readily solved by applying two-equation–two-unknown methods. We shall discuss two applications in detail.

Example 5

Diet

A dietitian in a hospital is to arrange a special diet comprised of two foods, M and N. Each ounce of food M contains 8 units of calcium and 2 units of iron. Each ounce of food N contains 5 units of calcium and 4 units of iron. How many ounces of foods M and N should be used to obtain a food mix that contains 74 units of calcium and 35 units of iron?

Solution　It is convenient to first summarize the quantities involved in a table:

	Food *M*	Food *N*	Total Needed
Calcium	8	5	74
Iron	2	4	35

Let　x = Number of ounces of food *M*

y = Number of ounces of food *N*

$$\begin{pmatrix} \text{Calcium in} \\ x \text{ oz of food } M \end{pmatrix} + \begin{pmatrix} \text{Calcium in} \\ y \text{ oz of food } N \end{pmatrix} = \begin{pmatrix} \text{Total calcium} \\ \text{needed} \end{pmatrix}$$

$$\begin{pmatrix} \text{Iron in } x \text{ oz} \\ \text{of food } M \end{pmatrix} + \begin{pmatrix} \text{Iron in } y \text{ oz} \\ \text{of food } N \end{pmatrix} = \begin{pmatrix} \text{Total iron} \\ \text{needed} \end{pmatrix}$$

$$8x \quad + \quad 5y \quad = \quad 74$$
$$2x \quad + \quad 4y \quad = \quad 35$$

Solve by elimination by addition:

$$8x + 5y = 74$$
$$\underline{-8x - 16y = -140}$$
$$-11y = -66$$
$$y = 6 \text{ oz of food } N$$

$$2x + 4(6) = 35$$
$$2x = 11$$
$$x = 5.5 \text{ oz of food } M$$

Check

$8x + 5y = 74$	$2x + 4y = 35$
$8(5.5) + 5(6) \stackrel{?}{=} 74$	$2(5.5) + 4(6) \stackrel{?}{=} 35$
$74 \stackrel{\checkmark}{=} 74$	$35 \stackrel{\checkmark}{=} 35$

Problem 5　Repeat Example 5 given that each ounce of food *M* contains 10 units of calcium and 4 units of iron, each ounce of food *N* contains 6 units of calcium and 4 units of iron, and the mix of *M* and *N* must contain 92 units of calcium and 44 units of iron.

Example 6

Supply and Demand　The quantity of a product that people are willing to buy during some period of time depends on its price. Generally, the higher the price, the less the demand; the lower the price, the greater the demand. Similarly, the quantity of a product that a supplier is willing to sell during some period of time also depends on the price. Generally, a supplier will be willing to supply more of a product at higher prices and less of a product at lower prices. The simplest supply and demand model is a linear model where the graphs of a demand equation and a supply equation are straight lines.

Suppose in a given city on a given day supply and demand equations for cherries are given by

$$p = -0.2q + 4 \qquad \text{Demand equation (consumer)}$$
$$p = 0.07q + 0.76 \qquad \text{Supply equation (supplier)}$$

where q represents the quantity in thousands of pounds and p represents the price in dollars. For example, we see that consumers will purchase 10 thousand pounds ($q = 10$) when the price is $p = -0.2(10) + 4 = \$2$ per pound. On the other hand, suppliers will be willing to supply 17.714 thousand pounds of cherries at \$2 per pound (solve $2 = 0.07q + 0.76$). Thus, at \$2 per pound the suppliers are willing to supply more cherries than consumers are willing to purchase. The supply exceeds the demand at that price and the price will come down. At what price will cherries stabilize for the day? That is, at what price will supply equal demand? This price, if it exists, is called the **equilibrium price,** and the quantity sold at that price is called the **equilibrium quantity.** How do we find these quantities? We solve the linear system

$$p = -0.2q + 4 \qquad \text{Demand equation}$$
$$p = 0.07q + 0.76 \qquad \text{Supply equation}$$

We solve this system using substitution (substituting $p = -0.2q + 4$ into the second equation).

$$-0.2q + 4 = 0.07q + 0.76$$
$$-0.27q = -3.24$$
$$\boldsymbol{q = 12 \textbf{ thousand pounds (equilibrium quantity)}}$$

Now substitute $q = 12$ back into either of the original equations in the system and solve for p (we choose the first equation):

$$p = -0.2(12) + 4$$

$\boldsymbol{p = \$1.60}$ **per pound (equilibrium price)**

These results are interpreted geometrically in the figure.

   Equilibrium quantity = 12 thousand pounds

   Equilibrium price = \$1.60 per pound

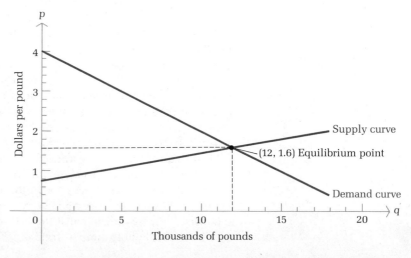

If the price was above the equilibrium price of $1.60 per pound, the supply would exceed the demand and the price would come down. If the price was below the equilibrium price of $1.60 per pound, the demand would exceed the supply and the price would rise. Thus, the price would reach equilibrium at $1.60. At this price, suppliers would supply 12 thousand pounds of cherries and consumers would purchase 12 thousand pounds.

Problem 6   Repeat Example 6 (including drawing the graph) given:

$$p = -0.1q + 3 \qquad \text{Demand equation}$$
$$p = \ \ 0.08q + 0.66 \qquad \text{Supply equation}$$

■ Systems in Three Variables

Any equation that can be written in the form

$$ax + by = c$$

where a, b, and c are constants (not both a and b zero) is called a **linear equation in two variables.** Similarly, any equation that can be written in the form

$$ax + by + cz = k$$

where a, b, c, and k are constants (not all a, b, and c zero) is called a **linear equation in three variables.** (A similar definition holds for a linear equation in four or more variables.)

Now that we know how to solve systems of linear equations in two variables, there is no reason to stop there. Systems of the form

$$
\begin{aligned}
a_1x + b_1y + c_1z &= k_1 \\
a_2x + b_2y + c_2z &= k_2 \\
a_3x + b_3y + c_3z &= k_3
\end{aligned}
\qquad (1)
$$

as well as higher-order systems are encountered frequently. In fact, systems of equations are so important in solving real-world problems that whole courses are devoted to this one topic. A triplet of numbers $x = x_0$, $y = y_0$, and $z = z_0$ [also written as an ordered triplet (x_0, y_0, z_0)] is a **solution** of system (1) if each equation is satisfied by this triplet. The set of all such ordered triplets of numbers is called the **solution set** of the system. If operations are performed on a system and the new system has the same solution set as the original, then both systems are said to be **equivalent.**

Linear equations in three variables represent planes in a three-dimensional space. Trying to visualize how three planes can intersect will give you insight as to what kind of solution sets are possible for system (1). Figure 1 shows several of the many ways in which three planes can intersect. It can be shown that system (1) will have exactly one solution, no solutions, or infinitely many solutions.

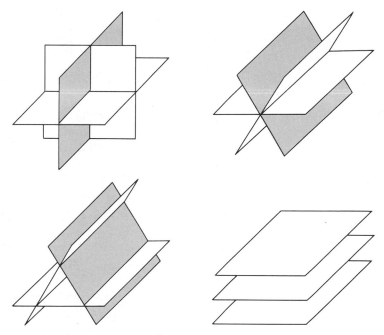

Figure 1 Three intersecting planes

In this section we will use an extension of the method of elimination discussed above to solve systems in the form of (1). In the next section we will consider techniques for solving linear systems that are more compatible with solving such systems with computers. In practice, most linear systems involving more than three variables are solved with the aid of a computer.

Steps in Solving Systems of Form (1)

1. Choose two equations from the system and eliminate one of the three variables using elimination by addition. The result is generally one equation in two unknowns.
2. Now eliminate the same variable from the unused equation and one of those used in step 1. We (generally) obtain another equation in two variables.
3. The two equations from steps 1 and 2 form a system of two equations and two unknowns. Solve as described in the earlier part of this section.
4. Substitute the solution from step 3 into any of the three original equations and solve for the third variable to complete the solution of the original system.

Example 7   Solve:

$$3x - 2y + 4z = \;\;6 \tag{2}$$
$$2x + 3y - 5z = -8 \tag{3}$$
$$5x - 4y + 3z = \;\;7 \tag{4}$$

Solution   **Step 1.**   We look at the coefficients of the variables and choose to eliminate y from equations (2) and (4) because of the convenient coefficients -2 and -4. Multiply equation (2) by -2 and add to equation (4):

$$\begin{aligned}
-6x + 4y - 8z &= -12 &&-2[\text{equation (2)}]\\
\underline{5x - 4y + 3z} &= \underline{\;\;7} &&\text{Equation (4)}\\
-x \qquad\;\; -5z &= \;-5 &&
\end{aligned} \tag{5}$$

Step 2.   Now we eliminate y (the same variable) from equations (2) and (3):

$$\begin{aligned}
9x - 6y + 12z &= \;\;18 &&3[\text{equation (2)}]\\
\underline{4x + 6y - 10z} &= \underline{-16} &&2[\text{equation (3)}]\\
13x \qquad\quad + 2z &= \;\;2 &&
\end{aligned} \tag{6}$$

Step 3.   From steps 1 and 2 we obtain the system

$$-x - 5z = -5 \tag{5}$$
$$13x + 2z = \;\;2 \tag{6}$$

[It has been shown that equations (5) and (6) along with (2), (3), or (4) form a system equivalent to the original system.] We solve system (5) and (6) as in the earlier part of this section:

$$\begin{aligned}
-13x - 65z &= -65 &&13[\text{equation (5)}]\\
\underline{13x + \;\;2z} &= \underline{\;\;2} &&\text{Equation (6)}\\
-63z &= -63 &&\\
z &= \;\;1 &&
\end{aligned}$$

Substitute $z = 1$ back into either equation (5) or (6) [we choose equation (5)] to find x:

$$\begin{aligned}
-x - \;\;5z &= -5 \tag{5}\\
-x - 5(1) &= -5\\
-x &= \;\;0\\
x &= \;\;0
\end{aligned}$$

Step 4.   Substitute $x = 0$ and $z = 1$ back into any of the three original equations [we choose equation (2)] to find y:

$$3x - 2y + 4z = 6 \qquad\qquad (2)$$
$$3(0) - 2y + 4(1) = 6$$
$$-2y + 4 = 6$$
$$-2y = 2$$
$$y = -1$$

Thus, the solution to the original system is $(0, -1, 1)$ or $x = 0$, $y = -1$, $z = 1$.

Check   To check the solution, we must check *each* equation in the original system:

$$3x - 2y + 4z = 6 \qquad\qquad 2x + 3y - 5z = -8$$
$$3(0) - 2(-1) + 4(1) \stackrel{?}{=} 6 \qquad 2(0) + 3(-1) - 5(1) \stackrel{?}{=} -8$$
$$6 \stackrel{\checkmark}{=} 6 \qquad\qquad\qquad -8 \stackrel{\checkmark}{=} -8$$

$$5x - 4y + 3z = 7$$
$$5(0) - 4(-1) + 3(1) \stackrel{?}{=} 7$$
$$7 \stackrel{\checkmark}{=} 7$$

Problem 7   Solve:

$$2x + 3y - 5z = -12$$
$$3x - 2y + 2z = 1$$
$$4x - 5y - 4z = -12$$

In the process described above, if we encounter an equation that states a contradiction, such as $0 = -2$, then we must conclude that the system has no solution (that is, the system is inconsistent). On the other hand, if one of the equations turns out to be $0 = 0$, the system has either infinitely many solutions or none. We must proceed further to determine which. Notice how this last result differs from the two-equation–two-unknown case. There, when we obtained $0 = 0$, we *knew* that there were infinitely many solutions. We shall have more to say about this in Section 7-3.

■ Applications

Now let us consider a real-world problem that leads to a system of three equations and three unknowns.

Example 8
Production Scheduling

A garment industry manufactures three shirt styles. Each style shirt requires the services of three departments as listed in the table on the next page. The cutting, sewing, and packaging departments have available a maximum of 1,160, 1,560, and 480 labor-hours per week, respectively. How many of each style shirt must be produced each week for the plant to operate at full capacity?

	Style *A*	Style *B*	Style *C*	Time Available
Cutting department	0.2 hr	0.4 hr	0.3 hr	1,160 hr
Sewing department	0.3 hr	0.5 hr	0.4 hr	1,560 hr
Packaging department	0.1 hr	0.2 hr	0.1 hr	480 hr

Solution   Let x = Number of style A produced per week

y = Number of style B produced per week

z = Number of style C produced per week

Then $0.2x + 0.4y + 0.3z = 1,160$ Cutting department

$0.3x + 0.5y + 0.4z = 1,560$ Sewing department

$0.1x + 0.2y + 0.1z = 480$ Packaging department

We can clear the system of decimals, if desired, by multiplying each side of each equation by 10. Thus,

$$2x + 4y + 3z = 11,600 \tag{7}$$

$$3x + 5y + 4z = 15,600 \tag{8}$$

$$x + 2y + z = 4,800 \tag{9}$$

Let us start by eliminating z from equations (7) and (9):

$$
\begin{array}{ll}
2x + 4y + 3z = 11,600 & \text{Equation (7)} \\
\underline{-3x - 6y - 3z = -14,400} & -3[\text{equation (9)}] \\
-x - 2y = -2,800 & \hspace{3em}(10)
\end{array}
$$

We now eliminate z from equations (8) and (9):

$$
\begin{array}{ll}
3x + 5y + 4z = 15,600 & \text{Equation (8)} \\
\underline{-4x - 8y - 4z = -19,200} & -4[\text{equation (9)}] \\
-x - 3y = -3,600 & \hspace{3em}(11)
\end{array}
$$

Equations (10) and (11) form a system of two equations and two unknowns:

$$-x - 2y = -2,800 \tag{10}$$

$$-x - 3y = -3,600 \tag{11}$$

We solve as in the earlier part of this section:

$$
\begin{array}{ll}
-x - 2y = -2,800 & \text{Equation (10)} \\
\underline{x + 3y = 3,600} & (-1)[\text{equation (11)}] \\
y = 800 &
\end{array}
$$

Substitute $y = 800$ into either (10) or (11) to find x:

$$-x - \quad 2y = -2,800 \qquad\qquad (10)$$
$$-x - 2(800) = -2,800$$
$$-x - \quad 1,600 = -2,800$$
$$-x = -1,200$$
$$x = \quad 1,200$$

Now use either (7), (8), or (9) to find z:

$$2x + \quad 4y + 3z = 11,600 \qquad\qquad (7)$$
$$2(1,200) + 4(800) + 3z = 11,600$$
$$2,400 + \quad 3,200 + 3z = 11,600$$
$$3z = \quad 6,000$$
$$z = \quad 2,000$$

Thus, each week, the company should produce 1,200 style *A* shirts, 800 style *B* shirts, and 2,000 style *C* shirts to operate at full capacity. The check of the solution is left to the reader.

Problem 8   Repeat Example 8 with the cutting, sewing, and packaging departments having available a maximum of 1,180, 1,560, and 510 labor-hours per week, respectively.

Answers to Matched Problems

1.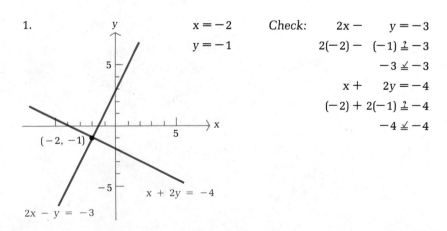

2. (A) $x = 2, y = 2$ (B)   Infinitely many solutions
(C)   No solution
3. $x = -2, y = 2$ 4. $x = 2, y = -1$
5.   6.5 oz of food *M*, 4.5 oz of food *N*

6. Equilibrium quantity = 13 thousand pounds

   Equilibrium price = $1.70 per pound

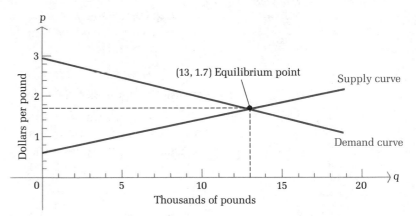

7. $x = -1, y = 0, z = 2$ 8. 900 style A; 1,300 style B; 1,600 style C

Exercise 7-1

A   *Solve by graphing.*

1. $x + y = 5$
   $x - y = 1$

2. $x - y = 2$
   $x + y = 6$

3. $3x - y = 2$
   $x + 2y = 10$

4. $3x - 2y = 12$
   $7x + 2y = 8$

5. $m + 2n = 4$
   $2m + 4n = -8$

6. $3u + 5v = 15$
   $6u + 10v = -30$

Solve using substitution.

7. $y = 2x - 3$
   $x + 2y = 14$

8. $y = x - 4$
   $x + 3y = 12$

9. $2x + y = 6$
   $x - y = -3$

10. $3x - y = 7$
    $2x + 3y = 1$

Solve using elimination by addition.

11. $3u - 2v = 12$
    $7u + 2v = 8$

12. $2x - 3y = -8$
    $5x + 3y = 1$

13. $2m - n = 10$
$\quad\;\; m - 2n = -4$

14. $2x + 3y = 1$
$\quad\;\; 3x - y = 7$

Solve using substitution or elimination by addition.

15. $9x - 3y = 24$
$\quad\;\; 11x + 2y = 1$

16. $4x + 3y = 26$
$\quad\;\; 3x - 11y = -7$

17. $2x - 3y = -2$
$\quad\;\; -4x + 6y = 7$

18. $3x - 6y = -9$
$\quad\;\; -2x + 4y = 12$

19. $3x + 8y = 4$
$\quad\;\; 15x + 10y = -10$

20. $7m + 12n = -1$
$\quad\;\; 5m - 3n = 7$

21. $-6x + 10y = -30$
$\quad\;\; 3x - 5y = 15$

22. $2x + 4y = -8$
$\quad\;\; x + 2y = 4$

23. $y = 0.07x$
$\quad\;\; y = 80 + 0.05x$

24. $y = 0.08x$
$\quad\;\; y = 100 + 0.04x$

B   *Solve using substitution or elimination by addition.*

25. $0.2x - 0.5y = 0.07$
$\quad\;\; 0.8x - 0.3y = 0.79$

26. $0.3u - 0.6v = 0.18$
$\quad\;\; 0.5u + 0.2v = 0.54$

27. $4y - z = -13$
$\quad\;\; 3y + 2z = 4$
$\quad\;\; 6x - 5y - 2z = 0$

28. $2x + z = -5$
$\quad\;\; x - 3z = -6$
$\quad\;\; 4x + 2y - z = -9$

29. $2x + y - z = 5$
$\quad\;\; x - 2y - 2z = 4$
$\quad\;\; 3x + 4y + 3z = 3$

30. $x - 3y + z = 4$
$\quad\;\; -x + 4y - 4z = 1$
$\quad\;\; 2x - y + 5z = -3$

31. $2a + 4b + 3c = 6$
$\quad\;\; a - 3b + 2c = -7$
$\quad\;\; -a + 2b - c = 5$

32. $3u - 2v + 3w = 11$
$\quad\;\; 2u + 3v - 2w = -5$
$\quad\;\; u + 4v - w = -5$

C   *Solve using substitution or elimination by addition.*

33. $2x - 3y + 3z = -15$
$\quad\;\; 3x + 2y - 5z = 19$
$\quad\;\; 5x - 4y - 2z = -2$

34. $3x - 2y - 4z = -8$
$\quad\;\; 4x + 3y - 5z = -5$
$\quad\;\; 6x - 5y + 2z = -17$

35. $x - 8y + 2z = -1$
$\quad\;\; x - 3y + z = 1$
$\quad\;\; 2x - 11y + 3z = 2$

36. $-x + 2y - z = -4$
$\quad\;\; 4x + y - 2z = 1$
$\quad\;\; x + y - z = -4$

Applications

Business & Economics

37. *Supply and demand.* Suppose the supply and demand equations for printed T-shirts in a resort town for a particular week are

$$p = 0.7q + 3 \qquad \text{Supply equation}$$
$$p = -1.7q + 15 \qquad \text{Demand equation}$$

where p is the price in dollars and q is the quantity in hundreds.

(A) Find the equilibrium price and quantity.

(B) Graph the two equations in the same coordinate system and identify the equilibrium point, supply curve, and demand curve.

38. *Supply and demand.* Repeat Problem 37 with the following supply and demand equations:

$$p = 0.4q + 3.2 \qquad \text{Supply equation}$$
$$p = -1.9q + 17 \qquad \text{Demand equation}$$

39. *Break-even analysis.* A small company manufactures portable home computers. The plant has fixed costs (leases, insurance, and so on) of $48,000 per month and variable costs (labor, materials, and so on) of $1,400 per unit produced. The computers are sold for $1,800 each. Thus, the cost and revenue equations are

$$C = 48{,}000 + 1{,}400x$$
$$R = 1{,}800x$$

where x is the total number of computers produced and sold each month, and C and R are, respectively, monthly costs and revenue in dollars.

(A) How many units must be manufactured and sold each month for the company to break even? (This is actually a three-equation–three-unknown problem with the third equation $R = C$. It can be solved by using the substitution method.)

(B) Graph both equations in the same coordinate system and show the break-even point. Interpret the regions between the lines to the left and to the right of the break-even point.

40. *Break-even analysis.* Repeat Problem 39 with the cost and revenue equations

$$C = 65{,}000 + 1{,}100$$
$$R = 1{,}600x$$

41. *Production scheduling.* A small manufacturing plant makes three types of inflatable boats: one-person, two-person, and four-person models. Each boat requires the services of three departments as listed in the table. The cutting, assembly, and packaging departments have available a maximum of 380, 330, and 120 labor-hours per week, respectively. How many boats of each type must be produced each week for the plant to operate at full capacity?

	One-Person Boat	Two-Person Boat	Four-Person Boat	
Cutting department	0.6 hr	1.0 hr	1.5 hr	380
Assembly department	0.6 hr	0.9 hr	1.2 hr	330
Packaging department	0.2 hr	0.3 hr	0.5 hr	120

42. *Production scheduling.* Repeat Problem 39 assuming the cutting, assembly, and packaging departments have available a maximum of 260, 234, and 82 labor-hours per week, respectively.

Life Sciences

43. *Nutrition.* Animals in an experiment are to be kept under a strict diet. Each animal is to receive, among other things, 20 grams of protein and 6 grams of fat. The laboratory technician is able to purchase two food mixes of the following compositions: Mix *A* has 10% protein and 6% fat; mix *B* has 20% protein and 2% fat. How many grams of each mix should be used to obtain the right diet for a single animal?

44. *Diet.* In an experiment involving mice, a zoologist needs a food mix that contains, among other things, 23 grams of protein, 6.2 grams of fat, and 16 grams of moisture. She has on hand mixes of the following compositions: Mix *A* contains 20% protein, 2% fat, and 15% moisture; mix *B* contains 10% protein, 6% fat, and 10% moisture; and mix *C* contains 15% protein, 5% fat, and 5% moisture. How many grams of each mix should be used to get the desired diet mix?

Social Sciences

45. *Psychology — approach and avoidance.* People often approach certain situations with "mixed emotions." For example, public speaking often brings forth the positive response of recognition and the negative response of failure. Which dominates? J. S. Brown, in an experiment on approach and avoidance, trained rats by feeding them from a goal box. Then the rats received mild electric shocks from the same goal box. This established an approach–avoidance conflict relative to the goal box. Using appropriate apparatus, Brown arrived at the following relationships:

$$p = -\tfrac{1}{5}d + 70$$
$$a = -\tfrac{4}{3}d + 230$$

$$30 \leqslant d \leqslant 175$$

Here *p* is the pull in grams toward the food goal box when the rat is placed *d* centimeters from it. The quantity *a* is the pull in grams away (avoidance) from the shock goal box when the rat is placed d centimeters from it.

(A) Graph the two equations above in the same coordinate system.
(B) Find d when $p = a$ by substitution.
(C) What do you think the rat would do when placed the distance d from the box found in part B?

(For additional discussion of this phenomenon, see J. S. Brown, "Gradients of Approach and Avoidance Responses and Their Relation to Motivation," *Journal of Comparative and Physiological Psychology*, 1948, 41:450–465.)

7-2 Systems of Linear Equations and Augmented Matrices—Introduction

- Introduction
- Augmented Matrices
- Solving Linear Systems

■ Introduction

Most linear systems of any consequence involve large numbers of equations and unknowns. These systems are solved with computers, since hand methods would be impractical (try solving even a five-equation–five-unknown problem and you will understand why). However, even if you have a computer facility to help solve a problem, it is still important for you to know how to formulate the problem so that it can be solved by a computer. In addition, it is helpful to have at least a general idea of how computers solve these problems. Finally, it is important for you to know how to interpret the results.

Even though the procedures and notation introduced in this and the next section are more involved than those used in the preceding section, it is important to keep in mind that our objective is not to find an efficient hand method for solving large-scale systems (there are none), but rather to find a process that generalizes readily for computer use. It turns out that you will receive an added bonus for your efforts, since several of the processes developed in this and the next section will be of considerable value in Sections 7-6, 7-7, 8-5, 8-6, and 8-7.

■ Augmented Matrices

In solving systems of equations by elimination, the coefficients of the variables and the constant terms played a central role. The process can be made more efficient for generalization and computer work by the introduction of a mathematical form called a *matrix*. A **matrix** is a rectangular array of numbers written within brackets. Some examples are

$$\begin{bmatrix} 3 & 5 \\ 0 & -2 \end{bmatrix} \qquad \begin{bmatrix} 2 \\ -3 \\ 0 \end{bmatrix} \qquad \begin{bmatrix} 1 & -1 & 0 & 5 \end{bmatrix}$$

$$\begin{bmatrix} -1 & 2 & -5 & 0 \\ 0 & 3 & 2 & 1 \end{bmatrix} \qquad \begin{bmatrix} 1 & 0 & 0 \\ 0 & 1 & 0 \\ 0 & 0 & 1 \end{bmatrix}$$

Each number in a matrix is called an **element** of the matrix.

Associated with the system

$$\begin{aligned} 2x - 3y &= 5 \\ x + 2y &= -3 \end{aligned} \tag{1}$$

is the *augmented matrix*

$$\begin{bmatrix} 2 & -3 & \bigm| & 5 \\ 1 & 2 & \bigm| & -3 \end{bmatrix}$$

which contains the essential parts of the system—namely, the coefficients of the variables and the constant terms. (The vertical bar is included only to separate the coefficients of the variables from the constant terms.)

For ease of generalization to the larger systems in the following sections, we are now going to change the notation for the variables in (1) to a subscript form (we could soon run out of letters, but we could not run out of subscripts). That is, in place of x and y, we will use x_1 and x_2 and (1) will be written as

$$\begin{aligned} 2x_1 - 3x_2 &= 5 \\ x_1 + 2x_2 &= -3 \end{aligned} \tag{2}$$

In general, associated with each linear system of the form

$$\begin{aligned} a_1x_1 + b_1x_2 &= k_1 \\ a_2x_1 + b_2x_2 &= k_2 \end{aligned} \tag{3}$$

where x_1 and x_2 are variables, is the **augmented matrix** of the system:

$$\begin{matrix} & \text{Column 1 } (C_1) \\ & \text{Column 2 } (C_2) \\ & \text{Column 3 } (C_3) \\ \begin{bmatrix} a_1 & b_1 & \bigm| & k_1 \\ a_2 & b_2 & \bigm| & k_2 \end{bmatrix} & \begin{matrix} \leftarrow \text{Row 1 } (R_1) \\ \leftarrow \text{Row 2 } (R_2) \end{matrix} \end{matrix} \tag{4}$$

This matrix contains the essential parts of system (3). Our objective is to learn how to manipulate augmented matrices in order to solve system (3), if a solution exists. The manipulative process is a direct outgrowth of the elimination process discussed in Section 7-1.

Recall that two linear systems are said to be **equivalent** if they have exactly the same solution set. How did we transform linear systems into equivalent linear systems? We used Theorem 1, which we restate here.

Theorem 1

A system of linear equations is transformed into an equivalent system if:

(A) Two equations are interchanged.
(B) An equation is multiplied by a nonzero constant.
(C) A constant multiple of another equation is added to a given equation.

Paralleling the discussion above, we say that two augmented matrices are **row-equivalent,** denoted by the symbol ~ placed between the two matrices, if they are augmented matrices of equivalent systems of equations. (Think about this.) How do we transform augmented matrices into row-equivalent matrices? We use Theorem 2, which is a direct consequence of Theorem 1:

Theorem 2

An augmented matrix is transformed into a row-equivalent matrix if:

(A) Two rows are interchanged ($R_i \leftrightarrow R_j$).
(B) A row is multiplied by a nonzero constant ($kR_i \rightarrow R_i$).
(C) A constant multiple of another row is added to a given row

   $(R_i + kR_j \rightarrow R_i)$.

Note: The arrow $\rightarrow$ means "replaces."

■ Solving Linear Systems

The use of Theorem 2 in solving systems in the form of (3) is best illustrated by examples.

Example 9   Solve using augmented matrix methods:

$$3x_1 + 4x_2 = 1$$
$$x_1 - 2x_2 = 7$$

(5)

Solution We start by writing the augmented matrix corresponding to (5)

$$\left[\begin{array}{cc|c} 3 & 4 & 1 \\ 1 & -2 & 7 \end{array}\right] \tag{6}$$

Our objective is to use row operations from Theorem 2 to try to transform (6) into the form

$$\left[\begin{array}{cc|c} 1 & 0 & m \\ 0 & 1 & n \end{array}\right] \tag{7}$$

where m and n are real numbers. The solution to system (5) will then be obvious, since matrix (7) will be the augmented matrix of the following system:

$$x_1 = m$$
$$ x_2 = n$$

We now proceed to use row operations to transform (6) into form (7).

Step 1. To get a 1 in the upper left corner, we interchange Rows 1 and 2 (Theorem 2A):

$$\left[\begin{array}{cc|c} 3 & 4 & 1 \\ 1 & -2 & 7 \end{array}\right] \quad \begin{array}{c} R_1 \leftrightarrow R_2 \\ \sim \end{array} \quad \left[\begin{array}{cc|c} 1 & -2 & 7 \\ 3 & 4 & 1 \end{array}\right]$$

Now you see why we wanted Theorem 1A!

Step 2. To get a 0 in the lower left corner, we multiply R_1 by (-3) and add to R_2 (Theorem 2C)—this changes R_2 but not R_1. Some people find it useful to write $(-3)R_1$ outside the matrix to help reduce errors in arithmetic, as shown:

$$\begin{array}{ccc} -3 & 6 & -21 \cdots \cdots \end{array}$$

$$\left[\begin{array}{cc|c} 1 & -2 & 7 \\ 3 & 4 & 1 \end{array}\right] \begin{array}{c} R_2 + (-3)R_1 \to R_2 \\ \sim \end{array} \left[\begin{array}{cc|c} 1 & -2 & 7 \\ 0 & 10 & -20 \end{array}\right]$$

Step 3. To get a 1 in the second row, second column, we multiply R_2 by $\frac{1}{10}$ (Theorem 2B):

$$\left[\begin{array}{cc|c} 1 & -2 & 7 \\ 0 & 10 & -20 \end{array}\right] \quad \begin{array}{c} \frac{1}{10}R_2 \to R_2 \\ \sim \end{array} \quad \left[\begin{array}{cc|c} 1 & -2 & 7 \\ 0 & 1 & -2 \end{array}\right]$$

Step 4. To get a 0 in the first row, second column, we multiply R_2 by 2 and add the result to R_1 (Theorem 2C)—this changes R_1 but not R_2:

$$\begin{array}{ccc} 0 & 2 & -4 \cdots \cdots \end{array}$$

$$\left[\begin{array}{cc|c} 1 & -2 & 7 \\ 0 & 1 & -2 \end{array}\right] \begin{array}{c} R_1 + 2R_2 \to R_1 \\ \sim \end{array} \left[\begin{array}{cc|c} 1 & 0 & 3 \\ 0 & 1 & -2 \end{array}\right]$$

We have accomplished our objective! The last matrix is the augmented matrix for the system

$$x_1 \quad = \quad 3$$
$$x_2 = -2$$

(8)

Since system (8) is equivalent to system (5), our starting system, we have solved (5); that is, $x_1 = 3$ and $x_2 = -2$.

Check $\quad 3x_1 + \quad 4x_2 = 1 \qquad x_1 - \quad 2x_2 = 7$

$$3(3) + 4(-2) \overset{?}{=} 1 \qquad 3 - 2(-2) \overset{?}{=} 7$$

$$9 - 8 \overset{\checkmark}{=} 1 \qquad\qquad 3 + 4 \overset{\checkmark}{=} 7$$

The above process is written more compactly as follows:

Step 1: Need a 1 here

$$\begin{bmatrix} 3 & 4 & | & 1 \\ 1 & -2 & | & 7 \end{bmatrix} \quad R_1 \leftrightarrow R_2$$

Step 2: Need a 0 here

$$\sim \begin{bmatrix} 1 & -2 & | & 7 \\ 3 & 4 & | & 1 \\ -3 & 6 & & -21 \end{bmatrix} \quad R_2 + (-3)R_1 \rightarrow R_2$$

Step 3: Need a 1 here

$$\sim \begin{bmatrix} 1 & -2 & | & 7 \\ 0 & 10 & | & -20 \end{bmatrix} \quad \tfrac{1}{10} R_2 \rightarrow R_2$$

Step 4: Need a 0 here

$$\sim \begin{bmatrix} 1 & -2 & | & 7 \\ 0 & 1 & | & -2 \\ 0 & 2 & & -4 \end{bmatrix} \quad R_1 + 2R_2 \rightarrow R_1$$

$$\sim \begin{bmatrix} 1 & 0 & | & 3 \\ 0 & 1 & | & -2 \end{bmatrix}$$

Therefore, $x_1 = 3$ and $x_2 = -2$.

Problem 9   Solve using augmented matrix methods:

$$2x_1 - \quad x_2 = -7$$
$$x_1 + 2x_2 = \quad 4$$

Example 10   Solve using augmented matrix methods:

$$2x_1 - 3x_2 = 6$$
$$3x_1 + 4x_2 = \tfrac{1}{2}$$

Solution

$$\text{Step 1:}\quad \text{Need a 1 here} \qquad \begin{bmatrix} 2 & -3 & \bigm| & 6 \\ 3 & 4 & \bigm| & \frac{1}{2} \end{bmatrix} \qquad \frac{1}{2}R_1 \rightarrow R_1$$

$$\text{Step 2:}\quad \text{Need a 0 here} \qquad \sim \begin{bmatrix} 1 & -\frac{3}{2} & \bigm| & 3 \\ 3 & 4 & \bigm| & \frac{1}{2} \end{bmatrix} \qquad R_2 + (-3)R_1 \rightarrow R_2$$

$$\begin{matrix} -3 & \frac{9}{2} & -9 \end{matrix}$$

$$\text{Step 3:}\quad \text{Need a 1 here} \qquad \sim \begin{bmatrix} 1 & -\frac{3}{2} & \bigm| & 3 \\ 0 & \frac{17}{2} & \bigm| & -\frac{17}{2} \end{bmatrix} \qquad \frac{2}{17}R_2 \rightarrow R_2$$

$$\text{Step 4:}\quad \text{Need a 0 here} \qquad \sim \begin{bmatrix} 1 & -\frac{3}{2} & \bigm| & 3 \\ 0 & 1 & \bigm| & -1 \end{bmatrix} \qquad R_1 + \frac{3}{2}R_2 \rightarrow R_1$$

$$\begin{matrix} 0 & \frac{3}{2} & -\frac{3}{2} \end{matrix}$$

$$\sim \begin{bmatrix} 1 & 0 & \bigm| & \frac{3}{2} \\ 0 & 1 & \bigm| & -1 \end{bmatrix}$$

Thus, $x_1 = \frac{3}{2}$ and $x_2 = -1$.

Problem 10 Solve using augmented matrix methods:

$$5x_1 - 2x_2 = 11$$
$$2x_1 + 3x_2 = \tfrac{5}{2}$$

Example 11 Solve using augmented matrix methods:

$$2x_1 - x_2 = 4$$
$$-6x_1 + 3x_2 = -12$$

Solution

$$\begin{bmatrix} 2 & -1 & \bigm| & 4 \\ -6 & 3 & \bigm| & -12 \end{bmatrix}$$

$\frac{1}{2}R_1 \rightarrow R_1$ (this produces a 1 in the upper left corner)

$\frac{1}{3}R_2 \rightarrow R_2$ (this simplifies R_2)

$$\sim \begin{bmatrix} 1 & -\frac{1}{2} & \bigm| & 2 \\ -2 & 1 & \bigm| & -4 \end{bmatrix}$$

$R_2 + 2R_1 \rightarrow R_2$ (this produces a 0 in the lower left corner)

$$\begin{matrix} 2 & -1 & 4 \end{matrix}$$

$$\sim \begin{bmatrix} 1 & -\frac{1}{2} & \bigm| & 2 \\ 0 & 0 & \bigm| & 0 \end{bmatrix}$$

The last matrix corresponds to the system

$$x_1 - \tfrac{1}{2}x_2 = 2$$
$$0x_1 + 0x_2 = 0$$

(10)

This system is equivalent to the original system. Geometrically, the graphs of the two original equations coincide and there are infinitely many solutions. In general, if we end up with a row of zeros in an augmented matrix for a two-equation–two-unknown system, the system is dependent and there are infinitely many solutions.

There are several ways of representing the infinitely many solutions to system (9). For example, solving the first equation in (10) for either variable in terms of the other (we solve for x_1 in terms of x_2), we obtain

$$x_1 = \tfrac{1}{2}x_2 + 2 \tag{11}$$

Thus, for any real number x_2,

$$(\tfrac{1}{2}x_2 + 2, \; x_2)$$

is a solution. Another way to represent the infinitely many solutions—a way that is convenient for the larger-scale systems we will be solving later in this chapter—is as follows: We choose another variable called a **parameter,** say t, and set the variable on the right of equation (11), x_2, equal to it. Then for t any real number,

$$x_1 = \tfrac{1}{2}t + 2$$
$$x_2 = t$$

represents a solution. For example, if $t = 8$, then

$$x_1 = \tfrac{1}{2}(8) + 2 = 6$$
$$x_2 = 8$$

That is, $(6, 8)$ is a solution of (9). If $t = -3$, then

$$x_1 = \tfrac{1}{2}(-3) + 2 = \tfrac{1}{2}$$
$$x_2 = -3$$

That is, $(\tfrac{1}{2}, -3)$ is a solution of (9). Other solutions can be obtained in a similar manner.

Problem 11   Solve using augmented matrix methods:

$$-2x_1 + 6x_2 = 6$$
$$3x_1 - 9x_2 = -9$$

Example 12   Solve using augmented matrix methods:

$$2x_1 + 6x_2 = -3$$
$$x_1 + 3x_2 = 2$$

Solution

$$\begin{bmatrix} 2 & 6 & | & -3 \\ 1 & 3 & | & 2 \end{bmatrix} \quad R_1 \leftrightarrow R_2$$

$$\sim \begin{bmatrix} 1 & 3 & | & 2 \\ 2 & 6 & | & -3 \end{bmatrix} \quad R_2 + (-2)R_1 \rightarrow R_2$$

$$-2 \quad -6 \qquad -4$$

$$\sim \begin{bmatrix} 1 & 3 & | & 2 \\ 0 & 0 & | & -7 \end{bmatrix} \quad R_2 \text{ implies the contradiction } 0 = -7$$

This is the augmented matrix of the system

$$x_1 + 3x_2 = 2$$
$$0x_1 + 0x_2 = -7$$

The second equation is not satisfied by any ordered pair of real numbers. Hence, the original system is inconsistent and has no solution—otherwise we have proved that $0 = -7$! Thus, if in a row of an augmented matrix we obtain all zeros to the left of the vertical bar and a nonzero number to the right, then the system is inconsistent and there are no solutions.

Problem 12 Solve using augmented matrix methods:

$$2x_1 - x_2 = 3$$
$$4x_1 - 2x_2 = -1$$

Summary

Form 1	Form 2	Form 3
A Unique Solution	Infinitely Many Solutions (Dependent)	No Solution (Inconsistent)

$$\begin{bmatrix} 1 & 0 & | & m \\ 0 & 1 & | & n \end{bmatrix} \qquad \begin{bmatrix} 1 & m & | & n \\ 0 & 0 & | & 0 \end{bmatrix} \qquad \begin{bmatrix} 1 & m & | & n \\ 0 & 0 & | & p \end{bmatrix}$$

$m, n, p.$ real numbers; $p \neq 0$

The process of solving systems of equations described in this section is referred to as **Gauss–Jordan elimination.** We will use this method to solve larger-scale systems in the next section, including systems where the number of equations and the number of variables are not the same.

Answers to
Matched Problems

9. $x_1 = -2, x_2 = 3$

10. $x_1 = 2, x_2 = -\frac{1}{2}$

11. The system is dependent. For t any real number,
$$x_1 = 3t - 3$$
$$x_2 = t$$
is a solution.

12. Inconsistent—no solution

Exercise 7-2

A   *Perform each of the indicated row operations on the following matrix:*

$$\begin{bmatrix} 1 & -3 & \bigm| & 2 \\ 4 & -6 & \bigm| & -8 \end{bmatrix}$$

1. $R_1 \leftrightarrow R_2$
2. $\frac{1}{2}R_2 \rightarrow R_2$
3. $-4R_1 \rightarrow R_1$
4. $-2R_1 \rightarrow R_1$
5. $2R_2 \rightarrow R_2$
6. $-1R_2 \rightarrow R_2$
7. $R_2 + (-4)R_1 \rightarrow R_2$
8. $R_1 + (-\frac{1}{2})R_2 \rightarrow R_1$
9. $R_2 + (-2)R_1 \rightarrow R_2$
10. $R_2 + (-3)R_1 \rightarrow R_2$
11. $R_2 + (-1)R_1 \rightarrow R_2$
12. $R_2 + (1)R_1 \rightarrow R_2$

Solve using augmented matrix methods.

13. $x_1 + x_2 = 5$
    $x_1 - x_2 = 1$

14. $x_1 - x_2 = 2$
    $x_1 + x_2 = 6$

B   *Solve using augmented matrix methods.*

15. $x_1 - 2x_2 = 1$
    $2x_1 - x_2 = 5$

16. $x_1 + 3x_2 = 1$
    $3x_1 - 2x_2 = 14$

17. $x_1 - 4x_2 = -2$
    $-2x_1 + x_2 = -3$

18. $x_1 - 3x_2 = -5$
    $-3x_1 - x_2 = 5$

19. $3x_1 - x_2 = 2$
    $x_1 + 2x_2 = 10$

20. $2x_1 + x_2 = 0$
    $x_1 - 2x_2 = -5$

21. $x_1 + 2x_2 = 4$
    $2x_1 + 4x_2 = -8$

22. $2x_1 - 3x_2 = -2$
    $-4x_1 + 6x_2 = 7$

23. $2x_1 + x_2 = 6$
    $x_1 - x_2 = -3$

24. $3x_1 - x_2 = -5$
    $x_1 + 3x_2 = 5$

25. $3x_1 - 6x_2 = -9$
    $-2x_1 + 4x_2 = 6$

26. $2x_1 - 4x_2 = -2$
    $-3x_1 + 6x_2 = 3$

27. $4x_1 - 2x_2 = 2$
    $-6x_1 + 3x_2 = -3$

28. $-6x_1 + 2x_2 = 4$
    $3x_1 - x_2 = -2$

C *Solve using augmented matrix methods.*

29. $3x_1 - x_2 = 7$
     $2x_1 + 3x_2 = 1$

30. $2x_1 - 3x_2 = -8$
     $5x_1 + 3x_2 = 1$

31. $3x_1 + 2x_2 = 4$
     $2x_1 - x_2 = 5$

32. $4x_1 + 3x_2 = 26$
     $3x_1 - 11x_2 = -7$

33. $0.2x_1 - 0.5x_2 = 0.07$
     $0.8x_1 - 0.3x_2 = 0.79$

34. $0.3x_1 - 0.6x_2 = 0.18$
     $0.5x_1 - 0.2x_2 = 0.54$

7-3 Gauss–Jordan Elimination

- Reduced Matrices
- Solving Systems by Gauss–Jordan Elimination
- Application

Now that you have had some experience with row operations on simple augmented matrices, we will consider systems involving more than two variables. In addition, we will not require that a system have the same number of equations as variables.

■ Reduced Matrices

Our objective is to start with the augmented matrix of a linear system and transform it by using row operations from Theorem 2 in the preceding section into a simple form where the solution can be read by inspection. The simple form so obtained is called the *reduced form*, and we define it as follows:

Reduced Matrix

A matrix is in **reduced form** if:

1. Each row consisting entirely of zeros is below any row having at least one nonzero element.
2. The leftmost nonzero element in each row is 1.
3. The column containing the leftmost 1 of a given row has zeros above and below the 1.
4. The leftmost 1 in any row is to the right of the leftmost 1 in the row above.

Example 13 The following matrices are in reduced form. Check each one carefully to convince yourself that the conditions in the definition are met.

$$\left[\begin{array}{cc|c} 1 & 0 & 2 \\ 0 & 1 & -3 \end{array}\right] \qquad \left[\begin{array}{ccc|c} 1 & 0 & 0 & 2 \\ 0 & 1 & 0 & -1 \\ 0 & 0 & 1 & 3 \end{array}\right]$$

$$\left[\begin{array}{cc|c} 1 & 0 & 3 \\ 0 & 1 & -1 \\ 0 & 0 & 0 \end{array}\right] \qquad \left[\begin{array}{cccc|c} 1 & 4 & 0 & 0 & -3 \\ 0 & 0 & 1 & 0 & 2 \\ 0 & 0 & 0 & 1 & 6 \end{array}\right]$$

$$\left[\begin{array}{ccc|c} 1 & 0 & 4 & 0 \\ 0 & 1 & 3 & 0 \\ 0 & 0 & 0 & 1 \end{array}\right]$$

Problem 13 The matrices below are not in reduced form. Indicate which condition in the definition is violated for each matrix.

(A) $\left[\begin{array}{cc|c} 1 & 0 & 2 \\ 0 & 3 & -6 \end{array}\right]$ 　(B) $\left[\begin{array}{ccc|c} 1 & 5 & 4 & 3 \\ 0 & 1 & 2 & -1 \\ 0 & 0 & 0 & 0 \end{array}\right]$

(C) $\left[\begin{array}{ccc|c} 0 & 1 & 2 & -3 \\ 1 & -2 & 3 & 0 \\ 0 & 0 & 1 & 2 \end{array}\right]$ 　(D) $\left[\begin{array}{ccc|c} 1 & 2 & 0 & 3 \\ 0 & 0 & 0 & 0 \\ 0 & 0 & 1 & 4 \end{array}\right]$

Example 14 Write the linear system corresponding to each reduced augmented matrix and solve.

(A) $\left[\begin{array}{ccc|c} 1 & 0 & 0 & 2 \\ 0 & 1 & 0 & -1 \\ 0 & 0 & 1 & 3 \end{array}\right]$ 　(B) $\left[\begin{array}{ccc|c} 1 & 0 & 4 & 0 \\ 0 & 1 & 3 & 0 \\ 0 & 0 & 0 & 1 \end{array}\right]$

(C) $\left[\begin{array}{ccc|c} 1 & 0 & 2 & -3 \\ 0 & 1 & -1 & 8 \\ 0 & 0 & 0 & 0 \end{array}\right]$ 　(D) $\left[\begin{array}{ccccc|c} 1 & 4 & 0 & 0 & 3 & -2 \\ 0 & 0 & 1 & 0 & -2 & 0 \\ 0 & 0 & 0 & 1 & 2 & 4 \end{array}\right]$

Solutions (A) $\begin{aligned} x_1 &= 2 \\ x_2 &= -1 \\ x_3 &= 3 \end{aligned}$

The solution is obvious: $x_1 = 2$, $x_2 = -1$, $x_3 = 3$.

(B) $\quad x_1 \qquad + 4x_3 = 0$

$\qquad\qquad x_2 + 3x_3 = 0$

$\quad 0x_1 + 0x_2 + 0x_3 = 1$

The last equation implies $0 = 1$, which is a contradiction. Hence, the system is inconsistent and has no solution.

(C) $\quad x_1 \quad + 2x_3 = -3 \qquad$ We disregard the equation corresponding to the

$\qquad\quad x_2 - \ x_3 = \quad 8 \qquad$ third row in the matrix, since it is satisfied by

$\qquad\qquad\qquad\qquad\qquad\qquad$ all values of x_1, x_2, and x_3

When a reduced system (a system corresponding to a reduced augmented matrix) has more variables than equations, the system is dependent and has infinitely many solutions. To represent these solutions, it is useful to divide the variables into two types: **basic variables** and **nonbasic variables.** To represent the infinitely many solutions to the system, we solve for the basic variables in terms of the nonbasic variables. This can be accomplished very easily if we **choose as basic variables the first variable (with a nonzero coefficient) in each equation of the reduced system.** Since each of these variables occurs in exactly one equation, it is easy to solve for each in terms of the other variables, the nonbasic variables. Returning to our original system, we choose x_1 and x_2 (the first variable in each equation) as basic variables and x_3 as a nonbasic variable. We then solve for the basic variables x_1 and x_2 in terms of the nonbasic variable x_3:

$\qquad x_1 = -2x_3 - 3$

$\qquad x_2 = x_3 + 8$

If we let $x_3 = t$, then for any real number t,

$\qquad x_1 = -2t - 3$

$\qquad x_2 = t + 8$

$\qquad x_3 = t$

is a solution. For example,

If $t = 0$, then	If $t = -2$, then
$x_1 = -2(0) - 3 = -3$	$x_1 = -2(-2) - 3 = 1$
$x_2 = 0 + 8 = 8$	$x_2 = -2 + 8 = 6$
$x_3 = 0$	$x_3 = -2$
is a solution.	is a solution.

(D) $\quad x_1 + 4x_2 \qquad + 3x_5 = -2$

$\qquad\qquad\quad x_3 \ - 2x_5 = \quad 0$

$\qquad\qquad\qquad x_4 + 2x_5 = \quad 4$

Solve for x_1, x_3, and x_4 (basic variables) in terms of x_2 and x_5 (nonbasic variables):

$$x_1 = -4x_2 - 3x_5 - 2$$
$$x_3 = 2x_5$$
$$x_4 = -2x_5 + 4$$

If we let $x_2 = s$ and $x_5 = t$, then for any real numbers s and t,

$$x_1 = -4s - 3t - 2$$
$$x_2 = s$$
$$x_3 = 2t$$
$$x_4 = -2t + 4$$
$$x_5 = t$$

is a solution. The system is dependent and has infinitely many solutions. Can you find two?

Problem 14   Write the linear system corresponding to each reduced augmented matrix and solve.

(A) $\begin{bmatrix} 1 & 0 & 0 & | & -5 \\ 0 & 1 & 0 & | & 3 \\ 0 & 0 & 1 & | & 6 \end{bmatrix}$ (B) $\begin{bmatrix} 1 & 2 & -3 & | & 0 \\ 0 & 0 & 0 & | & 1 \\ 0 & 0 & 0 & | & 0 \end{bmatrix}$

(C) $\begin{bmatrix} 1 & 0 & -2 & | & 4 \\ 0 & 1 & 3 & | & -2 \\ 0 & 0 & 0 & | & 0 \end{bmatrix}$ (D) $\begin{bmatrix} 1 & 0 & 3 & 2 & | & 5 \\ 0 & 1 & -2 & -1 & | & 3 \\ 0 & 0 & 0 & 0 & | & 0 \end{bmatrix}$

■ Solving Systems by Gauss–Jordan Elimination

We are now ready to outline the Gauss–Jordan elimination method for solving systems of linear equations. The method systematically transforms an augmented matrix into a reduced form from which we can write the solution to the original system by inspection, if a solution exists. The method will also reveal when a solution fails to exist (see Example 14B).

Example 15   Solve by Gauss–Jordan elimination:

$$2x_1 - 2x_2 + x_3 = 3$$
$$3x_1 + x_2 - x_3 = 7$$
$$x_1 - 3x_2 + 2x_3 = 0$$

Solution   Write the augmented matrix and follow the steps indicated at the right.

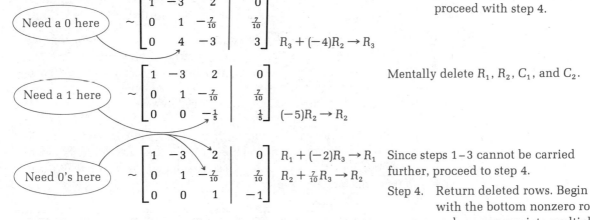

Need a 1 here

$$\begin{bmatrix} 2 & -2 & 1 & | & 3 \\ 3 & 1 & -1 & | & 7 \\ 1 & -3 & 2 & | & 0 \end{bmatrix} \quad R_1 \leftrightarrow R_3$$

Step 1. Choose leftmost nonzero column and get a 1 at the top.

Need 0's here

$$\sim \begin{bmatrix} 1 & -3 & 2 & | & 0 \\ 3 & 1 & -1 & | & 7 \\ 2 & -2 & 1 & | & 3 \end{bmatrix} \quad \begin{array}{l} R_2 + (-3)R_1 \to R_2 \\ R_3 + (-2)R_1 \to R_3 \end{array}$$

Step 2. Use multiples of the first row to get zeros below the 1 obtained in step 1.

Need a 1 here

$$\sim \begin{bmatrix} 1 & -3 & 2 & | & 0 \\ 0 & 10 & -7 & | & 7 \\ 0 & 4 & -3 & | & 3 \end{bmatrix} \quad \tfrac{1}{10}R_2 \to R_2$$

Step 3. Mentally delete R_1 and C_1, then repeat steps 1 and 2 with the **submatrix** (the matrix that remains after deleting the top row and first column). Continue the above process (steps 1–3) until it is not possible to go further; then proceed with step 4.

Need a 0 here

$$\sim \begin{bmatrix} 1 & -3 & 2 & | & 0 \\ 0 & 1 & -\tfrac{7}{10} & | & \tfrac{7}{10} \\ 0 & 4 & -3 & | & 3 \end{bmatrix} \quad R_3 + (-4)R_2 \to R_3$$

Need a 1 here

$$\sim \begin{bmatrix} 1 & -3 & 2 & | & 0 \\ 0 & 1 & -\tfrac{7}{10} & | & \tfrac{7}{10} \\ 0 & 0 & -\tfrac{1}{5} & | & \tfrac{1}{5} \end{bmatrix} \quad (-5)R_2 \to R_2$$

Mentally delete R_1, R_2, C_1, and C_2.

Need 0's here

$$\sim \begin{bmatrix} 1 & -3 & 2 & | & 0 \\ 0 & 1 & -\tfrac{7}{10} & | & \tfrac{7}{10} \\ 0 & 0 & 1 & | & -1 \end{bmatrix} \quad \begin{array}{l} R_1 + (-2)R_3 \to R_1 \\ R_2 + \tfrac{7}{10}R_3 \to R_2 \end{array}$$

Since steps 1–3 cannot be carried further, proceed to step 4.

Step 4. Return deleted rows. Begin with the bottom nonzero row and use appropriate multiples of it to get zeros above the leftmost 1. Continue the process, moving up row by row, until the matrix is in reduced form.

Need a 0 here

$$\sim \begin{bmatrix} 1 & -3 & 0 & | & 2 \\ 0 & 1 & 0 & | & 0 \\ 0 & 0 & 1 & | & -1 \end{bmatrix} \quad R_1 + 3R_2 \to R_1$$

$$\sim \begin{bmatrix} 1 & 0 & 0 & | & 2 \\ 0 & 1 & 0 & | & 0 \\ 0 & 0 & 1 & | & -1 \end{bmatrix}$$

The matrix is in reduced form, and we can write the solution to the original system by inspection.

Solution: $x_1 = 2$, $x_2 = 0$, $x_3 = -1$. It is left to the reader to check this solution.

Steps 1–4 outlined in the solution of Example 15 are referred to as *Gauss–Jordan elimination*. The steps are summarized in the box below for easy reference:

Gauss–Jordan Elimination

1. Choose the leftmost nonzero column and use appropriate row operations to get a 1 at the top.

2. Use multiples of the first row to get zeros in all places below the 1 obtained in step 1.

3. Delete (mentally) the top row and first column of the matrix. Repeat steps 1 and 2 with the **submatrix** (the matrix that remains after deleting the top row and first column). Continue this process (steps 1–3) until it is not possible to go further.

4. Consider the whole matrix obtained after mentally returning all the rows and columns to the matrix. Begin with the bottom nonzero row and use appropriate multiples of it to get zeros above the leftmost 1. Continue this process, moving up row by row, until the matrix is finally in reduced form.

Note: If at any point in the above process we obtain a row with all zeros to the left of the vertical line and a nonzero number to the right, we can stop, since we will have a contradiction ($0 = n$, $n \neq 0$). We can then conclude that the system has no solution.

Problem 15   Solve by Gauss–Jordan elimination:

$$3x_1 + x_2 - 2x_3 = 2$$
$$x_1 - 2x_2 + x_3 = 3$$
$$2x_1 - x_2 - 3x_3 = 3$$

Example 16   Solve by Gauss–Jordan elimination:

$$2x_1 - x_2 + 4x_3 = -2$$
$$3x_1 + 2x_2 - x_3 = 1$$

Solution

(Need a 1 here)

$$\begin{bmatrix} 2 & -1 & 4 & | & -2 \\ 3 & 2 & -1 & | & 1 \end{bmatrix} \quad \frac{1}{2}R_1 \rightarrow R_1$$

(Need a 0 here)

$$\sim \begin{bmatrix} 1 & -\frac{1}{2} & 2 & | & -1 \\ 3 & 2 & -1 & | & 1 \end{bmatrix} \quad R_2 + (-3)R_1 \rightarrow R_2$$

(Need a 1 here)

$$\sim \begin{bmatrix} 1 & -\frac{1}{2} & 2 & | & -1 \\ 0 & \frac{7}{2} & -7 & | & 4 \end{bmatrix} \quad \frac{2}{7}R_2 \rightarrow R_2$$

(Need a 0 here)

$$\sim \begin{bmatrix} 1 & -\frac{1}{2} & 2 & | & -1 \\ 0 & 1 & -2 & | & \frac{8}{7} \end{bmatrix} \quad R_1 + \frac{1}{2}R_2 \rightarrow R_1$$

$$\sim \begin{bmatrix} 1 & 0 & 1 & | & -\frac{3}{7} \\ 0 & 1 & -2 & | & \frac{8}{7} \end{bmatrix}$$

The matrix is now in reduced form. Write the corresponding system and the solution.

$$\begin{aligned} x_1 \quad + \quad x_3 &= -\tfrac{3}{7} \\ x_2 - 2x_3 &= \tfrac{8}{7} \end{aligned}$$

Solve for the basic variables x_1 and x_2 in terms of the nonbasic variable x_3:

$$\begin{aligned} x_1 &= -x_3 - \tfrac{3}{7} \\ x_2 &= 2x_3 + \tfrac{8}{7} \end{aligned}$$

If $x_3 = t$, then for t any real number,

$$\begin{aligned} x_1 &= -t - \tfrac{3}{7} \\ x_2 &= 2t + \tfrac{8}{7} \\ x_3 &= t \end{aligned}$$

is a solution.

Remark: In general, it can be proved that a system with more variables than equations cannot have a unique solution.

Problem 16 Solve by Gauss–Jordan elimination:

$$\begin{aligned} 3x_1 + 6x_2 - 3x_3 &= 2 \\ 2x_1 - x_2 + 2x_3 &= -1 \end{aligned}$$

Example 17 Solve by Gauss–Jordan elimination:

$$\begin{aligned} 2x_1 - x_2 &= -4 \\ x_1 + 2x_2 &= 3 \\ 3x_1 - x_2 &= -1 \end{aligned}$$

Solution

$$\begin{bmatrix} 2 & -1 & | & -4 \\ 1 & 2 & | & 3 \\ 3 & -1 & | & -1 \end{bmatrix} \quad R_1 \leftrightarrow R_2$$

$$\sim \begin{bmatrix} 1 & 2 & | & 3 \\ 2 & -1 & | & -4 \\ 3 & -1 & | & -1 \end{bmatrix} \quad \begin{array}{l} R_2 + (-2)R_1 \rightarrow R_2 \\ R_3 + (-3)R_1 \rightarrow R_3 \end{array}$$

$$\sim \begin{bmatrix} 1 & 2 & | & 3 \\ 0 & -5 & | & -10 \\ 0 & -7 & | & -10 \end{bmatrix} \quad -\tfrac{1}{5}R_2 \rightarrow R_2$$

$$\sim \begin{bmatrix} 1 & 2 & | & 3 \\ 0 & 1 & | & 2 \\ 0 & -7 & | & -10 \end{bmatrix} \quad R_3 + 7R_2 \rightarrow R_3$$

$$\sim \begin{bmatrix} 1 & 2 & | & 3 \\ 0 & 1 & | & 2 \\ 0 & 0 & | & 4 \end{bmatrix}$$

We stop the Gauss–Jordan elimination, even though the matrix is not in a reduced form, since the last row produces a contradiction.

The last row implies $0 = 4$, which is a contradiction; therefore, the system has no solution.

Problem 17   Solve by Gauss–Jordan elimination:

$$3x_1 + x_2 = 5$$
$$2x_1 + 3x_2 = 1$$
$$x_1 - x_2 = 3$$

■ Application

Example 18   A casting company produces three different bronze sculptures. The casting department has available a maximum of 350 labor-hours per week, and the finishing department has a maximum of 150 labor-hours available per week. Sculpture *A* requires 30 hours for casting and 10 hours for finishing; sculpture *B* requires 10 hours for casting and 10 hours for finishing; and sculpture *C* requires 10 hours for casting and 30 hours for finishing. If the plant is to operate at maximum capacity, how many of each sculpture should be produced each week?

Solution   First, we summarize the relevant manufacturing data in a table:

	Labor-Hours per Sculpture			Maximum Labor-Hours Available per Week
	A	B	C	
Casting department	30	10	10	350
Finishing department	10	10	30	150

Let x_1 = Number of sculpture A produced per week

x_2 = Number of sculpture B produced per week

x_3 = Number of sculpture C produced per week

Then $30x_1 + 10x_2 + 10x_3 = 350$ Casting department

$10x_1 + 10x_2 + 30x_3 = 150$ Finishing department

Now we can form the augmented matrix of the system and solve by using Gauss–Jordan elimination:

$$\begin{bmatrix} 30 & 10 & 10 & | & 350 \\ 10 & 10 & 30 & | & 150 \end{bmatrix} \quad \begin{matrix} \frac{1}{10}R_1 \rightarrow R_1 \\ \frac{1}{10}R_2 \rightarrow R_2 \end{matrix} \qquad \text{Simplify each row}$$

$$\sim \begin{bmatrix} 3 & 1 & 1 & | & 35 \\ 1 & 1 & 3 & | & 15 \end{bmatrix} \quad R_1 \leftrightarrow R_2$$

$$\sim \begin{bmatrix} 1 & 1 & 3 & | & 15 \\ 3 & 1 & 1 & | & 35 \end{bmatrix} \quad R_2 + (-3)R_1 \rightarrow R_2$$

$$\sim \begin{bmatrix} 1 & 1 & 3 & | & 15 \\ 0 & -2 & -8 & | & -10 \end{bmatrix} \quad -\frac{1}{2}R_2 \rightarrow R_2$$

$$\sim \begin{bmatrix} 1 & 1 & 3 & | & 15 \\ 0 & 1 & 4 & | & 5 \end{bmatrix} \quad R_1 + (-1)R_2 \rightarrow R_1$$

$$\sim \begin{bmatrix} 1 & 0 & -1 & | & 10 \\ 0 & 1 & 4 & | & 5 \end{bmatrix} \quad \begin{matrix}\text{Matrix is in reduced} \\ \text{form}\end{matrix}$$

$$\begin{array}{lll} x_1 \quad - \quad x_3 = 10 & \text{or} & x_1 = x_3 + 10 \\ x_2 + 4x_3 = 5 & & x_2 = -4x_3 + 5 \end{array}$$

Let $x_3 = t$. Then for t any real number,

$x_1 = t + 10$

$x_2 = -4t + 5$

$x_3 = t$

is a solution — or is it? We cannot produce a negative number of sculptures. If we also assume that we cannot produce a fractional number of sculptures, then t must be a nonnegative whole number. And because of the middle equation ($x_2 = -4t + 5$), t can only assume the values 0 and 1. Thus, for $t = 0$, we have $x_1 = 10$, $x_2 = 5$, $x_3 = 0$; and for $t = 1$, we have $x_1 = 11$, $x_2 = 1$, $x_3 = 1$. These are the only possible production schedules that utilize the full capacity of the plant.

Problem 18   Repeat Example 18 given a casting capacity of 400 labor-hours per week and a finishing capacity of 200 labor-hours per week.

Answers to Matched Problems

13. (A) Condition 2 is violated: The 3 in the second row should be a 1.
    (B) Condition 3 is violated: In the second column, the 5 should be a 0.
    (C) Condition 4 is violated: The leftmost 1 in the second row is not to the right of the leftmost 1 in the first row.
    (D) Condition 1 is violated: The all-zero second row should be at the bottom.

14. (A)
$$x_1 \qquad = -5$$
$$x_2 \quad = \quad 3$$
$$x_3 = \quad 6$$
Solution:
$$x_1 = -5, x_2 = 3, x_3 = 6$$

(B)
$$x_1 + 2x_2 - 3x_3 = 0$$
$$0x_1 + 0x_2 + 0x_3 = 1$$
$$0x_1 + 0x_2 + 0x_3 = 0$$
Inconsistent; no solution.

(C)
$$x_1 \quad - 2x_3 = \quad 4$$
$$x_2 + 3x_3 = -2$$
Dependent: let $x_3 = t$.

Then for any real t,
$$x_1 = 2t + 4$$
$$x_2 = -3t - 2$$
$$x_3 = t$$
is a solution.

(D)
$$x_1 \quad + 3x_3 + 2x_4 = 5$$
$$x_2 - 2x_3 - \quad x_4 = 3$$
Dependent: let $x_3 = s$ and $x_4 = t$.
Then for any real s and t,
$$x_1 = -3s - 2t + 5$$
$$x_2 = 2s + t + 3$$
$$x_3 = s$$
$$x_4 = t$$
is a solution.

15. $x_1 = 1, x_2 = -1, x_3 = 0$
16. $x_1 = -\frac{3}{5}t - \frac{4}{15}, x_2 = \frac{4}{5}t + \frac{7}{15}, x_3 = t$, t any real number
17. $x_1 = 2, x_2 = -1$
18. $x_1 = t + 10, x_2 = -4t + 10, x_3 = t$, where $t = 0, 1, 2$; that is, $(x_1, x_2, x_3) = (10, 10, 0)$, $(11, 6, 1)$, or $(12, 2, 2)$

Exercise 7-3

A   *Indicate whether each matrix is in reduced form.*

1. $\begin{bmatrix} 1 & 0 & | & 2 \\ 0 & 1 & | & -1 \end{bmatrix}$

2. $\begin{bmatrix} 0 & 1 & | & 2 \\ 1 & 0 & | & -1 \end{bmatrix}$

3. $\begin{bmatrix} 1 & 0 & 2 & | & 3 \\ 0 & 0 & 0 & | & 0 \\ 0 & 1 & -1 & | & 4 \end{bmatrix}$ **4.** $\begin{bmatrix} 1 & 0 & 0 & | & -2 \\ 0 & 1 & 0 & | & 0 \\ 0 & 0 & 1 & | & 1 \end{bmatrix}$

5. $\begin{bmatrix} 0 & 1 & 0 & | & 2 \\ 0 & 0 & 3 & | & -1 \\ 0 & 0 & 0 & | & 0 \end{bmatrix}$ **6.** $\begin{bmatrix} 1 & 3 & 0 & | & 0 \\ 0 & 0 & 1 & | & 0 \\ 0 & 0 & 0 & | & 1 \end{bmatrix}$

7. $\begin{bmatrix} 1 & 2 & 0 & 3 & | & 2 \\ 0 & 0 & 1 & -1 & | & 0 \end{bmatrix}$ **8.** $\begin{bmatrix} 0 & 1 & 2 & | & 1 \\ 1 & 0 & -3 & | & 2 \end{bmatrix}$

Write the linear system corresponding to each reduced augmented matrix and solve.

9. $\begin{bmatrix} 1 & 0 & 0 & | & -2 \\ 0 & 1 & 0 & | & 3 \\ 0 & 0 & 1 & | & 0 \end{bmatrix}$ **10.** $\begin{bmatrix} 1 & 0 & 0 & 0 & | & -2 \\ 0 & 1 & 0 & 0 & | & 0 \\ 0 & 0 & 1 & 0 & | & 1 \\ 0 & 0 & 0 & 1 & | & 3 \end{bmatrix}$

11. $\begin{bmatrix} 1 & 0 & -2 & | & 3 \\ 0 & 1 & 1 & | & -5 \\ 0 & 0 & 0 & | & 0 \end{bmatrix}$ **12.** $\begin{bmatrix} 1 & -2 & 0 & | & -3 \\ 0 & 0 & 1 & | & 5 \\ 0 & 0 & 0 & | & 0 \end{bmatrix}$

13. $\begin{bmatrix} 1 & 0 & | & 0 \\ 0 & 1 & | & 0 \\ 0 & 0 & | & 1 \end{bmatrix}$ **14.** $\begin{bmatrix} 1 & 0 & | & 5 \\ 0 & 1 & | & -3 \\ 0 & 0 & | & 0 \end{bmatrix}$

15. $\begin{bmatrix} 1 & -2 & 0 & -3 & | & -5 \\ 0 & 0 & 1 & 3 & | & 2 \end{bmatrix}$ **16.** $\begin{bmatrix} 1 & 0 & -2 & 3 & | & 4 \\ 0 & 1 & -1 & 2 & | & -1 \end{bmatrix}$

B   Use row operations to change each matrix to reduced form.

17. $\begin{bmatrix} 1 & 2 & | & -1 \\ 0 & 1 & | & 3 \end{bmatrix}$ **18.** $\begin{bmatrix} 1 & 3 & | & 1 \\ 0 & 2 & | & -4 \end{bmatrix}$

19. $\begin{bmatrix} 1 & 0 & -3 & | & 1 \\ 0 & 1 & 2 & | & 0 \\ 0 & 0 & 3 & | & -6 \end{bmatrix}$ **20.** $\begin{bmatrix} 1 & 0 & 4 & | & 0 \\ 0 & 1 & -3 & | & -1 \\ 0 & 0 & -2 & | & 2 \end{bmatrix}$

21. $\begin{bmatrix} 1 & 2 & -2 & | & -1 \\ 0 & 3 & -6 & | & 1 \\ 0 & -1 & 2 & | & -\frac{1}{3} \end{bmatrix}$ **22.** $\begin{bmatrix} 0 & -2 & 8 & | & 1 \\ 2 & -2 & 6 & | & -4 \\ 0 & -1 & 4 & | & \frac{1}{2} \end{bmatrix}$

Solve using Gauss–Jordan elimination.

23. $2x_1 + 4x_2 - 10x_3 = -2$
$3x_1 + 9x_2 - 21x_3 = 0$
$x_1 + 5x_2 - 12x_3 = 1$

24. $3x_1 + 5x_2 - x_3 = -7$
$x_1 + x_2 + x_3 = -1$
$2x_1 + 11x_3 = 7$

25. $3x_1 + 8x_2 - x_3 = -18$
$2x_1 + x_2 + 5x_3 = 8$
$2x_1 + 4x_2 + 2x_3 = -4$

26. $2x_1 + 7x_2 + 15x_3 = -12$
$4x_1 + 7x_2 + 13x_3 = -10$
$3x_1 + 6x_2 + 12x_3 = -9$

27. $2x_1 - x_2 - 3x_3 = 8$
$x_1 - 2x_2 = 7$

28. $2x_1 + 4x_2 - 6x_3 = 10$
$3x_1 + 3x_2 - 3x_3 = 6$

29. $2x_1 + 3x_2 - x_3 = 1$
$x_1 - 2x_2 + 2x_3 = -2$

30. $x_1 - 3x_2 + 2x_3 = -1$
$3x_1 + 2x_2 - x_3 = 2$

31. $2x_1 + 2x_2 = 2$
$x_1 + 2x_2 = 3$
$-3x_2 = -6$

32. $2x_1 - x_2 = 0$
$3x_1 + 2x_2 = 7$
$x_1 - x_2 = -1$

33. $2x_1 - x_2 = 0$
$3x_1 + 2x_2 = 7$
$x_1 - x_2 = -2$

34. $x_1 - 3x_2 = 5$
$2x_1 + x_2 = 3$
$x_1 - 2x_2 = 5$

35. $3x_1 - 4x_2 - x_3 = 1$
$2x_1 - 3x_2 + x_3 = 1$
$x_1 - 2x_2 + 3x_3 = 2$

36. $3x_1 + 7x_2 - x_3 = 11$
$x_1 + 2x_2 - x_3 = 3$
$2x_1 + 4x_2 - 2x_3 = 10$

C *Solve using Gauss–Jordan elimination*

37. $2x_1 - 3x_2 + 3x_3 = -15$
$3x_1 + 2x_2 - 5x_3 = 19$
$5x_1 - 4x_2 - 2x_3 = -2$

38. $3x_1 - 2x_2 - 4x_3 = -8$
$4x_1 + 3x_2 - 5x_3 = -5$
$6x_1 - 5x_2 + 2x_3 = -17$

39. $5x_1 - 3x_2 + 2x_3 = 13$
$2x_1 + 4x_2 - 3x_3 = -9$
$4x_1 - 2x_2 + 5x_3 = 13$

40. $4x_1 - 2x_2 + 3x_3 = 0$
$3x_1 - 5x_2 - 2x_3 = -12$
$2x_1 + 4x_2 - 3x_3 = -4$

41. $x_1 + 2x_2 - 4x_3 - x_4 = 7$
$2x_1 + 5x_2 - 9x_3 - 4x_4 = 16$
$x_1 + 5x_2 - 7x_3 - 7x_4 = 13$

42. $2x_1 + 4x_2 + 5x_3 + 4x_4 = 8$
$x_1 + 2x_2 + 2x_3 + x_4 = 3$

■

Applications

Solve all of the following problems using Gauss–Jordan elimination.

Business & Economics

43. *Production scheduling.* A small manufacturing plant makes three types of inflatable boats: one-person, two-person, and four-person models. Each boat requires the services of three departments, as listed in the table. The cutting, assembly, and packaging departments have available a maximum of 380, 330, and 120 labor-hours per week, respectively. How many boats of each type must be produced each week for the plant to operate at full capacity?

	One-Person Boat	Two-Person Boat	Four-Person Boat
Cutting department	0.5 hr	1.0 hr	1.5 hr *380*
Assembly department	0.6 hr	0.9 hr	1.2 hr *330.*
Packaging department	0.2 hr	0.3 hr	0.5 hr

44. *Production scheduling.* Repeat Problem 43 assuming the cutting, assembly, and packaging departments have available a maximum of 350, 330,and 115 labor-hours per week, respectively.

45. *Production scheduling.* Work Problem 43 assuming the packaging department is no longer used.

46. *Production scheduling.* Work Problem 44 assuming the packaging department is no longer used.

47. *Production scheduling.* Work Problem 43 assuming the four-person boat is no longer produced.

48. *Production scheduling.* Work Problem 44 assuming the four-person boat is no longer produced.

Life Sciences

49. *Nutrition.* A dietitian in a hospital is to arrange a special diet composed of three basic foods. The diet is to include exactly 340 units of calcium, 180 units of iron,and 220 units of vitamin A. The number of units per ounce of each special ingredient for each of the foods is indicated in the table. How many ounces of each food must be used to meet the diet requirements?

Units per Ounce			
	Food A	Food B	Food C
Calcium	30	10	20
Iron	10	10	20
Vitamin A	10	30	20

50. *Nutrition.* Repeat Problem 49 if the diet is to include exactly 400 units of calcium, 160 units of iron, and 240 units of vitamim A.

51. *Nutrition.* Solve Problem 49 with the assumption that food C is no longer available.

52. *Nutrition.* Solve Problem 50 with the assumption that food C is no longer available.

53. *Nutrition.* Solve Problem 49 with the assumption that the vitamin A requirement is deleted.

54. *Nutrition.* Solve Problem 50 with the assumption that the vitamin A requirement is deleted.

Social Sciences

55. *Sociology.* Two sociologists have grant money to study school busing in a particular city. They wish to conduct an opinion survey using 600 telephone contacts and 400 house contacts. Survey company *A* has personnel to do 30 telephone and 10 house contacts per hour; survey company *B* can handle 20 telephone and 20 house contacts per hour. How many hours should be scheduled for each firm to produce exactly the number of contacts needed?

56. *Sociology.* Repeat Problem 55 if 650 telephone contacts and 350 house contacts are needed.

7-4 Matrices—Addition and Multiplication by a Number

- Basic Definitions
- Sum and Difference
- Product of a Number *k* and a Matrix *M*
- Application

In the last two sections we introduced the important new idea of matrices. In this and the following sections, we shall develop this concept further.

■ Basic Definitions

Recall that we defined a **matrix** as any rectangular array of numbers enclosed within brackets. The **size** or **dimension of a matrix** is important to operations on matrices. We define an $m \times n$ **matrix** (read "m by n matrix") to be one with m rows and n columns. It is important to note that the number of rows is always given first. If a matrix has the same number of rows and columns, it is called a **square matrix.** A matrix with only one column is called a **column matrix,** and one with only one row is called a **row matrix.** These definitions are illustrated by the following:

$$
\begin{matrix}
3 \times 2 \\
\begin{bmatrix} -2 & 5 \\ 0 & -2 \\ 3 & 6 \end{bmatrix}
\end{matrix}
\qquad
\begin{matrix}
3 \times 3 \\
\begin{bmatrix} 0.5 & 0.2 & 1.0 \\ 0.0 & 0.3 & 0.5 \\ 0.7 & 0.0 & 0.2 \end{bmatrix} \\
\text{Square matrix}
\end{matrix}
\qquad
\begin{matrix}
4 \times 1 \\
\begin{bmatrix} 3 \\ -2 \\ 1 \\ 0 \end{bmatrix} \\
\text{Column matrix}
\end{matrix}
\qquad
\begin{matrix}
1 \times 4 \\
[2 \quad \tfrac{1}{2} \quad 0 \quad -\tfrac{2}{3}] \\
\text{Row matrix}
\end{matrix}
$$

Two matrices are **equal** if they have the same dimension and their corresponding elements are equal. For example,

$$
\begin{matrix}
2 \times 3 \\
\begin{bmatrix} a & b & c \\ d & e & f \end{bmatrix}
\end{matrix}
=
\begin{matrix}
2 \times 3 \\
\begin{bmatrix} u & v & w \\ x & y & z \end{bmatrix}
\end{matrix}
\quad \text{if and only if} \quad
\begin{matrix}
a = u & b = v & c = w \\
d = x & e = y & f = z
\end{matrix}
$$

■ Sum and Difference

The **sum of two matrices of the same dimension** is a matrix with elements that are the sum of the corresponding elements of the two given matrices. **Addition is not defined for matrices with different dimensions.**

Example 19   (A) $\begin{bmatrix} a & b \\ c & d \end{bmatrix} + \begin{bmatrix} w & x \\ y & z \end{bmatrix} = \begin{bmatrix} (a+w) & (b+x) \\ (c+y) & (d+z) \end{bmatrix}$

(B) $\begin{bmatrix} 2 & -3 & 0 \\ 1 & 2 & -5 \end{bmatrix} + \begin{bmatrix} 3 & 1 & 2 \\ -3 & 2 & 5 \end{bmatrix} = \begin{bmatrix} 5 & -2 & 2 \\ -2 & 4 & 0 \end{bmatrix}$

Problem 19   Add:

$$\begin{bmatrix} 3 & 2 \\ -1 & -1 \\ 0 & 3 \end{bmatrix} + \begin{bmatrix} -2 & 3 \\ 1 & -1 \\ 2 & -2 \end{bmatrix}$$

Because we add two matrices by adding their corresponding elements, it follows from the properties of real numbers that matrices of the same dimension are commutative and associative relative to addition. That is, if A, B, and C are matrices of the same dimension, then

$$A + B = B + A \qquad \text{Commutative}$$
$$(A + B) + C = A + (B + C) \qquad \text{Associative}$$

A matrix with elements that are all zeros is called a **zero matrix.** For example,

$$[0 \quad 0 \quad 0] \qquad \begin{bmatrix} 0 & 0 \\ 0 & 0 \end{bmatrix} \qquad \begin{bmatrix} 0 \\ 0 \\ 0 \\ 0 \end{bmatrix} \qquad \begin{bmatrix} 0 & 0 & 0 & 0 \\ 0 & 0 & 0 & 0 \\ 0 & 0 & 0 & 0 \end{bmatrix}$$

are zero matrices of different dimensions. [Note: "0" is often used to denote the zero matrix of an arbitrary dimension.] The **negative of a matrix M,** denoted by $-M$, is a matrix with elements that are the negatives of the elements in M. Thus, if

$$M = \begin{bmatrix} a & b \\ c & d \end{bmatrix} \qquad \text{then} \qquad -M = \begin{bmatrix} -a & -b \\ -c & -d \end{bmatrix}$$

Note that $M + (-M) = 0$ (a zero matrix).

If A and B are matrices of the same dimension, then we define **subtraction** as follows:

$$A - B = A + (-B)$$

Thus, to subtract matrix B from matrix A, we simply subtract corresponding elements.

Example 20

$$\begin{bmatrix} 3 & -2 \\ 5 & 0 \end{bmatrix} - \begin{bmatrix} -2 & 2 \\ 3 & 4 \end{bmatrix} = \begin{bmatrix} 3 & -2 \\ 5 & 0 \end{bmatrix} + \begin{bmatrix} 2 & -2 \\ -3 & -4 \end{bmatrix} = \begin{bmatrix} 5 & -4 \\ 2 & -4 \end{bmatrix}$$

Problem 20 Subtract: $[2 \quad -3 \quad 5] - [3 \quad -2 \quad 1]$

■ Product of a Number k and a Matrix M

Finally, the **product of a number k and a matrix M,** denoted by kM, is a matrix formed by multiplying each element of M by k. This definition is partly motivated by the fact that if M is a matrix, then we would like $M + M$ to equal $2M$.

Example 21

$$-2 \begin{bmatrix} 3 & -1 & 0 \\ -2 & 1 & 3 \\ 0 & -1 & -2 \end{bmatrix} = \begin{bmatrix} -6 & 2 & 0 \\ 4 & -2 & -6 \\ 0 & 2 & 4 \end{bmatrix}$$

Problem 21 Find: $10 \begin{bmatrix} 1.3 \\ 0.2 \\ 3.5 \end{bmatrix}$

■ Application

Example 22 Ms. Smith and Mr. Jones are salespeople in a new-car agency that sells only two models. August was the last month for this year's models, and next year's models were introduced in September. Gross dollar sales for each month are given in the following matrices:

| | August sales | | | September sales | |
	Compact	Luxury		Compact	Luxury
Ms. Smith	$18,000	$36,000	$= A$	$72,000	$144,000	$= B$
Mr. Jones	$36,000	0		$90,000	$108,000	

(For example, Ms. Smith had $18,000 in compact sales in August, and Mr. Jones had $108,000 in luxury car sales in September.)

(A) What was the combined dollar sales in August and September for each person and each model?
(B) What was the increase in dollar sales from August to September?
(C) If both salespeople receive 5% commissions on gross dollar sales, compute the commission for each person for each model sold in September.

Solutions (A) $A + B =$

$$\begin{array}{cc} \text{Compact} & \text{Luxury} \end{array}$$
$$\begin{bmatrix} \$90{,}000 & \$180{,}000 \\ \$126{,}000 & \$108{,}000 \end{bmatrix} \begin{array}{l} \text{Ms. Smith} \\ \text{Mr. Jones} \end{array}$$

(B) $B - A =$

$$\begin{array}{cc} \text{Compact} & \text{Luxury} \end{array}$$
$$\begin{bmatrix} \$54{,}000 & \$108{,}000 \\ \$54{,}000 & \$108{,}000 \end{bmatrix} \begin{array}{l} \text{Ms. Smith} \\ \text{Mr. Jones} \end{array}$$

(C) $0.05B =$

$$\begin{bmatrix} (0.05)(\$72{,}000) & (0.05)(\$144{,}000) \\ (0.05)(\$90{,}000) & (0.05)(\$108{,}000) \end{bmatrix}$$

$$= \begin{bmatrix} \$3{,}600 & \$7{,}200 \\ \$4{,}500 & \$5{,}400 \end{bmatrix} \begin{array}{l} \text{Ms. Smith} \\ \text{Mr. Jones} \end{array}$$

In Example 22 we chose a relatively simple example involving an agency with only two salespeople and two models. Consider the more realistic problem of an agency with nine models and perhaps seven salespeople — then you can begin to see the value of matrix methods.

Problem 22 Repeat Example 22 with

$$A = \begin{bmatrix} \$36{,}000 & \$36{,}000 \\ \$18{,}000 & \$36{,}000 \end{bmatrix} \quad \text{and} \quad B = \begin{bmatrix} \$90{,}000 & \$108{,}000 \\ \$72{,}000 & \$108{,}000 \end{bmatrix}$$

Answers to Matched Problems

19. $\begin{bmatrix} 1 & 5 \\ 0 & -2 \\ 2 & 1 \end{bmatrix}$ 20. $[-1 \quad -1 \quad 4]$ 21. $\begin{bmatrix} 13 \\ 2 \\ 35 \end{bmatrix}$

22. (A) $\begin{bmatrix} \$126{,}000 & \$144{,}000 \\ \$90{,}000 & \$144{,}000 \end{bmatrix}$ (B) $\begin{bmatrix} \$54{,}000 & \$72{,}000 \\ \$54{,}000 & \$72{,}000 \end{bmatrix}$

(C) $\begin{bmatrix} \$4{,}500 & \$5{,}400 \\ \$3{,}600 & \$5{,}400 \end{bmatrix}$

Exercise 7-4

A   *Problems 1–18 refer to the following matrices:*

$$A = \begin{bmatrix} 2 & -1 \\ 3 & 0 \end{bmatrix} \qquad B = \begin{bmatrix} -3 & 1 \\ 2 & -3 \end{bmatrix} \qquad C = \begin{bmatrix} 2 \\ -3 \\ 0 \end{bmatrix}$$

$$D = \begin{bmatrix} 1 \\ 3 \\ 5 \end{bmatrix} \qquad E = [-4 \quad 1 \quad 0 \quad -2] \qquad F = \begin{bmatrix} 2 & -3 \\ -2 & 0 \\ 1 & 2 \\ 3 & 5 \end{bmatrix}$$

1. What are the dimensions of B? Of E?
2. What are the dimensions of F? Of D?
3. What element is in the third row and second column of matrix F?
4. What element is in the second row and first column of matrix F?
5. Write a zero matrix of the same dimension as B.
6. Write a zero matrix of the same dimension as E.
7. Identify all column matrices.
8. Identify all row matrices.
9. Identify all square matrices.
10. How many additional columns would F have to have to be a square matrix?
11. Find $A + B$.
12. Find $C + D$.
13. Write the negative of matrix C.
14. Write the negative of matrix B.
15. Find $D - C$.
16. Find $A - A$.
17. Find 5B.
18. Find $-2E$.

B   *In Problems 19–26 perform the indicated operations.*

19. $\begin{bmatrix} 3 & -2 & 0 & 1 \\ 2 & -3 & -1 & 4 \\ 0 & 2 & -1 & 6 \end{bmatrix} + \begin{bmatrix} -2 & 5 & -1 & 0 \\ -3 & -2 & 8 & -2 \\ 4 & 6 & 1 & -8 \end{bmatrix}$

20. $\begin{bmatrix} 4 & -2 & 8 \\ 0 & -1 & -4 \\ -6 & 5 & 2 \\ 1 & 3 & -6 \end{bmatrix} + \begin{bmatrix} -6 & -2 & -3 \\ 5 & 2 & 4 \\ 8 & 3 & -4 \\ 1 & -5 & 0 \end{bmatrix}$

21. $\begin{bmatrix} 1.3 & 2.5 & -6.1 \\ 8.3 & -1.4 & 6.7 \end{bmatrix} - \begin{bmatrix} -4.1 & 1.8 & -4.3 \\ 0.7 & 2.6 & -1.2 \end{bmatrix}$

22. $\begin{bmatrix} 2.6 & 3.8 \\ -1.9 & 7.3 \\ 5.6 & -0.4 \end{bmatrix} - \begin{bmatrix} 4.8 & -2.1 \\ 3.2 & 5.9 \\ -1.5 & 2.2 \end{bmatrix}$ **23.** $1,000 \begin{bmatrix} 0.25 & 0.36 \\ 0.04 & 0.35 \end{bmatrix}$

24. $100 \begin{bmatrix} 0.32 & 0.05 & 0.17 \\ 0.22 & 0.03 & 0.21 \end{bmatrix}$

25. $0.08 \begin{bmatrix} 24{,}000 & 35{,}000 \\ 12{,}000 & 24{,}000 \end{bmatrix} + 0.03 \begin{bmatrix} 12{,}000 & 22{,}000 \\ 14{,}000 & 13{,}000 \end{bmatrix}$

26. $0.05 \begin{bmatrix} 430 & 212 \\ 210 & 165 \\ 435 & 315 \end{bmatrix} + 0.07 \begin{bmatrix} 234 & 436 \\ 160 & 212 \\ 410 & 136 \end{bmatrix}$

C **27.** Find a, b, c, and d so that

$$\begin{bmatrix} a & b \\ c & d \end{bmatrix} + \begin{bmatrix} 2 & -3 \\ 0 & 1 \end{bmatrix} = \begin{bmatrix} 1 & -2 \\ 3 & -4 \end{bmatrix}$$

28. Find w, x, y, and z so that

$$\begin{bmatrix} 4 & -2 \\ -3 & 0 \end{bmatrix} + \begin{bmatrix} w & x \\ y & z \end{bmatrix} = \begin{bmatrix} 2 & -3 \\ 0 & 5 \end{bmatrix}$$

29. Find x and y so that

$$\begin{bmatrix} 2x & 4 \\ -3 & 5x \end{bmatrix} + \begin{bmatrix} 3y & -2 \\ -2 & -y \end{bmatrix} = \begin{bmatrix} -5 & 2 \\ -5 & 13 \end{bmatrix}$$

30. Find x and y so that

$$\begin{bmatrix} 5 & 3x \\ 2x & -4 \end{bmatrix} + \begin{bmatrix} 1 & -4y \\ 7y & 4 \end{bmatrix} = \begin{bmatrix} 6 & -7 \\ 5 & 0 \end{bmatrix}$$

Applications

Business & Economics

31. *Cost analysis.* A company with two different plants manufactures guitars and banjos. Its production costs for each instrument are given in the following matrices:

	Plant X			Plant Y	
	Guitar	Banjo		Guitar	Banjo
Materials	$30	$25 $=A$		$36	$27 $=B$
Labor	$60	$80		$54	$74

Find $\frac{1}{2}(A + B)$, the average cost of production for the two plants.

Life Sciences

32. *Heredity.* Gregor Mendel (1822–1884), an Austrian monk and botanist, made discoveries that revolutionized the science of heredity. In one experiment, he crossed dihybrid yellow round peas (yellow and round are dominant characteristics; the peas also contained genes for the recessive characteristics green and wrinkled) and obtained 560 peas of the types indicated in the matrix:

$$\begin{array}{cc} & \text{Round} \quad \text{Wrinkled} \\ \begin{matrix} \text{Yellow} \\ \text{Green} \end{matrix} & \begin{bmatrix} 319 & 101 \\ 108 & 32 \end{bmatrix} = M \end{array}$$

Suppose he carried out a second experiment of the same type and obtained 640 peas of the types indicated in this matrix:

$$\begin{array}{cc} & \text{Round} \quad \text{Wrinkled} \\ \begin{matrix} \text{Yellow} \\ \text{Green} \end{matrix} & \begin{bmatrix} 370 & 124 \\ 110 & 36 \end{bmatrix} = N \end{array}$$

If the results of the two experiments are combined, write the resulting matrix $M + N$. Compute the percentage of the total number of peas (1,200) in each category of the combined results. [*Hints:* Compute $(1/1{,}200)(M + N)$.]

Social Sciences

33. *Psychology.* Two psychologists independently carried out studies on the relationship between height and aggressive behavior in women over 18 years of age. The results of the studies are summarized in the following matrices:

Professor Aldquist

$$\begin{array}{cc} & \text{Under 5 ft} \quad 5-5\tfrac{1}{2}\text{ ft} \quad \text{Over } 5\tfrac{1}{2}\text{ ft} \\ \begin{matrix} \text{Passive} \\ \text{Aggressive} \end{matrix} & \begin{bmatrix} 70 & 122 & 20 \\ 30 & 118 & 80 \end{bmatrix} = A \end{array}$$

Professor Kelley

$$\begin{array}{cc} & \text{Under 5 ft} \quad 5-5\tfrac{1}{2}\text{ ft} \quad \text{Over } 5\tfrac{1}{2}\text{ ft} \\ \begin{matrix} \text{Passive} \\ \text{Aggressive} \end{matrix} & \begin{bmatrix} 65 & 160 & 30 \\ 25 & 140 & 75 \end{bmatrix} = B \end{array}$$

The two psychologists decided to combine their results and publish a joint paper. Write the matrix $A + B$ illustrating their combined results. Compute the percentage of the total sample in each category of the combined study. [*Hint:* Compute $(\tfrac{1}{935})(A + B)$.]

7-5 Matrix Multiplication

- ■ Dot Product
- ■ Application
- ■ Matrix Product
- ■ Arithmetic of Matrix Products
- ■ Application

In this section, we are going to introduce two types of matrix multiplication that will seem rather strange at first. In spite of this apparent strangeness, these operations are well founded in the general theory of matrices and, as we will see, extremely useful in practical problems.

■ Dot Product

We start by defining the **dot product** of two special matrices, a $1 \times n$ row matrix and an $n \times 1$ column matrix:

$$[a_1 \quad a_2 \cdots a_n] \cdot \begin{bmatrix} b_1 \\ b_2 \\ \cdot \\ \cdot \\ \cdot \\ b_n \end{bmatrix} = a_1 b_1 + a_2 b_2 + \cdots + a_n b_n \qquad \text{A real number}$$

The dot product is a real number, not a matrix. The dot between the two matrices is important. If the dot is omitted, the multiplication is of another type, which we will consider below.

Example 23
$$[2 \quad -3 \quad 0] \cdot \begin{bmatrix} -5 \\ 2 \\ -2 \end{bmatrix} = (2)(-5) + (-3)(2) + (0)(-2)$$
$$= -10 - 6 + 0 = -16$$

Problem 23
$$[-1 \quad 0 \quad 3 \quad 2] \cdot \begin{bmatrix} 2 \\ 3 \\ 4 \\ -1 \end{bmatrix} = ?$$

■ Application

Example 24

A factory produces a slalom water ski that requires 4 labor-hours in the fabricating department and 1 labor-hour in the finishing department. Fabricating personnel receive $8 per hour and finishing personnel receive $6 per hour. Total labor cost per ski is given by the dot product:

$$[4 \quad 1] \cdot \begin{bmatrix} 8 \\ 6 \end{bmatrix} = (4)(8) + (1)(6) = 32 + 6 = \$38 \text{ per ski}$$

Problem 24

If the factory in Example 24 also produces a trick water ski that requires 6 labor-hours in the fabricating department and 1.5 labor-hours in the finishing department, write a dot product between appropriate row and column matrices that will give the total labor cost for this ski. Compute the cost.

■ Matrix Product

It is important to remember that the dot product of a row matrix and a column matrix is a real number and not a matrix. We now define a matrix product for certain matrices. First, the product of two matrices A and B is defined only if the number of columns of A is equal to the number of rows of B. If A is an $m \times p$ matrix and B is a $p \times n$ matrix, then the **matrix product** of A and B, denoted by AB (not BA), is an $m \times n$ matrix whose element in the ith row and jth column is the dot product of the ith row matrix of A and the jth column matrix of B.

It is important to check dimensions before starting the multiplication process. If matrix A has dimension $a \times b$ and matrix B has dimension $c \times d$, then if $b = c$, the product AB will exist and have dimension $a \times d$. This is shown schematically in Figure 2. The definition is not as complicated as it might seem at first. An example should help to clarify the process. For

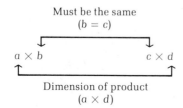

Must be the same
$(b = c)$

$a \times b \qquad c \times d$

Dimension of product
$(a \times d)$

Figure 2

$$A = \begin{bmatrix} 2 & 3 & -1 \\ -2 & 1 & 2 \end{bmatrix} \quad \text{and} \quad B = \begin{bmatrix} 1 & 3 \\ 2 & 0 \\ -1 & 2 \end{bmatrix}$$

A is 2×3, B is 3×2, and AB will be 2×2. The four dot products used to produce the four elements in AB (usually calculated mentally or with the aid of a hand calculator) are shown in the dashed box at the top of the next page. The shaded portions highlight the steps involved in computing the element in the first row and second column of the product matrix.

$$
\underset{2\times 3}{\begin{bmatrix} 2 & 3 & -1 \\ -2 & 1 & 2 \end{bmatrix}}
\begin{bmatrix} 1 & 3 \\ 2 & 0 \\ -1 & 2 \end{bmatrix}
=
\begin{bmatrix} [2 \;\; 3 \;\; -1]\cdot\begin{bmatrix}1\\2\\-1\end{bmatrix} & [2 \;\; 3 \;\; -1]\cdot\begin{bmatrix}3\\0\\2\end{bmatrix} \\[4pt] [-2 \;\; 1 \;\; 2]\cdot\begin{bmatrix}1\\2\\-1\end{bmatrix} & [-2 \;\; 1 \;\; 2]\cdot\begin{bmatrix}3\\0\\2\end{bmatrix} \end{bmatrix}
=
\underset{2\times 2}{\begin{bmatrix} 9 & 4 \\ -2 & -2 \end{bmatrix}}
$$

Example 25 Find the product AB, given

$$
A = \begin{bmatrix} 2 & 1 \\ 1 & 0 \\ -1 & 2 \end{bmatrix} \quad \text{and} \quad B = \begin{bmatrix} 1 & -1 & 0 & 1 \\ 2 & 1 & 2 & 0 \end{bmatrix}
$$

Solution A convenient way to carry out the multiplication is to arrange the matrices as shown below. The rows and columns in the product matrix will then be determined automatically.

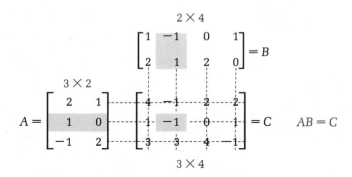

Problem 25 Find the product AB, given:

(A) $A = \begin{bmatrix} 1 & -1 & 0 \\ 2 & 1 & 2 \end{bmatrix}$ and $B = \begin{bmatrix} 2 & -1 \\ 1 & 0 \\ -1 & 2 \end{bmatrix}$

(B) $A = \begin{bmatrix} 3 & 2 & -1 \end{bmatrix}$ and $B = \begin{bmatrix} -2 \\ 0 \\ 3 \end{bmatrix}$

Note: This is a matrix product, not a dot product. The result is a matrix, not a real number.

■ Arithmetic of Matrix Products

Relative to addition, in the last section we noted that

$$A + B = B + A \qquad \text{Commutative}$$

$$A + (B + C) = (A + B) + C \qquad \text{Associative}$$

where A, B, and C are matrices of the same dimension. Do similar properties hold for multiplication? What about the commutative property? Let us compute AB and BA for

$$A = \begin{bmatrix} -3 & 5 \\ 2 & 0 \end{bmatrix} \quad \text{and} \quad B = \begin{bmatrix} 2 & -1 \\ 4 & 3 \end{bmatrix}$$

$$\begin{array}{cc}
2 \times 2 & 2 \times 2 \\
\begin{bmatrix} 2 & -1 \\ 4 & 3 \end{bmatrix} = B & \begin{bmatrix} -3 & 5 \\ 2 & 0 \end{bmatrix} = A
\end{array}$$

$$\begin{array}{cc}
2 \times 2 & 2 \times 2 \\
A = \begin{bmatrix} -3 & 5 \\ 2 & 0 \end{bmatrix} \begin{bmatrix} 14 & 18 \\ 4 & -2 \end{bmatrix} = C \qquad B = \begin{bmatrix} 2 & -1 \\ 4 & 3 \end{bmatrix} \begin{bmatrix} -8 & 10 \\ -6 & 20 \end{bmatrix} = D \\
2 \times 2 & 2 \times 2
\end{array}$$

$$AB = C \qquad\qquad BA = D$$

Thus, $AB \neq BA$. Only in some very special cases will matrix products commute; therefore, we must always be careful about the order in which matrix multiplication is performed—we cannot indiscriminately reverse order as in the arithmetic of real numbers. In fact, BA is often not even defined, even though AB is. Thus, $AB \neq BA$ in general.

Another important difference between matrix products and real number products is found in the zero property for real numbers:

For all real numbers a and b,

   $ab = 0$ **if and only if** $a = 0$ **or** $b = 0$ **(or both)**

For A and B matrices it is possible for $AB = 0$ and neither A nor B be 0. For example,

$$\begin{bmatrix} 2 & -6 \\ -3 & 9 \end{bmatrix} = B$$

$$A = \begin{bmatrix} 3 & 2 \\ 6 & 4 \end{bmatrix} \begin{bmatrix} 0 & 0 \\ 0 & 0 \end{bmatrix} = 0 \qquad \text{Zero matrix}$$

Here we see that $AB = 0$ and $A \neq 0$ and $B \neq 0$.

Matrix products do have some of the same types of properties as real number products. We state four important properties without proof. If all products and sums are defined for the indicated matrices A, B, and C, then for k a real number:

1. $(AB)C = A(BC)$ Associative property
2. $A(B + C) = AB + AC$ Left-hand distributive property
3. $(B + C)A = BA + CA$ Right-hand distributive property
4. $k(AB) = (kA)B = A(kB)$

Since matrix multiplication is not commutative, properties 2 and 3 must be listed as distinct properties.

■ Application

Example 26

Let us combine the time requirements for slalom and trick water skis discussed in Example 24 and Problem 24 into one matrix:

$$
\begin{array}{c}
\\
\text{Trick ski} \\
\text{Slalom ski}
\end{array}
\begin{array}{cc}
\text{Fabricating} & \text{Finishing} \\
\text{department} & \text{department} \\
\begin{bmatrix} 6 \text{ hr} & 1.5 \text{ hr} \\ 4 \text{ hr} & 1 \text{ hr} \end{bmatrix} = A
\end{array}
$$

Now suppose the company has two manufacturing plants X and Y in different parts of the country and that their hourly rates for each department are given in the following matrix:

$$
\begin{array}{c}
\\
\text{Fabricating department} \\
\text{Finishing department}
\end{array}
\begin{array}{cc}
\text{Plant X} & \text{Plant Y} \\
\begin{bmatrix} \$8 & \$7 \\ \$6 & \$4 \end{bmatrix} = B
\end{array}
$$

To find the total labor costs for each ski at each factory, we multiply A and B:

$$
AB = \overset{2 \times 2}{\begin{bmatrix} 6 & 1.5 \\ 4 & 1 \end{bmatrix}} \overset{2 \times 2}{\begin{bmatrix} 8 & 7 \\ 6 & 4 \end{bmatrix}} = \begin{bmatrix} \overset{X}{\$57} & \overset{Y}{\$48} \\ \$38 & \$32 \end{bmatrix} \begin{array}{l} \text{Trick ski} \\ \text{Slalom ski} \end{array}
$$

Notice that the dot product of the first row matrix of A and the first column matrix of B gives us the labor costs, \$57, for a trick ski manufactured at plant X; the dot product of the second row matrix of A and the second column matrix of B gives us the labor costs, \$32, for manufacturing a slalom ski at plant Y; and so on.

Example 26 is, of course, overly simplified. Companies manufacturing many different items in many different plants deal with matrices that have very large numbers of rows and columns.

Problem 26   Repeat Example 26 with

$$A = \begin{bmatrix} 7 \text{ hr} & 2 \text{ hr} \\ 5 \text{ hr} & 1.5 \text{ hr} \end{bmatrix} \quad \text{and} \quad B = \begin{bmatrix} \$10 & \$8 \\ \$6 & \$4 \end{bmatrix}$$

Answers to Matched Problems

23. 8

24. $[6 \quad 1.5] \cdot \begin{bmatrix} 8 \\ 6 \end{bmatrix} = \$57$

25. (A) $\begin{bmatrix} 1 & -1 \\ 3 & 2 \end{bmatrix}$ (B) $[-9]$

26. $\begin{matrix} & X & Y \\ \begin{bmatrix} \$82 & \$64 \\ \$59 & \$46 \end{bmatrix} & \begin{matrix} \text{Trick} \\ \text{Slalom} \end{matrix} \end{matrix}$

Exercise 7-5

A   *Find the dot products.*

1. $[2 \quad 4] \cdot \begin{bmatrix} 3 \\ 1 \end{bmatrix}$

2. $[3 \quad 1] \cdot \begin{bmatrix} 2 \\ 4 \end{bmatrix}$

3. $[-3 \quad 2] \cdot \begin{bmatrix} -1 \\ -2 \end{bmatrix}$

4. $[3 \quad -2] \cdot \begin{bmatrix} -4 \\ -1 \end{bmatrix}$

Find the matrix products.

5. $[2 \quad 5] \begin{bmatrix} 1 & -1 \\ 2 & 3 \end{bmatrix}$

6. $[1 \quad 3] \begin{bmatrix} 2 & 3 \\ 1 & -4 \end{bmatrix}$

7. $\begin{bmatrix} 3 & 4 \\ -1 & -2 \end{bmatrix} \begin{bmatrix} -1 \\ 2 \end{bmatrix}$

8. $\begin{bmatrix} -1 & 1 \\ 2 & -3 \end{bmatrix} \begin{bmatrix} 4 \\ -2 \end{bmatrix}$

9. $\begin{bmatrix} 2 & -3 \\ 1 & 2 \end{bmatrix} \begin{bmatrix} 1 & -1 \\ 0 & -2 \end{bmatrix}$

10. $\begin{bmatrix} -3 & 2 \\ 4 & -1 \end{bmatrix} \begin{bmatrix} -2 & 5 \\ -1 & 3 \end{bmatrix}$

11. $\begin{bmatrix} 1 & -1 \\ 0 & -2 \end{bmatrix} \begin{bmatrix} 2 & -3 \\ 1 & 2 \end{bmatrix}$
(Compare with Problem 9.)

12. $\begin{bmatrix} -2 & 5 \\ -1 & 3 \end{bmatrix} \begin{bmatrix} -3 & 2 \\ 4 & -1 \end{bmatrix}$
(Compare with Problem 10.)

13. $[5 \quad -2] \begin{bmatrix} -3 \\ -4 \end{bmatrix}$

14. $[-4 \quad 3] \begin{bmatrix} -2 \\ 1 \end{bmatrix}$

15. $\begin{bmatrix} -3 \\ -4 \end{bmatrix} [5 \quad -2]$

16. $\begin{bmatrix} -2 \\ 1 \end{bmatrix} [-4 \quad 3]$

B *Find the dot products.*

17. $[-1 \quad -2 \quad 2] \cdot \begin{bmatrix} 2 \\ -1 \\ 3 \end{bmatrix}$

18. $[-2 \quad 4 \quad 0] \cdot \begin{bmatrix} -1 \\ -3 \\ 2 \end{bmatrix}$

19. $[-1 \quad -3 \quad 0 \quad 5] \cdot \begin{bmatrix} 4 \\ -3 \\ -1 \\ 2 \end{bmatrix}$

20. $[-1 \quad 2 \quad 3 \quad -2] \cdot \begin{bmatrix} 3 \\ -2 \\ 0 \\ 4 \end{bmatrix}$

Find the matrix products.

21. $\begin{bmatrix} 2 & -1 & 1 \\ 1 & 3 & -2 \end{bmatrix} \begin{bmatrix} 1 & 3 \\ 0 & -1 \\ -2 & 2 \end{bmatrix}$

22. $\begin{bmatrix} -1 & -4 & 3 \\ 2 & 0 & 1 \end{bmatrix} \begin{bmatrix} 2 & -3 \\ 1 & 2 \\ 0 & -1 \end{bmatrix}$

23. $\begin{bmatrix} 1 & 3 \\ 0 & -1 \\ -2 & 2 \end{bmatrix} \begin{bmatrix} 2 & -1 & 1 \\ 1 & 3 & -2 \end{bmatrix}$

24. $\begin{bmatrix} 2 & -3 \\ 1 & 2 \\ 0 & -1 \end{bmatrix} \begin{bmatrix} -1 & -4 & 3 \\ 2 & 0 & 1 \end{bmatrix}$

25. $[3 \quad -2 \quad -4] \begin{bmatrix} 1 \\ 2 \\ -3 \end{bmatrix}$

26. $[1 \quad -2 \quad 2] \begin{bmatrix} 2 \\ -1 \\ 1 \end{bmatrix}$

27. $\begin{bmatrix} 1 \\ 2 \\ -3 \end{bmatrix} [3 \quad -2 \quad -4]$

28. $\begin{bmatrix} 2 \\ -1 \\ 1 \end{bmatrix} [1 \quad -2 \quad 2]$

29. $\begin{bmatrix} 2 & -1 & 3 & 0 \\ -3 & 4 & 2 & -1 \\ 0 & -2 & 1 & 4 \end{bmatrix} \begin{bmatrix} 2 & -3 & -2 \\ 1 & 0 & 1 \\ -1 & 2 & 0 \\ 2 & -2 & -3 \end{bmatrix}$

(Compare with Problem 30.)

30. $\begin{bmatrix} 2 & -3 & -2 \\ 1 & 0 & 1 \\ -1 & 2 & 0 \\ 2 & -2 & -3 \end{bmatrix} \begin{bmatrix} 2 & -1 & 3 & 0 \\ -3 & 4 & 2 & -1 \\ 0 & -2 & 1 & 4 \end{bmatrix}$

(Compare with Problem 29.)

C   **31.** $\begin{bmatrix} 2.1 & 3.2 & -1.1 \\ -0.8 & 5.7 & -4.3 \end{bmatrix} \begin{bmatrix} -4.5 & 3.7 \\ 1.1 & -2.6 \\ -2.0 & 4.3 \end{bmatrix}$

32. $\begin{bmatrix} 6.4 & 2.0 \\ -2.8 & 3.9 \\ -1.5 & -2.4 \end{bmatrix} \begin{bmatrix} -6.3 & 3.6 \\ -2.7 & 2.2 \end{bmatrix}$

In Problems 33–36 verify each statement by using the following matrices:

$$A = \begin{bmatrix} 1 & 2 \\ 0 & 1 \end{bmatrix} \qquad B = \begin{bmatrix} 1 & 1 \\ 2 & 3 \end{bmatrix} \qquad C = \begin{bmatrix} -3 & 1 \\ -1 & 2 \end{bmatrix}$$

33. $AB \neq BA$

34. $(AB)C = A(BC)$

35. $A(B + C) = AB + AC$

36. $(B + C)A = BA + CA$

Applications

Business & Economics

37. *Labor costs.* A company with manufacturing plants located in different parts of the country has labor-hour and wage requirements for the manufacturing of three types of inflatable boats as given in the following two matrices:

Labor-hours per boat

	Cutting department	Assembly department	Packaging department	
$M =$	0.6 hr	0.6 hr	0.2 hr	One-person boat
	1.0 hr	0.9 hr	0.3 hr	Two-person boat
	1.5 hr	1.2 hr.	0.4 hr	Four-person boat

Hourly wages

	Plant I	Plant II	
$N =$	$6	$7	Cutting department
	$8	$10	Assembly department
	$3	$4	Packaging department

(A) Find the labor costs for a one-person boat manufactured at plant I. That is, find the dot product

$$[0.6 \quad 0.6 \quad 0.2] \cdot \begin{bmatrix} 6 \\ 8 \\ 3 \end{bmatrix}$$

(B) Find the labor costs for a four-person boat manufactured at plant II. Set up a dot product as in part A and multiply.

(C)   What is the dimension of MN?

(D)   Find MN and interpret.

38.   *Inventory value.* A personal computer retail company sells five different computer models through three stores located in a large metropolitan area. The inventory of each model on hand in each store is
summarized in matrix M. Wholesale (W) and retail (R) values of each
model computer are summarized in matrix N.

Model

$$M = \begin{array}{c c} & \begin{array}{c c c c c} A & B & C & D & E \end{array} \\ \begin{bmatrix} 4 & 2 & 3 & 7 & 1 \\ 2 & 3 & 5 & 0 & 6 \\ 10 & 4 & 3 & 4 & 3 \end{bmatrix} & \begin{array}{c} \text{Store 1} \\ \text{Store 2} \\ \text{Store 3} \end{array} \end{array}$$

$$N = \begin{array}{c} \begin{array}{c c} W & R \end{array} \\ \begin{bmatrix} \$700 & \$840 \\ \$1,400 & \$1,800 \\ \$1,800 & \$2,400 \\ \$2,700 & \$3,300 \\ \$3,500 & \$4,900 \end{bmatrix} \begin{array}{c} A \\ B \\ C \\ D \\ E \end{array} \end{array}$$

(A)   What is the retail value of the inventory at store 2?

(B)   What is the wholesale value of the inventory at store 3?

(C)   Compute MN and interpret.

39.   (A)   Multiply M in Problem 38 by [1   1   1] and interpret.
(The multiplication only makes sense in one direction.)

(B)   Multiply MN in Problem 38 by [1   1   1] and interpret.
(The multiplication only makes sense in one direction.)

Life Sciences   40.   *Nutrition.* A nutritionist for a cereal company blends two cereals in
different mixes. The amounts of protein, carbohydrate, and fat (in
grams per ounce) in each cereal are given by matrix M. The amounts
of each cereal used in the three mixes are given by matrix N.

$$M = \begin{array}{c c} & \begin{array}{c c} \text{Cereal } A & \text{Cereal } B \end{array} \\ \begin{bmatrix} 4 \text{ g} & 2 \text{ g} \\ 20 \text{ g} & 16 \text{ g} \\ 3 \text{ g} & 1 \text{ g} \end{bmatrix} & \begin{array}{c} \text{Protein} \\ \text{Carbohydrate} \\ \text{Fat} \end{array} \end{array}$$

$$N = \begin{array}{c c} & \begin{array}{c c c} \text{Mix } X & \text{Mix } Y & \text{Mix } Z \end{array} \\ \begin{bmatrix} 15 \text{ oz} & 10 \text{ oz} & 5 \text{ oz} \\ 5 \text{ oz} & 10 \text{ oz} & 15 \text{ oz} \end{bmatrix} & \begin{array}{c} \text{Cereal } A \\ \text{Cereal } B \end{array} \end{array}$$

(A) Find the amount of protein in mix X by computing the dot product

$$[4 \quad 2] \cdot \begin{bmatrix} 15 \\ 5 \end{bmatrix}$$

(B) Find the amount of fat in mix Z. Set up a dot product as in part A and multiply.
(C) What is the dimension of MN?
(D) Find MN and interpret.
(E) Find $\frac{1}{20} MN$ and interpret.

Social Sciences   **41.** *Politics.* In a local California election, a group hired a public relations firm to promote its candidate in three ways: telephone calls, house calls, and letters. The cost per contact is given in matrix M:

$$M = \begin{bmatrix} \$0.40 \\ \$0.75 \\ \$0.25 \end{bmatrix} \begin{array}{l} \text{Telephone call} \\ \text{House call} \\ \text{Letter} \end{array}$$

Cost per contact

The number of contacts of each type made in two adjacent cities is given in matrix N:

$$N = \begin{bmatrix} 1,000 & 500 & 5,000 \\ 2,000 & 800 & 8,000 \end{bmatrix} \begin{array}{l} \text{Berkeley} \\ \text{Oakland} \end{array}$$

Telephone call   House call   Letter

(A) Find the total amount spent in Berkeley by computing the dot product

$$[1,000 \quad 500 \quad 5,000] \cdot \begin{bmatrix} \$0.40 \\ \$0.75 \\ \$0.25 \end{bmatrix}$$

(B) Find the total amount spent in Oakland by computing the dot product of appropriate matrices.
(C) Compute NM and interpret.
(D) Multiply N by the matrix $[1 \quad 1]$ and interpret.

7-6 Inverse of a Square Matrix; Matrix Equations

- Identity Matrix for Multiplication
- Inverse of a Square Matrix
- Matrix Equations
- Application

■ Identity Matrix for Multiplication

We know that

$$1a = a1 = a$$

for all real numbers a. The number 1 is called the **identity** for real number multiplication. Does the set of all matrices of a given dimension have an identity element for multiplication? The answer, in general, is no. However, the set of all **square matrices of order n** (dimension $n \times n$) does have an identity, and it is given as follows: The **identity element for multiplication** for the set of all square matrices of order n is the square matrix of order n, denoted by I, with 1's along the **main diagonal** (from the upper left corner to the lower right) and 0's elsewhere. For example,

$$\begin{bmatrix} 1 & 0 \\ 0 & 1 \end{bmatrix} \quad \text{and} \quad \begin{bmatrix} 1 & 0 & 0 \\ 0 & 1 & 0 \\ 0 & 0 & 1 \end{bmatrix}$$

are the identity matrices for all square matrices of order 2 and 3, respectively.

Example 27

$$\begin{bmatrix} 1 & 0 & 0 \\ 0 & 1 & 0 \\ 0 & 0 & 1 \end{bmatrix} \begin{bmatrix} a & b & c \\ d & e & f \\ g & h & i \end{bmatrix} = \begin{bmatrix} a & b & c \\ d & e & f \\ g & h & i \end{bmatrix}$$

$$= \begin{bmatrix} a & b & c \\ d & e & f \\ g & h & i \end{bmatrix} \begin{bmatrix} 1 & 0 & 0 \\ 0 & 1 & 0 \\ 0 & 0 & 1 \end{bmatrix}$$

Problem 27   Multiply: $\begin{bmatrix} 1 & 0 \\ 0 & 1 \end{bmatrix} \begin{bmatrix} 2 & -3 \\ 5 & 7 \end{bmatrix}$ and $\begin{bmatrix} 2 & -3 \\ 5 & 7 \end{bmatrix} \begin{bmatrix} 1 & 0 \\ 0 & 1 \end{bmatrix}$

In general, we can show that if M is a square matrix of order n and I is the identity matrix of order n, then

$$IM = MI = M$$

■ Inverse of a Square Matrix

In the set of real numbers, we know that for each real number a (except zero) there exists a real number a^{-1} such that

$$a^{-1}a = 1$$

The number a^{-1} is called the **inverse** of the number a relative to multiplication, or the **multiplicative inverse** of a. For example, 2^{-1} is the multiplicative inverse of 2, since $2^{-1} \cdot 2 = 1$. For each square matrix M, does there exist an inverse matrix M^{-1} such that the following relation is true?

$$M^{-1}M = MM^{-1} = I$$

If M^{-1} exists for a given matrix M, then M^{-1} is called the **inverse of M relative to multiplication.** Let us use this definition to find M^{-1} for

$$M = \begin{bmatrix} 2 & 3 \\ 1 & 2 \end{bmatrix}$$

We are looking for

$$M^{-1} = \begin{bmatrix} a & c \\ b & d \end{bmatrix}$$

such that

$$MM^{-1} = M^{-1}M = I$$

Thus, we write

$$\overset{M}{\begin{bmatrix} 2 & 3 \\ 1 & 2 \end{bmatrix}} \overset{M^{-1}}{\begin{bmatrix} a & c \\ b & d \end{bmatrix}} = \overset{I}{\begin{bmatrix} 1 & 0 \\ 0 & 1 \end{bmatrix}}$$

and try to find a, b, c, and d so that the product of M and M^{-1} is the identity matrix I. Multiplying M and M^{-1} on the left side, we obtain

$$\begin{bmatrix} (2a + 3b) & (2c + 3d) \\ (a + 2b) & (c + 2d) \end{bmatrix} = \begin{bmatrix} 1 & 0 \\ 0 & 1 \end{bmatrix}$$

which is true only if

$$\begin{aligned} 2a + 3b &= 1 & 2c + 3d &= 0 \\ a + 2b &= 0 & c + 2d &= 1 \end{aligned}$$

Solving these two systems, we find that $a = 2$, $b = -1$, $c = -3$, and $d = 2$. Thus,

$$M^{-1} = \begin{bmatrix} 2 & -3 \\ -1 & 2 \end{bmatrix}$$

as is easily checked:

$$\overset{M}{\begin{bmatrix} 2 & 3 \\ 1 & 2 \end{bmatrix}} \overset{M^{-1}}{\begin{bmatrix} 2 & -3 \\ -1 & 2 \end{bmatrix}} = \overset{I}{\begin{bmatrix} 1 & 0 \\ 0 & 1 \end{bmatrix}} = \overset{M^{-1}}{\begin{bmatrix} 2 & -3 \\ -1 & 2 \end{bmatrix}} \overset{M}{\begin{bmatrix} 2 & 3 \\ 1 & 2 \end{bmatrix}}$$

Inverses do not always exist for square matrices. For example, if

$$M = \begin{bmatrix} 2 & 1 \\ 4 & 2 \end{bmatrix}$$

then, proceeding as above, we are led to the systems

$$\begin{array}{ll} 2a + b = 1 & 2c + d = 0 \\ 4a + 2b = 0 & 4c + 2d = 1 \end{array}$$

These are both inconsistent and have no solution. Hence, M^{-1} does not exist.

Finding inverses (when they exist) leads to direct and simple solutions to many practical problems. At the end of this section, we shall show how inverses can be used to solve systems of linear equations.

The method outlined above for finding M^{-1}, if it exists, gets very involved for matrices of order larger than 2. Now that we know what we are looking for, we can introduce the idea of the augmented matrix (considered in Section 7-2) to make the process more efficient. For example, to find the inverse (if it exists) of

$$M = \begin{bmatrix} 1 & -1 & 1 \\ 0 & 2 & -1 \\ 2 & 3 & 0 \end{bmatrix}$$

we start as before and write

$$\overset{M}{\begin{bmatrix} 1 & -1 & 1 \\ 0 & 2 & -1 \\ 2 & 3 & 0 \end{bmatrix}} \overset{M^{-1}}{\begin{bmatrix} a & d & g \\ b & e & h \\ c & f & i \end{bmatrix}} = \overset{I}{\begin{bmatrix} 1 & 0 & 0 \\ 0 & 1 & 0 \\ 0 & 0 & 1 \end{bmatrix}}$$

which is true only if

$$\begin{array}{lll} a - b + c = 1 & d - e + f = 0 & g - h + i = 0 \\ 2b - c = 0 & 2e - f = 1 & 2h - i = 0 \\ 2a + 3b = 0 & 2d + 3e = 0 & 2g + 3h = 1 \end{array}$$

Now we write augmented matrices for each of the three systems:

$$\overset{First}{\left[\begin{array}{rrr|r} 1 & -1 & 1 & 1 \\ 0 & 2 & -1 & 0 \\ 2 & 3 & 0 & 0 \end{array}\right]} \quad \overset{Second}{\left[\begin{array}{rrr|r} 1 & -1 & 1 & 0 \\ 0 & 2 & -1 & 1 \\ 2 & 3 & 0 & 0 \end{array}\right]} \quad \overset{Third}{\left[\begin{array}{rrr|r} 1 & -1 & 1 & 0 \\ 0 & 2 & -1 & 0 \\ 2 & 3 & 0 & 1 \end{array}\right]}$$

Since each matrix to the left of the vertical bar is the same, exactly the same row operations can be used on each total matrix to transform it into a reduced form. We can speed up the process substantially by combining all three augmented matrices into the single augmented matrix form

$$\left[\begin{array}{ccc|ccc} 1 & -1 & 1 & 1 & 0 & 0 \\ 0 & 2 & -1 & 0 & 1 & 0 \\ 2 & 3 & 0 & 0 & 0 & 1 \end{array}\right] = [M|I] \tag{1}$$

We now try to perform row operations on matrix (1) until we obtain a row-equivalent matrix that looks like matrix (2):

$$\begin{array}{cc} I & B \\ \left[\begin{array}{ccc|ccc} 1 & 0 & 0 & a & d & g \\ 0 & 1 & 0 & b & e & h \\ 0 & 0 & 1 & c & f & i \end{array}\right] \end{array} \tag{2}$$

If this can be done, then the new matrix to the right of the vertical bar will be M^{-1}! Now let us try to transform (1) into a form like (2).

$$\begin{array}{cc} M & I \\ \left[\begin{array}{ccc|ccc} 1 & -1 & 1 & 1 & 0 & 0 \\ 0 & 2 & -1 & 0 & 1 & 0 \\ 2 & 3 & 0 & 0 & 0 & 1 \end{array}\right] & R_3 + (-2)R_1 \to R_3 \end{array}$$

$$\sim \left[\begin{array}{ccc|ccc} 1 & -1 & 1 & 1 & 0 & 0 \\ 0 & 2 & -1 & 0 & 1 & 0 \\ 0 & 5 & -2 & -2 & 0 & 1 \end{array}\right] \quad \tfrac{1}{2}R_2 \to R_2$$

$$\sim \left[\begin{array}{ccc|ccc} 1 & -1 & 1 & 1 & 0 & 0 \\ 0 & 1 & -\tfrac{1}{2} & 0 & \tfrac{1}{2} & 0 \\ 0 & 5 & -2 & -2 & 0 & 1 \end{array}\right] \quad R_3 + (-5)R_2 \to R_3$$

$$\sim \left[\begin{array}{ccc|ccc} 1 & -1 & 1 & 1 & 0 & 0 \\ 0 & 1 & -\tfrac{1}{2} & 0 & \tfrac{1}{2} & 0 \\ 0 & 0 & \tfrac{1}{2} & -2 & -\tfrac{5}{2} & 1 \end{array}\right] \quad 2R_3 \to R_3$$

$$\sim \left[\begin{array}{ccc|ccc} 1 & -1 & 1 & 1 & 0 & 0 \\ 0 & 1 & -\tfrac{1}{2} & 0 & \tfrac{1}{2} & 0 \\ 0 & 0 & 1 & -4 & -5 & 2 \end{array}\right] \quad \begin{array}{l} R_1 + (-1)R_3 \to R_1 \\ R_2 + \tfrac{1}{2}R_3 \to R_2 \end{array}$$

$$\sim \left[\begin{array}{ccc|ccc} 1 & -1 & 0 & 5 & 5 & -2 \\ 0 & 1 & 0 & -2 & -2 & 1 \\ 0 & 0 & 1 & -4 & -5 & 2 \end{array}\right] \quad R_1 + R_2 \to R_1$$

$$\sim \begin{matrix} & I & & & B \\ \begin{bmatrix} 1 & 0 & 0 \\ 0 & 1 & 0 \\ 0 & 0 & 1 \end{bmatrix} & & & \begin{bmatrix} 3 & 3 & -1 \\ -2 & -2 & 1 \\ -4 & -5 & 2 \end{bmatrix} \end{matrix}$$

Converting back to systems of equations equivalent to our three original systems (we don't have to do this step in practice), we have

$$\begin{aligned} a &= 3 & d &= 3 & g &= -1 \\ b &= -2 & e &= -2 & h &= 1 \\ c &= -4 & f &= -5 & i &= 2 \end{aligned}$$

And these are just the elements of M^{-1} that we are looking for! Hence,

$$M^{-1} = \begin{bmatrix} 3 & 3 & -1 \\ -2 & -2 & 1 \\ -4 & -5 & 2 \end{bmatrix}$$

Note that this is the matrix to the right of the vertical line in the last augmented matrix above. (You should check that $MM^{-1} = I$.)

Inverse of a Square Matrix M

If $[M|I]$ is transformed by row operations into $[I|B]$, then the resulting matrix B is M^{-1}. However, if we obtain all zeros in one or more rows to the left of the vertical line, then M^{-1} does not exist.

Example 28   Find M^{-1}, given: $M = \begin{bmatrix} 3 & -1 \\ -4 & 2 \end{bmatrix}$

Solution

$$\begin{bmatrix} 3 & -1 & | & 1 & 0 \\ -4 & 2 & | & 0 & 1 \end{bmatrix} \quad \tfrac{1}{3} R_1 \to R_1$$

$$\sim \begin{bmatrix} 1 & -\tfrac{1}{3} & | & \tfrac{1}{3} & 0 \\ -4 & 2 & | & 0 & 1 \end{bmatrix} \quad R_2 + 4 R_1 \to R_2$$

$$\sim \begin{bmatrix} 1 & -\tfrac{1}{3} & | & \tfrac{1}{3} & 0 \\ 0 & \tfrac{2}{3} & | & \tfrac{4}{3} & 1 \end{bmatrix} \quad \tfrac{2}{3} R_2 \to R_2$$

$$\sim \begin{bmatrix} 1 & -\tfrac{1}{3} & | & \tfrac{1}{3} & 0 \\ 0 & 1 & | & 2 & \tfrac{3}{2} \end{bmatrix} \quad R_1 + \tfrac{1}{3} R_2 \to R_1$$

$$\sim \begin{bmatrix} 1 & 0 & | & 1 & \tfrac{1}{2} \\ 0 & 1 & | & 2 & \tfrac{3}{2} \end{bmatrix}$$

Thus,

$$M^{-1} = \begin{bmatrix} 1 & \frac{1}{2} \\ 2 & \frac{3}{2} \end{bmatrix} = \frac{1}{2}\begin{bmatrix} 2 & 1 \\ 4 & 3 \end{bmatrix}$$

Check by showing that $M^{-1}M = I$

$$\frac{1}{2}\begin{bmatrix} 2 & 1 \\ 4 & 3 \end{bmatrix}\begin{bmatrix} 3 & -1 \\ -4 & 2 \end{bmatrix} = \frac{1}{2}\begin{bmatrix} 2 & 0 \\ 0 & 2 \end{bmatrix} = \begin{bmatrix} 1 & 0 \\ 0 & 1 \end{bmatrix} = I$$

Problem 28   Find M^{-1}, given: $M = \begin{bmatrix} 2 & -6 \\ 1 & -2 \end{bmatrix}$

Example 29   Find M^{-1}, given: $M = \begin{bmatrix} 2 & -4 \\ -3 & 6 \end{bmatrix}$

Solution

$$\begin{bmatrix} 2 & -4 & | & 1 & 0 \\ -3 & 6 & | & 0 & 1 \end{bmatrix} \sim \begin{bmatrix} 1 & -2 & | & \frac{1}{2} & 0 \\ -3 & 6 & | & 0 & 1 \end{bmatrix}$$

$$\sim \begin{bmatrix} 1 & -2 & | & \frac{1}{2} & 0 \\ 0 & 0 & | & \frac{3}{2} & 1 \end{bmatrix}$$

We have all zeros in the second row to the left of the vertical bar; therefore, the inverse does not exist.

Problem 29   Find M^{-1}, given: $M = \begin{bmatrix} -6 & 3 \\ -4 & 2 \end{bmatrix}$

Example 30   Find M^{-1}, given: $M = \begin{bmatrix} -1 & 2 & 0 \\ 3 & 2 & -1 \\ 4 & 0 & 3 \end{bmatrix}$

Solution

$$\begin{bmatrix} -1 & 2 & 0 & | & 1 & 0 & 0 \\ 3 & 2 & -1 & | & 0 & 1 & 0 \\ 4 & 0 & 3 & | & 0 & 0 & 1 \end{bmatrix}$$

$$\sim \begin{bmatrix} 1 & -2 & 0 & | & -1 & 0 & 0 \\ 3 & 2 & -1 & | & 0 & 1 & 0 \\ 4 & 0 & 3 & | & 0 & 0 & 1 \end{bmatrix}$$

$$\sim \begin{bmatrix} 1 & -2 & 0 & | & -1 & 0 & 0 \\ 0 & 8 & -1 & | & 3 & 1 & 0 \\ 0 & 8 & 3 & | & 4 & 0 & 1 \end{bmatrix}$$

$$\sim \begin{bmatrix} 1 & -2 & 0 & | & -1 & 0 & 0 \\ 0 & 1 & -\frac{1}{8} & | & \frac{3}{8} & \frac{1}{8} & 0 \\ 0 & 8 & 3 & | & 4 & 0 & 1 \end{bmatrix}$$

$$\sim \begin{bmatrix} 1 & -2 & 0 & | & -1 & 0 & 0 \\ 0 & 1 & -\frac{1}{8} & | & \frac{3}{8} & \frac{1}{8} & 0 \\ 0 & 0 & 4 & | & 1 & -1 & 1 \end{bmatrix}$$

$$\sim \begin{bmatrix} 1 & -2 & 0 & | & -1 & 0 & 0 \\ 0 & 1 & -\frac{1}{8} & | & \frac{3}{8} & \frac{1}{8} & 0 \\ 0 & 0 & 1 & | & \frac{1}{4} & -\frac{1}{4} & \frac{1}{4} \end{bmatrix}$$

$$\sim \begin{bmatrix} 1 & -2 & 0 & | & -1 & 0 & 0 \\ 0 & 1 & 0 & | & \frac{13}{32} & \frac{3}{32} & \frac{1}{32} \\ 0 & 0 & 1 & | & \frac{1}{4} & -\frac{1}{4} & \frac{1}{4} \end{bmatrix}$$

$$\sim \begin{bmatrix} 1 & 0 & 0 & | & -\frac{3}{16} & \frac{3}{16} & \frac{1}{16} \\ 0 & 1 & 0 & | & \frac{13}{32} & \frac{3}{32} & \frac{1}{32} \\ 0 & 0 & 1 & | & \frac{1}{4} & -\frac{1}{4} & \frac{1}{4} \end{bmatrix}$$

Thus,

$$M^{-1} = \begin{bmatrix} -\frac{3}{16} & \frac{3}{16} & \frac{1}{16} \\ \frac{13}{32} & \frac{3}{32} & \frac{1}{32} \\ \frac{1}{4} & -\frac{1}{4} & \frac{1}{4} \end{bmatrix} = \frac{1}{32} \begin{bmatrix} -6 & 6 & 2 \\ 13 & 3 & 1 \\ 8 & -8 & 8 \end{bmatrix}$$

Check $$\frac{1}{32} \begin{bmatrix} -6 & 6 & 2 \\ 13 & 3 & 1 \\ 8 & -8 & 8 \end{bmatrix} \begin{bmatrix} -1 & 2 & 0 \\ 3 & 2 & -1 \\ 4 & 0 & 3 \end{bmatrix} = \frac{1}{32} \begin{bmatrix} 32 & 0 & 0 \\ 0 & 32 & 0 \\ 0 & 0 & 32 \end{bmatrix}$$

$$= \begin{bmatrix} 1 & 0 & 0 \\ 0 & 1 & 0 \\ 0 & 0 & 1 \end{bmatrix}$$

Problem 30   Find M^{-1}, given: $M = \begin{bmatrix} 2 & -1 & 3 \\ 1 & 0 & 2 \\ 3 & 2 & 1 \end{bmatrix}$

- Matrix Equations

We will now show how certain systems of equations can be solved by using inverses of square matrices.

Example 31   Solve the system

$$-x_1 + 2x_2 \quad\quad = k_1$$
$$3x_1 + 2x_2 - \quad x_3 = k_2 \quad\quad\quad (3)$$
$$4x_1 \quad\quad + 3x_3 = k_3$$

for:

(A) $k_1 = 2, \quad k_2 = -1, \quad k_3 = 3$ (B) $k_1 = -1, \quad k_2 = 3, \quad k_3 = -2$
(C) $k_1 = 0, \quad k_2 = 2, \quad k_3 = 6$

Solutions   Once we obtain the inverse of the coefficient matrix

$$A = \begin{bmatrix} -1 & 2 & 0 \\ 3 & 2 & -1 \\ 4 & 0 & 3 \end{bmatrix}$$

we will be able to solve parts A, B, and C very easily. To see why, we convert system (3) into the following equivalent **matrix equation:**

$$\overset{A}{\begin{bmatrix} -1 & 2 & 0 \\ 3 & 2 & -1 \\ 4 & 0 & 3 \end{bmatrix}} \overset{X}{\begin{bmatrix} x_1 \\ x_2 \\ x_3 \end{bmatrix}} = \overset{B}{\begin{bmatrix} k_1 \\ k_2 \\ k_3 \end{bmatrix}} \quad\quad (4)$$

Now we see another important reason for defining matrix multiplication as it was defined. You should check that matrix equation (4) is equivalent to system (3) by multiplying the left side and then equating corresponding elements on the left with those on the right.

We are now interested in finding a column matrix X that will satisfy the matrix equation

$$AX = B$$

To solve this equation, we multiply both sides by A^{-1} (if it exists) to isolate X on the left side:

$$AX = B \quad\quad \text{Multiply both sides by } A^{-1}$$
$$A^{-1}(AX) = A^{-1}B \quad\quad \text{Use the associative property}$$
$$(A^{-1}A)X = A^{-1}B \quad\quad A^{-1}A = I$$
$$IX = A^{-1}B \quad\quad IX = X$$
$$X = A^{-1}B$$

The inverse of A was found in Example 30 to be

$$A^{-1} = \frac{1}{32} \begin{bmatrix} -6 & 6 & 2 \\ 13 & 3 & 1 \\ 8 & -8 & 8 \end{bmatrix}$$

Thus,

$$
\begin{matrix}
X & A^{-1} & B
\end{matrix}
$$

$$
\begin{bmatrix} x_1 \\ x_2 \\ x_3 \end{bmatrix} = \frac{1}{32} \begin{bmatrix} -6 & 6 & 2 \\ 13 & 3 & 1 \\ 8 & -8 & 8 \end{bmatrix} \begin{bmatrix} k_1 \\ k_2 \\ k_3 \end{bmatrix}
$$

To solve parts A, B, and C, we simply replace k_1, k_2, and k_3 with the given values and multiply.

(A) $\begin{bmatrix} x_1 \\ x_2 \\ x_3 \end{bmatrix} = \frac{1}{32} \begin{bmatrix} -6 & 6 & 2 \\ 13 & 3 & 1 \\ 8 & -8 & 8 \end{bmatrix} \begin{bmatrix} 2 \\ -1 \\ 3 \end{bmatrix} = \begin{bmatrix} -\frac{12}{32} \\ \frac{26}{32} \\ \frac{48}{32} \end{bmatrix}$

Thus, $x_1 = -\frac{3}{8}$, $x_2 = \frac{13}{16}$, and $x_3 = \frac{3}{2}$.

(B) $\begin{bmatrix} x_1 \\ x_2 \\ x_3 \end{bmatrix} = \frac{1}{32} \begin{bmatrix} -6 & 6 & 2 \\ 13 & 3 & 1 \\ 8 & -8 & 8 \end{bmatrix} \begin{bmatrix} -1 \\ 3 \\ 2 \end{bmatrix} = \begin{bmatrix} \frac{28}{32} \\ -\frac{2}{32} \\ -\frac{16}{32} \end{bmatrix}$

Thus, $x_1 = \frac{7}{8}$, $x_2 = -\frac{1}{16}$, and $x_3 = -\frac{1}{2}$.

(C) $\begin{bmatrix} x_1 \\ x_2 \\ x_3 \end{bmatrix} = \frac{1}{32} \begin{bmatrix} -6 & 6 & 2 \\ 13 & 3 & 1 \\ 8 & -8 & 8 \end{bmatrix} \begin{bmatrix} 0 \\ -2 \\ 6 \end{bmatrix} = \begin{bmatrix} 0 \\ 0 \\ \frac{64}{32} \end{bmatrix}$

Thus, $x_1 = 0$, $x_2 = 0$, and $x_3 = 2$.

Problem 31   Solve the system

$$
\begin{aligned}
2x_1 - x_2 + 3x_3 &= k_1 \\
x_1 + 2x_3 &= k_2 \\
3x_1 + 2x_2 + x_3 &= k_3
\end{aligned}
$$

by using the inverse of the coefficient matrix (found in Problem 30 above). Find the particular solutions of the system for:

(A) $k_1 = 2$, $k_2 = 0$, $k_3 = -3$ (B) $k_1 = -1$, $k_2 = 1$, $k_3 = 2$
(C) $k_1 = 3$, $k_2 = -3$, $k_3 = 0$

Computer programs are readily available for finding the inverse of square matrices. A great advantage of using an inverse to solve a system of linear equations is that once the inverse is found, it can be used to solve any new system formed by changing the constant terms. However, this method is not suited for cases where the number of equations and the number of unknowns are not the same. (Why?)

■

■ Application

The following application will illustrate the usefulness of the inverse method for solving systems of equations.

Example 32   An investment advisor currently has two types of investments available for clients; a conservative investment A that pays 10% per year and an investment B of higher risk that pays 20% per year. Clients may divide their investments between the two to achieve any total return desired between 10% and 20%. However, the higher the desired return, the higher the risk. How should each client listed in the table invest to achieve the indicated return?

	Client			
	1	2	3	k
Total investment	$20,000	$50,000	$10,000	k_1
Annual return desired	$ 2,400	$ 7,500	$ 1,300	k_2
	(12%)	(15%)	(13%)	

Solution   We will solve the problem for an arbitrary client k by using inverses. Then we will apply the result to the three specific clients.

Let $x_1 =$ Amount invested in A

$x_2 =$ Amount invested in B

Then

$$\begin{aligned} x_1 + x_2 &= k_1 \qquad \text{Total invested} \\ 0.1x_1 + 0.2x_2 &= k_2 \qquad \text{Total annual return desired} \end{aligned}$$

Write as a matrix equation:

$$\overset{A}{\begin{bmatrix} 1 & 1 \\ 0.1 & 0.2 \end{bmatrix}} \overset{X}{\begin{bmatrix} x_1 \\ x_2 \end{bmatrix}} = \overset{B}{\begin{bmatrix} k_1 \\ k_2 \end{bmatrix}}$$

If A^{-1} exists, then

$X = A^{-1}B$

We now find A^{-1} by starting with $[A|I]$ and proceeding as discussed earlier in this section:

$$\begin{bmatrix} 1 & 1 & | & 1 & 0 \\ 0.1 & 0.2 & | & 0 & 1 \end{bmatrix} \quad 10R_2 \rightarrow R_2$$

$$\sim \begin{bmatrix} 1 & 1 & | & 1 & 0 \\ 1 & 2 & | & 0 & 10 \end{bmatrix} \quad R_2 + (-1)R_1 \rightarrow R_2$$

$$\sim \begin{bmatrix} 1 & 1 & | & 1 & 0 \\ 0 & 1 & | & -1 & 10 \end{bmatrix} \quad R_1 + (-1)R_2 \rightarrow R_1$$

$$\sim \begin{bmatrix} 1 & 0 & | & 2 & -10 \\ 0 & 1 & | & -1 & 10 \end{bmatrix}$$

Thus,

$$A^{-1} = \begin{bmatrix} 2 & -10 \\ -1 & 10 \end{bmatrix} \quad Check: \quad \overset{A^{-1}}{\begin{bmatrix} 2 & -10 \\ -1 & 10 \end{bmatrix}} \overset{A}{\begin{bmatrix} 1 & 1 \\ 0.1 & 0.2 \end{bmatrix}} = \overset{I}{\begin{bmatrix} 1 & 0 \\ 0 & 1 \end{bmatrix}}$$

and

$$\overset{X}{\begin{bmatrix} x_1 \\ x_2 \end{bmatrix}} = \overset{A^{-1}}{\begin{bmatrix} 2 & -10 \\ -1 & 10 \end{bmatrix}} \overset{B}{\begin{bmatrix} k_1 \\ k_2 \end{bmatrix}}$$

To solve each client's investment problem, we replace k_1 and k_2 with appropriate values from the table and multiply by A^{-1}:

Client 1

$$\begin{bmatrix} x_1 \\ x_2 \end{bmatrix} = \begin{bmatrix} 2 & -10 \\ -1 & 10 \end{bmatrix} \begin{bmatrix} 20,000 \\ 2,400 \end{bmatrix} = \begin{bmatrix} 16,000 \\ 4,000 \end{bmatrix} \quad \text{Type A invest.}$$
$$\text{Type B invest}$$

Solution: $x_1 = \$16,000$ in A, $x_2 = \$4,000$ in B

Client 2

$$\begin{bmatrix} x_1 \\ x_2 \end{bmatrix} = \begin{bmatrix} 2 & -10 \\ -1 & 10 \end{bmatrix} \begin{bmatrix} 50,000 \\ 7,500 \end{bmatrix} = \begin{bmatrix} 25,000 \\ 25,000 \end{bmatrix}$$

Solution: $x_1 = \$25,000$ in A, $x_2 = \$25,000$ in B

Client 3

$$\begin{bmatrix} x_1 \\ x_2 \end{bmatrix} = \begin{bmatrix} 2 & -10 \\ -1 & 10 \end{bmatrix} \begin{bmatrix} 10,000 \\ 1,300 \end{bmatrix} = \begin{bmatrix} 7,000 \\ 3,000 \end{bmatrix}$$

Solution: $x_1 = \$7,000$ in A, $x_2 = \$3,000$ in B

Problem 32 Repeat Example 32 with investment A paying 8% and investment B paying 24%.

Answers to Matched Problems

27. $\begin{bmatrix} 2 & -3 \\ 5 & 7 \end{bmatrix}$

28. $\begin{bmatrix} -1 & 3 \\ -\frac{1}{2} & 1 \end{bmatrix} = \frac{1}{2} \begin{bmatrix} -2 & 6 \\ -1 & 2 \end{bmatrix}$

29. Does not exist

30. $M^{-1} = \frac{1}{7} \begin{bmatrix} 4 & -7 & 2 \\ -5 & 7 & 1 \\ -2 & 7 & -1 \end{bmatrix}$

31. (A) $x_1 = \frac{2}{7}$, $x_2 = -\frac{13}{7}$, $x_3 = -\frac{1}{7}$ (B) $x_1 = -1$, $x_2 = 2$, $x_3 = 1$
     (C) $x_1 = \frac{33}{7}$, $x_2 = -\frac{36}{7}$, $x_3 = -\frac{27}{7}$

32. $A^{-1} = \begin{bmatrix} 1.5 & -6.25 \\ -0.5 & 6.25 \end{bmatrix}$ Client 1: \$15,000 in A and \$5,000 in B
                                                          Client 2: \$28,125 in A and \$21,875 in B
                                                          Client 3: \$6,875 in A and \$3,125 in B

Exercise 7-6

A   *Perform the indicated operations.*

1. $\begin{bmatrix} 1 & 0 \\ 0 & 1 \end{bmatrix} \begin{bmatrix} 2 & -3 \\ 4 & 5 \end{bmatrix}$

2. $\begin{bmatrix} 2 & -3 \\ 4 & 5 \end{bmatrix} \begin{bmatrix} 1 & 0 \\ 0 & 1 \end{bmatrix}$

3. $\begin{bmatrix} 1 & 0 & 0 \\ 0 & 1 & 0 \\ 0 & 0 & 1 \end{bmatrix} \begin{bmatrix} -2 & 1 & 3 \\ 2 & 4 & -2 \\ 5 & 1 & 0 \end{bmatrix}$

4. $\begin{bmatrix} -2 & 1 & 3 \\ 2 & 4 & -2 \\ 5 & 1 & 0 \end{bmatrix} \begin{bmatrix} 1 & 0 & 0 \\ 0 & 1 & 0 \\ 0 & 0 & 1 \end{bmatrix}$

For each problem, show that the two matrices are inverses of each other by showing that their product is the identity matrix I.

5. $\begin{bmatrix} 3 & -4 \\ -2 & 3 \end{bmatrix}, \begin{bmatrix} 3 & 4 \\ 2 & 3 \end{bmatrix}$

6. $\begin{bmatrix} 5 & -7 \\ -2 & 3 \end{bmatrix}, \begin{bmatrix} 3 & 7 \\ 2 & 5 \end{bmatrix}$

7. $\begin{bmatrix} 1 & -1 & 1 \\ 0 & 2 & -1 \\ 2 & 3 & 0 \end{bmatrix}, \begin{bmatrix} 3 & 3 & -1 \\ -2 & -2 & 1 \\ -4 & -5 & 2 \end{bmatrix}$

8. $\begin{bmatrix} 3 & 3 & -1 \\ -2 & -2 & 1 \\ -4 & -5 & 2 \end{bmatrix}, \begin{bmatrix} 1 & -1 & 1 \\ 0 & 2 & -1 \\ 2 & 3 & 0 \end{bmatrix}$

Find x_1 and x_2.

9. $\begin{bmatrix} x_1 \\ x_2 \end{bmatrix} = \begin{bmatrix} 3 & -2 \\ 1 & 4 \end{bmatrix} \begin{bmatrix} -2 \\ 1 \end{bmatrix}$

10. $\begin{bmatrix} x_1 \\ x_2 \end{bmatrix} = \begin{bmatrix} -2 & 1 \\ -1 & 2 \end{bmatrix} \begin{bmatrix} 3 \\ -2 \end{bmatrix}$

11. $\begin{bmatrix} x_1 \\ x_2 \end{bmatrix} = \begin{bmatrix} -2 & 3 \\ 2 & -1 \end{bmatrix} \begin{bmatrix} 3 \\ 2 \end{bmatrix}$

12. $\begin{bmatrix} x_1 \\ x_2 \end{bmatrix} = \begin{bmatrix} 3 & -1 \\ 0 & 2 \end{bmatrix} \begin{bmatrix} -2 \\ 1 \end{bmatrix}$

B   Given M as indicated, find M^{-1} and show that $M^{-1}M = I$.

13. $\begin{bmatrix} 1 & 2 \\ 1 & 3 \end{bmatrix}$ 14. $\begin{bmatrix} 2 & 1 \\ 5 & 3 \end{bmatrix}$

15. $\begin{bmatrix} 1 & 3 \\ 2 & 7 \end{bmatrix}$ 16. $\begin{bmatrix} 2 & 1 \\ 1 & 1 \end{bmatrix}$

17. $\begin{bmatrix} 1 & -3 & 0 \\ 0 & 3 & 1 \\ 2 & -1 & 2 \end{bmatrix}$ 18. $\begin{bmatrix} 2 & 9 & 0 \\ 1 & 2 & 3 \\ 0 & -1 & 1 \end{bmatrix}$

19. $\begin{bmatrix} 1 & 1 & 0 \\ 0 & 3 & -1 \\ 1 & 0 & 1 \end{bmatrix}$ 20. $\begin{bmatrix} 1 & 0 & -1 \\ 2 & -1 & 0 \\ 1 & 1 & 1 \end{bmatrix}$

Write each system as a matrix equation and solve by using inverses.
[Note: The inverses were found in Problems 13–18.]

21. $x_1 + 2x_2 = k_1$
     $x_1 + 3x_2 = k_2$

(A) $k_1 = 1,\ k_2 = 3$
(B) $k_1 = 3,\ k_2 = 5$
(C) $k_1 = -2,\ k_2 = 1$

22. $2x_1 + x_2 = k_1$
     $5x_1 + 3x_2 = k_2$

(A) $k_1 = 2,\ k_2 = 13$
(B) $k_1 = 2,\ k_2 = 4$
(C) $k_1 = 1,\ k_2 = -3$

23. $x_1 + 3x_2 = k_1$
     $2x_1 + 7x_2 = k_2$

(A) $k_1 = 2,\ k_2 = -1$
(B) $k_1 = 1,\ k_2 = 0$
(C) $k_1 = 3,\ k_2 = -1$

24. $2x_1 + x_2 = k_1$
     $x_1 + x_2 = k_2$

(A) $k_1 = -1,\ k_2 = -2$
(B) $k_1 = 2,\ k_2 = 3$
(C) $k_1 = 2,\ k_2 = 0$

25. $x_1 - 3x_2 \qquad = k_1$
     $\qquad 3x_2 + x_3 = k_2$
     $2x_1 - x_2 + 2x_3 = k_3$

(A) $k_1 = 1,\ k_2 = 0,\ k_3 = 2$
(B) $k_1 = -1,\ k_2 = 1,\ k_3 = 0$
(C) $k_1 = 2,\ k_2 = -2,\ k_3 = 1$

26. $2x_1 + 9x_2 \qquad = k_1$
     $x_1 + 2x_2 + 3x_3 = k_2$
     $\quad - x_2 + x_3 = k_3$

(A) $k_1 = 0,\ k_2 = 2,\ k_3 = 1$
(B) $k_1 = -2,\ k_2 = 0,\ k_3 = 1$
(C) $k_1 = 3,\ k_2 = 1,\ k_3 = 0$

C Write each system as a matrix equation and solve using inverses.
[Note: The inverses were found in Problems 19 and 20.]

27. $x_1 + x_2 \qquad = k_1$

$\qquad 3x_2 - x_3 = k_2$

$x_1 \qquad + x_3 = k_3$

(A) $k_1 = 2$, $k_2 = 0$, $k_3 = 4$
(B) $k_1 = 0$, $k_2 = 4$, $k_3 = -2$
(C) $k_1 = 4$, $k_2 = 2$, $k_3 = 0$

28. $x_1 \qquad - x_3 = k_1$

$2x_1 - x_2 \qquad = k_2$

$x_1 + x_2 + x_3 = k_3$

(A) $k_1 = 4$, $k_2 = 8$, $k_3 = 0$
(B) $k_1 = 4$, $k_2 = 0$, $k_3 = -4$
(C) $k_1 = 0$, $k_2 = 8$, $k_3 = -8$

Show that the inverses of the following matrices do not exist:

29. $\begin{bmatrix} 3 & 9 \\ 2 & 6 \end{bmatrix}$
30. $\begin{bmatrix} 2 & -4 \\ -3 & 6 \end{bmatrix}$

31. $\begin{bmatrix} 2 & 1 & 1 \\ 1 & 1 & 0 \\ -1 & -1 & 0 \end{bmatrix}$
32. $\begin{bmatrix} 1 & -1 & 0 \\ 2 & -1 & 1 \\ 0 & 1 & 1 \end{bmatrix}$

33. Show that $(A^{-1})^{-1} = A$ for: $A = \begin{bmatrix} 3 & 4 \\ 2 & 3 \end{bmatrix}$

34. Show that $(AB)^{-1} = B^{-1}A^{-1}$ for:

$$A = \begin{bmatrix} 3 & 4 \\ 2 & 3 \end{bmatrix} \quad \text{and} \quad B = \begin{bmatrix} 3 & 7 \\ 2 & 5 \end{bmatrix}$$

Applications

Solve using systems of equations and inverses.

Business & Economics

35. *Resource allocation.* A concert hall has 10,000 seats. If tickets are $4 and $8, how many of each type of ticket should be sold (assuming all seats can be sold) to bring in each of the returns indicated in the table? Use decimals in computing the inverse.

	Concert		
	1	2	3
Tickets sold	10,000	10,000	10,000
Return required	$56,000	$60,000	$68,000

36. *Production scheduling.* Labor and material costs for manufacturing two guitar models are given in the table below:

Guitar Model	Labor Cost	Material Cost
A	$30	$20
B	$40	$30

If a total of $3,000 a week is allowed for labor and material, how many of each model should be produced each week to use exactly each of the allocations of the $3,000 indicated in the following table? Use decimals in computing the inverse.

	Weekly Allocation		
	1	2	3
Labor	$1,800	$1,750	$1,720
Material	$1,200	$1,250	$1,280

Life Sciences

37. *Diets.* A biologist has available two commercial food mixes containing the following percentages of protein and fat:

Mix	Protein (%)	Fat (%)
A	20	2
B	10	6

How many ounces of each mix should be used to prepare each of the diets listed in the following table?

	Diet		
	1	2	3
Protein	20 oz	10 oz	10 oz
Fat	6 oz	4 oz	6 oz

7-7 Leontief Input–Output Analysis (Optional)

- Introduction
- Two-Industry Model

Introduction

A very important application of matrices and their inverses is found in the relatively recently developed branch of applied mathematics called **input**

—**output analysis.** Wassily Leontief, the primary force behind these new developments, was awarded the Nobel Prize in economics in 1973 because of the significant impact his work had on economic planning for industrialized countries. Among other things, he conducted a comprehensive study of how 500 sectors of the American economy interacted with each other. Of course, large-scale computers played an important role in this analysis.

Our investigation will be more modest. In fact, we will start with an economy comprised of only two industries. From these humble beginnings, ideas and definitions will evolve that can be readily generalized for more realistic economies. Input–output analysis attempts to establish equilibrium conditions under which industries in an economy have just enough output to satisfy each other's demands in addition to final (outside) demands. Given the internal demands within the industries for each other's output, the problem is to determine output levels that will meet various levels of final (outside) demands.

■ Two-Industry Model

To make the problem concrete, let us start with a hypothetical economy with only two industries—electric company E and water company W. Outputs for both companies are measured in dollars. The heart of input–output analysis is a matrix, called the **technology matrix,** that expresses how each industry uses the other industries' outputs as well as its own for its own output. (In this case, the electric company will use both electricity and water as input for its own output, and the water company will use both electricity and water as input for its output.) Suppose that each dollar's worth of the electric company's output requires $0.10 of its own output and $0.30 of the water company's output, and each dollar's worth of the water company's output requires $0.40 of the electric company's output and $0.20 of its own output. These internal requirements can be conveniently summarized in a technology matrix:

$$
\begin{array}{cc}
& \begin{array}{cc} E & \quad W \end{array} \\
\begin{array}{c} E \\ W \end{array} & \left[\begin{array}{cc} 0.1 & 0.4 \\ 0.3 & 0.2 \end{array} \right] = M \quad \text{Technology matrix}
\end{array}
$$

Thus, the first column indicates that each dollar of the electric company's output requires $0.10 of its own output as input and $0.30 of the water company's output as input. The second column is interpreted similarly. The first row tells us that $0.10 of electricity is needed to produce a dollar's worth of electricity, and $0.40 of electricity is needed to produce a dollar's worth of water. The second row has a similar interpretation.

Now that we know how much of each output dollar must be used by each industry for input, we are ready to attack the main problem.

> **Basic Input–Output Problem**
>
> Given the internal demands for each industry's output, determine output levels for the various industries that will meet a given final (outside) level of demand as well as the internal demand.

Suppose the final demand (the demand from the outside sector) is

$d_1 = \$12$ million for electricity

$d_2 = \$6$ million for water

What dollar outputs, $\$x_1$ from the electric company and $\$x_2$ from the water company, are required to meet these final demands? Before we answer this question, we introduce the final demand matrix and the output matrix:

$$D = \begin{bmatrix} d_1 \\ d_2 \end{bmatrix} \quad \textbf{Final demand matrix}$$

$$X = \begin{bmatrix} x_1 \\ x_2 \end{bmatrix} \quad \textbf{Output matrix}$$

Given the technology matrix M and the final demand matrix D, the problem is to find the output matrix X. The following verbal equations (which summarize the discussion above) lead to a matrix equation and a solution to the problem:

$$\begin{pmatrix} \text{Total output} \\ \text{from } E \end{pmatrix} = \begin{pmatrix} \text{Input} \\ \text{required by } E \\ \text{from } E \end{pmatrix} + \begin{pmatrix} \text{Input} \\ \text{required by } W \\ \text{from } E \end{pmatrix} + \begin{pmatrix} \text{Final} \\ \text{(outside) demand} \\ \text{from } E \end{pmatrix}$$

$$\begin{pmatrix} \text{Total output} \\ \text{from } W \end{pmatrix} = \begin{pmatrix} \text{Input} \\ \text{required by } E \\ \text{from } W \end{pmatrix} + \begin{pmatrix} \text{Input} \\ \text{required by } W \\ \text{from } W \end{pmatrix} + \begin{pmatrix} \text{Final} \\ \text{(outside) demand} \\ \text{from } W \end{pmatrix}$$

Converted to symbolic forms, these equations become

$$\begin{aligned} x_1 &= 0.1x_1 + 0.4x_2 + d_1 \\ x_2 &= 0.3x_1 + 0.2x_2 + d_2 \end{aligned} \tag{1}$$

or

$$\begin{bmatrix} x_1 \\ x_2 \end{bmatrix} = \begin{bmatrix} 0.1 & 0.4 \\ 0.3 & 0.2 \end{bmatrix} \begin{bmatrix} x_1 \\ x_2 \end{bmatrix} + \begin{bmatrix} d_1 \\ d_2 \end{bmatrix}$$

or

$$X = MX + D \tag{2}$$

Now our problem is to solve this matrix equation for X. We proceed as in the preceding section:

$$X - MX = D$$
$$IX - MX = D$$
$$(I - M)X = D$$
$$X = (I - M)^{-1}D \qquad (3)$$

$$I = \begin{bmatrix} 1 & 0 \\ 0 & 1 \end{bmatrix}$$

assuming $I - M$ has an inverse. Since

$$I - M = \begin{bmatrix} 0.9 & -0.4 \\ -0.3 & 0.8 \end{bmatrix}$$

and

$$(I - M)^{-1} = \begin{bmatrix} \frac{4}{3} & \frac{2}{3} \\ \frac{1}{2} & \frac{3}{2} \end{bmatrix}$$ We convert decimals to fractions in this example to work with exact forms

we have

$$\begin{bmatrix} x_1 \\ x_2 \end{bmatrix} = \begin{bmatrix} \frac{4}{3} & \frac{2}{3} \\ \frac{1}{2} & \frac{3}{2} \end{bmatrix} \begin{bmatrix} d_1 \\ d_2 \end{bmatrix} \qquad (4)$$

$$= \begin{bmatrix} \frac{4}{3} & \frac{2}{3} \\ \frac{1}{2} & \frac{3}{2} \end{bmatrix} \begin{bmatrix} 12 \\ 6 \end{bmatrix}$$

$$= \begin{bmatrix} 20 \\ 15 \end{bmatrix}$$

Therefore, the electric company must have an output of $20 million and the water company an output of $15 million so that each company can meet both internal and final demands.

Actually, (4) solves the original problem for arbitrary final demands d_1 and d_2. This is very useful, since (4) gives a quick solution not only for the final demands stated but also to the original problem for various other projected final demands. If we had solved (1) by elimination, then we would have to start over for each new set of final demands.

Suppose in the original problem that the projected final demands 5 years from now were $d_1 = 18$ and $d_2 = 12$. Determine each company's output for this projection. We simply substitute these values into (4) and multiply:

$$\begin{bmatrix} x_1 \\ x_2 \end{bmatrix} = \begin{bmatrix} \frac{4}{3} & \frac{2}{3} \\ \frac{1}{2} & \frac{3}{2} \end{bmatrix} \begin{bmatrix} 18 \\ 12 \end{bmatrix}$$

$$= \begin{bmatrix} 32 \\ 27 \end{bmatrix}$$

We summarize these results for convenient reference.

Solution to a Two-Company Input–Output Problem

Given

| Technology | Output | Final demand |
| matrix* | matrix | matrix |

$$M = \begin{bmatrix} a_{11} & a_{12} \\ a_{21} & a_{22} \end{bmatrix} \qquad X = \begin{bmatrix} x_1 \\ x_2 \end{bmatrix} \qquad D = \begin{bmatrix} d_1 \\ d_2 \end{bmatrix}$$

The solution to the input–output matrix equation

| Total | Internal | Final |
| output | demand | demand |

$$X \;=\; MX \;+\; D \tag{2}$$

is

$$X = (I - M)^{-1}D \tag{3}$$

assuming $I - M$ has an inverse.

* We introduce the double subscript notation in matrix M for convenience of generalization. The first number in the subscript indicates the row containing the element and the second number indicates the column. Thus, a_{21} is in the second row and first column. For a larger matrix, a_{74} would be the element in the seventh row and fourth column.

Another consequence of expressing the input–output problem in matrix form (2) is that matrix equation (2) and its solution (3) are the same for a three-industry economy, a four-industry economy, or an economy with n industries (where n is any natural number), since the steps we took going from (2) to (3) hold for arbitrary matrices as long as they are dimensionally correct and $(I - M)^{-1}$ exists.

Exercise 7-7

A *Problems 1–6 pertain to the following input–output model: Assume an economy is based on two industrial sectors–agriculture (A) and energy (E). The technology matrix M and final demand matrices are (in billions of dollars)*

$$\begin{array}{cc} & A \quad\; E \\ \begin{array}{c} A \\ E \end{array} & \begin{bmatrix} 0.4 & 0.2 \\ 0.2 & 0.1 \end{bmatrix} = M \end{array} \qquad D_1 = \begin{bmatrix} 6 \\ 4 \end{bmatrix} \qquad D_2 = \begin{bmatrix} 8 \\ 5 \end{bmatrix} \qquad D_3 = \begin{bmatrix} 12 \\ 9 \end{bmatrix}$$

1. How much input from A and E are required to produce a dollar's worth of output for A?

2. How much input from A and E are required to produce a dollar's worth of output for E?
3. Find $I - A$ and $(I - A)^{-1}$.
4. Find the output for each sector that is needed to satisfy the final demand D_1.
5. Repeat Problem 4 for D_2.
6. Repeat Problem 4 for D_3.

B   *Problems 7–12 pertain to the following input–output model: Assume an economy is based on three industrial sectors–agriculture (A), building (B), and energy (E). The technology matrix M and final demand matrices are (in billions of dollars)*

$$
\begin{array}{c@{\qquad}c@{\qquad}c}
 & A & B & E
\end{array}
$$

$$
\begin{array}{c}
A \\ B \\ E
\end{array}
\begin{bmatrix}
0.422 & 0.100 & 0.266 \\
0.089 & 0.350 & 0.134 \\
0.134 & 0.100 & 0.334
\end{bmatrix} = M
$$

$$
D_1 = \begin{bmatrix} 4 \\ 3 \\ 2 \end{bmatrix} \qquad D_2 = \begin{bmatrix} 12 \\ 10 \\ 8 \end{bmatrix}
$$

7. How much input from A, B, and E are required to produce a dollar's worth of output for B?
8. How much of each of B's output dollar is required as input for each of the three sectors?
9. Show that: $I - M = \begin{bmatrix} 0.578 & -0.100 & -0.266 \\ -0.089 & 0.650 & -0.134 \\ -0.134 & -0.100 & 0.666 \end{bmatrix}$
10. Given

$$
(I - M)^{-1} = \begin{bmatrix} 2.006 & 0.446 & 0.891 \\ 0.368 & 1.670 & 0.482 \\ 0.458 & 0.340 & 1.752 \end{bmatrix}
$$

show that: $(I - M)^{-1}(I - M) \approx I$

[*Note:* Because of round-off errors, the results will not be exact.]
11. Use $(I - M)^{-1}$ in Problem 10 to find the output for each sector that is needed to satisfy the final demand D_1.
12. Repeat Problem 11 for D_2.

C   13. Find $(I - M)^{-1}$ given in Problem 10 using the procedure described in Section 7-6.

7-8 Chapter Review

Important Terms and Symbols

7-1 *Review: systems of linear equations.* graphing method, substitution method, elimination by addition, equivalent systems, inconsistent systems, dependent systems, parameter, equilibrium price, equilibrium quantity, linear equation in two variables, linear equation in three variables, solution of a system, solution set

7-2 *Systems of linear equations and augmented matrices—introduction.* matrix, element, augmented matrix, column, row, equivalent systems, row-equivalent matrices, row operations, $R_i \leftrightarrow R_j$, $kR_i \rightarrow R_i$, $R_i + kR_j \rightarrow R_i$

7-3 *Gauss–Jordan elimination.* reduced matrix, basic variables, nonbasic variables, submatrix, Gauss–Jordan elimination

7-4 *Matrices—addition and multiplication by a number.* size or dimension of a matrix, $m \times n$ matrix, square matrix, column matrix, row matrix, equal matrices, sum of two matrices, zero matrix, negative of a matrix M, subtraction of matrices, product of a number k and a matrix M

7-5 *Matrix multiplication.* dot product, matrix product, associative property, distributive properties.

7-6 *Inverse of a square matrix; matrix equations.* inverse of a number, multiplicative inverse of a number, identity matrix, multiplicative inverse of a matrix, matrix equation, M^{-1}

7-7 *Leontief input–output analysis (optional).* input–output analysis, technology matrix, final demand matrix, output matrix

Exercise 7-8 Chapter Review

Work through all the problems in this chapter review and check your answers in the back of the book. (Answers to all review problems are there.) Where weaknesses show up, review appropriate sections in the text. When you are satisfied that you know the material, take the practice test following this review.

A 1. Solve the following system by graphing:

$$2x - y = 4$$
$$x - 2y = -4$$

2. Solve the system in Problem 1 by substitution.

In Problems 3–11 perform the operations that are defined, given the following matrices:

$$A = \begin{bmatrix} 1 & 2 \\ 3 & 1 \end{bmatrix} \qquad B = \begin{bmatrix} 2 & 1 \\ 1 & 1 \end{bmatrix} \qquad C = \begin{bmatrix} 2 & 3 \end{bmatrix} \qquad D = \begin{bmatrix} 1 \\ 2 \end{bmatrix}$$

3. $A + B$
4. $B + D$
5. $A - 2B$
6. AB
7. AC
8. AD
9. DC
10. $C \cdot D$
11. $C + D$

12. Find the inverse of the matrix A given below by appropriate row operations on $[A \mid I]$. Show that $A^{-1}A = I$.

$$A = \begin{bmatrix} 3 & 2 \\ 4 & 3 \end{bmatrix}$$

13. Solve the following system by elimination by addition:

$$3x_1 + 2x_2 = 3$$
$$4x_1 + 3x_2 = 5$$

14. Solve the system in Problem 13 by performing appropriate row operations on the augmented matrix of the system.

15. Solve the system in Problem 13 by writing the system as a matrix equation and using the inverse of the coefficient matrix (see Problem 12). Also, solve the system if the constants 3 and 5 are replaced by 7 and 10, respectively. By 4 and 2, respectively.

B   *In Problems 16–21 perform the specified operations given the following matrices:*

$$A = \begin{bmatrix} 2 & -2 \\ 1 & 0 \\ 3 & 2 \end{bmatrix} \qquad B = \begin{bmatrix} -1 \\ 2 \\ 3 \end{bmatrix} \qquad C = \begin{bmatrix} 2 & 1 & 3 \end{bmatrix}$$

$$D = \begin{bmatrix} 3 & -2 & 1 \\ -1 & 1 & 2 \end{bmatrix} \qquad E = \begin{bmatrix} 3 & -4 \\ -1 & 0 \end{bmatrix}$$

16. $A + D$
17. $E + DA$
18. $DA - 3E$
19. $C \cdot B$
20. CB
21. $AD - BC$

22. Find the inverse of the matrix A given below by appropriate row operations on $[A \mid I]$. Show that $A^{-1}A = I$.

$$A = \begin{bmatrix} 1 & 2 & 3 \\ 2 & 3 & 4 \\ 1 & 2 & 1 \end{bmatrix}$$

23. Solve by Gauss–Jordan elimination:

(A) $x_1 + 2x_2 + 3x_3 = 1$ (B) $x_1 + 2x_2 - x_3 = 2$

  $2x_1 + 3x_2 + 4x_3 = 3$ $2x_1 + 3x_2 + x_3 = -3$

  $x_1 + 2x_2 + x_3 = 3$ $3x_1 + 5x_2 = -1$

24. Solve the system in Problem 23A by writing the system as a matrix equation and using the inverse of the coefficient matrix (see Problem 22). Also, solve the system if the constants 1, 3, and 3 are replaced by 0, 0, and -2, respectively. By -3, -4, and 1, respectively.

C **25.** Find the inverse of the matrix A given below. Show that $A^{-1}A = I$.

$$A = \begin{bmatrix} 4 & 5 & 6 \\ 4 & 5 & -6 \\ 1 & 1 & 1 \end{bmatrix}$$

26. Solve the system

$$0.04x_1 + 0.05x_2 + 0.06x_3 = 360$$
$$0.04x_1 + 0.05x_2 - 0.06x_3 = 120$$
$$x_1 + x_2 + x_3 = 7{,}000$$

by writing as a matrix equation and using the inverse of the coefficient matrix. (Before starting, multiply the first two equations by 100 to eliminate decimals. Also, see Problem 25.)

27. Solve Problem 26 by Gauss–Jordan elimination.

Applications

Business & Economics

28. *Resource allocation.* A mining company has two mines with ore compositions as given in the table. How many tons of each ore should be used to obtain 4.5 tons of nickel and 10 tons of copper? Set up a system of equations and solve using Gauss–Jordan elimination.

Ore	Nickel (%)	Copper (%)
A	1	2
B	2	5

29. (A) Set up Problem 28 as a matrix equation and solve using the inverse of the coefficient matrix.

(B) Solve Problem 28 (as in part A) if 2.3 tons of nickel and 5 tons of copper are needed.

30. *Material costs.* A metal foundry wishes to make two different bronze alloys. The quantities of copper, tin, and zinc needed are indicated in matrix M. The costs for these materials in dollars per pound from two suppliers is summarized in matrix N. The company must choose one supplier or the other.

$$M = \begin{matrix} & \text{Copper} & \text{Tin} & \text{Zinc} \\ & \begin{bmatrix} 4{,}800 \text{ lb} & 600 \text{ lb} & 300 \text{ lb} \\ 6{,}000 \text{ lb} & 1{,}400 \text{ lb} & 700 \text{ lb} \end{bmatrix} & & \begin{matrix} \text{Alloy 1} \\ \text{Alloy 2} \end{matrix} \end{matrix}$$

$$N = \begin{matrix} & \text{Supplier } A & \text{Supplier } B \\ & \begin{bmatrix} \$0.75 & \$0.70 \\ \$6.50 & \$6.70 \\ \$0.40 & \$0.50 \end{bmatrix} & & \begin{matrix} \text{Copper} \\ \text{Tin} \\ \text{Zinc} \end{matrix} \end{matrix}$$

(A) Find MN and interpret.   (B) Find [1 1]MN and interpret.

Practice Test: Chapter 7

1. Transform the following matrix into reduced form using row operations:

$$\left[\begin{array}{ccc|c} 2 & 1 & 1 & 8 \\ 1 & 2 & -4 & 7 \\ 1 & 1 & -1 & 5 \end{array} \right]$$

Solve each of the following systems using augmented matrices and Gauss–Jordan elimination:

2. $2x_1 + 4x_2 = 2$
   $3x_1 + 8x_2 = 1$

3. $4x_1 - 8x_2 = 8$
   $3x_1 - 4x_2 = 9$
   $2x_1 - 4x_2 = 2$

4. $2x_1 + x_2 + x_3 = 8$
   $x_1 + 2x_2 - 4x_3 = 7$
   $x_1 + x_2 - x_3 = 5$

In Problems 5–10 perform the indicated operations (if possible) given the following matrices:

$$A = \begin{bmatrix} 2 & -1 & 3 \\ -1 & 2 & 0 \end{bmatrix} \qquad B = \begin{bmatrix} 1 & -2 & -3 \end{bmatrix} \qquad C = \begin{bmatrix} 4 \\ -1 \\ 2 \end{bmatrix}$$

$$D = \begin{bmatrix} 2 & -3 \\ -1 & 2 \end{bmatrix} \qquad E = \begin{bmatrix} 1 & -2 & 0 \\ 3 & 1 & 2 \\ -1 & 0 & 1 \end{bmatrix}$$

5. $B \cdot C$ and BC
6. DA
7. AD
8. $CB + 2E$
9. $A + E$
10. $(CB)E$ and $C(BE)$

11. Convert the system

$$x_1 - x_2 + x_3 = k_1$$
$$2x_2 - x_3 = k_2$$
$$2x_1 + 3x_2 = k_3$$

into a matrix equation and solve using the inverse of the coefficient matrix for:

(A) $k_1 = 2$, $k_2 = 0$, $k_3 = -1$ (B) $k_1 = -1$, $k_2 = 2$, $k_3 = 0$

12. A person has $5,000 to invest, part at 5% and the rest at 10%. How much should be invested at each rate to yield $400 per year? Set up a system of equations and solve using augmented matrix methods.

13. Solve Problem 12 by using a matrix equation and the inverse of the coefficient matrix.

14. *Labor costs.* A company with manufacturing plants in California and Texas has labor-hour and wage requirements for the manufacturing of two inexpensive hand calculators as given in matrices M and N below:

Labor-Hours per Calculator

	Fabricating department	Assembly department	Packaging department	
$M =$	0.15 hr	0.10 hr	0.05 hr	Model A
	0.25 hr	0.20 hr	0.05 hr	Model B

Hourly Wages

	California plant	Texas plant	
	$15	$12	Fabricating department
$N =$	$12	$10	Assembly department
	$ 4	$ 4	Packaging department

(A) What is the labor cost for producing one model B calculator in California? Set up a dot product and multiply.

(B) Find MN and interpret.

Linear Inequalities and Linear Programming

Contents

In this chapter we will discuss linear inequalities in two and more variables; in addition, we will introduce a relatively new and powerful mathematical tool called *linear programming* that will be used to solve a variety of interesting practical problems. The row operations on matrices introduced in Chapter 7 will be particularly useful in Sections 8-5, 8-6, and 8-7.

8-1 Linear Inequalities in Two Variables

Having graphed linear equations such as

$$y = -2x + 3 \quad \text{and} \quad 2x - 3y = 12$$

we now turn to **graphing linear inequalities in two variables** such as

$$y \leq -2x + 3 \quad \text{and} \quad 2x - 3y > 12$$

Graphing inequalities of this type is almost as easy as graphing equations. The following discussion leads to a simple solution of the problem.

A vertical line divides a plane into left and right *half-planes*; a nonvertical line divides a plane into **upper** and **lower half-planes** (Fig. 1).

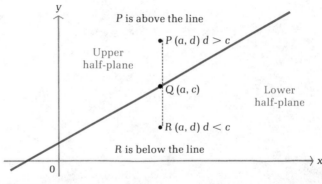

Figure 1

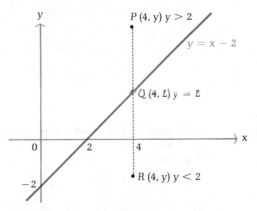

Figure 2

Let us compare the graphs of

$$y < x - 2 \qquad y = x - 2 \qquad y > x - 2$$

We start by graphing $y = x - 2$ (Fig. 2). It is clear from Figure 2 that for $x = 4$, any point P above Q will satisfy $y > 2$, and any point R below Q will satisfy $y < 2$. Since the same result holds for each x, we conclude that the graph of $y > x - 2$ is the upper half-plane determined by the graph of $y = x - 2$, and $y < x - 2$ is the lower half-plane.

To graph $y > x - 2$, we show the line $y = x - 2$ as a broken line (Fig. 3),

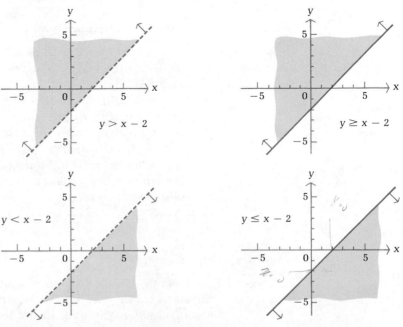

Figure 3

indicating that the line is not part of the graph. In graphing $y \geqslant x - 2$, we show the line $y = x - 2$ as a solid line, indicating that it is part of the graph. Four typical cases are illustrated in Figure 3 (on the preceding page).

The above discussion suggests the following theorem, which is stated without proof:

Theorem 1

The graph of the linear inequality

$$Ax + By < C \quad \text{or} \quad Ax + By > C$$

with $B \neq 0$ is either the upper half-plane or the lower half-plane (but not both) determined by the line $Ax + By = C$. If $B = 0$, the graph of

$$Ax < C \quad \text{or} \quad Ax > C$$

is either the left or right half-plane (but not both) as determined by the line $Ax = C$.

As a consequence of this theorem, we state a simple and fast mechanical procedure for graphing linear inequalities.

Procedure for Graphing Linear Inequalities

1. First graph $Ax + By = C$ as a broken line if equality is not included in the original statement or as a solid line if equality is included.
2. Choose a test point anywhere in the plane not on the line [the origin $(0, 0)$ often requires the least computation] and substitute the coordinates into the inequality.
3. The graph of the original inequality includes the half-plane containing the test point if the inequality is satisfied by that point or the half-plane not containing the test point if the inequality is not satisfied by that point.

Example 1   Graph $2x - 3y \leqslant 6$.

Solution   First graph the line $2x - 3y = 6$ as a solid line. Choose a convenient test point above or below the line. The origin $(0, 0)$ requires the least computation. So, substituting $(0, 0)$ into the inequality, we see that $2(0) - 3(0) \leqslant 6$ is true. Hence, the graph of the inequality is the upper half-plane.

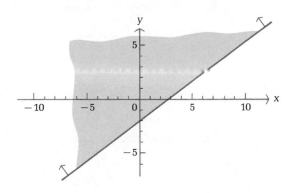

Problem 1   Graph $6x - 3y > 18$.

If you were asked to graph an inequality such as $y \leq 5$, the first question you would need to ask is "Is the graph to be done in a one- or two-dimensional coordinate system?" We have already graphed linear inequalities of this type in a one-dimensional coordinate system (see Section 1-3); now we turn to their graphs in a two-dimensional coordinate system.

In graphing $y \leq 5$ in an xy-coordinate system, we are actually asking for the graph of all ordered pairs of real numbers (x, y) such that $0x + y \leq 5$. Since x has a zero coefficient, it can be any real number in the ordered pair (x, y) as long as y is less than or equal to 5. Conclusion? The graph is the lower half-plane determined by the line $y = 5$.

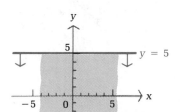

Example 2   Graph: (A) $x > -3$ (B) $-2 \leq y < 3$

Solutions   (A)

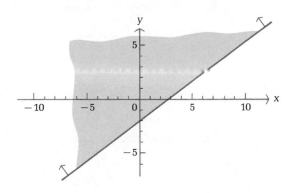

(B) Graphing $-2 \leq y < 3$ in an xy-coordinate system is the same as graphing $-2 \leqslant 0x + y < 3$. Thus, x in (x, y) can be any real number as long as y is between -2 and 3, including -2 but not 3.

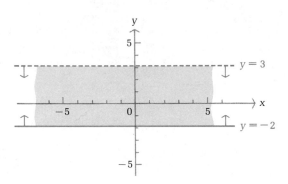

Problem 2   Graph: (A) $y < 4$ (B) $-3 < x \leqslant 3$

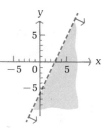

Answers to Matched Problems

1. Graph $6x - 3y = 18$ as a broken line (since equality is not included). Choosing the origin $(0, 0)$ as a test point, we see that $6(0) - 3(0) > 18$ is a false statement; thus, the lower half-plane (determined by $6x - 3y = 18$) is the graph of $6x - 3y > 18$.

2. (A)   (B)

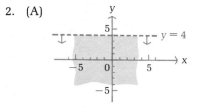

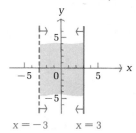

Exercise 8-1

Graph each inequality.

A   1. $y \leqslant x - 1$ 2. $y > x + 1$
        3. $3x - 2y > 6$ 4. $2x - 5y \leqslant 10$

        5. $y < \dfrac{x}{2} - 2$ 6. $y \geqslant \dfrac{x}{3} + 1$

        7. $3x - 8y > -24$ 8. $2x + 3y \leqslant -6$
        9. $x \geqslant -4$ 10. $y < 5$
        11. $-4 \leqslant y < 4$ 12. $0 \leqslant x < 6$

B

13. $10x + 2y \geqslant 84$

14. $8x + 4y \geqslant 120$

15. $0.9x + 1.8y \leqslant 864$

16. $0.8x + 1.2y \leqslant 672$

17. $0.01x + 0.03y \geqslant 3.9$

18. $0.03x + 0.04y \leqslant 36$

C *Graph each set. (Problems similar to 19–23 will be discussed in detail in the next section. See if you can figure these out now.)*

19. $\{(x, y) | -1 \leqslant x \leqslant 5 \quad \text{and} \quad 2 \leqslant y \leqslant 6\}$
20. $\{(x, y) | -2 \leqslant x \leqslant 2 \quad \text{and} \quad 0 \leqslant y \leqslant 5\}$
21. $\{(x, y) | x \geqslant 0, \quad y \geqslant 0, \quad \text{and} \quad 2x + 3y \leqslant 12\}$
22. $\{(x, y) | x \geqslant 0, \quad y \geqslant 0, \quad \text{and} \quad 3x + 8y \leqslant 24\}$
23. A manufacturer of water skis produces two models, a trick ski and a slalom ski. The relevant manufacturing data are given in the table. If x is the number of trick skis and y is the number of slalom skis produced per week, write a system of inequalities that indicates necessary restrictions on x and y. Graph this system showing the region of permissible values for x and y.

	Labor-Hours per Ski		Maximum Labor-Hours Available per Week
	Trick	Slalom	
Fabricating department	6	4	96
Finishing department	1	1	20

24. Repeat Problem 23 with 96 and 20 (in the table) replaced by 180 and 35, respectively.

8-2 Systems of Linear Inequalities in Two Variables

- Solving Systems of Linear Inequalities Graphically
- Special Definitions Pertaining to Solution Regions
- Application

■ Solving Systems of Linear Inequalities Graphically

We will now consider systems of linear inequalities such as

$$x \geqslant 0 \quad \text{and} \quad 6x + 2y \leqslant 36$$
$$y \geqslant 0 \qquad\qquad 2x + 4y \leqslant 32$$
$$x \leqslant 8 \qquad\qquad x \geqslant 0$$
$$y \leqslant 4 \qquad\qquad y \geqslant 0$$

We wish to **solve** such systems **graphically,** that is, to find the graph of all ordered pairs of real numbers (x, y) that simultaneously satisfy all inequalities in the system. The graph is called the **solution region** for the system. To find the solution region, we graph each inequality in the system and then take the intersection of all of the graphs.

Example 3   Solve the following linear system graphically:

$$x \geqslant 0$$
$$y \geqslant 0$$
$$x \leqslant 8$$
$$y \leqslant 4$$

Solution   Graph all the inequalities in the same coordinate system and shade in the region that satisfies all four inequalities — that is, the intersection of all four graphs. The coordinates of any point in the shaded region will simultaneously satisfy all the original inequalities.

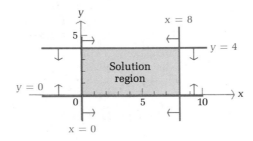

Problem 3   Solve the following linear system graphically:

$$x \geqslant 2$$
$$x \leqslant 6$$
$$y \leqslant 5$$
$$y \geqslant 2$$

Example 4   Solve the following linear system graphically:

$$6x + 2y \leqslant 36$$
$$2x + 4y \leqslant 32$$
$$x \geqslant 0$$
$$y \geqslant 0$$

Solution   Graph all the inequalities in the same coordinate system and shade in the intersection of all four graphs. The coordinates of any point in the shaded region of the accompanying figure specify a solution to the system. For example, (0, 0), (3, 4), and (2.35, 3.87) are three of infinitely many solutions, as can easily be checked.

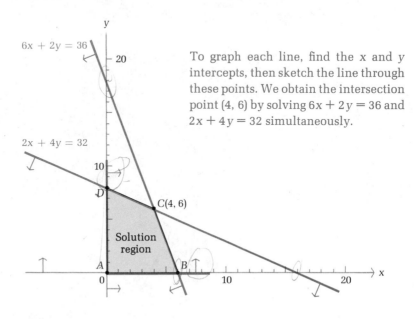

To graph each line, find the x and y intercepts, then sketch the line through these points. We obtain the intersection point (4, 6) by solving $6x + 2y = 36$ and $2x + 4y = 32$ simultaneously.

Problem 4   Solve by graphing:

$$3x + 2y \geqslant 24$$
$$x + 2y \geqslant 12$$
$$x \geqslant 0$$
$$y \geqslant 0$$

Example 5   Solve the system

$$2x + \ y \leqslant 20$$
$$10x + \ y \geqslant 36$$
$$2x + 5y \geqslant 36$$

graphically and find the coordinates of the intersection points of the boundary of the solution region.

Solution   The solution region is the intersection of the graphs of the three inequalities, as shown in the figure on the next page.

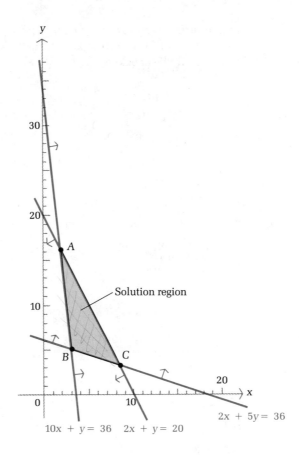

Coordinates of A   Solve the system

$$10x + y = 36$$
$$2x + y = 20$$

to obtain (2, 16).

Coordinates of B   Solve the system

$$10x + \ y = 36$$
$$2x + 5y = 36$$

to obtain (3, 6).

Coordinates of C   Solve the system

$$2x + \ y = 20$$
$$2x + 5y = 36$$

to obtain (8, 4).

Problem 5 Solve the system

$$3x + 4y \leqslant 48$$
$$x - y \leqslant 2$$
$$x \geqslant 4$$

graphically and find the coordinates of the intersection points of the boundary of the solution region.

■ Special Definitions Pertaining to Solution Regions

The following definitions pertain to the solution regions of systems of linear inequalities such as those found in the above examples and matched problems.

1. A solution region of a system of linear inequalities in two variables is **bounded** if it can be enclosed within a circle; if it cannot be enclosed within a circle, then it is **unbounded.** (The solution regions for Examples 4 and 5 are bounded; the solution region for Problem 4 is unbounded.)
2. A **corner point** of a solution region is the intersection of two boundary lines. [The corner points of the solution region in Example 5 are (2, 16), (3, 6), and (8, 4).]

That is enough new terminology for the moment. These definitions will be important to us in the next section. For now, let us introduce two applications that will be developed more fully as we proceed through this chapter.

■ Application

Example 6 A patient in a hospital is required to have at least 84 units of drug A and 120 units of drug B each day (assume that an overdosage of either drug is harmless). Each gram of substance M contains 10 units of drug A and 8 units of drug B, and each gram of substance N contains 2 units of drug A and 4 units of drug B. How many grams of substances M and N can be mixed to meet the minimum daily requirements?

Solution To clarify relationships, we summarize the information in the following table:

	Amount of Drug per Gram		Minimum Daily Requirement
	Substance M	Substance N	
Drug A	10 units	2 units	84 units
Drug B	8 units	4 units	120 units

Let $x =$ Number of grams of substance M used

   $y =$ Number of grams of substance N used

Then $10x =$ Number of units of drug A in x grams of substance M

   $2y =$ Number of units of drug A in y grams of substance N

   $8x =$ Number of units of drug B in x grams of substance M

   $4y =$ Number of units of drug B in y grams of substance N

The following conditions must be satisfied to meet daily requirements:

$$\begin{pmatrix} \text{Number of units of} \\ \text{drug } A \\ \text{in } x \text{ grams of substance } M \end{pmatrix} + \begin{pmatrix} \text{Number of units of} \\ \text{drug } A \\ \text{in } y \text{ grams of substance } N \end{pmatrix} \geqslant 84$$

$$\begin{pmatrix} \text{Number of units of} \\ \text{drug } B \\ \text{in } x \text{ grams of substance } M \end{pmatrix} + \begin{pmatrix} \text{Number of units of} \\ \text{drug } B \\ \text{in } y \text{ grams of substance } N \end{pmatrix} \geqslant 120$$

$$\begin{pmatrix} \text{Number of grams of} \\ \text{substance } M \text{ used} \end{pmatrix} \geqslant 0$$

$$\begin{pmatrix} \text{Number of grams of} \\ \text{substance } N \text{ used} \end{pmatrix} \geqslant 0$$

Converting these verbal statements into symbolic statements by using the variables x and y introduced above, we obtain the system of linear inequalities

$$10x + 2y \geqslant 84 \qquad \text{Drug } A \text{ restriction}$$
$$8x + 4y \geqslant 120 \qquad \text{Drug } B \text{ restriction}$$
$$x \geqslant 0 \qquad \text{Cannot use a negative amount of } M$$
$$y \geqslant 0 \qquad \text{Cannot use a negative amount of } N$$

Graphing this system of linear inequalities, we obtain the set of **feasible solutions,** or the **feasible region,** as shown in the following figure.* Thus, any point in the shaded area (including the straight line boundaries) will meet the daily requirements; any point outside the shaded area will not. (Note that the feasible region is unbounded.)

---

* For problems of this type and for linear programming problems in general (Sections 8-3 through 8-7), solution regions are often referred to as feasible regions.

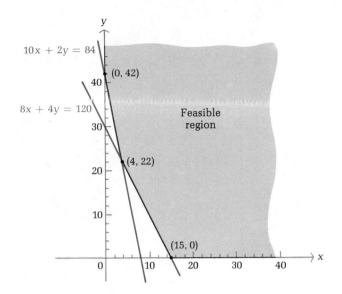

Problem 6

A manufacturing plant makes two types of inflatable boats, a two-person boat and a four-person boat. Each two-person boat requires 0.9 labor-hour in the cutting department and 0.8 labor-hour in the assembly department. Each four-person boat requires 1.8 labor-hours in the cutting department and 1.2 labor-hours in the assembly department. The maximum labor-hours available each month in the cutting and assembly departments are 864 and 672, respectively.

(A) Summarize this information in a table.
(B) If x two-person boats and y four-person boats are manufactured each month, write a system of linear inequalities that reflect the conditions indicated. Find the set of feasible solutions graphically.

Answers to
Matched Problems

3.

4.

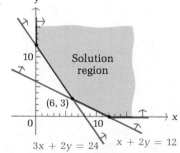

5.

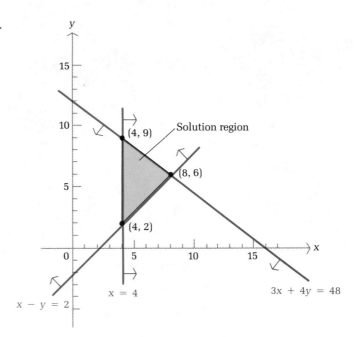

6. (A)

	Labor-Hours Required		Maximum Labor-Hours Available per Month
	Two-Person Boat	Four-Person Boat	
Cutting department	0.9	1.8	864
Assembly department	0.8	1.2	672

(B)

$$0.9x + 1.8y \leq 864$$
$$0.8x + 1.2y \leq 672$$
$$x \geq 0$$
$$y \geq 0$$

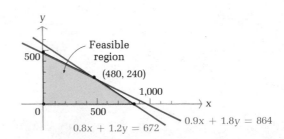

Exercise 8-2

A   *Solve the following linear systems graphically:*

1. $x \geqslant 0$
    $x \leqslant 7$
    $y \geqslant 0$
    $y \leqslant 4$

2. $y \geqslant 0$
    $x \leqslant 4$
    $y \geqslant 3$
    $y \leqslant 7$

3. $2x + 3y \leqslant 12$
    $x \geqslant 0$
    $y \geqslant 0$

4. $3x + 4y \leqslant 24$
    $x \geqslant 0$
    $y \geqslant 0$

5. $2x + y \leqslant 10$
    $x + 2y \leqslant 8$
    $x \geqslant 0$
    $y \geqslant 0$

6. $6x + 3y \leqslant 24$
    $3x + 6y \leqslant 30$
    $x \geqslant 0$
    $y \geqslant 0$

7. $2x + y \geqslant 10$
    $x + 2y \geqslant 8$
    $x \geqslant 0$
    $y \geqslant 0$

8. $4x + 3y \geqslant 24$
    $3x + 4y \geqslant 8$
    $x \geqslant 0$
    $y \geqslant 0$

B   *Solve the following systems graphically and indicate whether each solution set is bounded or unbounded. Find the coordinates of each corner.*

9. $2x + y \leqslant 10$
    $x + y \leqslant 7$
    $x + 2y \leqslant 12$
    $x \geqslant 0$
    $y \geqslant 0$

10. $3x + y \leqslant 21$
    $x + y \leqslant 9$
    $x + 3y \leqslant 21$
    $x \geqslant 0$
    $y \geqslant 0$

11. $2x + y \geqslant 16$
    $x + y \geqslant 12$
    $x + 2y \geqslant 14$
    $x \geqslant 0$
    $y \geqslant 0$

12. $3x + y \geqslant 24$
    $x + y \geqslant 16$
    $x + 3y \geqslant 30$
    $x \geqslant 0$
    $y \geqslant 0$

13. $x + 4y \leqslant 32$
    $3x + y \leqslant 30$
    $4x + 5y > 51$

14. $x + y < 11$
    $x + 5y \geqslant 15$
    $2x + y \geqslant 12$

15. $4x + 3y < 48$
    $2x + y \geqslant 24$
    $x < 9$

16. $2x + 3y > 24$
    $x + 3y < 15$
    $y > 4$

17. $x - y \leqslant 0$
    $2x - y \leqslant 4$
    $0 \leqslant x \leqslant 8$

18. $2x + 3y \geqslant 12$
    $-x + 3y \leqslant 3$
    $0 \leqslant y \leqslant 5$

C *Solve the following systems graphically and indicate whether each solution set is bounded or unbounded. Find the coordinates of each corner.*

19. $-x + 3y \geqslant 1$
    $5x - y \geqslant 9$
    $x + y \leqslant 9$
    $x \leqslant 5$

20. $x + y \leqslant 10$
    $5x + 3y \geqslant 15$
    $-2x + 3y \leqslant 15$
    $2x - 5y \leqslant 6$

21. $0.6x + 1.2y \leqslant 960$
    $0.03x + 0.04y \leqslant 36$
    $0.3x + 0.2y \leqslant 270$
    $x \geqslant 0$
    $y \geqslant 0$

22. $1.8x + 0.9y \geqslant 270$
    $0.3x + 0.2y \geqslant 54$
    $0.01x + 0.03y \geqslant 3.9$
    $x \geqslant 0$
    $y \geqslant 0$

Applications

Business & Economics

23. *Manufacturing—resource allocation.* A manufacturing company makes two types of water skis, a trick ski and a slalom ski. The trick ski requires 6 labor-hours for fabricating and 1 labor-hour for finishing. The slalom ski requires 4 labor-hours for fabricating and 1 labor-hour for finishing. The maximum labor-hours available per day for fabricating and finishing are 108 and 24, respectively. If x is the number of trick skis and y is the number of slalom skis produced per day, write a system of inequalities that indicates appropriate restraints on x and y. Find the set of feasible solutions graphically for the number of each type of ski that can be produced.

Life Sciences

24. *Nutrition.* A dietitian in a hospital is to arrange a special diet using two foods. Each ounce of food M contains 30 units of calcium, 10 units of iron, and 10 units of vitamin A. Each ounce of food N contains 10 units of calcium, 10 units of iron, and 30 units of vitamin A. The minimum requirements in the diet are 360 units of calcium, 160 units of iron, and 240 units of vitamin A. If x is the number of ounces of food M used and y is the number of ounces of food N used, write a system of linear inequalities that reflects the conditions indicated above. Find the set of feasible solutions graphically for the amount of each kind of food that can be used.

Social Sciences

25. *Psychology.* In an experiment on conditioning, a psychologist uses two types of Skinner (conditioning) boxes with mice and rats. Each mouse spends 10 minutes per day in box *A* and 20 minutes per day in box *B*. Each rat spends 20 minutes per day in box *A* and 10 minutes per day in box *B*. The total maximum time available per day is 800 minutes for box *A* and 640 minutes for box *B*. We are interested in the various numbers of mice and rats that can be used in the experiment under the conditions stated. If we let x be the number of mice used and y the number of rats used, write a system of inequalities that indicates appropriate restrictions on x and y. Find the set of feasible solutions graphically.

8-3 Linear Programming in Two Dimensions—A Geometric Approach

- A Maximization Problem
- A Minimization Problem
- Linear Program—A General Description
- A Maximum–Minimum Problem with Mixed Constraints
- Exceptional Situations

Several problems in the last section are related to a more general type of problem—a *linear programming problem*. Linear programming is a mathematical process that has been developed to help management in decision-making, and it has become one of the most widely used and best-known tools of management science. We will introduce this topic by considering a couple of examples in detail, using an intuitive geometric approach. Insight gained from this approach will prove invaluable when we later consider an algebraic approach that is less intuitive but necessary in solving most real-world problems.

> **Notation Change**
>
> For efficiency of generalization in later sections, we will now change variable notation from letters such as x and y to subscript forms such as x_1 and x_2.

A Maximization Problem

Example 7

A manufacturer of lightweight mountain tents makes a standard model and an expedition model. Each standard tent requires 1 labor-hour from the cutting department and 3 labor-hours from the assembly department. Each expedition tent requires 2 labor-hours from the cutting department and 4 labor-hours from the assembly department. The maximum labor-hours available per week in the cutting department and the assembly department are 32 and 84, respectively. In addition, the distributor, because of demand, will not take more than 12 expedition tents per week. If the company makes a profit of $50 on each standard tent and $80 on each expedition tent, how many tents of each type should be manufactured each week to maximize the total weekly profit?

Solution   This is an example of a *linear programming problem*. To see relationships more clearly, let us summarize the manufacturing requirements, objectives, and restrictions in table form (see Table 1).

Table 1

	Labor-Hours per Tent		Maximum Labor-Hours Available per Week
	Standard Model	Expedition Model	
Cutting department	1	2	32
Assembly department	3	4	84
Profit per tent	$50	$80	

In addition, as stated above, the weekly production of expedition tents cannot exceed 12.

We now proceed to formulate a mathematical model for the problem and then to solve it by using geometric methods.

Objective Function   The *objective* of management is to *decide* how many of each tent model should be produced each week so as to maximize profit. Let

x_1 = Number of standard tents produced per week
x_2 = Number of expedition tents produced per week
$\left.\right\}$ Decision variables

The following equation gives the total profit for x_1 standard tents and x_2 expedition tents manufactured each week, assuming all tents manufactured are sold:

$P = 50x_1 + 80x_2$ Objective function

Mathematically, the management needs to decide on values for the **decision variables** (x_1, x_2) that achieve its objective, that is, maximizing the **objective function** (profit) $P = 50x_1 + 80x_2$. It appears that the profit can be made as large as we like by manufacturing more and more tents — or can it?

Constraints

Any manufacturing company, no matter how large or small, has manufacturing limits imposed by available resources, plant capacity, demand, and so forth. These limits are referred to as constraints.

Cutting department constraint:

$$\begin{pmatrix} \text{Weekly cutting} \\ \text{time for } x_1 \\ \text{standard tents} \end{pmatrix} + \begin{pmatrix} \text{Weekly cutting} \\ \text{time for } x_2 \\ \text{expedition tents} \end{pmatrix} \leq \begin{pmatrix} \text{Maximum labor-} \\ \text{hours available} \\ \text{per week} \end{pmatrix}$$

$$1x_1 \qquad + \qquad 2x_2 \qquad \leq \qquad 32$$

Assembly department constraint:

$$\begin{pmatrix} \text{Weekly assembly} \\ \text{time for } x_1 \\ \text{standard tents} \end{pmatrix} + \begin{pmatrix} \text{Weekly assembly} \\ \text{time for } x_2 \\ \text{expedition tents} \end{pmatrix} \leq \begin{pmatrix} \text{Maximum labor-} \\ \text{hours available} \\ \text{per week} \end{pmatrix}$$

$$3x_1 \qquad + \qquad 4x_2 \qquad \leq \qquad 84$$

Demand constraints. The distributor will not take more then 12 expedition tents per week; thus,

$$x_2 \leq 12$$

Nonnegative constraints. It is not possible to manufacture a negative number of tents; thus, we have the **nonnegative constraints**

$$x_1 \geq 0$$
$$x_2 \geq 0$$

which we usually write in the form

$$x_1, x_2 \geq 0$$

Mathematical Model

We now have a **mathematical model** for the problem under consideration:

Maximize $\quad P = 50x_1 + 80x_2 \qquad$ Objective function

Subject to $\quad x_1 + 2x_2 \leq 32$

$\qquad\qquad\quad 3x_1 + 4x_2 \leq 84 \quad\Big\}\quad$ Problem constraints

$\qquad\qquad\qquad\quad x_2 \leq 12$

$\qquad\qquad\quad x_1, x_2 \geq 0 \qquad$ Nonnegative constraints

Graphical Solution

Solving the set of linear inequality constraints **graphically** (see the last section), we obtain the feasible region for production schedules (Fig. 4).

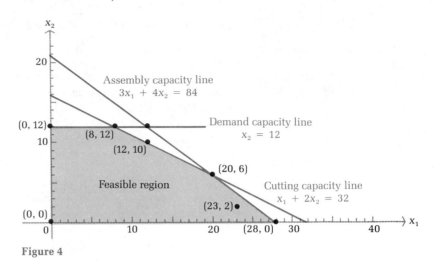

Figure 4

By choosing a production schedule (x_1, x_2) from the feasible region, a profit can be determined using the objective function

$$P = 50x_1 + 80x_2$$

For example, if $x_1 = 12$ and $x_2 = 10$, then the profit for the week would be

$$P = 50(12) + 80(10)$$
$$= \$1,400$$

Or if $x_1 = 23$ and $x_2 = 2$, then the profit for the week would be

$$P = 50(23) + 80(2)$$
$$= \$1,310$$

The question is, out of all possible production schedules (x_1, x_2) from the feasible region, which schedule(s) produces the maximum profit? Thus, we have a **maximization problem.** Since point-by-point checking is impossible (there are infinitely many points to check), we must find another way.

By assigning P in

$$P = 50x_1 + 80x_2$$

a particular value and plotting the resulting equation in Figure 4, we obtain a **constant-profit line (isoprofit line).** Every point in the feasible region on this line represents a production schedule that will produce the same profit. By doing this for a number of values for P, we obtain a family of constant-profit lines (Fig. 5) that are parallel to each other, since they all have the same slope. To see that they all have the same slope, we write $P = 50x_1 + 80x_2$ in the slope–intercept form

$$x_2 = -\frac{5}{8}x_1 + \frac{P}{80}$$

and note that for any profit P, the constant-profit line has slope $-\frac{5}{8}$. We also observe that as the profit P increases, the x_2 intercept $(P/80)$ increases, and the line moves away from the origin.

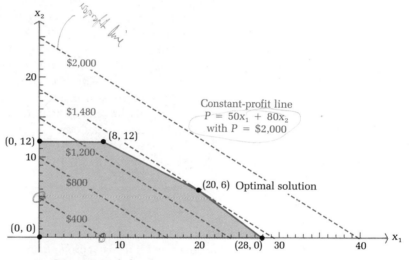

Figure 5 Constant-profit lines

Thus, the maximum profit occurs at a point where a constant-profit line is the farthest from the origin but still in contact with the feasible region. In this example, this occurs at (20, 6), as is seen in Figure 5. Thus, if the manufacturer makes 20 standard tents and 6 expedition tents per week, the profit will be maximized at

$$P = 50(20) + 80(6)$$
$$= \$1{,}480$$

The point (20, 6) is called an **optimal solution** to the problem, because it maximizes the objective (profit) function and is in the feasible region. In general, it appears that a maximum profit occurs at one of the corner points. We also note that the minimum profit $(P = 0)$ occurs at the corner point (0, 0).

Graphical Solution of a Maximization Problem

1. Form a mathematical model for the problem:
   a. Introduce decision variables and write a linear objective function.
   b. Write problem constraints using linear inequalities and/or equations.
   c. Write nonnegative constraints.
2. Graph the feasible region.
3. Draw a constant-profit line that passes through the feasible region.
4. Move parallel constant-profit lines toward higher profits (usually away from the origin) until they cannot be moved farther without leaving the feasible region.
5. The last line found in step 4 (if it exists) will pass through a corner of the feasible region. This corner represents an optimal solution.
6. Find the optimal solution (coordinates of the corner found in step 5) by simultaneously solving the constraint equations whose graphs pass through the corner point.
7. Evaluate the objective function at the optimal solution to find the maximum value.

Problem 7   We now convert the boat problem in the preceding section (Problem 6) into a linear programming problem. A manufacturing plant makes two types of inflatable boats, a two-person boat and a four-person boat. Each boat requires the services of two departments as listed in the table. In addition, the distribution will not take more than 750 two-person boats each month. How many boats of each type should be manufactured each month to maximize the profit? What is the maximum profit?

| | Labor-Hours Required | | Maximum Labor-Hours Available per Month |
	Two-Person Boat	Four-Person Boat	
Cutting department	0.9	1.8	864
Assembly department	0.8	1.2	672
Profit per boat	$25	$40	

■ A Minimization Problem

Example 8   We now convert the drug example in the preceding section (Example 6) into a linear programming problem. A patient in a hospital is required to have at least 84 units of drug D_1 and 120 units of drug D_2 each day. Two substances M and N contain each of these drugs; however, in addition, suppose both M and N contain an undesirable drug D_3. The relevant information is contained in the table. How many grams each of substances M and N should be mixed to meet the minimum daily requirement and at the same time minimize the intake of drug D_3? How many units of the undesirable drug D_3 will be in this mixture? (This is a **minimization problem**.)

	Amount of Drug per Gram		Minimum Daily Requirement
	Substance M	Substance N	
Drug D_1	10 units	2 units	84 units
Drug D_2	8 units	4 units	120 units
Drug D_3	3 units	1 unit	

Solution   Let x_1 = Number of grams of substance M used

x_2 = Number of grams of substance N used $\Big\}$ Decision variables

We form the linear objective function

$$C = 3x_1 + x_2$$

which gives the amount of the undesirable drug D_3 in x_1 grams of M and x_2 grams of N. Proceeding as in Example 7, we formulate the following mathematical model for the problem:

Minimize	$C = 3x_1 + x_2$	Objective function
Subject to	$10x_1 + 2x_2 \geq 84$	Drug D_1 constraint
	$8x_1 + 4x_2 \geq 120$	Drug D_2 constraint
	$x_1, x_2 \geq 0$	Nonnegative constraints

Solving the system of constraint inequalities graphically (as in the last section), we obtain the feasible region shown in Figure 6 (page 396).

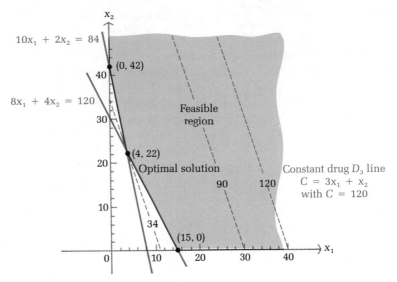

Figure 6

Proceeding as in Example 7, we see that the constant drug D_3 lines (found by assigning C in the objective function a constant value and graphing) produce the minimum value for C when the line is closest to the origin and still in contact with the feasible region. This takes place at the corner $(4, 22)$, which is the optimal solution to the problem. Thus, if we use 4 grams of substance M and 22 grams of substance N, we shall supply the minimum requirements for drugs D_1 and D_2 and minimize the intake of the undesirable drug D_3 at 34 units. Any other combination of M and N from the feasible region will not result in a smaller amount of the undesirable drug D_3. (Note that $C = 3x_1 + x_2$ has no maximum value in the feasible region, since x_1 and x_2 can be made as large as we like.)

The key steps used in solving a minimization problem by graphing are listed at the top of the next page.

Problem 8   A chicken farmer can buy a special food mix A at 20¢ per pound and a special food mix B at 40¢ per pound. Each pound of mix A contains 3,000 units of nutrient N_1 and 1,000 units of nutrient N_2; each pound of mix B contains 4,000 units of nutrient N_1 and 4,000 units of nutrient N_2. If the minimum daily requirements for the chickens collectively are 36,000 units of nutrient N_1 and 20,000 units of nutrient N_2, how many pounds of each food mix should be used each day to minimize daily food costs while meeting (or exceeding) the minimum daily nutrient requirements? What is the minimum daily cost?

Graphical Solution of a Minimization Problem

1. Form a mathematical model for the problem.
2. Graph the feasible region.
3. Draw a constant-cost line that passes through the feasible region.
4. Move parallel constant-cost lines toward lower costs (usually toward the origin) until they cannot be moved further without leaving the feasible region.
5. The last line found in step 4 (if it exists) will pass through a corner of the feasible region. This corner represents an optimal solution.
6. Find the optimal solution (coordinates of the corner point in step 5) by simultaneously solving the constraint equations whose graphs pass through the corner point.
7. Evaluate the objective function at the optimal solution to find the minimum value.

■ Linear Program — A General Description

The mathematical model for the tent problem in Example 7,

Maximize $P = 50x_1 + 80x_2$ Objective function

Subject to $x_1 + 2x_2 \leqslant 32$

$3x_1 + 4x_2 \leqslant 84$ Problem constraints

$x_2 \leqslant 12$

$x_1, x_2 \geqslant 0$ Nonnegative constraints

is an example of a linear programming problem or, simply, a linear program. The special feature that makes it a *linear* program is the fact that the objective function and the left sides of all the constraint statements are linear functions of the decision variables x_1 and x_2.

Linear Function

A **linear function** is any function defined by an equation of the form

$$z = c_1 x_1 + c_2 x_2 + \cdots + c_n x_n$$

where $x_1, x_2, \ldots, x_n$ are independent variables, z is a dependent variable, and $c_1, c_2, \ldots, c_n$ are constants not all zero.

In general, we describe a linear program as follows:

A Linear Program

All linear programs involve maximizing or minimizing a linear objective function of two or more decision variables subject to constraints in the form of linear inequalities or equations. All variables except the dependent variable for the objective function must be nonnegative.

The following theorem, which should seem reasonable after our earlier discussion, is fundamental to solving linear programming problems.

Theorem 2

Fundamental Theorem of Linear Programming

Let S be the feasible region in a linear programming problem and z the objective function.

(A) If S is bounded, the objective function z will have both a maximum and minimum value on S, and these will occur at corner points. [Thus, to find the maximum (minimum) value of an objective function, simply evaluate it at the corner points and choose the maximum (minimum) value.]

(B) If S is unbounded, the objective function z may not have a maximum or minimum. If it does, it will occur at a corner point.

■ A Maximum–Minimum Problem with Mixed Constraints

Example 9 (A) Maximize $P = 3x_1 + x_2$ (B) Minimize $C = x_1 + 6x_2$

P and C both subject to

$$2x_1 + x_2 \leq 20$$
$$10x_1 + x_2 \geq 36$$
$$2x_1 + 5x_2 \geq 36$$
$$x_1, x_2 \geq 0$$

Solutions

(A) S = Feasible region

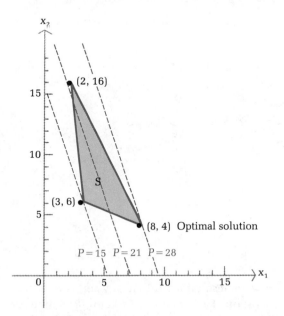

$P = 15$ $P = 21$ $P = 28$

Corner Point (x_1, x_2)	$P = 3x_1 + x_2$
(3, 6)	15
(2, 16)	22
(8, 4)	28

Optimal solution = (8, 4)
Max $P = 3(8) + (4) = 28$

(B) S = Feasible region

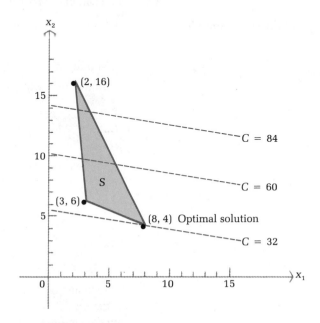

$C = 84$

$C = 60$

$C = 32$

Corner Point (x_1, x_2)	$C = x_1 + 6x_2$
(3, 6)	39
(2, 16)	99
(8, 4)	32

Optimal solution = (8, 4)
Min $C = 8 + 6(4) = 32$

Remarks 1. A common error is to assume that the maximum of an objective function occurs at the corner point of a feasible region that is farthest from the origin point and that the minimum occurs at the corner point closest to the origin. Example 9 shows that this is clearly not the case.
2. Also note that the optimal solution (8, 4) happens to be the same for both parts A and B. In fact, Min $C = 32$ is larger than Max $P = 28$. Conclusion: Optimal solutions and maxima and minima depend on the objective function under consideration as well as on the feasible region.

Problem 9 In Example 9, minimize part A and maximize part B given the same constraints.

■ Exceptional Situations

We complete this section by considering three exceptional situations that one can encounter when solving a linear programming problem.

Multiple Optimal Solutions   A linear programming problem may have more than one optimal solution. Consider the following:

Maximize $P = 3x_1 + 2x_2$
Subject to $3x_1 + 2x_2 \leq 22$
$x_1 \leq 6$
$x_2 \leq 5$
$x_1, x_2 \geq 0$

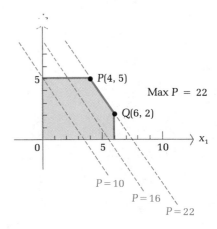

Notice that both P and Q are optimal solutions. In fact, any point on the line segment PQ is an optimal solution. This would be useful information for management, since it provides them with more choices for reaching their objective.

No Feasible Region   A linear programming problem may have no feasible region; that is, there may not be any points that simultaneously satisfy all constraints. Consider the following:

Maximize $P = 2x_1 + 3x_2$
Subject to $x_1 + x_2 \geq 8$
$x_1 + 2x_2 \leq 8$
$2x_1 + x_2 \leq 10$
$x_1, x_2 \geq 0$

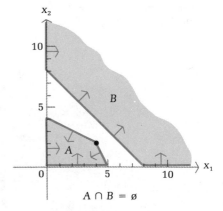

$A \cap B = \emptyset$

The intersection of the graphs of the constraint inequalities is the empty set; hence, the feasible region is empty. If this happens, then the problem should be reexamined to see if it has been formulated properly. If it has, then the management may have to reconsider items such as labor-hours, overtime, budget, and supplies allocated to the project in order to obtain a nonempty feasible region and a solution to the original problem.

Unbounded Objective Function

An objective function may increase (or decrease) without bound over a feasible region. Consider the following:

Maximize
$$P = 6x_1 + 3x_2$$

Subject to
$$-2x_1 + 3x_2 \leqslant 9$$
$$-x_1 + 3x_2 \leqslant 12$$
$$x_1, x_2 \geqslant 0$$

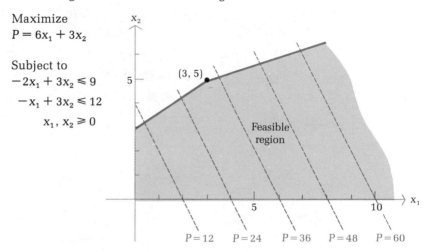

The farther constant-profit lines move from the origin, the larger P is. In fact, for any profit, however large, P can be made larger; hence, profit increases without bound. Before you rush out to buy stock in this company, consider the impossibility of such a feat. In an applied problem, you can assume a mistake was made in formulating the problem. Check to see if a mistake was made in writing a constraint statement or if a constraint was omitted.

The science of linear programming evolved mainly out of problems regarding supply allocations and nutrition during World War II. The geometric approach we have taken in this section serves well to make basic concepts clear, but it does not readily lend itself to problems with more than two variables. George B. Dantzig developed an algebraic approach in the 1940's, called the *simplex method*, that does generalize. We will consider this method in the following sections.

Answers to Matched Problems

7. 480 two-person boats, 240 four-person boats; Max $P = \$21,600$ per week
8. 8 lb of mix A, 3 lb of mix B; Min $C = \$2.80$ per day
9. (A) Optimal solution $= (3, 6)$, Min $P = 15$
    (B) Optimal solution $= (2, 16)$, Max $C = 98$

Exercise 8-3

A   *Solve the following linear programming problems:*

1. Maximize $P = 5x_1 + 5x_2$
    Subject to $2x_1 + x_2 \leqslant 10$
    $x_1 + 2x_2 \leqslant 8$
    $x_1, x_2 \geqslant 0$

2. Maximize $P = 3x_1 + 2x_2$
    Subject to $6x_1 + 3x_2 \leqslant 24$
    $3x_1 + 6x_2 \leqslant 30$
    $x_1, x_2 \geqslant 0$

3. Minimize and maximize
    $z = 2x_1 + 3x_2$
    Subject to $2x_1 + x_2 \geqslant 10$
    $x_1 + 2x_2 \geqslant 8$
    $x_1, x_2 \geqslant 0$

4. Minimize and maximize
    $z = 8x_1 + 7x_2$
    Subject to $4x_1 + 3x_2 \geqslant 24$
    $3x_1 + 4x_2 \geqslant 8$
    $x_1, x_2 \geqslant 0$

B   5. Maximize $P = 30x_1 + 40x_2$
    Subject to $2x_1 + x_2 \leqslant 10$
    $x_1 + x_2 \leqslant 7$
    $x_1 + 2x_2 \leqslant 12$
    $x_1, x_2 \geqslant 0$

6. Maximize $P = 20x_1 + 10x_2$
    Subject to $3x_1 + x_2 \leqslant 21$
    $x_1 + x_2 \leqslant 9$
    $x_1 + 3x_2 \leqslant 21$
    $x_1, x_2 \geqslant 0$

7. Minimize and maximize
    $z = 10x_1 + 30x_2$
    Subject to $2x_1 + x_2 \geqslant 16$
    $x_1 + x_2 \geqslant 12$
    $x_1 + 2x_2 \geqslant 14$
    $x_1, x_2 \geqslant 0$

8. Minimize and maximize
    $z = 400x_1 + 100x_2$
    Subject to $3x_1 + x_2 \geqslant 24$
    $x_1 + x_2 \geqslant 16$
    $x_1 + 3x_2 \geqslant 30$
    $x_1, x_2 \geqslant 0$

9. Minimize and maximize
    $P = 30x_1 + 10x_2$
    Subject to $2x_1 + 2x_2 \geqslant 4$
    $6x_1 + 4x_2 \leqslant 36$
    $2x_1 + x_2 \leqslant 10$
    $x_1, x_2 \geqslant 0$

10. Minimize and maximize
    $P = 2x_1 + x_2$
    Subject to $x_1 + x_2 \geqslant 2$
    $6x_1 + 4x_2 \leqslant 36$
    $4x_1 + 2x_2 \leqslant 20$
    $x_1, x_2 \geqslant 0$

11. Minimize and maximize
    $P = 3x_1 + 5x_2$
    Subject to $x_1 + 2x_2 \leqslant 6$
    $x_1 + x_2 \leqslant 4$
    $2x_1 + 3x_2 \geqslant 12$
    $x_1, x_2 \geqslant 0$

12. Minimize and maximize
    $P = -x_1 + 3x_2$
    Subject to $2x_1 - x_2 \geqslant 4$
    $-x_1 + 2x_2 \leqslant 4$
    $x_2 \leqslant 6$
    $x_1, x_2 \geqslant 0$

13. Minimize and maximize

$$P = 20x_1 + 10x_2$$

Subject to $\quad 2x_1 + 3x_2 \geqslant 30$

$2x_1 + x_2 \leqslant 26$

$2x_1 + 5x_2 \leqslant 34$

$x_1, x_2 \geqslant 0$

14. Minimize and maximize

$$P = 12x_1 + 14x_2$$

Subject to $\quad -2x_1 + x_2 \geqslant 6$

$x_1 + x_2 \leqslant 15$

$3x_1 - x_2 \geqslant 0$

$x_1, x_2 \geqslant 0$

15. Maximize $\quad P = 20x_1 + 30x_2$

Subject to $\quad 0.6x_1 + 1.2x_2 \leqslant 960$

$0.03x_1 + 0.04x_2 \leqslant 36$

$0.3x_1 + 0.2x_2 \leqslant 270$

$x_1, x_2 \geqslant 0$

16. Minimize $\quad C = 30x_1 + 10x_2$

Subject to $\quad 1.8x_1 + 0.9x_2 \geqslant 270$

$0.3x_1 + 0.2x_2 \geqslant 54$

$0.01x_1 + 0.03x_2 \geqslant 3.9$

$x_1, x_2 \geqslant 0$

17. The corner points for the system of inequalities

$$x_1 + 2x_2 \leqslant 10$$
$$3x_1 + x_2 \leqslant 15$$
$$x_1, x_2 \geqslant 0$$

are $O = (0, 0)$, $A = (0, 5)$, $B = (4, 3)$, and $C = (5, 0)$. If $P = ax_1 + bx_2$ and $a, b > 0$, determine conditions on a and b which will ensure that the maximum value of P occurs:

(A) Only at A (B) Only at B (C) Only at C
(D) At both A and B (E) At both B and C

18. The corner points for the system of inequalities

$$x_1 + 4x_2 \geqslant 30$$
$$3x_1 + x_2 \geqslant 24$$
$$x_1, x_2 \geqslant 0$$

are $A = (0, 24)$, $B = (6, 6)$, and $D = (30, 0)$. If $C = ax_1 + bx_2$ and $a, b > 0$, determine conditions on a and b which will ensure that the minimum value of C occurs:

(A) Only at A (B) Only at B (C) Only at D
(D) At both A and B (E) At both B and D

Applications

Business & Economics

19. *Manufacturing—resource allocation.* A manufacturing company makes two types of water skis, a trick ski and a slalom ski. The relevant manufacturing data are given in the table. How many of each type of ski should be manufactured each day to realize a maximum profit? What is the maximum profit?

	Labor-Hours per Ski		Maximum Labor-Hours Available per Day
	Trick Ski	Slalom Ski	
Fabricating department	6	4	108
Finishing department	1	1	24
Profit per ski	$40	$30	

20. *Investment.* An investor has $24,000 to invest in bonds of AAA and B qualities. The AAA bonds yield on the average 6% and the B bonds yield 10%. The investor's policy requires that she invest at least three times as much money in AAA bonds as in B bonds. How much should she invest in each type of bond to maximize her return? What is the maximum return?

21. *Pollution control.* Because of new federal regulations on pollution, a chemical plant introduced a new, more expensive process to supplement or replace an older process used in the production of a particular chemical. The older process emitted 15 grams of sulfur dioxide and 40 grams of particulate matter into the atmosphere for each gallon of chemical produced. The new process emits 5 grams of sulfur dioxide and 20 grams of particulate matter for each gallon produced. The company makes a profit of 30¢ per gallon and 20¢ per gallon on the old and new processes, respectively. If the government allows the plant to emit no more than 10,500 grams of sulfur dioxide and no more than 30,000 grams of particulate matter daily, how many gallons of the chemical should be produced by each process to maximize daily profit? What is the maximum profit?

Life Sciences

22. *Nutrition—people.* A dietitian in a hospital is to arrange a special diet composed of two foods, M and N. Each ounce of food M contains 30 units of calcium, 10 units of iron, 10 units of vitamin A, and 8 units of cholesterol. Each ounce of food N contains 10 units of calcium, 10 units of iron, 30 units of vitamin A, and 4 units of cholesterol. If the minimum daily requirements are 360 units of calcium, 160 units of

iron, and 240 units of vitamin A, how many ounces of each food should be used to meet the minimum requirements and at the same time minimize the cholesterol intake? What is the minimum cholesterol intake?

23. *Nutrition—plants.* A farmer can buy two types of plant food, mix A and mix B. Each cubic yard of mix A contains 20 pounds of phosphoric acid, 30 pounds of nitrogen, and 5 pounds of potash. Each cubic yard of mix B contains 10 pounds of phosphoric acid, 30 pounds of nitrogen, and 10 pounds of potash. The minimum monthly requirements are 460 pounds of phosphoric acid, 960 pounds of nitrogen, and 220 pounds of potash. If mix A costs $30 per cubic yard and mix B costs $35 per cubic yard, how many cubic yards of each mix should the farmer blend to meet the minimum monthly requirements at a minimal cost? What is this cost?

24. *Nutrition—animals.* A laboratory technician in a medical research center is asked to formulate a diet from two commercially packaged foods, food A and food B, for a group of animals. Each ounce of food A contains 8 units of fat, 16 units of carbohydrate, and 2 units of protein. Each ounce of food B contains 4 units of fat, 32 units of carbohydrate, and 8 units of protein. The minimum daily requirements are 176 units of fat, 1,024 units of carbohydrate, and 384 units of protein. If food A costs 5¢ per ounce and food B costs 5¢ per ounce, how many ounces of each food should be used to meet the minimum daily requirements at the least cost? What is the cost for this amount of food?

Social Sciences

25. *Psychology.* In an experiment on conditioning, a psychologist uses two types of Skinner boxes with mice and rats. The amount of time in minutes each mouse and each rat spends in each box per day is given in the table. What is the maximum number of mice and rats that can be used in this experiment? How many mice and how many rats produce this maximum?

	Time		Maximum Time Available per Day
	Mice	Rats	
Skinner box A	10 min	20 min	800 min
Skinner box B	20 min	10 min	640 min

$P = x_1 + x_2$

26. *Sociology.* A city council voted to conduct a study on inner-city community problems. A nearby university was contacted to provide sociologists and research assistants. Allocation of time and costs per

week are given in the table. How many sociologists and how many research assistants should be hired to minimize the cost and meet the weekly labor-hour requirements? What is the minimum weekly cost?

	Labor-Hours		Minimum Labor-Hours Needed per Week
	Sociologist	Research Assistant	
Fieldwork	10	30	180
Research center	30	10	140
Costs per week	$500	$300	

8-4 A Geometric Introduction to the Simplex Method

- Slack Variables
- Basic Feasible Solutions
- Basic Feasible Solutions and the Simplex Method

The geometric method of solving linear programming problems provided us with an overview of the subject and some useful terminology. But, practically speaking, the method is only useful for problems involving two decision variables and relatively few problem constraints. What happens when we need more decision variables or have many problem constraints? We use an algebraic approach called the *simplex method*. Using matrix methods and row operations, the simplex method is readily adapted to computer computation, and the method is commonly used to solve problems with hundreds and even thousands of variables.

The algebraic procedures utilized in the simplex method require the problem constraints to be written as equations rather than inequalities. This new form of the problem also prompts the use of some new terminology. We introduce this new form of a linear program and associated terminology through a simple example and an appropriate geometric interpretation. From this example we can illustrate what the simplex method does geometrically before we immerse ourselves in the algebraic details of the process.

■ Slack Variables

Let us return to the tent production problem in Example 7 from the last section. Recall the mathematical model for the problem:

Maximize $P = 50x_1 + 80x_2$ Objective function
Subject to $x_1 + 2x_2 \leqslant 32$ Cutting department constraint
$3x_1 + 4x_2 \leqslant 84$ Assembly department constraint   (1)
$x_2 \leqslant 12$ Demand constraint
$x_1, x_2 \geqslant 0$ Nonnegative constraints

where x_1 and x_2 are the number of standard and expedition tents, respectively, produced each week.

To take advantage of matrix methods in solving systems of equations (which is part of the algebraic process we will discuss in the next section), we convert the problem constraint inequalities in a linear program into a system of linear equations by using a simple device called a *slack variable*. In particular, to convert the system of problem constraint inequalities from (1)

$$x_1 + 2x_2 \leqslant 32$$
$$3x_1 + 4x_2 \leqslant 84$$ (2)
$$x_2 \leqslant 12$$

into a system of equations, we add nonnegative quantities s_1, s_2, and s_3 to the left members of (2) to obtain

$$x_1 + 2x_2 + s_1 \qquad\quad = 32$$
$$3x_1 + 4x_2 \quad\;\; + s_2 \quad\; = 84$$ (3)
$$x_2 \qquad\quad + s_3 = 12$$

The variables s_1, s_2, and s_3 are called **slack variables** because each makes up the difference (takes up the slack) between the left and right sides of the inequalities in (2). It is important to remember that **slack variables are nonnegative.**

Notice that system (3) has infinitely many solutions — just solve for s_1, s_2, and s_3 in terms of x_1 and x_2 and then assign x_1 and x_2 arbitrary values. Certain solutions to system (3) have an interesting relationship to the feasible region (see Fig. 7, page 408) for the original linear program (1).

In system (3), set any two variables equal to zero and solve, if possible, for the remaining three. The results of carrying out this project systematically are summarized in Table 2. Carefully compare the results in Table 2 with Figure 7.

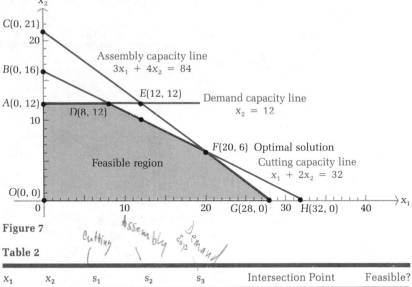

Figure 7

Table 2

x_1	x_2	s_1	s_2	s_3	Intersection Point	Feasible?
0	0	32	84	12	O	Yes
0	16	0	20	−4	B	No
0	21	−10	0	−9	C	No
0	12	8	36	0	A	Yes
32	0	0	−12	12	H	No
28	0	4	0	12	G	Yes
	0*			0*		No
20	6	0	0	6	F	Yes
8	12	0	12	0	D	Yes
12	12	−4	0	0	E	No

* Leads to an inconsistent system.

■ Basic Feasible Solutions

Table 2 contains interesting and useful information, and in conjunction with Figure 7, it leads to several important definitions.

The solutions in Table 2, obtained by setting two variables equal to zero and solving for the other three, are referred to as *basic solutions* of system (3). Each basic solution corresponds to an intersection point of two of the original constraint equations (including the nonnegativity constraints).

For example, the basic solution $x_1 = 0$, $x_2 = 21$, $s_1 = -10$, $s_2 = 0$, $s_3 = -9$ corresponds to the point $C(0, 21)$, the intersection of the assembly capacity line and the x_2 axis. Even though there is a technical distinction between a point in the plane (like C) and the corresponding basic solution to system (3), we will use the two concepts interchangeably to simplify our discussion. With this understanding, we can say that the basic solutions *are* the intersection points of the constraint equations taken two at a time.

Now note that the set of corner points (including the optimal solution) corresponds to a subset of the set of basic solutions. We refer to the basic solutions (O, A, D, F, and G) in Table 2, which correspond to corner points of the feasible region, as *basic feasible solutions* of system (3). In general,

---

Basic and Basic Feasible Solutions

Given a system of m linear equations with n variables, $n > m$, and assuming the system has infinitely many solutions, then the solution (if it exists) obtained by setting $n - m$ variables equal to zero and solving for the remaining m variables is called a **basic solution.** When a linear system is associated with a linear programming problem and a basic solution of the system has no negative values (that is, corresponds to a corner point of the feasible region), we refer to it as a **basic feasible solution.** The $n - m$ variables set equal to zero in obtaining a basic solution are called **nonbasic variables;** the m remaining variables are called **basic variables.**

Thus, in Table 2 when we set x_1 and x_2 equal to zero, then x_1 and x_2 become nonbasic variables and s_1, s_2, and s_3 become basic variables; if we set x_1 and s_1 equal to zero, then x_1 and s_1 become nonbasic variables and x_2, s_2, and s_3 become basic variables; and so on.

Looking again at Table 2, we notice some negative entries for some variables. **Any basic solution with a negative value for one or more variables is infeasible** (recall that all decision variables and slack variables must be nonnegative). These infeasible basic solutions correspond to points B, C, E, and H in Figure 7, which are outside of the feasible region. Thus, we see that we may subdivide basic solutions into two mutually exclusive sets: *basic feasible solutions* and *basic infeasible solutions.* The following important theorem [which is equivalent to the fundamental theorem (Theorem 2) in the preceding section] is stated without proof:

Theorem 3

If a linear program has an optimal solution, then it must be one (or more) of the basic feasible solutions.

Thus, to solve a linear programming problem (if a solution exists), we need only concern ourselves with basic feasible solutions (that is, corner points of the feasible region).

Let us carefully compare a couple of the basic feasible solutions in Table 2. If we choose the origin as a basic feasible solution, then we will not produce any tents, and the values of the slack variables represent 32 unused labor-hours in the cutting department, 84 unused labor-hours in the assembly department, and 12 units of unused demand for expedition tents. If we choose the basic feasible solution $D(8, 12)$, then we will produce

8 standard tents and 12 expedition tents. However, there will still be slack in the assembly department; that is, there will be 12 unused labor-hours in that department. There is no cutting department slack nor demand slack. Interpret the value of each slack variable at the optimal solution $F(20, 6)$.

The values of slack variables at optimal solutions provide management with useful information regarding resource utilization. For example, it would be useful to know after deciding on an optimal production schedule if there were any unused (slack) labor-hours in the cutting or assembly departments that could be utilized for other purposes.

▪ Basic Feasible Solutions and the Simplex Method

What does all of the above discussion have to do with the simplex method? The **simplex method** is an iterative (repetitive) algebraic procedure that moves automatically from one basic feasible solution to another, improving the situation each time until an optimal solution is reached (if it exists). Geometrically, the simplex method moves from one corner point of the feasible region (see Fig. 7) to another corner point, improving the situation each time until an optimal solution is reached (if it exists). With this background, we are now ready to discuss the details of the simplex method.

Exercise 8-4

A   **1.** Listed in the table below are all the basic solutions for the system

$$2x_1 + 3x_2 + s_1 \qquad = 24$$
$$4x_1 + 3x_2 \qquad + s_2 = 36$$

For each basic solution, identify the nonbasic variables and the basic variables, then classify each basic solution as feasible or not feasible.

	x_1	x_2	s_1	s_2
(A)	0	0	24	36
(B)	0	8	0	12
(C)	0	12	−12	0
(D)	12	0	0	−12
(E)	9	0	6	0
(F)	6	4	0	0

2. Repeat Problem 1 for the system

$$2x_1 + x_2 + s_1 \qquad = 30$$
$$x_1 + 5x_2 \qquad + s_2 = 60$$

whose basic solutions are given in the following table:

	x_1	x_2	s_1	s_2
(A)	0	0	30	60
(B)	0	30	0	−90
(C)	0	12	10	0
(D)	15	0	0	45
(E)	60	0	−90	0
(F)	10	10	0	0

3. Listed in the table below are all the possible choices of nonbasic variables for the system

$$2x_1 + x_2 + s_1 \qquad = 50$$
$$x_1 + 2x_2 \qquad + s_2 = 40$$

In each case, find the value of the basic variables and determine if the basic solution is feasible.

	x_1	x_2	s_1	s_2
(A)	0	0	?	?
(B)	0	?	0	?
(C)	0	?	?	0
(D)	?	0	0	?
(E)	?	0	?	0
(F)	?	?	0	0

4. Repeat Problem 3 for the system

$$x_1 + 2x_2 + s_1 \qquad = 12$$
$$3x_1 + 2x_2 \qquad + s_2 = 24$$

B *Graph the following systems of inequalities. Introduce slack variables to convert each system of inequalities to a system of equations and find all the basic solutions of the system. Construct a table (like Table 2) listing each basic solution, the corresponding point on the graph, and whether the basic solution is feasible.*

5. $x_1 + x_2 \leqslant 16$
    $2x_1 + x_2 \leqslant 20$
    $x_1, x_2 \geqslant 0$

6. $5x_1 + x_2 \leqslant 35$
    $4x_1 + x_2 \leqslant 32$
    $x_1, x_2 \geqslant 0$

7. $2x_1 + x_2 \leqslant 22$
    $x_1 + x_2 \leqslant 12$
    $x_1 + 2x_2 \leqslant 20$
    $x_1, x_2 \geqslant 0$

8. $4x_1 + x_2 \leqslant 28$
    $2x_1 + x_2 \leqslant 16$
    $x_1 + x_2 \leqslant 13$
    $x_1, x_2 \geqslant 0$

8-5   The Simplex Method: Maximization with ≤ Problem Constraints

- Standard Form of a Linear Program
- An Algebraic Introduction to the Simplex Method
- The Simplex Tableau and Method

In this section we will restrict our attention to maximization problems with ≤ problem constraints and nonnegative constants to the right of the inequality symbols.

■ Standard Form of a Linear Program

We start our discussion with the tent problem considered in the last two sections (Example 7). Recall the mathematical model for the problem:

$$
\begin{array}{lll}
\text{Maximize} & P = 50x_1 + 80x_2 & \text{Objective function} \\
\text{Subject to} & x_1 + 2x_2 \le 32 \\
& 3x_1 + 4x_2 \le 84 & \text{Problem constraints} \\
& x_2 \le 12 \\
& x_1, x_2 \ge 0 & \text{Nonnegative constraints}
\end{array} \tag{1}
$$

cutting *Assembly* *Demand of cap tents.*

Using (nonnegative) slack variables s_1, s_2, and s_3, we convert the problem constraint inequalities into equations and all of (1) into the following *standard* form:

$$
\begin{array}{ll}
\text{Maximize} & P = 50x_1 + 80x_2 \\
\text{Subject to} & x_1 + 2x_2 + s_1 \qquad\qquad = 32 \\
& 3x_1 + 4x_2 \qquad + s_2 \qquad = 84 \\
& x_2 \qquad\qquad + s_3 = 12 \\
& x_1, x_2, s_1, s_2, s_3 \ge 0
\end{array} \tag{2}
$$

For clarity, all variables with zero coefficients are omitted from each equation.

Standard Form

Whenever a linear programming problem is written in a form with all problem constraints expressed as equations, it is said to be written in **standard form.**

From our discussion in the last section, we know that out of the infinitely many solutions to the problem constraint equations

$$x_1 + 2x_2 + s_1 \qquad\qquad = 32$$
$$3x_1 + 4x_2 \qquad + s_2 \qquad = 84 \qquad\qquad (3)$$
$$x_2 \qquad\qquad + s_3 = 12$$

of the Constraints
Intersection

an optimal solution will be among the basic feasible solutions, which are the corner points of the feasible region. [Recall that a basic solution of (3) is found by setting $5 - 3 = 2$ variables equal to zero and solving for the other three. A basic solution is also feasible if none of the values in the solution are negative.]

■ An Algebraic Introduction to the Simplex Method

We will now discuss an algebraic method of moving from one basic feasible solution to another until the optimal solution is found. We will then streamline the process by introducing and using matrix methods. To start, we write system (2) in the following equivalent standard form:

$$x_1 + \;\; 2x_2 + s_1 \qquad\qquad\quad = 32$$
$$3x_1 + \;\; 4x_2 \qquad + s_2 \qquad\qquad = 84$$
$$x_2 \qquad\quad + s_3 \qquad = 12 \qquad\qquad (4)$$
$$-50x_1 - 80x_2 \qquad\qquad\quad + P = 0$$

The fourth equation is simply the objective function $P = 50x_1 + 80x_2$ written with all variable terms on the left. Our problem is to find a solution of (4) that maximizes P.

System (4) involves four equations and six variables. To find a basic solution to the system, we set $6 - 4 = 2$ variables equal to zero and solve for the other four variables. Any basic solution of (4) with P a basic variable (not one of the two variables set equal to zero) is also a basic solution of (3) if the P part of the solution is deleted. If the basic solution of (4) is also a feasible solution of (3) (the problem constraint system), we say it is a feasible solution of (4) (the problem constraint system with the objective equation added).

We must start our algebraic process with a basic feasible solution of (4) with P one of the basic variables. Our work will be somewhat easier if we start with an *obvious basic feasible solution*:

Obvious Basic Feasible Solution

When a linear program is written in standard form [as in (4)] with m equations and n variables, then choose m variables such that each occurs in one and only one equation and no two occur in the same equation. These m variables are the basic variables. (The objective function variable P will always be chosen as a basic variable.) The remaining $n - m$ variables (each usually occurring in more than one equation) are then nonbasic variables. The solution obtained by setting these $n - m$ nonbasic variables equal to zero and solving (by inspection) for the m basic variables will be referred to as an **obvious basic solution.** If no number in the solution is negative except possibly P, then the solution is an **obvious basic feasible solution.**

To obtain an obvious basic solution to (4), we choose s_1, s_2, s_3, and P as basic variables (since each occurs in exactly one equation and no two appear in the same equation). This leaves x_1 and x_2 (each occurring in more than one equation) as nonbasic variables. Setting the nonbasic variables equal to zero, a basic feasible solution to (4) can be obtained by inspection (thus the name "obvious basic feasible solution"):

$$
\begin{array}{rrrrrr}
x_1 & + \ 2\,x_2 & + \ s_1 & & & = 32 \\
3\,x_1 & + \ 4\,x_2 & & + \ s_2 & & = 84 \\
& x_2 & & & + \ s_3 & = 12 \\
-50\,x_1 & - \ 80\,x_2 & & & + \ P & = \ 0
\end{array}
$$

$$x_1 = 0, \quad x_2 = 0, \quad s_1 = 32, \quad s_2 = 84, \quad s_3 = 12, \quad P = 0$$

We would certainly expect a profit of zero if we do not produce any tents! We can improve the situation by increasing either x_1 or x_2 or both. Let us start by increasing the decision variable that contributes most to the profit for each unit increase in the variable. Referring to the objective function $P = 50x_1 + 80x_2$, we see that each unit increase in x_2 increases the profit by \$80, while each unit increase in x_1 increases the profit by only \$50, so we increase x_2 first. How much can we increase x_2 in (4), holding $x_1 = 0$, without causing s_1, s_2, or s_3 to become negative? (Remember that if any of the variables except P become negative, we no longer have a feasible solution.) To see how much x_2 can be increased, rewrite the first three equations in (4), with $x_1 = 0$, as follows:

$$s_1 = 32 - 2x_2$$
$$s_2 = 84 - 4x_2 \qquad (5)$$
$$s_3 = 12 - x_2$$

We can increase x_2 in the first equation to 16 without causing s_1 to become negative, to 21 in the second equation without causing s_2 to become negative, and to 12 in the third equation without causing s_3 to become negative. Thus, we can increase x_2 to 12 (the minimum of 16, 32, and 12) without causing *any* of the variables s_1, s_2, or s_3 to become negative.

So that $x_2 = 12$ can be read directly (by inspection) as part of an obvious basic feasible solution, we eliminate x_2 from all equations in (4) but the third (then x_2 will change from a nonbasic variable to a basic variable). To do this, we multiply the third equation by -2 and add it to the first equation; then we multiply the third equation by -4 and add it to the second equation; finally, we multiply the third equation by 80 and add it to the fourth equation. (The third equation is not changed in this process.) Completing these operations on (4), we obtain the following equivalent system:

$$
\begin{aligned}
x_1 + s_1 - 2s_3 &= 8 \\
3x_1 \qquad + s_2 - 4s_3 &= 36 \\
+ x_2 \qquad + s_3 &= 12 \\
-50x_1 \qquad + 80s_3 + P &= 960
\end{aligned}
\qquad (6)
$$

To obtain an obvious basic solution to (6), which variables should be basic and which nonbasic? Since each variable x_2, s_1, s_2, and P occurs in exactly one equation and no two appear in the same equation, they will be chosen as basic. Thus, x_1 and s_3 will be nonbasic. Assigning x_1 and s_3 zero values and solving (by inspection) for x_2, s_1, s_2, and P, we obtain the obvious basic feasible solution

$$
\begin{array}{ccccc}
& 0 & & 0 & \\
x_1 & + s_1 & - 2\;s_3 & & = 8 \\
3\;x_1 & & + s_2 - 4\;s_3 & & = 36 \\
& + x_2 & + s_3 & & = 12 \\
-50\;x_1 & & + 80\;s_3 & + P & = 960
\end{array}
$$

$$x_1 = 0, \quad x_2 = 12, \quad s_1 = 8, \quad s_2 = 36, \quad s_3 = 0, \quad P = \$960$$

Increasing x_2 to 12 has increased the profit to \$960, a marked improvement! But is this the best we can do? Rewriting the objective function (fourth equation) in (6) in the form

$$P = 50x_1 - 80s_3 + 960$$

we see that P can be increased still further if we can increase x_1, holding $s_3 = 0$, without making x_2, s_1, and s_2 negative. To see how far we can increase x_1 under these conditions, rewrite the first three equations in (6), with $s_3 = 0$, in the form

$$s_1 = 8 - x_1$$
$$s_2 = 36 - 3x_1$$
$$x_2 = 12$$

We can increase x_1 in the first equation to 8, in the second equation to 12, and in the third equation indefinitely without causing s_1, s_2, or s_3, respectively, to become negative. Thus, we can increase x_1 to 8 (the minimum of 8 and 12) without causing any of the variables s_1, s_2, or x_2 to become negative.

So that $x_1 = 8$ can be read by inspection as part of an obvious basic feasible solution, we eliminate x_1 from all equations in (6) but the first (then x_1 will change from a nonbasic variable to a basic variable). Notice that x_2 does not lose its status as a basic variable in the process. (Why?) Proceeding as above, we eliminate x_1 from all equations in (6) except the first to obtain the equivalent system:

$$
\begin{aligned}
x_1 \quad + \quad s_1 \quad - \quad 2s_3 \quad &= 8 \\
- \quad 3s_1 + s_2 + \quad 2s_3 \quad &= 12 \\
+ x_2 \quad\quad\quad + \quad s_3 \quad &= 12 \\
+ 50s_1 \quad - 20s_3 + P &= 1{,}360
\end{aligned}
\tag{7}
$$

which has the obvious basic feasible solution (s_1 and s_3 are nonbasic variables set equal to zero):

$$x_1 = 8, \quad x_2 = 12, \quad s_1 = 0, \quad s_2 = 12, \quad s_3 = 0, \quad P = \$1{,}360$$

Increasing x_1 to 8 has increased the profit to \$1,360, another marked improvement. Can we do any better? Rewriting the objective function in (7) in the form

$$P = -50s_1 + 20s_3 + 1{,}360$$

we see that P can be increased still further if we increase s_3, holding $s_1 = 0$, without making x_1, x_2, or s_2 negative. To see how far we can increase s_3 under these conditions, rewrite the first three equations in (7), with $s_1 = 0$, in the form

$$x_1 = 8 + 2s_3$$
$$s_2 = 12 - 2s_3$$
$$x_2 = 12 - s_3$$

We can increase s_3 in the first equation indefinitely, to 6 in the second equation, and to 12 in the third equation without causing x_1, s_2, or x_2, respectively, to become negative. Thus, we can increase s_3 to 6 (the minimum of 6 and 12) without causing any of the variables x_1, s_2, or x_2 to become negative.

So that $s_3 = 6$ can be read by inspection as part of an obvious basic feasible solution, we multiply the second equation in (7) by $\frac{1}{2}$ (so that the coefficient of s_3 is 1), then use the second equation to eliminate s_3 from the first, third, and fourth equations. In the process, s_3 will change from a nonbasic variable to a basic variable, while x_1 and x_2 do not lose their status as basic variables. (Why?) Carrying out the elimination, we obtain the equivalent system:

$$
\begin{aligned}
x_1 \quad - \quad 2s_1 + \quad s_2 \quad &= 20 \\
-1.5s_1 + 0.5s_2 + s_3 \quad &= 6 \\
+ x_2 + 1.5s_1 - 0.5s_2 \quad &= 6 \\
20s_1 + 10\,s_2 \quad + P &= \$1{,}480
\end{aligned}
\tag{8}
$$

which has the obvious basic feasible solution

$$x_1 = 20, \quad x_2 = 6, \quad s_1 = 0, \quad s_2 = 0, \quad s_3 = 6, \quad P = \$1{,}480$$

And P has been improved even further. Have we found the production schedule that maximizes P? To find out, we write the objective function in (8) in the form

$$P = -20s_1 - 10s_2 + \$1{,}480$$

and note that any increase in s_1 or s_2 will reduce P.

It can be shown that when this situation occurs, we have found the optimal solution. Hence, P is a maximum when 20 standard tents and 6 expedition tents are produced (as we found geometrically in Section 8-3). Since s_1 and s_2 are both zero, there are no labor-hours (slack) left in the cutting or assembly departments. However, there is a slack in demand, since $s_3 = 6$. That is, a weekly demand of 6 expedition tents is left unfilled.

Listing the obvious basic feasible solution we considered at each stage above in table form (Table 3) and comparing the results with the corners of

Table 3 Obvious Basic Feasible Solutions

x_1	x_2	s_1	s_2	s_3	P	Corner Point
0	0	32	84	12	$ 0	O
0	12	8	36	0	960	A
8	12	0	12	0	1,360	D
20	6	0	0	6	1,480	F

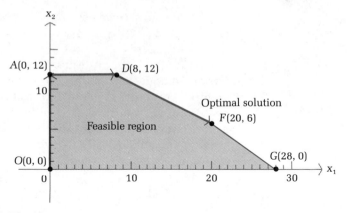

Figure 8

the feasible region discussed in the last section (Fig. 8), we see that the algebraic process moved from one corner point of the feasible region to another, improving P each time until the optimal solution $F(20, 6)$ was reached.

▪ The Simplex Tableau and Method

The above process can be made substantially more efficient by using the matrix methods discussed in the last chapter. We start with the original linear program:

$$\begin{aligned}
\text{Maximize} \quad & P = 50x_1 + 80x_2 && \text{Objective function} \\
\text{Subject to} \quad & \left.\begin{array}{l} x_1 + 2x_2 \leqslant 32 \\ 3x_1 + 4x_2 \leqslant 84 \\ x_2 \leqslant 12 \end{array}\right\} && \text{Problem constraints} \qquad\qquad (9) \\
& x_1, x_2 \geqslant 0 && \text{Nonnegative constraints}
\end{aligned}$$

Now we introduce slack variables $s_1 \geqslant 0$, $s_2 \geqslant 0$, and $s_3 \geqslant 0$ and convert (9) into the standard form:

$$\begin{aligned}
x_1 + \ 2x_2 + s_1 \qquad\qquad\qquad &= 32 \\
3x_1 + \ 4x_2 \qquad + s_2 \qquad\quad &= 84 \\
x_2 \qquad\quad + s_3 \quad &= 12 \\
-50x_1 - 80x_2 \qquad\qquad\quad + P &= 0 \\
x_1, x_2, s_1, s_2, s_3 \geqslant 0 &
\end{aligned} \qquad (10)$$

Let us again state our objective: Out of the infinitely many solutions to system (10), we are interested in finding a solution that maximizes the profit P by using matrix methods.

Our first step is to write the augmented matrix, called the **simplex tableau,** for system (10):

$$
\begin{array}{c}
x_1 \quad x_2 \quad s_1 \quad s_2 \quad s_3 \quad P \\
\left[
\begin{array}{cccccc|c}
1 & 2 & 1 & 0 & 0 & 0 & 32 \\
3 & 4 & 0 & 1 & 0 & 0 & 84 \\
0 & 1 & 0 & 0 & 1 & 0 & 12 \\
\hline
-50 & -80 & 0 & 0 & 0 & 1 & 0
\end{array}
\right]
\end{array}
\qquad (11)
$$

The row below the dashed line will always correspond to an objective function.

We now try to move from one obvious basic feasible solution to another (using basic row operations except for interchanging rows) until we find a solution that maximizes P. We start with the obvious basic feasible solution (x_1 and x_2 are the nonbasic variables set equal to zero):

$$
\begin{array}{c}
0 \qquad 0 \\
x_1 \quad x_2 \quad s_1 \quad s_2 \quad s_3 \quad P \\
\left[
\begin{array}{cccccc|c}
1 & 2 & 1 & 0 & 0 & 0 & 32 \\
3 & 4 & 0 & 1 & 0 & 0 & 84 \\
0 & 1 & 0 & 0 & 1 & 0 & 12 \\
\hline
-50 & -80 & 0 & 0 & 0 & 1 & 0
\end{array}
\right]
\end{array}
$$

$$x_1 = 0, \quad x_2 = 0, \quad s_1 = 32, \quad s_2 = 84, \quad s_3 = 12, \quad P = 0$$

We would like to transform (11) into a row-equivalent matrix that has a basic feasible solution with a larger value for P. Looking at the objective function $P = 50x_1 + 80x_2$, we observed that the change in P per unit increase in x_2 is greater than the change in P per unit increase in x_1. We can see this in (11) by looking for the most negative value in the fourth row, the objective function row. The column with the most negative value below the dashed line is called the **pivot column.** Now, how much can x_2 be increased when $x_1 = 0$ without causing s_1, s_2, or s_3 to become negative? We found this value by a process that is equivalent to dividing each *positive* element in the pivot column above the dashed line into the corresponding element in the last column and choosing the minimum value. Carrying out the calculations

$$\frac{32}{2} = 16, \qquad \frac{84}{4} = 21, \qquad \frac{12}{1} = 12$$

we see that the minimum quotient is 12, which is associated with the third row. This row is called the **pivot row.** The element in the pivot column (Column 2) and in the pivot row (Row 3) is called the **pivot element,** and we circle it:

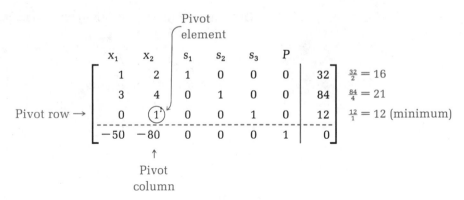

We will transform the nonbasic variable associated with the pivot column into a basic variable (by the suitable use of row operations) by transforming the pivot element into 1 (if it isn't already 1), then using the pivot element to transform the rest of the elements in the pivot column into 0's. This procedure is called a **pivot operation**. The **row operations** we can perform are the following two (we cannot interchange rows):

Permissible Row Operations

1. A row can be multiplied by a nonzero constant $(kR_i \rightarrow R_i)$.
2. A constant multiple of another row can be added to a given row $(R_i + kR_j \rightarrow R_i)$.

We now carry out the pivot operation:

$$
\begin{array}{c}
\begin{array}{cccccc} x_1 & x_2 & s_1 & s_2 & s_3 & P \end{array} \\
\left[\begin{array}{cccccc|c}
1 & 2 & 1 & 0 & 0 & 0 & 32 \\
3 & 4 & 0 & 1 & 0 & 0 & 84 \\
0 & 1 & 0 & 0 & 1 & 0 & 12 \\
\hline
-50 & -80 & 0 & 0 & 0 & 1 & 0
\end{array}\right]
\begin{array}{l}
R_1 + (-2)R_3 \rightarrow R_1 \\
R_2 + (-4)R_3 \rightarrow R_3 \\
\\
R_4 + 80R_3 \rightarrow R_4
\end{array}
\end{array}
$$

Since this is ① — No need to transform it first

$$
\sim
\left[\begin{array}{cccccc|c}
1 & 0 & 1 & 0 & -2 & 0 & 8 \\
3 & 0 & 0 & 1 & -4 & 0 & 36 \\
0 & 1 & 0 & 0 & 1 & 0 & 12 \\
\hline
-50 & 0 & 0 & 0 & 80 & 1 & 960
\end{array}\right]
$$

After completing the pivot operation, we write another obvious basic feasible solution (x_1 and s_3 are now nonbasic variables assigned zero values):

$$x_1 = 0 \quad x_2 = 12, \quad s_1 = 8, \quad s_2 = 36, \quad s_3 = 0, \quad P = \$960$$

This is an improvement over our earlier solution, but we can improve P still further since a negative number remains in the fourth row. (Write out the fourth row using variables to see why the negative number indicates that P can still be increased.)

We repeat the above sequence of steps using another pivot element. To locate the pivot element, we see that the pivot column is the first column in the matrix (since it contains the most negative element in the fourth row). To find the pivot row, divide each *positive* element in the pivot column above the dashed line into the corresponding element in the last column and choose the minimum value. The row corresponding to this value is the pivot row.

Pivot element

$$\text{Pivot row} \rightarrow \begin{bmatrix} \begin{array}{cccccc|c} x_1 & x_2 & s_1 & s_2 & s_3 & P & \\ \boxed{1} & 0 & 1 & 0 & -2 & 0 & 8 \\ 3 & 0 & 0 & 1 & -4 & 0 & 36 \\ 0 & 1 & 0 & 0 & 1 & 0 & 12 \\ \hdashline -50 & 0 & 0 & 0 & 80 & 1 & 960 \end{array} \end{bmatrix} \begin{array}{l} \frac{8}{1} = 8 \text{ (minimum)} \\ \\ \frac{36}{3} = 12 \\ \\ \end{array}$$

↑
Pivot
column

We can now use row operations to get 0's below the pivot element. (The nonbasic variable x_1 will then be transformed into a basic variable.)

$$\begin{bmatrix} \begin{array}{cccccc|c} x_1 & x_2 & s_1 & s_2 & s_3 & P & \\ \boxed{1} & 0 & 1 & 0 & -2 & 0 & 8 \\ 3 & 0 & 0 & 1 & -4 & 0 & 36 \\ 0 & 1 & 0 & 0 & 1 & 0 & 12 \\ \hdashline -50 & 0 & 0 & 0 & 80 & 1 & 960 \end{array} \end{bmatrix} \begin{array}{l} \\ \\ R_2 + (-3)R_1 \rightarrow R_2 \\ \\ R_4 + 50R_1 \rightarrow R_4 \end{array}$$

$$\sim \begin{bmatrix} \begin{array}{cccccc|c} 1 & 0 & 1 & 0 & -2 & 0 & 8 \\ 0 & 0 & -3 & 1 & 2 & 0 & 12 \\ 0 & 1 & 0 & 0 & 1 & 0 & 12 \\ \hdashline 0 & 0 & 50 & 0 & -20 & 1 & 1,360 \end{array} \end{bmatrix}$$

The obvious basic feasible solution is (choosing s_1 and s_3 as nonbasic variables set equal to zero):

$$x_1 = 8, \quad x_2 = 12, \quad s_1 = 0, \quad s_2 = 12, \quad s_3 = 0, \quad P = \$1,360$$

Since the fourth row in the matrix (the objective function row) still has a negative quantity -20, P can be improved still further. We find the pivot element as before and complete the pivot operation:

Pivot element

	x_1	x_2	s_1	s_2	s_3	P	
	1	0	1	0	-2	0	8
Pivot row →	0	0	-3	1	②	0	12
	0	1	0	0	1	0	12
	0	0	50	0	-20	1	1,360

$\frac{12}{2} = 6$ (minimum)

$\frac{12}{1} = 12$

↑
Pivot
column

Note: The minimum positive quotient is our only interest. To see why $8/(-2) = -4$ does not enter in, see the last part of the algebraic solution to this problem earlier in this section.

We now complete the pivot operation:

	x_1	x_2	s_1	s_2	s_3	P	
	1	0	1	0	-2	0	8
	0	0	-3	1	②	0	12
	0	1	0	0	1	0	12
	0	0	50	0	-20	1	1,360

$0.5R_2 \rightarrow R_2$

	x_1	x_2	s_1	s_2	s_3	P	
~	1	0	1	0	-2	0	8
	0	0	-1.5	0.5	1	0	6
	0	1	0	0	1	0	12
	0	0	50	0	-20	1	1,360

$R_1 + 2R_2 \rightarrow R_1$

$R_3 + (-1)R_2 \rightarrow R_3$

$R_4 + 20R_2 \rightarrow R_4$

	x_1	x_2	s_1	s_2	s_3	P	
~	1	0	-2	1	0	0	20
	0	0	-1.5	0.5	1	0	6
	0	1	1.5	-0.5	0	0	6
	0	0	20	10	0	1	1,480

The obvious basic feasible solution is

$$x_1 = 20, \quad x_2 = 6, \quad s_1 = 0, \quad s_2 = 0, \quad s_3 = 6, \quad P = \$1,480$$

Since the fourth row (the objective function row) has no more negative entries, P cannot be made larger by increasing any of the variables. This is seen more clearly by converting the fourth row back into the equation form

$$P = -20s_1 - 10s_2 + 1,480$$

So we are through, because any increase in s_1 or s_2 will reduce P.

Let us review the critical steps in the simplex method so that the process can be mechanized. The key idea in the matrix transformation centers on the selection of the pivot element, which we summarize here:

Selecting the Pivot Element

1. Locate the most negative element in the bottom row of the tableau. The column containing this element is the *pivot column*. If there is a tie for most negative, choose either.
2. Divide each *positive* element in the pivot column above the dashed line into the corresponding element in the last column. The *pivot row* is the row corresponding to the smallest quotient. If the pivot column above the dashed line has no positive elements, then there is no solution and we stop.
3. The *pivot* (or *pivot element*) is the element in the pivot column and in the pivot row. [*Note:* The pivot element is never in the bottom row.]

We now summarize the important parts of the simplex method.

Simplex Method — Key Steps for Maximization Problems

(Constraints involving two or more variables are of the ⩽ form with nonnegative constants on the right.)

1. Write the simplex tableau associated with a linear programming problem.
2. Determine the pivot element.
3. Use row transformations (except for interchanging rows) to convert the pivot element to 1 and all the other elements in the pivot column to 0.
4. Repeat steps 2 and 3 until all elements in the bottom row are nonnegative. When this occurs, we stop the process and read the optimal solution.

Example 10   Solve the following linear programming problem using the simplex method:

$$\text{Maximize} \quad P = 10x_1 + 5x_2$$
$$\text{Subject to} \quad 6x_1 + 2x_2 \leqslant 36$$
$$2x_1 + 4x_2 \leqslant 32$$
$$x_1, x_2 \geqslant 0$$

Solution   Introduce slack variables s_1 and s_2 and write in standard form:

$$6x_1 + 2x_2 + s_1 \qquad\quad = 36$$
$$2x_1 + 4x_2 \qquad + s_2 \quad = 32$$
$$-10x_1 - 5x_2 \qquad\quad + P = 0$$
$$x_1, x_2, s_1, s_2 \geqslant 0$$

Write the simplex tableau and identify the first pivot element:

$$
\begin{array}{ccccc}
x_1 & x_2 & s_1 & s_2 & P
\end{array}
$$
$$
\left[
\begin{array}{ccccc|c}
⑥ & 2 & 1 & 0 & 0 & 36 \\
2 & 4 & 0 & 1 & 0 & 32 \\
\hdashline
-10 & -5 & 0 & 0 & 1 & 0
\end{array}
\right]
\qquad
\begin{array}{l}
\frac{36}{6} = 6 \\[4pt]
\frac{32}{2} = 16
\end{array}
$$

Use row operations to pivot on 6:

$$
\begin{array}{ccccc}
x_1 & x_2 & s_1 & s_2 & P
\end{array}
$$
$$
\left[
\begin{array}{ccccc|c}
⑥ & 2 & 1 & 0 & 0 & 36 \\
2 & 4 & 0 & 1 & 0 & 32 \\
\hdashline
-10 & -5 & 0 & 0 & 1 & 0
\end{array}
\right]
\qquad \tfrac{1}{6}R_1 \rightarrow R_1
$$

$$
\sim
\left[
\begin{array}{ccccc|c}
1 & \frac{1}{3} & \frac{1}{6} & 0 & 0 & 6 \\
2 & 4 & 0 & 1 & 0 & 32 \\
\hdashline
-10 & -5 & 0 & 0 & 1 & 0
\end{array}
\right]
\qquad
\begin{array}{l}
R_2 + (-2)R_1 \rightarrow R_2 \\[4pt]
R_3 + 10R_1 \rightarrow R_3
\end{array}
$$

$$
\sim
\left[
\begin{array}{ccccc|c}
1 & \frac{1}{3} & \frac{1}{6} & 0 & 0 & 6 \\
0 & \frac{10}{3} & -\frac{1}{3} & 1 & 0 & 20 \\
\hdashline
0 & -\frac{5}{3} & \frac{5}{3} & 0 & 1 & 60
\end{array}
\right]
$$

Since there still is a negative element in the last row, we repeat the process by finding a new pivot element:

$$
\begin{array}{ccccc}
x_1 & x_2 & s_1 & s_2 & P
\end{array}
$$
$$
\left[
\begin{array}{ccccc|c}
1 & \frac{1}{3} & \frac{1}{6} & 0 & 0 & 6 \\
0 & ⑩\!\!\frac{10}{3} & -\frac{1}{3} & 1 & 0 & 20 \\
\hdashline
0 & -\frac{5}{3} & \frac{5}{3} & 0 & 1 & 60
\end{array}
\right]
\qquad
\begin{array}{l}
6 \div \frac{1}{3} = 18 \\[4pt]
20 \div \frac{10}{3} = 6
\end{array}
$$

Pivoting on $\frac{10}{3}$, we obtain

$$
\begin{array}{ccccc}
x_1 & x_2 & s_1 & s_2 & P
\end{array}
$$

$$
\left[
\begin{array}{ccccc|c}
1 & \frac{1}{3} & \frac{1}{6} & 0 & 0 & 6 \\
0 & \boxed{\frac{10}{3}} & -\frac{1}{3} & 1 & 0 & 20 \\
\hline
0 & -\frac{5}{3} & \frac{5}{3} & 0 & 1 & 60
\end{array}
\right]
\quad \frac{3}{10}R_2 \to R_2
$$

$$
\sim
\left[
\begin{array}{ccccc|c}
1 & \frac{1}{3} & \frac{1}{6} & 0 & 0 & 6 \\
0 & 1 & -\frac{1}{10} & \frac{3}{10} & 0 & 6 \\
\hline
0 & -\frac{5}{3} & \frac{5}{3} & 0 & 1 & 60
\end{array}
\right]
\quad
\begin{array}{l}
R_1 + (-\frac{1}{3})R_2 \to R_1 \\[6pt]
\\
R_3 + \frac{5}{3}R_2 \to R_3
\end{array}
$$

$$
\sim
\left[
\begin{array}{ccccc|c}
1 & 0 & \frac{1}{5} & -\frac{1}{10} & 0 & 4 \\
0 & 1 & -\frac{1}{10} & \frac{3}{10} & 0 & 6 \\
\hline
0 & 0 & \frac{3}{2} & \frac{1}{2} & 1 & 70
\end{array}
\right]
$$

Since all the elements in the last row are nonnegative, we stop and read the solution:

$$\text{Max } P = 70 \quad \text{at} \quad x_1 = 4, \quad x_2 = 6, \quad s_1 = 0, \quad s_2 = 0$$

(If this still is not clear, write the system of equations corresponding to the last matrix and see what happens to P when you try to increase s_1 or s_2.)

Problem 10   Solve the following linear programming problem using the simplex method:

$$\text{Maximize} \quad P = 2x_1 + x_2$$
$$\text{Subject to} \quad 4x_1 + x_2 \le 8$$
$$\phantom{\text{Subject to} \quad} 2x_1 + 2x_2 \le 10$$
$$\phantom{\text{Subject to} \quad} x_1, x_2 \ge 0$$

Example 11   Solve using the simplex method:

$$\text{Maximize} \quad P = 6x_1 + 3x_2$$
$$\text{Subject to} \quad -2x_1 + 3x_2 \le 9$$
$$\phantom{\text{Subject to} \quad} -x_1 + 3x_2 \le 12$$
$$\phantom{\text{Subject to} \quad} x_1, x_2 \ge 0$$

Solution   Write in standard form using the slack variables s_1 and s_2:

$$-2x_1 + 3x_2 + s_1 = 9$$
$$-x_1 + 3x_2 + s_2 = 12$$
$$-6x_1 - 3x_2 + P = 0$$

Write the simplex tableau and identify the first pivot element:

$$\begin{array}{ccccc} x_1 & x_2 & s_1 & s_2 & P \\ \end{array}$$

$$\left[\begin{array}{ccccc|c} -2 & 3 & 1 & 0 & 0 & 9 \\ -1 & 3 & 0 & 1 & 0 & 12 \\ \hline -6 & -3 & 0 & 0 & 1 & 0 \end{array}\right]$$

↑
Pivot column

Since both elements in the pivot column above the dashed line are negative, we conclude that there is no solution, and we stop. This was the example of an unbounded objective function discussed in Section 8-3. We include the graphical "solution" here for convenient reference.

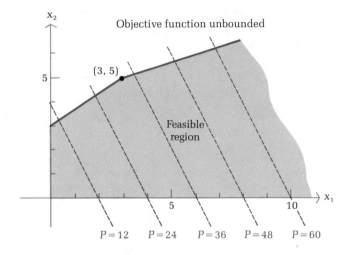

Problem 11   Solve using the simplex method:

Maximize $P = 2x_1 + 3x_2$

Subject to $-3x_1 + 4x_2 \leqslant 12$

$$x_2 \leqslant 9$$

$$x_1, x_2 \geqslant 0$$

Example 12   A farmer owns a 100 acre farm and plans to plant at most three crops. The seed for crops A, B, and C costs \$40, \$20, and \$30 per acre, respectively. A maximum of \$3,200 can be spent on seed. Crops A, B, and C require 1, 2, and 1 workdays per acre, respectively, and there are a maximum of 160 workdays available. If the farmer can make a profit of \$100 per acre on crop A, \$300 per acre on crop B, and \$200 per acre on crop C, how many acres of each crop should be planted to maximize profit?

Solution   Let

x_1 = Number of acres of crop A

x_2 = Number of acres of crop B

x_3 = Number of acres of crop C

P = Total profit

Then we have the following linear programming problem:

Maximize $P = 100x_1 + 300x_2 + 200x_3$

$$\left.\begin{array}{l} \text{Subject to} \quad x_1 + \quad x_2 + \quad x_3 \le 100 \\ \qquad\qquad 40x_1 + 20x_2 + 30x_3 \le 3{,}200 \\ \qquad\qquad x_1 + \quad 2x_2 + \quad x_3 \le 160 \end{array}\right\}$$ Problem constraints

$$x_1, x_2, x_3 \ge 0\}$$ Nonnegative constraints

Next, we introduce slack variables:

$$
\begin{array}{rcl}
x_1 + \quad x_2 + \quad x_3 + s_1 & & = 100 \\
40x_1 + 20x_2 + 30x_3 \quad + s_2 & & = 3{,}200 \\
x_1 + \quad 2x_2 + \quad x_3 \quad\quad + s_3 & & = 160 \\
-100x_1 - 300x_2 - 200x_3 \quad\quad\quad + P & = & 0
\end{array}
$$

$$x_1, x_2, x_3, s_1, s_2, s_3 \ge 0$$

Now we form the simplex tableau and solve using the technique described earlier:

x_1	x_2	x_3	s_1	s_2	s_3	P		
1	1	1	1	0	0	0	100	
40	20	30	0	1	0	0	3,200	
1	②	1	0	0	1	0	160	$0.5R_3 \to R_3$
−100	−300	−200	0	0	0	1	0	

x_1	x_2	x_3	s_1	s_2	s_3	P		
1	1	1	1	0	0	0	100	$R_1 + (-1)R_3 \to R_1$
40	20	30	0	1	0	0	3,200	$R_2 + (-20)R_3 \to R_2$
0.5	1	0.5	0	0	0.5	0	80	
−100	−300	−200	0	0	0	1	0	$R_4 + 300R_3 \to R_4$

x_1	x_2	x_3	s_1	s_2	s_3	P		
0.5	0	⓪.5	1	0	−0.5	0	20	$2R_1 \to R_1$
30	0	20	0	1	−10	0	1,600	
0.5	1	0.5	0	0	0.5	0	80	
50	0	−50	0	0	150	1	24,000	

$$\sim \begin{bmatrix} x_1 & x_2 & x_3 & s_1 & s_2 & s_3 & P & \\ 1 & 0 & 1 & 2 & 0 & -1 & 0 & 40 \\ 30 & 0 & 20 & 0 & 1 & -10 & 0 & 1{,}600 \\ 0.5 & 1 & 0.5 & 0 & 0 & 0.5 & 0 & 80 \\ \hline 50 & 0 & -50 & 0 & 0 & 150 & 1 & 24{,}000 \end{bmatrix} \begin{matrix} \\ R_2 + (-20)R_1 \to R_2 \\ R_3 + (-0.5)R_1 \to R_3 \\ R_4 + 50R_1 \to R_4 \end{matrix}$$

$$\sim \begin{bmatrix} 1 & 0 & 1 & 2 & 0 & -1 & 0 & 40 \\ 10 & 0 & 0 & -40 & 1 & 10 & 0 & 800 \\ 0 & 1 & 0 & -1 & 0 & 1 & 0 & 60 \\ \hline 100 & 0 & 0 & 100 & 0 & 100 & 1 & 26{,}000 \end{bmatrix}$$

All entries in the bottom row are nonnegative, and we can now read the optimal solution:

$$x_1 = 0, \quad x_2 = 60, \quad x_3 = 40, \quad s_1 = 0, \quad s_2 = 800, \quad s_3 = 0, \quad P = \$26{,}000$$

Thus, if the farmer plants 60 acres in crop B, 40 acres in crop C, and no crop A, the maximum profit of \$26,000 will be realized. The fact that $s_2 = 800$ tells us (look at the second row in the equations at the start) that this maximum profit is reached by using only \$2,400 of the \$3,200 available for seed; that is, we have a slack of \$800 that can be used for some other purpose.

Problem 12   Repeat Example 12 modified as follows:

	Investment per Acre			Maximum Available
	Crop A	Crop B	Crop C	
Seed cost	\$24	\$40	\$30	\$3,600
Workdays	1	2	2	160 workdays
Profit	\$140	\$200	\$160	

It is important to realize that in order to keep this introduction as simple as possible, we have purposely avoided certain degenerate cases that lead to difficulties. Discussion and resolution of these problems is left to a more advanced treatment of the subject.

Answers to Matched Problems

10. Max $P = 6$ when $x_1 = 1$ and $x_2 = 4$
11. No solution (unbounded objective function)
12. 40 acres of crop A, 60 acres of crop B, no crop C; Max $P = \$17{,}600$ (since $s = 240$, \$240 out of the \$3,600 will not be spent)

Exercise 8-5

A In Problems 1–4:

(A) Write the linear programming problem in standard form using slack variables.

(B) Write the simplex tableau and circle the first pivot.

(C) Use the simplex method to solve the problem.

1. Maximize $P = 15x_1 + 10x_2$
   Subject to $2x_1 + x_2 \leq 10$
   $\qquad\qquad x_1 + 2x_2 \leq 8$
   $\qquad\qquad x_1, x_2 \geq 0$

2. Maximize $P = 3x_1 + 2x_2$
   Subject to $6x_1 + 3x_2 \leq 24$
   $\qquad\qquad 3x_1 + 6x_2 \leq 30$
   $\qquad\qquad x_1, x_2 \geq 0$

3. Repeat Problem 1 with the objective function changed to $P = 30x_1 + x_2$.

4. Repeat Problem 2 with the objective function changed to $P = x_1 + 3x_2$.

B Solve the following linear programming problems using the simplex method:

5. Maximize $P = 30x_1 + 40x_2$
   Subject to $2x_1 + x_2 \leq 10$
   $\qquad\qquad x_1 + x_2 \leq 7$
   $\qquad\qquad x_1 + 2x_2 \leq 12$
   $\qquad\qquad x_1, x_2 \geq 0$

6. Maximize $P = 20x_1 + 10x_2$
   Subject to $3x_1 + x_2 \leq 21$
   $\qquad\qquad x_1 + x_2 \leq 9$
   $\qquad\qquad x_1 + 3x_2 \leq 21$
   $\qquad\qquad x_1, x_2 \geq 0$

7. Maximize $P = 2x_1 + 3x_2$
   Subject to $-2x_1 + x_2 \leq 2$
   $\qquad\qquad -x_1 + x_2 \leq 5$
   $\qquad\qquad\qquad x_2 \leq 6$
   $\qquad\qquad x_1, x_2 \geq 0$

8. Repeat Problem 7 with $P = -x_1 + 3x_2$.

9. Maximize $P = -x_1 + 2x_2$
   Subject to $-x_1 + x_2 \leq 2$
   $\qquad\qquad -x_1 + 3x_2 \leq 12$
   $\qquad\qquad x_1 - 4x_2 \leq 4$
   $\qquad\qquad x_1, x_2 \geq 0$

10. Repeat Problem 9 with $P = x_1 + 2x_2$.

11. Maximize
$$P = 5x_1 + 2x_2 - x_3$$
Subject to
$$x_1 + x_2 - x_3 \leqslant 10$$
$$2x_1 + 4x_2 + x_3 \leqslant 30$$
$$x_1, x_2, x_3 \geqslant 0$$

12. Maximize
$$P = 4x_1 - 3x_2 + 2x_3$$
Subject to
$$x_1 + 2x_2 - x_3 \leqslant 5$$
$$3x_1 + 2x_2 + 2x_3 \leqslant 22$$
$$x_1, x_2, x_3 \geqslant 0$$

C   **13.** Maximize
$$P = 20x_1 + 30x_2$$
Subject to
$$0.6x_1 + 1.2x_2 \leqslant 960$$
$$0.03x_1 + 0.04x_2 \leqslant 36$$
$$0.3x_1 + 0.2x_2 \leqslant 270$$
$$x_1, x_2 \geqslant 0$$

14. Repeat Problem 13 with
$$P = 20x_1 + 20x_2.$$

15. Maximize
$$P = x_1 + 2x_2 + 3x_3$$
Subject to
$$2x_1 + 2x_2 + 8x_3 \leqslant 600$$
$$x_1 + 3x_2 + 2x_3 \leqslant 600$$
$$3x_1 + 2x_2 + x_3 \leqslant 400$$
$$x_1, x_2, x_3 \geqslant 0$$

16. Maximize
$$P = 10x_1 + 50x_2 + 10x_3$$
Subject to
$$3x_1 + 3x_2 + 3x_3 \leqslant 66$$
$$6x_1 - 2x_2 + 4x_3 \leqslant 48$$
$$3x_1 + 6x_2 + 9x_3 \leqslant 108$$
$$x_1, x_2, x_3 \geqslant 0$$

In Problems 17 and 18, first solve the linear programming problem by the simplex method, keeping track of the obvious basic solution at each step. Then graph the feasible region and illustrate the path to the optimal solution determined by the simplex method.

17. Maximize $P = 2x_1 + 5x_2$
Subject to $x_1 + 2x_2 \leqslant 40$
$$x_1 + 3x_2 \leqslant 48$$
$$x_1 + 4x_2 \leqslant 60$$
$$x_2 \leqslant 14$$
$$x_1, x_2 \geqslant 0$$

18. Maximize $P = 5x_1 + 3x_2$
Subject to $5x_1 + 4x_2 \leqslant 100$
$$2x_1 + x_2 \leqslant 28$$
$$4x_1 + x_2 \leqslant 42$$
$$x_1 \leqslant 10$$
$$x_1, x_2 \geqslant 0$$

Applications

Formulate each of the following as a linear programming problem. Then solve the problem using the simplex method.

Business & Economics

19. *Manufacturing—resource allocation.* A company manufactures rackets for tennis, squash, and racketball. Each tennis racket requires 2 units of aluminum and 1 unit of nylon, each squash racket requires 1 unit of aluminum and 2 units of nylon, and each racketball racket requires 2 units of aluminum and 2 units of nylon. The company has 1,000 units of aluminum and 800 units of nylon. The profits on each tennis, squash and racketball racket are $7, $9, and $10, respectively. How many rackets of each type should the company manufacture in order to maximize its profit? What is the maximum profit?

20. *Manufacturing resource allocation.* Repeat Problem 19 under the additional assumption that the combined total number of rackets produced by the company cannot exceed 550.

21. *Investment.* An investor has $100,000 to invest in government bonds, mutual funds, and money market funds. The average yields for government bonds, mutual funds, and money market funds are 8%, 13%, and 15%, respectively. The investor's policy requires that the total amount invested in mutual and money market funds not exceed the amount invested in government bonds. How much should be invested in each type of investment in order to maximize the return? What is the maximum return?

22. *Investment.* Repeat Problem 21 under the additional assumption that no more than $30,000 can be invested in money market funds.

23. *Advertising.* A department store has $2,000 to spend on television advertising for a sale. An ad on a daytime show costs $100 and is viewed by 1,400 potential customers. An ad on a prime-time show costs $200 and is viewed by 2,400 potential customers. An ad on a late-night show costs $150 and is viewed by 1,800 potential customers. The television station will not accept a total of more than 15 ads in all three time periods. How many ads should be placed in each time period in order to maximize the number of potential customers who will see the ads? How many potential customers will see the ads?

24. *Advertising.* Repeat Problem 23 if the department store increases its advertising budget to $2,400 and requires that at least half of the ads be placed in prime-time shows.

Life Sciences   25. *Nutrition—animals.* The natural diet of a certain animal consists of three foods, A, B, and C. The number of units of calcium, iron, and protein in 1 gram of each food and the average daily intake are given in the table. A scientist wants to investigate the effect of increasing the protein in the animal's diet while not allowing the units of calcium and iron to exceed their average daily intakes. How many grams of each food should be used to maximize the amount of protein in the diet? What is the maximum amount of protein?

| | Units per Gram | | | Average Daily Intake |
	Food A	Food B	Food C	
Calcium	1	3	2	30
Iron	2	1	1	24
Protein	3	3	5	60

26. *Nutrition — animals.* Repeat Problem 25 if the scientist wants to maximize the daily calcium intake while not allowing the intake of iron or protein to exceed the average daily intake.

Social Sciences

27. *Opinion survey.* A political scientist has received a grant to fund a research project involving voting trends. The budget of the grant included $540 for conducting door-to-door interviews the day before an election. Undergraduate students, graduate students, and faculty members will be hired to conduct the interviews. Each undergraduate student will conduct 18 interviews and be paid $20. Each graduate student will conduct 25 interviews and be paid $30. Each faculty member will conduct 30 interviews and be paid $40. Due to limited transportation facilities, no more than 20 interviewers can be hired. How many undergraduate students, graduate students, and faculty members should be hired in order to maximize the number of interviews that will be conducted? What is the maximum number of interviews?

28. *Opinion survey.* Repeat Problem 27 if one of the requirements of the grant is that at least 50% of the interviewers be undergraduate students.

8-6   The Dual; Minimization with ≥ Problem Constraints

■ Formation of the Dual Problem
■ Solution of Minimization Problems
■ Application: A Transportation Problem

In the last section we restricted ourselves to maximization problems with ≤ problem constraints. Now we will consider minimization problems with ≥ problem constraints. These two types of problems turn out to be very closely related.

■ Formation of the Dual Problem

Associated with each minimization problem is a maximization problem called the **dual problem.** To illustrate the procedure for finding the dual problem, we will use Example 8 from Section 8-3. There we solved the following linear programming problem geometrically. Now we will form its dual, and later we will solve the problem using the dual and the simplex method.

Minimize $\quad C = 3x_1 + x_2$

Subject to $\quad 10x_1 + 2x_2 \geqslant 84$

$\qquad\qquad 8x_1 + 4x_2 \geqslant 120 \qquad\qquad\qquad (1)$

$\qquad\qquad\quad x_1, x_2 \geqslant 0$

The first step in forming the dual problem is to construct a matrix by using the problem constraints and the objective function written in the following form:

$$
\begin{array}{c}
10x_1 + 2x_2 \geqslant 84 \\
8x_1 + 4x_2 \geqslant 120 \\
3x_1 + \ x_2 = C
\end{array}
\qquad
A = \left[\begin{array}{cc|c}
10 & 2 & 84 \\
8 & 4 & 120 \\
3 & 1 & 1
\end{array}\right]
$$

Be careful not to confuse this matrix with the simplex tableau. We use a solid horizontal line in the matrix to help distinguish the dual matrix from the simplex tableau. No slack variables are involved in matrix A, and the coefficient of C is in the same column as the constants from the problem constraints.

Now we will form a second matrix B by using the rows of A as the columns of B. [Technically, B is called the *transpose of A*. In general, the **transpose** of a given matrix is formed by interchanging its rows and corresponding columns (first row with first column, second row with second column, and so on.)]

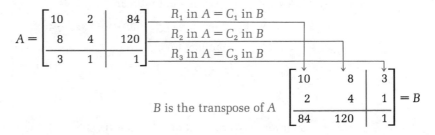

Finally, we use the rows of B to define a new linear programming problem. This new problem will always be a maximization problem with $\leqslant$ problem constraints. To avoid confusion, we shall use different variables in this new problem:

$$
\begin{array}{rl}
10y_1 + \ 8y_2 \leqslant 3 \\
2y_1 + \ 4y_2 \leqslant 1 \\
84y_1 + 120y_2 = P
\end{array}
\qquad
B = \begin{array}{c}
\begin{array}{ccc} y_1 & y_2 & \end{array} \\
\left[\begin{array}{cc|c}
10 & 8 & 3 \\
2 & 4 & 1 \\
84 & 120 & 1
\end{array}\right]
\end{array}
$$

The dual of the minimization problem (1) is:

Maximize $P = 84y_1 + 120y_2$

Subject to $10y_1 + 8y_2 \leqslant 3$

$2y_1 + 4y_2 \leqslant 1$

$y_1, y_2 \geqslant 0$

This procedure is summarized in the box below:

Formation of the Dual Problem

Given a minimization problem with $\geqslant$ problem constraints:

1. Use the coefficients of the problem constraints and the objective function to form a matrix A with the coefficients of the objective function in the last row.
2. Use the *rows* of the matrix A as the *columns* of a second matrix B (matrix B is the transpose of A).
3. Use the *rows* of B to form a maximization problem with $\leqslant$ problem constraints.

Example 13   Form the dual problem:

Minimize $C = 20x_1 + 12x_2 + 40x_3$

Subject to $x_1 + x_2 + 5x_3 \geqslant 20$

$2x_1 + x_2 + x_3 \geqslant 30$

$x_1, x_2, x_3 \geqslant 0$

Solution   Step 1.   Form the matrix A:

$$A = \begin{bmatrix} 1 & 1 & 5 & 20 \\ 2 & 1 & 1 & 30 \\ 20 & 12 & 40 & 1 \end{bmatrix}$$

Step 2.   Form the matrix B (the transpose of A):

$$B = \begin{bmatrix} 1 & 2 & 20 \\ 1 & 1 & 12 \\ 5 & 1 & 40 \\ 20 & 30 & 1 \end{bmatrix}$$

Step 3. State the dual problem:

$$\text{Maximize} \quad P = 20y_1 + 30y_2$$
$$\text{Subject to} \quad y_1 + 2y_2 \leqslant 20$$
$$y_1 + y_2 \leqslant 12$$
$$5y_1 + y_2 \leqslant 10$$
$$y_1, y_2 \geqslant 0$$

Problem 13 Find the dual problem:

$$\text{Minimize} \quad C = 16x_1 + 9x_2 + 21x_3$$
$$\text{Subject to} \quad x_1 + x_2 + 3x_3 \geqslant 16$$
$$2x_1 + x_2 + x_3 \geqslant 12$$
$$x_1, x_2, x_3 \geqslant 0$$

$16y_1 + 12y_2 = $ Dual, D.

$y_1 + 2y_2 \leq 16$

$y_1 + y_2 \leq 9$

$3y_1 + y \leq 21$

$$\begin{bmatrix} 1 & 1 & 3 & 16 \\ 2 & 1 & 1 & 12 \\ 16 & 9 & 21 & 1 \end{bmatrix}$$

$$\begin{bmatrix} 1 & 2 & 16 \\ 1 & 1 & 9 \\ 3 & 1 & 21 \\ 16 & 12 & \end{bmatrix}$$

■ **Solution of Minimization Problems**

The following theorem establishes the relationship between the solution of a minimization problem and the solution of its dual:

Theorem 4

> A minimization problem has a solution if and only if its dual problem has a solution. If a solution exists, then the optimal value of the minimization problem is the same as the optimal value of the dual problem.

In Section 8-5 we saw that the simplex method can be used to solve maximization problems with $\leqslant$ problem constraints and nonnegative constants on the right side of each problem constraint. When the dual of a minimization problem is formed, the coefficients of the objective function in the minimization problem (the last row in A) become the constants in the problem constraints in the dual problem (the last column in B). **Thus, a minimization problem whose objective function has nonnegative coefficients can be solved by applying the simplex method to the dual.** To illustrate this, let's return to Example 8 in Section 8-3, whose dual was found earlier in this section.

Original Problem	*Dual Problem*
Minimize $C = 3x_1 + x_2$	Maximize $P = 80y_1 + 120y_2$
Subject to $10x_1 + 2x_2 \geqslant 48$	Subject to $10y_1 + 8y_2 \leqslant 3$
$8x_1 + 4x_2 \geqslant 120$	$2y_1 + 4y_2 \leqslant 1$
$x_1, x_2 \geqslant 0$	$y_1, y_2 \geqslant 0$

$$\begin{bmatrix} 10 & 2 & 48 \\ 8 & 4 & 120 \\ 3 & 1 & 1 \end{bmatrix}$$

$8y_1 + y = P$

$10y_1 + 2y_2 \leq 48$

Now we will use the simplex method to solve the dual problem. For reasons that will become clear later, we will use the variables x_1 and x_2 from the original problem as the slack variables in the dual:

$$10y_1 + 8y_2 + x_1 = 3$$
$$2y_1 + 4y_2 + x_2 = 1$$
$$-84y_1 - 120y_2 + P = 0$$

$$
\begin{array}{ccccc}
y_1 & y_2 & x_1 & x_2 & P
\end{array}
$$

$$
\left[\begin{array}{ccccc|c}
10 & 8 & 1 & 0 & 0 & 3 \\
2 & \boxed{4} & 0 & 1 & 0 & 1 \\
\hline
-84 & -120 & 0 & 0 & 1 & 0
\end{array}\right] \quad \tfrac{1}{4}R_2 \to R_2
$$

$$
\sim \left[\begin{array}{ccccc|c}
10 & 8 & 1 & 0 & 0 & 3 \\
\tfrac{1}{2} & 1 & 0 & \tfrac{1}{4} & 0 & \tfrac{1}{4} \\
\hline
-84 & -120 & 0 & 0 & 1 & 0
\end{array}\right] \quad \begin{array}{l} R_1 + (-8)R_2 \to R_1 \\[6pt] R_3 + 120R_2 \to R_3 \end{array}
$$

$$
\sim \left[\begin{array}{ccccc|c}
\boxed{6} & 0 & 1 & -2 & 0 & 1 \\
\tfrac{1}{2} & 1 & 0 & \tfrac{1}{4} & 0 & \tfrac{1}{4} \\
\hline
-24 & 0 & 0 & 30 & 1 & 30
\end{array}\right] \quad \tfrac{1}{6}R_1 \to R_1
$$

$$
\sim \left[\begin{array}{ccccc|c}
1 & 0 & \tfrac{1}{6} & -\tfrac{1}{3} & 0 & \tfrac{1}{6} \\
\tfrac{1}{2} & 1 & 0 & \tfrac{1}{4} & 0 & \tfrac{1}{4} \\
\hline
-24 & 0 & 0 & 30 & 1 & 30
\end{array}\right] \quad \begin{array}{l} R_2 + (-\tfrac{1}{2})R_1 \to R_2 \\[6pt] R_3 + 24R_1 \to R_3 \end{array}
$$

$$
\sim \left[\begin{array}{ccccc|c}
1 & 0 & \tfrac{1}{6} & -\tfrac{1}{3} & 0 & \tfrac{1}{6} \\
0 & 1 & -\tfrac{1}{12} & \tfrac{5}{12} & 0 & \tfrac{1}{6} \\
\hline
0 & 0 & 4 & 22 & 1 & 34
\end{array}\right]
$$

Since all entries in the bottom row are nonnegative, the solution to the dual problem is

$$y_1 = \tfrac{1}{6}, \quad y_2 = \tfrac{1}{6}, \quad x_1 = 0, \quad x_2 = 0, \quad P = 34$$

and the maximum value of P is 34. According to Theorem 4, the minimum value of C must also be 34. This agrees with the solution we found in Section 8-3, where we used a geometric approach.

If we examine the geometric solution, we see that the minimum value of C occurred at $x_1 = 4$ and $x_2 = 22$, which are the numbers in the bottom row of the final simplex tableau. This is no accident. **The entire solution to the minimization problem can always be obtained from the bottom row of the final simplex tableau for the dual problem.** Thus, from the row

$$
\begin{array}{cc}
x_1 & x_2
\end{array}
$$
$$[0 \quad 0 \quad 4 \quad 22 \quad 1 \mid 34]$$

we can conclude that the solution to the minimization problem is

$$\text{Min } C = 34 \quad \text{at} \quad x_1 = 4, \quad x_2 = 22$$

Now we can see that using x_1 and x_2 as slack variables in the dual problem makes it easy to identify the solution of the original problem.

Solution of a Minimization Problem

(Problem constraints are of the $\geqslant$ form, and the coefficients of the objective function are nonnegative.)

1. Form the dual problem.
2. Use the simplex method to solve the dual problem.
3. Read the solution of the minimization problem from the bottom row of the final simplex tableau in step 2.

Example 14 Solve the following minimization problem by maximizing the dual (see Example 13).

$$\text{Minimize} \quad C = 20x_1 + 12x_2 + 40x_3$$
$$\text{Subject to} \quad x_1 + x_2 + 5x_3 \geqslant 20$$
$$2x_1 + x_2 + \ x_3 \geqslant 30$$
$$x_1, x_2, x_3 \geqslant 0$$

Solution From Example 13, the dual is

$$\text{Maximize} \quad P = 20y_1 + 30y_2$$
$$\text{Subject to} \quad y_1 + 2y_2 \leqslant 20$$
$$y_1 + \ y_2 \leqslant 12$$
$$5y_1 + \ y_2 \leqslant 40$$
$$y_1, y_2 \geqslant 0$$

Using x_1, x_2, and x_3 for slack variables, we obtain

$$y_1 + \ 2y_2 + x_1 \qquad\qquad = 20$$
$$y_1 + \ y_2 \qquad + x_2 \qquad = 12$$
$$5y_1 + \ y_2 \qquad\qquad + x_3 \ = 40$$
$$-20y_1 - 30y_2 \qquad\qquad + P = 0$$

Now we form the simplex tableau and solve the dual problem.

$$
\begin{array}{ccccccc}
y_1 & y_2 & x_1 & x_2 & x_3 & P & \\
\end{array}
$$

$$
\left[\begin{array}{cccccc|c}
1 & \boxed{2} & 1 & 0 & 0 & 0 & 20 \\
1 & 1 & 0 & 1 & 0 & 0 & 12 \\
5 & 1 & 0 & 0 & 1 & 0 & 40 \\
\hline
-20 & -30 & 0 & 0 & 0 & 1 & 0
\end{array}\right]
\quad \tfrac{1}{2}R_1 \to R_1
$$

$$
\sim
\left[\begin{array}{cccccc|c}
\tfrac{1}{2} & 1 & \tfrac{1}{2} & 0 & 0 & 0 & 10 \\
1 & 1 & 0 & 1 & 0 & 0 & 12 \\
5 & 1 & 0 & 0 & 1 & 0 & 40 \\
\hline
-20 & -30 & 0 & 0 & 0 & 1 & 0
\end{array}\right]
\quad\begin{array}{l}
R_2+(-1)R_1 \to R_2 \\
R_3+(-1)R_1 \to R_3 \\
R_4+30R_1 \to R_3
\end{array}
$$

$$
\sim
\left[\begin{array}{cccccc|c}
\tfrac{1}{2} & 1 & \tfrac{1}{2} & 0 & 0 & 0 & 10 \\
\boxed{\tfrac{1}{2}} & 0 & -\tfrac{1}{2} & 1 & 0 & 0 & 2 \\
\tfrac{9}{2} & 0 & -\tfrac{1}{2} & 0 & 1 & 0 & 30 \\
\hline
-5 & 0 & 15 & 0 & 0 & 1 & 300
\end{array}\right]
\quad 2R_2 \to R_2
$$

$$
\sim
\left[\begin{array}{cccccc|c}
\tfrac{1}{2} & 1 & \tfrac{1}{2} & 0 & 0 & 0 & 10 \\
1 & 0 & -1 & 2 & 0 & 0 & 4 \\
\tfrac{9}{2} & 0 & -\tfrac{1}{2} & 0 & 1 & 0 & 30 \\
\hline
-5 & 0 & 15 & 0 & 0 & 1 & 300
\end{array}\right]
\quad\begin{array}{l}
R_1+(-\tfrac{1}{2})R_2 \to R_1 \\
\\
R_3+(-\tfrac{9}{2})R_2 \to R_3 \\
R_4+5R_2 \to R_4
\end{array}
$$

$$
\sim
\left[\begin{array}{cccccc|c}
0 & 1 & 1 & -1 & 0 & 0 & 8 \\
1 & 0 & -1 & 2 & 0 & 0 & 4 \\
0 & 0 & 4 & -9 & 1 & 0 & 12 \\
\hline
0 & 0 & 10 & 10 & 0 & 1 & 320
\end{array}\right]
$$

From the bottom row of this tableau, we see that

$$\text{Min } C = 320 \quad \text{at} \quad x_1 = 10, \quad x_2 = 10, \quad \text{and} \quad x_3 = 0$$

Problem 14 Solve the following minimization problem by maximizing the dual (see Problem 13):

$$
\begin{aligned}
\text{Minimize} \quad & C = 16x_1 + 9x_2 + 21x_3 \\
\text{Subject to} \quad & x_1 + x_2 + 3x_3 \geqslant 16 \\
& 2x_1 + x_2 + x_3 \geqslant 12 \\
& x_1, x_2, x_3 \geqslant 0
\end{aligned}
$$

Example 15 Solve the following minimization problem by maximizing the dual:

Minimize $C = 5x_1 + 10x_2$

Subject to $x_1 - x_2 \geqslant 1$

$-x_1 + x_2 \geqslant 2$

$x_1, x_2 \geqslant 0$

Solution $A = \begin{bmatrix} 1 & -1 & | & 1 \\ -1 & 1 & | & 2 \\ \hline 5 & 10 & | & 1 \end{bmatrix}$ $B = \begin{bmatrix} 1 & -1 & | & 5 \\ -1 & 1 & | & 10 \\ \hline 1 & 2 & | & 1 \end{bmatrix}$

The dual problem is

Maximize $P = y_1 + 2y_2$

Subject to $y_1 - y_2 \leqslant 5$

$-y_1 + y_2 \leqslant 10$

$y_1, y_2 \geqslant 0$

Introduce slack variables x_1 and x_2:

$y_1 - y_2 + x_1 \qquad\qquad = 5$

$-y_1 + y_2 \qquad + x_2 \qquad = 10$

$-y_1 - 2y_2 \qquad\qquad + P = 0$

Form the simplex tableau and solve:

$$\begin{array}{ccccc} y_1 & y_2 & x_1 & x_2 & P \end{array}$$

$$\begin{bmatrix} 1 & -1 & 1 & 0 & 0 & | & 5 \\ -1 & \textcircled{1} & 0 & 1 & 0 & | & 10 \\ \hline -1 & -2 & 0 & 0 & 1 & | & 0 \end{bmatrix} \qquad \begin{array}{l} R_1 + R_2 \rightarrow R_1 \\ \\ R_3 + 2R_2 \rightarrow R_3 \end{array}$$

$$\sim \begin{bmatrix} 0 & 0 & 1 & 1 & 0 & | & 15 \\ -1 & 1 & 0 & 1 & 0 & | & 10 \\ \hline -3 & 0 & 0 & 2 & 1 & | & 20 \end{bmatrix} \qquad \begin{array}{l} \text{No positive elements} \\ \text{above dashed line in} \\ \text{pivot column} \end{array}$$

$\uparrow$
Pivot
column

The -3 in the bottom row indicates that Column 1 is the pivot column. Since no positive elements are in the pivot column above the dashed line, we are unable to select a pivot row. We stop the pivot operation and conclude that this maximization problem has no solution (see Section 8-5, page 423). Theorem 4 now implies that the original minimization problem has no solution. The graph of the inequalities in the minimization problem (see page 440) shows that the feasible region is empty; thus, it is not surprising that an optimal solution does not exist.

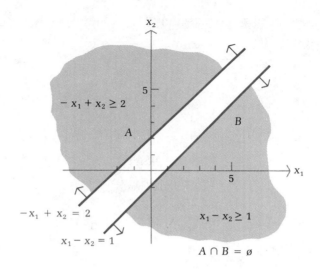

$A \cap B = \emptyset$

Problem 15   Solve the following minimization problem by maximizing the dual:

$$\text{Minimize} \quad C = 2x_1 + 3x_2$$
$$\text{Subject to} \quad x_1 - 2x_2 \geqslant 2$$
$$-x_1 + x_2 \geqslant 1$$
$$x_1, x_2 \geqslant 0$$

■ Application: A Transportation Problem

One of the first applications of linear programming was to the problem of minimizing the cost of transporting materials. Problems of this type are referred to as **transportation problems.**

Example 16   A computer manufacturing company has two assembly plants, plant A and plant B, and two distribution outlets, outlet I and outlet II. Plant A can assemble 700 computers a month, and plant B can assemble 900 computers a month. Outlet I must have at least 500 computers a month and outlet II must have at least 1,000 computers a month. Transportation costs for shipping one computer from each plant to each outlet are as follows: $6 from plant A to outlet I; $5 from plant A to outlet II; $4 from plant B to outlet I; $8 from plant B to outlet II. Find a shipping schedule that will minimize the total cost of shipping the computers from the assembly plants to the distribution outlets. What is this minimum cost?

Solution   First we summarize the relevant data in a table:

Assembly Plant	Distribution Outlet		Assembly Capacity
	I	II	
A	$6	$5	700
B	$4	$8	900
Minimum required	500	1,000	

In order to find a shipping schedule, we must determine the number of computers that should be shipped from each plant to each outlet. This will require the use of four decision variables:

x_1 = Number of computers shipped from plant A to outlet I

x_2 = Number of computers shipped from plant A to outlet II

x_3 = Number of computers shipped from plant B to outlet I

x_4 = Number of computers shipped from plant B to outlet II

The total number of computers shipped from plant A is $x_1 + x_2$. Since this cannot exceed the available number, we have

$x_1 + x_2 \leq 700$ Number shipped from plant A

Similarly, the total number shipped from plant B must satisfy

$x_3 + x_4 \leq 900$ Number shipped from plant B

The total number shipped to each outlet must satisfy

$x_1 + x_3 \geq 500$ Number shipped to outlet I

and

$x_2 + x_4 \geq 1,000$ Number shipped to outlet II

Using the shipping charges in the table, the total shipping charges are

$C = 6x_1 + 5x_2 + 4x_3 + 8x_4$

Thus, we must solve the following linear programming problem:

Minimize $C = 6x_1 + 5x_2 + 4x_3 + 8x_4$

Subject to $x_1 + x_2 \leq 700$ Available from A

$ x_3 + x_4 \leq 900$ Available from B

$x_1 + x_3 \geq 500$ Required at I

$ x_2 + x_4 \geq 1,000$ Required at II

$x_1, x_2, x_3, x_4 \geq 0$

Before we can solve this problem, we must multiply the first two constraints by -1 so that all the problem constraints are of the $\geqslant$ type. This will introduce negative constants into the minimization problem but not into the dual. Since the coefficients of C are nonnegative, the constants in the dual problem will be nonnegative, a requirement for the simplex method in Section 8-5. The problem can now be stated as

Minimize　$C = 6x_1 + 5x_2 + 4x_3 + 8x_4$

Subject to　$-x_1 - x_2 \qquad\qquad \geqslant -700$

$$-x_3 - x_4 \geqslant -900$$
$$x_1 \quad\;\; + x_3 \qquad \geqslant 500$$
$$x_2 \qquad + x_4 \geqslant 1{,}000$$
$$x_1, x_2, x_3, x_4 \geqslant 0$$

$$A = \begin{bmatrix} -1 & -1 & 0 & 0 & -700 \\ 0 & 0 & -1 & -1 & -900 \\ 1 & 0 & 1 & 0 & 500 \\ 0 & 1 & 0 & 1 & 1{,}000 \\ 6 & 5 & 4 & 8 & 1 \end{bmatrix}$$

$$B = \begin{bmatrix} -1 & 0 & 1 & 0 & 6 \\ -1 & 0 & 0 & 1 & 5 \\ 0 & -1 & 1 & 0 & 4 \\ 0 & -1 & 0 & 1 & 8 \\ -700 & -900 & 500 & 1{,}000 & 1 \end{bmatrix}$$ B is the transpose of A

The dual problem is

Maximize　$P = -700y_1 - 900y_2 + 500y_3 + 1{,}000y_4$

Subject to　$-y_1 \qquad + y_3 \qquad\quad\; \leqslant 6$

$$-y_1 \qquad\qquad + y_4 \leqslant 5$$
$$-y_2 + y_3 \qquad\quad\;\; \leqslant 4$$
$$-y_2 \qquad + y_4 \leqslant 8$$
$$y_1, y_2, y_3, y_4 \geqslant 0$$

Introduce slack variables x_1, x_2, x_3, and x_4:

$$-y_1 \qquad + \quad y_3 \qquad\qquad + x_1 \qquad\qquad\qquad = 6$$
$$-y_1 \qquad\qquad + \qquad y_4 \quad + x_2 \qquad\qquad = 5$$
$$-y_2 + \quad y_3 \qquad\qquad\qquad + x_3 \qquad = 4$$
$$-y_2 \qquad\qquad + \quad y_4 \qquad\qquad + x_4 \;\; = 8$$
$$700y_1 + 900y_2 - 500y_3 - 1{,}000y_4 \qquad\qquad\qquad + P = 0$$

Form the simplex tableau and solve:

$$
\begin{array}{c}
\begin{array}{ccccccccc}
y_1 & y_2 & y_3 & y_4 & x_1 & x_2 & x_3 & x_4 & P \\
\end{array} \\
\left[
\begin{array}{ccccccccc|c}
-1 & 0 & 1 & 0 & 1 & 0 & 0 & 0 & 0 & 6 \\
-1 & 0 & 0 & \boxed{1} & 0 & 1 & 0 & 0 & 0 & 5 \\
0 & -1 & 1 & 0 & 0 & 0 & 1 & 0 & 0 & 4 \\
0 & -1 & 0 & 1 & 0 & 0 & 0 & 1 & 0 & 8 \\
\hline
700 & 900 & -500 & -1{,}000 & 0 & 0 & 0 & 0 & 1 & 0 \\
\end{array}
\right]
\end{array}
$$

$R_4 + (-1)R_2 \rightarrow R_4$
$R_5 + 1{,}000R_2 \rightarrow R_5$

$$
\sim \left[
\begin{array}{ccccccccc|c}
-1 & 0 & 1 & 0 & 1 & 0 & 0 & 0 & 0 & 6 \\
-1 & 0 & 0 & 1 & 0 & 1 & 0 & 0 & 0 & 5 \\
0 & -1 & \boxed{1} & 0 & 0 & 0 & 1 & 0 & 0 & 4 \\
1 & -1 & 0 & 0 & 0 & -1 & 0 & 1 & 0 & 3 \\
\hline
-300 & 900 & -500 & 0 & 0 & 1{,}000 & 0 & 0 & 1 & 5{,}000 \\
\end{array}
\right]
$$

$R_1 + (-1)R_3 \rightarrow R_1$

$R_5 + 500R_3 \rightarrow R_5$

$$
\sim \left[
\begin{array}{ccccccccc|c}
-1 & 1 & 0 & 0 & 1 & 0 & -1 & 0 & 0 & 2 \\
-1 & 0 & 0 & 1 & 0 & 1 & 0 & 0 & 0 & 5 \\
0 & -1 & 1 & 0 & 0 & 0 & 1 & 0 & 0 & 4 \\
\boxed{1} & -1 & 0 & 0 & 0 & -1 & 0 & 1 & 0 & 3 \\
\hline
-300 & 400 & 0 & 0 & 0 & 1{,}000 & 500 & 0 & 1 & 7{,}000 \\
\end{array}
\right]
$$

$R_1 + R_4 \rightarrow R_1$
$R_2 + R_4 \rightarrow R_2$

$R_5 + 300R_4 \rightarrow R_5$

$$
\sim \left[
\begin{array}{ccccccccc|c}
0 & 0 & 0 & 0 & 1 & -1 & -1 & 1 & 0 & 5 \\
0 & -1 & 0 & 1 & 0 & 0 & 0 & 1 & 0 & 8 \\
0 & -1 & 1 & 0 & 0 & 0 & 1 & 0 & 0 & 4 \\
1 & -1 & 0 & 0 & 0 & -1 & 0 & 1 & 0 & 3 \\
\hline
0 & 100 & 0 & 0 & 0 & 700 & 500 & 300 & 1 & 7{,}900 \\
\end{array}
\right]
$$

From the bottom row of this tableau, we have

$$\text{Min } C = 7{,}900 \quad \text{at} \quad x_1 = 0, \quad x_2 = 700, \quad x_3 = 500, \quad x_4 = 300$$

The shipping schedule that minimizes the shipping charges is

700 from plant A to outlet II
500 from plant B to outlet I
300 from plant B to outlet II

The total shipping cost is \$7,900.

Problem 16   Repeat Example 16 if the shipping charge from plant A to outlet I is increased to \$7 and the shipping charge from plant B to outlet II is decreased to \$3.

Answers to
Matched Problems

13. Maximize $P = 16y_1 + 12y_2$

    Subject to $y_1 + 2y_2 \leqslant 16$

    $y_1 + y_2 \leqslant 9$

    $3y_1 + y_2 \leqslant 21$

    $y_1, y_2 \geqslant 0$

14. Min $C = 132$ at $x_1 = 0$, $x_2 = 10$, $x_3 = 2$

15. Dual problem:

    Maximize $P = 2y_1 + y_2$

    Subject to $y_1 - y_2 \leqslant 2$

    $-2y_1 + y_2 \leqslant 3$

    $y_1, y_2 \geqslant 0$

    No solution

16. 600 from plant A to outlet II, 500 from plant B to outlet I, 400 from plant B to outlet II; total shipping cost $6,200

Exercise 8-6

A   In Problems 1–8:

(A) Form the dual problem.

(B) Find the solution to the original problem by applying the simplex method to the dual problem.

1. Minimize $C = 9x_1 + 2x_2$

   Subject to $4x_1 + x_2 \geqslant 13$

   $3x_1 + x_2 \geqslant 12$

   $x_1, x_2 \geqslant 0$

2. Minimize $C = x_1 + 4x_2$

   Subject to $x_1 + 2x_2 \geqslant 5$

   $x_1 + 3x_2 \geqslant 6$

   $x_1, x_2 \geqslant 0$

3. Minimize $C = 7x_1 + 12x_2$

   Subject to $2x_1 + 3x_2 \geqslant 15$

   $x_1 + 2x_2 \geqslant 8$

   $x_1, x_2 \geqslant 0$

4. Minimize $C = 3x_1 + 5x_2$

   Subject to $2x_1 + 3x_2 \geqslant 7$

   $x_1 + 2x_2 \geqslant 4$

   $x_1, x_2 \geqslant 0$

5. Minimize $C = 11x_1 + 4x_2$

   Subject to $2x_1 + x_2 \geqslant 8$

   $-2x_1 + 3x_2 \geqslant 4$

   $x_1, x_2 \geqslant 0$

6. Minimize $C = 40x_1 + 10x_2$

   Subject to $2x_1 + x_2 \geqslant 12$

   $3x_1 - x_2 \geqslant 3$

   $x_1, x_2 \geqslant 0$

7. Minimize $C = 7x_1 + 9x_2$
   Subject to $-3x_1 + x_2 \geq 6$
   $x_1 - 2x_2 \geq 4$
   $x_1, x_2 \geq 0$

8. Minimize $C = 10x_1 + 15x_2$
   Subject to $-4x_1 + x_2 \geq 12$
   $12x_1 - 3x_2 \geq 10$
   $x_1, x_2 \geq 0$

B   *Solve the following linear programming problems by applying the simplex method to the dual problem.*

9. Minimize $C = 3x_1 + 9x_2$
   Subject to $2x_1 + x_2 \geq 8$
   $x_1 + 2x_2 \geq 8$
   $x_1, x_2 \geq 0$

10. Minimize $C = 2x_1 + x_2$
    Subject to $x_1 + x_2 \geq 8$
    $x_1 + 2x_2 \geq 4$
    $x_1, x_2 \geq 0$

11. Minimize $C = 7x_1 + 5x_2$
    Subject to $x_1 + x_2 \geq 4$
    $x_1 - 2x_2 \geq -8$
    $-2x_1 + x_2 \geq -8$
    $x_1, x_2 \geq 0$

12. Minimize $C = 10x_1 + 4x_2$
    Subject to $2x_1 + x_2 \geq 6$
    $x_1 - 4x_2 \geq -24$
    $-8x_1 + 5x_2 \geq -24$
    $x_1, x_2 \geq 0$

13. Minimize $C = 10x_1 + 30x_2$
    Subject to $2x_1 + x_2 \geq 16$
    $x_1 + x_2 \geq 12$
    $x_1 + 2x_2 \geq 14$
    $x_1, x_2 \geq 0$

14. Minimize $C = 40x_1 + 10x_2$
    Subject to $3x_1 + x_2 \geq 24$
    $x_1 + x_2 \geq 16$
    $x_1 + 3x_2 \geq 30$
    $x_1, x_2 \geq 0$

15. Minimize $C = 5x_1 + 7x_2$
    Subject to $x_1 \geq 4$
    $x_1 + x_2 \geq 8$
    $x_1 + 2x_2 \geq 10$
    $x_1, x_2 \geq 0$

16. Minimize $C = 4x_1 + 5x_2$
    Subject to $2x_1 + x_2 \geq 12$
    $x_1 + 2x_2 \geq 18$
    $x_2 \geq 6$
    $x_1, x_2 \geq 0$

17. Minimize
    $C = 10x_1 + 7x_2 + 12x_3$
    Subject to
    $x_1 + x_2 + 2x_3 \geq 7$
    $2x_1 + x_2 + x_3 \geq 4$
    $x_1, x_2, x_3 \geq 0$

18. Minimize
    $C = 18x_1 + 8x_2 + 20x_3$
    Subject to
    $x_1 + x_2 + 3x_3 \geq 6$
    $3x_1 + x_2 + x_3 \geq 9$
    $x_1, x_2, x_3 \geq 0$

19. Minimize
    $C = 5x_1 + 2x_2 + 2x_3$
    Subject to
    $x_1 - 4x_2 + x_3 \geq 6$
    $-x_1 + x_2 - 2x_3 \geq 4$
    $x_1, x_2, x_3 \geq 0$

20. Minimize
    $C = 6x_1 + 8x_2 + 3x_3$
    Subject to
    $-3x_1 - 2x_2 + x_3 \geq 4$
    $x_1 + x_2 - x_3 \geq 2$
    $x_1, x_2, x_3 \geq 0$

C   **21.** Minimize
$$C = 16x_1 + 8x_2 + 4x_3$$
Subject to
$$3x_1 + 2x_2 + 2x_3 \geqslant 16$$
$$4x_1 + 3x_2 + x_3 \geqslant 14$$
$$5x_1 + 3x_2 + x_3 \geqslant 12$$
$$x_1, x_2, x_3 \geqslant 0$$

22. Minimize
$$C = 6x_1 + 8x_2 + 12x_3$$
Subject to
$$x_1 + 3x_2 + 3x_3 \geqslant 6$$
$$x_1 + 5x_2 + 5x_3 \geqslant 4$$
$$2x_1 + 2x_2 + 3x_3 \geqslant 8$$
$$x_1, x_2, x_3 \geqslant 0$$

23. Minimize
$$C = 5x_1 + 4x_2 + 5x_3 + 6x_4$$
Subject to
$$x_1 + x_2 \qquad\quad \leqslant 12$$
$$x_3 + x_4 \leqslant 25$$
$$x_1 \qquad + x_3 \qquad \geqslant 20$$
$$x_2 \qquad + x_4 \geqslant 15$$
$$x_1, x_2, x_3, x_4 \geqslant 0$$

24. Repeat Problem 23 with
$$C = 4x_1 + 7x_2 + 5x_3 + 6x_4.$$

◼ ## Applications

Formulate each of the following as a linear programming problem. Then solve the problem by applying the simplex method to the dual problem.

Business & Economics

25. *Manufacturing—production scheduling.* A food processing company produces regular and deluxe ice cream at three plants. Per hour of operation, the plant in Cedarburg produces 20 gallons of regular ice cream and 10 gallons of deluxe ice cream, the Grafton plant 10 gallons of regular and 20 gallons of deluxe, and the West Bend plant 20 gallons of regular and 20 gallons of deluxe. It costs $70 per hour to operate the Cedarburg plant, $75 per hour to operate the Grafton plant, and $90 per hour to operate the West Bend plant. The company needs at least 300 gallons of regular ice cream and at least 200 gallons of deluxe ice cream each day. How many hours per day should each plant operate in order to produce the required amounts of ice cream and minimize the cost of production? What is the minimum production cost?

26. *Mining—production scheduling.* A mining company operates two mines, which produce three grades of ore. The West Summit mine can produce 4 tons of low-grade ore, 3 tons of medium-grade ore, and 2 tons of high-grade ore per hour of operation. The North Ridge mine can produce 1 ton of low-grade ore, 1 ton of medium-grade ore, and 2 tons of high-grade ore per hour of operation. It costs $1,000 per hour to operate the mine at West Summit and $300 per hour to operate the North Ridge mine. To satisfy existing orders, the company needs at least 48 tons of low-grade ore, 45 tons of medium-grade ore, and 31 tons of high-grade ore. How many hours should each mine be operated to supply the required amounts of ore and minimize the cost of production? What is the minimum production cost?

27. *Purchasing.* Acme Micros markets computers with single-sided and double-sided disk drives. The disk drives are supplied by two other companies, Associated Electronics and Digital Drives. Associated Electronics charges $250 for a single-sided disk drive and $350 for a double-sided disk drive. Digital Drives charges $290 for a single-sided disk drive and $320 for a double-sided disk drive. Each month Associated Electronics can supply at most 1,000 disk drives in any combination of single-sided and double-sided drives. The combined monthly total supplied by Digital Drives cannot exceed 2,000 disk drives. Acme Micros needs at least 1,200 single-sided drives and at least 1,600 double-sided drives each month. How many disk drives of each type should Acme Micros order from each supplier in order to meet its monthly demand and minimize the purchase cost? What is the minimum purchase cost?

28. *Transportation.* A feed company stores grain in elevators located in Ames, Iowa, and Bedford, Indiana. Each month the grain is shipped to processing plants in Columbia, Missouri, and Danville, Illinois. The monthly supply (in tons) of grain at each elevator, the monthly demand (in tons) at each processing plant, and the cost per ton for transporting the grain are given in the table. Determine a shipping schedule that will minimize the cost of transporting the grain. What is the minimum cost?

Originating Elevators	Shipping Cost per Ton		Supply in Tons
	Columbia	Danville	
Ames	$22	$38	700
Bedford	$46	$24	500
Demand in tons	400	600	

Life Sciences

29. *Nutrition — people.* A dietitian in a hospital is to arrange a special diet using three foods, *L*, *M*, and *N*. Each ounce of food *L* contains 20 units of calcium, 10 units of iron, 10 units of vitamin A, and 20 units of cholesterol. Each ounce of food *M* contains 10 units of calcium, 10 units of iron, 20 units of vitamin A, and 24 units of cholesterol. Each ounce of food *N* contains 10 units of calcium, 10 units of iron, 10 units of vitamin A, and 18 units of cholesterol. If the minimum daily requirements are 300 units of calcium, 200 units of iron, and 240 units of vitamin A, how many ounces of each food should be used to meet the minimum requirements and at the same time minimize the cholesterol intake? What is the minimum cholesterol intake?

30. *Nutrition — plants.* A farmer can buy three types of plant food, mix A, mix B, and mix C. Each cubic yard of mix A contains 20 pounds of phosphoric acid, 10 pounds of nitrogen, and 5 pounds of potash. Each cubic yard of mix B contains 10 pounds of phosphoric acid, 10 pounds of nitrogen, and 10 pounds of potash. Each cubic yard of mix C contains 20 pounds of phosphoric acid, 20 pounds of nitrogen, and 5 pounds of potash. The minimum monthly requirements are 480 pounds of phosphoric acid, 320 pounds of nitrogen, and 225 pounds of potash. If mix A costs $30 per cubic yard, mix B $36 per cubic yard, and mix C $39 per cubic yard, how many cubic yards of each mix should the farmer blend to meet the minimum monthly requirements at a minimal cost? What is the minimum cost?

Social Sciences

31. *Education — resource allocation.* A metropolitan school district has two high schools that are overcrowded and two that are under-enrolled. In order to balance the enrollment, the school board has decided to bus students from the overcrowded schools to the under-enrolled schools. North Division High School has 300 more students than it should have, and South Division High School has 500 more students than it should have. Central High School can accommodate 400 additional students and Washington High School can accommodate 500 additional students. The weekly cost of busing a student from North Division to Central is $5, from North Division to Washington is $2, from South Division to Central is $3, and from South Division to Washington is $4. Determine the number of students that should be bused from each of the overcrowded schools to each of the under-enrolled schools in order to balance the enrollment and minimize the cost of busing the students. What is the minimum cost?

32. *Education — resource allocation.* Repeat Problem 31 if the weekly cost of busing a student from North Division to Washington is $7 and all the other information remains the same.

8-7 Maximization and Minimization with Mixed Problem Constraints (Optional)

- An Introduction to the Big M Method
- The Big M Method
- Minimization by the Big M Method
- Summary of Methods of Solution
- Larger Problems — A Refinery Application

■ An Introduction to the Big M Method

In the preceding two sections, we have seen how to solve two types of linear programming problems: maximization problems with ⩽ problem constraints and nonnegative constants on the right side of each problem constraint, and minimization problems with $\geq$ problem constraints and nonnegative coefficients in the objective function. In this section we will present a generalized version of the simplex method that will solve both maximization and minimization problems with any combination of $\leq$, $\geq$, and $=$ problem constraints. The only requirement is that each problem constraint have a nonnegative constant on the right side.

To illustrate this new method, we will consider the following problem, which has one $\leq$ problem constraint and one $\geq$ problem constraint:

$$\begin{aligned} \text{Maximize} \quad & P = 2x_1 + x_2 \\ \text{Subject to} \quad & x_1 + x_2 \leq 10 \\ & -x_1 + x_2 \geq 2 \\ & x_1, x_2 \geq 0 \end{aligned} \tag{1}$$

To form an equation out of the first inequality, we introduce a nonnegative slack variable s_1, as before, and write:

$$x_1 + x_2 + s_1 = 10$$

How can we form an equation out of the second inequality? We introduce a second nonnegative variable s_2 and subtract it from the left side so that we can write

$$-x_1 + x_2 - s_2 = 2$$

The variable s_2 is called a **surplus variable** because it is the amount (surplus) by which the left side of the inequality exceeds the right side. **Surplus variables are always nonnegative quantities.**

We now express the linear programming problem (1) in the standard form.

$$\begin{aligned} x_1 + x_2 + s_1 \qquad\quad &= 10 \\ -x_1 + x_2 \qquad - s_2 \qquad &= 2 \\ -2x_1 - x_2 \qquad\qquad + P &= 0 \\ x_1, x_2, s_1, s_2 &\geq 0 \end{aligned} \tag{2}$$

The obvious basic solution (found by setting the nonbasic variables x_1 and x_2 equal to zero) is:

$$x_1 = 0, \quad x_2 = 0, \quad s_1 = 10, \quad s_2 = -2, \quad P = 0$$

This basic solution is not feasible. The surplus variable s_2 is negative, a violation of the nonnegative requirement for all variables except P. The simplex method works only when the obvious basic solution for each tableau is feasible, so we cannot solve this problem by writing the tableau for (2) and starting pivot operations.

In order to use the simplex method on a problem with mixed constraints, we must modify the problem. First, we introduce a second nonnegative variable a_1 in the equation involving the surplus variable s_2:

$$-x_1 + x_2 - s_2 + a_1 = 2$$

The variable a_1 is called an **artificial variable**, since it has no actual relationship to any of the variables in the original problem. **Artificial variables are always nonnegative quantities.** Next we add the term $-Ma_1$ to the objective function:

$$P = 2x_1 + x_2 - Ma_1$$

The number M is a very large positive constant whose value can be made as large as we wish. We now have a new problem, which we shall call the **modified problem:**

Maximize $P = 2x_1 + x_2 - Ma_1$

Subject to $x_1 + x_2 + s_1 \qquad = 10$

$\qquad\qquad -x_1 + x_2 \qquad - s_2 + a_1 = 2$ \hfill (3)

$\qquad\qquad x_1, x_2, s_1, s_2, a_1 \geqslant 0$

Rewriting (3) in the alternate standard form, we obtain:

$$x_1 + x_2 + s_1 \qquad\qquad\qquad = 10$$
$$-x_1 + x_2 \qquad - s_2 + \quad a_1 \qquad = 2 \qquad\qquad (4)$$
$$-2x_1 - x_2 \qquad\qquad\qquad + Ma_1 + P = 0$$

Again we see that the obvious basic solution is not feasible. (Setting the nonbasic variables x_1, x_2, and a_1 equal to zero, we see that s_2 is still negative.) To overcome this problem, we write the augmented matrix for (4) and proceed as follows:

x_1	x_2	s_1	s_2	a_1	P	
1	1	1	0	0	0	10
−1	1	0	−1	1	0	2
−2	−1	0	0	M	1	0

Let us eliminate M from the a_1 column so that a_1 will become a basic variable in an obvious basic solution. If the resulting obvious basic solution is also feasible, we can start pivot operations.

x_1	x_2	s_1	s_2	a_1	P	
1	1	1	0	0	0	10
−1	1	0	−1	1	0	2
−2	−1	0	0	M	1	0

$R_3 + (-M)R_2 \rightarrow R_3$

x_1	x_2	s_1	s_2	a_1	P	
1	1	1	0	0	0	10
−1	1	0	−1	1	0	2
$M-2$	$-M-1$	0	M	0	1	$-2M$

The obvious basic solution (setting the nonbasic variables x_1, x_2, and s_2 equal to zero) is

$$x_1 = 0, \quad x_2 = 0, \quad s_1 = 10, \quad s_2 = 0, \quad a_1 = 2, \quad P = -2M$$

This solution is also feasible, because all variables except P are nonnegative. Thus, we can commence with pivot operations,

The pivot column is determined by the most negative element in the bottom row of the tableau. Since M is a positive number, $-M-1$ is certainly a negative element. What about the element $M - 2$? Remember that M is a very large positive number. We will assume that M is so large that any expression of the form $M - k$ is positive. Thus, the only negative element in the bottom row is $-M - 1$.

$$
\begin{array}{cccccc}
x_1 & x_2 & s_1 & s_2 & a_1 & P
\end{array}
$$

Pivot row →
$$
\left[
\begin{array}{cccccc|c}
1 & 1 & 1 & 0 & 0 & 0 & 10 \\
-1 & \textcircled{1} & 0 & -1 & 1 & 0 & 2 \\
\hline
M-2 & -M-1 & 0 & M & 0 & 1 & -2M
\end{array}
\right]
\quad
\begin{array}{l}
R_1 + (-1)R_2 \rightarrow R_1 \\
\\
R_3 + (M+1)R_2 \rightarrow R_3
\end{array}
$$

$$\uparrow$$
Pivot
column

$$
\sim
\left[
\begin{array}{cccccc|c}
\textcircled{2} & 0 & 1 & 1 & -1 & 0 & 8 \\
-1 & 1 & 0 & -1 & 1 & 0 & 2 \\
\hline
-3 & 0 & 0 & -1 & M+1 & 1 & 2
\end{array}
\right]
\quad
\tfrac{1}{2}R_1 \rightarrow R_1
$$

$$
\sim
\left[
\begin{array}{cccccc|c}
1 & 0 & \tfrac{1}{2} & \tfrac{1}{2} & -\tfrac{1}{2} & 0 & 4 \\
-1 & 1 & 0 & -1 & 1 & 0 & 2 \\
\hline
-3 & 0 & 0 & -1 & M+1 & 1 & 2
\end{array}
\right]
\quad
\begin{array}{l}
R_2 + R_1 \rightarrow R_2 \\
R_3 + 3R_1 \rightarrow R_3
\end{array}
$$

$$
\sim
\left[
\begin{array}{cccccc|c}
1 & 0 & \tfrac{1}{2} & \tfrac{1}{2} & -\tfrac{1}{2} & 0 & 4 \\
0 & 1 & \tfrac{1}{2} & -\tfrac{1}{2} & \tfrac{1}{2} & 0 & 6 \\
\hline
0 & 0 & \tfrac{3}{2} & \tfrac{1}{2} & M-\tfrac{1}{2} & 1 & 14
\end{array}
\right]
$$

Since all the elements in the last row are nonnegative ($M - \tfrac{1}{2}$ is nonnegative because M is a very large positive number), we can stop and write the optimal solution:

$$\text{Max } P = 14 \quad \text{at} \quad x_1 = 4, \quad x_2 = 6, \quad s_1 = 0, \quad s_2 = 0, \quad a_1 = 0$$

This is an optimal solution to the modified problem (3). How is it related to the original problem (2)? Since $a_1 = 0$ in this solution,

$$x_1 = 4, \quad x_2 = 6, \quad s_1 = 0, \quad s_2 = 0, \quad P = 14 \tag{5}$$

is certainly a feasible solution for (2). [You can verify this by direct substitution into (2).] Surprisingly, it turns out that (5) is an optimal solution to the original problem. To see that this is true, suppose that we were able to find feasible values of x_1, x_2, s_1, and s_2 that satisfy the original system (2) and produce a value of $P > 14$. Then by using these same values in (3) along with $a_1 = 0$, we have found a feasible solution of (3) with $P > 14$. This contradicts the fact that $P = 14$ is the maximum value of P for the modified problem. Thus (5) is an optimal solution for the original problem.

As this example illustrates, if $a_1 = 0$ in an optimal solution for the modified problem, then deleting a_1 produces an optimal solution for the original problem. What happens if $a_1 \neq 0$ in the optimal solution for the modified problem? In this case, it can be shown that the original problem has no solution because its feasible set is empty.

In larger problems, each $\geq$ problem constraint will require the introduction of a surplus variable and an artificial variable. If one of the problem constraints is an equation rather than an inequality, there is no need to introduce a slack or surplus variable. However, each $=$ problem constraint will require the introduction of another artificial variable. Finally, each artificial variable must also be included in the objective function for the modified problem. The same constant M can be used for each artificial variable. Because of the role that the constant M plays in this approach, this method is often called the **big M method.**

■ The Big M Method

We now summarize the key steps used in the big M method and use them to solve several problems.

The Big M Method—Introducing Slack, Surplus, and Artificial Variables

1. If any problem constraints have negative constants on the right side, multiply both sides by -1 to obtain a constraint with a nonnegative constant. (If the constraint is an inequality, this will reverse the direction of the inequality.)
2. Introduce a slack variable in each $\leq$ constraint.
3. Introduce a surplus variable and an artificial variable in each $\geq$ constraint.
4. Introduce an artificial variable in each $=$ constraint.
5. For each artificial variable a_i, add $-Ma_i$ to the objective function. Use the same constant M for all artificial variables.

Example 17   Find the modified problem for the following linear programming problem. Do not attempt to solve the problem.

$$\text{Maximize} \quad P = 2x_1 + 5x_2 + 3x_3$$

$$\text{Subject to} \quad x_1 + 2x_2 - x_3 \leq 7$$

$$-x_1 + x_2 - 2x_3 \leq -5$$

$$x_1 + 4x_2 + 3x_3 \geq 1$$

$$2x_1 - x_2 + 4x_0 = 6$$

$$x_1, x_2, x_3 \geq 0$$

Solution   First, we multiply the second constraint by -1 to change -5 to 5:

$$(-1)(-x_2 + x_2 - 2x_3) \geq (-1)(-5)$$

$$x_1 - x_2 + 2x_3 \geq 5$$

Next, we introduce the slack, surplus, and artificial variables according to the rules stated in the box:

$$x_1 + 2x_2 - x_3 + s_1 \qquad\qquad\qquad = 7$$

$$x_1 - x_2 + 2x_3 \quad - s_2 + a_1 \qquad\qquad = 5$$

$$x_1 + 4x_2 + 3x_3 \qquad\qquad - s_3 + a_2 \quad = 1$$

$$2x_1 - x_2 + 4x_3 \qquad\qquad\qquad + a_3 = 6$$

Finally, we add $-Ma_1$, $-Ma_2$, and $-Ma_3$ to the objective function:

$$P = 2x_1 + 5x_2 + 3x_3 - Ma_1 - Ma_2 - Ma_3$$

The modified problem is

$$\text{Maximize} \quad P = 2x_1 + 5x_2 + 3x_3 - Ma_1 - Ma_2 - Ma_3$$

$$\text{Subject to} \quad x_1 + 2x_2 - x_3 + s_1 \qquad\qquad\qquad = 7$$

$$x_1 - x_2 + 2x_3 \quad - s_2 + a_1 \qquad\qquad = 5$$

$$x_1 + 4x_2 + 3x_3 \qquad\qquad - s_3 + a_2 \quad = 1$$

$$2x_1 - x_2 + 4x_3 \qquad\qquad\qquad + a_3 = 6$$

$$x_1, x_2, x_3, s_1, s_2, s_3, a_1, a_2, a_3 \geq 0$$

Problem 17   Repeat Example 17 for:

$$\text{Maximize} \quad P = 3x_1 - 2x_2 + x_3$$

$$\text{Subject to} \quad x_1 - 2x_2 + x_3 \geq 5$$

$$-x_1 - 3x_2 + 4x_3 \leq -10$$

$$2x_1 + 4x_2 + 5x_3 \leq 20$$

$$3x_1 - x_2 - x_3 = -15$$

$$x_1, x_2, x_3 \geq 0$$

> **The Big M Method—Solving the Problem**
>
> 1. Form the simplex tableau for the modified problem.
> 2. Use row operations to eliminate the M's in the bottom row of the simplex tableau in the columns corresponding to the artificial variables.
> 3. Solve the modified problem by the simplex method.
> 4. Relate the solution of the modified problem to the original problem.
>   a. If the modified problem has no solution, then the original problem has no solution.
>   b. If all artificial variables are zero in the solution to the modified problem, then delete the artificial variables to find a solution to the original problem.
>   c. If any artificial variables are nonzero in the solution to the modified problem, then the original problem has no solution.

Example 18   Solve the following linear programming problem using the big M method:

$$\text{Maximize} \quad P = x_1 - x_2 + 3x_3$$
$$\text{Subject to} \quad x_1 + x_2 \qquad \leqslant 20$$
$$x_1 \qquad + x_3 = 5$$
$$x_2 + x_3 \geqslant 10$$
$$x_1, x_2, x_3 \geqslant 0$$

Solution   State the modified problem:

$$\text{Maximize} \quad P = x_1 - x_2 + 3x_3 - Ma_1 - Ma_2$$
$$\text{Subject to} \quad x_1 + x_2 \qquad + s_1 \qquad\qquad = 20$$
$$x_1 \qquad + x_3 \qquad + a_1 \qquad = 5$$
$$x_2 + x_3 \qquad\qquad - s_2 + a_2 = 10$$
$$x_1, x_2, x_3, s_1, s_2, a_1, a_2 \geqslant 0$$

Write the simplex tableau for the modified problem.

$$
\begin{bmatrix}
x_1 & x_2 & x_3 & s_1 & a_1 & s_2 & a_2 & P & \\
1 & 1 & 0 & 1 & 0 & 0 & 0 & 0 & 20 \\
1 & 0 & 1 & 0 & 1 & 0 & 0 & 0 & 5 \\
0 & 1 & 1 & 0 & 0 & -1 & 1 & 0 & 10 \\
\hline
-1 & 1 & -3 & 0 & M & 0 & M & 1 & 0
\end{bmatrix}
$$

Eliminate M from the a_1 column

$R_4 + (-M)R_2 \to R_4$

$$\begin{array}{cccccccc} x_1 & x_2 & x_3 & s_1 & a_1 & s_2 & a_2 & P \end{array}$$

$$\sim \begin{bmatrix} 1 & 1 & 0 & 1 & 0 & 0 & (0) & 0 & | & 20 \\ 1 & 0 & 1 & 0 & 1 & 0 & 0 & 0 & | & 5 \\ 0 & 1 & 1 & 0 & 0 & -1 & 1 & 0 & | & 10 \\ \hline M & 1 & 1 & -M-3 & 0 & 0 & 0 & (M) & 1 & | & -5M \end{bmatrix}$$

Eliminate M from the a_2 column

$R_4 + (-M)R_3 \rightarrow R_4$

$$\sim \begin{bmatrix} 1 & 1 & 0 & 1 & 0 & 0 & 0 & 0 & | & 20 \\ 1 & 0 & 1 & 0 & 1 & 0 & 0 & 0 & | & 5 \\ 0 & 1 & 1 & 0 & 0 & -1 & 1 & 0 & | & 10 \\ \hline -M-1 & -M+1 & -2M-3 & 0 & 0 & M & 0 & 1 & | & -15M \end{bmatrix}$$

The obvious basic solution (setting the nonbasic variables x_1, x_2, x_3, and s_2 equal to zero) is

$$x_1 = 0, \quad x_2 = 0, \quad x_3 = 0, \quad s_1 = 20,$$
$$a_1 = 5, \quad s_2 = 0, \quad a_2 = 10, \quad P = -15M$$

Since this basic solution is feasible, we can commence with pivot operations to find the optimal solution. (It can be shown that except for some degenerate cases which we will not consider here, if the modified linear programming problem has a solution, then the obvious basic solution resulting after M has been eliminated from the artificial variable columns will be feasible. We can then perform pivot operations to find the optimal solution if it exists.)

$$\begin{array}{cccccccc} x_1 & x_2 & x_3 & s_1 & a_1 & s_2 & a_2 & P \end{array}$$

$$\begin{bmatrix} 1 & 1 & 0 & 1 & 0 & 0 & 0 & 0 & | & 20 \\ 1 & 0 & (1) & 0 & 1 & 0 & 0 & 0 & | & 5 \\ 0 & 1 & 1 & 0 & 0 & -1 & 1 & 0 & | & 10 \\ \hline -M-1 & -M+1 & -2M-3 & 0 & 0 & M & 0 & 1 & | & -15M \end{bmatrix}$$

$R_3 + (-1)R_2 \rightarrow R_3$

$R_4 + (2M+3)R_2 \rightarrow R_4$

$$\sim \begin{bmatrix} 1 & 1 & 0 & 1 & 0 & 0 & 0 & 0 & | & 20 \\ 1 & 0 & 1 & 0 & 1 & 0 & 0 & 0 & | & 5 \\ -1 & (1) & 0 & 0 & -1 & -1 & 1 & 0 & | & 5 \\ \hline M+2 & -M+1 & 0 & 0 & 2M+3 & M & 0 & 1 & | & -5M+15 \end{bmatrix}$$

$R_1 + (-1)R_3 \rightarrow R_1$

$R_4 + (M-1)R_3 \rightarrow R_4$

$$\sim \begin{bmatrix} 2 & 0 & 0 & 1 & 1 & 1 & -1 & 0 & | & 15 \\ 1 & 0 & 1 & 0 & 1 & 0 & 0 & 0 & | & 5 \\ -1 & 1 & 0 & 0 & -1 & -1 & 1 & 0 & | & 5 \\ \hline 3 & 0 & 0 & 0 & M+4 & 1 & M-1 & 1 & | & 10 \end{bmatrix}$$

Since the bottom row has no negative elements, we can stop and write the optimal solution:

$$x_1 = 0, \quad x_2 = 5, \quad x_3 = 5, \quad s_1 = 15, \quad a_1 = 0, \quad s_2 = 0, \quad a_2 = 0, \quad P = 10$$

Since $a_1 = 0$ and $a_2 = 0$, the solution to the original problem is

$$\text{Max } P = 10 \quad \text{at} \quad x_1 = 0, \quad x_2 = 5, \quad x_3 = 5$$

Problem 18   Solve the following linear programming problem using the big M method:

$$\text{Maximize} \quad P = x_1 + 4x_2 + 2x_3$$
$$\text{Subject to} \qquad x_2 + x_3 \leqslant 4$$
$$x_1 \qquad - x_3 = 6$$
$$x_1 - x_2 - x_3 \geqslant 1$$
$$x_1, x_2, x_3 \geqslant 0$$

Example 19   Solve the following linear programming problem using the big M method:

$$\text{Maximize} \quad P = 3x_1 + 5x_2$$
$$\text{Subject to} \quad 2x_1 + x_2 \leqslant 4$$
$$x_1 + 2x_2 \geqslant 10$$
$$x_1, x_2 \geqslant 0$$

Solution   Introducing slack, surplus, and artificial variables, we obtain the modified problem:

$$
\begin{aligned}
2x_1 + x_2 + s_1 & = 4 \\
x_1 + 2x_2 \quad - s_2 + a_1 & = 10 \qquad \text{Modified problem} \\
-3x_1 - 5x_2 \qquad\qquad + Ma_1 + P & = 0
\end{aligned}
$$

$$
\begin{bmatrix}
x_1 & x_2 & s_1 & s_2 & a_1 & P & \\
2 & 1 & 1 & 0 & 0 & 0 & 4 \\
1 & 2 & 0 & -1 & 1 & 0 & 10 \\
\hdashline
-3 & -5 & 0 & 0 & M & 1 & 0
\end{bmatrix}
$$

Simplex tableau: Eliminate M in the a_1 column, then begin pivot operations

$$R_3 + (-M)R_2 \to R_3$$

$$
\sim
\begin{bmatrix}
2 & \textcircled{1} & 1 & 0 & 0 & 0 & 4 \\
1 & 2 & 0 & -1 & 1 & 0 & 10 \\
\hdashline
-M-3 & -2M-5 & 0 & M & 0 & 1 & -10M
\end{bmatrix}
$$

Begin pivot operations:

$$R_2 + (-2)R_1 \to R_2$$
$$R_3 + (2M+5)R_1 \to R_3$$

$$
\sim
\begin{bmatrix}
2 & 1 & 1 & 0 & 0 & 0 & 4 \\
-3 & 0 & -2 & -1 & \textcircled{1} & 0 & 2 \\
\hdashline
3M+7 & 0 & 2M+5 & M & 0 & 1 & -2M+20
\end{bmatrix}
$$

The optimal solution of the modified problem is

$$x_1 = 0, \quad x_2 = 4, \quad s_1 = 0, \quad s_2 = 0, \quad a_1 = 2, \quad P = -2M + 20$$

Since a_1 is not zero, the original problem has no solution. The graph on the next page shows that the feasible region for the original problem is empty.

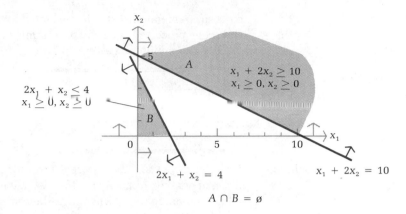

$$2x_1 + x_2 \leq 4$$
$$x_1 \geq 0, x_2 \geq 0$$

$$A \cap B = \emptyset$$

Problem 19   Solve the following linear programming problem using the big M method:

Maximize $P = 3x_1 + 2x_2$

Subject to $x_1 + 5x_2 \leq 5$

$\quad\quad\quad\quad 2x_1 + x_2 \geq 12$

$\quad\quad\quad\quad\quad\quad x_1, x_2 \geq 0$

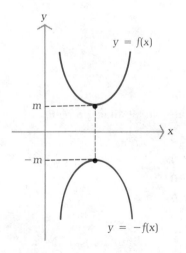

Figure 9

■ Minimization by the Big M Method

A minimization problem with $\geq$ problem constraints and nonnegative coefficients in the objective function can be solved by the dual method. How can we solve minimization problems that do not satisfy these two conditions? To minimize any objective function, we have only to maximize its negative. Figure 9 illustrates the fact that the minimum value of a function f occurs at the same point as the maximum value of the function $-f$. Furthermore, if m is the minimum value of f, then $-m$ is the maximum value of $-f$, and conversely. Thus, we can find the minimum value of a function f by finding the maximum value of $-f$ and then changing the sign of the maximum value.

Example 20   A small jewelry manufacturing company employs a person who is a highly skilled gem cutter, and it wishes to use this person at least 6 hours per day for this purpose. On the other hand, the polishing facilities can be used in any amounts up to 8 hours per day. The company specializes in three kinds of semiprecious gemstones, P, Q, and R. Relevant cutting, polishing, and cost requirements are listed in the table. How many gemstones of each type should be processed each day to minimize the cost of the finished stones? What is the minimum cost?

	P	Q	R
Cutting	2 hr	1 hr	1 hr
Polishing	1 hr	1 hr	2 hr
Cost per stone	$30	$30	$10

Solution   If we let x_1, x_2, and x_3 represent the number of type P, Q, and R stones finished per day, respectively, then we have the following linear programming problem to solve, where C is the cost of the stones:

Minimize $C = 30x_1 + 30x_2 + 10x_3$ Objective function

Subject to $\left.\begin{array}{l} 2x_1 + x_2 + x_3 \geqslant 6 \\ x_1 + x_2 + 2x_3 \leqslant 8 \end{array}\right\}$ Problem constraints

$x_1, x_2, x_3 \geqslant 0$ Nonnegative constraints

We convert this to a maximization problem by letting

$$P = -C = -30x_1 - 30x_2 - 10x_3$$

Thus, we get:

Maximize $P = -30x_1 - 30x_2 - 10x_3$

Subject to $2x_1 + x_2 + x_3 \geqslant 6$

$x_1 + x_2 + 2x_3 \leqslant 8$

$x_1, x_2, x_3 \geqslant 0$

and Min $C = -$Max P. To solve, we first state the modified problem:

$$2x_1 + x_2 + x_3 - s_1 + a_1 = 6$$
$$x_1 + x_2 + 2x_3 + s_2 = 8$$
$$30x_1 + 30x_2 + 10x_3 + Ma_1 + P = 0$$
$$x_1, x_2, x_3, s_1, s_2, a_1 \geqslant 0$$

$$\begin{array}{ccccccc|c} x_1 & x_2 & x_3 & s_1 & a_1 & s_2 & P & \\ \hline 2 & 1 & 1 & -1 & 1 & 0 & 0 & 6 \\ 1 & 1 & 2 & 0 & 0 & 1 & 0 & 8 \\ \hline 30 & 30 & 10 & 0 & M & 0 & 1 & 0 \end{array}$$

Eliminate M in the a_1 column

$R_3 + (-M)R_1 \rightarrow R_3$

Begin pivot operations. Assume M is so large that $-2M + 30, -M + 30,$ and $-M + 10$ are all negative.

$$\sim \begin{array}{ccccccc|c} \textcircled{2} & 1 & 1 & -1 & 1 & 0 & 0 & 6 \\ 1 & 1 & 2 & 0 & 0 & 1 & 0 & 8 \\ \hline -2M+30 & -M+30 & -M+10 & M & 0 & 0 & 1 & -6M \end{array}$$

$\frac{1}{2}R_1 \rightarrow R_1$

$$
\sim
\begin{bmatrix}
x_1 & x_2 & x_3 & s_1 & a_1 & s_2 & P & \\
1 & \frac{1}{2} & \frac{1}{2} & -\frac{1}{2} & \frac{1}{2} & 0 & 0 & 3 \\
1 & 1 & 2 & 0 & 0 & 1 & 0 & 8 \\
-2M+30 & -M+30 & -M+10 & M & 0 & 0 & 1 & -6M
\end{bmatrix}
\begin{matrix}
\\ R_2 + (-1)R_1 \to R_2 \\ R_3 + (2M-30)R_1 \to R_3
\end{matrix}
$$

$$
\sim
\begin{bmatrix}
1 & \frac{1}{2} & \frac{1}{2} & -\frac{1}{2} & \frac{1}{2} & 0 & 0 & 3 \\
0 & \frac{1}{2} & \boxed{\frac{3}{2}} & \frac{1}{2} & -\frac{1}{2} & 1 & 0 & 5 \\
0 & 15 & -5 & 15 & M-15 & 0 & 1 & -90
\end{bmatrix}
\begin{matrix}
\\ \frac{2}{3}R_2 \to R_2 \\
\end{matrix}
$$

$$
\sim
\begin{bmatrix}
1 & \frac{1}{2} & \frac{1}{2} & -\frac{1}{2} & \frac{1}{2} & 0 & 0 & 3 \\
0 & \frac{1}{3} & 1 & \frac{1}{3} & -\frac{1}{3} & \frac{2}{3} & 0 & \frac{10}{3} \\
0 & 15 & -5 & 15 & M-15 & 0 & 1 & -90
\end{bmatrix}
\begin{matrix}
R_1 + (-\frac{1}{2})R_2 \to R_1 \\ \\ R_3 + 5R_2 \to R_3
\end{matrix}
$$

$$
\sim
\begin{bmatrix}
1 & \frac{1}{3} & 0 & -\frac{2}{3} & \frac{2}{3} & -\frac{1}{3} & 0 & \frac{4}{3} \\
0 & \frac{1}{3} & 1 & \frac{1}{3} & -\frac{1}{3} & \frac{2}{3} & 0 & \frac{10}{3} \\
0 & \frac{50}{3} & 0 & \frac{50}{3} & M-\frac{50}{3} & \frac{10}{3} & 1 & -\frac{220}{3}
\end{bmatrix}
$$

Since the bottom row has no negative elements to the left of the vertical line, the optimal solution for the modified problem is

$$x_1 = \tfrac{4}{3}, \quad x_2 = 0, \quad x_3 = \tfrac{10}{3}, \quad s_1 = 0, \quad a_1 = 0, \quad s_2 = 0, \quad P = -\tfrac{220}{3}$$

Since $a_1 = 0$, deleting a_1 produces the solution to the original maximization problem and also to the minimization problem. Thus

$$\text{Min } C = -\text{Max } P = -\left(-\tfrac{220}{3}\right) = 73\tfrac{1}{3} \quad \text{at} \quad x_1 = \tfrac{4}{3}, \quad x_2 = 0, \quad x_3 = \tfrac{10}{3}$$

That is, a minimum cost of $\$73\frac{1}{3}$ for gemstones will be realized if $1\frac{1}{3}$ type A, no type B, and $3\frac{1}{3}$ type C stones are processed each day. The fractional values for production make sense if we think of them as average daily production figures.

Problem 20 Repeat Example 20 with $C = 40x_1 + 40x_2 + 10x_3$.

■ Summary of Methods of Solution

The big M method can be used to solve any minimization problem, including those that can also be solved by the dual method. However, when solving problems by hand, the dual method should be used whenever it is applicable. The following summary should help you select the proper method of solution:

Summary of Methods

Type of Problem	Method of Solution
1. Maximization, ≤ problem constraints, nonnegative constants on right side of problem constraints	Simplex method with slack variables
2. Minimization, ≥ problem constraints, nonnegative coefficients in objective function.	Form dual and solve by method 1
3. Maximization, mixed constraints, nonnegative constants on right side of problem constraints	Form modified problem with slack, surplus, and artificial variables and solve by the big M method
4. Minimization, mixed constraints, nonnegative constants on right side of problem constraints	Maximize negative of objective function by method 3

■ Larger Problems — A Refinery Application

Up to this point, all of the problems could be solved by hand. The simplex method will solve problems with a large number of variables and constraints; however, a computer is generally used to perform the actual pivot operations. As a final application, we will consider a problem that would require the use of a computer to complete the solution.

Example 21

A refinery produces two grades of gasoline, regular and premium, by blending together two components, A and B. Component A has an octane rating of 90 and costs $28 a barrel. Component B has an octane rating of 110 and costs $32 a barrel. The octane rating for regular gasoline must be at least 95, and the octane rating for premium must be at least 105. Regular gasoline sells for $34 a barrel and premium sells for $40 a barrel. Currently, the company has 30,000 barrels of component A and 20,000 barrels of component B. It also has orders for 20,000 barrels of regular and 10,000 barrels of premium that it must fill. Assuming that all the gasoline produced can be sold, determine the maximum possible profit.

Solution

First we will organize the information given in the problem in tabular form (Table 4).

Table 4

Component	Octane Rating	Cost	Available Supply
A	90	$28	30,000 barrels
B	110	$32	20,000 barrels

Grade	Minimum Octane Rating	Selling Price	Existing Orders
Regular	95	$34	20,000 barrels
Premium	105	$40	10,000 barrels

Let

x_1 = Number of barrels of component A used in regular gasoline

x_2 = Number of barrels of component A used in premium gasoline

x_3 = Number of barrels of component B used in regular gasoline

x_4 = Number of barrels of component B used in premium gasoline

The total amount of component A used is $x_1 + x_2$. This cannot exceed the available supply. Thus, one constraint is

$$x_1 + x_2 \leq 30,000$$

The corresponding inequality for component B is

$$x_3 + x_4 \leq 20,000$$

The amounts of regular and premium gasoline produced must be sufficient to meet the existing orders:

$$x_1 + x_3 \geq 20,000 \quad \text{Regular}$$
$$x_2 + x_4 \geq 10,000 \quad \text{Premium}$$

Now let's consider the octane ratings. The octane rating of a blend is simply the proportional average of the octane ratings of the components. Thus, the octane rating for regular gasoline is

$$90 \, \frac{x_1}{x_1 + x_3} + 110 \, \frac{x_3}{x_1 + x_3}$$

where $x_1 / (x_1 + x_3)$ is the percentage of component A used in regular gasoline and $x_3 / (x_1 + x_3)$ is the percentage of component B. The final octane rating of regular gasoline must be at least 95; thus,

$$90 \, \frac{x_1}{x_1 + x_3} + 110 \, \frac{x_3}{x_1 + x_3} \geq 95 \qquad \text{Multiply by } x_1 + x_3$$

$$90x_1 + 110x_3 \geq 95(x_1 + x_3) \qquad \text{Collect like terms on the left side}$$

$$-5x_1 + 15x_3 \geq 0$$

The corresponding inequality for premium gasoline is

$$90\,\frac{x_2}{x_2 + x_4} + 110\,\frac{x_4}{x_2 + x_4} \geqslant 105$$

$$90x_2 + 110x_4 \geqslant 105(x_2 + x_4)$$

$$-15x_2 + \;\;5x_4 \geqslant 0$$

The cost of the components used is

$$C = 28(x_1 + x_2) + 32(x_3 + x_4)$$

The revenue from selling all the gasoline is

$$R = 34(x_1 + x_3) + 40(x_2 + x_4)$$

and the profit is

$$
\begin{aligned}
P &= R - C \\
&= 34(x_1 + x_3) + 40(x_2 + x_4) - 28(x_1 + x_2) - 32(x_3 + x_4) \\
&= (34 - 28)x_1 + (40 - 28)x_2 + (34 - 32)x_3 + (40 - 32)x_4 \\
&= 6x_1 + 12x_2 + 2x_3 + 8x_4
\end{aligned}
$$

To find the maximum profit, we must solve the following linear programming problem:

Maximize
$$P = 6x_1 + 12x_2 + 2x_3 + 8x_4 \qquad\qquad \text{Profit}$$

Subject to

				Label
$x_1 +$	x_2		$\leqslant 30{,}000$	Available A
		$x_3 + \;x_4$	$\leqslant 20{,}000$	Available B
x_1		$+ \;\;x_3$	$\geqslant 20{,}000$	Required regular
	x_2	$+ \;\;x_4$	$\geqslant 10{,}000$	Required premium
$-5x_1$		$+ 15x_3$	$\geqslant 0$	Octane for regular
	$-15x_2$	$+ 5x_4$	$\geqslant 0$	Octane for premium

$$x_1, x_2, x_3, x_4 \geqslant 0$$

The tableau for this problem would have seven rows and sixteen columns. Solving this problem by hand is possible but would require considerable effort. Instead, we used a computer program to solve this problem. [The computer program can be found in the computer supplement for this text (see Preface).] The output from this problem is displayed in Table 5.

According to the output in Table 5, the refinery should blend 26,250 barrels of component A and 8,750 barrels of component B to produce 35,000 barrels of regular. They should blend 3,750 barrels of component A and 11,250 barrels of component B to produce 15,000 barrels of premium. This will result in a maximum profit of $310,000.

Table 5

Input to Program	Output from Program
| ```
NUMBER OF VARIABLES = 4
NUMBER OF CONSTRAINTS = 6
2 OF THE FORM <=
4 OF THE FORM >=
0 OF THE FORM =
``` | ```
DECISION VARIABLES
X1 = 26250
X2 = 3750
X3 = 8750
X4 = 11250
``` |
| ```
CONSTRAINTS
1 1 0 0 30000
``` | ```
SLACK VARIABLES
S1 = 0
S2 = 0
``` |
| ```
0 0 1 1 20000
``` | |
| ```
1 0 1 0 20000
``` | ```
SURPLUS VARIABLES
S3 = 15000
S4 = 5000
``` |
| ```
0 1 0 1 10000
``` | ```
S5 = 0
S6 = 0
``` |
| ```
-5 0 1 5 0 0
``` | |
| ```
0 -1 5 0 5 0
``` | ```
MAXIMUM VALUE OF OBJECTIVE FUNCTION
310000
``` |
| ```
OBJECTIVE FUNCTION
6 12 2 8
``` | |

Problem 21 Suppose the refinery in Example 21 has 35,000 barrels of component A, which costs \$25 a barrel, and 15,000 barrels of component B, which costs \$35 a barrel. If all the other information is unchanged, formulate a linear programming problem whose solution is the maximum profit. Do not attempt to solve the problem (unless you have access to a computer and a program that solves linear programming problems).

Answers to Matched Problems

17. Maximize $P = 3x_1 - 2x_2 + x_3 - Ma_1 - Ma_2 - Ma_3$

Subject to
$$x_1 - 2x_2 + x_3 - s_1 + a_1 = 5$$
$$x_1 + 3x_2 - 4x_3 - s_2 + a_2 = 10$$
$$2x_1 + 4x_2 + 5x_3 + s_3 = 20$$
$$-3x_1 + x_2 + x_3 + a_3 = 15$$
$$x_1, x_2, x_3, s_1, a_1, s_2, a_2, s_3, a_3 \geq 0$$

18. Max $P = 22$ at $x_1 = 6$, $x_2 = 4$, $x_3 = 0$
19. No solution (empty feasible region)
20. Min $C = \$86\frac{2}{3}$ at $x_1 = \frac{4}{3}$, $x_2 = 0$, $x_3 = \frac{10}{3}$
```

21. Maximize $P = 9x_1 + 15x_2 - x_3 + 5x_4$

    Subject to $\quad x_1 + \quad x_2 \qquad\qquad\qquad \leqslant 35{,}000$

    $\qquad\qquad\qquad\qquad\qquad x_3 + \quad x_4 \leqslant 15{,}000$

    $\qquad\qquad x_1 \qquad\qquad + \quad x_3 \qquad\qquad \geqslant 20{,}000$

    $\qquad\qquad\qquad\quad x_2 \qquad\qquad + \quad x_4 \geqslant 10{,}000$

    $\qquad -5x_1 \qquad\qquad + 15x_3 \qquad\qquad \geqslant 0$

    $\qquad\qquad\qquad - 15x_2 \qquad\qquad + 5x_4 \geqslant 0$

    $\qquad\qquad\qquad x_1, x_2, x_3, x_4 \geqslant 0$

Exercise 8-7

A   In Problems 1–8:

(A) Introduce slack, surplus, and artificial variables and form the modified problem.

(B) Write the augmented coefficient matrix for the modified problem and eliminate M from the columns of the artificial variables.

(C) Solve the modified problem using the simplex method.

1. Maximize $P = 5x_1 + 2x_2$

   Subject to $x_1 + 2x_2 \leqslant 12$

   $\qquad\qquad x_1 + \quad x_2 \geqslant 4$

   $\qquad\qquad\quad x_1, x_2 \geqslant 0$

2. Maximize $P = 3x_1 + 7x_2$

   Subject to $2x_1 + x_2 \leqslant 16$

   $\qquad\qquad x_1 + x_2 \geqslant 6$

   $\qquad\qquad\quad x_1, x_2 \geqslant 0$

3. Maximize $P = 3x_1 + 5x_2$

   Subject to $2x_1 + x_2 \leqslant 8$

   $\qquad\qquad x_1 + x_2 = 6$

   $\qquad\qquad\quad x_1, x_2 \geqslant 0$

4. Maximize $P = 4x_1 + 3x_2$

   Subject to $x_1 + 3x_2 \leqslant 24$

   $\qquad\qquad x_1 + \quad x_2 = 12$

   $\qquad\qquad\quad x_1, x_2 \geqslant 0$

5. Maximize $P = 4x_1 + 3x_2$

   Subject to $-x_1 + 2x_2 \leqslant 2$

   $\qquad\qquad x_1 + \quad x_2 \geqslant 4$

   $\qquad\qquad\quad x_1, x_2 \geqslant 0$

6. Maximize $P = 3x_1 + 4x_2$

   Subject to $x_1 - 2x_2 \leqslant 2$

   $\qquad\qquad x_1 + \quad x_2 \geqslant 5$

   $\qquad\qquad\quad x_1, x_2 \geqslant 0$

7. Maximize $P = 5x_1 + 10x_2$

   Subject to $\quad x_1 + \quad x_2 \leqslant 3$

   $\qquad\qquad 2x_1 + 3x_2 \geqslant 12$

   $\qquad\qquad\quad x_1, x_2 \geqslant 0$

8. Maximize $P = 4x_1 + 6x_2$

   Subject to $\quad x_1 + \quad x_2 \leqslant 2$

   $\qquad\qquad 3x_1 + 5x_2 \geqslant 15$

   $\qquad\qquad\quad x_1, x_2 \geqslant 0$

B *Use the big M method to solve the following problems:*

9. Minimize and maximize

$$P = 6x_1 - 2x_2$$

Subject to $\quad x_1 + x_2 \leq 10$

$ 3x_1 + 2x_2 \geq 24$

$ x_1 \ x_2 \geq 0$

10. Minimize and maximize

$$P = -4x_1 + 16x_2$$

Subject to $\quad 3x_1 + x_2 \leq 28$

$ x_1 + 2x_2 \geq 16$

$ x_1, x_2 \geq 0$

11. Maximize $\quad P = 2x_1 + 5x_2$

Subject to $\quad x_1 + 2x_2 \leq 18$

$ 2x_1 + x_2 \leq 21$

$ x_1 + x_2 \geq 10$

$ x_1, x_2 \geq 0$

12. Maximize $\quad P = 6x_1 + 2x_2$

Subject to $\quad x_1 + 2x_2 \leq 20$

$ 2x_1 + x_2 \leq 16$

$ x_1 + x_2 \geq 9$

$ x_1, x_2 \geq 0$

13. Maximize

$$P = 10x_1 + 12x_2 + 20x_3$$

Subject to

$$3x_1 + x_2 + 2x_3 \geq 12$$

$$x_1 - x_2 + 2x_3 = 6$$

$$x_1, x_2, x_3 \geq 0$$

14. Maximize

$$P = 5x_1 + 7x_2 + 9x_3$$

Subject to

$$x_1 - x_2 + x_3 \geq 20$$

$$2x_1 + x_2 + 3x_3 = 36$$

$$x_1, x_2, x_3 \geq 0$$

15. Minimize

$$C = -5x_1 - 12x_2 + 16x_3$$

Subject to

$$x_1 + 2x_2 + x_3 \leq 10$$

$$2x_1 + 3x_2 + x_3 \geq 6$$

$$2x_1 + x_2 - x_3 = 1$$

$$x_1, x_2, x_3 \geq 0$$

16. Minimize

$$C = -3x_1 + 15x_2 - 4x_3$$

Subject to

$$2x_1 + x_2 + 3x_3 \leq 24$$

$$x_1 + 2x_2 + x_3 \geq 6$$

$$x_1 - 3x_2 + x_3 = 2$$

$$x_1, x_2, x_3 \geq 0$$

C *Problems 17–24 are mixed. Some can be solved by the methods presented in Sections 8-5 and 8-6, while others must be solved by the big M method.*

17. Minimize

$$C = 10x_1 - 40x_2 - 5x_3$$

Subject to

$$x_1 + 3x_2 \leq 6$$

$$4x_2 + x_3 \leq 3$$

$$x_1, x_2, x_3 \geq 0$$

18. Maximize

$$P = 7x_1 - 5x_2 + 2x_3$$

Subject to

$$x_1 - 2x_2 + x_3 \geq -8$$

$$x_1 - x_2 + x_3 \leq 10$$

$$x_1, x_2, x_3 \geq 0$$

19. Maximize

$$P = -5x_1 + 10x_2 + 15x_3$$

Subject to

$$2x_1 + 3x_2 + x_3 \leqslant 24$$
$$x_1 - 2x_2 - 2x_3 \geqslant 1$$
$$x_1, x_2, x_3 \geqslant 0$$

20. Minimize

$$C = -5x_1 + 10x_2 + 15x_3$$

Subject to

$$2x_1 + 3x_2 + x_3 \leqslant 24$$
$$x_1 - 2x_2 - 2x_3 \geqslant 1$$
$$x_1, x_2, x_3 \geqslant 0$$

21. Minimize

$$C = 10x_1 + 40x_2 + 5x_3$$

Subject to

$$x_1 + 3x_2 \qquad \geqslant 6$$
$$4x_2 + x_3 \geqslant 3$$
$$x_1, x_2, x_3 \geqslant 0$$

22. Maximize

$$P = 8x_1 + 2x_2 - 10x_3$$

Subject to

$$x_1 + x_2 - 3x_3 \leqslant 6$$
$$4x_1 - x_2 + 2x_3 \leqslant -7$$
$$x_1, x_2, x_3 \geqslant 0$$

23. Maximize

$$P = 12x_1 + 9x_2 + 5x_3$$

Subject to

$$x_1 + 2x_2 + x_3 \leqslant 40$$
$$2x_1 + x_2 + 3x_3 \leqslant 60$$
$$x_1, x_2, x_3 \geqslant 0$$

24. Minimize

$$C = 10x_1 + 12x_2 + 28x_3$$

Subject to

$$4x_1 + 2x_2 + 3x_3 \geqslant 20$$
$$5x_1 - x_2 - 4x_3 \leqslant 10$$
$$x_1, x_2, x_3 \geqslant 0$$

■

Applications

In Problems 25–32, formulate each problem as a linear programming problem and solve by using the big M method.

Business & Economics

25. *Manufacturing—resource allocation.* An electronics company manufacturers two types of add-on memory modules for microcomputers, a 16K module and a 64K module. Each 16K module requires 10 minutes for assembly and 2 minutes for testing. Each 64K module requires 15 minutes for assembly and 4 minutes for testing. The company makes a profit of $18 on each 16K module and $30 on each 64K module. The assembly department can work a maximum of 1,500 minutes per day, and the testing department can work a maximum of 500 minutes a day. In order to satisfy current orders, the company must produce at least 50 16K modules per day. How many units of each module should the company manufacture each day in order to maximize the daily profit? What is the maximum profit?

26. *Manufacturing—resource allocation.* Repeat Problem 25 if the assembly department can work a maximum of 2,100 minutes daily.

27. *Advertising.* A company planning an advertising campaign to attract new customers wants to place a total of at most ten ads in three newspapers. Each ad in the *Sentinel* costs $200 and will be read by

2,000 people. Each ad in the *Journal* costs $200 and will be read by 500 people. Each ad in the *Tribune* costs $100 and will be read by 1,500 people. The company wants at least 16,000 people to read its ads. How many ads should it place in each paper in order to minimize the advertising costs? What is the minimum cost?

28. *Advertising.* Repeat Problem 27 if the *Tribune* is unable to accept more than 4 ads from the company.

Life Sciences

29. *Nutrition—people.* An individual on a high-protein, low-carbohydrate diet requires at least 100 units of protein and at most 24 units of carbohydrates daily. The diet will consist entirely of three special liquid diet foods, *A*, *B*, and *C*. The contents and cost of the diet foods are given in the table. How many bottles of each brand of diet food should be consumed daily in order to meet the protein and carbohydrate requirements at minimal cost? What is the minimum cost?

| | Units per Bottle | | |
	A	*B*	*C*
Protein	10	10	20
Carbohydrates	2	3	4
Cost per bottle	$0.60	$0.40	$0.90

30. *Nutrition—people.* Repeat Problem 29 if brand *C* liquid diet food costs $1.50 a bottle.

31. *Nutrition—plants.* A farmer can use three types of plant food, mix *A*, mix *B*, and mix *C*. The amounts (in pounds) of nitrogen, phosphoric acid, and potash in a cubic yard of each mix are given in the table. Tests performed on the soil in a large field indicate that the field needs at least 800 pounds of potash. The tests also indicate that no more than 700 pounds of phosphoric acid should be added to the field. The farmer plans to plant a crop that requires a great deal of nitrogen. How many cubic yards of each mix should he add to the field in order to satisfy the potash and phosphoric acid requirements and maximize the amount of nitrogen added? What is the maximum amount of nitrogen?

| | Pounds per Cubic Yard | | |
	A	*B*	*C*
Nitrogen	12	16	8
Phosphoric acid	12	8	16
Potash	16	8	16

32. *Nutrition—plants.* Repeat Problem 31 if the field should have no more than 1,000 pounds of phosphoric acid.

In Problems 33–39, formulate each problem as a linear programming problem. Do not solve the linear programming problem.

Business & Economics

33. *Manufacturing—production scheduling.* A company manufactures car and truck frames at plants in Milwaukee and Racine. The Milwaukee plant has a daily operating budget of $50,000 and can produce at most 300 frames daily in any combination. It costs $150 to manufacture a car frame and $200 to manufacture a truck frame at the Milwaukee plant. The Racine plant has a daily operating budget of $35,000, can produce a maximum combined total of 200 frames daily, and produces a car frame at a cost of $135 and a truck frame at a cost of $180. Based on past demand, the company wants to limit production to a maximum of 250 car frames and 350 truck frames per day. If the company realizes a profit of $50 on each car frame and $70 on each truck frame, how many frames of each type should be produced at each plant to maximize the daily profit?

34. *Finances—loan distributions.* A savings and loan company has $3 million to lend. The types of loans and annual returns offered by the company are given in the table. State laws require that at least 50% of the money loaned for mortgages must be for first mortgages and that at least 30% of the total amount loaned must be for either first or second mortgages. Company policy requires that the amount of signature and automobile loans cannot exceed 25% of the total amount loaned and that signature loans cannot exceed 15% of the total amount loaned. How much money should be allocated to each type of loan in order to maximize the company's return?

Type of Loan	Annual Return
Signature	18%
First mortgage	12%
Second mortgage	14%
Automobile	16%

35. *Blending—petroleum.* A refinery produces two grades of gasoline, regular and premium, by blending together three components, *A*, *B*, and *C*. Component *A* has an octane rating of 90 and costs $28 a barrel, component *B* has an octane rating of 100 and costs $30 a barrel, and component *C* has an octane rating of 110 and costs $34 a barrel. The octane rating for regular must be at least 95 and the octane rating for premium must be at least 105. Regular gasoline sells for $38 a barrel and premium sells for $46 a barrel. The company has 40,000 barrels of component *A*, 25,000 barrels of component *B*, and 15,000 barrels of component *C* and must produce at least 30,000 barrels of regular and 25,000 barrels of premium. How should they blend the components in order to maximize their profit?

36. *Blending — food processing.* A company produces two brands of trail mix, regular and deluxe, by mixing dried fruits, nuts, and cereal. The recipes for the mixes are given in the table. The company has 1,200 pounds of dried fruits, 750 pounds of nuts, and 1,500 pounds of cereal to be used in producing the mixes. The company makes a profit of $0.40 on each pound of regular mix and $0.60 on each pound of deluxe mix. How many pounds of each ingredient should be used in each mix in order to maximize the company's profit?

Type of Mix	Ingredients
Regular	At least 20% nuts
	At most 40% cereal
Deluxe	At least 30% nuts
	At most 25% cereal

Life Sciences

37. *Nutrition — people.* A dietitian in a hospital is to arrange a special diet using the foods L, M, and N. The table below gives the nutritional contents and the cost of one ounce of each food. The daily requirements for the diet are at least 400 units of calcium, at least 200 units of iron, at least 300 units of vitamin A, at most 150 units of cholesterol, and at most 900 calories. How many ounces of each food should be used in order to meet the requirements of the diet at minimal cost?

	Units per Ounce		
	L	M	N
Calcium	30	10	30
Iron	10	10	10
Vitamin A	10	30	20
Cholesterol	8	4	6
Calories	60	40	50
Cost per ounce	$0.40	$0.60	$0.80

38. *Nutrition — feed mixtures.* A farmer grows three crops, corn, oats, and soybeans, which he mixes together to feed his cows and pigs. At least 40% of the feed mix for the cows must be corn. The feed mix for the pigs must contain at least twice as much soybeans as corn. He has harvested 1,000 bushels of corn, 500 bushels of oats, and 1,000 bushels of soybeans. He needs 1,000 bushels of each feed mix for his livestock. The unused corn, oats, and soybeans can be sold for $4, $3.50, and $3.25 a bushel, respectively (thus, these amounts also represent the cost of the crops used to feed the livestock). How many bushels of each crop should be used in each feed mix in order to produce sufficient food for the livestock at minimal cost?

Social Sciences

39. *Education — resource allocation.* Three towns are forming a consolidated school district with two high schools. Each high school has a maximum capacity of 2,000 students. Town *A* has 500 high school students, town *B* has 1,200, and town *C* has 1,800. The weekly costs of transporting a student from each town to each school are given in the table. In order to keep the enrollment balanced, the school board has decided that each high school must enroll at least 40% of the total student population. Furthermore, no more than 60% of the students in any town should be sent to the same high school. How many students from each town should be enrolled in each school in order to meet these requirements and minimize the cost of transporting the students?

| | Weekly Transportation Cost per Student | |
	School I	School II
Town *A*	$4	$8
Town *B*	6	4
Town *C*	3	9

8-8   Chapter Review

Important Terms and Symbols

8-1 *Linear inequalities in two variables.* graph of a linear inequality in two variables, upper half-plane, lower half-plane

8-2 *Systems of linear inequalities in two variables.* graphical solution, solution region, bounded regions, unbounded regions, corner point, feasible solution, feasible region

8-3 *Linear programming in two dimensions — a geometric approach.* linear programming problem, decision variables, objective function, constraints, nonnegative constraints, mathematical model, graphical solution, maximization problem, constant-profit line, isoprofit line, optimal solution, minimization problem, linear function, multiple optimal solutions, no feasible region, unbounded objective function

8-4 *A geometric introduction to the simplex method.* slack variables, basic solution, basic feasible solution, nonbasic variables, basic variables, simplex method

8-5 *The simplex method: maximization with ≤ problem constraints.* standard form of a linear programming problem, obvious basic solution, obvious basic feasible solution, simplex tableau, pivot column, pivot row, pivot element, pivot operation, row operations, $kR_i \rightarrow R_i$, $R_i + kR_j \rightarrow R_i$

Exercise 8-8   Chapter Review

Work through all the problems in this chapter review and check your answers in the back of the book. (Answers to all review problems are there.) Where weaknesses show up, review appropriate sections in the text. When you are satisfied that you know the material, take the practice test following this review.

A

1. Solve the system of linear inequalities graphically:

$$3x_1 + x_2 \leq 9$$
$$2x_1 + 6x_2 \leq 18$$
$$x_1, x_2 \geq 0$$

2. Solve the linear programming problem geometrically:

Maximize $P = 6x_1 + 2x_2$
Subject to $2x_1 + x_2 \leq 8$
$$x_1 + 2x_2 \leq 10$$
$$x_1, x_2 \geq 0$$

3. Convert Problem 2 into a system of equations using slack variables.
4. Find all basic solutions for the system in Problem 3 and determine which basic solutions are feasible.
5. Write the simplex tableau for Problem 2 and circle the pivot element.
6. Solve Problem 2 using the simplex method.
7. Solve the linear programming problem geometrically:

Minimize $C = 5x_1 + 2x_2$
Subject to $x_1 + 3x_2 \geq 15$
$$2x_1 + x_2 \geq 20$$
$$x_1, x_2 \geq 0$$

8. Form the dual of Problem 7.
9. Solve Problem 8 by applying the simplex method to the dual problem.

B   **10.** Solve the linear programming problem geometrically:

$$\text{Maximize} \quad P = 3x_1 + 4x_2$$
$$\text{Subject to} \quad 2x_1 + 4x_2 \leqslant 24$$
$$3x_1 + 3x_2 \leqslant 21$$
$$4x_1 + 2x_2 \leqslant 20$$
$$x_1, x_2 \geqslant 0$$

11. Solve Problem 10 using the simplex method.

12. Solve the linear programming problem geometrically:

$$\text{Minimize} \quad C = 3x_1 + 8x_2$$
$$\text{Subject to} \quad x_1 + x_2 \geqslant 10$$
$$x_1 + 2x_2 \geqslant 15$$
$$x_2 \geqslant 3$$
$$x_1, x_2 \geqslant 0$$

13. Form the dual of Problem 12.

14. Solve Problem 13 by applying the simplex method.

Solve the following linear programming problems:

15. Maximize $P = 5x_1 + 3x_2 - 3x_3$

$$\text{Subject to} \quad x_1 - x_2 - 2x_3 \leqslant 3$$
$$2x_1 + 2x_2 - 5x_3 \leqslant 10$$
$$x_1, x_2, x_3 \geqslant 0$$

16. Maximize $P = 5x_1 + 3x_2 - 3x_3$

$$\text{Subject to} \quad x_1 - x_2 - 2x_3 \leqslant 3$$
$$x_1 + x_2 \qquad \leqslant 5$$
$$x_1, x_2, x_3 \geqslant 0$$

C   **17.** Minimize $C = 2x_1 + 3x_2$

$$\text{Subject to} \quad 2x_1 + x_2 \leqslant 20$$
$$2x_1 + x_2 \geqslant 10$$
$$x_1 + 2x_2 \geqslant 8$$
$$x_1, x_2 \geqslant 0$$

18. Minimize $C = 15x_1 + 12x_2 + 15x_3 + 18x_4$

$$\text{Subject to} \quad x_1 + x_2 \qquad\qquad \leqslant 240$$
$$x_3 + x_4 \leqslant 500$$
$$x_1 \qquad + x_3 \qquad \geqslant 400$$
$$x_2 \qquad + x_4 \geqslant 300$$
$$x_1, x_2, x_3, x_4 \geqslant 0$$

■

Applications

Business & Economics

19. *Manufacturing—resource allocation.* Set up (but do not solve) Problem 19 in Exercise 8-3 with the added restrictions that the fabricating department must operate at least 60 labor-hours per day and the finishing department at least 12 labor-hours per day.

(A) Write the linear programming problem using appropriate inequalities and objective function.

(B) Tranform part A into a system of equations using slack, surplus, and artificial variables.

(C) Write the simplex tableau for part B. Do not solve.

Formulate the following as linear programming problems but do not solve:

20. *Transportation—shipping schedule.* A company produces motors for washing machines at factory *A* and factory *B*. Factory *A* can produce 1,500 motors a month and factory *B* can produce 1,000 motors a month. The motors are then shipped to one of three plants, where the washing machines are assembled. In order to meet anticipated demand, plant *X* must assemble 500 washing machines a month, plant *Y* must assemble 700 washing machines a month, and plant *Z* must assemble 800 washing machines a month. The shipping charges for one motor are given in the table. Determine a shipping schedule that will minimize the cost of transporting the motors from the factories to the assembly plants.

	Shipping Charges		
	Plant *X*	Plant *Y*	Plant *Z*
Factory *A*	$5	$8	$12
Factory *B*	$9	$7	$ 6

Life Sciences

21. *Nutrition—animals.* A special diet for laboratory animals is to contain at least 300 units of vitamins, 200 units of minerals, and 900 calories. There are two feed mixes available, mix *A* and mix *B*. A gram of mix *A* contains 3 units of vitamins, 2 units of minerals, and 6 calories. A gram of mix *B* contains 4 units of vitamins, 5 units of minerals, and 10 calories. Mix *A* costs $0.02 per gram and mix *B* costs $0.04 per gram. How many grams of each mix should be used to satisfy the requirements of the diet at minimal cost?

Practice Test: Chapter 8

---

1. Solve the system of linear inequalities graphically:

   $$2x_1 + \ x_2 \leqslant 8$$
   $$2x_1 + 3x_2 \leqslant 12$$
   $$x_1, x_2 \geqslant 0$$

2. Convert the following maximization problem into a system of equations using slack variables:

   Maximize $P = 8x_1 + 10x_2$
   Subject to $2x_1 + \ x_2 \leqslant 8$
   $$2x_1 + 3x_2 \leqslant 12$$
   $$x_1, x_2 \geqslant 0$$

3. Solve Problem 2 by using the geometric method.
4. Write a simplex tableau for Problem 2 and circle the pivot element.
5. Solve Problem 2 by using the simplex method.
6. Solve the linear programming problem geometrically:

   Minimize $C = 8x_1 + 4x_2$
   Subject to $x_1 \qquad \geqslant 6$
   $$x_1 + x_2 \geqslant 10$$
   $$x_1, x_2 \geqslant 0$$

7. Form the dual of Problem 6.
8. Solve Problem 6 by applying the simplex method to the dual problem.
9. Find the obvious basic solution for each tableau. Determine whether the optimal solution has been reached, additional pivoting is required, or the problem has no solution.

(A)
x_1	x_2	s_1	s_2	P	
4	1	0	0	0	2
2	0	1	1	0	5
−2	0	3	0	1	12

(B)
−1	3	0	1	0	7
0	2	1	0	0	0
−2	1	0	0	1	22

(C)
1	−2	0	4	0	6
0	2	1	6	0	15
0	3	0	2	1	10

Formulate the following problems as linear programming problems but do not solve:

10. South Shore Sail Loft manufactures regular and competition sails. Each regular sail takes 2 hours to cut and 4 hours to sew. Each competition sail takes 3 hours to cut and 9 hours to sew. The Loft makes a profit of $100 on each regular sail and $200 on each competition sail. If there are 150 hours available in the cutting department and 360 hours available in the sewing department, how many sails of each type should the company manufacture in order to maximize their profit?

11. An individual requires 400 units of vitamin B and 800 units of vitamin C daily. The local drugstore carries two types of vitamin tablets, brand X and brand Y. A brand X tablet contains 75 units of vitamin B and 100 units of vitamin C and costs $0.05. A brand Y tablet contains 50 units of vitamin B and 200 units of vitamin C and costs $0.04. How many tablets of each brand should be taken in order to meet the daily vitamin requirements at minimal cost?

Probability

CHAPTER 9 — Contents

9-1 Introduction

Probability, like many branches of mathematics, evolved out of practical considerations. Girolamo Cardano (1501–1576), a gambler and physician, produced some of the best mathematics of his time, including a systematic analysis of gambling problems. In 1654 another gambler, the Chevalier de Méré, plagued with bad luck, approached the well-known French philosopher and mathematician Blaise Pascal (1623–1662) regarding certain dice problems. Pascal became interested in these problems, studied them, and discussed them with Pierre de Fermat (1601–1665), another French mathematician. Thus, out of the gaming rooms of western Europe probability was born.

In spite of this lowly birth, probability has matured into a highly respected and immensely useful branch of mathematics. It is used in practically every field. Probability can be thought of as the science of uncertainty. If, for example, a card is drawn from a deck of 52 cards, it is uncertain which card will be drawn. But suppose a card is drawn and replaced in the deck and a card is again drawn and replaced, and this action is repeated a large number of times. A particular card, say the ace of spades, will be drawn over the long run with a relative frequency that is approximately predictable. Probability theory is concerned with determining the long-run frequency of the occurrence of a given event.

How do we assign probabilities to events? There are two basic approaches to this problem, one theoretical and the other empirical. An example will illustrate the difference between the two approaches.

Suppose you were asked, "What is the probability of obtaining a 2 on a single throw of a die?" Using a *theoretical approach*, we would reason as follows: Since there are six *equally likely* ways the die can turn up (assuming the die is fair) and there is only one way a 2 can turn up, then the probability of obtaining a 2 is $\frac{1}{6}$. Here we have arrived at a probability assignment without even rolling a die once; we have used certain assumptions and a reasoning process.

What does the result have to do with reality? We would expect that in the long run (after rolling a die many times) the 2 would appear approximately $\frac{1}{6}$ of the time. With the *empirical approach*, we make no assumption about the equally likely ways in which the die can turn up. We simply set up an experiment and roll the die a large number of times. Then we compute the percentage of times the 2 appears and use this number as an estimate of the probability of obtaining a 2 on a single roll of the die. Each approach has advantages and drawbacks; these will be discussed in the following sections.

We will first consider the theoretical approach and develop procedures that will lead to the solution of a large variety of interesting problems. These procedures require counting the number of ways certain events can happen, and this is not always easy. However, powerful mathematical tools can assist us in this counting task. The development of these tools is the subject matter of the next two sections.

9-2 The Fundamental Principle of Counting

The best way to start this discussion is with an example.

Example 1   Suppose we spin a spinner that can land on four possible numbers, 1, 2, 3, or 4. We then flip a coin that can turn up either heads (H) or tails (T). What are the possible combined outcomes?

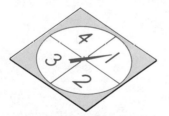

Solution   To solve this problem, let us use a **tree diagram:**

SPINNER OUTCOMES	COIN OUTCOMES	COMBINED OUTCOMES
1	H	(1, H)
	T	(1, T)
2	H	(2, H)
	T	(2, T)
3	H	(3, H)
	T	(3, T)
4	H	(4, H)
	T	(4, T)

Start

Thus, there are eight possible combined outcomes (there are four places the spinner can stop followed by two ways the coin can land).

Problem 1   Use a tree diagram to determine the possible combined outcomes of flipping a coin followed by spinning the dial in Example 1.

Now suppose you asked, "From the twenty-six letters in the alphabet, how many ways can three letters appear in a row on a license plate if no letter is repeated?" To try to count the possibilities using a tree diagram would be extremely tedious, to say the least. The following **fundamental principle of counting** will enable us to solve this problem easily; in addition, it forms the basis for several other counting devices that are developed in the next section:

Fundamental Principle of Counting

1. If two operations O_1 and O_2 are performed in order, with N_1 possible outcomes for the first operation and N_2 possible outcomes for the second operation, then there are

    $$N_1 \cdot N_2$$

    possible combined outcomes of the first operation followed by the second.

2. In general, if n operations $O_1, O_2, \ldots, O_n$ are performed in order, with possible number of outcomes $N_1, N_2, \ldots, N_n$, respectively, then there are

    $$N_1 \cdot N_2 \cdot \cdots \cdot N_n$$

    possible combined outcomes of the operations performed in the given order.

In Example 1, we see that there are four possible outcomes of spinning the dial (first operation) and two possible outcomes of flipping the coin (second operation); hence, by the fundamental principle of counting, there are $4 \cdot 2 = 8$ possible combined outcomes. Use the fundamental principle to solve Problem 1. [*Answer:* $2 \cdot 4 = 8$]

To answer the license plate question: There are twenty-six ways the first letter can be chosen; after a first letter is chosen, there are twenty-five ways a second letter can be chosen; and after two letters are chosen, there are twenty-four ways a third letter can be chosen. Hence, using the fundamental principle of counting, there are $26 \cdot 25 \cdot 24 = 15{,}600$ possible ways three letters can be chosen from the alphabet without repeats.

Example 2 Many colleges and universities are now using computer-assisted testing procedures. Suppose a screening test is to consist of five questions, and a computer stores five comparable questions for the first test question, eight for the second, six for the third, five for the fourth, and ten for the fifth. How many different five-question tests can the computer select? (Two tests are considered different if they differ in one or more questions.)

Solution O_1: Selecting the first question

N_1: 5 ways

O_2: Selecting the second question

N_2: 8 ways

O_3: Selecting the third question

N_3: 6 ways

O_4: Selecting the fourth question

N_4: 5 ways

O_5: Selecting the fifth question

N_5: 10 ways

Thus, the computer can generate

$5 \cdot 8 \cdot 6 \cdot 5 \cdot 10 = 12{,}000$ different tests

Problem 2 Each question on a multiple-choice test has five choices. If there are five such questions on a test, how many different response sheets are possible if only one choice is marked for each question?

Example 3 How many three-letter code words are possible using the first eight letters of the alphabet if:

(A) No letter can be repeated?   (B) Letters can be repeated?
(C) Adjacent letters cannot be alike?

Solutions To form three-letter code words from the eight letters available, we select a letter for the first position, one for the second position, and one for the third position. Altogether, there are three operations.

(A) O_1: Selecting the first letter

N_1: 8 ways

O_2: Selecting the second letter

N_2: 7 ways (since one letter has been used)

O_3: Selecting the third letter

N_3: 6 ways (since two letters have been used)

Thus, there are

$$8 \cdot 7 \cdot 6 = 336 \text{ possible code words}$$

(possible combined operations).

(B) O_1: Selecting the first letter

N_1: 8 ways

O_2: Selecting the second letter

N_2: 8 ways (repeats are allowed)

O_3: Selecting the third letter

N_3: 8 ways (repeats are allowed)

Thus, there are

$$8 \cdot 8 \cdot 8 = 8^3 = 512 \text{ possible code words}$$

(C) O_1: Selecting the first letter

N_1: 8 ways

O_2: Selecting the second letter

N_2: 7 ways (cannot be the same as the first)

O_3: Selecting the third letter

N_3: 7 ways (cannot be the same as the second letter, but can be the same as the first)

Thus, there are

$$8 \cdot 7 \cdot 7 = 392 \text{ possible code words}$$

Problem 3 How many four-letter code words are possible using the first ten letters of the alphabet under the three different conditions stated in Example 3?

Answers to Matched Problems 1.

2. 5^5 or 3,125

3. (A) $10 \cdot 9 \cdot 8 \cdot 7 = 5,040$ (B) $10 \cdot 10 \cdot 10 \cdot 10 = 10,000$

(C) $10 \cdot 9 \cdot 9 \cdot 9 = 7,290$

Exercise 9-2

A   *Solve by using a tree diagram.*

1.   In how many ways can two coins turn up if the combined outcome (H, T) is to be distinguished from (T, H)? [*Hint:*   Think of the problem in terms of two operations: (1) Flipping coin 1 with possible outcomes H or T and (2) flipping coin 2 with possible outcomes H or T.]

2.   How many two-letter "words" can be formed from the first three letters of the alphabet with no letter being used more than once?

3.   A coin is flipped with possible outcomes H or T; then a single die is rolled with possible outcomes 1, 2, 3, 4, 5, or 6. How many combined outcomes are there?

4.   In how many ways can three coins turn up if combined outcomes such as (H, T, H), (H, H, T), and (T, H, H) are to be considered different? [*Hint:*   See Problem 1 above.]

Solve the indicated problem from above by using the fundamental principle of counting.

5.   Problem 1   6.   Problem 2
7.   Problem 3   8.   Problem 4

B   9.   How many ways can two dice turn up if combined outcomes such as (3, 6) and (6, 3) are to be considered different?

10.   In how many ways can two coins and two dice turn up if combined outcomes such as (H, T, 2, 4), (H, T, 4, 2), (T, H, 2, 4), and (T, H, 4, 2) are to be considered different?

11.   In how many ways can a chairperson, a vice chairperson, and a secretary be selected from a committee of ten people? (One person can hold only one office.)

12.   How many ways can five people be seated in a row of five chairs? Ten people?

13.   In a sailboat race, how many different finishes among the first three places are possible for a ten-boat race? (Exclude ties.)

14.   In a long distance foot race, how many different finishes among the first five places are possible for a fifty-person race? (Exclude ties.)

15.   How many four-letter code words are possible from the first six letters of the alphabet with no letter repeated? Allowing letters to repeat?

16.   How many five-letter code words are possible from the first seven letters of the alphabet with no letters repeated? Allowing letters to repeat?

17.   How many different license plates are possible if each contains three letters followed by three digits? How many of these license plates contain no repeated letters and no repeated digits? [*Digits:*   0, 1, 2, 3, 4, 5, 6, 7, 8, 9]

18. How many five-digit ZIP code numbers are possible? How many of these numbers contain no repeated digits?

C **19.** How many three-letter code words are possible out of the alphabet if:

(A) No letter can be used more than once?
(B) Adjacent letters cannot be alike?

20. Each of two countries sends five delegates to a negotiating conference. A rectangular table is used with five chairs on each long side. If each country is assigned one long side of the table (operation 1), how many seating arrangements are possible?

Applications

Business & Economics

21. *Management selection.* A management selection service classifies its applicants (using tests and interviews) as high-IQ, middle-IQ, or low-IQ and as aggressive or passive. How many combined classifications are possible?

(A) Solve by using a tree diagram.
(B) Solve by using the fundamental principle of counting.

22. *Management selection.* A corporation plans to fill two different vice-president positions, V_1 and V_2, from the administrative officers in two of its manufacturing plants. Plant A has six officers and plant B has eight. How many ways can these two positions be filled if the V_1 position is to be filled from plant A and the V_2 position from plant B? If the selection is made without regard to plant?

23. *Product choice.* A particular new car model is available with five choices of color, three choices of transmission, four types of interior, and two types of engine. How many different cars of this model are possible?

24. *Security.* For security purposes, a company classifies each employee according to five hair colors, three eye colors, three weight categories, four height categories, and two sex categories. How many classifications are possible?

Life Sciences

25. *Medicine.* A medical researcher classifies subjects according to male or female, smoker or nonsmoker, and underweight, average weight, or overweight. How many combined classifications are possible?

(A) Solve by using a tree diagram.
(B) Solve by using the fundamental principle of counting.

26. *Family planning.* A couple is planning to have three children. How many boy–girl combinations are possible? Distinguish between combined outcomes such as (B, B, G)), (B, G, B), and (G, B, B).

(A) Solve by using a tree diagram.
(B) Solve by using the fundamental principle of counting.

Social Sciences

27. *Psychology—behavior modification.* In an experiment on the use of the drug Ritalin to modify behavior, a psychologist classified subjects according to four dosage levels of the drug, 0, 1, 2, and 3; male or female; and hyperactive (*H*), normal (*N*), and hypoactive (*L*). How many combined classifications are possible?

(A) Solve by using a tree diagram.

(B) Solve by using the fundamental principle of counting.

9-3 Permutations, Combinations, and Set Partitioning

- Factorial
- Permutations
- Combinations
- Set Partitioning

The fundamental principle of counting studied in the last section can be used to develop three additional devices for counting that are extremely useful in more complicated counting problems. All of these devices use a function called a *factorial function*, which we introduce first.

■ Factorial

In the last section, when using the fundamental principle of counting, we encountered expressions of the form

$$5 \cdot 4 \cdot 3 \cdot 2 \cdot 1 \qquad 50 \cdot 49 \cdot 48 \cdot 47 \cdot 46$$

where each natural number factor is decreased by one as we move from left to right. Forms of this type are encountered with such great frequency in certain types of counting problems that it is useful to express them in a concise notation. The product of the first n natural numbers is called **n factorial** and is denoted by $n!$. Also, we define **zero factorial** to be 1. Symbolically,

---

Factorial

For n a natural number,

$n! = n(n-1)(n-2) \cdot \cdots \cdot 2 \cdot 1$

$0! = 1$

$n! = n \cdot (n-1)!$

[*Note:* $n!$ appears on many hand calculators.]

Example 4   (A) $5! = 5 \cdot 4 \cdot 3 \cdot 2 \cdot 1 = 120$

(B) $\dfrac{7!}{6!} = \dfrac{7 \cdot \cancel{6!}}{\cancel{6!}} = 7$

(C) $\dfrac{8!}{5!} = \dfrac{8 \cdot 7 \cdot 6 \cdot \cancel{5!}}{\cancel{5!}} = 8 \cdot 7 \cdot 6 = 336$

(D) $\dfrac{52!}{5!47!} = \dfrac{52 \cdot 51 \cdot 50 \cdot 49 \cdot 48 \cdot \cancel{47!}}{5 \cdot 4 \cdot 3 \cdot 2 \cdot 1 \cdot \cancel{47!}} = 2{,}598{,}960$

Problem 4   Find: (A) $6!$ (B) $\dfrac{10!}{9!}$ (C) $\dfrac{10!}{7!}$ (D) $\dfrac{5!}{0!3!}$ (E) $\dfrac{20!}{3!17!}$

It is interesting and useful to note that $n!$ grows very rapidly. Compare the following:

$5! = 120$

$10! = 3{,}628{,}800$

$15! = 1{,}307{,}674{,}000{,}000$

Try $69!$, $70!$, and $71!$ on your calculator.

■ Permutations

Suppose four pictures are to be arranged from left to right on one wall of an art gallery. How many arrangements are possible? Using the fundamental principle of counting, there are four ways of selecting the first picture; after the first picture is selected, there are three ways of selecting the second picture; after the first two pictures are selected, there are two ways of selecting the third picture; and after the first three pictures are selected, there is only one way to select the fourth. Thus, the number of arrangements possible for the four pictures is

$$4 \cdot 3 \cdot 2 \cdot 1 = 4! \quad \text{or} \quad 24$$

In general, we refer to a particular arrangement or ordering of n objects as a **permutation** of the n objects. How many orderings (permutations) of n objects are there? From the reasoning above, there are n ways in which the first object can be chosen, there are $n - 1$ ways in which the second object can be chosen, and so on. Using the fundamental principle of counting, we have

---

Permutations of n Objects

Number of permutations of n objects $= n(n - 1) \cdot \cdots \cdot 2 \cdot 1 = n!$

Now suppose the museum director decides to use only two of the four available pictures on the wall arranged from left to right. How many arrangements of two pictures can be formed from the four? There are four ways the first picture can be selected; after selecting the first picture, there are three ways the second picture can be selected. Thus, the number of arrangements of two pictures from four pictures, denoted by $P_{4,2}$, is given by

$$P_{4,2} = 4 \cdot 3$$

or in terms of factorials, multiplying $4 \cdot 3$ by $2!/2!$, we have

$$P_{4,2} = 4 \cdot 3 = \frac{4 \cdot 3 \cdot 2!}{2!} = \frac{4!}{2!}$$

We write this last form for the purposes of generalization. Reasoning in the same way as in the example, we find that the **number of permutations of n objects taken r at a time**, $0 \leq r \leq n$, denoted by $P_{n,r}$, is given by

$$P_{n,r} = n(n-1)(n-2) \cdot \cdots \cdot (n-r+1)$$

Multiplying the right side by 1 in the form $(n-r)!/(n-r)!$, we obtain a factorial form for $P_{n,r}$:

$$P_{n,r} = n(n-1)(n-2) \cdot \cdots \cdot (n-r+1) \frac{(n-r)!}{(n-r)!}$$

But

$$n(n-1)(n-2) \cdot \cdots \cdot (n-r+1)(n-r)! = n!$$

Hence,

Permutation of n Objects Taken r at a Time

The number of permutations of n objects taken r at a time is given by*

$$P_{n,r} = \frac{n!}{(n-r)!} \qquad 0 \leq r \leq n$$

Note: $P_{n,n} = \frac{n!}{(n-n)!} = \frac{n!}{0!} = n!$ Permutations of n objects

---

Example 5   From a committee of eight people, in how many ways can we choose a chairperson and a vice-chairperson, assuming one person cannot hold more than one position?

* In place of the symbol $P_{n,r}$, one will also see P_r^n, $_nP_r$, and $P(n, r)$.

Solution   We are actually asking for the number of permutations of eight objects taken two at a time; that is, $P_{8,2}$:

$$P_{8,2} = \frac{8!}{(8-2)!} = \frac{8!}{6!} = \frac{8 \cdot 7 \cdot 6!}{6!} = 56$$

Problem 5   From a committee of ten people, in how many ways can we choose a chairperson, vice-chairperson, and a secretary, assuming one person cannot hold more than one position?

As we mentioned earlier, many hand calculators have an n! button, and some even have a $P_{n,r}$ button. The use of such a calculator will greatly facilitate many of the calculations in this and the following sections.

Example 6   Find the number of permutations of 25 objects taken 8 at a time. Compute the answer to four significant digits using a calculator.

Solution $P_{25,8} = \dfrac{25!}{(25-8)!} = \dfrac{25!}{17!} = 4.361 \times 10^{10}$

Problem 6   Find the number of permutations of 30 objects taken 4 at a time. Compute the answer exactly using a calculator.

▪ Combinations

Now suppose that an art museum owns eight paintings by a given artist and another art museum wishes to borrow three of these paintings for a special show. How many ways can three paintings be selected out of the eight available? Here the order does not matter. What we are actually interested in is how many three-object subsets can be formed from a set of eight objects. We call such a subset a **combination** of eight objects taken three at a time. The total number of such subsets (combinations) is denoted by the symbol

$$C_{8,3} \quad \text{or} \quad \binom{8}{3}$$

To find the number of combinations of eight objects taken three at a time, $C_{8,3}$, we make use of the formula for $P_{n,r}$ developed above and the fundamental principle of counting. We know that the number of permutations of eight objects taken three at a time is given by $P_{8,3}$, and we have a formula for computing this. Now suppose we think of $P_{8,3}$ in terms of two operations:

O_1:   Selecting a subset of three objects (paintings)

N_1: $C_{8,3}$ ways

O_2:   Arranging the subset in a given order

N_2:   3! ways

The combined operation, O_1 followed by O_2, produces a permutation of eight objects taken three at a time. Thus,

$$P_{8,3} = C_{8,3} \cdot 3!$$

To find $C_{8,3}$, the number of combinations of eight objects taken three at a time, we replace $P_{8,3}$ with $8!/(8-3)!$ and solve for $C_{8,3}$

$$\frac{8!}{(8-3)!} = C_{8,3} \cdot 3!$$

$$C_{8,3} = \frac{8!}{3!(8-3)!} = \frac{8 \cdot 7 \cdot 6 \cdot 5!}{3 \cdot 2 \cdot 1 \cdot 5!} = 56$$

Thus, the museum can make 56 choices in selecting three paintings from the eight available.

In general, reasoning in the same way as in the example, the number of **combinations of n objects taken r at a time**, $0 \leq r \leq n$, denoted by $C_{n,r}$, can be obtained by replacing $P_{n,r}$ with $n!/(n-r)!$ and solving $C_{n,r}$ in the relationship:

$$P_{n,r} = C_{n,r} \cdot r!$$

$$\frac{n!}{(n-r)!} = C_{n,r} \cdot r!$$

$$C_{n,r} = \frac{n!}{r!(n-r)!}$$

In summary,

Combinations of n Objects Taken r at a Time

The number of combinations of n objects taken r at a time is given by*

$$C_{n,r} = \binom{n}{r} = \frac{n!}{r!(n-r)!} \qquad 0 \leq r \leq n$$

If n and r are other than small numbers, a calculator with an $n!$ button will simplify the computation, and one with a $C_{n,r}$ button will simplify the computation even further.

Example 7   From a committee of eight people, in how many ways can we choose a subcommittee of two people?

Solution   Notice how this example differs from Example 4, where we asked in how many ways can we choose a chairperson and a vice-chairperson from a

---

* In place of the symbols $C_{n,r}$ and $\binom{n}{r}$, one will also see C_r^n, $_nC_r$, and $C(n, r)$.

committee of eight people. In Example 4, ordering matters; in choosing a two-person subcommittee the ordering does not matter. Thus, we are actually asking how many combinations of eight objects taken two at a time there are. The number is given by

$$C_{8,2} = \binom{8}{2} = \frac{8!}{2!(8-2)!} = \frac{8 \cdot 7 \cdot \cancel{6!}}{2 \cdot 1 \cdot \cancel{6!}} = 28$$

Problem 7   How many three-person subcommittees can be chosen from a committee of eight people?

Example 8   Find the number of combinations of 25 objects taken 8 at a time. Compute the answer to four significant digits using a calculator.

Solution $$C_{25,8} = \binom{25}{8} = \frac{25!}{8!(25-8)!} = \frac{25!}{8!17!} = 1.082 \times 10^6$$

Compare this result with that obtained in Example 6.

Problem 8   Find the number of combinations of 30 objects taken 4 at a time. Compute the answer exactly using a calculator.

Remember

In a permutation, order counts.

In a combination, order does not count.

Example 9   A company has seven senior and five junior officers. An ad hoc legislative committee is to be formed. In how many ways can a four-officer committee be formed so that it is composed of:

(A)   Any four officers?
(B)   Four senior officers?
(C)   Three senior officers and one junior officer?
(D)   Two senior and two junior officers?
(E)   At least two senior officers?

Solutions   (A)   Since there are a total of 12 officers in the company, the number of different four-member committees is

$$C_{12,4} = \frac{12!}{4!(12-4)!} = \frac{12!}{4!8!} = 495$$

(B)   If only senior officers can be on the committee, the number of different committees is

$$C_{7,4} = \frac{7!}{4!(7-4)!} = \frac{7!}{4!3!} = 35$$

(C) The three senior officers can be selected in $C_{7,3}$ ways, and the one junior officer can be selected in $C_{5,1}$ ways. Applying the fundamental principle of counting, the number of ways that three senior officers and one junior officer can be selected is

$$C_{7,3} \cdot C_{5,1} = \frac{7!}{3!(7-3)!} \cdot \frac{5!}{1!(5-1)!} = \frac{7!5!}{3!4!1!4!} = 175$$

(D) $$C_{7,2} \cdot C_{5,2} = \frac{7!}{2!(7-2)!} \cdot \frac{5!}{2!(5-2)!} = \frac{7!5!}{2!5!2!3!} = 210$$

(E) The committees with at least two senior officers can be divided into three disjoint collections:

1. Committees with four senior and zero junior officers
2. Committees with three senior officers and one junior officer
3. Committees with two senior and two junior officers

The number of committees of types 1, 2, and 3 were computed in parts B, C, and D, respectively. The total number of committees of all three types is the sum of these quantities

Type 1   Type 2   Type 3

$$C_{7,4} + C_{7,3}C_{5,1} + C_{7,2}C_{5,2} = 35 + 175 + 210 = 420$$

Problem 9 Given the information in Example 9, answer the following questions:

(A) How many committees with one senior and three junior officers can be formed?

(B) How many committees with four junior officers can be formed?

(C) How many committees with at least two junior officers can be formed?

■ Set Partitioning

The combination of n objects taken r at a time can be thought of in another way, a way that generalizes into something very useful. Let us return to the art museum that agreed to lend to another museum three paintings from the eight they own by a given artist. In choosing the three, they are actually **partitioning** (dividing) the set of eight paintings into two subsets: the subset containing the three paintings to be loaned and the subset containing the five paintings that will remain. Now let us suppose that in addition to the one museum wishing to borrow three paintings, a second museum wishes to borrow two paintings by the same artist. The museum owning the eight paintings is now confronted with the following problem: How to partition (divide) the set of eight paintings into three subsets, one containing three paintings, one containing two paintings, and one containing three paintings (the three left over). We now ask, "In how many ways can a set of eight objects be partitioned into three subsets with the first containing three objects, the second containing two objects, and the third containing three objects?" Again we call on the fundamental principle of counting. Think of the problem in terms of the following three operations:

O_1:   Select a subset with three paintings from eight paintings — $C_{8,3}$ ways

O_2:   Select a subset with two paintings from the five remaining paintings — $C_{5,2}$ ways

O_3:   Select a subset with three paintings from the three remaining paintings — $C_{3,3}$ ways

The combined operations O_1, O_2, and O_3 produce a *partition* of the set of eight paintings into three subsets as desired. Thus, the number of such partitions, denoted by $\begin{pmatrix} 8 \\ 3, 2, 3 \end{pmatrix}$, is given by

$$\begin{pmatrix} 8 \\ 3, 2, 3 \end{pmatrix} = C_{8,3} \cdot C_{5,2} \cdot C_{3,3}$$

$$= \frac{8!}{3!(8-3)!} \cdot \frac{5!}{2!(5-2)!} \cdot \frac{3!}{3!(3-3)!}$$

$$= \frac{8!}{3!5!} \cdot \frac{5!}{2!3!} \cdot \frac{3!}{3!0!}$$

$$= \frac{8!}{3!2!3!} \qquad \text{Compare this arrangement with the arrangement } \begin{pmatrix} 8 \\ 3, 2, 3 \end{pmatrix}$$

$$= 560$$

In general, to partition a set of n elements into k subsets such that r_1 elements are in the first subset, r_2 elements are in the second subset, . . . , r_k elements are in the kth subset, $r_1 + r_2 + \cdots + r_k = n$, we can think of the problem in terms of k operations as above and apply the fundamental principle of counting to obtain

$$\begin{pmatrix} n \\ r_1, r_2, r_3, \ldots, r_k \end{pmatrix} = C_{n,r_1} \cdot C_{n-r_1,r_2} \cdot C_{n-r_1-r_2,r_3} \cdots \cdots C_{n-r_1-r_2-\cdots-r_{k-1},r_k}$$

$$= \frac{n!}{r_1!(n-r_1)!} \cdot \frac{(n-r_1)!}{r_2!(n-r_1-r_2)!} \cdot \frac{(n-r_1-r_2)!}{r_3!(n-r_1-r_2-r_3)!}$$

$$\cdots \cdots \frac{(n-r_1-r_2-\cdots-r_{k-1})!}{r_k!(n-r_1-r_2-\cdots-r_k)!}$$

$$= \frac{n!}{r_1!r_2!\cdots\cdots r_k!} \qquad \text{Note: } (n-r_1-r_2-\cdots-r_k)! = 0! = 1$$

In summary,

Partition of *n* Elements into *k* Subsets

The number of partitions of a set with n elements into k subsets is given by

$$\binom{n}{r_1, r_2, \ldots, r_k} = \frac{n!}{r_1! r_2! \cdot \cdots \cdot r_k!}$$

where

r_i elements are in the ith subset

and

$$r_1 + r_2 + \cdots + r_k = n$$

Note: $\binom{n}{r_1, r_2} = C_{n,r_1} = C_{n,r_2}$ if $r_1 + r_2 = n$

Example 10

If four people are playing poker, how many deals are possible if each person receives five cards?

Solution

This is a partition problem. We are actually dividing (partitioning) the deck (set of 52 cards) into five subsets: four subsets correspond to the hands for four players, and the fifth subset is what is left over after dealing the four hands. We use a calculator to compute the following to four significant digits.

$$\binom{52}{5, 5, 5, 5, 32} = \frac{52!}{5!5!5!5!32!}$$

$$= \frac{52!}{(5!)^4 32!} \approx 1.478 \times 10^{24} \text{ deals}$$

Problem 10

If three people are playing cards and each is dealt 7 cards from a 52 card deck, how many deals are possible? Compute the answer to four significant digits using a calculator.

Answers to Matched Problems

4. (A) 720 (B) 10 (C) 720 (D) 20 (E) 1,140

5. $P_{10,3} = \dfrac{10!}{(10-3)!} = 720$ 6. $P_{30,4} = \dfrac{30!}{(30-4)!} = 657,720$

7. $C_{8,3} = \dfrac{8!}{3!(8-3)!} = 56$ 8. $C_{30,4} = \dfrac{30!}{4!(30-4)!} = 27,405$

9. (A) $C_{7,1}C_{5,3} = 70$ (B) $C_{5,4} = 5$
   (C) $C_{7,2}C_{5,2} + C_{7,1}C_{5,3} + C_{5,4} = 285$

10. $\dbinom{52}{7, 7, 7, 31} = \dfrac{52!}{7!7!7!31!} \approx 7.662 \times 10^{22}$

Exercise 9-3

A   *Evaluate.*

1. $4!$ 2. $6!$ 3. $\dfrac{9!}{8!}$ 4. $\dfrac{14!}{13!}$

5. $\dfrac{11!}{8!}$ 6. $\dfrac{14!}{12!}$ 7. $\dfrac{5!}{2!3!}$ 8. $\dfrac{6!}{4!2!}$

9. $\dfrac{7!}{4!(7-4)!}$ 10. $\dfrac{8!}{3!(8-3)!}$ 11. $\dfrac{7!}{7!(7-7)!}$ 12. $\dfrac{8!}{0!(8-0)!}$

13. $P_{5,3}$ 14. $P_{4,2}$ 15. $P_{52,4}$ 16. $P_{52,2}$

17. $C_{5,3}$ 18. $C_{4,2}$ 19. $C_{52,4}$ 20. $C_{52,2}$

21. $\dbinom{8}{5, 3}$ 22. $\dbinom{7}{2, 5}$ 23. $\dbinom{12}{2, 7, 3}$ 24. $\dbinom{14}{6, 3, 3, 2}$

Solve using permutation, combination, or partitioning formulas wherever possible.

25. A small combination lock on a suitcase has ten positions labeled with the ten digits 0, 1, 2, 3, 4, 5, 6, 7, 8, and 9. How many three-number opening combinations are possible, assuming no digit is used more than once?

26. A bookshelf has space for three books. Out of six different books available, how many arrangements can be made on the shelf?

27. There are ten teams in a conference. If each team is to play every other team exactly once, how many games must be scheduled?

28. Given seven points, no three of which are on a straight line, how many lines can be drawn joining two points at a time?

29. In how many different ways can six candidates for an office be listed on a ballot?

30. An ice cream parlor has 25 different flavors of ice cream. How many different two-scoop cones can they offer if 2 different flavors are to be used and the order of the scoops does not matter?

B   31. From a standard 52 card deck, how many 5 card hands will have all hearts?

32. From a standard 52 card deck, how many 5 card hands will have all face cards, including aces? Not including aces?

33. Seven paintings are left by a wealthy collector to three museums. If three are to go to one museum and two each to the other two, how many ways can they be distributed?

34. Thirteen rare books are left to two universities. If eight are to go to one and five to the other, how many ways can they be distributed?

35. A catering service offers eight appetizers, ten main courses, and seven desserts. A banquet chairperson is to select three appetizers, four main courses, and two desserts for a banquet. How many ways can this be done?

36. Three departments have 12, 15, and 18 members, respectively. If each department is to select a delegate and an alternate to represent the department at a conference, how many ways can this be done?

37. From a standard 52 card deck, how many 7 card hands have exactly 5 spades and 2 hearts?

38. How many 5 card hands will have 2 clubs and 3 hearts?

C   39. (A) How many 13 card bridge hands are possible from a standard 52 card deck?

   (B) If four people are playing cards and each is dealt 13 cards, how many different deals are possible from a 52 card deck?

40. (A) How many 7 card hands are possible from a standard 52 card deck?

   (B) If 5 people are playing cards and each is dealt 7 cards, how many different deals are possible from a 52 card deck?

Applications

Business & Economics

41. *Consumer testing.* From six known brands of cola, three are chosen at random for a consumer to identify. Assuming that the consumer guesses blindly, how many responses are possible?

42. *Contests.* In how many ways can ten finalists finish first, second, and third in a promotion contest?

43. *Personnel selection.* Six female and five male applicants have been successfully screened for five positions. In how many ways can the following compositions be selected?

   (A) Three females and two males
   (B) Four females and one male
   (C) Five females
   (D) Five people regardles of sex
   (E) At least four females

44. *Committee selection.* A 4-person grievance committee is to be selected

out of two departments, *A* and *B*, with 15 and 20 people, respectively. In how many ways can the following committees be selected?

(A) Three from *A* and one from *B*
(B) Two from *A* and two from *B*
(C) All from *A*
(D) Four people regardless of department
(E) At least three from department *A*

45. *Management.*

   (A) In how many ways can 4 accounts be assigned to four salespeople so that each receives 1 account?
   (B) Out of 6 accounts available, how many ways can 4 be selected and assigned to four salespeople so that each receives 1 account?
   (C) Out of 6 accounts, how many ways can 4 be selected and assigned to one salesperson?
   (D) In how many ways can 12 accounts be assigned to three salespeople, with 4 to the first, 3 to the second, and 5 to the third?

46. *Management.* A company has just completed a new office building and must make some decisions regarding office assignments.

   (A) In how many ways can 5 vice presidents be assigned to five offices?
   (B) In how many ways can 12 secretaries be assigned to three offices, 3 in the first, 4 in the second, and 5 in the third?
   (C) How many ways can 4 managers be assigned to two offices, two in each?

Life Sciences

47. *Medicine.* A prospective laboratory technician is given a test to identify blood types from eight standard classifications. If three different types are chosen at random for the identification test, how many responses are possible if the candidate guesses blindly?

48. *Medical research.* Because of limited funds, five research centers are to be chosen out of eight suitable ones for a study on heart disease. How many choices are possible?

Social Sciences

49. *Politics.* A nominating convention is to select a president and a vice-president from among four candidates. Campaign buttons, listing a president and a vice-president, are to be designed for each possible outcome before the convention. How many different kinds of buttons should be designed?

50. *Survey.* Twelve regions are to be divided among three trained surveyors, with five to the first, three to the second, and four to the third. In how many ways can this be done?

9-4 Experiments, Sample Spaces, and Probability of an Event

- Experiments
- Sample Spaces
- Events
- Probability of an Event
- Equally Likely Assumption

■ Experiments

Certain experiments in science produce the same results when performed repeatedly under exactly the same conditions. For example, when all other conditions are held constant, a given liquid will always freeze at the same temperature. Experiments of this type are called **deterministic** — the conditions of the experiment determine the outcome. There are also experiments that do not yield the same results no matter how carefully they are repeated under the same conditions. These experiments are called **random experiments.** Familiar examples of the latter are flipping coins, rolling dice, observing the sex of a newborn child, or observing the frequency of death in a certain age group. Probability theory is a branch of mathematics that has been developed to deal with outcomes of random experiments, both real and conceptual. In the work that follows, the word *experiment* will be used to mean a random experiment.

■ Sample Spaces

Consider the experiment "A child is born." What can we observe about this child? We might be interested in the day of the week the birth takes place; the sex of the child; the child's weight, height, eye color, or blood type; and so on. The list of possible outcomes of the experiment appears to be endless. In general, there is no unique method of analyzing all possible outcomes of an experiment. Therefore, before conducting an experiment, it is important to decide just what outcomes are of interest.

In the birth experiment, suppose we limit our interest to questions concerning the day of the week on which a birth takes place. Having decided what to observe, we make a list of outcomes of the experiment such that on each trial of the experiment one and only one of the results on the list will occur. Thus,

$$U = \{M, T, W, Th, F, S, Su\}$$

is an appropriate list for our interests (M represents the outcome "The child was born on Monday," and so on). Note that each birth will correspond to exactly one element in U. The set of outcomes U is called a *sample space* for the experiment. In general,

Sample Space

A set S is a **sample space** for an experiment (real or conceptual) if:

1. Each element of S is an outcome of the experiment.
2. Each outcome of the experiment corresponds to one and only one element of S.

Each element in the sample space is called a **sample point** or a **simple outcome.**

Notice that we did not include the outcome "The child was born on a weekend" in set U, since this outcome would occur if either S or Su occurs, violating the condition that one and only one of the outcomes in the sample space occurs on a given trial. The outcome "The child was born on a weekend" is called a *compound outcome.* In general, C is a **compound outcome** relative to a sample space S if there exist at least two simple outcomes in S that imply the occurrence of C. The outcome "The child was born on a weekday" is a compound outcome relative to the sample space U, since this outcome will occur if any of the simple outcomes in the set {M, T, W, Th, F} occurs. Of course, none of the outcomes in a sample space are compound outcomes relative to that space; that is why they are called simple outcomes.

Suppose we are interested in the sex of each child as well as the day of the week of the birth. Then we must refine the sample space U given above. A suitable new sample space for the experiment "A child is born," reflecting our additional interest, is

$$V = \{M-m, M-f, \dots , Su-m, Su-f\}$$

where M−m is the outcome "A male is born on Monday," and so on. Each simple outcome in the original sample space U is now a compound outcome in the new sample space V; that is, we will know that M has occurred if we know that either M−m or M−f has occurred.

Important Remark

There is no one correct sample space for a given experiment. We do require, however, that a set of outcomes satisfies both conditions in the definition above before it can be called a sample space. When specifying a sample space for an experiment, we include as much detail as is necessary to answer *all* questions of interest regarding the outcomes of the experiment. If in doubt, include more sample points rather than less.

Example 11   A nickel and a dime are tossed. How shall we identify a sample space for this experiment? There are a number of possibilities, depending on our interest. We shall consider three.

(A)   If we are interested in whether each coin falls heads (H) or tails (T), then, using a tree diagram, we can easily determine an appropriate sample space for the experiment:

NICKEL	DIME	COMBINED
OUTCOMES	OUTCOMES	OUTCOMES

Thus,

$$S_1 = \{HH, HT, TH, TT\}$$

and there are four sample points in the sample space.

(B)   If we are only interested in the number of heads that appear on a single toss of the two coins, then we can let

$$S_2 = \{0, 1, 2\}$$

and there are three sample points in the sample space.

(C)   If we are interested in whether the coins match (M) or do not match (D), then we can let

$$S_3 = \{M, D\}$$

and there are only two sample points in the sample space.

In Example 11, sample space S_1 contains more information than either S_2 or S_3. If we know which outcome has occurred in S_1, then we know which outcome has occurred in S_2 and S_3. However, the reverse is not true. In this sense, we say that **S_1 is a more fundamental sample space than either S_2 or S_3.**

Problem 11   An experiment consists of recording the boy–girl composition of two-child families.

(A)   What is an appropriate sample space if we are interested in the sex of each child in the order of their births? Draw a tree diagram.

(B)   What is an appropriate sample space if we are only interested in the number of girls in a family?

(C)   What is an appropriate sample space if we are only interested in whether the sexes are alike (A) or different (D)?

(D)   What is an appropriate sample space for all three interests expressed above?

A sample space may be **finite** or **infinite.** For example, a sample space for a single roll of a die might be

$$S = \{1, 2, 3, 4, 5, 6\}$$

This is a finite sample space, since there are only a finite number of outcomes of the experiment.

On the other hand, if we are interested in the number of rolls it takes for the die to turn up 5 for the first time, then an appropriate sample space would be the set of natural numbers

$$N = \{1, 2, 3, 4, \ldots \}$$

which is infinite. In this book, unless stated to the contrary, we will restrict our attention to finite sample spaces.

■ Events

We have now completed a first step in constructing a mathematical model for probability studies. That is, we have introduced a sample space, a set of sample points, as the mathematical counterpart of an experiment. The sample space becomes the universal set for all discussion pertaining to the experiment.

Now let us return to the two-coin problem in Example 11 and the sample space

$$S_1 = \{HH, HT, TH, TT\}$$

Suppose we are interested in the compound outcome "Exactly one head is up." Looking at S_1, we find that it will occur if either of the two simple outcomes HT or TH occurs. Thus, to say that the compound outcome "Exactly one head is up" occurs is the same as saying the experiment has an outcome in the set

$$E = \{HT, TH\}$$

which is a subset of the sample space S_1. We will call the subset E an *event.*

Event

In general, given a sample space S for an experiment, we define an **event E** to be any subset of S. We say that **an event E occurs** if any of the simple outcomes in E occurs. If an event E has only one element in it, it is called a **simple event;** if it has more than one element, it is called a **compound event.**

A second step in constructing a mathematical model for probability studies has now been completed by introducing an event as a subset of a sample space. In the coin example above, the event E, a subset of S_1, is the mathematical counterpart of the experimental outcome "Exactly one head is up" in the toss of a nickel and a dime.

Let us now consider a more complex experiment.

Example 12

Consider an experiment of rolling two dice. A convenient sample space that will enable us to answer many questions about interesting events is shown in Figure 1. Let S be the set of all ordered pairs in the table. The sample point $(3, 2)$ is to be distinguished from the sample point $(2, 3)$; the former indicates a 3 turned up on the first die and a 2 on the second, while the latter indicates that a 2 turned up on the first die and a 3 on the second.

SECOND DIE

(1, 1)	(1, 2)	(1, 3)	(1, 4)	(1, 5)	(1, 6)
(2, 1)	(2, 2)	(2, 3)	(2, 4)	(2, 5)	(2, 6)
(3, 1)	(3, 2)	(3, 3)	(3, 4)	(3, 5)	(3, 6)
(4, 1)	(4, 2)	(4, 3)	(4, 4)	(4, 5)	(4, 6)
(5, 1)	(5, 2)	(5, 3)	(5, 4)	(5, 5)	(5, 6)
(6, 1)	(6, 2)	(6, 3)	(6, 4)	(6, 5)	(6, 6)

FIRST DIE

Figure 1

Write down the event (subset of the sample space S) that corresponds to each of the following outcomes:

(A) A 7 turns up. (B) An 11 turns up.
(C) A sum less than 4 turns up. (D) A 12 turns up.

Solutions

(A) By "A 7 turns up," we mean that the sum of all dots on both turned-up faces is 7. This compound outcome corresponds to the event:

$$\{(6, 1), (5, 2), (4, 3), (3, 4), (5, 2), (1, 6)\}$$

Notice that there are 6 ways in which a 7 can be obtained out of the 36 simple outcomes in the sample space.

(B) "An 11 turns up" corresponds to the event

$$\{(6, 5), (5, 6)\}$$

(C) "A sum less than 4 turns up" corresponds to the event

$$\{(1, 1), (2, 1), (1, 2)\}$$

(D) "A 12 turns up" corresponds to the event

{(6, 6)}

which is a simple event.

Problem 12 Using the sample space in Example 12 (Figure 1), write down the events corresponding to the following outcomes:

(A) A 5 turns up.   (B) A prime number* greater than 7 turns up.

Informal Use of the Word *Event*

Informally, to facilitate discussion, we will often use *event* and *outcome of an experiment* interchangeably. Thus, in Example 12 we might use "the event 'An 11 turns up'" in place of "the outcome 'An 11 turns up,'" or even write

$E =$ An 11 turns up $= \{(6, 5), (5, 6)\}$

Technically speaking, as we said earlier, an event is the mathematical counterpart of an outcome of an experiment. Formally, we have

Real World	Mathematical Model
Experiment (real or conceptual)	Sample space (set S)
Outcome (simple or compound)	Event (subset of S) (simple or compound)

■ Probability of an Event

The next step in developing our mathematical model for probability studies is the introduction of a *probability function*. This is a function that assigns to an arbitrary event associated with a sample space a real number between 0 and 1, inclusive. Since an arbitrary event relative to a sample space S can be thought of as the union of simple events in S, we start by discussing ways in which probabilities are assigned to simple events in S. We will then use these results as building blocks in assigning probabilities for compound events.

---

* Recall that a *prime number* is a natural number greater than 1 that cannot be divided by any natural number other than itself or 1.

> **Probabilities for Simple Events**
>
> Given a sample space
>
> $$S = \{e_1, e_2, \ldots, e_n\}$$
>
> to each simple event* e_i we assign a real number, denoted by $P(e_i)$, that is called the **probability of the event e_i.** These numbers can be assigned in an arbitrary manner as long as the following two conditions are satisfied:
>
> 1. If e_i is a simple event, then $0 \leqslant P(e_i) \leqslant 1$.
> 2. $P(e_1) + P(e_2) + \cdots + P(e_n) = 1$; that is, the sum of the probabilities of all simple events in the sample space is 1.
>
> Any probability assignment that meets conditions 1 and 2 is said to be an **acceptable probability assignment.**

How specific acceptable probabilities are assigned to simple events is a question our mathematical theory does not answer. These assignments, however, are generally based on expected or actual long-run relative frequencies of the occurrences of the various simple events for a given experiment.

Example 13 Let an experiment be the flipping of a single coin, and let us choose a sample space S to be

$$S = \{H, T\}$$

Psychologically, if a coin appears to be "fair," we are inclined to assign probabilities to the simple events in S as follows:

$$P(H) = \tfrac{1}{2} \quad \text{and} \quad P(T) = \tfrac{1}{2}$$

thinking (since there are two ways a coin can land) that in the long run a head will turn up half the time and a tail will turn up half the time. These probability assignments are acceptable, since both conditions for acceptable probability assignments are satisfied:

1. $0 \leqslant P(H) \leqslant 1, \quad 0 \leqslant P(T) \leqslant 1$
2. $P(H) + P(T) = \tfrac{1}{2} + \tfrac{1}{2} = 1$

But there are other acceptable assignments. Maybe after flipping a coin 1,000 times, we find that the head turns up 376 times and the tail 624 times.

* Technically, we should write $\{e_i\}$, since there is a logical distinction between an element of a set and a subset consisting only of that element. But we will just keep this in mind and drop the braces for simple events to simplify the notation.

With this result, we might suspect that the coin is not fair and assign simple events in S the probabilities

$$P(H) = .376 \quad \text{and} \quad P(T) = .624$$

This is also an acceptable assignment. Which of the following are acceptable assignments?

(A) $P(H) = \frac{7}{8}$ and $P(T) = \frac{1}{8}$
(B) $P(H) = 1$ and $P(T) = 0$
(C) $P(H) = .6$ and $P(T) = .8$

Assignments A and B are acceptable, but C is not. The latter has a sum of 1.4, which violates condition 2.

It is important to keep in mind that out of the infinitely many possible acceptable probability assignments to simple events in a sample space, we are generally inclined to choose one assignment over another based on feelings, reasoning, or experimental results. In Example 13, we would probably choose

$$P(H) = .376 \quad \text{and} \quad P(T) = .624$$

if, in 1,000 tosses of a coin, a head turned up 376 times and a tail turned up 624 times. On the other hand, if we just pull a coin out of a pocket and flip it to see who pays for lunch, then we would probably be happy with

$$P(H) = \tfrac{1}{2} \quad \text{and} \quad P(T) = \tfrac{1}{2}$$

In neither case would we be happy with

$$P(H) = 1 \quad \text{and} \quad P(T) = 0$$

even though it is also an acceptable assignment. In short, the word *acceptable* has nothing to do with suitability. Mathematical probability theory simply requires the assignment to be acceptable.

Problem 13   A blank six-sided die is marked with a 1 on two sides, a 2 on two sides, and a 3 on the two remaining sides. If we choose a sample space for a single roll of the die to be

$$S = \{1, 2, 3\}$$

which of the following probability assignments to the simple events in S are acceptable?

(A) $P(1) = 1, \quad P(2) = -1, \quad P(3) = 1$
(B) $P(1) = \tfrac{1}{3}, \quad P(2) = \tfrac{2}{3}, \quad P(3) = 0$
(C) $P(1) = \tfrac{1}{3}, \quad P(2) = \tfrac{1}{3}, \quad P(3) = \tfrac{1}{3}$
(D) $P(1) = .35, \quad P(2) = .32, \quad P(3) = .33$

Given an acceptable probability assignment for simple events in a sample space S, how do we define the probability of an arbitrary event associated with S?

Probability of an Event E

Given an acceptable probability assignment for the simple events in a sample space S, we define the **probability of an arbitrary event E,** denoted by $P(E)$, as follows:

1. If E is the empty set, then $P(E) = 0$.
2. If E is a simple event, then $P(E)$ has already been assigned.
3. If E is the union of two or more simple events, then $P(E)$ is the sum of the probabilities of the simple events whose union is E.
4. If $E = S$, then $P(E) = P(S) = 1$ (this is a special case of 3).

Example 14   Let us return to Example 11, the tossing of a nickel and dime, with sample space

$S = \{HH, HT, TH, TT\}$

Since there are four simple outcomes and the coins are assumed to be fair, it would appear that each outcome would occur in the long run 25% of the time. Let us assign the same probability of $\frac{1}{4}$ to each simple event in S:

Simple Event e_i	HH	HT	TH	TT
$P(e_i)$	$\frac{1}{4}$	$\frac{1}{4}$	$\frac{1}{4}$	$\frac{1}{4}$

This is an acceptable assignment and is a reasonable assignment for ideal coins or coins close to ideal (perfectly balanced).

(A) What is the probability of getting one head (and one tail)?
(B) What is the probability of getting at least one head?
(C) What is the probability of getting three heads?

Solutions   (A) $E_1 = $ Getting one head

$= \{HT, TH\} = \{HT\} \cup \{TH\}$

$P(E_1) = P(HT) + P(TH) = \frac{1}{4} + \frac{1}{4} = \frac{1}{2}$

(B) $E_2 = $ Getting at least one head

$= \{HH, HT, TH\} = \{HH\} \cup \{HT\} \cup \{TH\}$

$P(E_2) = P(HH) + P(HT) + P(TH)$

$= \frac{1}{4} + \frac{1}{4} + \frac{1}{4} = \frac{3}{4}$

(C) E_3 = Getting three heads = $\varnothing$

   $P(\varnothing) = 0$

Let us summarize the key steps for finding probabilities of events:

Steps for Finding Probabilities of Events

1. Set up an appropriate sample space S for the experiment.
2. Assign acceptable probabilities to the simple events of S.
3. To obtain the probability of an arbitrary event E, add the probabilities of the simple events whose union is E.
4. If it is easier to find $P(E')$, then we can use $P(E) = 1 - P(E')$. (More will be said about this later in the chapter.) The event E' is the complement of E relative to S.

The function P defined in steps 2 and 3 is called a **probability function** with domain all possible events (subsets) in the sample space S and range a set of real numbers between 0 and 1, inclusive.

Problem 14   Suppose in Example 14 after flipping the nickel and dime 1,000 times, we find that HH turns up 273 times, HT 206 times, TH 312 times, and TT 209 times. On the basis of this evidence, we assign probabilities to simple events in S as follows:

Simple Event e_i	HH	HT	TH	TT
$P(e_i)$	.273	.206	.312	.209

This is an acceptable probability assignment. What are the probabilities of the following events?

(A) E_1 = Getting at least one tail   (B) E_2 = Getting two tails
(C) E_3 = Getting either a head or a tail

Example 14 and Problem 14 illustrate two important ways in which acceptable probability assignments are made for simple events in a sample space S.

1. *Theoretical.* The internal structure of an experiment is analyzed and assignments are made through a deductive process by using certain basic assumptions. (No coins are flipped or dice rolled ahead of time.) These assignments are used as an approximation of the actual probabilities. This is what we did in Example 14 above.
2. *Empirical.* Nothing is assumed about the internal structure of the

experiment; instead, we take a sample of all possible outcomes and use the relative frequency (percentage) of each simple event in the total sample as approximations of the actual probabilities.* This is what we did in Problem 14 above.

Each approach has its advantages in certain situations. For the rest of this section, we will emphasize the theoretical approach. In the next section we will consider the empirical approach in more detail.

■ Equally Likely Assumption

In tossing a nickel and dime (Example 14), we assigned the same probability, $\frac{1}{4}$, to each simple event in the sample space

S = {HH, HT, TH, TT}

By assigning the same probability to each simple event in S, we are actually making the assumption that each simple event is as likely to occur as any other. We refer to this as an **equally likely assumption.**

In general,

Probability of a Simple Event (Under Equally Likely Assumption)

If, in a sample space

$S = \{e_1, e_2, \ldots, e_n\}$

with n elements, we assume each simple event is as likely to occur as any other, then we assign the probability $1/n$ to each; that is,

$$P(e_i) = \frac{1}{n}$$

Under the equally likely assumption, we can develop a very useful formula for finding probabilities of arbitrary events associated with S. Consider the following example.

If a single die is rolled and we assume each face is as likely to come up as any other, then for the sample space

S = {1, 2, 3, 4, 5, 6}

we assign $\frac{1}{6}$ to each simple event, since there are six simple events. The

---

* The **actual probability of an event** is generally defined as the single fixed number (if it exists) that the relative frequency of the occurrence of the event approaches as an experiment is repeated without end.

probability of

E = Rolling a prime number

$= \{2, 3, 5\} = \{2\} \cup \{3\} \cup \{5\}$

is

$P(E) = P(2) + P(3) + P(5)$

$= \frac{1}{6} + \frac{1}{6} + \frac{1}{6} = \frac{3}{6} = \frac{1}{2}$

Notice the following:

$$\frac{3}{6} = \frac{\text{Number of elements in } E}{\text{Number of elements in } S}$$

Thus, under the assumption that each simple event is as likely to occur as any other, the computation of the probability of the occurrence of any event E in S is relatively easy. We simply count the number of elements in E and divide by the number of elements in the sample space S:

Probability of an Arbitrary Event E (Under Equally Likely Assumption)

If we assume each simple event in S is as likely to occur as any other, then the probability of an arbitrary event E in S is given by

$$P(E) = \frac{\text{Number of elements in } E}{\text{Number of elements in } S} = \frac{n(E)}{n(S)}$$

Example 15 If in rolling two dice we assume each simple event in the sample space shown in Figure 1 (page 501) is as likely as any other, find the probabilities of the following events (each event refers to the sum of the dots facing up on both dice):

(A) E_1 = A 7 turns up (B) E_2 = An 11 turns up
(C) E_3 = A sum less than 4 turns up (D) E_4 = A 12 turns up

Solutions Referring to Figure 1 (page 501), we see that

(A) $P(E_1) = \dfrac{n(E_1)}{n(S)} = \dfrac{6}{36} = \dfrac{1}{6}$

(B) $P(E_2) = \dfrac{n(E_2)}{n(S)} = \dfrac{2}{36} = \dfrac{1}{18}$

(C) $P(E_3) = \dfrac{n(E_3)}{n(S)} = \dfrac{3}{36} = \dfrac{1}{12}$

(D) $P(E_4) = \dfrac{n(E_4)}{n(S)} = \dfrac{1}{36}$

Problem 15 Under the conditions in Example 15, find the probabilities of the following events (each event refers to the sum of the dots facing up on both dice):

(A) $E_5 = $ A 5 turns up
(B) $E_6 = $ A prime number greater than 7 turns up

Example 16 The following questions pertain to the composition of a three-child family excluding multiple births.

(A) Under the assumption that a girl is as likely as a boy at each birth, select a sample space S such that all simple events can be assumed equally likely to occur.
(B) What is the probability of having three girls?
(C) What is the probability of having two boys and a girl in that order?
(D) What is the probability of having two boys and a girl in any order?

Solutions (A) A tree diagram is helpful in selecting a sample space S:

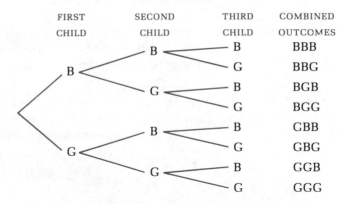

FIRST CHILD	SECOND CHILD	THIRD CHILD	COMBINED OUTCOMES
B	B	B	BBB
	B	G	BBG
	G	B	BGB
	G	G	BGG
G	B	B	GBB
	B	G	GBG
	G	B	GGB
	G	G	GGG

Under the assumption that a boy is as likely as a girl at each birth, each branch at the end is as likely as any other; hence, each combined outcome is as likely as any other. Thus, we let

$S = \{$BBB, BBG, BGB, BGG, GBB, GBG, GGB, GGG$\}$

(B) The event of having three girls is the simple event

$E = \{$GGG$\}$

Thus,

$$P(E) = \frac{n(E)}{n(S)} = \frac{1}{8}$$

(C) The event of having two boys and a girl in that order is the simple event

$E = \{$BBG$\}$

Thus,

$$P(E) = \frac{n(E)}{n(S)} = \frac{1}{8}$$

(D)   The event of having two boys and a girl in any order is

$$E = \{BBG, BGB, GBB\}$$

Thus,

$$P(E) = \frac{n(E)}{n(S)} = \frac{3}{8}$$

Problem 16   Using the sample space in Example 16, find the probability of having:

(A)   Three boys   (B)   At least two girls

We now turn to some examples that make use of the counting techniques developed in the last section.

Example 17   In drawing 5 cards from a 52 card deck without replacement (without replacing a drawn card before selecting the next card), what is the probability of getting five spades?

Solution   Let the sample space S be the set of all 5 card hands from a 52 card deck. Since the order in a hand does not matter, $n(S) = C_{52,5}$. The event $E =$ the set of all 5 card hands from 13 spades. Again, the order does not matter and $n(E) = C_{13,5}$. Thus, assuming each 5 card hand is as likely as any other,

$$P(E) = \frac{n(E)}{n(S)} = \frac{C_{13,5}}{C_{52,5}} = \frac{13!/(5!8!)}{52!/(5!47!)} = \frac{13!}{5!8!} \cdot \frac{5!47!}{52!} \approx .0005$$

Problem 17   In drawing 7 cards from a 52 card deck without replacement, what is the probability of getting seven hearts?

Example 18   The board of regents of a university is made up of 12 men and 16 women. If a committee of 6 is chosen at random, what is the probability that it will contain 3 men and 3 women?

Solution   Let $S =$ the set of all six-person committees out of 28 people:

$$n(S) = C_{28,6}$$

$E =$ the set of all six-person committees with 3 men and 3 women. Using the fundamental principle of counting,

$$n(E) = C_{12,3} C_{16,3}$$

Thus,

$$P(E) = \frac{n(E)}{n(S)} = \frac{C_{12,3} C_{16,3}}{C_{28,6}} \approx .327$$

Problem 18 What is the probability that the committee in Example 18 will have four men and two women?

Example 19 Four people are playing poker. In a single deal of 5 cards each from a standard 52 card deck, what is the probability that each player has a flush in a different suit? (A flush is 5 cards of the same suit.)

Solution Let the sample space S be the set of all four-person deals of 5 cards each from a 52 card deck. Using the partition formula (see Example 10 in Section 9-3), we find the number of elements in S is given by

$$n(S) = \begin{pmatrix} 52 \\ 5, 5, 5, 5, 32 \end{pmatrix}$$

We assume each deal is as likely as any other. The event E is the set of all four-person deals of 5 cards each from a 52 card deck such that each hand is a flush in a different suit. To find the number of elements in E, we utilize the fundamental principle of counting as follows:

O_1: Selecting a suit for the first hand—4 ways

O_2: Selecting 5 cards out of 13 possible in the suit—$C_{13,5}$ ways

O_3: Selecting a suit for the second hand—3 ways (three suits left after using one suit for the first hand)

O_4: Selecting 5 cards out of 13 possible in the suit—$C_{13,5}$ ways

O_5: Selecting a suit for the third hand—2 ways (two suits left after using two suits for first two hands)

O_6: Selecting 5 cards out of 13 possible in the suit—$C_{13,5}$ ways

O_7: Selecting a suit for the fourth hand—1 way (one suit left after using three suits for first three hands)

O_8: Selecting 5 cards out of 13 possible in the suit—$C_{13,5}$ ways

Thus, the number of elements in E is given by

$$n(E) = 4C_{13,5} \cdot 3C_{13,5} \cdot 2C_{13,5} \cdot 1C_{13,5}$$
$$= 4!(C_{13,5})^4$$

We can now compute the probability of E, under the equally likely assumption, to be

$$P(E) = \frac{n(E)}{n(S)} = \frac{4!(C_{13,5})^4}{\begin{pmatrix} 52 \\ 5, 5, 5, 5, 32 \end{pmatrix}}$$

$$= 4! \left(\frac{13!}{5!8!} \right)^4 \div \frac{52!}{5!5!5!5!32!}$$

$$= 4! \left(\frac{13!}{5!8!} \right)^4 \cdot \frac{(5!)^4 32!}{52!}$$

$$= 4.45 \times 10^{-11}$$

Problem 19   Repeat Example 19 for three people.

We should point out that there are many counting problems for which it is not possible to produce a simple formula that will yield the number of possible cases. In cases of this type, we usually revert back to tree diagrams and count branches.

Answers to Matched Problems

11. (A)

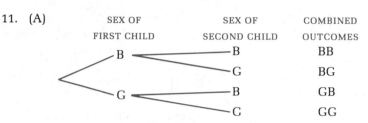

| SEX OF FIRST CHILD | SEX OF SECOND CHILD | COMBINED OUTCOMES |

Thus, $S_1 = \{BB, BG, GB, GG\}$.
(B) $S_2 = \{0, 1, 2\}$ (C) $S_3 = \{A, D\}$
(D) The sample space in part A.

12. (A) (4, 1), (3, 2), (2, 3), (1, 4)   (B) (6, 5), (5,6)
13. (A) Not acceptable   (B) Acceptable
    (C) Acceptable   (D) Acceptable
14. (A) .727   (B) .209   (C) 1
15. (A) $P(E_5) = \frac{1}{9}$ (B) $P(E_6) = \frac{1}{18}$

16. (A) $\frac{1}{8}$ (B) $\frac{1}{2}$ 17. $\dfrac{C_{13,7}}{C_{52,7}} \approx .000013$ 18. $\dfrac{C_{12,4}\,C_{16,2}}{C_{28,6}} \approx .158$

19. $4!(C_{13,5})^3 \div \begin{pmatrix} 52 \\ 5, 5, 5, 37 \end{pmatrix} \approx 1.51 \times 10^{-8}$

Exercise 9-4

A   A spinner is marked from 1 to 8, and each number is as likely to turn up as any other. An experiment consists of spinning the dial once. Problems 1–6 refer to this experiment.

1. Find a sample space S composed of equally likely simple events.
2. How many simple events are in S; that is, find $n(S)$.
3. What is the probability of obtaining a 7?
4. What is the probability of obtaining a 3?
5. What is the event E associated with obtaining an even number? What is the probability of obtaining an even number?
6. What is the event E associated with obtaining a number exactly divisible by 3? What is the probability of obtaining a number exactly divisible by 3?
7. What is the probability of having exactly one girl in a two-child family? (See Problem 11 in the text.)

8. What is the probability of getting exactly one head in tossing a coin twice? (See Example 11.)

B

9. How would you interpret $P(E) = 1$?
10. How would you interpret $P(E) = 0$?

An experiment consists of a coin being tossed three times in succession. Answer the questions in Problems 11–14 regarding this experiment.

11. Find a sample space composed of equally likely simple events. (See Example 16.)
12. Find the event E associated with exactly two heads occurring in three tosses of a coin. What is the probability of getting exactly two heads in three tosses?
13. Find the event E associated with at least two heads occurring. What is the probability of getting at least two heads?
14. Find the event E associated with at least one tail occurring. What is the probability of getting at least one tail?
15. A spinner can land on four different colors: red (R), green (G), yellow (Y), and blue (B). If we do not assume each color is as likely to turn up as any other, which of the probability assignments below have to be rejected, and why?

    (A) $P(R) = .15$, $P(G) = -.35$, $P(Y) = .50$, $P(B) = .70$
    (B) $P(R) = .32$, $P(G) = .28$, $P(Y) = .24$, $P(B) = .30$
    (C) $P(R) = .26$, $P(G) = .14$, $P(Y) = .30$, $P(B) = .30$

16. Using the probability assignments in Problem 15C, what is the probability that the spinner will not land on blue?
17. Using the probability assignments in Problem 15C, what is the probability that the spinner will land on red or yellow?
18. Using the probability assignments in Problem 15C, what is the probability that the spinner will not land on red or yellow?
19. Five thank-you notes are written and five envelopes are addressed. Accidentally, the notes are randomly inserted into the envelopes and mailed without checking the addresses. What is the probability that all notes will be inserted into the correct envelopes?
20. Six people check their coats in a checkroom. If all claim checks are lost and the six coats are randomly returned, what is the probability that all people will get their own coats back?

An experiment consists of rolling two fair dice and adding the dots on the two sides facing up. Using the sample space shown in Figure 1 and assuming each simple event is as likely as any other, find the probability of the sum of the dots in Problems 21–32:

21. Being 2
22. Being 10
23. Being 6
24. Being 8
25. Being less than 5
26. Being greater than 8

27. Not being 7 or 11

28. Not being 2, 4, or 6

29. Being 1

30. Not being 13

31. Being divisible by 3

32. Being divisible by 4

C   33. If four-digit numbers less than 5,000 are randomly formed from the digits 1, 3, 5, 7, and 9, what is the probability of forming a number divisible by 5? (Digits may be repeated; for example, 1,355 is acceptable.)

34. If four-letter code words are generated at random using the letters A, B, C, D, E, and F, what is the probability of forming a word without a vowel in it? (Letters may be repeated.)

An experiment consists of dealing 5 cards from a standard 52 card deck. In Problems 35–42 what is the probability of being dealt:

35. 5 cards, jacks through aces?

36. 5 cards, 2 through 10?

37. 4 aces?

38. Four of a kind?

39. Straight flush, ace high?

40. Straight flush, starting with 2?

41. 2 aces and 3 queens?

42. 2 kings and 3 aces?

Applications

Business & Economics

43. *Promotion.* From 12 known brands of beer, 4 are chosen at random for a participant to identify in a blind tasting. What is the probability that the four brands could be identified just by guessing?

44. *Consumer testing.* From six known brands of cola, three are chosen at random for a consumer to identify in a blind tasting. What is the probability that the three brands could be identified exactly by just guessing?

45. *Personnel selection.* Six female and 5 male applicants have been successfully screened for five positions. If the five positions are selected at random from the 11 finalists, what is the probability of selecting:

    (A) 3 females and 2 males?   (B) 4 females and 1 male?
    (C) 5 females?   (D) At least 4 females?

46. *Committee selection.* A four-person grievance committee is to be composed of employees in two departments A and B with 15 and 20 employees, respectively. If the 4 people are selected at random from the 35 people, what is the probability of selecting:

    (A) 3 from A and 1 from B?   (B) 2 from A and 2 from B?
    (C) All from A?   (D) At least 3 from A?

47. *Personnel selection.* A personnel director has selected, after final screening, seven equally qualified salespeople, three women and four

men. Three are to be selected at random and sent to New York, and two are to be selected at random and sent to Los Angeles. The remaining two people will not be hired at this time. What is the probability that three men go to New York and two women go to Los Angeles?

48. *Personnel selection.* Repeat Problem 47 if the pool of seven qualified salespeople is composed of four women and three men

Life Sciences   **49.** *Medicine.* A prospective laboratory technician is to be tested on identifying blood types from eight standard classifications. If three different samples are chosen at random from the eight types, what is the probability that the technician could identify these correctly by just guessing?

50. *Medical research.* Because of limited funds, five research centers are to be chosen out of eight suitable ones for a study on heart disease. If the selection is made at random, what is the probability that five particular regions will be chosen?

Social Sciences   **51.** *Membership selection.* A town council has 11 members, 6 Democrats and 5 Republicans.

(A) If the president and vice-president are selected at random, what is the probability that they are both Democrats?

(B) If a three-person committee is selected at random, what is the probability that a majority are Republicans?

9-5 Empirical Probability

- Theoretical versus Empirical Probability
- Statistics versus Probability Theory
- Law of Large Numbers

■ Theoretical versus Empirical Probability

In the last section we indicated that probability assignments are made for events in a sample space in two common ways, theoretical and empirical. Let us look at another example and compare the two approaches.

There are 20,000 students registered in a state university. Students are legally either state residents, out-of-state residents, or foreign residents. What is the probability that a student chosen at random is a state resident? An out-of-state resident? A foreign resident? How do we proceed to find these probabilities?

Theoretical Approach Suppose resident information is available in the registrar's office and can be obtained from a computer printout. Requesting the printout, we find

State residents (E_1) 12,000

Out-of-state residents (E_2) 5,000

Foreign residents (E_3) 3,000

$\overline{}$

20,000 $= N$

Looking at the total structure, we reason as follows: We choose the total register of registered students with resident status indicated as our sample space S. We assume one student is as likely to be chosen as another in a random sample of one. Thus, we assign the probability $\frac{1}{20,000}$ to each simple event in S. This is an acceptable assignment. Under the equally likely assumption,

$$P(E_1) = \frac{n(E_1)}{n(S)} = \frac{12,000}{20,000} = .60$$

$$P(E_2) = \frac{n(E_2)}{n(S)} = \frac{5,000}{20,000} = .25$$

$$P(E_3) = \frac{n(E_3)}{n(S)} = \frac{3,000}{20,000} = .15$$

Our approach here is analogous to that used in assigning a probability of $\frac{1}{4}$ to the drawing of a heart in a single draw of one card from a 52 card deck.

Empirical Approach Suppose residency status was not recorded during registration and the information is not available through the registrar. Not having the time, inclination, or money to interview each student, we choose a random sample of 200 students and find:

State residents   128

Out-of-state residents   47

Foreign residents   25

$\overline{}$

200 $= n$

It would be reasonable to say that

$$P(E_1) \approx \frac{128}{200} = .640$$

$$P(E_2) \approx \frac{47}{200} = .235$$

$$P(E_3) \approx \frac{25}{200} = .125$$

As we increase the sample size, our confidence in the probability assignments would likely increase. We refer to these probability assignments as

approximate empirical probabilities and use them to approximate the actual probabilities for the total population.

In general, if we conduct an experiment n times and an event E occurs with frequency $f(E)$, then the ratio $f(E)/n$ is called the **relative frequency** of the occurrence of event E in n trials. We define the **empirical probability** of E, denoted by $P(E)$, by the number (if it exists) that the relative frequency $f(E)/n$ approaches as n gets larger and larger. Of course, for any particular n, the relative frequency $f(E)/n$ is generally only approximately equal to $P(E)$. However, as n increases in size, we would expect the approximation to improve.

Empirical Probability Approximation

$$P(E) \approx \frac{\text{Frequency of occurrence of } E}{\text{Total number of trials}} = \frac{f(E)}{n}$$

(The larger n is, the better the approximation.)

If equally likely assumptions used to obtain theoretical probability assignments are actually warranted, then we would also expect corresponding approximate empirical probabilities to approach the theoretical ones as the number of trials n of actual experiments becomes very large.

Example 20 Two coins are tossed 1,000 times with the following frequencies of outcomes:

2 heads	200
1 head	560
0 heads	240

(A) Compute the approximate empirical probability for each type of outcome.

(B) Compute the theoretical probabilities for each type of outcome.

Solutions (A) $P(2 \text{ heads}) \approx \dfrac{200}{1,000} = .20$ (B) (See Example 14.)

$P(1 \text{ head}) \approx \dfrac{560}{1,000} = .56$ $P(2 \text{ heads}) = .25$

$P(0 \text{ heads}) \approx \dfrac{240}{1,000} = .24$ $P(1 \text{ head}) = .50$

$P(0 \text{ heads}) = .25$

Problem 20 One die is rolled 1,000 times with the following frequencies of outcomes:

1	180	4	138
2	140	5	175
3	152	6	215

(A) Calculate approximate empirical probabilities for each indicated outcome.

(B) Do the indicated outcomes seem equally likely?

(C) Assuming the indicated outcomes are equally likely, compute their theoretical probabilities.

Example 21

An insurance company selected 1,000 drivers at random in a particular city to determine a relationship between age and accidents. The data obtained are listed in Table 1. Compute the following approximate empirical probabilities for a driver chosen at random in the city:

(A) Of being under 20 years old **and** having three accidents in 1 year (E_1)

(B) Of being 30–39 years old **and** having one or more accidents in 1 year (E_2)

(C) Of having no accidents in 1 year (E_3)

(D) Of being under 20 years old **or** having three accidents in 1 year (E_4)

Table 1

Age	Accidents in One Year				
	0	1	2	3	Over 3
Under 20	50	62	53	35	20
20–29	64	93	67	40	36
30–39	82	68	32	14	4
40–49	38	32	20	7	3
Over 49	43	50	35	28	24

Solutions

(A) $P(E_1) \approx \dfrac{35}{1,000} = .035$

(B) $P(E_2) \approx \dfrac{68 + 32 + 14 + 4}{1,000} = .118$

(C) $P(E_3) \approx \dfrac{50 + 64 + 82 + 38 + 43}{1,000} = .277$

(D) $P(E_4) \approx \dfrac{50 + 62 + 53 + 35 + 20 + 40 + 14 + 7 + 28}{1,000} = .309$

Notice that in this type of problem, which is typical of many realistic problems, approximate empirical probabilities are the only type we can compute.

Problem 21

Referring to the results of the survey in Example 21, compute each of the following approximate empirical probabilities for a driver chosen at random in the city:

(A) Of being under 20 years old with no accidents in 1 year (E_1)

(B)   Of being 20–29 years old and having fewer than two accidents in 1 year (E_2)

(C)   Of not being over 49 years old (E_3)

Approximate empirical probabilities are often used to test theoretical probabilities. As we said before, equally likely assumptions may not be justified in reality. In addition to this use, there are many situations in which it is either very difficult or impossible to compute the theoretical probabilities for given events. For example, insurance companies use past experience to establish approximate empirical probabilities to predict the future, baseball teams use batting averages (approximate empirical probabilities based on past experience) to predict the future performance of a player, and pollsters use approximate empirical probabilities to predict outcomes of elections.

■ Statistics versus Probability Theory

We are now entering the area of mathematical statistics, which we will not pursue too far in this book. Mathematical statistics is a branch of mathematics that draws inferences about certain characteristics of a total population, called **population parameters,** based on corresponding characteristics of a random sample from the population. In general, a **population** is the set containing every element we are describing (all people in a school, all flashbulbs produced by a given company using a particular type of manufacturing process, all flips of a certain coin, or all rolls of a certain die). A **sample** is a subset of a population. The population size, if finite, is denoted by N; the sample size is denoted by n. [Except when the sample is a census (the whole population), n is less than N.]

Because samples are used to draw inferences about the total population, it is desirable that a sample be **representative** of the population, that is, that various population characteristics are proportionately represented in the sample. **Random samples** are those in which each element of the population has the same probability of being chosen for the sample. Much statistical theory is based on random samples.

Statistics starts with a known sample and proceeds to describe certain characteristics of the total population that are not known. [For example, in Example 21 the insurance company used the approximate empirical probability .035 (computed from the sample) as an approximation for the actual probability of a person drawn at random from the *total* population being under 20 years old and having three accidents in one year.]

Probability theory, on the other hand, starts with a known composition of a population and from this deduces the probable composition of a sample. [For example, knowing the composition of a standard deck of 52 cards, we can (assuming each 5 card hand has the same probability of being dealt as any other) deduce that the probability of being dealt a flush (5 cards of the same suit) is given by $4C_{13,5}/C_{52,5} = .00198$.] In short, statistics proceeds from a sample to the population, while probability theory proceeds from a population to a sample.

■ Law of Large Numbers

How does the approximate empirical probability of an event determined from a sample relate to the actual probability of the event relative to the total population? In mathematical statistics an important theorem, called the **law of large numbers** (or the **law of averages**), is proved. Informally, it states that the approximate empirical probability can be made as close to the actual probability as we please by making the sample sufficiently large.

For example, if we roll a fair die a large number of times, we would expect to get each number about (not exactly) $\frac{1}{6}$ of the time. The law of large numbers states (informally) the greater the number of times we roll a fair die, the closer the relative frequency of the occurrence of a given number will be to $\frac{1}{6}$ [or if the die is not fair (and no die can be absolutely fair), then the closer the relative frequency of the occurrence of a given number will be to the actual probability of the occurrence of that number].

Answers to Matched Problems

20. (A) $P(1) \approx .180$, $P(2) \approx .140$, $P(3) \approx .152$, $P(4) \approx .138$, $P(5) \approx .175$, $P(6) \approx .215$
    (B) No (C) $\frac{1}{6} \approx .167$ for each

21. (A) $P(E_1) \approx .05$ (B) $P(E_2) \approx .157$
    (C) $P(E_3) \approx .82$ or $P(E_3) = 1 - P(E_3') = 1 - .18 = .82$

Exercise 9-5

A

1. A ski jumper has jumped over 300 feet in 25 out of 250 jumps. What is the approximate empirical probability of the next jump being over 300 feet?

2. In a city there are 4,000 youths between 16 and 20 years old who drive cars. If 560 of them were involved in accidents last year, what is the approximate empirical probability of a youth in this age group being involved in an accident this year?

3. Out of 420 times at bat, a baseball player gets 189 hits. What is the approximate empirical probability that the player will get a hit next time at bat?

4. In a medical experiment, a new drug is found to help 2,400 out of 3,000 people. If a doctor prescribes the drug for a particular patient, what is the approximate empirical probability that the patient will be helped?

5. A thumbtack is tossed 1,000 times with the following outcome frequencies:

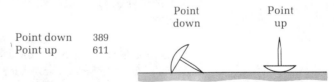

Point down   389
Point up   611

Compute the approximate empirical probability for each outcome. Does each outcome appear to be equally likely?

6. Toss a thumbtack 100 times and let it fall to the floor. Count the number of times it lands point down. What is the approximate empirical probability of the tack landing point down? Point up? (Actually, you can toss ten tacks at a time and count the total number pointing down in ten throws.)

B 7. A random sample of 10,000 two-child families excluding those with twins produced the following frequencies:

    2,351 families with two girls

    5,435 families with one girl

    2,214 families with no girls

(A) Compute the approximate empirical probability for each outcome.

(B) Compute the theoretical probability for each outcome assuming a boy is as likely as a girl at each birth.

8. If we multiply the probability of the occurrence of an event E by the total number of trials n, we obtain the **expected frequency** of the occurrence of E in n trials. Using the theoretical probabilities found in Problem 7B, compute the expected frequency of each outcome in Problem 7 from the sample of 10,000.

9. Three coins are flipped 1,000 times with the following frequencies of outcomes:

    3 heads   132

    2 heads   368

    1 head   380

    0 heads   120

(A) Compute the approximate empirical probabilities for each outcome.

(B) Compute the theoretical probability for each outcome, assuming fair coins.

(C) Compute the expected frequency for each outcome, assuming fair coins. (See Problem 8 above for a definition of expected frequency.)

10. Toss three coins 50 times and compute the approximate empirical probability for three heads, two heads, one head, and no heads, respectively.

C   11. If four fair coins are tossed 80 times, what is the expected frequency of four heads turning up? Three heads? Two heads? One head? No heads? (See Problem 8 above for a definition of expected frequency.)

12. Actually toss four coins 80 times and tabulate the frequencies of the outcomes indicated in Problem 11. What are the approximate empirical probabilities for these outcomes?

■

Applications

Business & Economics   13. *Market analysis.* A company selected 1,000 households at random and surveyed them to determine a relationship between income level and the number of television sets in a home.

Yearly Income	Televisions per Household				
	0	1	2	3	Above 3
Less than $6,000	0	40	51	11	0
$6,000–10,000	0	70	80	15	1
$10,000–20,000	2	112	130	80	12
$20,000–30,000	10	90	80	60	21
More than $30,000	30	32	28	25	20

Compute the approximate empirical probabilities:

(A) Of a household earning $6,000–10,000 per year **and** owning three television sets

(B) Of a household earning $10,000–20,000 per year **and** owning more than one television

(C) Of a household earning more than $30,000 per year **or** owning more than three television sets

(D) Of a household not owning zero television sets

14. *Market analysis.* Compute approximate empirical probabilities (from the sample results in Problem 13):

(A) Of a household earning $20,000–30,000 per year **and** owning no television sets.

(B) Of a household earning $6,000–20,000 per year **and** owning more than two television sets.

(C) Of a household earning less than \$10,000 per year **or** owning two television sets.

(D) Of a household not owning more than three television sets.

Life Sciences   **15.** *Genetics.* A particular type of flowering plant has the following possible colors:

Genes	Flowers
RR	Red
RW	Pink
WW	White

If two pink plants are crossed, the theoretical probabilities associated with each possible flower color are determined by the table:

		Pink-Flowered Plant	
		R	W
Pink-Flowered Plant	R	RR	RW
	W	WR	WW

$P(\text{Red}) = \frac{1}{4}$
$P(\text{Pink}) = \frac{1}{2}$
$P(\text{White}) = \frac{1}{4}$

In an experiment, 1,000 crosses were made with pink flowered plants with the following results:

Red	300
Pink	440
White	260

(A) What is the approximate empirical probability for each color?

(B) What is the expected number of plants with each color in the experiment, based on the theoretical probabilities?

Social Sciences   **16.** *Sociology.* One thousand women between the ages of 50 and 60 who had been married at least once were chosen at random. They were surveyed to determine a relationship between the age at which they were first married and the total number of marriages they had had to date.

First Marriage Age	Number of Marriages				
	1	2	3	4	Above 4
Under 18	44	88	25	12	7
18–20	82	70	30	14	8
21–25	130	110	30	10	4
26–30	95	84	12	6	3
Over 30	56	48	25	5	2

Compute the approximate empirical probabilities:

(A) Of a woman being 21–25 years old on her first marriage **and** having a total of three marriages

(B) Of a woman being 18–20 years old on her first marriage **and** having more than one marriage

(C) Of a woman being under 18 on her first marriage **or** having two marriages

(D) Of a woman not being over 30 on her first marriage

9-6   Union, Intersection, and Complement of Events

- Union and Intersection
- Complement of an Event
- Applications to Empirical Probability

Recall that in Section 9-4 we said that given a sample space

$$S = \{e_1, e_2, \ldots, e_n\}$$

any function P defined on S such that

$$0 \leq P(e_i) \leq 1 \qquad i = 1, 2, \ldots, n$$

and

$$P(e_1) + P(e_2) + \cdots + P(e_n) = 1$$

is called a *probability function*. In addition, we said that any subset of S is called an *event E*, and we defined the probability of E to be the sum of the probabilities of the simple events in E.

■ Union and Intersection

Let us start the discussion of union and intersection with an example.

Example 22   Consider the sample space of equally likely events for the rolling of a single fair die

$$S = \{1, 2, 3, 4, 5, 6\}$$

(A) What is the probability of rolling an even number **or** a 3?

(B) What is the probability of rolling a number that is odd **and** exactly divisible by 3?

(C) What is the probability of rolling a number that is odd **or** exactly divisible by 3?

Solutions   (A)   Let A be the event of rolling an even number, B the event of rolling a 3, and E the event of rolling an even number or a 3. Then

$$A = \{2, 4, 6\} \qquad B = \{3\} \qquad E = \{2, 3, 4, 6\}$$

Now let us look at events A, B, and E in the Venn diagrams:

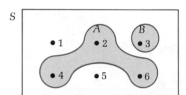 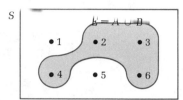

The event E of rolling an even number **or** a 3 is simply the union of the events A and B:

$$E = A \cup B = \{2, 3, 4, 6\}$$

Since this is an equally likely sample space,

$$P(E) = P(A \cup B) = \frac{n(A \cup B)}{n(S)} = \frac{4}{6} = \frac{2}{3}$$

(B) Let A be the event of rolling an odd number, B the event of rolling a number divisible by 3, and F the event of rolling a number that is odd **and** divisible by 3. Then

$$A = \{1, 3, 5\} \qquad B = \{3, 6\} \qquad F = \{3\}$$

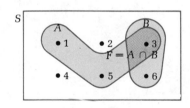

Look at the Venn diagram for events A, B, and F. We can see that the event F of rolling a number that is odd **and** exactly divisible by 3 is the *intersection* of the events A and B,

$$F = A \cap B = \{3\}$$

Thus, the probability of rolling a number that is odd **and** exactly divisible by 3 is

$$P(F) = P(A \cap B) = \frac{n(A \cap B)}{n(S)} = \frac{1}{6}$$

(C) Let A and B be the same events as in part B and let E be the event of rolling a number that is odd **or** divisible by 3. Then,

$$A = \{1, 3, 5\} \qquad B = \{3, 6\} \qquad E = \{1, 3, 5, 6\}$$

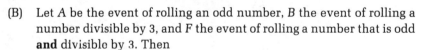

Once again, examining the Venn diagram shows that

$$E = A \cup B = \{1, 3, 5, 6\}$$

and

$$P(E) = P(A \cup B) = \frac{n(A \cup B)}{n(E)} = \frac{4}{6} = \frac{2}{3}$$

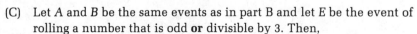

Problem 22   Use the sample space in Example 22 to answer the following:

(A)   What is the probability of rolling a number that is less than 3 **or** greater than 4?
(B)   What is the probability of rolling an odd number **and** a prime number?
(C)   What is the probability of rolling an odd number **or** a prime number?

In general, if A and B are two events in a sample space, the **event A or B** is defined to be the union of A and B and the **event A and B** is defined to be the intersection of A and B.

In this section we shall concentrate on the union of events and only consider simple cases of intersection.

Suppose

$$E = A \cup B$$

Can we find $P(E)$ in terms of A and B? The answer is almost yes, but we must be careful. There are two cases to be considered:

Case 1.   Events A and B are **mutually exclusive**; that is, $A \cap B = \varnothing$.
Case 2.   Events A and B are not mutually exclusive; that is, $A \cap B \neq \varnothing$.

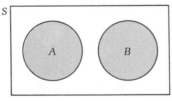

Case 1 $A \cap B = \varnothing$

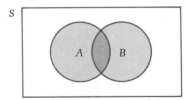

Case 2 $A \cap B \neq \varnothing$

In case 1, since $E = A \cup B$ and $A \cap B = \varnothing$, to find $P(E)$ we just add the sum of the probabilities of the elements in A to the sum of the probabilities of the elements in B. But this is the same as adding $P(A)$ to $P(B)$:

$$\text{If } A \cap B = \varnothing, \text{ then}$$
$$P(A \cup B) = P(A) + P(B) \qquad (1)$$

In case 2, if we simply added the probabilities of the elements in A to the probabilities of the elements in B, we would be adding some of the probabilities twice, namely those for elements that are in both A and B. To compensate for this double counting, we subtract $P(A \cap B)$ from equation

(1) to obtain the following:

If $A \cap B \neq \emptyset$, then

$$P(A \cup B) = P(A) + P(B) - P(A \cap B) \qquad (2)$$

[*Note:* Formula (2) holds for both cases. (Why?)]

To illustrate the difference between these two cases, let us return to Example 22. In Example 22A, we saw that

$$A = \{2, 4, 6\} \qquad B = \{3\} \qquad A \cup B = \{2, 3, 4, 6\}$$

Suppose we compute $P(A) + P(B)$ and $P(A \cup B)$ and compare them:

$$P(A) + P(B) = [P(2) + P(4) + P(6)] + P(3) = \tfrac{2}{3}$$
$$P(A \cup B) = P(2) + P(3) + P(4) + P(6) = \tfrac{2}{3}$$

As equation (1) indicates, $P(A \cup B) = P(A) + P(B)$ when $A \cap B = \emptyset$. Now consider Example 22B, where

$$A = \{1, 3, 5\} \qquad B = \{3, 6\} \qquad A \cup B = \{1, 3, 5, 6\} \qquad A \cap B = \{3\}$$

Once again, let us compare $P(A) + P(B)$ and $P(A \cup B)$:

$$P(A) + P(B) = [P(1) + P(3) + P(5)] + [P(3) + P(6)] = \frac{5}{6}$$

$$P(A \cup B) = P(1) + P(3) + P(5) + P(6) = \frac{4}{6}$$

Notice that $P(3) = P(A \cap B)$ shows up twice in the sum for $P(A) + P(B)$ but only once in the sum for $P(A \cup B)$. Thus, we must subtract $P(A \cap B)$ from $P(A) + P(B)$ when $A \cap B \neq \emptyset$.

Example 23 In the experiment of rolling two dice, use the equally likely sample space of ordered pairs shown in Figure 1 (Example 12, Section 9-4) to answer the following:

(A) What is the probability that a 7 or 11 turns up?
(B) What is the probability that both dice turn up the same or that a sum less than 5 turns up?

Solutions (A) If A is the event that a 7 turns up and B is the event that an 11 turns up, then the event that a 7 or 11 turns up is $A \cup B$ where

$$A = \{(1, 6), (2, 5), (3, 4), (4, 3), (5, 2), (6, 1)\}$$

and

$$B = \{(5, 6), (6, 5)\}$$

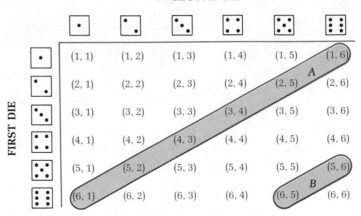

Figure 2

Since $A \cap B = \emptyset$ (see Fig. 2), we can use equation (1) to calculate $P(A \cup B)$:

$$P(A \cup B) = P(A) + P(B)$$
$$= \tfrac{6}{36} + \tfrac{2}{36} \qquad \text{In this equally likely sample space,}$$
$$= \tfrac{8}{36} \qquad\qquad n(A) = 6, \; n(B) = 2, \text{ and } n(S) = 36$$
$$= \tfrac{2}{9}$$

(B)   If A is the event that both dice turn up the same and B is the event that the sum is less than 5, then $A \cup B$ is the event that both dice turn up the same or the sum is less than 5.

$$A = \{(1, 1), (2, 2), (3, 3), (4, 4), (5, 5), (6, 6)\}$$
$$B = \{(1, 1), (1, 2), (1, 3), (2, 1), (2, 2), (3, 1)\}$$

Since $A \cap B = \{(1, 1), (2, 2)\}$ (see Fig. 3), A and B are not mutually exclusive and we must use equation (2):

$$P(A \cup B) = P(A) + P(B) - P(A \cap B)$$
$$= \tfrac{6}{36} + \tfrac{6}{36} - \tfrac{2}{36}$$
$$= \tfrac{10}{36}$$
$$= \tfrac{5}{18}$$

Problem 23   Use the sample space in Example 23 to answer the following:

(A)   What is the probability that a sum of 2 or a sum of 3 turns up?

(B)   What is the probability that both dice turn up the same or that a sum greater than 8 turns up?

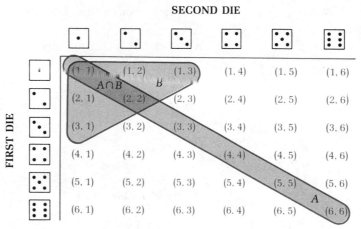

Figure 3

Equations (1) and (2) can also be used in sample spaces when the outcomes are not equally likely.

Example 24 A pair of dice are rolled 1,000 times with the following frequencies of outcomes:

Sum	2	3	4	5	6	7	8	9	10	11	12
Frequency	10	30	50	70	110	150	170	140	120	80	70

Use these frequencies to calculate the approximate empirical probabilities of the following events:

(A) The sum is a prime number or is exactly divisible by 4.
(B) The sum is an odd number or is exactly divisible by 3.

Solutions Dividing the frequency of occurrence of each outcome by 1,000 produces an approximate empirical probability function:

Simple Outcome e_i	2	3	4	5	6	7	8	9	10	11	12
$P(e_i)$	.01	.03	.05	.07	.11	.15	.17	.14	.12	.08	.07

(A) Let A be the event that the sum is a prime number and B the event that the sum is exactly divisible by 4. Then

$$A = \{2, 3, 5, 7, 11\} \quad \text{and} \quad P(A) = P(2) + P(3) + P(5) + P(7) + P(11)$$
$$= .01 + .03 + .07 + .15 + .08$$
$$= .34$$
$$B = \{4, 8, 12\} \qquad \text{and} \quad P(B) = P(4) + P(8) + P(12)$$
$$= .05 + .17 + .07$$
$$= .29$$

Since $A \cap B = \emptyset$, the probability that the sum is a prime number or exactly divisible by 4 is

$$P(A \cup B) = P(A) + P(B)$$
$$= .34 + .29$$
$$= .63$$

(B) Let A be the event that the sum is an odd number and B the event that the sum is exactly divisible by 3.

$$A = \{3, 5, 7, 9, 11\} \quad \text{and} \qquad P(A) = P(3) + P(5) + P(7) + P(9) + P(11)$$
$$= .03 + .07 + .15 + .14 + .08$$
$$= .47$$
$$B = \{3, 6, 9, 12\} \qquad \text{and} \qquad P(B) = P(3) + P(6) + P(9) + P(12)$$
$$= .03 + .11 + .14 + .07$$
$$= .35$$
$$A \cap B = \{3, 9\} \qquad \text{and} \quad P(A \cap B) = P(3) + P(9)$$
$$= .03 + .14$$
$$= .17$$

Using equation (2), the probability that the sum is an odd number or exactly divisible by 3 is

$$P(A \cup B) = P(A) + P(B) - P(A \cap B)$$
$$= .47 + .35 - .17$$
$$= .65$$

Problem 24   Use the empirical probability function in Example 24 to calculate the probability of the following events:

(A) The sum is less than 4 or greater than 9.
(B) The sum is even or exactly divisible by 5.

■ Complement of an Event

Suppose we divide a finite sample space

$$S = \{e_1, \ldots, e_n\}$$

into two subsets E and E' such that

$$E \cap E' = \varnothing$$

that is, E and E' are mutually exclusive, and

$$E \cup E' = S$$

Then E' is called the **complement of E** relative to S. Thus, E' contains all the elements of S that are not in E (Fig. 4). Furthermore,

$$P(S) = P(E \cup E')$$
$$= P(E) + P(E') = 1$$

Hence,

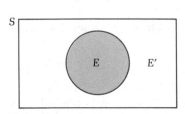

Figure 4

Complements
$P(E) = 1 - P(E')$
$P(E') = 1 - P(E)$

(3)

Example 25   If the probability of rain is .67, then the probability of no rain is $1 - .67 = .33$; if the probability of striking oil is .01, then the probability of not striking oil is .99; and so on.

Problem 25   If the probability of having at least one boy in a two-child family is .75, what is the probability of having no boys?

As was stated in Section 9-4, in finding $P(E)$, there are situations in which it is easier to find $P(E')$ first, and then use equations (3) to find $P(E)$. Consider the following example.

Example 26   What is the probability of getting at least one diamond in a 5 card hand dealt from a 52 card deck?

Solution S = Set of all 5 card hands

E = Set of all 5 card hands with at least 1 diamond

E' = Set of all 5 card hands with 0 diamonds

Since $P(E')$ is easier to compute than $P(E)$, we calculate it first and then use equations (3) to find $P(E)$:

$$P(E') = \frac{n(E')}{n(S)} = \frac{C_{39,5}}{C_{52,5}} \approx .22$$

Thus,

$$P(E) = 1 - P(E') \approx 1 - .22 = .78$$

Problem 26   What is the probability of getting at least one ace in a 5 card hand dealt from a 52 card deck?

Example 27
Birthday Problem   In a group of n people, what is the probability that at least two people have the same birthday (the same month and day excluding leap years)? (Make a guess for a class of 40 people, and check your guess with the conclusion of this example.)

Solution   If we form a list of the birthdays of all the people in the group, then we have a simple event in the sample space

S = Set of all lists of n birthdays

For any person in the group, we will assume that any birthday is as likely as any other, so that the simple events in S are equally likely. How many simple events are in the set S? Since any person could have any one of 365 birthdays (excluding leap years), the fundamental principle of counting implies that the number of simple events in S is

$$
\begin{array}{cccccc}
 & \text{1st} & \text{2nd} & \text{3rd} & & \text{nth} \\
 & \text{person} & \text{person} & \text{person} & & \text{person} \\
n(S) = & 365 & \cdot\ 365 & \cdot\ 365 & \cdots & 365 \\
 = & 365^n
\end{array}
$$

Now, let E be the event that at least two people in the group have the same birthday. Then E' is the event that no two people have the same birthday. The fundamental principle of counting can be used to determine the number of simple events in E':

$$
\begin{array}{cccccc}
 & \text{1st} & \text{2nd} & \text{3rd} & & \text{nth} \\
 & \text{person} & \text{person} & \text{person} & & \text{person} \\
n(E') = & 365 & \cdot\ 364 & \cdot\ 363 & \cdots & (366 - n)
\end{array}
$$

$$= \frac{365 \cdot 364 \cdot 363 \cdot \cdots \cdot (366 - n)(365 - n)(364 - n) \cdot \cdots \cdot 1}{(365 - n)(364 - n) \cdot \cdots \cdot 1}$$

Multiply numerator and denominator by $(365 - n)!$

$$= \frac{365!}{(365 - n)!}$$

Since we have assumed that S is an equally likely sample space,

$$P(E') = \frac{n(E')}{n(S)} = \frac{\dfrac{365!}{(365-n)!}}{365^n} = \frac{365!}{365^n(365-n)!}$$

Thus,

$$P(E) = 1 - P(E')$$

$$= 1 - \frac{365!}{365^n(365-n)!} \tag{4}$$

Equation (4) is valid for any n satisfying $1 \leqslant n \leqslant 365$. [What is $P(E)$ if $n > 365$?] For example, in a group of six people,

$$P(E) = 1 - \frac{365!}{(365)^6 359!}$$

$$= 1 - \frac{\cancel{365} \cdot 364 \cdot 363 \cdot 362 \cdot 361 \cdot 360 \cdot 359!}{\cancel{365} \cdot 365 \cdot 365 \cdot 365 \cdot 365 \cdot 365 \cdot 359!}$$

$$= .04$$

It is interesting to note that as the size of the group increases, $P(E)$ increases more rapidly than you might expect. Table 2 gives the value of $P(E)$ for selected values of n. Notice that for a group of only twenty-three people, the probability that two or more have the same birthday is greater than $\frac{1}{2}$.

Table 2 The Birthday Problem

Number of People in Group n	Probability That 2 or More Have Same Birthday $P(E)$
5	.027
10	.117
15	.253
20	.411
23	.507
30	.706
40	.891
50	.970
60	.994
70	.999

Problem 27   Use equation (4) to evaluate $P(E)$ for $n = 4$.

■ Applications to Empirical Probability

The following examples illustrate the application of the concepts discussed in this section to problems involving data from surveys of a randomly selected sample from a total population. In this situation, the distinction between theoretical and empirical probabilities is a subtle one. If we use the data to assign probabilities to events in the sample population, we are dealing with theoretical probabilities. If we use the same data to assign probabilities to events in the total population, then we are working with empirical probabilities. (See the discussion at the beginning of Section 9-5). Fortunately, the procedures for computing the probabilities are the same in either case, and all we must do is be careful to use the correct terminology. In the following discussions, we will use *empirical probability* to mean the probability of an event determined by a sample that is used to approximate the probability of the corresponding event in the total population.

Example 28   From a survey involving 1,000 people in a certain city, it was found that 500 people had tried a certain brand of diet cola, 600 had tried a certain brand of regular cola, and 200 had tried both brands. If a resident of the city is selected at random, what is the (empirical) probability that:

(A)   He or she has tried the diet or the regular cola?
(B)   He or she has tried one of the colas but not both?

Solutions   Let D be the event that a person has tried the diet cola and R the event that a person has tried the regular cola. The events D and R can be used to partition the residents of the city into four mutually exclusive subsets (a collection of subsets is mutually exclusive if the intersection of any two of them is the empty set):

$D \cap R$ = Set of people who have tried both colas

$D \cap R'$ = Set of people who have tried the diet cola but not the regular cola

$D' \cap R$ = Set of people who have tried the regular cola but not the diet cola

$D' \cap R'$ = Set of people who have not tried either cola

These sets are displayed in the Venn diagram in Figure 5.

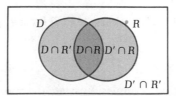

Figure 5 Total population

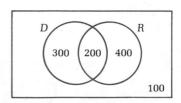

Figure 6 Sample population

The sample population of 1,000 residents is also partitioned into four mutually exclusive sets, with $n(D) = 500$, $n(R) = 600$, and $n(D \cap R) = 200$. By using a Venn diagram (Fig. 6), we can determine the number of sample points in the sets $D \cap R'$, $D' \cap R$, and $D' \cap R'$ (see Example 5 in Section 1-1). These frequencies can be conveniently displayed in a table:

		Regular R	No Regular R'	Total
Diet	D	200	300	500
No Diet	D'	400	100	500
	Total	600	400	1,000

Assuming that each sample point is equally likely, we form a probability table by dividing each entry in this table by 1,000, the total number surveyed. These are theoretical probabilities for the sample population, which we can use as empirical probabilities to approximate the corresponding probabilities for the total population.

		Regular R	No Regular R'	Total
Diet	D	.2	.3	.5
No Diet	D'	.4	.1	.5
	Total	.6	.4	1.0

Now we are ready to compute the required probabilities.

(A)   The event that a person has tried the diet or the regular cola is $D \cup R$.

$$P(D \cup R) = P(D) + P(R) - P(D \cap R)$$
$$= .5 + .6 - .2$$
$$= .9$$

(B)   The event that a person has tried one cola but not both is the event that the person has tried diet and not regular cola or has tried regular and not diet cola. In terms of sets, this is $(D \cap R') \cup (D' \cap R)$. Since $D \cap R'$ and $D' \cap R$ are mutually exclusive (look at the Venn diagram in Fig. 5),

$$P[(D \cap R') \cup (D' \cap R)] = P(D \cap R') + P(D' \cap R)$$
$$= .3 + .4$$
$$= .7$$

Problem 28　If a person is selected at random from the city in Example 28, what is the (empirical) probability that:

(A)　He or she has not tried either cola?

(B)　He or she has tried the diet cola or has not tried the regular cola?

Example 29　The data in the table were obtained by surveying 1,000 residents of a state concerning their political affiliations and their preferences in an upcoming gubernatorial election. If a resident of the state is selected at random, what is the (empirical) probability that the:

(A)　Resident is not affiliated with a political party or has no preference?

(B)　Resident is affiliated with a political party and prefers candidate A?

		Democrat D	**Republican** R	**Unaffiliated** U	Total
Candidate A	A	200	100	85	385
Candidate B	B	250	230	50	530
No Preference	N	50	20	15	85
	Total	500	350	150	1,000

Solutions　First, we form a table of empirical probabilities by dividing each entry in the above table by 1,000, the number of people surveyed.

		Democrat D	**Republican** R	**Unaffiliated** U	Total
Candidate A	A	.2	.1	.085	.385
Candidate B	B	.25	.23	.05	.53
No Preference	N	.05	.02	.015	.085
	Total	.5	.35	.15	1.000

(A)　$P(U \cup N) = P(U) + P(N) - P(U \cap N)$

$$= .15 + .085 - .015$$

$$= .22$$

(B)　A person affiliated with a political party who prefers candidate A must be a Democrat who prefers candidate A or a Republican who prefers candidate A. This is the event $(D \cap A) \cup (R \cap A)$. Since $D \cap A$ and $R \cap A$ are mutually exclusive,

$$P[(D \cap A) \cup (R \cap A)] = P(D \cap A) + P(R \cap A)$$

$$= .2 + .1$$

$$= .3$$

Problem 29   Use the data in the survey in Example 29 to find the (empirical) probability that a resident of the state selected at random is:

(A)   A Democrat or prefers candidate *B*
(B)   Not a Democrat and has no preference

Answers to
Matched Problems

22.   (A) $\frac{2}{3}$ (B) $\frac{1}{3}$ (C) $\frac{2}{3}$ 23.   (A) $\frac{1}{12}$ (B) $\frac{7}{18}$

24.   (A) .31   (B) .6   25.   .25

26. $1 - \dfrac{C_{48,5}}{C_{52,5}} \approx .341$ 27.   .016

28.   (A) $P(D' \cap R') = .1$ (B) $P(D \cup R') = .6$
29.   (A) $P(D \cup B) = .78$ (B) $P[(R \cap N) \cup (U \cap N)] = .035$

Exercise 9-6

A   1.   If a manufactured item has the probability of .003 of failing within 90 days, what is the probability that the item will not fail in that time period?
2.   If in a particular cross of two plants the probability that the flowers will be red is .25, what is the probability that they will not be red?

A spinner is numbered from 1 through 10, and each number is as likely to occur as any other. Use equation (1) or (2), indicating which is used, to compute the probability that in a single spin the dial will stop at:

3.   A number less than 3 or larger than 7
4.   A 2 or a number larger than 6
5.   An even number or a number divisible by 3
6.   An odd number or a number divisible by 3

Problems 7–18 refer to the Venn diagram for events A and B in an equally likely sample space S

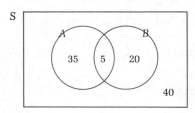

Find each of the following probabilities:

7. $P(A)$	8. $P(A')$	9. $P(B)$	10. $P(B')$
11. $P(A \cap B)$	12. $P(A \cap B')$	13. $P(A' \cap B)$	14. $P(A' \cap B')$
15. $P(A \cup B)$	16. $P(A \cup B')$	17. $P(A' \cup B)$	18. $P(A' \cup B')$

B   *Use the equally likely sample space in Example 23 and equation (1) or (2), indicating which is used, to compute the probability of the following events:*

19.   A sum of 5 or 6

20.   A sum of 9 or 10

21.   The number on the first die is a 1 or the number on the second die is a 1.

22.   The number on the first die is a 1 or the number on the second die is less than 3.

Use the sample space and probability function in Example 24 and equation (1) or (2), indicating which is used, to find the empirical probability of the following events:

23.   The sum is exactly divisible by 4 or exactly divisible by 5.

24.   The sum is odd or exactly divisible by 6.

25.   The sum is an odd number or a prime number.

26.   The sum is even or exactly divisible by 4.

In drawing single card from a deck of 52 cards, use equation (1), (2), or (3), indicating which is used, to determine the probability of drawing:

27.   A king or a queen

28.   A spade or a heart

29.   A king or a heart

30.   A 10 or a club

31.   A black card (spade or club) or an ace

32.   A heart or a number less than 3 (count an ace as 1)

33.   A card other than a king or ace

34.   A card other than a spade or king

Two spinners are each numbered from 1 to 4. On both spinners, each number is as likely to occur as any other. An experiment consists of spinning each dial once. Find the probability that:

35.   The dials stop at two numbers whose sum is 2 or 3.

36.   The dials stop at two numbers whose sum is 5 or 6.

37.   Both dials stop at the same number or the first dial stops at the number 2.

38.   The first dial stops at 1 or the second dial stops at 2.

Given a sample space S, an event E, and its complement E', we define

$$\text{Odds in favor of } E = \frac{P(E)}{P(E')} \qquad P(E') \neq 0$$

$$\text{Odds against } E = \frac{P(E')}{P(E)} \qquad P(E) \neq 0$$

For example, the odds in favor of rolling a 3 in a single roll of a fair die is $(\frac{1}{6})/(\frac{5}{6}) = \frac{1}{5}$, or 1 to 5; and the odds against rolling a 3 is 5 to 1. Compute the odds in favor of obtaining:

39.   A head in a single toss of a coin

40.   A number divisible by 3 in a single roll of a die

41. At least one head when a single coin is tossed three times

42. One head when a single coin is tossed twice

Compute the odds against obtaining:

43. A number greater than 4 in a single roll of a die

44. Two heads when a single coin is tossed twice

45. A 3 or an even number in a single roll of a die

46. An odd number or a number divisible by 3 in a single roll of a die

If the odds in favor of an event E are a to b, then $P(E) = a/(a + b)$. For example, if the odds in favor of a horse winning a race are 2 to 3, then the probability that the horse wins is $2/(2 + 3) = \frac{2}{5}$. Compute the probability of the event E if:

47. The odds in favor of E are 5 to 9.

48. The odds in favor of E are 4 to 3.

49. The odds in favor of E′ are 2 to 7.

50. The odds in favor of E′ are 11 to 9.

C **51.** In a group of n people ($n \le 12$), what is the probability that at least two of them have the same birth month? (Assume any birth month is as likely as any other.)

52. In a group of n people ($n \le 100$), each person is asked to select a number between 1 and 100, write the number on a slip of paper, and place the slip in a hat. What is the probability that at least two of the slips in the hat have the same number written on them?

53. If the odds in favor of an event E occurring are a to b, show that

$$P(E) = \frac{a}{a + b}$$

[*Hint:* Solve the equation $P(E)/P(E') = a/b$ for $P(E)$.]

54. If $P(E) = c/d$, show that the odds in favor of E occurring are c to $d - c$.

Applications

Business & Economics

55. *Market research.* From a survey involving 1,000 students at a large university, a market research company found that 750 students owned stereos, 450 owned cars, and 350 owned cars and stereos. If a student at the university is selected at random, what is the (empirical) probability that:

(A) The student owns either a car or a stereo?

(B) The student owns neither a car nor a stereo?

56. *Market research.* If a student at the university in Problem 55 is selected at random, what is the (empirical) probability that:

(A) The student does not own a car?

(B) The student owns a car but not a stereo?

57. *Insurance.* By examining the past driving records of drivers in a certain city, an insurance company has determined the (empirical) probabilities in the table below.

| | | Miles Driven per Year | | | |
		Less than 10,000, M_1	10,000 to 15,000 Inclusive, M_2	More than 15,000, M_3	Total
Accident	A	.05	.1	.15	.3
No Accident	A'	.15	.2	.35	.7
	Total	.2	.3	.5	1.0

If a driver in this city is selected at random, what is the probability that:

(A) He or she drives less than 10,000 miles per year or has an accident?

(B) He or she drives 10,000 or more miles per year and has no accidents?

58. *Insurance.* Use the (empirical) probabilities in Problem 57 to find the probability that a driver in the city selected at random:

(A) Drives more than 15,000 miles per year or has an accident

(B) Drives 15,000 or fewer miles per year and has an accident

59. *Manufacturing.* Manufacturers of a portable computer provide a 90-day limited warranty covering only the keyboard and the disk drive. Their records indicate that during the warranty period, 6% of their computers are returned because they have defective keyboards, 5% are returned because they have defective disk drives, and 1% are returned because both the keyboard and the disk drive are defective. What is the (empirical) probability that a computer will not be returned during the warranty period?

60. *Product testing.* In order to test a new car, an automobile manufacturer wants to select 4 employees to test drive the car for one year. If 12 management and 8 union employees volunteer to be test drivers and the selection is made at random, what is the probability that at least one union employee is selected?

Life Sciences

61. *Medicine.* In order to test a new drug for adverse reactions, the drug was administered to 1,000 test subjects with the following results: 60 subjects reported that their only adverse reaction was a loss of appetite, 90 subjects reported that their only adverse reaction was a loss of sleep, and 800 subjects reported no adverse reactions at all. If this drug is released for general use, what is the (empirical) probability that a person using the drug will suffer both a loss of appetite and a loss of sleep?

62. *Medicine.* Thirty animals are to be used in a medical experiment on diet deficiency: three male and seven female rhesus monkeys, six male and four female chimpanzees, and two male and eight female dogs. If one animal is selected at random, what is the probability of getting:

(A) A chimpanzee or a dog?

(B) A chimpanzee or a male?

(C) An animal other than a female monkey?

Social Sciences

63. *Sociology.* A group of five Blacks, five Asians, five Latinos, and five Whites were used in a study on racial influence in small group dynamics. If three people are chosen at random, what is the probability that at least one is Black? [*Hint:* See Example 26.]

64. *Political science.* In Example 29 suppose that candidate *A* is a Democrat and candidate *B* is a Republican. If a resident of the state is selected at random, what is the (empirical) probability that he or she is a member of a political party and prefers the candidate of the other party?

9-7 Chapter Review

Important Terms and Symbols

9-2 *The fundamental principle of counting.* tree diagram, fundamental principle of counting

9-3 *Permutations, combinations, and set partitioning.* n factorial, zero factorial, permutation, permutations of n objects, permutation of n objects taken r at a time, combination, combination of n objects taken r at a time, set partitioning,

$$n! = n(n-1)(n-2) \cdot \cdots \cdot 2 \cdot 1, \qquad P_{n,r} = \frac{n!}{(n-r)!},$$

$$C_{n,r} = \binom{n}{r} = \frac{n!}{r!(n-r)!}, \qquad \binom{n}{r_1, r_2, \ldots, r_n} = \frac{n!}{r_1!r_2! \cdots r_k!}$$

9-4 *Experiments, sample spaces, and probability of an event.* deterministic experiment, random experiment, sample space, sample point, simple outcome, compound outcome, finite sample space, infinite sample space, event, simple event, compound event, probability of an event, acceptable probability assignment, probability function, equally likely assumptions, $P(E)$

9-5 *Empirical probability.* approximate empirical probability, empirical probability, relative frequency, population parameters, population, sample, representative sample, random sample, law of large numbers (or law of averages), expected frequency, $P(E)$, $f(E)/n$

9-6 *Union, intersection, and complement of events.* event A **or** event B, event A **and** event B, mutually exclusive, complement of an event, $A \cup B$, $A \cap B$, $P(A \cup B) = P(A) + P(B)$ if $A \cap B = \varnothing$, $P(A \cup B) = P(A) + P(B) - P(A \cap B)$ if $A \cap B \neq \varnothing$, A', $P(A') = 1 - P(A)$

Exercise 9-7   Chapter Review

Work through all the problems in this chapter review and check your answers in the back of the book. (Answers to all review problems are there.) Where weaknesses show up, review appropriate sections in the text. When you are satisfied that you know the material, take the practice test following this review.

1. If one spinner can land on either red or green and a second spinner can land on the number 1, 2, 3, or 4, how many combined outcomes are possible? Solve by using a tree diagram.
2. Solve Problem 1 using the fundamental principle of counting.
3. Evaluate $C_{6,2}$ and $P_{6,2}$.
4. A spinner lands on R with probability .3, on G with probability .5, and on B with probability .2. What is an appropriate sample space S? Find the probability of the spinner landing on either R or G.
5. A drug has side effects for 50 out of 1,000 people in a test. What is the approximate empirical probability that a person using the drug will have side effects?
6. If A and B are events in an equally likely sample space S and $P(A) = .3$, $P(B) = .4$, and $P(A \cap B) = .1$, find:

    (A) $P(A')$ (B) $P(A \cup B)$

7. How many different five-child families are possible where the sex of each child in the order of their birth is taken into consideration [that is, birth sequences such as (B, G, G, B, B) and (G, B, G, B, B) produce different families]? How many families are possible if the order pattern is not taken into account?
8. How many seating arrangements are possible with six people and six chairs in a row? Solve by using the fundamental principle of counting.
9. Solve Problem 8 using permutations or combinations, whichever is applicable.
10. How many ways can eight people be divided into four two-player bridge teams?

11. In a single draw from a 52 card deck, what is the probability of drawing:

    (A) A jack or a queen?   (B) A jack or a spade?
    (C) A card other than an ace?

12. A pair of dice are rolled. The sample space is chosen as the set of all ordered pairs of integers taken from {1, 2, 3, 4, 5, 6}. What is the event A that corresponds to the sum being divisible by 4? What is the event B that corresponds to the sum being divisible by 6? What are $P(A)$, $P(B)$, $P(A \cap B)$, and $P(A \cup B)$?

13. Each letter of the first ten letters of the alphabet is printed on one of ten different cards. What is the probability of drawing the code word *dig* by drawing *d* on the first draw, *i* on the second draw, and *g* on the third draw? What is the probability of being dealt a three-card hand containing the letters *d*, *i*, and *g* in any order?

14. Two coins are flipped 1,000 times with the following frequencies:

    2 heads   210
    1 head   480
    0 heads   310

    (A) Compute the empirical probability for each outcome.
    (B) Compute the theoretical probability for each outcome.
    (C) Compute the expected frequency of each outcome, assuming fair coins.

15. *Market analysis.* From a survey of 100 residents of a city, it was found that 40 read the daily morning paper, 70 read the daily evening paper, and 30 read both papers. What is the (empirical) probability that a resident selected at random:

    (A) Reads a daily paper?
    (B) Does not read a daily paper?
    (C) Reads exactly one daily paper?

16. *Personnel selection.* A software development department consists of six women and four men.

    (A) How many ways can they select a chief programmer, a backup programmer, and a programming librarian?
    (B) If the positions in part A are selected by lottery, what is the probability that women are selected for all three positions?
    (C) How many ways can they select a team of three programmers to work on a particular project?
    (D) If the selections in part C are made by lottery, what is the probability that a majority of the team members will be women?

17. *Membership selection.* A mathematics department has 12 members. The department wants to form a curriculum committee with 5

members, an executive committee with 3 members, and a textbook selection committee with 4 members. Each faculty member will serve on one committee. How many ways can these committees be formed?

18. How many three-letter code words are possible using the first eight letters of the alphabet if no letter can be repeated? If letters can be repeated? If adjacent letters cannot be alike?

19. From a standard deck of 52 cards, how many 5 card hands have exactly 3 hearts and 2 clubs?

20. What is the probability of being dealt 5 clubs from a deck of 52 cards?

21. A person tells you that the following approximate empirical probabilities apply to the sample space $\{e_1, e_2, e_3, e_4\}$: $P(e_1) \approx .1$, $P(e_2) \approx -.2$, $P(e_3) \approx .6$, $P(e_4) \approx 2$. There are three reasons why P cannot be a probability function. Name them.

22. A group of ten people includes one married couple. If four people are selected at random, what is the probability that the married couple is selected?

23. If each of five people is asked to identify his or her favorite book from a list of ten best-sellers, what is the probability that at least two of them identify the same book?

Practice Test: Chapter 9

1. A single die is rolled and a coin is flipped. How many combined outcomes are possible? Solve:

   (A) By using a tree diagram
   (B) By using the fundamental principle of counting

2. Solve the following problems using $P_{n,r}$ or $C_{n,r}$:

   (A) How many three-digit opening combinations are possible on a combination lock with six digits if the digits cannot be repeated?
   (B) Five tennis players have made the finals. If each of the five players is to play every other player exactly once, how many games must be scheduled?

3. Why are the following probability assignments for the sample space $\{e_1, e_2, e_3, e_4\}$ not possible?

   $$P(e_1) = .3 \qquad P(e_2) = -.2 \qquad P(e_3) = 1.2 \qquad P(e_4) = .1$$

4. Betty and Bill are members of a 15-person ski club. If the president and treasurer are selected by lottery, what is the probability that Betty will be president and Bill will be treasurer? (A person cannot hold more than one office.)

5. From a standard deck of 52 cards, what is the probability of obtaining a 5 card hand:

   (A) Of all diamonds?   (B) Of 3 diamonds and 2 spades?

   Write answers in terms of $C_{n,r}$ or $P_{n,r}$; do not evaluate.

6. Three fair coins are tossed 1,000 times with the following frequencies of outcomes:

Number of Heads	0	1	2	3
Frequency	120	360	350	170

   What is the approximate empirical probability of obtaining two heads? What is the theoretical probability of obtaining two heads?

7. A spinning device has three numbers, 1, 2, and 3, each as likely to turn up as the other. If the device is spun twice, what is the probability that:

   (A) The same number turns up both times?
   (B) The sum of the numbers turning up is 5?

8. If three people are selected from a group of seven men and 3 women, what is the probability that at least one women is selected?

9. From a survey of 100 students in a school, it was found that 70 played video games at home, 60 played video games in an arcade, and 40 played video games both at home and in arcades. If a student in the school is selected at random, what is the (empirical) probability that:

   (A) The student plays video games at home or in arcades?
   (B) The student plays video games only at home?

10. A record company selected 1,000 persons at random and surveyed them to determine a relationship between age of purchaser and annual record album purchases.

		Albums Purchased Annually				
		0	1	2	Above 2	Total
	Under 12	60	70	30	10	170
	12–18	30	100	100	60	290
Age	19–25	70	110	120	30	330
	Over 25	100	50	40	20	210
	Total	260	330	290	120	1,000

    Find the empirical probability that a person selected at random:

    (A) Is over 25 and buys two albums annually
    (B) Is 12–18 years old and buys more than one album annually
    (C) Is 12–18 years old or buys more than one album annually

CALCULUS

III

The Derivative

CHAPTER 10   Contents

10-1   Introduction

How do algebra and calculus differ? The two words *static* and *dynamic* probably come as close as any in expressing the difference between the two disciplines. In algebra, we solve equations for a particular value of a variable—a static notion. In calculus, we are interested in how a change in one variable affects another variable—a dynamic notion.

Figure 1 illustrates three basic problems in calculus. It may surprise you to learn that all three problems—as different as they appear—are mathematically related. The solutions to these problems and the discovery of their relationship required the creation of a new kind of mathematics. Isaac Newton (1642–1727) of England and Gottfried Wilhelm von Leibniz (1646–1716) of Germany simultaneously and independently developed this new mathematics, called **the calculus**—it was an idea whose time had come.

In addition to solving the problems described in Figure 1, calculus will

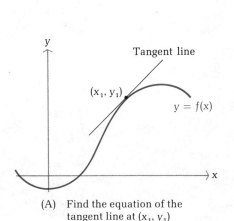

(A) Find the equation of the tangent line at (x_1, y_1) given $y = f(x)$

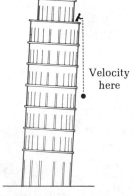

(B) Find the instantaneous velocity of a falling object

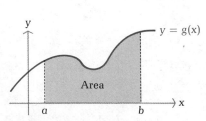

(C) Find the indicated area bounded by $y = g(x)$, $x = a$, $x = b$, and the x axis

Figure 1

enable us to solve many important problems. Until fairly recently, calculus was used primarily in the physical sciences, but now, people in many other disciplines are finding it a useful tool.

10-2 Limits and Continuity

- Limit
- One-Sided Limits
- Properties of Limits
- Continuity
- Application

Basic to the study of calculus are the concepts of *limit* and *continuity*. These concepts help us to describe, in a precise way, the behavior of $f(x)$ when x is close to but not equal to a particular value c. And as we will soon see, they are fundamental to the two main topics of calculus — *the derivative* and *the integral*. In our discussion, we will concentrate on concept development and understanding rather than on the formal details.

■ Limit

We introduce the concept of limit through a problem that goes back to early Grecian times. The problem concerns estimating the circumference of a circle using perimeters of regular polygons inscribed in the circle. Figure 2 illustrates three-sided, six-sided, and twelve-sided regular polygons inscribed in a circle. It appears that if we continue to double the number of sides of an inscribed regular polygon, we can make the perimeter as close to the circumference of the circle as we like. We say that the circumference C

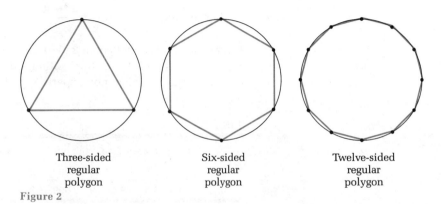

Three-sided	Six-sided	Twelve-sided
regular	regular	regular
polygon	polygon	polygon

Figure 2

of the circle is the "limit" of the perimeter of the inscribed regular polygon as the number of sides increases without bound. Archimedes, a Greek mathematician and inventor (287–212 B.C.), approximated the value of π as the "limit" of perimeters of inscribed regular polygons in a circle with diameter $D = 1$. (Recall that $C = \pi D$. If $D = 1$, then $\pi = C$.)

We now turn to another geometric example that will have far-reaching consequences in the whole development of calculus. Consider the graph of $f(x) = x^2$, a parabola, and the slope of the line through the point (2, 4) and another arbitrary point (x, x^2) on the graph (see Figure 3). A line through

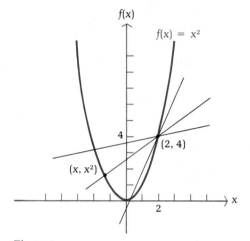

Figure 3

two points on a graph is called a **secant line.** The formula for the slope of the line passing through (x_1, y_1) and (x_2, y_2) is

$$m = \frac{y_2 - y_1}{x_2 - x_1} \qquad x_1 \neq x_2 \qquad \text{See Section 5-1.}$$

Thus, the slope of the secant line through (2, 4) and (x, x^2) is given by

$$\text{Slope of secant line} = m_s = \frac{x^2 - 4}{x - 2}$$

It is clear that x cannot equal 2 (0/0 is meaningless); but what happens to m_s when x approaches 2 from either side of 2? Let us investigate this question using a calculator experiment. Table 1 shows the secant line

Table 1

	x approaches 2 from the left $\rightarrow$ 2 $\leftarrow$ x approaches 2 from the right									
x	1.5	1.8	1.9	1.99	1.999 $\rightarrow$ 2 $\leftarrow$ 2.001	2.01	2.1	2.2	2.5
m_s	3.5	3.8	3.9	3.99	3.999 $\rightarrow$? $\leftarrow$ 4.001	4.01	4.1	4.2	4.5

slopes m_s for values of x approaching 2 from the left and for values of x approaching 2 from the right. It appears that m_s approaches 4 ($m_s \rightarrow 4$) as x approaches 2 ($x \rightarrow 2$) from either side of 2, and the closer x is to 2, the closer m_s will be to 4. We say that 4 is the "limit" of m_s as x approaches 2 and write

$$\lim_{x \to 2} \frac{x^2 - 4}{x - 2} = 4$$

As x approaches 2, $(x^2 - 4)/(x - 2)$ approaches 4, and it is this number 4 that we call the "limit," even though $(x^2 - 4)/(x - 2)$ is not defined at $x = 2$.

In Figure 3 we associate 4 with the slope of the "tangent line" to the graph at (2, 4). ("Tangent line" will be carefully defined in the next two sections.)

We now state an informal definition of **the limit of a function f as x approaches a number c.** A precise definition will not be needed for our discussion, but one is given in the footnote.*

Limit (Informal Definition)

We write

$$\lim_{x \to c} f(x) = L$$

if the functional value $f(x)$ is close to the single real number L whenever x is close to but not equal to c (on either side of c).

Some limits are easy to determine by guessing. For example, most people could guess that

$$\lim_{x \to 2} (x + 2) = 4 \qquad \text{x + 2 is close to 4 whenever x is close to 2 on either side of 2.}$$

But many people would have trouble (without the calculator experiment) guessing by inspection that

$$\lim_{x \to 2} \frac{x^2 - 4}{x - 2} = 4$$

---

* To make the informal definition of limit precise, the use of the word *close* must be made more precise. This is done as follows: We write $\lim_{x \to c} f(x) = L$ if for each $e > 0$, there exists a $d > 0$ such that $|f(x) - L| < e$ whenever $0 < |x - c| < d$. This definition is used to establish particular limits and to prove many useful properties of limits that will be helpful to us in finding particular limits. [Even though intuitive notions of limit existed for a long time, it was not until the nineteenth century that a precise definition was given by the German mathematician, Karl Weierstrass (1815–1897).]

With a little algebraic ingenuity, this result is obtained almost as easily as the preceding one. Factoring the numerator, we have

$$\frac{x^2 - 4}{x - 2} = \frac{(x - 2)(x + 2)}{(x - 2)} = x + 2 \qquad x \neq 2$$

Thus,

$$\lim_{x \to 2} \frac{x^2 - 4}{x - 2} = \lim_{x \to 2} (x + 2) = 4$$

Remember

A function f does not have to be defined at $x = c$ (but it can be) in order for a limit to exist as x approaches c. The function, however, must be defined on both sides of c.

■ One-Sided Limits

In our definition of limit,

$$\lim_{x \to c} f(x) = L$$

we require that $f(x)$ approach L as x approaches c from the left of c and from the right of c. There are many situations in which one-sided limits are useful. Symbolically, we use

$x \to c^-$ to mean "x approaches c from the left"

$x \to c^+$ to mean "x approaches c from the right"

One-Sided Limits (Informal Definition)

We write

$$\lim_{x \to c^-} f(x) = L \qquad \text{Left-hand limit}$$

if $f(x)$ is close to the single real number L whenever x is close to but not equal to c on the left of c. We write

$$\lim_{x \to c^+} f(x) = L \qquad \text{Right-hand limit}$$

if $f(x)$ is close to the single real number L whenever x is close to but not equal to c on the right of c.

Example 1   For $f(x) = |x|/x$, find

(A) $\lim\limits_{x \to 0^-} f(x)$ (B) $\lim\limits_{x \to 0^+} f(x)$ (C) $\lim\limits_{x \to 0} f(x)$

Solutions   We start by sketching a graph of f:

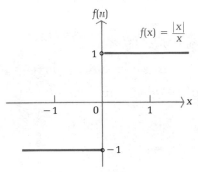

For $x > 0$, $|x|/x = x/x = 1$.
For $x < 0$, $|x|/x = -x/x = -1$.

(A) $\lim\limits_{x \to 0^-} f(x) = -1$ $f(x)$ is not only close to -1, but it is also equal to -1 for x to the left of 0.

(B) $\lim\limits_{x \to 0^+} f(x) = 1$ $f(x)$ is not only close to 1, but it is also equal to 1 for x to the right of 0.

(C) $\lim_{x \to 0} f(x)$ does not exist, since $f(x)$ is not close to a single fixed number whenever x is close to 0 on either side of 0.

If left- and right-hand limits exist and are not equal, or if either does not exist, then the ordinary limit does not exist. If both left- and right-hand limits exist and are equal, then the ordinary limit exists and has the same value. These observations are summarized in Theorem 1.

Theorem 1

For real numbers c and L,

$$\lim_{x \to c} f(x) = L$$

if and only if

$$\lim_{x \to c^-} f(x) = L = \lim_{x \to c^+} f(x)$$

Problem 1   Graph $f(x) = \sqrt{x}$, and from the graph determine

(A) $\lim\limits_{x \to 0^-} f(x)$ (B) $\lim\limits_{x \to 0^+} f(x)$ (C) $\lim\limits_{x \to 0} f(x)$

The domain of f is restricted so that $\sqrt{x}$ is real.

■ Properties of Limits

We now turn to some basic properties of limits that will enable us to evaluate limits of a rather large class of functions without resorting to geometric figures and graphs. We state some important properties without proof in Theorem 2.

Theorem 2

Properties of Limits

If k and c are constants, n is a positive integer, and

$$\lim_{x \to c} f(x) = A \qquad \lim_{x \to c} g(x) = B$$

then:

1. $\lim_{x \to c} k = k$

2. $\lim_{x \to c} kf(x) = k \lim_{x \to c} f(x) = kA$

3. $\lim_{x \to c} [f(x) \pm g(x)] = \lim_{x \to c} f(x) \pm \lim_{x \to c} g(x) = A \pm B$

4. If P is a polynomial function, then $\lim_{x \to c} P(x) = P(c)$

5. $\lim_{x \to c} [f(x) \cdot g(x)] = [\lim_{x \to c} f(x)][\lim_{x \to c} g(x)] = AB$

6. $\lim_{x \to c} \dfrac{f(x)}{g(x)} = \dfrac{\lim_{x \to c} f(x)}{\lim_{x \to c} g(x)} = \dfrac{A}{B} \qquad B \neq 0$

7. $\lim_{x \to c} \sqrt[n]{f(x)} = \sqrt[n]{\lim_{x \to c} f(x)} = \sqrt[n]{A}$

   (x is restricted to avoid even roots of negative numbers)

[Note: These properties also hold for one-sided limits.]

Example 2   Use the properties of limits to evaluate each limit.

(A) $\lim_{x \to 2} (3x^5 - 2x^2 + 3x - 1) = 3(2)^5 - 2(2)^2 + 3(2) - 1$ Use property 4.

$$= 93$$

(B) $\lim_{x \to 2} \sqrt{\dfrac{x^3 - 2x}{x^2 + 2}} = \sqrt{\lim_{x \to 2} \dfrac{x^3 - 2x}{x^2 + 2}}$

Use properties 7, 6, and 4. $(x^3 - 2x)/(x^2 + 2)$ is not negative for x close to 2 and $\lim_{x \to 2} (x^2 + 2) \neq 0$.

$$= \sqrt{\dfrac{2^3 - 2(2)}{2^2 + 2}}$$

$$= \sqrt{\dfrac{4}{6}} = \sqrt{\dfrac{2}{3}} \quad \text{or} \quad \dfrac{\sqrt{6}}{3}$$

Problem 2 Use the properties of limits to evaluate each limit.

(A) $\lim\limits_{x \to -2} (2x^4 - 3x^3 + 5)$ (B) $\lim\limits_{x \to -1} \sqrt{\dfrac{2x^4 + 1}{1 - x}}$

Example 3 Use the properties of limits and algebraic manipulation to find each limit, if it exists, for $f(x) = (x - 1)/(x^2 - 1)$.

(A) $\lim\limits_{x \to 3} f(x)$ (B) $\lim\limits_{x \to 1} f(x)$ (C) $\lim\limits_{x \to -1} f(x)$

Solutions (A) $\lim\limits_{x \to 3} \dfrac{x - 1}{x^2 - 1} = \dfrac{\lim_{x \to 3} (x - 1)}{\lim_{x \to 3} (x^2 - 1)}$ Use properties 6 and 4.

$= \dfrac{3 - 1}{3^2 - 1} = \dfrac{2}{8} = \dfrac{1}{4}$

(B) $\lim\limits_{x \to 1} \dfrac{x - 1}{x^2 - 1}$

We cannot use limit property 6, since $\lim_{x \to 1} (x^2 - 1) = 0$, so let us factor the denominator.

$= \lim\limits_{x \to 1} \dfrac{\overset{1}{\cancel{(x - 1)}}}{\underset{1}{\cancel{(x - 1)}}(x + 1)}$

We are interested in what $(x - 1)/[(x - 1)(x + 1)]$ approaches as x approaches (but does not equal) 1. The factor $(x - 1)$ cancels for any value of $x \neq 1$, and we can now use limit property 6 (as well as properties 1 and 4).

$= \lim\limits_{x \to 1} \dfrac{1}{x + 1}$

$= \dfrac{1}{2}$

(C) $\lim\limits_{x \to -1} \dfrac{x - 1}{x^2 - 1}$ Proceed as in part B.

$= \lim\limits_{x \to -1} \dfrac{\overset{1}{\cancel{(x - 1)}}}{\underset{1}{\cancel{(x - 1)}}(x + 1)}$

$= \lim\limits_{x \to -1} \dfrac{1}{x + 1}$

Does not exist

What does $1/(x + 1)$ approach as x approaches (but does not equal) -1? The denominator has a limit of 0, so limit property 6 cannot be used. As the denominator approaches 0, the fraction $1/(x + 1)$ can be made as large in absolute value as you like (see Table 2); hence, it does not approach any fixed number. The limit does not exist.

Table 2

	x approaches −1 from the left			→ −1 ←	x approaches −1 from the right			
x	−1.1 −1.01 −1.001 −1.0001			→ −1 ←	−0.9999 −0.999 −0.99 −0.9			
$f(x)$	−10 −100 −1,000 −10,000			→ ? ←	10,000 1,000 100 10			

Problem 3　Given $f(x) = (2 + x)/(4 - x^2)$, find

(A) $\lim\limits_{x \to -3} f(x)$　　(B) $\lim\limits_{x \to -2} f(x)$　　(C) $\lim\limits_{x \to 2} f(x)$

Example 4　For each of the following functions, find

$$\lim_{\Delta x \to 0} \frac{f(2 + \Delta x) - f(2)}{\Delta x}$$

(A) $f(x) = x^2 + 1$　　(B) $f(x) = \sqrt{x}$

[*Note:* Δx (read "delta x") represents a single real variable. We will discuss this new symbol in detail in the next section.]

Solutions　(A)
$$\frac{f(2 + \Delta x) - f(2)}{\Delta x} = \frac{[(2 + \Delta x)^2 + 1] - [2^2 + 1]}{\Delta x}$$

$$= \frac{4 + 4\Delta x + (\Delta x)^2 + 1 - 5}{\Delta x}$$

$$= \frac{4\Delta x + (\Delta x)^2}{\Delta x} = \frac{\Delta x(4 + \Delta x)}{\Delta x}$$

$$= 4 + \Delta x \qquad \Delta x \neq 0$$

Since property 6, the quotient property of limits, cannot be used here ($\lim_{\Delta x \to 3} \Delta x = 0$), we simplify the quotient first, then compute the limit, if possible.

Thus,

$$\lim_{\Delta x \to 0} \frac{f(2 + \Delta x) - f(2)}{\Delta x} = \lim_{\Delta x \to 0} (4 + \Delta x) = 4$$

(B)
$$\frac{f(2 + \Delta x) - f(2)}{\Delta x} = \frac{\sqrt{2 + \Delta x} - \sqrt{2}}{\Delta x}$$

$$= \frac{\sqrt{2 + \Delta x} - \sqrt{2}}{\Delta x} \cdot \frac{\sqrt{2 + \Delta x} + \sqrt{2}}{\sqrt{2 + \Delta x} + \sqrt{2}}$$

$$= \frac{2 + \Delta x - 2}{\Delta x(\sqrt{2 + \Delta x} + \sqrt{2})}$$

$$= \frac{1}{\sqrt{2 + \Delta x} + \sqrt{2}} \qquad \Delta x \neq 0$$

The quotient property of limits (property 6) cannot be used. (Why?) We try rationalizing the numerator, then use the limit properties.

Thus,

$$\lim_{\Delta x \to 0} \frac{\sqrt{2 + \Delta x} - \sqrt{2}}{\Delta x} = \lim_{\Delta x \to 0} \frac{1}{\sqrt{2 + \Delta x} + \sqrt{2}}$$

We can now use properties 6, 1, 7, and 4.

$$= \frac{1}{\sqrt{2} + \sqrt{2}}$$

$$= \frac{1}{2\sqrt{2}} \quad \text{or} \quad \frac{\sqrt{2}}{4}$$

Problem 4   For each of the following functions, find

$$\lim_{\Delta x \to 0} \frac{f(3 + \Delta x) - f(3)}{\Delta x}$$

(A) $f(x) = x^2 - x$ (B) $f(x) = \sqrt{x} + 1$

■ Continuity

Refer to Figure 4. Notice that some of the graphs are broken; that is, they cannot be drawn without lifting a pen off the paper. Informally, a function whose graph is broken (disconnected) at a certain point is said to be **discontinuous** at the point; if the graph is not broken at a point, then the function is said to be **continuous** at that point. A function is said to be continuous on an open interval* (a, b) if it is continuous (not broken) at each value in the interval. In Figure 4, functions f and g appear to be continuous for all x, while functions h and F are both discontinuous at $x = 0$.

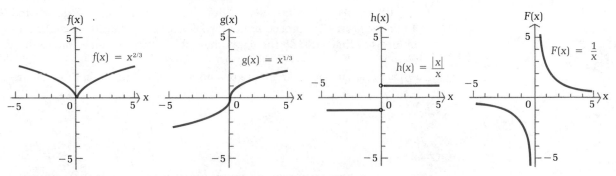

Figure 4

Most graphs of natural phenomena are continuous, whereas many graphs in business and economics have discontinuities. Figure 5A illus-

* (a, b) is an **open interval** (does not include either end point).
  $[a, b]$ is a **closed interval** (includes both end points).
  $(a, b]$ and $[a, b)$ are **half-open intervals** (include one end point and not the other).

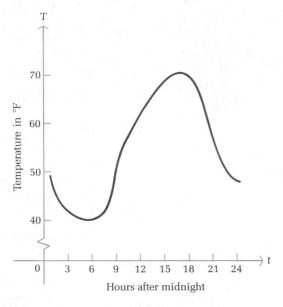

(A) Temperature for a 24-hour period
(continuous, natural behavior)

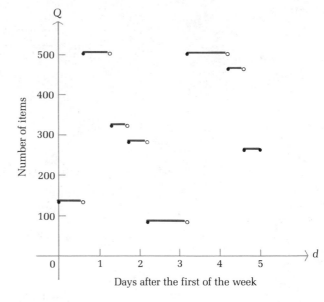

(B) Inventory in a warehouse during 1 week
(discontinuous, "unnatural" behavior)

Figure 5

trates continuous, natural behavior; Figure 5B illustrates discontinuous, "unnatural" behavior.

If we have a graph of a function, then it is usually easy to identify points of discontinuity. If a function is defined by an equation, how can we identify points of discontinuity without looking at its graph? Figure 6 and Table 3 suggest some procedures as well as a formal definition of continuity in terms of limits. Study the figure and table carefully before proceeding further.

The function shown in Figure 6 is not the type of function that you are

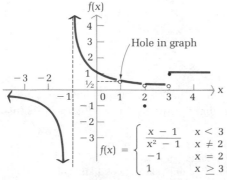

Figure 6

Table 3

c	$\lim_{x \to c} f(x)$	$f(c)$	**Graph**
-2	-1	-1	No break in graph
-1	Does not exist	Does not exist	Break in graph
0	1	1	No break in graph
1	$1/2$	Does not exist	Break in graph
2	$1/3$	-1	Break in graph
3	Does not exist	1	Break in graph
4	1	1	No break in graph

likely to encounter with great frequency. It was designed to illustrate most of the kinds of points of discontinuity exhibited by various types of functions. Looking at Table 3, we are led to the following precise definition of continuity:

Continuity

A function f is **continuous at the point $x = c$ if**

1. $\lim_{x \to c} f(x)$ exists.

2. $f(c)$ exists.

3. $\lim_{x \to c} f(x) = f(c)$.

A function is **continuous on the open interval (a, b)** if it is continuous at each point on the interval.

If one or more of the three conditions in the definition fails, then a function is **discontinuous** at $x = c$. Note that at least one of the conditions fails at each of the points $x = -1, 1, 2,$ and 3 in Figure 6; as you can see, these are the points of discontinuity for f.

We can talk about one-sided continuity as we talked about one-sided limits. For example, a function is said to be **continuous on the right** at $x = c$ if $\lim_{x \to c^+} f(x) = f(c)$ and **continuous on the left** at $x = c$ if $\lim_{x \to c^-} f(x) = f(c)$. For example, the function $f(x) = \sqrt{x}$ is continuous on the half-closed interval $[0, \infty)$, since

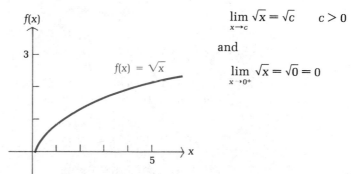

$$\lim_{x \to c} \sqrt{x} = \sqrt{c} \qquad c > 0$$

and

$$\lim_{x \to 0^+} \sqrt{x} = \sqrt{0} = 0$$

Functions have continuity properties similar to limit properties. For example, the sum, difference, product, and quotient of two continuous functions are continuous, except for values of x that make the denominator in a quotient 0. In addition, we state the following important theorem for polynomial and rational functions.*

* Rational functions are functions of the form $f(x)/g(x)$ where $f(x)$ and $g(x)$ are polynomials.

Theorem 3	**Continuity for Polynomial and Rational Functions**

Polynomials are continuous for all values of x. Rational functions are continuous for all values of x for which they are defined — that is, for all values of x except those which make a denominator 0.

Example 5   For what values of x are the following functions discontinuous?

(A) $f(x) = x^5 - 3x^2 + 1$ (B) $\dfrac{1}{x} + \dfrac{2x}{(x-3)(x+2)}$ (C) $\dfrac{x-1}{x^2 + 2x - 3}$

Solutions   (A) Continuous for all x
(B) Discontinuous at $x = -2, 0, 3$

(C) $\dfrac{x-1}{x^2 + 2x - 3} = \dfrac{x-1}{(x+3)(x-1)}$ Discontinuous at $x = -3, 1$

Problem 5   For what values of x are the following functions discontinuous?

(A) $\dfrac{x^2 - 5}{x(2x - 1)(x + 7)}$ (B) $3x^4 - 2x^3 + 3x^2 - x$ (C) $\dfrac{x^2 - 9}{2x^2 + 5x - 3}$

■ Application

Example 6
Compound interest
If \$1,000 is invested at 12% interest compounded quarterly, the amount in the account at the end of x months for a 1-year period is given by

$$F(x) = \begin{cases} \$1,000 & 0 \leqslant x < 3 \\ 1,030 & 3 \leqslant x < 6 \\ 1,061 & 6 \leqslant x < 9 \\ 1,093 & 9 \leqslant x < 12 \\ 1,126 & 12 = x \end{cases}$$

(A) Graph the function F.
(B) Find $\lim_{x \to 3^-} F(x)$, $\lim_{x \to 3^+} F(x)$, and $\lim_{x \to 3} F(x)$.
(C) Find $\lim_{x \to 6} F(x)$ and $F(6)$.
(D) Is F continuous at $x = 6$? At $x = 7$?

Solutions   (A)

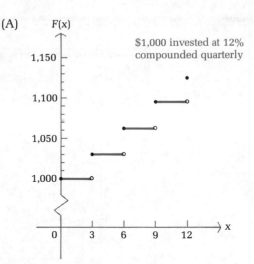

$F(x)$

$1,150$

$1,100$

$1,050$

$1,000$

0 3 6 9 12 x

$1,000 invested at 12% compounded quarterly

(B) $\lim_{x \to 3^-} F(x) = \$1,000$; $\lim_{x \to 3^+} F(x) = \$1,030$; $\lim_{x \to 3} F(x)$ does not exist

(C) $\lim_{x \to 6} F(x)$ does not exist; $F(6) = \$1,061$

(D)   No, since $\lim_{x \to 6} F(x) \neq F(6)$; yes, since $\lim_{x \to 7} F(x) = \$1,061 = F(7)$

Problem 6   Use the function F in Example 6.

(A)   Find $\lim_{x \to 9^-} F(x)$, $\lim_{x \to 9^+} F(x)$, and $\lim_{x \to 9} F(x)$.

(B)   Find $\lim_{x \to 0^+} F(x)$ and $F(0)$.

(C)   Is the function F continuous on the right at $x = 0$?

Answers to Matched Problems   1.

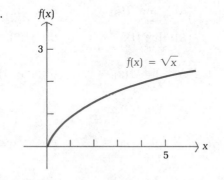

$f(x)$

3

$f(x) = \sqrt{x}$

5 x

(A) $\lim_{x \to 0^-} \sqrt{x}$ does not exist since values to the left of 0 are not in the domain of f.

(B) $\lim_{x \to 0^+} \sqrt{x} = 0$, since $\sqrt{x}$ approaches 0 as x approaches 0 from the right.

(C) $\lim_{x \to 0} \sqrt{x}$ does not exist, since $\lim_{x \to 0^-} \sqrt{x} \neq \lim_{x \to 0^+} \sqrt{x}$.

2.   (A)   61   (B) $\sqrt{3/2}$ or $\sqrt{6}/2$

3.   (A) $1/5$ (B) $1/4$ (C)   Does not exist

4.   (A)   5   (B) $1/(2\sqrt{3})$ or $\sqrt{3}/6$

5.   (A) $x = -7, 0, 1/2$ (B)   Continuous for all x (C) $x = -3, 1/2$

6.   (A) $\$1,061$; $\$1,093$; does not exist

     (B) $\$1,000$; $\$1,000$

     (C)   Yes

Exercise 10-2

A   *Problems 1–12 refer to the function f in the following graph. Use the graph to estimate limits.*

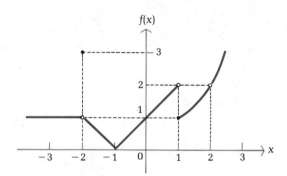

1. (A) $\lim\limits_{x \to 0^-} f(x)$ (B) $\lim\limits_{x \to 0^+} f(x)$ (C) $\lim\limits_{x \to 0} f(x)$

2. (A) $\lim\limits_{x \to -1^-} f(x)$ (B) $\lim\limits_{x \to -1^+} f(x)$ (C) $\lim\limits_{x \to -1} f(x)$

3. (A) $\lim\limits_{x \to 1^-} f(x)$ (B) $\lim\limits_{x \to 1^+} f(x)$ (C) $\lim\limits_{x \to 1} f(x)$

4. (A) $\lim\limits_{x \to 2^-} f(x)$ (B) $\lim\limits_{x \to 2^+} f(x)$ (C) $\lim\limits_{x \to 2} f(x)$

5. (A) $\lim\limits_{x \to -2^-} f(x)$ (B) $\lim\limits_{x \to -2^+} f(x)$ (C) $\lim\limits_{x \to -2} f(x)$

6. (A) $\lim\limits_{x \to 0.5^-} f(x)$ (B) $\lim\limits_{x \to 0.5^+} f(x)$ (C) $\lim\limits_{x \to 0.5} f(x)$

7. (A) $\lim\limits_{x \to 0} f(x)$ (B) $f(0) = ?$ (C) Is f continuous at $x = 0$?

8. (A) $\lim\limits_{x \to -1} f(x)$ (B) $f(-1) = ?$ (C) Is f continuous at $x = -1$?

9. (A) $\lim\limits_{x \to 1} f(x)$ (B) $f(1) = ?$ (C) Is f continuous at $x = 1$?

10. (A) $\lim\limits_{x \to 2} f(x)$ (B) $f(2) = ?$ (C) Is f continuous at $x = 2$?

11. (A) $\lim\limits_{x \to -2} f(x)$ (B) $f(-2) = ?$ (C) Is f continuous at $x = -2$?

12. (A) $\lim\limits_{x \to 0.5} f(x)$ (B) $f(0.5) = ?$ (C) Is f continuous at $x = 0.5$?

Find each limit.

13. $\lim\limits_{x \to 5} (2x^2 - 3)$ 14. $\lim\limits_{x \to 2} (x^2 - 8x + 2)$

15. $\lim\limits_{x\to 4}(x^2-5x)$

16. $\lim\limits_{x\to -2}(3x^3-9)$

17. $\lim\limits_{x\to 2}\dfrac{5x}{2+x^2}$

18. $\lim\limits_{x\to 10}\dfrac{2x+5}{3x-5}$

19. $\lim\limits_{x\to 2}(x+1)^3(2x-1)^2$

20. $\lim\limits_{x\to 3}(x+2)^2(2x-4)$

21. $\lim\limits_{x\to 0}\dfrac{x^2-3x}{x}$

22. $\lim\limits_{x\to 0}\dfrac{2x^2+5x}{x}$

Find points of discontinuity (if they exist) for each function.

23. $f(x)=2x-3$

24. $g(x)=3-5x$

25. $h(x)=\dfrac{2}{x-5}$

26. $k(x)=\dfrac{x}{x+3}$

27. $g(x)=\dfrac{x-5}{(x-3)(x+2)}$

28. $F(x)=\dfrac{1}{x(x+7)}$

B   *Problems 29–34 refer to the **greatest integer function**, which is denoted by* [x] *and is defined as follows:*

[x] = Greatest integer $\leqslant$ x

For example,

$[-3.6]$ = Greatest integer $\leqslant -3.6 = -4$

$[2]$ = Greatest integer $\leqslant 2 = 2$

$[2.5]$ = Greatest integer $\leqslant 2.5 = 2$

The graph of $f(x)=[x]$ *is as shown here:*

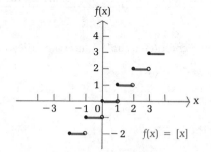

[x] = −2	for	−2 $\leq$ x < −1
[x] = −1	for	−1 $\leq$ x < 0
[x] = 0	for	0 $\leq$ x < 1
[x] = 1	for	1 $\leq$ x < 2
[x] = 2	for	2 $\leq$ x < 3

Find the indicated limits.

29. (A) $\lim\limits_{x\to 2^-}[x]$ (B) $\lim\limits_{x\to 2^+}[x]$ (C) $\lim\limits_{x\to 2}[x]$

30. (A) $\lim\limits_{x\to 0^-}[x]$ (B) $\lim\limits_{x\to 0^+}[x]$ (C) $\lim\limits_{x\to 0}[x]$

31. (A) $\lim\limits_{x\to 1.5^-}[x]$ (B) $\lim\limits_{x\to 1.5^+}[x]$ (C) $\lim\limits_{x\to 1.5}[x]$

32. (A) $\lim\limits_{x \to 2.6^-} [x]$ (B) $\lim\limits_{x \to 2.6^+} [x]$ (C) $\lim\limits_{x \to 2.6} [x]$

33. (A) $\lim\limits_{x \to 2} [x]$ (B) $[2] = ?$ (C) Is $[x]$ continuous at $x = 2$? At $x = 2.5$?

34. (A) $\lim\limits_{x \to 0} [x]$ (B) $[0] = ?$ (C) Is $[x]$ continuous at $x = 0$? At $x = 0.5$?

Find each limit, if it exists. (Use algebraic manipulation where necessary.)

35. $\lim\limits_{x \to 4} \sqrt[3]{x^2 - 3x}$

36. $\lim\limits_{x \to 2} \sqrt{x^2 + 2x}$

37. $\lim\limits_{x \to 4} \dfrac{\sqrt{x} - 2}{x^2 - 4}$

38. $\lim\limits_{x \to 25} \dfrac{5 - \sqrt{x}}{x + 25}$

39. $\lim\limits_{x \to -2} \dfrac{x^2 - x - 6}{x + 2}$

40. $\lim\limits_{x \to -4} \dfrac{2x^2 + 7x - 4}{x + 4}$

41. $\lim\limits_{x \to 3} \dfrac{x^2 - x - 6}{x^2 - 9}$

42. $\lim\limits_{x \to 2} \dfrac{x^2 + 2x - 8}{x^2 - 2x}$

43. $\lim\limits_{x \to 1} \dfrac{2x^3 - 3x + 2}{x^2 + x}$

44. $\lim\limits_{x \to 2} \dfrac{x^3 - x^2 + 1}{x - x^2}$

45. $\lim\limits_{x \to 3} \left(\dfrac{x}{x + 3} + \dfrac{x - 3}{x^2 - 9} \right)$

46. $\lim\limits_{x \to 2} \left(\dfrac{1}{x + 2} + \dfrac{x - 2}{x^2 - 4} \right)$

47. $\lim\limits_{x \to 4} \dfrac{\sqrt{x} - 2}{x - 4}$

48. $\lim\limits_{x \to 0} \dfrac{\sqrt{x + 4} - 2}{x}$

Complete the following table for each function in Problems 49–52:

x	0.9	0.99	0.999 $\to 1 \leftarrow$ 1.001	1.01	1.1
$f(x)$			$\to ? \leftarrow$		

From the completed table, guess the following (a calculator will be helpful for some):

(A) $\lim\limits_{x \to 1^-} f(x)$ (B) $\lim\limits_{x \to 1^+} f(x)$ (C) $\lim\limits_{x \to 1} f(x)$

49. $f(x) = \dfrac{|x - 1|}{x - 1}$

50. $f(x) = \dfrac{x - 1}{|x| - 1}$

51. $f(x) = \dfrac{x^3 - 1}{x - 1}$

52. $f(x) = \dfrac{x^4 - 1}{x - 1}$

For what values of x are the following functions continuous?

53. $F(x) = 2x^8 - 3x^4 + 5$

54. $h(x) = \dfrac{x^4 - 3x + 5}{x^2 + 2x}$

55. $g(x) = \sqrt{x - 5}$

56. $f(x) = \sqrt{3 - x}$

C *Compute*

$$\lim_{\Delta x \to 0} \frac{f(2 + \Delta x) - f(2)}{\Delta x}$$

for each function in Problems 57–66.

57. $f(x) = 3x + 1$ 58. $f(x) = 5x - 1$

59. $f(x) = x^2 + x$ 60. $f(x) = 2x^2 - 3$

61. $f(x) = -3$ 62. $f(x) = 2$

63. $f(x) = \dfrac{1}{x}$ 64. $f(x) = \dfrac{1}{x^2}$

65. $f(x) = \sqrt{x} + 5$ 66. $f(x) = \sqrt{x - 1}$

Find each limit.

67. $\displaystyle\lim_{x \to 2} \frac{x^3 - 8}{x - 2}$ 68. $\displaystyle\lim_{x \to 1} \frac{x^2 - 1}{x^3 - 1}$

69. $\displaystyle\lim_{x \to -2} \frac{x + 2}{x^3 + 8}$ 70. $\displaystyle\lim_{x \to -1^-} \frac{|x + 1|}{x + 1}$

71. $\displaystyle\lim_{x \to -1^+} \frac{|x + 1|}{x + 1}$ 72. $\displaystyle\lim_{x \to -1} \frac{|x + 1|}{x + 1}$

Applications

Business & Economics

73. *Postal rates.* First-class postage in 1983 was $.20 for the first ounce (or any fraction thereof) and $.17 for each additional ounce (or fraction thereof) up to 12 ounces. If $P(x)$ is the amount of postage for a letter weighing x ounces, then we can write

$$P(x) = \begin{cases} \$.20 & \text{for } 0 < x \leqslant 1 \\ \$.37 & \text{for } 1 < x \leqslant 2 \\ \$.54 & \text{for } 2 < x \leqslant 3 \\ \text{and so on} \end{cases}$$

(A) Graph P for $0 < x \leqslant 5$.

(B) Find $\lim_{x \to 2^-} P(x)$, $\lim_{x \to 2^+} P(x)$, and $\lim_{x \to 2} P(x)$.

(C) Find $\lim_{x \to 4} P(x)$ and $P(4)$.

(D) Is P continuous at $x = 4$? At $x = 4.5$?

74. *Telephone rates.* A person placing a station-to-station call on Saturday from San Francisco to New York is charged $.30 for the first minute (or any fraction thereof) and $.20 for each additional minute (or fraction thereof). If the length of a call is x minutes, then the long-distance charge $R(x)$ is

$$R(x) = \begin{cases} \$.30 & 0 < x \leqslant 1 \\ \$.50 & 1 < x \leqslant 2 \\ \$.70 & 2 < x \leqslant 3 \\ \text{and so on} \end{cases}$$

(A)   Graph R for $0 < x \leqslant 6$.
(B)   Find $\lim_{x \to 3^-} R(x)$, $\lim_{x \to 3^+} R(x)$, and $\lim_{x \to 3} R(x)$.
(C)   Find $\lim_{x \to 2} R(x)$ and $R(2)$.
(D)   Is R continuous at $x = 2$? At $x = 2.5$?

75.   *Compound interest.* If $1,000 is invested at 12% interest compounded quarterly, the amount of money in the account at the end of x months is given by

$$F(x) = 1,000(1.03)^{[x/3]}$$

where $[x/3] = $ (Greatest integer $\leqslant x/3$). (*Note:*   The greatest integer function is defined before Problem 29.)

(A)   Graph F for $0 \leqslant x \leqslant 12$.
(B)   For what values of x on the interval [0, 12] is F discontinuous?
(C)   Is F continuous on the right at $x = 9$? On the left at $x = 9$?

76.   *Compound interest.* Use the function F in Problem 75 to find the following:

(A) $\lim_{x \to 8^-} F(x)$, $\lim_{x \to 8^+} F(x)$, and $\lim_{x \to 8} F(x)$
(B) $\lim_{x \to 5} F(x)$ and $F(5)$
(C)   Is F continuous at $x = 5$?

Life Sciences

77.   *Animal supply.* A medical laboratory raises its own rabbits. The number of rabbits $N(t)$ available at any time t depends on the number of births and deaths. When a birth or death occurs, the function N generally has a discontinuity, as shown in the figure.

(A)   Where is the function N discontinuous?
(B) $\lim_{t \to t_5} N(t) = ?$, $N(t_5) = ?$
(C) $\lim_{t \to t_3} N(t) = ?$, $N(t_3) = ?$

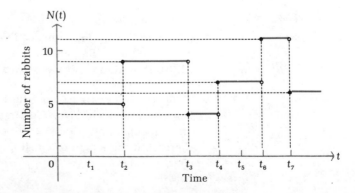

Social Sciences   **78.** *Learning.* The graph shown here might represent the history of a
particular person learning the material on limits and continuity in
this book. At time t_2, the student's mind goes blank during a quiz. At
time t_4, the instructor explains a concept particularly well, and sud-
denly, a big jump in understanding takes place.

(A) Where is the function p discontinuous?

(B) $\lim_{t \to t_1} p(t) = ?, \quad p(t_1) = ?$

(C) $\lim_{t \to t_2} p(t) = ?, \quad p(t_2) = ?$

(D) $\lim_{t \to t_4} p(t) = ?, \quad p(t_4) = ?$

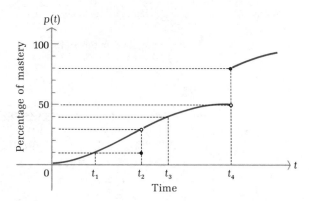

10-3   Increments, Tangent Lines, and Rates of Change

- Increments
- Slope and Tangent Line
- Average and Instantaneous Rates of Change

We will now use the concept of limit to solve two of the three basic
problems of calculus stated at the beginning of this chapter. The parts of
Figure 1 that we will concentrate on are repeated in Figure 7 (page 570).

■ Increments

Before pursuing these problems, we digress for a moment to introduce
increment notation. If we are given a function defined by $y = f(x)$ and the
independent variable x changes from x_1 to x_2, then the dependent variable
y will change from $y_1 = f(x_1)$ to $y_2 = f(x_2)$ (see Figure 8). Mathematically,
the change in x and the corresponding change in y, called **increments in x
and y,** respectively, are denoted by Δx and Δy (read "delta x" and "delta
y").

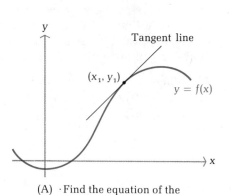

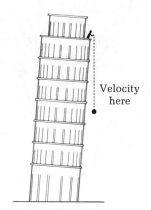

(A) · Find the equation of the
tangent line at (x_1, y_1)
given $y = f(x)$

(B) Find the instantaneous
velocity of a falling
object

Figure 7

Increments

For $y = f(x)$ (see Figure 8)

$$\Delta x = x_2 - x_1$$
$$x_2 = x_1 + \Delta x$$
$$\Delta y = y_2 - y_1$$
$$= f(x_2) - f(x_1)$$
$$= f(x_1 + \Delta x) - f(x_1)$$

Δy represents the change in y corresponding to a Δx change in x.

[*Note:* Δy depends on the function f, the input x, and the increment Δx.]

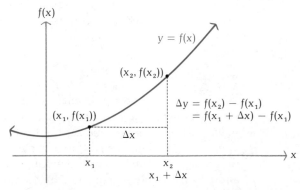

Figure 8

Example 7   Given the function

$$f(x) = \frac{x^2}{2}$$

(A) Find Δx, Δy, and $\Delta y/\Delta x$ for $x_1 = 1$ and $x_2 = 2$.
(B) Find

$$\frac{f(x_1 + \Delta x) - f(x_1)}{\Delta x}$$

for $x_1 = 1$ and $\Delta x = 2$.

Solutions   (A) $\Delta x = x_2 - x_1 = 2 - 1 = 1$
$\Delta y = f(x_2) - f(x_1)$

$$= f(2) - f(1) = \frac{4}{2} - \frac{1}{2} = \frac{3}{2}$$

$$\frac{\Delta y}{\Delta x} = \frac{f(x_2) - f(x_1)}{x_2 - x_1} = \frac{3/2}{1} = \frac{3}{2}$$

(B) $\dfrac{f(x_1 + \Delta x) - f(x_1)}{\Delta x} = \dfrac{f(1 + 2) - f(1)}{2}$

$$= \frac{f(3) - f(1)}{2} = \frac{(9/2) - (1/2)}{2} = \frac{4}{2} = 2$$

Problem 7   Given the function $f(x) = x^2 + 1$:

(A) Find Δx, Δy, and $\Delta y/\Delta x$ for $x_1 = 2$ and $x_2 = 3$.
(B) Find

$$\frac{f(x_1 + \Delta x) - f(x_1)}{\Delta x}$$

for $x_1 = 1$ and $\Delta x = 2$.

■ Slope and Tangent Line

From plane geometry, we know that a tangent to a circle is a line that passes through one and only one point on the circle; but how do we define and find a tangent line to a graph of a function at a point? The concept of the slope of a straight line (see Section 5-1) will play a central role in the process. If we pass a straight line through two points on the graph of $y = f(x)$, as in Figure 9 (next page), we obtain a secant line. Given the coordinates of the two points, we can find the slope of the secant line using the point–slope formula from Section 5-1. (This is exactly what we did in Figure 3 in the last section to motivate the concept of limit.)

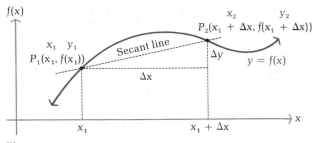

Figure 9

$$\textbf{Secant line slope} = \frac{y_2 - y_1}{x_2 - x_1} = \frac{f(x_1 + \Delta x) - f(x_1)}{\Delta x} = \frac{\Delta y}{\Delta x}$$

As we let Δx tend to 0, P_2 will approach P_1, and it appears that the secant lines will approach a limiting position and the secant slopes will approach a limiting value (see Figure 10). If they do, then we will call the line that the secant lines approach the *tangent line to the graph* at $(x_1, f(x_1))$, and the limiting slope will be the slope of the tangent line. This leads to the following definition of a tangent line:

Tangent Line

Given the graph of $y = f(x)$, then the **tangent line** at $(x_1, f(x_1))$ is the line that passes through this point with slope

$$\textbf{Tangent line slope} = \lim_{\Delta x \to 0} \frac{f(x_1 + \Delta x) - f(x_1)}{\Delta x} \qquad (1)$$

if the limit exists. The slope of the tangent line is also referred to as the **slope of the graph** at $(x_1, f(x_1))$. [Actually, in much of the work that follows, our main interest will be in the *slope of the graph of* $y = f(x)$ at $(x_1, f(x_1))$ rather than in the tangent line itself.]

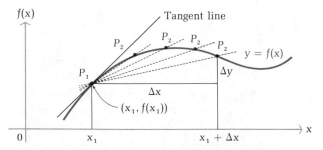

Figure 10 Dotted lines are secant lines for smaller and smaller Δx.

Example 8   Given $f(x) = x^2$, find the slope and equation of the tangent line at $x = 1$. Sketch the graph of f, the tangent line at $(1, f(1))$, and the secant line passing through $(1, f(1))$ and $(2, f(2))$.

Solution   First, we find the slope of the tangent line using equation (1).

$$\frac{f(1 + \Delta x) - f(1)}{\Delta x} = \frac{(1 + \Delta x)^2 - 1^2}{\Delta x}$$

We are computing the slope of a secant line passing through $(1, f(1))$

$$= \frac{1 + 2\Delta x + (\Delta x)^2 - 1}{\Delta x}$$

and $(1 + \Delta x, f(1 + \Delta x))$— see Figure 9.

$$= \frac{2\Delta x + (\Delta x)^2}{\Delta x}$$

$$= \frac{\Delta x(2 + \Delta x)}{\Delta x} = 2 + \Delta x \qquad \Delta x \neq 0$$

Tangent line slope $= \lim\limits_{\Delta x \to 0} \dfrac{f(1 + \Delta x) - f(1)}{\Delta x}$

This is also the slope of the graph of $f(x) = x^2$ at $(1, f(1))$.

$$= \lim\limits_{\Delta x \to 0} (2 + \Delta x) = 2$$

Now, to find the tangent line **equation,** we use the point–slope formula and substitute our known values:

$$y - y_1 = m(x - x_1)$$
$$x_1 = 1 \qquad y_1 = f(x_1) = f(1) = 1^2 = 1 \qquad m = 2$$

So,

$$y - 1 = 2(x - 1)$$
$$y - 1 = 2x - 2 \quad \text{or} \quad y = 2x - 1 \qquad \text{Tangent line equation}$$

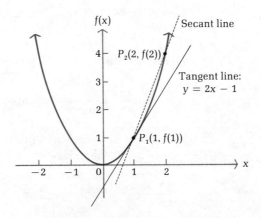

Problem 8   Find the equation of the tangent line for the graph of $f(x) = x^2$ at $x = 2$. Write the answer in the form $y = mx + b$.

■ Average and Instantaneous Rates of Change

We now show how increments and limits can be used to analyze rate problems. In the process, we will solve the second basic calculus problem we stated at the beginning of the chapter.

Example 9
Velocity

A small steel ball dropped from a tower will fall a distance of y feet in x seconds, as given approximately by the formula (from physics) $y = f(x) = 16x^2$. Let us determine the ball's position on a coordinate line at various times (see Figure 11). Our ultimate objective is to find the ball's *velocity* at a given instant, say, at the end of 2 seconds.

(A)   Find x_2 and Δy for $x_1 = 2$ and $\Delta x = 1$.
(B)   Find the average velocity for the time change in part A.
(C)   Find an expression for the average velocity from $x = 2$ to $x = 2 + \Delta x$, where Δx represents a small but arbitrary change in time and $\Delta x \neq 0$ (see Figure 11).
(D)   Find $\lim_{\Delta x \to 0} (\Delta y/\Delta x)$ using $\Delta y/\Delta x$ from part C.

Solutions   (A) $x_2 = x_1 + \Delta x = 2 + 1 = 3$

$$\Delta y = f(x_1 + \Delta x) - f(x_1)$$

$$= f(3) - f(2)$$

$$= 16(3^2) - 16(2^2)$$

$$= 144 - 64 = 80 \text{ ft} \qquad \text{Distance fallen from end of 2 seconds to end of 3 seconds (see Figure 11)}$$

(B)   Recall the formula $d = rt$, which can be written in the form

$$r = \frac{d}{t} = \frac{\text{Total distance}}{\text{Elapsed time}} = \text{Average rate}$$

For example, if a person drives from San Francisco to Los Angeles — a distance of about 420 miles — in 10 hours, then the average rate is

$$r = \frac{d}{t} = \frac{420}{10} = 42 \text{ miles per hour}$$

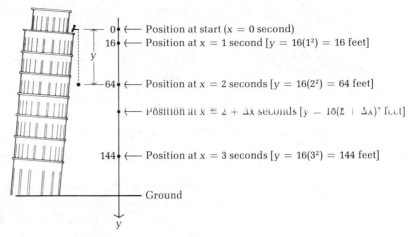

Figure 11 *Note:* Positive *y* direction is down.

Sometimes the person will be traveling faster and sometimes slower, but the *average rate* is 42 miles per hour. In our present problem, it is clear from Figure 11 that the ball is *accelerating* (falling faster and faster), but we can compute an average rate, or average velocity, just as we did for the trip from San Francisco to Los Angeles:

$$\textbf{Average velocity} = \frac{\text{Total distance}}{\text{Elapsed time}}$$

$$= \frac{\Delta y}{\Delta x} = \frac{f(3) - f(2)}{1} = \frac{80}{1} = 80 \text{ feet per second}$$

Thus, the average velocity from the end of 2 seconds to the end of 3 seconds is 80 feet per second.

(C)   Average velocity $= \dfrac{\Delta y}{\Delta x} = \dfrac{f(2 + \Delta x) - f(2)}{\Delta x}$ $\Delta x \neq 0$

$$= \frac{16(2 + \Delta x)^2 - 16(2^2)}{\Delta x}$$

$$= \frac{16[4 + 4\Delta x + (\Delta x)^2] - 64}{\Delta x}$$

$$= \frac{64 + 64\Delta x + 16(\Delta x)^2 - 64}{\Delta x}$$

$$= \frac{64\Delta x + 16(\Delta x)^2}{\Delta x} = \frac{\Delta x(64 + 16\Delta x)}{\Delta x} = 64 + 16\Delta x$$

Note that if $\Delta x = 1$, the average velocity is 80 feet per second; if $\Delta x = 0.5$, then the average velocity is 72 feet per second; if $\Delta x = 0.01$, then the average velocity is 64.16 feet per second; and so on. The smaller Δx gets, the closer the average velocity gets to 64 feet per second.

(D) $\lim\limits_{x \to 0} \dfrac{\Delta y}{\Delta x} = \lim\limits_{\Delta x \to 0} \dfrac{f(2 + \Delta x) - f(2)}{\Delta x}$

$\qquad\qquad = \lim\limits_{\Delta x \to 0} (64 + 16\Delta x)$

$\qquad\qquad = 64$ feet per second

We call 64 feet per second the **instantaneous velocity** at $x = 2$ seconds, and we have solved the second basic problem stated at the beginning of this chapter!

The discussion in Example 9 leads to the following general definitions of average rate and instantaneous rate:

Average and Instantaneous Rates

For $y = f(x)$

$$\textbf{Average rate} = \frac{\Delta y}{\Delta x} = \frac{f(x_2) - f(x_1)}{x_2 - x_1} = \frac{f(x_1 + \Delta x) - f(x_1)}{\Delta x}$$

$$\textbf{Instantaneous rate} = \lim_{\Delta x \to 0} \frac{\Delta y}{\Delta x} = \lim_{\Delta x \to 0} \frac{f(x_1 + \Delta x) - f(x_1)}{\Delta x}$$

if the limit exists

Problem 9

For the falling steel ball in Example 9, find:

(A) The average velocity from $x = 1$ to $x = 2$
(B) The average velocity from $x = 1$ to $x = 1 + \Delta x$
(C) The instantaneous velocity at $x = 1$

Now we consider a slightly different type of rate problem, but we will use the same approach as in Example 9.

Example 10
Supply

Suppose a produce grower is willing to supply crates of oranges according to the supply function illustrated in Figure 12 [$S(x) = 100x^2$]. At \$2 per crate, the supplier would be willing to supply $S(2) = 100(2^2) = 400$ crates of oranges; at \$4 per crate, the supplier would be willing to supply $S(4) = 100(4^2) = 1{,}600$ crates. As the price goes up, the supplier is willing to supply more oranges, just as we would expect.

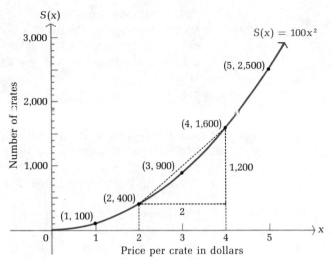

Figure 12

(A) What is the average rate of change in supply from $2 per crate to $4 per crate?

(B) What is the average rate of change in supply from $2 per crate to $(2 + \Delta x)$ per crate?

(C) What value does $\Delta S/\Delta x$ in part B approach as Δx tends to 0?

Solutions

(A) $$\frac{\Delta S}{\Delta x} = \frac{S(x_2) - S(x_1)}{x_2 - x_1}$$

$$= \frac{S(4) - S(2)}{4 - 2}$$

$$= \frac{1,600 - 400}{2} = \frac{1,200}{2} = 600 \text{ crates per dollar}$$

(B) $$\frac{\Delta S}{\Delta x} = \frac{S(2 + \Delta x) - S(2)}{\Delta x}$$

$$= \frac{100(2 + \Delta x)^2 - 100(2^2)}{\Delta x}$$

$$= \frac{100[4 + 4\Delta x + (\Delta x)^2] - 400}{\Delta x}$$

$$= \frac{400\Delta x + 100(\Delta x)^2}{\Delta x} = \frac{\Delta x(400 + 100\Delta x)}{\Delta x} = 400 + 100\Delta x$$

(C) $$\lim_{\Delta x \to 0} \frac{\Delta S}{\Delta x} = \lim_{\Delta x \to 0} (400 + 100\Delta x)$$

$$= 400 \text{ crates per dollar}$$

This is an "instantaneous" rate. It indicates the change in supply per unit change in dollars at the $2 price level. It approximates the actual change in supply, $S(3)-S(2) = 900 - 400 = 500$, for a price increase of 1 dollar at the $2 price level. We will say more about these concepts later.

Problem 10   For Example 10, find:

(A)   The average rate of change in supply from $1 per crate to $3 per crate.

(B)   The average rate of change in supply from $1 per crate to $(1 + \Delta x)$ per crate.

(C)   What value does $\Delta S/\Delta x$ in part B approach as Δx tends to 0?

Answers to
Matched Problems

7. (A) $\Delta x = 1, \Delta y = 5, \Delta y/\Delta x = 5$ (B) 4
8. $y = 4x - 4$
9. (A) 48 feet per second   (B) $32 + 16\Delta x$
    (C) 32 feet per second
10. (A) 400 crates per dollar   (B) $200 + 100\Delta x$
    (C) 200 crates per dollar

Exercise 10-3

In Problems 1–14 find the indicated quantities for $y = f(x) = 3x^2$.

A

1. Δx, Δy,   and $\Delta y/\Delta x$,   given $x_1 = 1$ and $x_2 = 4$

2. Δx, Δy,   and $\Delta y/\Delta x$,   given $x_1 = 2$ and $x_2 = 5$

3. $\dfrac{f(x_1 + \Delta x) - f(x_1)}{\Delta x}$,   given $x_1 = 1$ and $\Delta x = 2$

4. $\dfrac{f(x_1 + \Delta x) - f(x_1)}{\Delta x}$,   given $x_1 = 2$ and $\Delta x = 1$

5. $\dfrac{y_2 - y_1}{x_2 - x_1}$,   given $x_1 = 1$ and $x_2 = 3$

6. $\dfrac{y_2 - y_1}{x_2 - x_1}$,   given $x_1 = 2$ and $x_2 = 3$

7. $\dfrac{\Delta y}{\Delta x}$,   given $x_1 = 1$ and $x_2 = 3$

8. $\dfrac{\Delta y}{\Delta x}$,   given $x_1 = 2$ and $x_2 = 3$

B

9.   The average rate of change of y, for x changing from 1 to 4

10.   The average rate of change of y, for x changing from 2 to 5

11. (A) $\dfrac{f(2 + \Delta x) - f(2)}{\Delta x}$ (simplify)

    (B) What does the ratio in part A approach as Δx approaches 0?

12. (A) $\dfrac{f(3 + \Delta x) - f(3)}{\Delta x}$ (simplify)

    (B) What does the ratio in part A approach as Δx approaches 0?

13. (A) $\dfrac{f(4 + \Delta x) - f(4)}{\Delta x}$ (simplify)

    (B) What does the ratio in part A approach as Δx approaches 0?

14. (A) $\dfrac{f(5 + \Delta x) - f(5)}{\Delta x}$ (simplify)

    (B) What does the ratio in part A approach as Δx approaches 0?

Suppose an object moves along the y axis so that its location is $y = f(x) = x^2 + x$ at time x (y is in meters and x is in seconds). Find:

15. (A) The average velocity (the average rate of change of y) for x changing from 1 to 3 seconds
    (B) The average velocity for x changing from 1 to $(1 + \Delta x)$ seconds
    (C) The instantaneous velocity at $x = 1$

16. (A) The average velocity (the average rate of change of y) for x changing from 2 to 4 seconds
    (B) The average velocity for x changing from 2 to $(2 + \Delta x)$ seconds
    (C) The instantaneous velocity at $x = 2$

In Problems 17 and 18, find each of the following for the graph of $y = f(x) = x^2 + x$:

17. (A) The slope of the secant line joining $(1, f(1))$ and $(3, f(3))$
    (B) The slope of the secant line joining $(1, f(1))$ and $(1 + \Delta x, f(1 + \Delta x))$
    (C) The slope of the tangent line at $(1, f(1))$
    (D) The equation of the tangent line at $(1, f(1))$

18. (A) The slope of the secant line joining $(2, f(2))$ and $(4, f(4))$
    (B) The slope of the secant line joining $(2, f(2))$ and $(2 + \Delta x, f(2 + \Delta x))$
    (C) The slope of the tangent line at $(2, f(2))$
    (D) The equation of the tangent line at $(2, f(2))$

C 19. If an object moves on the x axis so that it is at $x = f(t) = t^2 - t$ at time t (t measured in seconds and x measured in meters), find the instantaneous velocity of the object at $t = 2$.

20. Find the equation of the tangent line for the graph of $y = x^2 - x$ at $x = 2$.

Applications

21. *Income.* The per capita income in the United States from 1969 to 1973 is given approximately in the table. Find the average rate of change of per capita income for a time change from:

(A)   1969 to 1971   (B)   1971 to 1973

Year	1969	1970	1971	1972	1973
Income	$3,700	$3,900	$4,100	$4,500	$5,000

22. *Demand function.* Suppose in a given grocery store people are willing to buy $D(x)$ pounds of chocolate candy per day at $x per pound, as given by the demand function

$$D(x) = 100 - x^2 \qquad \$1 \leqslant x \leqslant \$10$$

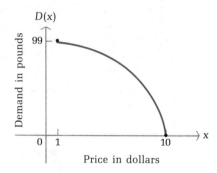

Note that as price goes up, demand goes down (see the figure).

(A)   Find the average rate of change in demand for a price change from $2 to $5; that is, find $\Delta y / \Delta x$ for $x_1 = 2$ and $x_2 = 5$.

(B)   Simplify:

$$\frac{D(2 + \Delta x) - D(2)}{\Delta x}$$

(C)   What does the ratio in part B approach as Δx approaches 0? [This is called "the instantaneous rate of change of $D(x)$ with respect to x at x = 2."]

23. *Medicine.* The area of a small (healing) wound in square millimeters, where time is measured in days, is given in the table.

Area	400	360	180	120	90	72	60
Days	0	1	2	3	4	5	6

Find the average rate of change of area for the time change from:

(A)   0 to 2 days   (B)   4 to 6 days

24. *Weight–height.* A formula relating the approximate weight of an average person and his or her height is

$$W(h) = 0.0005h^3$$

where $W(h)$ is in pounds and h is in inches.

(A)   Find the average rate of change of weight for a height change from 60 to 70 inches.

(B)   Simplify:

$$\frac{W(60 + \Delta h) - W(60)}{\Delta h}$$

(C)   What does the ratio in part B approach as Δh approaches 0? [This is called "the instantaneous rate of change of $W(h)$ with respect to h at $h = 60$."]

Social Sciences   25. *Illegitimate births.* The approximate numbers of illegitimate births per 1,000 live births in the United States from 1940 to 1970 are given in the table. Find the average rate of change of illegitimate births per 1,000 live births for the time change from:

(A)   1940 to 1945   (B)   1965 to 1970

Year	1940	1945	1950	1955	1960	1965	1970
Illegitimate Births Per 1,000 Live Births	38	41	40	47	54	80	120

26. *Learning.* A certain person learning to type has an achievement record given approximately by the function

$$N(t) = 60 \left(1 - \frac{2}{t} \right) \qquad 3 \le t \le 10$$

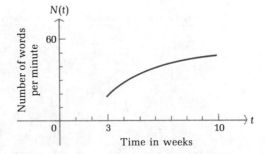

where $N(t)$ is in number of words per minute and t is in weeks. Find the average rate of change of the number of words per minute for the change in time from:

(A)   4 to 6 weeks   (B)   8 to 10 weeks

10-4   The Derivative

- The Derivative
- Slope Function and Tangent Lines
- Nonexistence of the Derivative
- Instantaneous Rates of Change
- Marginal Cost
- Summary

■ The Derivative

In the last section we found that the special limit

$$\lim_{\Delta x \to 0} \frac{f(x_1 + \Delta x) - f(x_1)}{\Delta x} \tag{1}$$

if it exists, gives us the slope of the tangent line to the graph of $y = f(x)$ at $(x_1, f(x_1))$. It also gives us the instantaneous rate of change of y per unit change in x at $x = x_1$. Formula (1) is of such basic importance to calculus and to the applications of calculus that we will give it a name and study it in detail. To keep formula (1) simple and general, we will drop the subscript on x_1 and think of the ratio

$$\frac{f(x + \Delta x) - f(x)}{\Delta x}$$

as a function of Δx, with x held fixed as we let Δx tend to 0. We are now ready to define one of the basic concepts in calculus, the *derivative*:

Derivative

For $y = f(x)$ we define the **derivative of f at x,** denoted by $f'(x)$, to be

$$f'(x) = \lim_{\Delta x \to 0} \frac{f(x + \Delta x) - f(x)}{\Delta x} \qquad \text{if the limit exists}$$

If $f'(x)$ exists, then f is said to be a **differentiable function** at x.

Thus, taking the derivative of a function f at x creates a new function f' that gives, among other things, the instantaneous rate of change of $y = f(x)$ and the slope of the tangent line to the graph of $y = f(x)$ for each x. The domain of f' is a subset of the domain of f, which will become clearer as we progress through this section.

Example 11 Find $f'(x)$, the derivative of f at x, for $f(x) = 4x - x^2$.

Solution To find $f'(x)$, we find

$$\lim_{\Delta x \to 0} \frac{f(x + \Delta x) - f(x)}{\Delta x}$$

To make the computation easier, we introduce a two-step process:

Step 1. Find $[f(x + \Delta x) - f(x)]/\Delta x$ and simplify.

$$\frac{f(x + \Delta x) - f(x)}{\Delta x} = \frac{[4(x + \Delta x) - (x + \Delta x)^2] - (4x - x^2)}{\Delta x}$$

$$= \frac{4x + 4\Delta x - x^2 - 2x\Delta x - (\Delta x)^2 - 4x + x^2}{\Delta x}$$

$$= \frac{4\Delta x - 2x\Delta x - (\Delta x)^2}{\Delta x}$$

$$= \frac{\Delta x}{\Delta x}(4 - 2x - \Delta x)$$

$$= 4 - 2x - \Delta x \qquad \Delta x \neq 0$$

Step 2. Find the limit of the result of step 1.

$$\lim_{\Delta x \to 0} \frac{f(x + \Delta x) - f(x)}{\Delta x} = \lim_{\Delta x \to 0} (4 - 2x - \Delta x)$$

$$= 4 - 2x$$

Thus, $f'(x) = 4 - 2x$.

Problem 11 Find $f'(x)$, the derivative of f at x, for $f(x) = 8x - 2x^2$.

Example 12 Find $f'(x)$, the derivative of f at x, for $f(x) = \sqrt{x} + 2$.

Solution To find $f'(x)$, we find

$$\lim_{\Delta x \to 0} \frac{f(x + \Delta x) - f(x)}{\Delta x}$$

We use the two-step process presented in Example 11.

Step 1. Find $[f(x + \Delta x) - f(x)]/\Delta x$ and simplify.

$$\frac{f(x + \Delta x) - f(x)}{\Delta x} = \frac{(\sqrt{x + \Delta x} + 2) - (\sqrt{x} + 2)}{\Delta x}$$

$$= \frac{\sqrt{x + \Delta x} + 2 - \sqrt{x} - 2}{\Delta x}$$

$$= \frac{\sqrt{x + \Delta x} - \sqrt{x}}{\Delta x}$$

Trying to apply the quotient property of limits, we find that $\lim_{\Delta x \to 0} \Delta x = 0$; hence, we cannot use it. We try rationalizing the numerator:

$$\frac{\sqrt{x + \Delta x} - \sqrt{x}}{\Delta x} \cdot \frac{\sqrt{x + \Delta x} + \sqrt{x}}{\sqrt{x + \Delta x} + \sqrt{x}} = \frac{x + \Delta x - x}{\Delta x(\sqrt{x + \Delta x} + \sqrt{x})}$$

$$= \frac{\Delta x}{\Delta x(\sqrt{x + \Delta x} + \sqrt{x})}$$

$$= \frac{1}{\sqrt{x + \Delta x} + \sqrt{x}} \qquad \Delta x \neq 0$$

Step 2. Find the limit of the result of step 1.

$$f'(x) = \lim_{\Delta x \to 0} \frac{\sqrt{x + \Delta x} - \sqrt{x}}{\Delta x}$$

$$= \lim_{\Delta x \to 0} \frac{1}{\sqrt{x + \Delta x} + \sqrt{x}} \qquad \Delta x \neq 0$$

$$= \frac{1}{\sqrt{x} + \sqrt{x}} = \frac{1}{2\sqrt{x}}$$

Note: The domain of $f(x) = \sqrt{x} + 2$ is $[0, \infty)$. Since $f'(0)$ is undefined, the domain of $f'(x) = 1/(2\sqrt{x})$ is $(0, \infty)$, a subset of the original domain.

Problem 12   Find $f'(x)$, the derivative of f at x, for $f(x) = x^{-1}$.

■ Slope Function and Tangent Lines

In the last section we defined the slope of the tangent line to the graph of $y = f(x)$ at $(x_1, f(x_1))$ to be

$$\lim_{\Delta x \to 0} \frac{f(x_1 + \Delta x) - f(x_1)}{\Delta x}$$

if the limit exists. This, of course, is $f'(x_1)$, the derivative of f at $x = x_1$. To find the equation of a tangent line to the graph of $y = f(x)$ at $(x_1, f(x_1))$, we use the point–slope form for the equation of a line, $y - y_1 = m(x - x_1)$, and the facts that $m = f'(x_1)$ and $y_1 = f(x_1)$ to obtain:

> **Tangent Line**
>
> The equation of the tangent line to the graph of $y = f(x)$ at $x = x_1$ is
>
> $$y - f(x_1) = f'(x_1)(x - x_1) \quad \text{if } f'(x_1) \text{ exists}$$

More generally,

$$f'(x) = \lim_{\Delta x \to 0} \frac{f(x + \Delta x) - f(x)}{\Delta x}$$

gives us the slope of the graph of f at *any* point $(x, f(x))$ on the graph of f for which the limit exists. In this context, we refer to f' as the **slope function** for the graph of f and we write

$$m(x) = f'(x) \qquad \text{Slope function}$$

Example 13   In Example 11 we started with the function specified by $f(x) = 4x - x^2$ and found the derivative of f at x to be $f'(x) = 4 - 2x$. Thus, the slope function for the graph of f is

$$m(x) = f'(x) = 4 - 2x$$

We will use this slope function in the following problems.

(A)   Find the slopes of the graph of f at $x = 1, 2,$ and 3.
(B)   Find the equations of the tangent lines at $x = 1, 2,$ and 3.
(C)   Sketch the tangent lines on the graph of $y = 4x - x^2$ at $x = 1, 2,$ and 3.

Solutions   (A) $m(1) = f'(1) = 4 - 2(1) = 2$
$\qquad m(2) = f'(2) = 4 - 2(2) = 0$
$\qquad m(3) = f'(3) = 4 - 2(3) = -2$

(B)   *Tangent Line at $x = 1$*   *Tangent Line at $x = 2$*

$$y - f(1) = f'(1)(x - 1) \qquad y - f(2) = f'(2)(x - 2)$$
$$y - 3 = 2(x - 1) \qquad\qquad y - 4 = 0(x - 2)$$
$$y - 3 = 2x - 2 \qquad\qquad\qquad y = 4$$
$$y = 2x + 1$$

Tangent Line at $x = 3$

$$y - f(3) = f'(3)(x - 3)$$
$$y - 3 = -2(x - 3)$$
$$y - 3 = -2x + 6$$
$$y = -2x + 9$$

(C)

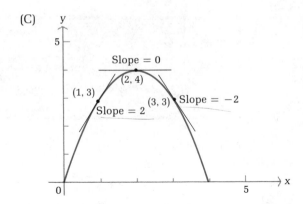

Observations

1. Slope is positive when curve is rising.
2. Slope is 0 at high point.
3. Slope is negative when curve is falling.

The observations in Example 13C will be very useful to us in Chapter 12, where we will consider the use of the derivative in graphing and the solution of maxima–minima problems.

Problem 13   In Problem 11 we started with the function specified by $f(x) = 8x - 2x^2$ and found the derivative of f at x to be $f'(x) = 8 - 4x$. Thus, the slope function for f is

$$m(x) = f'(x) = 8 - 4x$$

(A) Find the slopes of the graph of f at $x = 1, 2$, and 3.
(B) Find the equations of the tangent lines at $x = 1, 2$, and 3.
(C) Sketch the tangent lines on the graph of $y = 8x - 2x^2$ at $x = 1, 2$, and 3.

■ Nonexistence of the Derivative

The existence of a derivative at $x = a$ depends on the existence of a limit at $x = a$; that is, on the existence of

$$f'(a) = \lim_{\Delta x \to 0} \frac{f(a + \Delta x) - f(a)}{\Delta x}$$

If the limit does not exist at $x = a$, we say that the function f is **nondifferentiable at $x = a$ or $f'(a)$ does not exist.** Geometrically, a tangent line may not exist at a point, or a point may have a vertical tangent line (recall that the slope of a vertical line is not defined).

Given a function f, we now illustrate several common situations where its derivative will not exist. The derivative of f will not exist:

1.   Where f is not defined

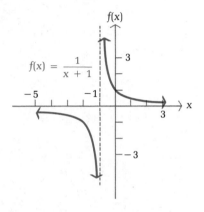

Since f is not defined at $x = -1$, the derivative of f does not exist at $x = -1$.

2.   Where f is defined but not continuous

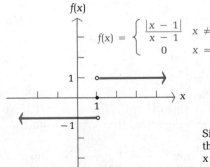

$$f(x) = \begin{cases} \dfrac{|x - 1|}{x - 1} & x \neq 1 \\ 0 & x = 1 \end{cases}$$

Since f is not continuous at $x = 1$, the derivative of f does not exist at $x = 1$.

3.   Where the graph of *f* has a vertical tangent

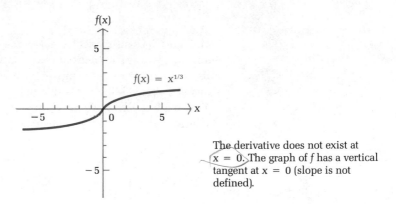

The derivative does not exist at
x = 0. The graph of *f* has a vertical
tangent at x = 0 (slope is not
defined).

4.   Where the graph of *f* is continuous but has a sharp corner (the curve is
not "smooth")

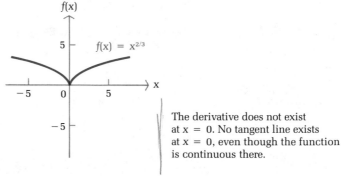

The derivative does not exist
at x = 0. No tangent line exists
at x = 0, even though the function
is continuous there.

Example 14   Show that $f'(0)$ does not exist for $f(x) = |x|$.

Solution $f'(0) = \lim_{\Delta x \to 0} \dfrac{f(0 + \Delta x) - f(0)}{\Delta x}$

$= \lim_{\Delta x \to 0} \dfrac{|0 + \Delta x| - |0|}{\Delta x}$

$= \lim_{\Delta x \to 0} \dfrac{|\Delta x|}{\Delta x}$

To show that $f'(0)$ does not exist, we compute left- and right-hand limits
and show that they are not equal:

$$\lim_{\Delta x \to 0^-} \frac{|\Delta x|}{\Delta x} = \lim_{\Delta x \to 0^-} \frac{-\Delta x}{\Delta x} = \lim_{\Delta x \to 0^-} -1 = -1$$

$$\lim_{\Delta x \to 0^+} \frac{|\Delta x|}{\Delta x} = \lim_{\Delta x \to 0^+} \frac{\Delta x}{\Delta x} = \lim_{\Delta x \to 0^+} 1 = 1$$

Geometrically, a tangent line cannot be defined at $x = 0$ because the graph has a sharp corner at this point:

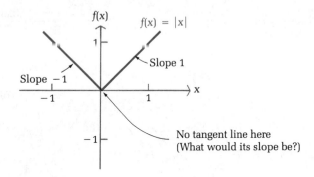

Problem 14   Show that $f'(0)$ does not exist for $f(x) = x^{1/3}$ (see illustration for situation 3).

If the derivative of f exists at $x = a$, what can we say about the function at $x = a$? We can prove that **if the derivative of f exists at $x = a$, then f must be continuous at $x = a$.** The converse of this statement is false. That is, if a function is continuous at $x = a$, then its derivative at $x = a$ may or may not exist (see illustrations for situations 3 and 4).

■ Instantaneous Rates of Change

From the definition of instantaneous rate of change of $f(x)$ at x given in Section 10-3, we see that the instantaneous rate of change is simply the derivative of f at x—that is, $f'(x)$.

Example 15   Refer to Example 9 in Section 10-3. Find a function that will give the instantaneous velocity, v, of the falling steel ball at any time x. Find the velocity at $x = 2$, 3, and 5 seconds.

Solution   Recall that the distance y (in feet) that the ball falls in x seconds is given by

$$y = f(x) = 16x^2$$

The instantaneous velocity function is $v = f'(x)$; thus,

$$v = f'(x) = \lim_{\Delta x \to 0} \frac{f(x + \Delta x) - f(x)}{\Delta x}$$

We will find $f'(x)$ using the two-step process described in Example 11.

Step 1. Find $[f(x + \Delta x) - f(x)]/\Delta x$ and simplify.

$$\frac{f(x + \Delta x) - f(x)}{\Delta x} = \frac{[16(x + \Delta x)^2] - (16x^2)}{\Delta x}$$

$$= \frac{16x^2 + 32x\Delta x + 16(\Delta x)^2 - 16x^2}{\Delta x}$$

$$= \frac{32x\Delta x + 16(\Delta x)^2}{\Delta x}$$

$$= \frac{\Delta x}{\Delta x}(32x + 16\Delta x) = 32x + 16\Delta x \qquad \Delta x \neq 0$$

Step 2. Find the limit of the result of step 1.

$$\lim_{\Delta x \to 0} \frac{f(x + \Delta x) - f(x)}{\Delta x} = \lim_{\Delta x \to 0}(32x + 16\Delta x)$$

$$= 32x$$

Thus,

$$v = f'(x) = 32x$$

The instantaneous velocities at $x = 2$, 3, and 5 seconds are

$f'(2) = 32(2) = 64$ feet per second

$f'(3) = 32(3) = 96$ feet per second

$f'(5) = 32(5) = 160$ feet per second

An instantaneous rate of 64 feet per second at the end of 2 seconds means that *if* the rate were to remain constant for the next second, the object would fall an additional 64 feet. If the object is accelerating or decelerating (that is, if the rate does not remain constant), then the instantaneous rate is an approximation of what actually happens during the next second.

Problem 15   A steel ball falls so that its distance y (in feet) at time x (in seconds) is given by $y = f(x) = 16x^2 - 4x$.

(A)   Find a function that will give the instantaneous velocity v at time x.
(B)   Find the velocity at $x = 2$, 4, and 6 seconds.

■ Marginal Cost

In business and economics one is often interested in the rate at which something is taking place. A manufacturer, for example, is not only interested in the total cost $C(x)$ at certain production levels x, but is also interested in the rate of change of costs at various production levels.

In economics the word **marginal** refers to a rate of change, that is, to a derivative. Thus, if

$C(x) =$ Total cost of producing x units during some unit of time then

$$C'(x) = \text{Marginal cost}$$

$$= \text{Rate of change in cost per unit change in production at an output level of x units}$$

Just as with instantaneous velocity, $C'(x)$ is an instantaneous rate. It indicates the change in cost for a 1 unit change in production at a production level of x units *if* the rate were to remain constant for the next unit change in production. If the rate does *not* remain constant, then the instantaneous rate is an approximation of what actually happens during the next unit change in production. Example 16 should help to clarify these ideas.

Example 16
Marginal Cost

Suppose the total cost $C(x)$ in thousands of dollars for manufacturing x sailboats per week is given by the function shown in the figure:

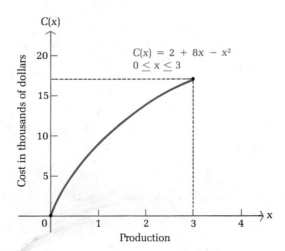

Find:

(A) The marginal cost at x
(B) The marginal cost at x = 1, 2, and 3 unit levels of production

Solutions

(A) Marginal cost at x is

$$C'(x) = \lim_{\Delta x \to 0} \frac{C(x + \Delta x) - C(x)}{\Delta x}$$

which we find using the two-step process discussed in Example 11 (steps omitted here).

Marginal cost $= C'(x) = 8 - 2x$

(B)   Marginal costs at $x = 1, 2$, and 3 unit levels of production are:

$$C'(1) = 8 - 2(1) = 6 \qquad \text{\$6,000 per unit increase in production}$$
$$C'(2) = 8 - 2(2) = 4 \qquad \text{\$4,000 per unit increase in production}$$
$$C'(3) = 8 - 2(3) = 2 \qquad \text{\$2,000 per unit increase in production}$$

Notice that, as production goes up, the marginal cost goes down, as we might expect.

Let us now look at the marginal cost at the 1 unit level of production and interpret the result geometrically:

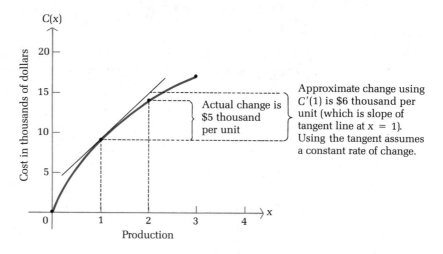

Problem 16   Repeat Example 16 with the cost function $C(x) = 3 + 10x - x^2$, $0 \leqslant x \leqslant 4$.

■ Summary

The concept of the derivative is a very powerful mathematical idea, and its applications are many and varied. In the next three sections we will develop formulas and general properties of derivatives that will enable us to find the derivatives of many functions without having to go through the (two-step) limiting process each time.

Answers to Matched Problems

11. $f'(x) = 8 - 4x$ 12. $f'(x) = -1/x^2$ or $-x^{-2}$
13. (A) $m(1) = 4$, $m(2) = 0$, $m(3) = -4$
     (B) At $x = 1$: $y = 4x + 2$; at $x = 2$: $y = 8$; at $x = 3$: $y = -4x + 18$

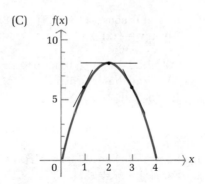

(C) $f(x)$

14. $\displaystyle \lim_{\Delta x \to 0} \frac{f(0 + \Delta x) - f(0)}{\Delta x} = \lim_{\Delta x \to 0} \frac{1}{(\Delta x)^{2/3}}$ does not exist $[1/(\Delta x)^{2/3}$ increases without bound as Δx approaches 0 from either side]

15. (A) $v = f'(x) = 32x - 4$
   (B) $f'(2) - 60$ feet per second, $f'(4) = 124$ feet per second, $f'(6) = 188$ feet per second

16. (A) Marginal cost $= C'(x) = 10 - 2x$
   (B) Marginal costs at 1, 2, and 3 unit levels of production are:

   $C'(1) = 8$ \$8,000 per unit increase

   $C'(2) = 6$ \$6,000 per unit increase

   $C'(3) = 4$ \$4,000 per unit increase

Exercise 10-4

A   In Problems 1–10, find $f'(x)$ for each indicated function; then find $f'(1)$, $f'(2)$, and $f'(3)$.

1. $f(x) = 2x - 3$
2. $f(x) = 4x + 3$
3. $f(x) = 6x - x^2$
4. $f(x) = 8x - x^2$

B

5. $f(x) = \dfrac{1}{x + 1}$
6. $f(x) = \dfrac{1}{x - 5}$
7. $f(x) = \sqrt{x} - 3$
8. $f(x) = 2 - \sqrt{x}$
9. $f(x) = x^{-2}$
10. $f(x) = \dfrac{1}{x^2}$

11. If an object moves along a line so that it is at $y = f(x) = 4x^2 - 2x$ at time x (in seconds), find the instantaneous velocity function $v = f'(x)$ and find the velocity at times 1, 3, and 5 seconds (y is measured in feet).

12. Repeat Problem 11 with $f(x) = 8x^2 - 4x$.
13. Given $y = f(x) = x^2$, $-3 \leqslant x \leqslant 3$:

   (A) Find the slope function $m = f'(x)$.
   (B) Find the slope of the tangent line to the graph of $y = x^2$ at $x = -2$, 0, and 2.
   (C) Find the equations of the tangents at $x = -2$, 0, and 2.
   (D) Sketch the tangent lines on the graph at $x = -2$, 0, and 2.

14. Repeat Problem 13 for $y = f(x) = x^2 + 1$, $-3 \leqslant x \leqslant 3$.

C 15. For $f(x) = x^3 + 2x$, find:

   (A) $f'(x)$ (B) $f'(1)$ and $f'(3)$

16. For $f(x) = x^2 - 3x^3$, find:

   (A) $f'(x)$ (B) $f'(1)$ and $f'(2)$

Applications

Business & Economics

17. *Marginal cost.* The total cost per day, $C(x)$ (in hundreds of dollars), for manufacturing x windsurfers is given by

$$C(x) = 3 + 10x - x^2 \qquad 0 \leqslant x \leqslant 4$$

   (A) Find the marginal cost at x.
   (B) Find the marginal cost at $x = 1, 3$, and 4 unit levels of production.

18. *Marginal cost.* Repeat Problem 17 for $C(x) = 5 + 12x - x^2$, $0 \leqslant x \leqslant 4$.

Life Sciences

19. *Negative growth.* A colony of bacteria was treated with a poison, and the number of survivors $N(t)$, in thousands, after t hours was found to be given approximately by

$$N(t) = t^2 - 8t + 16 \qquad 0 \leqslant t \leqslant 4$$

   (A) Find $N'(t)$.
   (B) Find the rate of change of the colony at $t = 1, 2$, and 3.

Social Sciences

20. *Learning.* A private foreign language school found that the average person learned $N(t)$ basic phrases in t continuous hours, as given approximately by

$$N(t) = 14t - t^2 \qquad 0 \leqslant t \leqslant 7$$

   (A) Find $N'(t)$.
   (B) Find the rate of learning at $t = 1, 3$, and 6 hours.

10-5   Derivatives of Constants, Power Forms, and Sums

- Derivative of a Constant
- Power Rule
- Derivative of a Constant Times a Function
- Derivatives of Sums and Differences
- Applications

In the last section we defined the derivative of f at x as

$$f'(x) = \lim_{\Delta x \to 0} \frac{f(x + \Delta x) - f(x)}{\Delta x}$$

(if the limit exists) and we used this definition and a two-step process to find the derivatives of a number of functions. In this and the next two sections we will develop some rules based on this definition that will enable us to determine the derivatives of a rather large class of functions without having to go through the two-step process each time.

Before starting on these rules, we list some symbols that are widely used to represent derivatives:

Derivative Notation

Given $y = f(x)$, then

$$f'(x) \qquad y' \qquad \frac{dy}{dx} \qquad D_x f(x)$$

all represent the derivative of f at x.

Each of these symbols for derivatives has its particular advantage in certain situations. All of them will become familiar to you after a little experience.

■ Derivative of a Constant

Suppose

$$f(x) = C \qquad C \text{ a constant} \qquad \text{A constant function}$$

Geometrically, the graph of $f(x) = C$ is a horizontal straight line with slope 0 (see Figure 13); hence, we would expect $D_x C = 0$. We will show that this is actually the case using the definition of the derivative and the two-step

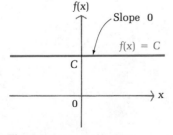

Figure 13

process introduced in the last section. We want to find

$$f'(x) = \lim_{\Delta x \to 0} \frac{f(x + \Delta x) - f(x)}{\Delta x} \qquad \text{Definition of } f'(x)$$

Step 1.

$$\frac{f(x + \Delta x) - f(x)}{\Delta x} = \frac{C - C}{\Delta x} = \frac{0}{\Delta x} = 0 \qquad \Delta x \neq 0$$

Step 2.

$$\lim_{\Delta x \to 0} 0 = 0$$

Thus,

$$D_x C = 0$$

And we conclude that **the derivative of any constant is 0.**

Derivative of a Constant

If $y = f(x) = C$, then

$$f'(x) = 0$$

Also, $y' = 0$, $dy/dx = 0$, and $D_x C = 0$.

Note: When we write $D_x C = 0$, we mean $D_x f(x) = 0$, where $f(x) = C$.

Example 17   (A) If $f(x) = 3$, then $f'(x) = 0$.   (B) If $y = -1.4$, then $y' = 0$.
(C) If $y = \pi$, then $dy/dx = 0$.   (D) $D_x(23) = 0$

Problem 17   Find:

(A) $f'(x)$ for $f(x) = -24$ (B) y' for $y = 12$
(C) dy/dx for $y = -\sqrt{7}$ (D) $D_x(-\pi)$

■ Power Rule

We are interested in finding

$$D_x x^n \qquad \text{for } n \text{ a positive integer}$$

Let us first find $D_x x^2$ and $D_x x^3$, and then generalize from these results.

For $f(x) = x^2$, we have

Step 1.

$$\frac{f(x + \Delta x) - f(x)}{\Delta x} = \frac{(x + \Delta x)^2 - x^2}{\Delta x}$$

$$= \frac{x^2 + 2x\Delta x + (\Delta x)^2 - x^2}{\Delta x}$$

$$= \frac{2x\Delta x + (\Delta x)^2}{\Delta x} \qquad \text{Factor out } \Delta x.$$

$$= \frac{\Delta x(2x + \Delta x)}{\Delta x} = 2x + \Delta x \qquad \Delta x \neq 0$$

Step 2.

$$\lim_{\Delta x \to 0} \frac{f(x + \Delta x) - f(x)}{\Delta x} = \lim_{\Delta x \to 0} (2x + \Delta x) = 2x$$

Thus,

$$D_x x^2 = 2x \tag{1}$$

Now, let $f(x) = x^3$. Then we have

Step 1.

$$\frac{f(x + \Delta x) - f(x)}{\Delta x} = \frac{(x + \Delta x)^3 - x^3}{\Delta x}$$

$$= \frac{x^3 + 3x^2\Delta x + 3x(\Delta x)^2 + (\Delta x)^3 - x^3}{\Delta x}$$

$$= \frac{3x^2\Delta x + 3x(\Delta x)^2 + (\Delta x)^3}{\Delta x} \qquad \text{Factor out } \Delta x.$$

$$= \frac{\Delta x[3x^2 + 3x\Delta x + (\Delta x)^2]}{\Delta x}$$

$$= 3x^2 + 3x\Delta x + (\Delta x)^2 \qquad \Delta x \neq 0$$

Step 2.

$$\lim_{\Delta x \to 0} \frac{f(x + \Delta x) - f(x)}{\Delta x} = \lim_{\Delta x \to 0} [3x^2 + 3x\Delta x + (\Delta x)^2] = 3x^2$$

Thus,

$$D_x x^3 = 3x^2 \tag{2}$$

Comparing equations (1) and (2) suggests $D_x x^4 = 4x^3$ and, in general,

$$D_x x^n = nx^{n-1} \qquad n \text{ a positive integer}$$

Let us see that this is the case.

Step 1.   If $f(x) = x^n$ (n a positive integer), then

$$\frac{f(x + \Delta x) - f(x)}{\Delta x} = \frac{(x + \Delta x)^n - x^n}{\Delta x}$$

To simplify we use the binomial formula (Appendix A):

$$(a + b)^n = a^n + na^{n-1}b + \frac{n(n-1)}{2} a^{n-2}b^2 + \cdots + b^n$$

Thus,

$$(x + \Delta x)^n = x^n + nx^{n-1}\Delta x + \frac{n(n-1)}{2} x^{n-2}(\Delta x)^2 + \cdots + (\Delta x)^n$$

and

$$\frac{(x + \Delta x)^n - x^n}{\Delta x}$$

$$= \frac{x^n + nx^{n-1}\Delta x + \frac{n(n-1)}{2} x^{n-2}(\Delta x)^2 + \cdots + (\Delta x)^n - x^n}{\Delta x}$$

$$= \frac{nx^{n-1}\Delta x + \frac{n(n-1)}{2} x^{n-2}(\Delta x)^2 + \cdots + (\Delta x)^n}{\Delta x} \qquad \text{Factor out } \Delta x.$$

$$= \frac{\Delta x \left[nx^{n-1} + \frac{n(n-1)}{2} x^{n-2}\Delta x + \cdots + (\Delta x)^{n-1} \right]}{\Delta x}$$

$$= nx^{n-1} + \frac{n(n-1)}{2} x^{n-2}\Delta x + \cdots + (\Delta x)^{n-1}$$

Step 2.

$$\lim_{\Delta x \to 0} \frac{f(x + \Delta x) - f(x)}{\Delta x} = \lim_{\Delta x \to 0} \left[nx^{n-1} + \frac{n(n-1)}{2} x^{n-2}\Delta x + \cdots + (\Delta x)^{n-1} \right]$$

$$= nx^{n-1}$$

and we conclude that

$$D_x x^n = nx^{n-1}$$

It can be shown that this formula holds for *any* real number n. We will assume this general result for the remainder of this book.

Power Rule

If $y = f(x) = x^n$, where n is a real number, then

$$f'(x) = nx^{n-1}$$

Example 18 (A) If $f(x) = x^5$, then $f'(x) = 5x^{5-1} = 5x^4$.

(B) If $y = \dfrac{1}{x^3} = x^{-3}$, then $y' = -3x^{-3-1} = -3x^{-4}$ or $\dfrac{-3}{x^4}$.

(C) If $y = x^{5/3}$, then $\dfrac{dy}{dx} = \dfrac{5}{3}\, x^{(5/3)-1} = \dfrac{5}{3}\, x^{2/3}$.

(D) $D_x \sqrt{x} = D_x x^{1/2} = \dfrac{1}{2}\, x^{(1/2)-1} = \dfrac{1}{2}\, x^{-1/2} = \dfrac{1}{2\sqrt{x}}$

Problem 18 Find:

(A) $f'(x)$ for $f(x) = x^3$ (B) y' for $y = x^{3/2}$

(C) $\dfrac{dy}{dx}$ for $y = \dfrac{1}{x^2}$ or x^{-2} (D) $D_x \dfrac{1}{\sqrt{x}}$

■ Derivative of a Constant Times a Function

Let

$$f(x) = ku(x)$$

where k is a constant and u is differentiable at x. Then we have

Step 1.

$$\frac{f(x + \Delta x) - f(x)}{\Delta x} = \frac{ku(x + \Delta x) - ku(x)}{\Delta x}$$

$$= k\left[\frac{u(x + \Delta x) - u(x)}{\Delta x}\right]$$

Step 2.

$$\lim_{\Delta x \to 0} \frac{f(x + \Delta x) - f(x)}{\Delta x}$$

$$= \lim_{\Delta x \to 0} k\left[\frac{u(x + \Delta x) - u(x)}{\Delta x}\right] \qquad \lim_{x \to c} kg(x) = k \lim_{x \to c} g(x)$$

$$= k \lim_{\Delta x \to 0} \frac{u(x + \Delta x) - u(x)}{\Delta x} \qquad \text{Definition of } u'(x)$$

$$= ku'(x)$$

Thus, **the derivative of a constant times a differentiable function is the constant times the derivative of the function.**

Constant Times a Function Rule

If $y = f(x) = ku(x)$, then

$\quad f'(x) = ku'(x)$

Also, $y' = ku'$, $dy/dx = k\,du/dx$, and $D_x ku(x) = kD_x u(x)$.

Example 19   (A) If $f(x) = 3x^2$, then $f'(x) \boxed{= 3 \cdot 2x^{2-1}} = 6x$.

(B) If $y = \dfrac{1}{2x^4} = \dfrac{1}{2}x^{-4}$, then $y' \boxed{= \dfrac{1}{2}(-4x^{-4-1})} = -2x^{-5}$ or $\dfrac{-2}{x^5}$.

(C) If $y = 8x^{3/2}$, then $\dfrac{dy}{dx} \boxed{= 8 \cdot \dfrac{3}{2}x^{(3/2)-1}} = 12x^{1/2}$ or $12\sqrt{x}$.

(D) $D_x\dfrac{4}{\sqrt{x^3}} = D_x\dfrac{4}{x^{3/2}} = D_x\, 4x^{-3/2} = 4\left[-\dfrac{3}{2}x^{(-3/2)-1}\right]$

$\qquad\qquad = -6x^{-5/2}$ or $-\dfrac{6}{\sqrt{x^5}}$

Problem 19   Find:

(A) $f'(x)$ for $f(x) = 4x^5$ (B) y' for $y = \dfrac{1}{3x^3}$

(C) $\dfrac{dy}{dx}$ for $y = 6x^{1/3}$ (D) $D_x\dfrac{9}{\sqrt[3]{x}}$

■ Derivatives of Sums and Differences

Let

$\quad f(x) = u(x) + v(x)$

and suppose that $u'(x)$ and $v'(x)$ exist. Then we can apply the two-step process as follows:

Step 1.

$$\frac{f(x + \Delta x) - f(x)}{\Delta x} = \frac{[u(x + \Delta x) + v(x + \Delta x)] - [u(x) + v(x)]}{\Delta x}$$

$$= \frac{u(x + \Delta x) + v(x + \Delta x) - u(x) - v(x)}{\Delta x}$$

$$= \frac{u(x + \Delta x) - u(x)}{\Delta x} + \frac{v(x + \Delta x) - v(x)}{\Delta x}$$

Step 2.

$$\lim_{\Delta x \to 0} \frac{f(x + \Delta x) - f(x)}{\Delta x}$$

$$= \lim_{\Delta x \to 0} \left[\frac{u(x + \Delta x) - u(x)}{\Delta x} + \frac{v(x + \Delta x) - v(x)}{\Delta x} \right]$$

$$\lim_{x \to c} [g(x) + h(x)] \quad \lim_{x \to c} g(x) + \lim_{x \to c} h(x)$$

$$= \lim_{\Delta x \to 0} \frac{u(x + \Delta x) - u(x)}{\Delta x} + \lim_{\Delta x \to 0} \frac{v(x + \Delta x) - v(x)}{\Delta x}$$

$$= u'(x) + v'(x)$$

So we see that **the derivative of the sum of two differentiable functions is the sum of the derivatives.** Similarly, we can show that **the derivative of the difference of two differentiable functions is the difference of the derivatives.** Together, we then have the sum and difference rule for differentiation.

Sum and Difference Rule

If $y = f(x) = u(x) \pm v(x)$, then

$$f'(x) = u'(x) \pm v'(x)$$

[Note: This rule generalizes to the sum and difference of any given number of functions.]

With this and the other rules stated previously, we will be able to compute the derivatives of all polynomials and a variety of other functions.

Example 20

(A) If $f(x) = 3x^2 + 2x$, then $f'(x)$ $= (3x^2)' + (2x)'$ $= 6x + 2$.

(B) If $y = 4 + 2x^3 - 3x^{-1}$, then y' $= (4)' + (2x^3)' - (3x^{-1})'$ $= 6x^2 + 3x^{-2}$.

(C) If $y = \sqrt[3]{x} - 3x$, then $\dfrac{dy}{dx} = \dfrac{d}{dx} x^{1/3} - \dfrac{d}{dx} 3x = \dfrac{1}{3} x^{-2/3} - 3$.

(D) $D_x \left(\dfrac{8}{\sqrt[4]{x}} + x^7 \right) = D_x 8x^{-1/4} + D_x x^7 = -2x^{-5/4} + 7x^6$

Problem 20

Find:

(A) $f'(x)$ for $f(x) = 3x^4 - 2x^3 + x^2 - 5x + 7$

(B) y' for $y = 3 - 7x^{-2}$

(C) $\dfrac{dy}{dx}$ for $y = 5x^3 - \sqrt[4]{x}$

(D) $D_x \left(\dfrac{6}{\sqrt[3]{x}} + \dfrac{2}{x^3} \right)$

■ Applications

Example 21
Instantaneous Velocity

The distance y in feet that a steel ball falls in x seconds is given by

$$y = f(x) = 16x^2$$

Find the instantaneous velocity function $v = f'(x)$. Find the velocity at $x = 1$ and $x = 6$ seconds.

Solution

$v = f'(x) = 16(2x^{2-1}) = 32x$

$f'(1) = 32(1) = 32$ feet per second

$f'(6) = 32(6) = 192$ feet per second

Problem 21

A steel ball falls so that its distance y in feet after x seconds is given by

$$y = f(x) = 16x^2 - 4x$$

(A)   Find the instantaneous velocity function.
(B)   Find the velocity at $x = 2$ and $x = 5$ seconds.

Example 22

(A)   Find the slope function $m = f'(x)$ for $y = f(x) = 4x - x^2$.
(B)   Find the slope of the tangent to the graph of $y = 4x - x^2$ at $x = 1, 2$, and 3.

Solutions

(A) $m = f'(x) = (4x)' - (x^2)' = 4 - 2x$
(B) $m_1 = f'(1) = 4 - 2(1) = 2$

$m_2 = f'(2) = 4 - 2(2) = 0$

$m_3 = f'(3) = 4 - 2(3) = -2$

Problem 22

Repeat Example 22 for $y = f(x) = 8x - 2x^2$.

Example 23
Marginal Cost

The total cost $C(x)$ in thousands of dollars for manufacturing x sailboats is given by

$$C(x) = 2 + 8x - x^2 \qquad 0 \leqslant x \leqslant 3$$

(A)   The marginal cost at a production level of x is

$$C'(x) \ \boxed{= (2)' + (8x)' - (x^2)'} \ = 8 - 2x$$

(B)   The marginal cost at $x = 1$ is

$$C'(1) = 8 - 2(1) = 6 \qquad \$6,000 \text{ per unit increase in production}$$

(C)   The marginal cost at $x = 3$ is

$$C'(3) = 8 - 2(3) - 2 \qquad \$2,000 \text{ per unit increase in production}$$

Problem 23   Repeat Example 23 with the cost function $C(x) = 3 + 10x - x^2$, $0 \leqslant x \leqslant 4$.

Answers to   17.   All are 0.
Matched Problems

18.   (A) $3x^2$ (B) $\dfrac{3}{2} x^{1/2}$

     (C) $-2x^{-3}$ (D) $-\dfrac{1}{2} x^{-3/2}$ or $\dfrac{-1}{2\sqrt{x^3}}$

19.   (A) $20x^4$ (B) $-x^{-4}$

     (C) $2x^{-2/3}$ or $\dfrac{2}{\sqrt[3]{x^2}}$ (D) $-3x^{-4/3}$ or $\dfrac{-3}{\sqrt[3]{x^4}}$

20.   (A) $12x^3 - 6x^2 + 2x - 5$ (B) $14x^{-3}$

     (C) $15x^2 - \dfrac{1}{4} x^{-3/4}$ (D) $-2x^{-4/3} - 6x^{-4}$

21.   (A) $v = 32x - 4$
     (B) $f'(2) = 60$ feet per second; $f'(5) = 156$ feet per second
22.   (A) $m = 8 - 4x$ (B) $m_1 = 4; m_2 = 0; m_3 = -4$
23.   (A)   Marginal cost $= C'(x) = 10 - 2x$
     (B) $C'(1) = 8$ $\$8,000$ per unit increase in production
     (C) $C'(3) = 4$ $\$4,000$ per unit increase in production

Exercise 10-5

Find each of the following:

A   1. $f'(x)$ for $f(x) = 12$ 2. $\dfrac{dy}{dx}$ for $y = -\sqrt{3}$

    3. $D_x 23$ 4. y' for $y = \pi$

    5. $\dfrac{dy}{dx}$ for $y = x^{12}$ 6. $D_x x^5$

    7. $f'(x)$ for $f(x) = x$ 8. y' for $y = x^7$

    9. $f'(x)$ for $f(x) = 2x^4$ 10. $\dfrac{dy}{dx}$ for $y = -3x$

11. $D_x\left(\dfrac{1}{3}x^6\right)$

12. y' for $y = \dfrac{1}{2}x^4$

B   **13.** $D_x(2x^{-5})$

14. y' for $y = -4x^{-1}$

15. $f'(x)$ for $f(x) = -3x^{1/3}$

16. $\dfrac{dy}{dx}$ for $y = -8x^{1/4}$

17. $\dfrac{dy}{dx}$ for $y = 3x^5 - 2x^3 + 5$

18. $f'(x)$ for $y = 2x^3 - 6x + 5$

19. $D_x(3x^{-4} + 2x^{-2})$

20. y' for $y = 2x^{-3} - 4x^{-1}$

21. $\dfrac{dy}{dx}$ for $y = \dfrac{3}{x^2}$

22. $f'(x)$ for $y = \dfrac{1}{x^4}$

23. $D_x(3x^{2/3} - 5x^{1/3})$

24. $D_x(8x^{3/4} + 4x^{-1/4})$

25. $D_x\left(\dfrac{3}{x^{3/5}} - \dfrac{6}{x^{1/2}}\right)$

26. $D_y\left(\dfrac{5}{y^{1/5}} - \dfrac{8}{y^{3/2}}\right)$

27. $D_x\dfrac{1}{\sqrt[3]{x}}$

28. y' for $y = \dfrac{10}{\sqrt[5]{x}}$

29. $\dfrac{dy}{dx}$ for $y = \dfrac{12}{\sqrt{x}} - 3x^{-2} + x$

30. $f'(x)$ for $f(x) = 2x^{-3} - \dfrac{6}{\sqrt[3]{x^2}} + 7$

31. Given the equation $y = f(x) = 6x - x^2$, find:

(A) The slope function $m = f'(x)$.
(B) The slope of the tangents to the graph at $x = 2$ and at $x = 4$.
(C) The value(s) of x such that the slope is 0.

32. Repeat Problem 31 for $y = f(x) = 2x^2 + 8x$.
33. Repeat Problem 31 for $y = f(x) = (1/3)x^3 - 3x^2 + 2$.
34. Repeat Problem 31 for $y = f(x) = 2x^3 - 3x^2 - 5$.
35. If an object moves along the y axis (marked in feet) so that its position at time x in seconds is given by $y = f(x) = 176x - 16x^2$, find:

(A) The instantaneous velocity function $v = f'(x)$
(B) The velocity at $x = 0$, 3, and 6 seconds
(C) The time(s) when $v = 0$

36. Repeat Problem 35 for $y = f(x) = 80x - 10x^2$.
37. Repeat Problem 35 for $y = f(x) = 10 + 40x - 5x^2$.
38. Repeat Problem 35 for $y = f(x) = -20 + 120x - 15x^2$.

C   **39.** $D_x\dfrac{x^4 - 3x^3 + 5}{x^2}$

40. y' for $y = \dfrac{2x^5 - 4x^3 + 2x}{x^3}$

41. $\dfrac{dy}{dx}$ for $y = \dfrac{\sqrt{x} - 6}{\sqrt{x^3}}$

42. $f'(x)$ for $f(x) = \dfrac{\sqrt[3]{x} + 3}{\sqrt[3]{x^2}}$

Applications

Business & Economics

43. *Advertising.* Using past records it is estimated that a company will sell $N(x)$ units of a product after spending $\$x$ thousand on advertising, as given by

$$N(x) = 60x - x^2 \qquad 5 \leqslant x \leqslant 30$$

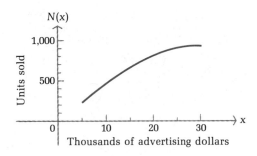

(A) Find $N'(x)$, the rate of change of sales per unit change in money spent on advertising at the $\$x$ thousand level.

(B) Find $N'(10)$ and $N'(20)$ and interpret.

44. *Marginal average cost.* (This topic is treated in detail in Section 11-5.) Economists often work with average costs — cost per unit output — rather than total costs. We would expect higher average costs, because of plant inefficiency, at low output levels and also at output levels near plant capacity. Therefore, we would expect the graph of an average cost function to be U-shaped. Suppose that for a given firm the total cost of producing x thousand units is given by

$$C(x) = x^3 - 6x^2 + 12x$$

Then the average cost $\overline{C}(x)$ is given by

$$\overline{C}(x) = \frac{C(x)}{x} = x^2 - 6x + 12$$

(A) Find the marginal average cost $\overline{C}'(x)$.

(B) Find the marginal average cost at $x = 2$, 3, and 4, and interpret.

Life Sciences

45. *Medicine.* A person x inches tall has a pulse rate of y beats per minute, as given approximately by

$$y = 590x^{-1/2} \qquad 30 \leqslant x \leqslant 75$$

What is the instantaneous rate of change of pulse rate at the:

(A) 36 inch level?   (B) 64 inch level?

46. *Ecology.* A coal-burning electrical generating plant emits sulfur dioxide into the surrounding air. The concentration $C(x)$ in parts per million is given approximately by

$$C(x) = \frac{0.1}{x^2}$$

where x is the distance from the plant in miles. Find the (instantaneous) rate of change of concentration at:
(A) $x = 1$ mile   (B) $x = 2$ miles

Social Sciences

47. *Learning.* Suppose a person learns y items in x hours, as given by

$$y = 50\sqrt{x} \qquad 0 \leqslant x \leqslant 9$$

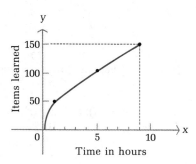

Find the rate of learning at the end of:

(A) 1 hour   (B) 9 hours

48. *Learning.* If a person learns y items in x hours, as given by

$$y = 21\sqrt[3]{x^2} \qquad 0 \leqslant x \leqslant 8$$

find the rate of learning at the end of:

(A) 1 hour   (B) 8 hours

10-6 Derivatives of Products and Quotients

- Derivatives of Products
- Derivatives of Quotients

The derivative rules discussed in the last section added substantially to our ability to compute and apply derivatives to many practical problems. In this and the next section we will add a few more rules that will increase this ability even further.

■ Derivatives of Products

In the last section we found that the derivative of a sum is the sum of the derivatives. Is the derivative of a product the product of the derivatives? Let us take a look at a simple example. Consider

$$f(x) = u(x)v(x) = (x^2 - 3x)(2x^3 - 1) \tag{1}$$

where $u(x) = x^2 - 3x$ and $v(x) = 2x^3 - 1$. The product of the derivatives is

$$u'(x)v'(x) = (2x - 3)6x^2 = 12x^3 - 18x^2 \tag{2}$$

To see if this is equal to the derivative of the product, we multiply the right side of (1) and use derivative formulas from the last section:

$$f(x) = (x^2 - 3x)(2x^3 - 1)$$
$$= 2x^5 - 6x^4 - x^2 + 3x$$

Thus,

$$f'(x) = 10x^4 - 24x^3 - 2x + 3 \tag{3}$$

Since (2) and (3) are not equal, we conclude that the derivative of a product is *not* the product of the derivatives. There is a product rule for derivatives, but it is slightly more complicated than you might expect. We will now derive the rule. Let

$$f(x) = u(x)v(x) \qquad \text{where} \qquad u'(x) \text{ and } v'(x) \text{ exist}$$

We will develop a derivative formula for $f'(x)$ in terms of $u'(x)$ and $v'(x)$. We proceed as follows:

$$f'(x) = \lim_{\Delta x \to 0} \frac{f(x + \Delta x) - f(x)}{\Delta x}$$

$$= \lim_{\Delta x \to 0} \frac{u(x + \Delta x)v(x + \Delta x) - u(x)v(x)}{\Delta x}$$

We now add 0 in a special form to the numerator. That is, we subtract and add $u(x + \Delta x)v(x)$ in the middle of the numerator to obtain

$$f'(x) = \lim_{\Delta x \to 0} \frac{u(x + \Delta x)v(x + \Delta x) - u(x + \Delta x)v(x) + u(x + \Delta x)v(x) - u(x)v(x)}{\Delta x}$$

$$= \lim_{\Delta x \to 0} \frac{u(x + \Delta x)[v(x + \Delta x) - v(x)] + v(x)[u(x + \Delta x) - u(x)]}{\Delta x}$$

$$= \lim_{\Delta x \to 0} \left[u(x + \Delta x) \frac{v(x + \Delta x) - v(x)}{\Delta x} + v(x) \frac{u(x + \Delta x) - u(x)}{\Delta x} \right]$$

$$= \lim_{\Delta x \to 0} u(x + \Delta x) \frac{v(x + \Delta x) - v(x)}{\Delta x} + \lim_{\Delta x \to 0} v(x) \frac{u(x + \Delta x) - u(x)}{\Delta x}$$

$$= u(x)v'(x) + v(x)u'(x) \qquad \text{Since } u'(x) \text{ exists, } u \text{ is continuous at } x$$
$$\text{and } \lim_{\Delta x \to 0} u(x + \Delta x) = u(x).$$

Thus, **the derivative of a product is the first times the derivative of the second plus the second times the derivative of the first.**

Product Rule

If $y = f(x) = u(x)v(x)$, then

$$f'(x) = u(x)v'(x) + v(x)u'(x)$$

Also,

$$y' = uv' + vu'$$

$$\frac{dy}{dx} = u\frac{dv}{dx} + v\frac{du}{dx}$$

$$D_x[u(x)v(x)] = u(x)D_xv(x) + v(x)D_xu(x)$$

Example 24   (A) Find $f'(x)$ for $f(x) = 2x^2(3x^4 - 2)$ two ways.

(B) Find $D_x[(x^2 - 2x + 1)(3x^3 + x - 5)]$.

(C) Find $D_x[(\sqrt[3]{x} + 2)(2\sqrt{x} - 1)]$.

Solutions   (A) Method I.   Use the product rule:

$$f'(x) = 2x^2(3x^4 - 2)' + (3x^4 - 2)(2x^2)' \qquad \text{First times derivative of}$$
$$= 2x^2(12x^3) + (3x^4 - 2)(4x) \qquad \text{second plus second times}$$
$$= 24x^5 + 12x^5 - 8x \qquad \text{derivative of first}$$
$$= 36x^5 - 8x$$

Method II.   Multiply first; then take derivatives:

$$f(x) = 2x^2(3x^4 - 2) = 6x^6 - 4x^2$$
$$f'(x) = 36x^5 - 8x$$

(B) $D_x[(x^2 - 2x + 1)(3x^3 + x - 5)]$
$$= (x^2 - 2x + 1)D_x(3x^3 + x - 5) + (3x^3 + x - 5)D_x(x^2 - 2x + 1)$$
$$= (x^2 - 2x + 1)(9x^2 + 1) + (3x^3 + x - 5)(2x - 2)$$
$$= 9x^4 - 18x^3 + 10x^2 - 2x + 1 + 6x^4 - 6x^3 + 2x^2 - 12x + 10$$
$$= 15x^4 - 24x^3 + 12x^2 - 14x + 11$$

How we write the final answer depends on what we want to do with it; we might have chosen to leave the answer in the unsimplified form two steps back for certain purposes.

(C) $D_x[(\sqrt[3]{x} + 2)(2\sqrt{x} - 1)]$

$\qquad = D_x[(x^{1/3} + 2)(2x^{1/2} - 1)]$ Change radicals to fractional exponent form.

$\qquad = (x^{1/3} + 2)D_x(2x^{1/2} - 1) + (2x^{1/2} - 1)D_x(x^{1/3} + 2)$

$\qquad = (x^{1/3} + 2)x^{-1/2} + (2x^{1/2} - 1)\dfrac{1}{3}x^{-2/3}$

$\qquad = \dfrac{\sqrt[3]{x} + 2}{\sqrt{x}} + \dfrac{2\sqrt{x} - 1}{3\sqrt[3]{x^2}}$

Problem 24   Find:

(A) $f'(x)$ for $f(x) = 3x^3(2x^2 - 3x + 1)$ two ways
(B) y' for $y = (2x^2 - 3x + 5)(x^2 + 3x + 1)$
(C) $D_x[(\sqrt{x} - 2)(\sqrt[3]{x^2} + 5)]$

■ Derivatives of Quotients

As is the case with a product, the derivative of a quotient is *not* the quotient of the derivatives.

Let

$$f(x) = \frac{u(x)}{v(x)} \qquad \text{where} \qquad u'(x) \text{ and } v'(x) \text{ exist}$$

Quotient Rule

If

$$y = f(x) = \frac{u(x)}{v(x)}$$

then

$$f'(x) = \frac{v(x)u'(x) - u(x)v'(x)}{[v(x)]^2}$$

Also,

$$y' = \frac{vu' - uv'}{v^2}$$

$$\frac{dy}{dx} = \frac{v(du/dx) - u(dv/dx)}{v^2}$$

$$D_x \frac{u(x)}{v(x)} = \frac{v(x)D_x u(x) - u(x)D_x v(x)}{[v(x)]^2}$$

Starting with the definition of a derivative, you can show that

$$f'(x) = \frac{v(x)u'(x) - u(x)v'(x)}{[v(x)]^2}$$

Thus, **the derivative of a quotient is the denominator times the derivative of the numerator minus the numerator times the derivative of the denominator, all over the denominator squared.**

Example 25   (A)   If

$$f(x) = \frac{x^2}{2x - 1}$$

find $f'(x)$.

(B)   Find

$$D_x \frac{x^2 - x}{x^3 + 1}$$

(C)   Find

$$D_x \frac{x^{1/2} - 3}{x^{1/2}}$$

by using the quotient rule and also by splitting the fraction into two fractions.

Solutions   (A) $f'(x) = \dfrac{(2x - 1)(x^2)' - x^2(2x - 1)'}{(2x - 1)^2}$

$\qquad\quad = \dfrac{(2x - 1)(2x) - x^2(2)}{(2x - 1)^2}$

$\qquad\quad = \dfrac{4x^2 - 2x - 2x^2}{(2x - 1)^2}$

$\qquad\quad = \dfrac{2x^2 - 2x}{(2x - 1)^2}$

The denominator times the derivative of the numerator minus the numerator times the derivative of the denominator, all over the square of the denominator

(B) $D_x \dfrac{x^2 - x}{x^3 + 1} = \dfrac{(x^3 + 1)D_x(x^2 - x) - (x^2 - x)D_x(x^3 + 1)}{(x^3 + 1)^2}$

$\qquad\qquad = \dfrac{(x^3 + 1)(2x - 1) - (x^2 - x)(3x^2)}{(x^3 + 1)^2}$

$\qquad\qquad = \dfrac{2x^4 - x^3 + 2x - 1 - 3x^4 + 3x^3}{(x^3 + 1)^2}$

$\qquad\qquad = \dfrac{-x^4 + 2x^3 + 2x - 1}{(x^3 + 1)^2}$

(C)   Method I.   Use the quotient rule:

$$D_x \frac{x^{1/2} - 3}{x^{1/2}} = \frac{x^{1/2}D_x(x^{1/2} - 3) - (x^{1/2} - 3)D_x x^{1/2}}{(x^{1/2})^2}$$

$$= \frac{x^{1/2}\left(\dfrac{1}{2}x^{-1/2}\right) - (x^{1/2} - 3)\dfrac{1}{2}x^{-1/2}}{x}$$

$$= \frac{\dfrac{1}{2} - \dfrac{1}{2} + \dfrac{3}{2}x^{-1/2}}{x}$$

$$= \frac{3}{2x(x^{1/2})} = \frac{3}{2x^{3/2}}$$

Method II.   Split into two fractions:

$$\frac{x^{1/2} - 3}{x^{1/2}} = \frac{x^{1/2}}{x^{1/2}} - \frac{3}{x^{1/2}} = 1 - 3x^{-1/2}$$

$$D_x(1 - 3x^{-1/2}) = 0 + \frac{3}{2}x^{-3/2} = \frac{3}{2x^{3/2}}$$

Comparing methods I and II, we see that it may sometimes pay to change an expression algebraically before blindly using a derivative formula.

Problem 25   Find:

(A) $f'(x)$ for $f(x) = \dfrac{2x}{x^2 + 3}$ (B) y' for $y = \dfrac{x^3 - 3x}{x^2 - 4}$

(C) $D_x \dfrac{2 + x^{1/3}}{x^{1/3}}$ two ways

Answers to Matched Problems

24.   (A) $30x^4 - 36x^3 + 9x^2$

(B) $(2x^2 - 3x + 5)(2x + 3) + (x^2 + 3x + 1)(4x - 3)$
$= 8x^3 + 9x^2 - 4x + 12$

(C) $(x^{1/2} - 2)\left(\dfrac{2}{3}x^{-1/3}\right) + (x^{2/3} + 5)\left(\dfrac{1}{2}x^{-1/2}\right)$

or $\dfrac{2(\sqrt{x} - 2)}{3\sqrt[3]{x}} + \dfrac{\sqrt[3]{x^2} + 5}{2\sqrt{x}}$

25.   (A) $\dfrac{(x^2 + 3)2 - (2x)(2x)}{(x^2 + 3)^2} = \dfrac{6 - 2x^2}{(x^2 + 3)^2}$

(B) $\dfrac{(x^2 - 4)(3x^2 - 3) - (x^3 - 3x)(2x)}{(x^2 - 4)^2} = \dfrac{x^4 - 9x^2 + 12}{(x^2 - 4)^2}$

(C) $\dfrac{-2}{3x^{4/3}}$

Exercise 10-6

A   For $f(x)$ as given, find $f'(x)$ and simplify.

1. $f(x) = 2x^3(x^2 - 2)$
2. $f(x) = 5x^2(x^3 + 2)$
3. $f(x) = (x - 3)(2x - 1)$
4. $f(x) = (3x + 2)(4x - 5)$

5. $f(x) = \dfrac{x}{x - 3}$
6. $f(x) = \dfrac{3x}{2x + 1}$

7. $f(x) = \dfrac{2x + 3}{x - 2}$
8. $f(x) = \dfrac{3x - 4}{2x + 3}$

9. $f(x) = (x^2 + 1)(2x - 3)$
10. $f(x) = (3x + 5)(x^2 - 3)$

11. $f(x) = \dfrac{x^2 + 1}{2x - 3}$
12. $f(x) = \dfrac{3x + 5}{x^2 - 3}$

B   Find each of the following (Problems 21–32 do not have to be simplified):

13. $f'(x)$ for $f(x) = (2x + 1)(x^2 - 3x)$
14. y' for $y = (x^3 + 2x^2)(3x - 1)$

15. $\dfrac{dy}{dx}$ for $y = (2x - x^2)(5x + 2)$

16. $D_x[(3 - x^3)(x^2 - x)]$

17. y' for $y = \dfrac{5x - 3}{x^2 + 2x}$
18. $f'(x)$ for $f(x) = \dfrac{3x^2}{2x - 1}$

19. $D_x \dfrac{x^2 - 3x + 1}{x^2 - 1}$
20. $\dfrac{dy}{dx}$ for $y = \dfrac{x^4 - x^3}{3x - 1}$

21. $f'(x)$ for $f(x) = (2x^4 - 3x^3 + x)(x^2 - x + 5)$

22. $\dfrac{dy}{dx}$ for $y = (x^2 - 3x + 1)(x^3 + 2x^2 - x)$

23. $D_x \dfrac{3x^2 - 2x + 3}{4x^2 + 5x - 1}$
24. y' for $y = \dfrac{x^3 - 3x + 4}{2x^2 + 3x - 2}$

25. $\dfrac{dy}{dx}$ for $y = 9x^{1/3}(x^3 + 5)$
26. $D_x[(4x^{1/2} - 1)(3x^{1/3} + 2)]$

27. $f'(x)$ for $f(x) = \dfrac{6\sqrt[3]{x}}{x^2 - 3}$
28. y' for $y = \dfrac{2\sqrt{x}}{x^2 - 3x + 1}$

C   29. $D_x \dfrac{x^3 - 2x^2}{\sqrt[3]{x^2}}$
30. $\dfrac{dy}{dx}$ for $y = \dfrac{x^2 - 3x + 1}{\sqrt[4]{x}}$

31. $f'(x)$ for $f(x) = \dfrac{(2x^2 - 1)(x^2 + 3)}{x^2 + 1}$

32. y' for $y = \dfrac{2x - 1}{(x^3 + 2)(x^2 - 3)}$

■ Applications

<table>
<tr><td>Business & Economics</td><td>**33.**</td><td>

Price–demand function. According to classical economic theory, the demand $d(x)$ for a commodity in a free market decreases as the price x increases. Suppose that the number $d(x)$ of transistor radios people are willing to buy per week in a given city at a price x is given by

</td></tr>
</table>

$$d(x) = \frac{50,000}{x^2 + 10x + 25} \qquad \$5 \le x \le \$15$$

(A) Find $d'(x)$, the rate of change of demand with respect to price change.
(B) Find $d'(5)$ and $d'(10)$.

Life Sciences   **34.** *Drug sensitivity.* One hour after x milligrams of a particular drug are given to a person, the change in body temperature $T(x)$ in degrees Fahrenheit is given approximately by

$$T(x) = x^2 \left(1 - \frac{x}{9}\right) \qquad 0 \le x \le 6$$

The rate at which T changes with respect to the size of the dosage x, $T'(x)$, is called the *sensitivity* of the body to the dosage.

(A) Find $T'(x)$, using the product rule.
(B) Find $T'(1)$, $T'(3)$, and $T'(6)$.

Social Sciences   **35.** *Learning.* In the early days of quantitative learning theory (around 1917), L. L. Thurstone found that a given person successfully accomplished $N(x)$ acts after x practice acts, as given by

$$N(x) = \frac{100x + 200}{x + 32}$$

(A) Find the rate of change of learning, $N'(x)$, with respect to the number of practice acts x.
(B) Find $N'(4)$ and $N'(68)$.

10-7 Chain Rule and General Power Rule

■ Composite Functions
■ Chain Rule
■ General Power Rule

Suppose you were asked to find the derivative of

$$h(x) = \sqrt{2x + 1}$$

We have developed formulas for computing the derivatives of square root functions and polynomial functions separately, but not in the indicated combination. In this section we will discuss one of the most important derivative rules of all—the **chain rule.** This rule will enable us to determine the derivatives of some fairly complicated functions in terms of derivatives of more elementary functions. The chain rule is used to compute derivatives of functions that are compositions of more elementary functions whose derivatives are known.

■ Composite Functions

Let us look at the given function h more closely:

$$h(x) = \sqrt{2x + 1}$$

Inside the radical is a first-degree polynomial that defines a linear function. So the function h is really a combination of a square root function and a linear function. To see this more clearly, let

$$y = f(u) = \sqrt{u}$$
$$u = g(x) = 2x + 1$$

Then we can express y as a function of x as follows:

$$y = f(u) = f[g(x)] = \sqrt{2x + 1} = h(x)$$

The function h is said to be the *composite* of the two simpler functions f and g. (Loosely speaking, we can think of h as a function of a function.) In general,

---

Composite Functions

A function h is a **composite** of functions f and g if

$$h(x) = f[g(x)]$$

The domain of h is the set of all numbers x such that x is in the domain of g and $g(x)$ is in the domain of f.

Example 26   Let $f(u) = u^{10}$ and $g(x) = 3x^4 - 1$. Find

(A) $f[g(x)]$ (B) $g[f(u)]$

Solutions   (A) $f[g(x)] = [g(x)]^{10} = (3x^4 - 1)^{10}$

(B) $g[f(u)] = 3[f(u)]^4 - 1 = 3(u^{10})^4 - 1 = 3u^{40} - 1$

Problem 26   Let $f(u) = \sqrt[3]{u}$ and $g(x) = x^2 - 3x + 1$. Find

(A) $f[g(x)]$ (B) $g[f(u)]$

Write answers using fractional exponents.

Example 27   Given:

$$\text{(A)}\quad y = \sqrt[4]{x^3 - 2x^2 + 1} \qquad \text{(B)}\quad y = \frac{1}{(x^2 - 1)^5}$$

Write each in the form $y = f(u) = u^n$ and $u = g(x)$ so that $y = f[g(x)]$.

Solutions   (A) $y = \sqrt[4]{x^3 - 2x^2 + 1} = (x^3 - 2x^2 + 1)^{1/4}$

Let

$$y = f(u) = u^{1/4}$$
$$u = g(x) = x^3 - 2x^2 + 1$$

Check $y = f[g(x)] = (x^3 - 2x^2 + 1)^{1/4} = \sqrt[4]{x^3 - 2x^2 + 1}$

(B) $y = \dfrac{1}{(x^2 - 1)^5} = (x^2 - 1)^{-5}$

Let

$$y = f(u) = u^{-5}$$
$$u = g(x) = x^2 - 1$$

Check $y = f[g(x)] = (x^2 - 1)^{-5} = \dfrac{1}{(x^2 - 1)^5}$

Problem 27   Given:

$$\text{(A)}\quad y = (x^{-1} + 3x^2)^{-2/5} \qquad \text{(B)}\quad y = \frac{1}{\sqrt{4 + \sqrt{x}}}$$

Write each in the form $y = f(u) = u^n$ and $u = g(x)$ so that $y = f[g(x)]$. Write all radicals in terms of fractional exponents.

▪ Chain Rule

The word "chain" comes from the fact that a function formed by composition (such as those in Example 26) involves a "chain" of functions — that is, "a function of a function." We now introduce the *chain rule*, which will enable us to compute the derivative of a composite function in terms of the derivatives of the functions making up the composition.

Suppose

$$y = h(x) = f[g(x)]$$

is a composite of f and g where

$$y = f(u) \quad \text{and} \quad u = g(x)$$

We would like to express the derivative dy/dx in terms of the derivatives of f and g. From the definition of a derivative we have

$$\frac{dy}{dx} = \lim_{\Delta x \to 0} \frac{h(x + \Delta x) - h(x)}{\Delta x}$$

$$= \lim_{\Delta x \to 0} \frac{\Delta y}{\Delta x} \tag{1}$$

Noting that

$$\frac{\Delta y}{\Delta x} = \frac{\Delta y}{\Delta u} \frac{\Delta u}{\Delta x} \tag{2}$$

we might be tempted to substitute (2) into (1) to obtain

$$\frac{dy}{dx} = \lim_{\Delta x \to 0} \frac{\Delta y}{\Delta u} \frac{\Delta u}{\Delta x}$$

and reason that $\Delta u \to 0$ as $\Delta x \to 0$ so that

$$\frac{dy}{dx} = \left(\lim_{\Delta u \to 0} \frac{\Delta y}{\Delta u} \right) \left(\lim_{\Delta x \to 0} \frac{\Delta u}{\Delta x} \right)$$

$$= \frac{dy}{du} \frac{du}{dx}$$

The result is correct under rather general conditions, and is called the *chain rule*, but our "derivation" is superficial, because it ignores a number of hidden problems. Since a formal proof of the **chain rule** is beyond the scope of this book, we simply state it as follows:

Chain Rule

If $y = f(u)$ and $u = g(x)$, define the composite function

$$y = h(x) = f[g(x)]$$

Then

$$\frac{dy}{dx} = \frac{dy}{du} \frac{du}{dx} \quad \text{provided } \frac{dy}{du} \text{ and } \frac{du}{dx} \text{ exist}$$

Example 28   Find dy/dx for $y = (x^2 - 2)^8$.

Solution   Let $y = u^8$ and $u = x^2 - 2$. Then

$$\frac{dy}{dx} = \frac{dy}{du}\frac{du}{dx}$$

$$= 8u^7(2x)$$

$$= 8(x^2 - 2)^7(2x) \qquad \text{Since } u = x^2 - 2$$

$$= 16x(x^2 - 2)^7$$

Gradually, you will want to be able to do most of these steps in your head and simply write

$$D_x[(x^2 - 2)^8] = 8(x^2 - 2)^7(2x)$$

$$= 16x(x^2 - 2)^7$$

Problem 28   Find dy/dx for $y = \sqrt{x^2 + 8x}$.

■ General Power Rule

Example 28 and Problem 28 are particular cases of the general power form

$$y = [g(x)]^n \quad \text{or} \quad y = u^n, \quad u = g(x)$$

a composite function form that occurs with great frequency. In fact, it occurs with sufficient frequency to warrant a special derivative formula as a special case of the chain rule. If $y = u^n$ and $u = g(x)$, then we can apply the chain rule to obtain

$$\frac{dy}{dx} = \frac{dy}{du}\frac{du}{dx}$$

$$= nu^{n-1}\frac{du}{dx}$$

or, equivalently,

$$\frac{dy}{dx} = n[g(x)]^{n-1}g'(x)$$

Thus, we have the following **general power rule:**

General Power Rule

For $y = [g(x)]^n$, n a real number,

$$\frac{dy}{dx} = n[g(x)]^{n-1}g'(x)$$

if $g'(x)$ exists. More compactly, if $u = u(x)$, then

$$D_x u^n = nu^{n-1}\frac{du}{dx}$$

The special power form of the chain rule will handle most of the problems we are considering now. Chapter 13 (on exponential and logarithmic functions) will require the use of the chain rule in its more general form. We conclude this section with a variety of examples using the general power rule.

Example 29   Find dy/dx, given

(A) $y = \dfrac{1}{\sqrt{4 + \sqrt{x}}}$ (B) $y = \left(\dfrac{1}{x^{-2} + 3x^{-1}}\right)^{10}$

Solutions   (A) $y = \dfrac{1}{\sqrt{4 + \sqrt{x}}}$ Write in the form u^n first.

$\qquad = \dfrac{1}{(4 + x^{1/2})^{1/2}}$

$\qquad = (4 + x^{1/2})^{-1/2}$ u^n form

$\dfrac{dy}{dx} = -\dfrac{1}{2}(4 + x^{1/2})^{(-1/2)-1}D_x(4 + x^{1/2})$ $nu^{n-1}\dfrac{du}{dx}$

$\qquad = -\dfrac{1}{2}(4 + x^{1/2})^{-3/2}\dfrac{1}{2}x^{-1/2}$

$\qquad = -\dfrac{x^{-1/2}}{4}(4 + x^{1/2})^{-3/2}$ or $\dfrac{-1}{4x^{1/2}(4 + x^{1/2})^{3/2}}$

(B) $y = \left(\dfrac{1}{x^{-2} + 3x^{-1}}\right)^{10}$ The inside could be treated as a quotient, but it is easier to proceed as indicated.

$\qquad = [(x^{-2} + 3x^{-1})^{-1}]^{10}$

$\qquad = (x^{-2} + 3x^{-1})^{-10}$ u^n form

$\dfrac{dy}{dx} = -10(x^{-2} + 3x^{-1})^{-10-1}D_x(x^{-2} + 3x^{-1})$ $nu^{n-1}\dfrac{du}{dx}$

$\qquad = -10(x^{-2} + 3x^{-1})^{-11}(-2x^{-3} - 3x^{-2})$

$\qquad$ or $\dfrac{-10(-2x^{-3} - 3x^{-2})}{(x^{-2} + 3x^{-1})^{11}}$

Problem 29   Find dy/dx, given:

(A) $y = \dfrac{1}{\sqrt[3]{x^3 - 9}}$ (B) $y = \left(\dfrac{1}{2x^{-4} + x}\right)^{-6}$

Example 30   Find dy/dx, given:

(A) $y = \dfrac{(3x - 5)^3}{(2x^2 + 1)^4}$ (B) $y = (3x^2 - 4)^3\sqrt{2x - 1}$

Solution (A) We use the quotient rule and the power rule:

$$\frac{dy}{dx} = \frac{(2x^2 + 1)^4 D_x(3x - 5)^3 - (3x - 5)^3 D_x(2x^2 + 1)^4}{[(2x^2 + 1)^4]^2}$$

$$= \frac{(2x^2 + 1)^4 3(3x - 5)^2 D_x(3x - 5) - (3x - 5)^3 4(2x^2 + 1)^3 D_x(2x^2 + 1)}{(2x^2 + 1)^8}$$

$$= \frac{(2x^2 + 1)^4 3(3x - 5)^2 3 - (3x - 5)^3 4(2x^2 + 1)^3 4x}{(2x^2 + 1)^8}$$

$$= \frac{9(2x^2 + 1)^4(3x - 5)^2 - 16x(3x - 5)^3(2x^2 + 1)^3}{(2x^2 + 1)^8}$$

To simplify, factor out the highest common powers of $(2x^2 + 1)$ and $(3x - 5)$, then reduce to lowest terms:

$$\frac{dy}{dx} = \frac{(2x^2 + 1)^3(3x - 5)^2[9(2x^2 + 1) - 16x(3x - 5)]}{(2x^2 + 1)^8}$$

$$= \frac{(3x - 5)^2(-30x^2 + 80x + 9)}{(2x^2 + 1)^5}$$

In general, derivatives should be simplified so that their uses will proceed more smoothly.

(B) $y = (3x^2 - 4)^3(2x - 1)^{1/2}$

Use the product and power rules.

$$\frac{dy}{dx} = (3x^2 - 4)^3 D_x(2x - 1)^{1/2} + (2x - 1)^{1/2} D_x(3x^2 - 4)^3$$

$$= (3x^2 - 4)^3 \frac{1}{2}(2x - 1)^{-1/2} D_x(2x - 1)$$

$$\quad + (2x - 1)^{1/2} 3(3x^2 - 4)^2 D_x(3x^2 - 4)$$

$$= (3x^2 - 4)^3 \frac{1}{2}(2x - 1)^{-1/2} 2 + (2x - 1)^{1/2} 3(3x^2 - 4)^2 6x$$

$$= \frac{(3x^2 - 4)^3}{(2x - 1)^{1/2}} + 18x(2x - 1)^{1/2}(3x^2 - 4)^2$$

$$= \frac{(3x^2 - 4)^3 + 18x(2x - 1)(3x^2 - 4)^2}{(2x - 1)^{1/2}}$$

To simplify further, factor out the highest common power of $(3x^2 - 4)$:

$$\frac{dy}{dx} = \frac{(3x^2 - 4)^2[(3x^2 - 4) + 18x(2x - 1)]}{(2x - 1)^{1/2}}$$

$$= \frac{(3x^2 - 4)^2(39x^2 - 18x - 4)}{(2x - 1)^{1/2}}$$

Problem 30   Find dy/dx and simplify.

(A) $y = \dfrac{(x^3 - 2)^4}{(6x + 1)^3}$ (B) $y = \sqrt[3]{2x^3 + 1}\,(x^3 - 4)^2$

Answers to   26.   (A) $f[g(x)] = (x^2 - 3x + 1)^{1/3}$
Matched Problems   (B) $g[f(u)] = u^{2/3} - 3u^{1/3} + 1$

27.   (A) $y = f(u) = u^{-2/5}$ and $g(x) = x^{-1} + 3x^2$
      (B) $y = f(u) = u^{-1/2}$ and $g(x) = 4 + x^{1/2}$

28. $(x^2 + 8x)^{-1/2}(x + 4)$

29.   (A) $-x^2(x^3 - 9)^{-4/3}$ or $\dfrac{-x^2}{(x^3 - 9)^{4/3}}$

      (B) $6(2x^{-4} + x)^5(-8x^{-5} + 1)$

30.   (A) $\dfrac{6(x^3 - 2)^3(9x^3 + 2x^2 + 6)}{(6x + 1)^4}$

      (B) $\dfrac{2x^2(x^3 - 4)(7x^3 - 1)}{(2x^3 + 1)^{2/3}}$

Exercise 10-7

A   *Write each composite function in the form $y = u^n$ and $u = g(x)$.*

1. $y = (2x + 5)^3$ 2. $y = (3x - 7)^5$
3. $y = (x^3 - x^2)^8$ 4. $y = (2x^2 - 3x + 1)^4$
5. $y = (x^3 + 3x)^{1/3}$ 6. $y = (x^2 - 6)^{3/2}$

Find dy/dx using the general power rule.

7. $y = (2x + 5)^3$ 8. $y = (3x - 7)^5$
9. $y = (x^3 - x^2)^8$ 10. $y = (2x^2 - 3x + 1)^4$
11. $y = (x^3 + 3x)^{1/3}$ 12. $y = (x^2 - 6)^{3/2}$

B   *Find dy/dx using the general power rule.*

13. $y = 3(x^2 - 2)^4$ 14. $y = 2(x^3 + 6)^5$
15. $y = 2(x^2 + 3x)^{-3}$ 16. $y = 3(x^3 + x^2)^{-2}$
17. $y = \sqrt{x^2 + 8}$ 18. $y = \sqrt[3]{3x - 7}$
19. $y = \sqrt[3]{3x + 4}$ 20. $y = \sqrt{2x - 5}$
21. $y = (x^2 - 4x + 2)^{1/2}$ 22. $y = (2x^2 + 2x - 3)^{1/2}$

23. $y = \dfrac{1}{2x + 4}$ 24. $y = \dfrac{1}{(x^2 - 3)^8}$

$\left[\text{Hint:}\quad y = \dfrac{1}{2x + 4} = (2x + 4)^{-1}\right]$

25. $y = \dfrac{1}{4x^2 - 4x + 1}$

26. $y = \dfrac{1}{2x^2 - 3x + 1}$

27. $y = \dfrac{4}{\sqrt{x^2 - 3x}}$

28. $y = \dfrac{3}{\sqrt[3]{x - x^2}}$

29. $y = \dfrac{1}{3 - \sqrt[3]{x}}$

30. $y = \dfrac{1}{2\sqrt{x} - 5}$

31. $y = \dfrac{4}{\sqrt{\sqrt{x} - 5}}$

32. $y = \dfrac{9}{\sqrt[3]{\sqrt[3]{x} + 2}}$

Find each derivative and simplify.

33. $D_x[3x(x^2 + 1)^3]$

34. $D_x[2x^2(x^3 - 3)^4]$

35. $D_x \dfrac{(x^3 - 7)^4}{2x^3}$

36. $D_x \dfrac{3x^2}{(x^2 + 5)^3}$

37. $D_x[(2x - 3)^2(2x^2 + 1)^3]$

38. $D_x[(x^2 - 1)^3(x^2 - 2)^2]$

39. $D_x[4x^2\sqrt{x^2 - 1}]$

40. $D_x[3x\sqrt{2x^2 + 3}]$

41. $D_x \dfrac{2x}{\sqrt{x - 3}}$

42. $D_x \dfrac{x^2}{\sqrt{x^2 + 1}}$

C *In Problems 43–44, find the derivative and simplify.*

43. $D_x\sqrt{(2x - 1)^3(x^2 + 3)^4}$

44. $D_x \sqrt{\dfrac{4x + 1}{2x^2 + 1}}$

45. Find the equation of the tangent line to the graph of

$$y = \frac{4}{2x^2 - 3x + 3} = 4(2x^2 - 3x + 3)^{-1}$$

at (1, 2), using the general power rule to find the slope. Write the answer in the form $y = mx + b$.

46. Find the equation of the tangent line to the graph of

$$y = \frac{6}{\sqrt{x^2 - 3x}} = 6(x^2 - 3x)^{-1/2}$$

at (4, 3), using the general power rule to find the slope. Write the answer in the form $Ax + By = C$, with A, B, and C integers and $A > 0$.

Applications

Business & Economics

47. *Marginal average cost.* A manufacturer of skis finds that the average cost $\overline{C}(x)$ per pair of skis at an output level of x thousand skis is

$$\overline{C}(x) = (2x - 8)^2 + 25$$

(A) Find the marginal average cost $\overline{C}'(x)$ using the general power rule.

(B) Find $\overline{C}'(2)$, $\overline{C}'(4)$, and $\overline{C}'(6)$.

48. *Compound interest.* If $100 is invested at an interest rate of i compounded semiannually, the amount in the account at the end of 5 years is given by

$$A = 100 \left(1 + \frac{1}{2} i\right)^{10}$$

Find dA/di.

Life Sciences

49. *Bacteria growth.* The number y of bacteria in a certain colony after x days is given approximately by

$$y = (3 \times 10^6)\left(1 - \frac{1}{\sqrt[3]{(x^2 - 1)^2}}\right)$$

Find dy/dx.

50. *Pollution.* A small lake in a resort area became contaminated with a harmful bacteria because of excessive septic tank seepage. After treating the lake with a bactericide, the Department of Public Health estimated the bacteria concentration (number per cubic centimeter) after t days to be given by

$$C(t) = 500(8 - t)^2 \qquad 0 \leq t \leq 7$$

(A)   Find $C'(t)$ using the general power rule.
(B)   Find $C'(1)$ and $C'(6)$, and interpret.

Social Sciences

51. *Learning.* In 1930, L. L. Thurstone developed the following formula to indicate how learning time T depends on the length of a list n:

$$T = f(n) = \frac{c}{k} n\sqrt{n - a}$$

where a, c, and k are empirical constants. Suppose for a particular person, time T in minutes for learning a list of length n is

$$T = f(n) = 2n\sqrt{n - 2}$$

(A)   Find dT/dn, the rate of change in time with respect to n.
(B)   Find $f'(11)$ and $f'(27)$, and interpret.

10-8   Chapter Review

Important Terms
and Symbols

10-2   *Limits and continuity.* secant line, limit, one-sided limits, left-hand limit, right-hand limit, properties of limits, continuity, discontinuous, continuous, open interval, closed interval, half-open interval, continuous at a point, continuous on an interval, continuous on the right, continuous on the left, $\lim_{x \to c} f(x) = L$, $\lim_{x \to c^-} f(x) = L$, $\lim_{x \to c^+} f(x) = L$

10-3   *Increments, tangent lines, and rates of change.* increments, slope, tangent line, slope of graph at a point, average rate of change, instantaneous rate of change, average velocity, instantaneous velocity, Δx, Δy, average rate $= \Delta y / \Delta x$, instantaneous rate $= \lim_{\Delta x \to 0} \Delta y / \Delta x$

10-4   *The derivative.* the derivative of f at x, slope function, tangent line, nonexistence of the derivative, nondifferentiable at $x = a$, instantaneous rates of change, marginal cost, $f'(x)$

10-5   *Derivatives of constants, power forms, and sums.* derivative notation, derivative of a constant, power rule, derivative of a constant times a function, derivatives of sums and differences, $f'(x)$, y', dy/dx, $D_x f(x)$

10-6   *Derivatives of products and quotients.* derivatives of products, derivatives of quotients

10-7   *Chain rule and general power rule.* composite function, chain rule, general power rule

Exercise 10-8   Chapter Review

Work through all the problems in this chapter review and check your answers in the back of the book. (Answers to all review problems are there.) Where weaknesses show up, review appropriate sections in the text. When you are satisfied that you know the material, take the practice test following this review.

A   In Problems 1–10 find $f'(x)$ for $f(x)$ as given.

1. $f(x) = 3x^4 - 2x^2 + 1$

2. $f(x) = 2x^{1/2} - 3x$

3. $f(x) = 5$

4. $f(x) = \dfrac{2}{3}$

5. $f(x) = (2x - 1)(3x + 2)$

6. $f(x) = (x^2 - 1)(x^3 - 3)$

7. $f(x) = \dfrac{2x}{x^2 + 2}$

8. $f(x) = \dfrac{1}{3x + 2}$

9. $f(x) = (2x - 3)^3$

10. $f(x) = (x^2 + 2)^{-2}$

B   In Problems 11–18 find the indicated derivatives.

11. $\dfrac{dy}{dx}$ for $y = 3x^4 - 2x^{-3} + 5$

12. y' for $y = (2x^2 - 3x + 2)(x^2 + 2x - 1)$

13. $f'(x)$ for $f(x) = \dfrac{2x - 3}{(x - 1)^2}$

14. y' for $y = 2\sqrt{x} + \dfrac{4}{\sqrt{x}}$

15. $D_x[(x^2 - 1)(2x + 1)^2]$

16. $D_x \sqrt[3]{x^3 - 5}$

17. $\dfrac{dy}{dx}$ for $y = \dfrac{1}{\sqrt[3]{3x^2 - 2}}$

18. $D_x \dfrac{(x^2 + 2)^4}{2x - 3}$

19. For $y = f(x) = x^2 + 4$, find:

(A) The slope of the graph at $x = 1$
(B) The equation of the tangent line at $x = 1$ in the form $y = mx + b$

20. For $y = f(x) = 10x - x^2$, find:

(A) The slope function
(B) The value(s) of x such that the slope is 0

21. If an object moves along the y axis (scale in feet) so that it is at $y = f(x) = 16x^2 - 4x$ at time x (in seconds), find:

(A) The instantaneous velocity function
(B) The velocity at time $x = 3$ seconds

22. An object moves along the y axis (scale in feet) so that at time x (in seconds) it is at $y = f(x) = 96x - 16x^2$. Find:

(A) The instantaneous velocity function
(B) The time(s) when the velocity is 0

Problems 23–24 refer to the function f described in the figure.

23. (A) $\lim\limits_{x \to 2^-} f(x) = ?$ (B) $\lim\limits_{x \to 2^+} f(x) = ?$ (C) $\lim\limits_{x \to 2} f(x) = ?$

(D) $f(2) = ?$ (E) Is f continuous at $x = 2$?

24. (A) $\lim\limits_{x \to 5^-} f(x) = ?$ (B) $\lim\limits_{x \to 5^+} f(x) = ?$ (C) $\lim\limits_{x \to 5} f(x) = ?$

(D) $f(5) = ?$ (E) Is f continuous at $x = 5$?

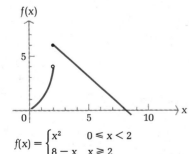

$f(x) = \begin{cases} x^2 & 0 \leqslant x < 2 \\ 8 - x & x \geqslant 2 \end{cases}$

In Problems 25–28 find points of discontinuity, if any exist.

25. $f(x) = 2x^2 - 3x + 1$

26. $f(x) = \dfrac{1}{x + 5}$

27. $f(x) = \dfrac{x - 3}{x^2 - x - 6}$

28. $f(x) = \sqrt{x - 3} \qquad x \geqslant 3$

In Problems 29–36 find each limit if it exists.

29. $\lim\limits_{x \to 3} \dfrac{2x - 3}{x + 5}$

30. $\lim\limits_{x \to 3} (2x^2 - x + 1)$

31. $\lim\limits_{x \to 0} \dfrac{2x}{3x^2 - 2x}$

32. $\lim\limits_{\Delta x \to 0} \dfrac{f(2 + \Delta x) - f(2)}{\Delta x}$

for $f(x) = x^2 + 4$

33. $\lim\limits_{x \to 3} \dfrac{x - 3}{x^2 - 9}$

34. $\lim\limits_{x \to -3} \dfrac{x - 3}{x^2 - 9}$

35. $\lim\limits_{x \to 7} \dfrac{\sqrt{x} - \sqrt{7}}{x - 7}$

36. $\lim\limits_{x \to -2} \sqrt{\dfrac{x^2 + 4}{2 - x}}$

In Problems 37–38 use the definition of the derivative to find $f'(x)$.

37. $f(x) = x^2 - x$

38. $f(x) = \sqrt{x} - 3$

C *Problems 39–41 refer to*

$$f(x) = \frac{2x^2 - 3x - 2}{3x^2 - 4x - 4}$$

39. (A) $\lim\limits_{x \to 2} f(x) = ?$ (B) $f(2) = ?$ (C) Is f continuous at $x = 2$?

40. (A) $\lim\limits_{x \to 0} f(x) = ?$ (B) $f(0) = ?$ (C) Is f continuous at $x = 0$?

41. Find all points of discontinuity for f.
42. Using the greatest integer function $[x] = ($Greatest integer $\leqslant x)$ find

(A) $\lim\limits_{x \to 3^-} [x]$ (B) $\lim\limits_{x \to 3^+} [x]$ (C) $\lim\limits_{x \to 3} [x]$

43. Find $D_x \sqrt[3]{\dfrac{(2x^3 + 1)^3}{(3x + 6)^2}}$ and simplify.

Applications

Business & Economics

44. *Marginal average cost.* Suppose a firm manufactures items having an average cost per item (in hundreds of dollars) given by

$$\overline{C}(x) = x^2 - 10x + 30$$

where x is the number of items manufactured.

(A) Find the marginal average cost $\overline{C}'(x)$.
(B) Find the marginal average cost at $x = 3$, 5, and 7, and interpret.

Life Sciences

45. *Pollution.* A sewage treatment plant disposes of its effluent through a pipeline that extends 1 mile toward the center of a large lake. The concentration of effluent $C(x)$, in parts per million, x meters from the end of the pipe is given approximately by

$$C(x) = 500(x + 1)^{-2}$$

What is the instantaneous rate of change of concentration at 9 meters? At 99 meters?

Social Sciences   46. *Learning.* If a person learns N items in t hours, as given by

$$N(t) = 20\sqrt{t}$$

find the rate of learning after:

(A) 1 hour   (B)   4 hours

Practice Test: Chapter 10

In Problems 1–4 find f'(x) for f(x) as given.

1. $f(x) = 3x^2 - 2x^{1/2} - 3$

2. $f(x) = (x^2 + 2)(2x - 3)$

3. $f(x) = \dfrac{3x^2 - 5}{x^2 + 1}$

4. $f(x) = (2x^3 - 3x + 1)^3$

In Problems 5–8, find the indicated derivative and simplify.

5. $D_x\left(\dfrac{3}{\sqrt[3]{x}} - \dfrac{2}{x} + x\right)$

6. $D_x[(x^2 - 1)^3(2x + 1)]$

7. $\dfrac{dy}{dx}$ for $y = \dfrac{1}{\sqrt[4]{2x^2 - 3}}$

8. $D_x \dfrac{\sqrt{2x - 1}}{(x^2 + 5)^4}$

9. Given $y = f(x) = 8x - x^2$. Find:

(A)   The slope function
(B)   The slope at $x = 2$
(C)   The equation of the tangent line at $x = 2$ in the form of $y = mx + b$
(D)   The value(s) of x that produces a slope of 0

10.   An object moves along the y axis (scale in feet) so that its position at time x (in seconds) is given by $y = f(x) = 20 + 80x - 10x^2$. Find:

(A)   The instantaneous velocity function
(B)   The velocity at $x = 3$ seconds
(C)   The time(s) when the velocity is 0

11.   (A)   Write the definition of the derivative of a function f at x.
(B)   Use the definition in part A to find $f'(x)$ for $f(x) = x - x^2$.

Problems 12–13 refer to the function f described in the figure.

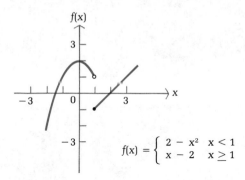

$$f(x) = \begin{cases} 2 - x^2 & x < 1 \\ x - 2 & x \geq 1 \end{cases}$$

12. (A) $\lim_{x \to 1^-} f(x) = ?$ (B) $\lim_{x \to 1^+} f(x) = ?$ (C) $\lim_{x \to 1} f(x) = ?$

    (D) $f(1) = ?$ (E) Is f continuous at $x = 1$?

13. (A) $\lim_{x \to 0^-} f(x) = ?$ (B) $\lim_{x \to 0^+} f(x) = ?$ (C) $\lim_{x \to 0} f(x) = ?$

    (D) $f(0) = ?$ (E) Is f continuous at $x = 0$?

Problems 14–15 refer to

$$f(x) = \frac{x - 4}{x^2 - 16}$$

14. (A) $\lim_{x \to 4} f(x) = ?$ (B) $f(4) = ?$

    (C) Is f continuous at $x = 4$?

15. (A) $\lim_{x \to 5} f(x) = ?$ (B) $f(5) = ?$

    (C) Is f continuous at $x = 5$?

16. The cost function $C(x)$ in dollars for manufacturing x video games per day is given by

$$C(x) = 1,000 + 15x - \frac{500}{\sqrt{2x + 1}}$$

    (A) Find the marginal cost function.

    (B) Find the marginal cost at $x = 12$.

Additional Derivative Topics

Contents

11-1 Implicit Differentiation

- Special Function Notation
- Implicit Differentiation

■ Special Function Notation

The equation

$$y = 2 - 3x^2 \tag{1}$$

defines a function f with y as a dependent variable and x as an independent variable. Using function notation, we would write

$$y = f(x) \qquad \text{or} \qquad f(x) = 2 - 3x^2$$

In order to reduce to a minimum the number of symbols involved in a discussion, we will often write equation (1) in the form

$$y = 2 - 3x^2 = y(x)$$

where y is *both* a dependent variable and a function symbol. This is a convenient notation and no harm is done as long as one is aware of the double role of y. Other examples are

$$x = 2t^2 - 3t + 1 = x(t)$$
$$z = \sqrt{u^2 - 3u} = z(u)$$
$$r = \frac{1}{(s^2 - 3s)^{2/3}} = r(s)$$

This type of notation will simplify much of the discussion and work that follows.

Until now we have considered functions involving only one independent variable. There is no reason to stop there. The concept can be generalized to functions involving two or more independent variables, and this will be done in detail in Chapter 16. For now, we will "borrow" the notation for a function involving two independent variables. For example,

$$F(x, y) = x^2 - 2xy + 3y^2 - 5$$

would specify a function F involving two independent variables.

■ Implicit Differentiation

Consider the equation

$$3x^2 + y - 2 = 0 \tag{2}$$

and the equation obtained by solving (2) for y in terms of x,

$$y = 2 - 3x^2 \tag{3}$$

Both equations define the same function using x as the independent variable and y as the dependent variable. For (1) we can write

$$y = f(x)$$

where

$$f(x) = 2 - 3x^2 \tag{4}$$

and we have an **explicit** (clearly stated) rule that enables us to determine y for each value of x. On the other hand, the y in equation (2) is the same y as in equation (3), and equation (2) **implicitly** gives (implies though does not plainly express) y as a function of x. Thus, we say that equations (3) and (4) define the function f explicitly and equation (2) defines f implicitly.

The direct use of an equation that defines a function implicitly to find the derivative of the dependent variable with respect to the independent variable is called **implicit differentiation.** Let us differentiate (2) implicitly and (3) directly, and compare results.

Starting with

$$3x^2 + y - 2 = 0$$

we think of y as a function of x — that is, $y = y(x)$ — and write

$$3x^2 + y(x) - 2 = 0$$

and differentiate both sides with respect to x:

$$D_x[3x^2 + y(x) - 2] = D_x 0$$
$$D_x 3x^2 + D_x y(x) - D_x 2 = 0$$
$$6x + y' - 0 = 0$$

Since y is a function of x, but is not explicitly given, we simply write $D_x y(x) = y'$ to indicate its derivative.

Now we solve for y':

$$y' = -6x$$

Note that we get the same result if we start with equation (3) and differentiate directly:

$$y = 2 - 3x^2$$
$$y' = -6x$$

Why are we interested in implicit differentiation? In general, why do we not solve for y in terms of x and differentiate directly? The answer is that there are many equations of the form

$$F(x, y) = 0 \tag{5}$$

that are either difficult or impossible to solve for y explicitly in terms of x (try it for $x^2y^5 - 3xy + 5 = 0$, for example), yet it can be shown that, under fairly general conditions on F, equation (5) will define one or more functions where y is a dependent variable and x is an independent variable. To find y' under these conditions, we differentiate (5) implicitly.

Example 1   Given

$$F(x, y) = x^2 + y^2 - 25 = 0 \tag{6}$$

find y' and the slope of the graph at x = 3.

Solution   We start with the graph of $x^2 + y^2 - 25 = 0$ (a circle) so that we can interpret our results geometrically:

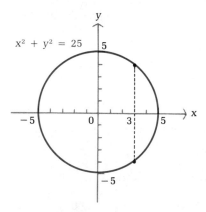

From the graph it is clear that equation (6) does not define a function. But with a suitable restriction on the variables, equation (6) can define two or more functions. For example, the upper half and the lower half of the circle each define a function. A point on each half-circle that corresponds to $x = 3$ is found by substituting $x = 3$ into (6) and solving for y:

$$x^2 + y^2 - 25 = 0$$
$$(3)^2 + y^2 = 25$$
$$y^2 = 16$$
$$y = \pm 4$$

Thus, the point $(3, 4)$ is on the upper half-circle and the point $(3, -4)$ is on the lower half-circle. We will use these results in a moment. We now differentiate (6) implicitly, treating y as a function of x; that is, $y = y(x)$.

$$x^2 + y^2 - 25 = 0$$

$$x^2 + [y(x)]^2 - 25 = 0$$
$$D_x\{x^2 + [y(x)]^2 - 25\} = D_x 0$$
$$D_x x^2 + D_x[y(x)]^2 - D_x 25 = 0 \qquad \text{Use the general power rule.}$$
$$2x + 2[y(x)]^{2-1} y'(x) - 0 = 0$$

$$2x + 2yy' = 0 \qquad \text{Solve for } y' \text{ in terms of } x \text{ and } y.$$

$$y' = -\frac{2x}{2y}$$

$$y' = -\frac{x}{y} \qquad \text{Leave the answer in terms of } x \text{ and } y.$$

We have found y' without first solving $x^2 + y^2 - 25 = 0$ for y in terms of x. And by leaving y' in terms of x and y, we can use $y' = -x/y$ to find y' for *any* point on the graph of $x^2 + y^2 - 25 = 0$ (except where $y = 0$). In particular, for $x = 3$, we found that $(3, 4)$ and $(3, -4)$ are on the graph; thus, the slope of the graph at $(3, 4)$ is

$$y'|_{(3,4)} = -\frac{3}{4} \qquad \text{The slope of the graph at } (3, 4)$$

and the slope at $(3, -4)$ is

$$y'|_{(3,-4)} = -\frac{3}{-4} = \frac{3}{4} \qquad \text{The slope of the graph at } (3, -4)$$

The symbol

$$y'|_{(a,b)}$$

is used to indicate that we are evaluating y' at $x = a$ and $y = b$.

The results are interpreted geometrically on the original graph as follows:

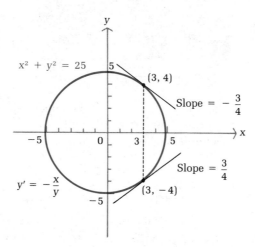

In Example 1 the fact that y' is given in terms of both x and y is not a great disadvantage. We have only to make certain that when we want **to evaluate y' for a particular value of x and y, say (x_0, y_0), the ordered pair must satisfy the original equation.**

Problem 1   Given $x^2 + y^2 - 169 = 0$, find y' by implicit differentiation and the slope of the graph when $x = 5$.

Example 2   Find the equation(s) of the tangent line(s) to the graph of

$$y - xy^2 + x^2 + 1 = 0 \qquad (7)$$

at the point(s) where $x = 1$.

Solution   We first find y when $x = 1$:

$$y - xy^2 + x^2 + 1 = 0$$
$$y - (1)y^2 + (1)^2 + 1 = 0$$
$$y - y^2 + 2 = 0$$
$$y^2 - y - 2 = 0$$
$$(y - 2)(y + 1) = 0$$
$$y = -1,\ 2$$

Thus, there are two points on the graph of (7) where $x = 1$; namely,

$$(1, -1) \quad \text{and} \quad (1, 2)$$

We next find the slope of the graph at these two points by differentiating (7) implicitly:

$$y - xy^2 + x^2 + 1 = 0$$
$$D_x\, y - D_x\, xy^2 + D_x\, x^2 + D_x\, 1 = D_x\, 0$$
$$y' - (x2yy' + y^2) + 2x = 0$$
$$y' - 2xyy' - y^2 + 2x = 0$$
$$y' - 2xyy' = y^2 - 2x$$
$$(1 - 2xy)y' = y^2 - 2x$$
$$y' = \frac{y^2 - 2x}{1 - 2xy}$$

Use the product rule for $D_x\, xy^2$.

Solve for y' by getting all terms involving y' on one side.

Now find the slope at each point:

$$y'|_{(1,-1)} = \frac{(-1)^2 - 2(1)}{1 - 2(1)(-1)} = \frac{1-2}{1+2} = \frac{-1}{3} = -\frac{1}{3}$$

$$y'|_{(1,2)} = \frac{2^2 - 2(1)}{1 - 2(1)(2)} = \frac{4-2}{1-4} = \frac{2}{-3} = -\frac{2}{3}$$

Equation of the tangent line at $(1, -1)$:

$$y - y_1 = m(x - x_1)$$
$$y + 1 = -\frac{1}{3}(x - 1)$$
$$y + 1 = -\frac{1}{3}x + \frac{1}{3}$$
$$y = -\frac{1}{3}x - \frac{2}{3}$$

Equation of the tangent line at $(1, 2)$:

$$y - y_1 = m(x - x_1)$$
$$y - 2 = -\frac{2}{3}(x - 1)$$
$$y - 2 = -\frac{2}{3}x + \frac{2}{3}$$
$$y = -\frac{2}{3}x + \frac{8}{3}$$

Problem 2 Repeat Example 2 for $x^2 + y^2 - xy - 7 = 0$ at $x = 1$.

Example 3 Find x' for $x = x(t)$ defined implicitly by

$$x^2 + 3t^2x - 10 = 0$$

and evaluate x' at $(t, x) = (1, 2)$.

Solution
$$x^2 + 3t^2x - 10 = 0$$
$$D_t\, x^2 + D_t\, 3t^2x - D_t\, 10 = D_t\, 0 \qquad \text{Remember, } x = x(t).$$
$$2xx' + 3t^2x' + x6t = 0 \qquad \text{Solve for } x'.$$
$$(2x + 3t^2)x' = -6tx$$
$$x' = \frac{-6tx}{2x + 3t^2}$$
$$x'|_{(1,\,2)} = \frac{-6(1)(2)}{2(2) + 3(1)^2} = \frac{-12}{7}$$

Problem 3 Find x' for $x = x(t)$ defined implicitly by

$$2x + 2t^3x^2 - 24 = 0$$

and evaluate x' at $(1, 3)$. Remember, x is the dependent variable and t is the independent variable.

Answers to Matched Problems

1. $y' = -x/y$; when $x = 5$, $y = \pm 12$, thus
$$y'|_{(5,\,12)} = -\frac{5}{12} \quad \text{and} \quad y'|_{(5,\,-12)} = \frac{5}{12}$$

2. $y' = \dfrac{y - 2x}{2y - x}$; $y = \dfrac{4}{5}x - \dfrac{14}{5}$, $y = \dfrac{1}{5}x + \dfrac{14}{5}$

3. $x' = \dfrac{-6t^2x^2}{2 + 4t^3x}$; $x'|_{(1,\,3)} = \dfrac{-27}{7}$

Exercise 11-1

In Problems 1–12, find y' without solving for y in terms of x (use implicit differentiation). Evaluate y' at the indicated point.

A
1. $y - 3x^2 + 5 = 0$, $(1, -2)$
3. $y^2 - 3x^2 + 8 = 0$, $(2, 2)$
5. $y^2 + y - x = 0$, $(2, 1)$

2. $3x^4 + y - 2 = 0$, $(1, -1)$
4. $3y^2 + 2x^3 - 14 = 0$, $(1, 2)$
6. $2y^3 + y^2 - x = 0$, $(-1, 1)$

B
7. $xy - 6 = 0$, $(2, 3)$
9. $2xy + y + 2 = 0$, $(-1, 2)$
11. $x^2y - 3x^2 - 4 = 0$, $(2, 4)$

8. $3xy - 2x - 2 = 0$, $(2, 1)$
10. $2y + xy - 1 = 0$, $(-1, 1)$
12. $2x^3y - x^3 + 5 = 0$, $(-1, 3)$

In Problems 13–14, find x' for $x = x(t)$ defined implicitly by the given equation. Evaluate x' at the indicated point.

13. $x^2 - t^2x + t^3 + 11 = 0$, $(-2, 1)$
14. $x^3 - tx^2 - 4 = 0$, $(-3, -2)$

Find the equation(s) of the tangent line(s) to the graphs of the indicated equations at the point(s) with abscissas as indicated.

15. $xy - x - 4 = 0$, $x = 2$

16. $3x + xy + 1 = 0$, $x = -1$

17. $y^2 - xy - 6 = 0$, $x = 1$

18. $xy^2 - y - 2 = 0$, $x = 1$

C *Find y' and the slope of the tangent line to the graph of each equation at the indicated point.*

19. $(1 + y)^3 + y = x + 7$, $(2, 1)$

20. $(y - 3)^4 - x = y$, $(-3, 4)$

21. $(x - 2y)^3 = 2y^2 - 3$, $(1, 1)$

22. $(2x - y)^4 - y^3 = 8$, $(-1, -2)$

23. $\sqrt{7 + y^2} - x^3 + 4 = 0$, $(2, 3)$

24. $6\sqrt{y^3 + 1} - 2x^{3/2} - 2 = 0$, $(4, 2)$

Applications

Business & Economics

For the demand equations in Problems 25–28, find the rate of change of p with respect to x by differentiating implicitly (x is the number of items that can be sold at a price of \$p).

25. $x = p^2 - 2p + 1{,}000$

26. $x = p^3 - 3p^2 + 200$

27. $x = \sqrt{10{,}000 - p^2}$

28. $x = \sqrt[3]{1{,}500 - p^3}$

Life Sciences

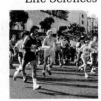

29. *Biophysics.* In biophysics, the equation

$$(L + m)(V + n) = k$$

is called the *fundamental equation of muscle contraction*, where m, n, and k are constants, and V is the velocity of the shortening of muscle fibers for a muscle subjected to a load of L. Find dL/dV using implicit differentiation.

11-2 Related Rates

In applications we often encounter two (or more) variables that are differentiable functions of time, say $x = x(t)$ and $y = y(t)$, but $x = x(t)$ and $y = y(t)$ may not be explicitly given. In addition, x and y may be related by an equation such as

$$x^2 + y^2 = 25 \tag{1}$$

Differentiating both sides of (1) with respect to t, we obtain

$$2x\frac{dx}{dt} + 2y\frac{dy}{dt} = 0 \qquad\qquad (2)$$

The derivatives dx/dt and dy/dt are related by equation (2); hence, they are referred to as **related rates.** If one of the rates and the value of one variable are both known, we can use equation (1) to find the value of the other variable and then we can use equation (2) to find the other rate. The following examples will illustrate how related rates can be used to solve certain types of practical problems.

Example 4   Suppose a point is moving on the graph of $x^2 + y^2 = 25$. When the point is at $(-3, 4)$ its x coordinate is increasing at the rate of 0.4 unit per second. How fast is the y coordinate changing at that moment?

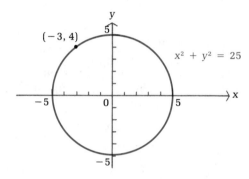

Solution   Since both x and y are changing with respect to time, we can think of each as a function of time:

$$x = x(t) \qquad \text{and} \qquad y = y(t)$$

but restricted so that

$$x^2 + y^2 = 25$$

Our problem is now to find dy/dt, given $x = -3$, $y = 4$, and $dx/dt = 0.4$. Implicitly differentiating both sides of (3) with respect to t, we have

$$x^2 + y^2 = 25$$

$$2x\frac{dx}{dt} + 2y\frac{dy}{dt} = 0 \qquad \text{Divide both sides by 2.}$$

$$x\frac{dx}{dt} + y\frac{dy}{dt} = 0 \qquad \text{Substitute } x = -3, y = 4, \text{ and } dx/dt = 0.4,$$
$$\text{and solve for } dy/dt.$$

$$(-3)(0.4) + 4\frac{dy}{dt} = 0$$

$$\frac{dy}{dt} = 0.3 \text{ unit per second}$$

Problem 4 A point is moving on the graph of $y^3 = x^2$. When the point is at $(-8, 4)$ its y coordinate is decreasing at 2 units per second. How fast is the x coordinate changing at that moment?

Example 5 Suppose two motor boats leave from the same point at the same time. If one travels north at 15 miles per hour and the other travels east at 20 miles per hour, how fast will the distance between them be changing after 2 hours?

Solution First, draw a picture.

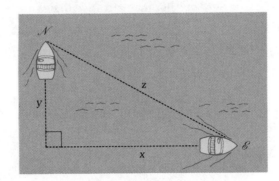

All variables, x, y, and z, are changing with time. Hence, they can be thought of as functions of time; $x = x(t)$, $y = y(t)$, and $z = z(t)$, given implicitly. It now makes sense to take derivatives of each variable with respect to time. From the Pythagorean theorem,

$$z^2 = x^2 + y^2 \tag{4}$$

We also know that

$$\frac{dx}{dt} = 20 \text{ miles per hour} \qquad \text{and} \qquad \frac{dy}{dt} = 15 \text{ miles per hour}$$

We would like to find dz/dt at the end of 2 hours; that is, when $x = 40$ miles and $y = 30$ miles. To do this we differentiate both sides of (4) with respect to t and solve for dz/dt:

$$2z\frac{dz}{dt} = 2x\frac{dx}{dt} + 2y\frac{dy}{dt} \tag{5}$$

We have everything we need except z. When $x = 40$ and $y = 30$, we find z from (4) to be 50. Substituting the known quantities into (5), we obtain

$$2(50)\frac{dz}{dt} = 2(40)(20) + 2(30)(15)$$

$$\frac{dz}{dt} = 25 \text{ miles per hour}$$

Problem 5 Repeat Example 5 for the situation at the end of 3 hours.

Suggestions for Solving Problems Involving Related Rates

1. Sketch a figure if helpful.
2. Identify all relevant variables, including those whose rates are given and those whose rates are to be found.
3. Find an equation connecting the variables in step 2.
4. Differentiate the equation implicitly and substitute in all given values.
5. Solve for the derivative that will give the unknown rate.

Example 6   Suppose that for a company manufacturing transistor radios, the cost, revenue, and profit equations are given by

$$C = 5,000 + 2x \qquad \text{Cost equation}$$

$$R = 10x - \frac{x^2}{1,000} \qquad \text{Revenue equation}$$

$$P = R - C \qquad \text{Profit equation}$$

where the production output in 1 week is x radios. If production is increasing at the rate of 500 radios per week when production is 2,000 radios, find the rate of increase in:

(A) Cost   (B) Revenue   (C) Profit

Solutions   If production x is a function of time (it must be since it is changing with respect to time), then C, R, and P must also be functions of time. They are implicitly (rather than explicitly) given. Letting t represent time in weeks, we differentiate both sides of each of the three equations above with respect to t, and then substitute $x = 2,000$ and $dx/dt = 500$ to find the desired rates.

(A) $C = 5,000 + 2x$ $\qquad$ Think $C = C(t)$ and $x = x(t)$.

$\dfrac{dC}{dt} = \dfrac{d}{dt}\,(5,000) + \dfrac{d}{dt}\,(2x)$ $\qquad$ Differentiate both sides with respect to t.

$\dfrac{dC}{dt} = 0 + 2\,\dfrac{dx}{dt} = 2\,\dfrac{dx}{dt}$

Since $dx/dt = 500$ when $x = 2,000$,

$\dfrac{dC}{dt} = 2(500) = \$1,000$ per week

Cost is increasing at a rate of \$1,000 per week.

(B) $R = 10x - \dfrac{x^2}{1,000}$

$\dfrac{dR}{dt} = \dfrac{d}{dt}(10x) - \dfrac{d}{dt}\dfrac{x^2}{1,000}$

$\dfrac{dR}{dt} = 10\dfrac{dx}{dt} - \dfrac{x}{500}\dfrac{dx}{dt}$

$\dfrac{dR}{dt} = \left(10 - \dfrac{x}{500}\right)\dfrac{dx}{dt}$

Since $dx/dt = 500$ when $x = 2,000$,

$\dfrac{dR}{dt} = \left(10 - \dfrac{2,000}{500}\right)(500) = \$3,000$ per week

Revenue is increasing at a rate of \$3,000 per week.

(C) $P = R - C$

$\dfrac{dP}{dt} = \dfrac{dR}{dt} - \dfrac{dC}{dt}$

$= \$3,000 - \$1,000$ Results from parts A and B

$= \$2,000$ per week

Profit is increasing at a rate of \$2,000 per week.

Problem 6   Repeat Example 6 for a production level of 6,000 radios per week.

Answers to   4. $\dfrac{dx}{dt} = 6$ units per second   5. $\dfrac{dz}{dt} = 25$ miles per hour
Matched Problems
6. (A) $dC/dt = \$1,000$ per week   (B) $dR/dt = -\$1,000$ per week
   (C) $dP/dt = -\$2,000$ per week

Exercise 11-2

A   In Problems 1–6 assume $x = x(t)$ and $y = y(t)$. Find the indicated rate, given the other information.

1. $y = 2x^2 - 1$, $dy/dt = ?$, $dx/dt = 2$ when $x = 30$
2. $y = 2x^{1/2} + 3$, $dy/dt = ?$, $dx/dt = 8$ when $x = 4$
3. $x^2 + y^2 = 25$, $dy/dt = ?$, $dx/dt = -3$ when $x = 3$ and $y = 4$
4. $y^2 + x = 11$, $dx/dt = ?$, $dy/dt = -2$ when $x = 2$ and $y = 3$
5. $x^2 + xy + 2 = 0$, $dy/dt = ?$, $dx/dt = -1$ when $x = 2$ and $y = -3$
6. $y^2 + xy - 3x = -3$, $dx/dt = ?$, $dy/dt = -2$ when $x = 1$ and $y = 0$

B   7. A point is moving on the graph of $xy = 36$. When the point is at $(4, 9)$, its x coordinate is increasing at 4 units per second. How fast is the y coordinate changing at that moment?

8. A point is moving on the graph of $4x^2 + 9y^2 = 36$. When the point is at $(3, 0)$, its y coordinate is decreasing at 2 units per second. How fast is its x coordinate changing at that moment?

9. A boat is being pulled toward a dock as indicated in the accompanying figure. If the rope is being pulled in at 3 feet per second, how fast is the distance between the dock and boat decreasing when it is 30 feet from the dock?

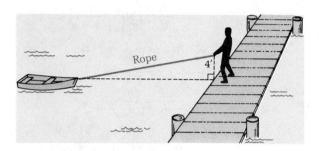

10. Refer to Problem 9. Suppose the distance between the boat and dock is decreasing at 3.05 feet per second. How fast is the rope being pulled in when the boat is 10 feet from the dock?

11. A rock is thrown into a still pond and causes a circular ripple. If the radius of the ripple is increasing at 2 feet per second, how fast is the area changing when the radius is 10 feet? (Use $A = \pi R^2$, $\pi \approx 3.14$.)

12. Refer to Problem 11. How fast is the circumference of a circular ripple changing when the radius is 10 feet? (Use $C = 2\pi R$, $\pi \approx 3.14$.)

13. The radius of a spherical balloon is increasing at the rate of 3 centimeters per minute. How fast is the volume changing when the radius is 10 centimeters? [Use $V = (4/3)\pi R^3$, $\pi \approx 3.14$.]

14. Refer to Problem 13. How fast is the surface area of the sphere increasing? (Use $S = 4\pi R^2$, $\pi \approx 3.14$.)

15. Boyle's law for enclosed gases states that if the volume is kept constant, then the pressure P and temperature T are related by the equation

$$\frac{P}{T} = k$$

where k is a constant. If the temperature is increasing at 3 degrees per hour, what is the rate of change of pressure when the temperature is 250° (Kelvin) and the pressure is 500 pounds per square inch?

16. Boyle's law for enclosed gases states that if the temperature is kept

constant, then the pressure P and volume V of the gas are related by the equation

$$VP = k$$

where k is a constant. If the volume is decreasing by 5 cubic inches per second, what is the rate of change of the pressure when the volume is 1,000 cubic inches and the pressure is 40 pounds per square inch?

17. A 10 foot ladder is placed against a vertical wall. Suppose the bottom slides away from the wall at a constant rate of 3 feet per second. How fast is the top sliding down the wall (negative rate) when the bottom is 6 feet from the wall? [*Hint:* Use the Pythagorean theorem: $a^2 + b^2 = c^2$, where c is the length of the hypotenuse of a right triangle and a and b are the lengths of the two shorter sides.]

18. A weather balloon is rising vertically at the rate of 5 meters per second. An observer is standing on the ground 300 meters from the point where the balloon was released. At what rate is the distance between the observer and the balloon changing when the balloon is 400 meters high?

C
19. A streetlight is on top of a 20 foot pole. A 5 foot tall person walks away from the pole at the rate of 5 feet per second. At what rate is the tip of the person's shadow moving away from the pole when he is 20 feet from the pole?

20. Refer to Problem 19. At what rate is the person's shadow growing when he is 20 feet from the pole?

Applications

Business & Economics

21. *Cost, revenue, and profit rates.* Suppose that for a company manufacturing hand calculators, the cost, revenue, and profit equations are given by

$$C = 90,000 + 30x$$

$$R = 300x - \frac{x^2}{30}$$

$$P = R - C$$

where the production output in 1 week is x calculators. If production is increasing at a rate of 500 calculators per week when production output is 6,000 calculators, find the rate of increase (decrease) in:

(A) Cost   (B) Revenue   (C) Profit

22. *Cost, revenue, and profit rates.* Repeat Problem 21 for

$$C = 72{,}000 + 60x$$

$$R = 200x - \frac{x^2}{30}$$

$$P = R - C$$

where production is increasing at a rate of 500 calculators per week at a production level of 1,500 calculators.

Life Sciences
23. *Pollution.* An oil tanker aground on a reef is leaking oil that forms a circular oil slick about 0.1 foot thick. To estimate the rate (in cubic feet per minute, dV/dt) at which the oil is leaking from the tanker, it was found that the radius of the slick was increasing at 0.32 foot per minute ($dR/dt = 0.32$) when the radius was 500 feet ($R = 500$). Find dV/dt, using $\pi \approx 3.14$.

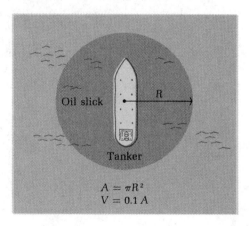

Social Sciences
24. *Learning.* A person who is new on an assembly line performs an operation in T minutes after x performances of the operation, as given by

$$T = 6\left(1 + \frac{1}{\sqrt{x}}\right)$$

If

$$\frac{dx}{dt} = 6 \text{ operations per hour}$$

where t is time in hours, find dT/dt after thirty-six performances of the operation.

11-3 Higher-Order Derivatives

- Higher-Order Derivatives for Explicitly Defined Functions
- Second-Order Derivatives for Implicitly Defined Functions

■ **Higher-Order Derivatives for Explicitly Defined Functions**

If we start with the function f defined by

$$f(x) = 3x^3 - 4x^2 - x + 1$$

and take the derivative, we obtain a new function f' defined by

$$f'(x) = 9x^2 - 8x - 1$$

Now if we take another derivative, called the **second derivative,** we obtain a new function f'' defined by

$$f''(x) = 18x - 8$$

And taking still another derivative produces the **third derivative** f''' defined by

$$f'''(x) = 18$$

and so on.

In general, the successive derivatives for a function f are denoted by

$$f', f'', f''', f^{(4)}, \ldots, f^{(n)}$$

It can be shown that domains of successive derivatives of a given function are subsets of the domain of the original function.

Example 7   Find f', f'', and f''' for $f(x) = 3x^{-1} + x^2$.

Solution $f'(x) = -3x^{-2} + 2x$ $f''(x) = 6x^{-3} + 2$ $f'''(x) = -18x^{-4}$

Problem 7   Find f', f'', and f''' for $f(x) = 2 - 3x^2 + x^{-2}$.

Along with the various other symbols for the first derivative that we considered earlier, we have corresponding symbols for higher-order derivatives. For example, if

$$y = f(x)$$

then

$$\frac{dy}{dx} = f'(x)$$

and the second derivative is given by

$$\frac{d}{dx}\left(\frac{dy}{dx}\right) = f''(x)$$

or, in short,

$$\frac{d^2y}{dx^2} = f''(x) \qquad \text{Note how the 2's are placed.}$$

Similarly,

$$\frac{d^3y}{dx^3} = f'''(x)$$

and so on. We summarize some of the more commonly used higher-derivative forms in the box.

Derivative Notation

For $y = f(x)$, we have the following:

First-derivative symbols

$$f'(x) \qquad \frac{dy}{dx} \qquad y' \qquad D_x f(x)$$

Second-derivative symbols

$$f''(x) \qquad \frac{d^2y}{dx^2} \qquad y'' \qquad D_x^2 f(x)$$

Third-derivative symbols

$$f'''(x) \qquad \frac{d^3y}{dx^3} \qquad y''' \qquad D_x^3 f(x)$$

Fourth-derivative symbols

$$f^{(4)}(x) \qquad \frac{d^4y}{dx^4} \qquad y^{(4)} \qquad D_x^4 f(x)$$

. . .

nth-derivative symbols

$$f^{(n)}(x) \qquad \frac{d^ny}{dx^n} \qquad y^{(n)} \qquad D_x^n f(x)$$

[Note: In the fourth derivative (and higher) we use $f^{(4)}(x)$ and $y^{(4)}$ to avoid confusion with powers represented by $f^4(x)$ and y^4.]

Example 8   If $y = 4x^{1/2}$, then

$$y' = 2x^{-1/2} \qquad \frac{dy}{dx} = 2x^{-1/2} \qquad D_x\, 4x^{1/2} = 2x^{-1/2}$$

$$y'' = -x^{-3/2} \qquad \frac{d^2y}{dx^2} = -x^{-3/2} \qquad D_x^2\, 4x^{1/2} = -x^{-3/2}$$

$$y''' = \frac{3}{2}x^{-5/2} \qquad \frac{d^3y}{dx^3} = \frac{3}{2}x^{-5/2} \qquad D_x^3\, 4x^{1/2} = \frac{3}{2}x^{5/2}$$

The domain of the original function is $[0, \infty)$, while the domain of each higher derivative is $(0, \infty)$, a subset of $[0, \infty)$.

Problem 8   If $y = 27x^{4/3}$, find:

(A) d^2y/dx^2 (B) $D_x^3\, 27x^{4/3}$ (C) $y^{(4)}$

Example 9   If

$$y = \frac{1}{\sqrt{2x-1}}$$

find y', y'', and y'''.

Solution

$$y = \frac{1}{\sqrt{2x-1}} = \frac{1}{(2x-1)^{1/2}} = (2x-1)^{-1/2}$$

$$y' = -\frac{1}{2}(2x-1)^{-3/2}2 = -(2x-1)^{-3/2}$$

$$y'' = \frac{3}{2}(2x-1)^{-5/2}2 = 3(2x-1)^{-5/2}$$

$$y''' = -\frac{15}{2}(2x-1)^{-7/2}2 = -15(2x-1)^{-7/2}$$

Problem 9   If

$$y = \frac{1}{(3x+2)^3}$$

find y', y'', and y'''.

■ Second-Order Derivatives for Implicitly Defined Functions

Suppose we have a function $y = y(x)$ defined implicitly by an equation of the form $F(x, y) = 0$. How can we find y'' without solving for y in terms of x? We will illustrate the process with an example.

Example 10   Find y'' for $y = y(x)$ defined implicitly by

$$x^2 + y^2 = 4 \tag{1}$$

Solution   Differentiate both sides with respect to x and solve for y'.

$$2x + 2yy' = 0$$
$$2yy' = -2x$$
$$y' = \frac{-x}{y} \tag{2}$$

Now differentiate both sides again with respect to x, thinking of $y = y(x)$, to obtain

$$y'' = \frac{y(-1) - (-x)y'}{y^2} \tag{3}$$

We are almost there! Substituting (2) into (3) we obtain

$$y'' = \frac{-y + x(-x/y)}{y^2}$$
$$= \frac{-y^2 - x^2}{y^3}$$
$$= \frac{-(x^2 + y^2)}{y^3}$$

Since $x^2 + y^2 = 4$ from our original equation, we obtain a further simplification:

$$y'' = \frac{-4}{y^3}$$

Problem 10   Find y'' for $y = y(x)$ defined implicitly by $3x^2 - y^2 = 9$.

In Chapter 12 we will see how second derivatives provide a useful tool in sketching graphs of equations and solving maxima–minima problems.

Answers to   7. $f'(x) = -6x - 2x^{-3}, f''(x) = -6 + 6x^{-4}, f'''(x) = -24x^{-5}$
Matched Problems   8. (A) $12x^{-2/3}$ (B) $-8x^{-5/3}$ (C) $(40/3)x^{-8/3}$
8. 9. $y' = -9(3x + 2)^{-4}, y'' = 108(3x + 2)^{-5}, y''' = -1,620(3x + 2)^{-6}$
10. $y'' = -27y^{-3}$

Exercise 11-3

A   *Find the indicated derivative for each function.*

1. $f''(x)$ for $f(x) = x^3 - 2x^2 - 1$
2. $g''(x)$ for $g(x) = x^4 - 3x^2 + 5$
3. $f'''(x)$ for $f(x) = 3x - 16x^2$

4. $g'''(x)$ for $g(x) = 1 - x - 2x^4$
5. d^2y/dx^2 for $y = 2x^5 - 3$
6. d^2y/dx^2 for $y = 3x^4 - 7x$
7. d^3y/dx^3 for $y = 120 - 30x^2$
8. d^3y/dx^3 for $y = 1 + 2x^2 - 4x^4$
9. $D_x^3(x^{-1})$
10. $D_x^3(x^{-2})$
11. $D_x^2(1 - 2x + x^3)$
12. $D_x^4(3x^2 - x^7)$

B *Find the indicated derivative for each function.*

13. $D_x^2(3x^{-1} + 2x^{-2} + 5)$
14. d^2y/dx^2 for $y = x^2 - \sqrt[3]{x}$
15. $y^{(4)}$ for $y = \sqrt{2x - 1}$
16. $f^{(4)}(x)$ for $f(x) = 27\sqrt[3]{x^2}$
17. $D_x^2(1 - 2x)^3$
18. $D_x^3(3 - x)^4$
19. y'' for $y = (x^2 - 1)^3$
20. y'' for $y = (x^2 + 4)^4$

21. $D_x^3 \dfrac{1}{\sqrt{3 - 2x}}$
22. $D_x^3 \dfrac{1}{(5 - 3x)^2}$

23. $f''(x)$ for $f(x) = (3x^2 - 1)^{4/3}$
24. $f''(x)$ for $f(x) = (2x^3 + 3)^{3/2}$

Use implicit differentiation to find y'' for each of the following:

25. $4x^2 - y^2 = 3$
26. $2x^3 - 3y^2 = 4$
27. $y^3 + x^2 = 7$
28. $3xy - x^2 = 2$

C 29. Find: $D_x^3 \dfrac{x}{2x - 1}$

30. Find y''' for $y = (2x - 1)(x^2 + 1)$.
31. Find y''' for $x^2 + y^2 = 4$.
32. Find y''' for $4x^2 - y^2 = 3$.

11-4 The Differential

- The Differential
- Approximations Using Differentials
- Differential Rules

■ The Differential

In Chapter 10 we introduced the concept of increment. Recall that for a function defined by

$$y = f(x)$$

we said that Δx represents a change in the independent variable x; that is,

$$\Delta x = x_2 - x_1 \qquad \text{or} \qquad x_2 = x_1 + \Delta x$$

And Δy represents the corresponding change in the dependent variable y; that is,

$$\Delta y = f(x_1 + \Delta x) - f(x_1)$$

We then defined the derivative of f at x_1 to be

$$\frac{dy}{dx} = \lim_{\Delta x \to 0} \frac{\Delta y}{\Delta x}$$

If the limit exists, then it follows that

$$\frac{\Delta y}{\Delta x} \approx \frac{dy}{dx} \quad \text{for small } \Delta x$$

or

$$\Delta y \approx \frac{dy}{dx} \Delta x \tag{1}$$

We used dy/dx as an alternate symbol for $f'(x)$. We will now give dy and dx special meaning, and we will show how dy can be used to approximate Δy. This turns out to be quite useful, since a number of practical problems require the computation of Δy, and we will be able to use the more readily computed dy. The symbols dy and dx are called **differentials** and are defined in the box.

Differentials

If $y = f(x)$ defines a differentiable function, then:

1. The **differential dx** of the independent variable x is an arbitrary real number.
2. The **differential dy** of the dependent variable y is defined as the product of $f'(x)$ and dx; that is,

   $$dy = f'(x)\, dx \tag{2}$$

   The differential dy is actually a function involving two independent variables, x and dx—a change in either one or both will affect dy.

Example 11 Find dy for $f(x) = x^2 + 3x$. Evaluate dy for $x = 2$ and $dx = 0.1$, for $x = 3$ and $dx = 0.1$, and for $x = 1$ and $dx = 0.02$.

Solution
$$dy = f'(x)\, dx$$
$$= (2x + 3)\, dx$$

When $x = 2$ and $dx = 0.1$,

$$dy = [2(2) + 3]0.1 = 0.7$$

When $x = 3$ and $dx = 0.1$,
$$dy = [2(3) + 3]0.1 = 0.9$$
When $x = 1$ and $dx = 0.02$,
$$dy = [2(1) + 3]0.02 = 0.1$$

Problem 11 Find dy for $f(x) = \sqrt{x} + 3$. Evaluate dy for $x = 4$ and $dx = 0.1$ for $x = 9$ and $dx = 0.12$, and for $x = 1$ and $dx = 0.01$.

If you compare the right-hand sides of (1) and (2) you will see what motivated the definition of dy. The differential concept has a very clear geometric interpretation, as is indicated in Figure 1 (study it carefully).

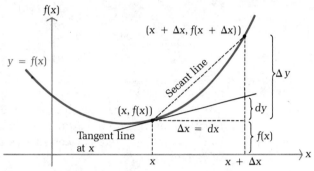

Figure 1

For small Δx, we have

Slope of secant line $\approx$ Slope of tangent line

$$\frac{\Delta y}{\Delta x} \approx \frac{dy}{dx}$$

$$\Delta y \approx dy = f'(x)\, dx$$

Example 12 Find Δy and dy for $f(x) = 6x - x^2$ when $x = 2$ and $\Delta x = dx = 0.1$.

Solution
$$\Delta y = f(x + \Delta x) - f(x)$$
$$= f(2.1) - f(2)$$
$$= [6(2.1) - (2.1)^2] - [6(2) - 2^2]$$
$$= 8.19 - 8$$
$$= 0.19$$

$$dy = f'(x)\, dx$$
$$= (6 - 2x)\, dx$$
$$= [6 - 2(2)](0.1)$$
$$= 0.2$$

Notice that dy and Δy differ by only 0.01 in this case.

Problem 12 Repeat Example 12 for $x = 4$ and $\Delta x = dx = 0.2$.

■ Approximations Using Differentials

Differentials provide a fast and convenient way of approximating certain quantities. The relationships given in the box can be used in this regard.

Differential Approximation

If $f'(x)$ exists, then for small Δx

$$\Delta y \approx dy$$

and

$$f(x + \Delta x) = f(x) + \Delta y$$
$$\approx f(x) + dy$$
$$= f(x) + f'(x)\,dx$$

We will use these relationships in the examples that follow. (Before proceeding, however, it should be mentioned that even though differentials can be used to approximate certain quantities, the error can be substantial in certain cases.)

Example 13
Weight–Height

A formula relating the approximate weight, W (in pounds), of an average person and their height, h (in inches), is

$$W = 0.0005h^3 \qquad 30 \leqslant h \leqslant 74$$

What is the approximate change in weight for a height increase from 40 to 42 inches?

Solution

We are actually interested in finding ΔW, the change in weight brought about by the change in height from 40 to 42 inches ($\Delta h = 2$). We will use the differential dW to approximate ΔW, since Δh is small. The problem is now to find dW for $h = 40$ and $dh = \Delta h = 2$.

$$W(h) = 0.0005h^3$$
$$dW = W'(h)\,dh$$
$$= 0.0015h^2\,dh$$
$$= 0.0015(40)^2(2)$$
$$= 4.8 \text{ pounds}$$

Thus, a child growing from 40 inches to 42 inches would expect to increase in weight by approximately 4.8 pounds. Notice that using the differential is somewhat easier than finding $\Delta W = W(42) - W(40)$.

Problem 13 Refer to Example 13. Approximate the change in weight resulting from a height increase from 70 to 72 inches.

Example 14
Cost–Revenue

A company manufactures and sells x transistor radios per week. If the weekly cost and revenue equations are

$$C(x) = 5,000 + 2x$$
$$R(x) = 10x - \frac{x^2}{1,000} \qquad 0 \leqslant x \leqslant 8,000$$

find the approximate changes in revenue and profit if production is increased from 2,000 to 2,010 units per week.

Solution We will approximate ΔR and ΔP with dR and dP, respectively, using $x = 2,000$ and $dx = \Delta x = 2,010 - 2,000 = 10$.

$$R(x) = 10x - \frac{x^2}{1,000}$$

$$dR = R'(x)\, dx$$
$$= \left(10 - \frac{x}{500}\right) dx$$
$$= \left(10 - \frac{2,000}{500}\right) 10$$
$$= \$60 \text{ per week}$$

$$P(x) = R(x) - C(x)$$
$$= 10x - \frac{x^2}{1,000} - 5,000 - 2x$$
$$= 8x - \frac{x^2}{1,000} - 5,000$$

$$dP = P'(x)\, dx$$
$$= \left(8 - \frac{x}{500}\right) dx$$
$$= \left(8 - \frac{2,000}{500}\right) 10$$
$$= \$40 \text{ per week}$$

Problem 14 Repeat Example 14 with production increasing from 6,000 to 6,010.

Comparing the results in Example 14 and Problem 14, we see that an increase in production results in a revenue and profit increase at the 2,000 production level, but a revenue and profit loss at the 6,000 production level.

Now we will consider a slightly different type of problem involving differential approximations.

Example 15   Use differentials to approximate $\sqrt[3]{27.54}$.

Solution   Even though the problem is trivial using a hand calculator, its solution using differentials will help increase the understanding of this concept. Form the function

$$y = f(x) = \sqrt[3]{x} = x^{1/3}$$

and note that we can compute $f(27)$ and $f'(27)$ exactly. Thus, if we let $x = 27$ and $dx = \Delta x = 0.54$ and use

$$f(x + \Delta x) = f(x) + \Delta y$$
$$\approx f(x) + dy$$
$$= f(x) + f'(x)\, dx$$

we will obtain an approximation for $f(27.54) = \sqrt[3]{27.54}$ that is easy to compute.

$$f(x + \Delta x) \approx f(x) + f'(x)\, dx$$

$$(x + \Delta x)^{1/3} \approx x^{1/3} + \frac{1}{3x^{2/3}}\, dx$$

$$(27 + 0.54)^{1/3} \approx 27^{1/3} + \frac{1}{3(27)^{2/3}}\, (0.54)$$

Thus,

$$\sqrt[3]{27.54} \approx 3 + \frac{0.54}{27} = 3.02 \qquad \text{(Calculator value} = 3.0199)$$

Problem 15   Use differentials to approximate $\sqrt{36.72}$.

■ Differential Rules

We close this section by listing a number of differential rules that will be of use to us in the next chapter. These rules follow directly from the definition of the differential and the derivative rules discussed earlier.

Differential Rules

If u and v are differentiable functions and c is a constant, then:

1. $dc = 0$
2. $du^n = nu^{n-1}\, du$
3. $d(u \pm v) = du \pm dv$
4. $d(uv) = u\, dv + v\, du$
5. $d\left(\dfrac{u}{v}\right) = \dfrac{v\, du - u\, dv}{v^2}$

To illustrate how these rules are established, we derive rule 4 as an example:

$$y = f(x) = u(x)v(x)$$

$$dy = f'(x)\, dx$$
$$= [u(x)v'(x) + v(x)u'(x)]\, dx$$
$$= u(x)v'(x)\, dx + v(x)u'(x)\, dx$$
$$= u\, dv + v\, du$$

Answers to Matched Problems

11. $dy = \dfrac{1}{2\sqrt{x}}\, dx$; 0.025, 0.02, 0.005

12. $\Delta y = -0.44,\ dy = -0.4$ 13.   14.7 pounds

14. $dR = -\$20$ per week, $dP = -\$40$ per week   15.   6.06

Exercise 11-4

A   *Find dy for each function.*

1. $y = 30 + 12x^2 - x^3$

2. $y = 200x - \dfrac{x^2}{30}$

3. $y = x^2\left(1 - \dfrac{x}{9}\right)$

4. $y = x^3(60 - x)$

5. $y = f(x) = \dfrac{590}{\sqrt{x}}$

6. $y = 52\sqrt{x}$

7. $y = 75\left(1 - \dfrac{2}{x}\right)$

8. $y = 100\left(x - \dfrac{4}{x^2}\right)$

B   *Evaluate dy and Δy for each function at the indicated values.*

9. $y = f(x) = x^2 - 3x + 2$, $x = 5$, $\Delta x = dx = 0.2$

10. $y = f(x) = 30 + 12x^2 - x^3$, $x = 2$, $\Delta x = dx = 0.1$

11. $y = f(x) = 75\left(1 - \dfrac{2}{x}\right)$, $x = 5$, $dx = \Delta x = 0.5$

12. $y = f(x) = 100\left(x - \dfrac{4}{x^2}\right)$, $x = 2$, $\Delta x = dx = 0.1$

Use differentials to approximate the indicated roots.

13. $\sqrt[4]{17}$

14. $\sqrt{83}$

15. $\sqrt[3]{28}$

16. $\sqrt[5]{34}$

17. A cube with sides 10 inches long is covered with a 0.2 inch thick coat of fiberglass. Use differentials to estimate the volume of the fiberglass shell.

18. A sphere with a radius of 5 centimeters is coated with ice 0.1 centimeter thick. Use differentials to estimate the volume of the ice [recall that $V = (4/3)\,\pi r^3$, $\pi \approx 3.14$].

C   19. Find dy if $y = \sqrt[3]{3x^2 - 2x + 1}$.

20. Find dy if $y = (2x^2 - 4)\sqrt{x + 2}$.

21. Find dy and Δy for $y = 52\sqrt{x}$, $x = 4$, and $\Delta x = dx = 0.3$.

22. Find dy and Δy for $y = 590/\sqrt{x}$, $x = 64$, and $\Delta x = dx = 1$.

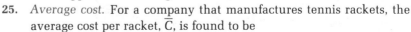

Applications

Use differential approximations in the following problems.

Business & Economics   23. *Advertising.* Using past records, it is estimated that a company will sell N units of a product after spending x thousand dollars in advertising, as given by

$$N = 60x - x^2 \qquad 5 \leqslant x \leqslant 30$$

Estimate the increase in sales that will result by increasing the advertising budget from \$10,000 to \$11,000. From \$20,000 to \$21,000.

24. *Price–demand.* Suppose in a grocery chain the daily demand in pounds for chocolate candy at \$x per pound is given by

$$D = 1,000 - 40x^2 \qquad 1 \leqslant x \leqslant 5$$

If the price is increased from \$3.00 per pound to \$3.20 per pound, what is the approximate change in demand?

25. *Average cost.* For a company that manufactures tennis rackets, the average cost per racket, $\overline{C}$, is found to be

$$\overline{C} = x^2 - 20x + 110 \qquad 6 \leqslant x \leqslant 14$$

where x is the number of rackets produced per hour. Approximate the change in cost per racket if production is increased from seven per hour to eight per hour. From twelve per hour to thirteen per hour.

26. *Revenue and profit.* A company manufactures and sells x televisions per month. If the cost and revenue equations are

$$C(x) = 72,000 + 60x$$

$$R(x) = 200x - \frac{x^2}{30} \qquad 0 \leqslant x \leqslant 6,000$$

find the approximate changes in revenue and profit if production is increased from 1,500 to 1,501. From 4,500 to 4,501.

Life Sciences   27. *Pulse rate.* The average pulse rate y in beats per minute of a healthy person x inches tall is given approximately by

$$y = \frac{590}{\sqrt{x}} \qquad 30 \leqslant x \leqslant 75$$

Approximate the change in pulse rate for a height change from 36 to 37 inches. From 64 to 65 inches.

28. *Measurement.* An egg of a particular bird is very nearly spherical. If the radius to the inside of the shell is 5 millimeters and the radius to the outside of the shell is 5.3 millimeters, approximately what is the volume of the shell? [Remember that $V = (4/3) \pi r^3$ and use $\pi \approx 3.14$.]

29. *Medicine.* A drug is given to a patient to dilate her arteries. If the radius of an artery is increased from 2 to 2.1 millimeters, approximately how much is a cross-sectional area increased? (Assume the cross-section of the artery is circular; $A = \pi r^2$ and $\pi \approx 3.14$.)

30. *Drug sensitivity.* One hour after x milligrams of a particular drug are given to a person, the change in body temperature T in degrees Fahrenheit is given by

$$T = x^2 \left(1 - \frac{x}{9}\right) \qquad 0 \leqslant x \leqslant 6$$

Approximate the changes in body temperature produced by the following changes in drug dosages:

(A) From 2 to 2.1 milligrams
(B) From 3 to 3.1 milligrams
(C) From 4 to 4.1 milligrams

Social Sciences

31. *Learning.* A particular person learning to type has an achievement record given approximately by

$$N = 75 \left(1 - \frac{2}{t}\right) \qquad 3 \leqslant t \leqslant 20$$

where N is the number of words per minute typed after t weeks of practice. What is the approximate improvement from 5 to 5.5 weeks of practice?

32. *Learning.* If a person learns y items in x hours, as given approximately by

$$y = 52\sqrt{x} \qquad 0 \leqslant x \leqslant 9$$

what is the approximate increase in the number of items learned when x changes from 1 to 1.1 hours? From 4 to 4.1 hours?

33. *Politics.* In a newly incorporated city it is estimated that the voting population (in thousands) will increase according to

$$N(t) = 30 + 12t^2 - t^3 \qquad 0 \leqslant t \leqslant 8$$

where t is time in years. Find the approximate change in votes for the following time changes:

(A) From 1 to 1.1 years
(B) From 4 to 4.1 years
(C) From 7 to 7.1 years

11-5   Marginal Analysis in Business and Economics

One important use of calculus in business and economics is in marginal analysis. We introduced the concept of marginal cost earlier. There is no reason to stop there. Economists also talk about **marginal revenue** and **marginal profit.** Recall that the word *marginal* refers to a rate of change — that is, a derivative. Thus, we define the following:

Marginal Cost, Revenue, and Profit

If x is the number of units of product produced in some time interval, then

$$\text{Total cost} = C(x)$$
$$\text{Marginal cost} = C'(x)$$
$$\text{Total revenue} = R(x)$$
$$\text{Marginal revenue} = R'(x)$$
$$\text{Total profit} = P(x) = R(x) - C(x)$$
$$\text{Marginal profit} = P'(x) = R'(x) - C'(x)$$
$$= (\text{Marginal revenue}) - (\text{Marginal cost})$$

In words, the marginal cost is the change in cost per unit change in production at a given output level; the marginal revenue is the change in revenue per unit change in production at a given output level; and the marginal profit is the change in profit per unit change in production at a given output level. Or, stated more simply, **the marginal cost, revenue, and profit represent the approximate changes in cost, revenue, and profit, respectively, that result from a unit increase in production** (see Figure 2).

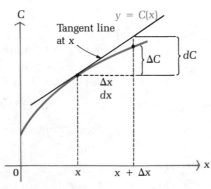

Figure 2

If $\Delta x = dx = 1$, then

$\Delta C =$ Actual change in cost C per unit change in output x at an output level of x units

Marginal cost

$dC = C'(x)\,dx = C'(x)$
    $=$ Approximate change in cost C per unit change in output x at an output level of x units (change up to tangent line)

We now present an example in market research to show how marginal cost, revenue, and profit are tied together.

Production Strategy

The market research department of a company recommends that the company manufacture and market a new transistor radio. After extensive surveys, the research department presents the following **demand equation:**

$$x = 10,000 - 1,000p \qquad x \text{ is demand at } \$p \text{ per radio} \qquad (1)$$

or

$$p = 10 - \frac{x}{1,000} \qquad\qquad (2)$$

where x is the number of radios retailers are likely to buy per week at $p per radio. Equation (2) is simply equation (1) solved for p in terms of x. Notice that as price goes up, demand goes down.

The financial department provides the following **cost equation:**

$$C(x) = 5,000 + 2x \qquad\qquad (3)$$

where $5,000 is the estimated fixed costs (tooling and overhead) and $2 is the estimated variable costs (cost per unit for materials, labor, marketing, transportation, storage, etc.).

The **marginal cost** is

$$C'(x) = 2$$

Since this is a constant, it costs an additional $2 to produce one more radio at all production levels.

The **revenue equation** [the amount of money $R(x)$ received by the company for manufacturing and selling x units at $p per unit] is

$$R(x) = (\text{Number of units sold})(\text{Price per unit})$$
$$= xp$$
$$= x\left(10 - \frac{x}{1,000}\right) \qquad \text{Using equation (2)} \qquad (4)$$
$$= 10x - \frac{x^2}{1,000}$$

The **marginal revenue** is

$$R'(x) = 10 - \frac{x}{500}$$

For production levels of x = 2,000, 5,000, and 7,000, we have

$$R'(2,000) = 6 \qquad R'(5,000) = 0 \qquad R'(7,000) = -4$$

This means that at production levels of 2,000, 5,000, and 7,000, the respective approximate changes in revenue per unit change in production are $6, $0, and −$4. That is, at the 2,000 output level revenue increases as production increases; at the 5,000 output level revenue does not change with a "small" change in production; and at the 7,000 output level revenue decreases with an increase in production. Figure 3 illustrates these results.

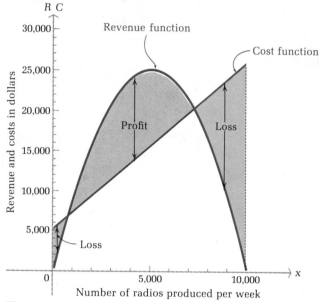

Figure 3

Finally, the **profit equation** is

$$P(x) = R(x) - C(x)$$
$$= \left(10x - \frac{x^2}{1,000} \right) - (5,000 + 2x)$$
$$= -\frac{x^2}{1,000} + 8x - 5,000$$

The **marginal profit** is

$$P'(x) = -\frac{x}{500} + 8$$

For production levels of 1,000, 4,000, and 6,000, we have

$$P'(1,000) = 6 \qquad P'(4,000) = 0 \qquad P'(6,000) = -4$$

This means that at production levels of 1,000, 4,000, and 6,000, the respective approximate changes in profit per unit change in production are $6, $0, and −$4. That is, at the 1,000 output level profit will be increased if production is increased; at the 4,000 output level profit does not change for "small" changes in production; and at the 6,000 output level profits will decrease if production is increased. It seems the best production level to produce a maximum profit is 4,000. [In the next chapter we will develop a systematic procedure for finding the production level (and, using the demand equation, the selling price) that will maximize profit.] This example warrants careful study, since a number of important ideas in economics and calculus are involved.

Sometimes it is desirable to carry out marginal analysis relative to **average cost (cost per unit), average revenue (revenue per unit), and average profit (profit per unit).** The relevant definitions are summarized in the following box:

Marginal Average Cost, Revenue, and Profit

If x is the number of units of a product produced in some time interval, then

$$\text{Average total cost} = \overline{C}(x) = \frac{C(x)}{x} \qquad \text{Cost per unit}$$

$$\text{Marginal average cost} = \overline{C}'(x)$$

$$\text{Average total revenue} = \overline{R}(x) = \frac{R(x)}{x} \qquad \text{Revenue per unit}$$

$$\text{Marginal average revenue} = \overline{R}'(x)$$

$$\text{Average total profit} = \overline{P}(x) = \frac{P(x)}{x} \qquad \text{Profit per unit}$$

$$\text{Marginal average profit} = \overline{P}'(x)$$

In the previous example, we have

$$\overline{C}(x) = \frac{C(x)}{x} = \frac{5,000 + 2x}{x} = \frac{5,000}{x} + 2 \qquad \text{As production goes up, cost per unit goes down.}$$

$$\overline{C}'(x) = -5,000x^{-2} = \frac{-5,000}{x^2}$$

$$\overline{C}'(100) = \frac{-5,000}{(100)^2} = -\$0.50 \qquad \text{Cost is decreasing at a rate of approximately 50¢ per unit at a production level of 100 units per week.}$$

$$\overline{C}'(1,000) = \frac{-5,000}{(1,000)^2} = -\$0.005 \qquad \text{Cost is decreasing at a rate of approximately 0.5¢ per unit at a production level of 1,000 units per week.}$$

Similar interpretations are given to $\overline{R}(x)$ and $\overline{R}'(x)$, and to $\overline{P}(x)$ and $\overline{P}'(x)$.

Exercise 11-5

Applications

Business & Economics

1. In the production strategy problem discussed in this section, suppose we have the demand equation

$$x = 6{,}000 - 30p \qquad \text{or} \qquad p = 200 - \frac{x}{30}$$

and the cost equation

$$C(x) = 72{,}000 + 60x$$

(A)   Find the marginal cost.
(B)   Find the revenue equation in terms of x.
(C)   Find the marginal revenue.
(D)   Find $R'(1{,}500)$ and $R'(4{,}500)$, and interpret.
(E)   Graph the cost function and the revenue function on the same coordinate system. Indicate regions of loss and profit. Use $0 \leqslant x \leqslant 6{,}000$.
(F)   Find the profit equation in terms of x.
(G)   Find the marginal profit.
(H)   Find $P'(1{,}500)$ and $P'(3{,}000)$, and interpret.

2. In the production strategy problem discussed in this section, suppose we have the demand equation

$$x = 9{,}000 - 30p \qquad \text{or} \qquad p = 300 - \frac{x}{30}$$

and the cost equation

$$C(x) = 90{,}000 + 30x$$

(A)   Find the marginal cost.
(B)   Find the revenue equation in terms of x.
(C)   Find the marginal revenue.
(D)   Find $R'(3{,}000)$ and $R'(6{,}000)$, and interpret.
(E)   Graph the cost function and the revenue function on the same coordinate system for $0 \leqslant x \leqslant 9{,}000$. Indicate regions of loss and profit.
(F)   Find the profit equation in terms of x.
(G)   Find the marginal profit.
(H)   Find $P'(1{,}500)$ and $P'(4{,}500)$, and interpret.

3. Referring to Problem 1, find:

    (A) $\overline{C}(x)$, $\overline{R}(x)$, and $\overline{P}(x)$
    (B) $\overline{C}'(x)$, $\overline{R}'(x)$, and $\overline{P}'(x)$
    (C) $\overline{P}'(1,000)$ and $\overline{P}'(6,000)$, and interpret

4. Referring to Problem 2, find:

    (A) $\overline{C}(x)$, $\overline{R}(x)$, and $\overline{P}(x)$
    (B) $\overline{C}'(x)$, $\overline{R}'(x)$, and $\overline{P}'(x)$
    (C) $\overline{P}'(1,000)$ and $\overline{P}'(2,000)$, and interpret

11-6 Chapter Review

Important Terms and Symbols

11-1 *Implicit differentiation.* special function notation, function explicitly defined, function implicitly defined, implicit differentiation, $y = f(x)$, $y = y(x)$, $F(x, y) = 0$

11-2 *Related rates.* related rates, $x = x(t)$, $y = y(t)$

11-3 *Higher-order derivatives.* $f''(x), f'''(x), f^{(4)}(x), \ldots, f^{(n)}(x), \ldots$;

$$\frac{d^2y}{dx^2}, \frac{d^3y}{dx^3}, \frac{d^4y}{dx^4}, \ldots, \frac{d^ny}{dx^n}, \ldots ;$$

$$y'', y''', y^{(4)}, \ldots, y^{(n)}, \ldots ;$$

$$D_x^2 f(x), D_x^3 f(x), D_x^4 f(x), \ldots, D_x^n f(x), \ldots$$

11-4 *The differential.* differential dx, differential $dy = f'(x)\, dx$, differential approximation, $\Delta y \approx dy$, $f(x + \Delta x) \approx f(x) + f'(x)\, dx$, differential rules, $dc = 0$,

$$du^n = nu^{n-1}\, du,$$

$$d(u \pm v) = du \pm dv,$$

$$d(uv) = u\, dv + v\, du,$$

$$d\left(\frac{u}{v}\right) = \frac{v\, du - u\, dv}{v^2}$$

11-5 *Marginal analysis in business and economics.* demand equation, cost equation, marginal cost, revenue equation, marginal revenue, profit equation, marginal profit, average cost, marginal average cost, average revenue, marginal average revenue, average profit, marginal average profit, $C'(x)$, $\overline{C}'(x)$, $R'(x)$, $\overline{R}'(x)$, $P'(x)$, $\overline{P}'(x)$

Exercise 11-6   Chapter Review

Work through all the problems in this chapter review and check your answers in the back of the book. (Answers to all review problems are there.) Where weaknesses show up, review appropriate sections in the text. When you are satisfied that you know the material, take the practice test following this review.

A

1. Find y' for $y = y(x)$ defined implicitly by

$$2y^2 - 3x^3 - 5 = 0$$

and evaluate at $(x, y) = (1, 2)$.

2. For $y = 3x^2 - 5$ where $x = x(t)$ and $y = y(t)$, find dy/dt if $dx/dt = 3$ when $x = 12$.

3. Find d^2y/dx^2 if $y = x^3 - \sqrt{x}$.

4. For $y = f(x) = (3x^2 - 7)^3$, find dy.

B

5. Find y' for $y = y(x)$ defined implicitly by

$$y^4 - xy - 4 = 0$$

and evaluate at $(x, y) = (6, 2)$.

6. Find x' for $x = x(t)$ defined implicitly by

$$x^3 - 2t^2x + 8 = 0$$

and evaluate at $(t, x) = (-2, 2)$.

7. Find y'' for $y = y(x)$ defined implicitly by $2x^2 - y^2 = 3$.

8. Find y'' for $y = (2x^2 - 3)^{7/4}$.

9. Find $D_x^3 \left(\dfrac{1}{\sqrt{5 - 4x}} \right)$.

10. Find dy and Δy for $f(x) = x^3 - 2x + 1$, $x = 5$, and $\Delta x = dx = 0.1$.

11. Approximate $\sqrt{17}$ using differentials.

12. A point is moving on the graph of $y^3 - xy - 2 = 0$ so that its y coordinate is decreasing at 2 units per second when $(x, y) = (5, -2)$. Find the rate of change of the x coordinate.

13. Water from a water heater is leaking onto a floor. A circular pool is created whose area is increasing at the rate of 24 square inches per minute. How fast is the radius R of the pool increasing when the radius is 12 inches? $(A = \pi R^2)$

C

14. Find y' for $y(x)$ defined implicitly by

$$\sqrt{5 - y^2} - 2x^2 + 6 = 0$$

Find the slope of the graph at $(2, 1)$.

15. Using implicit differentiation, find d^3y/dx^3 if $x^2 + y^2 = 6$.

16. Find dy and Δy for $y = (2/\sqrt{x}) + 8$, $x = 16$, and $\Delta x = dx = 0.2$.

■ Applications

Business & Economics 17. *Marginal analysis.* Let

$$p = 20 - x \quad \text{and} \quad C(x) = 2x + 56 \qquad 0 \leq x \leq 20$$

be the demand equation and the cost function, respectively, for a certain commodity.

(A) Find the marginal cost, average cost, and marginal average cost functions.
(B) Find the revenue, marginal revenue, average revenue, and marginal average revenue functions.
(C) Find the profit, marginal profit, average profit, and marginal average profit functions.
(D) Find the break-even point(s).
(E) Evaluate the marginal profit at $x = 7$, 9, and 11, and interpret.
(F) Graph $R = R(x)$ and $C = C(x)$ on the same axes and locate regions of profit and loss.

18. *Demand equation.* Given the demand equation

$$x = \sqrt{5,000 - 2p^3}$$

find the rate of change of p with respect to x by implicit differentiation (x is the number of items that can be sold at a price of $\$p$ per item).

19. *Rate of change of revenue.* A company is manufacturing a new video game and can sell all it manufactures. The revenue (in dollars) is given by

$$R = 36x - \frac{x^2}{20}$$

where the production output in 1 day is x games. If production is increasing at 10 games per day when production is 250 games per day, find the rate of increase in revenue.

Life Sciences 20. *Wound healing.* A circular wound on an arm is healing at the rate of 45 square millimeters per day (area of wound is decreasing at this rate). How fast is the radius R of the wound decreasing when $R = 15$ millimeters? ($A = \pi R^2$)

Social Sciences 21. *Learning.* A new worker on the production line performs an operation in T minutes after x performances of the operation, as given by

$$T = 2\left(1 + \frac{1}{x^{3/2}}\right)$$

If, after performing the operation nine times, the rate of improvement $dx/dt = 3$ operations per hour, find the rate of improvement in time dT/dt in performing each operation.

| Practice Test: Chapter 11 |

1. Find y' for $y = y(x)$ defined implicitly by $x^2 - 3xy + 4y^2 = 23$ and find the slope of the graph at $(-1, 2)$.
2. Find y'' for $y = y(x)$ defined implicitly by $x^2 + y^2 = 81$.
3. Find $D_x^3 \sqrt{1 - 2x}$.
4. Find y'' for $y = (5 - 3x^2)^{3/2}$.
5. Find dy and Δy for $f(x) = x^2 - 1$, $x = 2$, and $\Delta x = dx = 0.1$.
6. Approximate $\sqrt[3]{26}$ using differentials.
7. A point is moving on the graph of $y^2 - 4x^2 = 12$, so that its x coordinate is decreasing at 2 units per second when $(x, y) = (1, 4)$. Find the rate of change of the y coordinate.
8. A 17 foot ladder is placed against a wall. If the foot of the ladder is pushed toward the wall at 0.5 foot per second, how fast is the top rising when the foot of the ladder is 8 feet from the wall?
9. Let $p = 14 - x$ and $C(x) = 2x + 20$, $0 \le x \le 14$, be the demand equation and the cost function, respectively, for a certain commodity.

   (A) Find the revenue and profit functions.
   (B) Find the marginal profit, average profit, and marginal average profit functions.
   (C) Evaluate marginal profit at $x = 4$, 6, and 8, and interpret.
   (D) Find the break-even point(s).
   (E) Graph $R = R(x)$ and $C = C(x)$ on the same axes and locate regions of profit and loss.

Graphing and Optimization

CHAPTER 12   Contents

This chapter is concerned with two important applications of the derivative: sketching the graph of a function and solving optimization problems. The first three sections cover basic concepts that will be used in both of these applications. The last (optional) section shows how these basic concepts can be applied to a particular topic in economics.

12-1   Asymptotes; Limits at Infinity and Infinite Limits

- Limits at Infinity and Horizontal Asymptotes
- Infinite Limits and Vertical Asymptotes
- Application

In this section we take another look at limits. This time we are interested in limits as x increases or decreases without bound and limits at points where $f(x)$ is not defined.

■ Limits at Infinity and Horizontal Asymptotes

In Section 10-2 we said that the limit of $f(x)$ as x approaches a number c is the number L, written

$$\lim_{x \to c} f(x) = L$$

if the functional value $f(x)$ is close to the single real number L whenever x is close to but not equal to c on either side of c. In order to make an accurate sketch of the graph of a function, it will be helpful to know what happens to the functional value $f(x)$ as x assumes large positive values and large negative values. We will consider a specific example before we make any general statements.

Example 1   Consider the function

$$f(x) = \frac{2x^2}{1 + x^2}$$

which is defined for all real numbers. What happens to the functional value $f(x)$ as x assumes larger and larger positive values?

Solution   Let us investigate this question using a calculator. Table 1 shows the values of $f(x)$ for increasingly large values of x.

Table 1

x assumes larger and larger positive values							
x	0	5	10	20	50	100	500	1,000
$f(x)$	0	1.92	1.98	1.995	1.9992	1.9998	1.999992	1.999998

The calculations shown here suggest that as the values of x continue to increase, the functional value $f(x)$ approaches the number 2. It seems that we can make the functional value $f(x)$ come as close to 2 as we like by taking a sufficiently large value of x. If we use the symbol "$x \to \infty$" to indicate that x is increasing with no upper limit on its size, then we can write

$$\lim_{x \to \infty} \frac{2x^2}{1 + x^2} = 2$$

It is important to understand that the symbol ∞ does not represent an actual number that x is approaching, but is used to indicate only that the value of x is increasing with no upper limit on its size. In particular, the statement "$x = \infty$" is meaningless since ∞ is not a symbol for a real number. We will also use the statement "$x \to -\infty$" to indicate that the value of x is decreasing with no lower limit on its size. Since the function in this example is an even function [$f(-x) = f(x)$], the values in Table 1 also suggest that

$$\lim_{x \to -\infty} \frac{2x^2}{1 + x^2} = 2$$

Examining the graph of this function (see Figure 1) provides us with a geometric interpretation of these two limit statements. The graph of f is

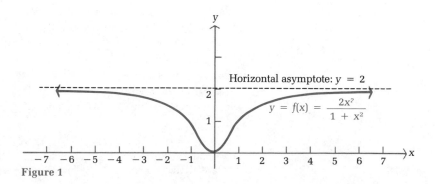

Figure 1

approaching the horizontal line with equation $y = 2$ as $x \to \infty$ and as $x \to -\infty$. The line $y = 2$ is called a **horizontal asymptote** of f.

Problem 1   Construct a table such as Table 1 in order to estimate (do not graph):

$$\lim_{x \to \infty} \frac{2}{1 + x^2}$$

We now state an informal* definition of **the limit of a function f as x approaches ∞ or $-\infty$** and the definition of a **horizontal asymptote.**

Limits at Infinity and Horizontal Asymptotes

We write

$$\lim_{x \to \infty} f(x) = L$$

if the functional value $f(x)$ is close to the single real number L whenever x is a very large positive number. We write

$$\lim_{x \to -\infty} f(x) = L$$

if the functional value $f(x)$ is close to the single real number L whenever x is a very large negative number. If either condition holds, then the line

$$y = L$$

is a **horizontal asymptote** of f.

Figure 2 illustrates several different possibilities for limits at infinity. Figures 2A and 2B both show that the existence of one of the limits at infinity does not imply the existence of the other. In Figure 2A, $\lim_{x \to -\infty} f(x)$ fails to exist because $f(x)$ is unbounded as $x \to -\infty$. In Figure 2B, $\lim_{x \to \infty} g(x)$ fails to exist because $g(x)$ oscillates as $x \to \infty$ and does not approach a single real number. Finally, Figure 2C shows that it is possible for a function to have two different horizontal asymptotes, one as $x \to \infty$ and a different one as $x \to -\infty$.

Now that we have a basic understanding of limits at infinity, how can we evaluate such a limit? Fortunately, all the limit properties we used in

---

* A more formal definition of $\lim_{x \to \infty} f(x) = L$ is as follows: Given any $e > 0$ (no matter how small), we can find a (large) positive number N such that $|f(x) - L| < e$ whenever $x > N$. A similar statement can be made for $\lim_{x \to -\infty} f(x)$.

(A) $\lim\limits_{x \to \infty} f(x) = L$

$\lim\limits_{x \to -\infty} f(x)$ does not exist

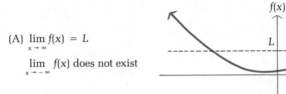

(B) $\lim\limits_{x \to -\infty} g(x) = L$

$\lim\limits_{x \to \infty} g(x)$ does not exist

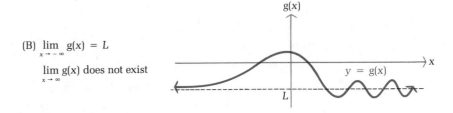

(C) $\lim\limits_{x \to \infty} h(x) = L_1$

$\lim\limits_{x \to -\infty} h(x) = L_2$

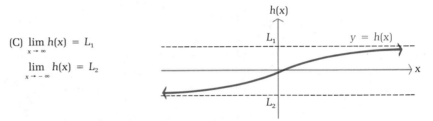

Figure 2 Limits at infinity

Section 10-2 (see the box on page 556) are valid if we replace the statement $x \to c$ with the statement $x \to \infty$ (or $x \to -\infty$). These properties, together with Theorem 1, will enable us to evaluate limits at infinity for many functions.

Theorem 1

> If p is a positive number, then
>
> $$\lim_{x \to \infty} \frac{1}{x^p} = 0 \quad \text{and} \quad \lim_{x \to -\infty} \frac{1}{x^p} = 0$$
>
> provided that x^p is defined for negative values of x.

Figure 3 on the next page illustrates the theorem for several values of p.

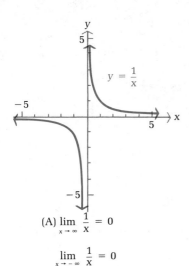

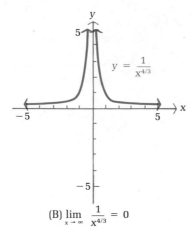

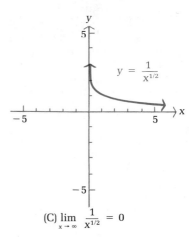

(A) $\lim\limits_{x \to \infty} \dfrac{1}{x} = 0$

$\lim\limits_{x \to -\infty} \dfrac{1}{x} = 0$

(B) $\lim\limits_{x \to \infty} \dfrac{1}{x^{4/3}} = 0$

$\lim\limits_{x \to -\infty} \dfrac{1}{x^{4/3}} = 0$

(C) $\lim\limits_{x \to \infty} \dfrac{1}{x^{1/2}} = 0$

$x^{1/2}$ is not defined for negative values of x

Figure 3 $\lim\limits_{x \to \infty} \dfrac{1}{x^p}$

Example 2   Find each limit.

(A) $\lim\limits_{x \to \infty} \left(3 + \dfrac{2}{x^{3/2}} \right)$ (B) $\lim\limits_{x \to \infty} \dfrac{5x + 4}{2x + 3}$

(C) $\lim\limits_{x \to -\infty} \dfrac{4x^2 + 3x + 2}{2x^3 + 5}$ (D) $\lim\limits_{x \to \infty} \dfrac{3x^4 + 6x}{x^2 + 4}$

Solutions   (A) $\lim\limits_{x \to \infty} \left(3 + \dfrac{2}{x^{3/2}} \right)$

$\lim\limits_{x \to \infty} [f(x) + g(x)] = \lim\limits_{x \to \infty} f(x) + \lim\limits_{x \to \infty} g(x)$

$= \lim\limits_{x \to \infty} 3 + \lim\limits_{x \to \infty} \dfrac{2}{x^{3/2}}$

$\lim\limits_{x \to \infty} k = k; \lim\limits_{x \to \infty} kf(x) = k \lim\limits_{x \to \infty} f(x)$

$= 3 + 2 \lim\limits_{x \to \infty} \dfrac{1}{x^{3/2}}$

$\lim\limits_{x \to \infty} \dfrac{1}{x^p} = 0 \quad \text{if } p > 0$

$= 3 + 2 \cdot 0$

$= 3$

(B) $\lim\limits_{x \to \infty} \dfrac{5x + 4}{2x + 3}$

$\neq \dfrac{\lim_{x \to \infty} 5x + 4}{\lim_{x \to \infty} 2x + 3}$ since $\lim\limits_{x \to \infty} (5x + 4)$

and $\lim\limits_{x \to \infty} (2x + 3)$ do not exist

$= \lim\limits_{x \to \infty} \dfrac{\dfrac{5x + 4}{x}}{\dfrac{2x + 3}{x}}$

We divide numerator and denominator by x in order to express the fraction in a form where the limits of the numerator and denominator do exist.

$$= \lim_{x \to \infty} \frac{5 + \dfrac{4}{x}}{2 + \dfrac{3}{x}}$$

$$= \frac{\lim\limits_{x \to \infty} \left(5 + \dfrac{4}{x}\right)}{\lim\limits_{x \to \infty} \left(2 + \dfrac{3}{x}\right)}$$

$$= \frac{5}{2}$$

$\lim\limits_{x \to \infty} \dfrac{f(x)}{g(x)} = \dfrac{\lim_{x \to \infty} f(x)}{\lim_{x \to \infty} g(x)}$, provided that

$\lim\limits_{x \to \infty} f(x)$ and $\lim\limits_{x \to \infty} g(x)$ both exist.

$\lim\limits_{x \to \infty} \left(5 + \dfrac{4}{x}\right) = 5 + 4 \cdot 0 = 5$ and

$\lim\limits_{x \to \infty} \left(2 + \dfrac{3}{x}\right) = 2 + 3 \cdot 0 = 2$ as in

part A

(C) $\lim\limits_{x \to -\infty} \dfrac{4x^2 + 3x + 2}{2x^3 + 5}$

Divide numerator and denominator by x^3, the highest power of x that occurs in the numerator or the denominator, simplify, and proceed as before.

$$= \lim_{x \to -\infty} \frac{\dfrac{4x^2 + 3x + 2}{x^3}}{\dfrac{2x^3 + 5}{x^3}}$$

$$= \lim_{x \to -\infty} \frac{\dfrac{4}{x} + \dfrac{3}{x^2} + \dfrac{2}{x^3}}{2 + \dfrac{5}{x^3}}$$

$$= \frac{\lim\limits_{x \to -\infty} \left(\dfrac{4}{x} + \dfrac{3}{x^2} + \dfrac{2}{x^3}\right)}{\lim\limits_{x \to -\infty} \left(2 + \dfrac{5}{x^3}\right)}$$

$$= \frac{0 + 0 + 0}{2 + 0} = \frac{0}{2} = 0$$

(D) $\lim\limits_{x \to \infty} \dfrac{3x^4 + 6x}{x^2 + 4}$

This time, divide numerator and denominator by x^4.

$$= \lim_{x \to \infty} \frac{\dfrac{3x^4 + 6x}{x^4}}{\dfrac{x^2 + 4}{x^4}}$$

$$= \lim_{x \to \infty} \frac{3 + \dfrac{6}{x^3}}{\dfrac{1}{x^2} + \dfrac{4}{x^4}}$$

Since the numerator of this fraction approaches 3 and the denominator approaches 0, the fraction can be made as large as you like; hence, the limit does not exist.

Does not exist

Problem 2   Find each limit.

(A) $\displaystyle\lim_{x \to -\infty} \left(2 - \frac{3}{x^{4/3}}\right)$ (B) $\displaystyle\lim_{x \to \infty} \frac{2x - 4}{3x + 5}$

(C) $\displaystyle\lim_{x \to \infty} \frac{3x^3 + 4}{2x^2 + 6}$ (D) $\displaystyle\lim_{x \to \infty} \frac{2x + 1}{x^2 + 4}$

The methods used to evaluate the limits in Example 2 can be applied to any rational function (ratio of two polynomials). The results are summarized in the box.

Limits at Infinity for Rational Functions

If

$$f(x) = \frac{p(x)}{q(x)}$$

where

$$p(x) = a_n x^n + a_{n-1} x^{n-1} + \cdots + a_0 \qquad a_n \neq 0$$

and

$$q(x) = b_m x^m + b_{m-1} x^{m-1} + \cdots + b_0 \qquad b_m \neq 0$$

then

$$\lim_{x \to \pm\infty} f(x) = \begin{cases} 0 & \text{if } n < m \qquad\qquad (1)\\[2mm] \dfrac{a_n}{b_m} & \text{if } n = m \qquad\qquad (2)\\[2mm] \text{Does not exist} & \text{if } n > m \qquad\qquad (3) \end{cases}$$

[*Note:*   If $n > m$, then the limit fails to exist because $f(x)$ is unbounded as $x \to \infty$ and as $x \to -\infty$.]

Thus, we see that the limit at infinity of a rational function can be determined by simply comparing the degree of the numerator and the degree of the denominator. Notice that the value of the limit is the same, whether $x \to \infty$ or $x \to -\infty$. This implies that a rational function can have at most one horizontal asymptote.

Example 3 Use equations (1), (2), and (3) to find each limit.

(A) $\lim\limits_{x\to\pm\infty} \dfrac{5x^2 + 3}{3x^2 + 4}$ (B) $\lim\limits_{x\to\pm\infty} \dfrac{2x^5 + 7}{6x^3 + 4}$ (C) $\lim\limits_{x\to\pm\infty} \dfrac{3x^4 + 9}{8x^6 + 5}$

Solutions (A) $\lim\limits_{x\to\pm\infty} \dfrac{5x^2 + 3}{3x^2 + 4} = \dfrac{5}{3}$ Use (2): $n = m = 2$, $a_n = 5$, $b_m = 3$

(B) $\lim\limits_{x\to\pm\infty} \dfrac{2x^5 + 7}{6x^3 + 4}$ Does not exist   Use (3): $n = 5$, $m = 3$, $n > m$

(C) $\lim\limits_{x\to\pm\infty} \dfrac{3x^4 + 9}{8x^6 + 5} = 0$ Use (1): $n = 4$, $m = 6$, $n < m$

Problem 3 Use equations (1), (2), and (3) in the box to find each limit.

(A) $\lim\limits_{x\to\pm\infty} \dfrac{4x^3 + 5}{2x^4 + 4}$ (B) $\lim\limits_{x\to\pm\infty} \dfrac{x^6 + 2}{x^5 + 4}$ (C) $\lim\limits_{x\to\pm\infty} \dfrac{4x^2 + 7}{9x^2 + 11}$

▪ Infinite Limits and Vertical Asymptotes

Now we turn our attention to another type of limit problem that is also related to asymptotes and curve sketching. In Section 10-2 when we encountered limits of the type

$$\lim_{x\to 2} \frac{x + 2}{x - 2}$$

we said that the limit did not exist because the numerator was approaching a nonzero number (in this case, the number 4) and the denominator was approaching 0. Hence, the fraction can be made as large in absolute value as you like. Table 2 illustrates this behavior.

Table 2 $f(x) = \dfrac{x + 2}{x - 2}$

	x approaches 2 from the left				$\to 2 \leftarrow$	x approaches 2 from the right			
x	1.9	1.99	1.999	1.9999	$\to 2 \leftarrow$	2.0001	2.001	2.01	2.1
$f(x)$	-39	-399	$-3{,}999$	$-39{,}999$	$\to ? \leftarrow$	40,001	4,001	401	41

The values in the table suggest that the value of $f(x) = (x + 2)/(x - 2)$ decreases with no lower limit as x approaches 2 from the left and increases with no upper limit as x approaches 2 from the right. Thus, we write

$$\lim_{x\to 2^-} \frac{x + 2}{x - 2} = -\infty \quad \text{and} \quad \lim_{x\to 2^+} \frac{x + 2}{x - 2} = \infty$$

As x approaches 2 from either side, the graph of $f(x)$ approaches the vertical line whose equation is $x = 2$. We call this line a **vertical asymptote** of f (see Figure 4). Notice that f also has a horizontal asymptote at $y = 1$.

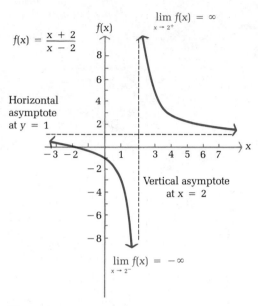

$$f(x) = \frac{x + 2}{x - 2}$$

Horizontal asymptote at $y = 1$

$$\lim_{x \to 2^+} f(x) = \infty$$

Vertical asymptote at $x = 2$

$$\lim_{x \to 2^-} f(x) = -\infty$$

Figure 4

Once again, it is important to understand that we are not using the symbol ∞ to represent the value of a limit, but to describe a certain type of behavior for a limit that does not exist. In general, when we write

$$\lim_{x \to c^+} f(x) = \infty$$

we mean that the functional value $f(x)$ is increasing without limit as x approaches c from the right. A similar statement can be made for x approaching c from the left and for $f(x)$ approaching $-\infty$. If

$$\lim_{x \to c^-} f(x) = \infty \qquad \text{and} \qquad \lim_{x \to c^+} f(x) = \infty$$

then we can write

$$\lim_{x \to c} f(x) = \infty$$

and if

$$\lim_{x \to c^-} f(x) = -\infty \qquad \text{and} \qquad \lim_{x \to c^+} f(x) = -\infty$$

then we can write

$$\lim_{x \to c} f(x) = -\infty$$

We can now state the general definition of a **vertical asymptote.**

Vertical Asymptotes

If

$$\lim_{x \to c^+} f(x) = \infty \quad (\text{or} -\infty)$$

or

$$\lim_{x \to c^-} f(x) = \infty \quad (\text{or} -\infty)$$

then the vertical line

$$x = c$$

is a **vertical asymptote** of f.

Refer again to the functions in Figure 3. Each of these functions has a vertical asymptote at $x = 0$. In each case, the vertical asymptote is due to the fact that as x approaches 0, the denominator of the function approaches 0 and the numerator does not. Theorem 2 formalizes this observation and provides a very important tool for locating vertical asymptotes.

Theorem 2

If

$$f(x) = \frac{p(x)}{q(x)}$$

where both p and q are continuous at $x = c$, $p(c) \neq 0$, and $q(c) = 0$, then f has a vertical asymptote at $x = c$.

This theorem is also true if p and q are both continuous on the right at $x = c$ or both continuous on the left at $x = c$.

Example 4 Find the vertical asymptotes of each function.

(A) $f(x) = \dfrac{x+1}{x(x-1)}$ (B) $f(x) = \dfrac{x+4}{x^2+6x+8}$

(C) $f(x) = \dfrac{x}{x^2+2}$ (D) $f(x) = \dfrac{1+x^2}{\sqrt{x+2}}$

Solutions (A) $f(x) = \dfrac{x+1}{x(x-1)} = \dfrac{p(x)}{q(x)}$

We let $p(x) = x + 1$ and $q(x) = x(x - 1)$ and note that both p and q are continuous for all values of x (polynomials are continuous everywhere). Since $q(0) = 0$ and $p(0) = 1 \neq 0$, f has a vertical asymptote at $x = 0$. Since $q(1) = 0$ and $p(1) = 2 \neq 0$, f also has a vertical asymptote at $x = 1$. There are no other values of x for which $q(x) = 0$, so these are all the vertical asymptotes of f. (At all other values of c, $\lim_{x \to c} f(x)$ exists.)

(B) $f(x) = \dfrac{x + 4}{x^2 + 6x + 8} = \dfrac{p(x)}{q(x)}$

We let $p(x) = x + 4$ and $q(x) = x^2 + 6x + 8$. First, we factor q to find the values of x that satisfy $q(x) = 0$:

$$q(x) = x^2 + 6x + 8$$
$$= (x + 4)(x + 2)$$

Thus, $q(x) = 0$ only for $x = -2$ and $x = -4$. Since $p(-2) = 2 \neq 0$, f has a vertical asymptote at $x = -2$. Since $p(-4) = 0$, Theorem 2 does not apply and we must evaluate $\lim_{x \to -4} f(x)$ to determine if $x = -4$ is a vertical asymptote:

$$\lim_{x \to -4} \frac{x + 4}{x^2 + 6x + 8} = \lim_{x \to -4} \frac{\overset{1}{\cancel{x + 4}}}{\underset{1}{\cancel{(x + 4)}}(x + 2)}$$

$$= \lim_{x \to -4} \frac{1}{x + 2}$$

$$= -\frac{1}{2}$$

Since this limit exists, f does not have a vertical asymptote at $x = -4$.

(C) $f(x) = \dfrac{x}{x^2 + 2} = \dfrac{p(x)}{q(x)}$

Let $p(x) = x$ and $q(x) = x^2 + 2$. Since

$$q(x) = x^2 + 2 \geqslant 2 > 0$$

for all values of x, $q(x)$ is never 0 and f has no vertical asymptotes.

(D) $f(x) = \dfrac{1 + x^2}{\sqrt{x + 2}} = \dfrac{p(x)}{q(x)}$

$p(x) = 1 + x^2$ is continuous for all x and $q(x) = \sqrt{x + 2}$ is continuous for $x > -2$ and continuous on the right at $x = -2$. Since $p(-2) = 5 \neq 0$ and $q(-2) = 0$, f has a vertical asymptote at $x = -2$.

Problem 4   Find the vertical asymptotes of each function.

(A) $f(x) = \dfrac{x^3}{x^2 - 1}$ (B) $f(x) = \dfrac{x^3}{x^2 + 1}$

(C) $f(x) = \dfrac{x + 1}{x^2 - 1}$ (D) $f(x) = \dfrac{x + 1}{\sqrt{x}}$

In order to use vertical asymptotes as an aid in sketching the graph of a function, it is not enough simply to locate all of the vertical asymptotes. In addition, we must determine the behavior of the graph as x approaches each asymptote from the left and the right. That is, we must evaluate $\lim_{x \to c^-} f(x)$ and $\lim_{x \to c^+} f(x)$ at each vertical asymptote $x = c$.

Example 5   Sketch a graph of f by first evaluating $\lim_{x \to c^-} f(x)$ and $\lim_{x \to c^+} f(x)$ at each vertical asymptote of

$$f(x) = \frac{x - 2}{x^2(x - 4)}$$

Then find any horizontal asymptotes and complete the graph using a calculator and point-by-point plotting.

Solution   Let $p(x) = x - 2$ and $q(x) = x^2(x - 4)$, and observe that $q(0) = 0$ and $q(4) = 0$. Since $p(0) = -2 \neq 0$ and $p(4) = 2 \neq 0$, f has vertical asymptotes at $x = 0$ and $x = 4$. It will be helpful to construct a sign chart for f relative to the real number line:

Sign of $(x - 2)$	$- \ - \ -$	$\mid$	$- \ - \ -$	$\mid$	$+ \ + \ +$	$\mid$	$+ \ + \ +$
Sign of x^2	$+ \ + \ +$	$\mid$	$+ \ + \ +$	$\mid$	$+ \ + \ +$	$\mid$	$+ \ + \ +$
Sign of $(x - 4)$	$- \ - \ -$	$\mid$	$- \ - \ -$	$\mid$	$- \ - \ -$	$\mid$	$+ \ + \ +$

$$\xrightarrow{} x$$

$$\quad\quad\quad 0 \quad\quad\quad 2 \quad\quad\quad 4$$

Sign of $f(x)$	$+ \ + \ +$	$\mid$	$+ \ + \ +$	$\mid$	$- \ - \ -$	$\mid$	$+ \ + \ +$

The sign chart indicates that $f(x) > 0$ for x close to and on the right of 4. Thus

$$\lim_{x \to 4^+} f(x) = \infty$$

If x is close to 4 and on the left, then $f(x) < 0$ and

$$\lim_{x \to 4^-} f(x) = -\infty$$

For x close to 0 on either side, $f(x) > 0$. Thus,

$$\lim_{x \to 0} f(x) = \infty$$

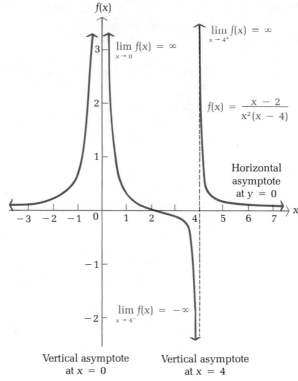

Vertical asymptote
at x = 0

Vertical asymptote
at x = 4

Figure 5

Since f is a rational function we can use (1) on page 674 to conclude that f has a horizontal asymptote at $y = 0$. We complete the graph (see Figure 5) using a calculator and point-by-point plotting for regions of uncertainty. (As we progress through this chapter, the aids to graphing that we will develop will tell us more and more about the shape of a graph and we will need less and less point-by-point plotting.)

Problem 5   Repeat Example 5 for $f(x) = \dfrac{4}{4 - x^2}$.

■ Application

Example 6
Average Cost

A company estimates that the fixed costs for manufacturing a new transistor radio are $5,000 and the cost per unit produced is $2 (see Section 11-5). The total cost of producing x radios is

$$C(x) = 5,000 + 2x$$

and the average cost per unit is

$$\overline{C}(x) = \frac{C(x)}{x} = \frac{5,000 + 2x}{x} = \frac{5,000}{x} + 2$$

Since

$$\lim_{x \to \infty} \overline{C}(x) = \lim_{x \to \infty} \left(\frac{5,000}{x} + 2 \right) = 2$$

the line $y = 2$ is a horizontal asymptote for the average cost function. Notice that 2 is also the cost per unit.

The function $\overline{C}(x)$ also has a vertical asymptote at $x = 0$. Since $C(x)$ is not defined for $x \le 0$, we need investigate only the right-hand limit at $x = 0$:

$$\lim_{x \to 0^+} \overline{C}(x) = \lim_{x \to 0^+} \left(\frac{5,000}{x} + 2 \right) = \infty$$

Figure 6 shows these results graphically.

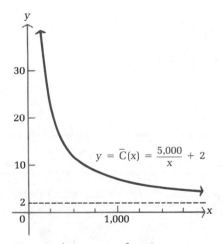

Figure 6 Average cost function

Problem 6 Refer to Example 6. Evaluate $\lim_{x \to \infty} \overline{C}(x)$ and $\lim_{x \to 0^+} \overline{C}(x)$ for the cost function $C(x) = 10,000 + 4x$.

Answers to Matched Problems

1.

x	0	5	10	20	50	100	500	1,000
$f(x)$	2	.08	.02	.005	.0008	.0002	.000008	.000002

$\lim_{x \to \infty} f(x) = 0$

2. (A) 2 (B) 2/3 (C) Does not exist (D) 0
3. (A) 0 (B) Does not exist (C) 4/9
4. (A) $x = -1$ and $x = 1$ (B) None (C) $x = 1$ (D) $x = 0$

5. Vertical asymptotes
    at $x = -2$ and $x = 2$

    $$\lim_{x \to -2^-} f(x) = -\infty$$

    $$\lim_{x \to -2^+} f(x) = \infty$$

    $$\lim_{x \to 2^-} f(x) = \infty$$

    $$\lim_{x \to 2^+} f(x) = -\infty$$

    Horizontal asymptote
    at $y = 0$

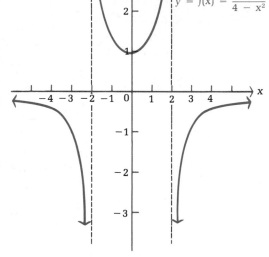

$$y = f(x) = \frac{4}{4 - x^2}$$

6. $\lim_{x \to \infty} \overline{C}(x) = 4;\ \lim_{x \to 0^+} \overline{C}(x) = \infty$

Exercise 12-1

A   *Use a calculator to evaluate each function at $x = 10, 100, 1{,}000$, and $10{,}000$. Use the results of these calculations to estimate $\lim_{x \to \infty} f(x)$.*

1. $f(x) = \dfrac{1}{x + 1}$

2. $f(x) = \dfrac{x}{x + 1}$

3. $f(x) = \dfrac{x^2}{x + 1}$

4. $f(x) = \dfrac{x + 1}{x^2}$

Use a calculator to evaluate each function at $x = 1.1, 1.01, 1.001$, and 1.0001, and at $.9, .99, .999$, and $.9999$. Use the results of your calculations to estimate $\lim_{x \to 1^+} f(x)$ and $\lim_{x \to 1^-} f(x)$.

5. $f(x) = \dfrac{1}{x - 1}$

6. $f(x) = \dfrac{1}{(1 - x)^{1/3}}$

7. $f(x) = \dfrac{1}{(x-1)^{2/3}}$

8. $f(x) = \dfrac{1}{(1-x)^{4/3}}$

Problems 9–12 refer to the following graph of $y = f(x)$:

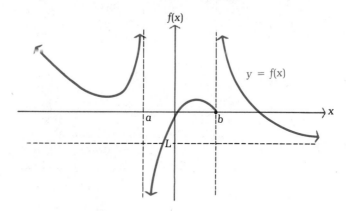

9. (A) $\lim\limits_{x \to -\infty} f(x) = ?$ (B) $\lim\limits_{x \to \infty} f(x) = ?$

10. (A) $\lim\limits_{x \to a^-} f(x) = ?$ (B) $\lim\limits_{x \to a^+} f(x) = ?$

11. (A) $\lim\limits_{x \to b^-} f(x) = ?$ (B) $\lim\limits_{x \to b^+} f(x) = ?$

12. (A) Where does f have horizontal asymptotes?
    (B) Where does f have vertical asymptotes?

B Evaluate the following limits.

13. $\lim\limits_{x \to \infty} \left(4 + \dfrac{2}{x} - \dfrac{3}{x^2}\right)$

14. $\lim\limits_{x \to -\infty} \left(3 - \dfrac{5}{x^3} + \dfrac{2}{x^4}\right)$

15. $\lim\limits_{x \to \infty} \dfrac{2x^3}{3x^3 + 5}$

16. $\lim\limits_{x \to \infty} \dfrac{4x^4}{9x^4 + 10}$

17. $\lim\limits_{x \to -\infty} \dfrac{2x^3}{4x^4 + 7}$

18. $\lim\limits_{x \to -\infty} \dfrac{3x^2}{x + 2}$

19. $\lim\limits_{x \to \infty} \dfrac{3x^3 + 5}{4x^2 + 2}$

20. $\lim\limits_{x \to \infty} \dfrac{7x^2}{x^5 + 7}$

Find the vertical asymptotes of each function.

21. $f(x) = \dfrac{x^2 + 1}{x^2 - 3x + 2}$

22. $f(x) = \dfrac{x^2 + 4}{x^2 - 3x - 10}$

23. $f(x) = \dfrac{x^2 - 1}{x^2 - 3x + 2}$

24. $f(x) = \dfrac{x^2 - 4}{x^2 - 3x - 10}$

25. $f(x) = \dfrac{1}{\sqrt{1 - x^2}}$

26. $f(x) = \dfrac{1}{\sqrt{9 - x^2}}$

For each function, find all horizontal and vertical asymptotes. Evaluate $\lim_{x \to c^+} f(x)$ and $\lim_{x \to c^-} f(x)$ at each vertical asymptote. Sketch the graph of the function. Use a calculator and point-by-point plotting in regions of uncertainty.

27. $f(x) = \dfrac{2x + 4}{x - 4}$

28. $f(x) = \dfrac{2 - x}{x + 2}$

29. $f(x) = \dfrac{x}{x^2 - 4}$

30. $f(x) = \dfrac{1}{x^2 - 4}$

C

31. Let $f(x) = \dfrac{x}{\sqrt{1 + x^2}}$.

   (A) Use a calculator to evaluate $f(x)$ at $x = 10, 100,$ and $1,000,$ and $x = -10, -100,$ and $-1,000.$
   (B) Evaluate $\lim_{x \to \infty} f(x)$ and $\lim_{x \to -\infty} f(x)$.
   (C) Sketch the graph of f.

32. Repeat Problem 31 for $f(x) = \dfrac{x}{\sqrt{4x^2 + 1}}$.

Each of the limits in Problems 33–36 is of the form

$$\lim_{x \to \infty} [f(x) - g(x)]$$

Evaluate each limit by first rationalizing the expression $[f(x) - g(x)]/1$. That is, multiply $[f(x) - g(x)]/1$ by $[f(x) + g(x)]/[f(x) + g(x)]$.

33. $\lim_{x \to \infty} (\sqrt{x + 1} - \sqrt{x})$

34. $\lim_{x \to \infty} (\sqrt{x^2 + 1} - x)$

35. $\lim_{x \to \infty} (\sqrt{x^2 + x} - x)$

36. $\lim_{x \to \infty} (\sqrt{x^2 + 4x} - x)$

Applications

Business & Economics

37. *Average cost.* The cost function for manufacturing x flashlights is

$$C(x) = 3,000 + 2.75x$$

   (A) Find $\overline{C}(x)$, the average cost function.
   (B) Evaluate $\lim_{x \to \infty} \overline{C}(x)$.
   (C) Evaluate $\lim_{x \to 0^+} \overline{C}(x)$.

38. *Average profit.* Suppose the flashlights in Problem 37 sell for $5.25 each and that the company manufactures and sells x flashlights.

   (A) Find the profit.
   (B) Find the average profit.
   (C) Find the limit of the average profit as x approaches infinity.

39. *Marginal cost.* The cost function for a publishing company is

$$C(x) = 10,000 + 12x + \frac{100}{x}$$

where x is the number of books produced in a single printing.

(A)   Find the marginal cost function.
(B)   Evaluate the limit of the marginal cost function as x approaches infinity.

40. *Advertising.* A company estimates that it will sell $N(x)$ units of a product after spending x thousand on advertising, as given by

$$N(x) = \frac{5,000x^2}{2.5x^2 + 4,000}$$

Evaluate the limit of $N(x)$ as x approaches infinity.

Life Sciences

41. *Pollution.* The bacteria concentration (number of bacteria per cubic centimeter) t days after a polluted lake is treated with a bactericide is given by

$$C(t) = \frac{50t^2 + 45,000}{t^2 + 225}$$

(A)   What was the concentration at the time the lake was initially treated?
(B)   What is the limit of the concentration as t approaches infinity?

42. *Animal population.* A biologist has estimated that the population of a certain species t years from now will be given by

$$P(t) = \frac{500t^2}{.5t^2 + 450}$$

What is the limit of $P(t)$ as t approaches infinity?

Social Sciences

43. *Learning.* A new worker on an assembly line performs an operation in T minutes after x performances of the operation, as given by

$$T = 6\left(1 + \frac{1}{\sqrt{x}}\right)$$

What is the limit of T as x approaches infinity?

12-2   First Derivative and Graphs

- Increasing and Decreasing Functions
- Critical Values and Local Extrema
- First-Derivative Test
- Application

Since the derivative is associated with the slope of the graph of a function at a point, we might expect that it is also associated with other properties of a graph. As we will see in this and the next section, the first and second derivatives can tell us a great deal about the shape of the graph of a function. In addition, this investigation will lead to methods for finding absolute maximum and minimum values for functions that do not require graphing. Companies can use these methods to find production levels that will minimize cost or maximize profit. Pharmacologists can use them to find levels of drug dosages that will produce maximum sensitivity to a drug. And so on.

■ Increasing and Decreasing Functions

Graphs of functions generally have rising or falling sections as we move from left to right. It would be an aid to graphing if we could figure out where these sections occur. Suppose the graph of a function f is as indicated in Figure 7. As we move from left to right, we see that on the interval (a, b) the graph of f is rising, $f(x)$ is increasing,* and the slope of the graph is positive $[f'(x) > 0]$. On the other hand, on the interval (b, c) the graph of f is falling, $f(x)$ is decreasing, and the slope of the graph is negative $[f'(x) < 0]$. At $x = b$ the graph of f changes direction (from rising to falling), $f(x)$ changes from increasing to decreasing, the slope of the graph is 0 $[f'(b) = 0]$, and the tangent line is horizontal.

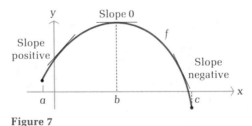

Figure 7

In general, we can prove that if $f'(x) > 0$ (is positive) on the interval (a, b), then $f(x)$ increases (↗) and the graph of f rises as we move from left to right over the interval; if $f'(x) < 0$ (is negative) on an interval (a, b), then $f(x)$ decreases (↘) and the graph of f falls as we move from left to right over the interval. We summarize these important results in the box.

---

* Formally, we say that $f(x)$ is *increasing* on an interval (a, b) if $f(x_2) > f(x_1)$ whenever $a < x_1 < x_2 < b$; f is decreasing on (a, b) if $f(x_2) < f(x_1)$ whenever $a < x_1 < x_2 < b$.

Increasing and Decreasing Functions

For the interval (a, b):

$f'(x)$	$f(x)$	Graph of f	Examples
$+$	Increases ↗	Rises ↗	
$-$	Decreases ↘	Falls ↘	

Example 7   Given $f(x) = x^3 - 3x^2 + 3$:

(A)   Which values of x correspond to horizontal tangents?
(B)   For which values of x is $f(x)$ increasing? Decreasing?
(C)   Sketch a graph of f. Add horizontal tangent lines.

Solutions   (A)   Take the derivative of f and determine which values of x make $f'(x) = 0$:

$$f'(x) = 3x^2 - 6x$$
$$= 3x(x - 2)$$

Now, $3x(x - 2) = 0$ if

$$3x = 0 \quad \text{or} \quad x - 2 = 0$$
$$x = 0 \quad \text{or} \quad x = 2$$

Thus, horizontal tangent lines exist at $x = 0$ and at $x = 2$.

(B)   Construct a sign chart for f' to determine which values of x make $f'(x) > 0$ and which values make $f'(x) < 0$.

Sign chart for $f'(x) = 3x(x - 2)$

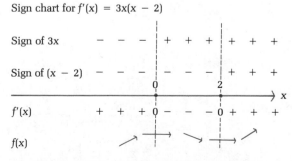

The last line of the sign chart indicates that f increases on $(-\infty, 0)$, has a horizontal tangent at $x = 0$, decreases on $(0, 2)$, has a horizontal tangent at $x = 2$, and increases on $(2, \infty)$. These facts are summarized in the table.

x	$f'(x)$	$f(x)$	Graph of f
$x < 0$	$+$	Increasing	Rising
$x = 0$	0		Horizontal tangent
$0 < x < 2$	$-$	Decreasing	Falling
$x = 2$	0		Horizontal tangent
$x > 2$	$+$	Increasing	Rising

(C) We sketch a graph of f using the information from part B and point-by-point plotting for regions of uncertainty. Notice how much we know about the shape of the graph before we sketch it and how few additional points we need for its sketch.

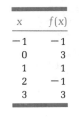

x	$f(x)$
-1	-1
0	3
1	1
2	-1
3	3

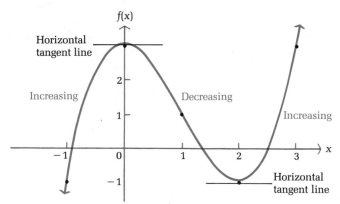

Problem 7   Repeat Example 7 for $f(x) = 3x^3 - 9x$.

Example 8   Given $f(x) = 5x^{2/3} - 2x^{5/3}$:

(A) Which values of x correspond to horizontal tangents? Where is f' not defined?

(B) For which values of x is $f(x)$ increasing? Decreasing?

(C) Sketch the graph of f.

Solutions   (A) $f'(x) = \dfrac{10}{3} x^{-1/3} - \dfrac{10}{3} x^{2/3}$ A term with a negative exponent often leads to a 0 in the denominator. Use algebraic operations to eliminate the negative exponents and combine the terms.

$$= \frac{10}{3}\left(\frac{1}{x^{1/3}} - x^{2/3}\right)$$

$$= \frac{10}{3}\left(\frac{1}{x^{1/3}} - \frac{x^{2/3}}{1} \cdot \frac{x^{1/3}}{x^{1/3}}\right)$$

$$= \frac{10}{3}\left(\frac{1-x}{x^{1/3}}\right)$$

Since $f'(1) = 0$, there is a horizontal tangent at $x = 1$. $f'(x)$ is not defined at $x = 0$.

(B) When you construct the sign chart for a fraction, remember to include the zeros for both the numerator and the denominator on the number line.

Sign chart for $f'(x) = \frac{10}{3}\left(\frac{1-x}{x^{1/3}}\right)$

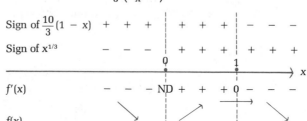

x	$f'(x)$	$f(x)$	Graph of f
x < 0	−	Decreasing	Falling
x = 0	Not defined	Is defined	Sharp corner
0 < x < 1	+	Increasing	Rising
x = 1	0		Horizontal tangent
x > 1	−	Decreasing	Falling

(C) We sketch a graph of f using the information from part B and point-by-point plotting for regions of uncertainty.

x	$f(x)$
−1	7
0	0
1	3
2.5	0

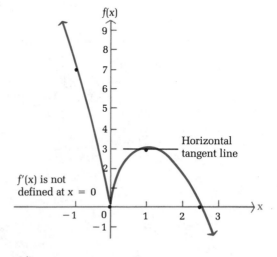

Problem 8 Given $f(x) = x^{5/3} - 5x^{2/3}$:

(A) Which values of x correspond to horizontal tangents? Where is f' not defined?

(B) For which values of x is $f(x)$ increasing? Decreasing?

(C) Sketch a graph of f.

Example 9   Given $f(x) = \dfrac{x-1}{x-2}$:

(A) Find the intervals where $f(x)$ is increasing and those where $f(x)$ is decreasing.

(B) Find the horizontal and vertical asymptotes of f.

(C) Sketch a graph of f.

Solutions   (A) $f'(x) = \dfrac{(x-2)(1) - (x-1)(1)}{(x-2)^2}$

$= \dfrac{x-2-x+1}{(x-2)^2}$

$= -\dfrac{1}{(x-2)^2}$

Notice that both f and f' are not defined at $x=2$. Although $f'(x) < 0$ for all other values of x, it would be incorrect to say that $f(x)$ is decreasing for all x except $x=2$. Instead, we must say that f is decreasing on $(-\infty, 2)$ and on $(2, \infty)$.

x	$f'(x)$	$f(x)$	Graph of f
$x < 2$	$-$	Decreasing	Falling
$x = 2$	ND	ND	
$x > 2$	$-$	Decreasing	Falling

(B)   Let $p(x) = x - 1$ and $q(x) = x - 2$. Since $q(2) = 0$ and $p(2) \neq 0$, f has a vertical asymptote at $x = 2$.

$\displaystyle\lim_{x \to 2^-} \frac{x-1}{x-2} = -\infty$ $x - 1 > 0$ and $x - 2 < 0$ for x near 2 and on the left.

$\displaystyle\lim_{x \to 2^+} \frac{x-1}{x-2} = \infty$ $x - 1 > 0$ and $x - 2 > 0$ for x near 2 and on the right.

Since p and q are polynomials of the same degree,

$\displaystyle\lim_{x \to \pm\infty} \frac{x-1}{x-2} = \frac{1}{1} = 1$ $n = m = 1$
   $a_n = 1 \qquad b_m = 1$

Thus, f has a horizontal asymptote at $y = 1$.

(C) We sketch a graph of f using the information from parts A and B and point-by-point plotting for regions of uncertainty.

x	$f(x)$
-1	$\dfrac{2}{3}$
0	$\dfrac{1}{2}$
1	0
3	2
4	$\dfrac{3}{2}$
5	$\dfrac{4}{3}$

Problem 9 Given $f(x) = \dfrac{2x}{1-x}$:

(A) Find the intervals where $f(x)$ is increasing and those where $f(x)$ is decreasing.

(B) Find the horizontal and vertical asymptotes of f.

(C) Sketch a graph of f.

■ Critical Values and Local Extrema

When the graph of a continuous function changes from rising to falling, a high point or *local maximum* occurs and when the graph changes from falling to rising, a low point or *local minimum* occurs. In Figure 8, high

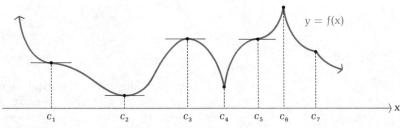

Figure 8

points occur at c_3 and c_6, and low points occur at c_2 and c_4. In general, we call a point $(c, f(c))$ a **local maximum** if there exists an interval (m, n) containing c such that

$$f(x) \leq f(c)$$

for all x in (m, n). A point $(c, f(c))$ is called a **local minimum** if there exists an interval (m, n) containing c such that

$$f(x) \geq f(c)$$

for all x in (m, n). Thus, in Figure 8 we see that local maxima occur at c_3 and c_6 and local minima occur at c_2 and c_4.

How can we locate local maxima and minima if we are given the equation for the function and not its graph? Figure 8 suggests an approach. It appears that local maxima and minima occur among those values of x such that $f'(x) = 0$ or $f'(x)$ does not exist — that is, among the values c_1, c_2, c_3, c_4, c_5, c_6, and c_7. [Recall from Section 10-4 that $f'(x)$ is not defined at sharp points or corners on a graph.] It is possible to prove the following theorem:

Theorem 3

> **Existence of Local Extrema**
>
> If f is a continuous function over the interval (a, b), then local maxima or minima, if they exist, must occur at values of x, called **critical values,** such that $f'(x) = 0$ or $f'(x)$ does not exist (is not defined).

Our strategy is now clear. We find all critical values for f and test each one to see if it is a local maximum, a local minimum, or neither. There are two derivative tests that can be used for this purpose. In this section we will discuss the first-derivative test, which works in all cases. In the next section we will discuss the second-derivative test, which is often easier to use but does not work in all cases.

■ First-Derivative Test

If $f'(x)$ exists on both sides of a critical value c, then the sign of $f'(x)$ can be used to determine if $f(c)$ is a local maximum, a local minimum, or neither. The various possibilities are summarized in the next box. Figure 9 illustrates several typical cases.

First-Derivative Test for Local Extrema

Let c be a critical value of $f[f'(c) = 0$ or $f'(c)$ is not defined, but $f(c)$ is defined].

Case 1

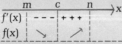

If $f'(x)$ changes from negative to positive at c, then $f(c)$ is a local minimum.

Case 2

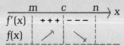

If $f'(x)$ changes from positive to negative at c, then $f(c)$ is a local maximum.

[Note: If $f'(x)$ does not change sign at c, then $f(c)$ is neither a local maximum nor a local minimum.]

$f'(c) = 0$
Horizontal tangent

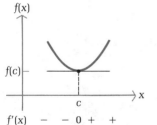

(A) $f(c)$ is a
    local minimum.

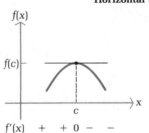

(B) $f(c)$ is a local
    maximum.

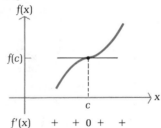

(C) $f(c)$ is neither
    a local maximum
    nor a local minimum.

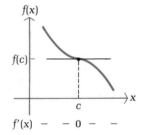

(D) $f(c)$ is neither a
    local maximum nor
    a local minimum.

$f'(c)$ is not defined
but $f(c)$ is defined

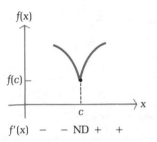

(E) $f(c)$ is a local
    minimum.

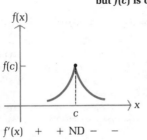

(F) $f(c)$ is a local
    maximum.

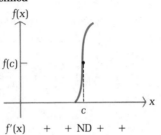

(G) $f(c)$ is neither a
    local maximum nor
    a local minimum.

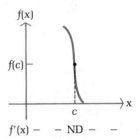

(H) $f(c)$ is neither a
    local maximum nor
    a local minimum.

Figure 9 Local extrema

Example 10 Given $f(x) = x^3 - 6x^2 + 9x + 1$:

(A) Find the critical values of f.
(B) Find the local maxima and minima.
(C) Sketch the graph of f.

Solutions (A) $$f'(x) = 3x^2 - 12x + 9$$
$$= 3(x^2 - 4x + 3)$$
$$= 3(x - 1)(x - 3)$$

Now, $f'(x) = 0$ if

$$x - 1 = 0 \quad \text{or} \quad x - 3 = 0$$
$$x = 1 \quad \text{or} \quad x = 3$$

Critical values are $x = 1$ and $x = 3$.

(B) The easiest way to apply the first-derivative test is to construct a sign chart for $f'(x)$:

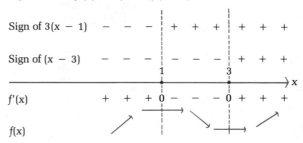

Sign chart for $f'(x) = 3(x - 1)(x - 3)$

Since $f'(x)$ changes from positive to negative at $x = 1$, $f(1) = 5$ is a local maximum. Since $f'(x)$ changes from negative to positive at $x = 3$, $f(3) = 1$ is a local minimum.

x	$f'(x)$	$f'(x)$	Graph of f
$x < 1$	$+$	Increasing	Rising
$x = 1$	0	Local maximum	Horizontal tangent
$1 < x < 3$	$-$	Decreasing	Falling
$x = 3$	0	Local minimum	Horizontal tangent
$3 < x$	$+$	Increasing	Rising

(C) We sketch a graph of f using the information from part B and point-by-point plotting.

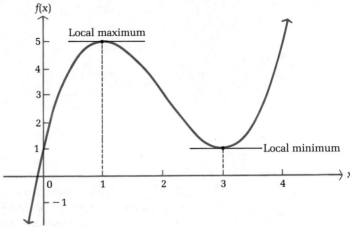

x	f(x)
0	1
1	5
2	3
3	1
4	5

Problem 10   Given $f(x) = x^3 - 9x^2 + 24x - 10$:

(A)   Find the critical values of f.
(B)   Find the local maxima and minima.
(C)   Sketch a graph of f.

Example 11   Find the local maxima and minima for each of the following functions:

(A) $f(x) = x^{4/3} + 4x^{1/3}$ (B) $f(x) = \dfrac{1 - 2x}{x^2}$

Solutions   (A) $f(x) = x^{4/3} + 4x^{1/3}$

$$f'(x) = \frac{4}{3} x^{1/3} + \frac{4}{3} x^{-2/3}$$

$$= \frac{4}{3} \left(x^{1/3} + \frac{1}{x^{2/3}} \right)$$

$$= \frac{4}{3} \left(\frac{x^{1/3}}{1} \cdot \frac{x^{2/3}}{x^{2/3}} + \frac{1}{x^{2/3}} \right)$$

$$= \frac{4}{3} \left(\frac{x + 1}{x^{2/3}} \right)$$

$f'(-1) = 0$ and $f'(0)$ is not defined. Thus, the critical values are $x = -1$ and $x = 0$.

Sign chart for $f'(x) = \frac{4}{3}\left(\frac{x+1}{x^{2/3}}\right)$

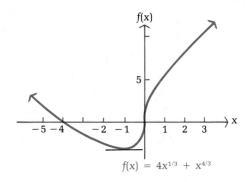

Since $f'(x)$ changes from negative to positive at $x = -1$, $f(-1) = -3$ is a local minimum. Since $f'(x)$ does not change sign at $x = 0$, $f(0) = 0$ is not a local extreme point, as shown in the figure.

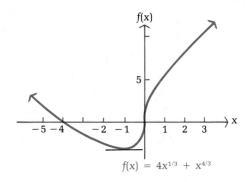

$$f(x) = 4x^{1/3} + x^{4/3}$$

(B) $f(x) = \dfrac{1-2x}{x^2}$.

$$f'(x) = \frac{x^2(-2) - (1-2x)2x}{(x^2)^2}$$

$$= \frac{-2x^2 - 2x + 4x^2}{x^4}$$

$$= \frac{2x^2 - 2x}{x^4}$$

$$= \frac{2x(x-1)}{x^4}$$

$$= \frac{2(x-1)}{x^3}$$

Since $f'(1) = 0$, $x = 1$ is a critical value. Even though $f'(0)$ does not exist, $x = 0$ is *not* a critical value of f. A critical value must be in the domain of the function and 0 is not in the domain of f—that is, $f(0)$

does not exist. Nevertheless, 0 must be included on the number line in the sign chart for $f'(x)$:

Sign chart for $f'(x) = \dfrac{2(x - 1)}{x^3}$

Sign of $2(x - 1)$ $-$ $-$ $-$ | $-$ $-$ $-$ | $+$ $+$ $+$

Sign of x^3 $-$ $-$ $-$ | $+$ $+$ $+$ | $+$ $+$ $+$

$$\begin{array}{c} & 0 & & 1 \end{array}$$

$f'(x)$ $+$ $+$ $+$ ND $-$ $-$ $-$ 0 $+$ $+$ $+$

$f(x)$ ↗   ND   ↘   ↗

The sign chart seems to indicate that there is a local maximum at $x = 0$. However, as we noted before, 0 is not in the domain of the function and it makes no sense to apply the first-derivative test at $x = 0$. In fact, since $\lim_{x \to 0} f(x) = \infty$, f has a vertical asymptote at $x = 0$. The test does apply at $x = 1$ and $f(1) = -1$ is a local minimum, as indicated in the figure.

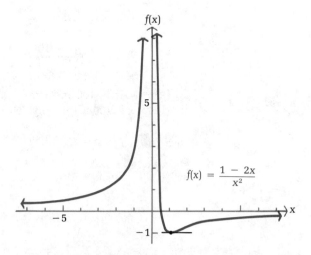

$$f(x) = \frac{1 - 2x}{x^2}$$

Notice that in Example 11A, $f(x)$ is defined at 0 but $f'(x)$ is not. Thus, $x = 0$ is a critical value. However, in part B, both $f(x)$ and $f'(x)$ are not defined at 0. Hence, $x = 0$ is not a critical value for this function. Be careful that you do not apply the first-derivative test to a value that is not a critical value.

Problem 11 Find the local maxima and minima for each of the following functions (do not graph):

(A) $f(x) = x^{5/3} + 5x^{2/3}$ (B) $f(x) = \dfrac{x+1}{x^2}$

■ Application

Example 12
Average Cost

Given the cost function $C(x) = 5{,}000 + (1/2)x^2$, where x is the number of units produced:

(A) Find the minimum average cost.
(B) Find the marginal cost function.
(C) Graph the average cost function and the marginal cost function on the same axes.

Solutions

(A) Let $\overline{C}(x) = \dfrac{C(x)}{x} = \dfrac{5{,}000}{x} + \dfrac{1}{2}x$, $x > 0$. Then

$$\overline{C}'(x) = -\dfrac{5{,}000}{x^2} + \dfrac{1}{2}$$

$$= -\dfrac{10{,}000}{2x^2} + \dfrac{x^2}{2x^2}$$

$$= \dfrac{x^2 - 10{,}000}{2x^2}$$

$$= \dfrac{(x - 100)(x + 100)}{2x^2}$$

$\overline{C}'(x) = 0$ at $x = 100$ and $x = -100$. Since the number of units must be positive, $x = 100$ is the only critical value.

Sign chart for $\overline{C}'(x) = \dfrac{(x - 100)(x + 100)}{2x^2}$

Sign of $(x - 100)$	$-\ -\ -$	$+\ +\ +$
Sign of $(x + 100)$	$+\ +\ +$	$+\ +\ +$
Sign of $2x^2$	$+\ +\ +$	$+\ +\ +$
	0	100
$\overline{C}'(x)$	$\mathrm{ND}\ -\ -\ -$	$0\ +\ +\ +$
$\overline{C}(x)$	ND $\searrow$	$\nearrow$

The first-derivative test implies that $\overline{C}(100) = 100$ is a local minimum.

An examination of the graph shows that $\overline{C}(x) > 100$ for all other values of x. Thus, *the minimum average cost is 100.*

(B) $C'(x) = x$

(C)

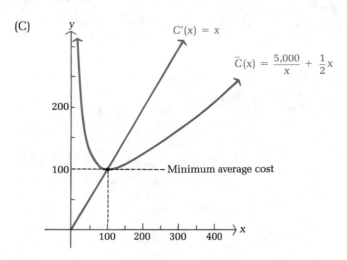

This graph illustrates an important principle in economics: The minimal average cost occurs when the average cost is equal to the marginal cost.

Problem 12 Given the cost function $C(x) = 1,600 + (1/4)x^2$:

(A) Find the minimum average cost.
(B) Find the marginal cost function.
(C) Graph the average cost function and the marginal cost function on the same axes.

Answers to Matched Problems

7. (A) $x = -1; x = 1$

(B) Increasing on $(-\infty, -1)$ and $(1, \infty)$

Decreasing on $(-1, 1)$

(C)

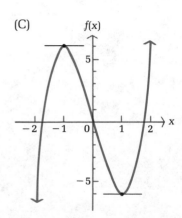

8. (A) Horizontal tangent at
x = 2

f'(0) does not exist

(B) Increasing on (−∞, 0)
and (2, ∞)

Decreasing on (0, 2)

(C) f(x)

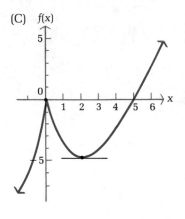

9. (A) Increasing on (−∞, 1)
and (1, ∞)

(B) Horizontal asymptote:
y = −2

Vertical asymptote: x = 1

(C) f(x)

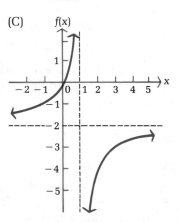

10. (A) Critical values: x = 2,
x = 4

(B) f(2) = 10 is a local
maximum

f(4) = 6 is a local
minimum

(C) f(x)

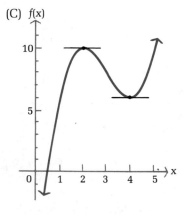

11. (A) f(−2) = (−2)^{5/3} + 5(−2)^{2/3} ≈ 4.8 is a local maximum

f(0) = 0 is a local minimum

(B) f(−2) = −1/4 is a local minimum

12. (A) Minimal average cost is 40 at x = 80

(B) C'(x) = (1/2)x

(C)

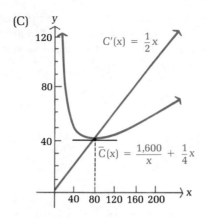

$$C'(x) = \frac{1}{2}x$$

$$\overline{C}(x) = \frac{1,600}{x} + \frac{1}{4}x$$

Exercise 12-2

A   Problems 1–6 refer to the following graph of y = f(x):

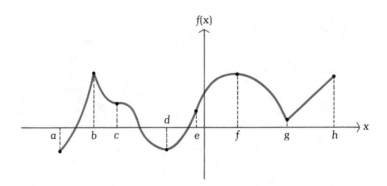

1. Identify the intervals over which $f(x)$ is increasing.

2. Identify the intervals over which $f(x)$ is decreasing.

3. Identify the points where $f'(x) = 0$.

4. Identify the points where $f'(x)$ does not exist.

5. Identify the points where f has a local maximum.

6. Identify the points where f has a local minimum.

In Problems 7–10 replace question marks in the tables with "Local maximum," "Local minimum," or "Neither," as appropriate. Assume f is continuous over (m, n) unless otherwise stated. (Sketching pictures may help you decide.)

7.

	$f'(c)$	$f'(x)$ (m, c)	$f'(x)$ (c, n)	$f(c)$
(A)	0	−	+	?
(B)	0	−	−	?

8.

	$f'(c)$	$f'(x)$ (m, c)	$f'(x)$ (c, n)	$f(c)$
(A)	0	+	−	?
(B)	0	+	+	?

9.

	$f'(c)$	$f(c)$	$f'(x)$ (m, c)	$f'(x)$ (c, n)	$f(c)$
(A)	Not defined	Defined	+	−	?
(B)	Not defined	Defined	+	+	?
(C)	Not defined	Not defined	−	+	?

10.

	$f'(c)$	$f(c)$	$f'(x)$ (m, c)	$f'(x)$ (c, n)	$f(c)$
(A)	Not defined	Defined	−	+	?
(B)	Not defined	Defined	−	−	?
(C)	Not defined	Not defined	+	−	?

In Problems 11–14 use the given information to make a rough sketch of a graph of $y = f(x)$. Assume that f is continuous on $(-\infty, \infty)$.

11.
$$f'(x) \quad + \quad + \quad + \quad 0 \quad + \quad + \quad + \quad 0 \quad - \quad - \quad - \quad \longrightarrow x$$
with -1 and 1 marked.

x	−2	−1	0	1	2
$f(x)$	−1	1	2	3	1

12.
$$f'(x) \quad + \quad + \quad + \quad 0 \quad - \quad - \quad - \quad 0 \quad - \quad - \quad - \quad \longrightarrow x$$
with -1 and 1 marked.

x	−2	−1	0	1	2
$f(x)$	1	3	2	1	−1

13.
$$f'(x) \quad - \quad - \quad - \quad 0 \quad + \quad + \quad + \quad ND \quad - \quad - \quad - \quad 0 \quad - \quad - \quad - \quad \longrightarrow x$$
with -1, 0, and 2 marked.

x	−2	−1	0	2	4
$f(x)$	2	1	2	1	0

14.
$$f'(x) \quad + \quad + \quad + \quad ND \quad + \quad + \quad + \quad 0 \quad - \quad - \quad - \quad 0 \quad + \quad + \quad + \quad \longrightarrow$$
with -1, 0, and 2 marked.

x	−2	−1	0	2	3
$f(x)$	−3	0	2	−1	0

B Determine the intervals over which the function is increasing and the intervals over which the function is decreasing. Find all local maxima and minima (do not graph).

15. $f(x) = 2x^2 - x^4$

16. $f(x) = 3x - x^3$

17. $f(x) = x^3 - 3x^2 - 24x + 7$

18. $f(x) = x^3 + 3x^2 - 9x + 5$

19. $f(x) = x(5 - x)^{2/3}$

20. $f(x) = x(4 - x)^{1/3}$

21. $f(x) = x + \dfrac{4}{x}$ 22. $f(x) = \dfrac{9}{x} + x$

23. $f(x) = 1 + \dfrac{1}{x} + \dfrac{1}{x^2}$ 24. $f(x) = 3 - \dfrac{4}{x} - \dfrac{2}{x^2}$

25. $f(x) = \sqrt{x}\,(x - 10)^2$ 26. $f(x) = x^2\sqrt{x + 5}$

27. $f(x) = \dfrac{x^2}{x - 2}$ 28. $f(x) = \dfrac{x^2}{x + 1}$

For each function, find the intervals over which the graph of f is rising and falling, locate horizontal tangents and points where f'(x) does not exist [but f(x) does exist], and sketch the graph.

29. $f(x) = 4 + 8x - x^2$ 30. $f(x) = 2x^2 - 8x + 9$
31. $f(x) = x^3 - 3x + 1$ 32. $f(x) = x^3 - 12x + 2$
33. $f(x) = (x - 2)^{2/3}$ 34. $f(x) = (x + 3)^{2/3}$
35. $f(x) = 2\sqrt{x} - x$ 36. $f(x) = x - 4\sqrt{x}$
37. $f(x) = 4x^{2/3} - x^{8/3}$ 38. $f(x) = (x - 1)^{8/3} - 4(x - 1)^{2/3}$

C *For each function, find the intervals over which the graph of f is rising and falling, locate horizontal tangents and points where f'(x) does not exist [but f(x) does exist], find horizontal and vertical asymptotes, and sketch the graph.*

39. $f(x) = \dfrac{x + 3}{x - 3}$ 40. $f(x) = \dfrac{2x - 4}{x + 2}$

41. $f(x) = x + \dfrac{1}{x}$ 42. $f(x) = x^2 + \dfrac{1}{x^2}$

43. $f(x) = 1 + \dfrac{1}{(x - 2)^2}$ 44. $f(x) = -1 + \dfrac{1}{(x + 1)^3}$

Applications

Business & Economics

45. *Average cost.* The cost of producing x units of a certain product is given by $C(x) = 1{,}000 + 5x + (1/10)x^2$.

   (A) Find the intervals where the average cost is decreasing and increasing.
   (B) Sketch the graph of the average cost function and the marginal cost function on the same axes.

46. *Average cost.* Repeat Problem 45 for $C(x) = 500 + 2x + (1/5)x^2$.

47. *Advertising.* A company estimates that it will sell N(x) units of a product after spending $x thousand on advertising, as given by

$$N(x) = -x^3 + 75x^2 - 1{,}200x + 15{,}000 \qquad 10 \leqslant x \leqslant 40$$

   (A) Determine when sales are increasing.

(B) Determine when the rate of change of sales is increasing. [*Hint:* Use $N''(x)$.]

48. *Profit function.* If the profit $P(x)$ in dollars for an output of x units is given by

$$P(x) = -\frac{x^2}{30} + 140x - 72{,}000 \qquad x \geqslant 0$$

find production levels for which P is increasing and levels for which P is decreasing.

Life Sciences

49. *Bacteria growth.* A colony of bacteria was treated with a slow-acting poison and the number of survivors $N(t)$, in thousands, t hours after the poison was administered was found to be given approximately by

$$N(t) = 2t^3 - 75t^2 + 600t + 2{,}000 \qquad 0 \leqslant t \leqslant 20$$

How long did the colony continue to grow after the drug was administered?

50. *Drug sensitivity.* One hour after x milligrams of a particular drug are given to a person, the change in body temperature $T(x)$ in degrees Fahrenheit is given by

$$T(x) = x^2 \left(1 - \frac{x}{9}\right) \qquad 0 \leqslant x \leqslant 6$$

The rate at which T changes with respect to the size of the dosage x, $T'(x)$, is called the *sensitivity* of the body to the dosage. For what values of x is $T'(x)$ increasing? Decreasing? [*Hint:* Use $T''(x)$.]

Social Sciences

51. *Learning.* The time T in minutes that it takes a particular person to learn a list of n items is

$$T = f(n) = 2n\sqrt{n - 12} \qquad n \geqslant 12$$

(A) When is T increasing?

(B) When is the rate of change of T increasing? [*Hint:* Use $f''(n)$.]

12-3 Second Derivative and Graphs

- Concavity
- Inflection Points
- Second-Derivative Test
- Application

In the preceding section we saw that the first derivative can be used to determine when a graph is rising and falling. Now we want to see what the second derivative can tell us about the shape of a graph.

■ Concavity

Consider the functions

$$f(x) = x^2 \quad \text{and} \quad g(x) = \sqrt{x}$$

for x in the interval (0, ∞). Since

$$f'(x) = 2x > 0 \quad \text{for } 0 < x < \infty$$

and

$$g'(x) = \frac{1}{2\sqrt{x}} > 0 \quad \text{for } 0 < x < \infty$$

both functions are increasing on (0, ∞).

   Notice the different shapes of the graphs of f and g (see Figure 10). Even though the graph of each function is rising and each graph starts at (0, 0) and goes through (1, 1), the graphs are quite dissimilar. The graph of f opens upward while the graph of g opens downward. We say that the graph of f is *concave upward* and the graph of g is *concave downward*. It will help us draw graphs if we can determine the concavity of the graph before we draw it. How can we find a mathematical formulation of concavity?

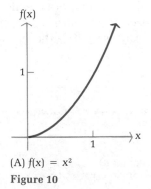

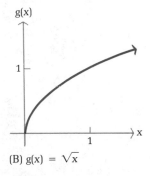

(A) $f(x) = x^2$ (B) $g(x) = \sqrt{x}$

Figure 10

   It will be instructive to examine the slopes of f and g at various points on their graphs (see Figure 11). There are two observations we can make about each graph. Looking at the graph of f in Figure 11A, we see that $f'(x)$ (the slope of the tangent line) is *increasing* and that the graph lies *above* each tangent line. Looking at Figure 11B, we see that $g'(x)$ is *decreasing* and that the graph lies *below* each tangent line. With these ideas in mind, we state the general definition of concavity: The graph of a function f is **concave upward (CU) on the interval (a, b)** if $f'(x)$ is *increasing* on (a, b) and is **concave downward (CD) on the interval (a, b)** if $f'(x)$ is *decreasing* on (a, b). Geometrically, the graph is concave upward on (a, b) if it lies above its tangent lines in (a, b) and is concave downward on (a, b) if it lies below its tangent lines in (a, b).

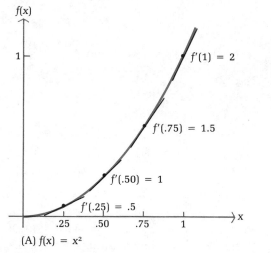

(A) $f(x) = x^2$

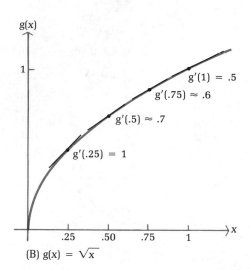

(B) $g(x) = \sqrt{x}$

Figure 11

How can we determine when $f'(x)$ is increasing or decreasing? In the last section we used the derivative of a function to find out when the function is increasing and decreasing. Thus, to determine when $f'(x)$ is increasing and decreasing, we can use $f''(x)$, the derivative of $f'(x)$. The results are summarized in the box.

Concavity

For the interval (a, b)

$f''(x)$	$f'(x)$	Graph of $y = f(x)$	Example
+	Increasing	Concave upward	⌣
−	Decreasing	Concave downward	⌢

Be careful not to confuse concavity with falling and rising. As Figure 12 illustrates, a graph that is concave upward on an interval may be falling, rising, or both falling and rising on that interval. A similar statement holds for a graph that is concave downward.

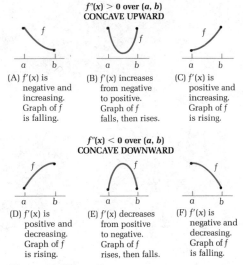

Figure 12 Concavity

Example 13   Find the intervals where each function is concave upward and the intervals where each function is concave downward. Sketch graphs of each function.

(A) $f(x) = x^3$ (B) $f(x) = (x-1)^{1/3}$

Solutions   (A)   To determine concavity, we must construct a sign chart for $f''(x)$.

$$f'(x) = 3x^2$$
$$f''(x) = 6x$$

Sign chart for $f''(x) = 6x$

Sign of $6x$ $-$ $-$ $-$ | $+$ $+$ $+$
                                            0
————————————————————————————————————→ x
$f''(x)$ $-$ $-$ $-$ 0 $+$ $+$ $+$

$f(x)$ CD | CU

Thus, the graph of f is concave downward on $(-\infty, 0)$ and concave upward on $(0, \infty)$. The graph of f (without going through other graphing details) is shown at the top of the next page.

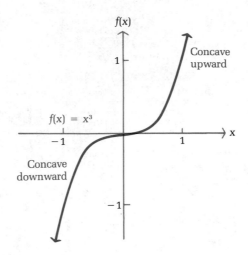

(B) $f'(x) = \dfrac{1}{3}(x-1)^{-2/3}$

$f''(x) = -\dfrac{2}{9}(x-1)^{-5/3}$

$= -\dfrac{2}{9(x-1)^{5/3}}$

Sign chart for $f''(x) = -\dfrac{2}{9(x-1)^{5/3}}$

Sign of $-\dfrac{2}{9(x-1)^{5/2}}$ +   +   + | −   −   −

                                                    |
                                                    1
―――――――――――――――――――――――――――――→ x
$f''(x)$ +   +   +   ND   −   −   −

$f(x)$ CU | CD

Thus, the graph of f is concave upward on $(-\infty, 1)$ and concave downward on $(1, \infty)$. The graph of f (without going through the other graphing details) is shown at the top of the next page.

Problem 13   Repeat Example 13 for the following functions:

(A) $f(x) = x - x^3$
(B) $f(x) = (x+2)^{5/3}$

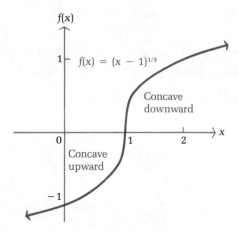

Each of the graphs in Example 13 has a point where the concavity changes. Such points are called *inflection points*.

■ Inflection Points

In general, an **inflection point** is a point on the graph of a function where the concavity changes (from upward to downward or from downward to upward). In order for the concavity to change at a point, $f''(x)$ must change sign at that point. Reasoning as we did in the previous section, we conclude that the inflection points must occur at points where $f''(x) = 0$ or $f''(x)$ does not exist [but $f(x)$ must exist]. Figure 13 illustrates several typical cases.

If $f'(c)$ exists, then the tangent line at an inflection point $(c, f(c))$ will always lie below the graph on the side that is concave upward and above the graph on the side that is concave downward (see Figures 13A, B, and C).

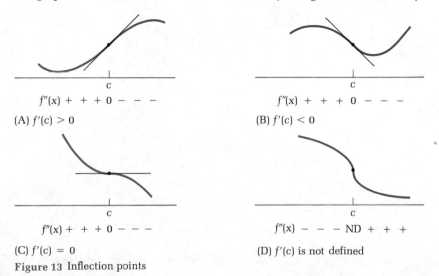

Figure 13 Inflection points

Example 14   Given $f(x) = x^4 - 2x^3 + 2$.

(A)   Find the intervals where f is concave upward. Concave downward.
(B)   Find the inflection points.
(C)   Graph f. Add tangent lines at all inflection points.

Solutions   (A) $f'(x) = 4x^3 - 6x^2$

$f''(x) = 12x^2 - 12x$

$= 12x(x-1)$

Now, $f''(x) = 0$ if

$x = 0$ or $x = 1$

Sign chart for $f''(x) = 12x(x-1)$

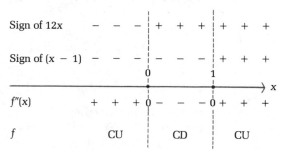

Thus, f is concave upward on $(-\infty, 0)$ and $(1, \infty)$ and concave downward on $(0, 1)$.

(B)   From the sign chart, we see that $f''(x)$ changes sign at $x = 0$ and $x = 1$; thus, f has inflection points at $x = 0$ and $x = 1$.

(C)

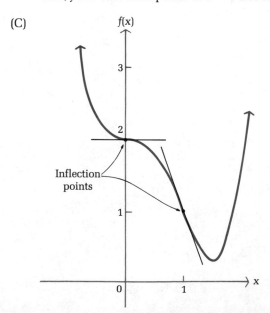

Problem 14   Given $f(x) = x^4 + 4x^3 + 10$.

    (A) Find intervals where f is concave upward. Concave downward.
    (B) Find the inflection points.
    (C) Graph f. Add tangent lines at all inflection points.

    The next example illustrates the same two important ideas that we discussed in the preceding section. That is,

1. The points where $f''(x) = 0$ or $f''(x)$ does not exist are only *possible* inflection points. The sign of $f''(x)$ must change at such a point in order for an inflection point to occur.
2. Numbers not in the domain of f must be included on the number line in the sign chart for $f''(x)$, but there cannot be an inflection point at a number where f is not defined.

Example 15   Find the inflection points (if any exist) for each of the following functions. Sketch a graph of each function.

    (A) $f(x) = (x - 2)^{4/3}$ (B) $f(x) = x + \dfrac{1}{x}$

Solutions   (A) $f'(x) = \dfrac{4}{3}(x - 2)^{1/3}$ Sign chart for $f''(x) = \dfrac{4}{9(x - 2)^{2/3}}$

$f''(x) = \dfrac{4}{9}(x - 2)^{-2/3}$ Sign of $\dfrac{4}{9(x - 2)^{2/3}}$ +   +   +   +   +   +

$= \dfrac{4}{9(x - 2)^{2/3}}$

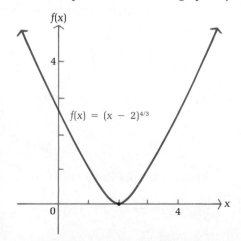

Since the second derivative does not change sign at $x = 2$, there is no inflection point at $x = 2$. The graph of f is shown here.

(B) $f'(x) = 1 - \dfrac{1}{x^2}$

$f''(x) = \dfrac{2}{x^3}$

Sign chart for $f''(x) = \dfrac{2}{x^3}$

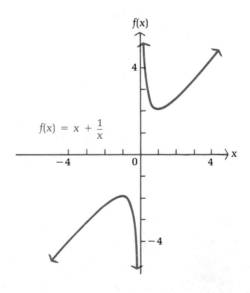

Sign of $\dfrac{2}{x^3}$ — — — $\vert$ + + +

                                   0

————————————————————○——————————→ x

$f''(x)$ — — — ND + + +

f CD   ND   CU

Even though the second derivative changes sign at $x = 0$, there is no inflection point at $x = 0$. The graph of f is given here.

$f(x) = x + \dfrac{1}{x}$

Problem 15   Repeat Example 15 for each of the following functions:

(A) $f(x) = x^4$ (B) $f(x) = \dfrac{1}{x^3}$

■ Second-Derivative Test

Now we want to see how the second derivative can be used to find local extrema. Suppose f is a function satisfying $f'(c) = 0$ and $f''(c) > 0$. First, note that if $f''(c) > 0$, then it follows from the properties of limits* that $f''(x) > 0$ in some interval (m, n) containing c. Thus, the graph of f must be concave upward in this interval. But this implies that $f'(x)$ is increasing in this interval. Since $f'(c) = 0$, $f'(x)$ must change from negative to positive at $x = c$ and $f(c)$ is a local minimum (see Figure 14). Reasoning in the same fashion, we conclude that if $f'(c) = 0$ and $f''(c) < 0$, then $f(c)$ is a local maximum. Of course, it is possible that both $f'(c) = 0$ and $f''(c) = 0$. In this case the second derivative cannot be used to determine the shape of the graph around $x = c$; $f(c)$ may be a local minimum, a local maximum, or neither.

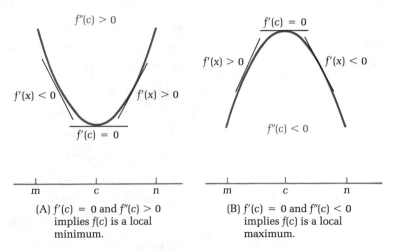

(A) $f'(c) = 0$ and $f''(c) > 0$
implies $f(c)$ is a local
minimum.

(B) $f'(c) = 0$ and $f''(c) < 0$
implies $f(c)$ is a local
maximum.

Figure 14 The second derivative and local extrema

The sign of the second derivative thus provides a simple test for identifying local maxima and minima. This test is most useful when we do not want to draw the graph of the function. If we are interested in drawing the graph and have already constructed the sign chart for $f'(x)$, then the first-derivative test can be used to identify the local extrema.

---

* Actually, we are assuming that $f''(x)$ is continuous in an interval containing c. It is very unlikely that we will encounter a function for which $f''(c)$ exists, but $f''(x)$ is not continuous in an interval containing c.

Second-Derivative Test for Local Maxima and Minima

$f'(c)$	$f''(c)$	$f(c)$	Example
0	+	Local minimum	$\smile$
0	−	Local maximum	$\frown$
0	0	Test fails	

Example 16 Find the local maxima and minima of each function. Use the second-derivative test when it applies.

(A) $f(x) = x^3 - 6x^2 + 9x + 1$ (B) $f(x) = (1/6)x^6 - 4x^5 + 25x^4$

Solutions (A) Take first and second derivatives and find critical values:

$$f(x) = x^3 - 6x^2 + 9x + 1$$
$$f'(x) = 3x^2 - 12x + 9 = 3(x - 1)(x - 3)$$
$$f''(x) = 6x - 12 = 6(x - 2)$$

Critical values are $x = 1, 3$.

$f''(1) = -6 < 0$ Therefore, $f(1)$ is a local maximum.

$f''(3) = 6 > 0$ Therefore, $f(3)$ is a local minimum.

(B) $f(x) = (1/6)x^6 - 4x^5 + 25x^4$
$$f'(x) = x^5 - 20x^4 + 100x^3 = x^3(x - 10)^2$$
$$f''(x) = 5x^4 - 80x^3 + 300x^2$$

Critical values are $x = 0$ and $x = 10$.

$f''(0) = 0$ The second-derivative test fails at both critical

$f''(10) = 0$ values, so the first-derivative test must be used.

Sign chart for $f'(x) = x^3(x - 10)^2$

| Sign of x^3 | − − − | + + + | + + + |

| Sign of $(x - 10)^2$ | + + + | + + + | + + + |

| | 0 | 10 | → x |

| $f'(x)$ | − − − 0 | + + + 0 | + + + |

| $f(x)$ | $\searrow$ | $\nearrow$ | $\nearrow$ |

Therefore, $f(0)$ is a local minimum and $f(10)$ is neither a local maximum nor minimum.

Problem 16   Find the local maxima and minima of each function. Use the second-derivative test when it applies.

(A) $f(x) = x^3 - 9x^2 + 24x - 10$ (B) $f(x) = 10x^6 - 24x^5 + 15x^4$

A common error is to assume that $f''(c) = 0$ implies that $f(c)$ is not a local extreme point. As Example 16B illustrates, if $f''(c) = 0$, then $f(c)$ may or may not be a local extreme point. The first-derivative test *must* be used whenever $f''(c) = 0$ [or $f''(c)$ does not exist].

■ Application

Example 17
Maximum Rate of Change

Using past records, a company estimates that it will sell $N(x)$ units of a product after spending \$$x$ thousand on advertising, as given by

$$N(x) = 2,000 - 2x^3 + 60x^2 - 450x \qquad 5 \leqslant x \leqslant 15$$

When is the rate of change of sales per unit (thousand dollars) change in advertising increasing? Decreasing? What is the maximum rate of change? Graph N and N' on the same axes and interpret.

Solution   The rate of change of sales per unit (thousand dollars) change in advertising expenditure is

$$N'(x) = -6x^2 + 120x - 450$$

We are interested in determining when $N'(x)$ is increasing and decreasing. This information can be obtained by examining the sign of $N''(x)$, the derivative of $N'(x)$:

$$N''(x) = -12x + 120 = 12(10 - x)$$

Since $N''(x) > 0$ for $5 < x < 10$ and $N''(x) < 0$ for $10 < x < 15$, $N'(x)$ is increasing on $(5, 10)$ and decreasing on $(10, 15)$. An examination of the graph of $N'(x)$ shows that the maximum rate of change is $N'(10) = 150$. (Refer to the figure at the top of the next page.) Graphing $N(x)$ on the same axes shows that the graph of $N(x)$ has an inflection point at $x = 10$. This point is often referred to as the **point of diminishing returns** since the rate of change of the number of units sold begins to decrease at this point.

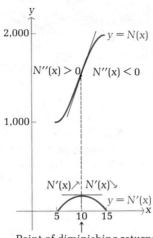

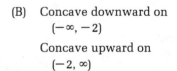

Problem 17   Repeat Example 17 for

$$N(x) = 5{,}000 - x^3 + 60x^2 - 900x \qquad 10 \leqslant x \leqslant 30$$

Answers to   13. (A)   Concave upward on   (B)   Concave downward on
Matched Problems $(-\infty, 0)$ $(-\infty, -2)$

Concave downward on   Concave upward on
$(0, \infty)$ $(-2, \infty)$

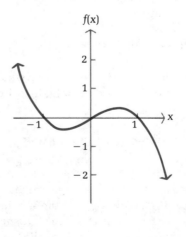

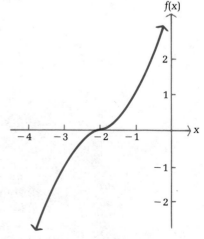

14. (A) Concave upward on $(-\infty, -2)$ and $(0, \infty)$;
    concave downward on $(-2, 0)$

    (B) Inflection points at $x = -2$ and $x = 0$

    (C)

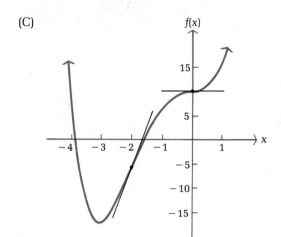

15. (A) No inflection point   (B) No inflection point

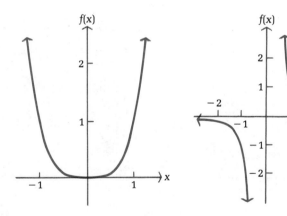

16. (A) $f(2)$ is a local maximum;
    $f(4)$ is a local minimum

    (B) $f(0)$ is a local minimum;
    $f(1)$ is neither a local maximum nor a local minimum

17. $N'(x)$ is increasing on $(10, 20)$, decreasing on $(20, 30)$; maximum rate of change is $N'(20) = 300$; $x = 20$ is point of diminishing returns

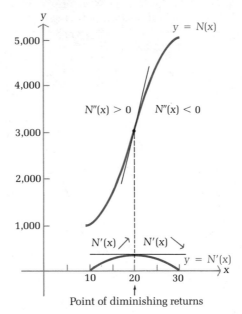

Exercise 12-3

A   *Problems 1–4 refer to the following graph of $y = f(x)$:*

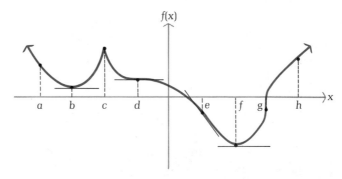

1. Identify intervals over which the graph of f is concave upward.
2. Identify intervals over which the graph of f is concave downward.
3. Identify inflection points.
4. Identify local extrema.

In Problems 5–6 replace question marks in the tables with "Local maximum," "Local minimum," "Neither," or "Test fails," as appropriate. As-

sume f is continuous over (m, n) unless otherwise stated. (Sketching pictures may help you decide.)

5.	$f'(c)$	$f''(c)$	$f(c)$
(A)	0	+	?
(B)	1	−	?
(C)	0	0	?

6.	$f'(c)$	$f''(c)$	$f(c)$
(A)	0	−	?
(B)	−1	+	?
(C)	−1	0	?

In Problems 7–10 use the given information to make a rough sketch of the graph of $y = f(x)$. Assume that f is continuous on $(-\infty, \infty)$.

7.

x	−4	−2	−1	0	2	4
$f(x)$	0	3	1.5	0	−1	−3

$f'(x)$: + + + 0 − − − 0 − − − (sign changes at −2 and 2)

$f''(x)$: − − − 0 + + + 0 − − − (sign changes at −1 and 2)

8.

x	−4	−2	−1	0	2	4
$f(x)$	0	−2	−1	0	1	3

$f'(x)$: − − − 0 + + + 0 + + + (−2 and 2)

$f''(x)$: + + + 0 − − − 0 + + + (−1 and 2)

9.

x	−3	0	1	2	4	5
$f(x)$	−4	0	2	1	−1	0

$f'(x)$: + + + ND + + + 0 − − − 0 + + + (0, 1, 4)

$f''(x)$: + + + ND − − − 0 + + + (0, 2)

10.

x	−4	−2	0	2	4	6
$f(x)$	0	3	0	−2	0	3

$f'(x)$: + + + 0 − − − ND − − − 0 + + + (−2, 0, 2)

$f''(x)$: − − − ND + + + 0 − − − (0, 4)

B   *Find all local maxima and minima using the second-derivative test whenever it applies (do not graph). If the second-derivative test fails, use the first-derivative test.*

11. $f(x) = 2x^2 - 8x + 6$

12. $f(x) = 6x - x^2 + 4$

13. $f(x) = 2x^3 - 3x^2 - 12x - 5$

14. $f(x) = 2x^3 + 3x^2 - 12x - 1$

15. $f(x) = 3 - x^3 + 3x^2 - 3x$

16. $f(x) = x^3 + 6x^2 + 12x + 2$

17. $f(x) = x^4 - 8x^2 + 10$

18. $f(x) = x^4 - 18x^2 + 50$

19. $f(x) = x^6 + 3x^4 + 2$

20. $f(x) = 4 - x^6 - 6x^4$

21. $f(x) = x + \dfrac{16}{x}$

22. $f(x) = x + \dfrac{25}{x}$

Find local maxima, local minima, and inflection points. Sketch the graph of each function. Include tangent lines at each local extreme point and inflection point.

23. $f(x) = x^3 - 6x^2 + 16$

24. $f(x) = x^3 - 9x^2 + 15x + 10$

25. $f(x) = x^3 + x + 2$

26. $f(x) = x^{1/3} + x + 2$

27. $f(x) = x^4 - 6x^2 + 7$

28. $f(x) = x^4 + 2x^2 - 3$

29. $f(x) = x^4 - 8x^3 + 18x^2 - 10$

30. $f(x) = x^4 - 4x^3$

31. $f(x) = (x + 3)\sqrt{x}$

32. $f(x) = 2x\sqrt{x - 3}$

33. $f(x) = 7x^{1/3} - x^{7/3}$

34. $f(x) = 7x^{4/3} - x^{7/3}$

C   *Find local maxima, local minima, inflection points, and asymptotes. Sketch the graph of each function. Include tangent lines at each local extreme point and inflection point.*

35. $f(x) = x - \dfrac{1}{x^3}$

36. $f(x) = 3x + \dfrac{1}{x^3}$

37. $f(x) = \dfrac{1}{1 + x^2}$

38. $f(x) = \dfrac{x^2}{1 + x^2}$

39. Given $f(x) = ax^2 + bx + c, \quad a \neq 0$.

   (A) When will $f(x)$ have a local maximum? What is the local maximum?

   (B) When will $f(x)$ have a local minimum? What is the local minimum?

40. Find the inflection points of each function.

   (A) $f(x) = ax^2 + bx + c, \quad a \neq 0$

   (B) $f(x) = ax^3 + bx^2 + cx + d, \quad a \neq 0$

■ ■

Applications

Business & Economics

41. *Long-run average cost.* The cost function of producing x units of a product is given by

$$C(x) = 260x - 10x^2 + .1x^3 \qquad x \geqslant 0$$

(A) Find the local extrema for the *average* cost function.

(B) Determine the intervals over which the graph of the average cost function is concave upward. Concave downward.

42. *Advertising.* A company estimates that it will sell N(x) units of a product after spending $x thousand on advertising, as given by

$$N(x) = -2x^3 + 90x^2 - 750x + 2{,}000 \qquad 5 \leqslant x \leqslant 25$$

(A) When is the rate of change of sales $N'(x)$ increasing? Decreasing?
(B) Find the inflection points for $N(x)$.
(C) Graph $N(x)$ and $N'(x)$ on the same axes.
(D) What is the maximum rate of change of sales?

Life Sciences

43. *Population growth — bacteria.* A drug that stimulates reproduction is introduced into a colony of bacteria. After t minutes, the number of bacteria is given approximately by

$$N(t) = 1{,}000 + 30t^2 - t^3 \qquad 0 \leqslant t \leqslant 20$$

(A) When is the rate of growth $N'(t)$ increasing? Decreasing?
(B) Find the inflection points for $N(t)$.
(C) Sketch the graph of $N(t)$ and $N'(t)$ on the same axes.
(D) What is the maximum rate of growth?

44. *Drug sensitivity.* One hour after x milligrams of a particular drug are given to a person, the change in body temperature $T(x)$ in degrees Fahrenheit is given by

$$T(x) = x^2 \left(1 - \frac{x}{9}\right) \qquad 0 \leqslant x \leqslant 6$$

The rate at which T changes with respect to the size of the dosage x, $T'(x)$, is called the *sensitivity* of the body to the dosage.

(A) When is $T'(x)$ increasing? Decreasing?
(B) Where does $T(x)$ have inflection points?
(C) Sketch the graph of $T(x)$ and $T'(x)$ on the same axes.
(D) What is the maximum value of $T'(x)$?

Social Sciences

45. *Learning.* The time T in minutes it takes a person to learn a list of length n is

$$T(n) = \frac{2}{25} n^3 - \frac{6}{5} n^2 + 6n \qquad 0 \leqslant n$$

(A) When is the rate of change of T with respect to the length of the list increasing? Decreasing?
(B) Where does the graph of T have inflection points? Graph T and T' on the same axes.
(C) What is the minimum value of $T'(n)$?

12-4   Curve Sketching

- A Graphing Strategy
- Using the Strategy

In this section we will apply, in a systematic way, all the graphing concepts discussed in the previous three sections as well as those discussed in Section 5-3. Before considering specific examples, we will outline a graphing strategy that you should find helpful in graphing many functions.

■ A Graphing Strategy

We now have powerful tools to determine the shape of a graph of a function, even before we plot any points. We can accurately sketch the graphs of many functions using these tools and point-by-point plotting as necessary (often, very little point-by-point plotting is necessary). We organize these tools in the graphing strategy summarized in the box on page 723.

■ Using the Strategy

Several examples will illustrate the use of the graphing strategy.

Example 18   Sketch the graph of $f(x) = 6x^{1/2} - x^{3/2} + 2$ using the graphing strategy.

Solution   First, notice that $f(x)$ is defined only for $x \geq 0$.

Step 1. $f'(x) = 3x^{-1/2} - \dfrac{3}{2} x^{1/2}$

$$= 3\left(\frac{1}{x^{1/2}} \cdot \frac{2}{2} - \frac{x^{1/2}}{2} \cdot \frac{x^{1/2}}{x^{1/2}}\right)$$

$$= \frac{3(2 - x)}{2x^{1/2}}$$

Sign chart for $f'(x) = 3(2 - x)/(2x^{1/2})$

A Graphing Strategy

[*Omit any of the following steps if procedures involved are too difficult or impossible (what may seem too difficult now, with a little practice, will become less so).*]

Step 1. **Use the first derivative.** Construct a sign chart for $f'(x)$, determine the intervals where $f(x)$ is increasing and decreasing, and find local maxima and minima.

Step 2. **Use the second derivative.** Construct a sign chart for $f''(x)$, determine the intervals where the graph of f is concave upward and downward, and find any inflection points.

Step 3. **Find horizontal and vertical asymptotes.** Find any horizontal asymptotes by calculating $\lim_{x\to\pm\infty} f(x)$. Find any vertical asymptotes by evaluating $\lim_{x\to c^+} f(x)$ and $\lim_{x\to c^-} f(x)$ at any value c where f is not defined.

Step 4. **Find intercepts.** Find the y intercept by evaluating $f(0)$, if it exists. Find x intercepts by solving the equation $f(x) = 0$ for x, if possible. This equation may be too difficult to solve and the x intercepts are omitted.

Step 5. **Determine symmetry.** The graph of f is symmetric with respect to the vertical axis if f is even — that is, if $f(x) = f(-x)$ for all x in the domain of f. The graph of f is symmetric with respect to the origin if f is odd — that is, if $f(-x) = -f(x)$ for all x in the domain of f.

Step 6. **Sketch the graph of f.** Draw asymptotes and locate intercepts, local maxima and minima, and inflection points. Sketch in what you know from steps 1–5. Use point-by-point plotting to complete the graph in regions of uncertainty.

Thus, $f(x)$ is increasing on $(0, 2)$ and decreasing on $(2, \infty)$. There is a local maximum at $x = 2$.

Step 2. $$f''(x) = -\frac{3}{2} x^{-3/2} - \frac{3}{4} x^{-1/2}$$

$$= 3\left(-\frac{1}{2x^{3/2}} \cdot \frac{2}{2} - \frac{1}{4x^{1/2}} \cdot \frac{x}{x}\right)$$

$$= \frac{-3(2 + x)}{4x^{3/2}}$$

Since $2 + x > 0$ and $x^{3/2} > 0$ for $x > 0$, $f''(x) < 0$ for $x > 0$. Thus, the graph of f is concave downward on $(0, \infty)$ and there are no inflection points.

Step 3.   There are no asymptotes.

Step 4. $f(0) = 2$ is the y intercept. Since $6x^{1/2} - x^{3/2} + 2 = 0$ cannot be solved easily, we will not find the x intercepts.

Step 5.   Symmetry with respect to the vertical axis or the origin is impossible since f is defined only for $x \geqslant 0$.

Step 6.

x	$f(x)$
0	2
1	7
2	7.7
4	6
9	−7

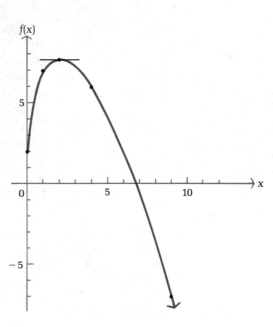

Problem 18   Sketch the graph of $f(x) = x^{3/2} - 3x^{1/2} - 1$ using the graphing strategy.

Example 19   Sketch the graph of $f(x) = (x^2 - 1)(x^2 - 7)$ using the graphing strategy.

Solution   Step 1.

$$f(x) = (x^2 - 1)(x^2 - 7) = x^4 - 8x^2 + 7$$

$$f'(x) = 4x^3 - 16x = 4x(x + 2)(x - 2)$$

Sign chart for $f'(x) = 4x(x + 2)(x - 2)$

Sign of $4x$ − − − | − − − | + + + | + + +

Sign of $(x + 2)$ − − − | + + + | + + + | + + +

Sign of $(x - 2)$ − − − | − − − | − − − | + + +

_____−2_____0_____2_____→ x

$f'(x)$ − − − 0 + + + 0 − − − 0 + + +

$f(x)$ ↘   ↗   ↘   ↗

$f(x)$ is increasing on $(-2, 0)$ and $(2, \infty)$, and decreasing on $(-\infty, -2)$ and $(0, 2)$. $f(0)$ is a local maximum. $f(-2)$ and $f(2)$ are local minima.

Step 2. $f''(x) = 12x^2 - 16 = 12\left(x - \dfrac{2\sqrt{3}}{3}\right)\left(x + \dfrac{2\sqrt{3}}{3}\right)$

Sign chart for $f''(x) = 12 (x - 2\sqrt{3}/3) (x + 2\sqrt{3}/3)$

Sign of $12 (x - 2\sqrt{3}/3)$ –   –   – $|$ –   –   – $|$ +   +   +

Sign of $(x + 2\sqrt{3}/3)$ –   –   – $|$ +   +   + $|$ +   +   +

$$\text{––} \quad -2\sqrt{3}/3 \qquad 2\sqrt{3}/3 \quad \longrightarrow x$$

$f''(x)$ +   +   +   0   –   –   –   0   +   +   +

$f(x)$ CU   CD   CU

$f(x)$ is concave upward on $(-\infty, -2\sqrt{3}/3)$ and $(2\sqrt{3}/3, \infty)$, and concave downward on $(-2\sqrt{3}/3, 2\sqrt{3}/3)$. f has inflection points at $x = -2\sqrt{3}/3$ and $x = 2\sqrt{3}/3$. Since the signs of both f' and f'' are related to the shape of the graph of f, it is helpful to combine the information from the two sign charts:

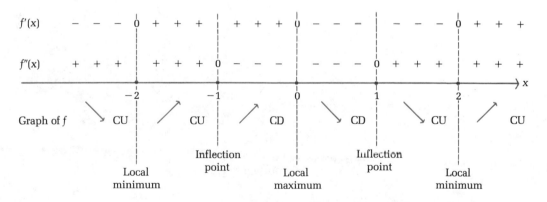

Step 3.   Since $f(x)$ is a polynomial, there are no horizontal or vertical asymptotes.

Step 4. $f(0) = 7$ is the y intercept.

$$f(x) = (x^2 - 1)(x^2 - 7)$$
$$= (x - 1)(x + 1)(x - \sqrt{7})(x + \sqrt{7})$$

The x intercepts are $x = -\sqrt{7}$, $x = -1$, $x = 1$, and $x = \sqrt{7}$.

Step 5. $f(-x) = (-x)^4 - 8(-x)^2 + 7$

$\qquad\qquad = x^4 - 8x^2 + 7$

$\qquad\qquad = f(x)$

Thus, f is an even function and its graph will be symmetric with respect to the y axis.

Step 6.

x	$f(x)$
$-\sqrt{7} \approx -2.6$	0
-2	-9
$-2\sqrt{3}/3 \approx -1.2$	-1.9
-1	0
0	7
1	0
$2\sqrt{3}/3 \approx 1.2$	-1.9
2	-9
$\sqrt{7} \approx 2.6$	0

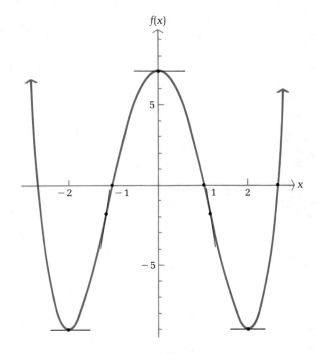

Problem 19   Sketch a graph of

$$f(x) = \frac{3}{5}x^5 - 4x^3$$

using the graphing strategy.

Example 20   Sketch a graph of

$$f(x) = \frac{x^2 - 1}{x^3}$$

using the graphing strategy.

Solution   Step 1. $f(x) = \dfrac{x^2 - 1}{x^3} = \dfrac{1}{x} - \dfrac{1}{x^3}$

$\qquad\qquad\qquad f'(x) = -\dfrac{1}{x^2} + \dfrac{3}{x^4} = \dfrac{3 - x^2}{x^4} = \dfrac{(\sqrt{3} - x)(\sqrt{3} + x)}{x^4}$

Sign chart for $f'(x) = (\sqrt{3} - x)(\sqrt{3} + x)/x^4$

Sign of $(\sqrt{3} - x)$ $+$ $+$ $+$ | $+$ $+$ $+$ | $+$ $+$ $+$ | $-$ $-$ $-$

Sign of $(\sqrt{3} + x)$ $-$ $-$ $-$ | $+$ $+$ $+$ | $+$ $+$ $+$ | $+$ $+$ $+$

Sign of x^4 $+$ $+$ $+$ | $+$ $+$ $+$ | $+$ $+$ $+$ | $+$ $+$ $+$

$$\quad\quad\quad\quad -\sqrt{3} \quad\quad\quad 0 \quad\quad\quad \sqrt{3} \quad\quad\quad\quad\quad\rightarrow x$$

$f'(x)$ $-$ $-$ $-$ 0 $+$ $+$ $+$ ND $+$ $+$ $+$ 0 $-$ $-$ $-$

$f(x)$ ↘   ↗   ND   ↗   ↘

Thus, $f(x)$ is increasing on $(-\sqrt{3}, 0)$ and $(0, \sqrt{3})$, and decreasing on $(-\infty, -\sqrt{3})$ and $(\sqrt{3}, \infty)$. The function has a local minimum at $x = -\sqrt{3}$ and a local maximum at $x = \sqrt{3}$.

Step 2. $f''(x) = \dfrac{2}{x^3} - \dfrac{12}{x^5} = \dfrac{2x^2 - 12}{x^5} = \dfrac{2(x + \sqrt{6})(x - \sqrt{6})}{x^5}$

Sign chart for $f''(x) = 2(x + \sqrt{6})(x - \sqrt{6})/x^5$

Sign of $2(x + \sqrt{6})$ $-$ $-$ $-$ | $+$ $+$ $+$ | $+$ $+$ $+$ | $+$ $+$ $+$

Sign of $(x - \sqrt{6})$ $-$ $-$ $-$ | $-$ $-$ $-$ | $-$ $-$ $-$ | $+$ $+$ $+$

Sign of x^5 $-$ $-$ $-$ | $-$ $-$ $-$ | $+$ $+$ $+$ | $+$ $+$ $+$

$$\quad\quad\quad\quad -\sqrt{6} \quad\quad\quad 0 \quad\quad\quad \sqrt{6} \quad\quad\quad\quad\quad\rightarrow x$$

$f''(x)$ $-$ $-$ $-$ 0 $+$ $+$ $+$ ND $-$ $-$ $-$ 0 $+$ $+$ $+$

$f(x)$ CD   CU   ND   CD   CU

Thus, the graph of f is concave upward on $(-\sqrt{6}, 0)$ and $(\sqrt{6}, \infty)$, and concave downward on $(-\infty, -\sqrt{6})$ and $(0, \sqrt{6})$. There are inflection points at $x = -\sqrt{6}$ and $x = \sqrt{6}$. The combined sign chart is shown at the top of the next page.

Step 3. $\displaystyle\lim_{x \to \pm\infty} f(x) = \lim_{x \to \pm\infty} \left(\frac{1}{x} - \frac{1}{x^3}\right) = 0$

The line $y = 0$ (the x axis) is a horizontal asymptote.

$$\lim_{x \to 0^-} \frac{x^2 - 1}{x^3} = +\infty \qquad \begin{array}{l} x^2 - 1 < 0 \text{ and } x^3 < 0 \text{ for } x \text{ close to and} \\ \text{on the left of } 0. \end{array}$$

$$\lim_{x \to 0^+} \frac{x^2 - 1}{x^3} = -\infty \qquad \begin{array}{l} x^2 - 1 < 0 \text{ and } x^3 > 0 \text{ for } x \text{ close to and} \\ \text{on the right of } 0. \end{array}$$

The line $x = 0$ (the y axis) is a vertical asymptote.

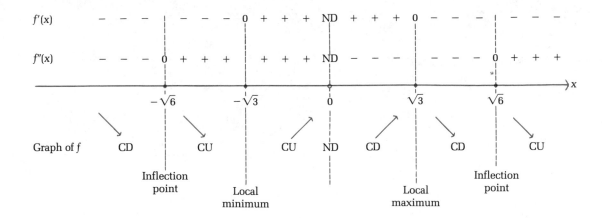

Step 4. Since $f(0)$ is not defined, there is no y intercept. Since $f(1) = 0$ and $f(-1) = 0$, the x intercepts are $x = -1$ and $x = 1$.

Step 5. $f(-x) = \dfrac{(-x)^2 - 1}{(-x)^3} = \dfrac{x^2 - 1}{-x^3} = -\dfrac{x^2 - 1}{x^3} = -f(x)$

Thus, f is an odd function and the graph of f is symmetric with respect to the origin.

Step 6.

x	$f(x)$
$-\sqrt{6} \approx -2.4$	$-.34$
$-\sqrt{3} \approx -1.7$	$-.38$
-1	0
1	0
$\sqrt{3} \approx 1.7$	$.38$
$\sqrt{6} \approx 2.4$	$.34$

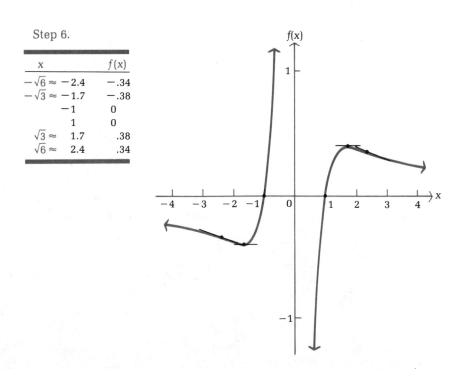

Problem 20   Sketch a graph of

$$f(x) = \frac{x^2 - 1}{x^4}$$

using the graphing strategy.

Answers to   18. Increasing on $(1, \infty)$
Matched Problems   Decreasing on $(0, 1)$

Local minimum at $x = 1$

Concave upward on $(0, \infty)$

y intercept: $f(0) = -1$

x	f(x)
0	-1
1	-3
4	1

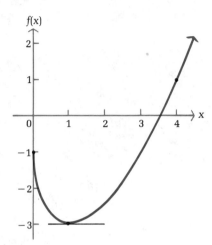

19. Increasing on $(-\infty, -2)$ and $(2, \infty)$, decreasing on $(-2, 0)$ and $(0, 2)$

Local maximum at $x = -2$ and local minimum at $x = 2$

Concave upward on $(-\sqrt{2}, 0)$ and $(\sqrt{2}, \infty)$, concave downward on $(-\infty, -\sqrt{2})$ and $(0, \sqrt{2})$

Inflection points at $x = -\sqrt{2}$, $x = 0$, and $x = \sqrt{2}$

$f(0) = 0$, $f(-2\sqrt{15}/3) = 0$, $f(2\sqrt{15}/3) = 0$

$f(-x) = -f(x)$; symmetry with respect to the origin

x	f(x)
$-\frac{2}{3}\sqrt{15} \approx -2.6$	0
-2.0	12.8
$-\sqrt{2} \approx -1.4$	7.9
0	0
$\sqrt{2} \approx 1.4$	-7.9
2.0	-12.8
$\frac{2}{3}\sqrt{15} \approx 2.6$	0

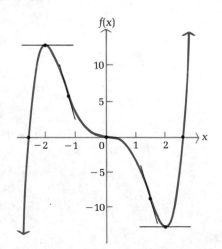

20. Increasing on $(-\infty, -\sqrt{2})$ and $(0, \sqrt{2})$; decreasing on $(-\sqrt{2}, 0)$ and $(\sqrt{2}, \infty)$

Local maxima at $x = -\sqrt{2}$ and $x = \sqrt{2}$

Concave upward on $(-\infty, -\sqrt{30}/3)$ and $(\sqrt{30}/3, \infty)$; concave downward on $(-\sqrt{30}/3, 0)$ and $(0, \sqrt{30}/3)$

Inflection points at $x = -\sqrt{30}/3$ and $x = \sqrt{30}/3$

Asymptotes: $\lim_{x \to \pm\infty} f(x) = 0$; horizontal asymptote at $y = 0$

$\lim_{x \to 0} f(x) = -\infty$; vertical asymptote at $x = 0$

$f(-1) = 0$ and $f(1) = 0$

$f(-x) = f(x)$; symmetry with respect to the y axis

x	$f(x)$
$-\sqrt{30}/3 \approx -1.8$	.21
$-\sqrt{2} \approx -1.4$	.25
-1	0
1	0
$\sqrt{2} \approx 1.4$	.25
$\sqrt{30}/3 \approx 1.8$	.21

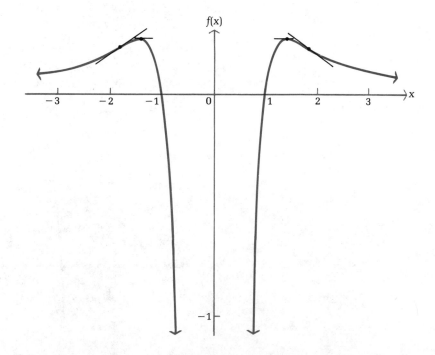

Exercise 12-4

A   *Problems 1–12 refer to the following graph of* $y = f(x)$:

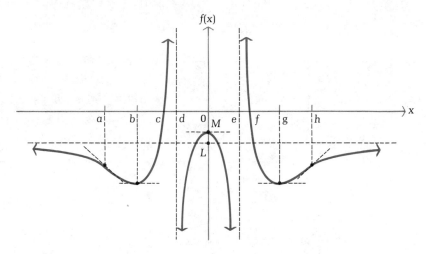

1. Identify the intervals over which $f(x)$ is increasing.
2. Identify the intervals over which $f(x)$ is decreasing.
3. Identify the points where $f(x)$ has a local maximum.
4. Identify the points where $f(x)$ has a local minimum.
5. Identify the intervals over which the graph of f is concave upward.
6. Identify the intervals over which the graph of f is concave downward.
7. Identify the inflection points.
8. Identify the horizontal asymptotes.
9. Identify the vertical asymptotes.
10. Identify the x intercepts.
11. Identify the y intercepts.
12. What type of symmetry does the graph exhibit?

B   *Use the given information to sketch a rough graph of* f.

13.

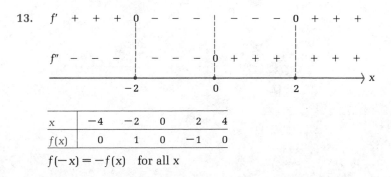

x	−4	−2	0	2	4
f(x)	0	1	0	−1	0

$f(-x) = -f(x)$ for all x

14.

```
f'   +   +   +   0   −   −   −   |   −   −   −   0   +   +   +   |   +   +   +   0   −   −   −
```

```
f''  −   −   −   |   −   −   −   0   +   +   +   |   +   +   +   0   −   −   −   |   −   −   −
```

$$\xrightarrow{\hspace{1cm}} x$$

$-2 \qquad -1 \qquad 0 \qquad 1 \qquad 2$

x	−4	−2	−1	0	1	2	4
f(x)	0	3	2	1	2	3	0

$f(-x) = f(x)$ for all x

15.

```
f'   −   −   −   |   −   −   −   0   +   +   +   ND   −   −   −   0   +   +   +   |   +   +   +
```

```
f''  −   −   −   0   +   +   +   |   +   +   +   ND   +   +   +   |   +   +   +   0   −   −   −
```

$$\xrightarrow{\hspace{1cm}} x$$

$-4 \qquad -2 \qquad 0 \qquad 2 \qquad 4$

x	−4	−2	0	2	4
f(x)	0	−2	0	−2	0

$f(-x) = f(x)$ for all x;
$\lim_{x \to \pm\infty} f(x) = 2$

16.

```
f'   +   +   +   |   +   +   +   0   −   −   −   ND   −   −   −   0   +   +   +   |   +   +   +
```

```
f''  +   +   +   0   −   −   −   |   −   −   −   ND   +   +   +   |   +   +   +   0   −   −   −
```

$$\xrightarrow{\hspace{1cm}} x$$

$-2 \qquad -1 \qquad 0 \qquad 1 \qquad 2$

x	−2	−1	0	1	2
f(x)	0	2	0	−2	0

$f(-x) = -f(x)$ for all x;
$\lim_{x \to -\infty} f(x) = -3$;
$\lim_{x \to \infty} f(x) = 3$

17.

```
f'   +   +   +   ND   +   +   +   0   −   −   −   |   −   −   −
```

```
f''  +   +   +   ND   −   −   −   |   −   −   −   0   +   +   +
```

$$\xrightarrow{\hspace{1cm}} x$$

$-2 \qquad 4 \qquad 6$

x	−4	0	4	6
f(x)	0	0	3	2

$\lim_{x \to -2^-} f(x) = \infty$;
$\lim_{x \to -2^+} f(x) = -\infty$;
$\lim_{x \to \infty} f(x) = 1$

18.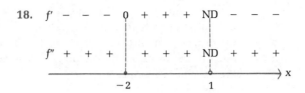

x	-4	-2	0	2
$f(x)$	0	-2	0	0

$\lim_{x \to 1^-} f(x) = \infty$;
$\lim_{x \to 1^+} f(x) = \infty$;
$\lim_{x \to \infty} f(x) = -2$

Sketch a graph of $y = f(x)$ using the graphing strategy.

19. $f(x) = x^2 - 6x + 5$ **20.** $f(x) = 3 + 2x - x^2$

21. $f(x) = x^3 - x$ **22.** $f(x) = x^3 + x$

23. $f(x) = (x^2 - 4)^2$ **24.** $f(x) = (x^2 - 1)(x^2 - 5)$

25. $f(x) = 2x^6 - 3x^5$ **26.** $f(x) = 3x^5 - 5x^4$

27. $f(x) = x - 4\sqrt{x}$ **28.** $f(x) = 3x - 2x^{3/2}$

29. $f(x) = x - 3x^{1/3}$ **30.** $f(x) = x - 3x^{2/3}$

C   **31.** $f(x) = \dfrac{x}{x - 2}$ **32.** $f(x) = \dfrac{2 + x}{3 - x}$

     33. $f(x) = \dfrac{x - 1}{x^2}$ **34.** $f(x) = \dfrac{x^2}{(x + 2)^2}$

     35. $f(x) = \dfrac{x^2 - 1}{x^2 + 3}$ **36.** $f(x) = \dfrac{9 - x^2}{x^2 + 3}$

     37. $f(x) = \dfrac{x}{\sqrt{x^2 + 1}}$ **38.** $f(x) = \dfrac{x}{\sqrt{x^2 - 1}}$

12-5   Optimization; Absolute Maxima and Minima

- Absolute Maxima and Minima
- Applications

We are now ready to consider one of the most important applications of the derivative, namely, the use of derivative to find the *absolute maximum* or *minimum* value of a function. As we mentioned earlier, an economist may be interested in the price or production level of a commodity that will bring a maximum profit; a doctor may be interested in the time it takes for a drug to reach its maximum concentration in the bloodstream after an injection; and a city planner might be interested in the location of heavy industry in a city to produce minimum pollution in residential and business areas. Before we launch an attack on problems of this type, we have to say a few words about the procedures needed to find absolute maximum and absolute minimum values of functions. We have most of the tools we need from the previous sections.

■ Absolute Maxima and Minima

First, what do we mean by *absolute maximum* and *absolute minimum*? We say that $f(c)$ is an **absolute maximum** of f if

$$f(c) \geq f(x)$$

for all x in the domain of f. Similarly, $f(c)$ is called an **absolute minimum** of f if

$$f(c) \leq f(x)$$

for all x in the domain of f. Figure 15 illustrates several typical examples.

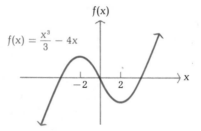

$f(x) = \dfrac{x^3}{3} - 4x$

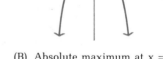

$f(x) = 4 - x^2$

(A) No absolute maximum or minimum
    One local maximum at $x = -2$
    One local minimum at $x = 2$

(B) Absolute maximum at $x = 0$
    No absolute minimum

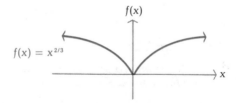

$f(x) = x^{2/3}$

(C) Absolute minimum at $x = 0$
    No absolute maximum

Figure 15

In many practical problems, the domain of a function is restricted because of practical or physical considerations. If the domain is restricted to some closed interval, as is often the case, then Theorem 4 can be proved. It is important to understand that the absolute maximum and minimum

Theorem 4	A function f continuous on a closed interval $[a, b]$ assumes both an absolute maximum and an absolute minimum on that interval.

depend on both the function f and the interval $[a, b]$ (see Figure 16). However, in all four cases illustrated in Figure 16, the absolute maximum and the absolute minimum both occur at a critical value or an end point. It can be proved that absolute extrema (if they exist) must always occur at critical values or end points. Thus, to find the absolute maximum or

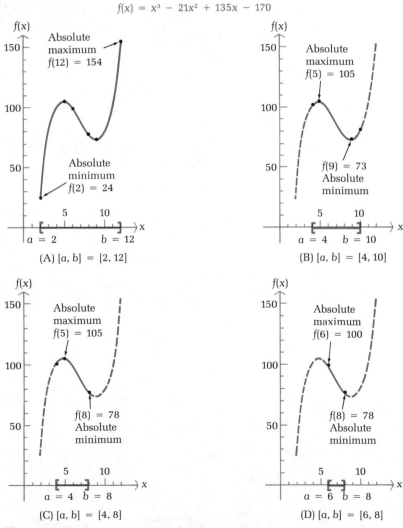

Figure 16 Absolute extrema on a closed interval

minimum value of a continuous function on a closed interval, we simply identify the end points and the critical values, evaluate each, and then choose the largest and smallest values out of this group.

Steps in Finding Absolute Maximum and Minimum Values of Continuous Functions

1. Check to make certain that f is continuous over $[a, b]$.
2. List end points and critical values: a, b, c_1, c_2, . . . , c_n.
3. Evaluate $f(a)$, $f(b)$, $f(c_1)$, $f(c_2)$, . . . , $f(c_n)$.
4. The absolute maximum $f(x)$ on $[a, b]$ is the largest of the values found in step 3.
5. The absolute minimum $f(x)$ on $[a, b]$ is the smallest of the values found in step 3.

Example 21   Find the absolute maximum and absolute minimum values of

$$f(x) = x^3 + 3x^2 - 9x - 7$$

on each of the following intervals:

(A) $[-6, 4]$ (B) $[-4, 2]$ (C) $[-2, 2]$

Solutions   (A)   The function is continuous for all values of x.

$$f'(x) = 3x^2 + 6x - 9 = 3(x - 1)(x + 3)$$

Thus, $x = -3$ and $x = 1$ are critical values. Evaluate f at the end points and critical values, -6, -3, 1, and 4, and choose the maximum and minimum from these:

$f(-6) = -61$ Absolute minimum
$f(-3) = 20$
$f(1) = -12$
$f(4) = 69$ Absolute maximum

(B)   Interval: $[-4, 2]$

x	$f(x)$	
-4	13	
-3	20	Absolute maximum
1	-12	Absolute minimum
2	-5	

(C) Interval: $[-2, 2]$

x	f(x)	
-2	15	Absolute maximum
1	-12	Absolute minimum
2	-5	

Problem 21   Find the absolute maximum and absolute minimum values of

$$f(x) = x^3 - 12x$$

on each of the following intervals:

(A) $[-5, 5]$ (B) $[-3, 3]$ (C) $[-3, 1]$

Now, suppose we want to find the absolute maximum or minimum value of a function that is continuous on an interval that is not closed. Since Theorem 4 no longer applies, we cannot be certain that the absolute maximum or minimum value exists. Figure 17 illustrates several ways that functions can fail to have absolute extrema.

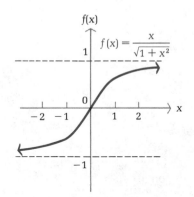

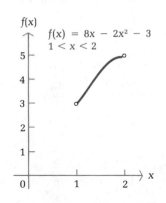

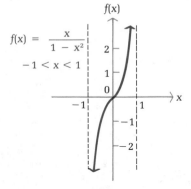

(A) No absolute extrema on $(-\infty, \infty)$:
$-1 < f(x) < 1$ for all x
$[f(x) \neq 1$ or -1 for any $x]$

(B) No absolute extrema on $(1, 2)$:
$3 < f(x) < 5$ for $x \in (1, 2)$
$[f(x) \neq 3$ or 5 for any $x \in (1, 2)]$

(C) No absolute extrema on $(-1, 1)$:
$\lim\limits_{x \to -1^+} f(x) = -\infty$ and $\lim\limits_{x \to 1^-} f(x) = \infty$

Figure 17 Functions with no absolute extrema

In general, the best procedure to follow when the interval is not a closed interval (that is, is not of the form $[a, b]$) is to sketch the graph of the function. However, there is one special case that occurs frequently in applications and that can be analyzed without drawing a graph. It often happens that f is continuous on an interval I and has only one critical value c in the interval I (here I can be any type of interval—open, closed, or half-closed). If this is the case and if $f''(c)$ exists, then we have the second-derivative test for absolute extrema given in the box on the next page.

Second-Derivative Test for Absolute Maximum and Minimum When f Is Continuous on an Interval I and c Is the Only Critical Value in I

$f'(c)$	$f''(c)$	$f(c)$	Example
0	+	Absolute minimum	
0	−	Absolute maximum	
0	0	Test fails	

Example 22   Find the absolute minimum value of

$$f(x) = x + \frac{4}{x}$$

on the interval $(0, \infty)$.

Solution $$f'(x) = 1 - \frac{4}{x^2} = \frac{x^2 - 4}{x^2} = \frac{(x - 2)(x + 2)}{x^2}$$

$$f''(x) = \frac{8}{x^3}$$

The only critical value in the interval $(0, \infty)$ is $x = 2$. Since $f''(2) = 1 > 0$, $f(2) = 4$ is the absolute minimum value of f on $(0, \infty)$.

Problem 22   Find the absolute maximum value of

$$f(x) = 12 - x - \frac{9}{x}$$

on the interval $(0, \infty)$.

■ Applications

Now we want to solve some applied problems that involve absolute extrema. Before beginning, we outline in the next box the steps to follow in solving this type of problem. The first step is the most difficult one. The techniques used to construct the model are best illustrated through a series of examples.

A Strategy for Solving Applied Optimization Problems

Step 1. Introduce variables and construct a mathematical model of the form

Maximize (or minimize) $f(x)$ on the interval I

Step 2. Find the absolute maximum (or minimum) value of $f(x)$ on the interval I and the value(s) of x where this occurs.

Step 3. Use the solution to the mathematical model to answer the questions asked in the application.

Example 23
Cost–Demand

A company manufactures and sells x transistor radios per week. If the weekly cost and demand equations are

$$C(x) = 5{,}000 + 2x$$

$$p = 10 - \frac{x}{1{,}000} \qquad 0 \leqslant x \leqslant 8{,}000$$

find for each week

(A) The maximum revenue

(B) The maximum profit, the production level that will realize the maximum profit, and the price that the company should charge for each radio

Solutions

(A) The revenue received for selling x radios at $\$p$ per radio is

$$R(x) = xp$$

$$= x\left(10 - \frac{x}{1{,}000}\right)$$

$$= 10x - \frac{x^2}{1{,}000}$$

Thus, the mathematical model is

$$\text{Maximize} \quad R(x) = 10x - \frac{x^2}{1{,}000} \qquad 0 \leqslant x \leqslant 8{,}000$$

$$R'(x) = 10 - \frac{x}{500}$$

$$10 - \frac{x}{500} = 0$$

$$x = 5{,}000 \qquad \text{Only critical value}$$

Use the second-derivative test for absolute extrema:

$$R''(x) = -\frac{1}{500} < 0 \qquad \text{for all } x$$

Thus, the maximum revenue is

$$\text{Max } R(x) = R(5,000) = \$25,000$$

(B)   Profit = Revenue − Cost

$$P(x) = R(x) - C(x)$$

$$= 10x - \frac{x^2}{1,000} - 5,000 - 2x$$

$$= 8x - \frac{x^2}{1,000} - 5,000$$

The mathematical model is

$$\text{Maximize} \quad P(x) = 8x - \frac{x^2}{1,000} - 5,000 \qquad 0 \leqslant x \leqslant 8,000$$

$$P'(x) = 8 - \frac{x}{500}$$

$$8 - \frac{x}{500} = 0$$

$$x = 4,000$$

$$P''(x) = -\frac{1}{500} < 0 \qquad \text{for all } x$$

Since $x = 4,000$ is the only critical value and $P''(x) < 0$,

$$\text{Max } P(x) = P(4,000) = \$11,000$$

Using the price–demand equation with $x = 4,000$, we find

$$p = 10 - \frac{4,000}{1,000} = \$6$$

Thus, a maximum profit of $11,000 per week is realized when 4,000 radios are produced weekly and sold for $6 each.

All the results in this example are illustrated in Figure 18. We also note that profit is maximum when

$$P'(x) = R'(x) - C'(x) = 0$$

that is, when the marginal revenue is equal to the marginal cost (the rate of increase in revenue is the same as the rate of increase in cost at the 4,000 output level—notice that the slopes of the two curves are the same at this point).

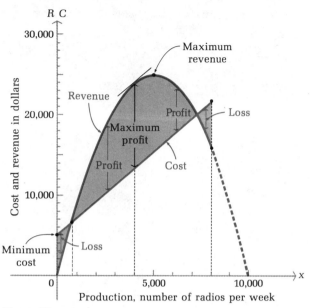

Figure 18

Problem 23   Repeat Example 23 for

$$C(x) = 90,000 + 30x$$

$$p = 300 - \frac{x}{30} \qquad 0 \leqslant x \leqslant 9,000$$

Example 24
Profit
In Example 23 the government has decided to tax the company $2 for each radio produced. Taking into account this additional cost, how many radios should the company manufacture each week in order to maximize its weekly profit? What is the maximum weekly profit? How much should it charge for the radios?

Solution   The tax of $2 per unit changes the company's cost equation:

$$C(x) = \text{Original cost} + \text{Tax}$$

$$= 5,000 + 2x + 2x$$

$$= 5,000 + 4x$$

The new profit function is

$$P(x) = R(x) - C(x)$$

$$= 10x - \frac{x^2}{1,000} - 5,000 - 4x$$

$$= 6x - \frac{x^2}{1,000} - 5,000$$

Thus, we must solve the following:

Maximize $P(x) = 6x - \dfrac{x^2}{1,000} - 5,000$ $0 \leqslant x \leqslant 8,000$

$P'(x) = 6 - \dfrac{x}{500}$

$6 - \dfrac{x}{500} = 0$

$x = 3,000$

$P''(x) = -\dfrac{1}{500} < 0$

Max $P(x) = P(3,000) = \$4,000$

Using the price–demand equation with $x = 3,000$, we find

$p = 10 - \dfrac{3,000}{1,000} = \$7$

Thus, the company's maximum profit is \$4,000 when 3,000 radios are produced and sold weekly at a price of \$7.

Even though the tax caused the company's cost to increase by \$2 per radio, the price that the company should charge to maximize its profit increases by only \$1. The company must absorb the other \$1 with a resulting decrease of \$7,000 in maximum profit.

Problem 24   Repeat Example 24 if

$C(x) = 90,000 + 30x$

$p = 300 - \dfrac{x}{30}$ $0 \leqslant x \leqslant 9,000$

and the government decides to tax the company \$20 for each unit produced. Compare the results with the results in Problem 23B.

Example 25
Maximize Yield

A walnut grower estimates from past records that if twenty trees are planted per acre, each tree will average 60 pounds of nuts per year. If for each additional tree planted per acre (up to fifteen) the average yield per tree drops 2 pounds, how many trees should be planted to maximize the yield per acre? What is the maximum yield?

Solution

Let x be the number of additional trees planted per acre. Then

$20 + x =$ Total number of trees per acre

$60 - 2x =$ Yield per tree

Yield per acre = (Total number of trees per acre)(Yield per tree)

$Y(x) = (20 + x)(60 - 2x)$

$= 1,200 + 20x - 2x^2$ $0 \leqslant x \leqslant 15$

Thus, we must solve the following:

Maximize $Y(x) = 1,200 + 20x - 2x^2$ $0 \leqslant x \leqslant 15$

$Y'(x) = 20 - 4x$

$20 - 4x = 0$

$x = 5$

$Y''(x) = -4 < 0$ for all x

Hence,

Max $Y(x) = Y(5) = 1,250$ pounds per acre

Thus, a maximum yield of 1,250 pounds of nuts per acre is realized if twenty-five trees are planted per acre.

Problem 25   Repeat Example 25 starting with thirty trees per acre and a reduction of 1 pound per tree for each additional tree planted.

Example 26   A farmer wants to construct a rectangular pen next to a barn 60 feet long,
Maximize Area   using all of the barn as part of one side of the pen. Find the dimensions of the pen with the largest area that the farmer can build if

(A)   160 feet of fencing material is available
(B)   260 feet of fencing material is available

Solutions   (A)   We begin by constructing and labeling a figure:

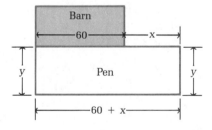

The area of the pen is

$A = (x + 60)y$

Before we can maximize the area, we must determine a relationship between x and y in order to express A as a function of one variable. In this case, x and y are related to the total amount of available fencing material:

$x + y + 60 + x + y = 160$

$2x + 2y = 100$

$y = 50 - x$

Thus,

$$A(x) = (x + 60)(50 - x)$$

Now we need to determine the permissible values of x. Since the farmer wants to use all of the barn as part of one side of the pen, x cannot be negative. Since y is the other dimension of the pen, y cannot be negative. Thus,

$$y = 50 - x \geqslant 0$$
$$50 \geqslant x$$

Thus, we must solve the following:

$$\text{Maximize} \quad A(x) = (x + 60)(50 - x) \qquad 0 \leqslant x \leqslant 50$$

$$A(x) = 3{,}000 - 10x - x^2$$
$$A'(x) = -10 - 2x$$
$$-10 - 2x = 0$$
$$x = -5$$

Since $x = -5$ is not in the interval [0, 50], there are no critical points in the interval. $A(x)$ is continuous on [0, 50], so the absolute maximum must occur at one of the end points.

$$A(0) = 3{,}000 \qquad \text{Maximum area}$$
$$A(50) = 0$$

If $x = 0$, then $y = 50$. Thus, the dimensions of the pen with largest area are 60 feet by 50 feet.

(B)   If there is 260 feet of fencing material available, then

$$x + y + x + 60 + y = 260$$
$$2x + 2y = 200$$
$$y = 100 - x$$

The model becomes

$$\text{Maximize} \quad A(x) = (x + 60)(100 - x) \qquad 0 \leqslant x \leqslant 100$$

$$A(x) = 6{,}000 + 40x - x^2$$
$$A'(x) = 40 - 2x$$
$$40 - 2x = 0$$
$$x = 20 \qquad \text{The only critical value}$$

$$A''(x) = -2 < 0$$
$$\text{Max } A(x) = A(20) = 6{,}400$$
$$y = 100 - 20 = 80$$

This time the dimensions of the pen with the largest area are 80 feet by 80 feet.

Problem 26 Repeat Example 26 if the barn is 80 feet long.

Example 27
Inventory Control

A record company anticipates that there will be a demand for 20,000 copies of a certain album during the following year. It costs the company $.50 to store a record for 1 year. Each time it must press additional records, it costs $200 to set up the equipment. How many records should the company press during each production run in order to minimize its total storage and set-up costs?

Solution

This type of problem is called an **inventory control problem.** One of the basic assumptions made in such problems is that the demand is uniform. For example, if there are 250 working days in a year, then the daily demand would be 20,000/250 = 800 records. The company could decide to produce all 20,000 records at the beginning of the year. This would certainly minimize the set-up costs, but would result in very large storage costs. At the other extreme, it could produce 800 records each day. This would minimize the storage costs, but would result in very large set-up costs. Somewhere between these two extremes is the optimal solution that will minimize the total storage and set-up costs. Let

x = Number of records pressed during each production run

y = Number of production runs

It is easy to see that the total set-up cost for the year is 200y, but what is the total storage cost? If the demand is uniform, then the number of records in storage between production runs will decrease from x to 0 and the average number in storage each day is x/2. This result is illustrated in the following figure:

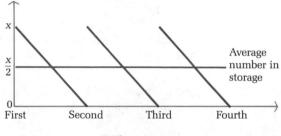

Number of records in storage

Production run

Since it costs $.50 to store one record for a year, the total storage cost is .5(x/2) = .25x and the total cost is

Total cost = Set-up cost + Storage cost

$$C = 200y + .25x$$

If the company produces x records in each of y production runs, then the total number of records produced is xy. Thus,

$$xy = 20,000$$

$$y = \frac{20,000}{x}$$

Certainly, x must be at least 1 and cannot exceed 20,000. Thus, we must solve the following:

$$\text{Minimize} \quad C(x) = 200\left(\frac{20,000}{x}\right) + .25x \qquad 1 \leqslant x \leqslant 20,000$$

$$C(x) = \frac{4,000,000}{x} + .25x$$

$$C'(x) = -\frac{4,000,000}{x^2} + .25$$

$$-\frac{4,000,000}{x^2} + .25 = 0$$

$$x^2 = \frac{4,000,000}{.25}$$

$$= 16,000,000$$

$$x = 4,000 \qquad -4,000 \text{ is not a critical value} \\ \text{since } 1 \leqslant x \leqslant 20,000$$

$$C''(x) = \frac{8,000,000}{x^3} > 0 \qquad \text{for } x \in (1, 20,000)$$

Thus,

$$\text{Min } C(x) = C(4,000) = 2,000$$

$$y = \frac{20,000}{4,000} = 5$$

The company will minimize its total cost by pressing 4,000 records five times during the year.

Problem 27   Repeat Example 27 if it costs $250 to set up a production run and $.40 to store one record for a year.

Answers to Matched Problems   21. (A) Absolute maximum: $f(5) = 65$; absolute minimum: $f(-5) = -65$
(B) Absolute maximum: $f(-2) = 16$; absolute minimum: $f(2) = -16$
(C) Absolute maximum: $f(-2) = 16$; absolute minimum: $f(1) = -11$
22. $f(3) = 6$

23. (A)   Max $R(x) = R(4,500) = \$675,000$
     (B)   Max $P(x) = P(4,050) = \$456,750$; $p = \$165$
24. Max $P(x) = P(3,750) = \$378,750$; $p = \$175$; price increases \$10, profit decreases \$78,000
25. Max $Y(x) = Y(15) = 2,025$ pounds per acre
26. (A)   80 feet by 40 feet   (B)   85 feet by 85 feet
27. Press 5,000 records four times during the year

Exercise 12-5

A   *Find the absolute maximum and absolute minimum, if either exists, for each function.*

1. $f(x) = x^2 - 4x + 5$
2. $f(x) = x^2 + 6x + 7$
3. $f(x) = 10 + 8x - x^2$
4. $f(x) = 6 - 8x - x^2$
5. $f(x) = 1 + x^{2/3}$
6. $f(x) = 2\sqrt{x} - x$

B   *Find the indicated extrema of each function.*

7. Absolute maximum value of $f(x) = 24 - 2x - 8/x$, $x > 0$
8. Absolute minimum value of $f(x) = 3x + 27/x$, $x > 0$
9. Absolute minimum value of $f(x) = 36 + x - 12x^{1/3}$, $x > 1$
10. Absolute maximum value of $f(x) = 9x^{2/3} - 2x + 3$, $x > 1$

Find the absolute maximum and minimum, if either exists, of each function on the indicated intervals.

11. $f(x) = x^3 - 6x^2 + 9x - 6$

    (A) $[-1, 5]$ (B) $[-1, 3]$ (C) $[2, 5]$

12. $f(x) = 2x^3 - 3x^2 - 12x + 24$

    (A) $[-3, 4]$ (B) $[-2, 3]$ (C) $[-2, 1]$

13. $f(x) = (x - 1)(x - 5)^3 + 1$

    (A) $[0, 3]$ (B) $[1, 7]$ (C) $[3, 6]$

14. $f(x) = x^4 - 8x^2 + 16$

    (A) $[-1, 3]$ (B) $[0, 2]$ (C) $[-3, 4]$

C   15. $f(x) = \dfrac{20x}{x^2 + 4}$

    (A) $(-\infty, \infty)$ (B) $[0, \infty)$ (C) $[1, \infty)$

16. $f(x) = \dfrac{3x^2}{x^2 + x + 1}$

(A) $(-\infty, \infty)$ (B) $[-1, \infty)$ (C) $[0, \infty)$

Preliminary Word Problems:

17. How would you divide a 10 inch line so that the product of the two lengths is a maximum?

18. What quantity should be added to 5 and subtracted from 5 in order to produce the maximum product of the results?

19. Find two numbers whose difference is 30 and whose product is a minimum.

20. Find two positive numbers whose sum is 60 and whose product is a maximum.

21. Find the dimensions of a rectangle with perimeter 100 centimeters that has maximum area. Find the maximum area.

22. Find the dimensions of a rectangle of area 225 square centimeters that has the least perimeter. What is the perimeter?

■

Applications

Business & Economics

23. *Average costs.* If the average manufacturing cost (in dollars) per pair of sunglasses is given by

$$\overline{C}(x) = x^2 - 6x + 12 \qquad 0 \leqslant x \leqslant 6$$

where x is the number (in thousands) of pairs manufactured, how many pairs of glasses should be manufactured to minimize the average cost per pair? What is the minimum average cost per pair?

24. *Maximum revenue and profit.* A company manufactures and sells x television sets per month. The monthly cost and demand equations are

$$C(x) = 72{,}000 + 60x$$

$$p = 200 - \frac{x}{30} \qquad 0 \leqslant x \leqslant 6{,}000$$

(A) Find the maximum revenue.

(B) Find the maximum profit, the production level that will realize the maximum profit, and the price the company should charge for each television set.

(C) If the government decides to tax the company \$5 for each set it produces, how many sets should the company manufacture each month in order to maximize its profit? What is the maximum profit? What should the company charge for each set?

25. *Car rental.* A car rental agency rents 100 cars per day at a rate of \$10 per day. For each \$1 increase in rate, five fewer cars are rented. At what rate should the cars be rented to produce the maximum income? What is the maximum income?

26. *Rental income.* A ninety room hotel in Las Vegas is filled to capacity every night at $25 a room. For each $1 increase in rent, three fewer rooms are rented. If each rented room costs $3 to service per day, how much should the management charge for each room to maximize gross profit? What is the maximum gross profit?

27. *Agriculture.* A commercial cherry grower estimates from past records that if thirty trees are planted per acre, each tree will yield an average of 50 pounds of cherries per season. If for each additional tree planted per acre (up to twenty) the average yield per tree is reduced by 1 pound, how many trees should be planted per acre to obtain the maximum yield per acre? What is the maximum yield?

28. *Agriculture.* A commercial pear grower must decide on the optimum time to have fruit picked and sold. If the pears are picked now, they will bring 30¢ per pound, with each tree yielding an average of 60 pounds of salable pears. If the average yield per tree increases 6 pounds per tree per week for the next 4 weeks, but the price drops 2¢ per pound per week, when should the pears be picked to realize the maximum return per tree? What is the maximum return?

29. *Manufacturing.* A candy box is to be made out of a piece of cardboard that measures 8 by 12 inches. Squares of equal size will be cut out of each corner, and then the ends and sides will be folded up to form a rectangular box. What size square should be cut from each corner to obtain a maximum volume?

30. *Packaging.* A parcel delivery service will deliver a package only if the length plus girth (distance around) does not exceed 108 inches.

    (A) Find the dimensions of a rectangular box with square ends that satisfies the delivery service's restriction and has maximum volume. What is the maximum volume?

    (B) Find the dimensions (radius and height) of a cylindrical container that meets the delivery service's requirement and has maximum volume. What is the maximum volume?

31. *Construction costs.* A fence is to be built to enclose a rectangular area of 800 square feet. The fence along three sides is to be made of material that costs $2 per foot. The material for the fourth side costs $6 per foot. Find the dimensions of the rectangle that will allow the most economical fence to be built.

32. *Construction costs.* The owner of a retail lumber store wants to construct a fence to enclose an outdoor storage area adjacent to the store as indicated in the accompanying figure. Find the dimensions that will enclose the largest area if:

    (A) 240 feet of fencing material is used.
    (B) 400 feet of fencing material is used.

33. *Inventory control.* A publishing company sells 50,000 copies of a certain book each year. It costs the company $1.00 to store a book for 1 year. Each time it must print additional copies, it costs the company $1,000 to set up the presses. How many books should the company produce during each printing in order to minimize its total storage and set-up costs?

34. *Operational costs.* The cost per hour for fuel for running a train is $v^2/4$ dollars, where v is the speed in miles per hour. (Note that the cost goes up as the square of the speed.) Other costs, including labor, are $300 per hour. How fast should the train travel on a 360 mile trip to minimize the total cost for the trip?

35. *Construction costs.* A freshwater pipeline is to be run from a source on the edge of a lake to a small resort community on an island 5 miles off-shore, as indicated in the figure.

(A)   If it costs 1.4 times as much to lay the pipe in the lake as it does on land, what should x be (in miles) to minimize the total cost of the project?

(B)   If it costs only 1.1 times as much to lay the pipe in the lake as it does on land, what should x be to minimize the total cost of the project? [*Note:*   Compare with Problem 40.]

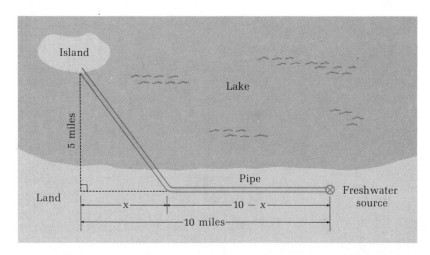

36. *Manufacturing costs.* A manufacturer wants to produce cans that will hold 12 ounces (approximately 22 cubic inches) in the form of a right circular cylinder. Find the dimensions (radius of an end and height) of the can that will use the smallest amount of material. Assume the circular ends are cut out of squares, with the corner portions wasted, and the sides are made from rectangles, with no waste.

Life Sciences   37. *Bacteria control.* A recreational swimming lake is treated periodically to control harmful bacteria growth. Suppose t days after a treatment, the concentration of bacteria per cubic centimeter is given by

$$C(t) = 30t^2 - 240t + 500 \qquad 0 \leqslant t \leqslant 8$$

How many days after a treatment will the concentration be minimal? What is the minimum concentration?

38. *Drug concentration.* The concentration $C(t)$ in milligrams per cubic centimeter of a particular drug in a patient's bloodstream is given by

$$C(t) = \frac{0.16t}{t^2 + 4t + 4}$$

where t is the number of hours after the drug is taken. How many hours after the drug is given will the concentration be maximum? What is the maximum concentration?

39. *Laboratory management.* A laboratory uses 500 white mice each year for experimental purposes. It costs $4.00 to feed a mouse for 1 year. Each time mice are ordered from a supplier, there is a service charge of $10 for processing the order. How many mice should be ordered each time in order to minimize the total cost of feeding the mice and of placing the orders for the mice?

40. *Bird flights.* Some birds tend to avoid flights over large bodies of water during daylight hours. It is speculated that more energy is required to fly over water than land because air generally rises over land and falls over water during the day. Suppose an adult bird with these tendencies is taken from its nesting area on the edge of a large lake to an island 5 miles off-shore and is then released (see the accompanying figure).

(A)   If it takes 1.4 times as much energy to fly over water as land, how far up-shore (x, in miles) should the bird head in order to minimize the total energy expended in returning to the nesting area?

(B)   If it takes only 1.1 times as much energy to fly over water as land, how far up-shore should the bird head in order to minimize the total energy expended in returning to the nesting area? [*Note:* Compare with Problem 35.]

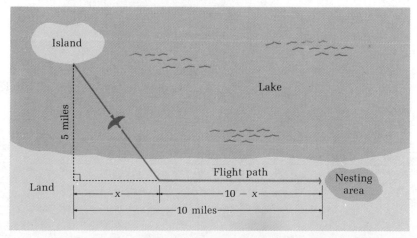

41. *Botany.* If it is known from past experiments that the height in feet of a given plant after t months is given approximately by

$$H(t) = 4t^{1/2} - 2t \qquad 0 \leqslant t \leqslant 2$$

how long, on the average, will it take a plant to reach its maximum height? What is the maximum height?

42. *Pollution.* Two heavy industrial areas are located 10 miles apart, as indicated in the figure. If the concentration of particulate matter in parts per million decreases as the reciprocal of the square of the distance from the source, and area A_1 emits eight times the particulate matter as A_2, then the concentration of particulate matter at any point between the two areas is given by

$$C(x) = \frac{8k}{x^2} + \frac{k}{(10 - x)^2} \qquad 0.5 \leqslant x \leqslant 9.5, \quad k > 0$$

How far from A_1 will the concentration of particulate matter be at a minimum?

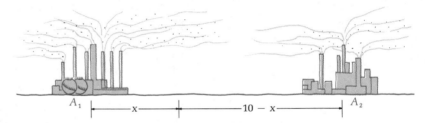

Social Sciences

43. *Politics.* In a newly incorporated city it is estimated that the voting population (in thousands) will increase according to

$$N(t) = 30 + 12t^2 - t^3 \qquad 0 \leqslant t \leqslant 8$$

where t is time in years. When will the rate of increase be most rapid?

44. *Learning.* A large grocery chain found that, on the average, a checker can memorize $P\%$ of a given price list in x continuous hours, as given approximately by

$$P(x) = 96x - 24x^2 \qquad 0 \leqslant x \leqslant 3$$

How long should a checker plan to take to memorize the maximum percentage? What is the maximum?

12-6   Elasticity of Demand (Optional)

- Price and Elasticity of Demand
- Revenue and Elasticity of Demand

■ Price and Elasticity of Demand

In this section we will study the effect that changes in price have on demand and revenue. Suppose the price $\$p$ and the demand x for a certain product are related by the *price–demand equation*:

$$x + 500p = 10,000 \tag{1}$$

In problems involving revenue, cost, and profit, it is customary to use the demand equation to express price as a function of demand. Since we are now interested in the effects that changes in price have on demand, it will be more convenient to express demand as a function of price. Solving (1) for x, we have

$$x = 10,000 - 500p \qquad \text{Demand as a function of price}$$
$$= 500(20 - p)$$

or

$$x = f(p) = 500(20 - p) \qquad 0 \leqslant p \leqslant 20 \tag{2}$$

Since x and p must be nonnegative quantities, we must restrict p so that $0 \leqslant p \leqslant 20$.

For most products, demand is assumed to be a decreasing function of price. That is, price increases result in lower demand and price decreases result in higher demand (see Figure 19). Suppose the price is changed by an

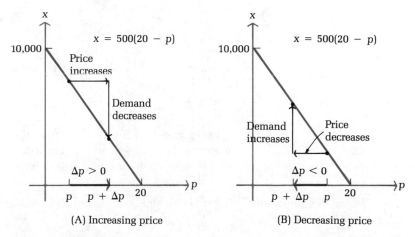

(A) Increasing price   (B) Decreasing price

Figure 19 Price and demand

amount Δp. Then the **relative change in price** and the **relative change in demand** are, respectively,

$$\frac{\Delta p}{p} \quad \text{and} \quad \frac{\Delta x}{x} = \frac{f(p + \Delta p) - f(p)}{f(p)}$$

Economists use the ratio

$$\frac{\dfrac{\Delta x}{x}}{\dfrac{\Delta p}{p}} = \frac{\text{Relative change in demand}}{\text{Relative change in price}} \tag{3}$$

to study the effect of price changes on demand. Economics texts that do not use calculus will call the expression in (3) the *elasticity of demand at price p*. However, this expression obviously depends on both p and Δp. Using calculus, we can let $\Delta p \to 0$ and obtain an expression for the point elasticity of demand at price p, denoted $E(p)$:

$$E(p) = \lim_{\Delta p \to 0} \frac{\dfrac{\Delta x}{x}}{\dfrac{\Delta p}{p}}$$

$$= \lim_{\Delta p \to 0} \frac{\dfrac{f(p + \Delta p) - f(p)}{f(p)}}{\dfrac{\Delta p}{p}}$$

$$= \lim_{\Delta p \to 0} \frac{f(p + \Delta p) - f(p)}{f(p)} \cdot \frac{p}{\Delta p}$$

$$= \frac{p}{f(p)} \lim_{\Delta p \to 0} \frac{f(p + \Delta p) - f(p)}{\Delta p}$$

$$= \frac{p}{f(p)} f'(p)$$

Thus, we define the **point elasticity of demand** to be

$$E(p) = \frac{pf'(p)}{f(p)}$$

Since p and $f(p)$ are always nonnegative quantities and $f'(p) \leq 0$ (remember, demand is assumed to be a decreasing function of price), $E(p) \leq 0$ for all values of p for which it is defined.

Example 28   If $x = f(p) = 500(20 - p)$, find $E(p)$ and calculate $E(p)$ at

(A) $p = \$4$ (B) $p = \$6$ (C) $p = \$10$

Solutions $$E(p) = \frac{pf'(p)}{f(p)}$$

$$= \frac{p(-500)}{500(20 - p)}$$

$$= \frac{-p}{20 - p} \qquad 0 \leqslant p < 20$$

(A) $E(4) = -\dfrac{4}{16} = -.25$

(B) $E(16) = -\dfrac{16}{4} = -4$

(C) $E(10) = -\dfrac{10}{10} = -1$

An economist would interpret the results in Example 28 as follows:

1. **$E(4) = -.25 > -1$.** At this price level ($p = 4$), a percentage change in price will result in a smaller percentage change in demand. For example, if the price is increased by 10%, then the demand will change by approximately

   $$-.25(10\%) = -2.5\%$$

   Since this change is negative, a 10% price increase will result in a 2.5% decrease in demand. On the other hand, a 10% price cut will result in a 2.5% increase in demand. Since the demand is not very sensitive to changes in price at this price level, we say that demand is **inelastic** when $E(p) > -1$.

2. **$E(16) = -4 < -1$.** At this price level ($p = 16$), a percentage change in price will result in a larger percentage change in demand. This time a 10% price increase will result in an approximate 40% decrease in demand, while a 10% price cut will result in an approximate 40% increase in demand. Since the demand is very sensitive to changes in price at this price level, we say that demand is **elastic** when $E(p) < -1$.

3. **$E(10) = -1$.** In this case, percentage changes in price will result in approximately equal percentage changes in demand. When $E(p) = -1$, we say that the demand has **unit elasticity.**

Problem 28 If $x = f(p) = 1{,}000(40 - p)$, find $E(p)$ and evaluate $E(p)$ at

(A) $p = \$8$ (B) $p = \$30$ (C) $p = \$20$

All the pertinent definitions are summarized in the box.

Point Elasticity of Demand

Let demand x and price p be related by the **price–demand equation**

$$x = f(p)$$

The **point elasticity of demand** is

$$E(p) = \frac{pf'(p)}{f(p)}$$

Demand is **inelastic** if $-1 < E(p) \leq 0$.

Demand is **elastic** if $E(p) < -1$.

Demand has **unit elasticity** if $E(p) = -1$.

Example 29
Price–Demand

Given $x = f(p) = 9{,}000 - 30p^2$:

(A) Determine the values of p for which demand is inelastic and the values for which it is elastic.

(B) Discuss the effect of a 10% price cut when $p = \$7$.

(C) Discuss the effect of a 10% price increase when $p = \$15$.

Solutions

(A) First, notice that

$$f(p) = 30(300 - p^2)$$
$$= 30(10\sqrt{3} - p)(10\sqrt{3} + p)$$

Since both p and $f(p)$ must be nonnegative, we must restrict p to

$$0 \leq p \leq 10\sqrt{3} \approx 17.3$$

$$E(p) = \frac{pf'(p)}{f(p)}$$

$$= \frac{p(-60p)}{9{,}000 - 30p^2}$$

$$= \frac{-2p^2}{300 - p^2} \qquad 0 \leq p < 10\sqrt{3}$$

The following observations will simplify our calculations:

Demand is inelastic: $E(p) > -1$ $E(p) + 1 > 0$

Demand is elastic: $E(p) < -1$ $E(p) + 1 < 0$

Thus, we can determine where demand is inelastic and where it is elastic by constructing a sign chart for $E(p) + 1$:

$$E(p) + 1 = \frac{-2p^2}{300 - p^2} + 1$$

$$= \frac{300 - 3p^2}{300 - p^2}$$

$$= \frac{3(10 - p)(10 + p)}{(10\sqrt{3} - p)(10\sqrt{3} + p)}$$

Sign of $3(10 - p)$ $+ \;\; + \;\; + \;$ $- \;\; - \;\; -$

Sign of $(10 + p)$ $+ \;\; + \;\; + \;$ $+ \;\; + \;\; +$

Sign of $(10\sqrt{3} - p)$ $+ \;\; + \;\; + \;$ $+ \;\; + \;\; +$

Sign of $(10\sqrt{3} + p)$ $+ \;\; + \;\; + \;$ $+ \;\; + \;\; +$

$$\quad 0 \qquad\qquad 10 \qquad\qquad 10\sqrt{3} \longrightarrow p$$

Sign of $[E(p) + 1]$ $+ \;\; + \;\; + \;\; 0$ $- \;\; - \;\; -$

Thus, demand is inelastic for $0 \leqslant p < 10$ and elastic for $10 < p < 10\sqrt{3}$.

(B) $E(7) = \dfrac{-2 \cdot 49}{300 - 49} \approx -.39$

Thus, a 10% price cut will result in a change in demand of approximately

$$-.39(-10\%) = 3.9\%$$

That is, the demand will increase approximately 3.9%.

(C) $E(15) = \dfrac{-2 \cdot 225}{300 - 225} = -6$

Thus, a 10% price increase will result in a change in demand of approximately

$$-6(10\%) = -60\%$$

That is, the demand will decrease approximately 60%.

Problem 29   Given $x = f(p) = 6{,}000 - 5p^2$:

(A)   Determine the values of p for which demand is inelastic and those for which it is elastic.

(B)   Discuss the effect of a 10% price increase when $p = \$10$.

(C)   Discuss the effect of a 10% price decrease when $p = \$25$.

■ Revenue and Elasticity of Demand

Now we want to see how revenue and elasticity of demand are related. We begin by considering an example.

Example 30
Price–Demand

Given the price–demand equation $x = f(p) = 500(20 - p), \quad 0 \leqslant p \leqslant 20$:

(A) Determine the values of p for which revenue is increasing and those for which revenue is decreasing.

(B) Determine the values of p for which demand is inelastic and those for which demand is elastic.

Solution

(A)   Revenue = (Price per unit)(Number of units)

$$R(p) = px$$
$$= 500p(20 - p)$$
$$= 10{,}000p - 500p^2 \qquad 0 \leqslant p \leqslant 20$$

$$R'(p) = 10{,}000 - 1{,}000p$$
$$= 1{,}000(10 - p)$$

The only critical value is $p = 10$. Thus,

$$R'(p) > 0 \quad \text{for } 0 < p < 10 \qquad \text{Increasing revenue}$$

and

$$R'(p) < 0 \quad \text{for } 10 < p < 20 \qquad \text{Decreasing revenue}$$

(B) $$E(p) = \frac{pf'(p)}{f(p)}$$

$$= \frac{-500p}{500(20 - p)}$$

$$= \frac{-p}{20 - p}$$

$$E(p) + 1 = -\frac{p}{20 - p} + 1$$

$$= \frac{20 - 2p}{20 - p}$$

$$= \frac{2(10 - p)}{20 - p}$$

Since the denominator is positive for $0 < p < 20$, we see that

$$E(p) + 1 > 0 \quad \text{for } 0 < p < 10 \qquad \text{Inelastic demand}$$

and

$$E(p) + 1 < 0 \quad \text{for } 10 < p < 20 \qquad \text{Elastic demand}$$

Problem 30   Repeat Example 30 for $x = f(p) = 1,000(40 - p)$, $0 \leqslant p \leqslant 40$.

Comparing the answers in Examples 30A and B, we see that revenue is increasing precisely when demand is inelastic and revenue is decreasing when demand is elastic. Is this always the case?

In general, let $x = f(p)$ be a demand function and let

$$R(p) = px = pf(p)$$

Then

$$R'(p) = pf'(p) + f(p)$$
$$= f(p)\left[\frac{pf'(p)}{f(p)} + 1\right]$$
$$= f(p)[E(p) + 1]$$

Since $x = f(p) > 0$, we conclude:

All are true or all are false	*All are true or all are false*
$R'(p) > 0$	$R'(p) < 0$
$E(p) + 1 > 0$	$E(p) + 1 < 0$
Demand is inelastic	Demand is elastic

Thus, if demand is inelastic, a price increase will increase revenue and a price cut will decrease revenue. On the other hand, if demand is elastic, then a price increase will decrease revenue and a price cut will increase revenue (see Figure 20 on the next page).

Example 31

A company can sell 4,500 pairs of sunglasses monthly when the price is $5.00. When the price of a pair of sunglasses is increased by 10%, the demand drops to 4,250 pairs a month. Assume that the demand equation is linear.

(A)   Find the point elasticity of demand at the new price level.
(B)   Approximate the change in demand if the price is increased by an additional 10%.
(C)   Will a second 10% price increase cause the revenue to increase or decrease?

Solutions   First, we must find the demand equation. Since we are given that the demand equation is linear, there must be constants a and b so that

$$x = a + bp \tag{4}$$

We know that $x = 4,500$ when $p = \$5.00$ and $x = 4,250$ when $p = 5 + (.1)5 = \$5.50$. Substituting these values into (4) produces a pair of equations

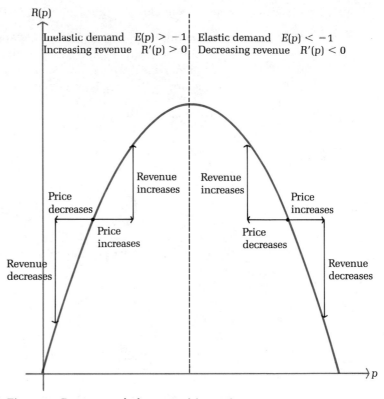

Figure 20 Revenue and elasticity of demand

that we can solve for a and b:

$4{,}500 = a + 5b$ Solve the first equation for a and substitute into
$4{,}250 = a + 5.5b$ the second equation.

$4{,}250 = (4{,}500 - 5b) + 5.5b$

$-250 = .5b$

$b = -500$ Substitute $b = -500$ into the first equation
               and solve for a.
$4{,}500 = a + 5(-500)$

$a = 7{,}000$

Thus, the demand equation is

$x = f(p) = 7{,}000 - 500p$ *Check:* $p = 5$: $x = 7{,}000 - 2{,}500 \overset{\checkmark}{=} 4{,}500$

                                            $p = 5.5$: $x = 7{,}000 - 2{,}750 \overset{\checkmark}{=} 4{,}250$

(A) $E(p) = \dfrac{pf'(p)}{f(p)}$

$= \dfrac{-500p}{7,000 - 500p}$

$= \dfrac{-p}{14 - p}$

$E(5.5) = \dfrac{-5.5}{14 - 5.5}$

$\approx -.65$ Point elasticity of demand at $p = 5.5$

(B)   At a price level of $5.50, a 10% increase in the price will result in a percentage change in demand of approximately

$E(p) \cdot 10\% \approx -.65(10\%) = -6.5\%$

Thus, the demand will decrease by

$0.065(4,250) \approx 276$

(C)   Since $E(5.5) \approx -.65 > -1$, demand is inelastic at this price level and an increase in price will increase revenue.

Problem 31   Repeat Example 31 if the demand drops from 4,500 to 3,250 when the price is increased from $5.00 to $5.50.

Answers to   28. $E(p) = -p/(40 - p)$;   (A) $E(8) = -.25$ (B) $E(30) = -3$
Matched Problems   (C) $E(20) = -1$

29.   (A)   Inelastic for $0 < p < 20$; elastic for $20 < p < 20\sqrt{3}$
      (B)   1.8% decrease in demand   (C)   22% increase in demand

30.   (A)   Revenue is increasing for $0 < p < 20$, decreasing for $20 < p < 40$
      (B)   Demand is inelastic for $0 < p < 20$, elastic for $20 < p < 40$

31.   (A) $E(5.5) \approx -4.2$ (B)   Changes by -42%; decreases by approxi-
      (C)   Revenue decreases   mately 1,365 pairs

Exercise 12-6

A   1.   Given the demand equation

$$p + \frac{1}{200}x = 30 \qquad 0 \leqslant p \leqslant 30$$

(A)   Express the demand x as a function of the price p.
(B)   Find the point elasticity of demand, $E(p)$.

(C)   What is the point elasticity of demand when $p = \$10$? If this price is increased by 10%, what is the approximate change in demand?

(D)   What is the point elasticity of demand when $p = \$25$? If this price is increased by 10%, what is the approximate change in demand?

(E)   What is the point elasticity of demand when $p = \$15$? If this price is increased by 10%, what is the approximate change in demand?

2. Given the demand equation $p + \dfrac{1}{100} x = 50$ $0 \leqslant p \leqslant 50$

(A)   Express the demand x as a function of the price p.
(B)   Find the point elasticity of demand, $E(p)$.
(C)   What is the point elasticity of demand when $p = \$10$? If this price is decreased by 5%, what is the approximate change in demand?
(D)   What is the point elasticity of demand when $p = \$45$? If this price is decreased by 5%, what is the approximate change in demand?
(E)   What is the point elasticity of demand when $p = \$25$? If this price is decreased by 5%, what is the approximate change in demand?

3. Given the demand equation $\dfrac{1}{50} x + p = 60$ $0 \leqslant p \leqslant 60$

(A)   Express the demand x as a function of the price p.
(B)   Express the revenue R as a function of the price p.
(C)   Find the point elasticity of demand, $E(p)$.
(D)   For which values of p is demand elastic? Inelastic?
(E)   For which values of p is revenue increasing? Decreasing?
(F)   If $p = \$10$ and the price is cut by 10%, will revenue increase or decrease?
(G)   If $p = \$40$ and the price is cut by 10%, will revenue increase or decrease?

4. Repeat Problem 3 for the demand equation

$\dfrac{1}{60} x + p = 50$ $0 \leqslant p \leqslant 50$

For each of the following demand equations, determine if demand is elastic, inelastic, or has unit elasticity at the indicated values of p.

5. $x = f(p) = 12{,}000 - 10p^2$

(A) $p = 10$ (B) $p = 20$ (C) $p = 30$

6. $x = f(p) = 1{,}875 - p^2$

(A) $p = 15$ (B) $p = 25$ (C) $p = 40$

7. $x = f(p) = 950 - 2p - \dfrac{1}{10}\,p^2$

   (A) $p = 30$ (B) $p = 50$ (C) $p = 70$

8. $x = f(p) = 875 - p - \dfrac{1}{20}\,p^2$

   (A) $p = 50$ (B) $p = 70$ (C) $p = 100$

B   For each of the following demand equations, find the values of p for which demand is elastic and the values for which demand is inelastic.

9. $x = f(p) = 10(p - 30)^2,\quad 0 \leqslant p \leqslant 30$
10. $x = f(p) = 5(p - 60)^2,\quad 0 \leqslant p \leqslant 60$
11. $f(p) = \sqrt{144 - 2p},\quad 0 \leqslant p \leqslant 72$
12. $f(p) = \sqrt{324 - 2p},\quad 0 \leqslant p \leqslant 162$
13. $x = f(p) = \sqrt{2{,}500 - 2p^2},\quad 0 \leqslant p \leqslant 25\sqrt{2}$
14. $x = f(p) = \sqrt{3{,}600 - 2p^2},\quad 0 \leqslant p \leqslant 30\sqrt{2}$

For each of the following demand equations, sketch the graph of the revenue function and indicate the regions of inelastic and elastic demand on the graph.

15. $x = f(p) = 20(10 - p),\quad 0 \leqslant p \leqslant 10$
16. $x = f(p) = 10(16 - p),\quad 0 \leqslant p \leqslant 16$
17. $x = f(p) = 40(p - 15)^2,\quad 0 \leqslant p \leqslant 15$
18. $x = f(p) = 10(p - 9)^2,\quad 0 \leqslant p \leqslant 9$
19. $x = f(p) = 30 - 10\sqrt{p},\quad 0 \leqslant p \leqslant 9$
20. $x = f(p) = 30 - 5\sqrt{p},\quad 0 \leqslant p \leqslant 36$

C   In Problems 21–24 use implicit differentiation to find the point elasticity of demand at the indicated values of x and p.

21. $x^{3/2} + 2px + p^3 = 4{,}000,\quad x = 100,\quad p = 10$
22. $2x^{3/2} + 4px + 10p^2 = 1{,}000,\quad x = 25,\quad p = 5$
23. $5x^3 + x^2p^2 + 20p^3 = 10{,}000,\quad x = 10,\quad p = 5$
24. $10\sqrt{x + 80} + 2x^2 + 4p^2 = 1{,}000,\quad x = 20,\quad p = 5$

In economics, it is common to use the demand x as the independent variable. If $p = g(x)$ is the demand equation, then it can be shown that the point elasticity of demand is given by

$$E(x) = \frac{g(x)}{xg'(x)}$$

Use this formula in Problems 25–28 to find the point elasticity of demand at the indicated value of x.

25. $p = g(x) = 50 - \dfrac{1}{10}\,x,\quad x = 200$

26. $p = g(x) = 30 - \dfrac{1}{20}\,x, \quad x = 400$

27. $p = g(x) = 50 - 2\sqrt{x}, \quad x = 400$

28. $p = g(x) = 20 - \sqrt{x}, \quad x = 100$

■

Applications

Business & Economics

29. *Revenue and elasticity.* The weekly demand for hamburgers sold by a chain of restaurants is 30,000 when the price of a hamburger is $2.00. A 10% price increase caused the weekly demand to drop to 28,000 hamburgers. Assume that the demand equation is linear.

(A)   Find the point elasticity of demand at the new price.
(B)   Approximate the change in demand if the price is increased by an additional 10% over the first 10% increase (see pages 754–755).
(C)   Will the second price increase cause the revenue to increase or decrease?

30. *Revenue and elasticity.* Repeat Problem 29 if the 10% price increase caused the weekly demand to drop to 24,000 hamburgers.

31. *Revenue and elasticity.* The weekly demand for a small personal computer is 2,000 when the price of a computer is $100. A 10% price cut caused the demand to increase to 2,100 computers per week. Assume that the demand equation is linear.

(A)   Find the point elasticity of demand at the new price.
(B)   Approximate the change in demand if the price is decreased by an additional 10% over the first 10% cut (see pages 754–755).
(C)   Will the second price decrease cause the revenue to increase or decrease?

32. *Revenue and elasticity.* Repeat Problem 31 if the 10% price cut caused the demand to increase to 2,400 computers per week.

12-7   Chapter Review

Important Terms
and Symbols

12-1   *Asymptotes; limits at infinity and infinite limits.* limit as x approaches ∞ or −∞, horizontal asymptote, limits at infinity for rational functions, infinite limits, vertical asymptote, one-sided limits at vertical asymptotes

Exercise 12-7   Chapter Review

Work through all the problems in this chapter review and check your answers in the back of the book. (Answers to all review problems are there.) Whcrc wcaknesscs show up, review appropriate sections in the text. When you are satisfied that you know the material, take the practice test following this review.

A   *Problems 1–8 refer to the following graph of $y = f(x)$:*

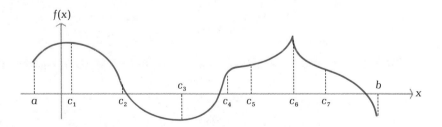

Identify the points or intervals on the x axis that produce the indicated behavior.

1. Graph of f is rising
2. $f'(x) < 0$
3. Graph of f is concave downward
4. Local minima

5.   Absolute maxima   6. $f'(x)$ appears to be 0
7. $f'(x)$ does not exist   8.   Inflection points
9.   Use the following information to sketch the graph of $y = f(x)$:

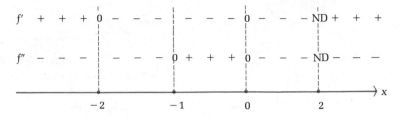

x	-3	-2	-1	0	2	3
$f(x)$	0	3	2	0	-3	0

Evaluate the following limits.

10. $\lim\limits_{x \to \infty} \left(3 + \dfrac{1}{x^{1/3}} + \dfrac{2}{x^3}\right)$ 11. $\lim\limits_{x \to \infty} \dfrac{2x^2 + 3}{3x^2 + 2}$

12. $\lim\limits_{x \to \infty} \dfrac{2x + 3}{3x^2 + 2}$ 13. $\lim\limits_{x \to \infty} \dfrac{2x^2 + 3}{3x + 2}$

Problems 14–19 refer to the function $y = f(x) = x^3 + 3x^2 - 24x - 3$.

14.   Identify critical values.
15.   Find intervals over which $f(x)$ is increasing. Decreasing.
16.   Find local maxima and minima.
17.   Find intervals over which the graph of f is concave upward. Concave downward.
18.   Identify inflection points.
19.   Graph f.

Problems 20–24 refer to the function $y = f(x) = 3x/(x + 2)$.

20.   Find horizontal asymptotes.
21.   Find vertical asymptotes.
22.   Find intervals over which $f(x)$ is increasing. Decreasing.
23.   Find intervals over which the graph of f is concave upward. Concave downward.
24.   Graph f.

Problems 25–30 refer to the function $y = f(x) = 3x^{1/3} - x + 2$.

25.   Identify critical values.
26.   Find intervals over which $f(x)$ is increasing. Decreasing.
27.   Find local maxima and minima.
28.   Find intervals over which the graph of f is concave upward. Concave downward.

29. Identify inflection points.

30. Graph f.

31. Find the absolute maximum and minimum for

$$y = f(x) = x^3 - 12x + 12 \qquad -3 \leqslant x \leqslant 5$$

32. Find the absolute minimum for

$$y = f(x) = x^2 + \frac{16}{x^2} \qquad x > 0$$

Find vertical asymptotes. Evaluate

$$\lim_{x \to c^+} f(x) \qquad and \qquad \lim_{x \to c^-} f(x)$$

at each vertical asymptote c.

33. $f(x) = \dfrac{x^2 - x - 2}{x^2 - 4x + 4}$
34. $f(x) = \dfrac{x^2 - 5x + 6}{x^2 - 3x + 2}$

C **35.** Find the absolute maximum for $f'(x)$ if

$$f(x) = 6x^2 - x^3 + 8$$

Graph f and f' on the same axes.

36. Sketch the graph of

$$f(x) = \frac{x^2 - 4}{x^2 - 1}$$

using the graphing strategy discussed in Section 12-4.

■

Applications

Business & Economics

37. *Profit.* The profit for a company manufacturing and selling x units per month is given by

$$P(x) = 150x - \frac{x^2}{40} - 50{,}000 \qquad 0 \leqslant x \leqslant 5{,}000$$

What production level will produce the maximum profit? What is the maximum profit?

38. *Average cost.* The total cost of producing x units per month is given by

$$C(x) = 4{,}000 + 10x + \frac{1}{10} x^2$$

Find the minimum average cost. Graph the average cost and the marginal cost functions on the same axes.

39. *Rental income.* A 100 room hotel in Fresno is filled to capacity every night at a rate of $20 per room. For each $1 increase in the nightly rate, two fewer rooms are rented. If each rented room costs $4 a day to

service, how much should the management charge per room in order to maximize gross profit?

40. *Inventory control.* A computer store sells 7,200 boxes of floppy discs annually. It costs the store $.20 to store a box of discs for 1 year. Each time it reorders discs, the store must pay a $5.00 service charge for processing the order. How many times during the year should the store order discs in order to minimize the total storage and reorder costs?

Life Sciences

41. *Bacteria control.* If t days after a treatment the bacteria count per cubic centimeter in a body of water is given by

$$C(t) = 20t^2 - 120t + 800 \qquad 0 \leqslant t \leqslant 9$$

in how many days will the count be a minimum?

Social Sciences

42. *Politics.* In a new suburb it is estimated that the number of registered voters will grow according to

$$N = 10 + 6t^2 - t^3 \qquad 0 \leqslant t \leqslant 5$$

where t is time in years and N is in thousands. When will the rate of increase be maximum?

Practice Test: Chapter 12

Problems 1–3 refer to the function $y = f(x) = 2x^3 - 9x^2 + 7$.

1. Find intervals over which $f(x)$ is increasing. Decreasing. Find all local maxima and minima.
2. Find intervals over which the graph of f is concave upward. Concave downward. Indicate the x coordinate(s) of any inflection point(s).
3. Sketch a graph of f.

Problems 4–7 refer to the function

$$y = f(x) = \frac{2x - 4}{x}$$

4. Find horizontal and vertical asymptotes.
5. Find intervals over which $f(x)$ is increasing. Decreasing.
6. Find intervals over which the graph of f is concave upward. Concave downward.
7. Sketch a graph of f.
8. Find all local maxima and minima for

$$f(x) = 2x - 3x^{2/3}$$

9. Find the absolute maximum and minimum for $f(x)$ in Problem 8 over the interval $[0, 8]$.

10. Find two positive numbers whose product is 400 and whose sum is a minimum. What is the minimum sum?

11. A cable television company has 3,600 subscribers in a city, each paying $10 per month for the service. A survey indicates that for each 50¢ reduction in rate, 300 more people will subscribe (and none of the original subscribers will be lost). What rate will maximize revenue? What is the maximum revenue and how many subscribers will produce this revenue?

Exponential and Logarithmic Functions

CHAPTER 13   Contents

Sections 13-1 and 13-2 provide a brief review of exponential and logarithmic functions without the use of calculus. If you have recently studied this material in an algebra or functions course, then you may go directly to Section 13-3. If you have forgotten some of the material, then a brief review should prove helpful.

13-1   Exponential Functions — A Review

- Exponential Functions
- Graphing an Exponential Function
- Typical Types of Exponential Graphs
- Base e
- Basic Exponential Properties

■ Exponential Functions

Until now we have considered mostly **algebraic functions** — that is, functions that can be defined using the algebraic operations of addition, subtraction, multiplication, division, powers, and roots. In no case has a variable been an exponent. In this and the next section we will consider two new kinds of functions that use variable exponents in their definitions.

To start, note that

$$f(x) = 2^x \qquad \text{and} \qquad g(x) = x^2$$

are not the same function. The function g is a quadratic function, which we have already discussed, and the function f is a new function, called an **exponential function.** In general, an exponential function is a function defined by the equation

$$f(x) = b^x \qquad b > 0, \quad b \neq 1$$

where b is a constant, called the **base,** and the exponent is a variable. The domain of f is the set of all real numbers. The range of f is the set of positive real numbers. We require the base b to be positive to avoid nonreal numbers such as $(-2)^{1/2}$.

▪ Graphing an Exponential Function

If asked to graph an exponential function such as

$$f(x) = 2^x$$

most students would not hesitate. They would likely make up a table by assigning integers to x, plot the resulting ordered pairs of numbers, and then join the plotted points with a smooth curve (see Figure 1). What has been overlooked? The exponent form 2^x has not been defined for *all* real numbers x. We assume 2^x is defined in such a way that if we plot $f(x) = 2^x$ for irrational values of x, the points will lie on the curve in Figure 1.

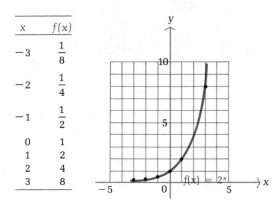

x	f(x)
−3	$\frac{1}{8}$
−2	$\frac{1}{4}$
−1	$\frac{1}{2}$
0	1
1	2
2	4
3	8

Figure 1

▪ Typical Types of Exponential Graphs

It is useful to compare the graphs of $y = 2^x$ and $y = (1/2)^x = 2^{-x}$ by plotting both on the same coordinate system (Figure 2A). Also, the graph of

$$f(x) = b^x \qquad b > 1 \qquad \text{(Figure 2B)}$$

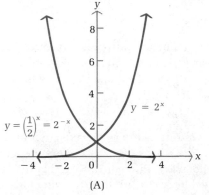

(A)

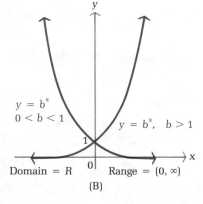

Domain = R   Range = (0, ∞)

(B)

Figure 2

will look very much like the graph of $y = 2^x$, and the graph of

$$f(x) = b^x \qquad 0 < b < 1 \quad \text{(Figure 2B)}$$

will look very much like the graph of $y = (1/2)^x$. [*Note:* In both cases, the x axis is a horizontal asymptote and the graphs will never touch it.]

Example 1   Graph $y = \left(\dfrac{1}{2}\right) 4^x$ for $-3 \leqslant x \leqslant 3$.

Solution

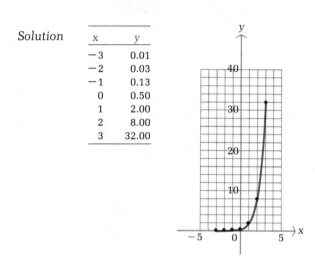

x	y
-3	0.01
-2	0.03
-1	0.13
0	0.50
1	2.00
2	8.00
3	32.00

Problem 1   Graph $y = \left(\dfrac{1}{2}\right) 4^{-x}$ for $-3 \leqslant x \leqslant 3$.

   A great variety of growth phenomena can be described by exponential functions, which is the reason such functions are often referred to as **growth functions.** They are used to describe the growth of money at compound interest; population growth of people, animals, and bacteria; radioactive decay (negative growth); and the growth of learning a skill such as typing or swimming relative to practice.

■ Base e

For introductory purposes, the bases 2 and 1/2 were convenient choices; however, a certain irrational number, denoted by e, is by far the most frequently used exponential base for both theoretical and practical purposes. In fact,

$$f(x) = e^x$$

is often referred to as *the* exponential function because of its widespread use. The reasons for the preference for e as a base will be explained in Sections 13-3 through 13-5. And at that time, it is shown that e is approximated by $(1 + 1/n)^n$ to any decimal accuracy desired by making n (an integer) sufficiently large. The irrational number e to eight decimal places is

$$e \approx 2.718\ 281\ 83$$

Since, for large n,

$$\left(1 + \frac{1}{n}\right)^n \approx e$$

we can raise each side to the xth power to obtain

$$\left(1 + \frac{1}{n}\right)^{nx} \approx e^x$$

Thus, for any x, e^x can be approximated as close as we like by making n (an integer) sufficiently large in $\left(1 + \dfrac{1}{n}\right)^{nx}$. Because of the importance of e^x and e^{-x}, tables for their evaluation are readily available. In fact, all scientific and financial calculators can evaluate these functions directly. A short table for evaluating e^x and e^{-x} is provided in Table I of Appendix B. The important constant e, along with two other important constants—$\sqrt{2}$ and π—are shown on the number line below:

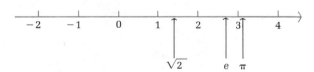

Example 2   Graph $y = 10e^{-0.5x}$, $-3 \leqslant x \leqslant 3$,   using a hand calculator or Table I of Appendix B.

Solution

x	y
−3	44.82
−2	27.18
−1	16.49
0	10.00
1	6.07
2	3.68
3	2.23

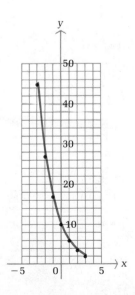

Problem 2   Graph $y = 10e^{0.5x}$, $-3 \leq x \leq 3$,   using a hand calculator or Table I of Appendix B.

■ Basic Exponential Properties

In Sections 3-1 and 3-3 we discussed five laws for integer and rational exponents. It can be shown that these laws also hold for irrational exponents. Thus, we now assume that all five laws of exponents hold for *any* real exponents as long as the involved bases are positive. In addition,

$$b^m = b^n \qquad \text{if and only if} \qquad m = n, \quad b > 0, \quad b \neq 1$$

Thus, if $2^{15} = 2^{3x}$, then $3x = 15$ and $x = 5$.

Answers to Matched Problems

1. $y = \left(\dfrac{1}{2}\right) 4^{-x}$

x	y
-3	32.00
-2	8.00
-1	2.00
0	0.50
1	0.13
2	0.03
3	0.01

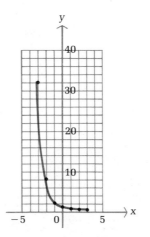

2. $y = 10e^{0.5x}$

x	y
-3	2.23
-2	3.68
-1	6.07
0	10.00
1	16.49
2	27.18
3	44.82

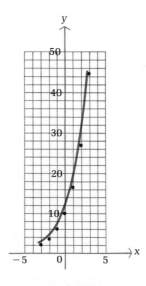

Exercise 13-1

A *Graph each equation for* $-3 \leqslant x \leqslant 3$. *Plot points using integers for x, and then join the points with a smooth curve.*

1. $y = 3^x$

2. $y = 10 \cdot 2^x$
[Note: $10 \cdot 2^x \neq 20^x$]

3. $y = \left(\dfrac{1}{3}\right)^x = 3^{-x}$

4. $y = 10 \cdot \left(\dfrac{1}{2}\right)^x = 10 \cdot 2^{-x}$

5. $y = 10 \cdot 3^x$

6. $y = 10 \cdot \left(\dfrac{1}{3}\right)^x = 10 \cdot 3^{-x}$

B *Graph each equation for* $-3 \leqslant x \leqslant 3$. *Use Table I of Appendix B or a calculator if the base is e.*

7. $y = 10 \cdot 2^{2x}$

8. $y = 10 \cdot 2^{-3x}$

9. $y = e^x$

10. $y = e^{-x}$

11. $y = 10e^{0.2x}$

12. $y = 100e^{0.1x}$

13. $y = 100e^{-0.1x}$

14. $y = 10e^{-0.2x}$

C 15. Graph $y = e^{-x^2}$ for $x = -1.5, -1.0, -0.5, 0, 0.5, 1.0, 1.5$, and then join these points with a smooth curve. Use Table I of Appendix B or a calculator. (This is a very important curve in probability and statistics.)

16. Graph $y = y_0 2^x$, where y_0 is the value of y when $x = 0$. (Express the vertical scale in terms of y_0.)

17. Graph $y = 2^x$ and $x = 2^y$ on the same coordinate system.

18. Graph $y = 10^x$ and $x = 10^y$ on the same coordinate system.

Applications

Business & Economics

19. *Exponential growth.* If we start with 2¢ and double the amount each day, we would have 2^n¢ after n days. Graph $f(n) = 2^n$ for $1 \leqslant n \leqslant 10$. (Label the vertical scale so that the graph will not go off the paper.)

20. *Compound interest.* If a certain amount of money P (the principal) is invested at $100r\%$ interest compounded annually, the amount of money (A) after t years is given by

$$A = P(1 + r)^t$$

Graph this equation for $P = \$100$, $r = 0.10$, and $0 \leqslant t \leqslant 6$. How much money would a person have after 10 years if no interest were withdrawn?

Life Sciences

21. *Bacteria growth.* A single cholera bacterium divides every $1/2$ hour to produce two complete cholera bacteria. If we start with 100 bacteria,

in t hours (assuming adequate food supply) we will have

$$A = 100 \cdot 2^{2t}$$

bacteria. Graph this equation for $0 \leq t \leq 5$.

22. *Ecology.* The atmospheric pressure (P, in pounds per square inch) may be calculated approximately from the formula

$$P = 14.7e^{-0.21h}$$

where h is the altitude above sea level in miles. Graph this equation for $0 \leq h \leq 12$.

Social Sciences

23. *Learning curves.* The performance record of a particular person learning to type is given approximately by

$$N = 100(1 - e^{-0.1t})$$

where N is the number of words typed per minute and t is the number of weeks of instruction. Graph this equation for $0 \leq t \leq 40$. What does N approach as t approaches ∞?

24. *Small group analysis.* After a lengthy investigation, sociologists Stephan and Mischler found that if the members of a discussion group of ten were ranked according to the number of times each participated, then the number of times, $N(k)$, the kth-ranked person participated was given approximately by

$$N(k) = N_1 e^{-0.11(k-1)} \qquad 1 \leq k \leq 10$$

where N_1 was the number of times the top-ranked person participated in the discussion. Graph the equation assuming $N_1 = 100$. [For a general discussion of this phenomenon, see J. S. Coleman, *Introduction to Mathematical Sociology* (London: The Free Press of Glencoe, 1964), pp. 28–31.]

13-2 Logarithmic Functions—A Review

- Definition of Logarithmic Functions
- From Logarithmic to Exponential Form and Vice Versa
- Properties of Logarithmic Functions
- Calculator Evaluation of Common and Natural Logarithms
- Application

Now we are ready to consider logarithmic functions, which are closely related to exponential functions.

■ Definition of Logarithmic Functions

If we start with an exponential function f defined by

$$y = 2^x \tag{1}$$

and interchange the variables, we obtain an equation that defines a new relation g defined by

$$x = 2^y \tag{2}$$

Any ordered pair of numbers that belongs to f will belong to g if we interchange the order of the components. For example, (3, 8) satisfies equation (1) and (8, 3) satisfies equation (2). Thus, the domain of f becomes the range of g and the range of f becomes the domain of g. Graphing f and g on the same coordinate system (Figure 3), we see that g is also a function. We call this new function the **logarithmic function with base 2,** and write

$$y = \log_2 x \quad \text{if and only if} \quad x = 2^y$$

Note that if we fold the paper along the dashed line $y = x$ in Figure 3, the two graphs match exactly.

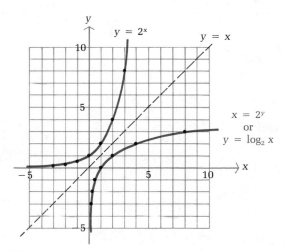

Exponential Function		Logarithmic Function	
x	$y = 2^x$	$x = 2^y$	y
-3	$1/8$	$1/8$	-3
-2	$1/4$	$1/4$	-2
-1	$1/2$	$1/2$	-1
0	1	1	0
1	2	2	1
2	4	4	2
3	8	8	3

$$\begin{bmatrix} \text{Ordered} \\ \text{pairs} \\ \text{reversed} \end{bmatrix}$$

Figure 3

In general, we define the logarithmic functions with base b as follows:

Logarithmic Function

$$y = \log_b x \quad \text{if and only if} \quad x = b^y \quad b > 0, \ b \neq 1$$

In words, **the logarithm of a number x to a base b ($b > 0$, $b \neq 1$) is the exponent to which b must be raised to equal x**. It is important to remember that $y = \log_b x$ and $x = b^y$ describe the same function, while $y = b^x$ is the related exponential function. Look at Figure 3 again.

Since the domain of an exponential function includes all real numbers and its range is the set of positive real numbers, the **domain** of a logarithmic function is the set of all positive real numbers and its **range** is the set of all real numbers. Remember that the logarithm of 0 or a negative number is not defined.

■ From Logarithmic to Exponential Form and Vice Versa

We now consider the matter of converting logarithmic forms to equivalent exponential forms and vice versa.

Example 3 Change from logarithmic form to exponential form.

(A) $\log_5 25 = 2$ is equivalent to $25 = 5^2$
(B) $\log_9 3 = 1/2$ is equivalent to $3 = 9^{1/2}$
(C) $\log_2 (1/4) = -2$ is equivalent to $1/4 = 2^{-2}$

Problem 3 Change to an equivalent exponential form.

(A) $\log_3 9 = 2$ (B) $\log_4 2 = 1/2$ (C) $\log_3 (1/9) = -2$

Example 4 Change from exponential form to logarithmic form.

(A) $64 = 4^3$ is equivalent to $\log_4 64 = 3$
(B) $6 = \sqrt{36}$ is equivalent to $\log_{36} 6 = 1/2$
(C) $1/8 = 2^{-3}$ is equivalent to $\log_2 (1/8) = -3$

Problem 4 Change to an equivalent logarithmic form.

(A) $49 = 7^2$ (B) $3 = \sqrt{9}$ (C) $1/3 = 3^{-1}$

Example 5 Find y, b, or x.

(A) $y = \log_4 16$ (B) $\log_2 x = -3$
(C) $y = \log_8 4$ (D) $\log_b 100 = 2$

Solutions (A) $y = \log_4 16$ is equivalent to $16 = 4^y$. Thus,

$$y = 2$$

(B) $\log_2 x = -3$ is equivalent to $x = 2^{-3}$. Thus,

$$x = \frac{1}{2^3} = \frac{1}{8}$$

(C) $y = \log_8 4$ is equivalent to

$$4 = 8^y \qquad \text{or} \qquad 2^2 = 2^{3y}$$

Thus,

$$3y = 2$$

$$y = \frac{2}{3}$$

(D) $\log_b 100 = 2$ is equivalent to $100 = b^2$. Thus,

$$b = 10 \qquad \text{Recall that } b \text{ cannot be negative.}$$

Problem 5 Find y, b, or x.

(A) $y = \log_9 27$ (B) $\log_3 x = -1$ (C) $\log_b 1{,}000 = 3$

Example 6 Graph $y = \log_2(x + 1)$ by converting to an equivalent exponential form first.

Solution Changing $y = \log_2(x + 1)$ to an equivalent exponential form, we have

$$x + 1 = 2^y \qquad \text{or} \qquad x = 2^y - 1$$

Even though x is the independent variable and y is the dependent variable, it is easier to assign y values and solve for x.

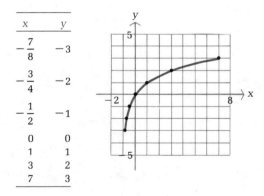

x	y
$-\dfrac{7}{8}$	-3
$-\dfrac{3}{4}$	-2
$-\dfrac{1}{2}$	-1
0	0
1	1
3	2
7	3

Problem 6 Graph $y = \log_3(x - 1)$ by converting to an equivalent exponential form first.

■ Properties of Logarithmic Functions

Logarithmic functions have several very useful properties that follow directly from their definitions. These properties will enable us to convert multiplication problems into addition problems, division problems into subtraction problems, and power and root problems into multiplication problems. We will also be able to solve exponential equations such as $2 = 1.06^n$.

Logarithmic Properties

$(b > 0, \quad b \neq 1, \quad M > 0, \quad N > 0)$

1. $\log_b b^x = x$
2. $\log_b MN = \log_b M + \log_b N$
3. $\log_b \dfrac{M}{N} = \log_b M - \log_b N$
4. $\log_b M^p = p \log_b M$
5. $\log_b M = \log_b N$ if and only if $M = N$
6. $\log_b 1 = 0$

The first property follows directly from the definition of a logarithmic function. Here, we will sketch a proof for property 2. The other properties are established in a similar way. Let

$$u = \log_b M \qquad \text{and} \qquad v = \log_b N$$

Or, in equivalent exponential form,

$$M = b^u \qquad \text{and} \qquad N = b^v$$

Now, see if you can provide reasons for each of the following steps:

$$\log_b MN = \log_b b^u b^v = \log_b b^{u+v} = u + v = \log_b M + \log_b N$$

Example 7   (A) $\log_b \dfrac{wx}{yz} \; = \log_b wx - \log_b yz$

$$= \log_b w + \log_b x - (\log_b y + \log_b z)$$

$$= \log_b w + \log_b x - \log_b y - \log_b z$$

(B) $\log_b (wx)^{3/5} \; = \dfrac{3}{5} \log_b wx$

$$= \dfrac{3}{5}(\log_b w + \log_b x)$$

Problem 7   Write in simpler logarithmic forms, as in Example 7.

(A) $\log_b \dfrac{R}{ST}$ (B) $\log_b \left(\dfrac{R}{S}\right)^{2/3}$

The following examples and problems, though somewhat artificial, will give you additional practice in using basic logarithmic properties.

Example 8 Find x so that

$$\frac{3}{2} \log_b 4 - \frac{2}{3} \log_b 8 + \log_b 2 = \log_b x$$

Solution

$$\frac{3}{2} \log_b 4 - \frac{2}{3} \log_b 8 + \log_b 2 = \log_b x$$

$$\log_b 4^{3/2} - \log_b 8^{2/3} + \log_b 2 = \log_b x \qquad \text{Property 4}$$

$$\log_b 8 - \log_b 4 + \log_b 2 = \log_b x$$

$$\log_b \frac{8 \cdot 2}{4} = \log_b x \qquad \text{Properties 2 and 3}$$

$$\log_b 4 = \log_b x$$

$$x = 4 \qquad \text{Property 5}$$

Problem 8 Find x so that

$$3 \log_b 2 + \frac{1}{2} \log_b 25 - \log_b 20 = \log_b x$$

Example 9 Solve $\log_{10} x + \log_{10}(x + 1) = \log_{10} 6$.

Solution

$$\log_{10} x + \log_{10}(x + 1) = \log_{10} 6$$

$$\log_{10} x(x + 1) = \log_{10} 6 \qquad \text{Property 2}$$

$$x(x + 1) = 6 \qquad \text{Property 5}$$

$$x^2 + x - 6 = 0 \qquad \text{Solve by factoring.}$$

$$(x + 3)(x - 2) = 0$$

$$x = -3, 2$$

We must exclude $x = -3$, since negative numbers are not in the domains of logarithmic functions; hence,

$$x = 2$$

is the only solution.

Problem 9 Solve $\log_3 x + \log_3(x - 3) = \log_3 10$.

■ Calculator Evaluation of Common and Natural Logarithms

Of all possible logarithmic bases, the base e and the base 10 are used almost exclusively. Before we can use logarithms in certain practical problems, we need to be able to approximate the logarithm of any number either to base 10 or to base e. And conversely, if we are given the logarithm of a number to base 10 or base e, we need to be able to approximate the number. Historically, tables such as Tables II and III of Appendix B were used for this purpose, but now with inexpensive scientific hand calculators readily available, most people will use a calculator, since it is faster and far more accurate.

Common logarithms (also called **Briggsian logarithms**) are logarithms with base 10. **Natural logarithms** (also called **Napierian logarithms**) are logarithms with base e. Most scientific calculators have a button labeled "log" (or "LOG") and a button labeled "ln" (or "LN"). The former represents a common (base 10) logarithm and the latter a natural (base e) logarithm. In fact, "log" and "ln" are both used extensively in mathematical literature, and whenever you see either used in this book without a base indicated they will be interpreted as follows:

Logarithmic Notation

$\log x = \log_{10} x$

$\ln x = \log_e x$

Finding the common or natural logarithm using a scientific calculator is very easy: you simply enter a number from the domain of the function and push the log or ln button.

Example 10   Use a scientific calculator to find each to six decimal places:

(A) $\log 3{,}184$ (B) $\ln 0.000\ 349$ (C) $\log(-3.24)$

Solutions

Enter	Press	Display
(A) 3184	$\boxed{\log}$	3.502973
(B) 0.000 349	$\boxed{\ln}$	−7.960439
(C) −3.24	$\boxed{\log}$	Error

An error is indicated in part C because -3.24 is not in the domain of the log function.

Problem 10   Use a scientific calculator to find each to six decimal places:

(A) $\log 0.013\ 529$ (B) $\ln 28.693\ 28$ (C) $\ln(-0.438)$

We now turn to the second problem to be discussed in this section: Given the logarithm of a number, find the number. We make direct use of the logarithmic–exponential relationships, which follow directly from the definition of logarithmic functions at the beginning of this section.

Logarithmic–Exponential Relationships

$\log x = y$ is equivalent to $x = 10^y$

$\ln x = y$ is equivalent to $x = e^y$

Example 11

Find x to three significant digits, given the indicated logarithms:

(A) $\log x = -9.315$ (B) $\ln x = 2.386$

Solutions

(A) $\log x = -9.315$ Change to equivalent exponential form.

      $x = 10^{-9.315}$

        $x = 4.84 \times 10^{-10}$ The answer is displayed in scientific notation in the calculator.

(B) $\ln x = 2.386$ Change to equivalent exponential form.

      $x = e^{2.386}$

      $x = 10.9$

Problem 11

Find x to four significant digits, given the indicated logarithms.

(A) $\ln x = -5.062$ (B) $\log x = 12.082\ 1$

■ Application

If P dollars are invested at $100i\%$ interest per period for n periods, and interest is paid to the account at the end of each period, then the amount of money in the account at the end of period n is given by

     $A = P(1 + i)^n$ Compound interest formula

The fact that interest paid to the account at the end of each period earns interest during the following periods is the reason this is called a **compound interest** formula.

Example 12
Doubling Time

How long (to the next whole year) will it take money to double if it is invested at 20% interest compounded annually?

Solution   Find n for $A = 2P$ and $i = 0.2$.

$$A = P(1 + i)^n$$

$$2P = P(1 + 0.2)^n$$

$1.2^n = 2$ Solve for n by taking the natural or common log of both sides.

$\ln 1.2^n = \ln 2$

$n \ln 1.2 = \ln 2$ Property 4

$n = \dfrac{\ln 2}{\ln 1.2}$ Use a calculator or a table.

$= 3.8$ years $\left[\textit{Note:} \quad \dfrac{\ln 2}{\ln 1.2} \neq \ln 2 - \ln 1.2 \right]$

≈ 4 years   To the next whole year

When interest is paid at the end of 3 years, the money will not be doubled; when paid at the end of 4 years, the money will be slightly more than doubled.

Problem 12   How long (to the next whole year) will it take money to double if it is invested at 13% interest compounded annually?

It is interesting and instructive to graph the doubling times for various

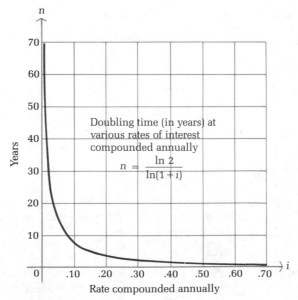

Figure 4

rates compounded annually. We proceed as follows:

$$A = P(1 + i)^n$$
$$2P = P(1 + i)^n$$
$$2 = (1 + i)^n$$
$$(1 + i)^n = 2$$
$$\ln(1 + i)^n = \ln 2$$
$$n \ln(1 + i) = \ln 2$$
$$n = \frac{\ln 2}{\ln(1 + i)}$$

Figure 4 shows the graph of this equation (doubling times in years) for interest rates compounded annually from 1% to 70%. Note the dramatic changes in doubling times from 1% to 20%.

Answers to
Matched Problems

3. (A) $9 = 3^2$ (B) $2 = 4^{1/2}$ (C) $1/9 = 3^{-2}$
4. (A) $\log_7 49 = 2$ (B) $\log_9 3 = 1/2$ (C) $\log_3 (1/3) = -1$
5. (A) $y = 3/2$ (B) $x = 1/3$ (C) $b = 10$
6. $y = \log_3(x - 1)$ is equivalent
   to $x = 3^y + 1$

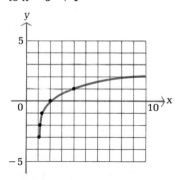

7. (A) $\log_b R - \log_b S - \log_b T$ (B) $(2/3)(\log_b R - \log_b S)$
8. $x = 2$
9. $x = 5$
10. (A) $-1.868\ 734$ (B) $3.356\ 663$ (C) Not defined
11. (A) 6.333×10^{-3} (B) 1.208×10^{12}
12. 6 years

Exercise 13-2

A *Rewrite in exponential form.*

1. $\log_3 27 = 3$ 2. $\log_2 32 = 5$
3. $\log_{10} 1 = 0$ 4. $\log_e 1 = 0$

5. $\log_4 8 = \dfrac{3}{2}$ **6.** $\log_9 27 = \dfrac{3}{2}$

Rewrite in logarithmic form.

7. $49 = 7^2$ **8.** $36 = 6^2$

9. $8 = 4^{3/2}$ **10.** $9 = 27^{2/3}$

11. $A = b^u$ **12.** $M = b^x$

Find each of the following:

13. $\log_{10} 10^3$ **14.** $\log_{10} 10^{-5}$

15. $\log_2 2^{-3}$ **16.** $\log_3 3^5$

17. $\log_{10} 1{,}000$ **18.** $\log_6 36$

Write in terms of simpler logarithmic forms as in Example 7.

19. $\log_b \dfrac{P}{Q}$ **20.** $\log_b FG$

21. $\log_b L^5$ **22.** $\log_b w^{15}$

23. $\log_b \dfrac{p}{qrs}$ **24.** $\log_b PQR$

B   *Find x, y, or b.*

25. $\log_3 x = 2$ **26.** $\log_2 x = 2$

27. $\log_7 49 = y$ **28.** $\log_3 27 = y$

29. $\log_b 10^{-4} = -4$ **30.** $\log_b e^{-2} = -2$

31. $\log_4 x = \dfrac{1}{2}$ **32.** $\log_{25} x = \dfrac{1}{2}$

33. $\log_{1/3} 9 = y$ **34.** $\log_{49} \dfrac{1}{7} = y$

35. $\log_b 1{,}000 = \dfrac{3}{2}$ **36.** $\log_b 4 = \dfrac{2}{3}$

Write in terms of simpler logarithmic forms going as far as you can with logarithmic properties (see Example 7).

37. $\log_b \dfrac{x^5}{y^3}$ **38.** $\log_b x^2 y^3$

39. $\log_b \sqrt[3]{N}$ **40.** $\log_b \sqrt[5]{Q}$

41. $\log_b x^2 \sqrt[3]{y}$ **42.** $\log_b \sqrt[3]{\dfrac{x^2}{y}}$

43. $\log_b (50 \cdot 2^{-0.2t})$ **44.** $\log_b (100 \cdot 1.06^t)$

45. $\log_b P(1 + r)^t$ **46.** $\log_e Ae^{-0.3t}$

47. $\log_e 100e^{-0.01t}$ **48.** $\log_{10} (67 \cdot 10^{-0.12x})$

(handwritten at left of 45.) $\log_b P + t\, \log_b (1 + r) \longrightarrow$

Find x.

49. $\log_b x = \dfrac{2}{3} \log_b 8 + \dfrac{1}{2} \log_b 9 - \log_b 6$

50. $\log_b x = \dfrac{2}{3} \log_b 27 + 2 \log_b 2 - \log_b 3$

51. $\log_b x = \dfrac{3}{2} \log_b 4 - \dfrac{2}{3} \log_b 8 + 2 \log_b 2$

52. $\log_b x = 3 \log_b 2 + \dfrac{1}{2} \log_b 25 - \log_b 20$

53. $\log_b x + \log_b (x - 4) = \log_b 21$
54. $\log_b (x + 2) + \log_b x = \log_b 24$
55. $\log_{10} (x - 1) - \log_{10} (x + 1) = 1$
56. $\log_{10} (x + 6) - \log_{10} (x - 3) = 1$

Graph by converting to exponential form first.

57. $y = \log_2 (x - 2)$ 　　　　　　　　**58.** $y = \log_3 (x + 2)$

In Problems 59 and 60, evaluate to five decimal places using a scientific calculator.

59. (A) $\log 3{,}527.2$ 　　(B) $\log 0.006\ 913\ 2$
　　　(C) $\ln 277.63$ 　　　(D) $\ln 0.040\ 883$

60. (A) $\log 72.604$ 　　(B) $\log 0.033\ 041$
　　　(C) $\ln 40{,}257$ 　　(D) $\ln 0.005\ 926\ 3$

In Problems 61 and 62, find x to four significant digits.

61. (A) $\log x = 3.128\ 5$ 　　(B) $\log x = -2.049\ 7$
　　　(C) $\ln x = 8.776\ 3$ 　　(D) $\ln x = -5.887\ 9$

62. (A) $\log x = 5.083\ 2$ 　　(B) $\log x = -3.157\ 7$
　　　(C) $\ln x = 10.133\ 6$ 　　(D) $\ln x = -4.328\ 1$

C　**63.** Find the logarithm of 1 for any permissible base.
　　64. Why is 1 not a suitable logarithmic base? [*Hint:*　Try to find $\log_1 8$.]
　　65. Write $\log_{10} y - \log_{10} c = 0.8x$ in an exponential form that is free of logarithms.
　　66. Write $\log_e x - \log_e 25 = 0.2t$ in an exponential form that is free of logarithms.

■

Applications

Business & Economics　**67.** *Doubling time.* How long (to the next whole year) will it take money to double if it is invested at 6% interest compounded annually?

68. *Doubling time.* How long (to the next whole year) will it take money to double if it is invested at 3% interest compounded annually?

69. *Tripling time.* Write a formula similar to the doubling time formula in Figure 4 for the tripling time of money invested at $100i$% interest compounded annually?

70. *Tripling time.* How long (to the next whole year) will it take money to triple if invested at 15% interest compounded annually?

Life Sciences

71. *Sound intensity—decibels.* Because of the extraordinary range of sensitivity of the human ear (a range of over 1,000 million millions to 1), it is helpful to use a logarithmic scale, rather than an absolute scale, to measure sound intensity over this range. The unit of measure is called the *decibel*, after the inventor of the telephone, Alexander Graham Bell. If we let N be the number of decibels, I the power of the sound in question (in watts per square centimeter), and I_0 the power of sound just below the threshold of hearing (approximately 10^{-16} watt per square centimeter), then

$$I = I_0 10^{N/10}$$

Show that this formula can be written in the form

$$N = 10 \log \frac{I}{I_0}$$

72. *Sound intensity—decibels.* Use the formula in Problem 71 (with $I_0 = 10^{-16}$ watt/cm²) to find the decibel ratings of the following sounds:

    (A) Whisper: 10^{-13} watt/cm²
    (B) Normal conversation: 3.16×10^{-10} watt/cm²
    (C) Heavy traffic: 10^{-8} watt/cm²
    (D) Jet plane with afterburner: 10^{-1} watt/cm²

Social Sciences

73. *World population.* If the world population is now 4 billion (4×10^9) people and if it continues to grow at 2% per year compounded annually, how long will it be before there is only 1 square yard of land per person? (The earth contains approximately 1.68×10^{14} square yards of land.)

74. *Archaeology—carbon-14 dating.* Cosmic-ray bombardment of the atmosphere produces neutrons, which in turn react with nitrogen to produce radioactive carbon-14. Radioactive carbon-14 enters all living tissues through carbon dioxide which is first absorbed by plants. As long as a plant or animal is alive, carbon-14 is maintained at a constant level in its tissues. Once dead, however, it ceases taking in carbon and the carbon-14 diminishes by radioactive decay according to the equation

$$A = A_0 e^{-0.000124t}$$

where t is time in years. Estimate the age of a skull uncovered in an archaeological site if 10% of the original amount of carbon-14 is still present. [*Hint:* Find t such that $A = 0.1A_0$.]

13-3 The Constant e and Continuous Compound Interest

- The Constant e
- Continuous Compound Interest

■ The Constant e

In the last two sections we introduced the special irrational number e as a particularly suitable base for both exponential and logarithmic functions. In this and the following sections we will see why this is so. We said earlier that e can be approximated as closely as we like by $[1 + (1/n)]^n$ by taking n sufficiently large. Now we will use the limit concept to formally define e as either of the following two limits:

The Number e

$$e = \lim_{n \to \infty} \left(1 + \frac{1}{n}\right)^n$$

or, alternately,

$$e = \lim_{s \to 0} (1 + s)^{1/s}$$

$$e = 2.718\ 281\ 8\ \ldots$$

We will use both these forms. [*Note:* If $s = 1/n$, then as $n \to \infty$, $s \to 0$.]

The proof that the indicated limits exist and represent an irrational number between 2 and 3 is not easy and is omitted here. Many people reason (incorrectly) that the limits are 1, since "$(1 + s)$ approaches 1 as $s \to 0$, and 1 to any power is 1." A little experimentation with a pocket calculator can convince you otherwise. Consider the table of values for s and $f(s) = (1 + s)^{1/s}$ and the graph shown in Figure 5 for s close to 0.

s approaches 0 from the left $\to 0 \leftarrow$ s approaches 0 from the right

s	-0.5	-0.2	-0.1	-0.01 $\to 0 \leftarrow$ 0.01		0.1	0.2	0.5
$(1+s)^{1/s}$	4.000 0	3.051 8	2.868 0	2.732 0 $\to e \leftarrow$ 2.704 8		2.593 7	2.488 3	2.250 0

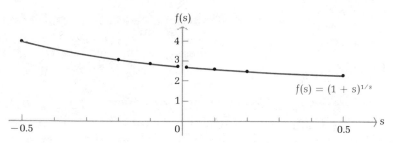

Figure 5

Compute some of the table values with a calculator yourself and also try several values of s even closer to 0. Note that the function is discontinuous at $s = 0$.

Exactly who discovered e is still being debated. It is named after the great mathematician Leonhard Euler (1707–1783), who computed e to twenty-three decimal places using $[1 + (1/n)]^n$.

■ Continuous Compound Interest

Now we will see how e appears quite naturally in the important application of compound interest. Let us start with simple interest, move on to compound interest, and then to continuous compound interest.

If a principal P is borrowed at an annual rate r, then after t years at simple interest the borrower will owe the lender an amount A given by

$$A = P + Prt = P(1 + rt) \qquad \text{Simple interest} \qquad (1)$$

On the other hand, if interest is compounded n times a year, then the borrower will owe the lender an amount A given by

$$A = P\left(1 + \frac{r}{n}\right)^{nt} \qquad \text{Compound interest} \qquad (2)$$

Suppose P, r, and t in (2) are held fixed and n is increased. Will the amount A increase without bound or will it tend to some limiting value?

Let us perform a calculator experiment before we attack the general limit problem. If $P = \$100$, $r = 0.06$, and $t = 2$ years, then

$$A = 100\left(1 + \frac{0.06}{n}\right)^{2n}$$

We compute A for several values of n in Table 1. The biggest gain appears in the first step; then the gains slow down as n increases. In fact, it appears that A might be tending to something close to \$112.75 as n gets larger and larger.

Now we turn back to the general problem for a moment. Keeping P, r, and t fixed in equation (2), we compute the following limit and observe an

Table 1

Compounding Frequency	n	$A = 100\left(1 + \dfrac{0.06}{n}\right)^{2n}$
Annually	1	$112.3600
Semiannually	2	112.5509
Quarterly	4	112.6493
Weekly	52	112.7419
Daily	365	112.7486
Hourly	8,760	112.7491

interesting and useful result:

$$\lim_{n \to \infty} P\left(1 + \frac{r}{n}\right)^{nt} = P \lim_{n \to \infty} \left(1 + \frac{r}{n}\right)^{(n/r)rt} \qquad \text{Insert } r/r \text{ in the exponent and let } s = r/n.$$

$$= P[\lim_{s \to 0}(1 + s)^{1/s}]^{rt} \qquad \lim_{s \to 0}(1 + s)^{1/s} = e$$

$$= Pe^{rt}$$

The resulting formula is called the **continuous compound interest formula,** a very important and widely used formula in business and economics.

Continuous Compound Interest

$A = Pe^{rt}$

where

$P = $ Principal

$r = $ Annual interest rate compounded continuously

$t = $ Time in years

$A = $ Amount at time t

Example 13   If $100 is invested at 6% interest compounded continuously, what amount will be in the account after 2 years?

Solution $A = Pe^{rt}$

$= 100e^{(0.06)(2)}$

$\approx \$112.7497$

(Compare this result with the values calculated in Table 1.)

Problem 13   What amount (to the nearest cent) will an account have after 5 years if $100 is invested at 8% interest compounded annually? Semiannually? Continuously?

Example 14   If $100 is invested at 12% interest compounded continuously, graph the amount in the account relative to time for a period of 10 years.

Solution   We want to graph

$$A = 100e^{0.12t} \qquad 0 \le t \le 10$$

We construct a table of values using a calculator or Table I of Appendix B, graph the points from the table, and join the points with a smooth curve.

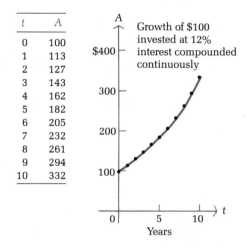

t	A
0	100
1	113
2	127
3	143
4	162
5	182
6	205
7	232
8	261
9	294
10	332

Growth of $100 invested at 12% interest compounded continuously

Problem 14   If $5,000 is invested at 20% interest compounded continuously, graph the amount in the account relative to time for a period of 10 years.

Example 15   How long will it take money to double if it is invested at 18% interest compounded continuously?

Solution   Starting with the continuous compound interest formula $A = Pe^{rt}$, we must solve for t given $A = 2P$ and $r = 0.18$.

$$2P = Pe^{0.18t} \qquad \text{Divide both sides by } P.$$
$$e^{0.18t} = 2 \qquad \text{Take natural logs of both sides.}$$
$$\ln e^{0.18t} = \ln 2 \qquad \text{Recall that } \log_b b^x = x.$$
$$0.18t = \ln 2$$
$$t = \frac{\ln 2}{0.18}$$
$$t = 3.85 \text{ years}$$

Problem 15   How long will it take money to triple if it is invested at 12% interest compounded continuously?

Answers to   13. $146.93; $148.02; $149.18
Matched Problems   14. $A = 5,000e^{0.2t}$

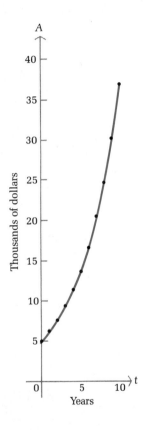

t	A
0	5,000
1	6,107
2	7,459
3	9,111
4	11,128
5	13,591
6	16,601
7	20,276
8	24,765
9	30,248
10	36,945

15.   9.16 years

Exercise 13-3

A   *Use a calculator or table to evaluate A to the nearest cent in Problems 1–2.*

1. $A = \$1,000e^{0.1t}$ for $t = 2, 5,$ and 8
2. $A = \$5,000e^{0.08t}$ for $t = 1, 4,$ and 10

B   In Problems 3 – 8 solve for t or r to two decimal places.

3. $2 = e^{0.06t}$ 4. $2 = e^{0.03t}$

5. $3 = e^{0.1t}$ 6. $3 = e^{0.25t}$

7. $2 = e^{5r}$ 8. $3 = e^{10r}$

C   In Problems 9 and 10 complete each table to five decimal places using a hand calculator.

9.

n	$(1 + 1/n)^n$
10	2.593 74
100	
1,000	
10,000	
100,000	
1,000,000	
10,000,000	
↓	↓
∞	$e = 2.718\ 281\ 8\ \ldots$

10.

s	$(1 + s)^{1/s}$
0.01	2.704 81
−0.01	
0.001	
−0.001	
0.000 1	
−0.000 1	
0.000 01	
−0.000 01	
↓	↓
0	$e = 2.718\ 281\ 8\ \ldots$

Applications

Business & Economics

11. *Continuous compound interest.* If $20,000 is invested at 12% interest compounded continuously, how much will it be worth in 8.5 years?

12. *Continuous compound interest.* Assume $1 had been invested at 4% interest compounded continuously at the birth of Christ. What would be the value of the account in solid gold earths in the year 2000? (Assume that the earth weighs approximately 2.11×10^{26} ounces and that gold will be worth $1,000 an ounce in the year 2000.) What would be the value of the account in dollars at simple interest?

13. *Present value.* A note will pay $20,000 at maturity 10 years from now. How much should you be willing to pay for the note now if money is worth 7% compounded continuously?

14. *Present value.* A note will pay $50,000 at maturity 5 years from now. How much should you be willing to pay for the note now if money is worth 8% compounded continuously?

15. *Doubling time.* How long will it take money to double if invested at 25% interest compounded continuously?

16. *Doubling time.* How long will it take money to double if invested at 5% interest compounded continuously?

17. *Doubling rate.* At what rate compounded continuously must money be invested to double in 5 years?

18. *Doubling rate.* At what rate compounded continuously must money be invested to double in 3 years?

19. *Doubling time.* It is instructive to look at doubling times for money invested at various rates of interest compounded continuously. Show that doubling time t at $100r\%$ interest compounded continuously is given by

$$t = \frac{\ln 2}{r}$$

20. *Doubling time.* Graph the doubling time equation from Problem 19 for $0 < r < 1.00$. Identify vertical and horizontal asymptotes.

Life Sciences   **21.** *World population.* A mathematical model for world population growth over short periods of time is given by

$$P = P_0 e^{rt}$$

where

$P_0 =$ Population at time $t = 0$

$r =$ Rate compounded continuously

$t =$ Time in years

$P =$ Population at time t

How long will it take the earth's population to double if it continues to grow at its current rate of 2% per year (compounded continuously)?

22. *World population.* Repeat Problem 21 under the assumption that the world population is growing at a rate of 1% per year compounded continuously.

23. *Population growth.* Some underdeveloped nations have population doubling times of 20 years. At what rate compounded continuously is the population growing? (Use the population growth model in Problem 21.)

24. *Population growth.* Some developed nations have population doubling times of 120 years. At what rate compounded continuously is the population growing? (Use the population growth model in Problem 21.)

Social Sciences   **25.** *World population.* If the world population is now 4 billion (4×10^9) people and if it continues to grow at 2% per year compounded continuously, how long will it be before there is only 1 square yard of land per person? (The earth has approximately 1.68×10^{14} square yards of land.)

13-4   Derivatives of Logarithmic Functions

- Derivatives of Logarithmic Functions
- Graph Properties of $y = \ln x$

■ Derivatives of Logarithmic Functions

We are now ready to derive a formula for the derivative of

$$f(x) = \log_b x \qquad b > 0, \quad b \neq 1, \quad x > 0$$

using the definition of the derivative

$$f'(x) = \lim_{\Delta x \to 0} \frac{f(x + \Delta x) - f(x)}{\Delta x}$$

and the two-step process discussed in Section 10-4.

Step 1.   Simplify the difference quotient first.

$$\frac{f(x + \Delta x) - f(x)}{\Delta x} = \frac{\log_b(x + \Delta x) - \log_b x}{\Delta x}$$

$$= \frac{1}{\Delta x} [\log_b(x + \Delta x) - \log_b x]$$

$$= \frac{1}{\Delta x} \log_b \frac{x + \Delta x}{x} \qquad \text{Property of logs}$$

$$= \frac{1}{x} \left(\frac{x}{\Delta x} \right) \log_b \left(1 + \frac{\Delta x}{x} \right) \qquad \text{Multiply by } \frac{x}{x} = 1.$$

$$= \frac{1}{x} \log_b \left(1 + \frac{\Delta x}{x} \right)^{x/\Delta x} \qquad \text{Property of logs}$$

Step 2.   Find the limit.

Let $s = \Delta x/x$. For x fixed, if $\Delta x \to 0$, then $s \to 0$. Thus,

$$D_x \log_b x = \lim_{\Delta x \to 0} \frac{f(x + \Delta x) - f(x)}{\Delta x}$$

$$= \lim_{\Delta x \to 0} \frac{1}{x} \log_b \left(1 + \frac{\Delta x}{x} \right)^{x/\Delta x} \qquad \text{Let } s = \Delta x/x.$$

$$= \lim_{s \to 0} \frac{1}{x} \log_b (1 + s)^{1/s}$$

$$= \frac{1}{x} \log_b [\lim_{s \to 0} (1 + s)^{1/s}] \qquad \begin{array}{l} \text{Properties of limits and} \\ \text{continuity of log functions} \end{array}$$

$$= \frac{1}{x} \log_b e \qquad \text{Definition of } e$$

Thus,

$$D_x \log_b x = \frac{1}{x} \log_b e \qquad (1)$$

This derivative formula takes on a particularly simple form for one particular base. Which base? Since $\log_b b = 1$ for any permissible base b, then

$$\log_e e = 1$$

Thus, for the natural logarithmic function

$$\ln x = \log_e x$$

we have

$$D_x \ln x = D_x \log_e x = \frac{1}{x} \log_e e = \frac{1}{x} \cdot 1 = \frac{1}{x} \qquad (2)$$

Now you see why we might want the complicated irrational number e as a base — of all possible bases, it provides the simplest derivative formula for logarithmic functions.

We will now see the power of the chain rule discussed in Section 10-7. Recall that if

$$y = f(u) \qquad \text{and} \qquad u = g(x)$$

then

$$\frac{dy}{dx} = \frac{dy}{du}\frac{du}{dx} \qquad \text{Chain rule}$$

In particular, if

$$y = \log_b u \qquad \text{and} \qquad u = u(x)$$

or

$$y = \ln u \qquad \text{and} \qquad u = u(x)$$

then

$$D_x \log_b u = \frac{1}{u} \log_b e \, D_x u$$

and

$$D_x \ln u = \frac{1}{u} D_x u$$

Let us summarize these results for convenient reference and then consider several examples. Formulas 1 and 2 in the box are used far more frequently than the others; hence, they will be given more attention in the examples and exercises that follow.

Derivatives of Logarithmic Functions

For $b > 0$ and $b \neq 1$:

1. $D_x \ln x = \dfrac{1}{x}$

2. $D_x \ln u = \dfrac{1}{u} D_x u$

3. $D_x \log_b x = \dfrac{1}{x} \log_b e$

4. $D_x \log_b u = \dfrac{1}{u} \log_b e \, D_x u$

Example 16   Differentiate.

(A) $D_x \ln (x^2 + 1)$ (B) $D_x (\ln x)^4$ (C) $D_x \ln x^4$

Solutions   (A) $\ln(x^2 + 1)$ is a composite function of the form

$$y = \ln u \qquad u = u(x) = x^2 + 1$$

Formula 2 applies; thus,

$$D_x \ln(x^2 + 1) = \frac{1}{x^2 + 1} D_x(x^2 + 1)$$

$$= \frac{2x}{x^2 + 1}$$

(B) $(\ln x)^4$ is a composite function of the form

$$y = u^p \qquad u = u(x) = \ln x$$

Hence, $D_x u^p = p u^{p-1} D_x u$, and

$$D_x(\ln x)^4 = 4(\ln x)^3 D_x \ln x \qquad \text{Power rule}$$

$$= 4(\ln x)^3 \left(\frac{1}{x}\right) \qquad \text{Formula 2}$$

$$= \frac{4(\ln x)^3}{x}$$

(C)   We work this problem two ways. The second method takes particular advantage of logarithmic properties.

Method I. $D_x \ln x^4 = \dfrac{1}{x^4} D_x x^4 = \dfrac{4x^3}{x^4} = \dfrac{4}{x}$

Method II. $D_x \ln x^4 = D_x(4 \ln x) = 4 D_x \ln x = \dfrac{4}{x}$

Problem 16   Differentiate.

(A) $D_x \ln(x^3 + 5)$ (B) $D_x(\ln x)^{-3}$ (C) $D_x \ln x^{-3}$

Example 17   Find:

$$D_x \ln \frac{x^5}{\sqrt{x+1}}$$

Solution   Using the chain rule directly results in a messy operation. (Try it.) Instead, we first take advantage of logarithmic properties to write

$$\ln \frac{x^5}{(x+1)^{1/2}} = \ln x^5 - \ln(x+1)^{1/2} = 5 \ln x - (1/2)\ln(x+1)$$

Then,

$$D_x \ln \frac{x^5}{(x+1)^{1/2}} = 5D_x \ln x - (1/2)D_x \ln (x+1)$$

$$= \frac{5}{x} - \frac{1}{2(x+1)}$$

Problem 17   Find $D_x \ln[(x-1)^2 \sqrt{x+2}]$. [*Hint:* Use logarithmic properties first.]

Example 18   Find $D_x[\ln(2x^2 - x)]^3$.

Solution   This problem involves two successive uses of the chain rule:

$$D_x[\ln(2x^2 - x)]^3 = 3[\ln(2x^2 - x)]^2 D_x \ln(2x^2 - x)$$

$$= 3[\ln(2x^2 - x)]^2 \frac{1}{2x^2 - x} [D_x(2x^2 - x)]$$

$$= 3[\ln(2x^2 - x)]^2 \frac{1}{2x^2 - x} (4x - 1)$$

$$= \frac{3(4x - 1)[\ln(2x^2 - x)]^2}{2x^2 - x}$$

Problem 18   Find $D_x \sqrt[3]{\ln(1 + x^3)}$

▪ Graph Properties of $y = \ln x$

Using techniques discussed in Chapter 12, we can use the first and second derivatives of $\ln x$ to give us useful information about the graph of $y = \ln x$. Using the derivative formulas given previously, we have

$$y = \ln x \qquad x > 0$$

$$y' = \frac{1}{x} = x^{-1}$$

$$y'' = -x^{-2} = \frac{-1}{x^2}$$

We see that the first derivative is positive for all x in the domain of ln x (all positive real numbers); hence, ln is an increasing function for *all* x > 0. We also see that the second derivative is negative for all x in the domain of ln x; hence, the graph of y = ln x is concave downward everywhere. It can also be shown that

$$\lim_{x \to 0^+} \ln x = -\infty$$

$$\lim_{x \to \infty} \ln x = \infty$$

Thus, the y axis is a vertical asymptote (there are no horizontal asymptotes) and ln x increases without bound as x → ∞, but ln x increases more slowly than x. The graph of y = ln x is shown in Figure 6.

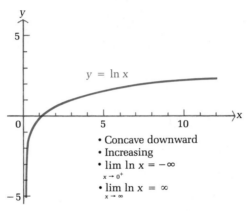

Figure 6

Answers to Matched Problems

16. (A) $\dfrac{3x^2}{x^3 + 5}$ (B) $\dfrac{-3(\ln x)^{-4}}{x}$ (C) $\dfrac{-3}{x}$

17. $\dfrac{2}{x - 1} + \dfrac{1}{2(x + 2)}$ 18. $\dfrac{x^2}{(1 + x^3)[\ln(1 + x^3)]^{2/3}}$

Exercise 13-4

A   *Find each derivative.*

1. $D_t \ln t$ 2. $D_z \ln z$
3. $D_x \ln(x - 3)$ 4. $D_w \ln(w + 100)$
5. $D_x 3 \ln(x - 1)$ 6. $D_x 5 \ln x$
7. $D_z(z^2 + 3 \ln z)$ 8. $D_t(2t^3 - 5 \ln t)$

9. $D_t(6\sqrt{t} - \ln t)$ 10. $D_z\left(\dfrac{2}{z^3} + 2 \ln z\right)$

B   *Find each derivative.*

11. $D_x \ln x^7$
12. $D_x \ln x^{-3}$
13. $D_x \ln \sqrt{x}$
14. $D_x \ln \sqrt[3]{x}$
15. $D_x(\sqrt{\ln x} + \ln \sqrt{x})$
16. $D_x[(\ln x)^5 - \ln x^5]$
17. $D_x \ln(x + 1)^4$
18. $D_x \ln(x + 1)^{-3}$
19. $D_t \ln(t^2 + 3t)$
20. $D_x \ln(x^3 - 3x^2)$
21. $D_x[2x^3 + \ln(x^2 + 1)]$
22. $D_x[\ln(x^2 - 5) + 4x^3]$

23. $D_x \dfrac{\ln x}{x^2}$
24. $D_x \dfrac{\ln 3x}{x^3}$

25. $D_x(x \ln x - x)$
26. $D_x(x^2 \ln x)$
27. $D_x[(x^2 + x) \ln(x^2 + x)]$
28. $D_x[(x^3 + x^2)\ln x]$

29. $D_x \dfrac{\ln x^2}{\ln x^4}$
30. $D_x \dfrac{\ln \sqrt{x}}{\ln x^3}$

31. $D_x \log_{10} x$
32. $D_x \log_2 x$

C   *Find each derivative in Problems 33–48.*

33. $D_x \log_2(3x^2 - 1)$
34. $D_x \log_2(1 - x^3)$
35. $D_x \ln(x^2 + 1)^{1/2}$
36. $D_x \ln(x^2 + 5)^4$
37. $D_x \log_{10}(3x^2 - 2x)$
38. $D_x \log_{10}(x^3 - 1)$

39. $D_x \ln \dfrac{(x - 1)^2}{(x + 1)^3}$
40. $D_x \ln \dfrac{\sqrt{x}}{(x + 1)^2}$

41. $D_x \ln[(x - 1)^2 \sqrt{x}]$
42. $D_x \ln(x^4 \sqrt{x - 1})$
43. $D_x[\ln(x^2 - 1)]^3$
44. $D_z[\ln(2 - z^2)]^5$

45. $D_x \dfrac{1}{\ln(1 + x^2)}$
46. $D_x \dfrac{1}{\ln(1 - x^3)}$

47. $D_x \sqrt[3]{\ln(1 - x^2)}$
48. $D_t \sqrt[5]{\ln(1 - t^5)}$

49. Show that $D_x \ln |x| = 1/x$, $x \neq 0$, by completing the following two cases:

Case 1. $x > 0$

$D_x \ln |x| = D_x \ln x =$

Case 2. $x < 0$

$D_x \ln |x| = D_x \ln (-x) =$

50. Use the results of Problem 49 and the chain rule to find $D_x \ln |x^2 - 1|$.

■

Applications

Business & Economics

51.   *Rate of change of doubling time.* In Section 13-2 we found that n, the doubling time of money invested at 100i% interest compounded an-

nually, is given by

$$n = \frac{\ln 2}{\ln(1 + i)}$$

Find dn/di.

52. *Rate of change of doubling time.* Using dn/di found in Problem 51 and a hand calculator, complete the table to two significant figures. (Remember that a unit change in i corresponds to 100%.)

i		dn/di
0.01	(1%)	$-6,900$
0.03	(3%)	
0.05	(5%)	
0.10	(10%)	
0.20	(20%)	
0.30	(30%)	
0.50	(50%)	
0.80	(80%)	
1.00	(100%)	

Compare the results with Figure 4 in Section 13-2.

Life Sciences

53. *Sound intensity — decibels.* If we let N be the number of decibels and I the power of sound in question (in watts per square centimeter), then N and I are related by

$$N = 10 \log_{10}(I \times 10^{16})$$

Find dN/dI.

13-5   Derivatives of Exponential Functions

- Derivatives of Exponential Functions
- Graph Properties of $y = e^x$ and $y = e^{-x}$

■ Derivatives of Exponential Functions

Recall from Section 13-1 that an exponential function is a function of the form

$$y = b^x \qquad b > 0, \quad b \neq 1 \tag{1}$$

To derive a derivative formula for exponential functions, instead of starting with the basic definition of a derivative, as we did in the last section, we can take advantage of the formulas derived in that section and use implicit differentiation (see Section 11-1).

We start by taking the natural logarithm of both sides of the equation in (1) to obtain

$$\ln y = \ln b^x$$
$$= x \ln b \qquad (2)$$

Now, thinking of y as a function of x,

$$y = y(x)$$

we differentiate both sides of (2) with respect to x:

$$D_x \ln y = D_x(x \ln b)$$

Using formula 2 from Section 13-4 and implicit differentiation, we arrive at

$$\frac{1}{y}\frac{dy}{dx} = \ln b$$

and we solve for dy/dx:

$$\frac{dy}{dx} = y \ln b$$

Recall that $y = b^x$ from equation (1), and we have

$$D_x b^x = b^x \ln b \qquad (3)$$

We ask, as before, for what number b will the derivative formula (3) be the simplest? If $b = e$, then (3) becomes

$$D_x e^x = e^x \ln e = e^x \cdot 1 = e^x$$

and we find that the derivative of the exponential function with base e is the function itself. Thus, all higher-order derivatives of e^x are e^x; that is,

$$D_x^n e^x = e^x \qquad (4)$$

for all natural numbers n.

If we have a composite function

$$y = e^u \qquad u = u(x)$$

or

$$y = b^u \qquad u = u(x)$$

then, using the chain rule, we obtain

$$D_x e^u = e^u \frac{du}{dx} \qquad (5)$$

and

$$D_x b^u = b^u \ln b \frac{du}{dx} \qquad (6)$$

We summarize these results for convenient reference. Then we will consider several examples. Formulas 1 and 2 in the box are used far more frequently than the others; hence, they will be given more attention in the examples and exercises that follow.

Derivatives of Exponential Functions

For $b > 0$ and $b \neq 1$:

1. $D_x e^x = e^x$
2. $D_x e^u = e^u D_x u$
3. $D_x b^x = b^x \ln b$
4. $D_x b^u = b^u \ln b \, D_x u$

Example 19   Differentiate.

(A) $D_x(2x^5 + 3e^x)$ (B) $D_x e^{2x-1}$ (C) $D_x e^{-x^2}$ (D) $D_x 3^{2x}$

Solutions   (A) $D_x(2x^5 + 3e^x) = D_x 2x^5 + D_x 3e^x$
$$= 2D_x x^5 + 3D_x e^x$$
$$= 10x^4 + 3e^x$$

(B) e^{2x-1} is a composite function of the form

$$y = e^u \qquad u = u(x) = 2x - 1$$

and formula 1 applies. Thus,

$$D_x e^{2x-1} = e^{2x-1} D_x(2x - 1)$$
$$= e^{2x-1}(2)$$
$$= 2e^{2x-1}$$

(C) e^{-x^2} is also a composite function of the form

$$y = e^u \qquad u = u(x) = -x^2$$

and, using formula 1, we obtain

$$D_x e^{-x^2} = e^{-x^2} D_x(-x^2)$$
$$= e^{-x^2}(-2x)$$
$$= -2xe^{-x^2}$$

(D) 3^{2x} is a composite function of the form b^u, $u = u(x) = 2x$; hence, formula 3 is used to obtain

$$D_x \, 3^{2x} = 3^{2x}(\ln 3)D_x(2x)$$
$$= 3^{2x}(\ln 3)(2)$$
$$= 2(3^{2x})(\ln 3) = 3^{2x} \ln 3^2 = 3^{2x} \ln 9$$

[Note: $2(3^{2x}) \neq 6^{2x}$ (Why?) and $D_x \, 3^{2x} \neq 2x \, 3^{2x-1}$ (Why?)]

Problem 19   Differentiate

(A) $D_t(2e^t - 3t)$ (B) $D_x \, e^{3x}$ (C) $D_x \, e^{x^2-x}$ (D) $D_x \, 10^{5x+2}$

Example 20   Find:

(A) $D_x(x - e^x \ln x)$ (B) $D_x \dfrac{1 - e^{-x}}{x^2 + 1}$

Solutions   (A) $D_x(x - e^x \ln x) = D_x \, x - D_x \, e^x \ln x$

$$= 1 - (e^x \, D_x \ln x + \ln x \, D_x \, e^x)$$

$$= 1 - \frac{e^x}{x} - e^x \ln x$$

(B) $D_x \dfrac{1 - e^{-x}}{x^2 + 1} = \dfrac{(x^2 + 1)D_x(1 - e^{-x}) - (1 - e^{-x})D_x(x^2 + 1)}{(x^2 + 1)^2}$

$$= \frac{e^{-x}(x^2 + 1) - 2x(1 - e^{-x})}{(x^2 + 1)^2}$$

Problem 20   Find:

(A) $D_x(x^2 e^x + \ln x)$ (B) $D_x \dfrac{e^{2x}}{x + 1}$

Example 21   Find $D_x \, e^{\sqrt{2x-3}}$.

Solution   This problem requires the use of the chain rule twice in succession:

$$D_x \, e^{\sqrt{2x-3}} = D_x \, e^{(2x-3)^{1/2}}$$

$$= e^{(2x-3)^{1/2}} D_x(2x - 3)^{1/2}$$

$$= e^{(2x-3)^{1/2}} \frac{1}{2}(2x - 3)^{-1/2} D_x(2x - 3)$$

$$= e^{(2x-3)^{1/2}} \frac{1}{2}(2x - 3)^{-1/2} \, 2$$

$$= \frac{e^{\sqrt{2x-3}}}{\sqrt{2x - 3}}$$

Problem 21   Find $D_x \, e^{(\ln x)^2}$.

■ Graph Properties of $y = e^x$ and $y = e^{-x}$

Using techniques discussed in Chapter 12, we can use the first and second derivatives of e^x and e^{-x} to give us useful information about the graphs of $y = e^x$ and $y = e^{-x}$. Using the derivative formulas given previously, we have

$$y = e^x$$
$$y' = e^x$$
$$y'' = e^x$$

Both the first and second derivatives are positive for all real x; hence, e^x is an increasing function and the graph of $y = e^x$ is concave upward every-where. In addition, it can be shown that

$$\lim_{x \to -\infty} e^x = 0$$
$$\lim_{x \to \infty} e^x = \infty$$

Thus, the x axis is a horizontal asymptote (there are no vertical asymptotes) and e^x increases without bound as $x \to \infty$, but e^x increases much more rapidly than x. The graph of $y = e^x$ is shown in Figure 7.

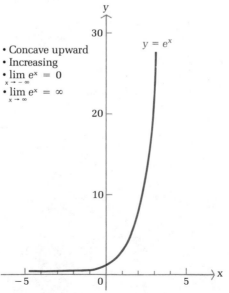

• Concave upward
• Increasing
• $\lim_{x \to -\infty} e^x = 0$
• $\lim_{x \to \infty} e^x = \infty$

$y = e^x$

Figure 7

We now use the same techniques to analyze the graph of e^{-x}.

$$y = e^{-x}$$

$$y' = D_x\, e^{-x} = e^{-x}\, D_x(-x) = -e^{-x}$$

$$y'' = D_x(-e^{-x}) = -e^{-x} D_x(-x) = e^{-x}$$

For all real values of x the first derivative is negative and the second derivative is positive. Hence, e^{-x} is a decreasing function and the graph of $y = e^{-x}$ is concave upward everywhere. In addition, it can be shown that

$$\lim_{x \to -\infty} e^{-x} = \infty \qquad \text{and} \qquad \lim_{x \to \infty} e^{-x} = 0$$

Thus, the x axis is a horizontal asymptote (there are no vertical asymptotes) and e^{-x} increases without bound as $x \to -\infty$. The graph of $y = e^{-x}$ is shown in Figure 8.

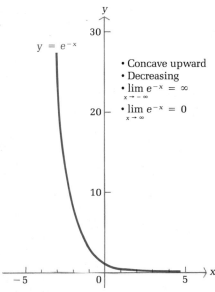

Figure 8

19. (A) $2e^t - 3$ (B) $3e^{3x}$ (C) $(2x - 1)e^{x^2-x}$

(D) $5(10^{5x+2})(\ln 10)$

20. (A) $x^2 e^x + 2x e^x + \dfrac{1}{x}$ (B) $\dfrac{e^{2x}(2x + 1)}{(x + 1)^2}$

21. $\dfrac{2e^{(\ln x)^2} \ln x}{x}$

Exercise 13-5

A   *Find each derivative.*

1. $D_t\, e^t$
2. $D_z\, e^z$
3. $D_x\, e^{8x}$
4. $D_x\, e^{5x-1}$
5. $D_x\, 3e^{2x}$
6. $D_y\, 2e^{3y}$
7. $D_t\, 2e^{-4t}$
8. $D_r\, 6e^{-3r}$
9. $D_x(\ln x + 2e^x)$
10. $D_x(3x^4 - e^{2x})$
11. $D_x(2e^{2x} - 3e^x + 5)$
12. $D_t(1 + e^{-t} - e^{-2t})$

B   *Find each derivative.*

13. $D_x\, e^{3x^2-2x}$
14. $D_x\, e^{x^3-3x^2+1}$
15. $D_x\, \dfrac{e^x - e^{-x}}{2}$
16. $D_x\, \dfrac{e^x + e^{-x}}{2}$
17. $D_x(e^{2x} - 3x^2 + 5)$
18. $D_x(2e^{3x} - 2e^{2x} + 5x)$
19. $D_x(xe^x)$
20. $D_x(x - 1)e^x$
21. $D_x\, 100e^{-0.03x}$
22. $D_t\, 1{,}000e^{0.06t}$
23. $D_x(e^{2x} - 1)^4$
24. $D_x(e^{x^2} + 3)^5$
25. $D_x\, 7^x$
26. $D_x\, 2^x$
27. $D_x[(e^x)^4 + e^{x^4}]$
28. $D_x(\sqrt{e^x} + e^{\sqrt{x}})$
29. $D_x\, \dfrac{e^{2x}}{x^2 + 1}$
30. $D_x\, \dfrac{e^{x+1}}{x + 1}$
31. $D_x(x^2 + 1)e^{-x}$
32. $D_x(1 - x)e^{2x}$
33. $D_x\, e^{-x} \ln x$
34. $D_x\, \dfrac{\ln x}{e^x + 1}$

C   *Find each derivative.*

35. $D_x\, e^{\sqrt[3]{3x+1}}$
36. $D_x\, e^{\sqrt{1-x^2}}$
37. $D_x\, xe^x \ln x$
38. $D_x\, \dfrac{e^x - e^{-x}}{e^x + e^{-x}}$
39. $D_x\, 10^{x^2+x}$
40. $D_x\, 8^{1-2x^2}$

Applications

Business & Economics

41. *Marginal analysis.* Suppose the price–demand equation for x units of a commodity is

$$p(x) = 100e^{-0.05x}$$

Then the revenue equation is

$$R(x) = xp(x) = x100e^{-0.05x}$$

Find the marginal revenue.

42. *Marginal analysis.* Suppose the price–supply equation for x units of a commodity is

$$p(x) = 10e^{0.05x}$$

where p is in dollars. Find the marginal price.

43. *Salvage value.* The salvage value S of a company airplane after t years is estimated to be given by

$$S(t) = 300,000e^{-0.1t}$$

What is the rate of depreciation in dollars per year after 1 year? 5 years? 10 years?

44. *Resale value.* The resale value R of a company car after t years is estimated to be given by

$$R(t) = 20,000e^{-0.15t}$$

What is the rate of depreciation in dollars per year after 1 year? 2 years? 3 years?

Life Sciences

45. *Bacterial growth.* A single cholera bacterium divides every 0.5 hour to produce two complete cholera bacteria. If we start with a colony of 5,000 bacteria, then after t hours there will be

$$A = 5,000 \cdot 2^{2t}$$

bacteria. Find $A'(t)$, $A'(5)$, and $A'(10)$. Compute numerical quantities to three significant digits.

46. *Ecology.* The atmospheric pressure P (in pounds per square inch) at x miles above sea level is given approximately by

$$P = 14.7e^{-0.21x}$$

What is the rate of change in pressure at $x = 1$? At $x = 5$? At $x = 10$?

Social Sciences

47. *Psychology — learning.* Suppose a particular person's history of learning to type is given by the equation

$$N = 80(1 - e^{-0.08t})$$

where N is the number of words per minute typed after t weeks of instruction. Find $N'(t)$, $N'(1)$, $N'(5)$, and $N'(20)$.

13-6 Chapter Review

Important Terms and Symbols

13-1 *Exponential functions—a review.* algebraic function, exponential function, graphs of exponential functions, base e, exponential properties, b^x, e^x

Exercise 13-6   Chapter Review

A   1.   Write $\log_{10} y = x$ in exponential form.

2.   Write $\log_b \dfrac{wx}{y}$ in terms of simpler logarithms.

Find the indicated derivatives in Problems 3–5.

3. $D_x(2 \ln x + 3e^x)$

4. $D_x e^{2x-3}$

5. y' for $y = \ln(2x^3 - 3x)$

B   6.   (A)   Find b: $\log_b 9 = 2$ (B)   Find x: $\log_4 x = -3/2$

7.   Write in terms of simpler logarithmic forms:

$$\log_b \frac{\sqrt[3]{x}}{uv^3}$$

8.   Write in terms of simpler logarithmic forms:

$$\log_b(100 \cdot 1.06^t)$$

9.   Graph $y = 100e^{-0.1x}$, $0 \leqslant x \leqslant 10$, using a calculator or a table.

10.   Find x: $\log_b x = 3 \log_b 2 - \dfrac{3}{2} \log_b 4 - \dfrac{1}{2} \log_b 36$

11.   Find x: $\log x + \log(x - 3) = 1$

12.   Evaluate to five decimal places using a scientific calculator:

(A) $\log 0.009\ 108\ 5$ (B) $\ln 9{,}843.3$

13.   Find x to four significant digits:

(A) $\log x = -3.805\ 5$ (B) $\ln x = 12.814\ 3$

14.   Find t: $240 = 80e^{0.12t}$

Find the indicated derivatives in Problems 15 – 19.

15. $D_x[2\sqrt{x} + \ln(x^3 + 1)]$

16. dy/dx for $y = e^{-2x} \ln 5x$

17. $D_x^2 e^{x^2}$

18. $D_x \ln \dfrac{(x^2 - x)^2}{x^3 + 1}$

19. $D_z(\sqrt{\ln z} - \ln \sqrt{z})$

C 20. Write $\ln y - \ln c = -0.2x$ in an exponential form free of logarithms.

Find the indicated derivatives in Problems 21 – 23.

21. y' for $y = 5^{x^2-1}$

22. $D_x \log_5(x^2 - x)$

23. $D_x \sqrt{\ln (x^2 + x)}$

Applications

Business & Economics

24. *Doubling time.* How long (to three significant digits) will it take money to double if it is invested at 5% interest compounded

   (A) Annually?   (B) Continuously?

25. *Continuous compound interest.* If $100 is invested at 10% interest compounded continuously, the amount (in dollars) at the end of t years is given by

$$A = 100e^{0.1t}$$

   Find $A'(t)$, $A'(1)$, and $A'(10)$. Compute numerical quantities to four significant digits.

26. *Marginal analysis.* If the price–demand equation for x units of a commodity is

$$p(x) = 1{,}000e^{-0.02x}$$

   then the revenue equation is

$$R(x) = xp(x) = 1{,}000xe^{-0.02x}$$

   Find the marginal revenue equation.

Practice Test: Chapter 13

Find x in Problems 1 – 3.

1. $\log_8 x = -\dfrac{2}{3}$

2. $\ln x + \ln(x - 3) = \ln 28$
3. $22 = 11e^{0.08x}$
4. Write in equivalent exponential form: $\ln y - \ln c = -0.03x$

Find the indicated derivatives in Problems 5–10.

5. $D_x[3e^{-x} - \ln(x + 1)]$
6. $D_x^2 e^{x^2-x}$
7. dy/dx for $y = \ln \dfrac{3x - 5}{(x^2 - 1)^3}$
8. y' for $y = \sqrt{\ln \sqrt{x}}$
9. $D_x 10^{x^2-1}$
10. $D_x \log_{10}(x^2 - x)$
11. Suppose the price–demand equation for x units of a commodity is

    $$p(x) = 10,000e^{-0.015x}$$

    (A) Find the revenue equation.
    (B) Find the marginal revenue equation.

12. Find, to three significant digits, the tripling time for money invested at 15% interest:

    (A) Compounded annually
    (B) Compounded continuously

Integration

CHAPTER 14   Contents

The last four chapters dealt with differential calculus. We now begin the development of the second main part of calculus, called *integral calculus*. Two types of integrals will be introduced, the *indefinite integral* and the *definite integral*; each is quite different from the other. But through the remarkable *fundamental theorem of calculus*, we will show that not only are the two integral forms intimately related, but both are intimately related to differentiation.

14-1   Antiderivatives and Indefinite Integrals

- Antiderivatives
- Indefinite Integrals
- Indefinite Integrals Involving Algebraic Functions
- Indefinite Integrals Involving Exponential and Logarithmic Functions
- Applications

■ Antiderivatives

Many operations in mathematics have reverses—compare addition and subtraction, multiplication and division, and powers and roots. The function $f(x) = (1/3) x^3$ has the derivative $f'(x) = x^2$. Reversing this process is referred to as *antidifferentiation*. Thus,

$$\frac{x^3}{3} \quad \text{is an antiderivative of} \quad x^2$$

since

$$D_x \left(\frac{x^3}{3} \right) = x^2$$

In general, we say that $F(x)$ is an **antiderivative** of $f(x)$ if

$$F'(x) = f(x)$$

Note that

$$D_x \left(\frac{x^3}{3} + 2 \right) = x^2 \qquad D_x \left(\frac{x^3}{3} - \pi \right) = x^2 \qquad D_x \left(\frac{x^3}{3} + \sqrt{5} \right) = x^2$$

Hence,

$$\frac{x^3}{3} + 2 \qquad \frac{x^3}{3} - \pi \qquad \frac{x^3}{3} + \sqrt{5}$$

are also antiderivatives of x^2, since each has x^2 as a derivative. In fact, it appears that

$$\frac{x^3}{3} + C$$

for any real number C, is an antiderivative of x^2, since

$$D_x \left(\frac{x^3}{3} + C \right) = x^2$$

Thus, antidifferentiation of a given function does not, in general, lead to a unique function, but to a whole set of functions.

Does the expression

$$\frac{x^3}{3} + C$$

with C any real number, include all antiderivatives of x^2? Theorem 1 (which we state without proof) indicates that the answer is yes.

Theorem 1

> If F and G are differentiable functions on the interval (a, b) and $F'(x) = G'(x)$, then $F(x) = G(x) + k$ for some constant k.

■ Indefinite Integrals

In words, Theorem 1 states that **if the derivatives of two functions are equal, then the functions differ by at most a constant.** We use the symbol

$$\int f(x)\, dx$$

called the **indefinite integral,** to represent all antiderivatives of $f(x)$, and we write

$$\int f(x)\, dx = F(x) + C \qquad \text{where } F'(x) = f(x)$$

that is, if $F(x)$ is any antiderivative of $f(x)$. The symbol $\int$ is called an **integral sign** and $f(x)$ is called the **integrand.** (We will have more to say about the symbol dx later.) The arbitrary constant C is called the **constant of integration.**

■ Indefinite Integrals Involving Algebraic Functions

Just as with differentiation, we can develop formulas and special properties that will enable us to find indefinite integrals of many frequently encountered functions. To start, we list some formulas that can be established using the definitions of antiderivative and indefinite integral, and the many properties of derivatives considered in Chapter 10.

Indefinite Integral Formulas and Properties

For k and C constants:

1. $\displaystyle \int k \, dx = kx + C$

2. $\displaystyle \int x^n \, dx = \frac{x^{n+1}}{n+1} + C \qquad n \neq -1$

3. $\displaystyle \int kf(x) \, dx = k \int f(x) \, dx$

4. $\displaystyle \int [f(x) \pm g(x)] \, dx = \int f(x) \, dx \pm \int g(x) \, dx$

We will establish formula 2 and property 3 here (the others may be shown to be true in a similar manner). To establish formula 2, we simply differentiate the right side to obtain the integrand on the left side. Thus,

$$D_x \left(\frac{x^{n+1}}{n+1} + C \right) = \frac{(n+1)x^n}{(n+1)} + 0 = x^n \qquad n \neq -1$$

(The case when $n = -1$ will be considered later in this section.) To establish property 3, let F be a function such that $F'(x) = f(x)$. Then

$$k \int f(x) \, dx = k \int F'(x) \, dx = k[F(x) + C_1] = kF(x) + kC_1$$

and since $(kF(x))' = kF'(x) = kf(x)$, we have

$$\int kf(x) \, dx = \int kF'(x) \, dx = kF(x) + C_2$$

But $kF(x) + kC_1$ and $kF(x) + C_2$ describe the same set of functions, since C_1 and C_2 are arbitrary real numbers. It is important to remember that property 3 states that a constant factor can be moved across an integral sign; a variable factor cannot be moved across an integral sign.

Correct   *Incorrect*

$$\int 5x^{1/2}\,dx = 5\int x^{1/2}\,dx \qquad \int xx^{1/2}\,dx = x\int x^{1/2}\,dx$$

Now let us put the formulas and properties to use.

Example 1 (A) $\displaystyle\int 5\,dx = 5x + C$

(B) $\displaystyle\int x^4\,dx = \frac{x^{4+1}}{4+1} + C = \frac{x^5}{5} + C$

(C) $\displaystyle\int 5x^7\,dx = 5\int x^7\,dx = 5\,\frac{x^8}{8} + C = \frac{5}{8}x^8 + C$

(D) $\displaystyle\int (4x^3 + 2x - 1)\,dx \;\begin{aligned}[t] &= \int 4x^3\,dx + \int 2x\,dx - \int dx \\[4pt] &= 4\int x^3\,dx + 2\int x\,dx - \int dx \\[4pt] &= \frac{4x^4}{4} + \frac{2x^2}{2} - x + C \end{aligned}$

Property 4 can be extended to the sum and difference of an arbitrary number of functions.

$$= x^4 + x^2 - x + C$$

(E) $\displaystyle\int \frac{3\,dx}{x^2} = \int 3x^{-2}\,dx = \frac{3x^{-2+1}}{-2+1} + C = -3x^{-1} + C$

(F) $\displaystyle\int 5\sqrt[3]{x^2}\,dx = 5\int x^{2/3}\,dx = 5\,\frac{x^{(2/3)+1}}{(2/3)+1} + C$

$$= 5\,\frac{x^{5/3}}{5/3} + C = 3x^{5/3} + C$$

To check any of these, we differentiate the final result to obtain the integrand in the original indefinite integral. When you evaluate an indefinite integral, do not forget to include the arbitrary constant C.

Problem 1 Find each of the following:

(A) $\displaystyle\int dx$ (B) $\displaystyle\int 3x^4\,dx$ (C) $\displaystyle\int (2x^5 - 3x^2 + 1)\,dx$

(D) $\displaystyle\int 4\sqrt[5]{x^3}\,dx$ (E) $\displaystyle\int \left(2x^{2/3} - \frac{3}{x^4}\right)dx$

Example 2   (A) $\displaystyle\int\left(\frac{2}{\sqrt[3]{x}}-6\sqrt{x}\right)dx = \int(2x^{-1/3}-6x^{1/2})\,dx$

$$= 2\int x^{-1/3}\,dx - 6\int x^{1/2}\,dx$$

$$= 2\frac{x^{(-1/3)+1}}{(-1/3)+1} - 6\frac{x^{(1/2)+1}}{(1/2)+1} + C$$

$$= 2\frac{x^{2/3}}{2/3} - 6\frac{x^{3/2}}{3/2} + C$$

$$= 3x^{2/3} - 4x^{3/2} + C$$

(B) $\displaystyle\int\frac{x^3-3}{\sqrt{x}}\,dx = \int\left(\frac{x^3}{x^{1/2}}-\frac{3}{x^{1/2}}\right)dx$

$$= \int(x^{5/2}-3x^{-1/2})\,dx$$

$$= \frac{x^{(5/2)+1}}{(5/2)+1} - 3\frac{x^{(-1/2)+1}}{(-1/2)+1} + C$$

$$= \frac{x^{7/2}}{7/2} - 3\frac{x^{1/2}}{1/2} + C$$

$$= \frac{2}{7}x^{7/2} - 6x^{1/2} + C$$

Problem 2   Find each indefinite integral.

(A) $\displaystyle\int\left(8\sqrt[3]{x}-\frac{6}{\sqrt{x}}\right)dx$ (B) $\displaystyle\int\frac{\sqrt{x}-8x^3}{x^2}\,dx$

▪ Indefinite Integrals Involving Exponential and Logarithmic Functions

The four indefinite integral formulas given in the next box follow immediately from the derivative formulas discussed in the last chapter. Because of the absolute value, formula 8 is the least obvious of the four and causes the most confusion among students. Let us show that

$$D_x \ln|x| = \frac{1}{x} \qquad x \neq 0$$

We consider two cases:

Case 1. $x > 0$

$$D_x \ln|x| = D_x \ln x \qquad \text{Since } |x| = x \text{ for } x > 0$$

$$= \frac{1}{x}$$

Indefinite Integral Formulas

5. $\displaystyle\int e^x \, dx = e^x + C$

6. $\displaystyle\int e^{ax} \, dx = \frac{1}{a} e^{ax} + C \qquad a \neq 0$

7. $\displaystyle\int \frac{dx}{x} = \ln x + C \qquad x > 0$

8. $\displaystyle\int \frac{dx}{x} = \ln|x| + C \qquad x \neq 0$

Note: Formula 7 is a special case of formula 8.

Case 2. $x < 0$

$$D_x \ln|x| = D_x \ln(-x) \qquad \text{Since } |x| = -x \text{ for } x < 0$$

$$= \frac{1}{-x} D_x(-x)$$

$$= \frac{-1}{-x} = \frac{1}{x}$$

Thus,

$$D_x \ln|x| = \frac{1}{x} \qquad x \neq 0$$

Hence,

$$\int \frac{1}{x} \, dx = \ln|x| + C \qquad x \neq 0$$

What about the indefinite integral of ln x? We postpone a discussion of $\int \ln x \, dx$ until Section 15-2, where we will be able to find it using a technique called *integration by parts*.

Example 3   (Λ) $\displaystyle\int \frac{e^x - e^{-x}}{2} \, dx \quad = \int \left(\frac{e^x}{2} - \frac{e^{-x}}{2} \right) dx$

$$= \frac{1}{2} \int e^x \, dx - \frac{1}{2} \int e^{-x} \, dx$$

$$= \frac{1}{2} e^x - \frac{1}{2}(-e^{-x}) + C$$

$$= \frac{1}{2} e^x + \frac{1}{2} e^{-x} + C$$

$$= \frac{e^x + e^{-x}}{2} + C$$

(B) $\displaystyle\int \frac{2x^5 - 3x}{x^2}\, dx = \int \left(\frac{2x^5}{x^2} - \frac{3x}{x^2}\right) dx$

$\displaystyle = \int \left(2x^3 - \frac{3}{x}\right) dx$

$\displaystyle = 2\int x^3 - 3\int \frac{1}{x} dx$

$\displaystyle = 2\frac{x^4}{4} - 3\,\ln|x| + C$

$\displaystyle = \frac{1}{2}x^4 - 3\,\ln|x| + C$

Problem 3   Find each indefinite integral.

(A) $\displaystyle\int \frac{3e^{5x} - 4}{e^{2x}}\, dx$ (B) $\displaystyle\int \frac{xe^x - 3}{x}\, dx$

Let us now consider some applications of the indefinite integral to see why we are interested in finding antiderivatives of functions.

■ Applications

Example 4   Find the equation of the curve that passes through (2, 5) if its slope is given by $dy/dx = 2x$ at any point x.

Solution   We are interested in finding a function $y = f(x)$ such that

$$\frac{dy}{dx} = 2x \tag{1}$$

and

$$y = 5 \qquad \text{when } x = 2 \tag{2}$$

If

$$\frac{dy}{dx} = 2x$$

then

$$y = \int 2x\, dx$$
$$= x^2 + C \tag{3}$$

Since $y = 5$ when $x = 2$, we determine the *particular* value of C so that

$$5 = 2^2 + C$$

Thus,

$$C = 1$$

and

$$y = x^2 + 1$$

is the particular antiderivative out of all those possible from (3) that satisfies both (1) and (2). See Figure 1.

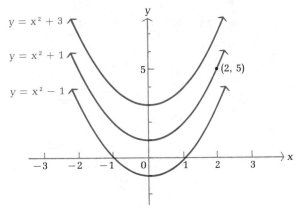

Figure 1 $y = x^2 + C$

Problem 4   Find the equation of the curve that passes through (2, 6) if the slope of the curve at any point x is given by $dy/dx = 3x^2$.

In certain situations it is easier to determine the rate at which something happens than how much of it has happened in a given length of time (e.g., population growth rates, business growth rates, rate of healing of a wound, rates of learning or forgetting). If a rate function (derivative) is given and we know the value of the dependent variable for a given value of the independent variable, then—if the rate function is not too complicated—we can often find the original function by integration.

Example 5
Cost Function   If the marginal cost of producing x units is given by

$$C'(x) = 3x^2 - 2x$$

and the fixed cost is \$2,000, find the cost function C(x) and the cost of producing twenty units.

Solution   Recall that marginal cost is the derivative of the cost function and that fixed cost is cost at a zero production level. Thus, the mathematical problem is to find C(x) given

$$C'(x) = 3x^2 - 2x \qquad C(0) = 2,000$$

We now find the indefinite integral of $3x^2 - 2x$ and determine the arbitrary integration constant using $C(0) = 2,000$.

$$C'(x) = 3x^2 - 2x$$

$$C(x) = \int (3x^2 - 2x)\,dx$$

$$= x^3 - x^2 + K \qquad \text{Since } C \text{ represents the cost, we use } K \text{ for the constant of integration.}$$

But

$$C(0) = 0^3 - 0^2 + K = 2,000$$

Thus,

$$K = 2,000$$

and the particular cost function is

$$C(x) = x^3 - x^2 + 2,000$$

We now find $C(20)$, the cost of producing twenty units:

$$C(20) = 20^3 - 20^2 + 2,000$$
$$= \$9,600$$

Problem 5   Find the revenue function $R(x)$ when the marginal revenue is

$$R'(x) = 400 - 0.4x$$

and no revenue results at a zero production level. What is the revenue at a production level of 1,000 units?

Example 6
Price–Demand   The marginal price $p'(x)$ at x units per month demand for a given model sailboat is given by

$$p'(x) = -500e^{-0.05x}$$

Find the price–demand equation if at a price $17,788 each the demand is 5 boats per month.

Solution $$p(x) = \int -500e^{-0.05x}\,dx$$

$$= -500 \int e^{-0.05x}\,dx$$

$$= -500\,\frac{e^{-0.05x}}{-0.05} + C$$

$$= 10,000e^{-0.05x} + C$$

We find C by noting that

$$p(5) = 10{,}000e^{-0.05(5)} + C = \$17{,}788$$
$$C = \$17{,}788 - 10{,}000e^{-0.25} \qquad \text{Use a calculator or a table.}$$
$$C = \$17{,}788 - 7{,}788$$
$$C = \$10{,}000$$

Thus,

$$p(x) = 10{,}000e^{-0.05x} + 10{,}000$$

Problem 6   The marginal price $p'(x)$ at x units per month supply for a given model sailboat is given by

$$p'(x) = 500e^{0.05x}$$

Find the price–supply equation if the supplier is willing to supply 10 boats per month at a price of $14,487 each.

Answers to
Matched Problems

1. (A) $x + C$ (B) $(3/5)x^5 + C$ (C) $x^6/3 - x^3 + x + C$
   (D) $(5/2)x^{8/5} + C$ (E) $(6/5)x^{5/3} + x^{-3} + C$
2. (A) $6x^{4/3} - 12x^{1/2} + C$ (B) $-2x^{-1/2} - 4x^2 + C$
3. (A) $e^{3x} + 2e^{-2x} + C$ (B) $e^x - 3\ln|x| + C$
4. $y = x^3 - 2$
5. $R(x) = 400x - 0.2x^2;$ $R(1{,}000) = \$200{,}000$
6. $p(x) = 10{,}000e^{0.05x} - 2{,}000$

Exercise 14-1

A   *Find each indefinite integral. (Check by differentiating.)*

1. $\displaystyle\int 7\, dx$

2. $\displaystyle\int \pi\, dx$

3. $\displaystyle\int x^6\, dx$

4. $\displaystyle\int x^3\, dx$

5. $\displaystyle\int 8t^3\, dt$

6. $\displaystyle\int 10t^4\, dt$

7. $\displaystyle\int (2u + 1)\, du$

8. $\displaystyle\int (1 - 2u)\, du$

9. $\displaystyle\int (3x^2 + 2x - 5)\, dx$

10. $\displaystyle\int (2 + 4x - 6x^2)\, dx$

11. $\displaystyle\int (s^4 - 8s^5)\, ds$

12. $\displaystyle\int (t^5 + 6t^3)\, dt$

13. $\int e^{3t} dt$

14. $\int e^{-2t} dt$

15. $\int 2z^{-1} dz$

16. $\int \frac{3}{s} ds$

Find all the antiderivatives for each derivative.

17. $\frac{dy}{dx} = 200x^4$

18. $\frac{dx}{dt} = 42t^5$

19. $\frac{dP}{dx} = 24 - 6x$

20. $\frac{dy}{dx} = 3x^2 - 4x^3$

21. $\frac{dy}{du} = 2u^5 - 3u^2 - 1$

22. $\frac{dA}{dt} = 3 - 12t^3 - 9t^5$

23. $\frac{dy}{dx} = e^{-x} + 3$

24. $\frac{dy}{dx} = x - e^{-x}$

25. $\frac{dx}{dt} = 5t^{-1} + 1$

26. $\frac{du}{dv} = \frac{4}{v} + \frac{v}{4}$

B   Find each indefinite integral. (Check by differentiation.)

27. $\int 6x^{1/2} dx$

28. $\int 8t^{1/3} dt$

29. $\int 8x^{-3} dx$

30. $\int 12u^{-4} du$

31. $\int \frac{du}{\sqrt{u}}$

32. $\int \frac{dt}{\sqrt[3]{t}}$

33. $\int \frac{dx}{4x^3}$

34. $\int \frac{6\,dm}{m^2}$

35. $\int \frac{du}{2u^5}$

36. $\int \frac{dy}{3y^4}$

37. $\int (3x^{1/2} - x^{-1/2}) dx$

38. $\int (4x^{1/3} + 2x^{-1/3}) dx$

39. $\int (10x^{2/3} - 8x^{1/3} - 2) dx$

40. $\int (6x^{-4} - 2x^{-3} + 1) dx$

41. $\int \left(3\sqrt{x} + \frac{2}{\sqrt{x}}\right) dx$

42. $\int \left(\frac{2}{\sqrt[3]{x}} - \sqrt[3]{x^2}\right) dx$

43. $\int \left(\sqrt[3]{x^2} - \frac{4}{x^3}\right) dx$

44. $\int \left(\frac{12}{x^5} - \frac{1}{\sqrt[3]{x^2}}\right) dx$

45. $\int \frac{e^{3x} - e^{-3x}}{2} dx$

46. $\int \frac{e^x + e^{-x}}{2} dx$

47. $\int (2z^{-3} + z^{-2} + z^{-1}) dz$

48. $\int (3x^{-2} - x^{-1}) dx$

In Problems 49–58, find the particular antiderivative of each derivative that satisfies the given condition.

49. $\dfrac{dy}{dx} = 2x - 3, \quad y(0) = 5$

50. $\dfrac{dy}{dx} = 5 - 4x, \quad y(0) = 20$

51. $C'(x) = 6x^2 - 4x, \quad C(0) = 3{,}000$

52. $R'(x) = 600 - 0.6x, \quad R(0) = 0$

53. $\dfrac{dx}{dt} = \dfrac{20}{\sqrt{t}}, \quad x(1) = 40$

54. $\dfrac{dR}{dt} = \dfrac{100}{t^2}, \quad R(1) = 400$

55. $\dfrac{dy}{dx} = 2x^{-2} + 3x^{-1} - 1, \quad y(1) = 0$

56. $\dfrac{dy}{dx} = 3x^{-1} + x^{-2}, \quad y(1) = 1$

57. $\dfrac{dx}{dt} = 4e^{-2t} - 3e^{-t} - 2, \quad x(0) = 1$

58. $\dfrac{dy}{dt} = 5e^{5t} - 4e^{4t} + e^{-t}, \quad y(0) = -1$

59. Find the equation of the curve that passes through (2, 3) if its slope is given by

$$\dfrac{dy}{dx} = 4x - 3$$

for each x.

60. Find the equation of the curve that passes through (1, 3) if its slope is given by

$$\dfrac{dy}{dx} = 12x^2 - 12x$$

for each x.

C Find each indefinite integral.

61. $\displaystyle\int \dfrac{2x^4 - x}{x^3}\, dx$

62. $\displaystyle\int \dfrac{x^{-1} - x^4}{x^2}\, dx$

63. $\displaystyle\int \dfrac{x^5 - 2x}{x^4}\, dx$

64. $\displaystyle\int \dfrac{1 - 3x^4}{x^2}\, dx$

65. $\displaystyle\int \dfrac{x^2 e^{2x} - 2x}{x^2}\, dx$

66. $\displaystyle\int \dfrac{x - e^x}{xe^x}\, dx$

Find the antiderivative of each of the derivatives that satisfies the given condition.

67. $\dfrac{dM}{dt} = \dfrac{\sqrt{t} - 1}{\sqrt{t}}$, $M(4) = 5$

68. $\dfrac{dR}{dx} = \dfrac{1 - \sqrt[3]{x^2}}{\sqrt[3]{x}}$, $R(8) = 4$

69. $\dfrac{dy}{dx} = \dfrac{5x + 2}{\sqrt[3]{x}}$, $y(1) = 0$

70. $\dfrac{dx}{dt} = \dfrac{\sqrt{t^3} - t}{\sqrt{t^3}}$, $x(9) = 4$

71. $p'(x) = -100e^{-0.05x}$, $p(0) = 3{,}000$

72. $p'(x) = 40e^{0.004x}$, $p(0) = 4{,}000$

■

Applications

Business & Economics

73. *Profit function.* If the marginal profit for producing x units is given by

$$P'(x) = 50 - 0.04x \qquad P(0) = 0$$

where $P(x)$ is the profit in dollars, find the profit function P and the profit on 100 units of production.

74. *Natural resources.* The world demand for wood is increasing. In 1975 the demand was 12.6 billion cubic feet, and the rate of increase in demand is given approximately by

$$d'(t) = 0.009t$$

where t is time in years after 1975. Noting that $d(0) = 12.6$, find $d(t)$. Also find $d(25)$, the demand in the year 2000.

75. *Price–demand equation.* The marginal price $p'(x)$ at x units demand per month for a given model water-skiing boat is given by

$$p'(x) = -200e^{-0.04x}$$

Find the price–demand equation if at a price of $6,094 each the demand is 12 boats per month.

76. *Price–supply equation.* The marginal price $p'(x)$ at x units per month supply for a given model water-skiing boat is given by

$$p'(x) = 200e^{0.04x}$$

Find the price–supply equation if the supplier is willing to supply 5 boats per month at a price of $5,107 each.

Life Sciences

77. *Weight–height.* The rate of change of an average person's weight with respect to their height h (in inches) is given approximately by

$$\frac{dW}{dh} = 0.0015h^2$$

Find $W(h)$ if $W(60) = 108$ pounds. Also find the average weight for a person who is 5 feet 10 inches tall.

78. *Wound healing.* If the area of a healing wound changes at a rate given approximately by

$$\frac{dA}{dt} = -4t^{-3} \qquad 1 \le t \le 10$$

where t is in days and $A(1) = 2$ square centimeters, what will the area of the wound be in 10 days?

Social Sciences

79. *Urban growth.* A suburban area of Chicago incorporated into a city. The growth rate t years after incorporation is estimated to be

$$\frac{dN}{dt} = 400 + 600\sqrt{t} \qquad 0 \le t \le 9$$

If the current population is 5,000, what will the population be 9 years from now?

80. *Learning.* In an experiment on memorizing vocabulary from a foreign language, it is found that the rate of learning during a study session increases and then decreases because of saturation. A typical rate might be given by

$$v'(t) = 0.04t - 0.0003t^2$$

where $v(t)$ is the amount of vocabulary learned after t minutes of study. Find $v(t)$ if $v(0) = 0$. Then find how many words are learned after 60 minutes of study.

14-2 Differential Equations—Growth and Decay

- Differential Equations
- Continuous Compound Interest Revisited
- Exponential Growth Law
- Population Growth, Radioactive Decay, Learning
- A Comparison of Exponential Growth Phenomena

■ Differential Equations

In the last section we considered equations of the form

$$\frac{dy}{dx} = 6x^2 - 4x \qquad p'(x) = -400e^{-0.04x}$$

These are examples of differential equations. In general an equation is a **differential equation** if it involves an unknown function (often denoted by *y*) and one or more of its derivatives. Other examples of differential equations are

$$\frac{dy}{dx} = ky \qquad y'' - xy' + x^2 = 5$$

Finding solutions to different types of differential equations (functions that satisfy the equation) is the subject matter for whole books and courses on the subject. Here we will consider only a few very special but very important types of equations that have immediate and significant application. We start by considering the problem of continuous compound interest from another point of view, which will enable us to generalize the concept and apply the results to problems from a number of different fields.

■ Continuous Compound Interest Revisited

Suppose we say that the amount of money *A* in an account grows at a rate proportional to the amount present, and the amount in the account at the start is *P*. Mathematically,

$$\frac{dA}{dt} = rA \qquad A(0) = P \qquad A, P > 0$$

where *r* is an appropriate constant. We would like to find a function $A = A(t)$ that satisfies these conditions. Using differentials, we can treat *dA* and *dt* as separate quantities and multiply both sides of the first equation by dt/A to obtain

$$\frac{dA}{A} = r\, dt$$

Now we integrate both sides,

$$\int \frac{dA}{A} = \int r\, dt$$

$$\ln |A| = rt + C \qquad |A| = A \quad \text{since } A > 0$$

$$\ln A = rt + C$$

and convert this last equation into the equivalent exponential form

$$A = e^{rt+C} \qquad \text{From Section 13-2, } y = \ln x \text{ if and only if } x = e^y$$

$$= e^C e^{rt} \qquad \text{Property of exponents: } b^m b^n = b^{m+n}$$

Since $A(0) = P$, we evaluate $A(t) = e^C e^{rt}$ at $t = 0$ and set it equal to *P*:

$$A(0) = e^C e^0 = e^C = P$$

Hence, $e^C = P$, and we can rewrite $A = e^C e^{rt}$ in the form

$$A = Pe^{rt}$$

This is the same continuous compound interest formula obtained in Section 13-3, where the principal P is invested at an annual rate of r compounded continuously for t years.

■ Exponential Growth Law

In general, if a quantity Q changes at a rate proportional to the amount present and $Q(0) = Q_0$, then proceeding in exactly the same way as above, we obtain the following:

Exponential Growth Law

If $\dfrac{dQ}{dt} = rQ$ and $Q(0) = Q_0$, then $Q = Q_0 e^{rt}$.

$Q_0 =$ Amount at $t = 0$

$r =$ Annual rate compounded continuously

$t =$ Time

$Q =$ Quantity at time t

Once we know that the rate of growth of something is proportional to the amount present, then we know it has exponential growth and we can use the results summarized in the box without having to solve the involved differential equation each time. The exponential growth law applies not only to money invested at interest compounded continuously, but also to many other types of problems—population growth, radioactive decay, natural resource depletion, and so on.

■ Population Growth, Radioactive Decay, Learning

The world population is growing at an ever-increasing rate, as illustrated in Figure 2 on the next page. **Population growth** over certain periods of time can often be approximated by the exponential growth law described above.

Example 7
Population Growth

India had a population of 500 million people in 1966 ($t = 0$) and a growth rate of 3% per year (which we will assume is compounded continuously). If P is the population in millions t years after 1966, and the same growth rate

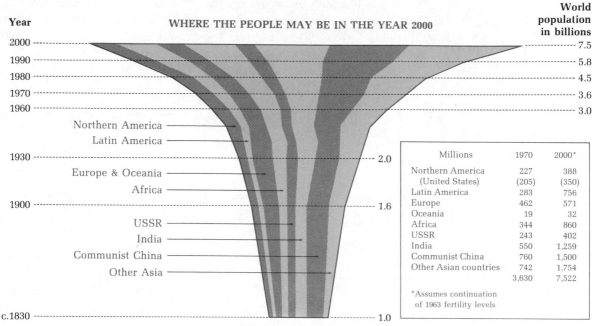

Figure 2 The population explosion. *Source:* United States State Department

continues, then

$$\frac{dP}{dt} = 0.03P \qquad P(0) = 500$$

Thus, using the exponential growth law, we obtain

$$P = 500e^{0.03t}$$

With this result, we can estimate the population of India in 1986 ($t = 20$) to be

$$P(20) = 500e^{0.03(20)}$$

$$\approx 911 \text{ million people}$$

Problem 7 Assuming the same rate of growth, what will India's population be in the year 2001?

Example 8 If the exponential growth law applies to Russia's population growth, at
Population Growth what rate compounded continuously will the population double over the next 100 years?

Solution The problem is to find r, given $P = 2P_0$ and $t = 100$:

$$P = P_0 e^{rt}$$
$$2P_0 = P_0 e^{100r}$$
$$2 = e^{100r} \qquad \text{Take ln of both sides and reverse equation.}$$
$$100r = \ln 2$$
$$r = \frac{\ln 2}{100}$$
$$\approx 0.0069 \quad \text{or} \quad 0.69\%$$

Problem 8 If the exponential growth law applies to population growth in Mexico, find the doubling time of the population if it continues to grow at 3.2% per year compounded continuously.

We now turn to another type of exponential growth — **radioactive decay,** a negative growth. In 1946, Willard Libby (who later received a Nobel Prize in chemistry) found that as long as a plant or animal is alive, radioactive carbon-14 is maintained at a constant level in its tissues. Once the plant or animal is dead, however, the radioactive carbon-14 diminishes by radioactive decay at a rate proportional to the amount present. Thus,

$$\frac{dQ}{dt} = rQ \qquad Q(0) = Q_0$$

and we have another example of the exponential growth law. The rate of decay for radioactive carbon-14 is found to be 0.000 123 8; thus, $r = -0.000\ 123\ 8$, since decay is negative growth.

Example 9
Archaeology
A piece of human bone was found at an archaeological site in Africa. If 10% of the original amount of radioactive carbon-14 was present, estimate the age of the bone.

Solution Using the exponential growth law for

$$\frac{dQ}{dt} = -0.000\ 123\ 8Q \qquad Q(0) = Q_0$$

we find that

$$Q = Q_0 e^{-0.0001238t}$$

and our problem is to find t so that $Q = 0.1Q_0$ (the amount of carbon-14 present now is 10% of the amount present, Q_0, at the death of the person). Thus,

$$0.1Q_0 = Q_0 e^{-0.0001238t}$$
$$0.1 = e^{-0.0001238t}$$
$$\ln 0.1 = \ln e^{-0.0001238t}$$
$$t = \frac{\ln 0.1}{-0.0001238} \approx 18{,}600 \text{ years}$$

Problem 9   Estimate the age of the bone in Example 9 if 50% of the original amount of carbon-14 is present.

In **learning** certain skills such as typing and swimming, a mathematical model often used is one that assumes there is a maximum skill attainable, say M, and the rate of improving is proportional to the difference between that achieved, y, and that attainable, M. Mathematically,

$$\frac{dy}{dt} = k(M - y) \qquad y(0) = 0$$

We solve this using the same technique that was used to obtain the exponential growth law. First, multiply both sides of the first equation by $dt/(M - y)$ to obtain

$$\frac{dy}{M - y} = k \, dt$$

and then integrate both sides:

$$\int \frac{dy}{M - y} = \int k \, dt$$

The indefinite integral on the left side can be verified by differentiation. We will have more to say about this type of integral in Section 15-1.

$$-\ln(M - y) = kt + C$$

Change this last equation to equivalent exponential form:

$$M - y = e^{-kt-C}$$
$$M - y = e^{-C}e^{-kt}$$
$$y = M - e^{-C}e^{-kt}$$

Now $y(0) = 0$; hence,

$$y(0) = M - e^{-C}e^0 = 0$$

Solving for e^{-C}, we obtain

$$e^{-C} = M$$

and our final solution is

$$y = M - Me^{-kt} = M(1 - e^{-kt})$$

Example 10
Learning

For a particular person who is learning to swim, it is found that the distance y (in feet) the person is able to swim in 1 minute after t hours of practice is given approximately by

$$y = 50(1 - e^{-0.04t})$$

What is the rate of improvement after 10 hours of practice?

Solution

$$y = 50 - 50e^{-0.04t}$$
$$y'(t) = 2e^{-0.04t}$$
$$y'(10) = 2e^{-0.04(10)} \approx 1.34 \text{ feet per hour of practice}$$

Problem 10   In Example 10, what is the rate of improvement after 50 hours of practice?

▪ A Comparison of Exponential Growth Phenomena

The graphs and equations given in Table 1 compare several widely used growth models. These are divided basically into two groups: unlimited growth and limited growth. Following each graph and equation is a short, incomplete list of areas in which the models are used. This only touches on

Table 1 Exponential Growth

Description	Model	Solution	Graph	Uses
Unlimited growth Rate of growth is proportional to the amount present.	$\dfrac{dy}{dt} = ky$ $k, t > 0$ $y(0) = c$	$y = ce^{kt}$		• Short-term population growth (people, bacteria, etc.) • Growth of money at continuous compound interest • Price–supply curves • Depletion of natural resources
Exponential decay Rate of growth is proportional to the amount present.	$\dfrac{dy}{dt} = -ky$ $k, t > 0$ $y(0) = c$	$y = ce^{-kt}$		• Radioactive decay • Light absorption in water • Price–demand curves • Atmospheric pressure (t is altitude)
Limited growth Rate of growth is proportional to the difference between the amount present and a fixed limit.	$\dfrac{dy}{dt} = k(M - y)$ $k, t > 0$ $y(0) = 0$	$y = c(1 - e^{-kt})$		• Learning • Sales fads (e.g., skateboards) • Depreciation of equipment • Company growth
Logistic growth Rate of growth is proportional to the amount present and to the difference between the amount present and a fixed amount.	$\dfrac{dy}{dt} = ky(M - y)$ $k, t > 0$ $y(0) = \dfrac{M}{1 + c}$	$y = \dfrac{M}{1 + ce^{-kMt}}$		• Learning • Long-term population growth • Epidemics • Sales of new products • Company growth

a subject that has been extensively developed and which you are likely to encounter in greater depth in the future.

7. 1,429 million people
8. Approximately 22 years
9. Approximately 5,600 years
10. Approximately 0.27 foot per hour

Exercise 14-2

Applications

Business & Economics

1. *Continuous compound interest.* Find the amount A in an account after t years if

$$\frac{dA}{dt} = 0.08A \quad \text{and} \quad A(0) = 1{,}000$$

2. *Continuous compound interest.* Find the amount A in an account after t years if

$$\frac{dA}{dt} = 0.12A \quad \text{and} \quad A(0) = 5{,}250$$

3. *Continuous compound interest.* Find the amount A in an account after t years if

$$\frac{dA}{dt} = rA \quad A(0) = 8{,}000 \quad A(2) = 9{,}020$$

4. *Continuous compound interest.* Find the amount A in an account after t years if

$$\frac{dA}{dt} = rA \quad A(0) = 5{,}000 \quad A(5) = 7{,}460$$

5. *Price–demand.* If the marginal price dp/dx at x units of demand per week is proportional to the price p, and if at $100 there is no weekly demand [$p(0) = 100$], and if at $77.88 there is a weekly demand of 5 units [$p(5) = 77.88$], find the price–demand equation.

6. *Price–supply.* If the marginal price dp/dx at x units of supply per day is proportional to the price p, and if at a price of $10 there is no daily supply [$p(0) = 10$], and if at a price of $12.84 there is a daily supply of 50 units [$p(50) = 12.84$], find the price–supply equation.

Kibbles
'n Bits

7. *Advertising.* A company is trying to expose a new product to as many people as possible through television advertising. Suppose the rate of exposure to new people is proportional to the number of those who have not seen the product out of L possible viewers. If no one is aware of the product at the start of the campaign and after 10 days 40% of L are aware of the product, solve

$$\frac{dN}{dt} = k(L - N) \qquad N(0) = 0 \quad N(10) = 0.4L$$

for $N = N(t)$, the number of people who are aware of the product after t days of advertising.

8. *Advertising.* Repeat Problem 7 for

$$\frac{dN}{dt} = k(L - N) \qquad N(0) = 0 \quad N(10) = 0.1L$$

Life Sciences

9. *Ecology.* For relatively clear bodies of water, light intensity is reduced according to

$$\frac{dI}{dx} = -kI \qquad I(0) = I_0$$

where I is the intensity of light at x feet below the surface. For the Sargasso Sea off the West Indies, $k = 0.009\ 42$. Find I in terms of x and find the depth at which the light is reduced to half of that at the surface.

10. *Blood pressure.* It can be shown under certain assumptions that blood pressure P in the largest artery in the human body (the aorta) changes between beats with respect to time t according to

$$\frac{dP}{dt} = -aP \qquad P(0) = P_0$$

where a is a constant. Find $P = P(t)$ that satisfies both conditions.

11. *Drug concentrations.* A single injection of a drug is administered to a patient. The amount Q in the body then decreases at a rate proportional to the amount present, and for this particular drug the rate is 4% per hour. Thus,

$$\frac{dQ}{dt} = -0.04Q \qquad Q(0) = Q_0$$

where t is time in hours. If the initial injection is 3 milliliters $[Q(0) = 3]$, find $Q = Q(t)$ that satisfies both conditions. How many milliliters of the drug are still in the body after 10 hours?

12. *Simple epidemic.* A community of 1,000 individuals is assumed to be homogeneously mixed. One individual who has just returned from another community has influenza. Assume the home community has not had influenza shots and all are susceptible. One mathematical model for an influenza epidemic assumes that influenza tends to spread at a rate in direct proportion to the number who have it, N, and to the number who have not contracted it, in this case, $1{,}000 - N$. Mathematically,

$$\frac{dN}{dt} = kN(1{,}000 - N) \qquad N(0) = 1$$

where N is the number of people who have contracted influenza after t days. For $k = 0.0004$, it can be shown that $N(t)$ is given by

$$N(t) = \frac{1{,}000}{1 + 999e^{-0.4t}}$$

See Table 1 (logistic growth) for the characteristic graph.

(A) How many people have contracted influenza after 10 days? After 20 days?

(B) How many days will it take until half the community has contracted influenza?

(C) Find $\lim_{t \to \infty} N(t)$.

Social Sciences

13. *Archaeology.* A skull from an ancient tomb was discovered and was found to have 5% of the original amount of radioactive carbon-14 present. Estimate the age of the skull. (See Example 9.)

14. *Learning.* For a particular person learning to type, it was found that the number of words per minute, N, the person was able to type after t hours of practice was given approximately by

$$N = 100(1 - e^{-0.02t})$$

See Table 1 (limited growth) for a characteristic graph. What is the rate of improvement after 10 hours of practice? After 40 hours of practice?

15. *Small group analysis.* In a study on small group dynamics, sociologists Stephan and Mischler found that, when the members of a discussion group of ten were ranked according to the number of times each participated, the number of times $N(k)$ the kth-ranked person participated was given approximately by

$$N(k) = N_1 e^{-0.11(k-1)} \qquad 1 \leqslant k \leqslant 10$$

where N_1 is the number of times the first-ranked person participated in the discussion. If, in a particular discussion group of ten people, $N_1 = 180$, estimate how many times the sixth-ranked person participated. The tenth-ranked person.

16. *Perception.* One of the oldest laws in mathematical psychology is the Weber–Fechner law (discovered in the middle of the nineteenth century). It concerns a person's sensed perception of various strengths

of stimulation involving weights, sound, light, shock, taste, and so on. One form of the law states that the rate of change of sensed sensation S with respect to stimulus R is inversely proportional to the strength of the stimulus R. Thus,

$$\frac{dS}{dR} = \frac{k}{R}$$

where k is a constant. If we let R_0 be the threshold level at which the stimulus R can be detected (the least amount of sound, light, weight, and so on that can be detected), then it is appropriate to write

$$S(R_0) = 0$$

Find a function S in terms of R that satisfies the above conditions.

17. *Rumor spread.* A group of 400 parents, relatives, and friends are waiting anxiously at Kennedy Airport for a student charter to return after a year in Europe. It is stormy and the plane is late. A particular parent thought he had heard that the plane's radio had gone out and related this news to some friends, who in turn passed it on to others, and so on. Sociologists have studied rumor propagation and have found that a rumor tends to spread at a rate in direct proportion to the number who have heard it, x, and to the number who have not, $P - x$, where P is the total population. Mathematically, for our case, $P = 400$ and

$$\frac{dx}{dt} = 0.001x(400 - x) \qquad x(0) = 1$$

where t is time in minutes. From this, it can be shown that

$$x(t) = \frac{400}{1 + 399e^{-0.4t}}$$

See Table 1 (logistic growth) for a characteristic graph.

(A) How many people have heard the rumor after 5 minutes? 20 minutes?

(B) Find $\lim_{t \to \infty} x(t)$.

14-3 General Power Rule

■ Introduction
■ General Power Rule
■ Common Errors
■ Remarks

■ Introduction

Just as the general power rule for differentiation

$$D_x\, u^n = n u^{n-1} \frac{du}{dx}$$

significantly increases the variety of functions we can differentiate (see Section 10-7), a corresponding power rule for integration will significantly increase the number of functions we can integrate. Let us start with several illustrations and generalize from the experience.

1. Since

$$D_x \frac{(x^2-1)^5}{5} = \frac{5(x^2-1)^4}{5}\, D_x(x^2-1) = (x^2-1)^4 2x$$

then

$$\int (x^2-1)^4 2x\, dx = \frac{(x^2-1)^5}{5} + C$$

2. Since

$$D_x \frac{(x-x^3)^{-4}}{-4} = \frac{-4(x-x^3)^{-5}}{-4}\, D_x(x-x^3)$$

$$= (x-x^3)^{-5}(1-3x^2)$$

then

$$\int (x-x^3)^{-5}(1-3x^2)\, dx = \frac{(x-x^3)^{-4}}{-4} + C$$

3. Since, for $u = u(x)$,

$$D_x \frac{u^{n+1}}{n+1} = \frac{(n+1)u^n}{n+1}\, D_x u = u^n \frac{du}{dx} \qquad n \neq -1$$

then

$$\int u^n \frac{du}{dx}\, dx = \frac{u^{n+1}}{n+1} + C \qquad n \neq -1$$

■ General Power Rule

The last illustration establishes the **general power rule for integration.** This rule is the inverse of the power rule for differentiation. We will illustrate its use with several examples.

> **General Power Rule**
>
> If $u = u(x)$ and $u'(x)$ exists, then for all real numbers $n(n \neq -1)$
>
> $$\int u^n \frac{du}{dx} \, dx = \frac{u^{n+1}}{n+1} + C$$

Example 11 (A) $\displaystyle\int 2x(x^2 + 5)^4 \, dx$

Note that if $u = x^2 + 5$, then $du/dx = 2x$.

$$\underset{u^n}{} \quad \underset{\frac{du}{dx}}{}$$

$$= \int (x^2 + 5)^4 (2x) \, dx$$

Write in $\displaystyle\int u^n \frac{du}{dx} \, dx$ form and apply the power rule.

$$= \frac{(x^2 + 5)^5}{5} + C$$

Check $\displaystyle D_x \frac{1}{5}(x^2 + 5)^5 = (x^2 + 5)^4 2x$

(B) $\displaystyle\int \frac{3x^2 + 1}{(x^3 + x)^2} \, dx$

Note that if $u = x^3 + x$, then $du/dx = 3x^2 + 1$.

$$\underset{u^n}{} \quad \underset{\frac{du}{dx}}{}$$

$$= \int (x^3 + x)^{-2}(3x^2 + 1) \, dx$$

Write in $\displaystyle\int u^n \frac{du}{dx} \, dx$ form and apply the power rule.

$$= \frac{(x^3 + x)^{-1}}{-1} + C$$

$$= -(x^3 + x)^{-1} + C$$

Check $D_x[-(x^3 + x)^{-1}] = (x^3 + x)^{-2}(3x^2 + 1)$

$$= \frac{3x^2 + 1}{(x^3 + x)^2}$$

Problem 11 Find: (A) $\displaystyle\int 3x^2(x^3 - 1)^2 \, dx$ (B) $\displaystyle\int \frac{e^x}{(e^x + 1)^3} \, dx$

If an integrand is within a constant factor of $u^n \dfrac{du}{dx}$, we can adjust the integral to achieve this form. Example 12 illustrates the process.

Example 12   Integrate

$$\text{(A)} \quad \int x^2\sqrt{x^3 - 10}\ dx \qquad \text{(B)} \quad \int \frac{x-1}{(x^2-2x)^3}\ dx$$

Solutions   (A) Rewrite in power form:

$$\int (x^3 - 10)^{1/2} x^2\ dx$$

If $u = x^3 - 10$, then $du/dx = 3x^2$. We are missing a factor of 3 in the integrand to have the form

$$\int u^n \frac{du}{dx}\ dx$$

(We must have this form *exactly* in order to apply the power rule.) Recalling that a constant factor can be moved across an integral sign, we proceed as follows:

$$\int (x^3 - 10)^{1/2} x^2\ dx = \int (x^3 - 10)^{1/2} \frac{3}{3}\ x^2\ dx$$

$$= \frac{1}{3} \int \underbrace{(x^3 - 10)^{1/2}}_{u^n}\ \underbrace{(3x^2)}_{\frac{du}{dx}}\ dx$$

$$= \frac{1}{3} \frac{(x^3 - 10)^{3/2}}{3/2} + C$$

$$= \frac{2}{9} (x^3 - 10)^{3/2} + C$$

Check $D_x \dfrac{2}{9} (x^3 - 10)^{3/2} = \dfrac{3}{2} \cdot \dfrac{2}{9} (x^3 - 10)^{1/2}(3x^2)$

$$= (x^3 - 10)^{1/2} x^2$$

(B) Rewrite in power form:

$$\int (x^2 - 2x)^{-3}(x - 1)\ dx$$

If $u = x^2 - 2x$, then $du/dx = 2x - 2 = 2(x - 1)$. Again, we are within a constant factor of having $u^n\ du/dx$. We adjust the integrand as in part A:

$$\int (x^2 - 2x)^{-3}(x - 1)\,dx = \int (x^2 - 2x)^{-3}\frac{2}{2}\,(x - 1)\,dx$$

$$= \frac{1}{2}\int \underbrace{(x^2 - 2x)^{-3}}_{u^n}\underbrace{2(x - 1)}_{\frac{du}{dx}}\,dx$$

$$= \frac{1}{2}\frac{(x^2 - 2x)^{-2}}{-2} + C$$

$$= -\frac{1}{4}(x^2 - 2x)^{-2} + C$$

Check $D_x\left[-\frac{1}{4}(x^2 - 2x)^{-2}\right] = \left(-\frac{1}{4}\right)(-2)(x^2 - 2x)^{-3}(2x - 2)$

$$= (x^2 - 2x)^{-3}(x - 1)$$

Problem 12   Integrate:

(A) $\displaystyle\int \sqrt[3]{3x + 5}\,dx$ (B) $\displaystyle\int \frac{x^2 + 1}{(x^3 + 3x)^4}\,dx$

Example 13   Solve the differential equation:

$$\frac{dy}{dt} = \frac{5t^3}{\sqrt[3]{(t^4 - 6)^2}}$$

Solution $\displaystyle y = \int \frac{5t^3}{(t^4 - 6)^{2/3}}\,dt$

$$= \int (t^4 - 6)^{-2/3}(5t^3)\,dt$$

If $u = t^4 - 6$, then $du/dt = 4t^3$. We need a 4 in place of the 5. We move the 5 across the integral sign and proceed as in Example 12:

$$y = 5\int (t^4 - 6)^{-2/3}t^3\,dt$$

$$= 5\int (t^4 - 6)^{-2/3}\frac{4}{4}\,t^3\,dt$$

$$= \frac{5}{4}\int \underbrace{(t^4 - 6)^{-2/3}}_{u^n}\underbrace{(4t^3)}_{\frac{du}{dt}}\,dt$$

$$= \frac{5}{4}\frac{(t^4 - 6)^{1/3}}{1/3} + C$$

$$= \frac{15}{4}(t^4 - 6)^{1/3} + C$$

Problem 13   Solve the differential equation:

$$\frac{dx}{dt} = \frac{7t^2}{(t^3 + 2)^5}$$

■ Common Errors

1. $$\int 2(x^2 - 3)^{3/2} \, dx = \int (x^2 - 3)^{3/2} 2 \, \frac{x}{x} \, dx$$

$$= \frac{1}{x} \int (x^2 - 3)^{3/2}(2x) \, dx$$

A variable cannot be moved across an integral sign! This integral requires techniques that are beyond the scope of this book.

2. $$\int \frac{2x^2}{(x^2 - 3)^2} \, dx = \int (x^2 - 3)^{-2} 2x^2 \, dx$$

$$= x \int (x^2 - 3)^{-2}(2x) \, dx$$

No, for the same reason as in illustration 1.

A constant factor can be moved back and forth across an integral sign, but a variable factor cannot.

Yes	No
$\int kf(x)\,dx = k \int f(x)\,dx$	$\int f(x)g(x)\,dx = f(x)\int g(x)\,dx$
(k a constant factor)	[$f(x)$ a variable factor]

■ Remarks

In this section we have touched on an integration technique that will be generalized in the next chapter. In fact, Chapter 15 covers several other commonly used techniques of integration as well. However, even with that chapter, our treatment will not be exhaustive.

Answers to Matched Problems

11.   (A) $\dfrac{1}{3}(x^3 - 1)^3 + C$ (B) $-\dfrac{1}{2}(e^x + 1)^{-2} + C$

12.   (A) $\dfrac{1}{4}(3x + 5)^{4/3} + C$ (B) $-\dfrac{1}{9}(x^3 + 3x)^{-3} + C$

13. $x = -\dfrac{7}{12}(t^3 + 2)^{-4} + C$

Exercise 14-3

Find each indefinite integral and check by differentiating the result.

A 1. $\int (x^2 - 4)^5 2x \, dx$ 2. $\int (x^3 + 1)^4 3x^2 \, dx$

3. $\int \sqrt{2x^2 - 1} \, 4x \, dx$ 4. $\int \sqrt[3]{2x^3 + 5} \, 6x^2 \, dx$

5. $\int (3x - 2)^7 \, dx$ 6. $\int (5x + 3)^9 \, dx$

B 7. $\int (x^2 + 3)^7 x \, dx$ 8. $\int (x^3 - 5)^4 x^2 \, dx$

9. $\int x\sqrt{3x^2 + 7} \, dx$ 10. $\int x^2\sqrt{2x^3 + 1} \, dx$

11. $\int \frac{x^3}{\sqrt{2x^4 + 3}} \, dx$ 12. $\int \frac{x^2}{\sqrt{4x^3 - 1}} \, dx$

13. $\int (x - 1)\sqrt{x^2 - 2x - 3} \, dx$ 14. $\int (x^3 - x)\sqrt{x^4 - 2x^2 + 7} \, dx$

15. $\int \frac{t}{(3t^2 + 1)^4} \, dt$ 16. $\int \frac{t^2}{(t^3 - 2)^5} \, dt$

17. $\int \frac{x^2}{\sqrt{4 - x^3}} \, dx$ 18. $\int \frac{x}{(5 - 2x^2)^5} \, dx$

19. $\int (e^x - 2x)^3(e^x - 2) \, dx$ 20. $\int (x^2 - e^x)^4(2x - e^x) \, dx$

C 21. $\int \frac{\sqrt{1 + \ln x}}{x} \, dx$ 22. $\int \frac{(\ln x)^4}{x} \, dx$

23. $\int \frac{x^3 + x}{(x^4 + 2x^2 + 1)^4} \, dx$ 24. $\int \frac{x^2 - 1}{\sqrt[3]{x^3 - 3x + 7}} \, dx$

Solve each differential equation.

25. $\dfrac{dx}{dt} = 7t^2\sqrt{t^3 + 5}$ 26. $\dfrac{dm}{dn} = 10n(n^2 - 8)^7$

27. $\dfrac{dy}{dt} = \dfrac{3t}{\sqrt{t^2 - 4}}$ 28. $\dfrac{dy}{dx} = \dfrac{5x^2}{(x^3 - 7)^4}$

29. $\dfrac{dp}{dx} = \dfrac{e^x + e^{-x}}{(e^x - e^{-x})^2}$ 30. $\dfrac{dm}{dt} = \dfrac{\ln(t - 5)}{t - 5}$

Applications

Business & Economics

31. *Revenue function.* If the marginal revenue in thousands of dollars of producing x units is given by

$$R'(x) = x(x^2 + 9)^{-1/2}$$

and no revenue results from a zero production level, find the revenue function $R(x)$. Find the revenue at a production level of four units.

Life Sciences

32. *Pollution.* An oil tanker aground on a reef is losing oil and producing an oil slick that is radiating outward at a rate given approximately by

$$\frac{dR}{dt} = \frac{60}{\sqrt{t+9}} \qquad t \geq 0$$

where R is the radius in feet of the circular slick after t minutes. Find the radius of the slick after 16 minutes if the radius is 0 when $t = 0$.

Social Sciences

33. *College enrollment.* The projected rate of increase in enrollment in a new college is estimated by

$$\frac{dE}{dt} = 5{,}000(t + 1)^{-3/2} \qquad t \geq 0$$

where $E(t)$ is the projected enrollment in t years. If enrollment is 2,000 when $t = 0$, find the projected enrollment 15 years from now.

14-4 Definite Integral

- Definite Integral
- Properties
- Applications

■ Definite Integral

We start this discussion with a simple example, out of which will evolve a new integral form, called the *definite integral*. Our approach in this section will be intuitive and informal; these concepts will be made more precise in Section 14-6.

Suppose a manufacturing company's marginal cost equation for a given product is given by

$$C'(x) = 2 - 0.2x \qquad 0 \leq x \leq 8$$

where the marginal cost is in thousands of dollars and production is x units per day. What is the total change in cost per day going from a production level of 2 units per day to 6 units per day? If $C = C(x)$ is the cost function, then

$$\left(\begin{array}{l}\text{Total net change in cost}\\ \text{between } x = 2 \text{ and } x = 6\end{array}\right) = C(6) - C(2) = C(x)|_2^6 \tag{1}$$

The special symbol $C(x)|_2^6$ is a convenient way of representing the center expression that will prove useful to us later.

To evaluate (1), we need to find the antiderivative of $C'(x)$, that is,

$$C(x) = \int (2 - 0.2x)\, dx = 2x - 0.1x^2 + K \tag{2}$$

Thus, we are within a constant of knowing the original marginal cost function. However, we do not need to know the constant K to solve the original problem (1). We compute $C(6) - C(2)$ for $C(x)$ found in (2):

$$C(6) - C(2) = [2(6) - 0.1(6)^2 + K] - [2(2) - 0.1(2)^2 + K]$$

$$= 12 - 3.6 + K - 4 + 0.4 - K$$

$$= \$4.8 \text{ thousand per day increase in costs for a production}$$
$$\text{increase from 2 to 6 units per day}$$

The unknown constant K canceled out! Thus, we conclude that any antiderivative of $C'(x) = 2 - 0.2x$ will do, since antiderivatives of a given function can differ by at most a constant (see Section 14-1). Thus, we really do not have to find the original cost function to solve the problem.

Since $C(x)$ is an antiderivative of $C'(x)$, the above discussion suggests the following notation:

$$C(6) - C(2) = C(x)|_2^6 = \int_2^6 C'(x)\, dx \tag{3}$$

The integral form on the right in (3) is called a *definite integral* — it represents the number found by evaluating an antiderivative of the integrand at 6 and 2 and taking the difference as indicated.

Definite Integral

The **definite integral** of a continuous function f over an interval from $x = a$ to $x = b$ is the net change of an antiderivative of f over the interval. Symbolically, if $F(x)$ is an antiderivative of $f(x)$, then

$$\int_a^b f(x)\, dx = F(x)|_a^b = F(b) - F(a) \qquad \text{where} \quad F'(x) = f(x)$$

Integrand: $f(x)$ **Upper limit:** b **Lower limit:** a

In Section 14-6 we will formally define a definite integral as a limit of a special sum. Then the relationship in the box turns out to be the most important theorem in calculus—the fundamental theorem of calculus. Our intent in this and the next section is to give you some intuitive experience with the definite integral concept and its use. You will then be better able to understand a formal definition and to appreciate the significance of the fundamental theorem.

Example 14   Evaluate $\int_{-1}^{2} (3x^2 - 2x)\, dx$.

Solution   We choose the simplest antiderivative of $(3x^2 - 2x)$, namely $(x^3 - x^2)$, since any antiderivative will do (see discussion at beginning of section).

$$\int_{-1}^{2} (3x^2 - 2x)\, dx = (x^3 - x^2)|_{-1}^{2}$$

$$= (2^3 - 2^2) - [(-1)^3 - (-1)^2] \qquad \text{Be careful of}$$
$$= 4 - (-2) = 6 \qquad\qquad\qquad \text{sign errors here.}$$

Problem 14   Evaluate $\int_{-2}^{2} (2x - 1)\, dx$.

Remark

Do not confuse a definite integral with an indefinite integral. The definite integral $\int_a^b f(x)\, dx$ is a real number; the indefinite integral $\int f(x)\, dx$ is a whole set of functions—all the antiderivatives of $f(x)$.

■ Properties

In the next box we state several useful properties of the definite integral. You will note that some of these parallel the properties for the indefinite integral listed in Section 14-1.

These properties are justified as follows: If $F'(x) = f(x)$, then

1. $\displaystyle\int_a^a f(x)\, dx = F(x)|_a^a = F(a) - F(a) = 0$

2. $\displaystyle\int_a^b f(x)\, dx = F(x)|_a^b = F(b) - F(a) = -[F(a) - F(b)] = -\int_b^a f(x)\, dx$

3. $\displaystyle\int_a^b Kf(x)\, dx = KF(x)|_a^b = KF(b) - KF(a) = K[F(b) - F(a)]$

$$= K \int_a^b f(x)\, dx$$

and so on.

Definite Integral Properties

1. $\displaystyle\int_a^a f(x)\,dx = 0$

2. $\displaystyle\int_a^b f(x)\,dx = -\int_b^a f(x)\,dx$

3. $\displaystyle\int_a^b Kf(x)\,dx = K\int_a^b f(x)\,dx$ K a constant

4. $\displaystyle\int_a^b [f(x) \pm g(x)]\,dx = \int_a^b f(x)\,dx \pm \int_a^b g(x)\,dx$

5. $\displaystyle\int_a^b f(x)\,dx = \int_a^c f(x)\,dx + \int_c^b f(x)\,dx$

Example 15   Evaluate $\displaystyle\int_0^1 [(2x-1)^3 + 2x]\,dx$.

Solution $\displaystyle\int_0^1 [(2x-1)^3 + 2x]\,dx$

$$= \int_0^1 (2x-1)^3\,dx + \int_0^1 2x\,dx$$

$$= \frac{1}{2}\int_0^1 \underset{\underset{u}{}}{(2x-1)^3}\,\overset{\overset{\frac{du}{dx}}{}}{2}\,dx + 2\int_0^1 x\,dx$$

$$= \frac{1}{2}\cdot\frac{(2x-1)^4}{4}\bigg|_0^1 + 2\cdot\frac{x^2}{2}\bigg|_0^1$$

$$= \left[\frac{(2\cdot1-1)^4}{8} - \frac{(2\cdot0-1)^4}{8}\right] + (1^2 - 0^2) \qquad \text{Be careful of sign orrors here.}$$

$$= \left[\frac{1^4}{8} - \frac{(-1)^4}{8}\right] + 1 = 1$$

Problem 15   Evaluate $\displaystyle\int_1^2 [3x^2 - x\sqrt{x^2-1}]\,dx$.

Example 16   Evaluate $\displaystyle\int_1^e \frac{\sqrt{\ln x}}{x}\,dx$.

Solution

$$\int_1^e \frac{\sqrt{\ln x}}{x}\,dx = \int_1^e \underbrace{(\ln x)^{1/2}}_{u^n}\underbrace{\frac{1}{x}}_{\frac{du}{dx}}\,dx$$

$$= \frac{2}{3}(\ln x)^{3/2}\,\Big|_1^e$$

$$= \frac{2}{3}(\ln e)^{3/2} - \frac{2}{3}(\ln 1)^{3/2}$$

$$= \frac{2}{3}\cdot 1 - \frac{2}{3}\cdot 0 = \frac{2}{3}$$

Problem 16   Evaluate $\displaystyle\int_0^{1.5} \frac{e^{2x} - e^{-2x}}{2}\,dx$ to four significant digits.

■ Applications

Example 17
Velocity

A steel ball is dropped from a tower. Its velocity t seconds later is $v(t) = 32t$ feet per second. How far will the ball fall from the end of 2 seconds to the end of 4 seconds?

Solution

The antiderivative of a velocity function is a positive function $s = s(t)$, and we are looking for $s(4) - s(2)$:

$$s(4) - s(2) = \int_2^4 32t\,dt = 16t^2\Big|_2^4$$

$$= 16\cdot 4^2 - 16\cdot 2^2 = 256 - 64 = 192\text{ feet}$$

Problem 17   Repeat Example 17 with $v(t) = 32t - 10$.

Example 18
Pollution

A large factory on the Mississippi River discharges pollutants into the river at a rate that is estimated by a water quality control agency to be

$$P'(t) = R(t) = t\sqrt{t^2 + 1} \qquad 0 \le t \le 5$$

where $P(t)$ is the total number of tons of pollutants discharged into the river after t years of operation. What quantity of pollutants will be discharged into the river during the first 3 years of operation?

Solution

$$P(3) - P(0) = \int_0^3 t\sqrt{t^2 + 1}\,dt$$

$$= \int_0^3 (t^2 + 1)^{1/2}t\,dt$$

$$= \frac{1}{2}\int_0^3 (t^2 + 1)^{1/2}2t\,dt$$

$$= \frac{1}{2} \cdot \frac{(t^2 + 1)^{3/2}}{3/2} \Big|_0^3$$

$$= \frac{1}{3} (t^2 + 1)^{3/2} \Big|_0^3$$

$$= \frac{1}{3} (3^2 + 1)^{3/2} - \frac{1}{3} (0^2 + 1)^{3/2}$$

$$= \frac{1}{3} (10^{3/2} - 1) \approx 10.2 \text{ tons}$$

Problem 18   Repeat Example 18 for the time interval from 3 to 5 years.

Answers to   14. -4
Matched Problems   15. $7 - \sqrt{3}$

16. $\dfrac{e^3 + e^{-3} - 2}{4} \approx 4.534$

17. 172 feet

18. $\dfrac{1}{3} (26^{3/2} - 10^{3/2}) \approx 33.7$ tons

Exercise 14-4

Evaluate.

A   1. $\displaystyle\int_2^3 2x \, dx$

2. $\displaystyle\int_1^2 3x^2 \, dx$

3. $\displaystyle\int_3^4 5 \, dx$

4. $\displaystyle\int_{12}^{20} dx$

5. $\displaystyle\int_1^3 (2x - 3) \, dx$

6. $\displaystyle\int_1^3 (6x + 5) \, dx$

7. $\displaystyle\int_0^4 (3x^2 - 4) \, dx$

8. $\displaystyle\int_0^2 (6x^2 - 2x) \, dx$

9. $\displaystyle\int_{-3}^4 (4 - x^2) \, dx$

10. $\displaystyle\int_{-1}^2 (x^2 - 4x) \, dx$

11. $\displaystyle\int_0^1 24x^{11} \, dx$

12. $\displaystyle\int_0^2 30x^5 \, dx$

13. $\displaystyle\int_0^1 e^{2x} \, dx$

14. $\displaystyle\int_{-1}^1 e^{5x} \, dx$

15. $\displaystyle\int_1^{3.5} 2x^{-1} \, dx$

16. $\displaystyle\int_1^2 \frac{dx}{x}$

B 17. $\int_1^2 (2x^{-2} - 3)\, dx$

18. $\int_1^2 (5 - 16x^{-3})\, dx$

19. $\int_1^4 3\sqrt{x}\, dx$

20. $\int_4^{25} \frac{2}{\sqrt{x}}\, dx$

21. $\int_2^3 12(x^2 - 4)^5 x\, dx$

22. $\int_0^1 32(x^2 + 1)^7 x\, dx$

23. $\int_1^9 \sqrt[3]{x - 1}\, dx$

24. $\int_{-1}^0 \sqrt[5]{x + 1}\, dx$

25. $\int_0^1 (e^{2x} - 2x)^2(e^{2x} - 1)\, dx$

26. $\int_0^1 \frac{2e^{4x} - 3}{e^{2x}}\, dx$

27. $\int_{-2}^{-1} (x^{-1} + 2x)\, dx$

28. $\int_{-3}^{-1} (-3x^{-2} + x^{-1})\, dx$

C 29. $\int_2^3 x\sqrt{2x^2 - 3}\, dx$

30. $\int_0^1 x\sqrt{3x^2 + 2}\, dx$

31. $\int_0^1 \frac{x - 1}{\sqrt[3]{x^2 - 2x + 3}}\, dx$

32. $\int_1^2 \frac{x + 1}{\sqrt{2x^2 + 4x - 2}}\, dx$

33. $\int_{-1}^1 \frac{e^x + e^{-x}}{(e^x - e^{-x})^2}\, dx$

34. $\int_6^7 \frac{\ln(t - 5)}{t - 5}\, dt$

■

Applications

Business & Economics

35. *Marginal analysis.* A company's marginal cost, revenue, and profit equations (in thousands of dollars per day) are:

$$\left. \begin{array}{l} C'(x) = 1 \\ R'(x) = 10 - 2x \\ P'(x) = R'(x) - C'(x) \end{array} \right\} 0 \leqslant x \leqslant 10$$

where x is the number of units produced per day. Find the change in

(A) Cost (B) Revenue (C) Profit

in going from a production level of 2 units per day to 4 units per day.

36. *Marginal analysis.* Repeat Problem 35 with $C'(x) = 2$ and $R'(x) = 12 - 2x$.

37. *Salvage value.* A new piece of industrial equipment will depreciate in value rapidly at first, then less rapidly as time goes on. Suppose the rate (in dollars per year) at which the book value of a new milling machine changes is given approximately by

$$V'(t) = f(t) = 500(t - 12) \qquad 0 \leqslant t \leqslant 10$$

where $V(t)$ is the value of the machine after t years. Find the total loss in value of the machine in the first 5 years. In the second 5 years. Set up appropriate integrals and solve.

38. *Maintenance costs.* Maintenance costs for an apartment house generally increase as the building gets older. From past records, a managerial service determines that the rate of increase in maintenance costs (in dollars per year) for a particular apartment complex is given approximately by

$$M'(x) = f(x) = 90x^2 + 5{,}000$$

where x is the age of the apartment in years and M(x) is the total (accumulated) cost of maintenance for x years. Write a definite integral that will give the total maintenance costs from 2 to 7 years after the apartment house was built, and evaluate it.

Life Sciences   **39.** *Pulse rate versus height.* The rate of change of an average person's pulse rate with respect to height is given approximately by

$$P'(x) = f(x) = -295x^{-3/2} \qquad 30 \leqslant x \leqslant 75$$

where x is height in inches. Find the total change in pulse rate for a child growing from 49 to 64 inches. Set up an appropriate definite integral and solve.

40. *Drug sensitivity.* One hour after x milligrams of a particular drug are given to a person, the rate of change of temperature in degrees Fahrenheit, T'(x), with respect to dosage x (called *sensitivity*) is given approximately by

$$T'(x) = 2x - \frac{x^2}{3} \qquad 0 \leqslant x \leqslant 6$$

What total change in temperature results from a dosage change from 0 to 2 milligrams? From 2 to 3 milligrams? Set up definite integrals and evaluate.

41. *Natural resource depletion.* The instantaneous rate of change of demand for wood in the United States since 1970 ($t = 0$) in billions of cubic feet per year is estimated to be given by

$$Q'(t) = 12 + 0.006t^2 \qquad 0 \leqslant t \leqslant 50$$

where Q(t) is the total amount of wood consumed in billions of cubic feet t years after 1970. How many billions of cubic feet of wood will be consumed from 1980 to 1990?

42. *Natural resource depletion.* Repeat Problem 41 for the time interval from 1990 to 2000.

Social Sciences   **43.** *Learning.* A person learns N items at a rate given approximately by

$$N'(t) = f(t) = \frac{25}{\sqrt{t}} \qquad 1 \leqslant t \leqslant 9$$

where t is the number of hours of continuous study. Use a definite integral to determine the total number of items learned from $t = 1$ to $t = 9$ hours of study.

14-5   Area and the Definite Integral

- Area under a Curve
- Area between Two Curves
- Signed Areas
- Consumers' and Producers' Surplus

■ Area under a Curve

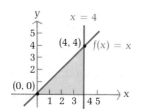

Figure 3

Consider the graph of $f(x) = x$ from $x = 0$ to $x = 4$ (Figure 3). We can easily compute the area of the triangle bounded by $f(x) = x$, the x axis $(y = 0)$, and the line $x = 4$, using the formula for the area of a triangle:

$$A = \frac{bh}{2} = \frac{4 \cdot 4}{2} = 8$$

Let us integrate $f(x) = x$ from $x = 0$ to $x = 4$:

$$\int_0^4 x \, dx = \frac{x^2}{2} \bigg|_0^4 = \frac{4^2}{2} - \frac{0^2}{2} = 8$$

We get the same result! It turns out that this is not a coincidence. In general, we can prove the following:

Area under a Curve

If f is continuous and $f(x) \geq 0$ over the interval $[a, b]$, then the area bounded by $y = f(x)$, the x axis $(y = 0)$, and the vertical lines $x = a$ and $x = b$ is given exactly by

$$A = \int_a^b f(x) \, dx$$

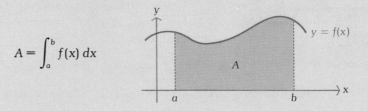

Let us see why the definite integral gives us the area exactly. Let $A(x)$ be the area under the graph of $y = f(x)$ from a to x, as indicated in Figure 4.

$A(x) = \text{Area from } a \text{ to } x$

$A(b) = \text{Area from } a \text{ to } b = A$

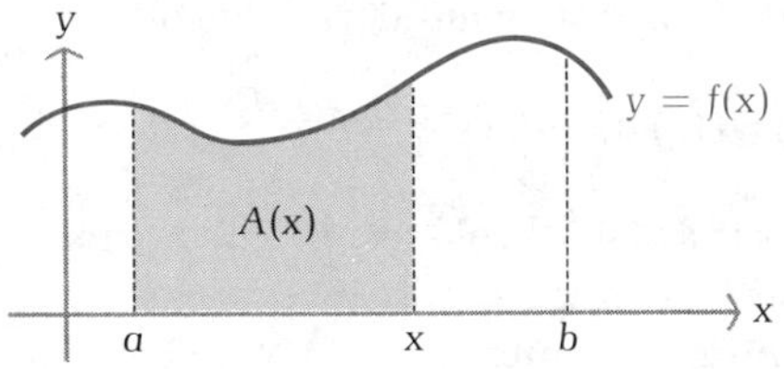

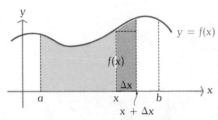

Figure 4

If we can show that $A(x)$ is an antiderivative of $f(x)$, then we can write

$$\int_a^b f(x)\,dx = A(x)|_a^b = A(b) - A(a)$$

$$= \begin{pmatrix} \text{Area from} \\ x = a \text{ to } x = b \end{pmatrix} - \begin{pmatrix} \text{Area from} \\ x = a \text{ to } x = a \end{pmatrix}$$

$$= A - 0 = A$$

To show that $A(x)$ is an antiderivative of $f(x)$—that is, $A'(x) = f(x)$—we use the definition of a derivative (Section 10-4) and write

$$A'(x) = \lim_{\Delta x \to 0} \frac{A(x + \Delta x) - A(x)}{\Delta x}$$

Geometrically, $A(x + \Delta x) - A(x)$ is the area from x to $x + \Delta x$ (see Figure 5). This area is given approximately by the area of the rectangle $\Delta x \cdot f(x)$, and the smaller Δx is, the better the approximation. Using

$$A(x + \Delta x) - A(x) \approx \Delta x \cdot f(x)$$

and dividing both sides by Δx, we obtain

$$\frac{A(x + \Delta x) - A(x)}{\Delta x} \approx f(x)$$

Now, if we let $\Delta x \to 0$, then the left side has $A'(x)$ as a limit, which is equal to the right side. Hence,

$$A'(x) = f(x)$$

Figure 5

that is, $A(x)$ is an antiderivative of $f(x)$. Thus,

$$\int_a^b f(x)\,dx = A(x)|_a^b = A(b) - A(a) = A - 0 = A$$

This is a remarkable result: The area under the graph of $y = f(x)$, $f(x) \geqslant 0$, can be obtained simply by evaluating the antiderivative of $f(x)$ at the end points of the interval $[a, b]$. We have now solved, at least in part, the third basic problem of calculus stated in Section 10-1.

Example 19   Find the area bounded by $f(x) = 6x - x^2$ and $y = 0$ for $1 \leqslant x \leqslant 4$.

Solution   We sketch a graph of the region first:

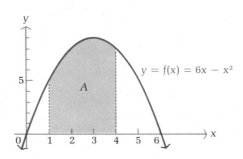

$$A = \int_1^4 (6x - x^2)\,dx = \left(3x^2 - \frac{x^3}{3} \right)\Big|_1^4$$

$$= \left[3(4)^2 - \frac{4^3}{3} \right] - \left[3(1)^2 - \frac{1^3}{3} \right]$$

$$= 48 - \frac{64}{3} - 3 + \frac{1}{3}$$

$$= 48 - 21 - 3 = 24$$

Problem 19   Find the area bounded by $f(x) = x^2 + 1$ and $y = 0$ for $-1 \leqslant x \leqslant 3$.

Example 20   Find the area between the curve $y = 1/t$ and the t axis from $t = 1$ to $t = 2$.

Solution

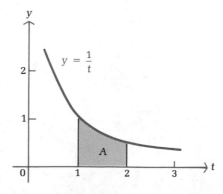

$$A = \int_1^2 \frac{1}{t}\, dt$$

$$= (\ln|t|) \Big|_1^2$$

$$= \ln 2 - \ln 1 = \ln 2$$

Problem 20 Find the area between the curve $y = 1/t$ and the t axis from $t = 1$ to $t = 3.5$.

Generalizing from the results of Example 20 and Problem 20, we can determine the area between the curve $y = 1/t$ and the t axis from $t = 1$ to $t = x$, $x > 0$.

$$A = \int_1^x \frac{dt}{t} = (\ln|t|) \Big|_1^x$$

$$= \ln x - \ln 1 \qquad |x| = x \text{ since } x > 0$$

$$= \ln x$$

Thus, $\ln x$ is exactly the area indicated in Figure 6.

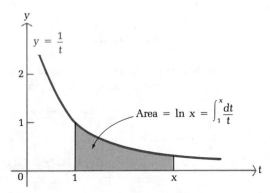

Figure 6

This is a significant result. In a more advanced treatment of logarithmic functions, $\ln x$ is defined by

$$\ln x = \int_1^x \frac{dt}{t} \qquad x > 0$$

and all of the basic logarithmic properties can be obtained from this definition. For example,

$$\ln 1 = \int_1^1 \frac{dt}{t} = 0$$

■ Area between Two Curves

Consider the area bounded by $y = f(x)$ and $y = g(x)$, $f(x) \geq g(x) \geq 0$, for $a \leq x \leq b$, as indicated in Figure 7.

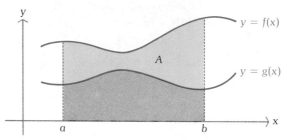

Figure 7

$$\begin{pmatrix} \text{Area } A \text{ between} \\ f(x) \text{ and } g(x) \end{pmatrix}$$

$$= \begin{pmatrix} \text{Area under} \\ f(x) \end{pmatrix} - \begin{pmatrix} \text{Area under} \\ g(x) \end{pmatrix} \qquad \begin{array}{l} \text{Areas are from} \\ x = a \text{ to } x = b \\ \text{above the } x \text{ axis} \end{array}$$

$$= \int_a^b f(x) \, dx - \int_a^b g(x) \, dx \qquad \begin{array}{l} \text{From definite in-} \\ \text{tegral property 4} \\ \text{(Section 14-4)} \end{array}$$

$$= \int_a^b [f(x) - g(x)] \, dx$$

It can be shown that the above result does not require $f(x)$ or $g(x)$ to remain positive over the interval $[a, b]$. A more general result is stated in the box:

Area between Two Curves

If f and g are continuous and $f(x) \geq g(x)$ over the interval $[a, b]$, then the area bounded by $y = f(x)$ and $y = g(x)$ for $a \leq x \leq b$ is given exactly by

$$A = \int_a^b [f(x) - g(x)] \, dx$$

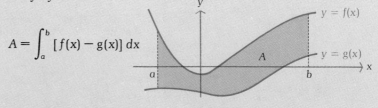

Example 21   Find the area bounded by $f(x) = (1/2)x + 3$, $g(x) = -x^2 + 1$, $x = -2$, and $x = 1$.

Solution   We first sketch the area, then set up and evaluate an appropriate definite integral.

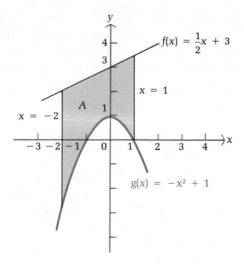

We observe from the graph that $f(x) \geqslant g(x)$ for $-2 \leqslant x \leqslant 1$, so

$$A = \int_{-2}^{1} [f(x) - g(x)] \, dx$$

$$= \int_{-2}^{1} [(x/2 + 3) - (-x^2 + 1)] \, dx$$

$$= \int_{-2}^{1} (x^2 + x/2 + 2) \, dx$$

$$= \left(\frac{x^3}{3} + \frac{x^2}{4} + 2x \right) \Big|_{-2}^{1}$$

$$= \left(\frac{1}{3} + \frac{1}{4} + 2 \right) - \left(\frac{-8}{3} + \frac{4}{4} - 4 \right)$$

$$= \frac{33}{4}$$

Problem 21   Find the area bounded by $f(x) = x^2 - 1$, $g(x) = -(1/2)x - 3$, $x = -1$, and $x = 2$.

Example 22   Find the area bounded by $f(x) = 5 - x^2$ and $g(x) = x^2 - 3$.

Solution

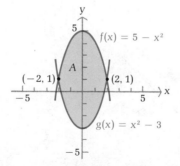

The two graphs are parabolas, one opening up and the other down, as shown in the figure. To find the points of intersection (hence, the upper and lower limits of integration), we solve $y = 5 - x^2$ and $y = x^2 - 3$ simultaneously by setting $5 - x^2$ equal to $x^2 - 3$ (substitution method):

$$5 - x^2 = x^2 - 3$$
$$2x^2 - 8 = 0$$
$$x^2 - 4 = 0$$
$$x = \pm 2$$

Thus,

$$A = \int_{-2}^{2} [(5 - x^2) - (x^2 - 3)] \, dx$$

$$= \int_{-2}^{2} (8 - 2x^2) \, dx$$

$$= \left(8x - \frac{2x^3}{3} \right) \Big|_{-2}^{2}$$

$$= \left[8(2) - \frac{2(2)^3}{3} \right] - \left[8(-2) - \frac{2(-2)^3}{3} \right]$$

$$= 16 - \frac{16}{3} + 16 - \frac{16}{3} = \frac{64}{3}$$

Problem 22   Find the area bounded by $f(x) = 6 - x^2$ and $g(x) = x$.

■ Signed Areas

Consider the area bounded by $f(x) = x$, the x axis $(y = 0)$, $x = -2$, and $x = 2$, as indicated in Figure 8. Integrating $f(x) = x$ from $x = -2$ to $x = 2$, we obtain

$$\int_{-2}^{2} x \, dx = \frac{x^2}{2} \Big|_{-2}^{2} = \frac{(2)^2}{2} - \frac{(-2)^2}{2} = 2 - 2 = 0$$

which is not the actual area indicated in Figure 8. But now consider the

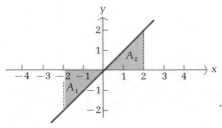

Figure 8

following two integrals:

$$\int_{-2}^{0} x \, dx = \frac{x^2}{2} \Big|_{-2}^{0} = \frac{0^2}{2} - \frac{(-2)^2}{2} = -2$$

$$\int_{0}^{2} x \, dx = \frac{x^2}{2} \Big|_{0}^{2} = \frac{2^2}{2} - \frac{0^2}{2} = 2$$

We interpret the results as **signed areas:** Area A_2 above the x axis is positive and area A_1 below the x axis is negative. The actual area can then be obtained by adding the absolute value of the negative area to the positive area:

$$\text{Total area} = \left| \int_{-2}^{0} x \, dx \right| + \int_{0}^{2} x \, dx = |-2| + 2 = 4$$

Note that the integral from -2 to 2 is the algebraic sum of the signed areas:

$$\int_{-2}^{2} x \, dx = \int_{-2}^{0} x \, dx + \int_{0}^{2} x \, dx = -2 + 2 = 0 \qquad \text{From definite integral property 5 (Section 14-4)}$$

We summarize these observations as follows:

Signed Areas and the Definite Integral

The **area** bounded by $y = f(x)$, the x axis $(y = 0)$, $x = a$, and $x = b$ is **positive** where the area is above the x axis and **negative** where the area is below the x axis. The definite integral of $f(x)$ from $x = a$ to $x = b$ can always be interpreted as the algebraic sum of these signed areas:

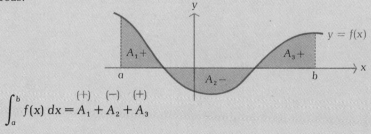

$$\int_{a}^{b} f(x) \, dx = A_1 + A_2 + A_3$$

If we want the **actual bounded area,** then we add the absolute value of each negative area to the sum of the positive areas.

Example 23 (A) Find the finite area bounded by $f(x) = 1 - x^2$ and $y = 0$, $0 \leqslant x \leqslant 2$.

(B) Find the definite integral of $f(x)$ from $x = 0$ to $x = 2$.

Solutions (A) We need to sketch a graph first to see if negative areas are involved.

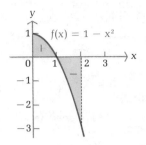

$$\text{Actual area} = \int_0^1 (1 - x^2)\, dx + \left| \int_1^2 (1 - x^2)\, dx \right|$$

$$= \frac{2}{3} + \left| -\frac{4}{3} \right| = \frac{2}{3} + \frac{4}{3} = 2$$

(B) $\displaystyle \int_0^2 (1 - x^2)\, dx = \left(x - \frac{x^3}{3} \right) \Big|_0^2 = 2 - \frac{8}{3} = -\frac{2}{3}$ This is the algebraic sum of the signed areas $2/3$ and $-4/3$.

Problem 23   (A)   Find the area bounded by $f(x) = x^2 - 1$, $y = 0$, $x = -1$, and $x = 2$.

(B)   Evaluate the definite integral of $f(x)$ from $x = -1$ to $x = 2$.

■ Consumers' and Producers' Surplus

If we graph the supply and demand functions $p = S(x)$ and $p = D(x)$ and locate the equilibrium point (a, b) (the point at which supply is equal to demand), then the area between $p = b$ and $p = D(x)$ from $x = 0$ to $x = a$ is called **consumers' surplus**. The area between $p = S(x)$ and $p = b$ is called **producers' surplus** (see Figure 9).

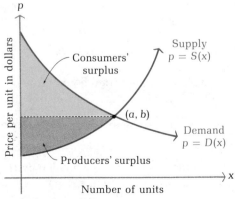

Figure 9

$$\text{Consumers' surplus} = \int_0^a [D(x) - b] \, dx$$

$$\text{Producers' surplus} = \int_0^a [b - S(x)] \, dx$$

In other words, if the price stabilizes at $b per unit, then there is still a demand by some people at higher prices, and people who are willing to pay a higher price benefit by only having to pay the equilibrium price. The total of these benefits over [0, a] is the consumers' surplus. On the other hand, there are still some producers who are willing to supply at a lower price, and these people benefit by receiving the equilibrium price. The total of these benefits for the producers over the interval [0, a] is the producers' surplus.

Example 24   Find the consumers' surplus for

$$p = D(x) = -\frac{x}{2} + 11 \quad \text{and} \quad p = S(x) = x + 2$$

Solution   Sketch a graph:

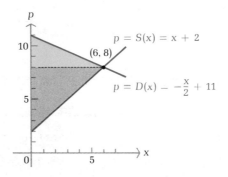

To find the equilibrium point, set $(x + 2)$ equal to $[(-x/2) + 11]$:

$$x + 2 = -\frac{x}{2} + 11$$

$$2x + 4 = -x + 22$$

$$3x = 18$$

$$x = 6$$

$$p = x + 2$$

$$= 6 + 2 = 8$$

Therefore, the equilibrium point is (6, 8), as shown in the figure. Now,

$$\text{Consumers' surplus} = \int_0^a [D(x) - b] \, dx$$

$$= \int_0^6 \left(-\frac{x}{2} + 11 - 8\right) dx$$

$$= \int_0^6 \left(-\frac{x}{2} + 3\right) dx$$

$$= \left(-\frac{x^2}{4} + 3x\right)\Big|_0^6 = 9$$

Problem 24   Find the producers' surplus for Example 24.

Answers to
Matched Problems

19. $A = \int_{-1}^3 (x^2 + 1) \, dx = \dfrac{40}{3}$

20. $\ln 3.5$

21. $A = \int_{-1}^2 [(x^2 - 1) - (-x/2 - 3)] \, dx = \dfrac{39}{4}$

22. $A = \int_{-3}^2 [(6 - x^2) - x] \, dx = \dfrac{125}{6}$

23. (A) $\dfrac{8}{3}$ (B) 0

24. 18

Exercise 14-5

Find the area bounded by the graphs of the indicated equations.

A
1. $y = 2x + 4$, $y = 0$, $1 \leqslant x \leqslant 3$
2. $y = -2x + 6$, $y = 0$, $0 \leqslant x \leqslant 2$
3. $y = 3x^2$, $y = 0$, $1 \leqslant x \leqslant 2$
4. $y = 4x^3$, $y = 0$, $1 \leqslant x \leqslant 2$
5. $y = x^2 + 2$, $y = 0$, $-1 \leqslant x \leqslant 0$
6. $y = 3x^2 + 1$, $y = 0$, $-2 \leqslant x \leqslant 0$
7. $y = 4 - x^2$, $y = 0$, $-1 \leqslant x \leqslant 2$
8. $y = 12 - 3x^2$, $y = 0$, $-2 \leqslant x \leqslant 1$
9. $y = e^x$, $y = 0$, $-1 \leqslant x \leqslant 2$
10. $y = e^{-x}$, $y = 0$, $-2 \leqslant x \leqslant 1$
11. $y = \dfrac{1}{t}$, $y = 0$, $0.5 \leqslant t \leqslant 1$
12. $y = \dfrac{1}{t}$, $y = 0$, $0.1 \leqslant t \leqslant 1$

B   13. $y = 12, \quad y = -2x + 8, \quad -1 \leqslant x \leqslant 2$

14. $y = 3, \quad y = 2x + 6, \quad -1 \leqslant x \leqslant 2$

15. $y = 3x^2, \quad y = 12$

16. $y = x^2, \quad y = 9$

17. $y = 4 - x^2, \quad y = -5$

18. $y = x^2 - 1, \quad y = 3$

19. $y = x^2 + 1, \quad y = 2x - 2, \quad -1 \leqslant x \leqslant 2$

20. $y = x^2 - 1, \quad y = x - 2, \quad -2 \leqslant x \leqslant 1$

21. $y = -x, \quad y = 0, \quad -2 \leqslant x \leqslant 1$

22. $y = -x + 1, \quad y = 0, \quad -1 \leqslant x \leqslant 2$

23. $y = e^{0.5x}, \quad y = \dfrac{-1}{x}, \quad 1 \leqslant x \leqslant 2$

24. $y = \dfrac{1}{x}, \quad y = -e^x, \quad 0.5 \leqslant x \leqslant 1$

C   25. $y = x^2 - 4, \quad y = 0, \quad 0 \leqslant x \leqslant 3$

26. $y = 4\sqrt[3]{x}, \quad y = 0, \quad -1 \leqslant x \leqslant 8$

27. $y = x^2 + 2x + 3, \quad y = 2x + 4$

28. $y = 8 + 4x - x^2, \quad y = x^2 - 2x$

Applications

Business & Economics

29. *Consumers' and producers' surplus.* Find the consumers' surplus and the producers' surplus for

$$p = D(x) = -\frac{x}{2} + 2$$

$$p = S(x) = \frac{x^2}{4}$$

30. *Consumers' and producers' surplus.* Find the consumers' surplus and the producers' surplus for

$$p = D(x) = 50 - x^2$$
$$p = S(x) = x^2 + 2x + 10$$

31. *Marginal analysis.* A company has a vending machine with the following marginal cost and revenue equations (in thousands of dollars per year):

$$C'(t) = 2$$
$$R'(t) = 12 - 2t \qquad 0 \leqslant t \leqslant 10$$

where $C(t)$ and $R(t)$ represent total accumulated costs and revenues, respectively, t years after the machine is put into use. The area between the graphs of the marginal equations for the time period such that $R'(t) \geqslant C'(t)$ represents the total accumulated profit for the useful life of the machine. What is the useful life of the machine and what is the total profit?

32. *Marginal analysis.* Repeat Problem 31 for $C'(t) = 0.5t + 2$ and $R'(t) = 10 - 0.5t$, $0 \leqslant t \leqslant 20$.

33. *Consumers' surplus.* Supply and demand functions are given by

$$p = D(x) = 100e^{-0.05x}$$

$$p = S(x) = 10e^{0.05x}$$

(A) Show that the equilibrium point is approximately (23.03, 31.62).
(B) Compute the consumers' surplus to two decimal places.

34. *Producers' surplus.* Compute the producers' surplus to two decimal places for Problem 33.

14-6 Definite Integral as a Limit of a Sum

- Rectangle Rule for Approximating Definite Integrals
- Definite Integral as a Limit of a Sum
- Recognizing a Definite Integral
- Average Value of a Continuous Function
- Volume of a Solid of Revolution (Optional)

Up to this point, in order to evaluate a definite integral

$$\int_a^b f(x) \, dx$$

we need to find an antiderivative of the function f so that we can write

$$\int_a^b f(x) \, dx = F(x) \Big|_a^b = F(b) - F(a) \qquad F'(x) = f(x)$$

But suppose we cannot find an antiderivative of f (it may not even exist in a convenient or closed form). For example, how would you evaluate the following?

$$\int_2^8 \sqrt{x^3 + 1} \, dx \qquad \int_1^5 \left(\frac{x}{x+1}\right)^3 dx \qquad \int_0^5 e^{-x^2} dx$$

We now introduce the *rectangle rule* for approximating definite integrals, and out of this discussion will evolve a new way of looking at definite integrals.

■ Rectangle Rule for Approximating Definite Integrals

In the last section we saw that any definite integral of a continuous function f over an interval $[a, b]$ can always be interpreted as the algebraic sum of

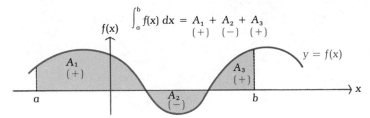

$$\int_a^b f(x)\, dx = A_1 + A_2 + A_3$$
$$(+)\quad(-)\quad(+)$$

Figure 10

the signed areas bounded by $y = f(x)$, $y = 0$, $x = a$, and $x = b$ (see Figure 10). What we need is a way of approximating such areas, given $y = f(x)$ and an interval $[a, b]$.

Let us start with a concrete example and generalize from the experience. We will start with a simple definite integral we can evaluate exactly:

$$\int_1^5 (x^2 + 3)\, dx = \left(\frac{x^3}{3} + 3x\right)\Big|_1^5$$

$$= \left[\frac{5^3}{3} + 3(5)\right] - \left[\frac{1^3}{3} + 3(1)\right]$$

$$= \left(\frac{125}{3} + 15\right) - \left(\frac{1}{3} + 3\right)$$

$$= \frac{160}{3} = 53\frac{1}{3}$$

This integral represents the area bounded by $y = x^2 + 3$, $y = 0$, $x = 1$, and $x = 5$, as indicated in Figure 11.

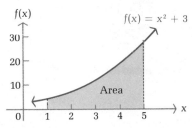

Figure 11

Since areas of rectangles are easy to compute, we cover the area in Figure 11 with rectangles so that the top of each rectangle has a point in common with the graph of $y = f(x)$. As our first approximation, we divide the interval $[1, 5]$ into two equal subintervals, each with length $(b - a)/2 =$

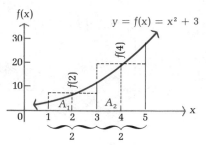

Figure 12

$(5-1)/2 = 2$, and use the midpoint of each subinterval to compute the altitude of the rectangle sitting on top of that subinterval (see Figure 12).

$$\int_1^5 (x^2 + 3)\,dx \approx A_1 + A_2$$
$$= f(2) \cdot 2 + f(4) \cdot 2$$
$$= 2[f(2) + f(4)]$$
$$= 2(7 + 19) = 52$$

This approximation is less than 3% off of the exact area we found above $(53\frac{1}{3})$.

Now let us divide the interval $[1, 5]$ into four equal subintervals, each of length $(b - a)/4 = (5 - 1)/4 = 1$, and use the midpoint* of each subinterval to compute the altitude of the rectangle corresponding to that subinterval (see Figure 13).

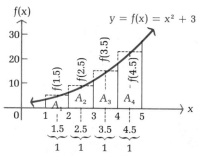

Figure 13

---

* We actually do not need to choose the midpoint of each subinterval; any point from each subinterval will do. The midpoint is often a convenient point to choose, because then the rectangle tops are usually above part of the graph and below part of the graph. This tends to cancel some of the error that occurs.

$$\int_1^5 (x^2 + 3)\, dx \approx A_1 + A_2 + A_3 + A_4$$

$$= f(1.5) \cdot 1 + f(2.5) \cdot 1 + f(3.5) \cdot 1 + f(4.5) \cdot 1$$
$$= f(1.5) + f(2.5) + f(3.5) + f(4.5)$$
$$= 5.25 + 9.25 + 15.25 + 23.25$$
$$= 53$$

Now we are less than 1% off of the exact area ($53\frac{1}{3}$).

We would expect the approximations to continue to improve as we use more and more rectangles with smaller and smaller bases. We now state the rectangle rule for approximating definite integrals of a continuous function f over the interval from $x = a$ to $x = b$.

Rectangle Rule

Divide the interval from $x = a$ to $x = b$ into n equal subintervals of length $\Delta x = (b - a)/n$. Let c_k be any point in the kth subinterval. Then

$$\int_a^b f(x)\, dx \approx f(c_1)\Delta x + f(c_2)\Delta x + \cdots + f(c_n)\Delta x$$

$$= \Delta x [\, f(c_1) + f(c_2) + \cdots + f(c_n)\,]$$

Example 25   Use the rectangle rule to approximate

$$\int_2^{10} \frac{x}{x + 1}\, dx$$

using $n = 4$ and c_k the midpoint of each subinterval. Compute the approximation to three significant digits.

Solution   **Step 1.** *Find Δx, the length of each subinterval.*

$$\Delta x = \frac{b - a}{n} = \frac{10 - 2}{4} = \frac{8}{4} = 2$$

Step 2. *Use the midpoint of each subinterval for c_k.*

Subintervals: [2, 4], [4, 6], [6, 8], [8, 10]
Midpoints: $c_1 = 3$, $c_2 = 5$, $c_3 = 7$, $c_4 = 9$

Step 3. *Use the rectangle rule with $n = 4$.*

$$\int_a^b f(x)\,dx \approx f(c_1)\Delta x + f(c_2)\Delta x + f(c_3)\Delta x + f(c_4)\Delta x$$
$$= \Delta x[f(c_1) + f(c_2) + f(c_3) + f(c_4)]$$
$$= 2[f(3) + f(5) + f(7) + f(9)]$$
$$= 2(0.750 + 0.833 + 0.875 + 0.900)$$
$$= 2(3.358) = 6.72 \qquad \text{To 3 significant digits}$$

Problem 25 Use the rectangle rule to approximate

$$\int_2^{14} \frac{x}{x-1}\,dx$$

using $n = 4$ and c_k the midpoint of each subinterval. Compute the approximation to three significant digits.

■ Definite Integral as a Limit of a Sum

In using the rectangle rule to approximate a definite integral, one might expect

$$\lim_{\Delta x \to 0} [f(c_1)\,\Delta x + f(c_2)\,\Delta x + \cdots + f(c_n)\,\Delta x] = \int_a^b f(x)\,dx$$

This idea motivates the formal definition of a definite integral that we referred to in Section 14-4.

Definition of a Definite Integral

Let f be a continuous function defined on the closed interval $[a, b]$, and let

1. $a = x_0 \leqslant x_1 \leqslant \cdots \leqslant x_{n-1} \leqslant x_n = b$
2. $\Delta x_k = x_k - x_{k-1}$ \quad for $k = 1, 2, \ldots, n$
3. $\Delta x_k \to 0$ \quad as $n \to \infty$
4. $x_{k-1} \leqslant c_k \leqslant x_k$ \quad for $k = 1, 2, \ldots, n$

Then

$$\int_a^b f(x)\,dx = \lim_{n \to \infty} [f(c_1)\Delta x_1 + f(c_2)\Delta x_2 + \cdots + f(c_n)\Delta x_n]$$

is called a **definite integral**.

In the definition of a definite integral, we divide the closed interval $[a, b]$ into n subintervals of arbitrary lengths in such a way that the length of each subinterval $\Delta x_k = x_k - x_{k-1}$ tends to 0 as n increases without bound. From each of the n subintervals we then select a point c_k.

Under the conditions stated in the definition, it can be shown that the limit always exists and it is a real number. The limit is independent of the nature of the subdivisions of $[a, b]$ as long as condition 3 holds, and it is independent of the choice of the c_k as long as condition 4 holds.

In a more formal treatment of the subject, we would then prove the remarkable **fundamental theorem of calculus,** which shows that the limit in the definition of a definite integral can be determined exactly by evaluating an antiderivative of $f(x)$, if it exists, at the end points of the interval $[a, b]$ and taking the difference.

Theorem 2

> **Fundamental Theorem of Calculus**
>
> Under the conditions stated in the definition of a definite integral
>
> (Definition)
>
> $$\int_a^b f(x)\, dx = \lim_{n \to \infty} [\, f(c_1)\Delta x_1 + f(c_2)\Delta x_2 + \cdots + f(c_n)\Delta x_n]$$
>
> (Theorem)
>
> $$= F(b) - F(a) \qquad \text{where } F'(x) = f(x)$$

Now we are free to evaluate a definite integral by using the fundamental theorem if an antiderivative of $f(x)$ can be found; otherwise, we can approximate it using the formal definition in the form of the rectangle rule.

■ Recognizing a Definite Integral

Recall that the derivative of a function f was defined by

$$f'(x) = \lim_{\Delta x \to 0} \frac{f(x + \Delta x) - f(x)}{\Delta x}$$

a form that is generally not easy to compute directly, but is easy to recognize in certain practical problems (slope, instantaneous velocity, rates of change, etc.). Once it is recognized that we are dealing with a derivative, we then proceed to try to compute it using derivative formulas and rules.

Similarly, evaluating a definite integral using the definition

$$\int_a^b f(x)\, dx = \lim_{n \to \infty} [f(c_1)\Delta x_1 + f(c_2)\Delta x_2 + \cdots + f(c_n)\Delta x_n] \tag{1}$$

is generally not easy; but the form on the right occurs naturally in many practical problems. We can use the fundamental theorem to evaluate the integral (once it is recognized) if an antiderivative can be found; otherwise, we will approximate it using the rectangle rule. We will now illustrate these points by finding the average value of a continuous function and the volume of a solid of revolution.

■ Average Value of a Continuous Function

Suppose the temperature T (in degrees Fahrenheit) in the middle of a small shallow lake from 8 AM ($t = 0$) to 6 PM ($t = 10$) during the month of May is

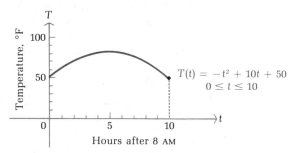

Figure 14

given approximately as shown in Figure 14. How can we compute the average temperature from 8 AM to 6 PM? We know that the average of a finite number of values

$$a_1, a_2, \ldots, a_n$$

is given by

$$\text{Average} = \frac{a_1 + a_2 + \cdots + a_n}{n}$$

But how can we handle a continuous function with infinitely many values? It would seem reasonable to divide the time interval [0, 10] into n equal subintervals, compute the temperature at a point in each subinterval, and then use the average of these values as an approximation of the average value of the continuous function $T = T(t)$ over [0, 10]. We would expect the approximations to improve as n increases. In fact, we would be inclined to define the limit of the average for n values as $n \to \infty$ as *the average value of T over* [0, 10], if the limit exists. This is exactly what we will do:

$$\left(\begin{matrix} \text{Average temperature} \\ \text{for } n \text{ values} \end{matrix} \right) = \frac{1}{n} [T(t_1) + T(t_2) + \cdots + T(t_n)] \qquad (2)$$

where t_k is a point in the kth subinterval. We will call the limit of (2) as $n \to \infty$ *the average temperature over the time interval* [0, 10].

Form (2) looks sort of like form (1), but we are missing the Δt_k. We take care of this by multiplying (2) by $(b - a)/(b - a)$, which will change the form of (2) without changing its value.

$$\frac{b-a}{b-a} \cdot \frac{1}{n} [T(t_1) + T(t_2) + \cdots + T(t_n)]$$

$$= \frac{1}{b-a} \cdot \frac{b-a}{n} [T(t_1) + T(t_2) + \cdots + T(t_n)]$$

$$= \frac{1}{b-a} \cdot \left[T(t_1) \frac{b-a}{n} + T(t_2) \frac{b-a}{n} + \cdots + T(t_n) \frac{b-a}{n} \right]$$

$$= \frac{1}{b-a} [T(t_1)\Delta t + T(t_2)\Delta t + \cdots + T(t_n)\Delta t]$$

Thus,

$$\left(\begin{array}{c} \text{Average temperature} \\ \text{over } [a, b] = [0, 10] \end{array} \right)$$

$$= \lim_{n \to \infty} \frac{1}{b-a} [T(t_1)\Delta t + T(t_2)\Delta t + \cdots + T(t_n)\Delta t]$$

$$= \frac{1}{b-a} \left\{ \lim_{n \to \infty} [T(t_1)\Delta t + T(t_2)\Delta t + \cdots + T(t_n)\Delta t] \right\}$$

Now the part in the braces is of form (1) — that is, a definite integral. Thus,

$$\left(\begin{array}{c} \text{Average temperature} \\ \text{over } [a, b] = [0, 10] \end{array} \right)$$

$$= \frac{1}{b-a} \int_a^b T(t) \, dt$$

$$= \frac{1}{10 - 0} \int_0^{10} (-t^2 + 10t + 50) \, dt \qquad \begin{array}{l} \text{We now evaluate the definite} \\ \text{integral using the fundamental} \\ \text{theorem.} \end{array}$$

$$= \frac{1}{10} \left(-\frac{t^3}{3} + 5t^2 + 50t \right) \Big|_0^{10}$$

$$= \frac{200}{3} \approx 67°F$$

In general, proceeding as above for an arbitrary continuous function f over an interval $[a, b]$, we obtain the general formula:

Average Value of a Continuous Function f over $[a, b]$

$$\frac{1}{b-a} \int_a^b f(x) \, dx$$

Example 26   Find the average value of $f(x) = x - 3x^2$ over the interval $[-1, 2]$.

Solution

$$\frac{1}{b-a}\int_a^b f(x)\,dx = \frac{1}{2-(-1)}\int_{-1}^{2}(x-3x^2)\,dx$$

$$= \frac{1}{3}\left(\frac{x^2}{2}-x^3\right)\Big|_{-1}^{2} = -\frac{5}{2}$$

Problem 26   Find the average value of $g(t) = 6t^2 - 2t$ over the interval $[-2, 3]$.

Example 27
Average Price

Given the demand function

$$p = D(x) = 100e^{-0.05x}$$

Find the average price (in dollars) over the demand interval $[0, 100]$.

Solution

$$\text{Average price} = \frac{1}{b-a}\int_a^b D(x)\,dx$$

$$= \frac{1}{100-0}\int_0^{100} 100e^{-0.05x}\,dx$$

$$= \frac{100}{100}\int_0^{100} e^{-0.05x}\,dx$$

$$= -20e^{-0.05x}\Big|_0^{100}$$

$$= 20(1-e^{-5}) \approx \$19.87$$

Problem 27   Given the supply equation

$$p = S(x) = 10e^{0.05x}$$

Find the average price (in dollars) over the supply interval $[0, 40]$.

■ Volume of a Solid of Revolution (Optional)

Let us consider another application in which expression (1) occurs naturally. Suppose we start out with the region bounded by the graphs of $y = f(x) = \sqrt{x}$, $y = 0$, and $x = 9$ (see Figure 15). This is the upper half of a parabola opening to the right.

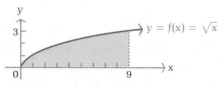

Figure 15

If we rotate the shaded area in Figure 15 around the x axis, we obtain a three-dimensional object called a **solid of revolution.** Figure 16 shows the result, which in this case is called a **paraboloid.** What is its volume?

Figure 16

Let us cover the region in Figure 15 with rectangles (as we did earlier in the section using the rectangle rule) and rotate the rectangles around the x axis (see Figure 17). We can then use the stacked cylinders to give an approximation of the volume—the more rectangles we use, the better the approximation.

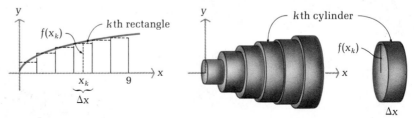

Figure 17

Volumes of cylinders are easy to compute:

$$V = (\text{Area of circular base})(\text{Height}) = \pi R^2 h$$

In terms of the kth cylinder in Figure 17, we have:

$$V_k = \pi[f(x_k)]^2 \Delta x$$

The volume of n cylinders is

$$\pi[f(x_1)]^2\Delta x + \pi[f(x_2)]^2\Delta x + \cdots + \pi[f(x_n)]^2\Delta x$$
$$= \pi\{[f(x_1)]^2\Delta x + [f(x_2)]^2\Delta x + \cdots + [f(x_n)]^2\Delta x\}$$

Again we recognize form (1) within the braces. And, from the fundamental theorem, in the limit we have a definite integral. Thus, the exact volume is given by

$$V = \pi \int_a^b [f(x)]^2\, dx$$

$$= \pi \int_0^9 (\sqrt{x})^2\, dx = \pi \int_0^9 x\, dx = \pi \frac{x^2}{2}\Big|_0^9 = \frac{81\pi}{2} \approx 127$$

In general, proceeding as above, we have:

Volume of a Solid of Revolution

The **volume of the solid of revolution** obtained by revolving the region bounded by the graphs of $y = f(x)$, $y = 0$, $x = a$, and $x = b$ about the x axis is given by

$$V = \pi \int_a^b [f(x)]^2\, dx$$

Example 28   Find the volume of the solid of revolution formed by rotating the region bounded by the graphs of $y = x^2$, $y = 0$, and $x = 2$ about the x axis.

Solution   Sketch a graph of the region first:

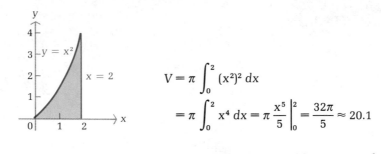

$$V = \pi \int_0^2 (x^2)^2 \, dx$$

$$= \pi \int_0^2 x^4 \, dx = \pi \frac{x^5}{5} \bigg|_0^2 = \frac{32\pi}{5} \approx 20.1$$

Problem 28   Find the volume of the solid of revolution formed by rotating the region bounded by the graphs of $y = x^2$, $y = 0$, $x = 1$, and $x = 2$ about the x axis.

Answers to   25. 14.4   26. 13
Matched Problems   27. $5(e^2 - 1) \approx \$31.95$ 28. $31\pi/5 \approx 19.5$

Exercise 14-6

For Problems 1–12:

(A) *Use the rectangle rule to approximate (to three significant digits) each definite integral for the indicated number of subintervals n. Choose c_k as the midpoint of each subinterval.*

(B) *Evaluate each integral exactly using an antiderivative. If an antiderivative cannot be found by methods we have considered, say so.*

A   1. $\displaystyle\int_1^5 3x^2 \, dx, \quad n = 2$ 2. $\displaystyle\int_2^6 x^2 \, dx, \quad n = 2$

   3. $\displaystyle\int_1^5 3x^2 \, dx, \quad n = 4$ 4. $\displaystyle\int_2^6 x^2 \, dx, \quad n = 4$

B   5. $\displaystyle\int_0^4 (4 - x^2) \, dx, \quad n = 2$ 6. $\displaystyle\int_0^4 (3x^2 - 12) \, dx, \quad n = 2$

   7. $\displaystyle\int_0^4 (4 - x^2) \, dx, \quad n = 4$ 8. $\displaystyle\int_0^4 (3x^2 - 12) \, dx, \quad n = 4$

   9. $\displaystyle\int_0^4 \left(\frac{x}{x+1}\right)^2 dx, \quad n = 2$ 10. $\displaystyle\int_1^7 \frac{1}{x} \, dx, \quad n = 3$

   11. $\displaystyle\int_0^4 \left(\frac{x}{x+1}\right)^2 dx, \quad n = 4$ 12. $\displaystyle\int_1^7 \frac{1}{x} \, dx, \quad n = 6$

Find the average value of each function over the indicated interval.

13. $f(x) = 500 - 50x$, $[0, 10]$
14. $g(x) = 2x + 7$, $[0, 5]$
15. $f(t) = 3t^2 - 2t$, $[-1, 2]$
16. $g(t) = 4t - 3t^2$, $[-2, 2]$
17. $f(x) = \sqrt[3]{x}$, $[1, 8]$
18. $g(x) = \sqrt{x + 1}$, $[3, 8]$
19. $f(x) = 4e^{-0.2x}$, $[0, 10]$
20. $f(x) = 64e^{0.08x}$, $[0, 10]$

(Optional) In Problems 21–26 find the volume of the solid of revolution formed by rotating the region bounded by the graphs of the indicated equations about the x axis. Express the answer in terms of π.

21. $y = \sqrt{3}\,x$, $y = 0$, $x = 1$, $x = 3$
22. $y = x + 1$, $y = 0$, $x = 1$, $x = 2$
23. $y = \sqrt{2x}$, $y = 0$, $x = 8$
24. $y = \sqrt{5}\,x^2$, $y = 0$, $x = 2$
25. $y = \sqrt{4 - x^2}$, $y = 0$
26. $y = \sqrt{9 - x^2}$, $y = 0$

C *Use the rectangle rule to approximate (to three significant figures) each quantity in Problems 27–30. Use $n = 4$ and c_k the midpoint of each subinterval. Problems 29 and 30 are optional.*

27. The average value of $f(x) = (x + 1)/(x^2 + 1)$ for $[-1, 1]$
28. The average value of $f(x) = x/(x + 1)$ for $[0, 4]$
29. The volume of the solid of revolution formed by rotating the region bounded by the graphs of $y = 1/\sqrt{x}$, $y = 0$, $x = 1$, and $x = 9$ about the x axis. Use $\pi \approx 3.14$.
30. The volume of the solid of revolution formed by rotating the region bounded by the graphs of $y = x/(x + 1)$, $y = 0$, and $x = 8$ about the x axis. Use $\pi \approx 3.14$.

Applications

Business & Economics

31. *Inventory.* A store orders 600 units of a product every 3 months. If the product is steadily depleted to zero by the end of each 3 months, the inventory on hand, I, at any time t during the year is illustrated as follows:

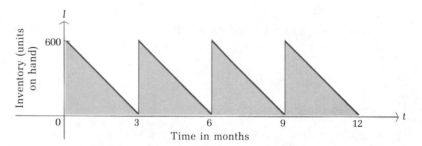

(A) Write an inventory function (assume it is continuous) for the first 3 months. [The graph is a straight line joining (0, 600) and (3, 0).]

(B)   What is the average number of units on hand for a 3 month period?

32.   Repeat Problem 31 with an order of 1,200 units every 4 months.

33.   *Cash reserves.* Suppose cash reserves (in thousands of dollars) are approximated by

$$C(x) = 1 + 12x - x^2 \qquad 0 \leqslant x \leqslant 12$$

where x is the number of months after the first of the year. What is the average cash reserve for the first quarter?

34.   Repeat Problem 33 for the second quarter.

35.   *Supply function.* Given the supply function

$$p = S(x) = 10(e^{0.02x} - 1)$$

Find the average price (in dollars) over the supply interval [0, 50].

36.   *Demand function.* Given the demand function

$$p = D(x) = \frac{1,000}{x}$$

Find the average price (in dollars) over the demand range [100, 600].

Life Sciences

37.   *Temperature.* If the temperature C(t) in an artificial habitat was made to change according to

$$C(t) = t^3 - 2t + 10 \qquad 0 \leqslant t \leqslant 2$$

(in degrees Celsius) over a 2 hour period, what is the average temperature over this period?

Social Sciences

38.   *Population composition.* Because of various factors (such as birth rate expansion, then contraction; family flights from urban areas; etc.), the number of children in a large city was found to increase and then decrease rather drastically. If the number of children over a 6 year period was found to be given approximately by

$$N(t) = -(1/4)t^2 + t + 4 \qquad 0 \leqslant t \leqslant 6$$

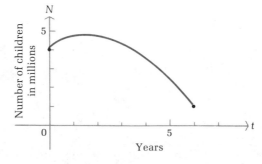

what was the average number of children in the city over the 6 year time period? [Assume N = N(t) is continuous.]

14-7 Chapter Review

Exercise 14-7 Chapter Review

Work through all the problems in this chapter review and check your answers in the back of the book. (Answers to all review problems are there.) Where weaknesses show up, review appropriate sections in the text. When you are satisfied that you know the material, take the practice test following this review.

A *Find each integral in Problems 1–6.*

1. $\displaystyle\int (3t^2 - 2t)\, dt$

2. $\displaystyle\int_2^5 (2x - 3)\, dx$

3. $\displaystyle\int (3t^{-2} - 3)\, dt$

4. $\displaystyle\int_1^4 x\, dx$

5. $\displaystyle\int e^{-0.5x}\, dx$

6. $\displaystyle\int_1^5 \frac{2}{u}\, du$

7. Find a function $y = f(x)$ that satisfies both conditions:

$$\frac{dy}{dx} = 3x^2 - 2 \qquad f(0) = 4$$

8. Find the area bounded by the graphs of $y = 3x^2 + 1$, $y = 0$, $x = -1$, and $x = 2$.

9. Approximate $\int_1^5 (x^2 + 1)\, dx$ using the rectangle rule with $n = 2$ and c_k the midpoint of the kth subinterval.

B   *Find each integral in Problems 10–15.*

10. $\int \sqrt[3]{6x-5}\, dx$

11. $\int_0^1 10(2x-1)^4\, dx$

12. $\int \left(\frac{2}{x^2} - \sqrt[3]{x^2} \right) dx$

13. $\int_0^4 \sqrt{x^2+4}\; x\, dx$

14. $\int (e^{-2x} + x^{-1})\, dx$

15. $\int_0^{10} 10e^{-0.02x}\, dx$

16. Find a function $y = f(x)$ that satisfies both conditions:

$$\frac{dy}{dx} = 3\sqrt{x} - x^{-2} \qquad f(1) = 5$$

17. Find the equation of the curve that passes through $(2, 10)$ if its slope is given by

$$\frac{dy}{dx} = 6x + 1$$

for each x.

18. Approximate $\int_{-2}^4 (x^2 - 4)\, dx$ using the rectangle rule with $n = 3$ and c_k the midpoint of the kth subinterval.

19. Find the average value of $f(x) = 3x^{1/2}$ over the interval $[1, 9]$.

C   20. Find the actual area bounded by the graphs of $y = x^2 - 4$, $y = 0$, $x = -2$, and $x = 4$.

Find each integral in Problems 21–23.

21. $\int_0^5 \sqrt[3]{x^2 - 2x}\, (x - 1)\, dx$

22. $\int \frac{\sqrt{x} - 2x^{-2}}{x}\, dx$

23. $\int \frac{\sqrt{x^3}e^{-2x} - \sqrt{x}}{\sqrt{x^3}}\, dx, \qquad x > 0$

24. Find a function $y = f(x)$ that satisfies both conditions:

$$\frac{dy}{dx} = x^2\sqrt{x^3 + 4} \qquad f(0) = 2$$

25. Solve the differential equation:

$$\frac{dN}{dt} = 0.06N, \qquad N(0) = 800, \qquad N > 0$$

26. Find the area bounded by the graphs of $y = 6 - x^2$, $y = x^2 - 2$, $x = 0$, and $x = 3$. Be careful!

27. Approximate the average value of $f(x) = 1/(x + 1)$ over the interval $[0, 4]$ using the rectangle rule with $n = 4$ and c_k the midpoint of the kth subinterval.

28. *Optional.* Find the volume of the solid of revolution formed by rotating the region bounded by the graphs of $y = 1/x$, $y = 0$, $x = 1$, and $x = 2$ about the x axis. State the answer in terms of π.

■ Applications

Business & Economics

29. *Profit function.* If the marginal profit for producing x units per day is given by

$$P'(x) = 100 - 0.02x \qquad P(0) = 0$$

where $P(x)$ is the profit in dollars, find the profit function P and the profit on ten units of production per day.

30. *Resource depletion.* An oil well starts out producing oil at the rate of 60,000 barrels of oil per year, but the production rate is expected to decrease by 4,000 barrels per year. Thus, if $P(t)$ is the total production (in thousands of barrels) in t years, then

$$P'(t) = f(t) = 60 - 4t \qquad 0 \le t \le 15$$

Write a definite integral that will give the total production after 15 years of operation. Evaluate it.

31. *Profit and production.* The weekly marginal profit for an output of x units is given approximately by

$$P'(x) = 150 - \frac{x}{10} \qquad 0 \le x \le 40$$

What is the total change in profit for a production change from ten units per week to forty units? Set up a definite integral and evaluate it.

32. *Inventory.* Suppose the inventory of a certain item t months after the first of the year is given approximately by

$$I(t) = 10 + 36t - 3t^2 \qquad 0 \le t \le 12$$

What is the average inventory for the second quarter of the year?

33. *Supply function.* Given the supply function

$$p = S(x) = 80(e^{0.05x} - 1)$$

find the average price (in dollars) over the supply interval $[0, 40]$.

Life Sciences

34. *Wound healing.* The area of a small, healing surface wound changes at a rate given approximately by

$$\frac{dA}{dt} = -5t^{-2} \qquad 1 \le t \le 5$$

where t is in days and $A(1) = 5$ square centimeters. What will the area of the wound be in 5 days?

35. *Height–weight relationship.* **For an average person, the rate of change of weight $W'(h)$ (in pounds) per unit change in height h (in inches) is given approximately by**

$$W'(h) = 0.0015h^2$$

What is the expected total change in weight in a child growing from 50 to 60 inches? Set up an appropriate definite integral and evaluate.

36. *Population growth.* **If a bacteria culture is growing at a rate given by**

$$N'(t) = 2{,}000e^{0.2t} \qquad N(0) = 10{,}000$$

where t is time in hours, find $N(t)$ and the number of bacteria after 10 hours.

Social Sciences

37. *School enrollment.* **The student enrollment in a new high school is expected to grow at a rate that is estimated to be**

$$\frac{dN}{dt} = 200 + 300t \qquad 0 \leqslant t \leqslant 4$$

where $N(t)$ is the number of students t years after opening. If the initial enrollment $(t = 0)$ is 2,000, what will be the enrollment 4 years from now?

38. *Politics.* **In a newly incorporated city, it is estimated that the rate of change of the voting population, $N'(t)$, with respect to time t in years is given by**

$$N'(t) = 12t - 3t^2 \qquad 0 \leqslant t \leqslant 4$$

where $N(t)$ is in thousands. What is the total increase in the voting population during the first 4 years? Set up an appropriate definite integral and evaluate.

Practice Test: Chapter 14

Find each integral in Problems 1–6.

1. $\displaystyle\int_{1}^{2} (5t^{-3} - t)\, dt$

2. $\displaystyle\int x^2\sqrt{x^3 + 9}\, dx$

3. $\displaystyle\int \frac{4 + x^4}{x^3}\, dx$

4. $\displaystyle\int_{1}^{5} \sqrt{2x - 1}\, dx$

5. $\displaystyle\int_{0}^{10} 4(e^{0.2t} - 1)\, dt$

6. $\displaystyle\int \frac{x^2 - x^3 e^{-0.1x}}{x^3}\, dx$

7. Find the equation of a function whose graph passes through the point (3, 10) and whose slope is given by

$$f'(x) = 6 - 2x$$

8. Find the area bounded by the graphs of $y = x^2$ and $y = \sqrt{x}$.
9. Find the finite area bounded by the graphs of $y = 1 - x^2$ and $y = 0$, $0 \leqslant x \leqslant 2$.
10. Approximate $\int_0^6 (x^2 - 4)\, dx$ using the rectangle rule with $n = 3$ and c_k the midpoint of the kth subinterval. (Calculate the approximation to three significant digits.) Also, evaluate the integral exactly.
11. Suppose the inventory of a certain item t months after the first of the year is given approximately by

$$I(t) = -2t + 36 \qquad 0 \leqslant t \leqslant 12$$

What is the average inventory for the second quarter of the year?
12. The instantaneous rate of change of production for a gold mine, in thousands of ounces of gold per year, is estimated to be given by

$$Q'(t) = 40 - 4t \qquad 0 \leqslant t \leqslant 10$$

where $Q(t)$ is the total quantity (in thousands of ounces) of gold produced after t years of operation. How much gold is produced during the first 2 years of operation? During the next 2 years?
13. Solve the differential equation:

$$\frac{dQ}{dt} = 0.12Q \quad Q(0) = 10{,}000 \quad Q > 0$$

Additional Integration Topics

15

CHAPTER 15   Contents

By now you should realize that finding antiderivatives is not as routine a process as finding derivatives. Indeed, it is not difficult to find functions whose antiderivatives cannot be expressed in terms of the elementary functions we are familiar with. The classic example of this case is $f(x) = e^{-x^2}$, an important function in statistics. Nevertheless, there are certain methods of integration that increase the number of functions we can integrate. We will now consider some of these methods.

15-1 Integration by Substitution

- Introduction
- Integration by Substitution
- Definite Integrals and Substitution
- Common Errors

■ Introduction

In Section 14-3 we saw that if an integrand is of the form

$$u^n \frac{du}{dx}$$

where $u = u(x)$ is a function of x, then we can use the generalized power rule to conclude that

$$\int u^n \frac{du}{dx} \, dx = \frac{u^{n+1}}{n+1} + C \qquad n \neq -1$$

For example,

$$\overset{u^n}{} \qquad \overset{\frac{du}{dx}}{}$$

$$\int (x^2 + 1)^{1/2} 2x \, dx = \frac{(x^2 + 1)^{3/2}}{3/2} + C \qquad \text{If } u = x^2 + 1, \text{ then } du/dx = 2x.$$

$$= \frac{2}{3} (x^2 + 1)^{3/2} + C$$

In this section we will see how to use the relationship $u = x^2 + 1$ to change the variable of integration from x to u. This technique, called **integration**

by substitution, is a very powerful tool that will enable us to evaluate a large number of indefinite integrals.

Recall from Section 11-4 that if $u = u(x)$, then the differential of u is

$$du = \frac{du}{dx}\, dx$$

Thus, for

$$u = x^2 + 1$$

the differential is

$$du = 2x\, dx$$

Substituting for u and du in $\int 2x(x^2 + 1)^{1/2}\, dx$, we have

$$\int \overset{u^{1/2}}{(x^2 + 1)^{1/2}}\overset{du}{2x\, dx} = \int u^{1/2}\, du \qquad \begin{array}{l}\text{We momentarily ``forget''}\\ \text{that u is a function of x and}\\ \text{treat u as if it were the}\\ \text{variable of integration.}\end{array}$$

$$= \frac{u^{3/2}}{3/2} + C \qquad \begin{array}{l}\text{Now we ``remember'' that}\\ \text{we started with } u = x^2 + 1.\end{array}$$

$$= \frac{2}{3}(x^2 + 1)^{3/2} + C$$

At first glance, it appears that we are actually making things more complicated but, as later examples will illustrate, making a substitution in order to change the variable in an indefinite integral can greatly simplify many problems. The important point is that, once the substitution has been made, we can treat u as the variable of integration and proceed to evaluate the simplified integral directly. We will now generalize this process of substitution.

■ Integration by Substitution

In general, if $u = u(x)$ and $du = (du/dx)\, dx$, then

$$\int f[\overset{u}{u(x)}]\overset{du}{\frac{du}{dx}\, dx} = \int f(u)\, du \qquad \begin{array}{l}\text{Regarding u as the variable of}\\ \text{integration, we try to find an}\\ \text{antiderivative } F(u) \text{ for } f(u).\end{array}$$

$$= F(u) + C \qquad \begin{array}{l}\text{Now we substitute } u = u(x) \text{ to}\\ \text{complete the process.}\end{array}$$

$$= F[u(x)] + C$$

This statement is easily verified by applying the chain rule to $F[u(x)] + C$:

$$D_x\{F[u(x)] + C\} = F'[u(x)]\frac{du}{dx} \qquad F' = f$$

$$= f[u(x)]\frac{du}{dx}$$

It is convenient to restate some of the basic integration formulas in terms of u and du.

Basic Integration Formulas

If F is an antiderivative of f, then

$$\int f(u)\, du = F(u) + C$$

In particular,

$$\int u^n\, du = \frac{u^{n+1}}{n+1} + C \qquad n \neq -1 \tag{1}$$

$$\int \frac{1}{u}\, du = \ln|u| + C \tag{2}$$

$$\int e^u\, du = e^u + C \tag{3}$$

These formulas are valid in each of the following cases:

1. u is the variable of integration
2. $u = u(x)$ is a function of x and

$$du = \frac{du}{dx}\, dx$$

Example 1   Use substitution to find the following indefinite integrals:

(A) $\displaystyle\int (2x+1)(x^2 + x + 5)^4\, dx$

(B) $\displaystyle\int \frac{x}{4 + x^2}\, dx$ (C) $\displaystyle\int x^2 e^{x^3}\, dx$

Solutions   (A) In selecting a substitution, you should begin by trying to find u so that du is a factor in the integrand. In this problem, if we let $u = x^2 + x + 5$, then $du = (2x+1)\, dx$ and

$$\int (2x + 1)(x^2 + x + 5)^4 \, dx$$

$$= \int \overset{u^4}{(x^2 + x + 5)^4}\,\overset{du}{(2x + 1)}\,dx \qquad \text{Substitution:}$$

$$\qquad\qquad\qquad u = x^2 + x + 5$$

$$\qquad\qquad\qquad du = (2x + 1)\,dx$$

$$= \int u^4 \, du \qquad\qquad\qquad \text{Use formula (1).}$$

$$= \frac{u^5}{5} + C \qquad\qquad\qquad \text{Substitute:}$$

$$\qquad\qquad\qquad u = x^2 + x + 5$$

$$= \frac{1}{5}(x^2 + x + 5)^5 + C \qquad \begin{array}{l}\text{Check by}\\ \text{differentiating.}\end{array}$$

Check $\qquad D_x\left[\dfrac{1}{5}(x^2 + x + 5)^5\right] = (x^2 + x + 5)^4(2x + 1)$

$$\overset{\frac{1}{u}}{}\quad \overset{\frac{1}{2}\,du}{}$$

(B) $\quad \displaystyle\int \frac{x}{4 + x^2}\,dx = \int \frac{1}{4 + x^2}\,x\,dx \qquad \text{Substitution:}$

$$\qquad\qquad\qquad u = 4 + x^2$$

$$\qquad\qquad\qquad du = 2x\,dx$$

$$\qquad\qquad\qquad \frac{1}{2}\,du = x\,dx$$

$$= \int \frac{1}{u}\frac{1}{2}\,du$$

$$= \frac{1}{2}\int \frac{1}{u}\,du \qquad\qquad \text{Use formula (2).}$$

$$= \frac{1}{2}\ln|u| + C \qquad\qquad \text{Substitute:}$$

$$\qquad\qquad\qquad u = 4 + x^2$$

$$= \frac{1}{2}\ln(4 + x^2) + C \qquad \begin{array}{l}\text{Absolute value signs}\\ \text{can be dropped, since}\\ 4 + x^2 > 0.\end{array}$$

Check $\qquad D_x\left[\dfrac{1}{2}\ln(4 + x^2)\right] = \dfrac{1}{2}\dfrac{1}{4 + x^2}\,2x$

$$= \frac{x}{4 + x^2}$$

$$e^u \frac{1}{3} \, du$$

(C) $\displaystyle\int x^2 e^{x^3} \, dx = \int e^{x^3} x^2 \, dx$ Substitution:

$$u = x^3$$
$$du = 3x^2 \, dx$$
$$\frac{1}{3} \, du = x^2 \, dx$$

$$= \int e^u \frac{1}{3} \, du$$

$$= \frac{1}{3} \int e^u \, du \qquad \text{Use formula (3).}$$

$$= \frac{1}{3} e^u + C \qquad \text{Substitute } u = x^3.$$

$$= \frac{1}{3} e^{x^3} + C \qquad \text{Check by differentiating.}$$

Check $\displaystyle D_x\left(\frac{1}{3} e^{x^3}\right) = \frac{1}{3} e^{x^3} \, 3x^2 = x^2 e^{x^3}$

Problem 1 Use substitution to find the following indefinite integrals:

(A) $\displaystyle\int (3x^2 + 2)(x^3 + 2x + 4)^6 \, dx$

(B) $\displaystyle\int xe^{x^2+5} \, dx$ (C) $\displaystyle\int \frac{x^2}{8 + x^3} \, dx$

We now summarize the steps we have been following.

Integration by Substitution

1. Select a substitution that appears to simplify the integrand. In particular, try to select u so that du is a factor in the integrand.
2. Express the integrand in terms of u and du, completely eliminating x and dx. In some cases, this will be easier to do if you first solve the equation $u = u(x)$ for x. (See Example 3.)
3. The integral should now be of the form

$$k \int f(u) \, du \qquad k \text{ a constant}$$

If possible, find an antiderivative for f. If you cannot find an antiderivative go back to step 1 and try a different substitution.
4. Substitute $u = u(x)$ in the antiderivative found in step 3 and express the answer in terms of the original variable.

Example 2   Find each of the following indefinite integrals:

$$\text{(A)} \quad \int \frac{e^x}{\sqrt{4 + e^x}} \, dx \qquad \text{(B)} \quad \int \frac{(\ln x)^2}{x} \, dx$$

$$\overset{u^{-1/2}}{} \qquad \overset{du}{}$$

Solutions   (A) $\displaystyle\int \frac{e^x}{\sqrt{4 + e^x}} \, dx = \int (4 + e^x)^{-1/2} e^x \, dx$ Substitution:

$$u = 4 + e^x$$
$$du = e^x \, dx$$

$$= \int u^{-1/2} \, du \qquad\qquad \text{Use formula (1).}$$

$$= \frac{u^{1/2}}{1/2} + C \qquad\qquad \text{Substitute } u = 4 + e^x.$$

$$= 2(4 + e^x)^{1/2} + C \qquad \text{Check by differentiating.}$$

Check $D_x[2(4 + e^x)^{1/2}] = 2 \cdot \dfrac{1}{2}\,(4 + e^x)^{-1/2} e^x$

$$= \frac{e^x}{\sqrt{4 + e^x}}$$

$$\overset{u^2}{} \qquad \overset{du}{}$$

(B) $\displaystyle\int \frac{(\ln x)^2}{x} \, dx = \int (\ln x)^2 \, \frac{1}{x} \, dx$ Substitution:

$$u = \ln x$$
$$du = \frac{1}{x} \, dx$$

$$= \int u^2 \, du \qquad\qquad \text{Use formula (1).}$$

$$= \frac{u^3}{3} + C \qquad\qquad \text{Substitute } u = \ln x.$$

$$= \frac{1}{3}\,(\ln x)^3 + C \qquad \text{Check by differentiating.}$$

Check $D_x\left[\dfrac{1}{3}\,(\ln x)^3\right] = \dfrac{1}{3} \cdot 3(\ln x)^2 \,\dfrac{1}{x}$

$$= \frac{(\ln x)^2}{x}$$

Problem 2   Find each of the following indefinite integrals:

$$\text{(A)} \quad \int \frac{e^x}{(5 + e^x)^2} \, dx \qquad \text{(B)} \quad \int \frac{\sqrt{\ln x}}{x} \, dx$$

Example 3   Find each of the following indefinite integrals:

(A) $\displaystyle\int \frac{x}{x+2}\, dx$ (B) $\displaystyle\int \frac{x}{\sqrt{x+2}}\, dx$

Solutions   (A)   No obvious substitution presents itself here. However, if we let $u = x + 2$, the integrand may simplify to something that we can integrate.

If $u = x + 2$, then $du = dx$ and

$$\int \frac{x}{x+2}\, dx = \int \frac{x}{u}\, du$$

To eliminate x in the numerator, we solve $u = x + 2$ for x:

$$u = x + 2$$
$$x = u - 2$$

$$= \int \frac{u-2}{u}\, du$$

$$= \int \left(1 - \frac{2}{u}\right) du$$

$$= u - 2\ln|u| + c \qquad \text{Substitute } u = x + 2.$$

$$= x + 2 - 2\ln|x + 2| + c \qquad \text{If } c \text{ is an arbitrary constant, so is } C = c + 2.$$

$$= x - 2\ln|x + 2| + C \qquad \text{Check by differentiating.}$$

Check $\displaystyle D_x(x - 2\ln|x + 2|) = 1 - 2 \cdot \frac{1}{x+2}$

$$= \frac{x+2}{x+2} - \frac{2}{x+2}$$

$$= \frac{x}{x+2}$$

(B)   We will let $u = \sqrt{x + 2}$ in the hope of simplifying the integrand. This time we will solve for x first and then determine the relationship between dx and du:

$$u = \sqrt{x + 2}$$
$$u^2 = x + 2$$
$$x = u^2 - 2 \qquad \text{If } x = x(u), \text{ then } dx = (dx/du)\, du.$$
$$dx = 2u\, du$$

Thus,

$$\int \frac{x}{\sqrt{x+2}}\,dx \quad = \quad \int \frac{u^2-2}{u}\,2u\,du$$

with substitutions: $x = u^2 - 2$, $dx = 2u\,du$, $\sqrt{x+2} = u$

$$= \int (2u^2 - 4)\,du$$

$$= \frac{2}{3}u^3 - 4u + C \qquad\qquad \text{Substitute } u = (x+2)^{1/2}.$$

$$= \frac{2}{3}(x+2)^{3/2} - 4(x+2)^{1/2} + C \qquad \text{Check by differentiating.}$$

Check
$$D_x\left[\frac{2}{3}(x+2)^{3/2} - 4(x+2)^{1/2}\right] = (x+2)^{1/2} - 2(x+2)^{-1/2}$$

$$= \frac{x+2}{(x+2)^{1/2}} - \frac{2}{(x+2)^{1/2}}$$

$$= \frac{x}{(x+2)^{1/2}}$$

Problem 3 Find each of the following indefinite integrals:

(A) $\displaystyle\int \frac{x+2}{x+1}\,dx$ (B) $\displaystyle\int \frac{x+2}{\sqrt{x+1}}\,dx$

■ **Definite Integrals and Substitution**

Example 4 illustrates two different methods for evaluating definite integrals by substitution.

Example 4 Evaluate

$$\int_0^2 \frac{x^2}{\sqrt{x^3+1}}\,dx$$

Solution   Method 1.   First find the indefinite integral:

$$\int \frac{x^2}{\sqrt{x^3+1}}\, dx = \int (x^3+1)^{-1/2}x^2\, dx \qquad \text{Substitution:}$$

$$u = x^3 + 1$$
$$du = 3x^2\, dx$$
$$\frac{1}{3}\, du = x^2\, dx$$

$$= \int u^{-1/2}\, \frac{1}{3}\, du$$

$$= \frac{1}{3}\frac{u^{1/2}}{1/2} + C$$

$$= \frac{2}{3}(x^3+1)^{1/2} + C$$

Now evaluate the definite integral.

$$\int_0^2 \frac{x^2}{\sqrt{x^3+1}}\, dx = \frac{2}{3}(x^3+1)^{1/2}\Big|_0^2$$

$$= \frac{2}{3}[(2)^3+1]^{1/2} - \frac{2}{3}[(0)^3+1]^{1/2}$$

$$= \frac{2}{3}(9)^{1/2} - \frac{2}{3}(1)^{1/2}$$

$$= 2 - \frac{2}{3} = \frac{4}{3}$$

Method 2.   Substitute directly in the definite integral, changing the limits of integration:

$$\overbrace{\phantom{\int_0^2 \frac{x^2}{\sqrt{x^3+1}}\, dx}}^{u=(2)^3+1}$$

$$\int_0^2 \frac{x^2}{\sqrt{x^3+1}}\, dx \quad = \int_1^9 u^{-1/2}\, \frac{1}{3}\, du \qquad \text{If } u = x^3+1, \text{ then}$$
$$\underbrace{\phantom{\int_0^2 \frac{x^2}{\sqrt{x^3+1}}\, dx}}_{u=(0)^3+1} \qquad\qquad\qquad x = 0 \text{ implies } u = 1$$
$$\text{and } x = 2 \text{ implies}$$
$$u = 9.$$

$$= \frac{2}{3}u^{1/2}\Big|_1^9$$

$$= \frac{2}{3}(9)^{1/2} - \frac{2}{3}(1)^{1/2}$$

$$= 2 - \frac{2}{3} = \frac{4}{3}$$

Problem 4   Evaluate

$$\int_1^3 \frac{x}{(x^2+1)^2}\,dx$$

Example 5   Find the area bounded by the graphs of $y = 5/(5-x)$ and $y = 0, 0 \le x \le 4$.

Solution   First we sketch a graph.

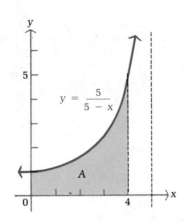

$$A = \int_0^4 \frac{5}{5-x}\,dx$$ Substitution: Limits:

$$u = 5 - x \qquad x = 0 \quad \text{implies} \quad u = 5$$
$$du = -dx \qquad x = 4 \quad \text{implies} \quad u = 1$$
$$dx = -du$$

$$= \int_5^1 \frac{5}{u}\,(-du)$$

$$= -5\,\ln|u|\,\Big|_5^1$$

$$= (-5\,\ln|1|) - (-5\,\ln|5|)$$

$$= 5\,\ln 5 \approx 8.05$$

Problem 5   Find the area bounded by the graphs of $y = 8/(6-x)$ and $y = 0, 2 \le x \le 5$.

■ **Common Errors**

1. $\displaystyle \int \frac{x^2}{x-1}\,dx = \int \frac{x^2}{u}\,du \qquad u = x - 1$
    $$du = dx$$

$$= x^2 \cancel{\int} \frac{1}{u}\,du$$

Remember that **only a constant can be moved across the integral sign.** Since u is now the variable of integration, it appears that x can be considered a constant. This is not correct, since x and u are related by the equation $u = x - 1$. You must substitute for x *wherever* it occurs in the integrand. The correct procedure is as follows:

$$\int \frac{x^2}{x - 1} \, dx = \int \frac{(u + 1)^2}{u} \, du \qquad \begin{array}{l} u = x - 1 \\ x = u + 1 \\ dx = du \end{array}$$

$$= \int \left(u + 2 + \frac{1}{u} \right) du$$

$$= \frac{1}{2} u^2 + 2u + \ln|u| + C$$

$$= \frac{1}{2} (x - 1)^2 + 2(x - 1) + \ln|x - 1| + C$$

2. $\displaystyle\int_1^9 \frac{1}{5 + \sqrt{x}} \, dx = \int \frac{1}{u} 2(u - 5) \, du \qquad \begin{array}{l} u = 5 + \sqrt{x} \\ x = (u - 5)^2 \\ dx = 2(u - 5) \, du \end{array}$

If a substitution is made in a definite integral, the limits of integration also must be changed. The new limits are determined by the particular substitution used in the integral. The correct procedure for this problem is as follows:

$$\int_1^9 \frac{1}{5 + \sqrt{x}} \, dx = \int_6^8 \frac{1}{u} 2(u - 5) \, du \qquad \begin{array}{l} u = 5 + \sqrt{x} \\ x = 1 \quad \text{implies} \quad u = 6 \\ x = 9 \quad \text{implies} \quad u = 8 \end{array}$$

$$= \int_6^8 \left(2 - \frac{10}{u} \right) du$$

$$= (2u - 10 \ln|u|) \Big|_6^8$$

$$= (16 - 10 \ln 8) - (12 - 10 \ln 6)$$

$$= 4 - 10 \ln 8 + 10 \ln 6 \approx 1.12$$

Answers to Matched Problems

1. (A) $\frac{1}{7}(x^3 + 2x + 4)^7 + C$ (B) $\frac{1}{2} e^{x^2+5} + C$

   (C) $\frac{1}{3} \ln|8 + x^3| + C$

2. (A) $-(5 + e^x)^{-1} + C$ (B) $\frac{2}{3}(\ln x)^{3/2} + C$

3. (A) $x + \ln|x + 1| + C$ (B) $\frac{2}{3}(x + 1)^{3/2} + 2(x + 1)^{1/2} + C$

4. $\frac{1}{5}$ 5. $8 \ln 4 \approx 11.1$

Exercise 15-1

A   Find each indefinite integral.

1. $\displaystyle \int x(x^2 + 9)^3 \, dx$

2. $\displaystyle \int x^2(x^3 + 9)^4 \, dx$

3. $\displaystyle \int \frac{1 + x}{4 + 2x + x^2} \, dx$

4. $\displaystyle \int \frac{x^2 - 1}{x^3 - 3x + 7} \, dx$

5. $\displaystyle \int (2x + 1)e^{x^2 + x + 1} \, dx$

6. $\displaystyle \int (x^2 + 2x)e^{x^3 + 3x^2} dx$

Evaluate each definite integral.

7. $\displaystyle \int_0^3 x\sqrt[3]{x^2 - 1} \, dx$

8. $\displaystyle \int_{-1}^2 x^2\sqrt{x^3 + 1} \, dx$

9. $\displaystyle \int_{-1}^1 \frac{1}{4x + 5} \, dx$

10. $\displaystyle \int_0^1 xe^{x^2 - 1} \, dx$

B   Find each indefinite integral.

11. $\displaystyle \int e^{2x}(1 + e^{2x})^3 \, dx$

12. $\displaystyle \int \frac{e^x}{1 + e^x} \, dx$

13. $\displaystyle \int \frac{(\ln x)^3}{x} \, dx$

14. $\displaystyle \int \frac{\ln(x + 4)}{x + 4} \, dx$

15. $\displaystyle \int x(x - 5)^4 \, dx$

16. $\displaystyle \int (x + 1)(x + 3)^5 \, dx$

17. $\displaystyle \int x\sqrt{4 + x} \, dx$

18. $\displaystyle \int x\sqrt[3]{2 - x} \, dx$

19. $\displaystyle \int \frac{x}{\sqrt{x + 3}} \, dx$

20. $\displaystyle \int \frac{x - 2}{\sqrt{4 - x}} \, dx$

21. $\displaystyle \int \frac{x}{x - 2} \, dx$

22. $\displaystyle \int \frac{x}{(x - 2)^2} \, dx$

23. $\displaystyle \int \frac{x^2}{(x - 2)} \, dx$

24. $\displaystyle \int \frac{x^2}{(x - 2)^2} \, dx$

Find the area bounded by the graphs of the indicated equations.

25. $y = \dfrac{8x}{x^2 + 4}$, $\quad y = 0$, $\quad 0 \leq x \leq 4$

26. $y = 4xe^{-x^2}$, $\quad y = 0$, $\quad 0 \leq x \leq 1$

27. $y = x\sqrt{9 - x^2}$, $\quad y = 0$, $\quad 0 \leq x \leq 3$

28. $y = \dfrac{x}{(5 - x^2)^2}$, $\quad y = 0$, $\quad 0 \leq x \leq 2$

29. $y = x\sqrt{2 - x}$, $\quad y = 0$, $\quad 0 \leq x \leq 2$

30. $y = \dfrac{x}{\sqrt{10 - x}}$, $\quad y = 0$, $\quad 6 \leq x \leq 9$

In Problems 31–34, find each indefinite integral two ways: first use the substitution $u = \sqrt{x - 1}$ and then use the substitution $u = x - 1$.

31. $\displaystyle\int \dfrac{x}{\sqrt{x - 1}}\, dx$

32. $\displaystyle\int x\sqrt{x - 1}\, dx$

33. $\displaystyle\int \dfrac{x^2}{\sqrt{x - 1}}\, dx$

34. $\displaystyle\int x^2\sqrt{x - 1}\, dx$

C *Find each indefinite integral.*

35. $\displaystyle\int \dfrac{\sqrt{x} - 2}{\sqrt{x} - 1}\, dx$

36. $\displaystyle\int \dfrac{1}{3 + \sqrt{x - 2}}\, dx$

37. $\displaystyle\int \dfrac{1}{\sqrt{x}(1 + \sqrt{x})}\, dx$

38. $\displaystyle\int \dfrac{1}{x + 2\sqrt{x}}\, dx$

39. $\displaystyle\int \dfrac{1}{x^2}\, e^{-1/x}\, dx$

40. $\displaystyle\int \dfrac{1}{x \ln x}\, dx$

41. Use the substitution $u = -x$ to show that if $f(x)$ is an odd function [that is, if $f(-x) = -f(x)$], then

$$\int_{-a}^{0} f(x)\, dx = - \int_{0}^{a} f(u)\, du$$

Then show that $\displaystyle\int_{-a}^{a} f(x)\, dx = 0$.

42. Use the substitution $u = -x$ to show that if $f(x)$ is an even function [that is, if $f(-x) = f(x)$], then

$$\int_{-a}^{0} f(x)\, dx = \int_{0}^{a} f(u)\, du$$

Then show that $\displaystyle\int_{-a}^{a} f(x)\, dx = 2\int_{0}^{a} f(x)\, dx$.

Applications

Business & Economics

43. *Price–demand equation.* The marginal price $p'(x)$ at x units per week for a certain style of designer jeans is given by

$$p'(x) = \frac{-300,000x}{(5,000 + x^2)^2}$$

At a price of $30 each, the weekly demand is 100. Find the price–demand equation.

44. *Consumers' surplus.* (Refer to Section 14-5.) Find the consumers' surplus for

$$p = D(x) = \frac{400 + 10x}{10 + x} \qquad p = S(x) = \frac{5}{2}x$$

45. *Marginal analysis.* The marginal cost and revenue equations (in thousands of dollars per year) for a coin-operated photocopying machine are given by

$$R'(t) = 5te^{-t^2}$$

$$C'(t) = \frac{1}{11}t$$

where t is time in years. The area between the graphs of the marginal equations for the time period such that $R'(t) \geq C'(t)$ represents the total accumulated profit for the useful life of the machine.

What is the useful life of the machine? What is the total profit?

46. *Cash reserves.* Suppose cash reserves (in thousands of dollars) are approximated by

$$C(x) = 1 + x\sqrt{12 - x} \qquad 0 \leq x \leq 12$$

where x is the number of months after the first of the year. What is the average cash reserve for the first quarter? The fourth quarter?

Life Sciences

47. *Pollution.* A contaminated lake is treated with a bactericide. The rate of decrease in harmful bacteria t days after the treatment is given by

$$\frac{dN}{dt} = -\frac{2,000t}{1 + t^2} \qquad 0 \leq t \leq 10$$

where $N(t)$ is the number of bacteria per milliliter of water. If the initial count was 5,000 bacteria per milliliter, find $N(t)$ and then find the bacteria count after 10 days.

48. *Medicine.* One hour after x milligrams of a particular drug are given to a person, the rate of change of temperature in degrees Fahrenheit, $T'(x)$, with respect to dosage x (called *sensitivity*) is given approximately by

$$T'(x) = \frac{1}{10}x\sqrt{9 - x} \qquad 0 \leq x \leq 9$$

What total change in temperature results from a dosage change from 0 to 5 milligrams? From 8 to 9 milligrams?

Social Sciences 49. *Learning.* A person learns N items at a rate given approximately by

$$N'(t) = \frac{15t}{\sqrt{1+t}} \qquad 0 \leqslant t \leqslant 10$$

where t is the number of hours of continuous study. Find the total number of items learned from $t = 0$ to $t = 8$ hours of study.

15-2 Integration by Parts

In Section 14-1 we said that we would return to the indefinite integral

$$\int \ln x \, dx$$

later, since none of the integration techniques considered up to that time could be used to find an antiderivative for ln x. We will now develop a very useful technique, called *integration by parts,* that will not only enable us to find the above integral, but also many others, including integrals such as

$$\int x \ln x \, dx \qquad \text{and} \qquad \int xe^x \, dx$$

The integration by parts technique is also used to derive many integration formulas that are tabulated in mathematical handbooks.

The method of integration by parts is based on the product formula for derivatives. If f and g are differentiable functions, then

$$D_x[f(x)g(x)] = f(x)g'(x) + g(x)f'(x)$$

which can be written in the equivalent form

$$f(x)g'(x) = D_x[f(x)g(x)] - g(x)f'(x)$$

Integrating both sides, we obtain

$$\int f(x)g'(x) \, dx = \int D_x[f(x)g(x)] \, dx - \int g(x)f'(x) \, dx$$

The first integral to the right of the equal sign is $f(x)g(x) + C$. (Why?) We will leave out the constant of integration for now, since we can add it after integrating the second integral to the right of the equal sign. So we have

$$\int f(x)g'(x) \, dx = f(x)g(x) - \int g(x)f'(x) \, dx$$

This last form can be transformed into a more convenient form by letting $u = f(x)$ and $v = g(x)$; then $du = f'(x) \, dx$ and $dv = g'(x) \, dx$. Making these substitutions, we obtain the **integration by parts formula:**

Integration by Parts Formula

$$\int u \, dv = uv - \int v \, du$$

This formula can be very useful when the integral on the left is difficult to integrate using standard formulas. If u and dv are chosen with care, then the integral on the right side may be easier to integrate than the one on the left. Several examples will demonstrate the use of the formula.

Example 6 Find $\int x \ln x \, dx$, $x > 0$, using integration by parts.

Solution First, write the integration by parts formula

$$\int u \, dv = uv - \int v \, du$$

Then try to identify u and dv in $\int x \ln x \, dx$ (this is the key step) so that when $\int u \, dv$ is written in the form $uv - \int v \, du$, the new integral will be easier to integrate.

Suppose we choose

$$u = x \quad \text{and} \quad dv = \ln x \, dx$$

Then

$$du = dx \qquad v = ?$$

We do not know an antiderivative of $\ln x$ yet, so we change our choice for u and dv to

$$u = \ln x \qquad dv = x \, dx$$

Then

$$du = \frac{1}{x} \, dx \qquad v = \frac{x^2}{2}$$

Any constant may be added to v (we choose 0 for simplicity). There are cases where it is convenient to add a constant other than 0, but in most cases 0 will do. The general arbitrary constant of integration will be added at the end of the process.

Using the chosen u, du, dv, and v in the integration by parts formula, we obtain

$$\int u \quad dv = u \quad v \; - \int v \quad du$$

$$\int (\ln x)x \, dx = (\ln x)\left(\frac{x^2}{2}\right) - \int \left(\frac{x^2}{2}\right)\frac{1}{x} \, dx$$

$$= \frac{x^2}{2} \ln x - \int \frac{x}{2} \, dx \qquad \text{This new integral is easy to integrate.}$$

$$= \frac{x^2}{2} \ln x - \frac{x^2}{4} + C$$

To check this result, show that

$$D_x\left(\frac{x^2}{2} \ln x - \frac{x^2}{4} + C\right) = x \ln x$$

which is the integrand in the original integral.

Problem 6   Find $\int x \ln 2x \, dx$.

Example 7   Find $\int xe^x \, dx$.

Solution   We write the integration by parts formula

$$\int u \, dv = uv - \int v \, du$$

and choose

$$u = e^x \qquad dv = x \, dx$$

Then

$$du = e^x \, dx \qquad v = \frac{x^2}{2}$$

and

$$\int u \quad dv = u \quad v \; - \int v \quad du$$

$$\int e^x x \, dx = e^x\left(\frac{x^2}{2}\right) - \int \left(\frac{x^2}{2}\right)e^x \, dx$$

$$= \frac{x^2}{2} e^x - \frac{1}{2} \int x^2 e^x \, dx \qquad \text{This new integral is more complicated than the original one.}$$

This time the integration by parts formula leads to a new integral that is more complicated than the one we started with. This does not mean that there is an error in our calculations or in the formula. It simply means that our first choice for u and dv did not change the original problem into one that we can solve. Thus, we must make a different selection. Suppose we choose

$$u = x \qquad dv = e^x \, dx$$

Then

$$du = dx \qquad v = e^x$$

and

$$\int u \, dv = uv - \int v \, du$$

$$\int xe^x \, dx = xe^x - \int e^x \, dx \qquad \text{This integral is}$$
$$\text{one we can evaluate.}$$

$$= xe^x - e^x + C$$

Problem 7   Find $\int xe^{2x} \, dx$.

Integration by Parts: Selection of *u* and *dv*

1. It must be possible to integrate dv (preferably by using standard formulas or simple substitutions).
2. The new integral, $\int v \, du$, should be simpler than the original integral, $\int u \, dv$.
3. For integrals involving $x^p(\ln x)^q$, try

    $$u = (\ln x)^q \qquad dv = x^p \, dx$$

4. For integrals involving $x^p e^{ax}$, try

    $$u = x^p \qquad dv = e^{ax} \, dx$$

Example 8   Find $\int x^2 e^{-x} \, dx$.

Solution   Following suggestion 4 in the box, we choose

$$u = x^2 \qquad dv = e^{-x} \, dx$$

Then

$$du = 2x \, dx \qquad v = -e^{-x}$$

and

$$\int u \ dv = u \ v - \int v \ du$$

$$\int ^-x^2 e^{-x} \ dx = x^2(-e^{-x}) - \int (-e^{-x})2x \ dx$$

$$= -x^2 e^{-x} + 2 \int xe^{-x} \ dx \tag{1}$$

The new integral is not one we can evaluate by standard formulas, but it is simpler than the original integral. Applying the integration by parts formula to it will produce an even simpler integral. For the integral $\int xe^{-x} \ dx$, we choose

$$u = x \qquad dv = e^{-x} \ dx$$

Then

$$du = dx \qquad v = -e^{-x}$$

and

$$\int u \ dv = u \ v - \int v \ du$$

$$\int xe^{-x} \ dx = x(-e^{-x}) - \int (-e^{-x}) \ dx$$

$$= -xe^{-x} + \int e^{-x} \ dx$$

$$= -xe^{-x} - e^{-x} \tag{2}$$

Substituting (2) into (1) and adding a constant of integration, we have

$$\int x^2 e^{-x} \ dx = -x^2 e^{-x} + 2(-xe^{-x} - e^{-x}) + C$$

$$= -x^2 e^{-x} - 2xe^{-x} - 2e^{-x} + C$$

Problem 8   Find $\int x^2 e^{2x} \ dx$.

Example 9   Find $\int_1^e \ln x \ dx$.

Solution   First, find $\int \ln x \ dx$; then return to the definite integral. Following suggestion 3 in the box (with $p = 0$), we choose

$$u = \ln x \qquad dv = dx$$

Then

$$du = \frac{1}{x} \ dx \qquad v = x$$

Hence,

$$\int \ln x \, dx = (\ln x)(x) - \int (x) \frac{1}{x} \, dx$$
$$= x \ln x - x + C$$

Thus,

$$\int_1^e \ln x \, dx = (x \ln x - x) \Big|_1^e$$
$$= (e \ln e - e) - (1 \ln 1 - 1)$$
$$= (e - e) - (0 - 1)$$
$$= 1$$

Problem 9   Find $\int_1^2 \ln 3x \, dx$.

Answers to Matched Problems

6. $\dfrac{x^2}{2} \ln 2x - \dfrac{x^2}{4} + C$ 7. $\dfrac{x}{2} e^{2x} - \dfrac{1}{4} e^{2x} + C$

8. $\dfrac{x^2}{2} e^{2x} - \dfrac{x}{2} e^{2x} + \dfrac{1}{4} e^{2x} + C$ 9. $2 \ln 6 - \ln 3 - 1 \approx 1.4849$

Exercise 15-2

A   *Integrate using integration by parts. Assume $x > 0$ whenever the natural log function is involved.*

1. $\displaystyle\int xe^{3x} \, dx$ 2. $\displaystyle\int xe^{4x} \, dx$

3. $\displaystyle\int x^2 \ln x \, dx$ 4. $\displaystyle\int x^3 \ln x \, dx$

B   *Problems 5–18 are mixed—some require integration by parts and others can be solved using techniques we have considered earlier. Integrate as indicated, assuming $x > 0$ whenever the natural log function is involved.*

5. $\displaystyle\int xe^{-x} \, dx$ 6. $\displaystyle\int (x-1)e^{-x} \, dx$

7. $\displaystyle\int xe^{x^2} \, dx$ 8. $\displaystyle\int xe^{-x^2} \, dx$

9. $\displaystyle\int_0^1 (x-3)e^x \, dx$ 10. $\displaystyle\int_0^2 (x+5)e^x \, dx$

11. $\displaystyle\int_1^3 \ln 2x \, dx$ 12. $\displaystyle\int_2^3 \ln 7x \, dx$

13. $\int \dfrac{2x}{x^2 + 1}\, dx$

14. $\int \dfrac{x^2}{x^3 + 5}\, dx$

15. $\int \dfrac{\ln x}{x}\, dx$

16. $\int \dfrac{e^x}{e^x + 1}\, dx$

17. $\int \sqrt{x}\, \ln x\, dx$

18. $\int \dfrac{\ln x}{\sqrt{x}}\, dx$

C   *Some of these problems may require using the integration by parts formula more than once. Assume $x > 0$ whenever the natural log function is involved.*

19. $\int x^2 e^x\, dx$

20. $\int x^3 e^x\, dx$

21. $\int x e^{ax}\, dx, \quad a \neq 0$

22. $\int \ln(ax)\, dx, \quad a > 0$

23. $\int_1^e \dfrac{\ln x}{x^2}\, dx$

24. $\int_1^2 x^3 e^{x^2}\, dx$

25. $\int (\ln x)^2\, dx$

26. $\int x(\ln x)^2\, dx$

27. $\int (\ln x)^3\, dx$

28. $\int x(\ln x)^3\, dx$

Problems 29–34 require both integration by parts and techniques we have considered earlier.

29. $\int e^{\sqrt{x}}\, dx$

30. $\int \sqrt{x}\, e^{\sqrt{x}}\, dx$

31. $\int x \ln(1 + x^2)\, dx$

32. $\int x \ln(1 + x)\, dx, \quad x > -1$

33. $\int \dfrac{\ln(1 + \sqrt{x})}{\sqrt{x}}\, dx$

34. $\int \ln(1 + \sqrt{x})\, dx$

Applications

Business & Economics

35. *Marginal profit.* If the marginal profit per year in millions of dollars is given by

$$P'(t) = 2t - te^{-t}$$

where t is time in years and the profit at time 0 is 0, find $P = P(t)$.

36. *Production.* An oil field is estimated to produce $R(t)$ thousand barrels of oil per month t months from now, as given by

$$R(t) = 10te^{-0.1t}$$

Estimate the total production in the first year of operation by use of an appropriate definite integral.

Life Sciences

37. *Pollution.* The concentration of particulate matter in parts per million *t* hours after a factory ceases operation for the day is given by

$$C(t) = \frac{20 \ln(t + 1)}{(t + 1)^2}$$

Find the average concentration for the time period from $t = 0$ to $t = 5$.

38. *Medicine.* After a person takes a pill, the drug contained in the pill is assimilated into the bloodstream. The rate of assimilation *t* minutes after taking the pill is

$$R(t) = te^{-0.2t}$$

Find the total amount of the drug that is assimilated into the bloodstream during the first 10 minutes after the pill is taken.

Social Sciences

39. *Politics.* The number of voters (in thousands) in a certain city is given by

$$N(t) = 20 + 4t - 5te^{-0.1t}$$

where *t* is the time in years. Find the average number of voters during the time period from $t = 0$ to $t = 5$.

15-3   Integration Using Tables

- Introduction
- Using a Table of Integrals
- Substitution and Integral Tables
- Application

■ Introduction

A **table of integrals** is a list of integration formulas that can be used to evaluate definite integrals. Individuals who must evaluate complicated integrals often refer to a table that may contain hundreds of formulas. Tables of this type can be found in mathematical handbooks; a short table illustrating the types of formulas found in more extensive tables is located inside the back cover of this book. These formulas have been derived by techniques we have not considered; however, it is possible to verify each formula by differentiating the right side.

You may notice some logical gaps in the list of formulas we have selected for this table. There are two reasons for this:

1. We have not included formulas for integrals that can be evaluated by the techniques we have already discussed. Thus, you will find formulas for $\int \sqrt{u^2 + a^2}\ du$ and $\int u^2\sqrt{u^2 + a^2}\ du$, but not for

$\int u\sqrt{u^2 + a^2}\, du$, since this last integral can be evaluated by making a simple substitution.

2. Many antiderivatives involve functions we have not considered. Thus, for example, a formula for $\int \sqrt{a^2 - u^2}\, du$ is not included in the table because the antiderivative of $\sqrt{a^2 - u^2}$ involves an inverse trigonometric function.

Even though our table is not very large, it will still permit us to evaluate many new indefinite integrals. We will now consider some examples that will illustrate the use of a table of integrals.

■ Using a Table of Integrals

Example 10   Use the Table of Integrals inside the back cover to find

$$\int \frac{x}{(2x + 5)(3x + 4)}\, dx$$

Solution   Since the integrand

$$f(x) = \frac{x}{(2x + 5)(3x + 4)}$$

is a rational function, we examine formulas 1–7 to determine if any of the integrands in these formulas has the same form as f. Comparing the integrand in formula 2 with f, we conclude that this formula can be used to evaluate $\int f(x)\, dx$. Letting $u = x$ and identifying the appropriate values for a, b, c, d, and $\Delta = bc - ad$, we have

$$a = 2 \qquad b = 5 \qquad c = 3 \qquad d = 4$$
$$\Delta = bc - ad = (5)(3) - (2)(4) = 7$$

$$\int \frac{u}{(au + b)(cu + d)}\, du = \frac{1}{\Delta}\left(\frac{b}{a}\ln|au + b| - \frac{d}{c}\ln|cu + d|\right) \qquad \text{Formula 2}$$

$$\int \frac{x}{(\underset{a}{2x} + \underset{b}{5})(\underset{c}{3x} + \underset{d}{4})}\, dx = \frac{1}{7}\left(\frac{5}{2}\ln|2x + 5| - \frac{4}{3}\ln|3x + 4|\right) + C$$

$$= \frac{5}{14}\ln|2x + 5| - \frac{4}{21}\ln|3x + 4| + C$$

Notice that the constant of integration C is not included in any of the formulas in the table. You must still include C in your antiderivatives.

Problem 10   Use the Table of Integrals inside the back cover to find

$$\int \frac{1}{(3x + 5)^2(x + 1)}\, dx$$

Example 11 Evaluate

$$\int_3^4 \frac{1}{x\sqrt{25 - x^2}}\, dx$$

Solution First we will use the Table of Integrals to find

$$\int \frac{1}{x\sqrt{25 - x^2}}\, dx$$

Since the integrand involves the expression $\sqrt{25 - x^2}$, we examine formulas 8–10 and select formula 8 with $a^2 = 25$ and $a = 5$.

$$\int \frac{1}{u\sqrt{a^2 - u^2}}\, du = -\frac{1}{a} \ln\left|\frac{a + \sqrt{a^2 - u^2}}{u}\right| \qquad \text{Formula 8}$$

$$\int \frac{1}{x\sqrt{25 - x^2}}\, dx = -\frac{1}{5} \ln\left|\frac{5 + \sqrt{25 - x^2}}{x}\right| + C$$

Thus,

$$\int_3^4 \frac{1}{x\sqrt{25 - x^2}}\, dx = -\frac{1}{5} \ln\left|\frac{5 + \sqrt{25 - x^2}}{x}\right|\Bigg|_3^4$$

$$= -\frac{1}{5} \ln\left|\frac{5 + 3}{4}\right| + \frac{1}{5} \ln\left|\frac{5 + 4}{3}\right|$$

$$= -\frac{1}{5} \ln 2 + \frac{1}{5} \ln 3 = \frac{1}{5} \ln 1.5 \approx .0811$$

Problem 11 Evaluate

$$\int_6^8 \frac{1}{x^2\sqrt{100 - x^2}}\, dx$$

■ Substitution and Integral Tables

As Examples 10 and 11 illustrate, if the integral we want to evaluate corresponds exactly to one in the table, then evaluating the indefinite integral consists of simply substituting the correct values of the constants into the formula. What happens if we cannot match an integral with one of the formulas in the table? In many cases, a substitution will change the given integral into one that appears in the table. The following examples illustrate several frequently used substitutions.

Example 12 Find

$$\int \frac{x^2}{\sqrt{16x^2 - 25}}\, dx$$

Solution   In order to relate this integral to one of the formulas involving $\sqrt{u^2 - a^2}$ (formulas 19–24), we observe that if $u = 4x$, then

$$u^2 = 16x^2 \quad \text{and} \quad \sqrt{16x^2 - 25} = \sqrt{u^2 - 25}$$

Thus, we will use the substitution $u = 4x$ to change this integral into one that appears in the table.

$$\int \frac{x^2}{\sqrt{16x^2 - 25}}\, dx = \int \frac{(1/16)\, u^2}{\sqrt{u^2 - 25}} \cdot \frac{1}{4}\, du$$

Substitution:

$u = 4x$

$$= \frac{1}{64} \int \frac{u^2}{\sqrt{u^2 - 25}}\, du \qquad x = \frac{1}{4} u$$

$$dx = \frac{1}{4}\, du$$

This last integral can be evaluated by using formula 23 with $a = 5$:

$$\int \frac{u^2}{\sqrt{u^2 - 25}}\, du = \frac{1}{2}\left(u\sqrt{u^2 - a^2} + a^2 \ln|u + \sqrt{u^2 - a^2}|\right) \qquad \text{Formula 23}$$

Thus,

$$\int \frac{x^2}{\sqrt{16x^2 - 25}}\, dx$$

$$= \frac{1}{64} \int \frac{u^2}{\sqrt{u^2 - 25}}\, du \qquad \qquad \text{Use formula 23 with } a = 5.$$

$$= \frac{1}{128}\left(u\sqrt{u^2 - 25} + 25\, \ln|u + \sqrt{u^2 - 25}|\right) + C \qquad \text{Substitute } u = 4x.$$

$$= \frac{1}{128}\left(4x\sqrt{16x^2 - 25} + 25\, \ln|4x + \sqrt{16x^2 - 25}|\right) + C$$

Problem 12   Find $\int \sqrt{9x^2 - 16}\, dx$.

Example 13   Find

$$\int \frac{x}{\sqrt{x^4 + 1}}\, dx$$

Solution   None of the formulas in the table involve fourth powers; however, if we let $u = x^2$, then

$$\sqrt{x^4 + 1} = \sqrt{u^2 + 1}$$

which does appear in formulas 11–18.

$$\int \frac{1}{\sqrt{x^4 + 1}}\, x\, dx = \int \frac{1}{\sqrt{u^2 + 1}} \frac{1}{2}\, du$$

Substitution:

$u = x^2$

$$= \frac{1}{2} \int \frac{1}{\sqrt{u^2 + 1}}\, du \qquad du = 2x\, dx$$

$$\frac{1}{2}\, du = x\, dx$$

We recognize the last integral as formula 15 with $a = 1$:

$$\int \frac{1}{\sqrt{u^2 + a^2}}\, du = \ln|u + \sqrt{u^2 + a^2}| \qquad \text{Formula 15}$$

Thus,

$$\int \frac{x}{\sqrt{x^4 + 1}}\, dx = \frac{1}{2} \int \frac{1}{\sqrt{u^2 + 1}}\, du \qquad \text{Use formula 15 with } a = 1.$$

$$= \frac{1}{2} \ln|u + \sqrt{u^2 + 1}| + C \qquad \text{Substitute } u = x^2.$$

$$= \frac{1}{2} \ln|x^2 + \sqrt{x^4 + 1}| + C$$

Problem 13   Find $\int x\sqrt{x^4 + 1}\, dx$.

Example 14   Evaluate

$$\int_5^{21} \frac{\sqrt{x + 4}}{x}\, dx$$

Solution   Since none of the formulas in the table involve $\sqrt{ax + b}$ (a more extensive table would contain formulas of this type), we first make a substitution to eliminate the square root:

$$\int_5^{21} \frac{\sqrt{x + 4}}{x}\, dx$$

$$= \int_3^5 \frac{u}{u^2 - 4}\, 2u\, du \qquad \text{Substitution:} \qquad \text{Limits:}$$

$$= 2 \int_3^5 \frac{u^2}{(u + 2)(u - 2)}\, du \qquad \begin{aligned} u &= \sqrt{x + 4} \\ x &= u^2 - 4 \\ dx &= 2u\, du \end{aligned} \qquad \begin{aligned} x &= 5 \text{ implies } u = 3 \\ x &= 21 \text{ implies } u = 5 \end{aligned}$$

Use formula 3 with $a = 1$, $b = 2$, $c = 1$, $d = -2$, and $\Delta = 4$:

$$\int \frac{u^2}{(au + b)(cu + d)}\, du$$

$$= \frac{1}{ac} u - \frac{1}{\Delta} \left(\frac{b^2}{a^2} \ln|au + b| - \frac{d^2}{c^2} \ln|cu + d| \right) \qquad \text{Formula 3}$$

Thus,

$$\int_5^{21} \frac{\sqrt{x + 4}}{x}\, dx = 2 \int_3^5 \frac{u^2}{(u + 2)(u - 2)}\, du \qquad \begin{aligned} &\text{Use formula 3 with } a = 1, \\ &b = 2, c = 1, d = -2, \text{ and} \\ &\Delta = 4. \end{aligned}$$

$$= 2 \left[u - \frac{1}{4} \left(\frac{4}{1} \ln|u + 2| - \frac{4}{1} \ln|u - 2| \right) \right] \Big|_3^5$$

$$= 2(5 - \ln|7| + \ln|3|) - 2(3 - \ln|5| + \ln|1|)$$

$$= 4 + 2 \ln \frac{15}{7} \approx 5.5243$$

Problem 14   Evaluate

$$\int_7^{40} \frac{1}{x\sqrt{x+9}}\, dx$$

▪ Application

One of the growth laws discussed in Section 14-2 was referred to as *logistic growth*. In this situation, the rate of growth of a quantity y is assumed to be proportional both to y and to the difference between y and a fixed upper limit M. Hence, y must satisfy the differential equation

$$\frac{dy}{dt} = ky(M - y) \qquad \text{Logistic growth equation}$$

Using the Table of Integrals, we can now solve this differential equation.

Example 15
Logistic Growth

Solve the differential equation

$$\frac{dy}{dt} = ky(M - y) \qquad y(0) = 1$$

Solution

$$\frac{dy}{dt} = ky(M - y) \qquad \begin{array}{l}\text{Multiply both sides of}\\ \text{this equation by } dt/[y(M - y)].\end{array}$$

$$\frac{dy}{y(M - y)} = k\, dt \qquad \begin{array}{l}\text{Integrate both sides of}\\ \text{this equation.}\end{array}$$

$$\int \frac{dy}{y(M - y)} = \int k\, dt$$

$$= kt + C \tag{1}$$

To evaluate

$$\int \frac{dy}{y(M - y)}$$

we use formula 1 with $a = 1$, $b = 0$, $c = -1$, $d = M$, and $\Delta = -M$.

$$\int \frac{1}{(au + b)(cu + d)}\, du = \frac{1}{\Delta} \ln \left|\frac{cu + d}{au + b}\right| \qquad \text{Formula 1}$$

$$\int \frac{dy}{y(M - y)} = -\frac{1}{M} \ln \left|\frac{M - y}{y}\right| \tag{2}$$

Substituting (2) into (1) and simplifying yields

$$\frac{M-y}{y} = e^{-Mkt-MC}$$ Now solve for y.

$$\frac{M-y}{y} = e^{-MC}e^{-Mkt}$$

$$M - y = ye^{-MC}e^{-Mkt}$$

$$y = \frac{M}{1 + e^{-MC}e^{-Mkt}}$$

Now $y(0) = 1$; hence,

$$y(0) = \frac{M}{1 + e^{-MC}} = 1$$

$$M = 1 + e^{-MC}$$

$$e^{-MC} = M - 1$$

Thus,

$$y = \frac{M}{1 + (M-1)e^{-Mkt}}$$ Solution to the logistic growth equation when $y(0) = 1$

Problem 15 In some biological applications, the logistic growth equation is written as

$$\frac{dy}{dt} = k\left(1 - \frac{y}{M}\right)y$$

Find the solution to this equation that satisfies $y(0) = 1$.

Answers to Matched Problems

10. $\dfrac{1}{2}\dfrac{1}{3x+5} + \dfrac{1}{4}\ln\left|\dfrac{x+1}{3x+5}\right| + C$ 11. $\dfrac{7}{1,200} \approx 0.0058$

12. $\dfrac{1}{6}(3x\sqrt{9x^2 - 16} - 16\ln|3x + \sqrt{9x^2 - 16}|) + C$

13. $\dfrac{1}{4}(x^2\sqrt{x^4 + 1} + \ln|x^2 + \sqrt{x^4 + 1}|) + C$

14. $\dfrac{1}{3}\ln 2.8 \approx 0.3432$ 15. $y = \dfrac{M}{1 + (M-1)e^{-kt}}$

Exercise 15-3

A Use the Table of Integrals inside the back cover of this book to find each indefinite integral.

1. $\displaystyle\int \frac{1}{x(1+x)}\,dx$

2. $\displaystyle\int \frac{1}{x^2(x+1)}\,dx$

3. $\int \dfrac{1}{(x+3)^2(2x+5)} \, dx$

4. $\int \dfrac{x}{(2x+5)^2(x+2)} \, dx$

5. $\int \dfrac{1}{x\sqrt{x^2+4}} \, dx$

6. $\int \dfrac{1}{x^2\sqrt{x^2-16}} \, dx$

7. $\int \dfrac{\sqrt{1-x^2}}{x} \, dx$

8. $\int \dfrac{x^2}{\sqrt{x^2+64}} \, dx$

Evaluate each definite integral. Use the Table of Integrals to find the anti-derivative.

9. $\displaystyle\int_0^7 \dfrac{1}{(x+3)(x+1)} \, dx$

10. $\displaystyle\int_0^7 \dfrac{x}{(x+3)(x+1)} \, dx$

11. $\displaystyle\int_0^4 \dfrac{1}{\sqrt{x^2+9}} \, dx$

12. $\displaystyle\int_4^5 \sqrt{x^2-16} \, dx$

B   *Use substitution techniques and the Table of Integrals to find each indefinite integral.*

13. $\int \dfrac{\sqrt{4x^2+1}}{x^2} \, dx$

14. $\int x^2\sqrt{9x^2-1} \, dx$

15. $\int \dfrac{x}{\sqrt{x^4-16}} \, dx$

16. $\int x\sqrt{x^4-16} \, dx$

17. $\int x^2\sqrt{x^6+4} \, dx$

18. $\int \dfrac{x^2}{\sqrt{x^6+4}} \, dx$

19. $\int \dfrac{\sqrt{x+16}}{x} \, dx$

20. $\int \dfrac{1}{x\sqrt{x+16}} \, dx$

21. $\int \dfrac{1}{x^3\sqrt{4-x^4}} \, dx$

22. $\int \dfrac{\sqrt{x^4+4}}{x} \, dx$

23. $\int \dfrac{1}{x^2\sqrt{x+1}} \, dx$

24. $\int \dfrac{1}{x(1+\sqrt{x})^2} \, dx$

C   *Problems 25–32 are mixed—some require the use of the Table of Integrals and others can be solved using techniques we have considered earlier.*

25. $\displaystyle\int_3^5 x\sqrt{x^2-9} \, dx$

26. $\displaystyle\int_3^5 x^2\sqrt{x^2-9} \, dx$

27. $\displaystyle\int_2^4 \dfrac{1}{(x^2-1)^2} \, dx$

28. $\displaystyle\int_2^4 \dfrac{x}{(x^2-1)^2} \, dx$

29. $\int \dfrac{x+1}{x^2+2x} \, dx$

30. $\int \dfrac{x+1}{x^2+x} \, dx$

31. $\int \dfrac{x+1}{x^2+3x} \, dx$

32. $\int \dfrac{x^2+1}{x^2+3x} \, dx$

■

Applications

Business & Economics

33. *Consumers' surplus.* (Refer to Section 14-5.) Find the consumers' surplus for

$$p = D(x) = \frac{360}{(x + 2)(x + 1)}$$

$$p = S(x) = \frac{5x}{x + 2}$$

34. *Marginal analysis.* The marginal cost and revenue equations (in thousands of dollars per year) for a vending machine are given by

$$R'(t) = \frac{25t}{(t + 1)(t + 6)}$$

$$C'(t) = \frac{1}{2} t$$

where t is time in years. The area between the graphs of the marginal equations for the time period such that $R'(t) \geqslant C'(t)$ represents the total accumulated profit for the useful life of the machine.

What is the useful life of the machine? What is the total profit?

Life Sciences

35. *Pollution.* An oil tanker aground on a reef is losing oil and producing an oil slick that is radiating outward at a rate given approximately by

$$\frac{dR}{dt} = \frac{100}{\sqrt{t^2 + 9}} \qquad t \geqslant 0$$

where R is the radius (in feet) of the circular slick after t minutes. Find the radius of the slick after 4 minutes if the radius is 0 when $t = 0$.

36. *Simple epidemic.* An influenza epidemic is spreading through a community of 1,000 people at a rate proportional both to the number of people who have been infected and to the number who have not been infected. If one individual was infected initially and 100 were infected 10 days later, how many will be infected after 20 days?

Social Sciences

37. *Learning.* A person learns N items at a rate given approximately by

$$N'(t) = \frac{60}{\sqrt{t^2 + 25}} \qquad t \geqslant 0$$

where t is the number of hours of continuous study. Determine the total number of items learned in the first 12 hours of continuous study.

15-4   Improper Integrals

- Improper Integrals
- Probability Density Functions

■ Improper Integrals

We are now going to consider an integral form that has wide application in probability studies as well as other areas. Earlier, when we introduced the idea of a definite integral,

$$\int_a^b f(x)\, dx \tag{1}$$

we required f to be continuous over a closed interval $[a, b]$. Now we are going to extend the meaning of (1) so that the interval $[a, b]$ may become infinite in length.

Let us investigate a particular example that will motivate several general definitions. What would be a reasonable interpretation for the following expression?

$$\int_1^\infty \frac{dx}{x^2}$$

Sketching a graph of $f(x) = 1/x^2$, $x \geq 1$ (see Figure 1), we note that for any fixed $b > 1$, $\int_1^b f(x)\, dx$ is the area between the curve $y = 1/x^2$, the x axis, $x = 1$, and $x = b$.

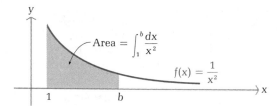

Figure 1

Let us see what happens when we let $b \to \infty$; that is, when we compute the following limit:

$$\lim_{b \to \infty} \int_1^b \frac{dx}{x^2} = \lim_{b \to \infty} \left[(-x^{-1}) \Big|_1^b \right]$$

$$= \lim_{b \to \infty} \left(-\frac{1}{b} + 1 \right) = 1$$

Did you expect this result? No matter how large b is taken, the area under the curve from $x = 1$ to $x = b$ never exceeds 1, and in the limit it is 1. This suggests that we write

$$\int_1^\infty \frac{dx}{x^2} = \lim_{b \to \infty} \int_1^b \frac{dx}{x^2} = 1$$

This integral is an example of an *improper integral*. In general, the forms

$$\int_{-\infty}^b f(x)\,dx \qquad \int_a^\infty f(x)\,dx \qquad \int_{-\infty}^\infty f(x)\,dx$$

where f is continuous over the indicated interval, are called **improper integrals.** (There are also other types of improper integrals that will not be considered here. These involve certain types of points of discontinuity within the interval of integration.) Each type of improper integral above is formally defined in the box:

Improper Integrals

If f is continuous over the indicated interval and the limit exists, then:

1. $\displaystyle \int_a^\infty f(x)\,dx = \lim_{b \to \infty} \int_a^b f(x)\,dx$

2. $\displaystyle \int_{-\infty}^b f(x)\,dx = \lim_{a \to -\infty} \int_a^b f(x)\,dx$

3. $\displaystyle \int_{-\infty}^\infty f(x)\,dx = \int_{-\infty}^c f(x)\,dx + \int_c^\infty f(x)\,dx$

where c is any point on $(-\infty, \infty)$, provided *both* improper integrals on the right exist.

If the indicated limit exists, then the improper integral is said to exist or **converge;** if the limit does not exist, then the improper integral is said not to exist or to **diverge** (and no value is assigned to it).

Example 16 Evaluate $\int_2^\infty dx/x$ if it converges.

Solution
$$\int_2^\infty \frac{dx}{x} = \lim_{b \to \infty} \int_2^b \frac{dx}{x}$$
$$= \lim_{b \to \infty} (\ln x)\Big|_2^b$$
$$= \lim_{b \to \infty} (\ln b - \ln 2)$$

Since $\ln b \to \infty$ as $b \to \infty$, the limit does not exist. Hence, the improper integral diverges.

Problem 16   Evaluate $\int_3^\infty dx/(x-1)^2$ if it converges.

Example 17   Evaluate $\int_{-\infty}^2 e^x \, dx$ if it converges.

Solution

$$\int_{-\infty}^2 e^x \, dx = \lim_{a \to -\infty} \int_a^2 e^x \, dx$$

$$= \lim_{a \to -\infty} (e^x|_a^2)$$

$$= \lim_{a \to -\infty} (e^2 - e^a) = e^2 - 0 = e^2 \qquad \text{The integral converges.}$$

Problem 17   Evaluate $\int_{-\infty}^{-1} x^{-2} \, dx$ if it converges.

Example 18   Evaluate

$$\int_{-\infty}^\infty \frac{2x}{(1+x^2)^2} \, dx$$

if it converges.

Solution

$$\int_{-\infty}^\infty \frac{2x}{(1+x^2)^2} \, dx = \int_{-\infty}^0 (1+x^2)^{-2} 2x \, dx + \int_0^\infty (1+x^2)^{-2} 2x \, dx$$

$$= \lim_{a \to -\infty} \int_a^0 (1+x^2)^{-2} 2x \, dx + \lim_{b \to \infty} \int_0^b (1+x^2)^{-2} 2x \, dx$$

$$= \lim_{a \to -\infty} \left[\frac{(1+x^2)^{-1}}{-1} \bigg|_a^0 \right] + \lim_{b \to \infty} \left[\frac{(1+x^2)^{-1}}{-1} \bigg|_0^b \right]$$

$$= \lim_{a \to -\infty} \left[-1 + \frac{1}{1+a^2} \right] + \lim_{b \to \infty} \left[-\frac{1}{1+b^2} + 1 \right]$$

$$= -1 + 1 = 0 \qquad \text{The integral converges.}$$

Problem 18   Evaluate $\int_{-\infty}^\infty dx/e^x$ if it converges.

Example 19
Oil Production   It is estimated that an oil well will produce $R(t)$ thousand barrels of oil per month t months from now, as given by

$$R(t) = 50e^{-0.05t} - 50e^{-0.1t}$$

Estimate the total amount of oil produced by this well.

Solution   The total amount of oil produced in T months of operation is

$$\int_0^T R(t) \, dt$$

At some point in time, the monthly production rate will become so low that it will no longer be economically feasible to operate the well. However, for the purpose of estimating the total production, it is convenient to assume that the well is operated indefinitely. Thus, the total amount of oil produced is

$$\int_0^\infty R(t)\ dt = \lim_{T\to\infty} \int_0^T R(t)\ dt$$

$$= \lim_{T\to\infty} \int_0^T (50e^{-0.05t} - 50e^{-0.1t})\ dt$$

$$= \lim_{T\to\infty} \left[(-1{,}000e^{-0.05t} + 500e^{-0.1t}) \Big|_0^T \right]$$

$$= \lim_{T\to\infty} (-1{,}000e^{-0.05T} + 500e^{-0.1T} + 500)$$

$$= 500\ \text{thousand barrels}$$

Problem 19   Find the total amount of oil produced by a well whose monthly production rate (in thousands of barrels) is given by

$$R(t) = 100e^{-0.1t} - 25e^{-0.2t}$$

■ Probability Density Functions

We will now take a brief look at the use of improper integrals relative to probability density functions. The approach will be intuitive and informal. Hopefully, when you next encounter these concepts in a more formal setting, you will have a better idea how calculus enters into the subject.

Suppose an experiment is designed in such a way that any real number x on the interval [a, b] is a possible outcome. For example, x may represent an IQ score, the height of a person in inches, or the life of a light bulb in hours.

In certain situations it is possible to find a function f with x as an independent variable that can be used to determine the probability that x will assume a value on a given subinterval of $(-\infty, \infty)$. Such a function, called a **probability density function,** must satisfy the following three conditions (see Figure 2 on the next page):

1. $f(x) \geqslant 0$ for all $x \in (-\infty, \infty)$
2. $\int_{-\infty}^\infty f(x)\ dx = 1$
3. If [c, d] is a subinterval of $(-\infty, \infty)$, then

$$\text{Probability}(c \leqslant x \leqslant d) = \int_c^d f(x)\ dx$$

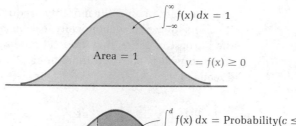

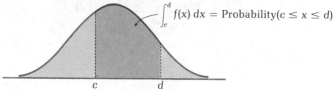

Figure 2

Example 20   A sailing club has a race over the same course twice a month. The races always start at 12 noon on Sunday, and the boats finish according to the probability density function (where x is hours after noon):

$$f(x) = \begin{cases} -\dfrac{x}{2} + 2 & 2 \leqslant x \leqslant 4 \\ 0 & \text{otherwise} \end{cases}$$

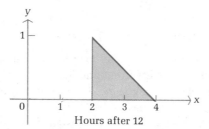

Note that

$$f(x) \geqslant 0$$

and

$$\int_{-\infty}^{\infty} f(x)\, dx = \int_{2}^{4} \left(-\frac{x}{2} + 2 \right) dx = \left(-\frac{x^2}{4} + 2x \right)\Bigg|_{2}^{4} = 1$$

The probability that a boat selected at random from the sailing fleet will finish between 2 and 3 hours after the start is given by

$$\text{Probability}(2 \leqslant x \leqslant 3) = \int_{2}^{3} \left(-\frac{x}{2} + 2 \right) dx$$

$$= \left(-\frac{x^2}{4} + 2x \right)\Bigg|_{2}^{3} = .75$$

which is the area under the curve from x = 2 to x = 3.

Problem 20   In Example 20, find the probability that a boat selected at random from the fleet will finish between 2:30 and 3:30 PM.

Example 21   Suppose the length of telephone calls (in minutes) in a public telephone booth has the probability density function

$$f(t) = \begin{cases} \dfrac{1}{4}\,e^{-t/4} & t \geqslant 0 \\ 0 & \text{otherwise} \end{cases}$$

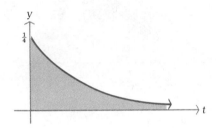

(A)   Compute $\int_{-\infty}^{\infty} f(t)\, dt$.

(B)   Determine the probability that a call selected at random will last between 2 and 3 minutes.

Solutions   (A) $\displaystyle \int_{-\infty}^{\infty} f(t)\, dt = \int_{-\infty}^{0} f(t)\, dt + \int_{0}^{\infty} f(t)\, dt$

$$= 0 + \int_{0}^{\infty} \frac{1}{4}\,e^{-t/4}\, dt$$

$$= \lim_{b \to \infty} \int_{0}^{b} \frac{1}{4}\,e^{-t/4}\, dt$$

$$= \lim_{b \to \infty} \left(-e^{-t/4} \Big|_{0}^{b} \right)$$

$$= \lim_{b \to \infty} (-e^{-b/4} + e^{0})$$

$$= \lim_{b \to \infty} \left(-\frac{1}{e^{b/4}} + 1 \right)$$

$$= 0 + 1 = 1$$

(B)   Probability$(2 \leqslant t \leqslant 3) = \displaystyle \int_{2}^{3} \frac{1}{4}\,e^{-t/4}\, dt$

$$= (-e^{-t/4}) \Big|_{2}^{3}$$

$$= -e^{-3/4} + e^{-1/2} \approx .13$$

Problem 21   In Example 21, find the probability that a call selected at random will last longer than 4 minutes.

The most important probability density function is the **normal probability density function** defined below and graphed in Figure 3.

$$f(x) = \frac{1}{\sigma\sqrt{2\pi}}\, e^{-(x-\mu)^2/2\sigma^2}$$

μ is the mean

σ is the standard deviation

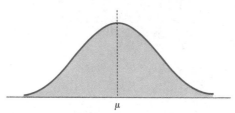

Figure 3   Normal curve

It can be shown, but not easily, that

$$\frac{1}{\sigma\sqrt{2\pi}} \int_{-\infty}^{\infty} e^{-(x-\mu)^2/2\sigma^2}\, dx = 1$$

Since $\int e^{-x^2}\, dx$ is nonintegrable in terms of elementary functions (that is, the antiderivative cannot be expressed as a finite combination of simple functions), probabilities such as

$$\text{Probability}(c \leqslant x \leqslant d) = \frac{1}{\sigma\sqrt{2\pi}} \int_{c}^{d} e^{-(x-\mu)^2/2\sigma^2}\, dx$$

are generally determined by making an appropriate substitution in the integrand and then using a table of areas under the standard normal curve (that is, the normal curve with $\mu = 0$ and $\sigma = 1$). Such tables are readily available in most mathematical handbooks. A table can be constructed by using the rectangle rule discussed in Section 14-6; however, digital computers that use refined techniques are generally used for this purpose. Some hand calculators have the capability of computing normal curve areas directly.

Answers to Matched Problems

16. $\dfrac{1}{2}$ 17. 1   18. Diverges

19. 875 thousand barrels   20. .5

21. $e^{-1} \approx .37$

Exercise 15-4

Find the value of each improper integral that converges.

A 1. $\displaystyle\int_1^\infty \frac{dx}{x^4}$ 2. $\displaystyle\int_1^\infty \frac{dx}{x^3}$

   3. $\displaystyle\int_0^\infty e^{-x/2}\, dx$ 4. $\displaystyle\int_0^\infty e^{-x}\, dx$

B 5. $\displaystyle\int_1^\infty \frac{dx}{\sqrt{x}}$ 6. $\displaystyle\int_1^\infty \frac{dx}{\sqrt[3]{x}}$

   7. $\displaystyle\int_0^\infty \frac{dx}{(x+1)^2}$ 8. $\displaystyle\int_0^\infty \frac{dx}{(x+1)^3}$

   9. $\displaystyle\int_0^\infty \frac{dx}{(x+1)^{2/3}}$ 10. $\displaystyle\int_0^\infty \frac{dx}{\sqrt{x+1}}$

   11. $\displaystyle\int_1^\infty \frac{dx}{x^{0.99}}$ 12. $\displaystyle\int_1^\infty \frac{dx}{x^{1.01}}$

   13. $0.3\displaystyle\int_0^\infty e^{-0.3x}\, dx$ 14. $0.01\displaystyle\int_0^\infty e^{-0.1x}\, dx$

   15. In Example 20, find the probability that a randomly selected boat will finish before 3:30 PM.
   16. In Example 20, find the probability that a randomly selected boat will finish after 2:30 PM.
   17. In Example 21, find the probability that a telephone call selected at random will last longer than 1 minute.
   18. In Example 21, find the probability that a telephone call selected at random will last less than 3 minutes.

C Find the value of each improper integral that converges. Note that $\lim_{x\to\infty} x^n e^{-x} = 0$ and $\lim_{x\to\infty} x^{-n} \ln x = 0$ for all positive integers n.

   19. $\displaystyle\int_0^\infty \frac{1}{k} e^{-x/k}\, dx,\ k>0$ 20. $\displaystyle\int_0^\infty xe^{-x}\, dx$

   21. $\displaystyle\int_{-\infty}^0 \frac{dx}{\sqrt{1-x}}$ 22. $\displaystyle\int_{-\infty}^\infty xe^{-x^2}\, dx$

   23. $\displaystyle\int_0^\infty (e^{-x} - e^{-2x})\, dx$ 24. $\displaystyle\int_0^\infty x^2 e^{-x}\, dx$

   25. $\displaystyle\int_1^\infty \frac{\ln x}{x}\, dx$ 26. $\displaystyle\int_1^\infty \frac{\ln x}{x^2}\, dx$

Applications

Business & Economics

27. *Production.* The monthly production of a natural gas well (in millions of cubic feet) t months from now is given by

$$R(t) = te^{-0.4t}$$

Assuming that the well is operated indefinitely, find the total production.

28. *Investment.* An investment will return $1,000e^{-0.125t}$ dollars per year t years from now. Assuming that the returns continue indefinitely, determine the total amount returned by this investment.

29. *Consumption.* The daily per capita use of water (in hundreds of gallons) for domestic purposes has a probability density function of the form

$$g(x) = \begin{cases} .05e^{-.05x} & x \geqslant 0 \\ 0 & \text{otherwise} \end{cases}$$

Find the probability that a person chosen at random will use at least 300 gallons of water per day.

30. *Warranty.* A manufacturer guarantees a product for 1 year. The time for failure of a new product after it is sold is given by the probability density function

$$f(t) = \begin{cases} .01e^{-.01t} & t \geqslant 0 \\ 0 & \text{otherwise} \end{cases}$$

where t is time in months. What is the probability that a buyer chosen at random will have a product failure during the warranty period?

Life Sciences

31. *Pollution.* It has been estimated that the seepage of toxic chemicals from a waste dump is $R(t)$ gallons per year t years from now, where

$$R(t) = \frac{500}{(1 + t)^2}$$

Assuming that this seepage continues indefinitely, find the total amount of toxic chemicals that seep from the dump.

32. *Drug assimilation.* When a person takes a drug, the body does not assimilate all of the drug. One way to determine the amount of the drug that is assimilated is to measure the rate at which the drug is eliminated from the body. If the rate of elimination of the drug (in milliliters per minute) is given by

$$R(t) = te^{-0.2t}$$

where t is the time in minutes since the drug was administered, how much of the drug is eliminated from the body?

33. *Medicine.* If the length of stay for people in a hospital has a probability

density function

$$g(t) = \begin{cases} .2e^{-.2t} & t \geqslant 0 \\ 0 & \text{otherwise} \end{cases}$$

where t is time in days, find the probability that a patient chosen at random will stay in the hospital less than 5 days.

34. *Medicine.* For a particular disease, the length of time in days for recovery has a probability density function of the form

$$R(t) = \begin{cases} .03e^{-.03t} & t \geqslant 0 \\ 0 & \text{otherwise} \end{cases}$$

For a randomly selected person who contracts this disease, what is the probability that he or she will take at least 7 days to recover?

35. *Politics.* In a particular election, the length of time each voter spent on campaigning for a candidate or issue was found to have a probability density function

$$F(x) = \begin{cases} \dfrac{1}{(x+1)^2} & x \geqslant 0 \\ 0 & \text{otherwise} \end{cases}$$

where x is time in minutes. For a voter chosen at random, what is the probability of his or her spending at least 9 minutes on the campaign?

36. *Psychology.* In an experiment on conditioning, pigeons were required to recognize on a light display one pattern of dots out of five possible patterns to receive a food pellet. After the ninth successful trial, it was found that the probability density function for the length of time in seconds until success on the tenth trial is given by

$$f(t) = \begin{cases} e^{-t} & t \geqslant 0 \\ 0 & \text{otherwise} \end{cases}$$

What is the probability that a pigeon selected at random from those having successfully completed nine trials will take two or more seconds to complete the tenth trial successfully?

15-5 Chapter Review

Important Terms and Symbols

15-1 *Integration by substitution.* substitution, substitution in definite integrals,

$$\int f[u(x)]u'(x)\, dx = \int f(u)\, du = F(u) + C = F[u(x)] + C$$
$$\underbrace{}_{u}\underbrace{}_{du} \qquad\qquad F'(u) = f(u)$$

$$\int u^n \, du = \frac{u^{n+1}}{n+1} + C \; (n \neq -1), \int \frac{1}{u} \, du = \ln |u| + C,$$

$$\int e^u \, du = e^u + C$$

15-2 *Integration by parts.* $\int u \, dv = uv - \int v \, du$

15-3 *Integration using tables.* Table of Integrals, substitution and integral tables

15-4 *Improper integrals.* improper integral, converge, diverge, probability density function, normal probability density function,
$\int_a^\infty f(x) \, dx = \lim_{b \to \infty} \int_a^b f(x) \, dx$, $\int_{-\infty}^b f(x) \, dx = \lim_{a \to -\infty} \int_a^b f(x) \, dx$,
$\int_{-\infty}^\infty f(x) \, dx = \int_{-\infty}^c f(x) \, dx + \int_c^\infty f(x) \, dx$

Exercise 15-5 Chapter Review

Work through all the problems in this chapter review and check your answers in the back of the book. (Answers to all review problems are there.) Where weaknesses show up, review appropriate sections in the text. When you are satisfied that you know the material, take the practice test following this review.

A *Evaluate the indicated integrals, if possible.*

1. $\displaystyle\int x\sqrt{1 + x^2} \, dx$

2. $\displaystyle\int x^2\sqrt{1 + x^2} \, dx$

3. $\displaystyle\int_0^3 \frac{x}{1 + x^2} \, dx$

4. $\displaystyle\int_0^\infty e^{-2x} \, dx$

5. $\displaystyle\int e^{x^2}x \, dx$

6. $\displaystyle\int_0^\infty \frac{1}{x + 1} \, dx$

B 7. $\displaystyle\int_0^1 xe^{-x} \, dx$

8. $\displaystyle\int x \ln x \, dx$

9. $\displaystyle\int \frac{e^{-x}}{e^{-x} + 3} \, dx$

10. $\displaystyle\int \frac{e^x}{(e^x + 2)^2} \, dx$

11. $\displaystyle\int \frac{x}{\sqrt{x + 9}} \, dx$

12. $\displaystyle\int \frac{x}{x + 9} \, dx$

13. $\displaystyle\int \frac{1}{x(x + 9)} \, dx$

14. $\displaystyle\int_0^\infty \frac{dx}{(x + 3)^2}$

15. $\displaystyle\int_{-\infty}^0 e^x \, dx$

16. $\displaystyle\int \frac{1}{\sqrt{9x^2 + 4}} \, dx$

17. Find the area bounded by the graphs of $y = \ln x$, $y = 0$, and $x = e$.

C 18. $\displaystyle\int \frac{(\ln x)^2}{x}\, dx$

19. $\displaystyle\int_0^\infty (x + 1)e^{-x}\, dx$

20. $\displaystyle\int x(\ln x)^2\, dx$

21. $\displaystyle\int_1^9 \frac{1}{4 + \sqrt{x}}\, dx$

22. $\displaystyle\int \frac{x^2}{\sqrt{x^6 - 16}}\, dx$

23. $\displaystyle\int_{-\infty}^\infty \frac{x}{(1 + x^2)^3}\, dx$

Applications

Business & Economics

24. *Consumers' and producers' surplus.* (Refer to Section 14-5.) Find the consumers' surplus and the producers' surplus for

$$p = D(x) = \frac{40}{\sqrt{x + 15}}$$

$$p = S(x) = \frac{4x}{\sqrt{x + 15}}$$

25. *Inventory.* Suppose the inventory of a certain item t months after the first of the year is given approximately by

$$I(t) = 4 + 5t\sqrt{12 - t} \qquad 0 \leqslant t \leqslant 12$$

What is the average inventory during the last nine months of the year?

26. *Production.* An oil field is estimated to produce $R(t)$ thousand barrels of oil per month t months from now, as given by

$$R(t) = 25te^{-0.05t}$$

How much oil is produced during the first 2 years of operation? If the well is operated indefinitely, what is the total amount of oil produced?

27. *Parts testing.* If in testing printed circuits for hand calculators, failures occur relative to time in hours according to the probability density function

$$F(t) = \begin{cases} .02e^{-.02t} & t \geqslant 0 \\ 0 & \text{otherwise} \end{cases}$$

what is the probability that a circuit chosen at random will fail in the first hour of testing?

Life Sciences

28. *Drug assimilation.* The rate at which the body eliminates a drug (in milliliters per hour) is given by

$$R(t) = \frac{20t}{(t + 1)^3}$$

where t is the number of hours since the drug was administered. How much of the drug is eliminated in the first hour after it was administered? What is the total amount of the drug that is eliminated by the body?

29. *Medicine*. For a particular doctor, the length of time in hours spent with a patient per office visit has the probability density function

$$f(t) = \begin{cases} \dfrac{4/3}{(t + 1)^2} & 0 \leqslant t \leqslant 3 \\ 0 & \text{otherwise} \end{cases}$$

What is the probability that the doctor will spend more than 1 hour with a randomly selected patient?

Social Sciences   30. *Politics*. The rate of change of the voting population of a city, $N'(t)$, with respect to time t in years is estimated to be

$$N'(t) = \frac{100t}{(1 + t^2)^2}$$

where $N(t)$ is in thousands. If $N(0)$ is the current voting population, how much will this population increase during the next 3 years? If the population continues to grow at this rate indefinitely, what is the total increase in the voting population?

31. *Psychology*. Rats were trained to go through a maze by rewarding them with a food pellet upon successful completion. After the seventh successful run, it was found that the probability density function for length of time in minutes until success on the eighth trial is given by

$$f(t) = \begin{cases} .5e^{-.5t} & t \geqslant 0 \\ 0 & \text{otherwise} \end{cases}$$

What is the probability that a rat selected at random after seven successful runs will take 2 or more minutes to complete the eighth run successfully?

Practice Test: Chapter 15

Evaluate the indicated integrals in Problems 1–9.

1. $\displaystyle\int \frac{x}{(x^2 + 3)^4}\, dx$ 2. $\displaystyle\int xe^{x^2}\, dx$

3. $\displaystyle\int_2^\infty e^{-8x}\, dx$ 4. $\displaystyle\int (x + 3)e^x\, dx$

5. $\int \dfrac{(\ln x)^7}{x}\, dx$

6. $\int x^7 \ln x\, dx$

7. $\int_{-2}^{2} x\sqrt{2 + x}\, dx$

8. $\int \dfrac{1}{x(x-1)}\, dx$

9. $\int (\ln x)^2\, dx$

10. Find the area bounded by the graphs of $y = \sqrt{x^2 + 16}$, $y = 0$, $x = 0$, and $x = 3$.

11. An oil well is estimated to produce $R(t)$ thousands of barrels of oil per year t years from now, as given by

$$R(t) = 10te^{-t}$$

How much oil is produced during the first year of operation? If the well is operated indefinitely, what is the total amount of oil produced?

Multivariable Calculus

16-1   Functions of Several Variables

- Functions of Two or More Independent Variables
- Examples of Functions of Several Variables
- Three-Dimensional Coordinate Systems

■ Functions of Two or More Independent Variables

In Section 5-2 we introduced the concept of a function with one independent variable. Now we will broaden the concept to include functions with more than one independent variable. We start with an example.

A small manufacturing company produces a standard type of surfboard and no other products. If fixed costs are $500 per week and variable costs are $70 per board produced, then the weekly cost function is given by

$$C(x) = 500 + 70x \tag{1}$$

where x is the number of boards produced per week. The cost function is a function of a single independent variable x. For each value of x from the domain of C there exists exactly one value of $C(x)$ in the range of C.

Now, suppose the company decides to add a high-performance competition board to its line. If the fixed costs for the competition board are $200 per week and the variable costs are $100 per board, then the cost function (1) must be modified to

$$C(x, y) = 700 + 70x + 100y \tag{2}$$

where $C(x, y)$ is the cost for weekly output of x standard boards and y competition boards. Equation (2) is an example of a function with two independent variables, x and y. Of course, as the company expands its product line even further, its weekly cost function must be modified to include more and more independent variables, one for each new product produced.

In general, an equation of the form

$$z = f(x, y)$$

will describe a **function of two independent variables** if for each ordered pair (x, y) from the domain of f there is one and only one value of z determined by $f(x, y)$ in the range of f. Unless otherwise stated, we will assume that the domain of a function specified by an equation of the form $z = f(x, y)$ is the set of all ordered pairs of real numbers (x, y) such that $f(x, y)$ is also a real number. It should be noted, however, that certain conditions in practical problems often lead to further restrictions of the domain of a function.

We can similarly define functions of three independent variables, $w = f(x, y, z)$; of four independent variables, $u = f(w, x, y, z)$; and so on. In this chapter, we will primarily concern ourselves with functions with two independent variables.

Example 1   For $C(x, y) = 700 + 70x + 100y$, find $C(10, 5)$.

Solution $C(10, 5) = 700 + 70(10) + 100(5) = \$1,900$

Problem 1   Find $C(20, 10)$ for the cost function in Example 1.

Example 2   For $f(x, y, z) = 2x^2 - 3xy + 3z + 1$, find $f(3, 0, -1)$.

Solution $f(3, 0, -1) = 2(3)^2 - 3(3)(0) + 3(-1) + 1$
$$= 18 - 0 - 3 + 1 = 16$$

Problem 2   Find $f(-2, 2, 3)$ for f in Example 2.

Example 3   The surfboard company discussed previously has determined that the demand equations for the two types of boards they produce are given by

$$p = 210 - 4x + y$$
$$q = 300 + x - 12y$$

where p is the price of the standard board, q is the price of the competition board, x is the weekly demand for standard boards, and y is the weekly demand for competition boards.

(A)   Find the weekly revenue function $R(x, y)$ and evaluate $R(20, 10)$.
(B)   If the weekly cost function is

$$C(x, y) = 700 + 70x + 100y$$

find the weekly profit function $P(x, y)$ and evaluate $P(20, 10)$.

Solution   (A)   Revenue = $\begin{pmatrix} \text{Demand for} \\ \text{standard} \\ \text{boards} \end{pmatrix} \times \begin{pmatrix} \text{Price of a} \\ \text{standard} \\ \text{board} \end{pmatrix}$

$\qquad\qquad\qquad + \begin{pmatrix} \text{Demand for} \\ \text{competition} \\ \text{boards} \end{pmatrix} \times \begin{pmatrix} \text{Price of a} \\ \text{competition} \\ \text{board} \end{pmatrix}$

$$R(x, y) = xp + yq$$
$$= x(210 - 4x + y) + y(300 + x - 12y)$$
$$= 210x + 300y - 4x^2 + 2xy - 12y^2$$

$$R(20, 10) = 210(20) + 300(10) - 4(20)^2 + 2(20)(10) - 12(10)^2$$
$$= \$4,800$$

(B)   Profit = Revenue − Cost

$$P(x, y) = R(x, y) - C(x, y)$$
$$= 210x + 300y - 4x^2 + 2xy - 12y^2 - 700 - 70x - 100y$$
$$= 140x + 200y - 4x^2 + 2xy - 12y^2 - 700$$

$$P(20, 10) = 140(20) + 200(10) - 4(20)^2 + 2(20)(10) - 12(10)^2 - 700$$
$$= \$1,700$$

Problem 3   Repeat Example 3 if the demand and cost equations are given by

$$p = 220 - 6x + y$$
$$q = 300 + 3x - 10y$$
$$C(x, y) = 40x + 80y + 1,000$$

■ Examples of Functions of Several Variables

A number of concepts we have already considered can be thought of in terms of functions of two or more variables. We list a few of these below.

Area of a
rectangle $A(x, y) = xy$

Volume of a
box $V(x, y, z) = xyz$

Volume of a
right circular
cylinder $V(r, h) = \pi r^2 h$

Simple interest	$A(P, r, t) = P(1 + rt)$	A = Amount
		P = Principal
		r = Annual rate
		t = Time in years

Compound interest	$A(P, r, t, n) = P\left(1 + \dfrac{r}{n}\right)^{nt}$	A = Amount
		P = Principal
		r = Annual rate
		t = Time in years
		n = Compound periods per year

IQ	$Q(M, C) = \dfrac{M}{C}(100)$	Q = IQ = Intelligence quotient
		M = MA = Mental age
		C = CA = Chronological age

Resistance for blood flow in a vessel	$R(L, r) = k\dfrac{L}{r^4}$	R = Resistance
		L = Length of vessel
		r = Radius of vessel
		k = Constant

Example 4 A company uses a box with a square base and an open top for one of its products (see the figure). If x is the length in inches of each side of the base and y is the height in inches, find the total amount of material $M(x, y)$ required to construct one of these boxes and evaluate $M(5, 10)$.

Solution

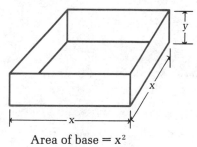

Area of base $= x^2$

Area of one side $= xy$

Total material $=$ (Area of base) $+$ 4(Area of one side)

$$M(x, y) = x^2 + 4xy$$

$$M(5, 10) = 5^2 + 4(5)(10)$$

$$= 225 \text{ square inches}$$

Problem 4 For the box in Example 4, find the volume $V(x, y)$ and evaluate $V(5, 10)$.

■ Three-Dimensional Coordinate Systems

We now take a brief look at some graphs of functions of two independent variables. Since functions of the form $z = f(x, y)$ involve two independent variables, x and y, and one dependent variable, z, we need a three-dimensional coordinate system for their graphs. We take three mutually perpendicular number lines intersecting at their origins to form a rectangular coordinate system in three-dimensional space (see Figure 1). In such a system, every ordered triplet of numbers (x, y, z) can be associated with a unique point, and conversely.

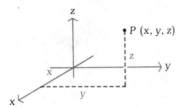

Figure 1 Rectangular coordinate system

Example 5   Locate $(-3, 5, 2)$ in a rectangular coordinate system.

Solution

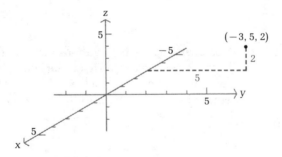

Problem 5   Find the coordinates of the corners A, C, G, and D of the rectangular box shown in the figure.

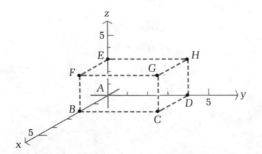

What does the graph of $z = x^2 + y^2$ look like? If we let $x = 0$ and graph $z = 0^2 + y^2 = y^2$ in the yz plane, we obtain a parabola; if we let $y = 0$ and graph $z = x^2 + 0^2 = x^2$ in the xz plane, we obtain another parabola. It can be shown that the graph of $z = x^2 + y^2$ is just one of these parabolas rotated around the z axis (see Figure 2). This cup-shaped figure is a *surface* and is called a **paraboloid.**

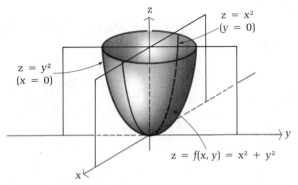

Figure 2 Paraboloid

In general, the graph of any function of the form $z = f(x, y)$ is called a **surface.** The graph of such a function is the graph of all ordered triplets of numbers (x, y, z) that satisfy the equation. Graphing functions of two independent variables is often a very difficult task, and the general process will not be dealt with in this book. We present only a few simple graphs to suggest extensions of earlier geometric interpretations of the derivative and local maxima and minima to functions of two variables. Note that $z = f(x, y) = x^2 + y^2$ appears (see Figure 2) to have a local minimum at $(x, y) = (0, 0)$. Figure 3 shows a local maximum at $(x, y) = (0, 0)$, and Figure

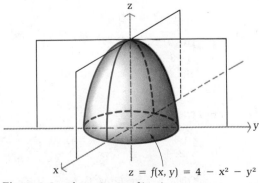

Figure 3 Local maximum: $f(0, 0) = 4$

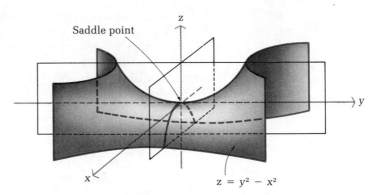

Figure 4 Saddle point at (0, 0, 0)

4 shows a point at $(x, y) = (0, 0)$, called a **saddle point,** which is neither a local minimum nor a local maximum. More will be said about local maxima and minima in Section 16-4.

Answers to
Matched Problems

1. \$3,100   2.   30

3. (A) $R(x, y) = 220x + 300y - 6x^2 + 4xy - 10y^2; R(20, 10) = \$4,800$
   (B) $P(x, y) = 180x + 220y - 6x^2 + 4xy - 10y^2 - 1,000;$
          $P(20, 10) = \$2,200$

4. $V(x, y) = x^2y; V(5, 10) = 250$ cubic inches
5. $A(0, 0, 0); C(2, 4, 0); G(2, 4, 3); D(0, 4, 0)$

Exercise 16-1

A   *For the functions*

$$f(x, y) = 10 + 2x - 3y \qquad g(x, y) = x^2 - 3y^2$$

find each of the following:

1. $f(0, 0)$
2. $f(2, 1)$
3. $f(-3, 1)$
4. $f(2, -7)$
5. $g(0, 0)$
6. $g(0, -1)$
7. $g(2, -1)$
8. $g(-1, 2)$

B   *Find each of the following:*

9. $A(2, 3)$ for $A(x, y) = xy$
10. $V(2, 4, 3)$ for $V(x, y, z) = xyz$

11. $Q(12, 8)$ for $Q(M, C) = \dfrac{M}{C}(100)$

12. $T(50, 17)$ for $T(V, x) = \dfrac{33V}{x + 33}$

13. $V(2, 4)$ for $V(r, h) = \pi r^2 h$

14. $S(4, 2)$ for $S(x, y) = 5x^2 y^3$

15. $R(1, 2)$ for $R(x, y) = -5x^2 + 6xy - 4y^2 + 200x + 300y$

16. $P(2, 2)$ for $P(x, y) = -x^2 + 2xy - 2y^2 - 4x + 12y + 5$

17. $R(6, 0.5)$ for $R(L, r) = 0.002\,\dfrac{L}{r^4}$

18. $L(2,000, 50)$ for $L(w, v) = (1.25 \times 10^{-5})wv^2$

19. $A(100, 0.06, 3)$ for $A(P, r, t) = P + Prt$

20. $A(10, 0.04, 3, 2)$ for $A(P, r, t, n) = P\left(1 + \dfrac{r}{n}\right)^{tn}$

21. $A(100, 0.08, 10)$ for $A(P, r, t) = Pe^{rt}$

22. $A(1,000, 0.06, 8)$ for $A(P, r, t) = Pe^{rt}$

C

23. For the function $f(x, y) = x^2 + 2y^2$, find:

$$\frac{f(x + \Delta x, y) - f(x, y)}{\Delta x}$$

24. For the function $f(x, y) = x^2 + 2y^2$, find:

$$\frac{f(x, y + \Delta y) - f(x, y)}{\Delta y}$$

25. For the function $f(x, y) = 2xy^2$, find:

$$\frac{f(x + \Delta x, y) - f(x, y)}{\Delta x}$$

26. For the function $f(x, y) = 2xy^2$, find:

$$\frac{f(x, y + \Delta y) - f(x, y)}{\Delta y}$$

27. Find the coordinates of E and F in the figure for Problem 5 in the text.

28. Find the coordinates of B and H in the figure for Problem 5 in the text.

Applications

Business & Economics

29. *Cost function.* A small manufacturing company produces two models of a surfboard: a standard model and a competition model. If the standard model is produced at a variable cost of $70 each, the competition model at a variable cost of $100 each, and the total fixed costs

per month are \$2,000, then the monthly cost function is given by

$$C(x, y) = 2,000 + 70x + 100y$$

where x and y are the numbers of standard and competition models produced per month, respectively. Find $C(20, 10)$, $C(50, 5)$, and $C(30, 30)$.

30. *Advertising and sales.* A company spends x thousand dollars per week on newspaper advertising and y thousand dollars per week on television advertising. Its weekly sales were found to be given by

$$S(x, y) = 5x^2y^3$$

Find $S(3, 2)$ and $S(2, 3)$.

31. *Revenue, cost, and profit functions.* A firm produces two types of calculators, x thousand of type A and y thousand of type B per year. The revenue and cost functions for the year are (in thousands of dollars)

$$R(x, y) = 14x + 20y$$
$$C(x, y) = x^2 - 2xy + 2y^2 + 12x + 16y + 5$$

Find $R(3, 5)$, $C(3, 5)$, and $P(3, 5)$.

32. *Revenue, cost, and profit functions.* A company manufactures ten-speed and three-speed bicycles. The weekly demand and cost equations are

$$p = 230 - 9x + y$$
$$q = 130 + x - 4y$$
$$C(x, y) = 200 + 80x + 30y$$

where \$$p$ is the price of a ten-speed bicycle, \$$q$ is the price of a three-speed bicycle, x is the weekly demand for ten-speed bicycles, y is the weekly demand for three-speed bicycles, and $C(x, y)$ is the cost function. Find the weekly revenue function $R(x, y)$ and the weekly profit function $P(x, y)$. Evaluate $R(10, 15)$ and $P(10, 15)$.

33. *Revenue function.* A supermarket sells two brands of coffee: brand A at \$$p$ per pound and brand B at \$$q$ per pound. The daily demand equations for brands A and B are, respectively,

$$x = 200 - 5p + 4q$$
$$y = 300 + 2p - 4q$$

(both in pounds). Find the daily revenue function $R(p, q)$. Evaluate $R(2, 3)$ and $R(3, 2)$.

34. *Package design.* The packaging department in a company has been asked to design a rectangular box with no top and a partition down the middle (see the accompanying figure). If x, y, and z are the dimensions in inches, find the total amount of material $M(x, y, z)$ used in constructing one of these boxes and evaluate $M(10, 12, 6)$.

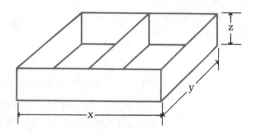

Life Sciences

35. *Marine biology.* In using scuba diving gear, a marine biologist estimates the time of a dive according to the equation

$$T(V, x) = \frac{33V}{x + 33}$$

where

   T = Time of dive in minutes

   V = Volume of air, at sea level pressure, compressed into tanks

   x = Depth of dive in feet

Find $T(70, 47)$ and $T(60, 27)$.

36. *Blood flow.* Poiseuille's law states that the resistance, R, for blood flowing in a blood vessel varies directly as the length of the vessel, L, and inversely as the fourth power of its radius, r. This relationship may be stated in equation form as follows:

$$R(L, r) = k \frac{L}{r^4} \qquad k \text{ a constant}$$

Find $R(8, 1)$ and $R(4, 0.2)$.

37. *Physical anthropology.* Anthropologists, in their study of race and human genetic groupings, often use an index called the *cephalic index*. The cephalic index, C, varies directly as the width, W, of the head, and inversely as the length, L, of the head (both viewed from the top). In terms of an equation,

$$C(W, L) = 100 \frac{W}{L}$$

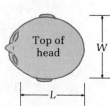

where

   W = Width in inches

   L = Length in inches

Find $C(6, 8)$ and $C(8.1, 9)$.

Social Sciences

38. *Safety research.* Under ideal conditions, if a person driving a car slams on the brakes and skids to a stop, the length of the skid marks (in feet) is given by the formula

$$L(w, v) = kwv^2$$

where

k = Constant

w = Weight of car in pounds

v = Speed of car in miles per hour

For $k = 0.0000133$, find $L(2,000, 40)$ and $L(3,000, 60)$.

39. *Psychology.* Intelligence quotient (IQ) is defined to be the ratio of the mental age (MA), as determined by certain tests, and the chronological age (CA), multiplied by 100. Stated as an equation,

$$Q(M, C) = \frac{M}{C} \cdot 100$$

where

Q = IQ

M = MA

C = CA

Find $Q(12, 10)$ and $Q(10, 12)$.

16-2 Partial Derivatives

■ Partial Derivatives
■ Higher-Order Partial Derivatives

■ Partial Derivatives

We know how to differentiate many kinds of functions of one independent variable and how to interpret the results. What about functions with two or more independent variables? Let us return to the surfboard example considered at the beginning of the chapter.

For the company producing only the standard board, the cost function was

$$C(x) = 500 + 70x$$

Differentiating with respect to x, we obtain the marginal cost function

$$C'(x) = 70$$

Since the marginal cost is constant, $70 is the change in cost for one unit increase in production at any output level.

For the company producing two boards, a standard model and a competition model, the cost function was

$$C(x, y) = 700 + 70x + 100y$$

Now suppose we differentiate with respect to x, holding y fixed, and denote this by $C_x(x, y)$; or we differentiate with respect to y, holding x fixed, and denote this by $C_y(x, y)$. Differentiating in this way, we obtain

$$C_x(x, y) = 70 \qquad C_y(x, y) = 100$$

Both these are called **partial derivatives** and, in this example, both represent marginal costs. The first is the change in cost due to one unit increase in production of the standard board with the production of the competition model held fixed. The second is the change in cost due to one unit increase in production of the competition board with the production of the standard board held fixed.

In general, if $z = f(x, y)$, then the **partial derivative of f with respect to x,** denoted by $\partial z / \partial x$, f_x, or $f_x(x, y)$, is defined by

$$\frac{\partial z}{\partial x} = \lim_{\Delta x \to 0} \frac{f(x + \Delta x, y) - f(x, y)}{\Delta x}$$

provided the limit exists. This is the ordinary derivative of f with respect to x, holding y constant. Thus, we are able to continue to use all the derivative rules and properties discussed in Chapters 10 and 11 for partials.

Similarly, the **partial derivative of f with respect to y,** denoted by $\partial z / \partial y$, f_y, or $f_y(x, y)$, is defined by

$$\frac{\partial z}{\partial y} = \lim_{\Delta y \to 0} \frac{f(x, y + \Delta y) - f(x, y)}{\Delta y}$$

which is the ordinary derivative with respect to y, holding x constant.

Parallel definitions and interpretations hold for functions with three or more independent variables.

Example 6 For $z = f(x, y) = 2x^2 - 3x^2y + 5y + 1$, find:

(A) $\dfrac{\partial z}{\partial x}$ (B) $f_x(2, 3)$

Solution (A) $z = 2x^2 - 3x^2y + 5y + 1$

Differentiating with respect to x, holding y constant (that is, treating y as a constant), we obtain

$$\frac{\partial z}{\partial x} = 4x - 6xy$$

(B) $f(x, y) = 2x^2 - 3x^2y + 5y + 1$

First differentiate with respect to x (part A) to obtain

$$f_x(x, y) = 4x - 6xy$$

Then evaluate at (2, 3). Thus,

$$f_x(2, 3) = 4(2) - 6(2)(3) = -28$$

Problem 6   For f in Example 6, find:

(A) $\dfrac{\partial z}{\partial y}$ (B) $f_y(2, 3)$

Example 7   For $z = f(x, y) = e^{x^2+y^2}$, find:

(A) $\dfrac{\partial z}{\partial x}$ (B) $f_y(2, 1)$

Solution   (A) Using the chain rule [thinking of $z = e^u$, $u = u(x)$; y is held constant], we obtain

$$\frac{\partial z}{\partial x} = e^{x^2+y^2} \frac{\partial(x^2 + y^2)}{\partial x}$$

$$= 2xe^{x^2+y^2}$$

(B) $f_y(x, y) = 2ye^{x^2+y^2}$

$$f_y(2, 1) = 2(1)e^{2^2+1^2}$$

$$= 2e^5$$

Problem 7   For $z = f(x, y) = (x^2 + 2xy)^5$, find:

(A) $\dfrac{\partial z}{\partial y}$ (B) $f_x(1, 0)$

Example 8
Profit   The profit function for the surfboard company in Example 3 in Section 16-1 was

$$P(x, y) = 140x + 200y - 4x^2 + 2xy - 12y^2 - 700$$

Find $P_x(15, 10)$ and $P_x(30, 10)$, and interpret.

Solution $P_x(x, y) = 140 - 8x + 2y$

$P_x(15, 10) = 140 - 8(15) + 2(10) = 40$

$P_x(30, 10) = 140 - 8(30) + 2(10) = -80$

At a production level of 15 standard and 10 competition boards per week, increasing the production of standard boards by one and holding the production of competition boards fixed at 10 will increase profit by approximately $40. At a production level of 30 standard and 10 competition boards per week, increasing the production of standard boards by one unit and holding the production of competition boards fixed at 10 will decrease profit by approximately $80.

Problem 8 For the profit function in Example 8, find $P_y(25, 10)$ and $P_y(25, 15)$, and interpret.

Partials have simple geometric interpretations, as indicated in Figure 5. If we hold x fixed, say $x = a$, then $f_y(a, y)$ is the slope of the curve obtained by intersecting the plane $x = a$ with the surface $z = f(x, y)$. A similar interpretation is given to $f_x(x, b)$.

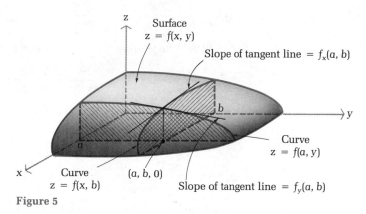

Figure 5

■ Higher-Order Partial Derivatives

Just as there are higher-order ordinary derivatives, there are higher-order partials, and we will be using some of these in Section 16-4 when we discuss local maxima and minima. The following second-order partials will be useful:

Second-Order Partials

If $z = f(x, y)$, then

$$\frac{\partial^2 z}{\partial x^2} = \frac{\partial}{\partial x}\left(\frac{\partial z}{\partial x}\right) = f_{xx}(x, y) = f_{xx}$$

$$\frac{\partial^2 z}{\partial x\, \partial y} = \frac{\partial}{\partial x}\left(\frac{\partial z}{\partial y}\right) = f_{yx}(x, y) = f_{yx}$$

$$\frac{\partial^2 z}{\partial y\, \partial x} = \frac{\partial}{\partial y}\left(\frac{\partial z}{\partial x}\right) = f_{xy}(x, y) = f_{xy}$$

$$\frac{\partial^2 z}{\partial y^2} = \frac{\partial}{\partial y}\left(\frac{\partial z}{\partial y}\right) = f_{yy}(x, y) = f_{yy}$$

In the mixed partial $\dfrac{\partial^2 z}{\partial x\,\partial y} = f_{yx}$, we start with $z = f(x, y)$ and first differentiate with respect to y (holding x constant). Then we differentiate with respect to x (holding y constant). What is the order of differentiation for $\dfrac{\partial^2 z}{\partial y\,\partial x} = f_{xy}$? It can be shown that for the functions we will consider, $f_{xy}(x, y) = f_{yx}(x, y)$.

Example 9   For $z = f(x, y) = 3x^2 - 2xy^3 + 1$, find:

(A) $\dfrac{\partial^2 z}{\partial x\,\partial y}, \dfrac{\partial^2 z}{\partial y\,\partial x}$ (B) $\dfrac{\partial^2 z}{\partial x^2}$ (C) $f_{yx}(2, 1)$

Solution   (A) First differentiate with respect to y and then with respect to x:

$$\frac{\partial z}{\partial y} = -6xy^2 \qquad \frac{\partial^2 z}{\partial x\,\partial y} = \frac{\partial}{\partial x}\left(\frac{\partial z}{\partial y}\right) = \frac{\partial}{\partial x}(-6xy^2) = -6y^2$$

First differentiate with respect to x and then with respect to y:

$$\frac{\partial z}{\partial x} = 6x - 2y^3 \qquad \frac{\partial^2 z}{\partial y\,\partial x} = \frac{\partial}{\partial y}\left(\frac{\partial z}{\partial x}\right) = \frac{\partial}{\partial y}(6x - 2y^3) = -6y^2$$

(B) Differentiate with respect to x twice:

$$\frac{\partial z}{\partial x} = 6x - 2y^3 \qquad \frac{\partial^2 z}{\partial x^2} = \frac{\partial}{\partial x}\left(\frac{\partial z}{\partial x}\right) = 6$$

(C) First find $f_{yx}(x, y)$. Then evaluate at $(2, 1)$. Again, remember that f_{yx} means to differentiate with respect to y first and then with respect to x. Thus,

$$f_y(x, y) = -6xy^2$$
$$f_{yx}(x, y) = -6y^2$$

and

$$f_{yx}(2, 1) = -6(1)^2 = -6$$

Problem 9   For the function in Example 9, find:

(A) $\dfrac{\partial^2 z}{\partial y\,\partial x}$ (B) $\dfrac{\partial^2 z}{\partial y^2}$ (C) $f_{xy}(2, 3)$ (D) $f_{yx}(2, 3)$

Answers to Matched Problems

6. (A) $\dfrac{\partial z}{\partial y} = -3x^2 + 5$ (B) $f_y(2, 3) = -7$

7. (A) $10x(x^2 + 2xy)^4$ (B) 10

8. $P_y(25, 10) = 10$: At a production level of $x = 25$ and $y = 10$, increasing y by one unit and holding x fixed at 25 will increase profit by approxi-

mately \$10; $P_y(25, 15) = -110$: At a production level of $x = 25$ and $y = 15$, increasing y by one unit and holding x fixed at 25 will decrease profit by approximately \$110.

9. (A) $-6y^2$ (B) $-12xy$ (C) -54 (D) -54

Exercise 16-2

A   For $z = f(x, y) = 10 + 3x + 2y$, find each of the following:

1. $\dfrac{\partial z}{\partial x}$

2. $\dfrac{\partial z}{\partial y}$

3. $f_y(1, 2)$

4. $f_x(1, 2)$

For $z = f(x, y) = 3x^2 - 2xy^2 + 1$, find each of the following:

5. $\dfrac{\partial z}{\partial y}$

6. $\dfrac{\partial z}{\partial x}$

7. $f_x(2, 3)$

8. $f_y(2, 3)$

For $S(x, y) = 5x^2y^3$, find each of the following:

9. $S_x(x, y)$
11. $S_y(2, 1)$

10. $S_y(x, y)$
12. $S_x(2, 1)$

B   For $C(x, y) = x^2 - 2xy + 2y^2 + 6x - 9y + 5$, find each of the following:

13. $C_x(x, y)$
15. $C_x(2, 2)$
17. $C_{xy}(x, y)$
19. $C_{xx}(x, y)$

14. $C_y(x, y)$
16. $C_y(2, 2)$
18. $C_{yx}(x, y)$
20. $C_{yy}(x, y)$

For $z = f(x, y) = e^{2x+3y}$, find each of the following:

21. $\dfrac{\partial z}{\partial x}$

22. $\dfrac{\partial z}{\partial y}$

23. $\dfrac{\partial^2 z}{\partial x\, \partial y}$

24. $\dfrac{\partial^2 z}{\partial y\, \partial x}$

25. $f_{xy}(1, 0)$
27. $f_{xx}(0, 1)$

26. $f_{yx}(0, 1)$
28. $f_{yy}(1, 0)$

Find $f_x(x, y)$ and $f_y(x, y)$ for each function f given by:

29. $f(x, y) = (x^2 - y^3)^3$
31. $f(x, y) = (3x^2y - 1)^4$
33. $f(x, y) = \ln(x^2 + y^2)$
35. $f(x, y) = y^2 e^{xy^2}$

30. $f(x, y) = \sqrt{2x - y^2}$
32. $f(x, y) = (3 + 2xy^2)^3$
34. $f(x, y) = \ln(2x - 3y)$
36. $f(x, y) = x^3 e^{x^2y}$

37. $f(x, y) = \dfrac{x^2 - y^2}{x^2 + y^2}$

38. $f(x, y) = \dfrac{2x^2 y}{x^2 + y^2}$

Find $f_{xx}(x, y)$, $f_{xy}(x, y)$, $f_{yx}(x, y)$, and $f_{yy}(x, y)$ for each function f given by:

39. $f(x, y) = x^2 y^2 + x^3 + y$

40. $f(x, y) = x^3 y^3 + x + y^2$

41. $f(x, y) = \dfrac{x}{y} - \dfrac{y}{x}$

42. $f(x, y) = \dfrac{x^2}{y} - \dfrac{y^2}{x}$

43. $f(x, y) = xe^{xy}$

44. $f(x, y) = x \ln(xy)$

C

45. For

$$P(x, y) = -x^2 + 2xy - 2y^2 - 4x + 12y - 5$$

find values of x and y such that

$$P_x(x, y) = 0 \quad \text{and} \quad P_y(x, y) = 0$$

simultaneously.

46. For

$$C(x, y) = 2x^2 + 2xy + 3y^2 - 16x - 18y + 54$$

find values of x and y such that

$$C_x(x, y) = 0 \quad \text{and} \quad C_y(x, y) = 0$$

simultaneously.

In Problems 47–48, show that the function f satisfies $f_{xx}(x, y) + f_{yy}(x, y) = 0$.

47. $f(x, y) = \ln(x^2 + y^2)$
48. $f(x, y) = x^3 - 3xy^2$
49. For $f(x, y) = x^2 + 2y^2$, find:

(A) $\lim\limits_{\Delta x \to 0} \dfrac{f(x + \Delta x, y) - f(x, y)}{\Delta x}$ (B) $\lim\limits_{\Delta y \to 0} \dfrac{f(x, y + \Delta y) - f(x, y)}{\Delta y}$

50. For $f(x, y) = 2xy^2$, find:

(A) $\lim\limits_{\Delta x \to 0} \dfrac{f(x + \Delta x, y) - f(x, y)}{\Delta x}$ (B) $\lim\limits_{\Delta y \to 0} \dfrac{f(x, y + \Delta y) - f(x, y)}{\Delta y}$

■

Applications

Business & Economics

51. *Cost function.* The cost function for the surfboard company in Problem 29 in Exercise 16-1 was

$$C(x, y) = 2{,}000 + 70x + 100y$$

Find $C_x(x, y)$ and $C_y(x, y)$, and interpret.
52. *Advertising and sales.* A company spends x thousand dollars per week

on newspaper advertising and y thousand dollars per week on television advertising. Its weekly sales were found to be given by

$$S(x, y) = 5x^2y^3$$

Find $S_x(3, 2)$ and $S_y(3, 2)$, and interpret.

53. *Profit function.* A firm produces two types of calculators, x thousand of type A and y thousand of type B per year. The revenue and cost functions for the year are (in thousands of dollars)

$$R(x, y) = 14x + 20y$$
$$C(x, y) = x^2 - 2xy + 2y^2 + 12x + 16y + 5$$

Find $P_x(1, 2)$ and $P_y(1, 2)$, and interpret.

54. *Revenue and profit functions.* A company manufactures ten-speed and three-speed bicycles. The weekly demand and cost functions are

$$p = 230 - 9x + y$$
$$q = 130 + x - 4y$$
$$C(x, y) = 200 + 80x + 30y$$

where $\$p$ is the price of a ten-speed bicycle, $\$q$ is the price of a three-speed bicycle, x is the weekly demand for ten-speed bicycles, y is the weekly demand for three-speed bicycles, and $C(x, y)$ is the cost function. Find $R_x(10, 5)$ and $P_x(10, 5)$, and interpret.

55. *Demand equations.* A supermarket sells two brands of coffee, brand A at $\$p$ per pound and brand B at $\$q$ per pound. The daily demand equations for brands A and B are, respectively,

$$x = 200 - 5p + 4q$$
$$y = 300 + 2p - 4q$$

Find $\partial x/\partial p$ and $\partial y/\partial p$, and interpret.

56. *Marginal productivity.* A company has determined that its productivity (units per employee per week) is given approximately by

$$z(x, y) = 50xy - x^2 - 3y^2$$

where x is the size of the labor force in thousands and y is the amount of capital investment in millions of dollars.

(A) Determine the marginal productivity of labor when $x = 5$ and $y = 4$. Interpret.

(B) Determine the marginal productivity of capital when $x = 5$ and $y = 4$. Interpret.

Life Sciences

57. *Marine biology.* In using scuba diving gear, a marine biologist estimates the time of a dive according to the equation

$$T(V, x) = \frac{33V}{x + 33}$$

where

T = Time of dive in minutes

V = Volume of air, at sea level pressure, compressed into tanks

x = Depth of dive in feet

Find $T_V(70, 47)$ and $T_x(70, 47)$, and interpret.

58. *Blood flow.* Poiseuille's law states that the resistance, R, for blood flowing in a blood vessel varies directly as the length of the vessel, L, and inversely as the fourth power of its radius, r. This relationship may be stated in equation form as follows:

$$R(L, r) = k \frac{L}{r^4} \qquad k \text{ a constant}$$

Find $R_L(4, 0.2)$ and $R_r(4, 0.2)$, and interpret.

59. *Physical anthropology.* Anthropologists, in their study of race and human genetic groupings, often use an index called the *cephalic index.* The cephalic index, C, varies directly as the width, W, of the head, and inversely as the length, L, of the head (both viewed from the top). In terms of an equation,

$$C(W, L) = 100 \frac{W}{L}$$

where

W = Width in inches

L = Length in inches

Find $C_W(6, 8)$ and $C_L(6, 8)$, and interpret.

60. *Safety research.* Under ideal conditions, if a person driving a car slams on the brakes and skids to a stop, the length of the skid marks (in feet) is given by the formula

$$L(w, v) = kwv^2$$

where

k = Constant

w = Weight of car in pounds

v = Speed of car in miles per hour

For $k = 0.0000133$, find $L_w(2,500, 60)$ and $L_v(2,500, 60)$, and interpret.

61. *Psychology.* Intelligence quotient (IQ) is defined to be the ratio of the mental age (MA), as determined by certain tests, and the chronological age (CA), multiplied by 100. Stated as an equation,

$$Q(M, C) = \frac{M}{C} \cdot 100$$

where

$$Q = IQ$$
$$M = MA$$
$$C = CA$$

Find $Q_M(12, 10)$ and $Q_C(12, 10)$, and interpret.

16-3 Total Differentials and Their Applications

■ The Total Differential
■ Approximations Using Differentials

■ The Total Differential

Recall (Section 11-4) that for a function defined by

$$y = f(x)$$

the differential dx of the *independent variable* x is another independent variable, which can be viewed as Δx, the change in x. The differential dy of the *dependent variable* y is given by $dy = f'(x)\, dx$. Thus, the differential of a function with one independent variable is a function with *two* independent variables, x and dx. How can the differential concept be extended to functions with two or more independent variables?

Suppose $z = f(x, y)$ is a function with the independent variables x and y. We define the **total differential** of the dependent variable z to be

$$dz = f_x(x, y)\, dx + f_y(x, y)\, dy$$

Notice that dz is a function of *four* variables: the independent variables x and y, and their differentials dx and dy.

Example 10   Find dz for $f(x, y) = x^2y^3$. Evaluate dz for:

(A) $x = 2$, $y = -1$, $dx = 0.1$,   and $dy = 0.2$
(B) $x = 1$, $y = 2$, $dx = -0.1$,   and $dy = 0.05$
(C) $x = -2$, $y = 1$, $dx = 0.3$,   and $dy = -0.1$

Solution   Since $f_x(x, y) = 2xy^3$ and $f_y(x, y) = 3x^2y^2$,

$$dz = f_x(x, y)\, dx + f_y(x, y)\, dy$$
$$= 2xy^3\, dx + 3x^2y^2\, dy$$

(A)   When $x = 2$, $y = -1$, $dx = 0.1$, and $dy = 0.2$,

$$dz = 2(2)(-1)^3(0.1) + 3(2)^2(-1)^2(0.2) = 2$$

(B)   When $x = 1$, $y = 2$, $dx = -0.1$, and $dy = 0.05$,

$$dz = 2(1)(2)^3(-0.1) + 3(1)^2(2)^2(0.05) = -1$$

(C)   When $x = -2$, $y = 1$, $dx = 0.3$, and $dy = -0.1$,

$$dz = 2(-2)(1)^3(0.3) + 3(-2)^2(1)^2(-0.1) = -2.4$$

Problem 10   Find dz for $f(x, y) = xy^2 + x^2$. Evaluate dz for:

(A) $x = 3$, $y = 1$, $dx = 0.05$,   and $dy = -0.1$
(B) $x = -2$, $y = 2$, $dx = 0.2$,   and $dy = 0.1$
(C) $x = 1$, $y = -2$, $dx = 0.1$,   and $dy = -0.04$

If $w = f(x, y, z)$, then the **total differential** is

$$dw = f_x(x, y, z)\, dx + f_y(x, y, z)\, dy + f_z(x, y, z)\, dz$$

This time, dw is a function of six independent variables: the original independent variables x, y, and z, and their differentials dx, dy, and dz. Generalizations to functions with more than three independent variables follow the same pattern.

Example 11   Find dw for $f(x, y, z) = xyz^2$. Evaluate dw for:

(A) $x = 2$, $y = 3$, $z = -1$, $dx = 0.1$, $dy = -0.2$,   and $dz = 0.05$
(B) $x = 1$, $y = -2$, $z = 0$, $dx = -0.1$, $dy = 0.1$,   and $dz = 0$
(C) $x = -1$, $y = 1$, $z = 2$, $dx = 0.2$, $dy = 0.3$,   and $dz = -0.4$

Solution   Since $f_x(x, y, z) = yz^2$, $f_y(x, y, z) = xz^2$, and $f_z(x, y, z) = 2xyz$,

$$dw = yz^2\, dx + xz^2\, dy + 2xyz\, dz$$

(A)   When $x = 2$, $y = 3$, $z = -1$, $dx = 0.1$, $dy = -0.2$, and $dz = 0.05$,

$$dw = (3)(-1)^2(0.1) + (2)(-1)^2(-0.2) + 2(2)(3)(-1)(0.05) = -0.7$$

(B)   When $x = 1$, $y = -2$, $z = 0$, $dx = -0.1$, $dy = 0.1$, and $dz = 0$,

$$dw = (-2)(0)^2(-0.1) + (1)(0)^2(0.1) + 2(1)(-2)(0)(0) = 0$$

(C)   When $x = -1$, $y = 1$, $z = 2$, $dx = 0.2$, $dy = 0.3$, and $dz = -0.4$,

$$dw = (1)(2)^2(0.2) + (-1)(2)^2(0.3) + 2(-1)(1)(2)(-0.4) = 1.2$$

Problem 11   Find dw for $f(x, y, z) = xy + yz + zx$. Evaluate dw for:

(A) $x = 1$, $y = 1$, $z = 1$, $dx = 0.1$, $dy = 0.1$,   and $dz = 0.1$
(B) $x = 2$, $y = 2$, $z = -2$, $dx = 0.5$, $dy = 0.5$,   and $dz = 0$
(C) $x = 4$, $y = -3$, $z = 1$, $dx = 0.1$, $dy = -0.2$,   and $dz = 0.4$

■ Approximations Using Differentials

If $z = f(x, y)$ and Δx and Δy represent the changes in the independent variables x and y, then the corresponding change in the dependent variable z is given exactly by

$$\Delta z = f(x + \Delta x, y + \Delta y) - f(x, y)$$

For small values of Δx and Δy, the differential dz can be used to approximate the change Δz.

Example 12   Find Δz and dz for $f(x, y) = x^2 + y^2$ when $x = 3$, $y = 4$, $\Delta x = dx = 0.01$, and $\Delta y = dy = -0.02$.

Solution

$$\Delta z = f(x + \Delta x, y + \Delta y) - f(x, y)$$
$$= f(3.01, 3.98) - f(3, 4)$$
$$= [(3.01)^2 + (3.98)^2] - [3^2 + 4^2]$$
$$= 24.9005 - 25$$
$$= -0.0995 \longleftarrow$$
$$dz = f_x(x, y)\,dx + f_y(x, y)\,dy$$
$$= 2x\,dx + 2y\,dy$$
$$= 2(3)(0.01) + 2(4)(-0.02)$$
$$= -0.1 \longleftarrow$$

Note that dz is a good approximation for Δz, the exact change in z, and dz was easier to calculate

Problem 12   Repeat Example 12 for $x = 2$, $y = 5$, $\Delta x = dx = -0.01$, and $\Delta y = dy = 0.05$.

In addition to approximating Δz, the differential can also be used to approximate $f(x + \Delta x, y + \Delta y)$. These approximations are summarized in the box and illustrated in the examples that follow.

Differential Approximation

If $f_x(x, y)$ and $f_y(x, y)$ exist, then for small Δx and Δy,

$$\Delta z \approx dz$$

and

$$f(x + \Delta x, y + \Delta y) = f(x, y) + \Delta z$$
$$\approx f(x, y) + dz$$
$$= f(x, y) + f_x(x, y)\,dx + f_y(x, y)\,dy$$

Example 13
Cost

Suppose the cost equation for a company producing standard and competition surfboards is

$$C(x, y) = 700 + 70x^{3/2} + 100y^{3/2} - 20x^{1/2}y^{1/2}$$

where x is the number of standard boards produced and y is the number of competition boards produced.

(A)   What is the cost of producing 100 boards of each type?

(B)   What is the approximate change in the cost if one fewer standard and two more competition boards are produced? Approximate the change using differentials.

Solution   (A) $C(100, 100) = 700 + 70(100)^{3/2} + 100(100)^{3/2} - 20(100)^{1/2}(100)^{1/2}$

$$= 700 + 70,000 + 100,000 - 2,000$$

$$= \$168,700$$

(B)   We will use dC to approximate ΔC as x changes from 100 to 99 and y changes from 100 to 102. We must evaluate dC for $x = 100$, $y = 100$, $dx = \Delta x = -1$, and $dy = \Delta y = 2$:

$$\Delta C \approx dC$$

$$= C_x(x, y)\, dx + C_y(x, y)\, dy$$

$$= [105x^{1/2} - 10x^{-1/2}y^{1/2}]\, dx + [150y^{1/2} - 10x^{1/2}y^{-1/2}]\, dy$$

$$= [105(100)^{1/2} - 10(100)^{-1/2}(100)^{1/2}](-1)$$

$$\qquad\qquad + [150(100)^{1/2} - 10(100)^{1/2}(100)^{-1/2}](2)$$

$$= -1,040 + 2,980$$

$$= \$1,940$$

Thus, decreasing the production of standard boards by one and increasing the production of competition boards by two will increase the cost by approximately $1,940.

Problem 13   For the cost function in Example 13:

(A)   What is the cost of producing 25 standard boards and 100 competition boards?

(B)   What is the approximate change in the cost if three more standard boards and five fewer competition boards are produced?

Example 14   Approximate the hypotenuse of a right triangle with legs of length 6.02 and 7.97 inches.

Solution   If x and y are the lengths of the legs of a right triangle, then from the Pythagorean theorem we find the hypotenuse z to be

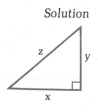

$$z = f(x, y) = \sqrt{x^2 + y^2}$$

We could use a calculator to compute the value of $f(6.02, 7.97)$ directly, however, our purpose here is to illustrate the use of the differential to approximate the value of a function. Thus, we will proceed as though a calculator is not available. This means that we must select values of x and y that satisfy two conditions: First, they must be near 6.02 and 7.97; and second, we must be able to evaluate $\sqrt{x^2 + y^2}$ without using a calculator. Since

$$\sqrt{6^2 + 8^2} = \sqrt{36 + 64} = \sqrt{100} = 10$$

$x = 6$ and $y = 8$ satisfy both of these conditions. So, we let $x = 6$, $y = 8$, $dx = \Delta x = 0.02$, and $dy = \Delta y = -0.03$, and then we use

$$f(x + \Delta x, y + \Delta y) = f(x, y) + \Delta z$$
$$\approx f(x, y) + dz$$
$$= f(x, y) + f_x(x, y)\, dx + f_y(x, y)\, dy$$

Now we can obtain an approximation to $f(6.02, 7.97)$ that we can evaluate by hand:

$$f(x + \Delta x, y + \Delta y) \approx f(x, y) + f_x(x, y)\, dx + f_y(x, y)\, dy$$

$$\sqrt{(x + \Delta x)^2 + (y + \Delta y)^2} \approx \sqrt{x^2 + y^2} + \frac{x}{\sqrt{x^2 + y^2}}\, dx + \frac{y}{\sqrt{x^2 + y^2}}\, dy$$

$$\sqrt{(6 + 0.02)^2 + [8 + (-0.03)]^2} \approx \sqrt{6^2 + 8^2} + \frac{6}{\sqrt{6^2 + 8^2}}\,(0.02) + \frac{8}{\sqrt{6^2 + 8^2}}\,(-0.03)$$

$$\sqrt{(6.02)^2 + (7.97)^2} \approx 10 + 0.012 - 0.024 = 9.988$$

Problem 14   Approximate the hypotenuse of a right triangle with legs of length 2.95 and 4.02.

Answers to Matched Problems

10. $dz = (y^2 + 2x)\, dx + 2xy\, dy$:   (A) -0.25 (B) -0.8 (C) 0.76
11. $dw = (y + z)\, dx + (x + z)\, dy + (y + x)\, dz$:   (A) 0.6 (B) 0
     (C) -0.8
12. $\Delta z = 0.4626$, $dz = 0.46$ 13. (A) $\$108{,}450$ (C) $-\$5{,}960$
14. 4.986

Exercise 16-3

A   Find dz for each function.

1. $z = x^2 + y^2$ 2. $z = 2x + xy + 3y$
3. $z = x^4 y^3$ 4. $z = \sqrt{2x + 6y}$
5. $z = \sqrt{x} + \dfrac{5}{\sqrt{y}}$ 6. $z = x\sqrt{1 + y}$

Find dw for each function.

7. $w = x^3 + y^3 + z^3$ 8. $w = xy^2 z^3$
9. $w = xy + 2xz + 3yz$ 10. $w = \sqrt{2x + 3y - z}$

B   Evaluate dz and Δz for each function at the indicated values.

11. $z = f(x, y) = x^2 - 2xy + y^2$, $x = 3$, $y = 1$, $\Delta x = dx = 0.1$,
     $\Delta y = dy = 0.2$
12. $z = f(x, y) = 2x^2 + xy - 3y^2$, $x = 2$, $y = 4$, $\Delta x = dx = 0.1$,
     $\Delta y = dy = 0.05$

13. $z = f(x, y) = 100\left(3 - \dfrac{x}{y}\right)$, $\quad x = 2$, $\quad y = 1$, $\quad \Delta x = dx = 0.05$,

$\Delta y = dy = 0.1$

14. $z = f(x, y) = 50\left(1 + \dfrac{x^2}{y}\right)$, $\quad x = 3$, $\quad y = 9$, $\quad \Delta x = dx = -0.1$,

$\Delta y = dy = 0.2$

In Problems 15–18 evaluate dw and Δw for each function at the indicated values.

15. $w = f(x, y, z) = x^2 + yz$, $\quad x = 2$, $\quad y = 3$, $\quad z = 5$, $\quad \Delta x = dx = 0.1$,
    $\Delta y = dy = 0.2$, $\quad \Delta z = dz = 0.1$

16. $w = f(x, y, z) = 2xz + y^2 - z^2$, $\quad x = 4$, $\quad y = 2$, $\quad z = 3$,
    $\Delta x = dx = 0.2$, $\quad \Delta y = dy = 0.1$, $\quad \Delta z = dz = -0.1$

17. $w = f(x, y, z) = \dfrac{10x + 20y}{z}$, $\quad x = 4$, $\quad y = 3$, $\quad z = 5$,

    $\Delta x = dx = 0.05$, $\quad \Delta y = dy = -0.05$, $\quad \Delta z = dz = 0.1$

18. $w = f(x, y, z) = 50\left(x + \dfrac{1}{y} + \dfrac{1}{z^2}\right)$, $\quad x = 2$, $\quad y = 2$, $\quad z = 1$,

    $\Delta x = dx = 0.2$, $\quad \Delta y = dy = 0.1$, $\quad \Delta z = dz = 0.1$

19. Approximate the hypotenuse of a right triangle with legs of length 3.1 and 3.9 inches.

20. Approximate the hypotenuse of a right triangle with legs of length 4.95 and 12.02 inches.

21. A can in the shape of a right circular cylinder with radius 5 inches and height 10 inches is coated with ice 0.1 inch thick. Use differentials to approximate the volume of the ice ($V = \pi r^2 h$).

22. A box with edges of length 10, 15, and 20 centimeters is covered with a 1 centimeter thick coat of fiberglass. Use differentials to approximate the volume of the fiberglass shell.

23. A plastic box is to be constructed with a square base and an open top. The plastic material used in construction is 0.1 centimeter thick. The inside dimensions of the box are 10 by 10 by 5 centimeters. Use differentials to approximate the volume of the plastic required for one box.

24. The surface area of a right circular cone with radius r and altitude h is given by

$$S = \pi r \sqrt{r^2 + h^2}$$

Use differentials to approximate the change in S when r changes from 6 to 6.1 inches and h changes from 8 to 8.05 inches.

C 25. Find dz if $z = xye^{x^2 + y^2}$.
   26. Find dz if $z = x \ln(xy) + y \ln(xy)$.

27. Find dw if $w = xyze^{xyz}$.

28. Find dw if $w = xy \ln(xz) + yz \ln(xy)$.

Applications

Business & Economics

29. *Cost function.* A microcomputer company manufactures two types of computers, model I and model II. The cost in thousands of dollars of producing x model I's and y model II's per month is given by

$$C(x, y) = x + 2y - \frac{1}{10}\sqrt{x^2 + y^2}$$

Currently, the company manufactures 30 model I computers and 40 model II computers each month. Use differentials to approximate the change in the cost function if the company decides to produce 5 more model I and 3 more model II computers each month.

30. *Advertising and sales.* A company spends x thousand dollars per week on newspaper advertising and y thousand dollars per week on television advertising. Its weekly sales were found to be given by

$$S(x, y) = 5x^2y^3$$

Use differentials to approximate the change in sales if the amount spent on newspaper advertising is increased from \$3,000 to \$3,100 per week and the amount spent on television advertising is increased from \$2,000 to \$2,200 per week.

31. *Revenue function.* A supermarket sells two brands of coffee: brand A at \$x per pound and brand B at \$y per pound. The daily demand equations for brands A and B are, respectively,

$$u = 200 - 5x + 4y$$

$$v = 300 - 4y + 2x$$

(both in pounds). Thus, the daily revenue equation is

$$R(x, y) = xu + yv$$
$$= x(200 - 5x + 4y) + y(300 - 4y + 2x)$$
$$= -5x^2 + 6xy - 4y^2 + 200x + 300y$$

Use differentials to approximate the change in revenue if the price of brand A is increased from \$2.00 to \$2.10 per pound and the price of brand B is decreased from \$3.00 to \$2.95 per pound.

32. *Marginal productivity.* A company has determined that its productivity (units per employee per week) is given approximately by

$$z(x, y) = 50xy - x^2 - 3y^2$$

where x is the size of the labor force in thousands and y is the amount of capital investment in millions of dollars. The current labor force is

5,000 workers. The current capital investment is $4 million. Use differentials to approximate the change in productivity if both the labor force and the capital investment are increased by 10%.

Life Sciences   33.   *Blood flow.* Poiseuille's law states that the resistance, R, for blood flowing in a blood vessel varies directly as the length of the vessel, L, and inversely as the fourth power of its radius, r. This relationship may be stated in equation form as follows:

$$R(L, r) = k\,\frac{L}{r^4} \qquad k \text{ a constant}$$

Use differentials to approximate the change in the resistance if the length of the vessel decreases from 8 to 7.5 centimeters and the radius decreases from 1 to 0.95 centimeter.

34.   *Drug concentration.* The concentration of a drug in the bloodstream after having been injected into a vein is given by

$$C(x, y) = \frac{1}{1 + \sqrt{x^2 + y^2}}$$

where x is the time passed since the injection and y is the distance from the point of injection. Use differentials to approximate the concentration C(3.1, 4.1).

Social Sciences   35.   *Safety research.* Under ideal conditions, if a person driving a car slams on the brakes and skids to a stop, the length of the skid marks (in feet) is given by the formula

$$L(w, v) = kwv^2$$

where

$k = \text{Constant}$

$w = \text{Weight of car in pounds}$

$v = \text{Speed of car in miles per hour}$

For $k = 0.000\ 013\ 3$, use differentials to approximate the change in the length of the skid marks if the weight of the car is increased from 2,000 to 2,200 pounds and the speed is increased from 40 to 45 miles per hour.

36.   *Psychology.* Intelligence quotient (IQ) is defined to be the ratio of the mental age (MA), as determined by certain tests, and the chronological age (CA) multiplied by 100. Stated as an equation,

$$Q(M, C) = \frac{M}{C} \cdot 100$$

where

$$Q = IQ$$
$$M = MA$$
$$C = CA$$

Use differentials to approximate the change in IQ as a person's mental age changes from 12 to 12.5 and chronological age changes from 10 to 11.

16-4 Maxima and Minima

We are now ready to undertake a brief but useful analysis of local maxima and minima for functions of the type $z = f(x, y)$. Basically, we are going to extend the second-derivative test developed for functions of a single independent variable. To start, we assume that all second-order partials exist for the function f in some circular region in the xy plane. This guarantees that the surface $z = f(x, y)$ has no sharp points, breaks, or ruptures. In other words, we are dealing only with smooth surfaces with no edges (like the edge of a box); or breaks (like an earthquake fault); or sharp points (like the bottom point of a golf tee). See Figure 6.

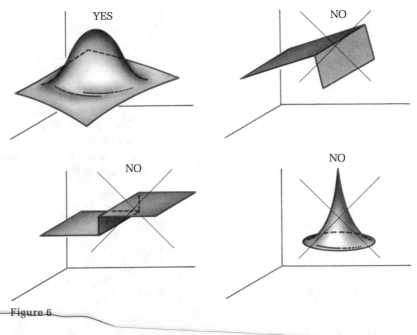

Figure 6

In addition, we will not concern ourselves with boundary points or absolute maxima–minima theory. In spite of these restrictions, the procedure we are now going to describe will help us solve a large number of useful problems.

What does it mean for $f(a, b)$ to be a local maximum or a local minimum? We say that $f(a, b)$ **is a local maximum** if there exists a circular region in the domain of f with (a, b) as the center, such that

$$f(a, b) \geqslant f(x, y)$$

for all (x, y) in the region. Similarly, we say that $f(a, b)$ **is a local minimum** if there exists a circular region in the domain of f with (a, b) as the center, such that

$$f(a, b) \leqslant f(x, y)$$

for all (x, y) in the region. In Section 16-1, Figure 2 illustrates a local minimum, Figure 3 illustrates a local maximum, and Figure 4 illustrates a saddle point, which is neither.

What happens to $f_x(a, b)$ and $f_y(a, b)$ if $f(a, b)$ is a local minimum or a local maximum and the partials of f exist in a circular region containing (a, b)? Figure 7 suggests that $f_x(a, b) = 0$ and $f_y(a, b) = 0$, since the tangents to the indicated curves are horizontal. Theorem 1 indicates that our intuitive reasoning is correct.

Theorem 1

> Let $f(a, b)$ be an extreme (a local maximum or a local minimum) for the function f. If both f_x and f_y exist at (a, b), then
>
> $$f_x(a, b) = 0 \qquad \text{and} \qquad f_y(a, b) = 0 \tag{1}$$

The converse of this theorem is false; that is, if $f_x(a, b) = 0$ and $f_y(a, b) = 0$, then $f(a, b)$ may or may not be a local extreme — the point $(a, b, f(a, b))$ may be a saddle point, for example.

Theorem 1 gives us what are called *necessary* (but not *sufficient*) conditions for $f(a, b)$ to be a local extreme. We thus find all points (a, b) such that

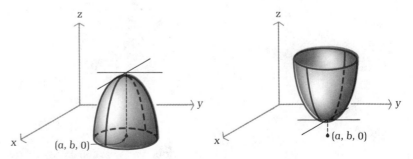

Figure 7

$f_x(a, b) = 0$ and $f_y(a, b) = 0$ and test these further to determine whether $f(a, b)$ is a local extreme or a saddle point. Points (a, b) such that (1) holds are called **critical points.** The next theorem, using second-derivative tests, gives us *sufficient* conditions for a local point to produce a local extreme or a saddle point. As was the case with Theorem 1, we state this theorem without proof.

Theorem 2

Second-Derivative Test for Local Extrema

If:

1. $z = f(x, y)$
2. $f_x(a, b) = 0$ and $f_y(a, b) = 0$ [(a, b) is a critical point]
3. All second-order partials of f exist in some circular region containing (a, b) as a center
4. $A = f_{xx}(a, b)$, $B = f_{xy}(a, b)$, $C = f_{yy}(a, b)$

Then:

1. If $AC - B^2 > 0$ and $A < 0$, then $f(a, b)$ is a local maximum.
2. If $AC - B^2 > 0$ and $A > 0$, then $f(a, b)$ is a local minimum.
3. If $AC - B^2 < 0$, then f has a saddle point at (a, b).
4. If $AC - B^2 = 0$, the test fails.

Let us consider a few examples.

Example 15   Use Theorem 2 to find local extrema for

$$f(x, y) = -x^2 - y^2 + 6x + 8y - 21$$

Solution   Step 1.   *Find critical points.* Find (x, y) such that $f_x(x, y) = 0$ and $f_y(x, y) = 0$, simultaneously.

$$f_x(x, y) = -2x + 6 = 0$$
$$x = 3$$

$$f_y(x, y) = -2y + 8 = 0$$
$$y = 4$$

The only critical point is $(a, b) = (3, 4)$.

Step 2.   *Compute $A = f_{xx}(3, 4)$, $B = f_{xy}(3, 4)$, and $C = f_{yy}(3, 4)$.*

$$f_{xx}(x, y) = -2, \quad \text{thus} \quad A = f_{xx}(3, 4) = -2$$
$$f_{xy}(x, y) = 0, \quad \text{thus} \quad B = f_{xy}(3, 4) = 0$$
$$f_{yy}(x, y) = -2, \quad \text{thus} \quad C = f_{yy}(3, 4) = -2$$

Step 3. *Evaluate $AC - B^2$ and try to classify the critical point $(3, 4)$ using Theorem 2.*

$$AC - B^2 = (-2)(-2) - (0)^2 = 4 > 0 \quad \text{and} \quad A = -2 < 0$$

Therefore, case 1 in Theorem 2 holds. That is, $f(3, 4) = 4$ is a local maximum.

Problem 15 Use Theorem 2 to find local extrema for

$$f(x, y) = x^2 + y^2 - 10x - 2y + 36$$

Example 16 Use Theorem 2 to find local extrema for

$$f(x, y) = x^3 + y^3 - 6xy$$

Solution **Step 1.** *Find critical points.*

$$f_x(x, y) = 3x^2 - 6y = 0 \qquad \text{Solve for } y.$$
$$6y = 3x^2$$
$$y = \frac{1}{2} x^2 \qquad\qquad\qquad (2)$$

$$f_y(x, y) = 3y^2 - 6x = 0$$
$$3y^2 = 6x \qquad \text{Use (2) to eliminate } y.$$
$$3 \left(\frac{1}{2} x^2 \right)^2 = 6x$$
$$\frac{3}{4} x^4 = 6x \qquad \text{Solve for } x.$$
$$3x^4 - 24x = 0$$
$$3x(x^3 - 8) = 0$$

$$x = 0 \quad \text{or} \quad x = 2$$
$$y = 0 \qquad\qquad y = \frac{1}{2}(2)^2 = 2$$

The critical points are $(0, 0)$ and $(2, 2)$. Since there are two critical points, steps 2 and 3 must be performed twice.

Test $(0, 0)$ **Step 2.** *Compute $A = f_{xx}(0, 0)$, $B = f_{xy}(0, 0)$, and $C = f_{yy}(0, 0)$.*

$$f_{xx}(x, y) = 6x, \quad \text{thus} \quad A = f_{xx}(0, 0) = \quad 0$$
$$f_{xy}(x, y) = -6, \quad \text{thus} \quad B = f_{xy}(0, 0) = -6$$
$$f_{yy}(x, y) = 6y, \quad \text{thus} \quad C = f_{yy}(0, 0) = \quad 0$$

Step 3. *Evaluate $AC - B^2$ and try to classify the critical point $(0, 0)$ using Theorem 2.*

$$AC - B^2 = (0)(0) - (-6)^2 = -36 < 0$$

Therefore, case 3 in Theorem 2 applies. That is, f has a saddle point at $(0, 0)$.

Now we will consider the second critical point, $(2, 2)$.

Test (2, 2) **Step 2.** *Compute $A = f_{xx}(2, 2)$, $B = f_{xy}(2, 2)$, and $C = f_{yy}(2, 2)$.*

$$f_{xx}(x, y) = 6x, \quad \text{thus} \quad A = f_{xx}(2, 2) = 12$$
$$f_{xy}(x, y) = -6, \quad \text{thus} \quad B = f_{xy}(2, 2) = -6$$
$$f_{yy}(x, y) = 6y, \quad \text{thus} \quad C = f_{yy}(2, 2) = 12$$

Step 3. *Evaluate $AC - B^2$ and try to classify the critical point $(2, 2)$ using Theorem 2.*

$$AC - B^2 = (12)(12) - (-6)^2 = 108 > 0 \quad \text{and} \quad A = 12 > 0$$

Thus, case 2 in Theorem 2 applies and $f(2, 2) = -8$ is a local minimum.

Problem 16 Use Theorem 2 to find local extrema for

$$f(x, y) = x^3 + y^2 - 6xy$$

Example 17 Suppose the surfboard company discussed earlier has developed the yearly
Profit profit equation

$$P(x, y) = -2x^2 + 2xy - y^2 + 10x - 4y + 107$$

where x is the number (in thousands) of standard surfboards produced per year, y is the number (in thousands) of competition surfboards produced per year, and P is profit (in thousands of dollars). How many of each type of board should be produced per year to realize a maximum profit? What is the maximum profit?

Solution **Step 1.** *Find critical points.*

$$P_x(x, y) = -4x + 2y + 10 = 0$$
$$P_y(x, y) = 2x - 2y - 4 = 0$$

Solving this system, we obtain $(3, 1)$ as the only critical point.

Step 2. *Compute $A = P_{xx}(3, 1)$, $B = P_{xy}(3, 1)$, and $C = P_{yy}(3, 1)$.*

$$P_{xx}(x, y) = -4, \quad \text{thus} \quad A = P_{xx}(3, 1) = -4$$
$$P_{xy}(x, y) = 2, \quad \text{thus} \quad B = P_{xy}(3, 1) = 2$$
$$P_{yy}(x, y) = -2, \quad \text{thus} \quad C = P_{yy}(3, 1) = -2$$

Step 3. *Evaluate $AC - B^2$ and try to classify the critical point $(3, 1)$ using Theorem 2.*

$$AC - B^2 = (-4)(-2) - (2)^2 = 8 - 4 = 4 > 0$$
$$A = -4 < 0$$

Therefore, case 1 in Theorem 2 applies. That is, $P(3, 1) = \$120,000$ is a local maximum. This is obtained by producing 3,000 standard boards and 1,000 competition boards per year.

Problem 17   Repeat Example 17 with

$$P(x, y) = -2x^2 + 4xy - 3y^2 + 4x - 2y + 77$$

Example 18   The packaging department in a company has been asked to design a
Package Design   rectangular box with no top and a partition down the middle. The box must have a volume of 48 cubic inches. Find the dimensions that will minimize the amount of material used to construct the box.

Solution

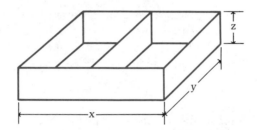

The amount of material used in constructing this box is

$$
\begin{array}{ccc}
& \text{Front} & \text{Sides} \\
& \text{and} & \text{and} \\
\text{Base} & \text{back} & \text{partition}
\end{array}
$$

$$M = \;\; xy \;\; + \;\; 2xz \;\; + \;\; 3yz \tag{3}$$

The volume of the box is

$$V = xyz = 48 \tag{4}$$

Since Theorem 2 applies only to functions with two independent variables, we must use (4) to eliminate one of the variables in (3).

$$M = xy + 2xz + 3yz \qquad\qquad \text{Substitute } z = \frac{48}{xy}$$

$$= xy + 2x\left(\frac{48}{xy}\right) + 3y\left(\frac{48}{xy}\right)$$

$$= xy + \frac{96}{y} + \frac{144}{x}$$

Thus, we must find the minimum value of

$$M(x, y) = xy + \frac{96}{y} + \frac{144}{x}$$

Step 1. *Find critical points.*

$$M_x(x, y) = y - \frac{144}{x^2} = 0$$

$$y = \frac{144}{x^2} \tag{5}$$

$$M_y(x, y) = x - \frac{96}{y^2} = 0$$

$$x = \frac{96}{y^2} \qquad \text{Solve for } y^2.$$

$$y^2 = \frac{96}{x} \qquad \text{Use (5) to eliminate } y \text{ and solve for } x.$$

$$\left(\frac{144}{x^2}\right)^2 = \frac{96}{x}$$

$$\frac{20{,}736}{x^4} = \frac{96}{x}$$

$$x^3 = \frac{20{,}736}{96} = 216$$

$$x = 6 \qquad \text{Use (5) to find } y.$$

$$y = \frac{144}{36} = 4$$

Step 2. *Compute* $A = M_{xx}(6, 4)$, $B = M_{xy}(6, 4)$, *and* $C = M_{yy}(6, 4)$.

$$M_{xx}(x, y) = \frac{288}{x^3}, \quad \text{thus} \quad A = M_{xx}(6, 4) = \frac{288}{216} = \frac{4}{3}$$

$$M_{xy}(x, y) = \quad 1, \quad \text{thus} \quad B = M_{xy}(6, 4) = \quad 1$$

$$M_{yy}(x, y) = \frac{192}{y^3}, \quad \text{thus} \quad C = M_{yy}(6, 4) = \frac{192}{64} = 3$$

Step 3. *Evaluate* $AC - B^2$ *and try to classify the critical point* (6, 4) *using Theorem 2.*

$$AC - B^2 = \left(\frac{4}{3}\right)(3) - (1)^2 = 3 > 0 \quad \text{and} \quad A = \frac{4}{3} > 0$$

Therefore, case 2 in Theorem 2 applies; $M(x, y)$ has a local minimum at (6, 4). If $x = 6$ and $y = 4$, then

$$z = \frac{48}{xy} = \frac{48}{6(4)} = 2$$

Thus, the dimensions that will require the minimum amount of material are 6 inches by 4 inches by 2 inches.

Problem 18 If the box in Example 18 must have a volume of 384 cubic inches, find the dimensions that will require the least amount of material.

Answers to
Matched Problems

15. $f(5, 1) = 10$ is a local minimum
16. f has a saddle point at $(0, 0)$; $f(6, 18) = -108$ is a local minimum
17. Local maximum for $x = 2$ and $y = 1$; $P(2, 1) = \$80,000$
18. 12 inches by 8 inches by 4 inches

Exercise 16-4

Find local extrema using Theorem 2.

A
1. $f(x, y) = 6 - x^2 - 4x - y^2$
2. $f(x, y) = 3 - x^2 - y^2 + 6y$
3. $f(x, y) = x^2 + y^2 + 2x - 6y + 14$
4. $f(x, y) = x^2 + y^2 - 4x + 6y + 23$

B
5. $f(x, y) = xy + 2x - 3y - 2$
6. $f(x, y) = x^2 - y^2 + 2x + 6y - 4$
7. $f(x, y) = -3x^2 + 2xy - 2y^2 + 14x + 2y + 10$
8. $f(x, y) = -x^2 + xy - 2y^2 + x + 10y - 5$
9. $f(x, y) = 2x^2 - 2xy + 3y^2 - 4x - 8y + 20$
10. $f(x, y) = 2x^2 - xy + y^2 - x - 5y + 8$

C
11. $f(x, y) = e^{xy}$
12. $f(x, y) = x^2y - xy^2$
13. $f(x, y) = x^3 + y^3 - 3xy$
14. $f(x, y) = 2y^3 - 6xy - x^2$
15. $f(x, y) = 2x^4 + y^2 - 12xy$
16. $f(x, y) = 16xy - x^4 - 2y^2$
17. $f(x, y) = x^3 - 3xy^2 + 6y^2$
18. $f(x, y) = 2x^2 - 2x^2y + 6y^3$

■

Applications

Business & Economics

19. *Product mix for maximum profit.* A firm produces two types of calculators, x thousand of type A and y thousand of type B per year. If the revenue and cost equations for the year are (in millions of dollars)

$$R(x, y) = 2x + 3y$$
$$C(x, y) = x^2 - 2xy + 2y^2 + 6x - 9y + 5$$

find how many of each type of calculator should be produced per year to maximize profit. What is the maximum profit?

20. *Automation–labor mix for minimum cost.* The annual labor and automated equipment cost (in millions of dollars) for a company's production of television sets is given by

$$C(x, y) = 2x^2 + 2xy + 3y^2 - 16x - 18y + 54$$

where x is the amount spent per year on labor and y is the amount spent per year on automated equipment (both in millions of dollars). Determine how much should be spent on each per year to minimize this cost. What is the minimum cost?

21. *Research–advertising mix for maximum profit.* A pocket calculator company has developed the profit equation

$$P(x, y) = -3x^2 + 3xy - y^2 + 12x - 5y + 17$$

where x is the amount spent per year on research and development and y is the amount spent per year on advertising (all units are in millions of dollars). How much should be spent in each area per year to maximize profit? What is the maximum profit for this budget?

22. *Minimum material.* A rectangular box with no top is to be made to hold 32 cubic inches. What should its dimensions be in order to use the least amount of material in its construction?

23. *Minimum material.* A rectangular box with no top and two parallel partitions (see accompanying figure) is to be made to hold 64 cubic inches. Find the dimensions that will require the least amount of material.

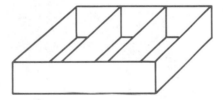

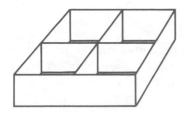

24. *Minimum material.* A rectangular box with no top and two intersecting partitions (see accompanying figure) is to be made to hold 72 cubic inches. What should its dimensions be in order to use the least amount of material in its construction?

25. *Maximum volume.* A mailing service states that a rectangular package shall have the sum of the length and girth not to exceed 120 inches (see the figure). What are the dimensions of the largest (in volume) mailing carton that can be constructed meeting these restrictions?

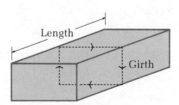

Length   Girth

16-5   Maxima and Minima Using Lagrange Multipliers

- Functions of Two Independent Variables
- Functions of Three Independent Variables

■ Functions of Two Independent Variables

We will now consider a particularly powerful method of solving a certain class of maxima–minima problems. The method is due to Joseph Louis Lagrange (1736–1813), an eminent eighteenth century French mathematician, and it is called the **method of Lagrange multipliers.** We introduce the method through an example; then we will formalize the discussion in the form of a theorem.

A rancher wants to construct two feeding pens of the same size along an existing fence (see Figure 8). If 720 feet of fencing are available, how long should x and y be in order to obtain the maximum total area? What is the maximum area?

Existing fence

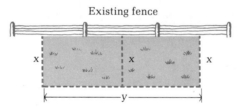

Figure 8

The total area is given by

$$f(x, y) = xy$$

which can be made as large as we like providing there are no restrictions on x and y. But there are restrictions on x and y, since we have only 720 feet of fencing. That is, x and y must be chosen so that

$$3x + y = 720$$

This restriction on x and y, also called a **constraint,** leads to the following maxima–minima problem:

Maximize $f(x, y) = xy$ (1)

Subject to $3x + y = 720$ or $3x + y - 720 = 0$ (2)

This problem is a special case of a general class of problems of the form

Maximize (or minimize) $z = f(x, y)$ (3)

Subject to $g(x, y) = 0$ (4)

Of course, we could try to solve (4) for y in terms of x, or for x in terms of y, then substitute the result into (3), and use methods developed in Section 12-5 for functions of a single variable. But what if (4) were more complicated than (2), and solving for one variable in terms of the other was either very difficult or impossible? In the method of Lagrange multipliers we work with $g(x, y)$ directly and avoid having to solve (4) for one variable in terms of the other. In addition, the method generalizes to functions of arbitrarily many variables subject to one or more constraints.

Now, to the method. We form a new function F, using functions f and g in (3) and (4), as follows:

$$F(x, y, \lambda) = f(x, y) + \lambda g(x, y) \tag{5}$$

where λ (lambda) is called a **Lagrange multiplier.** Theorem 3 forms the basis for the method.

Theorem 3

> The relative maxima and minima of the function $z = f(x, y)$ subject to the constraint $g(x, y) = 0$ will be among those points (x_0, y_0) for which (x_0, y_0, λ_0) is a solution to the system
>
> $F_x(x, y, \lambda) = 0$
> $F_y(x, y, \lambda) = 0$
> $F_\lambda(x, y, \lambda) = 0$
>
> where $F(x, y, \lambda) = f(x, y) + \lambda g(x, y)$, provided all the partial derivatives exist.

We now solve the fence problem using the method of Lagrange multipliers.

Step 1. *Formulate the problem in the form of equations (3) and (4).*

      Maximize $f(x, y) = xy$
      Subject to $g(x, y) = 3x + y - 720 = 0$

Step 2. *Form the function F, introducing the Lagrange multiplier λ.*

$$F(x, y, \lambda) = f(x, y) + \lambda g(x, y)$$
$$= xy + \lambda(3x + y - 720)$$

Step 3. *Solve the system $F_x = 0$, $F_y = 0$, $F_\lambda = 0$.* (Solutions are called **critical points** for F.)

$$F_x = y + 3\lambda = 0$$
$$F_y = x + \lambda = 0$$
$$F_\lambda = 3x + y - 720 = 0$$

From the first two equations, we see that

$$y = -3\lambda$$
$$x = -\lambda$$

Substitute these values for x and y into the third equation and solve for λ.

$$-3\lambda - 3\lambda = 720$$
$$-6\lambda = 720$$
$$\lambda = -120$$

Thus,

$$y = -3(-120) = 360 \text{ feet}$$
$$x = -(-120) = 120 \text{ feet}$$

Step 4. *Test the critical points for maxima and minima.* The function F has only one critical point at $(120, 360, -120)$, and since $f(x, y) = xy$ has a minimum at $(0, 0)$, we conclude that $(120, 360)$ produces a maximum for f. Hence,

$$\text{Max } f(x, y) = f(120, 360)$$
$$= (120)(360)$$
$$= 43,200 \text{ square feet}$$

Method of Lagrange Multipliers — Key Steps

1. Formulate the problem in the form

   Maximize (or minimize) $z = f(x, y)$
   Subject to $g(x, y) = 0$

2. Form the function F:

   $$F(x, y, \lambda) = f(x, y) + \lambda g(x, y)$$

3. Find the critical points for F; that is, solve the system

   $$F_x(x, y, \lambda) = 0$$
   $$F_y(x, y, \lambda) = 0$$
   $$F_\lambda(x, y, \lambda) = 0$$

4. Evaluate $z = f(x, y)$ at each point (x_0, y_0) such that (x_0, y_0, λ_0) satisfies the system in step 3. The maximum or minimum value of $f(x, y)$ will be among these values in the problems we consider.

Example 19   Minimize $f(x, y) = x^2 + y^2$ subject to $x + y = 10$.

Solution   **Step 1.**   Minimize $f(x, y) = x^2 + y^2$

Subject to $g(x, y) = x + y - 10 = 0$

Step 2. $F(x, y, \lambda) = x^2 + y^2 + \lambda(x + y - 10)$

Step 3. $F_x = 2x + \lambda = 0$

$F_y = 2y + \lambda = 0$

$F_\lambda = x + y - 10 = 0$

From the first two equations,

$$x = -\frac{\lambda}{2} \qquad y = -\frac{\lambda}{2}$$

Substituting these into the third equation, we obtain

$$-\frac{\lambda}{2} - \frac{\lambda}{2} = 10$$

$$-\lambda = 10$$

$$\lambda = -10$$

The critical point is $(5, 5, -10)$.

Step 4. $f(5, 5) = 5^2 + 5^2 = 50$

Checking other points on the line $x + y = 10$ near $(5, 5)$, we see that this is a minimum. (See Figure 9.)

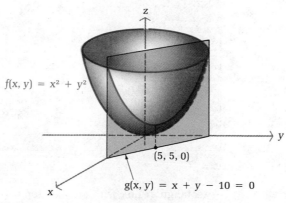

Figure 9

Problem 19   Maximize $f(x, y) = 25 - x^2 - y^2$ subject to $x + y = 4$. (See Figure 10.)

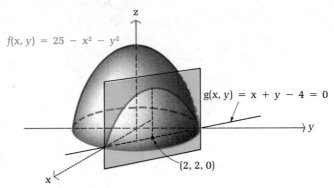

$f(x, y) = 25 - x^2 - y^2$

z

$g(x, y) = x + y - 4 = 0$

y

$(2, 2, 0)$

x

Figure 10

■ Functions of Three Independent Variables

We have indicated that the method of Lagrange multipliers can be extended to functions with arbitrarily many independent variables with one or more constraints. We state a theorem for functions with three independent variables and one constraint and consider an example that will demonstrate the advantage of the method of Lagrange multipliers over the method used in Section 16-4.

Theorem 4

The relative maxima and minima of the function $w = f(x, y, z)$ subject to the constraint $g(x, y, z) = 0$ will be among the set of points (x_0, y_0, z_0) for which $(x_0, y_0, z_0, \lambda_0)$ is a solution to the system

$$F_x(x, y, z, \lambda) = 0$$
$$F_y(x, y, z, \lambda) = 0$$
$$F_z(x, y, z, \lambda) = 0$$
$$F_\lambda(x, y, z, \lambda) = 0$$

where $F(x, y, z, \lambda) = f(x, y, z) + \lambda g(x, y, z)$, provided all the partial derivatives exist.

Example 20
Package Design

A rectangular box with an open top and one partition is to be constructed from 162 square inches of cardboard. Find the dimensions that will result in a box with the largest possible volume.

Solution

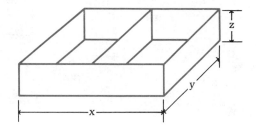

We must maximize

$$V(x, y, z) = xyz$$

subject to the constraint that the amount of material used is 162 square inches. Thus, x, y, and z must satisfy

$$xy + 2xz + 3yz = 162$$

Step 1.   Maximize $V(x, y, z) = xyz$

            Subject to $g(x, y, z) = xy + 2xz + 3yz - 162 = 0$

Step 2. $F(x, y, z, \lambda) = xyz + \lambda(xy + 2xz + 3yz - 162)$

Step 3. $F_x = yz + \lambda(y + 2z) = 0$

          $F_y = xz + \lambda(x + 3z) = 0$

          $F_z = xy + \lambda(2x + 3y) = 0$

          $F_\lambda = xy + 2xz + 3yz - 162 = 0$

From the first two equations, we can write

$$\lambda = \frac{-yz}{y + 2z} \qquad \lambda = \frac{-xz}{x + 3z}$$

Eliminating λ, we have

$$\frac{-yz}{y + 2z} = \frac{-xz}{x + 3z}$$

$$-xyz - 3yz^2 = -xyz - 2xz^2 \qquad \text{We can assume } z \neq 0.$$

$$3yz^2 = 2xz^2$$

$$3y = 2x$$

$$x = \frac{3}{2}y$$

From the second and third equations,

$$\lambda = \frac{-xz}{x + 3z} \qquad \lambda = \frac{-xy}{2x + 3y}$$

Eliminating λ, we have

$$\frac{-xz}{x+3z} = \frac{-xy}{2x+3y}$$

$$-2x^2z - 3xyz = -x^2y - 3xyz$$

$$2x^2z = x^2y \qquad \text{We can assume } x \neq 0.$$

$$2z = y$$

$$z = \frac{1}{2}\,y$$

Substituting $x = \frac{3}{2}y$ and $z = \frac{1}{2}y$ in the fourth equation, we have

$$\left(\frac{3}{2}\,y\right)y + 2\left(\frac{3}{2}\,y\right)\left(\frac{1}{2}\,y\right) + 3y\left(\frac{1}{2}\,y\right) - 162 = 0$$

$$\frac{3}{2}\,y^2 + \frac{3}{2}\,y^2 + \frac{3}{2}\,y^2 = 162$$

$$y^2 = 36 \qquad \text{We can assume}$$

$$y = 6 \qquad\qquad y > 0.$$

$$x = \frac{3}{2}\,(6) = 9 \qquad \text{Using } x = \frac{3}{2}\,y$$

$$z = \frac{1}{2}\,(6) = 3 \qquad \text{Using } z = \frac{1}{2}\,y$$

Since $(9, 6, 3)$ is the only critical point with x, y, and z all positive, the dimensions of the box with maximum volume are 9 inches by 6 inches by 3 inches.

Problem 20   Find the dimensions of the box of the type described in Example 20 with the largest volume that can be constructed from 288 square inches of cardboard.

Suppose we had decided to solve Example 20 by the method used in Section 16-4. First we would have to solve the material constraint for one of the variables, say z:

$$z = \frac{162 - xy}{2x + 3y}$$

Then we would eliminate z in the volume function and maximize

$$V(x, y) = xy\,\frac{162 - xy}{2x + 3y}$$

Using the method of Lagrange multipliers allows us to avoid the formidable task of finding the partial derivatives of V.

Answers to
Matched Problems

19.   Max $f(x, y) = f(2, 2) = 17$ (see Figure 10)

20.   12 inches by 8 inches by 4 inches

Exercise 16-5

Use the method of Lagrange multipliers in the following problems:

A
1. Maximize $f(x, y) = 2xy$
   Subject to $x + y = 6$

2. Minimize $f(x, y) = 6xy$
   Subject to $y - x = 6$

3. Minimize $f(x, y) = x^2 + y^2$
   Subject to $3x + 4y = 25$

4. Maximize $f(x, y) = 25 - x^2 - y^2$
   Subject to $2x + y = 10$

B
5. Find the maximum and minimum of $f(x, y) = 2xy$ subject to $x^2 + y^2 = 18$.

6. Find the maximum and minimum of $f(x, y) = x^2 - y^2$ subject to $x^2 + y^2 = 25$.

7. Maximize the product of two numbers if their sum must be 10.

8. Minimize the product of two numbers if their difference must be 10.

C
9. Minimize $f(x, y, z) = x^2 + y^2 + z^2$
   Subject to $2x - y + 3z = -28$

10. Maximize $f(x, y, z) = xyz$
    Subject to $2x + y + 2z = 120$

11. Maximize and minimize $f(x, y, z) = x + y + z$
    Subject to $x^2 + y^2 + z^2 = 12$

12. Maximize and minimize $f(x, y, z) = 2x + 4y + 4z$
    Subject to $x^2 + y^2 + z^2 = 9$

▪ Applications

Business & Economics

13. *Budgeting for least cost.* A manufacturing company produces two models of a television set, x units of model A and y units of model B per week, at a cost in dollars of

$$C(x, y) = 6x^2 + 12y^2$$

If it is necessary (because of shipping considerations) that

$$x + y = 90$$

how many of each type of set should be manufactured per week to minimize cost? What is the minimum cost?

14. *Budgeting for maximum production.* A manufacturing firm has budgeted $60,000 per month for labor and materials. If x thousand dollars is spent on labor and y thousand dollars is spent on materials, and if the monthly output in units is given by

$$N(x, y) = 4xy - 8x$$

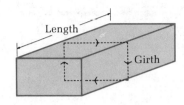

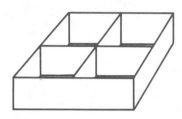

how should the $60,000 be allocated to labor and materials in order to maximize N? What is the maximum N?

15. *Maximum volume.* A mailing service states that a rectangular package shall have the sum of the length and girth not to exceed 120 inches (see the figure). What are the dimensions of the largest (in volume) mailing carton that can be constructed meeting these restrictions?

16. *Maximum volume.* A rectangular box with no top is to be constructed from 192 square inches of cardboard. Find the dimensions that will maximize the volume.

17. *Maximum volume.* A rectangular box with no top and two intersecting partitions is to be constructed from 192 square inches of cardboard (see accompanying figure). Find the dimensions that will maximize the volume.

18. *Scheduling production for least cost.* A company manufactures mattresses at three different plants. The cost functions for each plant are as follows:

Plant A: Cost of producing x mattresses is

$$1,000 + 50x - \frac{1}{20}x^2$$

Plant B: Cost of producing y mattresses is

$$1,200 + 60y - \frac{1}{10}y^2$$

Plant C: Cost of producing z mattresses is

$$800 + 40z - \frac{1}{30}z^2$$

The company must produce a total of 1,850 mattresses at the three plants. How many mattresses should it produce at each plant in order to minimize the total production cost? What is the minimum cost?

Life Sciences   19. *Agriculture.* Three pens of the same size are to be built along an existing fence (see the figure). If 400 feet of fencing are available, what length should x and y be to produce the maximum total area? What is the maximum area?

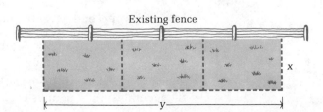

20. *Diet and minimum cost.* A group of guinea pigs is to receive 25,600 calories per week. Two available foods produce 200xy calories for a mixture of x kilograms of type M food and y kilograms of type N food. If type M costs $1 per kilogram and type N costs $2 per kilogram, how much of each type of food should be used to minimize weekly food costs? What is the minimum cost? [*Note:* $x \geq 0$, $y \geq 0$]

16-6   Method of Least Squares

- Least Squares Approximation
- Applications

■ Least Squares Approximation

In this section we will use the optimization techniques discussed in Section 16-4 to find the equation of a line which is a "best" approximation to a set of points in a rectangular coordinate system. This very popular method is known as **least squares approximation** or **linear regression.** Let us begin by considering a specific case.

A manufacturer wants to approximate the cost function for a product. The value of the cost function has been determined for certain levels of production, as listed in the table:

Number of Units x, in hundreds	**Cost** y, in thousands of dollars
2	4
5	6
6	7
9	8

Although these points do not all lie on a line (see Figure 11, page 978), they are very close to being linear. The manufacturer would like to approximate the cost function by a linear function; that is, determine values m and d so that the line

$$y = mx + d$$

is, in some sense, the "best" approximation to the cost function.

What do we mean by "best"? Since the line $y = mx + d$ will not go through all four points, it is reasonable to examine the differences between the y coordinates of the points listed in the table and the y coordinates of

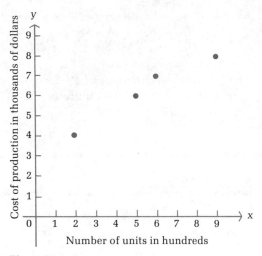

Figure 11

the corresponding points on the line. Each of these differences is called the **residual** at that point (see Figure 12). For example, at $x = 2$ the point from the table is $(2, 4)$ and the point on the line is $(2, 2m + d)$, so the residual is

$$4 - (2m + d) = 4 - 2m - d$$

All the residuals are listed in the table below:

x	y	$mx + d$	Residual
2	4	$2m + d$	$4 - 2m - d$
5	6	$5m + d$	$6 - 5m - d$
6	7	$6m + d$	$7 - 6m - d$
9	8	$9m + d$	$8 - 9m - d$

Our criterion for the "best" approximation is the following: Determine the values of m and d that *minimize* the sum of the squares of the residuals. The resulting line is called the **least squares line** or the **regression line.** To this end, we minimize

$$F(m, d) = (4 - 2m - d)^2 + (6 - 5m - d)^2 + (7 - 6m - d)^2$$
$$+ (8 - 9m - d)^2$$

Step 1.   *Find critical points.*

$$F_m(m, d) = 2(4 - 2m - d)(-2) + 2(6 - 5m - d)(-5)$$
$$+ 2(7 - 6m - d)(-6) + 2(8 - 9m - d)(-9)$$
$$= -304 + 292m + 44d = 0$$

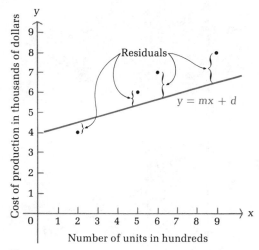

Figure 12

$$F_d(m, d) = 2(4 - 2m - d)(-1) + 2(6 - 5m - d)(-1)$$
$$+ 2(7 - 6m - d)(-1) + 2(8 - 9m - d)(-1)$$
$$= -50 + 44m + 8d = 0$$

Solving the system

$$-304 + 292m + 44d = 0$$
$$-50 + 44m + 8d = 0$$

we obtain $(m, d) = (0.58, 3.06)$ as the only critical point.

Step 2. *Compute* $A = F_{mm}(m, d)$, $B = F_{md}(m, d)$, *and* $C = F_{dd}(m, d)$.

$$F_{mm}(m, d) = 292, \quad \text{thus} \quad A = F_{mm}(0.58, 3.06) = 292$$
$$F_{md}(m, d) = 44, \quad \text{thus} \quad B = F_{md}(0.58, 3.06) = 44$$
$$F_{dd}(m, d) = 8, \quad \text{thus} \quad C = F_{dd}(0.58, 3.06) = 8$$

Step 3. *Evaluate* $AC - B^2$ *and try to classify the critical point* (m, d) *using Theorem 2 in Section 16-4.*

$$AC - B^2 = (292)(8) - (44)^2 = 400 > 0$$
$$A = 292 > 0$$

Therefore, case 2 in Theorem 2 applies, and $F(m, d)$ has a local minimum at the critical point $(0.58, 3.06)$.

Thus, the least squares line for the given data is

$y = 0.58x + 3.06$ Least squares line

Note that the sum of the squares of the residuals is minimized for this choice of m and d (see Figure 13, page 980).

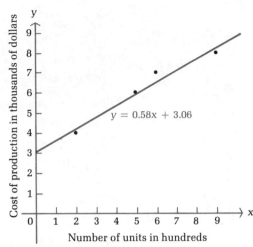

Figure 13

This linear function can now be used by the manufacturer to estimate any of the quantities normally associated with the cost function—such as costs, marginal costs, average costs, and so on. For example, the cost of producing 2,000 units is approximately

$$y = (0.58)(20) + 3.06 = 14.66 \quad \text{or} \quad \$14,660$$

The marginal cost function is

$$\frac{dy}{dx} = 0.58$$

The average cost function is

$$\bar{y} = \frac{0.58x + 3.06}{x}$$

In general, if we are given a set of n points (x_1, y_1), (x_2, y_2), . . . , (x_n, y_n), then it can be shown that the coefficients m and d of the least squares line $y = mx + d$ must satisfy the system of equations

$$\left(\sum_{k=1}^{n} x_k\right)m + nd = \sum_{k=1}^{n} y_k \tag{1}$$

$$\left(\sum_{k=1}^{n} x_k^2\right)m + \left(\sum_{k=1}^{n} x_k\right)d = \sum_{k=1}^{n} x_k y_k \tag{2}$$

Using the notation

$$\bar{x} = \frac{1}{n}\sum_{k=1}^{n} x_k \qquad \text{Average of the } x \text{ coordinates}$$

$$\bar{y} = \frac{1}{n}\sum_{k=1}^{n} y_k \qquad \text{Average of the } y \text{ coordinates}$$

to simplify the form of equations (1) and (2) and solving
produces the formulas given in the box.

Least Squares Approximation

For a set of n points $(x_1, y_1), (x_2, y_2), \ldots, (x_n, y_n)$, the coefficients m and d of the least squares line

$$y = mx + d$$

are given by the formulas

$$m = \frac{\displaystyle\sum_{k=1}^{n} x_k y_k - n\bar{x}\bar{y}}{\displaystyle\sum_{k=1}^{n} x_k^2 - n\bar{x}^2} \tag{3}$$

$$d = \bar{y} - \bar{x}m \tag{4}$$

where

$$\bar{x} = \frac{1}{n}\sum_{k=1}^{n} x_k \qquad \text{Average of the } x \text{ coordinates}$$

$$\bar{y} = \frac{1}{n}\sum_{k=1}^{n} y_k \qquad \text{Average of the } y \text{ coordinates}$$

Since the value of m is used in equation (4) to compute the value of d, the value of m must always be computed first. Notice that equation (4) implies that the point $(\bar{x}, \bar{y})$ is always on the least squares line.

■ Applications

Example 21
Educational Testing

The table lists the midterm and final examination scores for ten students in a calculus course.

Midterm	Final
49	61
53	47
67	72
71	76
74	68
78	77
83	81
85	79
91	93
99	99

(A) Find the least squares line for the data given in the table.

(B) Use the least squares line to predict the final examination score for a student who scored 95 on the midterm examination.

(C) Graph the data and the least squares line on the same set of axes.

Solution

(A) A table is a convenient way to compute all the sums in the formulas for m and d:

x_k	y_k	$x_k y_k$	x_k^2
49	61	2,989	2,401
53	47	2,491	2,809
67	72	4,824	4,489
71	76	5,396	5,041
74	68	5,032	5,476
78	77	6,006	6,084
83	81	6,723	6,889
85	79	6,715	7,225
91	93	8,463	8,281
99	99	9,801	9,801
Totals 750	753	58,440	58,496

Thus,

$$\bar{x} = \frac{1}{10} \sum_{k=1}^{10} x_k = \frac{1}{10}(750) = 75.0$$

$$\bar{y} = \frac{1}{10} \sum_{k=1}^{10} y_k = \frac{1}{10}(753) = 75.3$$

$$\sum_{k=1}^{10} x_k y_k = 58,440$$

$$\sum_{k=1}^{10} x_k^2 = 58,496$$

Substituting the appropriate values in equation (3),

$$m = \frac{\sum\limits_{k=1}^{n} x_k y_k - n\bar{x}\bar{y}}{\sum\limits_{k=1}^{n} x_k^2 - n\bar{x}^2}$$

$$= \frac{58,440 - 10(75.0)(75.3)}{58,496 - 10(75.0)^2} = \frac{1,965}{2,246} \approx 0.875$$

Then, using equation (4),

$$d = \bar{y} - \bar{x}m$$

$$\approx 75.3 - (75.0)(0.875) \approx 9.68$$

The least squares line is given (approximately) by

$$y = 0.875x + 9.68$$

(B)   If x = 95, then the predicted score on the final examination is

$$y = 0.875(95) + 9.68$$

$$\approx 93 \qquad \text{Assuming that the score}$$
$$\text{must be an integer}$$

(C)

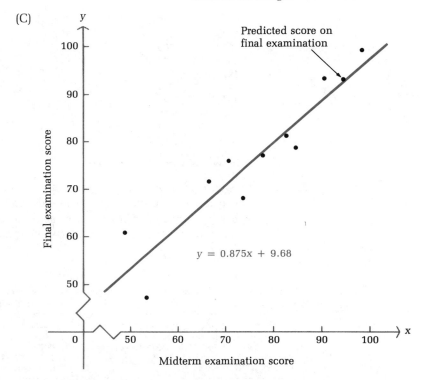

Problem 21   Repeat Example 21 for the following scores:

Midterm	Final
54	50
60	66
75	80
76	68
78	71
84	80
88	95
89	85
97	94
99	86

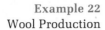

Example 22
Wool Production

Table 1 lists the annual production of wool throughout the world for the years 1970–1980. Use the data in the table to predict the worldwide wool production for 1981.

Table 1
World Wool Production

Year	Millions of Pounds
1970	6,107
1971	5,972
1972	5,560
1973	5,474
1974	5,769
1975	5,911
1976	5,827
1977	5,838
1978	5,983
1979	6,168
1980	6,285

Solution

Solving this problem by hand is certainly possible, but would require considerable effort. Instead, we used a computer to perform the necessary computations. (The program we used can be found in the computer supplement for this text. See the Preface.) The computer output is listed in Table 2.

Table 2

Input to Program	Output from Program
11 DATA POINTS HAVE BEEN ENTERED.	<------- LEAST SQUARES LINE ------->
DO YOU WANT TO SEE THE POINTS (Y/N)?Y	SLOPE: M = 33.9
	Y INTERCEPT: D = 5729.96
DATA POINTS	EQUATION: Y = 33.9 X + 5729.96
-------------------	-----------------------------------
0 6107	
1 5972	TO COMPUTE AN ESTIMATED VALUE OF Y,
2 5560	ENTER AN X VALUE. ENTER 999 TO STOP.
3 5474	?11
4 5769	
5 5911	X = 11 Y = 6102.85
6 5827	
7 5838	ENTER AN X VALUE. ENTER 999 TO STOP.
8 5983	?999
9 6168	
10 6285	

PRESS RETURN TO CONTINUE?	

Notice that we used $x = 0$ for 1970, $x = 1$ for 1971, and so on. Examining the computer output in Table 2, we see that the least squares line is

$$y = 33.9x + 5{,}729.96$$

and the estimated worldwide wool production in 1981 is 6,102.85 million pounds.

Problem 22   Use the least squares line in Example 22 to estimate the worldwide wool production in 1982.

Answers to   21.   (A) $y = 0.85x + 9.47$ (B)   90.2
Matched Problems   (C)

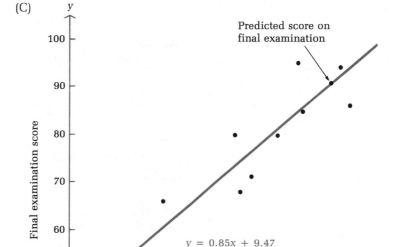

22.   6,136.76 million pounds

Exercise 16-6

A   *Find the least squares line. Graph the data and the least squares line.*

1.	x	y
	1	1
	2	3
	3	4
	4	3

2.	x	y
	1	−2
	2	−1
	3	3
	4	5

3.	x	y
	1	8
	2	5
	3	4
	4	0

4.

x	y
1	20
2	14
3	11
4	3

5.

x	y
1	3
2	4
3	5
4	6

6.

x	y
1	2
2	3
3	3
4	2

B   *Find the least squares line and use it to estimate y for the indicated value of x.*

7.

x	y
0	10
5	22
10	31
15	46
20	51

Estimate y when x = 25.

8.

x	y
−5	60
0	50
5	30
10	20
15	15

Estimate y when x = 20.

9.

x	y
−1	14
1	12
3	8
5	6
7	5

Estimate y when x = 2.

10.

x	y
2	−4
6	0
10	8
14	12
18	14

Estimate y for x = 15.

11.

x	y
0.5	25
2	22
3.5	21
5	21
6.5	18
9.5	12
11	11
12.5	8
14	5
15.5	1

Estimate y for x = 8.

12.

x	y
0	−15
2	−9
4	−7
6	−7
8	−1
12	11
14	13
16	19
18	25
20	33

Estimate y for x = 10.

C   **13.**   The method of least squares can be generalized to curves other than straight lines. To find the coefficients of the parabola

$$y = ax^2 + bx + c$$

that is the "best" fit for the points (1, 2), (2, 1), (3, 1), and (4, 3), minimize the sum of the squares of the residuals

$$F(a, b, c) = (a + b + c - 2)^2 + (4a + 2b + c - 1)^2$$
$$+ (9a + 3b + c - 1)^2 + (16a + 4b + c - 3)^2$$

by solving the system

$$F_a(a, b, c) = 0 \qquad F_b(a, b, c) = 0 \qquad F_c(a, b, c) = 0$$

for a, b, and c. Graph the points and the parabola.

14. Repeat Problem 13 for the points $(-1, -2)$, $(0, 1)$, $(1, 2)$, and $(2, 0)$.

■

Applications

Business and Economics

15. *Cost.* The cost y in thousands of dollars for producing x units of a product at various times in the past is given in the table.

x	y
10	5
12	6
15	7
18	8
20	9

(A) Find the least squares line for the data.

(B) Use the least squares line to estimate the cost of producing 25 units.

16. *Advertising and sales.* A company spends x thousand dollars on advertising each month and has y thousand dollars in monthly sales. The data in the table were obtained by examining the past history of the company.

x	y
4	100
5	120
6	150
7	190
8	240

(A) Find the least squares line for the data.

(B) Use the least squares line to estimate the sales if $10,000 is spent on advertising.

17. *Price–demand.* The price x in cents and the demand y in thousands of units for a certain item at various times in the past are given in the table at the top of the next page.

x	y
10	120
15	120
20	130
25	125
30	135

(A) Find the least squares line for the data.

(B) Use the least squares line to estimate the demand and the revenue if the price is 40¢.

18. *Profit.* A company's annual profits in millions of dollars from 1970 to 1980 are listed in the table.

Year	Profit
1970	1.2
1971	1.4
1972	1.6
1973	1.8
1974	2.1
1975	2.4
1976	2.9
1977	3.3
1978	3.4
1979	3.5
1980	3.6

(A) Find the least squares line for the data.

(B) Use the least squares line to estimate the profit in 1985.

Life Sciences   19. *Air pollution.* The amounts of air pollution in parts per million in a large city at certain times of day are listed in the table. [*Note:*   Count 12 noon as 0; then 8 AM is −4, 3 PM is 3, and so on.]

Time	Pollution
8 AM	20
10 AM	47
1 PM	82
3 PM	107
4 PM	114

(A) Find the least squares line for the data.

(B) Use the least squares line to estimate the pollution at noon.

20. *Spread of disease.* A virus that affects dogs and other small mammals is spreading through a community. The table lists the number of cases (in thousands) each year from 1978 to 1982.

Year	Cases
1978	10
1979	14
1980	17
1981	19
1982	20

(A) Find the least squares line for the data.

(B) Use the least squares line to estimate the number of cases expected in 1987.

Social Sciences

21. *Learning.* The table lists the number of weeks of instruction in typing and the average number of words per minute typed for a group of students.

Weeks of Practice	Words per Minute
1	20
2	28
3	50
4	45
5	62

(A) Find the least squares line for the data.

(B) Estimate the number of weeks of practice required to be able to type 100 words per minute.

22. *Education.* The table lists the high school grade-point averages of ten students and their college grade-point averages after one semester of college.

High School GPA	College GPA
2.0	1.5
2.2	1.5
2.4	1.6
2.7	1.8
2.9	2.1
3.0	2.3
3.1	2.5
3.3	2.9
3.4	3.2
3.7	3.5

(A) Find the least squares line for the data.

(B) Estimate the college GPA for a student with a high school GPA of 3.5.

(C) Estimate the high school GPA necessary for a college GPA of 2.7.

16-7 Double Integrals over Rectangular Regions

- Introduction
- Definition of the Double Integral
- Average Value over Rectangular Regions
- Volume and Double Integrals

■ Introduction

We have generalized the concept of differentiation to functions with two or more independent variables. How can we do the same with integration and how can we interpret the results? Let us first look at the operation of antidifferentiation. We can antidifferentiate a function of two or more variables with respect to one of the variables by treating all the other variables as though they were constants. Thus, this operation is the reverse operation of partial differentiation, just as ordinary antidifferentiation is the reverse operation of ordinary differentiation. We write $\int f(x, y)\, dx$ to indicate that we are to antidifferentiate $f(x, y)$ with respect to x, holding y fixed; we write $\int f(x, y)\, dy$ to indicate that we are to antidifferentiate $f(x, y)$ with respect to y, holding x fixed.

Example 23 Evaluate:

(A) $\displaystyle\int (6xy^2 + 3x^2)\, dy$ (B) $\displaystyle\int (6xy^2 + 3x^2)\, dx$

Solution   (A) Treating x as a constant and using the properties of antidifferentiation from Section 14-1, we have

$$\int (6xy^2 + 3x^2)\, dy = \int 6xy^2\, dy + \int 3x^2\, dy$$

$$= 6x \int y^2\, dy + 3x^2 \int dy$$

$$= 6x \left(\frac{y^3}{3}\right) + 3x^2(y) + C(x)$$

$$= 2xy^3 + 3x^2y + C(x)$$

The dy tells us we are looking for the antiderivative of $(6xy^2 + 3x^2)$ with respect to y only, holding x constant.

Notice that the constant of integration can actually be *any function of x alone*, since, for any such function, $\partial/\partial y\, [C(x)] = 0$. We can verify that our answer is correct by using partial differentiation:

$$\frac{\partial}{\partial y}\, [2xy^3 + 3x^2y + C(x)] = 6xy^2 + 3x^2 + 0$$

$$= 6xy^2 + 3x^2$$

(B)   Now we treat y as a constant:

$$\int (6xy^2 + 3x^2)\, dx = \int 6xy^2\, dx + \int 3x^2\, dx$$

$$= 6y^2 \int x\, dx + 3 \int x^2\, dx$$

$$= 6y^2 \left(\frac{x^2}{2}\right) + 3 \left(\frac{x^3}{3}\right) + E(y)$$

$$= 3x^2y^2 + x^3 + E(y)$$

This time the antiderivative contains an arbitrary function $E(y)$ of y alone.

Check $$\frac{\partial}{\partial x}\,[3x^2y^2 + x^3 + E(y)] = 6xy^2 + 3x^2 + 0$$

$$= 6xy^2 + 3x^2$$

Problem 23   Evaluate:

(A) $\displaystyle\int (4xy + 12x^2y^3)\, dy$ (B) $\displaystyle\int (4xy + 12x^2y^3)\, dx$

Now that we have extended the concept of antidifferentiation to functions with two variables, we can also evaluate definite integrals of the form

$$\int_a^b f(x,\, y)\, dx \quad\quad \text{or} \quad\quad \int_c^d f(x,\, y)\, dy$$

Example 24   Evaluate, substituting the limits of integration in y if dy is used and in x if dx is used:

(A) $\displaystyle\int_0^2 (6xy^2 + 3x^2)\, dy$ (B) $\displaystyle\int_0^1 (6xy^2 + 3x^2)\, dx$

Solution   (A)   From Example 23A, we know that $\int (6xy^2 + 3x^2)\, dy = 2xy^3 + 3x^2y + C(x)$. According to the definition of the definite integral for a function of one variable, we can use any antiderivative to evaluate the definite integral. Thus, choosing $C(x) = 0$, we have

$$\int_0^2 (6xy^2 + 3x^2)\, dy = (2xy^3 + 3x^2y)\Big|_{y=0}^{y=2}$$

$$= [2x(2)^3 + 3x^2(2)] - [2x(0)^3 + 3x^2(0)]$$

$$= 16x + 6x^2$$

(B)   From Example 23B, we know that $\int (6xy^2 + 3x^2)\, dx = 3x^2y^2 + x^3 + E(y)$. Thus, choosing $E(y) = 0$, we have

$$\int_0^1 (6xy^2 + 3x^2)\, dx = (3x^2y^2 + x^3)\Big|_{x=0}^{x=1}$$

$$= [3y^2(1)^2 + (1)^3] - [3y^2(0)^2 + (0)^3]$$

$$= 3y^2 + 1$$

Problem 24   Evaluate:

(A) $\displaystyle\int_0^1 (4xy + 12x^2y^3)\, dy$ (B) $\displaystyle\int_0^3 (4xy + 12x^2y^3)\, dx$

Notice that integrating and evaluating a definite integral, with integrand $f(x, y)$, with respect to y produces a function of x alone (or a constant). Likewise, integrating and evaluating a definite integral, with integrand $f(x, y)$, with respect to x produces a function of y alone (or a constant). Each of these results, involving at most one variable, can now be used as an integrand in a second definite integral.

Example 25   Evaluate:

(A) $\displaystyle\int_0^1 \left[\int_0^2 (6xy^2 + 3x^2)\, dy\right] dx$ (B) $\displaystyle\int_0^2 \left[\int_0^1 (6xy^2 + 3x^2)\, dx\right] dy$

Solution   (A)   Example 24A showed that

$$\int_0^2 (6xy^2 + 3x^2)\, dy = 16x + 6x^2$$

Thus,

$$\int_0^1 \left[\int_0^2 (6xy^2 + 3x^2)\, dy\right] dx = \int_0^1 (16x + 6x^2)\, dx$$

$$= (8x^2 + 2x^3)\Big|_{x=0}^{x=1}$$

$$= [8(1)^2 + 2(1)^3] - [8(0)^2 + 2(0)^3]$$

$$= 10$$

(B)   Example 24B showed that

$$\int_0^1 (6xy^2 + 3x^2)\, dx = 3y^2 + 1$$

Thus,

$$\int_0^2 \left[\int_0^1 (6xy^2 + 3x^2) \, dx \right] dy = \int_0^2 (3y^2 + 1) \, dy$$

$$= (y^3 + y) \Big|_{y=0}^{y=2}$$

$$= [(2)^3 + 2] - [(0)^3 + 0]$$

$$= 10$$

Problem 25 Evaluate:

(A) $\displaystyle\int_0^3 \left[\int_0^1 (4xy + 12x^2y^3) \, dy \right] dx$

(B) $\displaystyle\int_0^1 \left[\int_0^3 (4xy + 12x^2y^3) \, dx \right] dy$

■ Definition of the Double Integral

Notice that the answers in Examples 25A and 25B are identical. This is not an accident. In fact, it is this property that enables us to define the **double integral.**

Double Integral

The double integral of a function $f(x, y)$ over a rectangle $R = \{(x, y) | a \leqslant x \leqslant b, \ c \leqslant y \leqslant d\}$ is

$$\iint_R f(x, y) \, dA$$

$$= \int_a^b \left[\int_c^d f(x, y) \, dy \right] dx$$

$$= \int_c^d \left[\int_a^b f(x, y) \, dx \right] dy$$

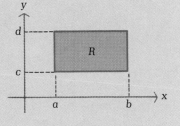

In the double integral $\iint_R f(x, y) \, dA$, $f(x, y)$ is called the **integrand** and R is called the **region of integration.** The expression dA indicates that this is an integral over a two-dimensional region. The integrals

$$\int_a^b \left[\int_c^d f(x, y) \, dy \right] dx \qquad \text{and} \qquad \int_c^d \left[\int_a^b f(x, y) \, dx \right] dy$$

are referred to as **iterated integrals** (the brackets are often omitted), and the order in which dx and dy are written indicates the order of integration. This is not the most general definition of the double integral over a rectangular region; however, it is equivalent to the general definition for all the functions we will consider.

Example 26   Evaluate $\iint_R (x + y)\, dA$ over $R = \{(x, y)|1 \leqslant x \leqslant 3, \ -1 \leqslant y \leqslant 2\}$.

Solution   We can choose either order of iteration. As a check, we will evaluate the integral both ways:

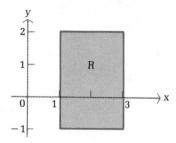

$$\iint_R (x + y)\, dA = \int_1^3 \int_{-1}^2 (x + y)\, dy\, dx$$

$$= \int_1^3 \left[\left(xy + \frac{y^2}{2} \right) \Big|_{y=-1}^{y=2} \right] dx$$

$$= \int_1^3 \left[(2x + 2) - \left(-x + \frac{1}{2} \right) \right] dx$$

$$= \int_1^3 \left(3x + \frac{3}{2} \right) dx$$

$$= \left(\frac{3}{2} x^2 + \frac{3}{2} x \right) \Big|_{x=1}^{x=3}$$

$$= \left(\frac{27}{2} + \frac{9}{2} \right) - \left(\frac{3}{2} + \frac{3}{2} \right)$$

$$= (18) - (3) = 15$$

$$\iint_R (x + y)\, dA = \int_{-1}^2 \int_1^3 (x + y)\, dx\, dy$$

$$= \int_{-1}^2 \left[\left(\frac{x^2}{2} + xy \right) \Big|_{x=1}^{x=3} \right] dy$$

$$= \int_{-1}^2 \left[\left(\frac{9}{2} + 3y \right) - \left(\frac{1}{2} + y \right) \right] dy$$

$$= \int_{-1}^2 (4 + 2y)\, dy$$

$$= (4y + y^2) \Big|_{y=-1}^{y=2}$$

$$= (8 + 4) - (-4 + 1)$$

$$= (12) - (-3) = 15$$

Problem 26   Evaluate both ways:

$$\iint\limits_{R} (2x - y)\, dA \quad \text{over } R = \{(x, y)|-1 \leqslant x \leqslant 5, \quad 2 \leqslant y \leqslant 4\}$$

Example 27   Evaluate:

$$\iint\limits_{R} 2xe^{x^2+y}\, dA \quad \text{over } R = \{(x, y)|0 \leqslant x \leqslant 1, \quad -1 \leqslant y \leqslant 1\}$$

Solution

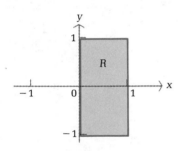

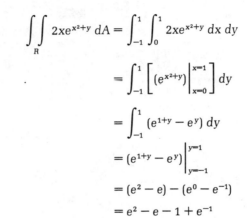

$$\iint\limits_{R} 2xe^{x^2+y}\, dA = \int_{-1}^{1} \int_{0}^{1} 2xe^{x^2+y}\, dx\, dy$$

$$= \int_{-1}^{1} \left[(e^{x^2+y}) \Big|_{x=0}^{x=1} \right] dy$$

$$= \int_{-1}^{1} (e^{1+y} - e^{y})\, dy$$

$$= (e^{1+y} - e^{y}) \Big|_{y=-1}^{y=1}$$

$$= (e^2 - e) - (e^0 - e^{-1})$$

$$= e^2 - e - 1 + e^{-1}$$

Problem 27   Evaluate:

$$\iint\limits_{R} \frac{x}{y^2} e^{x/y}\, dA \quad \text{over } R = \{(x, y)|0 \leqslant x \leqslant 1, \quad 1 \leqslant y \leqslant 2\}$$

■ Average Value over Rectangular Regions

In Section 14-6 the average value of a function $f(x)$ over an interval $[a, b]$ was defined as

$$\frac{1}{b - a} \int_{a}^{b} f(x)\, dx$$

This definition is easily extended to functions of two variables over rectangular regions, as shown in the box on the next page. Notice that the denominator in the expression given in the box, $(b - a)(d - c)$, is simply the area of the rectangle R.

Average Value over Rectangular Regions

The **average value** of the function $f(x, y)$ over the rectangle $R = \{(x, y) | a \leqslant x \leqslant b, \ c \leqslant y \leqslant d\}$ is

$$\frac{1}{(b - a)(d - c)} \iint_R f(x, y) \, dA$$

Example 28 Find the average value of $f(x, y) = 4 - \frac{1}{2}x - \frac{1}{2}y$ over the rectangle $R = \{(x, y) | 0 \leqslant x \leqslant 2, \ 0 \leqslant y \leqslant 2\}$.

Solution

$$\frac{1}{(b - a)(d - c)} \iint_R f(x, y) \, dA = \frac{1}{(2 - 0)(2 - 0)} \iint_R \left(4 - \frac{1}{2}x - \frac{1}{2}y \right) dA$$

$$= \frac{1}{4} \int_0^2 \int_0^2 \left(4 - \frac{1}{2}x - \frac{1}{2}y \right) dy \, dx$$

$$= \frac{1}{4} \int_0^2 \left[\left(4y - \frac{1}{2}xy - \frac{1}{4}y^2 \right) \Big|_{y=0}^{y=2} \right] dx$$

$$= \frac{1}{4} \int_0^2 (7 - x) \, dx$$

$$= \frac{1}{4} \left(7x - \frac{1}{2}x^2 \right) \Big|_{x=0}^{x=2}$$

$$= \frac{1}{4} (12) = 3$$

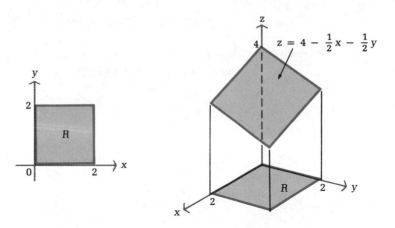

Problem 28 Find the average value of $f(x, y) = x + 2y$ over the rectangle $R = \{(x, y)|0 \le x \le 2, \quad 0 \le y \le 1\}$.

■ Volume and Double Integrals

One application of the definite integral of a function with one variable is the calculation of areas, so it is not surprising that the definite integral of a function of two variables can be used to calculate volumes of solids.

Volume under a Surface

If $f(x, y) \ge 0$ over a rectangle R, $R = \{(x, y)|a \le x \le b, \quad c \le y \le d\}$, then the volume of the solid formed by graphing f over the rectangle R is given by

$$V = \iint\limits_{R} f(x, y) \, dA$$

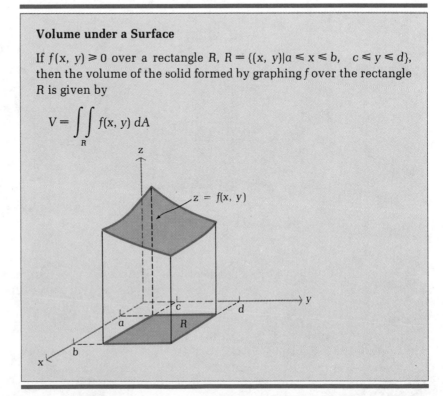

A proof of the statement in the box is left to a more advanced text.

Example 29   Find the volume of the solid under the graph of $f(x, y) = 1 + x^2 + y^2$ over the rectangle $R = \{(x, y)|0 \leqslant x \leqslant 1, \quad 0 \leqslant y \leqslant 1\}$.

Solution

$$V = \int\!\!\int_R (1 + x^2 + y^2)\, dA$$

$$= \int_0^1 \int_0^1 (1 + x^2 + y^2)\, dx\, dy$$

$$= \int_0^1 \left[\left(x + \frac{1}{3} x^3 + xy^2\right)\Big|_{x=0}^{x=1}\right] dy$$

$$= \int_0^1 \left(\frac{4}{3} + y^2\right) dy$$

$$= \left(\frac{4}{3} y + \frac{1}{3} y^3\right)\Big|_{y=0}^{y=1} = \frac{5}{3} \text{ cubic units}$$

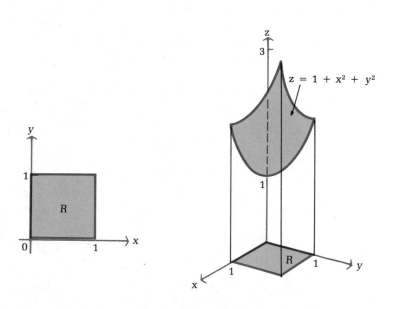

Problem 29   Find the volume of the solid under the graph of $f(x, y) = 1 + x + y$ over the rectangle $R = \{(x, y)|0 \leqslant x \leqslant 1, \quad 0 \leqslant y \leqslant 2\}$.

Answers to
Matched Problems

23. (A) $2xy^2 + 3x^2y^4 + C(x)$ (B) $2x^2y + 4x^3y^3 + E(y)$
24. (A) $2x + 3x^2$ (B) $18y + 108y^3$ 25. (A) 36   (B) 36
26. 12   27. $e - 2e^{1/2} + 1$ 28. 2   29. 5 cubic units

Exercise 16-7

A   *Find each antiderivative. Then use the antiderivative to evaluate the definite integral.*

1. (A) $\displaystyle\int 12x^2y^3\,dy$ (B) $\displaystyle\int_0^1 12x^2y^3\,dy$

2. (A) $\displaystyle\int 12x^2y^3\,dx$ (B) $\displaystyle\int_{-1}^2 12x^2y^3\,dx$

3. (A) $\displaystyle\int (4x+6y+5)\,dx$ (B) $\displaystyle\int_{-2}^3 (4x+6y+5)\,dx$

4. (A) $\displaystyle\int (4x+6y+5)\,dy$ (B) $\displaystyle\int_1^4 (4x+6y+5)\,dy$

5. (A) $\displaystyle\int \frac{x}{\sqrt{y+x^2}}\,dx$ (B) $\displaystyle\int_0^2 \frac{x}{\sqrt{y+x^2}}\,dx$

6. (A) $\displaystyle\int \frac{x}{\sqrt{y+x^2}}\,dy$ (B) $\displaystyle\int_1^5 \frac{x}{\sqrt{y+x^2}}\,dy$

B   *Evaluate each iterated integral. (See the indicated problem for the evaluation of the inner integral.)*

7. $\displaystyle\int_{-1}^2 \int_0^1 12x^2y^3\,dy\,dx$

    (see Problem 1)

8. $\displaystyle\int_0^1 \int_{-1}^2 12x^2y^3\,dx\,dy$

    (see Problem 2)

9. $\displaystyle\int_1^4 \int_{-2}^3 (4x+6y+5)\,dx\,dy$

    (see Problem 3)

10. $\displaystyle\int_{-2}^3 \int_1^4 (4x+6y+5)\,dy\,dx$

    (see Problem 4)

11. $\displaystyle\int_1^5 \int_0^2 \frac{x}{\sqrt{y+x^2}}\,dx\,dy$

    (see Problem 5)

12. $\displaystyle\int_0^2 \int_1^5 \frac{x}{\sqrt{y+x^2}}\,dy\,dx$

    (see Problem 6)

Use both orders of iteration to evaluate each double integral.

13. $\displaystyle\iint_R xy\,dA;\quad R=\{(x,y)|0\leqslant x\leqslant 2,\ \ 0\leqslant y\leqslant 4\}$

14. $\displaystyle\iint_R \sqrt{xy}\,dA;\quad R=\{(x,y)|1\leqslant x\leqslant 4,\ \ 1\leqslant y\leqslant 9\}$

15. $\displaystyle\iint_R (x+y)^5\,dA;\quad R=\{(x,y)|-1\leqslant x\leqslant 1,\ \ 1\leqslant y\leqslant 2\}$

16. $\displaystyle\iint\limits_{R} xe^y \, dA; \quad R = \{(x, y)|-2 \leqslant x \leqslant 3, \quad 0 \leqslant y \leqslant 2\}$

Find the average value of each function over the given rectangle.

17. $f(x, y) = (x + y)^2; \quad R = \{(x, y)|1 \leqslant x \leqslant 5, \quad -1 \leqslant y \leqslant 1\}$

18. $f(x, y) = x^2 + y^2; \quad R = \{(x, y)|-1 \leqslant x \leqslant 2, \quad 1 \leqslant y \leqslant 4\}$

19. $f(x, y) = \dfrac{x}{y}; \quad R = \{(x, y)|1 \leqslant x \leqslant 4, \quad 2 \leqslant y \leqslant 7\}$

20. $f(x, y) = x^2 y^3; \quad R = \{(x, y)|-1 \leqslant x \leqslant 1, \quad 0 \leqslant y \leqslant 2\}$

Find the volume of the solid under the graph of each function over the given rectangle.

21. $f(x, y) = 2 - x^2 - y^2; \quad R = \{(x, y)|0 \leqslant x \leqslant 1, \quad 0 \leqslant y \leqslant 1\}$

22. $f(x, y) = 5 - x; \quad R = \{(x, y)|0 \leqslant x \leqslant 5, \quad 0 \leqslant y \leqslant 5\}$

23. $f(x, y) = 4 - y^2; \quad R = \{(x, y)|0 \leqslant x \leqslant 2, \quad 0 \leqslant y \leqslant 2\}$

24. $f(x, y) = e^{-x-y}; \quad R = \{(x, y)|0 \leqslant x \leqslant 1, \quad 0 \leqslant y \leqslant 1\}$

C   *Evaluate each double integral. Select the order of integration carefully—each problem is easy to do one way and difficult the other.*

25. $\displaystyle\iint\limits_{R} xe^{xy} \, dA; \quad R = \{(x, y)|0 \leqslant x \leqslant 1, \quad 1 \leqslant y \leqslant 2\}$

26. $\displaystyle\iint\limits_{R} xye^{x^2 y} \, dA; \quad R = \{(x, y)|0 \leqslant x \leqslant 1, \quad 1 \leqslant y \leqslant 2\}$

27. $\displaystyle\iint\limits_{R} \dfrac{2y + 3xy^2}{1 + x^2} \, dA; \quad R = \{(x, y)|0 \leqslant x \leqslant 1, \quad -1 \leqslant y \leqslant 1\}$

28. $\displaystyle\iint\limits_{R} \dfrac{2x + 2y}{1 + 4y + y^2} \, dA; \quad R = \{(x, y)|1 \leqslant x \leqslant 3, \quad 0 \leqslant y \leqslant 1\}$

Applications

Business & Economics

29. *Economics—multiplier principle.* Suppose Congress enacts a one-time-only 10% tax rebate that is expected to infuse y billion dollars, $5 \leqslant y \leqslant 7$, into the economy. If every individual and corporation is expected to spend a proportion x, $0.6 \leqslant x \leqslant 0.8$, of each dollar received, then by the **multiplier principle** in economics (using the sum of an infinite geometric progression), the total amount of spending S (in billions of dollars) generated by this tax rebate is given by

$$S(x, y) = \frac{y}{1 - x}$$

What is the average total amount of spending for the indicated ranges of the values of x and y? Set up a double integral and evaluate.

30. *Economics — multiplier principle.* Repeat Problem 29 if $6 \leqslant y \leqslant 10$ and $0.7 \leqslant x \leqslant 0.9$.

31. *Economics — Cobb–Douglas production function.* If an industry invests x thousand labor-hours, $10 \leqslant x \leqslant 20$, and y million dollars, $1 \leqslant y \leqslant 2$, in the production of N thousand units of a certain item, then N is given by

$$N(x, y) = x^{0.75}y^{0.25}$$

Functions of this form are called **Cobb–Douglas production functions** and are used extensively in economics. What is the average number of units produced for the indicated ranges of x and y? Set up a double integral and evaluate.

32. *Economics — Cobb–Douglas production function.* Repeat Problem 31 for

$$N(x, y) = x^{0.5}y^{0.5}$$

where $10 \leqslant x \leqslant 30$ and $1 \leqslant y \leqslant 3$.

Life Sciences

33. *Population distribution.* In order to study the population distribution of a certain species of insects, a biologist has constructed an artificial habitat in the shape of a rectangle 16 feet long and 12 feet wide. The only food available to the insects in this habitat is located at its center. The biologist has determined that the concentration C of insects per square foot at a point d units from the food supply (see the figure) is given approximately by

$$C = 10 - \tfrac{1}{10}d^2$$

What is the average concentration of insects throughout the habitat? Express C as a function of x and y, set up a double integral, and evaluate.

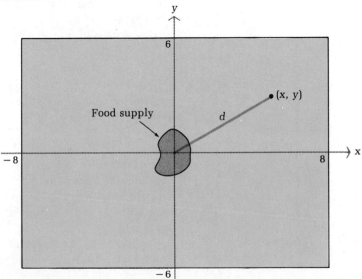

34. *Population distribution.* Repeat Problem 33 for a square habitat that measures 12 feet on each side, where the insect concentration is given by

$$C = 8 - \tfrac{1}{10}d^2$$

35. *Pollution.* A heavy industrial plant located in the center of a small town emits particulate matter into the atmosphere. Suppose the concentration of particulate matter in parts per million at a point d miles from the plant is given by

$$C = 100 - 15d^2$$

If the boundaries of the town form a rectangle 4 miles long and 2 miles wide, what is the average concentration of particulate matter throughout the city? Express C as a function of x and y, set up a double integral, and evaluate.

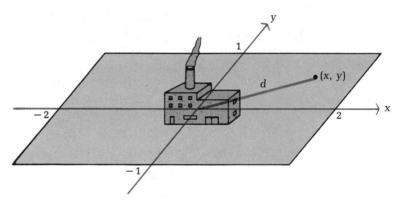

36. *Pollution.* Repeat Problem 35 if the boundaries of the town form a rectangle 8 miles long and 4 miles wide and the concentration of particulate matter is given by

$$C = 100 - 3d^2$$

Social Sciences

37. *Safety research.* Under ideal conditions, if a person driving a car slams on the brakes and skids to a stop, the length of the skid marks (in feet) is given by the formula

$$L = 0.000\ 013\ 3xy^2$$

where x is the weight of the car in pounds and y is the speed of the car in miles per hour. What is the average length of the skid marks for cars weighing between 2,000 and 3,000 pounds and traveling at speeds between 50 and 60 miles per hour? Set up a double integral and evaluate.

38. *Safety research.* Repeat Problem 37 for cars weighing between 2,000 and 2,500 pounds and traveling at speeds between 40 and 50 miles per hour.

39. *Psychology.* The intelligence quotient Q for an individual with mental age x and chronological age y is given by

$$Q(x, y) = 100\,\frac{x}{y}$$

In a group of sixth graders, the mental age varies between 8 and 16 years and the chronological age varies between 10 and 12 years. What is the average intelligence quotient for this group? Set up a double integral and evaluate.

40. *Psychology.* Repeat Problem 39 for a group with mental ages between 6 and 14 years and chronological ages between 8 and 10 years.

16-8   Double Integrals over More General Regions

- Regular Regions
- Double Integrals over Regular Regions
- Reversing the Order of Integration
- Volume and Double Integrals

In this section we will extend the concept of double integration to nonrectangular regions. We begin with an example and some new terminology.

■ Regular Regions

Let R be the region graphed in Figure 14. We can describe R with the following inequalities:

$$R = \{(x, y)\,|\,x \leqslant y \leqslant 6x - x^2, \quad 0 \leqslant x \leqslant 5\}$$

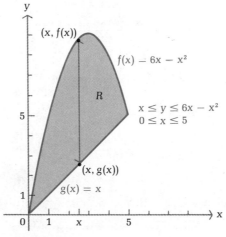

Figure 14

The region R can be viewed as a union of vertical line segments. For each x in the interval $[0, 5]$, the line segment from the point $(x, g(x))$ to the point $(x, f(x))$ lies in the region R. Any region that can be covered by vertical line segments in this manner is called a *regular x region*.

Now consider the region S in Figure 15. This is *not* a regular x region, but it can be described with inequalities:

$$S = \{(x, y) | y^2 \leq x \leq y + 2, \quad -1 \leq y \leq 2\}$$

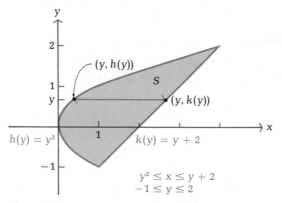

Figure 15

The region S can be viewed as a union of horizontal line segments going from the graph of $h(y) = y^2$ to the graph of $k(y) = y + 2$ on the interval $[-1, 2]$. Regions that can be described in this manner are called *regular y regions*. In general, *regular regions* are defined as follows:

Regular Regions

A region R in the xy plane is a **regular x region** if there exist functions $f(x)$ and $g(x)$ and numbers a and b so that

$$R = \{(x, y) | g(x) \leq y \leq f(x), \quad a \leq x \leq b\}$$

A region R is a **regular y region** if there exist functions $h(y)$ and $k(y)$ and numbers c and d so that

$$R = \{(x, y) | h(y) \leq x \leq k(y), \quad c \leq y \leq d\}$$

See Figure 16 for a geometric interpretation.

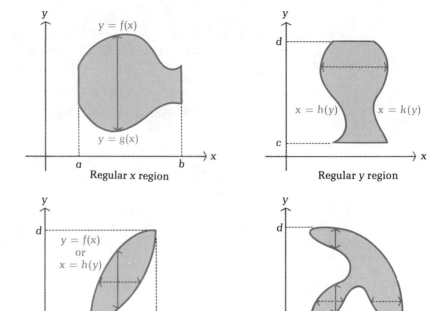

Figure 16

Example 30   The region R is bounded by the graphs of $y = 4 - x^2$ and $y = x - 2$, $x \geqslant 0$, and the y axis. Graph R and describe R as a regular x region, a regular y region, both, or neither. If possible, represent R in terms of set notation and double inequalities.

Solution

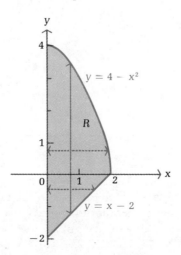

As the solid line in the figure indicates, R can be covered by vertical line segments which go from the graph of $y = x - 2$ to the graph of $y = 4 - x^2$. Thus, R is a regular x region. In terms of set notation and double inequalities, we can write

$$R = \{(x, y)|x - 2 \leqslant y \leqslant 4 - x^2, \quad 0 \leqslant x \leqslant 2\}$$

On the other hand, a horizontal line passing through a point in the interval $[-2, 0]$ on the y axis will intersect R in a line segment which goes from the y axis to the graph of $y = x - 2$, while one that passes through a point in the interval $[0, 4]$ on the y axis goes from the y axis to the graph of $y = 4 - x^2$. Two such segments are shown as dashed lines in the figure. Thus, the region is not a regular y region.

Problem 30   Repeat Example 30 for the region R bounded by the graphs of $x = 6 - y$, $x = y^2$, $y \geqslant 0$, and the x axis, as shown in the figure.

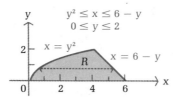

Example 31   The region R is bounded by the graphs of $x + y^2 = 9$ and $x + 3y = 9$. Graph R and describe R as a regular x region, a regular y region, both, or neither. If possible, represent R using set notation and double inequalities.

Solution

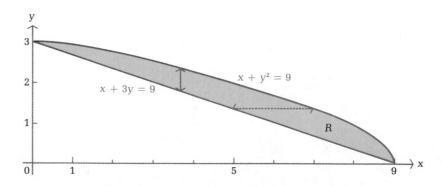

Test for x region.   Region R can be covered by vertical line segments which go from the graph of $x + 3y = 9$ to the graph of $x + y^2 = 9$. Thus, R is a regular x region. In order to describe R with inequalities, we must solve each equation for y in terms of x:

$$x + 3y = 9$$
$$3y = 9 - x$$
$$y = 3 - \tfrac{1}{3}x$$

$$x + y^2 = 9$$
$$y^2 = 9 - x$$
$$y = \sqrt{9 - x} \qquad$$ We use the positive square root, since the graph is in the first quadrant.

Thus,

$$R = \{(x, y) | 3 - \tfrac{1}{3}x \leqslant y \leqslant \sqrt{9 - x}, \quad 0 \leqslant x \leqslant 9\}$$

Test for y region. Since region R can also be covered by horizontal line segments (dashed line in the figure) which go from the graph of $x + 3y = 9$ to the graph of $x + y^2 = 9$, it is a regular y region. Now we must solve each equation for x in terms of y:

$$x + 3y = 9 \qquad\qquad x + y^2 = 9$$
$$x = 9 - 3y \qquad\qquad x = 9 - y^2$$

Thus,

$$R = \{(x, y) | 9 - 3y \leqslant x \leqslant 9 - y^2, \quad 0 \leqslant y \leqslant 3\}$$

Problem 31 Repeat Example 31 for the region bounded by the graphs of $2y - x = 4$ and $y^2 - x = 4$, as shown in the figure.

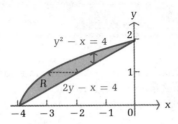

■ Double Integrals over Regular Regions

Now we want to extend the definition of double integration to include regular x regions and regular y regions.

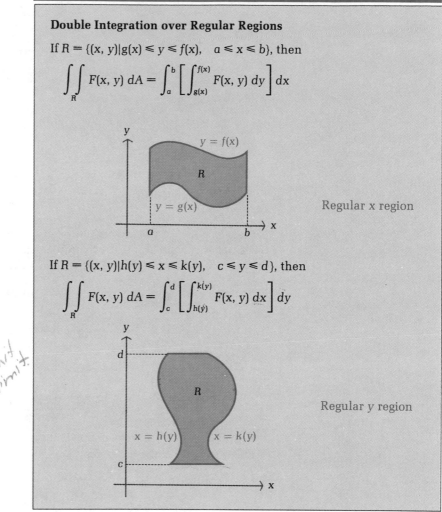

Double Integration over Regular Regions

If $R = \{(x, y) | g(x) \leq y \leq f(x), \quad a \leq x \leq b\}$, then

$$\iint\limits_{R} F(x, y)\, dA = \int_{a}^{b} \left[\int_{g(x)}^{f(x)} F(x, y)\, dy \right] dx$$

Regular x region

If $R = \{(x, y) | h(y) \leq x \leq k(y), \quad c \leq y \leq d\}$, then

$$\iint\limits_{R} F(x, y)\, dA = \int_{c}^{d} \left[\int_{h(y)}^{k(y)} F(x, y)\, dx \right] dy$$

Regular y region

(handwritten margin note:) if x changing — left is lower limit, — right upper limit.

Notice that the order of integration now depends on the nature of the region R. If R is a regular x region, we integrate with respect to y first, while if R is a regular y region, we integrate with respect to x first.

It is also important to note that the variable limits of integration (when present) are always on the inner integral, and the constant limits of integration are always on the outer integral.

Example 32   Evaluate $\iint_{R} 2xy\, dA$, where R is the region bounded by the graphs of $y = -x$ and $y = x^2$, $x \geq 0$, and the graph of $x = 1$.

Solution From the graph we can see that R is a regular x region described by

$$R = \{(x, y)| -x \leqslant y \leqslant x^2, \quad 0 \leqslant x \leqslant 1\}$$

Thus,

$$\iint\limits_R 2xy \, dA = \int_0^1 \left(\int_{-x}^{x^2} 2xy \, dy \right) dx$$

$$= \int_0^1 \left(xy^2 \, \Big|_{y=-x}^{y=x^2} \right) dx$$

$$= \int_0^1 [x(x^2)^2 - x(-x)^2] \, dx$$

$$= \int_0^1 (x^5 - x^3) \, dx$$

$$= \left(\frac{x^6}{6} - \frac{x^4}{4} \right) \Big|_{x=0}^{x=1}$$

$$= (\tfrac{1}{6} - \tfrac{1}{4}) - (0 - 0) = -\tfrac{1}{12}$$

Problem 32 Evaluate $\iint_R 3xy^2 \, dA$, where R is the region in Example 32.

Example 33 Evaluate $\iint_R (2x + y) \, dA$, where R is the region bounded by the graphs of $y = \sqrt{x}$, $x + y = 2$, and $y = 0$.

Solution From the graph we can see that R is a regular y region. After solving each equation for x, we can write

$$R = \{(x, y)| y^2 \leqslant x \leqslant 2 - y, \quad 0 \leqslant y \leqslant 1\}$$

Thus,

$$\iint\limits_R (2x + y) \, dA = \int_0^1 \left[\int_{y^2}^{2-y} (2x + y) \, dx \right] dy$$

$$= \int_0^1 \left[(x^2 + yx) \, \Big|_{x=y^2}^{x=2-y} \right] dy$$

$$= \int_0^1 \{[(2 - y)^2 + y(2 - y)] - [(y^2)^2 + y(y^2)]\} \, dy$$

$$= \int_0^1 (4 - 2y - y^3 - y^4) \, dy$$

$$= (4y - y^2 - \tfrac{1}{4}y^4 - \tfrac{1}{5}y^5) \, \Big|_{y=0}^{y=1}$$

$$= (4 - 1 - \tfrac{1}{4} - \tfrac{1}{5}) - 0 = \tfrac{51}{20}$$

Problem 33 Evaluate $\iint_R (y - 4x) \, dA$, where R is the region in Example 33.

Example 34 The region R is bounded by the graphs of $y = \sqrt{x}$ and $y = \tfrac{1}{2}x$. Evaluate $\iint_R 4xy^3 \, dA$ two different ways.

Solution

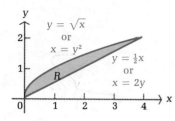

Region R is both a regular x region and a regular y region:

$$R = \{(x, y)|\tfrac{1}{2}x \leqslant y \leqslant \sqrt{x}, \quad 0 \leqslant x \leqslant 4\} \qquad \text{Regular x region}$$

$$R = \{(x, y)|y^2 \leqslant x \leqslant 2y, \quad 0 \leqslant y \leqslant 2\} \qquad \text{Regular y region}$$

Using the first representation (a regular x region), we obtain

$$\iint\limits_R 4xy^3\, dA = \int_0^4 \left(\int_{(1/2)x}^{\sqrt{x}} 4xy^3\, dy \right) dx$$

$$= \int_0^4 \left(xy^4 \Big|_{y=(1/2)x}^{y=\sqrt{x}} \right) dx$$

$$= \int_0^4 [x(\sqrt{x})^4 - x(\tfrac{1}{2}x)^4]\, dx$$

$$= \int_0^4 (x^3 - \tfrac{1}{16}x^5)\, dx$$

$$= (\tfrac{1}{4}x^4 - \tfrac{1}{96}x^6) \Big|_{x=0}^{x=4}$$

$$= (64 - \tfrac{128}{3}) - 0 = \tfrac{64}{3}$$

Using the second representation (a regular y region), we obtain

$$\iint\limits_R 4xy^3\, dA = \int_0^2 \left(\int_{y^2}^{2y} 4xy^3\, dx \right) dy$$

$$= \int_0^2 \left(2x^2y^3 \Big|_{x=y^2}^{x=2y} \right) dy$$

$$= \int_0^2 [2(2y)^2y^3 - 2(y^2)^2y^3]\, dy$$

$$= \int_0^2 (8y^5 - 2y^7)\, dy$$

$$= (\tfrac{4}{3}y^6 - \tfrac{1}{4}y^8) \Big|_{y=0}^{y=2}$$

$$= (\tfrac{256}{3} - 64) - 0 = \tfrac{64}{3}$$

Problem 34

The region R is bounded by the graphs of $y = x$ and $y = \tfrac{1}{2}x^2$. Evaluate $\iint_R 4x^3y\, dA$ two different ways.

■ Reversing the Order of Integration

Example 34 shows that

$$\iint\limits_R 4xy^3\, dA = \int_0^4 \left(\int_{(1/2)x}^{\sqrt{x}} 4xy^3\, dy \right) dx = \int_0^2 \left(\int_{y^2}^{2y} 4xy^3\, dx \right) dy$$

In general, if R is both a regular x region and a regular y region, the two iterated integrals are equal. In rectangular regions, reversing the order of integration in an iterated integral was a simple matter. As Example 34 illustrates, the process is more complicated in nonrectangular regions. The next example illustrates how to start with an iterated integral and reverse the order of integration. Since we are interested in the reversal process and not in the value of either integral, the integrand will not be specified.

Example 35 Reverse the order of integration in $\int_1^3 \left[\int_0^{x-1} f(x, y) \, dy \right] dx$.

Solution The order of integration indicates that the region of integration is a regular x region:

$$R = \{(x, y) | 0 \leqslant y \leqslant x - 1, \quad 1 \leqslant x \leqslant 3\}$$

Graph region R to determine whether it is also a regular y region. The graph shows that R is also a regular y region, and we can write

$$R = \{(x, y) | y + 1 \leqslant x \leqslant 3, \quad 0 \leqslant y \leqslant 2\}$$

Thus,

$$\int_1^3 \left[\int_0^{x-1} f(x, y) \, dy \right] dx = \int_0^2 \left[\int_{y+1}^3 f(x, y) \, dx \right] dy$$

Problem 35 Reverse the order of integration in $\int_2^4 \left[\int_0^{4-x} f(x, y) \, dy \right] dx$.

■ Volume and Double Integrals

In Section 16-7 we used the double integral to calculate the volume of a solid with a rectangular base. In general, if a solid can be described by the graph of a positive function $f(x, y)$ over a regular region R (not necessarily a rectangle), then the double integral of the function f over the region R still represents the volume of the corresponding solid.

Example 36 The region R is bounded by the graphs of $x + y = 1$, $y = 0$, and $x = 0$. Find the volume of the solid under the graph of $z = 1 - x - y$ over the region R.

Solution

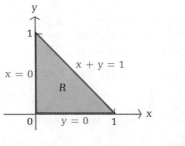

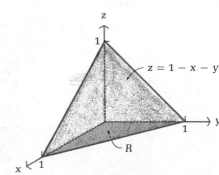

The graph of R indicates that R is a regular x region and can be described by

$$R = \{(x, y)|0 \leqslant y \leqslant 1 - x, \quad 0 \leqslant x \leqslant 1\}$$

Thus, the volume of the solid is

$$
\begin{aligned}
V = \iint\limits_{R} (1 - x - y) \, dA &= \int_0^1 \left[\int_0^{1-x} (1 - x - y) \, dy \right] dx \\
&= \int_0^1 \left[(y - xy - \tfrac{1}{2}y^2) \, \Big|_{y=0}^{y=1-x} \right] dx \\
&= \int_0^1 [(1 - x) - x(1 - x) - \tfrac{1}{2}(1 - x)^2] \, dx \\
&= \int_0^1 (\tfrac{1}{2} - x + \tfrac{1}{2}x^2) \, dx \\
&= (\tfrac{1}{2}x - \tfrac{1}{2}x^2 + \tfrac{1}{6}x^3) \, \Big|_{x=0}^{x=1} \\
&= (\tfrac{1}{2} - \tfrac{1}{2} + \tfrac{1}{6}) - 0 = \tfrac{1}{6}
\end{aligned}
$$

Problem 36 The region R is bounded by the graphs of $y + 2x = 2$, $y = 0$, and $x = 0$. Find the volume of the solid under the graph of $z = 2 - 2x - y$ over the region R. [*Hint:* Sketch the region first—the solid does not have to be sketched.]

Answers to Matched Problems

30. $R = \{(x, y)|y^2 \leqslant x \leqslant 6 - y, \quad 0 \leqslant y \leqslant 2\}$ is a regular y region; R is not a regular x region

31. R is both a regular x region and a regular y region;

$$R = \{(x, y)|\tfrac{1}{2}x + 2 \leqslant y \leqslant \sqrt{x + 4}, \quad -4 \leqslant x \leqslant 0\}$$
$$= \{(x, y)|y^2 - 4 \leqslant x \leqslant 2y - 4, \quad 0 \leqslant y \leqslant 2\}$$

32. $\tfrac{13}{40}$ 33. $-\tfrac{77}{20}$ 34. $\tfrac{16}{3}$ 35. $\displaystyle\int_0^2 \int_2^{4-y} f(x, y) \, dx \, dy$ 36. $\tfrac{2}{3}$

Exercise 16-8

A *Graph the region R bounded by the graphs of the equations. Express R in terms of set notation and double inequalities that describe R as a regular x region, a regular y region, or both.*

1. $y = 4 - x^2, \quad y = 0, \quad 0 \leqslant x \leqslant 2$
2. $y = x^2, \quad y = 9, \quad 0 \leqslant x \leqslant 3$
3. $y = x^3, \quad y = 12 - 2x, \quad x = 0$
4. $y = 5 - x, \quad y = 1 + x, \quad y = 0$
5. $y^2 = 2x, \quad y = x - 4$
6. $y = 4 + 3x - x^2, \quad x + y = 4$

Evaluate each integral.

7. $\displaystyle\int_0^1 \int_0^x (x+y)\,dy\,dx$

8. $\displaystyle\int_0^2 \int_0^y xy\,dx\,dy$

9. $\displaystyle\int_0^1 \int_{y^3}^{\sqrt{y}} (2x+y)\,dx\,dy$

10. $\displaystyle\int_1^4 \int_x^{x^2} (x^2+2y)\,dy\,dx$

B   *Use the description of the region R to evaluate the indicated integral.*

11. $\displaystyle\iint_R (x^2+y^2)\,dA;\quad R=\{(x,y)|0\leqslant y\leqslant 2x,\ \ 0\leqslant x\leqslant 2\}$

12. $\displaystyle\iint_R 2x^2y\,dA;\quad R=\{(x,y)|0\leqslant y\leqslant 9-x^2,\ \ -3\leqslant x\leqslant 3\}$

13. $\displaystyle\iint_R (x+y-2)^3\,dA;\quad R=\{(x,y)|0\leqslant x\leqslant y+2,\ \ 0\leqslant y\leqslant 1\}$

14. $\displaystyle\iint_R (2x+3y)\,dA;\quad R=\{(x,y)|y^2-4\leqslant x\leqslant 4-2y,\ \ 0\leqslant y\leqslant 2\}$

15. $\displaystyle\iint_R e^{x+y}\,dA;\quad R=\{(x,y)|-x\leqslant y\leqslant x,\ \ 0\leqslant x\leqslant 2\}$

16. $\displaystyle\iint_R \frac{x}{\sqrt{x^2+y^2}}\,dA;\quad R=\{(x,y)|0\leqslant x\leqslant \sqrt{4y-y^2},\ \ 0\leqslant y\leqslant 2\}$

Graph the region R bounded by the graphs of the indicated equations. Describe R in set notation with double inequalities and evaluate the indicated integral.

17. $y=x+1,\ \ y=0,\ \ x=0,\ \ x=1;\quad \displaystyle\iint_R \sqrt{1+x+y}\,dA$

18. $y=x^2,\ \ y=\sqrt{x};\quad \displaystyle\iint_R 12xy\,dA$

19. $y=4x-x^2,\ \ y=0;\quad \displaystyle\iint_R \sqrt{y+x^2}\,dA$

20. $x=1+3y,\ \ x=1-y,\ \ y=1;\quad \displaystyle\iint_R (x+y+1)^3\,dA$

21. $y=1-\sqrt{x},\ \ y=1+\sqrt{x},\ \ x=4;\quad \displaystyle\iint_R x(y-1)^2\,dA$

22. $y=\tfrac{1}{2}x,\ \ y=6-x,\ \ y=1;\quad \displaystyle\iint_R \frac{1}{x+y}\,dA$

$2y=x\qquad y-6=-x$
$\qquad\qquad\quad x=6-y$

Evaluate each integral. Graph the region of integration, reverse the order of integration, and then evaluate the integral with the order reversed.

23. $\displaystyle\int_0^3 \int_0^{3-x} (x + 2y)\, dy\, dx$

24. $\displaystyle\int_0^2 \int_0^y (y - x)^4\, dx\, dy$

25. $\displaystyle\int_0^1 \int_0^{1-x^2} x\sqrt{y}\, dy\, dx$

26. $\displaystyle\int_0^2 \int_{x^3}^{4x} (1 + 2y)\, dy\, dx$

27. $\displaystyle\int_0^4 \int_{x/4}^{\sqrt{x}/2} x\, dy\, dx$

28. $\displaystyle\int_0^4 \int_{y^2/4}^{2\sqrt{y}} (1 + 2xy)\, dx\, dy$

Find the volume of the solid under the graph of $f(x, y)$ over the region R bounded by the graphs of the indicated equations. Sketch the region R — the solid does not have to be sketched.

29. $f(x, y) = 4 - x - y$;   R is bounded by the graphs of $x + y = 4$, $y = 0$, $x = 0$

30. $f(x, y) = (x - y)^2$;   R is the region bounded by the graphs of $y = x$, $y = 2$, $x = 0$

31. $f(x, y) = 4$;   R is the region bounded by the graphs of $y = 1 - x^2$ and $y = 0$ for $0 \leqslant x \leqslant 1$

32. $f(x, y) = 4xy$;   R is the region bounded by the graphs of $y = \sqrt{1 - x^2}$ and $y = 0$ for $0 \leqslant x \leqslant 1$

C   *Reverse the order of integration for each integral. Evaluate the integral with the order reversed. Do not attempt to evaluate the integral in the original form.*

33. $\displaystyle\int_0^2 \int_{x^2}^4 \frac{4x}{1 + y^2}\, dy\, dx$

34. $\displaystyle\int_0^1 \int_y^1 \sqrt{1 - x^2}\, dx\, dy$

35. $\displaystyle\int_0^1 \int_{y^2}^1 4ye^{x^2}\, dx\, dy$

36. $\displaystyle\int_0^4 \int_{\sqrt{x}}^2 \sqrt{3x + y^2}\, dy\, dx$

16-9   Chapter Review

Important Terms and Symbols

16-1   *Functions of several variables.* functions of two independent variables, functions of several independent variables, surface, paraboloid, saddle point, $z = f(x, y)$, $w = f(x, y, z)$

16-2   *Partial derivatives.* partial derivative of f with respect to x, partial derivative of f with respect to y, second-order partials,

$$\frac{\partial z}{\partial x}, \frac{\partial z}{\partial y}, f_x(x, y), f_y(x, y), \quad \frac{\partial^2 z}{\partial x^2} = f_{xx}(x, y), \quad \frac{\partial^2 z}{\partial x\, \partial y} = f_{yx}(x, y),$$

$$\frac{\partial^2 z}{\partial y\, \partial x} = f_{xy}(x, y), \quad \frac{\partial^2 z}{\partial y^2} = f_{yy}(x, y)$$

16-3 *Total differentials and their applications.* total differential of $z = f(x, y)$, $dz = f_x(x, y) \, dx + f_y(x, y) \, dy$, total differential of $w = f(x, y, z)$, $dw = f_x(x, y, z) \, dx + f_y(x, y, z) \, dy + f_z(x, y, z) \, dz$, differential approximation, $\Delta z = f(x + \Delta x, y + \Delta y) - f(x, y) \approx f_x(x, y) \, dx + f_y(x, y) \, dy = dz$

16-4 *Maxima and minima.* local maximum, local minimum, critical point, second-derivative test

16-5 *Maxima and minima using Lagrange multipliers.* constraint, Lagrange multiplier, method of Lagrange multipliers for functions of two variables, method of Lagrange multipliers for functions of three variables

16-6 *Method of least squares.* least squares approximation, linear regression, residual, least squares line, regression line, estimation, approximation

16-7 *Double integrals over rectangular regions.* double integral, iterated integral, average value over rectangular regions, volume under a surface,

$$\iint_R f(x, y) \, dA = \int_a^b \left[\int_c^d f(x, y) \, dy \right] dx = \int_c^d \left[\int_a^b f(x, y) \, dx \right] dy$$

16-8 *Double integrals over more general regions.* regular x region, regular y region, reversing the order of integration, volume under a surface,

$$\int_a^b \left[\int_{g(x)}^{f(x)} F(x, y) \, dy \right] dx, \quad \int_c^d \left[\int_{h(y)}^{k(y)} F(x, y) \, dx \right] dy$$

Exercise 16-9 Chapter Review

Work through all the problems in this chapter review and check your answers in the back of the book. (Answers to all review problems are there.) Where weaknesses show up, review appropriate sections in the text. When you are satisfied that you know the material, take the practice test following this review.

A
1. For $f(x, y) = 2{,}000 + 40x + 70y$, find $f(5, 10)$, $f_x(x, y)$, and $f_y(x, y)$.
2. For $z = x^3 y^2$, find $\partial^2 z / \partial x^2$ and $\partial^2 z / \partial x \, \partial y$.
3. For $z = 2x + 3y$, find dz.   4. For $z = x^4 y^3$, find dz.

5. Evaluate: $\displaystyle\int (6xy^2 + 4y) \, dy$ 6. Evaluate: $\displaystyle\int (6xy^2 + 4y) \, dx$

7. Evaluate: $\displaystyle\int_0^1 \int_0^1 4xy \, dy \, dx$ 8. Evaluate: $\displaystyle\int_0^1 \int_0^x 4xy \, dy \, dx$

B
9. For $f(x, y) = 3x^2 - 2xy + y^2 - 2x + 3y - 7$, find $f(2, 3)$, $f_y(x, y)$, and $f_y(2, 3)$.

10. For $f(x, y) = -4x^2 + 4xy - 3y^2 + 4x + 10y + 81$, find

$$[f_{xx}(2, 3)][f_{yy}(2, 3)] - [f_{xy}(2, 3)]^2$$

11. Find Δz and dz for $z = f(x, y) = x^4 + y^4$, $x = 1$, $y = 2$, $\Delta x = dx = 0.1$, and $\Delta y = dy = 0.2$.

12. Use the least squares line for the data in the table to estimate y when $x = 10$.

x	y
2	12
4	10
6	7
8	3

13. For $R = \{(x, y)|-1 \leqslant x \leqslant 1, \quad 1 \leqslant y \leqslant 2\}$, evaluate the following two ways:

$$\iint\limits_{R} (4x + 6y) \, dA$$

14. Evaluate $\displaystyle\iint\limits_{R} (x + y)^3 \, dA$ for

$$R = \{(x, y)|0 \leqslant x \leqslant y + 1, \quad 0 \leqslant y \leqslant 3\}$$

C

15. For $f(x, y) = e^{x^2 + 2y}$, find f_x, f_y, and f_{xy}.

16. For $f(x, y) = (x^2 + y^2)^5$, find f_x and f_{xy}.

17. Use differentials to approximate the hypotenuse of a right triangle with legs 7.1 and 24.05 inches long, respectively.

18. Find all critical points and test for extrema for

$$f(x, y) = x^3 - 12x + y^2 - 6y$$

19. Find the least squares line for the data in the table:

x	y	x	y
10	50	60	80
20	45	70	85
30	50	80	90
40	55	90	90
50	65	100	110

20. Find the average value of $f(x, y) = x^{2/3}y^{1/3}$ over the rectangle

$$R = \{(x, y)|-8 \leqslant x \leqslant 8, \quad 0 \leqslant y \leqslant 27\}$$

21. Find the volume of the solid under the graph of $z = x + y$ over the region bounded by the graphs of $y = \sqrt{1 - x^2}$ and $y = 0$ for $0 \leqslant x \leqslant 1$.

22. Evaluate $\displaystyle\int_0^1 \int_0^{\sqrt{1-x^2}} y \, dy \, dx$. Then evaluate the integral obtained by reversing the order of integration.

Applications

Business & Economics   23. *Maximizing profit.* A company produces x units of product A and y units of product B (both in hundreds per month). The monthly profit equation (in thousands of dollars) is found to be

$$P(x, y) = -4x^2 + 4xy - 3y^2 + 4x + 10y + 81$$

(A) Find $P_x(1, 3)$ and interpret.
(B) How many of each product should be produced each month to maximize profit? What is the maximum profit?

24. *Minimizing material.* A rectangular box with no top and six compartments (see the figure) is to have a volume of 96 cubic inches. Find the dimensions that will require the least amount of material.

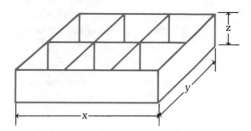

25. *Profit.* A company's annual profit (in millions of dollars) over a 5 year period is given in the table. Use the least squares line to estimate the profit for the sixth year.

Year	Profit
1	2
2	2.5
3	3.1
4	4.2
5	4.3

26. *Economics—Cobb–Douglas production function.* The Cobb–Douglas production function for an industry is

$$N(x, y) = x^{0.8}y^{0.2}$$

where x is the number of labor-hours (in thousands) and y is the amount of money (in millions) invested in the production of N thousand units of a certain item. If $10 \leq x \leq 12$ and $1 \leq y \leq 3$, find the average number of units produced. Set up a definite integral and evaluate.

Life Sciences

27. *Marine biology.* The function used for timing dives with scuba gear is

$$T(V, x) = \frac{33V}{x + 33}$$

where T is the time of the dive in minutes, V is the volume of air (at sea level pressure) compressed into tanks, and x is the depth of the dive in feet. Find $T_x(70, 17)$ and interpret.

28. *Blood flow.* In Poiseuille's law,

$$R(L, r) = k\frac{L}{r^4}$$

where R is the resistance for blood flow, L is the length of the blood vessel, r is the radius of the blood vessel, and k is a constant. Use differentials to approximate the change in R if L increases from 10 to 10.1 centimeters and r decreases from 0.5 to 0.45 centimeter.

Social Sciences

29. *Sociology.* Joseph Cavanaugh, a sociologist, found that the number of long-distance telephone calls, n, between two cities in a given period of time varied (approximately) jointly as the populations P_1 and P_2 of the two cities, and varied inversely as the distance, d, between the two cities. In terms of an equation for a time period of 1 week,

$$n(P_1, P_2, d) = 0.001 \, \frac{P_1 P_2}{d}$$

Find $n(100{,}000, \; 50{,}000, \; 100)$.

30. *Education.* At the beginning of the semester, students in a foreign language course are given a proficiency exam. The same exam is given at the end of the semester. The results for five students are given in the table. Use the least squares line to estimate the score on the second exam for a student who scored 40 on the first exam.

First Exam	Second Exam
30	60
50	75
60	80
70	85
90	90

Practice Test: Chapter 16

1. For $f(x, y) = 2x^2 y + y + 1$, find:

(A) $f(1, 2)$ (B) $f_x(1, 2)$

2. For $f(x, y) = (x^2 y^3 - 2x)^4$, find $f_x(x, y)$ and $f_y(x, y)$.

3. For $z = x^3 y^4$, find dz.

4. Find dz and Δz for $z = f(x, y) = x^2 + y^2$, $x = 1$, $y = 2$, $\Delta x = dx = 0.2$, and $\Delta y = dy = 0.1$.

5. For $z = x^3 - 2x^2 y + y^2$, find $\partial^2 z / \partial y \, \partial x$.

6. For $f(x, y) = e^{x^2 y}$, find $f_{yx}(x, y)$.

7. Evaluate: $\displaystyle\int_0^2 \int_0^3 (6x + 4y) \, dy \, dx$

8. The daily revenue equation for two commodities is

$$R(x, y) = 10(-3x^2 + 2xy - 5y^2 + 250x + 200y)$$

where x and y are the unit prices of the commodities in dollars. Find $R_x(30, 20)$ and $R_y(20, 30)$.

9. Find all local extrema for $f(x, y) = x^4 + 8x^2 + y^2 - 4y$.

10. Find the volume of the solid under the graph of $f(x, y) = 1 - x - y$ over the region bounded by the graphs of $y = 1 - x$, $y = 0$, and $x = 0$.

11. The cost y in thousands of dollars of producing x units of a certain product is given in the table for various levels of production. Use the least squares line to estimate the cost of producing 40 units.

x	y
20	40
25	45
30	51
35	56

Additional Probability Topics

Contents

In this chapter we return to the study of probability. In particular, we will see how calculus is used to extend the concepts of probability to experiments with an infinite number of possible outcomes.

17-1   Random Variable, Probability Distribution, and Expectation

- Random Variable
- Probability Distribution
- Expected Value
- Mean and Standard Deviation

■ Random Variable

When performing a random experiment, a sample space S is selected in such a way that all probability problems of interest relative to the experiment can be solved. In many situations, we may not be interested in each simple event in the sample space S but in some numerical value associated with the event. For example, if three coins are tossed, we may be interested in the number of heads that turn up rather than in the particular pattern that turns up. Or, in selecting a random sample of students, we may be interested in the proportion that are women rather than which particular students are women. And, in the same way, a craps player is usually interested in the sum of the dots showing on the faces of a pair of dice rather than the pattern of dots on each face.

In each of these examples, we have a rule that assigns to each simple event in a sample space S a single real number. Mathematically speaking, we are dealing with a function. Historically, this particular type of function has been called a *random variable*.

Random Variable

A **random variable** is a function that assigns a numerical value to each simple event in a sample space S.

The term *random variable* is an unfortunate choice, since it is neither random nor a variable—it is a function with a numerical value and it is defined on a sample space. But the terminology has stuck and is now standard, so we will have to live with it. Capital letters, such as X, are used to represent random variables.

Let us consider the experiment of tossing three coins. A sample space S of equally likely simple events is indicated in the first column of Table 1. The second column indicates the number of heads corresponding to a simple event. And the last column indicates the probability of each simple event occurring.

Table 1 Tossing Three Coins

S		Number of Heads $X(e_i)$	Probability $P(e_i)$
e_1:	TTT	0	$\frac{1}{8}$
e_2:	TTH	1	$\frac{1}{8}$
e_3:	THT	1	$\frac{1}{8}$
e_4:	HTT	1	$\frac{1}{8}$
e_5:	THH	2	$\frac{1}{8}$
e_6:	HTH	2	$\frac{1}{8}$
e_7:	HHT	2	$\frac{1}{8}$
e_8:	HHH	3	$\frac{1}{8}$

The random variable X (a function) associates exactly one of the numbers 0, 1, 2, or 3 with each simple event. For example, $X(e_1) = 0$, $X(e_2) = 1$, $X(e_3) = 1$, and so on.

▪ Probability Distribution

We are interested in the probability of the occurrence of each image value of X; that is, in the probability of the occurrence of zero heads, one head, two heads, or three heads in the single toss of three coins. We indicate this probability by

$p(x)$ where $x \in \{0, 1, 2, 3\}$

The function p is called the **probability function* of the random variable X.** What is $p(2)$, the probability of getting exactly two heads on the single toss of three coins? Exactly two heads occur if any of the simple events

e_5: THH e_6: HTH e_7: HHT

occurs. Adding the probabilities for these simple events (see Table 1), we obtain $p(2) = \frac{1}{8} + \frac{1}{8} + \frac{1}{8} = \frac{3}{8}$.

---

* Formally, the probability function p of the random variable X is defined by $p(x) = P(\{e_i \in S | X(e_i) = x\})$, which, because of its cumbersome nature, is usually simplified to $p(x) = P(X = x)$ or, simply, $p(x)$. We will use the simplified notation.

Proceeding similarly for $p(0)$, $p(1)$, and $p(3)$, we obtain the results in Table 2 and Figure 1. The table is called a *probability distribution for the random variable X* and the graph is called a *histogram*.

Table 2 Probability Distribution

Number of Heads x	0	1	2	3
Probability $p(x)$	$\dfrac{1}{8}$	$\dfrac{3}{8}$	$\dfrac{3}{8}$	$\dfrac{1}{8}$

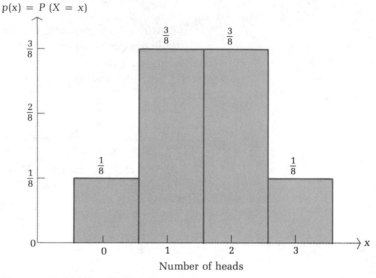

Figure 1 Probability distribution

Note from Table 2 or Figure 1 that

1. $0 \leqslant p(x) \leqslant 1$, $\quad x \in \{0, 1, 2, 3\}$
2. $p(0) + p(1) + p(2) + p(3) = \frac{1}{8} + \frac{3}{8} + \frac{3}{8} + \frac{1}{8} = 1$

These are general properties of any probability distribution of a random variable X associated with a finite sample space.

Probability Distribution of a Random Variable

A probability function $P(X = x) = p(x)$ is a **probability distribution of the random variable X** if

1. $0 \leqslant p(x) \leqslant 1$, $\quad x \in \{x_1, x_2, \ldots, x_m\}$
2. $p(x_1) + p(x_2) + \cdots + p(x_m) = 1$

where $\{x_1, x_2, \ldots, x_m\}$ are the (range) values of X. See Figure 2.

■ Expected Value

Suppose the experiment of tossing three coins was repeated a large number of times. What would be the average number of heads per toss (the total number of heads in all tosses divided by the total number of tosses)? Consulting the probability distribution in Table 2 or Figure 1, we see that

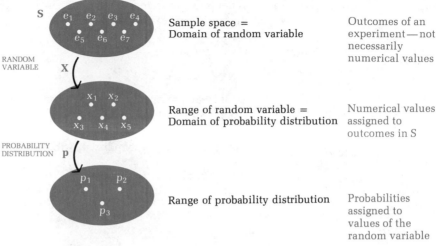

RANDOM
VARIABLE   **X**

PROBABILITY
DISTRIBUTION   **p**

Sample space =
Domain of random variable

Outcomes of an
experiment — not
necessarily
numerical values

Range of random variable =
Domain of probability distribution

Numerical values
assigned to
outcomes in S

Range of probability distribution

Probabilities
assigned to
values of the
random variable

Figure 2 Probability distribution of a random variable for a finite sample space

we would expect to toss zero heads one-eighth of the time, one head three-eighths of the time, two heads three-eighths of the time, and three heads one-eighth of the time. Thus, in the long run, we would expect the average number of heads per toss of the three coins, or the *expected value* $E(X)$, to be given by

$$E(X) = 0(\tfrac{1}{8}) + 1(\tfrac{3}{8}) + 2(\tfrac{3}{8}) + 3(\tfrac{1}{8}) = \tfrac{12}{8} = 1.5$$

It is important to note that the expected value is not a value that will necessarily occur in a single experiment (1.5 heads cannot occur in the toss of three coins), but it is an average of what occurs over a large number of experiments. Sometimes, we will toss more than 1.5 heads and sometimes less, but if the experiment is repeated many times, the average number of heads per experiment should approach 1.5.

We make the above discussion precise with the definition of expected value given in the box.

Expected Value of a Random Variable X

Given the probability distribution for the random variable X:

x_i	x_1	x_2	$\cdots$	x_m
p_i	p_1	p_2	$\cdots$	p_m

where $p_i = p(x_i)$, we define the **expected value of X,** denoted by $E(X)$, by the formula

$$E(X) = x_1 p_1 + x_2 p_2 + \cdots + x_m p_m$$

Example 1   What is the expected value (long-run average) of the number of dots facing up for the roll of a single die?

Solution   If we choose

$$S = \{1, 2, 3, 4, 5, 6\}$$

as our sample space, then each simple event is a numerical outcome reflecting our interest, and each is equally likely. The random variable X in this case is just the identity function (each number is associated with itself). Thus, the probability distribution for X is

x_i	1	2	3	4	5	6
p_i	$\frac{1}{6}$	$\frac{1}{6}$	$\frac{1}{6}$	$\frac{1}{6}$	$\frac{1}{6}$	$\frac{1}{6}$

Hence,

$$E(X) = 1(\tfrac{1}{6}) + 2(\tfrac{1}{6}) + 3(\tfrac{1}{6}) + 4(\tfrac{1}{6}) + 5(\tfrac{1}{6}) + 6(\tfrac{1}{6})$$
$$= \tfrac{21}{6} = 3.5$$

Problem 1   Suppose the die in Example 1 is not fair and we obtain (empirically) the following probability distribution for X:

x_i	1	2	3	4	5	6	
p_i	.14	.13	.18	.20	.11	.24	[*Note:* Sum = 1]

What is the expected value of X?

Example 2   A spinner device is numbered from 0 to 5, and each of the six numbers is as likely to come up as any other. A player who bets \$1 on any given number wins \$4 (and gets the bet back) if the pointer comes to rest on the chosen number; otherwise, the \$1 bet is lost. What is the expected value of the game (long-run average gain or loss per game)?

Solution   The sample space of equally likely events is

$$S = \{0, 1, 2, 3, 4, 5\}$$

Each sample point occurs with a probability of $\tfrac{1}{6}$. The random variable X assigns \$4 to the winning number and $-$\$1 to each of the remaining numbers. Thus, the probability distribution for X, called a **payoff table,** is as shown in the margin. The probability of winning \$4 is $\tfrac{1}{6}$ and of losing \$1 is $\tfrac{5}{6}$. We can now compute the expected value of the game:

$$E(X) = \$4(\tfrac{1}{6}) + (-\$1)(\tfrac{5}{6}) = -\$\tfrac{1}{6} \approx -\$0.1667 \approx -17\cent \text{ per game}$$

Thus, in the long run the player will lose an average of about 17¢ per game.

Payoff Table
(Probability
Distribution for X)

x_i	\$4	$-$\$1
p_i	$\frac{1}{6}$	$\frac{5}{6}$

In general, a game is said to be **fair** if $E(X) = 0$. The game in Example 2 is not fair—the "house" has an advantage, on the average, of about 17¢ per game.

Problem 2   Repeat Example 2 with the player winning $5 instead of $4 if the chosen number turns up. The loss is still $1 if any other number turns up. Is this now a fair game?

Example 3   Suppose you are interested in insuring a car stereo system for $500 against theft. An insurance company charges a premium of $60 for coverage for 1 year, claiming an empirically determined probability of .1 that the system will be stolen some time during the year. What is your expected gain or loss if you take out this insurance?

Solution   This is actually a game of chance in which your stake is $60. You have a .1 chance of winning $440 ($500 minus your stake of $60) and a .9 chance of losing your stake of $60. What is the expected value of this game? We form a payoff table (the probability distribution for X):

Payoff Table

x_i	$440	$-$60
p_i	.1	.9

Then we compute the expected value as follows:

$$E(X) = (\$440)(.1) + (-\$60)(.9) = -\$10$$

This means that if you insure with this company over many years and the circumstances remain the same, you would have an average net loss of $10 per year.

Problem 3   Find the expected value in Example 3 from the insurance company's point of view.

■ Mean and Standard Deviation

Since the expected value of a random variable represents the long-run average of repeated experiments, it is often referred to as the *mean*. Traditionally, the greek letter μ is used to denote the mean. Thus,

$$\mu = E(X) = x_1 p_2 + x_2 p_2 + \cdots + x_m p_m$$

is the **mean of the random variable X.** Geometrically, the mean, in some sense, is the center of the values of X and is often referred to as a *measure of central tendency*.

Another numerical quantity that is used to describe the properties of a random variable is the *standard deviation*. This quantity gives a *measure of the dispersion, or spread,* of the random variable X about the mean μ.

Standard Deviation of a Random Variable X

Given the probability distribution for the random variable X:

x_i	x_1	x_2	$\cdots$	x_m
p_i	p_1	p_2	$\cdots$	p_m

and the mean

$$\mu = x_1 p_1 + x_2 p_2 + \cdots + x_m p_m$$

we define the **variance of X,** denoted by $V(X)$, by the formula

$$V(X) = (x_1 - \mu)^2 p_1 + (x_2 - \mu)^2 p_2 + \cdots + (x_m - \mu)^2 p_m$$

and the **standard deviation of X,** denoted by σ, by the formula

$$\sigma = \sqrt{V(X)}$$

In other words, the variance is the expected value of the squares of the distances from each value of X to the mean. Standard deviation is defined using the square root so that it will be expressed in the same units as the values of X.

Returning to the coin tossing experiment, we have already shown that $\mu = E(X) = 1.5$. Thus,

$$V(X) = (0 - 1.5)^2(\tfrac{1}{8}) + (1 - 1.5)^2(\tfrac{3}{8}) + (2 - 1.5)^2(\tfrac{3}{8}) + (3 - 1.5)^2(\tfrac{1}{8})$$
$$= .75$$

and

$$\sigma = \sqrt{V(X)} = \sqrt{.75} \approx .866$$

Example 4 Find the variance and standard deviation for the random variable in Example 1.

Solution
$$\mu = E(X) = 3.5$$
$$V(X) = (1 - 3.5)^2(\tfrac{1}{6}) + (2 - 3.5)^2(\tfrac{1}{6}) + (3 - 3.5)^2(\tfrac{1}{6})$$
$$\qquad\qquad + (4 - 3.5)^2(\tfrac{1}{6}) + (5 - 3.5)^2(\tfrac{1}{6}) + (6 - 3.5)^2(\tfrac{1}{6})$$
$$= \tfrac{35}{12}$$
$$\sigma = \sqrt{V(X)} = \sqrt{\tfrac{35}{12}} \approx 1.708$$

Problem 4 Find the variance and standard deviation for the random variable in Problem 1.

The standard deviation is often used to compare different probability distributions. Figure 3 (parts A – D) gives four different probability distributions and their graphs. Notice the relationship between the standard deviation σ and the dispersion of the probability distribution about the mean. The tighter the cluster of the probability distribution about the mean, the smaller the standard deviation.

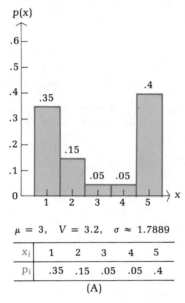

$\mu = 3, \quad V = 3.2, \quad \sigma \approx 1.7889$

x_i	1	2	3	4	5
p_i	.35	.15	.05	.05	.4

(A)

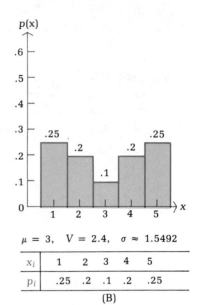

$\mu = 3, \quad V = 2.4, \quad \sigma \approx 1.5492$

x_i	1	2	3	4	5
p_i	.25	.2	.1	.2	.25

(B)

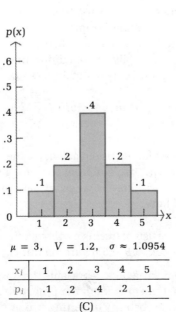

$\mu = 3, \quad V = 1.2, \quad \sigma \approx 1.0954$

x_i	1	2	3	4	5
p_i	.1	.2	.4	.2	.1

(C)

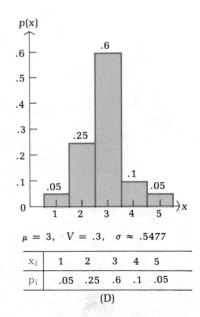

$\mu = 3, \quad V = .3, \quad \sigma \approx .5477$

x_i	1	2	3	4	5
p_i	.05	.25	.6	.1	.05

(D)

Figure 3

1. $E(X) = 3.73$
2. $E(X) = \$0$; the game is fair
3. $E(X) = (-\$440)(.1) + (\$60)(.9) = \$10$; this amount, of course, is necessary to cover expenses and profit
4. $V(X) = 2.9571$; $\sigma \approx 1.720$

Exercise 17-1

A   In Problems 1–4 graph the probability distribution and find the mean and standard deviation of the random variable X.

1.

x_i	-2	-1	0	1	2
p_i	.1	.2	.4	.2	.1

2.

x_i	-2	-1	0	1	2
p_i	.1	.1	.2	.2	.4

3.

x_i	-2	-1	0	1	2
p_i	.5	.2	.1	.1	.1

4.

x_i	-2	-1	0	1	2
p_i	.3	.1	.1	.1	.4

A spinner is marked from 1 to 10 and each number is as likely to turn up as any other. Problems 5–10 refer to this experiment.

5. Find a sample space S consisting of equally likely simple events.
6. The random variable X represents the number that turns up on the spinner. Find the probability distribution for X.
7. What is the probability of obtaining an even number?
8. What is the probability of obtaining a number that is exactly divisible by 3?
9. What is the expected value of X?
10. What is the standard deviation of X?

B   11. In tossing two fair coins once, what is the expected number of heads?
12. In a family with two children, excluding multiple births and assuming a boy is as likely as a girl at each birth, what is the expected number of boys?

An experiment consists of tossing a coin four times in succession. Answer the questions in Problems 13–18 regarding this experiment.

13. The random variable X represents the number of heads that occur in four tosses. Find and graph the probability distribution of X.
14. What is the probability of getting two or more heads?

15. What is the probability of getting an even number of heads?
16. What is the probability of getting more heads than tails?
17. Find the expected number of heads.
18. Find the standard deviation of X.

An experiment consists of rolling two fair dice. Answer the questions in Problems 19–24 regarding this experiment.

19. The random variable X represents the sum of the dots on the two up faces of the dice. Find and graph the probability distribution of X.
20. What is the probability that the sum is 7 or 11?
21. What is the probability that the sum is less than 6?
22. What is the probability that the sum is an even number?
23. Find the expected value of the sum of the dots.
24. Find the standard deviation of X.

C

25. After you pay $4 to play a game, a single fair die is rolled and you are paid back the number of dollars equal to the number of dots facing up. For example, if 5 dots turn up, $5 is returned to you for a net gain of $1. If 1 dot turns up, $1 is returned to you for a net gain, or payoff, of $-$3; and so on. If X is the random variable that represents net gain, or payoff, what is the expected value of X?
26. Repeat Problem 25 with the same game costing $3.50 for each play.
27. A player tosses two coins and wins $3 if two heads appear and $1 if one head appears, but loses $6 if two tails appear. If X is the random variable representing the player's net gain, what is the expected value of X?
28. Repeat Problem 27 if the player wins $5 if two heads appear and $2 if one head appears, but loses $7 if two tails appear.
29. Roulette wheels in the United States generally have thirty-eight equally spaced slots numbered 00, 0, 1, 2, . . . , 36. A player who bets $1 on any given number wins $35 (and gets the bet back) if the ball comes to rest on the chosen number; otherwise, the $1 bet is lost. If X is the random variable that represents the player's net gain, what is the expected value of X?
30. In roulette (see Problem 29) the numbers from 1 to 36 are evenly divided between red and black. A player who bets $1 on black, wins $1 (and gets the bet back) if the ball comes to rest on black; otherwise (if the ball lands on red, 0, or 00), the $1 bet is lost. If X is the random variable that represents the player's net gain, what is the expected value of X?

Applications

Business & Economics

31. *Insurance.* The annual premium for a $5,000 insurance policy against the theft of a painting is $150. If the probability that the painting will

be stolen during the year is .01, what is your expected gain or loss if you take out this insurance?

32. *Insurance.* Repeat Problem 31 from the point of view of the insurance company.

33. *Decision analysis.* After careful testing and analysis, an oil company is considering drilling in one of two different sites. It is estimated that site A will net $30 million if successful (probability .2) and lose $3 million if not (probability .8); site B will net $70 million if successful (probability .1) and lose $4 million if not (probability .9). Based on the expected return from each site, which site should the company choose?

34. *Decision analysis.* Repeat Problem 33, assuming additional analysis caused the estimated probability of success in field B to be changed from .1 to .11.

Life Sciences   35. *Genetics.* Suppose that at each birth having a girl is not as likely as having a boy, and that the probability assignments for the number of boys in a family with three children is approximated from past records to be:

Number of Boys

x_i	0	1	2	3
p_i	.12	.36	.38	.14

What is the expected number of boys in a three-child family?

36. *Genetics.* A pink-flowering plant is of genotype RW. If two such plants are crossed, we obtain a red plant (RR) with probability .25, a pink plant (RW) with probability .50, and a white plant (WW) with probability .25:

Number of W Genes
Present

x_i	0	1	2
p_i	.25	.50	.25

What is the expected number of W genes present in a crossing of this type?

Social Sciences   37. *Politics.* A money drive is organized by a campaign committee for a candidate running for public office. Two approaches are considered:

A_1 — A general mailing with a followup mailing

A_2 — Door-to-door solicitation with followup telephone calls

From campaign records of previous committees, average donations and their corresponding probabilities are estimated to be:

A_1		A_2	
x_1 (return per person)	p_i	x_i (return per person)	p_i
$10	.3	$15	.3
$ 5	.2	$ 3	.1
$ 0	.5	$ 0	.6
	1.0		1.0

Which course of action should be taken based on expected return?

17-2   Binomial Distributions

- Bernoulli Trials
- Binomial Theorem
- Binomial Distribution
- Application

■ Bernoulli Trials

If we toss a coin, either a head occurs or it does not. If we roll a die, either a 3 shows or it fails to show. If you are vaccinated for smallpox, either you contract smallpox or you do not. What do all these situations have in common? All can be classified as experiments with two possible outcomes, each the complement of the other. An experiment for which there are only two possible outcomes, E or E', is called a **Bernoulli experiment** or **trial,** after Jacob Bernoulli (1654–1705), a Swiss scientist and mathematician who was one of the first people to study systematically the probability problems related to a two-outcome experiment.

In a Bernoulli experiment or trial, it is customary to refer to one of the two outcomes as a success S and to the other as a failure F. If we designate the probability of success by

$$P(S) = p$$

then the probability of failure will be

$$P(F) = 1 - p = q \qquad \text{Note:} \quad p + q = 1$$

Example 5   We roll a fair die and ask for the probability of a 6 turning up. This can be viewed as a Bernoulli trial by identifying a success with a 6 turning up and a failure with any of the other numbers turning up. Thus,

$$p = \tfrac{1}{6} \qquad \text{and} \qquad q = 1 - \tfrac{1}{6} = \tfrac{5}{6}$$

Problem 5   Identify p and q for a single roll of a fair die where a success is a number divisible by 3 turning up.

Now, suppose a Bernoulli experiment is repeated five times. How can we compute the probability of the outcome SSFFS? In order to answer this question, we must make two basic assumptions about the trials in a sequence of Bernoulli experiments. First, we will assume that the probability of a success remains the same from trial to trial. Second, we will assume that the trials are **independent;** that is, the outcome of one trial has no effect on the outcome of any of the other trials. With these two assumptions, it can be shown that *the probability of a sequence of events is equal to the product of the probability of each event in the sequence.* Thus,

$$P(SSFFS) = P(S)P(S)P(F)P(F)P(S)$$
$$= ppqqp = p^3q^2$$

Example 6   If we roll a fair die five times and identify a success in a single roll with a 1 turning up, what is the probability of the sequence SFFSS occurring?

Solution $p = \frac{1}{6}$ $q = 1 - p = \frac{5}{6}$

$P(SFFSS) = pqqpp = p^3q^2$

$= (\frac{1}{6})^3(\frac{5}{6})^2 \approx .003$

Problem 6   In Example 6, find the probability of the outcome FSSSF.

In simple Bernoulli experiments, such as tossing a coin or rolling a die, it seems very reasonable to assume that the trials are independent. In more complicated situations, it can be very difficult to determine whether the trials are actually independent. We will assume that all the Bernoulli experiments we consider in this book have independent trials.

In general, we define a sequence of Bernoulli trials as follows:

Bernoulli Trials

A sequence of experiments is called a **sequence of Bernoulli trials** if:

1.   Only two outcomes are possible on each trial.
2.   The probability of success p for each trial is a constant (probability of failure is then $q = 1 - p$).
3.   All trials are independent.

In most applications involving sequences of Bernoulli trials, we will be interested in the number of successes, rather than in a specific outcome of the form SSFFS. If X is the random variable associated with the number of successes in a sequence of Bernoulli trials, we would like to find the probability distribution for X. Since this probability distribution is closely related to the binomial theorem, it will be helpful if we first review this important theorem. (A more detailed discussion of this theorem can be found in Appendix A-3.)

■ Binomial Theorem

To start, let us calculate directly the first five natural number powers of $(a + b)^x$:

$(a + b)^1 = a + b$

$(a + b)^2 = a^2 + 2ab + b^2$

$(a + b)^3 = a^3 + 3a^2b + 3ab^2 + b^3$

$(a + b)^4 = a^4 + 4a^3b + 6a^2b^2 + 4ab^3 + b^4$

$(a + b)^5 = a^5 + 5a^4b + 10a^3b^2 + 10a^2b^3 + 5ab^4 + b^5$

In general, it can be shown that a binomial expansion is given by the formula in the **binomial theorem:**

Theorem 1

Binomial Theorem

For x a natural number,

$$(a + b)^x = \binom{x}{0} a^x + \binom{x}{1} a^{x-1}b + \binom{x}{2} a^{x-2}b^2 + \cdots + \binom{x}{x} b^x$$

Example 7 Use the binomial theorem to expand $(p + q)^3$.

Solution
$$(p + q) = \binom{3}{0} p^3 + \binom{3}{1} p^2q + \binom{3}{2} pq^2 + \binom{3}{3} q^3$$
$$= p^3 + 3p^2q + 3pq^2 + q^3$$

Problem 7 Use the binomial theorem to expand $(p + q)^4$.

Example 8 Use the binomial theorem to find the fifth term in the expansion of $(p + q)^6$.

Solution The fifth term is given by

$$\binom{6}{4} p^2q^4 = \frac{6!}{4!(6 - 4)!} p^2q^4 = 15p^2q^4$$

Problem 8 Use the binomial theorem to find the third term in the expansion of $(p + q)^7$.

■ Binomial Distribution

Let us consider a sequence of three Bernoulli trials. Let the random variable X_3 represent the number of successes in three trials. Thus, the random variable X_3 can assume the value of 0, 1, 2, or 3. We are interested in the probability distribution for this random variable.

Which outcomes of an experiment consisting of a sequence of three Bernoulli trials lead to the random variable values 0, 1, 2, and 3, and what are the probabilities associated with these values? Table 3 answers these questions completely.

Table 3

Simple Event	Probability of Simple Event	X_3 x successes in 3 trials	$P(X_3 = x)$
SSS	$ppp = p^3$	3	p^3
SSF	$ppq = p^2q$		
SFS	$pqp = p^2q$	2	$3p^2q$
FSS	$qpp = p^2q$		
SFF	$pqq = pq^2$		
FSF	$qpq = pq^2$	1	$3pq^2$
FFS	$qqp = pq^2$		
FFF	$qqq = q^3$	0	q^3

The terms in the last column of Table 3 are the terms in the binomial expansion of $(p + q)^3$; see Example 7. The last two columns in the table provide a probability distribution for the random variable X_3. Note that both conditions for a probability distribution are met:

1. $0 \leqslant P(X_3 = x) \leqslant 1, \qquad x \in \{0, 1, 2, 3\}$
2. $1 = 1^3 = (p + q)^3$ Recall that $p + q = 1$

$$= \binom{3}{0} p^3 + \binom{3}{1} p^2q + \binom{3}{2} pq^2 + \binom{3}{3} q^3$$

$$= p^3 + 3p^2q + 3pq^2 + q^3$$

$$= P(X_3 = 3) + P(X_3 = 2) + P(X_3 = 1) + P(X_3 = 0)$$

In the general case, let X_n be the random variable associated with the number of successes in a sequence of n Bernoulli trials. Reasoning in the same way as we did for X_3, we see that each value of the probability distribution for X_n is a term in the binomial expansion of $(p + q)^n$. For this reason, a sequence of Bernoulli trials is often referred to as a **binomial experiment** and the probability distribution for the random variable associated with the number of successes is called a **binomial distribution.** In terms of a formula, we have:

Theorem 2

Binomial Distribution

$p(x) = P(X_n = x) = P(x \text{ successes in } n \text{ trials})$

$$= \binom{n}{x} p^x q^{n-x} \qquad x \in \{0, 1, 2, \ldots, n\} \qquad (1)$$

Example 9   If a fair coin is tossed four times, what is the probability of tossing:

(A)   Exactly two heads?   (B)   At least two heads?

Solutions (A) Use equation (1) with $n = 4$, $x = 2$, $p = \frac{1}{2}$, and $q = \frac{1}{2}$:

$$p(2) = P(X_4 = 2)$$

$$= \binom{4}{2} \left(\frac{1}{2}\right)^2 \left(\frac{1}{2}\right)^2$$

$$= \frac{4!}{2!2!} \left(\frac{1}{2}\right)^4 = .375$$

(B) Notice how this problem differs from part A. Here we have

$$P(X_4 \geqslant 2) = p(2) + p(3) + p(4)$$

$$= \binom{4}{2} \left(\frac{1}{2}\right)^2 \left(\frac{1}{2}\right)^2 + \binom{4}{3} \left(\frac{1}{2}\right)^3 \left(\frac{1}{2}\right)^1 + \binom{4}{4} \left(\frac{1}{2}\right)^4 \left(\frac{1}{2}\right)^0$$

$$= \frac{4!}{2!2!} \left(\frac{1}{2}\right)^4 + \frac{4!}{3!1!} \left(\frac{1}{2}\right)^4 + \frac{4!}{4!0!} \left(\frac{1}{2}\right)^4$$

$$= .375 + .25 + .0625 = .6875$$

Problem 9 If a fair coin is tossed four times, what is the probability of tossing:

(A) Exactly one head? (B) At most one head?

Example 10 Suppose a fair die is rolled three times and a success on a single roll is considered to be rolling a number divisible by 3.

(A) Write the probability function for the binomial distribution.
(B) Construct a table for this binomial distribution.
(C) Draw a histogram for this theoretical distribution.

Solutions (A) $p = \frac{1}{3}$ Since there are two numbers out of six that are divisible by 3

$$q = 1 - p = \frac{2}{3}$$
$$n = 3$$

Hence,

$$P(x \text{ successes in 3 trials}) = \binom{3}{x} \left(\frac{1}{3}\right)^x \left(\frac{2}{3}\right)^{3-x}$$

(B)

x	p(x)	
0	$\binom{3}{0} \left(\frac{1}{3}\right)^0 \left(\frac{2}{3}\right)^3$	$\approx$ .30
1	$\binom{3}{1} \left(\frac{1}{3}\right)^1 \left(\frac{2}{3}\right)^2$	$\approx$ .44
2	$\binom{3}{2} \left(\frac{1}{3}\right)^2 \left(\frac{2}{3}\right)^1$	$\approx$ .22
3	$\binom{3}{3} \left(\frac{1}{3}\right)^3 \left(\frac{2}{3}\right)^0$	$\approx$ .04
		1.00

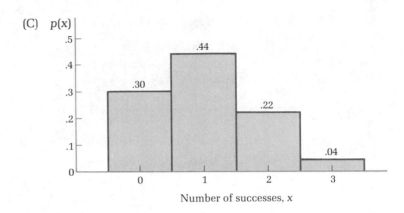

(C) $p(x)$

Number of successes, x

If we actually performed the binomial experiment described in Example 10 a large number of times with a fair die, we would find that we would roll no number divisible by 3 in three rolls of a die about 30% of the time, one number divisible by 3 in three rolls about 44% of the time, two numbers divisible by 3 in three rolls about 22% of the time, and three numbers divisible by 3 in three rolls only 4% of the time. Note that the sum of all the probabilities is 1, as it should be.

Problem 10   Repeat Example 10 where the binomial experiment consists of two rolls of a die instead of three rolls.

We close our discussion of binomial distributions by stating (without proof) formulas for the mean and standard deviation of the random variable associated with the distribution.

Theorem 3

> **Mean and Standard Deviation (Random Variable in a Binomial Distribution)**
>
> Mean: $\mu = np$
> Standard deviation: $\sigma = \sqrt{npq}$

Example 11   Compute the mean and standard deviation for the random variable in Example 10.

Solution $n = 3$ $p = \frac{1}{3}$ $q = 1 - \frac{1}{3} = \frac{2}{3}$

$\mu = np = 3(\frac{1}{3}) = 1$ $\sigma = \sqrt{npq} = \sqrt{3(\frac{1}{3})(\frac{2}{3})} \approx .82$

Problem 11   Compute the mean and standard deviation for the random variable in Problem 10 above.

■ Application

Binomial experiments are associated with a wide variety of practical problems: industrial sampling, drug testing, genetics, epidemics, medical diagnosis, opinion polls, analysis of social phenomena, qualifying tests, and so on. Several types of applications are included in Exercise 17-2. We will now consider one application in detail.

Example 12   The probability of recovering after a particular type of operation is .5. Let us investigate the binomial distribution involving eight patients undergoing this operation.

(A)   Write the function defining this distribution.
(B)   Construct a table for the distribution.
(C)   Construct a histogram for the distribution.
(D)   Find the mean and standard deviation for the distribution.

Solutions   (A) $p = .5$ $q = 1 - p = .5$ $n = 8$

Hence, letting a recovery be a success,

$$p(x) = P(\text{Exactly } x \text{ successes in 8 trials}) = \binom{8}{x}(.5)^x(.5)^{8-x}$$

$$= \binom{8}{x}(.5)^8$$

(B)

x	p(x)
0	$\binom{8}{0}(.5)^8 \approx .004$
1	$\binom{8}{1}(.5)^8 \approx .031$
2	$\binom{8}{2}(.5)^8 \approx .109$
3	$\binom{8}{3}(.5)^8 \approx .219$
4	$\binom{8}{4}(.5)^8 \approx .273$
5	$\binom{8}{5}(.5)^8 \approx .219$
6	$\binom{8}{6}(.5)^8 \approx .109$
7	$\binom{8}{7}(.5)^8 \approx .031$
8	$\binom{8}{8}(.5)^8 \approx \underline{.004}$
	$.999 \approx 1$

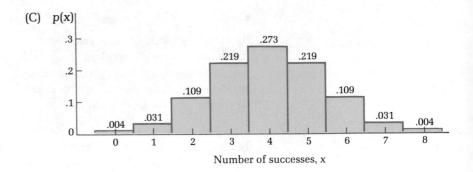

(C)

(D) $\mu = np = 8(.5) = 4$ $\sigma = \sqrt{npq} = \sqrt{8(.5)(.5)} = 1.41$

Problem 12   Repeat Example 12 for four patients.

Answers to   5. $p = \frac{1}{3}, q = \frac{2}{3}$ 6. $p^3 q^2 = (\frac{1}{6})^3 (\frac{5}{6})^2 \approx .003$
Matched Problems

7. $\binom{4}{0} p^4 + \binom{4}{1} p^3 q + \binom{4}{2} p^2 q^2 + \binom{4}{3} pq^3 + \binom{4}{4} q^4$

   $= p^4 + 4p^3 q + 6p^2 q^2 + 4pq^3 + q^4$

8. $\binom{7}{2} p^5 q^2 = 21 p^5 q^2$

9.   (A) $p(1) = \binom{4}{1} \left(\frac{1}{2} \right)^1 \left(\frac{1}{2} \right)^3 = .25$

   (B) $P(X_4 \leq 1) = p(0) + p(1)$

   $= \binom{4}{0} \left(\frac{1}{2} \right)^0 \left(\frac{1}{2} \right)^4 + \binom{4}{1} \left(\frac{1}{2} \right)^1 \left(\frac{1}{2} \right)^3 = .3125$

10.   (A) $p(x) = P(x \text{ successes in 2 trials}) = \binom{2}{x} \left(\frac{1}{3} \right)^x \left(\frac{2}{3} \right)^{2-x}, \quad x \in \{0, 1, 2\}$

(B)

x	p(x)
0	$\frac{4}{9} \approx .44$
1	$\frac{4}{9} \approx .44$
2	$\frac{1}{9} \approx .12$

(C)   p(x)

Number of successes, x

11. $\mu = .67;$ $\sigma = .67$

12. (A) $p(x) = P(\text{Exactly } x \text{ successes in 4 trials}) = \binom{4}{x}(.5)^4$

(B)

x	p(x)
0	.06
1	.25
2	.38
3	.25
4	.06
	1.00

(C)

Number of successes, x

(D) $\mu = 2;$ $\sigma = 1$

Exercise 17-2

A   A fair coin is tossed three times. What is the probability of obtaining:

1. Exactly two heads?
2. Exactly one head?
3. At least two heads?
4. At least one head?

A fair die is rolled four times. What is the probability of rolling:

5. Exactly three 2's?
6. Exactly two 3's?
7. At least one 6?
8. At least one 4?

B   Construct a histogram for each of the binomial distributions in Problems 9–14. Compute the mean and standard deviation for each distribution.

9. $p(x) = \binom{2}{x}(.3)^x(.7)^{2-x}$

10. $p(x) = \binom{2}{x}(.7)^x(.3)^{2-x}$

11. $p(x) = \binom{4}{x}(.5)^x(.5)^{4-x}$

12. $p(x) = \binom{6}{x}(.6)^x(.4)^{6-x}$

13. $p(x) = \binom{8}{x}(.3)^x(.7)^{8-x}$

14. $p(x) = \binom{8}{x}(.7)^x(.3)^{8-x}$

C   15.   Toss a coin three times or toss three coins simultaneously, and record the number of heads. Repeat the binomial experiment 100 times and compare your relative frequency distribution with the theoretical probability distribution.

16. Roll a die three times or roll three dice simultaneously, and record the number of 5's that occur. Repeat the binomial experiment 100 times and compare your relative frequency distribution with the theoretical probability distribution.

A coin is loaded so that the probability of a head occurring on a single toss is $\frac{3}{4}$. In five tosses of the coin, what is the probability of getting:

17. All heads or all tails?
18. Exactly two heads or exactly two tails?

■

Applications

Business & Economics

19. *Quality control.* A manufacturing process produces on the average 5 defective items out of 100. To control quality, each day a random sample of six completed items is selected and inspected. If a success on a single trial (inspection of one item) is finding the item defective, then the inspection of each of the 6 items in the sample constitutes a binomial experiment, which has a binomial distribution.

   (A) Write the function defining the distribution.
   (B) Construct a table for the distribution.
   (C) Draw a histogram.
   (D) Compute the mean and standard deviation.

20. *Quality control.* Refer to Problem 19. If the sample of six items that are inspected contains two or more defective items, then the whole day's output is inspected and the manufacturing process is reviewed. What is the probability of this happening, assuming that the process is still producing 5 defective items out of 100?

21. *Guarantees.* A manufacturing process produces, on the average, 3% defective items. The company ships ten items in each box and wishes to guarantee no more than one defective item per box. If this guarantee accompanies each box, what is the probability that the box will fail to satisfy the guarantee?

22. *Management training.* Each year a company selects 20 employees for a management training program given at a nearby university. On the average, 70% of those sent complete the course. Compute the mean and standard deviation for the number of employees who complete the program each year.

Life Sciences

23. *Epidemics.* If the probability of a person contracting influenza on exposure is .6, consider the binomial distribution for a family of six that has been exposed.

(A) Write the function defining the distribution.
(B) Construct a table for the distribution.
(C) Draw a histogram.
(D) Compute the mean and standard deviation.

24. *Epidemics.* Refer to Problem 23. Out of a family of six exposed to the virus, what is the probability that:

(A) No one will contract the disease?
(B) All will contract the disease?
(C) Exactly two will contract the disease?
(D) At least two will contract the disease?

25. *Genetics.* The probability that brown-eyed parents, both with the recessive gene for blue, will have a child with brown eyes is .75. If such parents have five children, what is the probability that they will have:

(A) All blue-eyed children?
(B) Exactly three children with brown eyes?
(C) At least three children with brown eyes?

26. *Side effects of drugs.* The probability that a given drug will produce a serious side effect in a person using the drug is .02. In the binomial distribution for 450 people using the drug, what are the mean and standard deviation?

Social Sciences   27. *Testing.* A multiple-choice test is given with five choices for each of five questions. Answering each of the five questions by guessing constitutes a binomial experiment with an associated binomial distribution.

(A) Write the function defining the distribution.
(B) Construct a table for the distribution.
(C) Draw a histogram.
(D) Compute the mean and standard deviation.

28. *Testing.* Refer to Problem 27. What is the probability of passing the test with a grade of 60% or better just by guessing?

29. *Opinion polls.* An opinion poll based on a small sample can be unrepresentative of the population. To see why, let us assume that 40% of the electorate favors a certain candidate. If a random sample of seven are asked their preference, what is the probability that a majority will favor the candidate?

30. *Sociology.* The probability that a marriage will end in divorce within 10 years is .4. What are the mean and standard deviation for the binomial distribution involving 1,000 marriages?

17-3   Continuous Random Variables

■ Continuous Random Variable
■ Probability Density Function
■ Comparing Probability Distribution Functions and Probability Density
  Functions
■ Cumulative Probability Distribution Function

■ Continuous Random Variable

All the random variables we considered in the preceding section assumed
one of a finite number of possible values. But in many experiments we use
random variables that can assume any one of an infinite number of possible
values. For example, we may be interested in the life expectancy of a
circuit in a computer, the time it takes a rat to find its way through a maze,
or the amount of a certain drug present in an individual's bloodstream. In
experiments of this type, the set of all possible outcomes forms an interval
on the real line. Thus, the life expectancy of a computer's circuit could be
any value in the interval $[0, \infty)$, and the transit time of a rat in a maze might
always lie in the interval $[5, 60]$. A random variable associated with this
kind of experiment is usually called *continuous*.

Continuous Random Variable

A **continuous random variable** is a random variable whose set of
possible values (range) is an interval on the real line. This interval
may be open or closed, and it may be bounded or unbounded.

The term *continuous* is not used in the same sense here as it was used in
Section 10-2. In this case, it refers to the fact that the values of the random
variable form a continuous set of numbers, such as $[0, \infty)$, rather than a
discrete set, such as $\{0, 1, 2, 3\}$. In fact, random variables of the type we
considered in the preceding two sections are often called **discrete random
variables** to emphasize the difference between them and continuous random variables.

■ Probability Density Function

In order to work with continuous random variables, we must have some
way of defining the probability of an event. Since there are an infinite

number of possible outcomes, we cannot define the probability of each outcome by means of a table. Instead, we introduce a new function that is used to compute probabilities. For convenience in stating definitions and formulas, we will assume that the value of a continuous random variable can be any real number; that is, the range is $(-\infty, \infty)$.

Probability Density Function

The function $f(x)$ is a **probability density function** for a continuous random variable X if:

1. $f(x) \geq 0$ for all $x \in (-\infty, \infty)$
2. $\int_{-\infty}^{\infty} f(x)\, dx = 1$
3. The probability that X lies in the interval $[c, d]$ is given by

$$P(c \leq X \leq d) = \int_{c}^{d} f(x)\, dx$$

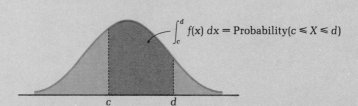

Range of $X = (-\infty, \infty) = $ Domain of f

Example 13 Let: $f(x) = \begin{cases} 12x^2 - 12x^3 & 0 \leq x \leq 1 \\ 0 & \text{otherwise} \end{cases}$

(A) Verify that f satisfies the first two conditions for a probability density function.

(B) Compute $P(\frac{1}{4} \leq X \leq \frac{3}{4})$, $P(X \leq \frac{1}{2})$, $P(X \geq \frac{2}{3})$, and $P(X = \frac{1}{3})$.

Solutions (A) For $0 \leq x \leq 1$, we have $f(x) = 12x^2 - 12x^3 = 12x^2(1 - x) \geq 0$. Since $f(x) = 0$ for all other values of x, it follows that $f(x) \geq 0$ for all x. Also,

$$\int_{-\infty}^{\infty} f(x)\, dx = \int_{0}^{1} (12x^2 - 12x^3)\, dx = (4x^3 - 3x^4)\Big|_{0}^{1} = (4 - 3) - (0) = 1$$

(B) $P(\frac{1}{4} \leqslant X \leqslant \frac{3}{4}) = \displaystyle\int_{1/4}^{3/4} f(x)\, dx$

$= \displaystyle\int_{1/4}^{3/4} (12x^2 - 12x^3)\, dx$

$= (4x^3 - 3x^4)\Big|_{1/4}^{3/4}$

$= \frac{189}{256} - \frac{13}{256} = \frac{11}{16}$

$P(X \leqslant \frac{1}{2}) = \displaystyle\int_{-\infty}^{1/2} f(x)\, dx$ Note that $f(x) = 0$ for $x < 0$.

$= \displaystyle\int_{0}^{1/2} (12x^2 - 12x^3)\, dx$

$= (4x^3 - 3x^4)\Big|_{0}^{1/2}$

$= \frac{5}{16}$

$P(X \geqslant \frac{2}{3}) = \displaystyle\int_{2/3}^{\infty} f(x)\, dx$ Note that $f(x) = 0$ for $x > 1$.

$= \displaystyle\int_{2/3}^{1} (12x^2 - 12x^3)\, dx$

$= (4x^3 - 3x^4)\Big|_{2/3}^{1}$

$= 1 - \frac{16}{27} = \frac{11}{27}$

$P(X = \frac{1}{3}) = \displaystyle\int_{1/3}^{1/3} f(x)\, dx = 0$ Property 1, page 849

Problem 13   Let: $f(x) = \begin{cases} 6x - 6x^2 & 0 \leqslant x \leqslant 1 \\ 0 & \text{otherwise} \end{cases}$

(A)   Verify that f satisfies the first two conditions for a probability density function.

(B)   Compute $P(\frac{1}{3} \leqslant X \leqslant \frac{2}{3})$, $P(X \leqslant \frac{1}{5})$, $P(X \geqslant \frac{1}{2})$, and $P(X = \frac{1}{4})$.

■ Comparing Probability Distribution Functions and Probability Density Functions

The last probability in Example 13 illustrates a fundamental difference between discrete and continuous random variables. In the discrete case, there is a *probability distribution* $p(x)$ that gives the probability of each possible value of the random variable. Thus, if c is one of the values of the random variable, then $P(X = c) = p(c)$. In the continuous case, the *integral*

*of the probability density function f(x) gives the probability that the out-
come lies in a certain interval.* If *c* is any real number, then the probability
that the outcome is *exactly c* is

$$P(X = c) = P(c \leqslant X \leqslant c) = \int_c^c f(x)\, dx = 0$$

Thus, $P(X = c) = 0$ for *any* number *c* and, since $f(c)$ is certainly not 0 for all
values of *c*, we see that $f(x)$ does not play the same role for a continuous
random variable as $p(x)$ does for a discrete random variable.

The fact that $P(X = c) = 0$ also implies that excluding either end point
from an interval does not change the probability that the random variable
lies in that interval; that is,

$$P(a < X < b) = P(a < X \leqslant b) = P(a \leqslant X < b) = P(a \leqslant X \leqslant b)$$

$$= \int_a^b f(x)\, dx$$

Example 14   Use the probability density function in Example 13 to compute
$P(.1 < X \leqslant .2)$ and $P(X > .9)$.

Solution $$P(.1 < X \leqslant .2) = \int_{.1}^{.2} f(x)\, dx \qquad\qquad f(x) = \begin{cases} 12x^2 - 12x^3 & 0 \leqslant x \leqslant 1 \\ 0 & \text{otherwise} \end{cases}$$

$$= \int_{.1}^{.2} (12x^2 - 12x^3)\, dx$$

$$= (4x^3 - 3x^4)\Big|_{.1}^{.2}$$

$$= .0272 - .0037 = .0235$$

$$P(X > .9) = \int_{.9}^{\infty} f(x)\, dx$$

$$= \int_{.9}^{1} (12x^2 - 12x^3)\, dx$$

$$= (4x^3 - 3x^4)\Big|_{.9}^{1}$$

$$= 1 - .9477 = .0523$$

Problem 14   Use the probability density function in Problem 13 to compute
$P(.2 \leqslant X < .4)$ and $P(X < .8)$.

If $f(x)$ is the probability density function in Example 13, notice that
$f(\frac{2}{3}) = \frac{16}{9} > 1$. Thus, a probability density function can assume values larger
than 1. This illustrates another difference between probability density
functions and probability distribution functions. In terms of inequalities, a
probability distribution function must always satisfy $0 \leqslant p(x) \leqslant 1$, while a

probability density function need only satisfy $f(x) \geqslant 0$. Despite these differences, we shall see that there are many similarities in the application of probability distribution functions and probability density functions.

Example 15
Shelf-Life

The shelf-life (in months) of a certain drug is a continuous random variable with probability density function

$$f(x) = \begin{cases} 50/(x+50)^2 & x \geqslant 0 \\ 0 & \text{otherwise} \end{cases}$$

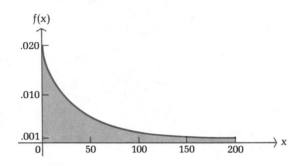

Find the probability that the drug has a shelf-life of:

(A)   Between 10 and 20 months   (B)   At most 30 months
(C)   Over 25 months

Solutions   (A) $P(10 \leqslant X \leqslant 20) = \displaystyle\int_{10}^{20} f(x)\, dx = \int_{10}^{20} \frac{50}{(x+50)^2}\, dx = \frac{-50}{x+50}\Big|_{10}^{20}$

$$= \left(-\tfrac{50}{70}\right) - \left(-\tfrac{50}{60}\right) = \tfrac{5}{42}$$

(B) $P(X \leqslant 30) = \displaystyle\int_{-\infty}^{30} f(x)\, dx = \int_{0}^{30} \frac{50}{(x+50)^2}\, dx = \frac{-50}{x+50}\Big|_{0}^{30}$

$$= \left(-\tfrac{50}{80}\right) - (-1) = \tfrac{3}{8}$$

(C) $P(X > 25) = \displaystyle\int_{25}^{\infty} f(x)\, dx = \int_{25}^{\infty} \frac{50}{(x+50)^2}\, dx$

$$= \lim_{R\to\infty} \int_{25}^{R} \frac{50}{(x+50)^2}\, dx = \lim_{R\to\infty} \frac{-50}{x+50}\Big|_{25}^{R}$$

$$= \lim_{R\to\infty} \left(-\frac{50}{R+50} + \frac{50}{75}\right)$$

$$= \tfrac{2}{3}$$

Problem 15   In Example 15 find the probability that the drug has a shelf-life of:

(A)   Between 50 and 100 months   (B)   At most 20 months
(C)   Over 10 months

▪ Cumulative Probability Distribution Function

Each time we compute the probability for a continuous random variable, we must find the antiderivative of the probability density function. This antiderivative is used so often that it is convenient to give it a name.

Cumulative Probability Distribution Function

If f is a probability density function, then the associated **cumulative probability distribution function** F is defined by

$$F(x) = P(X \leq x) = \int_{-\infty}^{x} f(t)\, dt$$

Furthermore,

$$P(c \leq X \leq d) = F(d) - F(c)$$

Figure 4 gives a geometric interpretation of these ideas.

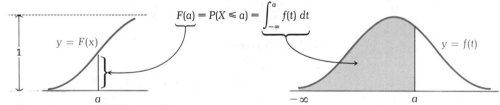

(A)   Cumulative probability distribution function   (B)   Probability density function

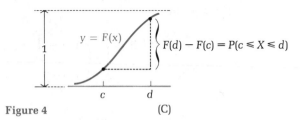

Figure 4   (C)

Notice that $F(x) = \int_{-\infty}^{x} f(t)\, dt$ is a function of x, the upper limit of integration, not t, the variable in the integrand. We state some important properties of cumulative probability distribution functions in Theorem 4 (next page). These properties follow directly from the fact that $F(x)$ can be interpreted geometrically as the area under the graph of $y = f(t)$ from $-\infty$ to x (see Section 14-5).

Theorem 4

> **Properties of Cumulative Probability Distribution Functions**
>
> If f is a probability density function and
>
> $$F(x) = \int_{-\infty}^{x} f(t)\, dt$$
>
> is the associated cumulative probability distribution function, then:
>
> 1. $F'(x) = f(x)$ wherever f is continuous
> 2. $0 \leqslant F(x) \leqslant 1, \quad -\infty < x < \infty$
> 3. $F(x)$ is nondecreasing on $(-\infty, \infty)^*$

Example 16   Find the cumulative probability distribution function for the probability density function in Example 13, and use it to compute $P(.1 \leqslant X \leqslant .9)$.

Solution   If $x < 0$, then

$$F(x) = \int_{-\infty}^{x} f(t)\, dt \qquad f(x) = \begin{cases} 12x^2 - 12x^3 & 0 \leqslant x \leqslant 1 \\ 0 & \text{otherwise} \end{cases}$$

$$= \int_{-\infty}^{x} 0\, dt = 0$$

If $0 \leqslant x \leqslant 1$, then

$$F(x) = \int_{-\infty}^{x} f(t)\, dt = \int_{-\infty}^{0} f(t)\, dt + \int_{0}^{x} f(t)\, dt$$

$$= 0 + \int_{0}^{x} (12t^2 - 12t^3)\, dt = (4t^3 - 3t^4)\Big|_{0}^{x}$$

$$= 4x^3 - 3x^4$$

If $x > 1$, then

$$F(x) = \int_{-\infty}^{x} f(t)\, dt = \int_{-\infty}^{0} f(t)\, dt + \int_{0}^{1} f(t)\, dt + \int_{1}^{x} f(t)\, dt$$

$$= 0 + 1 + 0 = 1$$

Thus,

$$F(x) = \begin{cases} 0 & x < 0 \\ 4x^3 - 3x^4 & 0 \leqslant x \leqslant 1 \\ 1 & x > 1 \end{cases}$$

And

$$P(.1 \leqslant X \leqslant .9) = F(.9) - F(.1) = .9477 - .0037 = .944$$

* A function $F(x)$ is nondecreasing on (a, b) if $F(x_1) \leqslant F(x_2)$ for $a < x_1 < x_2 < b$.

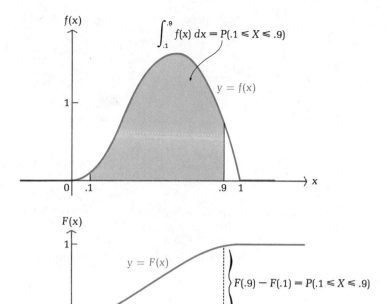

Problem 16   Find the cumulative probability distribution function for the probability density function in Problem 13, and use it to compute $P(.3 \le X \le .7)$.

Example 17
Shelf-Life

Returning to the discussion of the shelf-life of a drug in Example 15, suppose a pharmacist wants to be 95% certain that the drug is still good when it is sold. How long is it safe to leave the drug on the shelf?

Solution

Let x be the number of months the drug has been on the shelf when it is sold. The probability that the shelf-life of the drug is less than the number of months it has been sitting on the shelf is $P(0 \le X \le x)$. The pharmacist wants this probability to be .05. Thus, we must solve the equation $P(0 \le X \le x) = .05$ for x. First, we will find the cumulative probability distribution function F. For $x < 0$, we see that $F(x) = 0$. For $x \ge 0$,

$$F(x) = \int_0^x \frac{50}{(50+t)^2}\, dt = \frac{-50}{50+t}\Big|_0^x = \frac{-50}{50+x} - (-1) = 1 - \frac{50}{50+x}$$

$$= \frac{x}{50+x}$$

Thus,

$$F(x) = \begin{cases} 0 & x < 0 \\ x/(50+x) & x \ge 0 \end{cases}$$

Now, to solve the equation $P(0 \leqslant X \leqslant x) = .05$, we solve

$$F(x) - F(0) = .05 \qquad F(0) = 0$$

$$\frac{x}{50 + x} = .05$$

$$x = 2.5 + .05x$$

$$.95x = 2.5$$

$$x \approx 2.6$$

If the drug is sold during the first 2.6 months it is on the shelf, then the probability that it is still good is .95.

Problem 17   Repeat Example 17 if the pharmacist wants the probability that the drug is still good to be .99.

Answers to   13. (B) $\frac{13}{27}; \frac{13}{125}; \frac{1}{2}; 0$ 14. .248; .896
Matched Problems   15. (A) $\frac{1}{6}$ (B) $\frac{2}{7}$ (C) $\frac{5}{6}$

16. $F(x) = \begin{cases} 0 & x < 0 \\ 3x^2 - 2x^3 & 0 \leqslant x \leqslant 1 \\ 1 & x > 1 \end{cases}$ $P(.3 \leqslant X \leqslant .7) = .568$

17. Approx. $\frac{1}{2}$ month or 15 days

Exercise 17-3

A   *Problems 1–12 refer to the continuous random variable X with probability density function*

$$f(x) = \begin{cases} \frac{1}{8}x & 0 \leqslant x \leqslant 4 \\ 0 & \text{otherwise} \end{cases}$$

1. Graph f and verify that f satisfies the first two conditions for a probability density function.
2. Find $P(1 \leqslant X \leqslant 3)$ and illustrate with a graph.
3. Find $P(2 < X < 3)$.   4. Find $P(X \leqslant 2)$.
5. Find $P(X > 3)$.   6. Find $P(X = 1)$.
7. Find $P(X > 5)$.   8. Find $P(X < 5)$.
9. Find and graph the associated cumulative probability distribution function.
10. Use the associated cumulative probability distribution function to find $P(2 \leqslant X \leqslant 4)$ and illustrate with a graph.
11. Use the associated cumulative probability distribution function to find $P(0 < X < 2)$ and illustrate with a graph.
12. Use the associated cumulative probability distribution function to find the value of x that satisfies $P(0 \leqslant X \leqslant x) = \frac{1}{2}$.

B   *Problems 13–20 refer to the continuous random variable X with probability density function*

$$f(x) = \begin{cases} 2/(1+x)^3 & x \geqslant 0 \\ 0 & \text{otherwise} \end{cases}$$

13. Graph f and verify that f satisfies the first two conditions for a probability density function.
14. Find $P(1 \leqslant X \leqslant 4)$ and illustrate with a graph.
15. Find $P(X > 3)$.   16. Find $P(X \leqslant 2)$.
17. Find $P(X = 1)$.   18. Find $P(X > -1)$.
19. Find and graph the associated cumulative probability distribution function.
20. Use the associated cumulative probability distribution function to find the value of x that satisfies $P(0 \leqslant X \leqslant x) = \frac{3}{4}$.
21. Find the associated cumulative probability distribution function for

$$f(x) = \begin{cases} \frac{3}{2}x - \frac{3}{4}x^2 & 0 \leqslant x \leqslant 2 \\ 0 & \text{otherwise} \end{cases}$$

   Graph both functions (on separate sets of axes).
22. Repeat Problem 21 for

$$f(x) = \begin{cases} e^{-x} & x \geqslant 0 \\ 0 & \text{otherwise} \end{cases}$$

In Problems 23–26 find the associated cumulative probability function, and use it to find the indicated probability.

23. Find $P(1 \leqslant X \leqslant 2)$ for   24. Find $P(1 \leqslant X \leqslant 2)$ for

$$f(x) = \begin{cases} \ln x & 1 \leqslant x \leqslant e \\ 0 & \text{otherwise} \end{cases} \qquad f(x) = \begin{cases} 3x/(8\sqrt{1+x}) & 0 \leqslant x \leqslant 3 \\ 0 & \text{otherwise} \end{cases}$$

25. Find $P(X \leqslant 1)$ for   26. Find $P(X \geqslant e)$ for

$$f(x) = \begin{cases} xe^{-x} & x \geqslant 0 \\ 0 & \text{otherwise} \end{cases} \qquad f(x) = \begin{cases} (\ln x)/x^2 & x \geqslant 1 \\ 0 & \text{otherwise} \end{cases}$$

C   *In Problems 27–30 F(x) is the cumulative probability distribution function for a continuous random variable X. Find the probability density function f(x) associated with each F(x).*

27. $F(x) = \begin{cases} 0 & x < 0 \\ x^2 & 0 \leqslant x \leqslant 1 \\ 1 & x > 1 \end{cases}$ 28. $F(x) = \begin{cases} 0 & x < 1 \\ \frac{1}{2}x - \frac{1}{2} & 1 \leqslant x \leqslant 3 \\ 1 & x > 3 \end{cases}$

29. $F(x) = \begin{cases} 0 & x < 0 \\ 6x^2 - 8x^3 + 3x^4 & 0 \leqslant x \leqslant 1 \\ 1 & x > 1 \end{cases}$

30. $F(x) = \begin{cases} 1 - (1/x^3) & x \geqslant 1 \\ 0 & \text{otherwise} \end{cases}$

In Problems 31–34, find the associated cumulative distribution function.

31. $f(x) = \begin{cases} x & 0 \leqslant x \leqslant 1 \\ 2-x & 1 < x \leqslant 2 \\ 0 & \text{otherwise} \end{cases}$

32. $f(x) = \begin{cases} \frac{1}{4} & 0 \leqslant x \leqslant 1 \\ \frac{1}{2} & 1 < x \leqslant 2 \\ \frac{1}{4} & 2 < x \leqslant 3 \\ 0 & \text{otherwise} \end{cases}$

33. $f(x) = \begin{cases} |x| & -1 \leqslant x \leqslant 1 \\ 0 & \text{otherwise} \end{cases}$

34. $f(x) = \begin{cases} |x| + \frac{1}{2}x & -1 \leqslant x \leqslant 1 \\ 0 & \text{otherwise} \end{cases}$

Applications

Business & Economics

35. *Time-sharing.* In a computer time-sharing network, the time it takes (in seconds) to respond to a user's request is a continuous random variable with probability density given by

$$f(x) = \begin{cases} \frac{1}{10}e^{-x/10} & x \geqslant 0 \\ 0 & \text{otherwise} \end{cases}$$

(A) What is the probability that the computer responds within 1 second?

(B) What is the probability that a user must wait over 4 seconds for a response?

36. *Gasoline consumption.* The daily demand for gasoline (in millions of gallons) in a certain area is a continuous random variable with probability density given by

$$f(x) = \begin{cases} \frac{1}{4}xe^{-x/2} & x \geqslant 0 \\ 0 & \text{otherwise} \end{cases}$$

(A) What is the probability that no more than 1 million gallons are demanded?

(B) What is the probability that 2 million gallons will not be sufficient to meet the daily demand?

37. *Demand.* The weekly demand for hamburger (in thousands of pounds) for a chain of supermarkets is a continuous random variable with probability density given by

$$f(x) = \begin{cases} 0.003x\sqrt{100 - x^2} & 0 \leqslant x \leqslant 10 \\ 0 & \text{otherwise} \end{cases}$$

(A) What is the probability that more than 4,000 pounds of hamburger are demanded?

(B) The manager of the meat department orders 8,000 pounds of hamburger. What is the probability that the demand will not exceed this amount?

(C) The manager wants the probability that the demand does not exceed the amount ordered to be .9. How much hamburger meat should be ordered?

38. *Demand.* Repeat Problem 37 if

$$f(x) = \begin{cases} \frac{1}{2,500}x^2(1,000 - x^3)^{1/3} & 0 \leqslant x \leqslant 10 \\ 0 & \text{otherwise} \end{cases}$$

Life Sciences

39. *Life expectancy.* The life expectancy (in minutes) of a certain microscopic organism is a continuous random variable with probability density function given by

$$f(x) = \begin{cases} \frac{1}{5,000}(10x^3 - x^4) & 0 \leqslant x \leqslant 10 \\ 0 & \text{otherwise} \end{cases}$$

(A) What is the probability that an organism lives for at least 7 minutes?

(B) What is the probability that an organism lives for at most 5 minutes?

40. *Shelf-life.* The shelf-life (in days) of a perishable drug is a continuous random variable with probability density function given by

$$f(x) = \begin{cases} 200x/(100 + x^2)^2 & x \geqslant 0 \\ 0 & \text{otherwise} \end{cases}$$

(A) What is the probability that the drug has a shelf-life of at most 10 days?

(B) What is the probability that the shelf-life exceeds 15 days?

(C) If the user wants the probability that the drug is still good to be .8, when is the last time it should be used?

Social Sciences

41. *Learning.* The number of words per minute a beginner can type after 1 week of practice is a continuous random variable with probability density given by

$$f(x) = \begin{cases} \frac{1}{400}xe^{-x/20} & x \geqslant 0 \\ 0 & \text{otherwise} \end{cases}$$

(A) What is the probability that a beginner can type at least 30 words per minute after 1 week of practice?

(B) What is the probability that a beginner can type at least 80 words per minute after 1 week of practice?

42. *Learning.* The number of hours it takes a chimpanzee to learn a new task is a continuous random variable with probability density given by

$$f(x) = \begin{cases} \frac{4}{9}x^2 - \frac{4}{27}x^3 & 0 \leqslant x \leqslant 3 \\ 0 & \text{otherwise} \end{cases}$$

(A) What is the probability that the chimpanzee learns the task in the first hour?

(B) What is the probability that the chimpanzee does not learn the task in the first 2 hours?

17-4   Expected Value, Standard Deviation, and Median of Continuous Random Variables

μ - mean

- ■ Expected Value and Standard Deviation
- ■ Alternate Formula for Variance
- ■ Median

■ Expected Value and Standard Deviation

In Section 17-1 we discussed the expected value, variance, and standard deviation for discrete random variables. The formulas for these quantities can be generalized to the continuous case by replacing the finite summation operation with integration. Compare the formulas below with those in Section 17-1.

Expected Value and Standard Deviation for a Continuous Random Variable

Let $f(x)$ be the probability density function for a continuous random variable X. The **expected value, or mean, of X** is

$$\mu = E(X) = \int_{-\infty}^{\infty} xf(x)\,dx$$

The **variance** is

$$V(X) = \int_{-\infty}^{\infty} (x - \mu)^2 f(x)\,dx$$

and the **standard deviation** is

$$\sigma = \sqrt{V(X)}$$

The standard deviation of a continuous random variable measures the dispersion of the probability density function about the mean, just as in the discrete case. This is illustrated in Figure 5. The probability density function in Figure 5A has a standard deviation of 1. Most of the area under the curve is near the mean. In Figure 5B, the standard deviation is four times as large, and the area under the graph is much more spread out.

Example 18   Find the mean, variance, and standard deviation for

$$f(x) = \begin{cases} 12x^2 - 12x^3 & 0 \leq x \leq 1 \\ 0 & \text{otherwise} \end{cases}$$

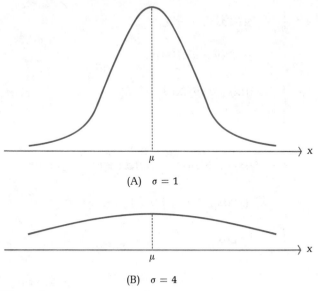

(A) $\sigma = 1$

(B) $\sigma = 4$

Figure 5

Solution

$$\mu = E(X) = \int_{-\infty}^{\infty} xf(x)\,dx = \int_{0}^{1} x(12x^2 - 12x^3)\,dx = \int_{0}^{1}(12x^3 - 12x^4)\,dx$$

$$= (3x^4 - \tfrac{12}{5}x^5)\Big|_{0}^{1} = \tfrac{3}{5}$$

$$V(X) = \int_{-\infty}^{\infty}(x-\mu)^2 f(x)\,dx = \int_{0}^{1}(x-\tfrac{3}{5})^2(12x^2 - 12x^3)\,dx$$

$$= \int_{0}^{1}(x^2 - \tfrac{6}{5}x + \tfrac{9}{25})(12x^2 - 12x^3)\,dx$$

$$= \int_{0}^{1}(\tfrac{108}{25}x^2 - \tfrac{468}{25}x^3 + \tfrac{132}{5}x^4 - 12x^5)\,dx$$

$$= (\tfrac{36}{25}x^3 - \tfrac{117}{25}x^4 + \tfrac{132}{25}x^5 - 2x^6)\Big|_{0}^{1} = \tfrac{1}{25}$$

$$\sigma = \sqrt{V(X)} = \sqrt{\tfrac{1}{25}} = \tfrac{1}{5}$$

Problem 18 Find the mean, variance, and standard deviation for

$$f(x) = \begin{cases} 6x - 6x^2 & 0 \leqslant x \leqslant 1 \\ 0 & \text{otherwise} \end{cases}$$

■ Alternate Formula for Variance

The term $(x - \mu)^2$ in the formula for $V(X)$ introduces some complicated algebraic manipulations in the evaluation of the integral. We can use the properties of the definite integral to simplify this formula. Thus,

$$V(X) = \int_{-\infty}^{\infty} (x - \mu)^2 f(x)\, dx \qquad\qquad \text{Expand } (x - \mu)^2.$$

$$= \int_{-\infty}^{\infty} (x^2 - 2x\mu + \mu^2) f(x)\, dx \qquad\qquad \text{Multiply by } f(x).$$

$$= \int_{-\infty}^{\infty} [x^2 f(x) - 2x\mu f(x) + \mu^2 f(x)]\, dx \qquad \text{Use property 4, page 849.}$$

$$= \int_{-\infty}^{\infty} x^2 f(x)\, dx - \int_{-\infty}^{\infty} 2x\mu f(x)\, dx + \int_{-\infty}^{\infty} \mu^2 f(x)\, dx \qquad \text{Use property 3, page 849.}$$

$$= \int_{-\infty}^{\infty} x^2 f(x)\, dx - 2\mu \int_{-\infty}^{\infty} x f(x)\, dx + \mu^2 \int_{-\infty}^{\infty} f(x)\, dx \qquad \int_{-\infty}^{\infty} x f(x)\, dx = \mu, \quad \int_{-\infty}^{\infty} f(x)\, dx = 1$$

$$= \int_{-\infty}^{\infty} x^2 f(x)\, dx - 2\mu(\mu) + \mu^2(1)$$

$$= \int_{-\infty}^{\infty} x^2 f(x)\, dx - \mu^2$$

In general, it will be easier to evaluate $\int_{-\infty}^{\infty} x^2 f(x)\, dx$ than to evaluate $\int_{-\infty}^{\infty} (x - \mu)^2 f(x)\, dx$.

Theorem 5

> **Alternate Formula for Variance**
>
> $$V(X) = \int_{-\infty}^{\infty} x^2 f(x)\, dx - \mu^2$$

Example 19   Use the alternate formula for variance (Theorem 5) to compute the variance in Example 18.

Solution   From Example 18, we have $\mu = \int_{-\infty}^{\infty} x f(x)\, dx = \frac{3}{5}$.

$$\int_{-\infty}^{\infty} x^2 f(x)\, dx = \int_0^1 x^2 (12x^2 - 12x^3)\, dx \qquad f(x) = \begin{cases} 12x^2 - 12x^3 & 0 \le x \le 1 \\ 0 & \text{otherwise} \end{cases}$$

$$= \int_0^1 (12x^4 - 12x^5)\, dx$$

$$= \left(\tfrac{12}{5} x^5 - \tfrac{12}{6} x^6 \right) \Big|_0^1 = \tfrac{2}{5}$$

$$V(X) = \int_{-\infty}^{\infty} x^2 f(x)\, dx - \mu^2 = \tfrac{2}{5} - \left(\tfrac{3}{5}\right)^2 = \tfrac{1}{25}$$

Problem 19   Use the alternate formula for variance (Theorem 5) to compute the variance in Problem 18.

Example 20   Find the mean, variance, and standard deviation for

$$f(x) = \begin{cases} 3/x^4 & x \geq 1 \\ 0 & \text{otherwise} \end{cases}$$

Solution

$$\mu = \int_{-\infty}^{\infty} xf(x)\,dx = \int_1^{\infty} x\frac{3}{x^4}\,dx = \lim_{R \to \infty} \int_1^R \frac{3}{x^3}\,dx$$

$$= \lim_{R \to \infty} \left[-\frac{3}{2}\left(\frac{1}{x^2}\right) \right]\Big|_1^R = \lim_{R \to \infty} \left[-\frac{3}{2}\left(\frac{1}{R^2}\right) + \frac{3}{2} \right] = \frac{3}{2}$$

$$\int_{-\infty}^{\infty} x^2 f(x)\,dx = \int_1^{\infty} x^2 \frac{3}{x^4}\,dx = \lim_{R \to \infty} \int_1^R \frac{3}{x^2}\,dx = \lim_{R \to \infty} \left(-\frac{3}{x} \right)\Big|_1^R$$

$$= \lim_{R \to \infty} \left(-\frac{3}{R} + 3 \right) = 3$$

$$V(X) = \int_{-\infty}^{\infty} x^2 f(x)\,dx - \mu^2 = 3 - \left(\frac{3}{2} \right)^2 = \frac{3}{4}$$

$$\sigma = \sqrt{V(X)} = \sqrt{\tfrac{3}{4}} = \frac{\sqrt{3}}{2} \approx .8660$$

Problem 20   Find the mean, variance, and standard deviation for

$$f(x) = \begin{cases} 24/x^4 & x \geq 2 \\ 0 & \text{otherwise} \end{cases}$$

Example 21
Life Expectancy

The life expectancy (in hours) for a particular brand of light bulbs is a continuous random variable with probability density function

$$f(x) = \begin{cases} \frac{1}{100} - \frac{1}{20,000}x & 0 \leq x \leq 200 \\ 0 & \text{otherwise} \end{cases}$$

(A) What is the average life expectancy of one of these light bulbs?
(B) What is the probability that a bulb will last longer than this average?

Solution

(A) Since the value of this random variable is the number of hours a bulb lasts, the average life expectancy is just the expected value of the random variable. Thus,

$$E(X) = \int_{-\infty}^{\infty} xf(x)\,dx = \int_0^{200} x\left(\tfrac{1}{100} - \tfrac{1}{20,000}x\right) dx$$

$$= \int_0^{200} \left(\tfrac{1}{100}x - \tfrac{1}{20,000}x^2\right) dx = \left(\tfrac{1}{200}x^2 - \tfrac{1}{60,000}x^3\right)\Big|_0^{200}$$

$$= \tfrac{200}{3} \quad \text{or} \quad 66\tfrac{2}{3} \text{ hours}$$

(B)   The probability that a bulb lasts longer than $66\frac{2}{3}$ hours is

$$P(X > \tfrac{200}{3}) = \int_{200/3}^{\infty} f(x)\,dx = \int_{200/3}^{200} (\tfrac{1}{100} - \tfrac{1}{20,000}x)$$

$$= (\tfrac{1}{100}x - \tfrac{1}{40,000}x^2)\Big|_{200/3}^{200} = 1 - \tfrac{5}{9} = \tfrac{4}{9}$$

Problem 21   Repeat Example 21 if the probability density function is

$$f(x) = \begin{cases} \tfrac{1}{200} - \tfrac{1}{90,000}x & 0 \le x \le 300 \\ 0 & \text{otherwise} \end{cases}$$

■ Median

Another measurement often used to describe the properties of a random variable is the median. The **median** is the value of the random variable that divides the area under the graph of the probability density function into two equal parts (see Figure 6). If x_m is the median, then x_m must satisfy

$$P(X \le x_m) = \tfrac{1}{2}$$

Generally, this equation is solved by first finding the cumulative probability distribution function.

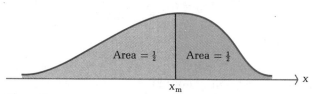

Figure 6

Example 22   Find the median of the continuous random variable with probability density function

$$f(x) = \begin{cases} 3/x^4 & x \ge 1 \\ 0 & \text{otherwise} \end{cases}$$

Solution   Step 1.   *Find the cumulative probability distribution function. For $x < 1$, we have $F(x) = 0$. If $x \ge 1$, then*

$$F(x) = \int_{-\infty}^{x} f(t)\,dt = \int_{1}^{x} \frac{3}{t^4}\,dt = -\frac{1}{t^3}\Big|_{1}^{x} = -\frac{1}{x^3} + 1 = 1 - \frac{1}{x^3}$$

Step 2.   *Solve the equation $P(X \le x_m) = \tfrac{1}{2}$ for x_m.*

$$F(x_m) = P(X \le x_m)$$

$$1 - \frac{1}{x_m^3} = \frac{1}{2}$$

$$\frac{1}{2} = \frac{1}{x_m^3}$$

$$x_m^3 = 2$$

$$x_m = \sqrt[3]{2}$$

Thus, the median is $\sqrt[3]{2} \approx 1.26$.

Problem 22   Find the median of the continuous random variable with probability density function

$$f(x) = \begin{cases} 24/x^4 & x \geqslant 2 \\ 0 & \text{otherwise} \end{cases}$$

Example 23
Life Expectancy

In Example 21, find the median life expectancy of a light bulb.

Solution   **Step 1.** *Find the cumulative probability distribution function.* If $x < 0$, we have $F(x) = 0$. If $0 \leqslant x \leqslant 200$, then

$$F(x) = \int_{-\infty}^{x} f(t)\, dt \qquad f(x) = \begin{cases} \frac{1}{100} - \frac{1}{20,000}x & 0 \leqslant x \leqslant 200 \\ 0 & \text{otherwise} \end{cases}$$

$$= \int_{0}^{x} (\tfrac{1}{100} - \tfrac{1}{20,000}t)\, dt$$

$$= (\tfrac{1}{100}t - \tfrac{1}{40,000}t^2) \Big|_{0}^{x}$$

$$= \tfrac{1}{100}x - \tfrac{1}{40,000}x^2$$

If $x > 200$, then

$$F(x) = \int_{-\infty}^{x} f(t)\, dt = \int_{-\infty}^{0} f(t)\, dt + \int_{0}^{200} f(t)\, dt + \int_{200}^{x} f(t)\, dt$$

$$= 0 + 1 + 0 = 1$$

Thus,

$$F(x) = \begin{cases} 0 & x < 0 \\ \tfrac{1}{100}x - \tfrac{1}{40,000}x^2 & 0 \leqslant x \leqslant 200 \\ 1 & x > 200 \end{cases}$$

Step 2.   *Solve the equation $P(X \leqslant x_m) = \frac{1}{2}$ for x_m.*

$$F(x_m) = P(X \leqslant x_m) = \tfrac{1}{2}$$

$$\tfrac{1}{100}x_m - \tfrac{1}{40,000}x_m^2 = \tfrac{1}{2} \qquad \text{The solution must occur for}$$

$$x_m^2 - 400x_m + 20,000 = 0 \qquad 0 \leqslant x_m \leqslant 200.$$

This quadratic equation has two solutions, $200 + 100\sqrt{2}$ and $200 - 100\sqrt{2}$. Since x_m must lie in the interval $[0, 200]$, the second root is the correct answer.

Thus, the median life expectancy is $200 - 100\sqrt{2} \approx 58.58$ hours.

$\frac{2}{100} - \frac{1}{200}$

$a + bx$

$\frac{1}{100}$

$Ans \frac{200}{100} \cdot 4$

Problem 23   In Problem 21, find the median life expectancy of a light bulb.

If you compare Examples 21 and 23, and Examples 20 and 22, you will see that the mean and the median generally are not equal (see Figure 7).

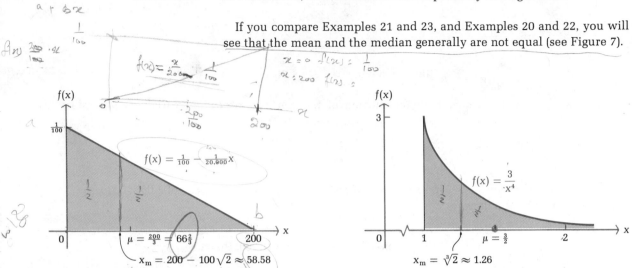

$f(x) = \frac{x}{200}$, $\frac{1}{100}$

$x = 0 \quad f(x) = \frac{1}{100}$
$x = 200 \quad f(x) =$

$f(x) = \frac{1}{100} - \frac{1}{20,900}x$

$\mu = \frac{200}{3} = 66\frac{2}{3}$

$x_m = 200 - 100\sqrt{2} \approx 58.58$

$f(x) = \frac{3}{x^4}$

$\mu = \frac{3}{2}$

$x_m = \sqrt[3]{2} \approx 1.26$

Figure 7

Answers to Matched Problems

18. $\mu = \frac{1}{2}$; $V(X) = \frac{1}{20}$; $\sigma \approx .2236$ 19. $\frac{1}{20}$
20. $\mu = 3$; $V(X) = 3$; $\sigma = \sqrt{3} \approx 1.732$ 21. (A) 125 hours   (B) $\frac{133}{288}$
22. $x_m = \sqrt[3]{16} = 2\sqrt[3]{2} \approx 2.52$ 23. $x_m = 450 - 150\sqrt{5} \approx 114.59$ hours

---

Exercise 17-4

A   Problems 1–8 refer to the random variable X with probability density function

$$f(x) = \begin{cases} \frac{1}{2}x & 0 \leq x \leq 2 \\ 0 & \text{otherwise} \end{cases}$$

1. Find the mean.
2. Find the variance.
3. Find the standard deviation.
4. Find the probability that the random variable is less than the mean.
5. Find the probability that the random variable is within 1 standard deviation of the mean (between $\mu - \sigma$ and $\mu + \sigma$).
6. Find the associated cumulative probability distribution function.
7. Find the median.
8. Find the probability that the random variable lies between the median and the mean.

B   Problems 9–16 refer to the continuous random variable X with probability density function

$$f(x) = \begin{cases} 4/x^5 & x \geq 1 \\ 0 & \text{otherwise} \end{cases}$$

   9. Find the mean.
   10. Find the variance.
   11. Find the standard deviation.
   12. Find the probability that the random variable is greater than the mean.
   13. Find the probability that the random variable is within 2 standard deviations of the mean (between $\mu - 2\sigma$ and $\mu + 2\sigma$).
   14. Find the associated cumulative probability density function.
   15. Find the median.
   16. Find the probability that the random variable is between the mean and the median.

In Problems 17–20 find the mean, variance, and standard deviation of the continuous random variable with the indicated probability density function.

17. $f(x) = \begin{cases} \frac{1}{3} & 2 \leq x \leq 5 \\ 0 & \text{otherwise} \end{cases}$ 18. $f(x) = \begin{cases} 3x^2 & 0 \leq x \leq 1 \\ 0 & \text{otherwise} \end{cases}$

19. $f(x) = \begin{cases} 1/(2\sqrt{1+x}) & 0 \leq x \leq 3 \\ 0 & \text{otherwise} \end{cases}$ 20. $f(x) = \begin{cases} \ln x & 1 \leq x \leq e \\ 0 & \text{otherwise} \end{cases}$

In Problems 21–24 find the median of the continuous random variable with the indicated probability density function.

21. $f(x) = \begin{cases} 1/x & 1 \leq x \leq e \\ 0 & \text{otherwise} \end{cases}$ 22. $f(x) = \begin{cases} 2/(1+x)^2 & 0 \leq x \leq 1 \\ 0 & \text{otherwise} \end{cases}$

23. $f(x) = \begin{cases} 1/(1+x)^2 & x \geq 0 \\ 0 & \text{otherwise} \end{cases}$ 24. $f(x) = \begin{cases} e^{-x} & x \geq 0 \\ 0 & \text{otherwise} \end{cases}$

C   In Problems 25 and 26 f(x) is a continuous probability density function with mean μ and standard deviation σ; a and b are constants. Evaluate each integral, expressing the result in terms of a, b, μ, and σ.

25. $\displaystyle\int_{-\infty}^{\infty} (ax + b)f(x)\, dx$ 26. $\displaystyle\int_{-\infty}^{\infty} (x - a)^2 f(x)\, dx$

In Problems 27 and 28 f(x) is an even continuous probability density function [f(−x) = f(x) for all x].

27. Find the mean.   28. Find the median.

The **quartile points** for a probability density function are the values x_1, x_2, x_3 that divide the area under the graph of the function into four equal parts.

Find the quartile points for the probability density functions in Problems 29 and 30.

29. $f(x) = \begin{cases} \frac{1}{2}x & 0 \leqslant x \leqslant 2 \\ 0 & \text{otherwise} \end{cases}$ **30.** $f(x) = \begin{cases} 1/(1+x)^2 & x \geqslant 0 \\ 0 & \text{otherwise} \end{cases}$

Applications

Business & Economics

31. *Profit.* A building contractor's profit (in thousands of dollars) on each unit in a subdivision is a continuous random variable with probability density given by

$$f(x) = \begin{cases} \frac{1}{8}(10 - x) & 6 \leqslant x \leqslant 10 \\ 0 & \text{otherwise} \end{cases}$$

(A)   What is the contractor's expected profit?
(B)   What is the median profit?

32. *Product life.* The life expectancy (in years) of an automobile battery is a continuous random variable with probability density given by

$$f(x) = \begin{cases} \frac{1}{2}e^{-x/2} & x \geqslant 0 \\ 0 & \text{otherwise} \end{cases}$$

Find the median life expectancy.

33. *Water consumption.* The daily consumption of water (in millions of gallons) in a small city is a continuous random variable with probability density given by

$$f(x) = \begin{cases} 1/(1 + x^2)^{3/2} & x \geqslant 0 \\ 0 & \text{otherwise} \end{cases}$$

Find the expected daily consumption.

34. *Demand.* The weekly demand for hamburger (in thousands of pounds) for a chain of supermarkets is a continuous random variable with probability density given by

$$f(x) = \begin{cases} 0.003x\sqrt{100 - x^2} & 0 \leqslant x \leqslant 10 \\ 0 & \text{otherwise} \end{cases}$$

Find the median demand for hamburger meat.

Life Sciences

35. *Life expectancy.* The life expectancy of a certain microscopic organism (in minutes) is a continuous random variable with probability density function given by

$$f(x) = \begin{cases} \frac{1}{5,000}(10x^3 - x^4) & 0 \leqslant x \leqslant 10 \\ 0 & \text{otherwise} \end{cases}$$

Find the mean life expectancy of one of these organisms.

36. *Shelf-life.* The shelf-life (in days) of a perishable drug is a continuous random variable with probability density given by

$$f(x) = \begin{cases} (200x)/(100 + x^2)^2 & x \geqslant 0 \\ 0 & \text{otherwise} \end{cases}$$

Find the median shelf-life of this drug.

Social Sciences

37. *Learning.* The number of hours it takes a chimpanzee to learn a new task is a continuous random variable with probability density given by

$$f(x) = \begin{cases} \frac{4}{9}x^2 - \frac{4}{27}x^3 & 0 \leqslant x \leqslant 3 \\ 0 & \text{otherwise} \end{cases}$$

What is the expected number of hours it will take a chimpanzee to learn the task?

17-5   Uniform, Beta, and Exponental Distributions

- Uniform Distribution
- Beta Distribution
- Exponential Distribution

In this section we will examine several important probability density functions. In actual practice, we do not usually construct a probability density function for each experiment. Instead, we try to select a known probability density function that seems to give a reasonable description of the experiment. Thus, it is important to be familiar with the properties and applications of a variety of probability density functions.

■ Uniform Distribution

To begin, suppose the outcome of an experiment can be any number in a certain finite interval $[a, b]$. If we believe that the probability of the outcome lying in a small interval of fixed length is independent of the location of this small interval within $[a, b]$, then we say that the continuous random variable for this experiment is **uniformly distributed** on the interval $[a, b]$. The **uniform probability density function** is

$$f(x) = \begin{cases} \dfrac{1}{b - a} & a \leqslant x \leqslant b \\ 0 & \text{otherwise} \end{cases}$$

Since $f(x) \geq 0$ and

$$\int_{-\infty}^{\infty} f(x)\, dx = \int_{a}^{b} \frac{1}{b-a}\, dx = \frac{x}{b-a}\Big|_{a}^{b} = \frac{b}{b-a} - \frac{a}{b-a} = 1$$

f satisfies the necessary conditions for a probability density function.

If F is the associated cumulative probability distribution function, then for $x < a$, $F(x) = 0$. For $a \leq x \leq b$, we have

$$F(x) = \int_{-\infty}^{x} f(t)\, dt = \int_{a}^{x} \frac{1}{b-a}\, dt = \frac{t}{b-a}\Big|_{a}^{x}$$

$$= \frac{x}{b-a} - \frac{a}{b-a} = \frac{x-a}{b-a}$$

For $x > b$, $F(x) = 1$.

Now we calculate the mean, median, and standard deviation for the uniform probability density function:

Mean

$$\mu = \int_{-\infty}^{\infty} xf(x)\, dx = \int_{a}^{b} \frac{x}{b-a}\, dx = \frac{x^2}{2(b-a)}\Big|_{a}^{b} = \frac{b^2}{2(b-a)} - \frac{a^2}{2(b-a)}$$

$$= \frac{b^2 - a^2}{2(b-a)} = \frac{(b-a)(b+a)}{2(b-a)} = \frac{1}{2}(a+b)$$

Median

$$F(x_m) = P(X \leq x_m)$$

$$\frac{x_m - a}{b-a} = \frac{1}{2}$$

$$x_m - a = \frac{1}{2}(b-a)$$

$$x_m = a + \frac{1}{2}(b-a) = \frac{1}{2}(a+b)$$

Standard Deviation

$$\int_{-\infty}^{\infty} x^2 f(x)\, dx = \int_{a}^{b} \frac{x^2}{b-a}\, dx = \frac{x^3}{3(b-a)}\Big|_{a}^{b} = \frac{b^3}{3(b-a)} - \frac{a^3}{3(b-a)}$$

$$= \frac{b^3 - a^3}{3(b-a)} = \frac{(b-a)(b^2 + ab + a^2)}{3(b-a)} = \frac{1}{3}(b^2 + ab + a^2)$$

$$V(X) = \int_{-\infty}^{\infty} x^2 f(x)\, dx - \mu^2 = \frac{1}{3}(b^2 + ab + a^2) - \left[\frac{1}{2}(a+b)\right]^2$$

$$= \frac{1}{3}(b^2 + ab + a^2) - \frac{1}{4}(a^2 + 2ab + b^2)$$

$$= \frac{4(b^2 + ab + a^2) - 3(a^2 + 2ab + b^2)}{12}$$

$$= \frac{b^2 - 2ab + a^2}{12} = \frac{1}{12}(b - a)^2$$

$$\sigma = \sqrt{V(X)} = \sqrt{\tfrac{1}{12}(b-a)^2} = \frac{1}{\sqrt{12}}(b - a)$$

These properties are summarized in the box.

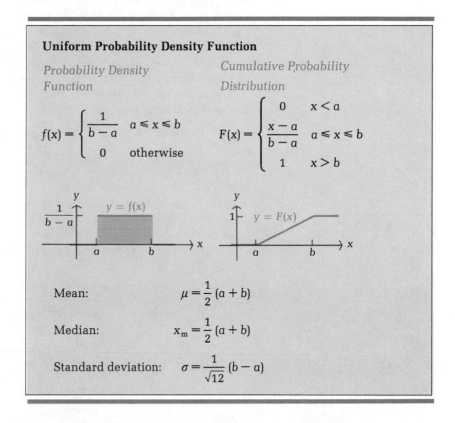

Uniform Probability Density Function

Probability Density Function

$$f(x) = \begin{cases} \dfrac{1}{b - a} & a \le x \le b \\ 0 & \text{otherwise} \end{cases}$$

Cumulative Probability Distribution

$$F(x) = \begin{cases} 0 & x < a \\ \dfrac{x - a}{b - a} & a \le x \le b \\ 1 & x > b \end{cases}$$

Mean: $\mu = \dfrac{1}{2}(a + b)$

Median: $x_m = \dfrac{1}{2}(a + b)$

Standard deviation: $\sigma = \dfrac{1}{\sqrt{12}}(b - a)$

Example 24
Electrical Current

Standard electrical current is uniformly distributed between 110 and 120 volts. What is the probability that the current is between 113 and 118 volts?

Solution

Since we are told that the current is uniformly distributed on the interval [110, 120], we choose the uniform probability density function

$$f(x) = \begin{cases} \frac{1}{10} & 110 \le x \le 120 \\ 0 & \text{otherwise} \end{cases}$$

Then

$$P(113 \leqslant X \leqslant 118) = \int_{113}^{118} \frac{1}{10} \, dx = \frac{x}{10} \Big|_{113}^{118} = \frac{118}{10} - \frac{113}{10} = \frac{1}{2}$$

Problem 24   In Example 24, what is the probability that the current is at least 116 volts?

■ Beta Distribution

In many applications, the outcomes of an experiment are expressed in terms of fractions or percentages. For example, $\frac{9}{10}$ or 90% of the students entering a certain college successfully complete the freshman year, $\frac{1}{5}$ or 20% of fast-food restaurants fail to show a profit during their first year of operation, and so on. In order to work with outcomes expressed as fractions or percentages, it is necessary to use a probability density function whose values lie in the interval [0, 1]. One special probability density function which is often used in this situation is the *beta probability density function*.

A continuous random variable has a **beta distribution*** and is referred to as a **beta random variable** if its probability density function is the **beta probability density function**

$$f(x) = \begin{cases} (\beta + 1)(\beta + 2)x^{\beta}(1 - x) & 0 \leqslant x \leqslant 1 \\ 0 & \text{otherwise} \end{cases}$$

where β is a constant, $\beta \geqslant 0$. The value of β is usually determined by examining the results of a particular experiment. The values of a beta random variable can be expressed as fractions or percentages; however, percentages should be converted to fractions before performing calculations involving a beta random variable.

First, we show that f satisfies the requirements for a probability density function:

$$f(x) = (\beta + 1)(\beta + 2)x^{\beta}(1 - x) \geqslant 0 \qquad 0 \leqslant x \leqslant 1$$

$$\int_{-\infty}^{\infty} f(x) \, dx = \int_{0}^{1} (\beta + 1)(\beta + 2)x^{\beta}(1 - x) \, dx$$

$$= \int_{0}^{1} (\beta + 1)(\beta + 2)(x^{\beta} - x^{\beta+1}) \, dx$$

$$= (\beta + 1)(\beta + 2) \left(\frac{x^{\beta+1}}{\beta + 1} - \frac{x^{\beta+2}}{\beta + 2} \right) \Big|_{0}^{1}$$

$$= (\beta + 1)(\beta + 2) \left(\frac{1}{\beta + 1} - \frac{1}{\beta + 2} \right)$$

$$= (\beta + 2) - (\beta + 1) = 1$$

Thus, f is a probability density function.

* There is a more general definition of a beta distribution, but we will not consider it here.

If $F(x)$ is the associated cumulative probability distribution function, then for $x < 0$, $F(x) = 0$. For $0 \leqslant x \leqslant 1$, we have

$$F(x) = \int_{-\infty}^{x} f(t)\, dt = \int_{0}^{x} (\beta + 1)(\beta + 2)t^{\beta}(1 - t)\, dt$$

$$= (\beta + 1)(\beta + 2)\left(\frac{t^{\beta+1}}{\beta+1} - \frac{t^{\beta+2}}{\beta+2}\right)\Big|_{0}^{x}$$

$$= (\beta + 1)(\beta + 2)\left(\frac{x^{\beta+1}}{\beta+1} - \frac{x^{\beta+2}}{\beta+2}\right)$$

$$= (\beta + 2)x^{\beta+1} - (\beta + 1)x^{\beta+2}$$

And for $x > 1$,

$$F(x) = 1$$

In general, it is not possible to solve the equation $F(x_m) = \frac{1}{2}$ for x_m. Thus, we will not discuss the median of a beta random variable. By straightforward (but tedious) integration we can show that

$$\mu = \frac{\beta + 1}{\beta + 3}$$

and

$$\sigma = \sqrt{\frac{2(\beta + 1)}{(\beta + 4)(\beta + 3)^2}}$$

The calculations are not included here. The above results are summarized below.

Beta Probability Density Function

$$f(x) = \begin{cases} (\beta + 1)(\beta + 2)x^{\beta}(1 - x) & 0 \leqslant x \leqslant 1 \\ 0 & \text{otherwise} \end{cases} \qquad \text{where } \beta \geqslant 0$$

$$F(x) = \begin{cases} 0 & x < 0 \\ (\beta + 2)x^{\beta+1} - (\beta + 1)x^{\beta+2} & 0 \leqslant x \leqslant 1 \\ 1 & x > 1 \end{cases}$$

Mean: $$\mu = \frac{\beta + 1}{\beta + 3}$$

Standard deviation: $$\sigma = \sqrt{\frac{2(\beta + 1)}{(\beta + 4)(\beta + 3)^2}}$$

Figure 8 (page 1070) shows the graphs of $f(x)$ for some typical values of β.

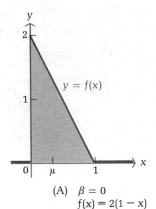

(A) $\beta = 0$
$f(x) = 2(1 - x)$

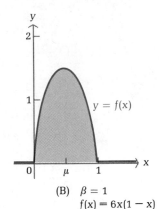

(B) $\beta = 1$
$f(x) = 6x(1 - x)$

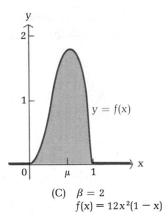

(C) $\beta = 2$
$f(x) = 12x^2(1 - x)$

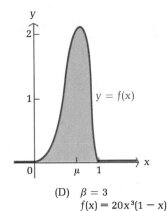

(D) $\beta = 3$
$f(x) = 20x^3(1 - x)$

Figure 8

Example 25
Income Tax

The annual percentage of correct income tax forms filed with the Internal Revenue Service is a beta random variable with $\beta = 8$.

(A)   What is the probability that at least half the returns filed are correct?
(B)   What is the expected percentage of correct returns?

Solutions

Substituting $\beta = 8$ in the definition of the beta probability density function, we have

$$f(x) = \begin{cases} 90x^8(1 - x) & 0 \leqslant x \leqslant 1 \\ 0 & \text{otherwise} \end{cases}$$

(A) $P(X \geqslant \frac{1}{2}) = \int_{1/2}^{1} 90x^8(1 - x)\, dx = \int_{1/2}^{1} (90x^8 - 90x^9)\, dx$

$$= (10x^9 - 9x^{10}) \Big|_{1/2}^{1} = 1 - \frac{11}{2^{10}} \approx .989$$

(B) $\mu = E(X) = \dfrac{\beta + 1}{\beta + 3} = \dfrac{8 + 1}{8 + 3} = \dfrac{9}{11} \approx .818$

Thus, we expect approximately 82% of the returns to be correct.

Problem 25 In Example 25, what is the probability that at least 90% of the returns are correct?

Example 26 A psychologist is studying the learning abilities of children in a certain age
Learning group. She has determined that on the average 75% of the children can learn to perform a particular task in 5 minutes. She believes that the percentage of children that can learn the task in 5 minutes is a continuous beta random variable. What is an appropriate value of β?

Solution Since the average percentage of children that learned the task is 75% and the mean for any beta distribution is $(\beta + 1)/(\beta + 3)$, the value of β must satisfy

$$\frac{\beta + 1}{\beta + 3} = .75 \qquad \text{Convert 75\% to the decimal fraction .75.}$$

Solving this equation, we obtain $\beta = 5$.

Problem 26 In Example 26, what is the probability that at least 75% of the children will learn the task in 5 minutes?

■ Exponential Distribution

A continuous random variable has an **exponential distribution** and is referred to as an **exponential random variable** if its probability density function is the **exponential probability density function**

$$f(x) = \begin{cases} (1/\lambda)e^{-x/\lambda} & x \geq 0 \\ 0 & \text{otherwise} \end{cases}$$

where λ is a positive constant. Exponential random variables are used in a variety of applications, including studies of the length of telephone conversations, the time customers spend waiting in line at a bank, and the life expectancy of a machine part.

Since $f(x) \geq 0$ and

$$\int_{-\infty}^{\infty} f(x)\, dx = \int_{0}^{\infty} \frac{1}{\lambda} e^{-x/\lambda}\, dx = \lim_{R \to \infty} \int_{0}^{R} \frac{1}{\lambda} e^{-x/\lambda}\, dx = \lim_{R \to \infty} (-e^{-x/\lambda}) \Big|_{0}^{R}$$
$$= \lim_{R \to \infty} (-e^{-R/\lambda} + 1) = 1$$

f satisfies the conditions for a probability density function. If F is the cumulative distribution function, we see that $F(x) = 0$ for $x < 0$. For $x \geq 0$, we have

$$F(x) = \int_{-\infty}^{x} f(t)\, dt = \int_{0}^{x} \frac{1}{\lambda} e^{-t/\lambda}\, dt = -e^{-t/\lambda} \Big|_{0}^{x} = 1 - e^{-x/\lambda}$$

Median

$$F(x_m) = P(X \leq x_m) = \frac{1}{2}$$

$$1 - e^{-x_m/\lambda} = \frac{1}{2}$$

$$\frac{1}{2} = e^{-x_m/\lambda}$$

$$\ln \frac{1}{2} = -\frac{x_m}{\lambda}$$

$$x_m = -\lambda \ln \frac{1}{2} = \lambda \ln 2 \qquad \text{Note:} \quad \ln \frac{1}{2} = -\ln 2$$

Integration by parts can be used to show that $\mu = \lambda$ and $\sigma = \lambda$. The calculations are not included here. The above results are summarized below.

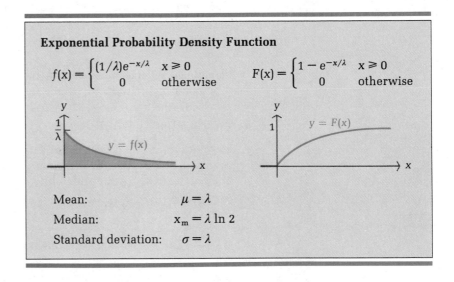

Exponential Probability Density Function

$$f(x) = \begin{cases} (1/\lambda)e^{-x/\lambda} & x \geq 0 \\ 0 & \text{otherwise} \end{cases} \qquad F(x) = \begin{cases} 1 - e^{-x/\lambda} & x \geq 0 \\ 0 & \text{otherwise} \end{cases}$$

Mean: $\qquad\qquad \mu = \lambda$

Median: $\qquad\quad\; x_m = \lambda \ln 2$

Standard deviation: $\quad \sigma = \lambda$

Example 27
Inventory

The number of units of a certain item sold each week in a chain of department stores is an exponential random variable. The average number of items sold each week is 10,000. What is the probability that 15,000 or more units will be sold in one week?

Solution

If we let x represent the number of units sold (in thousands), then the appropriate probability density function is

$$f(x) = \begin{cases} \frac{1}{10}e^{-x/10} & x \geq 0 \\ 0 & \text{otherwise} \end{cases}$$

Thus,

$$P(X \geq 15) = \int_{15}^{\infty} \tfrac{1}{10} e^{-x/10} \, dx = \lim_{R \to \infty} (-e^{-x/10}) \Big|_{15}^{R} = \lim_{R \to \infty} (-e^{-R/10} + e^{-15/10})$$

$$= e^{-1.5} \approx .223$$

Problem 27 In Example 27, what is the probability that at most 5,000 units will be sold?

Answers to 24. $\tfrac{2}{5}$ 25. .264   26. .555   27. $1 - e^{-0.5} \approx .393$
Matched Problems

Exercise 17-5

A   *Problems 1–4 refer to a continuous random variable X that is uniformly distributed on the interval [0, 2].*

1. Find the probability density function for X.
2. Find the associated cumulative probability distribution function for X.
3. Find the mean, median, and standard deviation.
4. Find the probability that the random variable is within 1 standard deviation of the mean (between $\mu - \sigma$ and $\mu + \sigma$).

Problems 5–8 refer to a beta random variable X with $\beta = 3$.

5. Find the probability density function for X.
6. Find the cumulative probability distribution function for X.
7. Find the mean and standard deviation.
8. Find the probability that the random variable is within 1 standard deviation of the mean (between $\mu - \sigma$ and $\mu + \sigma$).

Problems 9–12 refer to an exponential random variable X with $\lambda = \tfrac{1}{2}$.

9. Find the probability density function for X.
10. Find the cumulative probability distribution function for X.
11. Find the mean, median, and standard deviation.
12. Find the probability that the random variable is within 1 standard deviation of the mean (between $\mu - \sigma$ and $\mu + \sigma$).

B   *Problems 13–16 refer to a beta random variable with $\beta = \tfrac{1}{2}$.*

13. Find the probability density function for X.
14. Find the cumulative probability distribution function for X.
15. Find the mean and standard deviation.
16. Find the probability that the random variable is within 1 standard deviation of the mean (between $\mu - \sigma$ and $\mu + \sigma$).

Problems 17–20 refer to a beta random variable X with mean $\mu = .4$.

17. Find β.
18. Find the probability density function.
19. Find the cumulative probability distribution function.
20. Find the standard deviation.

Problems 21–24 refer to an exponential random variable X with median $x_m = 2$.

21. Find λ.
22. Find the probability density function.
23. Find the cumulative probability distribution function.
24. Find the mean and standard deviation.

C 25. Compute the following probabilities for an exponential random variable with mean λ:

  (A) $P(0 \leqslant X \leqslant \lambda)$ (B) $P(0 \leqslant X \leqslant 2\lambda)$ (C) $P(0 \leqslant X \leqslant 3\lambda)$

26. The point where a probability density function assumes its maximum value is sometimes referred to as the **mode.** Find the mode of the probability density function

$$f(x) = \begin{cases} (\beta + 1)(\beta + 2)x^\beta(1 - x) & 0 \leqslant x \leqslant 1 \\ 0 & \text{otherwise} \end{cases}$$

27. If the random variable X is the time at which a device malfunctions, then the failure rate is given by

$$\frac{f(x)}{1 - F(x)}$$

where $f(x)$ is the probability density function and $F(x)$ is the cumulative probability distribution function for X. Find the failure rate for:

  (A) A uniform random variable
  (B) An exponential random variable

28. A continuous random variable X is said to have a *Pareto distribution* if its probability density function is given by

$$f(x) = \begin{cases} p/x^{p+1} & x \geqslant 1 \\ 0 & \text{otherwise} \end{cases}$$

where p is a constant, $p > 0$.

  (A) Find the mean. What restrictions must you place on p?
  (B) Find the variance and standard deviation. What restrictions must you place on p?
  (C) Find the median.

Applications

Business & Economics

29. *Waiting time.* The time (in minutes) applicants must wait for an officer to give them a driver's examination is uniformly distributed on the interval [0, 40]. What is the probability that an applicant must wait more than 25 minutes?

30. *Business failures.* The percentage of computer hobby stores that fail during the first year of operation is a beta random variable with $\beta = 4$.

    (A) What is the expected percentage of failures?

    (B) What is the probability that over 50% of the stores fail during the first year?

31. *Absenteeism.* The percentage of assembly line workers that are absent one Monday each month is a beta random variable. The mean percentage is 50%.

    (A) What is the appropriate value of β?

    (B) What is the probability that no more than 75% of the workers will be absent on one Monday each month?

32. *Waiting time.* The waiting time (in minutes) for customers at a drive-in bank is an exponential random variable. The average (mean) time a customer waits is 4 minutes. What is the probability that a customer waits more than 5 minutes?

33. *Communication.* The length of time for telephone conversations (in minutes) is exponentially distributed. The average (mean) length of a conversation is 3 minutes. What is the probability that a conversation lasts less than 2 minutes?

34. *Component failure.* The life expectancy (in years) of a component in a microcomputer is an exponential random variable. Half the components fail in the first 3 years. The company that manufactures the component offers a 1 year warranty. What is the probability that a component will fail during the warranty period?

Life Sciences

35. *Nutrition.* The percentage of the daily requirement of vitamin D present in an 8 ounce serving of milk is a beta random variable with $\beta = .2$.

    (A) What is the expected percentage of vitamin D per serving?

    (B) What is the probability that a serving contains at least 50% of the daily requirement?

36. *Medicine.* A scientist is measuring the percentage of a drug present in the bloodstream 10 minutes after an injection. The results indicate that the percentage of the drug present is a beta random variable with mean $\mu = .75$.

    (A) What is the value of β?

    (B) What is the probability that no more than 25% of the drug is present 10 minutes after an injection?

37. *Survival time.* The time of death (in years) after patients have contracted a certain disease is exponentially distributed. The probability that a patient dies within 1 year is .3.

    (A) What is the expected time of death?
    (B) What is the probability that a patient survives longer than the expected time of death?

38. *Survival time.* Repeat Problem 37 if the probability that a patient dies within 1 year is .5.

Social Sciences   39. *Education.* The percentage of entering freshmen that complete the first year of college is a beta random variable with $\beta = 17$.

    (A) What is the expected percentage of students that complete the first year?
    (B) What is the probability that more than 95% of the students complete the first year?

40. *Psychology.* The time (in seconds) it takes rats to find their way through a maze is exponentially distributed. The average (mean) time is 30 seconds. What is the probability that it takes a rat over 1 minute to find a path through the maze?

17-6 Normal Distributions

- Normal Probability Density Functions
- The Standard Normal Curve
- Areas under Arbitrary Normal Curves
- Normal Distribution Approximation of a Binomial Distribution

We will now consider the most important of all probability density functions, the *normal probability density function*. This function is at the heart of a great deal of statistical theory, and it is also a useful tool in its own right for solving problems. We will see that the normal probability density function can also be used to provide a good approximation to the binomial distribution.

■ Normal Probability Density Functions

A continuous random variable X has a **normal distribution** and is referred to as a **normal random variable** if its probability density function is the **normal probability density function**

$$f(x) = \frac{1}{\sigma\sqrt{2\pi}} e^{-(x-\mu)^2/2\sigma^2}$$

where μ is any constant and σ is any positive constant. It can be shown, but not easily, that

$$\int_{-\infty}^{\infty} f(x)\ dx = 1$$

$$E(X) = \int_{-\infty}^{\infty} xf(x)\ dx = \mu$$

and

$$V(X) = \int_{-\infty}^{\infty} (x - \mu)^2 f(x)\ dx = \sigma^2$$

Thus, μ is the mean of the normal probability density function and σ is the standard deviation. The graph of $f(x)$ is always a bell-shaped curve called a **normal curve.** Figure 9 illustrates three normal curves for different values of μ and σ.

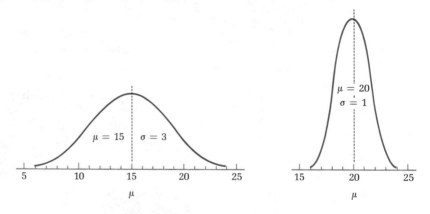

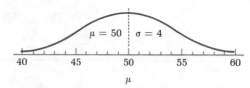

Figure 9 Normal probability distributions

The standard deviation measures the dispersion of the normal probability density function about the mean — a small standard deviation indicates a tight clustering about the mean and thus a tall, narrow curve; a large standard deviation indicates a large deviation from the mean and thus a broad, flat curve. Notice that each of the normal curves in Figure 9 is symmetric about a vertical line through the mean. This is true for any normal curve. Thus, the line $x = \mu$ divides the region under a normal curve into two regions with equal area. Since the total area under a normal curve is always 1, the area of each of these regions is .5. This implies that the median of a normal random variable is always equal to the mean (see Figure 10).

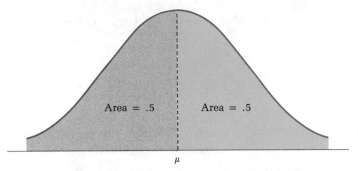

Figure 10 The mean and median of a normal random variable

The properties of the normal probability density function are summarized in the box for ease of reference.

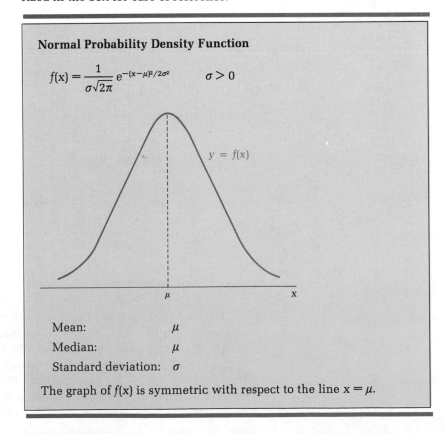

Normal Probability Density Function

$$f(x) = \frac{1}{\sigma\sqrt{2\pi}} \, e^{-(x-\mu)^2/2\sigma^2} \qquad \sigma > 0$$

$y = f(x)$

Mean: μ

Median: μ

Standard deviation: σ

The graph of $f(x)$ is symmetric with respect to the line $x = \mu$.

The cumulative distribution function for a normal random variable is given formally by

$$F(x) = \frac{1}{\sigma\sqrt{2\pi}} \int_{-\infty}^{x} e^{-(t-\mu)^2/2\sigma^2} \, dt$$

It is not possible to express $F(x)$ as a finite combination of the functions we are familiar with. Furthermore, we cannot use antidifferentiation to evaluate probabilities such as

$$P(c \leq X \leq d) = \frac{1}{\sigma\sqrt{2\pi}} \int_{c}^{d} e^{-(x-\mu)^2/2\sigma^2} \, dx$$

Instead, we will use a table to approximate probabilities of this type. Fortunately, we can use the same table for all normal probability density functions, irrespective of the values of μ and σ. It is a remarkable fact that the area under a normal curve between the mean and a given number of standard deviations to the right (or left) of μ is the same regardless of the values of μ and σ (see Figure 11).

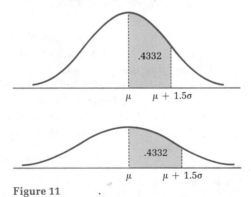

Figure 11

■ The Standard Normal Curve

It is convenient to relate the area under an arbitrary normal curve to the area under a particular normal curve called the *standard normal curve*.

Standard Normal Curve

The normal random variable Z with mean $\mu = 0$ and standard deviation $\sigma = 1$ is called the **standard normal random variable** and its graph is called the **standard normal curve**.

Table IV in the back of the book gives the area under the standard normal curve from 0 to z for values of z in the range $0 \leq z \leq 3.89$. The values in this table, together with the familiar properties of area under a curve, can be used to compute probabilities involving the standard normal random variable.

Example 28   Use Table IV to compute the following probabilities for the standard normal random variable Z:

(A) $P(0 \leq Z \leq .88)$ (B) $P(Z \leq 1.45)$ (C) $P(.3 \leq Z \leq 2.73)$

Solutions   (A) From Table IV, the area under the standard normal curve from $z = 0$ to $z = .88$ is .3106. Thus,

$$P(0 \leq Z \leq .88) = .3106$$

(B)

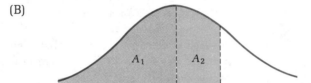

$P(Z \leq 1.45)$ is the area under the standard normal curve over the interval $(-\infty, 1.45]$. Since Table IV only gives the area over intervals of the form $[0, z_0]$, we must divide this region into two parts. Let A_1 be the area of the region over the interval $(-\infty, 0]$ and let A_2 be the area of the region over the interval $[0, 1.45]$ (see the figure). The median of the standard normal random variable is 0; thus, $A_1 = .5$. From Table IV, $A_2 = .4265$. Adding these, we have

$$P(Z \leq 1.45) = A_1 + A_2 = .5 + .4265 = .9265$$

(C) This time we let A_1 be the area of the region from 0 to 2.73 and let A_2 be the area of the region from 0 to .3 (see the figures). Using the appropriate values from Table IV, we have

$$P(.3 \leq Z \leq 2.73) = A_1 - A_2 = .4968 - .1179 = .3789$$

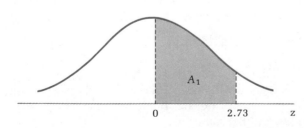

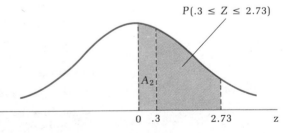

Problem 28   Use Table IV to find the following probabilities for the standard normal random variable Z:

(A) $P(0 \leq Z \leq 2.15)$ (B) $P(Z \leq .75)$ (C) $P(.7 \leq Z \leq 3.2)$

■ Areas under Arbitrary Normal Curves

Now that we have seen how to use Table IV to determine probabilities involving the standard normal random variable, we want to consider the more general case. Theorem 6 relates areas under any normal curve to corresponding areas under the standard normal curve. This will enable us to use Table IV to find areas under any normal curve, regardless of the values of μ and σ.

Theorem 6

If X is a normal random variable with mean μ and standard deviation σ and

$$z_i = \frac{x_i - \mu}{\sigma} \qquad i = 1, 2 \tag{1}$$

then

$$P(x_1 \leq X \leq x_2) = P(z_1 \leq Z \leq z_2) \tag{2}$$
$$P(x_1 \leq X) = P(z_1 \leq Z) \tag{3}$$
$$P(X \leq x_2) = P(Z \leq z_2) \tag{4}$$

Example 29

Scholastic Aptitude Test scores are normally distributed with a mean of 500 and a standard deviation of 100. What percentage of the students taking the test can be expected to score between 500 and 670?

Solution

Since the total area under a normal curve is 1, the percentage of students that can be expected to score between 500 and 670 on the test is the same as the area under the curve from 500 to 670 (see the figure).

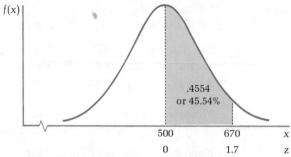

Scholastic Aptitude Test scores

If X is the random variable associated with a student's score on the test, then we must find

$$P(500 \leq X \leq 670)$$

First we use equation (1) in Theorem 6 to find the corresponding z values:

$$z_1 = \frac{x_1 - \mu}{\sigma} = \frac{500 - 500}{100} = 0$$

$$z_2 = \frac{x_2 - \mu}{\sigma} = \frac{670 - 500}{100} = 1.7$$

Next we use equation (2) in Theorem 6 to write

$$P(500 \leq X \leq 670) = P(0 \leq Z \leq 1.7)$$

Finally, we use Table IV to find the area under the standard normal curve from $z = 0$ to $z = 1.7$. This area is .4554. Thus,

$$P(500 \leq X \leq 670) = P(0 \leq Z \leq 1.7)$$

$$= .4554$$

and we conclude that 45.54% of the students can be expected to score between 500 and 670 on the SAT.

Problem 29   Refer to Example 29. What percentage of the students can be expected to score between 500 and 750?

Example 30   Refer to Example 29. From all high school students taking the Scholastic Aptitude Test, what is the probability that a student chosen at random scores between 380 and 500 on the test?

Solution   The corresponding z values are

$$z_1 = \frac{x_1 - \mu}{\sigma} = \frac{380 - 500}{100} = -1.2$$

$$z_2 = \frac{x_2 - \mu}{\sigma} = \frac{500 - 500}{100} = 0$$

Thus,

$$P(380 \leq X \leq 500) = P(-1.2 \leq Z \leq 0)$$

Table IV does not include negative values of z, but because normal curves are symmetric with respect to a vertical line through the mean, we simply use the absolute value of z in Table IV (see the figure at the top of page 1083).

The area under the standard normal curve from $z = -1.2$ to $z = 0$ is the same as the area from $z = 0$ to $z = 1.2$, which is .3849. Thus,

$$P(380 \leq X \leq 500) = P(-1.2 \leq Z \leq 0) \qquad \text{Theorem 6}$$

$$= P(0 \leq Z \leq 1.2) \qquad \text{Symmetry property of the normal curve}$$

$$= .3849 \qquad \text{Table IV}$$

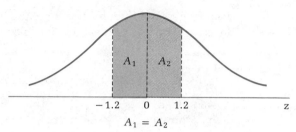

$A_1 = A_2$

Area under the standard normal curve for negative z

Problem 30   Refer to Example 29. What is the probability that a student selected at random scores between 400 and 500 on the test?

■ Normal Distribution Approximation of a Binomial Distribution

If we take the histogram for a binomial distribution, say, the one we drew for Example 12 in Section 17-2 ($n = 8$, $p = .5$), and join the midpoints of the top of each rectangle with a smooth curve, we obtain the bell-shaped curve in Figure 12. The mean of this binomial distribution is 4 and the standard

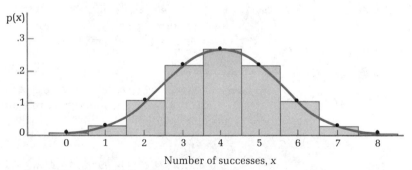

Figure 12 The binomial distribution and a bell-shaped curve

deviation is approximately 1.41. The normal curve with mean 4 and standard deviation 1.41 approximates the bell-shaped curve in Figure 12 and can be used to approximate probabilities involving this binomial distribution, although the results may not be very accurate. In general, the accuracy of the approximation of a binomial distribution by a normal curve depends on the values of p, q, and n. A normal curve is always symmetric with respect to a vertical line through the mean, whereas a binomial distribution is symmetric only if p is equal to (or nearly equal to) q. The more p and q differ from each other, the worse the approximation. The following rule-of-thumb states conditions under which it is reasonably safe to use a normal curve to approximate a binomial distribution:

Rule-of-Thumb Test for Approximating a Binomial Distribution

A binomial distribution with n trials and probability of success p can be approximated by the normal curve with

$$\mu = np \qquad \sigma = \sqrt{npq}$$

provided both np and nq are greater than 10.

Example 31
Quality Control

A company manufactures 50,000 ballpoint pens each day. The manufacturing process produces 50 defective pens per 1,000 on the average. A random sample of 400 pens is selected from each day's production and tested. What is the probability that in the sample there are:

(A)   At least 14 and no more than 25 defective pens?
(B)   33 or more defective pens?

Solutions

First, we use the rule-of-thumb test to determine whether it is appropriate to use a normal curve approximation for this binomial distribution. Using $n = 400$, $p = .05$, and $q = .95$, we see that

$$np = 400(.05) = 20 \qquad \text{and} \qquad nq = 400(.95) = 380$$

Since both of these values are greater than 10, we can use the rule-of-thumb test. We can approximate this binomial distribution with the normal curve with

$$\mu = np = 20$$
$$\sigma = \sqrt{npq} = \sqrt{(400)(.05)(.95)} \approx 4.36$$

(A)   To approximate the probability that the number of defective pens is at least 14 and not more than 25, we want to determine

$$P(14 \leqslant X \leqslant 25)$$

where X is the random variable associated with the number of defective pens in a sample. Using $\mu = 20$ and $\sigma = 4.36$, the corresponding z values are

$$z_1 = \frac{x_1 - \mu}{\sigma} = \frac{14 - 20}{4.36} \approx -1.38$$

$$z_2 = \frac{x_2 - \mu}{\sigma} = \frac{25 - 20}{4.36} \approx 1.15$$

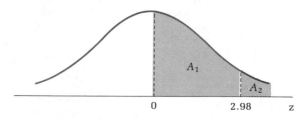

Using Theorem 6 and Table IV (see the figure) we have

$$P(14 \leqslant X \leqslant 25) = P(-1.38 \leqslant Z \leqslant 1.15)$$
$$= P(-1.38 \leqslant Z \leqslant 0) + P(0 \leqslant Z \leqslant 1.15)$$
$$= .4162 + .3749 = .7911$$

Therefore, the approximate probability that the number of defective pens in the sample is at least 14 and not more than 25 is .7911.

(B) Let

$$z = \frac{x - \mu}{\sigma} = \frac{33 - 20}{4.36} \approx 2.98$$

Then

$$P(33 \leqslant X) = P(2.98 \leqslant Z)$$

Since the total area under a normal curve from the mean on is .5, we find the area A_1 from Table IV and subtract it from .5 to obtain A_2 (see the figure):

$$P(2.98 \leqslant Z) = A_2 = .5 - A_1 = .5 - .4986 = .0014$$

Thus, the approximate probability of finding 33 or more defective pens in the sample is .0014. If a random sample of 400 included more than 33 defective pens, then the management would conclude that either a rare event has happened and the manufacturing process is still producing only 50 defective pens per 1,000 on the average, or something is wrong with the manufacturing process and it is producing more than 50 defective pens per 1,000 on the average. The company might very well have a policy of checking the manufacturing process whenever 33 or more defective pens are found in a sample rather than believing a rare event has happened and that the manufacturing process is still all right.

Problem 31   Suppose in Example 31 that the manufacturing process produces 40 defective pens per 1,000 on the average. What is the approximate probability that in the sample of pens there are:

(A)   At least 10 and no more than 20 defective pens?
(B)   27 or more defective pens?

Answers to   28. (A) .4842   (B) .7734   (C) .2413   29. 49.38%
Matched Problems   30. .3413   31. (A) .7831   (B) .0025

Exercise 17-6

A   *Use Table IV to find the area under the standard normal curve from 0 to the indicated value of z.*

1. 1   2. 2   3. −3   4. −1
5. .9   6. −1.7   7. 2.47   8. −1.96

Given a normal distribution with mean 50 and standard deviation 10, use Theorem 6 and Table IV to find the area under this normal curve from the mean to the indicated measurement.

9. 65   10. 75   11. 83   12. 79
13. 45   14. 38   15. 42   16. 26

B   *In Problems 17–24 find the indicated probability for the standard normal random variable Z.*

17. $P(-1.7 \leqslant Z \leqslant .6)$ 18. $P(-.4 \leqslant Z \leqslant 2)$ 19. $P(.45 \leqslant Z \leqslant 2.25)$
20. $P(1 \leqslant Z \leqslant 2.75)$ 21. $P(Z \geqslant .75)$ 22. $P(Z \leqslant -1.5)$
23. $P(Z \leqslant 1.88)$ 24. $P(Z \geqslant -.66)$

Given a normal random variable X with mean 70 and standard deviation 8, find the indicated probabilities.

25. $P(60 \leqslant X \leqslant 80)$ 26. $P(50 \leqslant X \leqslant 90)$ 27. $P(62 \leqslant X \leqslant 74)$
28. $P(66 \leqslant X \leqslant 78)$ 29. $P(X \geqslant 88)$ 30. $P(X \geqslant 90)$
31. $P(X \leqslant 60)$ 32. $P(X \leqslant 56)$

A binomial experiment consists of 500 trials with the probability of success for each trial .4. Thus,

$$\mu = np = 200 \quad and \quad \sigma = \sqrt{npq} = \sqrt{500(.4)(.6)} \approx 11$$

What is the probability of obtaining the number of successes indicated in Problems 33–38? Approximate with a normal curve.

33. 185–215   34. 180–220   35. 200 or more
36. 200 or less   37. 225 or more   38. 175 or less

Applications

Assume normal distributions are warranted in these problems.

Business & Economics

39. *Sales.* Salespeople for a business machine company have average annual sales of $200,000, with a standard deviation of $20,000. What percentage of the salespeople would be expected to make annual sales of $240,000 or more?

40. *Guarantees.* The average lifetime for a car battery of a certain brand is 170 weeks, with a standard deviation of 10. If the company guarantees the battery for 3 years, what is the probability that a battery will be returned during the warranty period?

41. *Quality control.* A manufacturing process produces a critical part of average length 100 millimeters, with a standard deviation of 2 millimeters. All parts deviating by more than 5 millimeters from the mean must be rejected. What is the probability that a part will be rejected?

42. *Labor relations.* A union representative claims 60% of the union membership will vote in favor of a particular settlement. A random sample of 100 members is polled, and out of these 47 favor the settlement. What is the approximate probability of 47 or less in a sample of 100 favoring the settlement when 60% of all the membership favor the settlement? Conclusion? [*Hint:* This problem involves a binomial distribution with $n = 100$ and $p = .6$.]

Life Sciences

43. *Medicine.* The average healing time of a certain type of incision is 240 hours, with standard deviation of 20 hours. What is the probability that an incision of this type will heal in 8 days or less?

44. *Agriculture.* The average height of a hay crop is 38 inches, with a standard deviation of 1.5 inches. What percentage of the crop will be 40 inches or more?

45. *Genetics.* In a two-child family, the probability that both children are girls is approximately .25. In a random sample of 1,000 two-child families, what is the approximate probability that 220 or fewer will have two girls? [*Hint:* This problem involves a binomial distribution with $n = 1,000$ and $p = .25$.]

46. *Genetics.* In Problem 45, what is the approximate probability of the number of families with two girls in the sample being at least 225 and not more than 275?

Social Sciences

47. *Testing.* Scholastic Aptitude Tests are scaled so that the mean score is 500 and the standard deviation is 100. What percentage of the students taking this test should score 700 or more?

48. *Politics.* Candidate Harkins claims a private poll shows that she will receive 52% of the vote for governor. Her opponent, Mankey, secures the services of another pollster, who finds that 470 out of a random sample of 1,000 registered voters favor Harkins. If Harkins' claim is correct, what is the probability that only 470 or fewer will favor her in a random sample of 1,000? Conclusion? [*Hint;* This problem involves a binomial distribution with $n = 1,000$ and $p = .52$.]

49. *Grading on a curve.* An instructor grades on a curve by assuming the grades on a test are normally distributed. If the average grade is 70 and the standard deviation is 8, find the test scores for each grade interval if the instructor wishes to assign grades as follows: 10% A's, 20% B's, 40% C's, 20% D's, and 10% F's.

50. *Psychology.* A test devised to measure aggressive–passive personalities was standardized on a large group of people. The scores were normally distributed with a mean of 50 and a standard deviation of 10. If we want to designate the highest 10% as aggressive, the next 20% as moderately aggressive, the middle 40% as average, the next 20% as moderately passive, and the lowest 10% as passive, what ranges of scores will be covered by these five designations?

17-7   Chapter Review

Important Terms
and Symbols

17-1 *Random variable, probability distribution, and expectation.* sample space, simple event, random variable, probability function, probability distribution of a random variable, expected value, payoff table, fair game, mean, variance, standard deviation, $E(X)$, μ, $V(X)$, σ

17-2 *Binomial distributions.* Bernoulli trial, independent trials, binomial theorem, binomial distribution, binomial experiment, $P(x \text{ successes in } n \text{ trials}) = \binom{n}{x} p^x q^{n-x}$, $x \in \{0, 1, \ldots, n\}$, mean $= \mu = np$, standard deviation $= \sigma = \sqrt{npq}$

17-3 *Continuous random variables.* continuous random variable, discrete random variable, probability density function, cumulative probability distribution function, $P(c \leq X \leq d)$, $F(x) = P(X \leq x)$

17-4 *Expected value, standard deviation, and median of continuous random variables.* expected value, mean, variance, standard deviation, alternate formula for variance, median, $E(X)$, μ, $V(X)$, σ, x_m

17-5 *Uniform, beta, and exponential distributions.* uniform probability density function, beta probability density function, exponential probability density function

17-6 *Normal distributions.* normal distribution, normal random variable, normal probability density function, normal curve, standard normal random variable, standard normal curve, table of areas under the standard normal curve, approximating binomial distributions with normal distributions

Exercise 17-7 Chapter Review

Work through all the problems in this chapter review and check your answers in the back of the book. (Answers to all review problems are there.) Where weaknesses show up, review appropriate sections in the text. When you are satisfied that you know the material, take the practice test following this review.

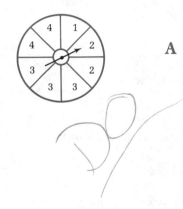

A   A spinner can land on any one of eight different sectors, and each sector is as likely to turn up as any other. The sectors are numbered in the figure. An experiment consists of spinning the dial once and recording the number in the sector that the spinner lands on. Problems 1–4 refer to this experiment.

1. Find a sample space and probability distribution for this experiment.
2. What is the probability that the spinner stops on an even-numbered sector?
3. Find the expected value, variance, and standard deviation for the probability distribution found in Problem 1.
4. After paying $3 to play, you spin the dial and are paid back the number of dollars corresponding to the number on the sector where the spinner stopped. What is the expected value of this game?

5. (A) Draw a histogram for the binomial distribution

$$p(x) = \binom{3}{x} (.4)^x (.6)^{3-x}$$

   (B) What are the mean and standard deviation?

Problems 6–9 refer to the continuous random variable X with probability density function.

$$f(x) = \begin{cases} 1 - \tfrac{1}{2}x & 0 \leqslant x \leqslant 2 \\ 0 & \text{otherwise} \end{cases}$$

6. Find $P(0 \leqslant X \leqslant 1)$ and illustrate with a graph.
7. Find the mean, variance, and standard deviation.
8. Find and graph the associated cumulative probability distribution function.
9. Find the median.
10. If Z is the standard normal random variable, find $P(0 \leqslant Z \leqslant 2.5)$.
11. If X is a normal random variable with a mean of 100 and a standard deviation of 10, find $P(100 \leqslant X \leqslant 118)$.

B   12. (A) Construct a histogram for the binomial distribution

$$p(x) = \binom{6}{x} (.5)^x (.5)^{6-x}$$

   (B) What are the mean and standard deviation?

Problems 13–16 refer to the continuous random variable X with probability density function

$$f(x) = \begin{cases} \frac{5}{2}x^{-7/2} & x \geqslant 1 \\ 0 & \text{otherwise} \end{cases}$$

13. Find $P(1 \leqslant X \leqslant 4)$ and illustrate with a graph.
14. Find the mean, variance, and standard deviation.
15. Find and graph the associated cumulative probability distribution function.
16. Find the median.

Problems 17–20 refer to a beta random variable X with $\beta = 5$.

17. Find and graph the probability density function.
18. Find $P(\frac{1}{4} \leqslant X \leqslant \frac{3}{4})$.
19. Find and graph the associated cumulative probability distribution function.
20. Find the mean and standard deviation.

Problems 21–24 refer to an exponentially distributed random variable X.

21. If $P(4 \leqslant X) = e^{-2}$, find the probability density function.
22. Find $P(0 \leqslant X \leqslant 2)$.
23. Find the associated cumulative probability distribution function.
24. Find the mean, standard deviation, and median.
25. What are the mean and standard deviation for a binomial distribution with $p = .6$ and $n = 1,000$?
26. If the probability of success in a single trial of a binomial experiment with 1,000 trials is .6, what is the probability of obtaining between 550 and 650 successes in 1,000 trials?
27. Given a normal distribution with mean 50 and standard deviation 6, find the area under the normal curve:

    (A) Between 41 and 62   (B) From 59 on

C 28. If X is a beta random variable with mean $\mu = .8$, what is the value of β?
    29. Find the mean and the median of the continuous random variable with probability density function

$$f(x) = \begin{cases} 50/(x+5)^3 & x \geqslant 0 \\ 0 & \text{otherwise} \end{cases}$$

30. If $f(x)$ is a continuous probability density function with mean μ and standard deviation σ and a, b, and c are constants, evaluate the integral given below. Express the result in terms of μ, σ, a, b, and c.

$$\int_{-\infty}^{\infty} (ax^2 + bx + c)f(x)\, dx$$

Applications

Business & Economics

31. *Quality control.* A manufacturing process produces, on the average, six defective items out of 100. To control quality, each day a sample of ten completed items is selected at random and is inspected. If the sample produces more than two defective items, then the whole day's output is inspected and the manufacturing process is reviewed. What is the probability of this happening, assuming that the process is still producing 6% defective items?

32. *Demand.* The manager of a movie theater has determined that the weekly demand for popcorn (in pounds) is a continuous random variable with probability density function

$$f(x) = \begin{cases} \frac{1}{50}(1 - .01x) & 0 \le x \le 100 \\ 0 & \text{otherwise} \end{cases}$$

  (A) If the manager has 50 pounds of popcorn on hand at the beginning of the week, what is the probability that this will be enough to meet the weekly demand?

  (B) If the manager wants the probability that the supply on hand exceeds the weekly demand to be .96, how much popcorn must be on hand at the beginning of the week?

33. *Credit applications.* The percentage of applications for a national credit card that are processed on the same day they are received is a beta random variable with $\beta = 1$.

  (A) What is the probability that at least 20% of the applications received are processed the same day they arrive?

  (B) What is the expected percentage of applications processed the same day they arrive?

34. *Computer failure.* A computer manufacturer has determined that the time between failures for its computers is an exponentially distributed random variable with a mean failure time of 4,000 hours. Suppose a particular computer has just been repaired.

  (A) What is the probability that the computer operates for the next 4,000 hours without a failure?

  (B) What is the probability that the computer fails in the next 1,000 hours?

35. *Radial tire failure.* The life expectancy (in miles) of a certain brand of radial tire is a normal random variable with a mean of 35,000 and a standard deviation of 5,000. What is the probability that a tire fails during the first 25,000 miles of use?

Life Sciences

36. *Medicine.* The shelf-life (in months) of a certain drug is a continuous random variable with probability density function

$$f(x) = \begin{cases} 10/(x+10)^2 & x \geqslant 0 \\ 0 & \text{otherwise} \end{cases}$$

(A) What is the probability that the drug is still usable after 5 months?

(B) What is the median shelf-life?

37. *Life expectancy.* The life expectancy (in months) after dogs have contracted a certain disease is an exponentially distributed random variable. The probability of surviving more than 1 month is e^{-2}. After contracting this disease:

(A) What is the probability of surviving more than 2 months?

(B) What is the mean life expectancy?

38. *Harmful side effects of drugs.* A drug causes harmful side effects in 25% of the patients treated with the drug. If the drug is administered to 100 patients, what is the probability that 30 or more of these patients will suffer from the side effects?

Social Sciences

39. *Testing.* The percentage of correct answers on a college entrance examination is a beta random variable. The mean score is 75%. What is the probability that a student answers over 50% of the questions correctly?

40. *Testing.* The IQ scores for 6-year-old children in a certain area are normally distributed with a mean of 108 and a standard deviation of 12. What percentage of the children can be expected to have IQ scores of 135 or more?

Practice Test: Chapter 17

1. Find the mean, variance, and standard deviation for the discrete random variable X with probability distribution

x_i	1	2	3	4	5
p_i	.1	.3	.2	.1	.3

2. Find the mean, variance, and standard deviation for the continuous random variable X with probability density function

$$f(x) = \begin{cases} \frac{1}{4}(1+x) & 0 \leqslant x \leqslant 2 \\ 0 & \text{otherwise} \end{cases}$$

3. If X is a continuous random variable with probability density function

$$f(x) = \begin{cases} 10/(9x^2) & 1 \leqslant x \leqslant 10 \\ 0 & \text{otherwise} \end{cases}$$

find $P(1 \leqslant X \leqslant 5)$ and illustrate this with a graph.

4. Find the associated cumulative probability distribution function $F(x)$ and the median for the random variable in Problem 3. Graph $F(x)$ and locate the median on your graph.

5. For a binomial distribution with $p = 0.3$ and $n = 200$, compute:

   (A) The mean   (B) The standard deviation

6. In Problem 5, what is the probability of obtaining between 50 and 70 successes in 200 trials?

7. Given a normal distribution with mean 100 and standard deviation 10, find the area under the normal curve:

   (A) Between 92 and 108   (B) From 115 on

8. The life expectancy of a certain brand of light bulbs (in hundreds of hours) is an exponential random variable with a mean life expectancy of 500 hours. What is the probability that a bulb lasts over 500 hours?

9. The percentage of completed Social Security application forms that contain errors is a beta random variable with mean $\mu = .4$. What is the appropriate value of β?

10. The daily demand for doughnuts in a chain of bakeries (in hundreds of dozens) is a continuous random variable with probability density function

$$f(x) = \begin{cases} \frac{1}{8}(6 - x) & 2 \leqslant x \leqslant 6 \\ 0 & \text{otherwise} \end{cases}$$

What is the expected demand and the median demand?

Special Topics

A

A-1   Arithmetic Progressions

- Arithmetic Progressions — Definitions
- Special Formulas
- Application

Arithmetic Progressions – Definitions

Consider the sequence of numbers

   1, 4, 7, 10, 13, . . .

Assuming the pattern continues, can you guess what the next two numbers are? If you guessed 16 and 19, you have observed that each number after the first can be obtained from the preceding one by adding 3 to it. This is an example of an *arithmetic progression*. In general,

---

Arithmetic Progression

A sequence of numbers

   $a_1, a_2, a_3, \ldots, a_n, \ldots$

is called an **arithmetic progression** if there is a constant d, called the **common difference**, such that

$$a_n - a_{n-1} = d$$

That is,

$$a_n = a_{n-1} + d \quad \text{for every} \quad n > 1 \tag{1}$$

Example 1   Which sequence of numbers is an arithmetic progression and what is its common difference?

(A) 2, 4, 8, 10, . . .   (B) 3, 8, 13, 18, . . .

Solution   Sequence A does not have a common difference, since $4 - 2 = 2$ and $8 - 4 = 4$; hence, it is not an arithmetic progression. Sequence B is an arithmetic progression, since the difference between any two successive terms is 5, the common difference, and each number after the first one can be obtained by adding 5 to the preceding number.

Problem 1   Which sequence of numbers is an arithmetic progression, and what is its common difference?

(A) 15, 13, 11, 9, . . .   (B) 3, 9, 27, 81, . . .

■ Special Formulas

Arithmetic progressions have a number of convenient properties. For example, it is easy to derive formulas for the nth term in terms of n and the sum of any number of consecutive terms. To obtain a formula for the nth term of an arithmetic progression, we note that if a_1 is the first term and d is the common difference, then

$$a_2 = a_1 + d$$
$$a_3 = a_2 + d = (a_1 + d) + d = a_1 + 2d$$
$$a_4 = a_3 + d = (a_1 + 2d) + d = a_1 + 3d$$

This suggests that

$$a_n = a_1 + (n - 1)d \qquad \text{for all} \quad n > 1 \qquad\qquad (2)$$

Example 2   Find the twenty-first term in the arithmetic progression 3, 8, 13, 18, . . .

Solution   Find the common difference d and use formula (2):

$$d = 5, \qquad n = 21, \qquad a_1 = 3$$

Thus

$$a_{21} = 3 + (21 - 1)5$$
$$= 103$$

Problem 2   Find the fifty-first term in the arithmetic progression 15, 13, 11, 9, . . .

We now derive two simple and very useful formulas for the sum of n consecutive terms of an arithmetic progression. Let

$$S_n = a_1 + a_2 + \cdots + a_{n-1} + a_n$$

be the sum of n terms of an arithmetic progression with common difference d. Then,

$$S_n = a_1 + (a_1 + d) + \cdots + [a_1 + (n-2)d] + [a_1 + (n-1)d]$$

Reversing the order of the sum, we obtain

$$S_n = [a_1 + (n-1)d] + [a_1 + (n-2)d] + \cdots + (a_1 + d) + a_1$$

Something interesting happens if we combine these last two equations by addition (adding corresponding terms on the right sides):

$$2S_n = [2a_1 + (n-1)d] + [2a_1 + (n-1)d] + \cdots$$
$$+ [2a_1 + (n-1)d] + [2a_1 + (n-1)d]$$

All the terms on the right side are the same, and there are n of them. Thus,

$$2S_n = n[2a_1 + (n-1)d]$$

and

$$S_n = \frac{n}{2}[2a_1 + (n-1)d] \tag{3}$$

Replacing

$$[a_1 + (n-1)d] \qquad \text{in} \qquad \frac{n}{2}[a_1 + a_1 + (n-1)d]$$

by a_n from equation (2), we can obtain a second useful formula for the sum:

$$S_n = \frac{n}{2}(a_1 + a_n) \tag{4}$$

Example 3   Find the sum of the first 30 terms in the arithmetic progression 3, 8, 13, 18, . . .

Solution   Use (3) with $n = 30$, $a_1 = 3$, and $d = 5$:

$$S_{30} = \frac{30}{2}[2 \cdot 3 + (30 - 1)5]$$

$$= 2{,}265$$

Problem 3   Find the sum of the first 40 terms in the arithmetic progression 15, 13, 11, 9, . . .

Example 4   Find the sum of all the even numbers between 31 and 87.

Solution   First, find n using (2):

$$a_n = a_1 + (n - 1)d$$

$$86 = 32 + (n - 1)2$$

$$n = 28$$

Now find S_{28} using (4):

$$S_n = \frac{n}{2}(a_1 + a_n)$$

$$S_{28} = \frac{28}{2}(32 + 86)$$

$$= 1{,}652$$

Problem 4   Find the sum of all the odd numbers between 24 and 208.

▪ Application

Example 5   A person borrows \$3,600 and agrees to repay the loan in monthly install-ments over a 3 year period. The agreement is to pay 1% of the unpaid balance each month for using the money and \$100 each month to reduce the loan. What is the total cost of the loan over the 3 year period?

Solution   Let us look at the problem relative to a time line:

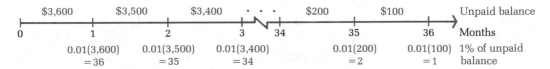

The total cost of the loan is

$$1 + 2 + \cdots + 34 + 35 + 36$$

The terms form an arithmetic progression with $n = 36$, $a_1 = 1$, and $a_{36} = 36$, so we can use (4):

$$S_n = \frac{n}{2}(a_1 + a_n)$$

$$S_{36} = \frac{36}{2}(1 + 36) = \$666$$

And we conclude that the total cost of the loan over the 3 year period is \$666.

Problem 5   Repeat Example 5 with a loan of $6,000 over a 5 year period.

Answers to   1. Sequence A; $d = -2$ 2. -85 3. -960
Matched Problems   4. 10,672   5. $1,830

Exercise A-1

A   1. Determine which of the following are arithmetic progressions. Find the common difference d and the next two terms for those progressions.

   (A) 5, 8, 11, . . .   (B) 4, 8, 16, . . .
   (C) $-2, -4, -8, . . .$ (D) $8, -2, -12, . . .$

2. Repeat Problem 1 for:

   (A) 11, 16, 21, . . .   (B) 16, 8, 4, . . .
   (C) $2, -3, -8, . . .$ (D) $-1, -2, -4, . . .$

Let $a_1, a_2, a_3, . . . , a_n, . . .$ be an arithmetic progression and S_n be the sum of the first n terms. In Problems 3–8 find the indicated quantities.

3. $a_1 = 7$, $d = 4$, $a_2 = ?$, $a_3 = ?$
4. $a_1 = -2$, $d = -3$, $a_2 = ?$, $a_3 = ?$

B   5. $a_1 = 2$, $d = 4$, $a_{21} = ?$, $S_{31} = ?$
6. $a_1 = 8$, $d = -10$, $a_{15} = ?$, $S_{23} = ?$
7. $a_1 = 18$, $a_{20} = 75$, $S_{20} = ?$
8. $a_1 = 203$, $a_{30} = 261$, $S_{30} = ?$
9. Find $f(1) + f(2) + f(3) + \cdots + f(50)$ if $f(x) = 2x - 3$.
10. Find $g(1) + g(2) + g(3) + \cdots + g(100)$ if $g(t) = 18 - 3t$.
11. Find the sum of all the odd integers between 12 and 68.
12. Find the sum of all the even integers between 23 and 97.

C   13. Show that the sum of the first n odd positive integers is n^2, using appropriate formulas from this section.
14. Show that the sum of the first n positive even integers is $n + n^2$, using formulas in this section.

Applications

Business & Economics   15. You are confronted with two job offers. Firm A will start you at $24,000 per year and guarantees you a $900 raise each year for 10 years. Firm B will start you at $22,000 per year but guarantees you a $1,300 raise each year for 10 years. Over the 10 year period, what is the total amount each firm will pay you?

16. In Problem 15, what would be your annual salary in each firm for the tenth year?

17. *Loan repayment.* If you borrow $4,800 and repay the loan by paying $200 per month to reduce the loan and 1% of the unpaid balance each month for the use of the money, what is the total cost of the loan over 24 months?

18. *Loan repayment.* Repeat Problem 17 replacing 1% with 1.5%.

A-2 Geometric Progressions

■ Geometric Progressions
■ Special Formulas
■ Infinite Geometric Progressions

■ Geometric Progressions

Consider the sequence of numbers

  2, 6, 18, 54, . . .

Assuming the pattern continues, can you guess what the next two numbers are? If you guessed 162 and 486, you have observed that each number after the first can be obtained from the preceding one by multiplying it by 3. This is an example of a *geometric progression.* In general,

---

Geometric Progression

A sequence of numbers

  $a_1, a_2, a_3, \ldots, a_n, \ldots$

is called a **geometric progression** if there exists a nonzero constant r, called a **common ratio,** such that

$$\frac{a_n}{a_{n-1}} = r$$

That is,

  $a_n = ra_{n-1}$ for every $n \geq 1$ (1)

Example 6 Which sequence of numbers is a geometric progression and what is its common ratio?

(A) 5, 3, 1, −1, . . .   (B) 1, 2, 4, 8, . . .

Solution Sequence A does not have a common ratio, since $3 \div 5 \neq 1 \div 3$; hence, it is not a geometric progression. Sequence B is a geometric progression, since the ratio of any two successive terms (the second divided by the first) is the constant 2, the common ratio, and each number after the first can be obtained by multiplying the preceding number by 2.

Problem 6 Which sequence of numbers is a geometric progression and what is its common ratio?

(A) $4, -2, 1, -\frac{1}{2}, \ldots$ (B) $2, 4, 6, 8, \ldots$

■ Special Formulas

Like arithmetic progressions, geometric progressions have several useful properties. It is easy to derive formulas for the nth term in terms of n and for the sum of any number of consecutive terms. To obtain a formula for the nth term of a geometric progression, we note that if a_1 is the first term and r is the common ratio, then

$$a_2 = ra_1$$
$$a_3 = ra_2 = r(ra_1) = r^2 a_1 = a_1 r^2$$
$$a_4 = ra_3 = r(r^2 a_1) = r^3 a_1 = a_1 r^3$$

This suggests that

$$a_n = a_1 r^{n-1} \qquad \text{for all} \quad n > 1 \tag{2}$$

Example 7 Find the eighth term in the geometric progression $\frac{1}{2}, \frac{1}{4}, \frac{1}{8}, \ldots$.

Solution Find the common ratio r and use formula (2):

$$r = \tfrac{1}{2}, \qquad n = 8, \qquad a_1 = \tfrac{1}{2}$$

Thus,

$$a_8 = (\tfrac{1}{2})(\tfrac{1}{2})^{8-1}$$
$$= \tfrac{1}{256}$$

Problem 7 Find the seventh term in the geometric progression $\frac{1}{32}, -\frac{1}{16}, \frac{1}{8}, \ldots$.

Example 8 If the first and tenth terms of a geometric progression are 2 and 4, respectively, find the common ratio r.

Solution $a_n = a_1 r^{n-1}$

$4 = 2 \cdot r^{10-1}$

$2 = r^9$

$r = 2^{1/9} \approx 1.08$ Use a calculator or logarithms

Problem 8   If the first and eighth terms of a geometric progression are 1,000 and 2,000, respectively, find the common ratio r.

We now derive two very useful formulas for the sum of n consecutive terms of a geometric progression. Let

$$a_1, a_1 r, a_1 r^2, \ldots, a_1 r^{n-2}, a_1 r^{n-1}$$

be n terms of a geometric progression. Their sum is

$$S_n = a_1 + a_1 r + a_1 r^2 + \cdots + a_1 r^{n-2} + a_1 r^{n-1}$$

If we multiply both sides by r, we obtain

$$r S_n = a_1 r + a_1 r^2 + a_1 r^3 + \cdots + a_1 r^{n-1} + a_1 r^n$$

Now combine these last two equations by subtraction to obtain

$$r S_n - S_n = (a_1 r + a_1 r^2 + a_1 r^3 + \cdots + a_1 r^{n-1} + a_1 r^n)$$
$$- (a_1 + a_1 r + a_1 r^2 + \cdots + a_1 r^{n-2} + a_1 r^{n-1})$$

$$(r - 1)S_n = a_1 r^n - a_1$$

Notice how many terms drop out on the right side. Hence,

$$S_n = \frac{a_1(r^n - 1)}{r - 1} \qquad r \neq 1 \qquad\qquad (3)$$

Since $a_n = a_1 r^{n-1}$, or $r a_n = a_1 r^n$, formula (3) can also be written in the form

$$S_n = \frac{r a_n - a_1}{r - 1} \qquad r \neq 1 \qquad\qquad (4)$$

Example 9   Find the sum of the first ten terms of the geometric progression
1, 1.05, 1.05², . . .

Solution   Use formula (3) with $a_1 = 1$, $r = 1.05$, and $n = 10$:

$$S_n = \frac{a_1(r^n - 1)}{r - 1}$$

$$S_{10} = \frac{1(1.05^{10} - 1)}{1.05 - 1}$$

$$\approx \frac{0.6289}{0.05} \approx 12.58$$

Problem 9   Find the sum of the first eight terms of the geometric progression
100, 100(1.08), 100(1.08)², . . .

■ Infinite Geometric Progressions

Given a geometric progression, what happens to the sum S_n of the first n
terms as n increases without stopping? To answer this question, let us write
formula (3) in the form

$$S_n = \frac{a_1 r^n}{r - 1} - \frac{a_1}{r - 1}$$

It is possible to show that if $|r| < 1$ (that is, $-1 < r < 1$), then r^n will tend to
zero as n increases. (See what happens, for example, if you let $r = \frac{1}{2}$ and
then increase n.) Thus, the first term above will tend to zero and S_n can be
made as close as we please to the second term, $-a_1/(r - 1)$ [which can be
written as $a_1/(1 - r)$], by taking n sufficiently large. Thus, if the common
ratio r is between -1 and 1, we define the sum of an infinite geometric
progression to be

$$S_\infty = \frac{a_1}{1 - r} \qquad |r| < 1 \qquad\qquad (5)$$

If $r \le -1$ or $r \ge 1$, then an infinite geometric progression has no sum.

Example 10   The government has decided on a tax rebate program to stimulate the
economy. Suppose you receive $600 and that you spend 80% of this, and
that each of the people who receive what you spend also spend 80% of what
they receive, and this process continues without end. According to the
multiplier doctrine in economics, the effect of your $600 tax rebate on the
economy is multiplied many times. What is the total amount spent if the
process continues as indicated?

Solution We need to find the sum of an infinite geometric progression with the first amount spent being $a_1 = (.08)(\$600) = \$480$ and $r = 0.8$. Using formula (5), we obtain

$$S_\infty = \frac{a_1}{1 - r}$$

$$= \frac{\$480}{1 - 0.8}$$

$$= \$2,400$$

Thus, assuming the process continues as indicated, we would expect the $600 tax rebate to result in about $2,400 of spending.

Problem 10 Repeat Example 10 with a tax rebate of $1,000.

Answers to Matched Problems

6. Sequence A; $r = -\frac{1}{2}$ 7. 2   8. Approximately 1.104
9. 1,063.66   10. $4,000

Exercise A-2

A

1. Determine which of the following are geometric progressions. Find the common ratio r and the next two terms for those that are:

   (A) $1, -2, 4, \ldots$ (B) $7, 6, 5, \ldots$ (C) $2, 1, \frac{1}{2}, \ldots$
   (D) $2, -4, 6, \ldots$

2. Repeat Problem 1 for:

   (A) $4, -1, -6, \ldots$ (B) $15, 5, \frac{5}{3}, \ldots$ (C) $\frac{1}{4}, -\frac{1}{2}, 1, \ldots$
   (D) $\frac{1}{2}, \frac{2}{3}, \frac{3}{4}, \ldots$

Let $a_1, a_2, a_3, \ldots, a_n, \ldots$ be a geometric progression and S_n be the sum of the first n terms. In Problems 3–12 find the indicated quantities. Use logarithms or a calculator as needed.

3. $a_1 = 3$, $r = -2$, $a_2 = ?$, $a_3 = ?$, $a_4 = ?$
4. $a_1 = 32$, $r = -\frac{1}{2}$, $a_2 = ?$, $a_3 = ?$, $a_4 = ?$
5. $a_1 = 1$, $a_7 = 729$, $r = -3$, $S_7 = ?$
6. $a_1 = 3$, $a_7 = 2,187$, $r = 3$, $S_7 = ?$

B

7. $a_1 = 100$, $r = 1.08$, $a_{10} = ?$
8. $a_1 = 240$, $r = 1.06$, $a_{12} = ?$
9. $a_1 = 100$, $a_9 = 200$, $r = ?$
10. $a_1 = 100$, $a_{10} = 300$, $r = ?$
11. $a_1 = 500$, $r = 0.6$, $S_{10} = ?$, $S_\infty = ?$
12. $a_1 = 8,000$, $r = 0.4$, $S_{10} = ?$, $S_\infty = ?$

13. Find the sum of each infinite geometric progression (if it exists).

(A) 2, 4, 8, . . .   (B) 2, $-\frac{1}{2}$, $\frac{1}{8}$, . . .

14. Repeat Problem 13 for:

(A) 16, 4, 1, . . .   (B) 1, -3, 9, . . .

C **15.** Find $f(1) + f(2) + \cdots + f(10)$ if $f(x) = (\frac{1}{2})^x$.

16. Find $g(1) + g(2) + \cdots + g(10)$ if $g(x) = 2^x$.

◼

Applications

Business & Economics

17. *Economy stimulation.* The government, through a subsidy program, distributes $5,000,000. If we assume each individual or agency spends 70% of what is received, and 70% of this is spent, and so on, how much total increase in spending results from this government action? (Let $a_1 = \$3,500,000$.)

18. *Economy stimulation.* Repeat Problem 17 using $10,000,000 as the amount distributed and 80%.

19. *Cost-of-living adjustment.* If the cost-of-living index increased 5% for each of the past 10 years and you had a salary agreement that increased your salary by the same percentage each year, what would your present salary be if you had a $20,000 per year salary 10 years ago? What would be your total earnings in the past 10 years? [*Hint:* $r = 1.05$.]

20. *Depreciation.* In *straight-line depreciation*, an asset less its salvage value at the end of its useful life is depreciated (for tax purposes) in equal annual amounts over its useful life. Thus, a $100,000 company airplane with a salvage value of $20,000 at the end of 10 years would be depreciated at $8,000 per year for each of the 10 years.

Since certain assets, such as airplanes, cars, and so on, depreciate more rapidly during the early years of their useful life, several methods of depreciation that take this into consideration are available to the taxpayer. One such method is called the *method of declining balance.* The rate used cannot exceed double that used for straight-line depreciation (ignoring salvage value) and is applied to the remaining value of an asset after the previous year's depreciation has been deducted. In our airplane example, the annual rate of straight-line depreciation over the 10 year period is 10%. Let us assume we can double this rate for the method of declining balance. At some point before the salvage value is reached (taxpayer's choice), we must switch over to the straight-line method to depreciate the final amount of the asset.

The table on the next page illustrates the two methods of depreciation for the company airplane.

	Straight-Line		Declining Balance	
Year end	Amount depreciated	Asset value	Amount depreciated	Asset value
0	$ 0	$100,000	$ 0	$100,000
1	0.1(80,000) = 8,000	92,000	0.2(100,000) = 20,000	80,000
2	0.1(80,000) = 8,000	84,000	0.2(80,000) = 16,000	64,000
3	0.1(80,000) = 8,000	76,000	0.2(64,000) = 12,800	51,200
⋮	⋮	⋮	⋮	⋮
7	0.1(80,000) = 8,000	44,000	0.2(26,214) = 5,243	20,972
8	0.1(80,000) = 8,000	36,000	$\frac{972}{3}$ = 324	20,648
9	0.1(80,000) = 8,000	28,000	$\frac{972}{3}$ = 324	20,324
10	0.1(80,000) = 8,000	20,000	$\frac{972}{3}$ = 324	20,000

Shift to straight-line, otherwise next entry will drop below salvage value

Arithmetic progression

Geometric progressions above dashed line

(A) For the declining balance, find the sum of the depreciation amounts above the dashed line using formula (4) and then add the entries below the line to this result.

(B) Repeat part A using formula (3).

(C) Find the asset value under declining balance at the end of the fifth year using formula (2).

(D) Find the asset value under straight-line at the end of the fifth year using formula (2) in Section A-1.

A-3 The Binomial Formula

- Factorial
- Binomial Theorem—Development

The binomial form

$$(a + b)^n$$

where n is a natural number, appears more frequently than you might expect. The coefficients in the expansion play an important role in probability studies. The binomial formula, which we will informally derive, enables us to expand $(a + b)^n$ directly for n any natural number. Since the formula involves **factorials,** we digress for a moment here to introduce this important concept.

■ Factorial

For n a natural number, **n factorial**—denoted by **n!**—is the product of the first n natural numbers. **Zero factorial** is defined to be one. Symbolically,

$$n! = n(n-1) \cdot \cdots \cdot 2 \cdot 1$$
$$1! = 1$$
$$0! = 1$$

It is also useful to note that

$$n! = n \cdot (n-1)!$$

Example 11   Evaluate each.

(A) $5! = 5 \cdot 4 \cdot 3 \cdot 2 \cdot 1 = 120$ (B) $\dfrac{8!}{7!} = \dfrac{8 \cdot 7!}{7!} = 8$

(C) $\dfrac{10!}{7!} = \dfrac{10 \cdot 9 \cdot 8 \cdot 7!}{7!} = 720$

Problem 11   Evaluate each: (A) $4!$ (B) $\dfrac{7!}{6!}$ (C) $\dfrac{8!}{5!}$

A special formula involving factorials is

$$\binom{n}{r} = \frac{n!}{r!(n-r)!} \qquad n \geqslant r \geqslant 0$$

Example 12   (A) $\dbinom{9}{2} = \dfrac{9!}{2!(9-2)!} = \dfrac{9!}{2!7!} = \dfrac{9 \cdot 8 \cdot 7!}{2 \cdot 7!} = 36$

(B) $\dbinom{5}{5} = \dfrac{5!}{5!(5-5)!} = \dfrac{5!}{5!0!} = \dfrac{5!}{5!} = 1$

Problem 12   Find: (A) $\dbinom{5}{2}$ (B) $\dbinom{6}{0}$

■ Binomial Theorem — Development

Let us expand $(a + b)^n$ for several values of n to see if we can observe a pattern that leads to a general formula for the expansion for any natural number n:

$$(a + b)^1 = a + b$$

$$(a + b)^2 = a^2 + 2ab + b^2$$

$$(a + b)^3 = a^3 + 3a^2b + 3ab^2 + b^3$$

$$(a + b)^4 = a^4 + 4a^3b + 6a^2b^2 + 4ab^3 + b^4$$

$$(a + b)^5 = a^5 + 5a^4b + 10a^3b^2 + 10a^2b^3 + 5ab^4 + b^5$$

Observations

1. The expansion of $(a + b)^n$ has $(n + 1)$ terms.
2. The power of a decreases by 1 for each term as we move from left to right.
3. The power of b increases by 1 for each term as we move from left to right.
4. In each term the sum of the powers of a and b always equals n.
5. Starting with a given term, we can get the coefficient of the next term by multiplying the coefficient of the given term by the exponent of a and dividing by the number that represents the position of the term in the series of terms. For example, in the expansion of $(a + b)^4$, above, the coefficient of the third term is found from the second term by multiplying 4 and 3, and then dividing by 2 [that is, the coefficient of the third term $= (4 \cdot 3)/2 = 6$].

We now postulate these same properties for the general case:

$$(a + b)^n = a^n + \frac{n}{1} a^{n-1}b + \frac{n(n-1)}{1 \cdot 2} a^{n-2}b^2 + \frac{n(n-1)(n-2)}{1 \cdot 2 \cdot 3} a^{n-3}b^3 + \cdots + b^n$$

$$= \frac{n!}{0!(n-0)!} a^n + \frac{n!}{1!(n-1)!} a^{n-1}b + \frac{n!}{2!(n-2)!} a^{n-2}b^2 + \frac{n!}{3!(n-3)!} a^{n-3}b^3 + \cdots + \frac{n!}{n!(n-n)!} b^n$$

$$= \binom{n}{0} a_n + \binom{n}{1} a^{n-1}b + \binom{n}{2} a^{n-2}b^2 + \binom{n}{3} a^{n-3}b^3 + \cdots + \binom{n}{n} b^n$$

And we are led to the formula in the binomial theorem (a formal proof requires mathematical induction, which is beyond the scope of this book):

Binomial Theorem

For all natural numbers n,

$$(a + b)^n = \binom{n}{0} a_n + \binom{n}{1} a^{n-1}b + \binom{n}{2} a^{n-2}b^2$$

$$+ \binom{n}{3} a^{n-3}b^3 + \cdots + \binom{n}{n} b^n$$

Example 13   Use the binomial formula to expand $(u + v)^6$.

Solution

$$(u + v)^6 = \binom{6}{0} u^6 + \binom{6}{1} u^5v + \binom{6}{2} u^4v^2 + \binom{6}{3} u^3v^3 + \binom{6}{4} u^2v^4 + \binom{6}{5} uv^5 + \binom{6}{6} v^6$$

$$= u^6 + 6u^5v + 15u^4v^2 + 20u^3v^3 + 15u^2v^4 + 6uv^5 + v^6$$

Problem 13   Use the binomial formula to expand $(x + 2)^5$.

Example 14   Use the binomial formula to find the sixth term in the expansion of $(x - 1)^{18}$.

Solution   Sixth term $= \binom{18}{5} x^{13}(-1)^5$

$$= \frac{18!}{5!(18 - 5)!} x^{13}(-1)$$

$$= -8,568x^{13}$$

Problem 14   Use the binomial formula to find the fourth term in the expansion of $(x - 2)^{20}$.

Answers to Matched Problems

11. (A) 24   (B) 7   (C) 336   12. (A) 10   (B) 1

13. $x^5 + 5x^4 \cdot 2 + 10x^3 \cdot 2^2 + 10x^2 \cdot 2^3 + 5x \cdot 2^4 + 2^5$

$$= x^5 + 10x^4 + 40x^3 + 80x^2 + 80x + 32$$

14. $-9,120x^{17}$

Exercise A-3

A   *Evaluate.*

1. 6!

2. 7!

3. $\dfrac{10!}{9!}$

4. $\dfrac{20!}{19!}$

5. $\dfrac{12!}{9!}$

6. $\dfrac{10!}{6!}$

7. $\dfrac{5!}{2!3!}$

8. $\dfrac{7!}{3!4!}$

9. $\dfrac{6!}{5!(6-5)!}$

10. $\dfrac{7!}{4!(7-4)!}$

11. $\dfrac{20!}{3!17!}$

12. $\dfrac{52!}{50!2!}$

B   *Evaluate.*

13. $\dbinom{5}{3}$

14. $\dbinom{7}{3}$

15. $\dbinom{6}{5}$

16. $\dbinom{7}{4}$

17. $\dbinom{5}{0}$

18. $\dbinom{5}{5}$

19. $\dbinom{18}{15}$

20. $\dbinom{18}{3}$

Expand each expression using the binomial formula.

21. $(a+b)^4$

22. $(m+n)^5$

23. $(x-1)^6$

24. $(u-2)^5$

25. $(2a-b)^5$

26. $(x-2y)^5$

Find the indicated term in each expansion.

27. $(x-1)^{18}$,   fifth term

28. $(x-3)^{20}$,   third term

29. $(p+q)^{15}$,   seventh term

30. $(p+q)^{15}$,   thirteenth term

31. $(2x+y)^{12}$,   eleventh term

32. $(2x+y)^{12}$,   third term

C   33.   Show that: $\dbinom{n}{0}=\dbinom{n}{n}$

34.   Show that: $\dbinom{n}{r}=\dbinom{n}{n-r}$

35.   The triangle below is called **Pascal's triangle.** Can you guess what the next two rows at the bottom are? Compare these numbers with the coefficients of binomial expansions.

Tables

B

APPENDIX B   Contents

Table I Exponential Functions (e^x and e^{-x})

x	e^x	e^{-x}	x	e^x	e^{-x}	x	e^x	e^{-x}
0.00	1.0000	1.00 000	0.50	1.6487	0.60 653	1.00	2.7183	0.36 788
0.01	1.0101	0.99 005	0.51	1.6653	0.60 050	1.01	2.7456	0.36 422
0.02	1.0202	0.98 020	0.52	1.6820	0.59 452	1.02	2.7732	0.36 059
0.03	1.0305	0.97 045	0.53	1.6989	0.58 860	1.03	2.8011	0.35 701
0.04	1.0408	0.96 079	0.54	1.7160	0.58 275	1.04	2.8292	0.35 345
0.05	1.0513	0.95 123	0.55	1.7333	0.57 695	1.05	2.8577	0.34 994
0.06	1.0618	0.94 176	0.56	1.7507	0.57 121	1.06	2.8864	0.34 646
0.07	1.0725	0.93 239	0.57	1.7683	0.56 553	1.07	2.9154	0.34 301
0.08	1.0833	0.92 312	0.58	1.7860	0.55 990	1.08	2.9447	0.33 960
0.09	1.0942	0.91 393	0.59	1.8040	0.55 433	1.09	2.9743	0.33 622
0.10	1.1052	0.90 484	0.60	1.8221	0.54 881	1.10	3.0042	0.33 287
0.11	1.1163	0.89 583	0.61	1.8404	0.54 335	1.11	3.0344	0.32 956
0.12	1.1275	0.88 692	0.62	1.8589	0.53 794	1.12	3.0649	0.32 628
0.13	1.1388	0.87 810	0.63	1.8776	0.53 259	1.13	3.0957	0.32 303
0.14	1.1503	0.86 936	0.64	1.8965	0.52 729	1.14	3.1268	0.31 982
0.15	1.1618	0.86 071	0.65	1.9155	0.52 205	1.15	3.1582	0.31 664
0.16	1.1735	0.85 214	0.66	1.9348	0.51 685	1.16	3.1899	0.31 349
0.17	1.1853	0.84 366	0.67	1.9542	0.51 171	1.17	3.2220	0.31 037
0.18	1.1972	0.83 527	0.68	1.9739	0.50 662	1.18	3.2544	0.30 728
0.19	1.2092	0.82 696	0.69	1.9937	0.50 158	1.19	3.2871	0.30 422
0.20	1.2214	0.81 873	0.70	2.0138	0.49 659	1.20	3.3201	0.30 119
0.21	1.2337	0.81 058	0.71	2.0340	0.49 164	1.21	3.3535	0.29 820
0.22	1.2461	0.80 252	0.72	2.0544	0.48 675	1.22	3.3872	0.29 523
0.23	1.2586	0.79 453	0.73	2.0751	0.48 191	1.23	3.4212	0.29 229
0.24	1.2712	0.78 663	0.74	2.0959	0.47 711	1.24	3.4556	0.28 938
0.25	1.2840	0.77 880	0.75	2.1170	0.47 237	1.25	3.4903	0.28 650
0.26	1.2969	0.77 105	0.76	2.1383	0.46 767	1.26	3.5254	0.28 365
0.27	1.3100	0.76 338	0.77	2.1598	0.46 301	1.27	3.5609	0.28 083
0.28	1.3231	0.75 578	0.78	2.1815	0.45 841	1.28	3.5966	0.27 804
0.29	1.3364	0.74 826	0.79	2.2034	0.45 384	1.29	3.6328	0.27 527
0.30	1.3499	0.74 082	0.80	2.2255	0.44 933	1.30	3.6693	0.27 253
0.31	1.3634	0.73 345	0.81	2.2479	0.44 486	1.31	3.7062	0.26 982
0.32	1.3771	0.72 615	0.82	2.2705	0.44 043	1.32	3.7434	0.26 714
0.33	1.3910	0.71 892	0.83	2.2933	0.43 605	1.33	3.7810	0.26 448
0.34	1.4049	0.71 177	0.84	2.3164	0.43 171	1.34	3.8190	0.26 185
0.35	1.4191	0.70 469	0.85	2.3396	0.42 741	1.35	3.8574	0.25 924
0.36	1.4333	0.69 768	0.86	2.3632	0.42 316	1.36	3.8962	0.25 666
0.37	1.4477	0.69 073	0.87	2.3869	0.41 895	1.37	3.9354	0.25 411
0.38	1.4623	0.68 386	0.88	2.4109	0.41 478	1.38	3.9749	0.25 158
0.39	1.4770	0.67 706	0.89	2.4351	0.41 066	1.39	4.0149	0.24 908
0.40	1.4918	0.67 032	0.90	2.4596	0.40 657	1.40	4.0552	0.24 660
0.41	1.5068	0.66 365	0.91	2.4843	0.40 252	1.41	4.0960	0.24 414
0.42	1.5220	0.65 705	0.92	2.5093	0.39 852	1.42	4.1371	0.24 171
0.43	1.5373	0.65 051	0.93	2.5345	0.39 455	1.43	4.1787	0.23 931
0.44	1.5527	0.64 404	0.94	2.5600	0.39 063	1.44	4.2207	0.23 693
0.45	1.5683	0.63 763	0.95	2.5857	0.38 674	1.45	4.2631	0.23 457
0.46	1.5841	0.63 128	0.96	2.6117	0.38 289	1.46	4.3060	0.23 224
0.47	1.6000	0.62 500	0.97	2.6379	0.37 908	1.47	4.3492	0.22 993
0.48	1.6161	0.61 878	0.98	2.6645	0.37 531	1.48	4.3939	0.22 764
0.49	1.6323	0.61 263	0.99	2.6912	0.37 158	1.49	4.4371	0.22 537
0.50	1.6487	0.60 653	1.00	2.7183	0.36 788	1.50	4.4817	0.22 313

x	e^x	e^{-x}	x	e^x	e^{-x}	x	e^x	e^{-x}
1.50	4.4817	0.22 313	2.00	7.3891	0.13 534	2.50	12.182	0.082 085
1.51	4.5267	0.22 091	2.01	7.4633	0.13 399	2.51	12.305	0.081 268
1.52	4.5722	0.21 871	2.02	7.5383	0.13 266	2.52	12.429	0.080 460
1.53	4.6182	0.21 654	2.03	7.6141	0.13 134	2.53	12.554	0.079 659
1.54	4.6646	0.21 438	2.04	7.6906	0.13 003	2.54	12.680	0.078 866
1.55	4.7115	0.21 225	2.05	7.7679	0.12 873	2.55	12.807	0.078 082
1.56	4.7588	0.21 014	2.06	7.8460	0.12 745	2.56	12.936	0.077 305
1.57	4.8066	0.20 805	2.07	7.9248	0.12 619	2.57	13.066	0.076 536
1.58	4.8550	0.20 598	2.08	8.0045	0.12 493	2.58	13.197	0.075 774
1.59	4.9037	0.20 393	2.09	8.0849	0.12 369	2.59	13.330	0.075 020
1.60	4.9530	0.20 190	2.10	8.1662	0.12 246	2.60	13.464	0.074 274
1.61	5.0028	0.19 989	2.11	8.2482	0.12 124	2.61	13.599	0.073 535
1.62	5.0531	0.19 790	2.12	8.3311	0.12 003	2.62	13.736	0.072 803
1.63	5.1039	0.19 593	2.13	8.4149	0.11 884	2.63	13.874	0.072 078
1.64	5.1552	0.19 398	2.14	8.4994	0.11 765	2.64	14.013	0.071 361
1.65	5.2070	0.19 205	2.15	8.5849	0.11 648	2.65	14.154	0.070 651
1.66	5.2593	0.19 014	2.16	8.6711	0.11 533	2.66	14.296	0.069 948
1.67	5.3122	0.18 825	2.17	8.7583	0.11 418	2.67	14.440	0.069 252
1.68	5.3656	0.18 637	2.18	8.8463	0.11 304	2.68	14.585	0.068 563
1.69	5.4195	0.18 452	2.19	8.9352	0.11 192	2.69	14.732	0.067 881
1.70	5.4739	0.18 268	2.20	9.0250	0.11 080	2.70	14.880	0.067 206
1.71	5.5290	0.18 087	2.21	9.1157	0.10 970	2.71	15.029	0.066 537
1.72	5.5845	0.17 907	2.22	9.2073	0.10 861	2.72	15.180	0.065 875
1.73	5.6407	0.17 728	2.23	9.2999	0.10 753	2.73	15.333	0.065 219
1.74	5.6973	0.17 552	2.24	9.3933	0.10 646	2.74	15.487	0.064 570
1.75	5.7546	0.17 377	2.25	9.4877	0.10 540	2.75	15.643	0.063 928
1.76	5.8124	0.17 204	2.26	9.5831	0.10 435	2.76	15.800	0.063 292
1.77	5.8709	0.17 033	2.27	9.6794	0.10 331	2.77	15.959	0.062 662
1.78	5.9299	0.16 864	2.28	9.7767	0.10 228	2.78	16.119	0.062 039
1.79	5.9895	0.16 696	2.29	9.8749	0.10 127	2.79	16.281	0.061 421
1.80	6.0496	0.16 530	2.30	9.9742	0.10 026	2.80	16.445	0.060 810
1.81	6.1104	0.16 365	2.31	10.074	0.099 261	2.81	16.610	0.060 205
1.82	6.1719	0.16 203	2.32	10.176	0.098 274	2.82	16.777	0.059 606
1.83	6.2339	0.16 041	2.33	10.278	0.097 296	2.83	16.945	0.059 013
1.84	6.2965	0.15 882	2.34	10.381	0.096 328	2.84	17.116	0.058 426
1.85	6.3598	0.15 724	2.35	10.486	0.095 369	2.85	17.288	0.057 844
1.86	6.4237	0.15 567	2.36	10.591	0.094 420	2.86	17.462	0.057 269
1.87	6.4883	0.15 412	2.37	10.697	0.093 481	2.87	17.637	0.056 699
1.88	6.5535	0.15 259	2.38	10.805	0.092 551	2.88	17.814	0.056 135
1.89	6.6194	0.15 107	2.39	10.913	0.091 630	2.89	17.993	0.055 576
1.90	6.6859	0.14 957	2.40	11.023	0.090 718	2.90	18.174	0.055 023
1.91	6.7531	0.14 808	2.41	11.134	0.089 815	2.91	18.357	0.054 476
1.92	6.8210	0.14 661	2.42	11.246	0.088 922	2.92	18.541	0.053 934
1.93	6.8895	0.14 515	2.43	11.359	0.088 037	2.93	18.728	0.053 397
1.94	6.9588	0.14 370	2.44	11.473	0.087 161	2.94	18.916	0.052 866
1.95	7.0287	0.14 227	2.45	11.588	0.086 294	2.95	19.106	0.052 340
1.96	7.0993	0.14 086	2.46	11.705	0.085 435	2.96	19.298	0.051 819
1.97	7.1707	0.13 946	2.47	11.822	0.084 585	2.97	19.492	0.051 303
1.98	7.2427	0.13 807	2.48	11.941	0.083 743	2.98	19.688	0.050 793
1.99	7.3155	0.13 670	2.49	12.061	0.082 910	2.99	19.886	0.050 287
2.00	7.3891	0.13 534	2.50	12.182	0.082 085	3.00	20.086	0.049 787

Table I (Continued)

x	e^x	e^{-x}	x	e^x	e^{-x}	x	e^x	e^{-x}
3.00	20.086	0.049 787	3.50	33.115	0.030 197	4.00	54.598	0.018 316
3.01	20.287	0.049 292	3.51	33.448	0.029 897	4.01	55.147	0.018 133
3.02	20.491	0.048 801	3.52	33.784	0.029 599	4.02	55.701	0.017 953
3.03	20.697	0.048 316	3.53	34.124	0.029 305	4.03	56.261	0.017 774
3.04	20.905	0.047 835	3.54	34.467	0.029 013	4.04	56.826	0.017 597
3.05	21.115	0.047 359	3.55	34.813	0.028 725	4.05	57.397	0.017 422
3.05	21.328	0.046 888	3.56	35.163	0.028 439	4.06	57.974	0.017 249
3.07	21.542	0.046 421	3.57	35.517	0.028 156	4.07	58.557	0.017 077
3.08	21.758	0.045 959	3.58	35.874	0.027 876	4.08	59.145	0.016 907
3.09	21.977	0.045 502	3.59	36.234	0.027 598	4.09	59.740	0.016 739
3.10	22.198	0.045 049	3.60	36.598	0.027 324	4.10	60.340	0.016 573
3.11	22.421	0.044 601	3.61	36.966	0.027 052	4.11	60.947	0.016 408
3.12	22.646	0.044 157	3.62	37.338	0.026 783	4.12	61.559	0.016 245
3.13	22.874	0.043 718	3.63	37.713	0.026 516	4.13	62.178	0.016 083
3.14	23.104	0.043 283	3.64	38.092	0.026 252	4.14	62.803	0.015 923
3.15	23.336	0.042 852	3.65	38.475	0.025 991	4.15	63.434	0.015 764
3.16	23.571	0.042 426	3.66	38.861	0.025 733	4.16	64.072	0.015 608
3.17	23.807	0.042 004	3.67	39.252	0.025 476	4.17	64.715	0.015 452
3.18	24.047	0.041 586	3.68	39.646	0.025 223	4.18	65.366	0.015 299
3.19	24.288	0.041 172	3.69	40.045	0.024 972	4.19	66.023	0.015 146
3.20	24.533	0.040 762	3.70	40.447	0.024 724	4.20	66.686	0.014 996
3.21	24.779	0.040 357	3.71	40.854	0.024 478	4.21	67.357	0.014 846
3.22	25.028	0.039 955	3.72	41.264	0.024 234	4.22	68.033	0.014 699
3.23	25.280	0.039 557	3.73	41.679	0.023 993	4.23	68.717	0.014 552
3.24	25.534	0.039 164	3.74	42.098	0.023 754	4.24	69.408	0.014 408
3.25	25.790	0.038 774	3.75	42.521	0.023 518	4.25	70.105	0.014 264
3.26	26.050	0.038 388	3.76	42.948	0.023 284	4.26	70.810	0.014 122
3.27	26.311	0.038 006	3.77	43.380	0.023 052	4.27	71.522	0.013 982
3.28	26.576	0.037 628	3.78	43.816	0.022 823	4.28	72.240	0.013 843
3.29	26.843	0.037 254	3.79	44.256	0.022 596	4.29	72.966	0.013 705
3.30	27.113	0.036 883	3.80	44.701	0.022 371	4.30	73.700	0.013 569
3.31	27.385	0.036 516	3.81	45.150	0.022 148	4.31	74.440	0.013 434
3.32	27.660	0.036 153	3.82	45.604	0.021 928	4.32	75.189	0.013 300
3.33	27.938	0.035 793	3.83	46.063	0.021 710	4.33	75.944	0.013 168
3.34	28.219	0.035 437	3.84	46.525	0.021 494	4.34	76.708	0.013 037
3.35	28.503	0.035 084	3.85	46.993	0.021 280	4.35	77.478	0.012 907
3.36	28.789	0.034 735	3.86	47.465	0.021 068	4.36	78.257	0.012 778
3.37	29.079	0.034 390	3.87	47.942	0.020 858	4.37	79.044	0.012 651
3.38	29.371	0.034 047	3.88	48.424	0.020 651	4.38	79.838	0.012 525
3.39	29.666	0.033 709	3.89	48.911	0.020 445	4.39	80.640	0.012 401
3.40	29.964	0.033 373	3.90	49.402	0.020 242	4.40	81.451	0.012 277
3.41	30.265	0.033 041	3.91	49.899	0.020 041	4.41	82.269	0.012 155
3.42	30.569	0.032 712	3.92	50.400	0.019 841	4.42	83.096	0.012 034
3.43	30.877	0.032 387	3.93	50.907	0.019 644	4.43	83.931	0.011 914
3.44	31.187	0.032 065	3.94	51.419	0.019 448	4.44	84.775	0.011 796
3.45	31.500	0.031 746	3.95	51.935	0.019 255	4.45	85.627	0.011 679
3.46	31.817	0.031 430	3.96	52.457	0.019 063	4.46	86.488	0.011 562
3.47	32.137	0.031 117	3.97	52.985	0.018 873	4.47	87.357	0.011 447
3.48	32.460	0.030 807	3.98	53.517	0.018 686	4.48	88.235	0.011 333
3.49	32.786	0.030 501	3.99	54.055	0.018 500	4.49	89.121	0.011 221
3.50	33.115	0.030 197	4.00	54.598	0.018 316	4.50	90.017	0.011 109

x	e^x	e^{-x}	x	e^x	e^{-x}	x	e^x	e^{-x}
4.50	90.017	0.011 109	5.00	148.41	0.006 7379	7.50	1,808.0	0.000 5531
4.51	90.922	0.010 998	5.05	156.02	0.006 4093	7.55	1,900.7	0.000 5261
4.52	91.836	0.010 889	5.10	164.02	0.006 0967	7.60	1,998.2	0.000 5005
4.53	92.759	0.010 781	5.15	172.43	0.005 7994	7.65	2,100.6	0.000 4760
4.54	93.691	0.010 673	5.20	181.27	0.005 5166	7.70	2,208.3	0.000 4528
4.55	94.632	0.010 567	5.25	190.57	0.005 2475	7.75	2,321.6	0.000 4307
4.56	95.583	0.010 462	5.30	200.34	0.004 9916	7.80	2,440.6	0.000 4097
4.57	96.544	0.010 358	5.35	210.61	0.004 7482	7.85	2,565.7	0.000 3898
4.58	97.514	0.010 255	5.40	221.41	0.004 5166	7.90	2,697.3	0.000 3707
4.59	98.494	0.010 153	5.45	232.76	0.004 2963	7.95	2,835.6	0.000 3527
4.60	99.484	0.010 052	5.50	244.69	0.004 0868	8.00	2,981.0	0.000 3355
4.61	100.48	0.009 9518	5.55	257.24	0.003 8875	8.05	3,133.8	0.000 3191
4.62	101.49	0.009 8528	5.60	270.43	0.003 6979	8.10	3,294.5	0.000 3035
4.63	102.51	0.009 7548	5.65	284.29	0.003 5175	8.15	3,463.4	0.000 2887
4.64	103.54	0.009 6577	5.70	298.87	0.003 3460	8.20	3,641.0	0.000 2747
4.65	104.58	0.009 5616	5.75	314.19	0.003 1828	8.25	3,827.6	0.000 2613
4.66	105.64	0.009 4665	5.80	330.30	0.003 0276	8.30	4,023.9	0.000 2485
4.67	106.70	0.009 3723	5.85	347.23	0.002 8799	8.35	4,230.2	0.000 2364
4.68	107.77	0.009 2790	5.90	365.04	0.002 7394	8.40	4,447.1	0.000 2249
4.69	108.85	0.009 1867	5.95	383.75	0.002 6058	8.45	4,675.1	0.000 2139
4.70	109.95	0.009 0953	6.00	403.43	0.002 4788	8.50	4,914.8	0.000 2035
4.71	111.05	0.009 0048	6.05	424.11	0.002 3579	8.55	5,166.8	0.000 1935
4.72	112.17	0.008 9152	6.10	445.86	0.002 2429	8.60	5,431.7	0.000 1841
4.73	113.30	0.008 8265	6.15	468.72	0.002 1335	8.65	5,710.1	0.000 1751
4.74	114.43	0.008 7386	6.20	492.75	0.002 2094	8.70	6,002.9	0.000 1666
4.75	115.58	0.008 6517	6.25	518.01	0.001 9305	8.75	6,310.7	0.000 1585
4.76	116.75	0.008 5656	6.30	544.57	0.001 8363	8.80	6,634.2	0.000 1507
4.77	117.92	0.008 4804	6.35	572.49	0.001 7467	8.85	6,974.4	0.000 1434
4.78	119.10	0.008 3960	6.40	601.85	0.001 6616	8.90	7,332.0	0.000 1364
4.79	120.30	0.008 3125	6.45	632.70	0.001 5805	8.95	7,707.9	0.000 1297
4.80	121.51	0.008 2297	6.50	665.14	0.001 5034	9.00	8,103.1	0.000 1234
4.81	122.73	0.008 1479	6.55	699.24	0.001 4301	9.05	8,518.5	0.000 1174
4.82	123.97	0.008 0668	6.60	735.10	0.001 3604	9.10	8,955.3	0.000 1117
4.83	125.21	0.007 9865	6.65	772.78	0.001 2940	9.15	9,414.4	0.000 1062
4.84	126.47	0.007 9071	6.70	812.41	0.001 2309	9.20	9,897.1	0.000 1010
4.85	127.74	0.007 8284	6.75	854.06	0.001 1709	9.25	10,405	0.000 0961
4.86	129.02	0.007 7505	6.80	897.85	0.001 1138	9.30	10,938	0.000 0914
4.87	130.32	0.007 6734	6.85	943.88	0.001 0595	9.35	11,499	0.000 0870
4.88	131.63	0.007 5970	6.90	992.27	0.001 0078	9.40	12,088	0.000 0827
4.89	132.95	0.007 5214	6.95	1,043.1	0.000 9586	9.45	12,708	0.000 0787
4.90	134.29	0.007 4466	7.00	1,096.6	0.000 9119	9.50	13,360	0.000 0749
4.91	135.64	0.007 3725	7.05	1,152.9	0.000 8674	9.55	14,045	0.000 0712
4.92	137.00	0.007 2991	7.10	1,212.0	0.000 8251	9.60	14,765	0.000 0677
4.93	138.38	0.007 2265	7.15	1,274.1	0.000 7849	9.65	15,522	0.000 0644
4.94	139.77	0.007 1546	7.20	1,339.4	0.000 7466	9.70	16,318	0.000 0613
4.95	141.17	0.007 0834	7.25	1,408.1	0.000 7102	9.75	17,154	0.000 0583
4.96	142.59	0.007 0129	7.30	1,480.3	0.000 6755	9.80	18,034	0.000 0555
4.97	144.03	0.006 9431	7.35	1,556.2	0.000 6426	9.85	18,958	0.000 0527
4.98	145.47	0.006 8741	7.40	1,636.0	0.000 6113	9.90	19,930	0.000 0502
4.99	146.94	0.006 8057	7.45	1,719.9	0.000 5814	9.95	20,952	0.000 0477
5.00	148.41	0.006 7379	7.50	1,808.0	0.000 5531	10.00	22,026	0.000 0454

N	0	1	2	3	4	5	6	7	8	9
1.0	0.0000	0.004321	0.008600	0.01284	0.01703	0.02119	0.02531	0.02938	0.03342	0.03743
1.1	0.04139	0.04532	0.04922	0.05308	0.05690	0.06070	0.06446	0.06819	0.07188	0.07555
1.2	0.07918	0.08279	0.08636	0.08991	0.09342	0.09691	0.1004	0.1038	0.1072	0.1106
1.3	0.1139	0.1173	0.1206	0.1239	0.1271	0.1303	0.1335	0.1367	0.1399	0.1430
1.4	0.1461	0.1492	0.1523	0.1553	0.1584	0.1614	0.1644	0.1673	0.1703	0.1732
1.5	0.1761	0.1790	0.1818	0.1847	0.1875	0.1903	0.1931	0.1959	0.1987	0.2014
1.6	0.2041	0.2068	0.2095	0.2122	0.2148	0.2175	0.2201	0.2227	0.2253	0.2279
1.7	0.2304	0.2330	0.2355	0.2380	0.2405	0.2430	0.2455	0.2480	0.2504	0.2529
1.8	0.2553	0.2577	0.2601	0.2625	0.2648	0.2673	0.2695	0.2718	0.2742	0.2765
1.9	0.2788	0.2810	0.2833	0.2856	0.2878	0.2900	0.2923	0.2945	0.2967	0.2989
2.0	0.3010	0.3032	0.3054	0.3075	0.3096	0.3118	0.3139	0.3160	0.3181	0.3201
2.1	0.3222	0.3243	0.3263	0.3284	0.3304	0.3324	0.3345	0.3365	0.3385	0.3404
2.2	0.3424	0.3444	0.3464	0.3483	0.3502	0.3522	0.3541	0.3560	0.3579	0.3598
2.3	0.3617	0.3636	0.3655	0.3674	0.3692	0.3711	0.3729	0.3747	0.3766	0.3784
2.4	0.3802	0.3820	0.3838	0.3856	0.3874	0.3892	0.3909	0.3927	0.3945	0.3962
2.5	0.3979	0.3997	0.4014	0.4031	0.4048	0.4065	0.4082	0.4099	0.4116	0.4133
2.6	0.4150	0.4166	0.4183	0.4200	0.4216	0.4232	0.4249	0.4265	0.4281	0.4298
2.7	0.4314	0.4330	0.4346	0.4362	0.4378	0.4393	0.4409	0.4425	0.4440	0.4456
2.8	0.4472	0.4487	0.4502	0.4518	0.4533	0.4548	0.4564	0.4579	0.4594	0.4609
2.9	0.4624	0.4639	0.4654	0.4669	0.4683	0.4698	0.4713	0.4728	0.4742	0.4757
3.0	0.4771	0.4786	0.4800	0.4814	0.4829	0.4843	0.4857	0.4871	0.4886	0.4900
3.1	0.4914	0.4928	0.4942	0.4955	0.4969	0.4983	0.4997	0.5011	0.5024	0.5038
3.2	0.5051	0.5065	0.5079	0.5092	0.5105	0.5119	0.5132	0.5145	0.5159	0.5172
3.3	0.5185	0.5198	0.5211	0.5224	0.5237	0.5250	0.5263	0.5276	0.5289	0.5302
3.4	0.5315	0.5328	0.5340	0.5353	0.5366	0.5378	0.5391	0.5403	0.5416	0.5428
3.5	0.5441	0.5453	0.5465	0.5478	0.5490	0.5502	0.5514	0.5527	0.5539	0.5551
3.6	0.5563	0.5575	0.5587	0.5599	0.5611	0.5623	0.5635	0.5647	0.5658	0.5670
3.7	0.5682	0.5694	0.5705	0.5717	0.5729	0.5740	0.5752	0.5763	0.5775	0.5786
3.8	0.5798	0.5809	0.5821	0.5832	0.5843	0.5855	0.5866	0.5877	0.5888	0.5899
3.9	0.5911	0.5922	0.5933	0.5944	0.5955	0.5966	0.5977	0.5988	0.5999	0.6010
4.0	0.6021	0.6031	0.6042	0.6053	0.6064	0.6075	0.6085	0.6096	0.6107	0.6117
4.1	0.6128	0.6138	0.6149	0.6160	0.6170	0.6180	0.6191	0.6201	0.6212	0.6222
4.2	0.6232	0.6243	0.6253	0.6263	0.6274	0.6284	0.6294	0.6304	0.6314	0.6325
4.3	0.6335	0.6345	0.6355	0.6365	0.6375	0.6385	0.6395	0.6405	0.6415	0.6425
4.4	0.6435	0.6444	0.6454	0.6464	0.6474	0.6484	0.6493	0.6503	0.6513	0.6522
4.5	0.6532	0.6542	0.6551	0.6561	0.6571	0.6580	0.6590	0.6599	0.6609	0.6618
4.6	0.6628	0.6637	0.6646	0.6656	0.6665	0.6675	0.6684	0.6693	0.6702	0.6712
4.7	0.6721	0.6730	0.6739	0.6749	0.6758	0.6767	0.6776	0.6785	0.6794	0.6803
4.8	0.6812	0.6821	0.6830	0.6839	0.6848	0.6857	0.6866	0.6875	0.6884	0.6893
4.9	0.6902	0.6911	0.6920	0.6928	0.6937	0.6946	0.6955	0.6964	0.6972	0.6981
5.0	0.6990	0.6998	0.7007	0.7016	0.7024	0.7033	0.7042	0.7050	0.7059	0.7067
5.1	0.7076	0.7084	0.7093	0.7101	0.7110	0.7118	0.7126	0.7135	0.7143	0.7152
5.2	0.7160	0.7168	0.7177	0.7185	0.7193	0.7202	0.7210	0.7218	0.7226	0.7235
5.3	0.7243	0.7251	0.7259	0.7267	0.7275	0.7284	0.7292	0.7300	0.7308	0.7316
5.4	0.7324	0.7332	0.7340	0.7348	0.7356	0.7364	0.7372	0.7380	0.7388	0.7396

N	0	1	2	3	4	5	6	7	8	9
5.5	0.7404	0.7412	0.7419	0.7427	0.7435	0.7443	0.7451	0.7459	0.7466	0.7474
5.6	0.7482	0.7490	0.7497	0.7505	0.7513	0.7520	0.7528	0.7536	0.7543	0.7551
5.7	0.7559	0.7566	0.7574	0.7582	0.7589	0.7597	0.7604	0.7612	0.7619	0.7627
5.8	0.7634	0.7642	0.7649	0.7657	0.7664	0.7672	0.7679	0.7686	0.7694	0.7701
5.9	0.7709	0.7716	0.7723	0.7731	0.7738	0.7745	0.7752	0.7760	0.7767	0.7774
6.0	0.7782	0.7789	0.7796	0.7803	0.7810	0.7818	0.7825	0.7832	0.7839	0.7846
6.1	0.7853	0.7860	0.7868	0.7875	0.7882	0.7889	0.7896	0.7903	0.7910	0.7917
6.2	0.7924	0.7931	0.7938	0.7945	0.7952	0.7959	0.7966	0.7973	0.7980	0.7987
6.3	0.7993	0.8000	0.8007	0.8014	0.8021	0.8028	0.8035	0.8041	0.8048	0.8055
6.4	0.8062	0.8069	0.8075	0.8082	0.8089	0.8096	0.8102	0.8109	0.8116	0.8122
6.5	0.8129	0.8136	0.8142	0.8149	0.8156	0.8162	0.8169	0.8176	0.8182	0.8189
6.6	0.8195	0.8202	0.8209	0.8215	0.8222	0.8228	0.8235	0.8241	0.8248	0.8254
6.7	0.8261	0.8267	0.8274	0.8280	0.8287	0.8293	0.8299	0.8306	0.8312	0.8319
6.8	0.8325	0.8331	0.8338	0.8344	0.8351	0.8357	0.8363	0.8370	0.8376	0.8382
6.9	0.8388	0.8395	0.8401	0.8407	0.8414	0.8420	0.8426	0.8432	0.8439	0.8445
7.0	0.8451	0.8457	0.8463	0.8470	0.8476	0.8482	0.8488	0.8494	0.8500	0.8506
7.1	0.8513	0.8519	0.8525	0.8531	0.8537	0.8543	0.8549	0.8555	0.8561	0.8567
7.2	0.8573	0.8579	0.8585	0.8591	0.8597	0.8603	0.8609	0.8615	0.8621	0.8627
7.3	0.8633	0.8639	0.8645	0.8651	0.8657	0.8663	0.8669	0.8675	0.8681	0.8686
7.4	0.8692	0.8698	0.8704	0.8710	0.8716	0.8722	0.8727	0.8733	0.8739	0.8745
7.5	0.8751	0.8756	0.8762	0.8768	0.8774	0.8779	0.8785	0.8791	0.8797	0.8802
7.6	0.8808	0.8814	0.8820	0.8825	0.8831	0.8837	0.8842	0.8848	0.8854	0.8859
7.7	0.8865	0.8871	0.8876	0.8882	0.8887	0.8893	0.8899	0.8904	0.8910	0.8915
7.8	0.8921	0.8927	0.8932	0.8938	0.8943	0.8949	0.8954	0.8960	0.8965	0.8971
7.9	0.8976	0.8982	0.8987	0.8993	0.8998	0.9004	0.9009	0.9015	0.9020	0.9025
8.0	0.9031	0.9036	0.9042	0.9047	0.9053	0.9058	0.9063	0.9069	0.9074	0.9079
8.1	0.9085	0.9090	0.9096	0.9101	0.9106	0.9112	0.9117	0.9122	0.9128	0.9133
8.2	0.9138	0.9143	0.9149	0.9154	0.9159	0.9165	0.9170	0.9175	0.9180	0.9186
8.3	0.9191	0.9196	0.9201	0.9206	0.9212	0.9217	0.9222	0.9227	0.9232	0.9238
8.4	0.9243	0.9248	0.9253	0.9258	0.9263	0.9269	0.9274	0.9279	0.9284	0.9289
8.5	0.9294	0.9299	0.9304	0.9309	0.9315	0.9320	0.9325	0.9330	0.9335	0.9340
8.6	0.9345	0.9350	0.9355	0.9360	0.9365	0.9370	0.9375	0.9380	0.9385	0.9390
8.7	0.9395	0.9400	0.9405	0.9410	0.9415	0.9420	0.9425	0.9430	0.9435	0.9440
8.8	0.9445	0.9450	0.9455	0.9460	0.9465	0.9469	0.9474	0.9479	0.9484	0.9489
8.9	0.9494	0.9499	0.9504	0.9509	0.9513	0.9518	0.9523	0.9528	0.9533	0.9538
9.0	0.9542	0.9547	0.9552	0.9557	0.9562	0.9566	0.9571	0.9576	0.9581	0.9586
9.1	0.9590	0.9595	0.9600	0.9605	0.9609	0.9614	0.9619	0.9624	0.9628	0.9633
9.2	0.9638	0.9643	0.9647	0.9652	0.9657	0.9661	0.9666	0.9671	0.9675	0.9680
9.3	0.9685	0.9689	0.9694	0.9699	0.9703	0.9708	0.9713	0.9717	0.9722	0.9727
9.4	0.9731	0.9736	0.9741	0.9745	0.9750	0.9754	0.9759	0.9763	0.9768	0.9773
9.5	0.9777	0.9782	0.9786	0.9791	0.9795	0.9800	0.9805	0.9809	0.9814	0.9818
9.6	0.9823	0.9827	0.9832	0.9836	0.9841	0.9845	0.9850	0.9854	0.9859	0.9863
9.7	0.9868	0.9872	0.9877	0.9881	0.9886	0.9890	0.9894	0.9899	0.9903	0.9908
9.8	0.9912	0.9917	0.9921	0.9926	0.9930	0.9934	0.9939	0.9943	0.9948	0.9952
9.9	0.9956	0.9961	0.9965	0.9969	0.9974	0.9978	0.9983	0.9987	0.9991	0.9996

Table III Natural Logarithms (ln $N = \log_e N$)

ln 10 = 2.3026	5 ln 10 = 11.5130	9 ln 10 = 20.7233
2 ln 10 = 4.6052	6 ln 10 = 13.8155	10 ln 10 = 23.0259
3 ln 10 = 6.9078	7 ln 10 = 16.1181	
4 ln 10 = 9.2103	8 ln 10 = 18.4207	

N	.00	.01	.02	.03	.04	.05	.06	.07	.08	.09
1.0	0.0000	0.0100	0.0198	0.0296	0.0392	0.0488	0.0583	0.0677	0.0770	0.0862
1.1	0.0953	0.1044	0.1133	0.1222	0.1310	0.1398	0.1484	0.1570	0.1655	0.1740
1.2	0.1823	0.1906	0.1989	0.2070	0.2151	0.2231	0.2311	0.2390	0.2469	0.2546
1.3	0.2624	0.2700	0.2776	0.2852	0.2927	0.3001	0.3075	0.3148	0.3221	0.3293
1.4	0.3365	0.3436	0.3507	0.3577	0.3646	0.3716	0.3784	0.3853	0.3920	0.3988
1.5	0.4055	0.4121	0.4187	0.4253	0.4318	0.4383	0.4447	0.4511	0.4574	0.4637
1.6	0.4700	0.4762	0.4824	0.4886	0.4947	0.5008	0.5068	0.5128	0.5188	0.5247
1.7	0.5306	0.5365	0.5423	0.5481	0.5539	0.5596	0.5653	0.5710	0.5766	0.5822
1.8	0.5878	0.5933	0.5988	0.6043	0.6098	0.6152	0.6206	0.6259	0.6313	0.6366
1.9	0.6419	0.6471	0.6523	0.6575	0.6627	0.6678	0.6729	0.6780	0.6831	0.6881
2.0	0.6931	0.6981	0.7031	0.7080	0.7129	0.7178	0.7227	0.7275	0.7324	0.7372
2.1	0.7419	0.7467	0.7514	0.7561	0.7608	0.7655	0.7701	0.7747	0.7793	0.7839
2.2	0.7885	0.7930	0.7975	0.8020	0.8065	0.8109	0.8154	0.8198	0.8242	0.8286
2.3	0.8329	0.8372	0.8416	0.8459	0.8502	0.8544	0.8587	0.8629	0.8671	0.8713
2.4	0.8755	0.8796	0.8838	0.8879	0.8920	0.8961	0.9002	0.9042	0.9083	0.9123
2.5	0.9163	0.9203	0.9243	0.9282	0.9322	0.9361	0.9400	0.9439	0.9478	0.9517
2.6	0.9555	0.9594	0.9632	0.9670	0.9708	0.9746	0.9783	0.9821	0.9858	0.9895
2.7	0.9933	0.9969	1.0006	1.0043	1.0080	1.0116	1.0152	1.0188	1.0225	1.0260
2.8	1.0296	1.0332	1.0367	1.0403	1.0438	1.0473	1.0508	1.0543	1.0578	1.0613
2.9	1.0647	1.0682	1.0716	1.0750	1.0784	1.0818	1.0852	1.0886	1.0919	1.0953
3.0	1.0986	1.1019	1.1053	1.1086	1.1119	1.1151	1.1184	1.1217	1.1249	1.1282
3.1	1.1314	1.1346	1.1378	1.1410	1.1442	1.1474	1.1506	1.1537	1.1569	1.1600
3.2	1.1632	1.1663	1.1694	1.1725	1.1756	1.1787	1.1817	1.1848	1.1878	1.1909
3.3	1.1939	1.1969	1.2000	1.2030	1.2060	1.2090	1.2119	1.2149	1.2179	1.2208
3.4	1.2238	1.2267	1.2296	1.2326	1.2355	1.2384	1.2413	1.2442	1.2470	1.2499
3.5	1.2528	1.2556	1.2585	1.2613	1.2641	1.2669	1.2698	1.2726	1.2754	1.2782
3.6	1.2809	1.2837	1.2865	1.2892	1.2920	1.2947	1.2975	1.3002	1.3029	1.3056
3.7	1.3083	1.3110	1.3137	1.3164	1.3191	1.3218	1.3244	1.3271	1.3297	1.3324
3.8	1.3350	1.3376	1.3403	1.3429	1.3455	1.3481	1.3507	1.3533	1.3558	1.3584
3.9	1.3610	1.3635	1.3661	1.3686	1.3712	1.3737	1.3762	1.3788	1.3813	1.3838
4.0	1.3863	1.3888	1.3913	1.3938	1.3962	1.3987	1.4012	1.4036	1.4061	1.4085
4.1	1.4110	1.4134	1.4159	1.4183	1.4207	1.4231	1.4255	1.4279	1.4303	1.4327
4.2	1.4351	1.4375	1.4398	1.4422	1.4446	1.4469	1.4493	1.4516	1.4540	1.4563
4.3	1.4586	1.4609	1.4633	1.4656	1.4679	1.4702	1.4725	1.4748	1.4770	1.4793
4.4	1.4816	1.4839	1.4861	1.4884	1.4907	1.4929	1.4951	1.4974	1.4996	1.5019
4.5	1.5041	1.5063	1.5085	1.5107	1.5129	1.5151	1.5173	1.5195	1.5217	1.5239
4.6	1.5261	1.5282	1.5304	1.5326	1.5347	1.5369	1.5390	1.5412	1.5433	1.5454
4.7	1.5476	1.5497	1.5518	1.5539	1.5560	1.5581	1.5602	1.5623	1.5644	1.5665
4.8	1.5686	1.5707	1.5728	1.5748	1.5769	1.5790	1.5810	1.5831	1.5851	1.5872
4.9	1.5892	1.5913	1.5933	1.5953	1.5974	1.5994	1.6014	1.6034	1.6054	1.6074
5.0	1.6094	1.6114	1.6134	1.6154	1.6174	1.6194	1.6214	1.6233	1.6253	1.6273
5.1	1.6292	1.6312	1.6332	1.6351	1.6371	1.6390	1.6409	1.6429	1.6448	1.6467
5.2	1.6487	1.6506	1.6525	1.6544	1.6563	1.6582	1.6601	1.6620	1.6639	1.6658
5.3	1.6677	1.6696	1.6715	1.6734	1.6752	1.6771	1.6790	1.6808	1.6827	1.6845
5.4	1.6864	1.6882	1.6901	1.6919	1.6938	1.6956	1.6974	1.6993	1.7011	1.7029

Note: ln 35, 200 = ln (3.52 × 10⁴) = ln 3.52 + 4 ln 10
$\qquad$ ln 0.00864 = ln (8.64 × 10⁻³) = ln 8.64 − 3 ln 10

N	.00	.01	.02	.03	.04	.05	.06	.07	.08	.09
5.5	1.7047	1.7066	1.7084	1.7102	1.7120	1.7138	1.7156	1.7174	1.7192	1.7210
5.6	1.7228	1.7246	1.7263	1.7281	1.7299	1.7317	1.7334	1.7352	1.7370	1.7387
5.7	1.7405	1.7422	1.7440	1.7457	1.7475	1.7492	1.7509	1.7527	1.7544	1.7561
5.8	1.7579	1.7596	1.7613	1.7630	1.7647	1.7664	1.7681	1.7699	1.7716	1.7733
5.9	1.7750	1.7766	1.7783	1.7800	1.7817	1.7834	1.7851	1.7867	1.7884	1.7901
6.0	1.7918	1.7934	1.7951	1.7967	1.7984	1.8001	1.8017	1.8034	1.8050	1.8066
6.1	1.8083	1.8099	1.8116	1.8132	1.8148	1.8165	1.8181	1.8197	1.8213	1.8229
6.2	1.8245	1.8262	1.8278	1.8294	1.8310	1.8326	1.8342	1.8358	1.8374	1.8390
6.3	1.8405	1.8421	1.8437	1.8453	1.8469	1.8485	1.8500	1.8516	1.8532	1.8547
6.4	1.8563	1.8579	1.8594	1.8610	1.8625	1.8641	1.8656	1.8672	1.8687	1.8703
6.5	1.8718	1.8733	1.8749	1.8764	1.8779	1.8795	1.8810	1.8825	1.8840	1.8856
6.6	1.8871	1.8886	1.8901	1.8916	1.8931	1.8946	1.8961	1.8976	1.8991	1.9006
6.7	1.9021	1.9036	1.9051	1.9066	1.9081	1.9095	1.9110	1.9125	1.9140	1.9155
6.8	1.9169	1.9184	1.9199	1.9213	1.9228	1.9242	1.9257	1.9272	1.9286	1.9301
6.9	1.9315	1.9330	1.9344	1.9359	1.9373	1.9387	1.9402	1.9416	1.9430	1.9445
7.0	1.9459	1.9473	1.9488	1.9502	1.9516	1.9530	1.9544	1.9559	1.9573	1.9587
7.1	1.9601	1.9615	1.9629	1.9643	1.9657	1.9671	1.9685	1.9699	1.9713	1.9727
7.2	1.9741	1.9755	1.9769	1.9782	1.9796	1.9810	1.9824	1.9838	1.9851	1.9865
7.3	1.9879	1.9892	1.9906	1.9920	1.9933	1.9947	1.9961	1.9974	1.9988	2.0001
7.4	2.0015	2.0028	2.0042	2.0055	2.0069	2.0082	2.0096	2.0109	2.0122	2.0136
7.5	2.0149	2.0162	2.0176	2.0189	2.0202	2.0215	2.0229	2.0242	2.0255	2.0268
7.6	2.0281	2.0295	2.0308	2.0321	2.0334	2.0347	2.0360	2.0373	2.0386	2.0399
7.7	2.0412	2.0425	2.0438	2.0451	2.0464	2.0477	2.0490	2.0503	2.0516	2.0528
7.8	2.0541	2.0554	2.0567	2.0580	2.0592	2.0605	2.0618	2.0631	2.0643	2.0656
7.9	2.0669	2.0681	2.0694	2.0707	2.0719	2.0732	2.0744	2.0757	2.0769	2.0782
8.0	2.0794	2.0807	2.0819	2.0832	2.0844	2.0857	2.0869	2.0882	2.0894	2.0906
8.1	2.0919	2.0931	2.0943	2.0956	2.0968	2.0980	2.0992	2.1005	2.1017	2.1029
8.2	2.1041	2.1054	2.1066	2.1078	2.1090	2.1102	2.1114	2.1126	2.1138	2.1150
8.3	2.1163	2.1175	2.1187	2.1199	2.1211	2.1223	2.1235	2.1247	2.1258	2.1270
8.4	2.1282	2.1294	2.1306	2.1318	2.1330	2.1342	2.1353	2.1365	2.1377	2.1389
8.5	2.1401	2.1412	2.1424	2.1436	2.1448	2.1459	2.1471	2.1483	2.1494	2.1506
8.6	2.1518	2.1529	2.1541	2.1552	2.1564	2.1576	2.1587	2.1599	2.1610	2.1622
8.7	2.1633	2.1645	2.1656	2.1668	2.1679	2.1691	2.1702	2.1713	2.1725	2.1736
8.8	2.1748	2.1759	2.1770	2.1782	2.1793	2.1804	2.1815	2.1827	2.1838	2.1849
8.9	2.1861	2.1872	2.1883	2.1894	2.1905	2.1917	2.1928	2.1939	2.1950	2.1961
9.0	2.1972	2.1983	2.1994	2.2006	2.2017	2.2028	2.2039	2.2050	2.2061	2.2072
9.1	2.2083	2.2094	2.2105	2.2116	2.2127	2.2138	2.2148	2.2159	2.2170	2.2181
9.2	2.2192	2.2203	2.2214	2.2225	2.2235	2.2246	2.2257	2.2268	2.2279	2.2289
9.3	2.2300	2.2311	2.2322	2.2332	2.2343	2.2354	2.2364	2.2375	2.2386	2.2396
9.4	2.2407	2.2418	2.2428	2.2439	2.2450	2.2460	2.2471	2.2481	2.2492	2.2502
9.5	2.2513	2.2523	2.2534	2.2544	2.2555	2.2565	2.2576	2.2586	2.2597	2.2607
9.6	2.2618	2.2628	2.2638	2.2649	2.2659	2.2670	2.2680	2.2690	2.2701	2.2711
9.7	2.2721	2.2732	2.2742	2.2752	2.2762	2.2773	2.2783	2.2793	2.2803	2.2814
9.8	2.2824	2.2834	2.2844	2.2854	2.2865	2.2875	2.2885	2.2895	2.2905	2.2915
9.9	2.2925	2.2935	2.2946	2.2956	2.2966	2.2976	2.2986	2.2996	2.3006	2.3016

Table IV Areas under the Standard Normal Curve

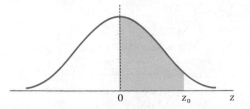

A represents the area between $z = 0$ and $z = z_0$, $z_0 \geq 0$

z	A	z	A	z	A	z	A
0.00	0.0000	0.30	0.1179	0.60	0.2258	0.90	0.3159
0.01	0.0040	0.31	0.1217	0.61	0.2291	0.91	0.3186
0.02	0.0080	0.32	0.1255	0.62	0.2324	0.92	0.3212
0.03	0.0120	0.33	0.1293	0.63	0.2357	0.93	0.3238
0.04	0.0160	0.34	0.1331	0.64	0.2389	0.94	0.3264
0.05	0.0199	0.35	0.1368	0.65	0.2422	0.95	0.3289
0.06	0.0239	0.36	0.1406	0.66	0.2454	0.96	0.3315
0.07	0.0279	0.37	0.1443	0.67	0.2486	0.97	0.3340
0.08	0.0319	0.38	0.1480	0.68	0.2518	0.98	0.3365
0.09	0.0359	0.39	0.1517	0.69	0.2549	0.99	0.3389
0.10	0.0398	0.40	0.1554	0.70	0.2580	1.00	0.3413
0.11	0.0438	0.41	0.1591	0.71	0.2612	1.01	0.3438
0.12	0.0478	0.42	0.1628	0.72	0.2642	1.02	0.3461
0.13	0.0517	0.43	0.1664	0.73	0.2673	1.03	0.3485
0.14	0.0557	0.44	0.1700	0.74	0.2704	1.04	0.3508
0.15	0.0596	0.45	0.1736	0.75	0.2734	1.05	0.3531
0.16	0.0636	0.46	0.1772	0.76	0.2764	1.06	0.3554
0.17	0.0675	0.47	0.1808	0.77	0.2794	1.07	0.3577
0.18	0.0714	0.48	0.1844	0.78	0.2823	1.08	0.3599
0.19	0.0754	0.49	0.1879	0.79	0.2852	1.09	0.3621
0.20	0.0793	0.50	0.1915	0.80	0.2881	1.10	0.3643
0.21	0.0832	0.51	0.1950	0.81	0.2910	1.11	0.3665
0.22	0.0871	0.52	0.1985	0.82	0.2939	1.12	0.3686
0.23	0.0910	0.53	0.2019	0.83	0.2967	1.13	0.3708
0.24	0.0948	0.54	0.2054	0.84	0.2996	1.14	0.3729
0.25	0.0987	0.55	0.2088	0.85	0.3023	1.15	0.3749
0.26	0.1026	0.56	0.2123	0.86	0.3051	1.16	0.3770
0.27	0.1064	0.57	0.2157	0.87	0.3079	1.17	0.3790
0.28	0.1103	0.58	0.2190	0.88	0.3106	1.18	0.3810
0.29	0.1141	0.59	0.2224	0.89	0.3133	1.19	0.3830

z	A	z	A	z	A	z	A
1.20	0.3849	1.55	0.4394	1.90	0.4713	2.25	0.4878
1.21	0.3869	1.56	0.4406	1.91	0.4719	2.26	0.4881
1.22	0.3888	1.57	0.4418	1.92	0.4726	2.27	0.4884
1.23	0.3907	1.58	0.4430	1.93	0.4732	2.28	0.4887
1.24	0.3925	1.59	0.4441	1.94	0.4738	2.29	0.4890
1.25	0.3944	1.60	0.4452	1.95	0.4744	2.30	0.4893
1.26	0.3962	1.61	0.4463	1.96	0.4750	2.31	0.4896
1.27	0.3980	1.62	0.4474	1.97	0.4756	2.32	0.4898
1.28	0.3997	1.63	0.4485	1.98	0.4762	2.33	0.4901
1.29	0.4015	1.64	0.4495	1.99	0.4767	2.34	0.4904
1.30	0.4032	1.65	0.4505	2.00	0.4773	2.35	0.4906
1.31	0.4049	1.66	0.4515	2.01	0.4778	2.36	0.4909
1.32	0.4066	1.67	0.4525	2.02	0.4783	2.37	0.4911
1.33	0.4082	1.68	0.4535	2.03	0.4788	2.38	0.4913
1.34	0.4099	1.69	0.4545	2.04	0.4793	2.39	0.4916
1.35	0.4115	1.70	0.4554	2.05	0.4798	2.40	0.4918
1.36	0.4131	1.71	0.4564	2.06	0.4803	2.41	0.4920
1.37	0.4147	1.72	0.4573	2.07	0.4808	2.42	0.4922
1.38	0.4162	1.73	0.4582	2.08	0.4812	2.43	0.4925
1.39	0.4177	1.74	0.4591	2.09	0.4817	2.44	0.4927
1.40	0.4192	1.75	0.4599	2.10	0.4821	2.45	0.4929
1.41	0.4207	1.76	0.4608	2.11	0.4826	2.46	0.4931
1.42	0.4222	1.77	0.4616	2.12	0.4830	2.47	0.4932
1.43	0.4236	1.78	0.4625	2.13	0.4834	2.48	0.4934
1.44	0.4251	1.79	0.4633	2.14	0.4838	2.49	0.4936
1.45	0.4265	1.80	0.4641	2.15	0.4842	2.50	0.4938
1.46	0.4279	1.81	0.4649	2.16	0.4846	2.51	0.4940
1.47	0.4292	1.82	0.4656	2.17	0.4850	2.52	0.4941
1.48	0.4306	1.83	0.4664	2.18	0.4854	2.53	0.4943
1.49	0.4319	1.84	0.4671	2.19	0.4857	2.54	0.4945
1.50	0.4332	1.85	0.4678	2.20	0.4861	2.55	0.4946
1.51	0.4345	1.86	0.4686	2.21	0.4865	2.56	0.4948
1.52	0.4357	1.87	0.4693	2.22	0.4868	2.57	0.4949
1.53	0.4370	1.88	0.4700	2.23	0.4871	2.58	0.4951
1.54	0.4382	1.89	0.4706	2.24	0.4875	2.59	0.4952

Table IV **(Continued)**

z	A	z	A	z	A	z	A
2.60	0.4953	2.95	0.4984	3.30	0.4995	3.65	0.4999
2.61	0.4955	2.96	0.4985	3.31	0.4995	3.66	0.4999
2.62	0.4956	2.97	0.4985	3.32	0.4996	3.67	0.4999
2.63	0.4957	2.98	0.4986	3.33	0.4996	3.68	0.4999
2.64	0.4959	2.99	0.4986	3.34	0.4996	3.69	0.4999
2.65	0.4960	3.00	0.4987	3.35	0.4996	3.70	0.4999
2.66	0.4961	3.01	0.4987	3.36	0.4996	3.71	0.4999
2.67	0.4962	3.02	0.4987	3.37	0.4996	3.72	0.4999
2.68	0.4963	3.03	0.4988	3.38	0.4996	3.73	0.4999
2.69	0.4964	3.04	0.4988	3.39	0.4997	3.74	0.4999
2.70	0.4965	3.05	0.4989	3.40	0.4997	3.75	0.4999
2.71	0.4966	3.06	0.4989	3.41	0.4997	3.76	0.4999
2.72	0.4967	3.07	0.4989	3.42	0.4997	3.77	0.4999
2.73	0.4968	3.08	0.4990	3.43	0.4997	3.78	0.4999
2.74	0.4969	3.09	0.4990	3.44	0.4997	3.79	0.4999
2.75	0.4970	3.10	0.4990	3.45	0.4997	3.80	0.4999
2.76	0.4971	3.11	0.4991	3.46	0.4997	3.81	0.4999
2.77	0.4972	3.12	0.4991	3.47	0.4997	3.82	0.4999
2.78	0.4973	3.13	0.4991	3.48	0.4998	3.83	0.4999
2.79	0.4974	3.14	0.4992	3.49	0.4998	3.84	0.4999
2.80	0.4974	3.15	0.4992	3.50	0.4998	3.85	0.4999
2.81	0.4975	3.16	0.4992	3.51	0.4998	3.86	0.4999
2.82	0.4976	3.17	0.4992	3.52	0.4998	3.87	0.5000
2.83	0.4977	3.18	0.4993	3.53	0.4998	3.88	0.5000
2.84	0.4977	3.19	0.4993	3.54	0.4998	3.89	0.5000
2.85	0.4978	3.20	0.4993	3.55	0.4998		
2.86	0.4979	3.21	0.4993	3.56	0.4998		
2.87	0.4980	3.22	0.4994	3.57	0.4998		
2.88	0.4980	3.23	0.4994	3.58	0.4998		
2.89	0.4981	3.24	0.4994	3.59	0.4998		
2.90	0.4981	3.25	0.4994	3.60	0.4998		
2.91	0.4982	3.26	0.4994	3.61	0.4999		
2.92	0.4983	3.27	0.4995	3.62	0.4999		
2.93	0.4983	3.28	0.4995	3.63	0.4999		
2.94	0.4984	3.29	0.4995	3.64	0.4999		

Table V Mathematics of Finance

A29

			$i = 0.0025\ (\frac{1}{4}\%)$				
n	$(1+i)^n$	$s_{\overline{n}\rvert i}$	$a_{\overline{n}\rvert i}$	n	$(1+i)^n$	$s_{\overline{n}\rvert i}$	$a_{\overline{n}\rvert i}$
1	1.002 500	1.000 000	0.997 506	51	1.135 804	54.321 654	47.826 604
2	1.005 006	2.002 500	1.992 525	52	1.138 644	55.457 459	48.704 842
3	1.007 519	3.007 506	2.985 062	53	1.141 490	56.596 102	49.580 890
4	1.010 038	4.015 025	3.975 124	54	1.144 344	57.737 593	50.454 753
5	1.012 563	5.025 063	4.962 718	55	1.147 205	58.881 936	51.326 437
6	1.015 094	6.037 625	5.947 848	56	1.150 073	60.029 141	52.195 947
7	1.017 632	7.052 719	6.930 522	57	1.152 948	61.179 214	53.063 288
8	1.020 176	8.070 351	7.910 745	58	1.155 830	62.332 162	53.928 467
9	1.022 726	9.090 527	8.888 524	59	1.158 720	63.487 993	54.791 489
10	1.025 283	10.113 253	9.863 864	60	1.161 617	64.646 713	55.652 358
11	1.027 846	11.138 536	10.836 772	61	1.164 521	65.808 329	56.511 080
12	1.030 416	12.166 383	11.807 254	62	1.167 432	66.972 850	57.367 661
13	1.032 992	13.196 799	12.775 316	63	1.170 351	68.140 282	58.222 106
14	1.035 574	14.229 791	13.740 963	64	1.173 277	69.310 633	59.074 420
15	1.038 163	15.265 365	14.704 203	65	1.176 210	70.483 910	59.924 608
16	1.040 759	16.303 529	15.665 040	66	1.179 150	71.660 119	60.772 676
17	1.043 361	17.344 287	16.623 481	67	1.182 098	72.839 270	61.618 630
18	1.045 969	18.387 648	17.579 533	68	1.185 053	74.021 368	62.462 474
19	1.048 584	19.433 617	18.533 200	69	1.188 016	75.206 421	63.304 213
20	1.051 205	20.482 201	19.484 488	70	1.190 986	76.394 437	64.143 853
21	1.053 834	21.533 407	20.433 405	71	1.193 964	77.585 423	64.981 400
22	1.056 468	22.587 240	21.379 955	72	1.196 948	78.779 387	65.816 858
23	1.059 109	23.643 708	22.324 145	73	1.199 941	79.976 335	66.650 232
24	1.061 757	24.702 818	23.265 980	74	1.202 941	81.176 276	67.481 528
25	1.064 411	25.764 575	24.205 466	75	1.205 948	82.379 217	68.310 751
26	1.067 072	26.828 986	25.142 609	76	1.208 963	83.585 165	69.137 907
27	1.069 740	27.896 059	26.077 416	77	1.211 985	84.794 128	69.962 999
28	1.072 414	28.965 799	27.009 891	78	1.215 015	86.006 113	70.786 034
29	1.075 096	30.038 213	27.940 041	79	1.218 053	87.221 129	71.607 017
30	1.077 783	31.113 309	28.867 871	80	1.221 098	88.439 181	72.425 952
31	1.080 478	32.191 092	29.793 388	81	1.224 151	89.660 279	73.242 845
32	1.083 179	33.271 570	30.716 596	82	1.227 211	90.884 430	74.057 700
33	1.085 887	34.354 749	31.637 503	83	1.230 279	92.111 641	74.870 524
34	1.088 602	35.440 636	32.556 112	84	1.233 355	93.341 920	75.681 321
35	1.091 323	36.529 237	33.472 431	85	1.236 438	94.575 275	76.490 095
36	1.094 051	37.620 560	34.386 465	86	1.239 529	95.811 713	77.296 853
37	1.096 787	38.714 612	35.298 220	87	1.242 628	97.051 242	78.101 599
38	1.099 528	39.811 398	36.207 700	88	1.245 735	98.293 871	78.904 339
39	1.102 277	40.910 927	37.114 913	89	1.248 849	99.539 605	79.705 076
40	1.105 033	42.013 204	38.019 863	90	1.251 971	100.788 454	80.503 816
41	1.107 796	43.118 237	38.922 557	91	1.255 101	102.040 425	81.300 565
42	1.110 565	44.226 033	39.822 999	92	1.258 239	103.295 526	82.095 327
43	1.113 341	45.336 598	40.721 196	93	1.261 384	104.553 765	82.888 106
44	1.116 125	46.449 939	41.617 154	94	1.264 538	105.815 150	83.678 909
45	1.118 915	47.566 064	42.510 876	95	1.267 699	107.079 688	84.467 740
46	1.121 712	48.684 979	43.402 370	96	1.270 868	108.347 387	85.254 603
47	1.124 517	49.806 692	44.291 641	97	1.274 046	109.618 255	86.039 504
48	1.127 328	50.931 208	45.178 695	98	1.277 231	110.892 301	86.822 448
49	1.130 146	52.058 536	46.063 536	99	1.280 424	112.169 532	87.603 440
50	1.132 972	53.188 683	46.946 170	100	1.283 625	113.449 956	88.382 483

Table V (Continued)

		$i = 0.005$ (½%)					
n	$(1 + i)^n$	$s_{\overline{n}i}$	$a_{\overline{n}i}$	n	$(1 + i)^n$	$s_{\overline{n}i}$	$a_{\overline{n}i}$
1	1.005 000	1.000 000	0.995 025	51	1.289 642	57.928 389	44.918 195
2	1.010 025	2.005 000	1.985 099	52	1.296 090	59.218 031	45.689 747
3	1.015 075	3.015 025	2.970 248	53	1.302 571	60.514 121	46.457 459
4	1.020 151	4.030 100	3.950 496	54	1.309 083	61.816 692	47.221 353
5	1.025 251	5.050 250	4.925 866	55	1.315 629	63.125 775	47.981 445
6	1.030 378	6.075 502	5.896 384	56	1.322 207	64.441 404	48.737 757
7	1.035 529	7.105 879	6.862 074	57	1.328 818	65.763 611	49.490 305
8	1.040 707	8.141 409	7.822 959	58	1.335 462	67.092 429	50.239 110
9	1.045 911	9.182 116	8.779 064	59	1.342 139	68.427 891	50.984 189
10	1.051 140	10.228 026	9.730 412	60	1.348 850	69.770 031	51.725 561
11	1.056 396	11.279 167	10.677 027	61	1.355 594	71.118 881	52.463 245
12	1.061 678	12.335 562	11.618 932	62	1.362 372	72.474 475	53.197 258
13	1.066 986	13.397 240	12.556 151	63	1.369 184	73.836 847	53.927 620
14	1.072 321	14.464 226	13.488 708	64	1.376 030	75.206 032	54.654 348
15	1.077 683	15.536 548	14.416 625	65	1.382 910	76.582 062	55.377 461
16	1.083 071	16.614 230	15.339 925	66	1.389 825	77.964 972	56.096 976
17	1.088 487	17.697 301	16.258 632	67	1.396 774	79.354 797	56.812 912
18	1.093 929	18.785 788	17.172 768	68	1.403 758	80.751 571	57.525 285
19	1.099 399	19.879 717	18.082 356	69	1.410 777	82.155 329	58.234 115
20	1.104 896	20.979 115	18.987 419	70	1.417 831	83.566 105	58.939 418
21	1.110 420	22.084 011	19.887 979	71	1.424 920	84.983 936	59.641 212
22	1.115 972	23.194 431	20.784 059	72	1.432 044	86.408 856	60.339 514
23	1.121 552	24.310 403	21.675 681	73	1.439 204	87.840 900	61.034 342
24	1.127 160	25.431 955	22.562 866	74	1.446 401	89.280 104	61.725 714
25	1.132 796	26.559 115	23.445 638	75	1.453 633	90.726 505	62.413 645
26	1.138 460	27.691 911	24.324 018	76	1.460 901	92.180 138	63.098 155
27	1.144 152	28.830 370	25.198 028	77	1.468 205	93.641 038	63.779 258
28	1.149 873	29.974 522	26.067 689	78	1.475 546	95.109 243	64.456 974
29	1.155 622	31.124 395	26.933 024	79	1.482 924	96.584 790	65.131 317
30	1.161 400	32.280 017	27.794 054	80	1.490 339	98.067 714	65.802 305
31	1.167 207	33.441 417	28.650 800	81	1.497 790	99.558 052	66.469 956
32	1.173 043	34.608 624	29.503 284	82	1.505 279	101.055 842	67.134 284
33	1.178 908	35.781 667	30.351 526	83	1.512 806	102.561 122	67.795 308
34	1.184 803	36.960 575	31.195 548	84	1.520 370	104.073 927	68.453 042
35	1.190 727	38.145 378	32.035 371	85	1.527 971	105.594 297	69.107 505
36	1.196 681	39.336 105	32.871 016	86	1.535 611	107.122 268	69.758 711
37	1.202 664	40.532 785	33.702 504	87	1.543 289	108.657 880	70.406 678
38	1.208 677	41.735 449	34.529 854	88	1.551 006	110.201 169	71.051 421
39	1.214 721	42.944 127	35.353 089	89	1.558 761	111.752 175	71.692 956
40	1.220 794	44.158 847	36.172 228	90	1.566 555	113.310 936	72.331 300
41	1.226 898	45.379 642	36.987 291	91	1.574 387	114.877 490	72.966 467
42	1.233 033	46.606 540	37.798 300	92	1.582 259	116.451 878	73.598 475
43	1.239 198	47.839 572	38.605 274	93	1.590 171	118.034 137	74.227 338
44	1.245 394	49.078 770	39.408 232	94	1.598 121	119.624 308	74.853 073
45	1.251 621	50.324 164	40.207 196	95	1.606 112	121.222 430	75.475 694
46	1.257 879	51.575 785	41.002 185	96	1.614 143	122.828 542	76.095 218
47	1.264 168	52.833 664	41.793 219	97	1.622 213	124.442 684	76.711 660
48	1.270 489	54.097 832	42.580 318	98	1.630 324	126.064 898	77.325 035
49	1.276 842	55.368 321	43.363 500	99	1.638 476	127.695 222	77.935 358
50	1.283 226	56.645 163	44.142 786	100	1.646 668	129.333 698	78.542 645

		$i = 0.0075\ (\sqrt[3]{4}\%)$					
n	$(1+i)^n$	$s_{\overline{n}i}$	$a_{\overline{n}i}$	n	$(1+i)^n$	$s_{\overline{n}i}$	$a_{\overline{n}i}$
1	1.007 500	1.000 000	0.992 556	51	1.463 854	61.847 214	42.249 575
2	1.015 056	2.007 500	1.977 723	52	1.474 833	63.311 068	42.927 618
3	1.022 669	3.022 556	2.955 556	53	1.485 894	64.785 901	43.600 614
4	1.030 339	4.045 225	3.926 110	54	1.497 038	66.271 796	44.268 599
5	1.038 067	5.075 565	4.889 440	55	1.508 266	67.768 834	44.931 612
6	1.045 852	6.113 631	5.845 598	56	1.519 578	69.277 100	45.589 689
7	1.053 696	7.159 484	6.794 638	57	1.530 975	70.796 679	46.242 868
8	1.061 599	8.213 180	7.736 613	58	1.542 457	72.327 659	46.891 184
9	1.069 561	9.274 779	8.671 576	59	1.554 026	73.870 111	47.534 674
10	1.077 583	10.344 339	9.599 580	60	1.565 681	75.424 137	48.173 374
11	1.085 664	11.421 922	10.520 675	61	1.577 424	76.989 818	48.807 319
12	1.093 807	12.507 586	11.434 913	62	1.589 254	78.567 242	49.436 545
13	1.102 010	13.601 393	12.342 345	63	1.601 174	80.156 496	50.061 086
14	1.110 276	14.703 404	13.243 022	64	1.613 183	81.757 670	50.680 979
15	1.118 603	15.813 679	14.136 995	65	1.625 281	83.370 852	51.296 257
16	1.126 992	16.932 282	15.024 313	66	1.637 471	84.996 134	51.906 955
17	1.135 445	18.059 274	15.905 025	67	1.649 752	86.633 605	52.513 107
18	1.143 960	19.194 718	16.779 181	68	1.662 125	88.283 356	53.114 746
19	1.152 540	20.338 679	17.646 830	69	1.674 591	89.945 482	53.711 907
20	1.161 184	21.491 219	18.508 020	70	1.687 151	91.620 073	54.304 622
21	1.169 893	22.652 403	19.362 799	71	1.699 804	93.307 223	54.892 925
22	1.178 667	23.822 296	20.211 215	72	1.712 553	95.007 028	55.476 849
23	1.187 507	25.000 963	21.053 315	73	1.725 397	96.719 580	56.056 426
24	1.196 414	26.188 471	21.889 146	74	1.738 337	98.444 977	56.631 688
25	1.205 387	27.384 884	22.718 755	75	1.751 375	100.183 314	57.202 668
26	1.214 427	28.590 271	23.542 189	76	1.764 510	101.934 689	57.769 397
27	1.223 535	29.804 698	24.359 493	77	1.777 744	103.699 199	58.331 908
28	1.232 712	31.028 233	25.170 713	78	1.791 077	105.476 943	58.890 231
29	1.241 957	32.260 945	25.975 893	79	1.804 510	107.268 021	59.444 398
30	1.251 272	33.502 902	26.775 080	80	1.818 044	109.072 531	59.994 440
31	1.260 656	34.754 174	27.568 318	81	1.831 679	110.890 575	60.540 387
32	1.270 111	36.014 830	28.355 650	82	1.845 417	112.722 254	61.082 270
33	1.279 637	37.284 941	29.137 122	83	1.859 258	114.567 671	61.620 119
34	1.289 234	38.564 578	29.912 776	84	1.873 202	116.426 928	62.153 965
35	1.298 904	39.853 813	30.682 656	85	1.887 251	118.300 130	62.683 836
36	1.308 645	41.152 716	31.446 805	86	1.901 405	120.187 381	63.209 763
37	1.318 460	42.461 361	32.205 266	87	1.915 666	122.088 787	63.731 774
38	1.328 349	43.779 822	32.958 080	88	1.930 033	124.004 453	64.249 900
39	1.338 311	45.108 170	33.705 290	89	1.944 509	125.934 486	64.764 169
40	1.348 349	46.446 482	34.446 938	90	1.959 092	127.878 995	65.274 609
41	1.358 461	47.794 830	35.183 065	91	1.973 786	129.838 087	65.781 250
42	1.368 650	49.153 291	35.913 713	92	1.988 589	131.811 873	66.284 119
43	1.378 915	50.521 941	36.638 921	93	2.003 503	133.800 462	66.783 245
44	1.389 256	51.900 856	37.358 730	94	2.018 530	135.803 965	67.278 655
45	1.399 676	53.290 112	38.073 181	95	2.033 669	137.822 495	67.770 377
46	1.410 173	54.689 788	38.782 314	96	2.048 921	139.856 164	68.258 439
47	1.420 750	56.099 961	39.486 168	97	2.064 288	141.905 085	68.742 867
48	1.431 405	57.520 711	40.184 782	98	2.079 770	143.969 373	69.223 689
49	1.442 141	58.952 116	40.878 195	99	2.095 369	146.049 143	69.700 932
50	1.452 957	60.394 257	41.566 447	100	2.111 084	148.144 512	70.174 623

Table V (Continued)

			$i = 0.01$ (1%)				
n	$(1 + i)^n$	$s_{\overline{n}\,i}$	$a_{\overline{n}\,i}$	n	$(1 + i)^n$	$s_{\overline{n}\,i}$	$a_{\overline{n}\,i}$
1	1.010 000	1.000 000	0.990 099	51	1.661 078	66.107 814	39.798 136
2	1.020 100	2.010 000	1.970 395	52	1.677 689	67.768 892	40.394 194
3	1.030 301	3.030 100	2.940 985	53	1.694 466	69.446 581	40.984 351
4	1.040 604	4.060 401	3.901 966	54	1.711 410	71.141 047	41.568 664
5	1.051 010	5.101 005	4.853 431	55	1.728 525	72.852 457	42.147 192
6	1.061 520	6.152 015	5.795 476	56	1.745 810	74.580 982	42.719 992
7	1.072 135	7.213 535	6.728 195	57	1.763 268	76.326 792	43.287 121
8	1.082 857	8.285 671	7.651 678	58	1.780 901	78.090 060	43.848 635
9	1.093 685	9.368 527	8.566 018	59	1.798 710	79.870 960	44.404 589
10	1.104 622	10.462 213	9.471 305	60	1.816 697	81.669 670	44.955 038
11	1.115 668	11.566 835	10.367 628	61	1.834 864	83.486 367	45.500 038
12	1.126 825	12.682 503	11.255 077	62	1.853 212	85.321 230	46.039 642
13	1.138 093	13.809 328	12.133 740	63	1.871 744	87.174 443	46.573 903
14	1.149 474	14.947 421	13.003 703	64	1.890 462	89.046 187	47.102 874
15	1.160 969	16.096 896	13.865 053	65	1.909 366	90.936 649	47.626 608
16	1.172 579	17.257 864	14.717 874	66	1.928 460	92.846 015	48.145 156
17	1.184 304	18.430 443	15.562 251	67	1.947 745	94.774 475	48.658 570
18	1.196 147	19.614 748	16.398 269	68	1.967 222	96.722 220	49.166 901
19	1.208 109	20.810 895	17.226 008	69	1.986 894	98.689 442	49.670 199
20	1.220 190	22.019 004	18.045 553	70	2.006 763	100.676 337	50.168 514
21	1.232 392	23.239 194	18.856 983	71	2.026 831	102.683 100	50.661 895
22	1.244 716	24.471 586	19.660 379	72	2.047 099	104.709 931	51.150 391
23	1.257 163	25.716 302	20.455 821	73	2.067 570	106.757 031	51.634 051
24	1.269 735	26.973 465	21.243 387	74	2.088 246	108.824 601	52.112 922
25	1.282 432	28.243 200	22.023 156	75	2.109 128	110.912 847	52.587 051
26	1.295 256	29.525 632	22.795 204	76	2.130 220	113.021 975	53.056 486
27	1.308 209	30.820 888	23.559 608	77	2.151 522	115.152 195	53.521 274
28	1.321 291	32.129 097	24.316 443	78	2.173 037	117.303 717	53.981 459
29	1.334 504	33.450 388	25.065 785	79	2.194 768	119.476 754	54.437 088
30	1.347 849	34.784 892	25.807 708	80	2.216 715	121.671 522	54.888 206
31	1.361 327	36.132 740	26.542 285	81	2.238 882	123.888 237	55.334 858
32	1.374 941	37.494 068	27.269 589	82	2.261 271	126.127 119	55.777 087
33	1.388 690	38.869 009	27.989 693	83	2.283 884	128.388 391	56.214 937
34	1.402 577	40.257 699	28.702 666	84	2.306 723	130.672 274	56.648 453
35	1.416 603	41.660 276	29.408 580	85	2.329 790	132.978 997	57.077 676
36	1.430 769	43.076 878	30.107 505	86	2.353 088	135.308 787	57.502 650
37	1.445 076	44.507 647	30.799 510	87	2.376 619	137.661 875	57.923 415
38	1.459 527	45.952 724	31.484 663	88	2.400 385	140.038 494	58.340 015
39	1.474 123	47.412 251	32.163 033	89	2.424 389	142.438 879	58.752 490
40	1.488 864	48.886 373	32.834 686	90	2.448 633	144.863 267	59.160 881
41	1.503 752	50.375 237	33.499 689	91	2.473 119	147.311 900	59.565 229
42	1.518 790	51.878 989	34.158 108	92	2.497 850	149.785 019	59.965 573
43	1.533 978	53.397 779	34.810 008	93	2.522 829	152.282 869	60.361 954
44	1.549 318	54.931 757	35.455 454	94	2.548 057	154.805 698	60.754 410
45	1.564 811	56.481 075	36.094 508	95	2.573 538	157.353 755	61.142 980
46	1.580 459	58.045 885	36.727 236	96	2.599 273	159.927 293	61.527 703
47	1.596 263	59.626 344	37.353 699	97	2.625 266	162.526 565	61.908 617
48	1.612 226	61.222 608	37.973 959	98	2.651 518	165.151 831	62.285 759
49	1.628 348	62.834 834	38.588 079	99	2.678 033	167.803 349	62.659 168
50	1.644 632	64.463 182	39.196 118	100	2.704 814	170.481 383	63.028 879

		$i = 0.0125\ (1\frac{1}{4}\%)$					
n	$(1 + i)^n$	$s_{\overline{n}i}$	$a_{\overline{n}i}$	n	$(1 + i)^n$	$s_{\overline{n}i}$	$a_{\overline{n}i}$
1	1.012 500	1.000 000	0.987 654	51	1.884 285	70.742 812	37.543 581
2	1.025 156	2.012 500	1.963 115	52	1.907 839	72.627 097	38.067 734
3	1.037 971	3.037 656	2.926 534	53	1.931 687	74.534 936	38.585 417
4	1.050 945	4.075 627	3.878 058	54	1.955 833	76.466 623	39.096 708
5	1.064 082	5.126 572	4.817 835	55	1.980 281	78.422 456	39.601 687
6	1.077 383	6.190 654	5.746 010	56	2.005 034	80.402 737	40.100 431
7	1.090 850	7.268 038	6.662 726	57	2.030 097	82.407 771	40.593 019
8	1.104 486	8.358 888	7.568 124	58	2.055 473	84.437 868	41.079 524
9	1.118 292	9.463 374	8.462 345	59	2.081 167	86.493 341	41.560 024
10	1.132 271	10.581 666	9.345 526	60	2.107 181	88.574 508	42.034 592
11	1.146 424	11.713 937	10.217 803	61	2.133 521	90.681 689	42.503 300
12	1.160 755	12.860 361	11.079 312	62	2.160 190	92.815 210	42.966 223
13	1.175 264	14.021 116	11.930 185	63	2.187 193	94.975 400	43.423 430
14	1.189 955	15.196 380	12.770 553	64	2.214 532	97.162 593	43.874 992
15	1.204 829	16.386 335	13.600 546	65	2.242 214	99.377 125	44.320 980
16	1.219 890	17.591 164	14.420 292	66	2.270 242	101.619 339	44.761 462
17	1.235 138	18.811 053	15.229 918	67	2.298 620	103.889 581	45.196 506
18	1.250 577	20.046 192	16.029 549	68	2.327 353	106.188 201	45.626 178
19	1.266 210	21.296 769	16.819 308	69	2.356 444	108.515 553	46.050 547
20	1.282 037	22.562 979	17.599 316	70	2.385 900	110.871 998	46.469 676
21	1.298 063	23.845 016	18.369 695	71	2.415 724	113.257 898	46.883 630
22	1.314 288	25.143 078	19.130 563	72	2.445 920	115.673 621	47.292 474
23	1.330 717	26.457 367	19.882 037	73	2.476 494	118.119 542	47.696 271
24	1.347 351	27.788 084	20.624 235	74	2.507 450	120.596 036	48.095 082
25	1.364 193	29.135 435	21.357 269	75	2.538 794	123.103 486	48.488 970
26	1.381 245	30.499 628	22.081 253	76	2.570 528	125.642 280	48.877 995
27	1.398 511	31.880 873	22.796 299	77	2.602 660	128.212 809	49.262 218
28	1.415 992	33.279 384	23.502 518	78	2.635 193	130.815 469	49.641 696
29	1.433 692	34.695 377	24.200 018	79	2.668 133	133.450 662	50.016 490
30	1.451 613	36.129 069	24.888 906	80	2.701 485	136.118 795	50.386 657
31	1.469 759	37.580 682	25.569 290	81	2.735 254	138.820 280	50.752 254
32	1.488 131	39.050 441	26.241 274	82	2.769 444	141.555 534	51.113 337
33	1.506 732	40.538 571	26.904 962	83	2.804 062	144.324 978	51.469 963
34	1.525 566	42.045 303	27.560 456	84	2.839 113	147.129 040	51.822 185
35	1.544 636	43.570 870	28.207 858	85	2.874 602	149.968 153	52.170 060
36	1.563 944	45.115 506	28.847 267	86	2.910 534	152.842 755	52.513 639
37	1.583 493	46.679 449	29.478 783	87	2.946 916	155.753 289	52.852 977
38	1.603 287	48.292 642	30.102 501	88	2.983 753	158.700 206	53.188 125
39	1.623 328	49.886 229	30.718 520	89	3.021 049	161.683 958	53.519 136
40	1.643 619	51.489 557	31.326 933	90	3.058 813	164.705 008	53.846 060
41	1.664 165	53.133 177	31.927 835	91	3.097 048	167.763 820	54.168 948
42	1.684 967	54.797 341	32.521 319	92	3.135 761	170.860 868	54.487 850
43	1.706 029	56.482 308	33.107 475	93	3.174 958	173.996 629	54.802 815
44	1.727 354	58.188 337	33.686 395	94	3.214 645	177.171 587	55.113 892
45	1.748 946	59.915 691	34.258 168	95	3.254 828	180.386 232	55.421 127
46	1.770 808	61.664 637	34.822 882	96	3.295 513	183.641 059	55.724 570
47	1.792 943	63.435 445	35.380 624	97	3.336 707	186.936 573	56.024 267
48	1.815 355	65.228 388	35.931 481	98	3.378 416	190.273 280	56.320 264
49	1.838 047	67.043 743	36.475 537	99	3.420 646	193.651 696	56.612 606
50	1.861 022	68.881 790	37.012 876	100	3.463 404	197.072 342	56.901 339

Table V (Continued)

		$i = 0.015\ (1\frac{1}{2}\%)$					
n	$(1+i)^n$	$s_{\overline{n}|i}$	$a_{\overline{n}|i}$	n	$(1+i)^n$	$s_{\overline{n}|i}$	$a_{\overline{n}|i}$
1	1.015 000	1.000 000	0.985 222	51	2.136 821	75.788 070	35.467 673
2	1.030 225	2.015 000	1.955 883	52	2.168 873	77.924 892	35.928 742
3	1.045 678	3.045 225	2.912 200	53	2.201 406	80.093 765	36.382 997
4	1.061 364	4.090 903	3.854 385	54	2.234 428	82.295 171	36.830 539
5	1.077 284	5.152 267	4.782 645	55	2.267 944	84.529 599	37.271 467
6	1.093 443	6.229 551	5.697 187	56	2.301 963	86.797 543	37.705 879
7	1.109 845	7.322 994	6.598 214	57	2.336 493	89.099 506	38.133 871
8	1.126 493	8.432 839	7.485 925	58	2.371 540	91.435 999	38.555 538
9	1.143 390	9.559 332	8.360 517	59	2.407 113	93.807 539	38.970 973
10	1.160 541	10.702 722	9.222 185	60	2.443 220	96.214 652	39.380 269
11	1.177 949	11.863 262	10.071 118	61	2.479 868	98.657 871	39.783 516
12	1.195 618	13.041 211	10.907 505	62	2.517 066	101.137 740	40.180 804
13	1.213 552	14.236 830	11.731 532	63	2.554 822	103.654 806	40.572 221
14	1.231 756	15.450 382	12.543 382	64	2.593 144	106.209 628	40.957 853
15	1.250 232	16.682 138	13.343 233	65	2.632 042	108.802 772	41.337 786
16	1.268 986	17.932 370	14.131 264	66	2.671 522	111.434 814	41.712 105
17	1.288 020	19.201 355	14.907 649	67	2.711 595	114.106 336	42.080 891
18	1.307 341	20.489 376	15.672 561	68	2.752 269	116.817 931	42.444 228
19	1.326 951	21.796 716	16.426 168	69	2.793 553	119.570 200	42.802 195
20	1.346 855	23.123 667	17.168 639	70	2.835 456	122.363 753	43.154 872
21	1.367 058	24.470 522	17.900 137	71	2.877 988	125.199 209	43.502 337
22	1.387 564	25.837 580	18.620 824	72	2.921 158	128.077 197	43.844 667
23	1.408 377	27.225 144	19.330 861	73	2.964 975	130.998 355	44.181 938
24	1.429 503	28.633 521	20.030 405	74	3.009 450	133.963 331	44.514 224
25	1.450 945	30.063 024	20.719 611	75	3.054 592	136.972 781	44.841 600
26	1.472 710	31.513 969	21.398 632	76	3.100 411	140.027 372	45.164 138
27	1.494 800	32.986 678	22.067 617	77	3.146 917	143.127 783	45.481 910
28	1.517 222	34.481 479	22.726 717	78	3.194 120	146.274 700	45.794 985
29	1.539 981	35.998 701	23.376 076	79	3.242 032	149.468 820	46.103 433
30	1.563 080	37.538 681	24.015 838	80	3.290 663	152.710 852	46.407 323
31	1.586 526	39.101 762	24.646 146	81	3.340 023	156.001 515	46.706 723
32	1.610 324	40.688 288	25.267 139	82	3.390 123	159.341 536	47.001 697
33	1.634 479	42.298 612	25.878 954	83	3.440 975	162.731 661	47.292 313
34	1.658 996	43.933 092	26.481 728	84	3.492 590	166.172 636	47.578 633
35	1.683 881	45.592 088	27.075 595	85	3.544 978	169.665 226	47.860 722
36	1.709 140	47.275 969	27.660 684	86	3.598 153	173.210 204	48.138 643
37	1.734 777	48.985 109	28.237 127	87	3.652 125	176.808 357	48.412 456
38	1.760 798	50.719 885	28.805 052	88	3.706 907	180.460 482	48.682 222
39	1.787 210	52.480 684	29.364 583	89	3.762 511	184.167 390	48.948 002
40	1.814 018	54.267 894	29.915 845	90	3.818 949	187.929 900	49.209 855
41	1.841 229	56.081 912	30.458 961	91	3.876 233	191.748 849	49.467 837
42	1.868 847	57.923 141	30.994 050	92	3.934 376	195.625 082	49.722 007
43	1.896 880	59.791 988	31.521 232	93	3.993 392	199.559 458	49.972 421
44	1.925 333	61.688 868	32.040 622	94	4.053 293	203.552 850	50.219 134
45	1.954 213	63.614 201	32.552 337	95	4.114 092	207.606 142	50.462 201
46	1.983 526	65.568 414	33.056 490	96	4.175 804	211.720 235	50.701 675
47	2.013 279	67.551 940	33.553 192	97	4.238 441	215.896 038	50.937 611
48	2.043 478	69.565 219	34.042 554	98	4.302 017	220.134 479	51.170 060
49	2.074 130	71.608 698	34.524 683	99	4.366 547	224.436 496	51.399 074
50	2.105 242	73.682 828	34.999 688	100	4.432 046	228.803 043	51.624 704

		$i = 0.0175\ (1\tfrac{3}{4}\%)$					
n	$(1+i)^n$	$s_{\overline{n}i}$	$a_{\overline{n}i}$	n	$(1+i)^n$	$s_{\overline{n}i}$	$a_{\overline{n}i}$
1	1.017 500	1.000 000	0.982 801	51	2.422 453	81.283 014	33.554 014
2	1.035 306	2.017 500	1.948 699	52	2.464 846	83.705 466	33.959 719
3	1.053 424	3.052 806	2.897 984	53	2.507 980	86.170 312	34.358 446
4	1.071 859	4.106 230	3.830 943	54	2.551 870	88.678 292	34.750 316
5	1.090 617	5.178 089	4.747 855	55	2.596 528	91.230 163	35.135 446
6	1.109 702	6.268 706	5.648 998	56	2.641 967	93.826 690	35.513 951
7	1.129 122	7.378 408	6.534 641	57	2.688 202	96.468 658	35.885 947
8	1.148 882	8.507 530	7.405 053	58	2.735 245	99.156 859	36.251 545
9	1.168 987	9.656 412	8.260 494	59	2.783 112	101.892 104	36.610 855
10	1.189 444	10.825 399	9.101 223	60	2.831 816	104.675 216	36.963 986
11	1.210 260	12.014 844	9.927 492	61	2.881 373	107.507 032	37.311 042
12	1.231 439	13.225 104	10.739 550	62	2.931 797	110.388 405	37.652 130
13	1.252 990	14.456 543	11.537 641	63	2.983 104	113.320 202	37.987 351
14	1.274 917	15.709 533	12.322 006	64	3.034 308	116.303 306	38.316 807
15	1.297 228	16.984 449	13.092 880	65	3.088 426	119.338 614	38.640 597
16	1.319 929	18.281 677	13.850 497	66	3.142 473	122.427 039	38.958 817
17	1.343 028	19.601 607	14.595 083	67	3.197 466	125.569 513	39.271 565
18	1.366 531	20.944 635	15.326 863	68	3.253 422	128.766 979	39.578 934
19	1.390 445	22.311 166	16.046 057	69	3.310 357	132.020 401	39.881 016
20	1.414 778	23.701 611	16.752 881	70	3.368 288	135.330 758	40.177 903
21	1.439 537	25.116 389	17.447 549	71	3.427 233	138.699 047	40.469 683
22	1.464 729	26.555 926	18.130 269	72	3.487 210	142.126 280	40.756 445
23	1.490 361	28.020 655	18.801 248	73	3.548 236	145.613 490	41.038 276
24	1.516 443	29.511 016	19.460 686	74	3.610 330	149.161 726	41.315 259
25	1.542 981	31.027 459	20.108 782	75	3.673 511	152.772 056	41.587 478
26	1.569 983	32.570 440	20.745 732	76	3.737 797	156.445 567	41.855 015
27	1.597 457	34.140 422	21.371 726	77	3.803 209	160.183 364	42.117 951
28	1.625 413	35.737 880	21.986 955	78	3.869 765	163.986 573	42.376 364
29	1.653 858	37.363 293	22.591 602	79	3.937 486	167.856 338	42.630 334
30	1.682 800	39.017 150	23.185 849	80	4.006 392	171.793 824	42.879 935
31	1.712 249	40.699 950	23.769 876	81	4.076 504	175.800 216	43.125 243
32	1.742 213	42.412 200	24.343 859	82	4.147 843	179.876 720	43.366 332
33	1.772 702	44.154 413	24.907 970	83	4.220 430	184.024 563	43.603 275
34	1.803 725	45.927 115	25.462 378	84	4.294 287	188.244 992	43.836 142
35	1.835 290	47.730 840	26.007 251	85	4.369 437	192.539 280	44.065 005
36	1.867 407	49.566 129	26.542 753	86	4.445 903	196.908 717	44.289 931
37	1.900 087	51.433 537	27.069 045	87	4.523 706	201.354 620	44.510 989
38	1.933 338	53.333 624	27.586 285	88	4.602 871	205.878 326	44.728 244
39	1.967 172	55.266 962	28.094 629	89	4.683 421	210.481 196	44.941 764
40	2.001 597	57.234 134	28.594 230	90	4.765 381	215.164 617	45.151 610
41	2.036 625	59.235 731	29.085 238	91	4.848 775	219.929 998	44.357 848
42	2.072 266	61.272 357	29.567 801	92	4.933 629	224.778 773	45.560 539
43	2.108 531	63.344 623	30.042 065	93	5.019 967	229.712 401	45.759 743
44	2.145 430	65.453 154	30.508 172	94	5.107 816	234.732 368	45.955 521
45	2.182 975	67.598 584	30.966 263	95	5.197 203	239.840 185	46.147 933
46	2.221 177	69.781 559	31.416 474	96	5.288 154	245.037 388	46.337 035
47	2.260 048	72.002 736	31.858 943	97	5.380 697	250.325 542	46.522 884
48	2.299 599	74.262 784	32.293 801	98	5.474 859	255.706 239	46.705 537
49	2.339 842	76.562 383	32.721 181	99	5.570 669	261.181 099	46.885 049
50	2.380 789	78.902 225	33.141 209	100	5.668 156	266.751 768	47.061 473

Table V (Continued)

		$i = 0.02$ (2%)					
n	$(1 + i)^n$	$s_{\overline{n}|i}$	$a_{\overline{n}|i}$	n	$(1 + i)^n$	$s_{\overline{n}|i}$	$a_{\overline{n}|i}$
1	1.020 000	1.000 000	0.980 392	51	2.745 420	87.270 989	31.787 849
2	1.040 400	2.020 000	1.941 561	52	2.800 328	90.016 409	32.144 950
3	1.061 208	3.060 400	2.883 883	53	2.856 335	92.816 737	32.495 049
4	1.082 432	4.121 608	3.807 729	54	2.913 461	95.673 072	32.838 283
5	1.104 081	5.204 040	4.713 460	55	2.971 731	98.586 534	33.174 788
6	1.126 162	6.308 121	5.601 431	56	3.031 165	101.558 264	33.504 694
7	1.148 686	7.434 283	6.471 991	57	3.091 789	104.589 430	33.828 131
8	1.171 659	8.582 969	7.325 481	58	3.153 624	107.681 218	34.145 226
9	1.195 093	9.754 628	8.162 237	59	3.216 697	110.834 843	34.456 104
10	1.218 994	10.949 721	8.982 585	60	3.281 031	114.051 539	34.760 887
11	1.243 374	12.168 715	9.786 848	61	3.346 651	117.332 570	35.059 693
12	1.268 242	13.412 090	10.575 341	62	3.413 584	120.679 222	35.352 640
13	1.293 607	14.680 331	11.348 374	63	3.481 856	124.092 806	35.639 843
14	1.319 479	15.973 938	12.106 249	64	3.551 493	127.574 662	35.921 415
15	1.345 868	17.293 417	12.849 264	65	3.622 523	131.126 155	36.197 466
16	1.372 786	18.639 285	13.577 709	66	3.694 974	134.748 679	36.468 103
17	1.400 241	20.012 071	14.291 872	67	3.768 873	138.443 652	36.733 435
18	1.428 246	21.412 312	14.992 031	68	3.844 251	142.212 525	36.993 564
19	1.456 811	22.840 559	15.678 462	69	3.921 136	146.056 776	37.248 592
20	1.485 947	24.297 370	16.351 433	70	3.999 558	149.977 911	37.498 619
21	1.515 666	25.783 317	17.011 209	71	4.079 549	153.977 469	37.743 744
22	1.545 980	27.298 984	17.658 048	72	4.161 140	158.057 019	37.984 063
23	1.576 899	28.844 963	18.292 204	73	4.244 363	162.218 159	38.219 670
24	1.608 437	30.421 862	18.913 926	74	4.329 250	166.462 522	38.450 657
25	1.640 606	32.030 300	19.523 456	75	4.415 835	170.791 773	38.677 114
26	1.673 418	33.670 906	20.121 036	76	4.504 152	175.207 608	38.899 132
27	1.706 886	35.344 324	20.706 898	77	4.594 235	179.711 760	39.116 796
28	1.741 024	37.051 210	21.281 272	78	4.686 120	184.305 996	39.330 192
29	1.775 845	38.792 235	21.844 385	79	4.779 842	188.992 115	39.539 404
30	1.811 362	40.568 079	22.396 456	80	4.875 439	193.771 958	39.744 514
31	1.847 589	42.379 441	22.937 702	81	4.972 948	198.647 397	39.945 602
32	1.884 541	44.227 030	23.468 335	82	5.072 407	203.620 345	40.142 747
33	1.922 231	46.111 570	23.988 564	83	5.173 855	208.692 752	40.336 026
34	1.960 676	48.033 802	24.498 592	84	5.277 332	213.866 607	40.525 516
35	1.999 890	49.994 478	24.998 619	85	5.382 879	219.143 939	40.711 290
36	2.039 887	51.994 367	25.488 842	86	5.490 536	224.526 818	40.893 422
37	2.080 685	54.034 255	25.969 453	87	5.600 347	230.017 354	41.071 982
38	2.122 299	56.114 940	26.440 641	88	5.712 354	235.617 701	41.247 041
39	2.164 745	58.237 238	26.902 589	89	5.826 601	241.330 055	41.418 668
40	2.208 040	60.401 983	27.355 479	90	5.943 133	247.156 656	41.586 929
41	2.252 200	62.610 023	27.799 489	91	6.061 996	253.099 789	41.751 891
42	2.297 244	64.862 223	28.234 794	92	6.183 236	259.161 785	41.913 619
43	2.343 189	67.159 468	28.661 562	93	6.306 900	265.345 021	42.072 175
44	2.390 053	69.502 657	29.079 963	94	6.433 038	271.651 921	42.227 623
45	2.437 854	71.892 710	29.490 159	95	6.561 699	278.084 960	42.380 023
46	2.486 611	74.330 564	29.892 314	96	6.692 933	284.646 659	42.529 434
47	2.536 344	76.817 176	30.286 582	97	6.826 792	291.339 592	42.675 916
48	2.587 070	79.353 519	30.673 120	98	6.963 328	298.166 384	42.819 525
49	2.638 812	81.940 590	31.052 078	99	7.102 594	305.129 712	42.960 319
50	2.691 588	84.579 401	31.423 606	100	7.244 646	312.232 306	43.098 352

$i = 0.0225\ (2\frac{1}{4}\%)$						
n	$(1 + i)^n$	$s_{\overline{n}i}$	$a_{\overline{n}i}$	n	$(1 + i)^n$	$s_{\overline{n}i}$	$a_{\overline{n}i}$

n	$(1 + i)^n$	$s_{\overline{n}i}$	$a_{\overline{n}i}$	n	$(1 + i)^n$	$s_{\overline{n}i}$	$a_{\overline{n}i}$
1	1.022 500	1.000 000	0.977 995	51	3.110 492	93.799 664	30.155 889
2	1.045 506	2.022 500	1.934 470	52	3.180 479	96.910 157	30.470 307
3	1.069 030·	3.068 006	2.869 897	53	3.252 039	100.090 635	30.777 806
4	1.093 083	4.137 036	3.784 740	54	3.325 210	103.342 674	31.078 539
5	1.117 678	5.230 120	4.679 453	55	3.400 027	106.667 885	31.372 654
6	1.142 825	6.347 797	5.554 477	56	3.476 528	110.067 912	31.660 298
7	1.168 539	7.490 623	6.410 246	57	3.554 750	113.544 440	31.941 611
8	1.194 831	8.659 162	7.247 185	58	3.634 732	117.099 190	32.216 735
9	1.221 715	9.853 993	8.065 706	59	3.716 513	120.733 922	32.485 804
10	1.249 203	11.075 708	8.866 216	60	3.800 135	124.450 435	32.748 953
11	1.277 311	12.324 911	9.649 111	61	3.885 638	128.250 570	33.006 311
12	1.306 050	13.602 222	10.414 779	62	3.973 065	132.136 208	33.258 006
13	1.335 436	14.908 272	11.163 598	63	4.062 459	136.109 272	33.504 162
14	1.365 483	16.243 708	11.895 939	64	4.153 864	140.171 731	33.744 902
15	1.396 207	17.609 191	12.612 166	65	4.247 326	144.325 595	33.980 344
16	1.427 621	19.005 398	13.312 631	66	4.342 891	148.572 920	34.210 605
17	1.459 743	20.433 020	13.997 683	67	4.440 606	152.915 811	34.435 800
18	1.492 587	21.892 763	14.667 661	68	4.540 519	157.356 417	34.656 039
19	1.526 170	23.385 350	15.322 896	69	4.642 681	161.896 937	34.871 432
20	1.560 509	24.911 520	15.963 712	70	4.747 141	166.539 618	35.082 085
21	1.595 621	26.472 029	16.590 428	71	4.853 952	171.286 759	35.288 103
22	1.631 522	28.067 650	17.203 352	72	4.963 166	176.140 711	35.489 587
23	1.668 231	29.699 172	17.802 790	73	5.074 837	181.103 877	35.686 638
24	1.705 767	31.367 403	18.389 036	74	5.189 021	186.178 714	35.879 352
25	1.744 146	33.073 170	18.962 383	75	5.305 774	191.367 735	36.067 826
26	1.783 390	34.817 316	19.523 113	76	5.425 154	196.673 509	36.252 153
27	1.823 516	36.600 706	20.071 504	77	5.547 220	202.098 663	36.432 423
28	1.864 545	38.424 222	20.607 828	78	5.672 032	207.645 883	36.608 727
29	1.906 497	40.288 767	21.132 350	79	5.799 653	213.317 916	36.781 151
30	1.949 393	42.195 264	21.645 330	80	5.930 145	219.117 569	36.949 781
31	1.993 255	44.144 657	22.147 022	81	6.063 574	225.047 714	37.114 700
32	2.038 103	46.137 912	22.637 674	82	6.200 004	231.111 288	37.275 990
33	2.083 960	48.176 015	23.117 530	83	6.339 504	237.311 292	37.433 731
34	2.130 849	50.259 976	23.586 826	84	6.482 143	243.650 796	37.588 001
35	2.178 794	52.390 825	24.045 796	85	6.627 991	250.132 939	37.738 877
36	2.227 816	54.569 619	24.494 666	86	6.777 121	256.760 930	37.886 432
37	2.277 942	56.797 435	24.933 658	87	6.929 606	263.538 051	38.030 740
38	2.329 196	59.075 377	25.362 991	88	7.085 522	270.467 657	38.171 873
39	2.381 603	61.404 573	25.782 876	89	7.244 947	277.553 179	38.309 900
40	2.435 189	63.786 176	26.193 522	90	7.407 958	284.798 126	38.444 890
41	2.489 981	66.221 365	26.595 132	91	7.574 637	292.206 083	38.576 910
42	2.546 005	68.711 346	26.987 904	92	7.745 066	299.780 720	38.706 024
43	2.603 290	71.257 351	27.372 033	93	7.919 330	307.525 786	38.832 298
44	2.661 864	73.860 642	27.747 710	94	8.097 515	315.445 117	38.955 792
45	2.721 756	76.522 506	28.115 120	95	8.279 709	323.542 632	39.076 569
46	2.782 996	79.244 262	28.474 444	96	8.466 003	331.822 341	39.194 689
47	2.845 613	82.027 258	28.825 863	97	8.656 488	340.288 344	39.310 209
48	2.909 640	84.872 872	29.169 548	98	8.851 259	348.944 831	39.423 187
49	2.975 107	87.782 511	29.505 670	99	9.050 412	357.796 090	39.533 680
50	3.042 046	90.757 618	29.834 396	100	9.254 046	366.846 502	39.641 741

Table V (Continued)

		$i = 0.025\ (2\frac{1}{2}\%)$					
n	$(1+i)^n$	$s_{\overline{n}\rvert i}$	$a_{\overline{n}\rvert i}$	n	$(1+i)^n$	$s_{\overline{n}\rvert i}$	$a_{\overline{n}\rvert i}$
1	1.025 000	1.000 000	0.975 610	51	3.523 036	100.921 458	28.646 158
2	1.050 625	2.025 000	1.927 424	52	3.611 112	104.444 494	28.923 081
3	1.076 891	3.075 625	2.856 024	53	3.701 390	108.055 606	29.193 249
4	1.103 813	4.152 516	3.761 974	54	3.793 925	111.756 996	29.456 829
5	1.131 408	5.256 329	4.645 828	55	3.888 773	115.550 921	29.713 979
6	1.159 693	6.387 737	5.508 125	56	3.985 992	119.439 694	29.964 858
7	1.188 686	7.547 430	6.349 391	57	4.085 642	123.425 687	30.209 617
8	1.218 403	8.736 116	7.170 137	58	4.187 783	127.511 329	30.448 407
9	1.248 863	9.954 519	7.970 866	59	4.292 478	131.699 112	30.681 373
10	1.280 085	11.203 382	8.752 064	60	4.399 790	135.991 590	30.908 656
11	1.312 087	12.483 466	9.514 209	61	4.509 784	140.391 380	31.130 397
12	1.344 889	13.795 553	10.257 765	62	4.622 529	144.901 164	31.346 728
13	1.378 511	15.140 442	10.983 185	63	4.738 092	149.523 693	31.557 784
14	1.412 974	16.518 953	11.690 912	64	4.856 545	154.261 786	31.763 691
15	1.448 298	17.931 927	12.381 378	65	4.977 958	159.118 330	31.964 577
16	1.484 506	19.380 225	13.055 003	66	5.102 407	164.096 289	32.160 563
17	1.521 618	20.864 730	13.712 198	67	5.229 967	169.198 696	32.351 769
18	1.559 659	22.386 349	14.353 364	68	5.360 717	174.428 663	32.538 311
19	1.598 650	23.946 007	14.978 891	69	5.494 734	179.789 380	32.720 303
20	1.638 616	25.544 658	15.589 162	70	5.632 103	185.284 114	32.897 857
21	1.679 582	27.183 274	16.184 549	71	5.772 905	190.916 217	33.071 080
22	1.721 571	28.862 856	16.765 413	72	5.917 228	196.689 122	33.240 078
23	1.764 611	30.584 427	17.332 110	73	6.065 159	202.606 351	33.404 954
24	1.808 726	32.349 038	17.884 986	74	6.216 788	208.671 509	33.565 809
25	1.853 944	34.157 764	18.424 376	75	6.372 207	214.888 297	33.722 740
26	1.900 293	36.011 708	18.950 611	76	6.531 513	221.260 504	33.875 844
27	1.947 800	37.912 001	19.464 011	77	6.694 800	227.792 017	34.025 214
28	1.996 495	39.859 801	19.964 889	78	6.862 170	234.486 818	34.170 940
29	2.046 407	41.856 296	20.453 550	79	7.033 725	241.348 988	34.313 113
30	2.097 568	43.902 703	20.930 293	80	7.209 568	248.382 713	34.451 817
31	2.150 007	46.000 271	21.395 407	81	7.389 807	255.592 280	34.587 139
32	2.203 757	48.150 278	21.849 178	82	7.574 552	262.982 087	34.719 160
33	2.258 851	50.354 034	22.291 881	83	7.763 916	270.556 640	34.847 961
34	2.315 322	52.612 885	22.723 786	84	7.958 014	278.320 556	34.973 620
35	2.373 205	54.928 207	23.145 157	85	8.156 964	286.278 569	35.096 215
36	2.432 535	57.301 413	23.556 251	86	8.360 888	294.435 534	35.215 819
37	2.493 349	59.733 948	23.957 318	87	8.569 911	302.796 422	35.332 507
38	2.555 682	62.227 297	24.348 603	88	8.784 158	311.366 333	35.446 348
39	2.619 574	64.782 979	24.730 344	89	9.003 762	320.150 491	35.557 413
40	2.685 064	67.402 554	25.102 775	90	9.228 856	329.154 253	35.665 768
41	2.752 190	70.087 617	25.466 122	91	9.459 578	338.383 110	35.771 481
42	2.820 995	72.839 808	25.820 607	92	9.696 067	347.842 687	35.874 616
43	2.891 520	75.660 803	26.166 446	93	9.938 469	357.538 755	35.975 235
44	2.963 808	78.552 323	26.503 849	94	10.186 931	367.477 223	36.073 400
45	3.037 903	81.516 131	26.833 024	95	10.441 604	377.664 154	36.169 171
46	3.113 851	84.554 034	27.154 170	96	10.702 644	388.105 758	36.262 606
47	3.191 697	87.667 885	27.467 483	97	10.970 210	398.808 402	36.353 762
48	3.271 490	90.859 582	27.773 154	98	11.244 465	409.778 612	36.442 694
49	3.353 277	94.131 072	28.071 369	99	11.525 577	421.023 077	36.529 458
50	3.437 109	97.484 349	28.362 312	100	11.813 716	432.548 654	36.614 105

		$i = 0.03$ (3%)					
n	$(1 + i)^n$	$s_{\overline{n}|i}$	$a_{\overline{n}|i}$	n	$(1 + i)^n$	$s_{\overline{n}|i}$	$a_{\overline{n}|i}$
1	1.030 000	1.000 000	0.970 874	51	4.515 423	117.180 773	25.951 227
2	1.060 900	2.030 000	1.913 470	52	4.650 886	121.696 197	26.166 240
3	1.092 727	3.090 900	2.828 611	53	4.790 412	126.347 082	26.374 990
4	1.125 509	4.183 627	3.717 098	54	4.934 125	131.137 495	26.577 660
5	1.159 274	5.309 136	4.579 707	55	5.082 149	136.071 620	26.774 428
6	1.194 052	6.468 410	5.417 191	56	5.234 613	141.153 768	26.965 464
7	1.229 874	7.662 462	6.230 283	57	5.391 651	146.388 381	27.150 936
8	1.266 770	8.892 336	7.019 692	58	5.553 401	151.780 033	27.331 005
9	1.304 773	10.159 106	7.786 109	59	5.720 003	157.333 434	27.505 831
10	1.343 916	11.463 879	8.530 203	60	5.891 603	163.053 437	27.675 564
11	1.384 234	12.807 796	9.252 624	61	6.068 351	168.945 040	27.840 353
12	1.425 761	14.192 030	9.954 004	62	6.250 402	175.013 391	28.000 343
13	1.468 534	15.617 790	10.634 955	63	6.437 914	181.263 793	28.155 673
14	1.512 590	17.086 324	11.296 073	64	6.631 051	187.701 707	28.306 478
15	1.557 967	18.598 914	11.937 935	65	6.829 983	194.332 758	28.452 892
16	1.604 706	20.156 881	12.561 102	66	7.034 882	201.162 741	28.595 040
17	1.652 848	21.761 588	13.166 118	67	7.245 929	208.197 623	28.733 049
18	1.702 433	23.414 435	13.753 513	68	7.463 307	215.443 551	28.867 038
19	1.753 506	25.116 868	14.323 799	69	7.687 206	222.906 858	28.997 124
20	1.806 111	26.870 374	14.877 475	70	7.917 822	230.594 064	29.123 421
21	1.860 295	28.676 486	15.415 024	71	8.155 357	238.511 886	29.246 040
22	1.916 103	30.536 780	15.936 917	72	8.400 017	246.667 242	29.365 088
23	1.973 587	32.452 884	16.443 608	73	8.652 016	255.067 259	29.480 668
24	2.032 794	34.426 470	16.935 542	74	8.911 578	263.719 277	29.592 881
25	2.093 778	36.459 264	17.413 148	75	9.178 926	272.630 856	29.701 826
26	2.156 591	38.553 042	17.876 842	76	9.454 293	281.809 781	29.807 598
27	2.221 289	40.709 634	18.327 031	77	9.737 922	291.264 075	29.910 290
28	2.287 928	42.930 923	18.764 108	78	10.030 060	301.001 997	30.009 990
29	2.356 566	45.218 850	19.188 455	79	10.330 962	311.032 057	30.106 786
30	2.427 262	47.575 416	19.600 441	80	10.640 891	321.363 019	30.200 763
31	2.500 080	50.002 678	20.000 428	81	10.960 117	332.003 909	30.292 003
32	2.575 083	52.502 759	20.388 766	82	11.288 921	342.964 026	30.380 586
33	2.652 335	55.077 841	20.765 792	83	11.627 588	354.252 947	30.466 588
34	2.731 905	57.730 177	21.131 837	84	11.976 416	365.880 536	30.550 086
35	2.813 862	60.462 082	21.487 220	85	12.335 709	377.856 952	30.631 151
36	2.898 278	63.275 944	21.832 252	86	12.705 780	390.192 660	30.709 855
37	2.985 227	66.174 223	22.167 235	87	13.086 953	402.898 440	30.786 267
38	3.074 783	69.159 449	22.492 462	88	13.479 562	415.985 393	30.860 454
39	3.167 027	72.234 233	22.808 215	89	13.883 949	429.464 955	30.932 479
40	3.262 038	75.401 260	23.114 772	90	14.300 467	443.348 904	31.002 407
41	3.359 899	78.663 298	23.412 400	91	14.729 481	457.649 371	31.070 298
42	3.460 696	82.023 196	23.701 359	92	15.171 366	472.378 852	31.136 212
43	3.564 517	85.483 892	23.981 902	93	15.626 507	487.550 217	31.200 206
44	3.671 452	89.048 409	24.254 274	94	16.095 302	503.176 724	31.262 336
45	3.781 596	92.719 861	24.518 713	95	16.578 161	519.272 026	31.322 656
46	3.895 044	96.501 457	24.775 449	96	17.075 506	535.850 186	31.381 219
47	4.011 895	100.396 501	25.024 708	97	17.587 771	552.925 692	31.438 077
48	4.132 252	104.408 396	25.266 707	98	18.115 404	570.513 463	31.493 279
49	4.256 219	108.540 648	25.501 657	99	18.658 866	588.628 867	31.546 872
50	4.383 906	112.796 867	25.729 764	100	19.218 632	607.287 733	31.598 905

Table V (Continued)

$i = 0.035 (3\frac{1}{2}\%)$						
n	$(1 + i)^n$	$s_{\overline{n}|i}$	$a_{\overline{n}|i}$	n	$(1 + i)^n$	$s_{\overline{n}|i}$	$a_{\overline{n}|i}$

n	$(1 + i)^n$	$s_{\overline{n}|i}$	$a_{\overline{n}|i}$	n	$(1 + i)^n$	$s_{\overline{n}|i}$	$a_{\overline{n}|i}$
1	1.035 000	1.000 000	0.966 184	51	5.780 399	136.582 837	23.628 616
2	1.071 225	2.035 000	1.899 694	52	5.982 713	142.363 236	23.795 765
3	1.108 718	3.106 225	2.801 637	53	6.192 108	148.345 950	23.957 260
4	1.147 523	4.214 943	3.673 079	54	6.408 832	154.538 058	24.113 295
5	1.187 686	5.362 466	4.515 052	55	6.633 141	160.946 890	24.264 053
6	1.229 255	6.550 152	5.328 553	56	6.865 301	167.580 031	24.409 713
7	1.272 279	7.779 408	6.114 544	57	7.105 587	174.445 332	24.550 448
8	1.316 809	9.051 687	6.873 956	58	7.354 282	181.550 919	24.686 423
9	1.362 897	10.368 496	7.607 687	59	7.611 682	188.905 201	24.817 800
10	1.410 599	11.731 393	8.316 605	60	7.878 091	196.516 883	24.944 734
11	1.459 970	13.141 992	9.001 551	61	8.153 824	204.394 974	25.067 376
12	1.511 069	14.601 962	9.663 334	62	8.439 208	212.548 798	25.185 870
13	1.563 956	16.113 030	10.302 738	63	8.734 580	220.988 006	25.300 358
14	1.618 695	17.676 986	10.920 520	64	9.040 291	229.722 586	25.410 974
15	1.675 349	19.295 681	11.517 411	65	9.356 701	238.762 876	25.517 849
16	1.733 986	20.971 030	12.094 117	66	9.684 185	248.119 577	25.621 110
17	1.794 676	22.705 016	12.651 321	67	10.023 132	257.803 762	25.720 880
18	1.857 489	24.499 691	13.189 682	68	10.373 941	267.826 894	25.817 275
19	1.922 501	26.357 180	13.709 837	69	10.737 029	278.200 835	25.910 411
20	1.989 789	28.279 682	14.212 403	70	11.112 825	288.937 865	26.000 397
21	2.059 431	30.269 471	14.697 974	71	11.501 774	300.050 690	26.087 340
22	2.131 512	32.328 902	15.167 125	72	11.904 336	311.552 464	26.171 343
23	2.206 114	34.460 414	15.620 410	73	12.320 988	323.456 800	26.252 505
24	2.283 328	36.666 528	16.058 368	74	12.752 223	335.777 788	26.330 923
25	2.363 245	38.949 857	16.481 515	75	13.198 550	348.530 011	26.406 689
26	2.445 959	41.313 102	16.890 352	76	13.660 500	361.728 561	26.479 892
27	2.531 567	43.759 060	17.285 365	77	14.138 617	375.389 061	26.550 621
28	2.620 172	46.290 627	17.667 019	78	14.633 469	389.527 678	26.618 957
29	2.711 878	48.910 799	18.035 767	79	15.145 640	404.161 147	26.684 983
30	2.806 794	51.622 677	18.392 045	80	15.675 738	419.306 787	26.748 776
31	2.905 031	54.429 471	18.736 276	81	16.224 388	434.982 524	26.810 411
32	3.006 708	57.334 502	19.068 865	82	16.792 242	451.206 913	26.869 963
33	3.111 942	60.341 210	19.390 208	83	17.379 970	467.999 155	26.927 500
34	3.220 860	63.453 152	19.700 684	84	17.988 269	485.379 125	26.983 092
35	3.333 590	66.674 013	20.000 661	85	18.617 859	503.367 394	27.036 804
36	3.450 266	70.007 603	20.290 494	86	19.269 484	521.985 253	27.088 699
37	3.571 025	73.457 869	20.570 525	87	19.943 916	541.254 737	27.138 840
38	3.696 011	77.028 895	20.841 087	88	20.641 953	561.198 653	27.187 285
39	3.825 372	80.724 906	21.102 500	89	21.364 421	581.840 606	27.234 092
40	3.959 260	84.550 278	21.355 072	90	22.112 176	603.205 027	27.279 316
41	4.097 834	88.509 537	21.599 104	91	22.886 102	625.317 203	27.323 010
42	4.241 258	92.607 371	21.834 883	92	23.687 116	648.203 305	27.365 227
43	4.389 702	96.848 629	22.062 689	93	24.516 165	671.890 421	27.406 017
44	4.543 342	101.238 331	22.282 791	94	25.374 230	696.406 585	27.445 427
45	4.702 359	105.781 673	22.495 450	95	26.262 329	721.780 816	27.483 504
46	4.866 941	110.484 031	22.700 918	96	27.181 510	748.043 145	27.520 294
47	5.037 284	115.350 973	22.899 438	97	28.132 863	775.224 655	27.555 839
48	5.213 589	120.388 257	23.091 244	98	29.117 513	803.357 517	27.590 183
49	5.396 065	125.601 846	23.276 564	99	30.136 626	832.475 031	27.623 .366
50	5.584 927	130.997 910	23.455 618	100	31.191 408	662.611 657	27.653 425

$i = 0.04\ (4\%)$						
n	$(1+i)^n$	$s_{\overline{n}|i}$	$a_{\overline{n}|i}$	n	$(1+i)^n$	$s_{\overline{n}|i}$	$a_{\overline{n}|i}$

n	$(1+i)^n$	$s_{\overline{n}|i}$	$a_{\overline{n}|i}$	n	$(1+i)^n$	$s_{\overline{n}|i}$	$a_{\overline{n}|i}$
1	1.040 000	1.000 000	0.961 538	51	7.390 951	159.773 767	21.617 485
2	1.081 600	2.040 000	1.886 095	52	7.686 589	167.164 718	21.747 582
3	1.124 864	3.121 600	2.775 091	53	7.994 052	174.851 306	21.872 675
4	1.169 859	4.246 464	3.629 895	54	8.313 814	182.845 359	21.992 957
5	1.216 653	5.416 323	4.451 822	55	8.646 367	191.159 173	22.108 612
6	1.265 319	6.632 975	5.242 137	56	8.992 222	109.805 540	22.219 819
7	1.315 932	7.898 294	6.002 055	57	9.351 910	208.797 762	22.326 749
8	1.368 569	9.214 226	6.732 745	58	9.725 987	218.149 672	22.429 567
9	1.423 312	10.582 795	7.435 332	59	10.115 026	227.875 659	22.528 430
10	1.480 244	12.006 107	8.110 896	60	10.519 627	237.990 685	22.623 490
11	1.539 454	13.486 351	8.760 477	61	10.940 413	248.510 312	22.714 894
12	1.601 032	15.025 805	9.385 074	62	11.378 029	259.450 725	22.802 783
13	1.665 074	16.626 838	9.985 648	63	11.833 150	270.828 754	22.887 291
14	1.731 676	18.291 911	10.563 123	64	12.306 476	282.661 904	22.968 549
15	1.800 944	20.023 588	11.118 387	65	12.798 735	294.968 380	23.046 682
16	1.872 981	21.824 531	11.632 296	66	13.310 685	307.767 116	23.121 810
17	1.947 900	23.697 512	12.165 669	67	13.843 112	321.077 800	23.194 048
18	2.025 817	25.645 413	12.659 297	68	14.396 836	334.920 912	23.263 507
19	2.106 849	27.671 229	13.133 939	69	14.972 710	349.317 749	23.330 296
20	2.191 123	29.778 079	13.590 326	70	15.571 618	364.290 459	23.394 515
21	2.278 768	31.969 202	14.029 160	71	16.194 483	379.862 077	23.456 264
22	2.369 919	34.247 970	14.451 115	72	16.842 262	396.056 560	23.515 639
23	2.464 716	36.617 889	14.856 842	73	17.515 953	412.898 823	23.572 730
24	2.563 304	39.082 604	15.246 963	74	18.216 591	430.414 776	23.627 625
25	2.665 836	41.645 908	15.622 080	75	18.945 255	448.631 367	23.680 408
26	2.772 470	44.311 745	15.982 769	76	19.703 065	467.576 621	23.731 162
27	2.883 369	47.084 214	16.329 586	77	20.491 187	487.279 686	23.779 963
28	2.998 703	49.967 583	16.663 063	78	21.310 835	507.770 873	23.826 688
29	3.118 651	52.966 286	16.983 715	79	22.163 268	529.081 708	23.872 008
30	3.243 398	56.084 938	17.292 033	80	23.049 799	551.244 977	23.915 392
31	3.373 133	59.328 335	17.588 494	81	23.971 791	574.294 776	23.957 108
32	3.508 059	62.701 469	17.873 552	82	24.930 663	598.266 567	23.997 219
33	3.648 381	66.209 527	18.147 646	83	25.927 889	623.197 230	24.035 787
34	3.794 316	69.857 909	18.411 198	84	26.965 005	649.125 119	24.072 872
35	3.946 089	73.652 225	18.664 613	85	28.043 605	676.090 123	24.108 531
36	4.103 933	77.598 314	18.908 282	86	29.165 349	704.133 728	24.142 818
37	4.268 090	81.702 246	19.142 579	87	30.331 963	733.299 078	24.175 787
38	4.438 813	85.970 336	19.367 864	88	31.545 242	763.631 041	24.207 487
39	4.616 366	90.409 150	19.584 485	89	32.807 051	795.176 282	24.237 969
40	4.801 021	95.025 516	19.792 774	90	34.119 333	827.983 334	24.267 276
41	4.993 061	99.826 536	19.993 052	91	35.484 107	862.102 667	24.295 459
42	5.192 784	104.819 598	20.185 627	92	36.903 471	897.586 774	24.322 557
43	5.400 495	110.012 382	20.370 795	93	38.379 610	934.490 244	24.348 612
44	5.616 515	115.412 877	20.548 841	94	39.914 794	972.869 854	24.373 666
45	5.841 176	121.029 392	20.720 040	95	41.511 386	1012.784 648	24.397 756
46	6.074 823	126.870 568	20.884 654	96	43.171 841	1054.296 034	24.420 919
47	6.317 816	132.945 390	21.042 936	97	44.898 715	1097.467 876	24.443 191
48	6.570 528	139.263 206	21.195 131	98	46.694 664	1142.366 591	24.464 607
49	6.833 349	145.833 734	21.341 472	99	48.562 450	1189.061 254	24.485 199
50	7.106 683	152.667 084	21.482 185	100	50.504 948	1237.623 705	24.504 999

Table V (Continued)

			$i = 0.045 \, (4\frac{1}{2}\%)$				
n	$(1 + i)^n$	$s_{\overline{n}|i}$	$a_{\overline{n}|i}$	n	$(1 + i)^n$	$s_{\overline{n}|i}$	$a_{\overline{n}|i}$
1	1.045 000	1.000 000	0.956 938	51	9.439 105	187.535 665	19.867 950
2	1.092 025	2.045 000	1.872 668	52	9.863 865	196.974 770	19.969 330
3	1.141 166	3.137 025	2.748 964	53	10.307 739	206.838 634	20.066 345
4	1.192 519	4.278 191	3.587 526	54	10.771 587	217.146 373	20.159 181
5	1.246 182	5.470 710	4.389 977	55	11.256 308	227.917 959	20.248 021
6	1.302 260	6.716 892	5.157 872	56	11.762 842	239.174 268	20.333 034
7	1.360 862	8.019 152	5.892 701	57	12.292 170	250.937 110	20.414 387
8	1.422 101	9.380 014	6.595 886	58	12.845 318	263.229 280	20.492 236
9	1.486 095	10.802 114	7.268 790	59	13.423 357	276.074 597	20.566 733
10	1.552 969	12.288 209	7.912 718	60	14.027 408	289.497 954	20.638 022
11	1.622 853	13.841 179	8.528 917	61	14.658 641	303.525 362	20.706 241
12	1.695 881	15.464 032	9.118 581	62	15.318 280	318.184 031	20.771 523
13	1.772 196	17.159 913	9.682 852	63	16.007 603	333.502 283	20.833 993
14	1.851 945	18.932 109	10.222 825	64	16.727 945	349.509 868	20.893 773
15	1.935 282	20.784 054	10.739 546	65	17.480 702	366.237 831	20.950 979
16	2.022 370	22.719 337	11.234 015	66	18.267 334	383.718 533	21.005 722
17	2.113 377	24.741 707	11.707 191	67	19.089 364	401.985 867	21.058 107
18	2.208 479	26.855 084	12.159 992	68	19.948 385	421.075 231	21.108 236
19	2.307 860	29.063 562	12.593 294	69	20.846 063	441.023 617	21.156 207
20	2.411 714	31.371 423	13.007 936	70	21.784 136	461.869 680	21.202 112
21	2.520 241	33.783 137	13.404 724	71	22.764 422	483.653 815	21.246 040
22	2.633 652	36.303 378	13.784 425	72	23.788 821	506.418 237	21.288 077
23	2.752 166	38.937 030	14.147 775	73	24.859 318	530.207 057	21.328 303
24	2.876 014	41.689 196	14.495 478	74	25.977 987	555.066 375	21.366 797
25	3.005 434	44.565 210	14.828 209	75	27.146 996	581.044 362	21.403 634
26	3.140 679	47.570 645	15.146 611	76	28.368 611	608.191 358	21.438 884
27	3.282 010	50.711 324	15.451 303	77	29.645 199	636.559 969	21.472 616
28	3.429 700	53.993 333	15.742 874	78	30.979 233	666.205 168	21.504 896
29	3.584 036	57.423 033	16.021 889	79	32.373 298	697.184 401	21.535 785
30	3.745 318	61.007 070	16.288 889	80	33.830 096	729.557 699	21.565 345
31	3.913 857	64.752 388	16.544 391	81	35.352 451	763.387 795	21.593 632
32	4.089 981	68.666 245	16.788 891	82	36.943 311	798.740 246	21.620 700
33	4.274 030	72.756 226	17.022 862	83	38.605 760	835.683 557	21.646 603
34	4.466 362	77.030 256	17.246 758	84	40.343 019	874.289 317	21.671 390
35	4.667 348	81.496 618	17.461 012	85	42.158 455	914.632 336	21.695 110
36	4.877 378	86.163 966	17.666 041	86	44.055 586	956.790 791	21.717 809
37	5.096 860	91.041 344	17.862 240	87	46.038 087	1000.846 377	21.739 530
38	5.326 219	96.138 205	18.049 990	88	48.109 801	1046.884 464	21.760 316
39	5.565 899	101.464 424	18.229 656	89	50.274 742	1094.994 265	21.780 207
40	5.816 365	107.030 323	18.401 584	90	52.537 105	1145.269 007	21.799 241
41	6.078 101	112.846 688	18.566 109	91	54.901 275	1197.806 112	21.817 455
42	6.351 615	118.924 789	18.723 550	92	57.371 832	1252.707 387	21.834 885
43	6.637 438	125.276 404	18.874 210	93	59.953 565	1310.079 219	21.851 565
44	6.936 123	131.913 842	19.018 383	94	62.651 475	1370.032 784	21.867 526
45	7.248 248	138.849 965	19.156 347	95	65.470 792	1432.684 259	21.882 800
46	7.574 420	146.098 214	19.288 371	96	68.416 977	1498.155 051	21.897 417
47	7.915 268	153.672 633	19.414 709	97	71.495 741	1566.572 028	21.911 403
48	8.271 456	161.587 902	19.535 607	98	74.713 050	1638.067 770	21.924 788
49	8.643 671	169.859 357	19.651 298	99	78.075 137	1712.780 819	21.937 596
50	9.032 636	178.503 028	19.762 008	100	81.588 518	1790.855 956	21.949 853

	$i = 0.05$ (5%)				$i = 0.06$ (6%)		
n	$(1 + i)^n$	$s_{\overline{n}|i}$	$a_{\overline{n}|i}$	n	$(1 + i)^n$	$s_{\overline{n}|i}$	$a_{\overline{n}|i}$
1	1.050 000	1.000 000	0.952 381	1	1.060 000	1.000 000	0.943 396
2	1.102 500	2.050 000	1.859 410	2	1.123 600	2.060 000	1.833 393
3	1.157 625	3.152 500	2.723 248	3	1.191 016	3.183 600	2.673 012
4	1.215 506	4.310 125	3.545 951	4	1.262 477	4.374 616	3.465 106
5	1.276 282	5.525 631	4.329 477	5	1.338 226	5.637 093	4.212 364
6	1.340 096	6.801 913	5.075 692	6	1.418 519	6.975 319	4.917 324
7	1.407 100	8.142 008	5.786 373	7	1.503 630	8.393 838	5.582 381
8	1.477 455	9.549 109	6.463 213	8	1.593 848	9.897 468	6.209 794
9	1.551 328	11.026 564	7.107 822	9	1.689 479	11.491 316	6.801 692
10	1.628 895	12.577 893	7.721 735	10	1.790 848	13.180 795	7.360 087
11	1.710 339	14.206 787	8.306 414	11	1.898 299	14.971 643	7.886 875
12	1.795 856	15.917 127	8.863 252	12	2.012 196	16.869 941	8.383 844
13	1.885 649	17.712 983	9.393 573	13	2.132 928	18.882 138	8.852 683
14	1.979 932	19.598 632	9.898 641	14	2.260 904	21.015 066	9.294 984
15	2.078 928	21.578 564	10.379 658	15	2.396 558	23.275 970	9.712 249
16	2.182 875	23.657 492	10.837 770	16	2.540 352	25.672 528	10.105 895
17	2.292 018	25.040 366	11.274 066	17	2.692 773	28.212 880	10.477 260
18	2.406 619	28.132 385	11.689 587	18	2.854 339	30.905 653	10.827 603
19	2.526 950	30.539 004	12.085 321	19	3.025 600	33.759 992	11.158 116
20	2.653 298	33.065 954	12.462 210	20	3.207 135	36.785 591	11.469 921
21	2.785 963	35.719 252	12.821 153	21	3.399 564	39.992 727	11.764 077
22	2.925 261	38.505 214	13.163 003	22	3.603 537	43.392 290	12.041 582
23	3.071 524	41.430 475	13.488 574	23	3.819 750	46.995 828	12.303 379
24	3.225 100	44.501 999	13.798 642	24	4.048 935	50.815 577	12.550 358
25	3.386 355	47.727 099	14.093 945	25	4.291 871	54.864 512	12.783 356
26	3.555 673	51.113 454	14.375 185	26	4.549 383	59.156 383	13.003 166
27	3.733 456	54.669 126	14.643 034	27	4.822 346	63.705 766	13.210 534
28	3.920 129	58.402 583	14.898 127	28	5.111 687	68.528 112	13.406 164
29	4.116 136	62.322 712	15.141 074	29	5.418 388	73.639 798	13.590 721
30	4.321 942	66.438 848	15.372 451	30	5.743 491	79.058 186	13.764 831
31	4.538 039	70.760 790	15.592 810	31	6.088 101	84.801 677	13.929 086
32	4.764 941	75.298 829	15.802 677	32	6.453 387	90.889 778	14.084 043
33	5.003 189	80.063 771	16.002 549	33	6.840 590	97.343 165	14.230 230
34	5.253 348	85.066 959	16.192 904	34	7.251 025	104.183 755	14.368 141
35	5.516 015	90.320 307	16.374 194	35	7.686 087	111.434 780	14.498 246
36	5.791 816	95.836 323	16.546 852	36	8.147 252	119.120 867	14.620 987
37	6.081 407	101.628 139	16.711 287	37	8.636 087	127.268 119	14.736 780
38	6.385 477	107.709 546	16.867 893	38	9.154 252	135.904 206	14.846 019
39	6.704 751	114.095 023	17.017 041	39	9.703 507	145.058 458	14.949 075
40	7.039 989	120.799 774	17.159 086	40	10.285 718	154.761 966	15.046 297
41	7.391 988	127.839 763	17.294 368	41	10.902 861	165.047 684	15.138 016
42	7.761 588	135.231 751	17.423 208	42	11.557 033	175.950 545	15.224 543
43	8.149 667	142.993 339	17.545 912	43	12.250 455	187.507 577	15.306 173
44	8.557 150	151.143 006	17.662 773	44	12.985 482	199.758 032	15.383 182
45	8.985 008	159.700 156	17.774 070	45	13.764 611	212.743 514	15.455 832
46	9.434 258	168.685 164	17.880 066	46	14.590 487	226.508 125	15.524 370
47	9.905 971	178.119 422	17.981 016	47	15.465 917	241.098 612	15.589 028
48	10.401 270	188.025 393	18.077 158	48	16.393 872	256.564 529	15.650 027
49	10.921 333	198.426 663	18.168 722	49	17.377 504	272.958 401	15.707 572
50	11.467 400	209.347 996	18.255 925	50	18.420 154	290.335 905	15.761 861

· Table V **(Continued)**

	i = 0.07 (7%)				i = 0.08 (8%)		
n	$(1 + i)^n$	$s_{\overline{n}i}$	$a_{\overline{n}i}$	n	$(1 + i)^n$	$s_{\overline{n}i}$	$a_{\overline{n}i}$
1	1.070 000	1.000 000	0.934 579	1	1.080 000	1.000 000	0.925 925
2	1.144 900	2.070 000	1.808 018	2	1.166 400	2.080 000	1.783 265
3	1.225 043	3.214 900	2.624 316	3	1.259 712	3.246 400	2.577 097
4	1.310 796	4.439 943	3.387 211	4	1.360 489	4.506 112	3.312 127
5	1.402 552	5.750 739	4.100 197	5	1.469 328	5.866 601	3.992 710
6	1.500 730	7.153 291	4.766 540	6	1.586 874	7.335 929	4.622 880
7	1.605 781	8.654 021	5.389 289	7	1.713 824	8.922 803	5.206 370
8	1.718 186	10.259 803	5.971 299	8	1.850 930	10.636 628	5.746 639
9	1.838 459	11.977 989	6.515 232	9	1.999 005	12.487 558	6.246 888
10	1.967 151	13.816 448	7.023 582	10	2.158 925	14.486 562	6.710 081
11	2.104 852	15.783 599	7.498 674	11	2.331 639	16.645 487	7.138 964
12	2.252 192	17.888 451	7.942 686	12	2.518 170	18.977 126	7.536 078
13	2.409 845	20.140 643	8.357 651	13	2.719 624	21.495 297	7.903 776
14	2.578 534	22.550 488	8.745 468	14	2.937 194	24.214 920	8.244 237
15	2.759 032	25.129 022	9.107 914	15	3.172 169	27.152 114	8.559 479
16	2.952 164	27.888 054	9.446 649	16	3.425 943	30.324 283	8.851 369
17	3.158 815	30.840 217	9.763 223	17	3.700 018	33.750 226	9.121 638
18	3.379 932	33.999 033	10.059 087	18	3.996 019	37.450 244	9.371 887
19	3.616 528	37.378 965	10.335 595	19	4.315 701	41.446 263	9.603 599
20	3.869 684	40.995 492	10.594 014	20	4.660 957	45.761 964	9.818 147
21	4.140 562	44.865 177	10.835 527	21	5.033 834	50.422 921	10.016 803
22	4.430 402	49.005 739	11.061 240	22	5.436 540	55.456 755	10.200 744
23	4.740 530	53.436 141	11.272 187	23	5.871 464	60.893 296	10.371 059
24	5.072 367	58.176 671	11.469 334	24	6.341 181	66.764 759	10.528 758
25	5.427 433	63.249 038	11.653 583	25	6.848 475	73.105 940	10.674 776
26	5.807 353	68.676 470	11.825 779	26	7.396 353	79.954 415	10.809 978
27	6.213 868	74.483 823	11.986 709	27	7.988 061	87.350 768	10.935 165
28	6.648 838	80.697 691	12.137 111	28	8.627 106	95.338 830	11.051 078
29	7.114 257	87.346 529	12.277 674	29	9.317 275	103.965 936	11.158 406
30	7.612 255	94.460 786	12.409 041	30	10.062 657	113.283 211	11.257 783
31	8.145 113	102.073 041	12.531 814	31	10.867 669	123.345 868	11.349 799
32	8.715 271	110.218 154	12.646 555	32	11.737 083	134.213 537	11.434 999
33	9.325 340	118.933 425	12.753 790	33	12.676 050	145.950 620	11.513 888
34	9.978 114	128.258 765	12.854 009	34	13.690 134	158.626 670	11.586 934
35	10.676 581	138.236 878	12.947 672	35	14.785 344	172.316 804	11.654 568
36	11.423 942	148.913 460	13.035 208	36	15.968 172	187.102 148	11.717 193
37	12.223 618	160.337 402	13.117 017	37	17.245 626	203.070 320	11.775 179
38	13.079 271	172.561 020	13.193 473	38	18.625 276	220.315 945	11.828 869
39	13.994 820	185.640 292	13.264 928	39	20.115 298	238.941 221	11.878 582
40	14.974 458	199.635 112	13.331 709	40	21.724 522	259.056 519	11.924 613
41	16.022 670	214.609 570	13.394 120	41	23.462 483	280.781 040	11.967 235
42	17.144 257	230.632 240	13.452 449	42	25.339 482	304.243 523	12.006 699
43	18.344 355	247.776 496	13.506 962	43	27.366 640	329.583 005	12.043 240
44	19.628 460	266.120 851	13.557 908	44	29.555 972	356.949 646	12.077 074
45	21.002 452	285.749 311	13.605 522	45	31.920 449	386.505 617	12.108 402
46	22.472 623	306.751 763	13.650 020	46	34.474 085	418.426 067	12.137 409
47	24.045 707	329.224 386	13.691 608	47	37.232 012	452.900 152	12.164 267
48	25.728 907	353.270 093	13.730 474	48	40.210 573	490.132 164	12.189 136
49	27.529 930	378.999 000	13.766 799	49	43.427 419	530.342 737	12.212 163
50	29.457 025	406.528 929	13.800 746	50	46.901 613	573.770 156	12.233 485

Answers

Exercise 1-1

1. T **3.** T **5.** T **7.** T **9.** {1, 2, 3, 4, 5} **11.** {3, 4} **13.** ∅ **15.** {2} **17.** {−7, 7} **19.** {1, 3, 5, 7, 9} **21.** $A' = \{1, 5\}$
23. 40 **25.** 60 **27.** 60 **29.** 20 **31.** 95 **33.** 40 **35.** (A) {1, 2, 3, 4, 6} (B) {1, 2, 3, 4, 6} **37.** {1, 2, 3, 4, 6} **39.** Yes
41. Yes **43.** Yes **45.** (A) 2 (B) 4 (C) 8; 2^n **47.** 800 **49.** 200 **51.** 200 **53.** 800 **55.** 200 **57.** 200 **59.** 6
61. A+, AB+ **63.** A−, A+, B+, AB−, AB+, O+ **65.** O+, O− **67.** B−, B+
69. Everybody in the clique relates to each other.

Exercise 1-2

1. T **3.** F **5.** T **7.** Commutative **9.** Associative **11.** Commutative **13.** Associative **15.** 35x **17.** $t + 13$
19. $3x + 2y + 16$ **21.** 5 **23.** $-\frac{15}{2}$ **25.** $\frac{1}{8}$ **27.** $-\frac{7}{5}$ **29.** Commutative **31.** Associative **33.** Commutative
35. Associative **37.** $p + q + r + 15$ **39.** 40xy **41.** 60mnp **43.** $18x + 27$ **45.** $21u + 28v$ **47.** $am + an$ **49.** $5(3u + 5v)$
51. $5(2m + 1)$ **53.** $8(4x + 3y)$ **55.** $a(h + k)$ **57.** $\frac{3}{13}$ **59.** $-\frac{4}{15}$ **61.** F; $4 - 3 \neq 3 - 4$, for example **63.** T
65. F; $(8 - 6) - 4 \neq 8 - (6 - 4)$, for example **67.** T

Exercise 1-3

1. $-5 > -30$ **3.** $x \geq -6$ **5.** 8 is greater than -8 **7.** x is greater than or equal to 8 **9.** < **11.** > **13.** < **15.** <

17. < **19.** > **21.** > **23.** $(-\infty, 5]$; **25.** $(-5, \infty)$; **27.** $(-2, 3)$;

29. $[-5, -1]$; **31.** $(2, 8]$; **33.** $[-7, -2)$; **35.** $x > 5$;

37. $x \leq 4$; **39.** $-2 \leq x \leq 5$; **41.** $-7 < x < -2$;

43. $-2 \leq x < 2$; **45.** $2 < x \leq 10$; **47.** $x \leq 8$; $(-\infty, 8]$ **49.** $x > -6$; $(-6, \infty)$

51. $-3 \leq x \leq 9$; $[-3, 9]$ **53.** $-5 < x \leq 15$; $(-5, 15]$ **55.** $x \leq -3$ or $x > 5$ **57.** $x \leq -5$ or $x \geq 0$

59. $-5 \leq x \leq 5$; **61.** $-2 < x \leq 5$; **63.** $0 < x < 4$;

65. **67.** **69.** **71.** $\$14{,}000 \leq S \leq \$18{,}500$ **73.** $50°F \leq T \leq 70°F$

75. $n \geq 40$ **77.** Long-headed, $C < 75$; intermediate, $75 \leq C \leq 80$; round-headed, $C > 80$.

Exercise 1-4

1. −20 **3.** −7 **5.** 9 **7.** −7 **9.** −5 **11.** 20 **13.** −10 **15.** −21 **17.** −4 **19.** 63 **21.** −4 **23.** 0 **25.** Undefined
27. 48 **29.** −22 **31.** 0 **33.** 0 **35.** 0 **37.** −4 **39.** −15 **41.** sometimes **43.** sometimes **45.** −4 **47.** −31 **49.** −12
51. 3 **53.** 77 **55.** −53 **57.** 0 **59.** −48 **61.** 0 **63.** −40 **65.** 3 **67.** 8 **69.** \$28 **71.** 2,555 **73.** 187

Exercise 1-5

1. 7 **3.** 5 **5.** 18 **7.** 3 **9.** 8 **11.** 6 **13.** $\dfrac{b^5}{c^5}$ **15.** 4 **17.** 4 **19.** 3 **21.** 9 **23.** 1 **25.** $28y^5$ **27.** $2w^2$ **29.** $\dfrac{2}{5m^2}$ **31.** u^9v^9

33. $\dfrac{p^7}{q^7}$ **35.** m^{24} **37.** $70x^9$ **39.** x^8y^{12} **41.** $8x^6$ **43.** $\dfrac{a^{12}}{b^6}$ **45.** $\dfrac{3y^5}{2x^2}$ **47.** $27u^6v^9w^3$ **49.** $16a^6b^4$ **51.** $4x^{12}y^4$ **53.** $\dfrac{m^6}{n^6}$ **55.** $\dfrac{1}{y^2}$

57. −1 **59.** $36x^7y^6z^3$ **61.** $-108u^{22}v^{18}$ **63.** $\dfrac{2t^2u^9}{v^2}$

Exercise 1-6   Chapter Review

1. (A) F (B) T (C) F (D) T **2.** (A) T (B) F (C) T **3.** Commutative **4.** Associative **5.** Associative
6. Commutative **7.** $20y$ **8.** $42w$ **9.** $y + 18$ **10.** $2x + 11$ **11.** (A) 16 (B) −9 **12.** (A) $\frac{7}{2}$ (B) $-\frac{1}{18}$ **13.** $x \geqslant 3$
14. $-13 < -5$ **15.** 20 is greater than 7 **16.** x is less than or equal to −2 **17.** > **18.** < **19.** 3 **20.** −5 **21.** −11
22. −7 **23.** 0 **24.** Undefined **25.** −3 **26.** 24 **27.** 5 **28.** 4 **29.** 11 **30.** 10 **31.** $42m^7$ **32.** $8x^6y^3$ **33.** $\dfrac{a^{18}}{b^6}$ **34.** $\dfrac{a^2}{4}$

35. $\dfrac{3}{n^2}$ **36.** $\dfrac{8x^6}{27y^9}$ **37.** (A) {1, 2, 3, 4} (B) {2, 3} **38.** Associative **39.** Commutative **40.** Commutative
41. Associative **42.** $6x + 10y$ **43.** $20a + 5b$ **44.** $hm + km$ **45.** $pq + pr$ **46.** $9(r + s)$ **47.** $5(3x + 1)$ **48.** $8(3a + 2b)$

49. $k(m + n)$ **50.** $(-\infty, -8)$; x **51.** $[2, \infty)$; x **52.** $[-5, 5)$; x
53. $(8, 15)$; x **54.** $x \leqslant 5$; x **55.** $x > -3$; x **56.** $-4 \leqslant x \leqslant 3$; x
57. $5 < x \leqslant 15$; x **58.** $x < 8$ **59.** $-6 < x \leqslant 6$ **60.** −72 **61.** 3 **62.** 26 **63.** −3 **64.** 10 **65.** 5 **66.** 50
67. 85 **68.** 10 **69.** 20 **70.** 55 **71.** $6x^7y^5$ **72.** $\dfrac{2x^2}{3y^2}$ **73.** $-8x^9y^3z^6$ **74.** $\dfrac{v^{12}}{u^{12}}$ **75.** (A) 90 (B) 45

76. (A) 28 (B) 5 (C) 4 (D) 10 **77.** (A) $\frac{5}{3}$ (B) $-\dfrac{3}{23}$ **78.** $-5 < x < 3$; x
79. $-1 \leqslant x \leqslant 1$; x **80.** x **81.** x **82.** −4 **83.** 1 **84.** $\dfrac{y^{11}}{2}$ **85.** $-72m^{12}n^{15}$

Practice Test: Chapter 1

1. (A) {2, 4, 5, 6} (B) {5} (C) {8} (D) {2, 4} **2.** (A) F (B) T (C) T (D) T
3. (A) Commutative (B) Associative (C) Commutative (D) Associative **4.** (A) $9(4m + 3n)$ (B) $6(3x + 1)$
5. (A) $60xyz$ (B) $a + b + c + 14$ **6.** (A) −33 (B) 0 (C) −2 **7.** (A) 60 (B) −1

8. (A) $(-\infty, -5]$; x (B) $[-3, 5)$; x

9. (A) $x > 3$; $\xrightarrow{\quad}x$ (B) $-2 < x \leqslant 4$; $\xrightarrow{\quad}x$ **10.** (A) $\xleftrightarrow{\quad}x$ (B) $\xrightarrow{\quad}x$

11. (A) $\dfrac{9y^5}{5x^3}$ (B) $\dfrac{4}{v^2}$ **12.** (A) 850 (B) 350

Chapter 2

Exercise 2-1

1. 1 **3.** 1 **5.** $11u - v$ **7.** $-2m - 5n$ **9.** $-30a^5$ **11.** $15m^3n^5$ **13.** $10x^3 - 6x^2 + 4x$ **15.** $-13a + 19b$ **17.** $12a - 5$
19. $13x^2 - 7x + 7$ **21.** $5x + 1$ **23.** $3t^2 - 3t - 10$ **25.** $-4z^2 + 2z - 9$ **27.** $4x + y$ **29.** $-5x + 6y$ **31.** $-x - 10$
33. $-18m + 20n$ **35.** $a^2 - 7ab + 2b^2$ **37.** $15u^4v - 10u^3v^2 + 20u^2v^3$ **39.** $x^3 - y^3$ **41.** $12x^3 + 5x^2y - 11xy^2 - 6y^3$
43. $6x^4 + 2x^3 - 5x^2 + 4x - 1$ **45.** $3x^4 + 8x^3 - 2x^2 - 5x + 1$ **47.** $6x^3 + 5x^2 + 3x - 13$ **49.** $y^2 + 2y - 35$
51. $10x^2 + 3x - 1$ **53.** $20r^2 - 23r + 6$ **55.** $2x^2 - 7x + 3$ **57.** $9a^2 - 4b^2$ **59.** $4x^2 + 12x + 9$ **61.** $25x^2 - 40xy + 16y^2$
63. $-11x - 19y$ **65.** $67x - 38$ **67.** $-x + 27$ **69.** $-7x^2 - x - 16$ **71.** $4x^3 - 14x^2 + 8x - 6$ **73.** $x^3 + 3x^2y + 3xy^2 + y^3$
75. $x^4 - 4x^3y + 6x^2y^2 - 4xy^3 + y^4$

Exercise 2-2

1. $2a(3a - 4)$ **3.** $7u^2v^2(u - 2v)$ **5.** $(x + 2)(x + 5)$ **7.** $(a - 5)(3a - 2)$ **9.** $(4y - 3)(5y - 1)$ **11.** $3x(2x^2 - 3x + 5)$
13. $2uv(4u^2 + 3u - 7)$ **15.** $3x - 3$ **17.** $x - 4$ **19.** $(x - 1)(5x + 3)$ **21.** $(x - 4)(2x - 1)$ **23.** $(x - 1)(5x + 3)$
25. $(x - 4)(2x - 1)$ **27.** $(2x - y)(2x - 3y)$ **29.** $(2u - 3v)(u - 2v)$ **31.** $(x + 2y)(x + 3y)$ **33.** $(u - 3)(u - 4)$
35. Does not factor **37.** $(u - 2v)(u + 5v)$ **39.** Does not factor **41.** $(x - 3)(x + 3)$ **43.** $(a + 2b)^2$ **45.** $(x - 5y)(x + 4y)$
47. $(2x - 3)(x - 2)$ **49.** $(2x - 3y)(3x - 2y)$ **51.** Does not factor **53.** $(5u + 3)(5u - 2)$ **55.** $(2u + 3v)(2u - v)$
57. Does not factor **59.** $(5w - 2)(5w + 2)$ **61.** $(5u - 3v)^2$ **63.** $x(x - 3)(x + 3)$ **65.** $10w(w - 5)^2$ **67.** $6x(x - 3)(x + 4)$
69. $5uv(u - 2v)(u + 3v)$ **71.** $3m^2n(m - n)(m + 3n)$ **73.** $(x - 2)(x^2 + 2x + 4)$ **75.** $(x + 3)(x^2 - 3x + 9)$

Exercise 2-3

1. $\dfrac{3x}{4y^3}$ **3.** $\dfrac{x + 3}{x^2}$ **5.** $\dfrac{5(2u - 1)}{6u^4}$ **7.** $\dfrac{2}{x - 3}$ **9.** $35x^3y$ **11.** $3x^2y^3$ **13.** $\dfrac{u}{v}$ **15.** $\dfrac{u^2}{vw}$ **17.** $\dfrac{25x^2}{3y}$ **19.** $\dfrac{5x}{3z}$ **21.** $\dfrac{x}{wy}$ **23.** $-\dfrac{15a^2}{8b^2}$

25. $\dfrac{2x + 1}{2x}$ **27.** $\dfrac{1}{x + 3}$ **29.** v **31.** $\dfrac{x + 5y}{x - 5y}$ **33.** $\dfrac{x - 3}{3x^2}$ **35.** $\dfrac{3(x - 1)}{2x}$ **37.** $3x(x - 3)$ or $3x^2 - 9x$

39. $(x + 2y)^2$ or $x^2 + 4xy + 4y^2$ **41.** $9xz(x - 2)$ **43.** $\dfrac{b}{3}$ **45.** $\dfrac{2z^2}{3}$ **47.** $\dfrac{3(x - 4)}{y^2}$ **49.** $\dfrac{4x}{x - 5}$ **51.** $\dfrac{x - 4}{4}$ **53.** $\dfrac{1}{a}$ **55.** $\dfrac{y + 2}{y + 4}$

57. $\dfrac{1}{6z(z - 1)}$ **59.** $\dfrac{2x - y}{3x + y}$ **61.** $-\dfrac{10a}{3b^3c}$ **63.** $\dfrac{2}{x^2y}$ **65.** $\dfrac{2m^2(m - 4n)}{3n(m + 5n)}$ **67.** $\dfrac{x - y}{2x + y}$ **69.** $\dfrac{x + y}{x - y}$ **71.** $\dfrac{(x - y)^2}{y^2(x + y)}$

Exercise 2-4

1. $\dfrac{m}{2pq}$ **3.** $\dfrac{x + 2}{4y}$ **5.** 2 **7.** $\dfrac{1}{z + 5}$ **9.** $12cd$ **11.** $18x^2$ **13.** $(x - 1)(x + 1)$ **15.** $3y(y + 2)$ **17.** $\dfrac{3u - 2w}{9v}$ **19.** $\dfrac{7y - 2}{y}$

21. $\dfrac{8ad - 3bc}{12dc}$ **23.** $\dfrac{20 + 48y - 27y^2}{36y^2}$ **25.** $\dfrac{8z - 6}{(z + 3)(z - 2)}$ **27.** $\dfrac{17u + 15}{5u(u - 5)}$ **29.** $24m^2n^3$ **31.** $3(n - 1)(n - 2)$

33. $(x + 3)(x - 3)^2$ **35.** $4m^2(m - 2)^2$ **37.** $\dfrac{96u^2v^2 - 20u^3 + 27v^3}{36u^3v^3}$ **39.** $\dfrac{15s^2 + 2r + 24s}{12r^2s^3}$ **41.** $\dfrac{-3x + 16}{5(x - 2)(x - 3)}$

43. $\dfrac{8x + 4}{(x - 2)^2(x + 2)}$ **45.** $\dfrac{5m^2 + 16m - 20}{4m^2(m - 2)^2}$ **47.** $\dfrac{3x^2 + 6x - 6}{x + 3}$ **49.** $\dfrac{16}{(x - 4)(x + 4)}$ **51.** $\dfrac{-x + 3}{(2x - 1)(2x + 1)}$

53. $\dfrac{3x^2 + y^2}{(x - y)^2(x + y)^2}$ **55.** $\dfrac{3x^2 - 2x - 14}{(x - 3)(x + 4)(x - 2)}$ **57.** $\dfrac{3}{(x - 2)(x - 3)(x + 2)}$

Exercise 2-5   Chapter Review

1. $7x^2 - x - 4$ **2.** $-3x^2 - 11$ **3.** $3r + 4s$ **4.** $-4u + 11$ **5.** $-15x^3y^8$ **6.** $2t^3 - 4t^2 + 6t$ **7.** $7uv(2u^2 - v)$

8. $6x^2(2x^2 - 4x + 1)$ **9.** $7u^2v(u^2 - 3uv + 5v^2)$ **10.** $(3x - 5)(2x + 3)$ **11.** $(2a + 3)(5a - 1)$ **12.** $\dfrac{3b^5}{2a^2}$ **13.** $\dfrac{3y - 2}{4}$ **14.** $\dfrac{c^2}{b^3}$

15. $\dfrac{10u^2}{9w}$ **16.** $\dfrac{15z^2 - 14z + 27}{18z^2}$ **17.** $\dfrac{17x + 9}{4x(x - 3)}$ **18.** $8a^2 + 9ab + 12b^2$ **19.** $12r^4s - 9r^3s^2 + 21r^2s^3$

20. $6x^3 - 13x^2 + 8x - 3$ **21.** $2y^3 - 9y^2 + 11y - 3$ **22.** $2x^4 - 5x^3 + 5x^2 + 11x - 10$ **23.** $21t - 24$ **24.** $28a - 16b$

25. $24u - 10v$ **26.** $5x^2 - 16x + 3$ **27.** $8u^2 + 14uv + 3v^2$ **28.** $3r^2 + 4rs - 15s^2$ **29.** $49x^2 - 9y^2$

30. $9m^2 + 24mn + 16n^2$ **31.** $4a^2 - 4ab + b^2$ **32.** $5x^3 + 6x^2 - 9x - 10$ **33.** $-2x^3 + 7x^2 + 4x - 13$ **34.** $(z - 2)(5z + 2)$

35. $(3w + 2)(2w - 3)$ **36.** $(2u - v)(3u - v)$ **37.** $(a - 2b)(2a + b)$ **38.** $(x - 3)(x + 4)$ **39.** Does not factor

40. $(a - 4b)(a + 6b)$ **41.** $(2x - 3)(2x - 5)$ **42.** $(5a - b)(a + 7b)$ **43.** $(5x - y)^2$ **44.** $w^2(w - 9)(w + 9)$ **45.** $u^2(u - 5)^2$

46. $5y(y - 3)(y + 4)$ **47.** $7mn(m - 2n)(m + 4n)$ **48.** $\dfrac{z}{z + 4}$ **49.** $\dfrac{x + 4}{4x}$ **50.** $\dfrac{4a - b}{4a + b}$ **51.** $\dfrac{4(y + 2)}{3y}$ **52.** $\dfrac{a}{4}$ **53.** $\dfrac{2}{x^2}$

54. $\dfrac{x + 3}{7}$ **55.** $\dfrac{u - 4}{5}$ **56.** $\dfrac{-3y + 5}{(y + 1)(y - 3)}$ **57.** $\dfrac{4z - 6}{5(z + 2)(z - 4)}$ **58.** $\dfrac{2u - 4v}{(u - v)(u + v)^2}$ **59.** $\dfrac{6x^2 - 15x + 3}{2x - 3}$

60. $-x^2 + 17x - 11$ **61.** $-5x + 15$ **62.** $3m^2n(m - n)(m + 3n)$ **63.** $(2x - 3y)^2(2x + 3y)^2$ **64.** $\dfrac{x(x + 4y)}{2(x - 5y)}$ **65.** $\dfrac{x - 2}{x + 3}$

66. $\dfrac{a - 2b}{a - 4b}$ **67.** $\dfrac{u - 5}{u + 4}$ **68.** $\dfrac{x^2 + 2x + 5}{(x + 2)(x + 1)(x - 3)}$ **69.** $\dfrac{5x + 12}{(x - 2)(x - 3)(x + 3)}$

Practice Test: Chapter 2

1. $4x^3 + 5x^2 + 4x - 1$ **2.** $x^2 - 4xy - y^2$ **3.** $6u^3 - 13u^2v + 8uv^2 - 3v^3$ **4.** $(4x - 3y)(x + 4y)$ **5.** $5m^2(2m + 1)(m - 3)$

6. $(3a - b)(2a - 3)$ **7.** $\dfrac{x}{x - 3}$ **8.** $\dfrac{2u + 3v}{2u - 3v}$ **9.** $10x + 8y$ **10.** $\dfrac{2}{3b}$ **11.** $\dfrac{9x^2}{x + 4}$ **12.** $\dfrac{2x^2 - 5xy - 6y^2}{x + 2y}$

13. $\dfrac{10m - 10}{4m^2(m - 2)(m + 5)}$

Chapter 3

Exercise 3-1

1. 1 **3.** $\frac{1}{8}$ **5.** 27 **7.** $\frac{8}{27}$ **9.** $\dfrac{1}{u^9}$ **11.** z^3 **13.** 10^5 **15.** w^4 **17.** 1 **19.** 10^9 **21.** $\dfrac{1}{w^4}$ **23.** $\dfrac{1}{10^8}$ **25.** x^8 **27.** $\dfrac{1}{w^{12}}$ **29.** $\dfrac{3^4}{2^5}$

31. $\dfrac{v^6}{u^{12}}$ **33.** 1 **35.** 10^{22} **37.** a^{16} **39.** $\dfrac{n^6}{25m^4}$ **41.** $\dfrac{36u^4}{v^2}$ **43.** $\dfrac{3x^2}{7y}$ **45.** $\dfrac{4n^5}{5m^9}$ **47.** $\dfrac{1}{y^3}$ **49.** $\dfrac{x^6}{y^{21}}$ **51.** $\dfrac{m^8}{4n^{10}}$ **53.** $\dfrac{x^5}{8y^4}$

55. $\dfrac{1+x^6}{x^3}$ **57.** $\dfrac{1}{r^2-s^2}$ **59.** $\dfrac{3}{10}$ **61.** $\dfrac{1}{xy}$ **63.** $\dfrac{y^{18}z^{12}}{x^{24}}$ **65.** $\dfrac{9y^6}{x^6}$ **67.** $\dfrac{a^2b^2}{a^2+b^2}$ **69.** $\dfrac{10^4}{11}$ or $\dfrac{10,000}{11}$ **71.** 4
73. \$384,000; \$1,944,000; \$6,561,000 **75.** 5,120; 8,000; 15,625 **77.** 250,000; 160,000; 102,400

Exercise 3-2

1. 7.6×10^1 **3.** 8.6×10^4 **5.** 9.4×10^{-2} **7.** 2.9×10^{-7} **9.** 5.29×10^{10} **11.** 6.84×10^{-11} **13.** 3,700 **15.** 80
17. 0.000 8 **19.** 0.082 **21.** 2,800,000,000 **23.** 0.000 000 000 000 64 **25.** 9.29×10^7 **27.** 7×10^{10} **29.** 8.49×10^{-10}
31. 3,670,000,000 **33.** \$2,630,000,000,000 **35.** 0.000 000 000 000 000 000 000 000 000 91 **37.** 2.94×10^7
39. 3×10^2 **41.** 8×10^{-3} **43.** 2.79×10^{-5} **45.** 1.68×10^2; 168 **47.** 2×10^{-9}; 0.000 000 002 **49.** 6×10^4; 60,000
51. 6×10^{-5}; 0.000 06 **53.** $\dfrac{1.82\times10^{11}}{2.21\times10^8}=8.24\times10^2$ or \$824 per person **55.** 9×10^4 or 90,000 barrels
57. 1.5×10^{10} or 15,000,000,000 **59.** $(1.5\times10^3)(4.2\times10^7)=\$6.3\times10^{10}$ or \$63,000,000,000

Exercise 3-3

1. $10,-10$ **3.** $10,-10$ **5.** None **7.** 7 **9.** Not real **11.** -9 **13.** 3 **15.** -4 **17.** 27 **19.** 4 **21.** x **23.** $u^{2/5}$ **25.** a^4
27. rs^3 **29.** $\dfrac{u^3}{v^4}$ **31.** x^4y^6 **33.** $\frac{27}{64}$ **35.** $\frac{1}{7}$ **37.** $\frac{1}{64}$ **39.** $\frac{1}{9}$ **41.** $30x^{7/9}$ **43.** $\dfrac{64b}{a^4}$ **45.** $a^{1/10}$ **47.** $m^{3/2}$ **49.** $\dfrac{y^8}{x^{10}}$ **51.** $\dfrac{2u^2}{v^3}$
53. $\dfrac{v^3}{2u^2}$ **55.** $\dfrac{2}{x^{2/5}}$ **57.** $\dfrac{4b}{3a^3}$ **59.** $15x-20x^{17/5}$ **61.** $m-n$ **63.** $u+2u^{1/2}v^{1/2}+v$ **65.** $a^{1/2}b^{1/2}-\dfrac{1}{a^{1/2}b^{1/2}}$ **67.** $x-\dfrac{1}{y}$
69. $a+2+\dfrac{1}{a}$ **71.** \$512,000; \$800,000; \$1,562,500 **73.** 20; 640; 10,240 **75.** 1,920,000; 4,320,000; 14,580,000

Exercise 3-4

1. $\sqrt{15}$ **3.** $\sqrt[7]{x^3}$ **5.** $8\sqrt[3]{m^2}$ **7.** $\sqrt[5]{(5y)^4}$ **9.** $\sqrt[3]{(6a^3b^2)^2}$ **11.** $\sqrt{x-y}$ **13.** $x^{1/7}$ **15.** $z^{3/5}$ **17.** $(3x^2y^3)^{2/3}$ **19.** $(a^2+b^2)^{1/2}$ **21.** $7z$
23. $12u^2v^3$ **25.** $3\sqrt{3}$ **27.** $5x\sqrt{2x}$ **29.** $4u^2v^3\sqrt{3v}$ **31.** $\dfrac{2\sqrt{a}}{a}$ **33.** $\dfrac{\sqrt{6xy}}{3y}$ **35.** $8b^2\sqrt{3ab}$ **37.** $\sqrt[3]{x^2}$ **39.** $\sqrt[8]{x^3}$ **41.** $\dfrac{1}{\sqrt[7]{y^4}}$
43. $\dfrac{1}{\sqrt[4]{(3r^2s^3)^3}}$ **45.** $\sqrt{x}+\sqrt[3]{x}$ **47.** $\dfrac{1}{\sqrt{a}}+\dfrac{1}{\sqrt{b}}$ **49.** $-3(3x^3y)^{1/4}$ **51.** $(u^2-v^2)^{3/4}$ **53.** $5y^{-2/5}$ **55.** $xy^{-1/2}-yx^{-1/2}$ **57.** $2u^2v^3$
59. $2m^2n^3\sqrt[3]{3m}$ **61.** $2x^2y^3z\sqrt[5]{2y^2z^4}$ **63.** $4ab^2\sqrt[3]{a^2}$ **65.** $\dfrac{2a^3}{3b}$ **67.** $\dfrac{\sqrt[3]{9x}}{3x}$ **69.** $3x\sqrt[3]{5x^2y}$ **71.** $2\sqrt{a^2+b^2}$ **73.** $\dfrac{\sqrt{x^2-y^2}}{x+y}$
75. $\sqrt[3]{(a-b)^2}$ **77.** (A) $7x$ (B) x **79.** (A) $-x$ (B) $9x$

Exercise 3-5

1. $-5\sqrt{x}$ **3.** $-4\sqrt{7}+3\sqrt{3}$ **5.** $\sqrt{3}$ **7.** $5\sqrt{2}$ **9.** $5-3\sqrt{5}$ **11.** $u+3\sqrt{u}$ **13.** $7\sqrt{x}-x$ **15.** $10\sqrt{3}-3\sqrt{10}$ **17.** -4
19. $27+10\sqrt{2}$ **21.** $w-4$ **23.** $\sqrt{5}+2$ **25.** $\dfrac{4-\sqrt{6}}{2}$ **27.** $\dfrac{3\sqrt{2}-2\sqrt{3}}{2}$ **29.** $\dfrac{z+5\sqrt{z}}{z-25}$ **31.** $3\sqrt{2x}$ **33.** $9\sqrt{2}$ **35.** $10x\sqrt{2x}$
37. $2z\sqrt[3]{3}$ **39.** $\dfrac{10\sqrt{3}}{3}$ **41.** $\dfrac{9\sqrt[5]{5}}{5}$ **43.** $\dfrac{10\sqrt{6mn}}{3}$ **45.** $x+2\sqrt{x}-15$ **47.** $14+5\sqrt{3}$ **49.** $6a-\sqrt{a}-15$ **51.** $-6+11\sqrt{6}$
53. $m-n$ **55.** $m+\sqrt[3]{m^2n^2}-\sqrt[3]{mn}-n$ **57.** 0 **59.** $\dfrac{-7+3\sqrt{5}}{2}$ **61.** $5-2\sqrt{6}$ **63.** $\dfrac{x+4\sqrt{x}+4}{x-4}$ **65.** $\dfrac{4-\sqrt{6}}{5}$ **67.** $\dfrac{2\sqrt{6}}{3}$
69. $\dfrac{13xy\sqrt{2xy}}{12}$ **71.** $x-3\sqrt[5]{x^3y^3}+2\sqrt[5]{x^2y^2}-6y$ **73.** $\dfrac{20u-27\sqrt{uv}+9v}{16u-9v}$

Exercise 3-6   Chapter Review

1. 1 **2.** $\frac{25}{36}$ **3.** 125 **4.** $-\frac{1}{2}$ **5.** Not real **6.** 81 **7.** 1 **8.** $\frac{1}{m^3}$ **9.** r^8 **10.** $\frac{1}{u^2}$ **11.** $\frac{x^6}{y^4}$ **12.** $\frac{1}{a^{1/7}}$ **13.** $\frac{x^2}{y^3}$ **14.** u^5v^6

15. (A) 5.3×10^{10} (B) 4.9×10^{-6} **16.** (A) 38,000,000 (B) 0.000 057 **17.** (A) $\sqrt[6]{(7z)^5}$ (B) $4\sqrt[4]{w^3}$

18. (A) $(2x^2y)^{3/5}$ (B) $(m^2 - n^2)^{1/2}$ **19.** $10xy^3$ **20.** $6x^2\sqrt{2x}$ **21.** $4x^2y^4\sqrt{2x}$ **22.** $\frac{\sqrt{14xy}}{7y}$ **23.** $\frac{\sqrt{15uv}}{5v}$ **24.** $4a^2\sqrt{7ab}$

25. $\sqrt[5]{x^2}$ **26.** $\sqrt[10]{y^3}$ **27.** $-3\sqrt{5} - 2\sqrt{3}$ **28.** $\sqrt{3}$ **29.** $\sqrt{35} - 3\sqrt{7}$ **30.** $3\sqrt{2} - \sqrt{6}$ **31.** $9 + 4\sqrt{5}$ **32.** $3\sqrt{11} + 9$ **33.** $\frac{3\sqrt{2} - 2\sqrt{3}}{2}$

34. $\frac{1}{8}$ **35.** 2 **36.** $\frac{1}{16}$ **37.** $\frac{a^8}{49b^6}$ **38.** $\frac{27u^6}{v^{12}}$ **39.** $\frac{1}{10^6}$ **40.** $\frac{5y^5}{7x^3}$ **41.** $\frac{m^{12}}{8n^9}$ **42.** $\frac{1}{125v^7}$ **43.** $\frac{v^7 - 1}{v^3}$ **44.** $\frac{1}{ab}$ **45.** $\frac{256v^5}{u^6}$

46. $\frac{3m^2}{n^3}$ **47.** $t^{1/12}$ **48.** $\frac{1}{a^{1/30}}$ **49.** $\frac{2}{x^{2/9}}$ **50.** $\frac{u^4}{2v^3}$ **51.** $15x - 6$ **52.** $3x - 10x^{1/2}y^{1/2} + 3y$

53. (A) 5.24×10^8 (B) 5.83×10^{-4} (C) 8.32×10^8 (D) 5.29×10^{-3} **54.** 2×10^{-3}; 0.002 **55.** (A) $-5\sqrt[3]{y^2}$ (B) $\sqrt[3]{a} - \frac{1}{\sqrt[3]{a}}$

56. (A) $-6x(2xy^2)^{3/4}$ (B) $3w^{-5/6}$ **57.** $5x^3y^2$ **58.** $2xy^4\sqrt[3]{4x^2}$ **59.** $2xy\sqrt[4]{2y}$ **60.** $\frac{5u^4}{3v^3}$ **61.** $\frac{\sqrt[3]{4x}}{2x}$ **62.** $3m\sqrt[3]{25mn^2}$ **63.** $11\sqrt{5}$

64. $7x\sqrt{3x}$ **65.** $\frac{9\sqrt{14}}{14}$ **66.** $\frac{3\sqrt{2uv}}{2}$ **67.** $7z\sqrt[3]{2z}$ **68.** $6 - \sqrt{10}$ **69.** $10x + 13\sqrt{xy} - 3y$ **70.** $x - \sqrt[3]{x^2y^2} + \sqrt[3]{xy} - y$

71. $\frac{13 - 5\sqrt{7}}{3}$ **72.** $\frac{6x + 2\sqrt{xy}}{9x - y}$ **73.** $\frac{6 - \sqrt{6}}{6}$ **74.** 0 **75.** $\frac{uv}{u + v}$ **76.** $\frac{2}{w} - 3w - 5$ **77.** 6 **78.** $2\sqrt{4x^2 + 1}$ **79.** $\frac{\sqrt{4x^2 - 1}}{2x - 1}$

80. $4mn\sqrt[4]{3mn^2}$ **81.** $\frac{37\sqrt{35}}{35}$ **82.** $\frac{6a - 5\sqrt{ab} - 6b}{9a - 4b}$ **83.** (A) $2x$ (B) $-8x$

Practice Test: Chapter 3

1. $\frac{25y^6}{36x^4}$ **2.** $\frac{3a^3}{b^2}$ **3.** $\frac{(p + q)^2}{pq}$ **4.** 3×10^{-6} **5.** (A) $-5y(3x^2y)^{3/7}$ (B) $\frac{3}{\sqrt[3]{(x - y)^2}}$ **6.** $5xy^4z^5\sqrt[3]{2x^2z}$ **7.** $\frac{m\sqrt{15n}}{5n}$ **8.** $5b\sqrt[3]{5a^2b}$

9. $11a\sqrt[3]{5a^2}$ **10.** $-2\sqrt[3]{2}$ **11.** $-\frac{\sqrt{6}}{2}$ **12.** $\frac{3}{x} - 4x - 11$

Chapter 4

Exercise 4-1

1. 3 **3.** 0 **5.** 2 **7.** No solution **9.** $\frac{85}{2}$ **11.** All real numbers **13.** -6 **15.** 24 **17.** 76 **19.** 70 **21.** 30 **23.** 4 **25.** 3
27. $-\frac{5}{4}$ **29.** -1 **31.** $\frac{32}{5}$ **33.** No solution **35.** All real numbers **37.** $-\frac{1}{2}$ **39.** -3 **41.** 25,000 **43.** 120 **45.** 5
47. No solution **49.** -5 **51.** 3 **53.** -2 **55.** -2 **57.** $\frac{8}{5}$ **59.** $6,500 at 11%; $3,500 at 18%
61. First painting $6,000; second painting $9,000 **63.** $7,200 at 10%; $4,800 at 15%
65. (A) 800 pounds (B) 1,400 pounds **67.** 24,000 gallons **69.** 5,000 trout **71.** 12.6 years

Exercise 4-2

1. $x \geq 3$ **3.** $x \geq -2$ **5.** $x < -20$ **7.** $x \leq 21$ **9.** $x \leq -5$ **11.** $3 \leq x \leq 11$ **13.** $-4 \leq x < 7$ **15.** $6 \leq x < 15$

17. $-4 < x < 3$ **19.** $-15 < x \leq 12$ **21.** $-5 \leq x < 6$ **23.** $x \leq 6$;   **25.** $y \leq -9$;

27. $-3 < x \leq 4$;   **29.** $-2 \leq x \leq 3$;   **31.** $x < -5$;   **33.** $u < 3$;

35. $x \leq 150$;   **37.** $-2 < x \leq 4$;   **39.** $131 \leq F \leq 185$;

41. $50 \leq C \leq 95$;   **43.** $x \geq \frac{3}{2}$;   **45.** $-8 \leq x \leq 8$;   **47.** $x > \frac{31}{24}$;

49. \$7,500 or more   **51.** (A) More than 225 (B) At least 275   **53.** At least 3 pounds   **55.** $9 \leq MA \leq 18$

Exercise 4-3

1. ± 6 **3.** No solution   **5.** $\pm 2\sqrt{2}$ **7.** $\pm \frac{4}{3}$ **9.** $0, 7$ **11.** $-2, 4$ **13.** Cannot be factored using integers   **15.** $0, \frac{4}{5}$

17. $\frac{3}{2}, -1$ **19.** $-3 \pm 2\sqrt{3}$ **21.** $\dfrac{3 \pm \sqrt{15}}{2}$ **23.** $-5 \pm \sqrt{33}$ **25.** $\dfrac{5 \pm \sqrt{73}}{6}$ **27.** $-3, -4$ **29.** $\pm \dfrac{\sqrt{5}}{2}$ **31.** $\pm \dfrac{\sqrt{13}}{4}$ **33.** $-1, 11$

35. $\frac{2}{3}, \frac{3}{2}$ **37.** $1, -\frac{1}{2}$ **39.** $\frac{3}{2}, -1$ **41.** $\dfrac{1 \pm \sqrt{7}}{3}$ **43.** $\dfrac{-1 \pm \sqrt{5}}{2}$ **45.** No solution   **47.** ± 4 **49.** $\dfrac{5 \pm \sqrt{37}}{2}$ **51.** $5, -7$

53. $\dfrac{-5 \pm \sqrt{57}}{4}$ **55.** $-\frac{3}{2}, 5$ **57.** No solution   **59.** $\dfrac{-6 \pm \sqrt{15}}{7}$ **61.** $2a, 3a$ **63.** $\dfrac{-b \pm \sqrt{b^2 - 4ac}}{2a}$ **65.** $-m \pm \sqrt{m^2 - n}$

67. $3{,}000$ **69.** 50 or 150   **71.** $P = -4x^2 + 800x - 30{,}000$; 60 or 140   **73.** \$55   **75.** 40

77. 2 hours after the drug is taken   **79.** \$100

Exercise 4-4

1. $-3 < x < 4$; $(-3, 4)$

3. $x \leq -3$ or $x \geq 4$; $(-\infty, -3] \cup [4, \infty)$

5. $-5 < x < 2$; $(-5, 2)$

7. $x < 3$ or $x > 7$; $(-\infty, 3) \cup (7, \infty)$

9. $0 \leq x \leq 8$; $[0, 8]$

11. $-5 \leq x \leq 0$; $[-5, 0]$

13. $x < -2$ or $x > 2$; $(-\infty, -2) \cup (2, \infty)$

15. True for all real numbers. The graph is the whole real number line.

17. $-1 \leq x \leq 1$ or $x \geq 5$; $[-1, 1] \cup [5, \infty)$

19. $x < -5$ or $3 < x < 5$; $(-\infty, -5) \cup (3, 5)$

21. $-4 < x \leq 2$; $(-4, 2]$

23. $-5 \leq x \leq 0$ or $x > 3$; $[-5, 0] \cup (3, \infty)$

25. $x \leq -4$ or $x \geq 1$; $(-\infty, -4] \cup (1, \infty)$

27. $x < 0$ or $x > \frac{1}{4}$; $(-\infty, 0) \cup (\frac{1}{4}, \infty)$

29. $x < -3$ or $x \geq 3$; $(-\infty, -3) \cup [3, \infty)$

31. $-4 < x \leq \frac{3}{2}$; $(-4, \frac{3}{2}]$

33. $-1 < x < 2$ or $x \geq 5$; $(-1, 2) \cup [5, \infty)$

35. No solution

37. $x > 4$; $(4, \infty)$

39. $-2 \leq x \leq -\frac{1}{2}$ or $\frac{1}{2} \leq x \leq 2$; $[-2, -\frac{1}{2}] \cup [\frac{1}{2}, 2]$

41. (A) Profit: $\$4 < p < \$7$ (B) Loss: $\$0 \leq p < \$4$ or $p > \$7$ (C) Break-even: $p = \$4$ or $p = \$7$

43. $25 \leq p \leq 35$ **45.** $t < 1$ or $t > 9$

Exercise 4-5

1. $r = \dfrac{d}{t}$ **3.** $r = \dfrac{C}{2\pi}$ **5.** $d = \dfrac{C}{\pi}$ **7.** $b = \dfrac{V}{ac}$ **9.** $x = -\dfrac{b}{a}$ **11.** $A = \dfrac{3V}{b}$ **13.** $a = \dfrac{b}{m}$ **15.** $l = \dfrac{P - 2w}{2}$ **17.** $x = \dfrac{y - b}{m}$

19. $y = \dfrac{3x + 15}{5}$ **21.** $y = \dfrac{-Ax - C}{B}$ **23.** $W = \dfrac{CL}{100}$ **25.** $h = \dfrac{2A}{a + b}$ **27.** $d = \dfrac{bc}{a}$ **29.** $x = \dfrac{d - b}{a - c}$ **31.** $P = \dfrac{kT}{V}$

33. $F = \frac{9}{5}C + 32$ **35.** $R = \dfrac{R_1 R_2}{R_1 + R_2}$ **37.** $P_2 = \dfrac{P_1 V_1 T_2}{T_1 V_2}$ **39.** $d = \dfrac{a_n - a_1}{n - 1}$ **41.** $r = \sqrt{\dfrac{A}{\pi}}$ **43.** $i = \sqrt{\dfrac{A}{P}} - 1$ **45.** $x = \dfrac{y + 3}{2y - 4}$

47. $x = \dfrac{-m \pm \sqrt{m^2 - 4n}}{2}$ **49.** $M = \dfrac{P}{1 - dt}$ **51.** $d = 33\left(\dfrac{P}{15} - 1\right)$

Exercise 4-6   Chapter Review

1. $-\frac{5}{6}$ **2.** All real numbers **3.** No solution **4.** 14 **5.** 800 **6.** 42 **7.** 12 **8.** $x < -4$;

9. $2 \leqslant x \leqslant 12$; **10.** $-2 < x \leqslant 3$; **11.** $-5 \leqslant x \leqslant 10$; **12.** $\pm \frac{7}{4}$

13. No real solution **14.** $0, -9$ **15.** $-5, 7$ **16.** 5 **17.** $0, -\frac{7}{10}$ **18.** $\frac{1}{3}, -\frac{1}{2}$ **19.** $2 \pm 2\sqrt{3}$ **20.** $\dfrac{2 \pm \sqrt{22}}{2}$ **21.** $2 \pm \sqrt{2}$

22. $\dfrac{3 \pm 2\sqrt{6}}{3}$ **23.** $-3 < x < 2$; **24.** $x < -3$ or $x > 4$;

25. $x \leqslant -3$ or $x \geqslant 0$; **26.** $-2 \leqslant x \leqslant 7$; **27.** $p = \dfrac{k}{v}$ **28.** $R = \dfrac{W}{I^2}$ **29.** -3 **30.** 5

31. 15,000 **32.** All real numbers **33.** No solution **34.** 8 **35.** $m < 3$; **36.** $-3 \leqslant x < 4$;

37. $-2 < x < 3$; **38.** $-3 \leqslant x \leqslant 3$; **39.** $x > 4$; **40.** $z < 4$;

41. $x \leqslant 2,000$; **42.** $-30 \leqslant C \leqslant 30$; **43.** $\dfrac{-1 \pm \sqrt{7}}{4}$ **44.** $\frac{1}{7}, -3$ **45.** ± 3 **46.** $\dfrac{-5 \pm \sqrt{29}}{2}$

47. $-2, 4$ **48.** $\dfrac{7 \pm \sqrt{65}}{2}$ **49.** No real solution **50.** $\dfrac{-2 \pm \sqrt{19}}{5}$ **51.** $x \leqslant -7$ or $x \geqslant 7$;

52. $\frac{1}{2} < x < 5$; **53.** $-2 < x < 5$; **54.** $-3 < x < 0$ or $x > 3$;

55. $y = \dfrac{5x + 30}{3}$ **56.** $a = \dfrac{2s}{t^2}$ **57.** 4 **58.** $-45 \leqslant C \leqslant 65$ **59.** $\dfrac{-7 \pm \sqrt{15}}{3}$ **60.** $1 < x < 3$ **61.** $t = \sqrt{\dfrac{2s}{a}}$ **62.** $-3a, 4a$

63. \$10,000 at 15%; \$50,000 at 9% **64.** \$40,000 or more **65.** $40 < x < 100$ **66.** (A) 50 or 90 (B) $60 \leqslant x \leqslant 80$
67. $18 \leqslant r \leqslant 23$ **68.** 1 and 3 days **69.** $12 \leqslant MA \leqslant 16$

Practice Test: Chapter 4

1. 1,500 **2.** -3 **3.** $x > -12$; **4.** $-2 \leqslant x < 8$; **5.** $\frac{1}{3}, 2$ **6.** $\dfrac{1 \pm \sqrt{13}}{3}$ **7.** $\dfrac{5 \pm \sqrt{15}}{2}$

8. $x \leqslant -4$ or $x \geqslant 5$; **9.** $-3 < x \leqslant 5$; **10.** $B = \dfrac{Ap}{1 + q}$

11. \$18,000 at 10%; \$12,000 at 15% **12.** $60 \leqslant x \leqslant 90$

Chapter 5

Exercise 5-1

1.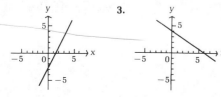

3.

5. Slope $= 2$; y intercept $= -3$
7. Slope $= -\frac{2}{3}$; y intercept $= 2$
9. $y = -2x + 4$
11. $y = -\frac{3}{5}x + 3$

13.

15.

17.

19. $y = -3x + 5$, $m = -3$
21. $y = -\frac{2}{3}x + 4$, $m = -\frac{2}{3}$
23. $y + 1 = -3(x - 4)$, $y = -3x + 11$
25. $y + 5 = \frac{2}{3}(x + 6)$, $y = \frac{2}{3}x - 1$
27. $\frac{1}{3}$ **29.** $-\frac{1}{5}$

31. $(y - 3) = \frac{1}{3}(x - 1)$, $x - 3y = -8$ **33.** $(y + 2) = -\frac{1}{5}(x + 5)$, $x + 5y = -15$ **35.** $x = 3$, $y = -5$ **37.** $x = -1$, $y = -3$
39. $y = -\frac{1}{2}x + 4$ **41.** $y = -\frac{1}{2}x + 1$ **43.** $y = \frac{1}{2}x$

45.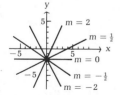

47. $x = 2$ **49.** $y = 3$ **51.** (A) 130; 220 (B) 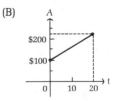 (C) 6

53. (A) 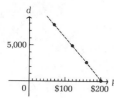 (B) $d = -60p + 12{,}000$ **55.** $0.2x + 0.1y = 20$

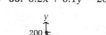

57. (A) 64 grams; 35 grams (B)  (C) $-\frac{1}{5}$

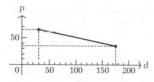

Exercise 5-2

1. Function **3.** Not a function **5.** Function **7.** Function **9.** Not a function **11.** Function **13.** Function $(x \neq 1)$
15. Function **17.** Not a function; when $x = 4$, $y = \pm 2$ **19.** Not a function; when $x = 0$, $y = 0, 1$ **21.** Function
23. Function **25.** 4 **27.** -5 **29.** -6 **31.** -2 **33.** -12 **35.** -1 **37.** -6 **39.** 12 **41.** $\frac{3}{4}$
43. Domain $= \{1, 2, 3\}$; range $= \{1, 2, 3\}$; not a function
45. Domain $= \{-1, 0, 1, 2, 3, 4\}$; range $= \{-2, -1, 0, 1, 2\}$; function

47. Domain = {0, 1, 2, 3}; range = {0, 2, 4, 6}; function   **49.** Domain = {0, 1, 4}; range = {−2, −1, 0, 1, 2}; not a function

51. 13   **53.** −3   **55.** 5   **57.** $\sqrt{2}$ **59.** $e^2 - e$ **61.** $\sqrt{u}$ **63.** $(2 + h)^2 - (2 + h) = h^2 + 3h + 2$ **65.** $2(a + h) + 1 = 2a + 2h + 1$

67. $\dfrac{[2(2 + h) + 1] - [2(2) + 1]}{h} = 2$ **69.** $\dfrac{[(2 + h)^2 - (2 + h)] - [2^2 - 2]}{h} = 3 + h$ **71.** All nonnegative real numbers

73. All real numbers x except x = −3, 5   **75.** All real numbers x such that x ⩾ 1

77. All real numbers x except x = −2, 3   **79.** (A)   1   (B)   0   (C)   2   (D)   6   **81.** C(x) = 4x

83. (A) $V(x) = x(8 - 2x)(12 - 2x)$ (B)   Domain = 0 < x < 4 = (0, 4)   **85.** $C(F) = \tfrac{5}{9}(F - 32)$ **87.** IQ = 100(MA/12)

(C)

x	V(x)
1	60
2	64
3	36

Exercise 5-3

1.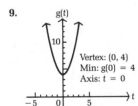
Slope: −2
y intercept: 4

3.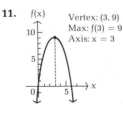
Slope: −$\frac{2}{3}$
y intercept: 4

5.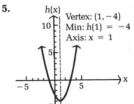
Vertex: (1, −4)
Min: h(1) = −4
Axis: x = 1

7.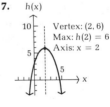
Vertex: (2, 6)
Max: h(2) = 6
Axis: x = 2

9.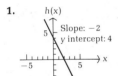
Vertex: (0, 4)
Min: g(0) = 4
Axis: t = 0

11.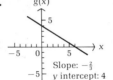
Vertex: (3, 9)
Max: f(3) = 9
Axis: x = 3

13.
Vertex: (4, 4)
Max: f(4) = 4
Axis: x = 4

15.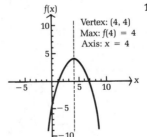
Vertex: (−2.5, 8.25)
Max: h(−2.5) = 8.25
Axis: x = −2.5

17.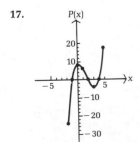

19.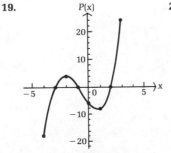

21. (A) f   (B) p   (C) g   **23.** (A) f   (B) g   (C) h

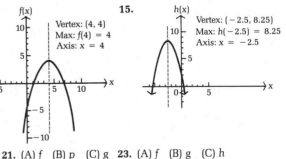

25.

Even

27.

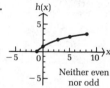

Neither even
nor odd

29.

Even

31.

Odd

33.

Neither even
nor odd

35.

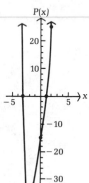

37.

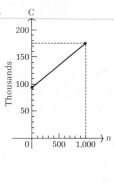

39.

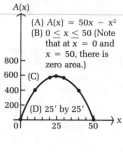

(A) $A(x) = 50x - x^2$
(B) $0 \leq x \leq 50$ (Note
that at $x = 0$ and
$x = 50$, there is
zero area.)
(C)
(D) 25' by 25'

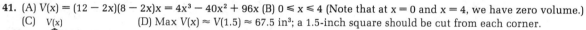

41. (A) $V(x) = (12 - 2x)(8 - 2x)x = 4x^3 - 40x^2 + 96x$ (B) $0 \leqslant x \leqslant 4$ (Note that at $x = 0$ and $x = 4$, we have zero volume.)
(C) $V(x)$ (D) Max $V(x) \approx V(1.5) \approx 67.5$ in³; a 1.5-inch square should be cut from each corner.

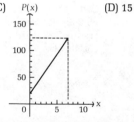

43. (A) $C = 360,000 - 900p$ (B) $R = xp = (9,000 - 30p)p = 9,000p - 30p^2$ (C)
(D) $42, $288

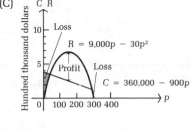

45. (A) $P(x) = 15x + 20$ (B) 95 (C) $P(x)$ (D) 15

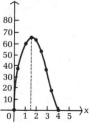

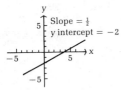

Exercise 5-4   Chapter Review

1.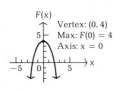
Slope = ½
y intercept = −2

2. $y = \frac{1}{2}x + 1$

3.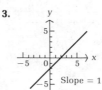
Slope = 1

4. −2

5.
Axis
x = 4
Min f(x):
f(4) = −2
(4, −2)
Vertex

6.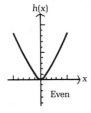
F(x)
Vertex: (0, 4)
Max: F(0) = 4
Axis: x = 0

7. $x + 2y = 4$; slope $= -\frac{1}{2}$

8.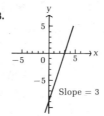
Slope = 3

9.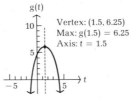
x = −5
y = 2

10. x = 4   **11.** (A) A function
(B) Not a function

12. (A) 4   (B) $\frac{2}{9}$ **13.** (A) All real numbers, except 3
(B) All real numbers greater than or equal to 1

14. 2   **15.**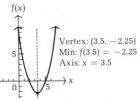
f(x)
Vertex: (3.5, −2.25)
Min: f(3.5) = −2.25
Axis: x = 3.5

16.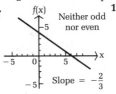
g(t)
Vertex: (1.5, 6.25)
Max: g(1.5) = 6.25
Axis: t = 1.5

17.
f(x)
Neither odd
nor even
Slope $= -\frac{2}{3}$

18.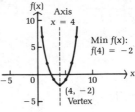
g(x)
Odd

19.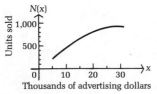
h(x)
Even

20.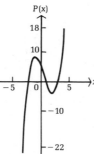
P(x)
$P(x) = [(x − 2)x − 5]x + 6$

21.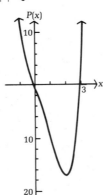
P(x)
$P(x) = [(x − 2)x^2 − 8]x − 1$

22. (A) $R = \frac{8}{5}C$ (B) $168

23.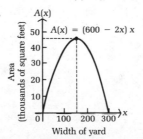
N(x)
Units sold
1,000
500
0   10   20   30
Thousands of advertising dollars

24.
A(x)
$A(x) = (600 − 2x)x$
Area (thousands of square feet)
50
40
30
20
10
0   100   200   300
Width of yard

A maximum area of
45,000 square feet
results for a yard
150 feet by 300 feet.

Practice Test: Chapter 5

1.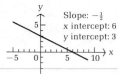
Slope: $-\frac{1}{2}$
x intercept: 6
y intercept: 3

2. $y = -\frac{3}{2}x + 2$ **3.** $x - 2y = 8$ **4.**
$x = 2$
$y = -3$

5. -9 **6.** Domain of f: R
Domain of g: all R,
except $x = 2$

7. (A) A function (B) Not a function   **8.**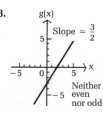
$g(x)$
Slope $= \frac{3}{2}$
Neither even nor odd

9.
$f(x)$
Even

10.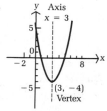
y Axis
$x = 3$
$(3, -4)$
Vertex
Min $f(x) = f(3) = -4$

11. $f(x) = [(x - 3)x - 1]x + 3$

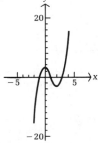

12. (A) $V = -1{,}800t + 20{,}000$ (B) \$9,200

13. Maximum revenue is \$18,000 at a price of \$3.

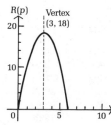
$R(p)$
Vertex
$(3, 18)$

14. (A) $A(x) = x(20 - 2x)$ (B) $[0,10]$

(C)

x	A(x)
2	32
4	48
5	50
6	48
8	32

Chapter 6

Exercise 6-1

1. $I = \$20$ **3.** $D = \$40$ **5.** $r = 0.08$ or 8% **7.** $d = 0.10$ or 10% **9.** $A = \$112$ **11.** $P = \$888.89$ **13.** $P = \$7{,}000$
15. $M = \$2{,}444.44$ **17.** $r = I/Pt$ **19.** $M = D/dt$ **21.** $P = A/(1 + rt)$ **23.** $A = \$560$ **25.** $A = \$10{,}500$
27. $P = \$440, D = \$60, M = \$500$ **29.** $P = \$2{,}678.57$ **31.** $P = \$9{,}523.81$ **33.** $r = 0.222\ 2$ or 22.22%
35. $M = \$3{,}000, D = \$600$ **37.** $r = 15.44\%$ **39.** \$5,060

Exercise 6-2

1. $A = \$112.68$　**3.** $A = \$3,433.50$　**5.** $A = \$2,980.68$　**7.** $\$2,419.99$　**9.** $P = \$7,351.04$　**11.** $n \approx 11.9$　**13.** $n \approx 55.5$
15. (A)　$\$126.25; \$26.25$　　(B)　$\$126.90; \$26.90$　　(C)　$\$127.05; \$27.05$　**17.** (A)　$\$7,147.51$　　(B)　$\$10,217.39$
19. (A)　$\$2,208.04$　　(B)　$\$4,875.44$　**21.** (A)　$\$6,755.64$　　(B)　$\$4,563.87$　**23.** (A)　$\$3,197.05$　　(B)　$\$2,044.22$
25. $\$6,729.71$　**27.** (A)　10.38%　　(B)　12.68%　**29.** (A)　7 years　　(B)　6 years
31. (A)　$\$8,243.05$　　(B)　$\$13,589.57$
33. (A)　$\$3,516.83$　　(B)　$\$3,908.37$　　(C)　$\$4,133.40$　　(D)　$\$4,296.10$　　(E)　$\$4,378.72$　**35.** 10.52%　**37.** 8.45%
39. (A)　8.67 years　　(B)　6.93 years　　(C)　5.78 years

Exercise 6-3

1. $S = \$13,435.19$　**3.** $S = \$60,401.98$　**5.** $R = \$123.47$　**7.** $R = \$310.62$　**9.** $n = 17$
11. Value: $\$30,200.99$; interest: $\$10,200.99$　**13.** $\$20,931.01$　**15.** $\$331.46$　**17.** $\$625.28$
19. First year: $\$50.76$; second year: $\$168.09$; third year: $\$296.42$　**21.** Value: $\$30,383.01$; interest: $\$18,383.01$
23. $\$59,987.37$　**25.** 20 years

Exercise 6-4

1. $P = \$3,458.41$　**3.** $P = \$4,606.09$　**5.** $R = \$199.29$　**7.** $R = \$586.01$　**9.** $n = 29$　**11.** $\$109,421.92$
13. $\$14,064.67; \$16,800.00$　**15.** (A)　$\$36.59$ per month; $\$58.62$ interest　　(B)　$\$38.28$ per month; $\$89.04$ interest
17. $\$273.69$ per month; $\$7,705.68$ interest

19. Amortization Schedule

Payment Number	Payment	Interest	Unpaid Balance Reduction	Unpaid Balance
0				$5,000.00
1	$ 758.05	$ 225.00	$ 533.05	4,466.95
2	758.05	201.01	557.04	3,909.91
3	758.05	175.95	582.10	3,327.81
4	758.05	149.75	608.30	2,719.51
5	758.05	122.38	635.67	2,083.84
6	758.05	93.77	664.28	1,419.56
7	758.05	63.88	694.17	725.39
8	758.05	32.64	725.39	0.00
Total	$6,064.38	$1,064.38	$5,000.00	

21. First year interest $= \$625.07$;
second year interest $= \$400.91$;
third year interest $= \$148.46$
23. $\$85,846.38; \$128,153.62$
25. $\$143.85$ per month; $\$904.80$
27. Monthly payment $R = \$841.39$
(A)　$\$70,952.33$　　(B)　$\$55,909.02$
(C)　$\$36,813.32$
29. (A)　Monthly payment $R = \$395.04$;
total interest $= \$64,809.60$
(B)　114 months or 9.5 years;
interest saved $= \$38,375.04$

Exercise 6-5　Chapter Review

1. $A = \$104.50$　**2.** $P = \$800$　**3.** $t = 0.75$ year or 9 months　**4.** $r = 0.06$ or 6%　**5.** $P = \$4,250.00$　**6.** $M = \$4,444.44$
7. $d = 0.12$ or 12%　**8.** $t = 1$ year　**9.** $A = \$1,393.68$　**10.** $P = \$3,193.50$　**11.** $S = \$69,770.03$　**12.** $R = \$115.00$
13. $P = \$33,944.27$　**14.** $R = \$166.07$　**15.** $n \approx 16$　**16.** $n \approx 41$　**17.** $\$3,350.00; \$350.00$　**18.** $\$2,650.00; \$3,000; \$350$
19. $\$4,290.96$　**20.** $\$9,043.63$　**21.** $\$12,282.85$　**22.** $\$33.70$ per month; $\$133.80$　**23.** $\$27,355.48$　**24.** $\$526.28$ per month
25. 9.38%　**26.** $\$5,106.25; \$5,375$　**27.** 70 months or 5 years and 10 months

28. Amortization Schedule

Payment Number	Payment	Interest	Unpaid Balance Reduction	Unpaid Balance
0				$1,000.00
1	$ 265.82	$25.00	$ 240.82	759.18
2	265.82	18.98	246.84	512.34
3	265.82	12.81	253.01	259.33
4	265.82	6.18	259.33	0.00
Total	$1,063.27	$63.27	$1,000.00	

29. (A) Approximately 14.93% compounded monthly   (B) $6,275.16
30. (A) $1,435.63   (B) $74,397.48   (C) $11,625.04
31. (A) $115,573.86   (B) $359.64   (C) $171,228.80
32. 43

Practice Test: Chapter 6

1. $9,422.24   **2.** $463.69   **3.** $15,084.83   **4.** $188.00 per month; $2,624.00 interest   **5.** $119,233.54
6. $10,481.25; $10,750   **7.** 8.24%   **8.** 35 quarters or 8 years and 9 months   **9.** 17 quarters or 4.25 years   **10.** $6,697.11

Chapter 7

Exercise 7-1

1. $x = 3, y = 2$ **3.** $x = 2, y = 4$ **5.** No solution (parallel lines)   **7.** $x = 4, y = 5$ **9.** $x = 1, y = 4$ **11.** $u = 2, v = -3$
13. $m = 8, n = 6$ **15.** $x = 1, y = -5$ **17.** No solution (inconsistent)   **19.** $x = -\frac{4}{3}, y = 1$
21. Infinitely many solutions (dependent)   **23.** $x = 4,000, y = 280$ **25.** $x = 1.1, y = 0.3$ **27.** $x = 0, y = -2, z = 5$
29. $x = 2, y = 0, z = -1$ **31.** $a = -1, b = 2, c = 0$ **33.** $x = 0, y = 2, z = -3$
35. No solution (inconsistent)
37. (A) Equilibrium price = $6.50, equilibrium quantity = 500
(B)

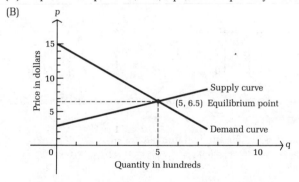

39. (A) For $x = 120$ units, $C = \$216,000 = R$

(B)

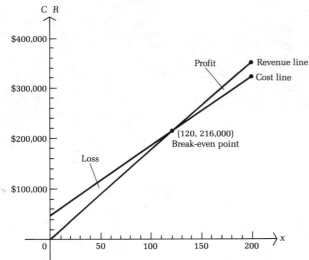

41. 50 one-person boats, 200 two-person boats, 100 four-person boats **43.** Mix A: 80 grams; mix B: 60 grams
45. (A) **(B)** $d = 141$ centimeters (approximately) **(C)** Vacillate

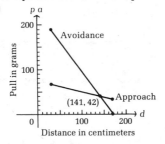

Exercise 7-2

1. $\begin{bmatrix} 4 & -6 & | & -8 \\ 1 & -3 & | & 2 \end{bmatrix}$ **3.** $\begin{bmatrix} -4 & 12 & | & -8 \\ 4 & -6 & | & -8 \end{bmatrix}$ **5.** $\begin{bmatrix} 1 & -3 & | & 2 \\ 8 & -12 & | & -16 \end{bmatrix}$ **7.** $\begin{bmatrix} 1 & -3 & | & 2 \\ 0 & 6 & | & -16 \end{bmatrix}$ **9.** $\begin{bmatrix} 1 & -3 & | & 2 \\ 2 & 0 & | & -12 \end{bmatrix}$

11. $\begin{bmatrix} 1 & -3 & | & 2 \\ 3 & -3 & | & -10 \end{bmatrix}$ **13.** $x_1 = 3, x_2 = 2$ **15.** $x_1 = 3, x_2 = 1$ **17.** $x_1 = 2, x_2 = 1$ **19.** $x_1 = 2, x_2 = 4$ **21.** No solution

23. $x_1 = 1, x_2 = 4$ **25.** Infinitely many solutions: $x_2 = s, x_1 = 2s - 3$ for any real number s
27. Infinitely many solutions: $x_2 = s, x_1 = \frac{1}{2}s + \frac{1}{2}$ for any real number s **29.** $x_1 = 2, x_2 = -1$ **31.** $x_1 = 2, x_2 = -1$
33. $x_1 = 1.1, x_2 = 0.3$

Exercise 7-3

1. Yes **3.** No **5.** No **7.** Yes **9.** $x_1 = -2, x_2 = 3, x_3 = 0$ **11.** $x_1 = 2t + 3$ **13.** No solution

$x_2 = -t - 5$

$x_3 = t$

t any real number

15. $x_1 = 2s + 3t - 5$ **17.** $\begin{bmatrix} 1 & 0 & | & -7 \\ 0 & 1 & | & 3 \end{bmatrix}$ **19.** $\begin{bmatrix} 1 & 0 & 0 & | & -5 \\ 0 & 1 & 0 & | & 4 \\ 0 & 0 & 1 & | & -2 \end{bmatrix}$ **21.** $\begin{bmatrix} 1 & 0 & 2 & | & -\frac{5}{3} \\ 0 & 1 & -2 & | & \frac{1}{3} \\ 0 & 0 & 0 & | & 0 \end{bmatrix}$

$x_2 = s$

$x_3 = -3t + 2$

$x_4 = t$

s and t any real numbers

23. $x_1 = -2, x_2 = 3, x_3 = 1$ **25.** $x_1 = 0, x_2 = -2, x_3 = 2$ **27.** $x_1 = 2t + 3$ **29.** $x_1 = (-4t - 4)/7$

$x_2 = t - 2$ $x_2 = (5t + 5)/7$

$x_3 = t$ $x_3 = t$

t any real number t any real number

31. $x_1 = -1, x_2 = 2$ **33.** No solution **35.** No solution **37.** $x_1 = 0, x_2 = 2, x_3 = -3$ **39.** $x_1 = 1, x_2 = -2, x_3 = 1$

41. $x_1 = 2s - 3t + 3$ **43.** 20 one-person boats, 220 two-person boats, 100 four-person boats

$x_2 = s + 2t + 2$

$x_3 = s$

$x_4 = t$

s and t any real numbers

45. $(t - 80)$ one-person boats, $(-2t + 420)$ two-person boats, t four-person boats, $80 \leqslant t \leqslant 210$, t an integer

47. No solution; no production schedule will use all the labor-hours in all departments

49. 8 ounces food A, 2 ounces food B, 4 ounces food C **51.** No solution

53. 8 ounces food A, $(-2t + 10)$ ounces food B, t ounces food C, $0 \leqslant t \leqslant 5$

55. Company A: 10 hours; company B: 15 hours

 Exercise 7-4

1. $2 \times 2; 1 \times 4$ **3.** 2 **5.** $\begin{bmatrix} 0 & 0 \\ 0 & 0 \end{bmatrix}$ **7.** C, D **9.** A, B **11.** $\begin{bmatrix} -1 & 0 \\ 5 & -3 \end{bmatrix}$ **13.** $\begin{bmatrix} -2 \\ 3 \\ 0 \end{bmatrix}$ **15.** $\begin{bmatrix} -1 \\ 6 \\ 5 \end{bmatrix}$ **17.** $\begin{bmatrix} -15 & 5 \\ 10 & -15 \end{bmatrix}$

19. $\begin{bmatrix} 1 & 3 & -1 & 1 \\ -1 & -5 & 7 & 2 \\ 4 & 8 & 0 & -2 \end{bmatrix}$ **21.** $\begin{bmatrix} 5.4 & 0.7 & -1.8 \\ 7.6 & -4.0 & 7.9 \end{bmatrix}$ **23.** $\begin{bmatrix} 250 & 360 \\ 40 & 350 \end{bmatrix}$ **25.** $\begin{bmatrix} 2,280 & 3,460 \\ 1,380 & 2,310 \end{bmatrix}$

27. $a = -1, b = 1, c = 3, d = -5$ **29.** $x = 2, y = -3$ **31.** $\begin{matrix} \text{Guitar} & \text{Banjo} \\ \begin{bmatrix} \$33 & \$26 \\ \$57 & \$77 \end{bmatrix} & \begin{matrix} \text{Materials} \\ \text{Labor} \end{matrix} \end{matrix}$

33. $\begin{bmatrix} 135 & 282 & 50 \\ 55 & 258 & 155 \end{bmatrix}; \begin{bmatrix} 0.14 & 0.30 & 0.05 \\ 0.06 & 0.28 & 0.17 \end{bmatrix}$

Exercise 7-5

1. 10 **3.** -1 **5.** $[12 \quad 13]$ **7.** $\begin{bmatrix} 5 \\ -3 \end{bmatrix}$ **9.** $\begin{bmatrix} 2 & 4 \\ 1 & -5 \end{bmatrix}$ **11.** $\begin{bmatrix} 1 & -5 \\ -2 & -4 \end{bmatrix}$ **13.** $[-7]$ **15.** $\begin{bmatrix} -15 & 6 \\ -20 & 8 \end{bmatrix}$ **17.** 6 **19.** 15

21. $\begin{bmatrix} 0 & 9 \\ 5 & -4 \end{bmatrix}$ **23.** $\begin{bmatrix} 5 & 8 & -5 \\ -1 & -3 & 2 \\ -2 & 8 & -6 \end{bmatrix}$ **25.** $[11]$ **27.** $\begin{bmatrix} 3 & -2 & -4 \\ 6 & -4 & -8 \\ -9 & 6 & 12 \end{bmatrix}$ **29.** $\begin{bmatrix} 0 & 0 & -5 \\ -6 & 15 & 13 \\ 5 & -6 & -14 \end{bmatrix}$ **31.** $\begin{bmatrix} -3.73 & -5.28 \\ 18.47 & -36.27 \end{bmatrix}$

33. $AB = \begin{bmatrix} 5 & 7 \\ 2 & 3 \end{bmatrix}$, $BA = \begin{bmatrix} 1 & 3 \\ 2 & 7 \end{bmatrix}$ **35.** Both sides equal $\begin{bmatrix} 0 & 12 \\ 1 & 5 \end{bmatrix}$

37. (A) \$9 per boat (B) $[1.5 \quad 1.2 \quad 0.4] \cdot \begin{bmatrix} 7 \\ 10 \\ 4 \end{bmatrix} = \$24.10$ (C) 3×2 (D)

	I	II	
	\$9.00	\$11.00	One-person
	\$14.10	\$17.20	Two-person
	\$19.80	\$24.10	Four-person

Labor costs per boat at each plant

39. (A)

A	B	C	D	E
[16	9	11	11	10],

which is the combined inventory in all three stores

(B)

W	R
[\$108,300	\$141,340],

which is the total wholesale and retail values of the total inventory in all three stores

41. (A) \$2,025 (B) $[2,000 \quad 800 \quad 8,000] \cdot \begin{bmatrix} \$0.40 \\ \$0.75 \\ \$0.25 \end{bmatrix} = \$3,400$ (C) $\begin{bmatrix} \$2,025 \\ \$3,400 \end{bmatrix}$ Berkeley
Oakland

Cost per
town

(D)

Telephone	House	Letter
[3,000	1,300	13,000]

Number of each type of contact made

Exercise 7-6

1. $\begin{bmatrix} 2 & -3 \\ 4 & 5 \end{bmatrix}$ **3.** $\begin{bmatrix} -2 & 1 & 3 \\ 2 & 4 & -2 \\ 5 & 1 & 0 \end{bmatrix}$ **9.** $x_1 = -8, x_2 = 2$ **11.** $x_1 = 0, x_2 = 4$ **13.** $\begin{bmatrix} 3 & -2 \\ -1 & 1 \end{bmatrix}$ **15.** $\begin{bmatrix} 7 & -3 \\ -2 & 1 \end{bmatrix}$

17. $\begin{bmatrix} 7 & 6 & -3 \\ 2 & 2 & -1 \\ -6 & -5 & 3 \end{bmatrix}$ **19.** $\frac{1}{2}\begin{bmatrix} 3 & -1 & -1 \\ -1 & 1 & 1 \\ -3 & 1 & 3 \end{bmatrix}$ **21.** (A) $x_1 = -3, x_2 = 2$ **23.** (A) $x_1 = 17, x_2 = -5$

(B) $x_1 = -1, x_2 = 2$ (B) $x_1 = 7, x_2 = -2$

(C) $x_1 = -8, x_2 = 3$ (C) $x_1 = 24, x_2 = -7$

25. (A) $x_1 = 1, x_2 = 0, x_3 = 0$ **27.** (A) $x_1 = 1, x_2 = 1, x_3 = 3$

(B) $x_1 = -1, x_2 = 0, x_3 = 1$ (B) $x_1 = -1, x_2 = 1, x_3 = -1$

(C) $x_1 = -1, x_2 = -1, x_3 = 1$ (C) $x_1 = 5, x_2 = -1, x_3 = -5$

35. Concert 1: 6,000 $4 tickets and 4,000 $8 tickets; Concert 2: 5,000 $4 tickets and 5,000 $8 tickets; Concert 3: 3,000 $4 tickets and 7,000 $8 tickets

37. Diet 1: 60 ounces mix A and 80 ounces mix B; Diet 2: 20 ounces mix A and 60 ounces mix B; Diet 3: 0 ounces mix A and 100 ounces mix B

Exercise 7-7

1. 40¢ from A, 20¢ from E **3.** $\begin{bmatrix} 0.6 & -0.2 \\ -0.2 & 0.9 \end{bmatrix}, \begin{bmatrix} 1.8 & 0.4 \\ 0.4 & 1.2 \end{bmatrix}$ **5.** $X = \begin{bmatrix} x_1 \\ x_2 \end{bmatrix} = \begin{bmatrix} 16.4 \\ 9.2 \end{bmatrix}$

7. 10¢ from A, 35¢ from B, 10¢ from E **11.** $X = \begin{bmatrix} x_1 \\ x_2 \\ x_3 \end{bmatrix} = \begin{bmatrix} 11.14 \\ 7.45 \\ 6.36 \end{bmatrix}$

Exercise 7-8   Chapter Review

1. $x = 4, y = 4$ **2.** $x = 4, y = 4$ **3.** $\begin{bmatrix} 3 & 3 \\ 4 & 2 \end{bmatrix}$ **4.** Not defined **5.** $\begin{bmatrix} -3 & 0 \\ 1 & -1 \end{bmatrix}$ **6.** $\begin{bmatrix} 4 & 3 \\ 7 & 4 \end{bmatrix}$ **7.** Not defined **8.** $\begin{bmatrix} 5 \\ 5 \end{bmatrix}$

9. $\begin{bmatrix} 2 & 3 \\ 4 & 6 \end{bmatrix}$ **10.** 8 (a real number) **11.** Not defined **12.** $\begin{bmatrix} 3 & -2 \\ -4 & 3 \end{bmatrix}$ **13.** $x_1 = -1, x_2 = 3$ **14.** $x_1 = -1, x_2 = 3$

15. $x_1 = -1, x_2 = 3; x_1 = 1, x_2 = 2; x_1 = 8, x_2 = -10$ **16.** Not defined **17.** $\begin{bmatrix} 10 & -8 \\ 4 & 6 \end{bmatrix}$ **18.** $\begin{bmatrix} -2 & 8 \\ 8 & 6 \end{bmatrix}$

19. 9 (a real number) **20.** [9](a matrix) **21.** $\begin{bmatrix} 10 & -5 & 1 \\ -1 & -4 & -5 \\ 1 & -7 & -2 \end{bmatrix}$ **22.** $\begin{bmatrix} -\frac{5}{2} & 2 & -\frac{1}{2} \\ 1 & -1 & 1 \\ \frac{1}{2} & 0 & -\frac{1}{2} \end{bmatrix}$ **23.** (A) $x_1 = 2, x_2 = 1, x_3 = -1$

(B) $x_1 = -5t - 12$

$x_2 = 3t + 7$

$x_3 = t$

t any real number

24. $x_1 = 2, x_2 = 1, x_3 = -1; x_1 = 1, x_2 = -2, x_3 = 1; x_1 = -1, x_2 = 2, x_3 = -2$ **25.** $\begin{bmatrix} -\frac{11}{12} & -\frac{1}{12} & 5 \\ \frac{10}{12} & \frac{2}{12} & -4 \\ \frac{1}{12} & -\frac{1}{12} & 0 \end{bmatrix}$

26. $x_1 = 1,000, x_2 = 4,000, x_3 = 2,000$ **27.** $x_1 = 1,000, x_2 = 4,000, x_3 = 2,000$ **28.** $0.01x_1 + 0.02x_2 = 4.5$

$0.02x_1 + 0.05x_2 = 10$

$x_1 = 250$ tons of ore A

$x_2 = 100$ tons of ore B

29. (A) $\begin{matrix} X \\ \begin{bmatrix} x_1 \\ x_2 \end{bmatrix} \end{matrix} = \begin{matrix} A^{-1} \\ \begin{bmatrix} 500 & -200 \\ -200 & 100 \end{bmatrix} \end{matrix} \begin{bmatrix} 4.5 \\ 10 \end{bmatrix} = \begin{bmatrix} 250 \\ 100 \end{bmatrix}$

$x_1 = 250$ tons of ore A

$x_2 = 100$ tons of ore B

(B) $\begin{matrix} X \\ \begin{bmatrix} x_1 \\ x_2 \end{bmatrix} \end{matrix} = \begin{matrix} A^{-1} \\ \begin{bmatrix} 500 & -200 \\ -200 & 100 \end{bmatrix} \end{matrix} \begin{bmatrix} 2.3 \\ 5 \end{bmatrix} = \begin{bmatrix} 150 \\ 40 \end{bmatrix}$

$x_1 = 150$ tons of ore A

$x_2 = 40$ tons of ore B

30. (A) $MN = \begin{matrix} \text{Supplier } A \quad \text{Supplier } B \\ \begin{bmatrix} \$7,620 & \$7,530 \\ \$13,880 & \$13,930 \end{bmatrix} \end{matrix} \begin{matrix} \text{Alloy 1} \\ \text{Alloy 2} \end{matrix}$

Cost of each alloy from each supplier

(B) $\begin{matrix} \text{Supplier } A \quad \text{Supplier } B \\ [\$21,500 \qquad \$21,460] \end{matrix}$

Total cost for both alloys from each supplier

Practice Test: Chapter 7

1. $\begin{bmatrix} 1 & 0 & 2 & | & 3 \\ 0 & 1 & -3 & | & 2 \\ 0 & 0 & 0 & | & 0 \end{bmatrix}$ **2.** $x_1 = 3, x_2 = -1$ **3.** No solution   **4.** $x_1 = -2t + 3$ **5.** $B \cdot C = 0, BC = [0]$
$x_2 = 3t + 2$
$x_3 = t$
t any real number

6. $\begin{bmatrix} 7 & -8 & 6 \\ -4 & 5 & -3 \end{bmatrix}$ **7.** Not defined   **8.** $\begin{bmatrix} 6 & -12 & -12 \\ 5 & 4 & 7 \\ 0 & -4 & -4 \end{bmatrix}$ **9.** Not defined   **10.** Both $\begin{bmatrix} -8 & -16 & -28 \\ 2 & 4 & 7 \\ -4 & -8 & -14 \end{bmatrix}$

11. (A) $x_1 = 7, x_2 = -5, x_3 = -10$ (B) $x_1 = 3, x_2 = -2, x_3 = -6$
12. \$2,000 at 5%, \$3,000 at 10%   **13.** \$2,000 at 5%, \$3,000 at 10%

14. (A) $[0.25 \quad 0.20 \quad 0.05] \cdot \begin{bmatrix} 15 \\ 12 \\ 4 \end{bmatrix} = \$6.35$ (B) $\begin{array}{cc} \text{Calif.} & \text{Texas} \\ \begin{bmatrix} \$3.65 & \$3.00 \\ \$6.35 & \$5.20 \end{bmatrix} & \begin{array}{l} \text{Model } A \\ \text{Model } B \end{array} \end{array}$
Total labor costs for each model
at each plant

Chapter 8

Exercise 8-1

1.

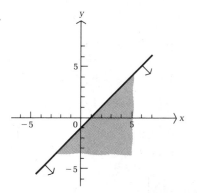

3.

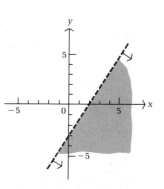

5.

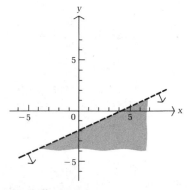

7.

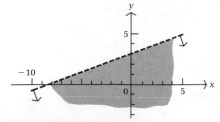

9.

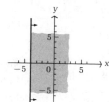

11.

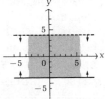

13.

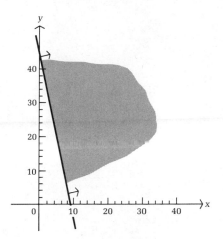

15.

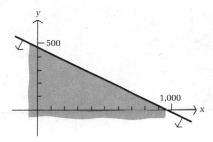

17.

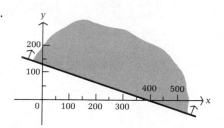

19.

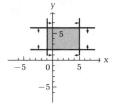

21.

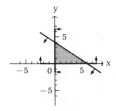

23.

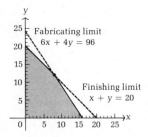

Fabricating limit
$6x + 4y = 96$

Finishing limit
$x + y = 20$

Exercise 8-2

1.

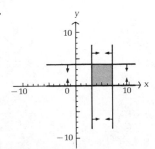

3.

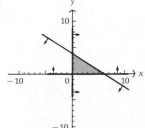

5.

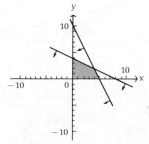

7.

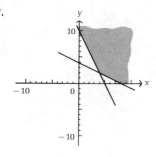

9.

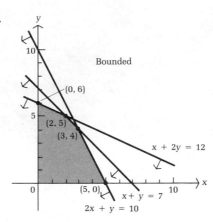

Bounded

$(0, 6)$

$(2, 5)$

$(3, 4)$

$x + 2y = 12$

$(5, 0)$

$x + y = 7$

$2x + y = 10$

11.

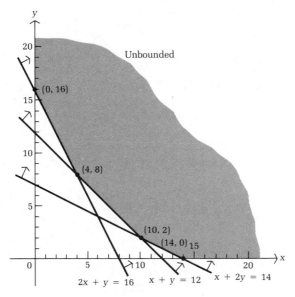

Unbounded

$(0, 16)$

$(4, 8)$

$(10, 2)$

$(14, 0)$

$2x + y = 16$ $x + y = 12$ $x + 2y = 14$

13.

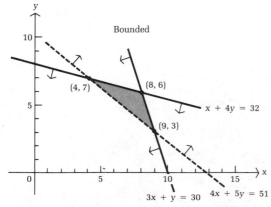

Bounded

$(4, 7)$

$(8, 6)$

$x + 4y = 32$

$(9, 3)$

$3x + y = 30$ $4x + 5y = 51$

15.

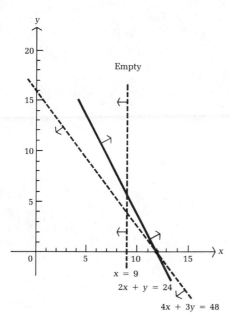

Empty

$x = 9$

$2x + y = 24$

$4x + 3y = 48$

17.

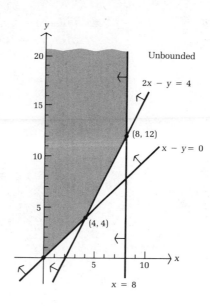

Unbounded

$2x - y = 4$

$(8, 12)$

$x - y = 0$

$(4, 4)$

$x = 8$

19.

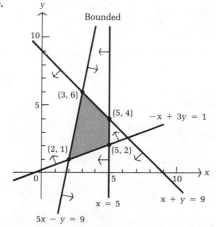

Bounded

$(3, 6)$

$(5, 4)$

$-x + 3y = 1$

$(2, 1)$

$(5, 2)$

$x = 5$

$x + y = 9$

$5x - y = 9$

21.

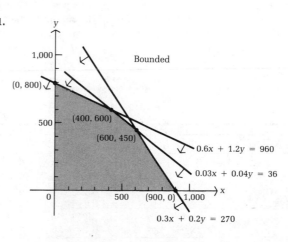

Bounded

$(0, 800)$

$(400, 600)$

$(600, 450)$

$0.6x + 1.2y = 960$

$0.03x + 0.04y = 36$

$(900, 0)$

$0.3x + 0.2y = 270$

23. $6x + 4y \leqslant 108$

$\ x + \ y \leqslant 24$

$\ x \geqslant 0$

$\ y \geqslant 0$

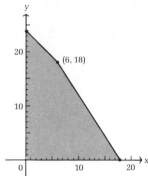

25. $10x + 20y \leqslant 800$

$\ 20x + 10y \leqslant 640$

$\ x \geqslant 0$

$\ y \geqslant 0$

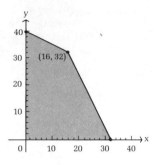

Exercise 8-3

1. Max $P = 30$ at $x_1 = 4$ and $x_2 = 2$ **3.** Min $z = 14$ at $x_1 = 4$ and $x_2 = 2$; no maximum

5. Max $P = 260$ at $x_1 = 2$ and $x_2 = 5$ **7.** Min $z = 140$ at $x_1 = 14$ and $x_2 = 0$; no maximum

9. Min $P = 20$ at $x_1 = 0$ and $x_2 = 2$; Max $P = 150$ at $x_1 = 5$ and $x_2 = 0$ **11.** Feasible region empty, no optimal solutions

13. Min $P = 140$ at $x_1 = 3$ and $x_2 = 8$; Max $P = 260$ at $x_1 = 8$ and $x_2 = 10$, at $x_1 = 12$ and $x_2 = 2$,

or at any point on the line segment from $(8, 10)$ to $(12, 2)$

15. Max $P = 26{,}000$ at $x_1 = 400$ and $x_2 = 600$

17. (A) $2a < b$ (B) $\frac{1}{3}a < b < 2a$ (C) $b < \frac{1}{3}a$ (D) $b = 2a$ (E) $b = \frac{1}{3}a$

19. 6 trick, 18 slalom; $780   **21.** 1,500 gallons by new process, none by old process; maximum profit $300

23. 20 cubic yards of A, 12 cubic yards of B; $1,020   **25.** 48; 16 mice, 32 rats

Exercise 8-4

1.

	Nonbasic	Basic	Feasible?
(A)	x_1, x_2	s_1, s_2	Yes
(B)	x_1, s_1	x_2, s_2	Yes
(C)	x_1, s_2	x_2, s_1	No
(D)	x_2, s_1	x_1, s_2	No
(E)	x_2, s_2	x_1, s_1	Yes
(F)	s_1, s_2	x_1, x_2	Yes

3.

	x_1	x_2	s_1	s_2	Feasible?
(A)	0	0	50	40	Yes
(B)	0	50	0	-60	No
(C)	0	20	30	0	Yes
(D)	25	0	0	15	Yes
(E)	40	0	-30	0	No
(F)	20	10	0	0	Yes

5.

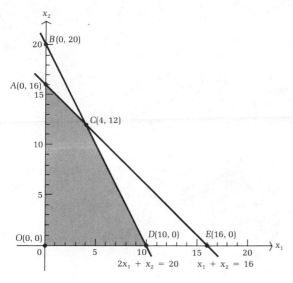

$$x_1 + x_2 + s_1 \quad = 16$$
$$2x_1 + x_2 \quad + s_2 = 20$$

x_1	x_2	s_1	s_2	Intersection Point	Feasible?
0	0	16	20	O	Yes
0	16	0	4	A	Yes
0	20	−4	0	B	No
16	0	0	−12	E	No
10	0	6	0	D	Yes
4	12	0	0	C	Yes

7.

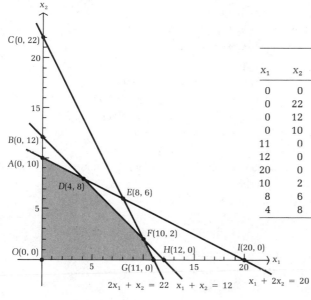

$$2x_1 + x_2 + s_1 \quad = 22$$
$$x_1 + x_2 \quad + s_2 \quad = 12$$
$$x_1 + 2x_2 \quad + s_3 = 20$$

x_1	x_2	s_1	s_2	s_3	Intersection Point	Feasible?
0	0	22	12	20	O	Yes
0	22	0	−10	−24	C	No
0	12	10	0	−4	B	No
0	10	12	2	0	A	Yes
11	0	0	1	9	G	Yes
12	0	−2	0	8	H	No
20	0	−18	−8	0	I	No
10	2	0	0	6	F	Yes
8	6	0	−2	0	E	No
4	8	6	0	0	D	Yes

Exercise 8-5

1. (A)
$$2x_1 + x_2 + s_1 \quad = 10$$
$$x_1 + 2x_2 \quad + s_2 \quad = 8$$
$$-15x_1 - 10x_2 \quad + P = 0$$

(B)
$$\begin{bmatrix} 2 & 1 & 1 & 0 & 0 & | & 10 \\ 1 & 2 & 0 & 1 & 0 & | & 8 \\ -15 & -10 & 0 & 0 & 1 & | & 0 \end{bmatrix}$$

(C) Max $P = 80$ at $x_1 = 4$ and $x_2 = 2$

3. (A)
$$2x_1 + x_2 + s_1 \qquad\qquad = 10$$
$$x_1 + 2x_2 \qquad + s_2 \qquad = 8$$
$$-30x_1 - x_2 \qquad\qquad + P = 0$$

(B)
$$\begin{bmatrix} \textcircled{2} & 1 & 1 & 0 & 0 & | & 10 \\ 1 & 2 & 0 & 1 & 0 & | & 8 \\ \hline -30 & -1 & 0 & 0 & 1 & | & 0 \end{bmatrix}$$

(C) Max $P = 150$ at $x_1 = 5$ and $x_2 = 0$
5. Max $P = 260$ at $x_1 = 2$ and $x_2 = 5$ **7.** No optimal solution exists **9.** Max $P = 7$ at $x_1 = 3$ and $x_2 = 5$
11. Max $P = \frac{190}{3}$ at $x_1 = \frac{40}{3}$, $x_2 = 0$, and $x_3 = \frac{10}{3}$ **13.** Max $P = 26{,}000$ at $x_1 = 400$ and $x_2 = 600$
15. Max $P = 450$ at $x_1 = 0$, $x_2 = 180$, and $x_3 = 30$ **17.** Max $P = 88$ at $x_1 = 24$ and $x_2 = 8$

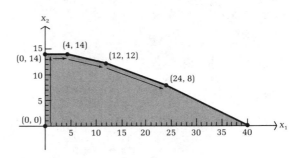

19. Let x_1 = Number of tennis rackets
x_2 = Number of squash rackets
x_3 = Number of racketball rackets

Maximize $P = 7x_1 + 9x_2 + 10x_3$
Subject to $2x_1 + x_2 + 2x_3 \leq 1{,}000$
$\qquad\qquad x_1 + 2x_2 + 2x_3 \leq 800$
$\qquad\qquad\qquad x_1, x_2, x_3 \geq 0$

400 tennis rackets, 200 squash rackets, and 0 racketball rackets; maximum profit \$4,600

21. Let x_1 = Amount invested in government bonds
x_2 = Amount invested in mutual funds
x_3 = Amount invested in money market funds

Maximize $P = 0.08x_1 + 0.13x_2 + 0.15x_3$
Subject to $x_1 + x_2 + x_3 \leq 100{,}000$
$\qquad\qquad -x_1 + x_2 + x_3 \leq 0$
$\qquad\qquad\qquad x_1, x_2, x_3 \geq 0$

\$50,000 in government bonds, \$0 in mutual funds, and \$50,000 in money market funds; maximum return \$11,500

23. Let x_1 = Number of ads placed in daytime shows
x_2 = Number of ads placed in prime-time shows
x_3 = Number of ads placed in late-night shows

Maximize $P = 1{,}400x_1 + 2{,}400x_2 + 1{,}800x_3$
Subject to $x_1 + x_2 + x_3 \leq 15$
$\qquad\qquad 100x_1 + 200x_2 + 150x_3 \leq 2{,}000$
$\qquad\qquad\qquad x_1, x_2, x_3 \geq 0$

10 daytime ads, 5 prime-time ads, and 0 late-night ads; maximum number of potential customers 26,000

25. Let x_1 = Number of grams of food A
x_2 = Number of grams of food B
x_3 = Number of grams of food C

Maximize $P = 3x_1 + 3x_2 + 5x_3$
Subject to $x_1 + 3x_2 + 2x_3 \leq 30$
$\qquad\qquad 2x_1 + x_2 + x_3 \leq 24$
$\qquad\qquad\qquad x_1, x_2, x_3 \geq 0$

6 grams of food A, 0 grams of food B, and 12 grams of food C; maximum protein 78 units

27. Let x_1 = Number of undergraduate students
x_2 = Number of graduate students
x_3 = Number of faculty members

Maximize $P = 18x_1 + 25x_2 + 30x_3$
Subject to $x_1 + x_2 + x_3 \leq 20$
$\qquad\qquad 20x_1 + 30x_2 + 40x_3 \leq 540$
$\qquad\qquad\qquad x_1, x_2, x_3 \geq 0$

6 undergraduate and 14 graduate students, 0 faculty members; maximum number of interviews 458

Exercise 8-6

1. (A) Maximize $P = 13y_1 + 12y_2$
Subject to $4y_1 + 3y_2 \leqslant 9$
$y_1 + y_2 \leqslant 2$
$y_1, y_2 \geqslant 0$
(B) Min $C = 26$ at $x_1 = 0$ and $x_2 = 13$

3. (A) Maximize $P = 15y_1 + 8y_2$
Subject to $2y_1 + y_2 \leqslant 7$
$3y_1 + 2y_2 \leqslant 12$
$y_1, y_2 \geqslant 0$
(B) Min $C = 54$ at $x_1 = 6$ and $x_2 = 1$

5. (A) Maximize $P = 8y_1 + 4y_2$
Subject to $2y_1 - 2y_2 \leqslant 11$
$y_1 + 3y_2 \leqslant 4$
$y_1, y_2 \geqslant 0$
(B) Min $C = 32$ at $x_1 = 0$ and $x_2 = 8$

7. (A) Maximize $P = 6y_1 + 4y_2$
Subject to $-3y_1 + y_2 \leqslant 7$
$y_1 - 2y_2 \leqslant 9$
$y_1, y_2 \geqslant 0$
(B) No optimal solution exists

9. Min $C = 24$ at $x_1 = 8$ and $x_2 = 0$ **11.** Min $C = 20$ at $x_1 = 0$ and $x_2 = 4$ **13.** Min $C = 140$ at $x_1 = 14$ and $x_2 = 0$
15. Min $C = 44$ at $x_1 = 6$ and $x_2 = 2$ **17.** Min $C = 43$ at $x_1 = 0$, $x_2 = 1$, and $x_3 = 3$ **19.** No optimal solution exists
21. Min $C = 44$ at $x_1 = 0$, $x_2 = 3$, and $x_3 = 5$ **23.** Min $C = 166$ at $x_1 = 0$, $x_2 = 12$, $x_3 = 20$, and $x_4 = 3$

25. Let $x_1 =$ Number of hours the Cedarburg plant is operated
$x_2 =$ Number of hours the Grafton plant is operated
$x_3 =$ Number of hours the West Bend plant is operated

Minimize $C = 70x_1 + 75x_2 + 90x_3$
Subject to $20x_1 + 10x_2 + 20x_3 \geqslant 300$
$10x_1 + 20x_2 + 20x_3 \geqslant 200$
$x_1, x_2, x_3 \geqslant 0$

Cedarburg plant 10 hours per day, West Bend plant 5 hours per day, Grafton plant not used; $1,150

27. Let $x_1 =$ Number of single-sided drives ordered from Associated Electronics
$x_2 =$ Number of double-sided drives ordered from Associated Electronics
$x_3 =$ Number of single-sided drives ordered from Digital Drives
$x_4 =$ Number of double-sided drives ordered from Digital Drives

Minimize $C = 250x_1 + 350x_2 + 290x_3 + 320x_4$
Subject to $x_1 + x_2 \leqslant 1,000$
$x_3 + x_4 \leqslant 2,000$
$x_1 + x_3 \geqslant 1,200$
$x_2 + x_4 \geqslant 1,600$
$x_1, x_2, x_3, x_4 \geqslant 0$
1,000 single-sided drives from Associated Electronics, 200 single-sided and 1,600 double-sided drives from Digital Drives; $820,000

29. Let $x_1 =$ Number of ounces of food L
$x_2 =$ Number of ounces of food M
$x_3 =$ Number of ounces of food N

Minimize $C = 20x_1 + 24x_2 + 18x_3$
Subject to $20x_1 + 10x_2 + 10x_3 \geqslant 300$
$10x_1 + 10x_2 + 10x_3 \geqslant 200$
$10x_1 + 20x_2 + 10x_3 \geqslant 240$
$x_1, x_2, x_3 \geqslant 0$

10 ounces of L, 4 ounces of M, 6 ounces of N; 404 units

31. Let x_1 = Number of students bused from North Division to Central

x_2 = Number of students bused from North Division to Washington

x_3 = Number of students bused from South Division to Central

x_4 = Number of students bused from South Division to Washington

Minimize $C = 5x_1 + 2x_2 + 3x_3 + 4x_4$

Subject to $x_1 + x_2 \qquad\qquad \geqslant 300$

$\qquad\qquad\qquad x_3 + x_4 \geqslant 500$

$\qquad x_1 \qquad + x_3 \qquad \leqslant 400$

$\qquad\qquad x_2 \qquad + x_4 \leqslant 500$

$\qquad x_1, x_2, x_3, x_4 \geqslant 0$

300 students bused from North Division to Washington, 400 from South Division to Central High, and 100 from South Division to Washington; $2,200

Exercise 8-7

1. (A) Maximize $P = 5x_1 + 2x_2 - Ma_1$

Subject to $x_1 + 2x_2 + s_1 \qquad\qquad = 12$

$\qquad x_1 + x_2 \qquad - s_2 + a_1 = 4$

$\qquad x_1, x_2, s_1, s_2, a_1 \geqslant 0$

(B)

x_1	x_2	s_1	s_2	a_1	P	
1	2	1	0	0	0	12
1	1	0	-1	1	0	4
$-M-5$	$-M-2$	0	M	0	1	$-4M$

(C) Max $P = 60$ at $x_1 = 12$ and $x_2 = 0$

3. (A) Maximize $P = 3x_1 + 5x_2 - Ma_1$

Subject to $2x_1 + x_2 + s_1 \qquad = 8$

$\qquad x_1 + x_2 \qquad + a_1 = 6$

$\qquad x_1, x_2, s_1, a_1 \geqslant 0$

(B)

x_1	x_2	s_1	a_1	P	
2	1	1	0	0	8
1	1	0	1	0	6
$-M-3$	$-M-5$	0	0	1	$-6M$

(C) Max $P = 30$ at $x_1 = 0$ and $x_2 = 6$

5. (A) Maximize $P = 4x_1 + 3x_2 - Ma_1$

Subject to $-x_1 + 2x_2 + s_1 \qquad = 2$

$\qquad x_1 + x_2 \qquad - s_2 + a_1 = 4$

$\qquad x_1, x_2, s_1, s_2, a_1 \geqslant 0$

(B)

x_1	x_2	s_1	s_2	a_1	P	
-1	2	1	0	0	0	2
1	1	0	-1	1	0	4
$-M-4$	$-M-3$	0	M	0	1	$-4M$

(C) No optimal solution exists

7. (A) Maximize $P = 5x_1 + 10x_2 - Ma_1$

Subject to $x_1 + x_2 + s_1 \qquad\qquad = 3$

$\qquad 2x_1 + 3x_2 \qquad - s_2 + a_1 = 12$

$\qquad x_1, x_2, s_1, s_2, a_1 \geqslant 0$

(B)

x_1	x_2	s_1	s_2	a_1	P	
1	1	1	0	0	0	3
2	3	0	-1	1	0	12
$-2M-5$	$-3M-10$	0	M	0	1	$-12M$

(C) No optimal solution exists

9. Min $P = 12$ at $x_1 = 4$ and $x_2 = 6$; Max $P = 60$ at $x_1 = 10$ and $x_2 = 0$ **11.** Max $P = 44$ at $x_1 = 2$ and $x_2 = 8$

13. No optimal solution exists **15.** Min $C = -9$ at $x_1 = 0$, $x_2 = \frac{7}{4}$, and $x_3 = \frac{3}{4}$

17. Min $C = -30$ at $x_1 = 0$, $x_2 = \frac{3}{4}$, and $x_3 = 0$ **19.** Max $P = 17$ at $x_1 = \frac{49}{5}$, $x_2 = 0$, and $x_3 = \frac{22}{5}$

21. Min $C = \frac{135}{2}$ at $x_1 = \frac{15}{4}$, $x_2 = \frac{3}{4}$, and $x_3 = 0$ **23.** Max $P = 380$ at $x_1 = \frac{80}{3}$, $x_2 = \frac{20}{3}$, and $x_3 = 0$

25. Let x_1 = Number of 16K modules manufactured daily

x_2 = Number of 64K modules manufactured daily

Maximize $P = 18x_1 + 30x_2$

Subject to $10x_1 + 15x_2 \leqslant 1,500$

$\qquad 2x_1 + 4x_2 \leqslant 500$

$\qquad x_1 \qquad \geqslant 50$

$\qquad x_1, x_2 \geqslant 0$

Average daily production: 50 16K modules and $66\frac{2}{3}$ 64K modules; $2,900

27. Let x_1 = Number of ads placed in the *Sentinel*   Minimize $C = 200x_1 + 200x_2 + 100x_3$
    x_2 = Number of ads placed in the *Journal*   Subject to $x_1 + \quad x_2 + \quad\quad x_3 \leqslant 10$
    x_3 = Number of ads placed in the *Tribune* $2{,}000x_1 + 500x_2 + 1{,}500x_3 \geqslant 16{,}000$
                $x_1, x_2, x_3 \geqslant 0$

2 ads in the *Sentinel*, 0 ads in the *Journal*, 8 ads in the *Tribune*; $1,200

29. Let x_1 = Number of bottles of brand *A*   Minimize $C = 0.6x_1 + 0.4x_2 + 0.9x_3$
    x_2 = Number of bottles of brand *B*   Subject to $10x_1 + 10x_2 + 20x_3 \geqslant 100$
    x_3 = Number of bottles of brand *C* $?x_1 + ?x_2 + 1x_3 \geqslant ?4$
                $x_1, x_2, x_3 \geqslant 0$

0 bottles of *A*, 4 bottles of *B*, 3 bottles of *C*; $4.30

31. Let x_1 = Number of cubic yards of mix *A*   Maximize $P = 12x_1 + 16x_2 + 8x_3$
    x_2 = Number of cubic yards of mix *B*   Subject to $12x_1 + 8x_2 + 16x_3 \leqslant 700$
    x_3 = Number of cubic yards of mix *C* $16x_1 + 8x_2 + 16x_3 \geqslant 800$
                $x_1, x_2, x_3 \geqslant 0$

25 cubic yards of *A*, 50 cubic yards of *B*, 0 cubic yards of *C*; 1,100 pounds

33. Let x_1 = Number of car frames produced at the Milwaukee plant
    x_2 = Number of truck frames produced at the Milwaukee plant
    x_3 = Number of car frames produced at the Racine plant
    x_4 = Number of truck frames produced at the Racine plant
Maximize $P = 50x_1 + 70x_2 + 50x_3 + 70x_4$
Subject to $x_1 \quad\quad + \quad x_3 \quad\quad \leqslant 250$
            $x_2 \quad\quad + \quad x_4 \leqslant 350$
         $x_1 + \quad x_2 \quad\quad\quad \leqslant 300$
                $x_3 + \quad x_4 \leqslant 200$
      $150x_1 + 200x_2 \quad\quad\quad \leqslant 50{,}000$
            $135x_3 + 180x_4 \leqslant 35{,}000$
                $x_1, x_2, x_3, x_4 \geqslant 0$

35. Let x_1 = Number of barrels of *A* used in regular gasoline
    x_2 = Number of barrels of *A* used in premium gasoline
    x_3 = Number of barrels of *B* used in regular gasoline
    x_4 = Number of barrels of *B* used in premium gasoline
    x_5 = Number of barrels of *C* used in regular gasoline
    x_6 = Number of barrels of *C* used in premium gasoline
Maximize $P = 10x_1 + 18x_2 + 8x_3 + 16x_4 + 4x_5 + 12x_6$
Subject to $x_1 + \quad x_2 \quad\quad\quad\quad\quad\quad \leqslant 40{,}000$
            $x_3 + \quad x_4 \quad\quad\quad\quad \leqslant 25{,}000$
                $x_5 + \quad x_6 \leqslant 15{,}000$
         $x_1 \quad\quad + x_3 \quad\quad + \quad x_5 \quad\quad \geqslant 30{,}000$
            $x_2 \quad\quad + x_4 \quad\quad + \quad x_6 \geqslant 25{,}000$
      $-5x_1 \quad\quad + 5x_3 \quad\quad + 15x_5 \quad\quad \geqslant 0$
         $-15x_2 \quad\quad - 5x_4 \quad\quad + 5x_6 \geqslant 0$
            $x_1, x_2, x_3, x_4, x_5, x_6 \geqslant 0$

37. Let x_1 = Number of ounces of food L Minimize $C = 0.4x_1 + 0.6x_2 + 0.8x_3$
x_2 = Number of ounces of food M Subject to $30x_1 + 10x_2 + 30x_3 \geqslant 400$
x_3 = Number of ounces of food N $10x_1 + 10x_2 + 10x_3 \geqslant 200$
$$10x_1 + 30x_2 + 20x_3 \geqslant 300$$
$$8x_1 + 4x_2 + 6x_3 \leqslant 150$$
$$60x_1 + 40x_2 + 50x_3 \leqslant 900$$
$$x_1, x_2, x_3 \geqslant 0$$

39. Let x_1 = Number of students from town A enrolled in school I
x_2 = Number of students from town A enrolled in school II
x_3 = Number of students from town B enrolled in school I
x_4 = Number of students from town B enrolled in school II
x_5 = Number of students from town C enrolled in school I
x_6 = Number of students from town C enrolled in school II

Minimize $C = 4x_1 + 8x_2 + 6x_3 + 4x_4 + 3x_5 + 9x_6$
Subject to $x_1 + x_2 \qquad\qquad\qquad = 500$
$$x_3 + x_4 \qquad\qquad = 1{,}200$$
$$x_5 + x_6 = 1{,}800$$
$$x_1 \quad + x_3 \quad + x_5 \quad \leqslant 2{,}000$$
$$x_2 \quad + x_4 \quad + x_6 \leqslant 2{,}000$$
$$x_1 \quad + x_3 \quad + x_5 \quad \geqslant 1{,}400$$
$$x_2 \quad + x_4 \quad + x_6 \geqslant 1{,}400$$
$$x_1 \qquad\qquad\qquad \leqslant 300$$
$$x_2 \qquad\qquad\qquad \leqslant 300$$
$$x_3 \qquad\qquad \leqslant 720$$
$$x_4 \qquad\qquad \leqslant 720$$
$$x_5 \qquad \leqslant 1{,}080$$
$$x_6 \leqslant 1{,}080$$
$$x_1, x_2, x_3, x_4, x_5, x_6 \geqslant 0$$

Exercise 8-8 Chapter Review

1.

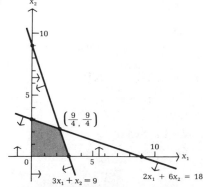

$3x_1 + x_2 = 9$
$2x_1 + 6x_2 = 18$

2. Max $P = 24$ at $x_1 = 4$ and $x_2 = 0$

3. $2x_1 + x_2 + s_1 \qquad\qquad = 8$
$$x_1 + 2x_2 \quad + s_2 \quad = 10$$
$$-6x_1 - 2x_2 \qquad\qquad + P = 0$$

4.

x_1	x_2	s_1	s_2	Feasible?
0	0	8	10	Yes
0	8	0	−6	No
0	5	3	0	Yes
4	0	0	6	Yes
10	0	−12	0	No
2	4	0	0	Yes

5.

$$\begin{array}{ccccc} x_1 & x_2 & s_1 & s_2 & P \end{array}$$

$$\left[\begin{array}{ccccc|c} \textcircled{2} & 1 & 1 & 0 & 0 & 8 \\ 1 & 2 & 0 & 1 & 0 & 10 \\ \hline -6 & -2 & 0 & 0 & 1 & 0 \end{array}\right]$$

6. Max $P = 24$ at $x_1 = 4$ and $x_2 = 0$

7. Min $C = 40$ at $x_1 = 0$ and $x_2 = 20$

8. Maximize $P = 15y_1 + 20y_2$

Subject to $y_1 + 2y_2 \leqslant 5$

$3y_1 + y_2 \leqslant 2$

$y_1, y_2 \geqslant 0$

9. Min $C = 40$ at $x_1 = 0$ and $x_2 = 20$

10. Max $P = 26$ at $x_1 = 2$ and $x_2 = 5$

11. Max $P = 26$ at $x_1 = 2$ and $x_2 = 5$

12. Min $C = 51$ at $x_1 = 9$ and $x_2 = 3$

13. Maximize $P = 10y_1 + 15y_2 + 3y_3$

Subject to $y_1 + y_2 \leqslant 3$

$y_1 + 2y_2 + y_3 \leqslant 8$

$y_1, y_2, y_3 \geqslant 0$

14. Min $C = 51$ at $x_1 = 9$ and $x_2 = 3$

15. No optimal solution exists

16. Max $P = 23$ at $x_1 = 4$, $x_2 = 1$, and $x_3 = 0$

17. Min $C = 14$ at $x_1 = 4$ and $x_2 = 2$ **18.** Min $C = 9,960$ at $x_1 = 0$, $x_2 = 240$, $x_3 = 400$, and $x_4 = 60$

19. (A) Maximize $P = 40x_1 + 30x_2$

Subject to $6x_1 + 4x_2 \geqslant 60$

$6x_1 + 4x_2 \leqslant 108$

$x_1 + x_2 \geqslant 12$

$x_1 + x_2 \leqslant 24$

$x_1, x_2 \geqslant 0$

(B)

$$6x_1 + 4x_2 - s_1 + a_1 = 60$$
$$6x_1 + 4x_2 + s_2 = 108$$
$$x_1 + x_2 - s_3 + a_2 = 12$$
$$x_1 + x_2 + s_4 = 24$$
$$-40x_1 - 30x_2 + Ma_1 + Ma_2 + P = 0$$
$$x_1, x_2, s_1, s_2, s_3, a_1, a_2 \geqslant 0$$

(C)

$$\begin{array}{cccccccc} x_1 & x_2 & s_1 & a_1 & s_2 & s_3 & a_2 & P \end{array}$$

$$\left[\begin{array}{cccccccc|c} 6 & 4 & -1 & 1 & 0 & 0 & 0 & 0 & 60 \\ 6 & 4 & 0 & 0 & 1 & 0 & 0 & 0 & 108 \\ 1 & 1 & 0 & 0 & 0 & -1 & 1 & 0 & 12 \\ 1 & 1 & 0 & 0 & 0 & 0 & 0 & 0 & 24 \\ \hline -40 & -30 & 0 & M & 0 & 0 & M & 1 & 0 \end{array}\right]$$

20. Let $x_1 = $ Number of motors shipped from factory A to plant X

$x_2 = $ Number of motors shipped from factory A to plant Y

$x_3 = $ Number of motors shipped from factory A to plant Z

$x_4 = $ Number of motors shipped from factory B to plant X

$x_5 = $ Number of motors shipped from factory B to plant Y

$x_6 = $ Number of motors shipped from factory B to plant Z

Minimize $C = 5x_1 + 8x_2 + 12x_3 + 9x_4 + 7x_5 + 6x_6$

Subject to $x_1 + x_2 + x_3 \leqslant 1,500$

$x_4 + x_5 + x_6 \leqslant 1,000$

$x_1 + x_4 \geqslant 500$

$x_2 + x_5 \geqslant 700$

$x_3 + x_6 \geqslant 800$

$x_1, x_2, x_3, x_4, x_5, x_6 \geqslant 0$

21. Let x_1 = Number of grams of mix A
x_2 = Number of grams of mix B

Minimize $C = 0.02x_1 + 0.04x_2$
Subject to $3x_1 + 4x_2 \geq 300$
$2x_1 + 5x_2 \geq 200$
$6x_1 + 10x_2 \geq 900$
$x_1, x_2 \geq 0$

Practice Test: Chapter 8

1.

x_2

(0, 8)

10

$2x_1 + x_2 = 8$

(0, 4) (3, 2)

(6, 0)

0 (4, 0) 10 x_1

$2x_1 + 3x_2 = 12$

2.
$$2x_1 + x_2 + s_1 = 8$$
$$2x_1 + 3x_2 + s_2 = 12$$
$$-8x_1 - 10x_2 + P = 0$$
$$x_1, x_2, s_1, s_2 \geq 0$$

3. Max $P = 44$ at $x_1 = 3$ and $x_2 = 2$

4.

x_1	x_2	s_1	s_2	P	
2	1	1	0	0	8
2	③	0	1	0	12
−8	−10	0	0	1	0

5. Max $P = 44$ at $x_1 = 3$ and $x_2 = 2$ **6.** Min $C = 64$ at $x_1 = 6$ and $x_2 = 4$

7. Maximize $P = 6y_1 + 10y_2$
Subject to $y_1 + y_2 \leq 8$
$y_2 \leq 4$
$y_1, y_2 \geq 0$

8. Min $C = 64$ at $x_1 = 6$ and $x_2 = 4$

9. (A) $x_1 = 0, x_2 = 2, s_1 = 0, s_2 = 5, P = 12$; additional pivoting required
  (B) $x_1 = 0, x_2 = 0, s_1 = 0, s_2 = 7, P = 22$; no optimal solution exists
  (C) $x_1 = 6, x_2 = 0, s_1 = 15, s_2 = 0, P = 10$; optimal solution

10. Let x_1 = Number of regular sails
x_2 = Number of competition sails

Maximize $P = 100x_1 + 200x_2$
Subject to $2x_1 + 3x_2 \leq 150$
$4x_1 + 9x_2 \leq 360$
$x_1, x_2 \geq 0$

11. Let x_1 = Number of brand X tablets
x_2 = Number of brand Y tablets

Minimize $C = 0.05x_1 + 0.04x_2$
Subject to $75x_1 + 50x_2 \geq 400$
$100x_1 + 200x_2 \geq 800$
$x_1, x_2 \geq 0$

Exercise 9-2

1. Four ways:

3. Twelve combined outcomes: **5.** $2 \cdot 2 = 4$ **7.** $2 \cdot 6 = 12$

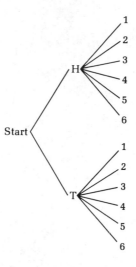

9. $6 \cdot 6 = 36$ **11.** $10 \cdot 9 \cdot 8 = 720$ **13.** $10 \cdot 9 \cdot 8 = 720$ **15.** $6 \cdot 5 \cdot 4 \cdot 3 = 360$; $6 \cdot 6 \cdot 6 \cdot 6 = 1,296$
17. $26 \cdot 26 \cdot 26 \cdot 10 \cdot 10 \cdot 10 = 17,576,000$; $26 \cdot 25 \cdot 24 \cdot 10 \cdot 9 \cdot 8 = 11,232,000$
19. (A) $26 \cdot 25 \cdot 24 = 15,600$ (B) $26 \cdot 25 \cdot 25 = 16,250$ **21.** (A) Six combined outcomes: (B) $3 \cdot 2 = 6$ **23.** $5 \cdot 4 \cdot 3 \cdot 2 = 120$

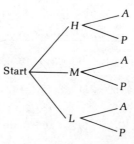

25. (A) Twelve classifications: (B) $2 \cdot 2 \cdot 3 = 12$ **27.** (A) Twenty-four classifications: (B) $4 \cdot 2 \cdot 3 = 24$

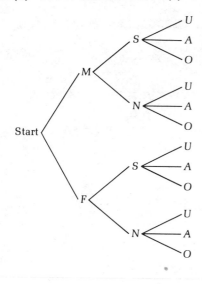

Exercise 9-3

1. 24 **3.** 9 **5.** 990 **7.** 10 **9.** 35 **11.** 1 **13.** 60 **15.** 6,497,400 **17.** 10 **19.** 270,725 **21.** 56 **23.** 7,920

25. $P_{10,3} = 720$ **27.** $C_{10,2} = 45$ **29.** $6! = 720$ **31.** $C_{13,5} = 1,287$ **33.** $\begin{pmatrix} 7 \\ 3, 2, 2 \end{pmatrix} = 210$ **35.** $C_{8,3}C_{10,4}C_{7,2} = 246,960$

37. $C_{13,5}C_{13,2} = 100,386$ **39.** (A) $C_{52,13} = 6.35 \times 10^{11}$ (B) $\begin{pmatrix} 52 \\ 13, 13, 13, 13 \end{pmatrix} = 5.36 \times 10^{28}$ **41.** $P_{6,3} = 120$

43. (A) $C_{6,3}C_{5,2} = 200$ (B) $C_{6,4}C_{5,1} = 75$ (C) $C_{6,5} = 6$ (D) $C_{11,5} = 462$ (E) $C_{6,4}C_{5,1} + C_{6,5} = 81$

45. (A) $P_{4,4} = 4! = 24$ (B) $P_{6,4} = 360$ (C) $C_{6,4} = \binom{6}{4} = 15$ (D) $\binom{12}{4,\,3,\,5} = 27{,}720$ **47.** 336 **49.** $P_{4,2} = 12$

Exercise 9-4

1. $S = \{1, 2, 3, 4, 5, 6, 7, 8\}$ **3.** $\frac{1}{8}$ **5.** $E = \{2, 4, 6, 8\}$; $P(E) = \frac{4}{8} = \frac{1}{2}$ **7.** $\frac{1}{2}$ **9.** Occurrence of E is certain
11. $\{(H, H, H), (H, H, T), (H, T, H), (H, T, T), (T, H, H), (T, H, T), (T, T, H), (T, T, T)\}$
13. $E = \{(H, H, T), (H, T, H), (T, H, H), (H, H, H)\}$; $\frac{1}{2}$
15. (A) No probability can be negative (B) $P(R) + P(G) + P(Y) + P(B) \neq 1$ **17.** $P(R) + P(Y) = .56$
19. $1/P_{5,5} = 1/5! = .008\ 33$ **21.** $\frac{1}{36}$ **23.** $\frac{5}{36}$ **25.** $\frac{1}{6}$ **27.** $\frac{7}{9}$ **29.** 0 **31.** $\frac{1}{3}$ **33.** $(2 \cdot 5 \cdot 5 \cdot 1)/(2 \cdot 5 \cdot 5 \cdot 5) = .2$
35. $C_{16,5}/C_{52,5} \approx .001\ 68$ **37.** $48/C_{52,5} \approx .000\ 018\ 5$ **39.** $4/C_{52,5} \approx .000\ 001\ 5$ **41.** $C_{4,2}C_{4,3}/C_{52,5} \approx .000\ 009$
43. $1/P_{12,4} = .000\ 084\ 2$
45. (A) $C_{6,3}C_{5,2}/C_{11,5} = .433$ (B) $C_{6,4}C_{5,1}/C_{11,5} = .162$ (C) $C_{6,5}/C_{11,5} = .013$ (D) $[C_{6,4}C_{5,1} + C_{6,5}]/C_{11,5} = .175$
47. $\dfrac{C_{4,3}C_{3,2}}{\binom{7}{3,\,2,\,2}} = .0571$ **49.** $1/P_{8,3} = \frac{1}{336} \approx .003$ **51.** (A) $C_{6,2}/C_{11,2} = .273$ (B) $[C_{5,2}C_{6,1} + C_{5,3}]/C_{11,3} = .424$

Exercise 9-5

1. .1 **3.** .45 **5.** P(Point down) = .389, P(Point up) = .611; no
7. (A) P(2 girls) $\approx$ .2351, P(1 girl) $\approx$ .5435, P(0 girls) $\approx$ .2214 (B) P(2 girls) = .25, P(1 girl) = .50, P(0 girls) = .25
9. (A) P(3 heads) $\approx$ .132, P(2 heads) $\approx$ .368, P(1 head) $\approx$ .38, P(0 heads) $\approx$ .12
   (B) P(3 heads) $\approx$ .125, P(2 heads) $\approx$ .375, P(1 head) $\approx$ .375, P(0 heads) $\approx$ .125
   (C) 3 heads, 125; 2 heads, 375; 1 head, 375; 0 heads, 125
11. 4 heads, 5; 3 heads, 20; 2 heads, 30; 1 head, 20; 0 heads, 5 **13.** (A) .015 (B) .222 (C) .169 (D) .958
15. (A) P(Red) = .3, P(Pink) = .44, P(White) = .26 (B) 250 red, 500 pink, 250 white

Exercise 9-6

1. .997 **3.** (1), $\frac{1}{2}$ **5.** (2), $\frac{7}{10}$ **7.** .4 **9.** .25 **11.** .05 **13.** .2 **15.** .6 **17.** .65 **19.** $\frac{1}{4}$ **21.** $\frac{11}{36}$ **23.** .48 **25.** .48 **27.** (1), $\frac{2}{13}$
29. (2), $\frac{4}{13}$ **31.** (2), $\frac{7}{13}$ **33.** (1) and (3), $\frac{11}{13}$ **35.** $\frac{3}{16}$ **37.** $\frac{7}{16}$ **39.** 1 to 1 **41.** 7 to 1 **43.** 2 to 1 **45.** 1 to 2 **47.** $\frac{5}{14}$ **49.** $\frac{7}{9}$

51. $P(E) = 1 - \dfrac{12!}{(12 - n)!\,12^n}$ **55.** (A) $P(C \cup S) = P(C) + P(S) - P(C \cap S) = .45 + .75 - .35 = .85$ (B) $P(C' \cap S') = .15$
57. (A) $P(M_1 \cup A) = P(M_1) + P(A) - P(M_1 \cap A) = .2 + .3 - .05 = .45$
   (B) $P[(M_2 \cap A') + (M_3 \cap A')] = P(M_2 \cap A') + P(M_3 \cap A') = .2 + .35 = .55$
59. $P(K' \cap D') = .9$ **61.** $P(A \cap S) = 50/1{,}000 = .05$ **63.** $1 - C_{15,3}/C_{20,3} \approx .6$

Exercise 9-7 Chapter Review

1. Eight combined outcomes: **2.** 8 **3.** 15, 30 **4.** {R, G, B}; P(R or G) = .8 **5.** .05 **6.** (A) .7 (B) .6 **7.** $2^5 = 32$; 6

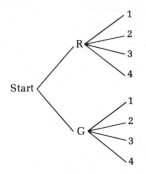

8. $6 \cdot 5 \cdot 4 \cdot 3 \cdot 2 \cdot 1 = 720$ **9.** $P_{(6,6)} = 6! = 720$ **10.** $\binom{8}{2,\,2,\,2,\,2} = 2{,}520$

11. (A) $\frac{2}{13}$ (B) $\frac{4}{13}$ (C) $\frac{12}{13}$

12. $A = \{(1, 3), (2, 2), (3, 1), (2, 6), (3, 5), (4, 4), (5, 3), (6, 2), (6, 6)\}$;
$B = \{(1, 5), (2, 4), (3, 3), (4, 2), (5, 1), (6, 6)\}$; $P(A) = \frac{1}{4}$, $P(B) = \frac{1}{6}$, $P(A \cap B) = \frac{1}{36}$,
$P(A \cup B) = \frac{7}{18}$

13. $1/P_{10,3} \approx .0014$; $1/C_{10,3} \approx .0083$

14. (A) P(2 heads) = .21, P(1 head) = .48, P(0 heads) = .31
(B) P(2 heads) = .25, P(1 head) = .50, P(0 heads) = .25
(C) 2 heads, 250; 1 head, 500; 0 heads, 250

15. (A) $P(M \cup E) = .8$ (B) $P[(M \cup E)'] = .2$ (C) $P[(M \cap E') \cup (M' \cap E)] = .5$

16. (A) $P_{10,3} = 720$ (B) $P_{6,3}/P_{10,3} = \frac{1}{6}$ (C) $C_{10,3} = 120$
(D) $(C_{6,3} + C_{6,2} \cdot C_{4,1})/C_{10,3} = \frac{2}{3}$

17. $\binom{12}{5,\,3,\,4} = 27{,}720$ **18.** 336; 512; 392 **19.** $C_{13,3}C_{13,2} = 22{,}308$ **20.** $C_{13,5}/C_{52,5} \approx .0005$

21. (1) Probability of an event cannot be negative; (2) sum of probabilities of simple events must be 1; (3) probability of an event cannot be greater than 1

22. $C_{8,2}/C_{10,4} = \frac{2}{15}$ **23.** $1 - 10!/(5!10^5) \approx .70$

Practice Test: Chapter 9

1. (A) Twelve combined outcomes: (B) $6 \cdot 2 = 12$ **2.** (A) $P_{6,3} = 120$ (B) $C_{5,2} = 10$

3. Probability of an event cannot be negative; probability of an event cannot be greater than 1; sum of probabilities of all simple events must be 1

4. $1/P_{15,2} \approx .0048$ **5.** (A) $C_{13,5}/C_{52,5}$ (B) $C_{13,3} \cdot C_{13,2}/C_{52,5}$

6. .350; $\frac{3}{8} = .375$ **7.** (A) $\frac{1}{3}$ (B) $\frac{2}{9}$ **8.** $1 - C_{7,3}/C_{10,3} = \frac{17}{24}$

9. (A) $P(H \cup A) = P(H) + P(A) - P(H \cap A) = .7 + .6 - .4 = .9$
(B) $P(H \cap A') = .3$

10. (A) .04 (B) .16 (C) .54

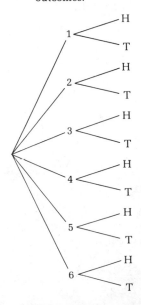

Chapter 10

Exercise 10-2

1. (A) 1 (B) 1 (C) 1 **3.** (A) 2 (B) 1 (C) Does not exist **5.** (A) 1 (B) 1 (C) 1 **7.** (A) 1 (B) 1 (C) Yes **9.** (A) Does not exist (B) 1
(C) No **11.** (A) 1 (B) 3 (C) No **13.** 47 **15.** −4 **17.** 5/3 **19.** 243 **21.** −3 **23.** None **25.** $x = 5$ **27.** $x = -2, 3$
29. (A) 1 (B) 2 (C) Does not exist **31.** (A) 1 (B) 1 (C) 1 **33.** (A) Does not exist (B) 2 (C) No; yes **35.** $\sqrt[3]{4}$ **37.** 0 **39.** −5
41. 5/6 **43.** 1/2 **45.** 2/3 **47.** 1/4 **49.**

x	0.9	0.99	0.999	→1←	1.001	1.01	1.1
$f(x)$	−1	−1	−1	→?←	1	1	1

(A) −1 (B) 1 (C) Does not exist

51.

x	0.9	0.99	0.999	→1←	1.001	1.01	1.1
$f(x)$	2.71	2.97	2.997	→?←	3.003	3.03	3.31

(A) 3 (B) 3 (C) 3 **53.** All x **55.** $x \geqslant 5$ **57.** 3 **59.** 5 **61.** 0

63. −1/4 **65.** $1/(2\sqrt{2})$ or $\sqrt{2}/4$ **67.** 12 **69.** 1/12 **71.** 1 **73.** (A)

(B) \$.37; \$.54; does not exist
(C) Does not exist; \$.71
(D) No; yes

75. (A)

(B) 3, 6, 9, 12 (C) Yes; no (see Example 6)

77. (A) t_2, t_3, t_4, t_6, t_7 (B) 7, 7 (C) Does not exist; 4

Exercise 10-3

1. $\Delta x = 3$; $\Delta y = 45$; $\Delta y / \Delta x = 15$ **3.** 12 **5.** 12 **7.** 12 **9.** 15 **11.** (A) $12 + 3\Delta x$ (B) 12 **13.** (A) $24 + 3\Delta x$ (B) 24
15. (A) 5 meters per second (B) $3 + \Delta x$ meters per second (C) 3 meters per second **17.** (A) 5 (B) $3 + \Delta x$ (C) 3
(D) $y = 3x - 1$ **19.** 3 meters per second **21.** (A) \$200 per year (B) \$450 per year **23.** (A) −110 square millimeters
per day (B) −15 square millimeters per day **25.** (A) 0.6 birth per year (B) 8 births per year

Exercise 10-4

1. $f'(x) = 2$; $f'(1) = f'(2) = f'(3) = 2$ **3.** $f'(x) = 6 - 2x$; $f'(1) = 4$, $f'(2) = 2$, $f'(3) = 0$ **5.** $f'(x) = -1/(x+1)^2$; $f'(1) = -1/4$,
$f'(2) = -1/9$, $f'(3) = -1/16$ **7.** $f'(x) = 1/(2\sqrt{x})$; $f'(1) = 1/2$, $f'(2) = 1/(2\sqrt{2})$, $f'(3) = 1/(2\sqrt{3})$ **9.** $f'(x) = -2/x^3$;
$f'(1) = -2$, $f'(2) = -1/4$, $f'(3) = -2/27$ **11.** $v = f'(x) = 8x - 2$; $f'(1) = 6$ feet per second; $f'(3) = 22$ feet per second;

$f'(5) = 38$ feet per second   **13.** (A) $m = f'(x) = 2x$ (B) $m_1 = f'(-2) = -4$; $m_2 = f'(0) = 0$; $m_3 = f'(2) = 4$ (C) $y = -4x - 4$; $y = 0$; $y = 4x - 4$ (D)

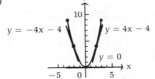

15. (A) $f'(x) = 3x^2 + 2$ (B) $f'(1) = 5$; $f'(3) = 29$

17. (A) $C'(x) = 10 - 2x$ (B) $C'(1) = \$8$ hundred per unit increase; $C'(3) = \$4$ hundred per unit increase; $C'(4) = \$2$ hundred per unit increase   **19.** (A) $N'(t) = 2t - 8$ (B) $N'(1) = -6$ thousand per hour; $N'(2) = -4$ thousand per hour; $N'(3) = -2$ thousand per hour [*Note:* A negative rate indicates the population is decreasing.]

Exercise 10-5

1. 0   **3.** 0   **5.** $12x^{11}$ **7.** 1   **9.** $8x^3$ **11.** $2x^5$ **13.** $-10x^{-6}$ **15.** $-x^{-2/3}$ **17.** $15x^4 - 6x^2$ **19.** $-12x^{-5} - 4x^{-3}$ **21.** $-6x^{-3}$
23. $2x^{-1/3} - (5/3)x^{-2/3}$ **25.** $-(9/5)x^{-8/5} + 3x^{-3/2}$ **27.** $-(1/3)x^{-4/3}$ **29.** $-6x^{-3/2} + 6x^{-3} + 1$ **31.** (A) $m = 6 - 2x$ (B) 2; -2
(C) $x = 3$ **33.** (A) $m = x^2 - 6x$ (B) -8; -8 (C) $x = 0, 6$ **35.** (A) $v = 176 - 32x$ (B) 176 feet per second; 80 feet per second;
-16 feet per second (C) $x = 5.5$ seconds   **37.** (A) $v = 40 - 10x$ (B) 40 feet per second; 10 feet per second; -20 feet
per second (C) $x = 4$ seconds   **39.** $2x - 3 - 10x^{-3}$ **41.** $-x^{-2} + 9x^{-5/2}$ **43.** (A) $N'(x) = 60 - 2x$ (B) $N'(10) = 40$ (at the
$\$10,000$ level of advertising, there would be an approximate increase of 40 units of sales per $\$1,000$ increase in
advertising); $N'(20) = 20$ (at the $\$20,000$ level of advertising, there would be an approximate increase of only 20 units of
sales per $\$1,000$ increase in advertising); the effect of advertising levels off as the amount spent increases.
45. (A) -1.37 beats per minute (B) -0.58 beat per minute   **47.** (A) 25 items per hour (B) 8.33 items per hour

Exercise 10-6

1. $2x^3(2x) + (x^2 - 2)(6x^2) = 10x^4 - 12x^2$ **3.** $(x - 3)(2) + (2x - 1)(1) = 4x - 7$ **5.** $\dfrac{(x - 3)(1) - x(1)}{(x - 3)^2} = \dfrac{-3}{(x - 3)^2}$

7. $\dfrac{(x - 2)(2) - (2x + 3)(1)}{(x - 2)^2} = \dfrac{-7}{(x - 2)^2}$ **9.** $(x^2 + 1)(2) + (2x - 3)(2x) = 6x^2 - 6x + 2$

11. $\dfrac{(2x - 3)(2x) - (x^2 + 1)(2)}{(2x - 3)^2} = \dfrac{2x^2 - 6x - 2}{(2x - 3)^2}$ **13.** $(2x + 1)(2x - 3) + (x^2 - 3x)(2) = 6x^2 - 10x - 3$

15. $(2x - x^2)(5) + (5x + 2)(2 - 2x) = -15x^2 + 16x + 4$ **17.** $\dfrac{(x^2 + 2x)(5) - (5x - 3)(2x + 2)}{(x^2 + 2x)^2} = \dfrac{-5x^2 + 6x + 6}{(x^2 + 2x)^2}$

19. $\dfrac{(x^2 - 1)(2x - 3) - (x^2 - 3x + 1)(2x)}{(x^2 - 1)^2} = \dfrac{3x^2 - 4x + 3}{(x^2 - 1)^2}$ **21.** $(2x^4 - 3x^3 + x)(2x - 1) + (x^2 - x + 5)(8x^3 - 9x^2 + 1)$

23. $\dfrac{(4x^2 + 5x - 1)(6x - 2) - (3x^2 - 2x + 3)(8x + 5)}{(4x^2 + 5x - 1)^2}$ **25.** $9x^{1/3}(3x^2) + (x^3 + 5)(3x^{-2/3})$ **27.** $\dfrac{(x^2 - 3)(2x^{-2/3}) - 6x^{1/3}(2x)}{(x^2 - 3)^2}$

29. $x^{-2/3}(3x^2 - 4x) + (x^3 - 2x^2)[(-2/3)x^{-5/3}]$ **31.** $\dfrac{(x^2 + 1)[(2x^2 - 1)(2x) + (x^2 + 3)(4x)] - (2x^2 - 1)(x^2 + 3)(2x)}{(x^2 + 1)^2}$

33. (A) $d'(x) = \dfrac{-50,000(2x + 10)}{(x^2 + 10x + 25)^2} = \dfrac{-100,000}{(x + 5)^3}$ (B) $d'(5) = -100$ radios per $\$1$ increase in price; $d'(10) = -30$ radios per $\$1$

increase in price   **35.** (A) $N'(x) = \dfrac{(x + 32)(100) - (100x + 200)}{(x + 32)^2} = \dfrac{3,000}{(x + 32)^2}$ (B) $N'(4) = 2.31$; $N'(68) = 0.30$

Exercise 10-7

1. $y = u^3$, $u = 2x + 5$ **3.** $y = u^8$, $u = x^3 - x^2$ **5.** $y = u^{1/3}$, $u = x^3 + 3x$ **7.** $6(2x + 5)^2$ **9.** $8(x^3 - x^2)^7(3x^2 - 2x)$
11. $(x^3 + 3x)^{-2/3}(x^2 + 1)$ **13.** $24x(x^2 - 2)^3$ **15.** $-6(x^2 + 3x)^{-4}(2x + 3)$ **17.** $x(x^2 + 8)^{-1/2}$ **19.** $(3x + 4)^{-2/3}$

21. $(1/2)(x^2 - 4x + 2)^{-1/2}(2x - 4) = (x - 2)/(x^2 - 4x + 2)^{1/2}$ **23.** $(-1)(2x + 4)^{-2}(2) = -2/(2x + 4)^2$

25. $(-1)(4x^2 - 4x + 1)^{-2}(8x - 4) = -4/(2x - 1)^3$ **27.** $-2(x^2 - 3x)^{-3/2}(2x - 3) = \dfrac{-2(2x - 3)}{(x^2 - 3x)^{3/2}}$

29. $-(3 - x^{1/3})^{-2}\left(-\dfrac{1}{3}x^{-2/3}\right) = \dfrac{1}{3(3 - x^{1/3})^2 x^{2/3}}$ **31.** $-(x^{1/2} - 5)^{-3/2}x^{-1/2} = \dfrac{-1}{(x^{1/2} - 5)^{3/2}x^{1/2}}$

33. $18x^2(x^2 + 1)^2 + 3(x^2 + 1)^3 = 3(x^2 + 1)^2(7x^2 + 1)$ **35.** $\dfrac{2x^3 4(x^3 - 7)^3 3x^2 - (x^3 - 7)^4 6x^2}{4x^6} = \dfrac{3(x^3 - 7)^3(3x^3 + 7)}{2x^4}$

37. $(2x - 3)^2[3(2x^2 + 1)^2(4x)] + (2x^2 + 1)^3[2(2x - 3)(2)] = 4(2x^2 + 1)^2(2x - 3)(8x^2 - 9x + 1)$

39. $4x^2[(1/2)(x^2 - 1)^{-1/2}(2x)] + (x^2 - 1)^{1/2}(8x) = (12x^3 - 8x)/\sqrt{x^2 - 1}$ **41.** $\dfrac{(x - 3)^{1/2}(2) - 2x[(1/2)(x - 3)^{-1/2}]}{x - 3} = \dfrac{x - 6}{(x - 3)^{3/2}}$

43. $(2x - 1)^{1/2}(x^2 + 3)(11x^2 - 4x + 9)$ **45.** $y = -x + 3$ **47.** (A) $\overline{C}'(x) = 2(2x - 8)2 = 8x - 32$ (B) $\overline{C}'(2) = -16; \overline{C}'(4) = 0;$
$\overline{C}'(6) = 16.$ An increase in production at the 2,000 level will reduce costs; at the 4,000 level, no increase or decrease will

occur; and at the 6,000 level, an increase in production will increase the costs. **49.** $\dfrac{(4 \times 10^6)x}{(x^2 - 1)^{5/3}}$

51. (A) $f'(n) = n(n - 2)^{-1/2} + 2(n - 2)^{1/2} = \dfrac{3n - 4}{(n - 2)^{1/2}}$ (B) $f'(11) = 29/3$ (rate of learning is 29/3 units per minute at the
$n = 11$ level); $f'(27) = 77/5$ (rate of learning is 77/5 units per minute at the $n = 27$ level)

Exercise 10-8 Chapter Review

1. $12x^3 - 4x$ **2.** $x^{-1/2} - 3 = (1/x^{1/2}) - 3$ **3.** 0 **4.** 0 **5.** $(2x - 1)(3) + (3x + 2)(2) = 12x + 1$

6. $(x^2 - 1)(3x^2) + (x^3 - 3)(2x) = 5x^4 - 3x^2 - 6x$ **7.** $\dfrac{(x^2 + 2)2 - 2x(2x)}{(x^2 + 2)^2} = \dfrac{4 - 2x^2}{(x^2 + 2)^2}$ **8.** $(-1)(3x + 2)^{-2}3 = -3/(3x + 2)^2$

9. $3(2x - 3)^2 2 = 6(2x - 3)^2$ **10.** $-2(x^2 + 2)^{-3}2x = -4x/(x^2 + 2)^3$ **11.** $12x^3 + 6x^{-4}$

12. $(2x^2 - 3x + 2)(2x + 2) + (x^2 + 2x - 1)(4x - 3) = 8x^3 + 3x^2 - 12x + 7$ **13.** $\dfrac{(x - 1)^2 2 - (2x - 3)2(x - 1)}{(x - 1)^4} = \dfrac{4 - 2x}{(x - 1)^3}$

14. $x^{-1/2} - 2x^{-3/2} = \dfrac{1}{\sqrt{x}} - \dfrac{2}{\sqrt{x^3}}$ **15.** $(x^2 - 1)[2(2x + 1)2] + (2x + 1)^2(2x) = 2(2x + 1)(4x^2 + x - 2)$

16. $(1/3)(x^3 - 5)^{-2/3}3x^2 = \dfrac{x^2}{\sqrt[3]{(x^3 - 5)^2}}$ **17.** $-(1/3)(3x^2 - 2)^{-4/3}6x = \dfrac{-2x}{\sqrt[3]{(3x^2 - 2)^4}}$

18. $\dfrac{(2x - 3)4(x^2 + 2)^3 2x - (x^2 + 2)^4 2}{(2x - 3)^2} = \dfrac{2(x^2 + 2)^3(7x^2 - 12x - 2)}{(2x - 3)^2}$ **19.** (A) $m = f'(1) = 2$ (B) $y = 2x + 3$

20. (A) $m = f'(x) = 10 - 2x$ (B) $x = 5$ **21.** (A) $v = f'(x) = 32x - 4$ (B) $f'(3) = 92$ feet per second
22. (A) $v = f'(x) = 96 - 32x$ (B) $x = 3$ seconds **23.** (A) 4 (B) 6 (C) Does not exist (D) 6 (E) No **24.** (A) 3 (B) 3 (C) 3 (D) 3
(E) Yes **25.** None **26.** $x = -5$ **27.** $x = -2, 3$ **28.** None **29.** $[2(3) - 3]/(3 + 5) = 3/8$ **30.** $2(3^2) - 3 + 1 = 16$ **31.** -1
32. 4 **33.** 1/6 **34.** Does not exist **35.** $1/(2\sqrt{7})$ **36.** $\sqrt{2}$ **37.** $2x - 1$ **38.** $1/(2\sqrt{x})$ **39.** (A) 5/8 (B) Does not exist
(C) No **40.** (A) 1/2 (B) 1/2 (C) Yes **41.** $x = -2/3, 2$ **42.** (A) 2 (B) 3 (C) Does not exist
43. $(14x^3 + 36x^2 - 2)/[(3x + 6)^{5/3}]$ **44.** (A) $\overline{C}'(x) = 2x - 10$ (B) $\overline{C}'(3) = -4$ (average cost per unit is decreasing at
approximately \$400 per unit as production increases at an output level of 3 units); $\overline{C}'(5) = 0$ (average cost per unit does
not change for a small change in production at an output level of 5 units); $\overline{C}'(7) = 4$ (average cost per unit is increasing
at approximately \$400 per unit as production increases at an output level of 7 units) **45.** $C'(9) = -1$ part per million
per meter; $C'(99) = -0.001$ part per million per meter **46.** (A) 10 items per hour (B) 5 items per hour

Practice Test: Chapter 10

1. $6x - x^{-1/2} = 6x - (1/x^{1/2})$ **2.** $(x^2 + 2)(2) + (2x - 3)(2x) = 6x^2 - 6x + 4$ **3.** $\dfrac{(x^2 + 1)(6x) - (3x^2 - 5)(2x)}{(x^2 + 1)^2} = \dfrac{16x}{(x^2 + 1)^2}$

4. $3(2x^3 - 3x + 1)^2(6x^2 - 3)$ **5.** $-x^{-4/3} + 2x^{-2} + 1$ **6.** $(x^2 - 1)^3(2) + (2x + 1)[3(x^2 - 1)^2 2x] = 2(x^2 - 1)^2(7x^2 + 3x - 1)$

7. $-(1/4)(2x^2 - 3)^{-5/4}4x = -x/\sqrt[4]{(2x^2 - 3)^5}$ **8.** $\dfrac{(x^2 + 5)^4(1/2)(2x - 1)^{-1/2}2 - (2x - 1)^{1/2}4(x^2 + 5)^3 2x}{(x^2 + 5)^8} = \dfrac{-15x^2 + 8x + 5}{(2x - 1)^{1/2}(x^2 + 5)^5}$

9. (A) $m = f'(x) = 8 - 2x$ (B) 4 (C) $y = 4x + 4$ (D) $x = 4$ **10.** (A) $v = f'(x) = 80 - 20x$ (B) 20 feet per second
(C) $x = 4$ seconds   **11.** (A) $f'(x) = \lim\limits_{\Delta x \to 0} \dfrac{f(x + \Delta x) - f(x)}{\Delta x}$ (B) $1 - 2x$ **12.** (A) 1 (B) -1 (C) Does not exist (D) -1 (E) No

13. (A) 2 (B) 2 (C) 2 (D) 2 (E) Yes   **14.** (A) 1/8 (B) Not defined (C) No   **15.** (A) 1/9 (B) 1/9 (C) Yes

16. (A) $C'(x) = 15 + \dfrac{500}{(2x + 1)^{3/2}}$; (B) $C'(12) = \$19$

Chapter 11

Exercise 11-1

1. $y' = 6x$; 6   **3.** $y' = 3x/y$; 3   **5.** $y' = 1/(2y + 1)$; 1/3   **7.** $y' = -y/x$; $-3/2$ **9.** $y' = -2y/(2x + 1)$; 4
11. $y' = (6 - 2y)/x$; -1 **13.** $x' = (2tx - 3t^2)/(2x - t^2)$; 8   **15.** $y = -x + 5$ **17.** $y = (2/5)x - 12/5$; $y = (3/5)x + 12/5$
19. $y' = 1/[3(1 + y)^2 + 1]$; 1/13   **21.** $y' = 3(x - 2y)^2/[6(x - 2y)^2 + 4y]$; 3/10   **23.** $y' = 3x^2(7 + y^2)^{1/2}/y$; 16
25. $p' = 1/(2p - 2)$ **27.** $p' = -\sqrt{10{,}000 - p^2}/p$ **29.** $dL/dV = -(L + m)/(V + n)$

Exercise 11-2

1. 240   **3.** 9/4   **5.** 1/2   **7.** Decreasing at 9 units per second   **9.** Approximately -3.03 feet per second
11. $dA/dt \approx 126$ square feet per second   **13.** 3,768 cubic centimeters per minute   **15.** 6 pounds per square inch
per hour   **17.** $-9/4$ feet per second   **19.** 20/3 feet per second   **21.** (A) $dC/dt = \$15{,}000$ per week (B) $dR/dt = -\$50{,}000$
per week (C) $dP/dt = -\$65{,}000$ per week   **23.** Approximately 100 cubic feet per minute

Exercise 11-3

1. $6x - 4$ **3.** 0   **5.** $40x^3$ **7.** 0   **9.** $-6x^{-4}$ **11.** $6x$ **13.** $6x^{-3} + 12x^{-4}$ **15.** $-15(2x - 1)^{-7/2}$ **17.** $24(1 - 2x)$
19. $24x^2(x^2 - 1) + 6(x^2 - 1)^2 = 6(x^2 - 1)(5x^2 - 1)$ **21.** $15(3 - 2x)^{-7/2}$
23. $16x^2(3x^2 - 1)^{-2/3} + 8(3x^2 - 1)^{1/3} = (40x^2 - 8)/(3x^2 - 1)^{2/3}$ **25.** $-12/y^3$ **27.** $-(6y^3 + 8x^2)/(9y^5)$ **29.** $-24(2x - 1)^{-4}$
31. $-12x/y^5$

Exercise 11-4

1. $dy = (24x - 3x^2)\, dx$ **3.** $dy = \left(2x - \dfrac{x^2}{3}\right) dx$ **5.** $dy = -\dfrac{295}{x^{3/2}}\, dx$ **7.** $dy = \dfrac{150}{x^2}\, dx$ **9.** $dy = 1.4$, $\Delta y = 1.44$

11. $dy = 3$, $\Delta y = 2.73$ **13.** 2.03   **15.** 3.04   **17.** 120 cubic inches   **19.** $dy = \dfrac{6x - 2}{3(3x^2 - 2x + 1)^{2/3}}\, dx$

21. $dy = 3.9$, $\Delta y = 3.83$ **23.** 40 unit increase; 20 unit increase   **25.** $-\$6$, $\$4$ **27.** -1.37 per minute; -0.58 per minute
29. 1.26 square millimeters   **31.** 3 words per minute   **33.** (A) 2,100 increase (B) 4,800 increase (C) 2,100 increase

Exercise 11-5

1. (A) $C'(x) = 60$ (B) $R(x) = xp(x) = 200x - (x^2/30)$ (C) $R'(x) = 200 - (x/15)$ (D) $R'(1,500) = 100$ (revenue is increasing at $100 per unit increase in production at the 1,500 output level); $R'(4,500) = -100$ (revenue is decreasing $100 per unit increase at the 4,500 output level) (E) (F) $P(x) = R(x) - C(x) = -(x^2/30) + 140x - 72,000$

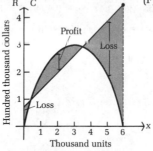

(G) $P'(x) = -(x/15) + 140$ (H) $P(1,500) = 40$ (profit is increasing at approximately $40 per unit increase in production at the 1,500 output level); $P'(3,000) = -60$ (profit is decreasing at approximately $60 per unit increase in production at the 3,000 output level)
3. (A) $\overline{C}(x) = (72,000/x) + 60; \overline{R}(x) = xp/x = 200 - (x/30); \overline{P}(x) = \overline{R}(x) - \overline{C}(x) = 140 - (x/30) - (72,000/x)$
(B) $\overline{C}'(x) = -72,000/x^2; \overline{R}'(x) = -1/30; \overline{P}'(x) = -1/30 + 72,000/x^2$
(C) $\overline{P}'(1,000) = \$0.039$ (profit is increasing at a rate of approximately 3.9¢ per unit at an output level of 1,000 units per week); $\overline{P}'(6,000) = -\$0.031$ (profit is decreasing at a rate of approximately 3.1¢ per unit at an output level of 6,000 units per week)

Exercise 11-6   Chapter Review

1. $y' = 9x^2/(4y); 9/8$ **2.** $\dfrac{dy}{dt} = 216$ **3.** $\dfrac{d^2y}{dx^2} = 6x + (1/4)x^{-3/2}$ **4.** $dy = 18x(3x^2 - 7)^2\,dx$ **5.** $y' = y/(4y^3 - x); 1/13$
6. $x' = 4tx/(3x^2 - 2t^2); -4$ **7.** $y'' = -2(2x^2 - y^2)/y^3 = -6/y^3$
8. $21x^2(2x^2 - 3)^{-1/4} + 7(2x^2 - 3)^{3/4} = (35x^2 - 21)/(2x^2 - 3)^{1/4}$ **9.** $120(5 - 4x)^{-7/2}$ **10.** $dy = 7.3, \Delta y = 7.45$ **11.** 4.13
12. 7 units per second   **13.** $\dfrac{dR}{dt} = 1/\pi \approx 0.318$ inch per minute   **14.** $y' = -4x(5 - y^2)^{1/2}/y; -16$ **15.** $\dfrac{d^3y}{dx^3} = \dfrac{-18x}{y^5}$
16. $dy = -0.0031, \Delta y = -0.0031$ **17.** (A) $C'(x) = 2; \overline{C}(x) = 2 + 56x^{-1}; \overline{C}'(x) = -56x^{-2}$ (B) $R(x) = xp = 20x - x^2;$
$R'(x) = 20 - 2x; \overline{R}(x) = 20 - x; \overline{R}'(x) = -1$ (C) $P(x) = R(x) - C(x) = 18x - x^2 - 56; P'(x) = 18 - 2x; \overline{P}(x) = 18 - x - 56x^{-1};$
$\overline{P}'(x) = -1 + 56x^{-2}$ (D) Solving $R(x) = C(x)$, we find break-even points at $x = 4, 14$. (E) $P'(7) = 4$ (increasing production increases profit); $P'(9) = 0$ (stable); $P'(11) = -4$ (increasing production decreases profit) (F)

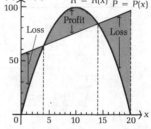

18. $p' = -(5,000 - 2p^3)^{1/2}/3p^2$ **19.** $\dfrac{dR}{dt} = \$110$ per day   **20.** $\dfrac{dR}{dt} = -\dfrac{3}{2\pi} \approx 0.477$ millimeters per day

21. $\dfrac{dT}{dt} = -\dfrac{1}{27} \approx -0.037$ minute per operation per hour

Practice Test: Chapter 11

1. $y' = (3y - 2x)/(8y - 3x)$; $8/19$ **2.** $-(y^2 + x^2)/y^3 = -81/y^3$ **3.** $-3(1 - 2x)^{-5/2}$

4. $27x^2(5 - 3x^2)^{-1/2} - 9(5 - 3x^2)^{1/2} = \dfrac{54x^2 - 45}{(5 - 3x^2)^{1/2}}$ **5.** $\Delta y = 0.41$, $dy = 0.4$ **6.** $3 - 1/27 \approx 2.96$ **7.** $\dfrac{dy}{dt} = -2$ units

per second   **8.** 0.27 foot per second   **9.** (A) $R(x) = xp = 14x - x^2$; $P(x) = R(x) - C(x) = -x^2 + 12x - 20$
(B) $P'(x) = -2x + 12$; $\overline{P}(x) = P(x)/x = -x + 12 - 20x^{-1}$; $\overline{P}'(x) = -1 + 20x^{-2}$ (C) $P'(4) = 4$ (increasing production increases
profit); $P'(6) = 0$ (stable); $P'(8) = -4$ (increasing production decreases profit) (D) $x = 2, 10$ (E)

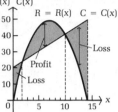

Chapter 12

Exercise 12-1

1.

x	10	100	1,000	10,000
$f(x)$	.091	.0099	.000999	.0000999

; $\lim\limits_{x \to \infty} f(x) = 0$

3.

x	10	100	1,000	10,000
$f(x)$	9.09	99.01	999.001	9,999.0001

; $\lim\limits_{x \to \infty} f(x)$ does not exist

5.

x	1.1	1.01	1.001	1.0001
$f(x)$	10	100	1,000	10,000

;

x	.9	.99	.999	.9999
$f(x)$	−10	−100	−1,000	−10,000

; $\lim\limits_{x \to 1^+} f(x) = \infty$, $\lim\limits_{x \to 1^-} f(x) = -\infty$

7.

x	1.1	1.01	1.001	1.0001
$f(x)$	4.64	21.5	100	464.2

;

x	.9	.99	.999	.9999
$f(x)$	4.64	21.5	100	464.2

; $\lim\limits_{x \to 1^+} f(x) = \infty$, $\lim\limits_{x \to 1^-} f(x) = \infty$

9. (A) Does not exist (B) L **11.** (A) 0 (B) ∞ **13.** 4   **15.** 2/3   **17.** 0   **19.** Does not exist   **21.** $x = 1$ and $x = 2$ **23.** $x = 2$

25. $x = -1$, $x = 1$ **27.** Horizontal asymptote at $y = 2$; vertical asymptote at $x = 4$; $\lim_{x \to 4^+} f(x) = \infty$; $\lim_{x \to 4^-} f(x) = -\infty$

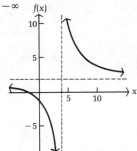

29. Horizontal asymptote at $y = 0$; vertical asymptotes at $x = -2$ and $x = 2$; $\lim_{x \to -2^+} f(x) = \infty$; $\lim_{x \to -2^-} f(x) = -\infty$; $\lim_{x \to 2^+} f(x) = \infty$; $\lim_{x \to 2^-} f(x) = -\infty$

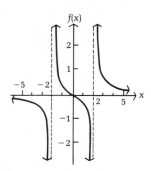

31. (A)

x	10	100	1,000
$f(x)$	.995	.99995	.9999995

;

x	-10	-100	$-1,000$
$f(x)$	$-.995$	$-.99995$	$-.9999995$

(B) $\lim_{x \to \infty} f(x) = 1$, $\lim_{x \to -\infty} f(x) = -1$

(C)

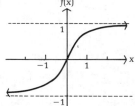

33. 0 **35.** $1/2$ **37.** (A) $\overline{C}(x) = 3{,}000/x + 2.75$ (B) 2.75 (C) ∞

39. (A) $C'(x) = 12 - 100/x^2$ (B) 12 **41.** (A) $C(0) = 200$ (B) 50 **43.** 6

Exercise 12-2

1. (a, b); (d, e); (e, f); (g, h) **3.** c, d, f **5.** b, f **7.** (A) Local minimum (B) Neither **9.** (A) Local maximum (B) Neither (C) Neither **11.**

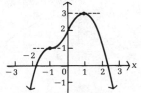

13.

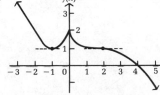

15. Increasing on $(-\infty, -1)$, $(0, 1)$; decreasing on $(-1, 0)$, $(1, \infty)$; $f(-1) = 1$ is a local maximum; $f(0) = 0$ is a local minimum; $f(1) = 1$ is a local maximum

17. Increasing on $(-\infty, -2)$, $(4, \infty)$; decreasing on $(-2, 4)$; $f(-2) = 35$ is a local maximum; $f(4) = -73$ is a local minimum   **19.** Increasing on $(-\infty, 3)$, $(5, \infty)$; decreasing on $(3, 5)$; $f(3) = 3(2^{2/3}) \approx 4.8$ is a local maximum; $f(5) = 0$ is a local minimum   **21.** Increasing on $(-\infty, -2)$, $(2, \infty)$; decreasing on $(-2, 0)$, $(0, 2)$; $f(-2) = -4$ is a local maximum; $f(2) = 4$ is a local minimum   **23.** Increasing on $(-2, 0)$; decreasing on $(-\infty, -2)$, $(0, \infty)$; $f(-2) = 3/4$ is a local minimum   **25.** Increasing on $(0, 2)$, $(10, \infty)$; decreasing on $(2, 10)$; $f(2) = 64\sqrt{2} \approx 90.5$ is a local maximum; $f(10) = 0$ is a local minimum   **27.** Increasing on $(-\infty, 0)$, $(4, \infty)$; decreasing on $(0, 2)$, $(2, 4)$; $f(0) = 0$ is a local maximum; $f(4) = 8$ is a local minimum   **29.** Rising on $(-\infty, 4)$; falling on $(4, \infty)$; horizontal tangent at $x = 4$

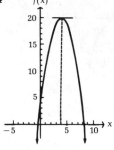

31. Rising on $(-\infty, -1)$, $(1, \infty)$; falling on $(-1, 1)$; horizontal tangents at $x = -1, 1$

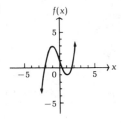

33. Falling on $(-\infty, 2)$; rising on $(2, \infty)$; $f'(2)$ does not exist

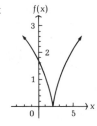

35. Rising on $(0, 1)$; falling on $(1, \infty)$; horizontal tangent at $x = 1$; $f'(0)$ does not exist

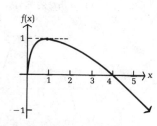

37. Rising on $(-\infty, -1)$ and $(0, 1)$; falling on $(-1, 0)$ and $(1, \infty)$; horizontal tangents at $x = 1, -1$; $f'(0)$ does not exist

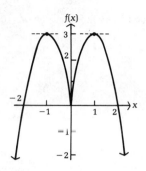

39. Falling on $(-\infty, 3)$ and $(3, \infty)$; horizontal asymptote $y = 1$; vertical asymptote $x = 3$

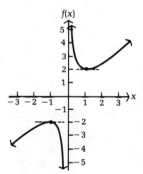

41. Rising on $(-\infty, -1)$ and $(1, \infty)$; falling on $(-1, 0)$ and $(0, 1)$; horizontal tangents at $x = -1, 1$; vertical asymptote $x = 0$

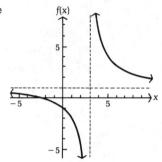

43. Rising on $(-\infty, 2)$; falling on $(2, \infty)$; horizontal asymptote $y = 1$; vertical asymptote $x = 2$

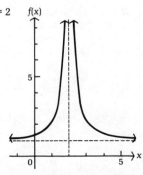

45. (A) Decreasing on (0, 100); increasing on (100, ∞) (B)

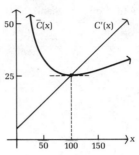

47. (A) 10 < x < 40 (B) 10 < x < 25 **49.** 5 hours   **51.** (A) (12, ∞) (B) (16, ∞)

Exercise 12-3

1. (a, c), (c, d), (e, g) **3.** d, e, g **5.** (A) Local minimum (B) Neither (C) Test fails **7.**

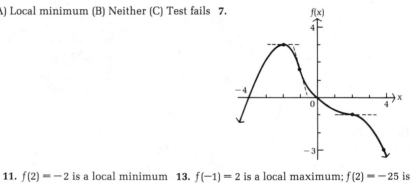

9.

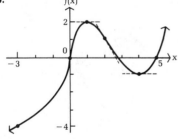

11. $f(2) = -2$ is a local minimum **13.** $f(-1) = 2$ is a local maximum; $f(2) = -25$ is a local minimum

15. No local extrema **17.** $f(-2) = -6$ is a local minimum; $f(0) = 10$ is a local maximum; $f(2) = -6$ is a local minimum
19. $f(0) = 2$ is a local minimum **21.** $f(-4) = -8$ is a local maximum; $f(4) = 8$ is a local minimum **23.** Local maximum at $x = 0$; local minimum at $x = 4$; inflection point at $x = 2$

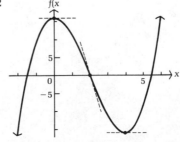

25. Inflection point at $x = 0$

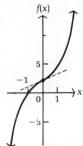

27. Local minima at $x = -\sqrt{3}$ and $x = \sqrt{3}$;
local maximum at $x = 0$; inflection points at $x = -1$ and $x = 1$

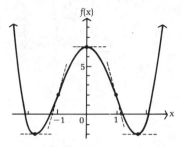

29. Local minimum at $x = 0$;
inflection points at $x = 1$ and $x = 3$

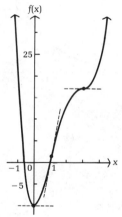

31. Inflection point at $x = 1$

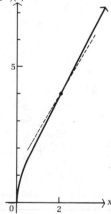

33. Local minimum at $x = -1$; local maximum
at $x = 1$; inflection point at $x = 0$

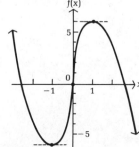

35. Vertical asymptote at $x = 0$

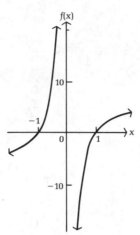

37. Local maximum at $x = 0$; inflection points at $x = -\sqrt{3}/3$ and $\sqrt{3}/3$; horizontal asymptote at $y = 0$

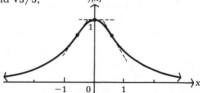

39. (A) If $a < 0$, $f\left(-\dfrac{b}{2a}\right) = \dfrac{4ac - b^2}{4a}$ is a local maximum (B) If $a > 0$, $f\left(-\dfrac{b}{2a}\right) = \dfrac{4ac - b^2}{4a}$ is a local minimum

41. (A) Local minimum at $x = 50$ (B) Concave upward on $(0, \infty)$

43. (A) Increasing on $(0, 10)$; decreasing on $(10, 20)$ (B) Inflection point at $t = 10$ (C)
(D) $N'(10) = 300$

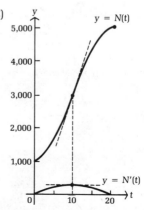

45. (A) Increasing on $(5, \infty)$; decreasing on $(0, 5)$ (B) Inflection point at $n = 5$
(C) $T'(5) = 0$

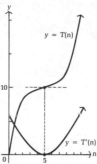

Exercise 12-4

1. $(b, d), (d, 0), (g, \infty)$ **3.** $x = 0$ **5.** $(a, d), (e, h)$ **7.** $x = a, x = h$ **9.** $x = d, x = e$ **11.** $y = M$
13. 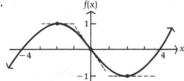   **15.** 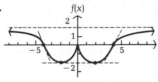   **17.**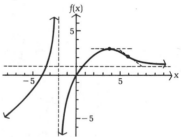

19. Decreasing on $(-\infty, 3)$; increasing on $(3, \infty)$; local minimum at $x = 3$; concave upward on $(-\infty, \infty)$; $f(1) = 0$; $f(5) = 0$;
$f(0) = 5$

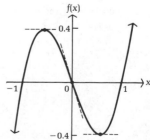

21. Increasing on $(-\infty, -\sqrt{3}/3)$ and $(\sqrt{3}/3, \infty)$; decreasing on $(-\sqrt{3}/3, \sqrt{3}/3)$; local maximum at $x = -\sqrt{3}/3$; local minimum at $x = \sqrt{3}/3$; concave downward on $(-\infty, 0)$; concave upward on $(0, \infty)$; inflection point at $x = 0$; $f(0) = 0$; $f(1) = 0$; $f(-1) = 0$; $f(-x) = -f(x)$

23. Decreasing on $(-\infty, -2)$ and $(0, 2)$; increasing on $(-2, 0)$ and $(2, \infty)$; local minima at $x = -2, 2$; local maximum at $x = 0$; concave upward on $(-\infty, -2\sqrt{3}/3)$ and $(2\sqrt{3}/3, \infty)$; concave downward on $(-2\sqrt{3}/3, 2\sqrt{3}/3)$; inflection points at $x = -2\sqrt{3}/3, 2\sqrt{3}/3$; $f(-2) = 0$; $f(2) = 0$; $f(0) = 16$; $f(-x) = f(x)$

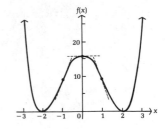

25. Decreasing on $(-\infty, 0)$ and $(0, 1.25)$; increasing on $(1.25, \infty)$; local minimum at $x = 1.25$; concave upward on $(-\infty, 0)$ and $(1, \infty)$; concave downward on $(0, 1)$; inflection points at $x = 0, 1$; $f(0) = 0$; $f(1.5) = 0$

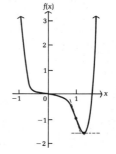

27. Decreasing on $(0, 4)$; increasing on $(4, \infty)$; local minimum at $x = 4$; concave upward on $(0, \infty)$; $f(0) = 0$; $f(16) = 0$

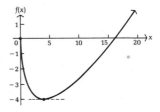

29. Increasing on $(-\infty, -1)$ and $(1, \infty)$; decreasing on $(-1, 0)$ and $(0, 1)$; local maximum at $x = -1$; local minimum at $x = 1$; concave downward on $(-\infty, 0)$; concave upward on $(0, \infty)$; inflection point at $x = 0$; $f(-3\sqrt{3}) = 0$; $f(3\sqrt{3}) = 0$; $f(0) = 0$; $f(-x) = -f(x)$

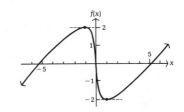

31. Decreasing on $(-\infty, 2)$ and $(2, \infty)$; concave downward on $(-\infty, 2)$; concave upward on $(2, \infty)$; $f(0) = 0$; horizontal asymptote at $y = 1$; vertical asymptote at $x = 2$

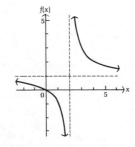

33. Decreasing on $(-\infty, 0)$ and $(2, \infty)$; increasing on $(0, 2)$; local maximum at $x = 2$; concave downward on $(-\infty, 0)$ and $(0, 3)$; concave upward on $(3, \infty)$; inflection point at $x = 3$; $f(1) = 0$; horizontal asymptote at $y = 0$; vertical asymptote at $x = 0$

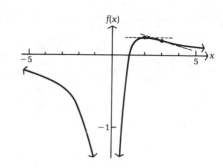

35. Decreasing on $(-\infty, 0)$; increasing on $(0, \infty)$; local minimum at $x = 0$; concave downward on $(-\infty, -1)$ and $(1, \infty)$; concave upward on $(-1, 1)$; inflection points at $x = 1, -1$; $f(-1) = 0$; $f(1) = 0$; $f(0) = -1/3$; $f(-x) = f(x)$; horizontal asymptote at $y = 1$

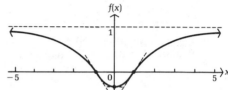

37. Increasing on $(-\infty, \infty)$; concave upward on $(-\infty, 0)$; concave downward on $(0, \infty)$; inflection point at $x = 0$; $f(0) = 0$; $f(-x) = -f(x)$; horizontal asymptotes at $y = -1, 1$

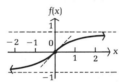

Exercise 12-5

1. Min $f(x) = f(2) = 1$; no maximum   **3.** Max $f(x) = f(4) = 26$; no minimum   **5.** Min $f(x) = f(0) = 1$; no maximum
7. Max $f(x) = f(2) = 16$ **9.** Min $f(x) = f(8) = 20$ **11.** (A) Max $f(x) = f(5) = 14$; min $f(x) = f(-1) = -22$
(B) Max $f(x) = f(1) = -2$; min $f(x) = f(-1) = -22$ (C) Max $f(x) = f(5) = 14$; min $f(x) = f(3) = -6$
13. (A) Max $f(x) = f(0) = 126$; min $f(x) = f(2) = -26$ (B) Max $f(x) = f(7) = 49$; min $f(x) = f(2) = -26$
(C) Max $f(x) = f(6) = 6$; min $f(x) = f(3) = -15$ **15.** (A) Max $f(x) = f(2) = 5$; min $f(x) = f(-2) = -5$
(B) Max $f(x) = f(2) = 5$; min $f(x) = f(0) = 0$ (C) Max $f(x) = f(2) = 5$; no minimum   **17.** Exactly in half   **19.** 15 and -15
21. A square of side 25 centimeters; max area = 625 square centimeters   **23.** 3,000 pairs; $3.00 per pair   **25.** $15
per day; $1,125 per day   **27.** 40 trees; 1,600 pounds   **29.** $(10 - 2\sqrt{7})/3 = 1.57$ inch squares   **31.** 20 feet by 40 feet (with
the expensive side being one of the short sides)   **33.** 10,000 books in 5 printings   **35.** (A) $x = 5.1$ miles (B) $x = 10$ miles
37. 4 days; 20 bacteria per cubic centimeter   **39.** 50 mice per order   **41.** 1 month; 2 feet   **43.** 4 years from now

Exercise 12-6

1. (A) $x = f(p) = 6,000 - 200p$ (B) $E(p) = -p/(30 - p)$ (C) $E(10) = -.5$; 5% decrease in demand (D) $E(25) = -5$; 50%
decrease in demand (E) $E(15) = -1$; 10% decrease in demand   **3.** (A) $x = f(p) = 3,000 - 50p$ (B) $R(p) = 3,000p - 50p^2$
(C) $E(p) = -p/(60 - p)$ (D) Elastic on $(30, 60)$; inelastic on $(0, 30)$ (E) Increasing on $(0, 30)$; decreasing on $(30, 60)$
(F) Decrease (G) Increase   **5.** $E(p) = -2p^2/(1,200 - p^2)$; (A) $E(10) = -2/11$ (inelastic) (B) $E(20) = -1$ (unit elasticity)
(C) $E(30) = -6$ (elastic)   **7.** $E(p) = -(20p + 2p^2)/(9,500 - 20p - p^2)$; (A) $E(30) = -\dfrac{3}{10}$ (inelastic) (B) $E(50) = -1$ (unit
elasticity) (C) $E(70) = -\dfrac{7}{2}$ (elastic)   **9.** Elastic on $(10, 30)$; inelastic on $(0, 10)$ **11.** Elastic on $(48, 72)$; inelastic on $(0, 48)$

13. Elastic on $(25, 25\sqrt{2})$; inelastic on $(0, 25)$ **15.**

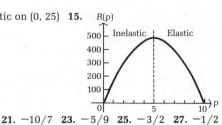

17.

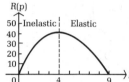

19.

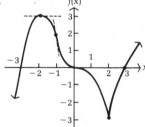

21. $-10/7$ **23.** $-5/9$ **25.** $-3/2$ **27.** $-1/2$

29. (A) $E(2.2) = -.8$ (B) Demand decreases by approximately 8% to 25,760 (C) Revenue increases
31. (A) $E(90) = -3/7$ (B) Demand increases by approximately 4% to 2,184 (C) Revenue decreases

Exercise 12-7　Chapter Review

1. $(a, c_1), (c_3, c_5), (c_5, c_6)$ **2.** $(c_1, c_3), (c_6, b)$ **3.** $(a, c_2), (c_4, c_5), (c_7, b)$ **4.** c_3 **5.** c_6 **6.** c_1, c_3, c_5 **7.** c_6 **8.** c_2, c_4, c_5, c_7
9. 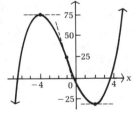 **10.** 3 **11.** $2/3$ **12.** 0 **13.** Does not exist **14.** $-4, 2$

15. Increasing on $(-\infty, -4)$ and $(2, \infty)$; decreasing on $(-4, 2)$ **16.** Local maximum at $x = -4$; local minimum at $x = 2$
17. Concave upward on $(-1, \infty)$; concave downward on $(-\infty, -1)$ **19.**
18. Inflection point at $x = -1$

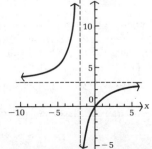

20. Horizontal asymptote at $y = 3$ **21.** Vertical asymptote at $x = -2$ **22.** Increasing on $(-\infty, -2)$ and $(-2, \infty)$
23. Concave upward on $(-\infty, -2)$; concave downward on $(-2, \infty)$ **24.** **25.** $-1, 0, 1$

26. Increasing on $(-1, 0)$ and $(0, 1)$; decreasing on $(-\infty, -1)$ and $(1, \infty)$ **30.**

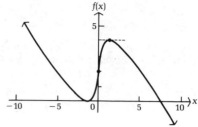

27. Local maximum at $x = 1$; local minimum at $x = -1$

28. Concave upward on $(-\infty, 0)$; concave downward on $(0, \infty)$

29. Inflection point at $x = 0$

31. Min $f(x) = f(2) = -4$; max $f(x) = f(5) = 77$ **32.** Min $f(x) = f(2) = 8$ **33.** Vertical asymptote at $x = 2$; $\lim_{x \to 2^+} f(x) = \infty$; $\lim_{x \to 2^-} f(x) = -\infty$ **34.** Vertical asymptote at $x = 1$; $\lim_{x \to 1^+} f(x) = -\infty$; $\lim_{x \to 1^-} f(x) = \infty$

35. Max $f'(x) = f'(2) = 12$

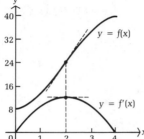

36. Decreasing on $(-\infty, -1)$ and $(-1, 0)$; increasing on $(0, 1)$ and $(1, \infty)$; local minimum at $x = 0$; concave downward on $(-\infty, -1)$ and $(1, \infty)$; concave upward on $(-1, 1)$; horizontal asymptote at $y = 1$; vertical asymptotes at $x = -1, 1$; $f(0) = 4, f(-2) = 0, f(2) = 0; f(-x) = f(x)$

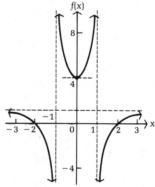

37. Max $P(x) = P(3,000) = \$175,000$ **38.** Min $\overline{C}(x) = \overline{C}(200) = 50$

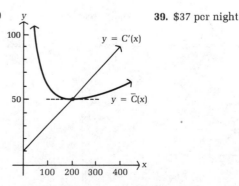

39. $37 per night

40. 12 orders per year   **41.** 3 days   **42.** In 2 years

Practice Test: Chapter 12

1. Increasing on $(-\infty, 0)$ and $(3, \infty)$; decreasing on $(0, 3)$; local maximum at $x = 0$; local minimum at $x = 3$ **2.** Concave upward on $(3/2, \infty)$; concave downward on $(-\infty, 3/2)$; inflection point at $x = 3/2$ **3.**

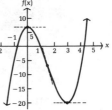

4. Horizontal asymptote at $y = 2$; vertical asymptote at $x = 0$ **5.** Increasing on $(-\infty, 0)$ and $(0, \infty)$ **6.** Concave upward on $(-\infty, 0)$; concave downward on $(0, \infty)$ **7.**   **8.** Local maximum at $x = 0$; local minimum at $x = 1$

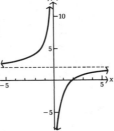

9. Min $f(x) = f(1) = -1$, max $f(x) = f(8) = 4$ **10.** Each number is 20; minimum sum is 40   **11.** Let $R(x) = (3,600 + 300x)(10 - 0.5x)$ where x is the number of 50¢ decreases. Max $R(x) = R(4) = \$38,400$ per month with 4,800 subscribers at a rate of $8 per month.

Chapter 13

Exercise 13-1

1.

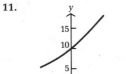

3.

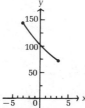

5.

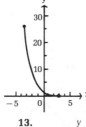

7.

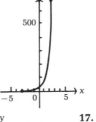

9.

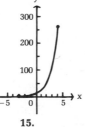

11.

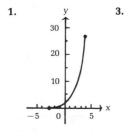

13.

15.

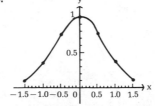

17.

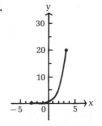

19. 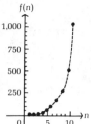 **21.** 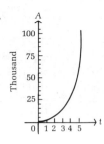 **23.**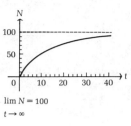

$\lim_{t \to \infty} N = 100$

Exercise 13-2

1. $27 = 3^3$ **3.** $10^0 = 1$ **5.** $8 = 4^{3/2}$ **7.** $\log_7 49 = 2$ **9.** $\log_4 8 = 3/2$ **11.** $\log_b A = u$ **13.** 3 **15.** -3 **17.** 3
19. $\log_b P - \log_b Q$ **21.** $5 \log_b L$ **23.** $\log_b p - \log_b q - \log_b r - \log_b s$ **25.** $x = 9$ **27.** $y = 2$ **29.** $b = 10$ **31.** $x = 2$
33. $y = -2$ **35.** $b = 100$ **37.** $5 \log_b x - 3 \log_b y$ **39.** $(1/3)\log_b N$ **41.** $2 \log_b x + (1/3)\log_b y$ **43.** $\log_b 50 - 0.2t \log_b 2$
45. $\log_b P + t \log_b(1 + r)$ **47.** $\log_e 100 - 0.01t$ **49.** $x = 2$ **51.** $x = 8$ **53.** $x = 7$ **55.** No solution
57. 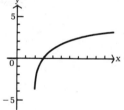   **59.** (A) 3.547 43 (B) $-2.160\ 32$ (C) 5.626 29 (D) $-3.197\ 04$

61. (A) 1,344 (B) 0.008 919 (C) 6,479 (D) 0.002 773 **63.** $\log_b 1 = 0, b > 0, b \neq 1$ **65.** $y = c10^{0.8x}$ **67.** 12 years
69. $n = \dfrac{\ln 3}{\ln(1 + i)}$ **73.** Approximately 538 years

---

Exercise 13-3

1. \$1,221.40; \$1,648.72; \$2,225.54 **3.** 11.55 **5.** 10.99 **7.** 0.14 **9.**

n	$(1 + 1/n)^n$
10	2.593 74
100	2.704 81
1,000	2.716 92
10,000	2.718 15
100,000	2.718 27
1,000,000	2.718 28
10,000,000	2.718 28
↓	↓
∞	$e = 2.718\ 281\ 8 \ldots$

11. \$55,463.90

13. \$9,931.71 **15.** 2.77 years **17.** 13.86% **19.** $A = Pe^{rt}$; $2P = Pe^{rt}$; $e^{rt} = 2$; $\ln e^{rt} = \ln 2$; $rt = \ln 2$; $t = (\ln 2)/r$
21. 34.66 years **23.** 3.47% **25.** Approximately 538 years

Exercise 13-4

1. $1/t$ **3.** $1/(x-3)$ **5.** $3/(x-1)$ **7.** $2z+(3/z)$ **9.** $3/(t^{1/2}) - 1/t = (3\sqrt{t}-1)/t$ **11.** $7/x$ **13.** $1/(2x)$
15. $1/(2x\sqrt{\ln x}) + 1/(2x)$ **17.** $4/(x+1)$ **19.** $(2t+3)/(t^2+3t)$ **21.** $6x^2+2x/(x^2+1)$ **23.** $(1-2\ln x)/x^3$ **25.** $\ln x$
27. $(2x+1)+(2x+1)\ln(x^2+x) = (2x+1)[1+\ln(x^2+x)]$ **29.** 0 **31.** $(\log_{10} e)/x$ **33.** $(6x)(\log_2 e)/(3x^2-1)$
35. $x/(x^2+1)$ **37.** $(6x-2)(\log_{10} e)/(3x^2-2x)$ **39.** $2/(x-1)-3/(x+1)$ **41.** $2/(x-1)+1/(2x)$
43. $6x[\ln(x^2-1)]^2/(x^2-1)$ **45.** $-2x/\{(1+x^2)[\ln(1+x^2)]^2\}$ **47.** $-2x/(3(1-x^2)[\ln(1-x^2)]^{2/3})$ **49.** Case 1 $(x>0)$:
$D_x \ln|x| = D_x \ln x = 1/x$; case 2 $(x<0)$: $D_x \ln|x| = D_x \ln(-x) = D_x(-x)/(-x) = (-1)/(-x) = 1/x$; conclusion:
$D_x \ln|x| = 1/x$, $x \neq 0$ **51.** $dn/di = (-\ln 2)/\{(1+i)[\ln(1+i)]^2\}$ **53.** $dN/dI = (10\log_{10} e)/I$

Exercise 13-5

1. e^t **3.** $8e^{8x}$ **5.** $6e^{2x}$ **7.** $-8e^{-4t}$ **9.** $1/x + 2e^x$ **11.** $4e^{2x}-3e^x$ **13.** $(6x-2)e^{3x^2-2x}$ **15.** $(e^x+e^{-x})/2$ **17.** $2e^{2x}-6x$
19. xe^x+e^x **21.** $-3e^{-0.03x}$ **23.** $4(e^{2x}-1)^3(2e^{2x}) = 8e^{2x}(e^{2x}-1)^3$ **25.** $7^x \ln 7$ **27.** $4e^{4x}+4x^3 e^{x^4}$
29. $2e^{2x}(x^2-x+1)/[(x^2+1)^2]$ **31.** $(x^2+1)(-e^{-x}) + e^{-x}(2x) = e^{-x}(2x-x^2-1)$ **33.** $(e^{-x}/x) - e^{-x}\ln x$
35. $e^{\sqrt[3]{3x+1}}/\sqrt[3]{(3x+1)^2}$ **37.** $e^x + xe^x \ln x + e^x \ln x$ **39.** $(2x+1)(10^{x^2+x})(\ln 10)$ **41.** $R'(x) = (100-5x)e^{-0.05x}$
43. $-\$27,145$ per year; $-\$18,196$ per year; $-\$11,036$ per year **45.** $A'(t) = 10,000 \cdot 2^{2t} \ln 2$; $A'(5) = 7.10 \times 10^6$ bacteria
per hour; $A'(10) = 7.27 \times 10^9$ bacteria per hour **47.** $N'(t) = 6.4e^{-0.08t}$; $N'(1) = 5.9$ words per minute per week; $N'(5) = 4.3$
words per minute per week; $N'(20) = 1.3$ words per minute per week

Exercise 13-6   Chapter Review

1. $y = 10^x$ **2.** $\log_b w + \log_b x - \log_b y$ **3.** $2/x + 3e^x$ **4.** $2e^{2x-3}$ **5.** $(6x^2-3)/(2x^3-3x)$ **6.** (A) $b=3$ (B) $x = \dfrac{1}{8}$

7. $\dfrac{1}{3}\log_b x - \log_b u - 3\log_b v$ **8.** $\log_b 100 + t\log_b 1.06$ **9.** **10.** $x = \dfrac{1}{6}$ **11.** $x = 5$

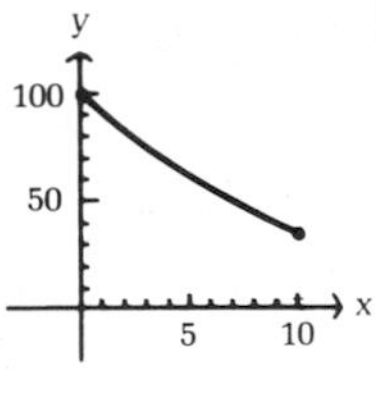

(graph: y-axis marked 100 and 50, x-axis marked 5 and 10, showing a decreasing curve)

12. (A) $-2.040\ 55$ (B) $9.194\ 55$ **13.** (A) $0.000\ 156\ 5$ (B) $367,400$ **14.** $t = 9.15$ **15.** $1/\sqrt{x} + 3x^2/(x^3+1)$
16. $e^{-2x}/x - 2e^{-2x}\ln 5x$ **17.** $(4x^2+2)e^{x^2}$ **18.** $2(2x-1)/(x^2-x) - 3x^2/(x^3+1)$ **19.** $1/(2z\sqrt{\ln z}) - 1/(2z)$
20. $y = ce^{-0.2x}$ **21.** $2x5^{x^2-1}\ln 5$ **22.** $(2x-1)(\log_5 e)/(x^2-x)$ **23.** $(2x+1)/[2(x^2+x)\sqrt{\ln(x^2+x)}]$ **24.** (A) 14.2 years
(B) 13.9 years **25.** $A'(t) = 10e^{0.1t}$; $A'(1) = \$11.05$ per year; $A'(10) = \$27.18$ per year **26.** $R'(x) = (1,000-20x)e^{-0.02x}$

Practice Test: Chapter 13

1. $x = 1/4$ **2.** $x = 7$ **3.** $x = 8.66$ **4.** $y = ce^{-0.03x}$ **5.** $-3e^{-x} - 1/(x+1)$ **6.** $[(2x-1)^2+2]e^{x^2-x} = (4x^2-4x+3)e^{x^2-x}$
7. $3/(3x-5) - 6x/(x^2-1)$ **8.** $1/(4x\sqrt{\ln \sqrt{x}})$ **9.** $(2x)(10^{x^2-1})(\ln 10)$ **10.** $(2x-1)(\log_{10} e)/(x^2-x)$
11. (A) $R(x) = xp(x) = 10,000xe^{-0.015x}$ (B) $R'(x) = (10,000-150x)e^{-0.015x}$ **12.** (A) 7.86 years (B) 7.32 years

Chapter 14

Exercise 14-1

1. $7x + C$ **3.** $(x^7/7) + C$ **5.** $2t^4 + C$ **7.** $u^2 + u + C$ **9.** $x^3 + x^2 - 5x + C$ **11.** $(s^5/5) - \dfrac{4}{3}s^6 + C$ **13.** $\dfrac{1}{3}e^{3t} + C$

15. $2\ln|z| + C$ **17.** $y = 40x^5 + C$ **19.** $P = 24x - 3x^2 + C$ **21.** $y = \dfrac{1}{3}u^6 - u^3 - u + C$ **23.** $y = -e^{-x} + 3x + C$

25. $x = 5\ln|t| + t + C$ **27.** $4x^{3/2} + C$ **29.** $-4x^{-2} + C$ **31.** $2\sqrt{u} + C$ **33.** $-(x^{-2}/8) + C$ **35.** $-(u^{-4}/8) + C$

37. $2x^{3/2} - 2x^{1/2} + C$ **39.** $6x^{5/3} - 6x^{4/3} - 2x + C$ **41.** $2x^{3/2} + 4x^{1/2} + C$ **43.** $\dfrac{3}{5}x^{5/3} + 2x^{-2} + C$ **45.** $\dfrac{e^{3x} + e^{-3x}}{6} + C$

47. $-z^{-2} - z^{-1} + \ln|z| + C$ **49.** $y = x^2 - 3x + 5$ **51.** $C(x) = 2x^3 - 2x^2 + 3{,}000$ **53.** $x = 40\sqrt{t}$

55. $y = -2x^{-1} + 3\ln|x| - x + 3$ **57.** $x = -2e^{-2t} + 3e^{-t} - 2t$ **59.** $y = 2x^2 - 3x + 1$ **61.** $x^2 + x^{-1} + C$

63. $\dfrac{1}{2}x^2 + x^{-2} + C$ **65.** $\dfrac{1}{2}e^{2x} - 2\ln|x| + C$ **67.** $M = t - 2\sqrt{t} + 5$ **69.** $y = 3x^{5/3} + 3x^{2/3} - 6$

71. $p(x) = 2{,}000e^{-0.05x} + 1{,}000$ **73.** $P(x) = 50x - 0.02x^2$; $P(100) = \$4{,}800$ **75.** $p(x) = 5{,}000e^{-0.04x} + 3{,}000$

77. $W(h) = 0.0005h^3$; $W(70) = 171.5$ pounds **79.** 19,400

Exercise 14-2

1. $A = 1{,}000e^{0.08t}$ **3.** $A = 8{,}000e^{0.06t}$ **5.** $p(x) = 100e^{-0.05x}$ **7.** $N = L(1 - e^{-0.051t})$ **9.** $I = I_0 e^{-0.00942x}$; $x \approx 74$ feet

11. $Q = 3e^{-0.04t}$; $Q(10) = 2.01$ milliliters **13.** 24,200 years (approximately) **15.** 104 times; 67 times **17.** (A) 7 people; 353 people (B) 400

Exercise 14-3

1. $\dfrac{(x^2 - 4)^6}{6} + C$ **3.** $\dfrac{2}{3}(2x^2 - 1)^{3/2} + C$ **5.** $\dfrac{(3x - 2)^8}{24} + C$ **7.** $\dfrac{(x^2 + 3)^8}{16} + C$ **9.** $\dfrac{(3x^2 + 7)^{3/2}}{9} + C$ **11.** $\dfrac{(2x^4 + 3)^{1/2}}{4} + C$

13. $\dfrac{1}{3}(x^2 - 2x - 3)^{3/2} + C$ **15.** $-\dfrac{1}{18}(3t^2 + 1)^{-3} + C$ **17.** $-\dfrac{2}{3}(4 - x^3)^{1/2} + C$ **19.** $\dfrac{1}{4}(e^x - 2x)^4 + C$

21. $\dfrac{2}{3}(1 + \ln x)^{3/2} + C$ **23.** $\dfrac{-1}{12(x^4 + 2x^2 + 1)^3} + C$ **25.** $x = \dfrac{14}{9}(t^3 + 5)^{3/2} + C$ **27.** $y = 3(t^2 - 4)^{1/2} + C$

29. $p = -(e^x - e^{-x})^{-1} + C$ **31.** $R(x) = \sqrt{x^2 + 9} - 3$; $R(4) = \$2{,}000$ **33.** $E(t) = 12{,}000 - \dfrac{10{,}000}{\sqrt{t + 1}}$; $E(15) = 9{,}500$ students

Exercise 14-4

1. 5 **3.** 5 **5.** 2 **7.** 48 **9.** $-\dfrac{7}{3}$ **11.** 2 **13.** $\dfrac{1}{2}(e^2 - 1)$ **15.** $2\ln 3.5$ **17.** -2 **19.** 14 **21.** $5^6 = 15{,}625$ **23.** 12

25. $\dfrac{1}{6}[(e^2 - 2)^3] - 1$ **27.** $-3 - \ln 2$ **29.** $\dfrac{1}{6}(15^{3/2} - 5^{3/2})$ **31.** $\dfrac{3}{4}(2^{2/3} - 3^{2/3})$ **33.** $\dfrac{2}{e^{-1} - e} = \dfrac{2e}{1 - e^2}$

35. (A) $C(4) - C(2) = \displaystyle\int_2^4 1\,dx = \$2$ thousand per day (B) $R(4) - R(2) = \displaystyle\int_2^4 (10 - 2x)dx = \$8$ thousand per day

(C) $P(4) - P(2) = \displaystyle\int_2^4 [R'(x) - C'(x)]dx = \$6$ thousand per day

37. $\int_0^5 500(t - 12)dt = -\$23,750;$ $\int_5^{10} 500(t - 12)dt = -\$11,250$ **39.** $\int_{49}^{64} -295x^{-3/2}dx \approx -10.5$ beats per minute

41. $\int_{10}^{20} (12 + 0.006t^2)dt = 134$ billion cubic feet **43.** $\int_1^9 \frac{25}{\sqrt{t}} dt = 100$ items

Exercise 14-5

1. 16 **3.** 7 **5.** $\frac{7}{3}$ **7.** 9 **9.** $e^2 - e^{-1}$ **11.** $-\ln 0.5$ **13.** 15 **15.** 32 **17.** 36 **19.** 9 **21.** $\frac{5}{2}$ **23.** $2e + \ln 2 - 2e^{0.5}$ **25.** $\frac{23}{3}$

27. $\frac{4}{3}$ **29.** Consumers' surplus $= 1$; producers' surplus $= \frac{4}{3}$ **31.** 5 years; \$25,000 **33.** (A) Solve $100e^{-0.05x} = 10e^{0.05x}$

(B) $\int_0^{23.03} (100e^{-0.05x} - 31.62)dx \approx 639.47$

Exercise 14-6

1. (A) 120 (B) 124 **3.** (A) 123 (B) 124 **5.** (A) -4 (B) -5.33 **7.** (A) -5 (B) -5.33 **9.** (A) 1.63 (B) Not possible at this time **11.** (A) 1.59 (B) Not possible at this time **13.** 250 **15.** 2 **17.** $45/28 \approx 1.61$ **19.** $2(1 - e^{-2}) \approx 1.73$ **21.** 26π

23. 64π **25.** $32\pi/3$ **27.** 0.791 **29.** 6.54 **31.** (A) $I = -200t + 600$ (B) $\frac{1}{3}\int_0^3 (-200t + 600)dt = 300$ **33.** \$16,000

35. \$7.18 **37.** $10°C$

Exercise 14-7   Chapter Review

1. $t^3 - t^2 + C$ **2.** 12 **3.** $-3t^{-1} - 3t + C$ **4.** $15/2$ **5.** $-2e^{-0.5x} + C$ **6.** $2 \ln 5$ **7.** $y = f(x) = x^3 - 2x + 4$ **8.** 12

9. $2(5 + 17) = 44$ **10.** $\frac{1}{8}(6x - 5)^{4/3} + C$ **11.** 2 **12.** $-2x^{-1} - \frac{3}{5}x^{5/3} + C$ **13.** $(20^{3/2} - 8)/3$ **14.** $-\frac{1}{2}e^{-2x} + \ln |x| + C$

15. $-500(e^{-0.2} - 1) \approx 90.63$ **16.** $y = f(x) = 2x^{3/2} + x^{-1} + 2$ **17.** $y = 3x^2 + x - 4$ **18.** -2 **19.** $\frac{13}{2}$

20. $\left|\int_{-2}^2 (x^2 - 4)dx\right| + \int_2^4 (x^2 - 4)dx = 64/3$ **21.** $\frac{3}{8}(15)^{4/3}$ **22.** $2x^{1/2} + x^{-2} + C$ **23.** $-\frac{1}{2}e^{-2x} - \ln x + C$

24. $y = \frac{2}{9}(x^3 + 4)^{3/2} + \frac{2}{9}$ **25.** $N = 800e^{0.06t}$ **26.** $\frac{46}{3}$ **27.** $\frac{1}{4}[f(0.5) + f(1.5) + f(2.5) + f(3.5)] = 0.394$ **28.** $\pi/2$

29. $P(x) = 100x - 0.01x^2$; $P(10) = \$999$ **30.** $\int_0^{15} (60 - 4t)dt = 450$ thousand barrels **31.** $\int_{10}^{40} \left(150 - \frac{x}{10}\right) dx = \$4,425$

32. 109 items **33.** $40(e^2 - 3) \approx \$175.56$ **34.** 1 square centimeter **35.** $\int_{50}^{60} 0.0015h^2 dh = 45.5$ pounds

36. $N(t) = 10,000e^{0.2t}$; $N(10) = 10,000e^2 \approx 73,890$ **37.** 5,200 **38.** $\int_0^4 (12t - 3t^2)dt = 32$ thousand

Practice Test: Chapter 14

1. $3/8$ **2.** $(2/9)(x^3 + 9)^{3/2} + C$ **3.** $-2x^{-2} + (x^2/2) + C$ **4.** $26/3$ **5.** $20(e^2 - 3) \approx 87.8$ **6.** $\ln |x| + 10e^{-0.1x} + C$

7. $f(x) = 1 + 6x - x^2$ **8.** $\frac{1}{3}$ **9.** 2 **10.** 46; 48 **11.** 27 **12.** $\int_0^2 (40 - 4t)dt = 72$ thousand ounces;

$\int_2^4 (40 - 4t)dt = 56$ thousand ounces **13.** $Q = 10,000e^{0.12t}$

Chapter 15

Exercise 15-1

1. $\frac{1}{8}(x^2+9)^4+C$ **3.** $\frac{1}{2}\ln(4+2x+x^2)+C$ **5.** $e^{x^2+x+1}+C$ **7.** $\frac{45}{8}$ **9.** $\frac{1}{4}\ln 9$ **11.** $\frac{1}{8}(1+e^{2x})^4+C$ **13.** $\frac{1}{4}(\ln x)^4+C$

15. $\frac{1}{6}(x-5)^6+(x-5)^5+C$ **17.** $\frac{2}{5}(4+x)^{5/2}-\frac{8}{3}(4+x)^{3/2}+C$ **19.** $\frac{2}{3}(x+3)^{3/2}-6(x+3)^{1/2}+C$

21. $x-2+2\ln|x-2|+c=x+2\ln|x-2|+C$ **23.** $\frac{1}{2}(x-2)^2+4(x-2)+4\ln|x-2|+C$ **25.** $4\ln 5$ **27.** 9

29. $16\sqrt{2}/15$ **31.** $\frac{2}{3}(x-1)^{3/2}+2(x-1)^{1/2}+C$ **33.** $\frac{2}{5}(x-1)^{5/2}+\frac{4}{3}(x-1)^{3/2}+2(x-1)^{1/2}+C$

35. $(\sqrt{x}-1)^2-2\ln|\sqrt{x}-1|+C$ **37.** $2\ln|1+\sqrt{x}|+C$ **39.** $e^{-1/x}+C$ **43.** $p(x)=150{,}000/(5{,}000+x^2)+20$ **45.** Useful life $=\sqrt{\ln 55}\approx 2$ years; total profit $\approx \$2{,}272$ **47.** $N(t)=5{,}000-1{,}000\ln(1+t^2)$; $N(10)\approx 385$ **49.** 200

Exercise 15-2

1. $\frac{1}{3}xe^{3x}-\frac{1}{9}e^{3x}+C$ **3.** $\frac{x^3}{3}\ln x-\frac{x^3}{9}+C$ **5.** $-xe^{-x}-e^{-x}+C$ **7.** $\frac{1}{2}e^{x^2}+C$ **9.** $(xe^x-4e^x)\Big|_0^1=-3e+4\approx -4.1548$

11. $(x\ln 2x-x)\Big|_1^3=(3\ln 6-3)-(\ln 2-1)\approx 2.6821$ **13.** $\ln(x^2+1)+C$ **15.** $\frac{(\ln x)^2}{2}+C$

17. $\frac{2}{3}x^{3/2}\ln x-\frac{4}{9}x^{3/2}+C$ **19.** $(x^2-2x+2)e^x+C$ **21.** $\frac{xe^{ax}}{a}-\frac{e^{ax}}{a^2}+C$ **23.** $\left(-\frac{\ln x}{x}-\frac{1}{x}\right)\Big|_1^e=-\frac{2}{e}+1\approx 0.2642$

25. $x(\ln x)^2-2x\ln x+2x+C$ **27.** $x(\ln x)^3-3x(\ln x)^2+6x\ln x-6x+C$ **29.** $2\sqrt{x}\,e^{\sqrt{x}}-2e^{\sqrt{x}}+C$

31. $\frac{1}{2}(1+x^2)\ln(1+x^2)-\frac{1}{2}(1+x^2)+c=\frac{1}{2}(1+x^2)\ln(1+x^2)-\frac{1}{2}x^2+C$

33. $2(1+\sqrt{x})\ln(1+\sqrt{x})-2(1+\sqrt{x})+c=2(1+\sqrt{x})\ln(1+\sqrt{x})-2\sqrt{x}+C$ **35.** $P(t)=t^2+te^{-t}+e^{-t}-1$

37. $(10-2\ln 6)/3\approx 2.1388$ parts per million **39.** 20,980

Exercise 15-3

1. $-\ln|(x+1)/x|+C$ **3.** $1/(x+3)+2\ln|(2x+5)/(x+3)|+C$ **5.** $\frac{1}{2}\ln|x/(2+\sqrt{x^2+4})|+C$

7. $\sqrt{1-x^2}-\ln|(1+\sqrt{1-x^2})/x|+C$ **9.** $\frac{1}{2}\ln 2.4\approx 0.4377$ **11.** $\ln 3\approx 1.0986$

13. $-\sqrt{4x^2+1}/x+2\ln|2x+\sqrt{4x^2+1}|+C$ **15.** $\frac{1}{2}\ln|x^2+\sqrt{x^4-16}|+C$ **17.** $\frac{1}{6}(x^3\sqrt{x^6+4}+4\ln|x^3+\sqrt{x^6+4}|)+C$

19. $2\sqrt{x+16}-4\ln|\sqrt{x+16}+4|+4\ln|\sqrt{x+16}-4|+C=2\sqrt{x+16}-8\ln|(4+\sqrt{x+16})/\sqrt{x}|+c$

21. $\sqrt{4-x^4}/(8x^2)+C$ **23.** $\frac{1}{2}\ln|(\sqrt{x+1}+1)/(\sqrt{x+1}-1)|-\sqrt{x+1}/x+C$ **25.** $\frac{64}{3}$ **27.** $\frac{1}{5}+\frac{1}{4}\ln\frac{5}{9}\approx 0.0531$

29. $\frac{1}{2}\ln|x^2+2x|+C$ **31.** $\frac{2}{3}\ln|x+3|+\frac{1}{3}\ln|x|+C$ **33.** $360\ln(1.8)-32\approx 179.6$ **35.** $100\ln 3\approx 110$ feet

37. $60\ln 5\approx 97$

Exercise 15-4

1. $1/3$ **3.** 2 **5.** Diverges **7.** 1 **9.** Diverges **11.** Diverges **13.** 1 **15.** $\int_{2}^{3.5}\left(-\dfrac{x}{2}+2\right)dx \approx .94$

17. $\dfrac{1}{4}\int_{1}^{\infty}e^{-t/4}\,dt \approx .78$ **19.** 1 **21.** Diverges **23.** $\dfrac{1}{2}$ **25.** Diverges **27.** 6.25 million cubic feet

29. $.05\int_{3}^{\infty}e^{-.05x}\,dx \approx .86$ **31.** 500 gallons **33.** $.2\int_{0}^{5}e^{-.2t}\,dt \approx .63$ **35.** $\int_{9}^{\infty}\dfrac{dx}{(x+1)^2}=.1$

Exercise 15-5   Chapter Review

1. $\dfrac{1}{3}(1+x^2)^{3/2}+C$ **2.** $\dfrac{1}{8}x(2x^2+1)\sqrt{x^2+1}-\dfrac{1}{8}\ln|x+\sqrt{x^2+1}|+C$ **3.** $\dfrac{1}{2}\ln 10$ **4.** $\dfrac{1}{2}$ **5.** $\dfrac{1}{2}e^{x^2}+C$ **6.** Diverges

7. $-2e^{-1}+1\approx 0.2642$ **8.** $\dfrac{x^2}{2}\ln x-\dfrac{x^2}{4}+C$ **9.** $-\ln(e^{-x}+3)+C$ **10.** $-(e^x+2)^{-1}+C$

11. $\dfrac{2}{3}(x+9)^{3/2}-18(x+9)^{1/2}+C$ **12.** $x-9\ln|x+9|+C$ **13.** $-\dfrac{1}{9}\ln\left|\dfrac{x+9}{x}\right|+C$ **14.** $\dfrac{1}{3}$ **15.** 1

16. $\dfrac{1}{3}\ln|3x+\sqrt{9x^2+4}|+C$ **17.** 1 **18.** $\dfrac{(\ln x)^3}{3}+C$ **19.** 2 **20.** $\dfrac{1}{2}x^2(\ln x)^2-\dfrac{1}{2}x^2\ln x+\dfrac{1}{4}x^2+C$ **21.** $4-8\ln 1.4$

22. $\dfrac{1}{3}\ln|x^3+\sqrt{x^6-16}|+C$ **23.** 0 **24.** Consumers' surplus $=320-80\sqrt{15}\approx 10.2$; producers'

surplus $=1{,}040/3-80\sqrt{15}\approx 36.8$ **25.** 70 **26.** 3,374 thousand barrels; 10,000 thousand barrels

27. $.02\int_{0}^{1}e^{-.02t}\,dt\approx .02$ **28.** 2.5 milliliters; 10 milliliters **29.** $\int_{1}^{\infty}f(t)dt=\int_{1}^{3}f(t)dt=1/3$ **30.** 45 thousand;

50 thousand **31.** $.5\int_{2}^{\infty}e^{-.5t}\,dt\approx .37$

Practice Test: Chapter 15

1. $-1/6(x^2+3)^3+C$ **2.** $\dfrac{1}{2}e^{x^2}+C$ **3.** $\dfrac{1}{8}e^{-16}$ **4.** $(x+2)e^x+C$ **5.** $(\ln x)^8/8+C$ **6.** $\dfrac{1}{8}x^8\ln x-\dfrac{1}{64}x^8+C$ **7.** $\dfrac{32}{15}$

8. $\ln[(x-1)/x]+C$ **9.** $x(\ln x)^2-2x\ln x+2x+C$ **10.** $7.5+8\ln 2\approx 13.05$ **11.** 2,642 barrels; 10,000 barrels

---

Chapter 16

Exercise 16-1

1. 10 **3.** 1 **5.** 0 **7.** 1 **9.** 6 **11.** 150 **13.** 16π **15.** 791 **17.** 0.192 **19.** 118 **21.** $100e^{0.8}\approx 222.55$ **23.** $2x+\Delta x$
25. $2y^2$ **27.** $E(0,0,3)$; $F(2,0,3)$ **29.** \$4,400; \$6,000; \$7,100 **31.** \$142 thousand; \$150 thousand; \$8 thousand loss
33. $R(p,q)=-5p^2+6pq-4q^2+200p+300q$; $R(2,3)=\$1{,}280$; $R(3,2)=\$1{,}175$ **35.** $T(70,47)\approx 29$ minutes;
$T(60,27)=33$ minutes **37.** $C(6,8)=75$; $C(8.1,9)=90$ **39.** $Q(12,10)=120$; $Q(10,12)\approx 83$

Exercise 16-2

1. 3 **3.** 2 **5.** $-4xy$ **7.** -6 **9.** $10xy^3$ **11.** 60 **13.** $2x - 2y + 6$ **15.** 6 **17.** -2 **19.** 2 **21.** $2e^{2x+3y}$ **23.** $6e^{2x+3y}$
25. $6e^2$ **27.** $4e^3$ **29.** $f_x(x, y) = 6x(x^2 - y^3)^2; f_y(x, y) = -9y^2(x^2 - y^3)^2$ **31.** $f_x(x, y) = 24xy(3x^2y - 1)^3;$
$f_y(x, y) = 12x^2(3x^2y - 1)^3$ **33.** $f_x(x, y) = 2x/(x^2 + y^2); f_y(x, y) = 2y/(x^2 + y^2)$ **35.** $f_x(x, y) = y^4e^{xy^2};$
$f_y(x, y) = 2xy^3e^{xy^2} + 2ye^{xy^2}$ **37.** $f_x(x, y) = 4xy^2/(x^2 + y^2)^2; f_y(x, y) = -4x^2y/(x^2 + y^2)^2$ **39.** $f_{xx}(x, y) = 2y^2 + 6x;$
$f_{xy}(x, y) = 4xy = f_{yx}(x, y); f_{yy}(x, y) = 2x^2$ **41.** $f_{xx}(x, y) = -2y/x^3; f_{xy}(x, y) = -1/y^2 + 1/x^2 = f_{yx}(x, y); f_{yy}(x, y) = 2x/y^3$
43. $f_{xx}(x, y) = (2y + xy^2)e^{xy}; f_{xy}(x, y) = (2x + x^2y)e^{xy} = f_{yx}(x, y); f_{yy}(x, y) = x^3e^{xy}$ **45.** $x = 2$ and $y = 4$
47. $f_{xx}(x, y) + f_{yy}(x, y) = (2y^2 - 2x^2)/(x^2 + y^2)^2 + (2x^2 - 2y^2)/(x^2 + y^2)^2 = 0$ **49.** (A) $2x$ (B) $4y$ **51.** $C_x(x, y) = 70$:
Increasing x by one unit and holding y fixed will increase costs by \$70 at any production level; $C_y(x, y) = 100$: Increasing
y by one unit and holding x fixed will increase costs by \$100 at any production level **53.** $P_x(1, 2) = 4$: Profit will
increase approximately \$4 thousand per 1,000 increase in production of type A calculator at the (1, 2) output level;
$P_y(1, 2) = -2$: Profit will decrease approximately \$2 thousand per 1,000 increase in production of type B calculator at
the (1, 2) output level **55.** $\partial x/\partial p = -5$: A \$1 increase in the price of brand A will decrease the demand for brand A by
5 pounds at any price level $(p, q); \partial y/\partial p = 2$: A \$1 increase in the price of brand A will increase the demand for brand
B by 2 pounds at any price level (p, q) **57.** $T_V(70, 47) \approx 0.41$ minute per unit increase in volume of air when $V = 70$
cubic feet and $x = 47$ feet; $T_x(70, 47) \approx -0.36$ minute per unit increase in depth when $V = 70$ cubic feet and $x = 47$
feet **59.** $C_W(6, 8) = 12.5$: Index increases approximately 12.5 units for 1 inch increase in width of the head (length held
fixed) when $W = 6$ and $L = 8; C_L(6, 8) = -9.38$: Index decreases approximately 9.38 units for 1 inch increase in length
(width held fixed) when $W = 6$ and $L = 8$ **61.** $Q_M(12, 10) = 10$: IQ increases approximately 10 points for 1 year increase
of mental age (chronological age held fixed) when $M = 12$ and $C = 10; Q_C(12, 10) = -12$: IQ decreases 12 points for 1
year increase in chronological age (mental age held fixed) when $M = 12$ and $C = 10$

Exercise 16-3

1. $2x\,dx + 2y\,dy$ **3.** $4x^3y^3\,dx + 3x^4y^2\,dy$ **5.** $\frac{1}{2}x^{-1/2}\,dx - \frac{5}{2}y^{-3/2}\,dy$ **7.** $3x^2\,dx + 3y^2\,dy + 3z^2\,dz$
9. $(y + 2z)\,dx + (x + 3z)\,dy + (2x + 3y)\,dz$ **11.** $dz = -0.4, \Delta z = -0.39$ **13.** $dz = 15, \Delta z = 13.636\ 364$
15. $dw = 1.7, \Delta w = 1.73$ **17.** $dw = -0.5, \Delta w = -0.490\ 196$ **19.** 4.98 inches **21.** $15\pi \approx 47.12$ cubic inches
23. 30 cubic centimeters **25.** $dz = e^{x^2+y^2}[(y + 2x^2y)\,dx + (x + 2xy^2)\,dy]$
27. $dw = (1 + xyz)e^{xyz}(yz\,dx + xz\,dy + xy\,dz)$ **29.** \$10,460 **31.** \$5.40 **33.** 1.1k **35.** 14.896

Exercise 16-4

1. $f(-2, 0) = 10$ is a local maximum **3.** $f(-1, 3) = 4$ is a local minimum **5.** f has a saddle point at $(3, -2)$
7. $f(3, 2) = 33$ is a local maximum **9.** $f(2, 2) = 8$ is a local minimum **11.** f has a saddle point at $(0, 0)$ **13.** f has a
saddle point at $(0, 0); f(1, 1) = -1$ is a local minimum **15.** f has a saddle point at $(0, 0); f(3, 18) = -162$ and
$f(-3, -18) = -162$ are local minima **17.** The test fails at $(0, 0); f$ has saddle points at $(2, 2)$ and $(2, -2)$ **19.** 2,000
type A and 4,000 type B; Max $P = P(2, 4) = \$15$ million **21.** \$3 million on research and development and \$2 million on
advertising; Max $P = P(3, 2) = \$30$ million **23.** 8 inches by 4 inches by 2 inches **25.** 20 inches by 20 inches by 40 inches

Exercise 16-5

1. Max $f(x, y) = f(3, 3) = 18$ **3.** Min $f(x, y) = f(3, 4) = 25$ **5.** Max $f(x, y) = f(3, 3) = f(-3, -3) = 18$;
min $f(x, y) = f(3, -3) = f(-3, 3) = -18$ **7.** Maximum product is 25 when each number is 5
9. Min $f(x, y, z) = f(-4, 2, -6) = 56$ **11.** Max $f(x, y, z) = f(2, 2, 2) = 6$; min $f(x, y, z) = f(-2, -2, -2) = -6$ **13.** 60 of
model A and 30 of model B will yield a minimum cost of \$32,400 per week **15.** A maximum volume of 16,000 cubic
inches occurs for a box 40 inches by 20 inches by 20 inches. **17.** 8 inches by 8 inches by 8/3 inches **19.** $x = 50$ feet
and $y = 200$ feet; maximum area is 10,000 square feet

Exercise 16-6

1. $y = 0.7x + 1$

3. 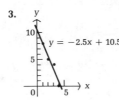 $y = -2.5x + 10.5$

5. $y = x + 2$

7. $y = 2.12x + 10.8$; $y = 63.8$ when $x = 25$ **9.** $y = -1.2x + 12.6$; $y = 10.2$ when $x = 2$
11. $y = -1.53x + 26.67$; $y = 14.4$ when $x = 8$
13. $y = 0.75x^2 - 3.45x + 4.75$ **15.** (A) $y = 0.382x + 1.265$ (B) $10,815

17. (A) $y = 0.7x + 112$ (B) Demand, 140,000 units; revenue, $56,000
19. (A) $y = 11.9x + 69.2$ (B) 69.2 parts per million **21.** (A) $y = 10.1x + 10.7$ (B) 9 weeks

Exercise 16-7

1. (A) $3x^2y^4 + C(x)$ (B) $3x^2$ **3.** (A) $2x^2 + 6xy + 5x + E(y)$ (B) $35 + 30y$ **5.** (A) $\sqrt{y + x^2} + E(y)$ (B) $\sqrt{y + 4} - \sqrt{y}$ **7.** 9

9. 330 **11.** $(56 - 20\sqrt{5})/3$ **13.** 16 **15.** 49 **17.** $\frac{1}{8}\int_1^5\int_{-1}^1 (x + y)^2 \, dy \, dx = \frac{32}{3}$

19. $\frac{1}{15}\int_1^4\int_2^7 (x/y) \, dy \, dx = \frac{1}{2}\ln\frac{7}{2} \approx 0.626$ 4 **21.** $\frac{4}{3}$ **23.** $\frac{32}{3}$ **25.** $\int_0^1\int_1^2 xe^{xy} \, dy \, dx = \frac{1}{2} + \frac{1}{2}e^2 - e$

27. $\int_0^1\int_{-1}^1 \frac{2y + 3xy^2}{1 + x^2} \, dy \, dx = \ln 2$ **29.** $\frac{1}{0.4}\int_{0.6}^{0.8}\int_5^7 \frac{y}{1 - x} \, dy \, dx = 30\ln 2 \approx 20.8 billion

31. $\frac{1}{10}\int_{10}^{20}\int_1^2 x^{0.75}y^{0.25} \, dy \, dx = \frac{8}{175}(2^{1.25} - 1)(20^{1.75} - 10^{1.75}) \approx 8.375$ or 8,375 items

33. $\frac{1}{192}\int_{-8}^8\int_{-6}^6 [10 - \frac{1}{10}(x^2 + y^2)] \, dy \, dx = \frac{20}{3}$ insects per square foot **35.** $\frac{1}{8}\int_{-2}^2\int_{-1}^1 [100 - 15(x^2 + y^2)] \, dy \, dx = 75$ parts
per million **37.** $\frac{1}{10,000}\int_{2,000}^{3,000}\int_{50}^{60} 0.0000133xy^2 \, dy \, dx \approx 100.86$ feet **39.** $\frac{1}{16}\int_8^{16}\int_{10}^{12} 100\frac{x}{y} \, dy \, dx = 600\ln 1.2 \approx 109.4$

Exercise 16-8

1. $R = \{(x, y) | 0 \leq y \leq 4 - x^2, \ 0 \leq x \leq 2\}$ **3.** $R = \{(x, y) | x^3 \leq y \leq 12 - 2x, \ 0 \leq x \leq 2\}$
    $= \{(x, y) | 0 \leq x \leq \sqrt{4 - y}, \ 0 \leq y \leq 4\}$

R is both a
regular x-region
and a regular
y-region
$y = 4 - x^2$

 $y = 12 - 2x$
R is a regular
x-region
$y = x^3$

5. $R = \{(x, y) | \frac{1}{2}y^2 \leqslant x \leqslant y + 4, \ -2 \leqslant y \leqslant 4\}$ **7.** $\frac{1}{2}$ **9.** $\frac{39}{70}$ **11.** $\frac{56}{3}$ **13.** $-\frac{3}{4}$ **15.** $\frac{1}{2}e^4 - \frac{5}{2}$

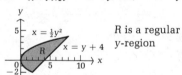

$x = \frac{1}{2}y^2$

$x = y + 4$

R is a regular
y-region

17. $R = \{(x, y) | 0 \leqslant y \leqslant x + 1, \ 0 \leqslant x \leqslant 1\}$

$\int_0^1 \int_0^{x+1} \sqrt{1 + x + y} \, dy \, dx = (68 - 24\sqrt{2})/15$

$y = x + 1$

19. $R = \{(x, y) | 0 \leqslant y \leqslant 4x - x^2, \ 0 \leqslant x \leqslant 4\}$

$\int_0^4 \int_0^{4x-x^2} \sqrt{y + x^2} \, dy \, dx = \frac{128}{5}$

$y = 4x - x^2$

21. $R = \{(x, y) | 1 - \sqrt{x} \leqslant y \leqslant 1 + \sqrt{x}, \ 0 \leqslant x \leqslant 4\}$ **23.** $\int_0^3 \int_0^{3-y} (x + 2y) \, dx \, dy = \frac{27}{2}$

$\int_0^4 \int_{1-\sqrt{x}}^{1+\sqrt{x}} x(y - 1)^2 \, dy \, dx = \frac{512}{21}$

$y = 1 + \sqrt{x}$

$y = 1 - \sqrt{x}$

$y = 3 - x$
or
$x = 3 - y$

25. $\int_0^1 \int_0^{\sqrt{1-y}} x\sqrt{y} \, dx \, dy = \frac{2}{15}$ **27.** $\int_0^1 \int_{4y^2}^{4y} x \, dx \, dy = \frac{16}{15}$

$y = 1 - x^2$
or
$x = \sqrt{1 - y}$

$y = \dfrac{\sqrt{x}}{2}$ or $x = 4y^2$

$y = \dfrac{x}{4}$ or $x = 4y$

29. $\int_0^4 \int_0^{4-x} (4 - x - y) \, dy \, dx = \frac{32}{3}$ **31.** $\int_0^1 \int_0^{1-x^2} 4 \, dy \, dx = \frac{8}{3}$

$y = 4 - x$

$y = 1 - x^2$

33. $\int_0^4 \int_0^{\sqrt{y}} \dfrac{4x}{1 + y^2} \, dx \, dy = \ln 17$ **35.** $\int_0^1 \int_0^{\sqrt{x}} 4ye^{x^2} \, dy \, dx = e - 1$

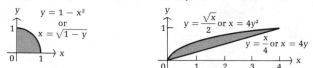

Exercise 16-9   Chapter Review

1. $f(5, 10) = 2,900; f_x(x, y) = 40; f_y(x, y) = 70$ **2.** $\partial^2 z/\partial x^2 = 6xy^2; \partial^2 z/\partial x\,\partial y = 6x^2 y$ **3.** $dz = 2\,dx + 3\,dy$
4. $dz = 4x^3 y^3\,dx + 3x^4 y^2\,dy$ **5.** $2xy^3 + 2y^2 + C(x)$ **6.** $3x^2 y^2 + 4xy + E(y)$ **7.** 1   **8.** $\frac{1}{2}$
9. $f(2, 3) = 7; f_y(x, y) = -2x + 2y + 3; f_y(2, 3) = 5$ **10.** $(-8)(-6) - 4^2 = 32$
11. $\Delta z = f(1.1, 2.2) - f(1, 2) - 7.8897; dz = 4(1)^3(0.1) + 4(2)^3(0.2) = 6.8$

12. $y = -1.5x + 15.5; y = 0.5$ when $x = 10$ **13.** 18   **14.** $\int_0^3 \int_0^{y+1} (x + y)^3\,dx\,dy = 408$

15. $f_x(x, y) = 2xe^{x^2+2y}; f_y(x, y) = 2e^{x^2+2y}; f_{xy}(x, y) = 4xe^{x^2+2y}$
16. $f_x(x, y) = 10x(x^2 + y^2)^4; f_{xy}(x, y) = 80xy(x^2 + y^2)^3$ **17.** 25.076 inches

18. $f(2, 3) = -25$ is a local minimum; f has a saddle point at $(-2, 3)$ **19.** $y = \frac{116}{165}x + \frac{100}{3}$ **20.** $\frac{27}{5}$

21. $\int_0^1 \int_0^{\sqrt{1-x^2}} (x + y)\,dy\,dx = \frac{2}{3}$ **22.** $\int_0^1 \int_0^{\sqrt{1-y^2}} y\,dx\,dy = \frac{1}{3}$ **23.** (A) $P_x(1, 3) = 8$; profit will increase \$8,000 for 100 units
increase in product A if production of product B is held fixed at an output level of (1, 3) (B) For 200 units of A and 300
units of B, $P(2, 3) = \$100$ thousand is a local maximum   **24.** 8 inches by 6 inches by 2 inches   **25.** $y = 0.63x + 1.33$;
profit in sixth year is \$5.11 million   **26.** $\frac{1}{4} \int_{10}^{12} \int_1^3 x^{0.8} y^{0.2}\,dy\,dx \approx 7.764$ or   7,764 units   **27.** $T_x(70, 17) = -0.924$
minute per foot increase in depth when $V = 70$ cubic feet and $x = 17$ feet   **28.** 65.6k   **29.** 50,000   **30.** $y = \frac{1}{2}x + 48$;
$y = 68$ when $x = 40$

Practice Test: Chapter 16

1. (A) 7 (B) 8   **2.** $f_x(x, y) = 4(x^2 y^3 - 2x)^3(2xy^3 - 2); f_y(x, y) = 12x^2 y^2(x^2 y^3 - 2x)^3$ **3.** $dz = 3x^2 y^4\,dx + 4x^3 y^3\,dy$
4. $\Delta z = 0.85; dz = 0.8$ **5.** $-4x$ **6.** $2x^3 ye^{x^2 y} + 2xe^{x^2 y}$ **7.** 72   **8.** $R_x(30, 20) = \$1,100; R_y(20, 30) = -\$600$
9. $f(0, 2) = -4$ is a local minimum   **10.** $\int_0^1 \int_0^{1-x} (1 - x - y)\,dy\,dx = \frac{1}{6}$
11. $y = 1.08x + 18.3$; cost of producing 40 units is \$61,500

Chapter 17

Exercise 17-1

1. $\mu = 0$
$\sigma = 1.095\ 445\ 1$

3. $\mu = -.9$
$\sigma = 1.374\ 772\ 7$

5. $S = \{1, 2, 3, 4, 5, 6, 7, 8, 9, 10\}$
7. .5   **9.** 5.5   **11.** 1

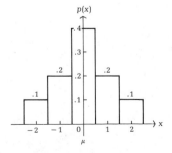

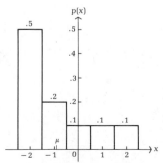

13.

x_i	0	1	2	3	4
p_i	$\frac{1}{16}$	$\frac{4}{16}$	$\frac{6}{16}$	$\frac{4}{16}$	$\frac{1}{16}$

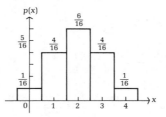

15. $\frac{1}{2}$ **17.** 2

19.

x_i	2	3	4	5	6	7	8	9	10	11	12
p_i	$\frac{1}{36}$	$\frac{2}{36}$	$\frac{3}{36}$	$\frac{4}{36}$	$\frac{5}{36}$	$\frac{6}{36}$	$\frac{5}{36}$	$\frac{4}{36}$	$\frac{3}{36}$	$\frac{2}{36}$	$\frac{1}{36}$

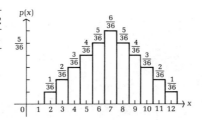

21. $\frac{5}{18}$ **23.** 7   **25.** $-\$0.50$ **27.** $-\$0.25$ **29.** $-\$0.052\ 631\ 58$

31.

x_i	4,850	-150
p_i	.01	.99

33. Site A, with $E(X) = \$3.6$ million   **35.** 1.54

$E(X) = -100$

37. A_2 is better, since for A_1, $E(X) = \$4$, and for A_2, $E(X) = \$4.80$

Exercise 17-2

1. $\binom{3}{2}(.5)^2(.5) = .375$ **3.** $\binom{3}{2}(.5)^2(.5) + \binom{3}{3}(.5)^3(.5)^0 = .500$ **5.** $\binom{4}{3}(\frac{1}{6})^3(\frac{5}{6}) \approx .0154$ **7.** $1 - \left[\binom{4}{0}(\frac{1}{6})^0(\frac{5}{6})^4\right] \approx .518$

9. $\mu = .6, \sigma = .65$ **11.** $\mu = 2, \sigma = 1$ **13.** $\mu = 2.4, \sigma = 1.3$

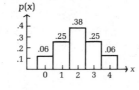
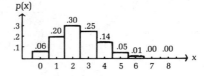

15. The theoretical probability distribution is given by $p(x) = \binom{3}{x}(.5)^x(.5)^{3-x} = \binom{3}{x}(.5)^3$ **17.** .238

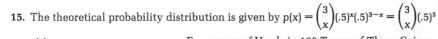

Frequency of Heads in 100 Tosses of Three Coins

Number of Heads	Theoretical Frequency	Actual Frequency
0	12.5	(List your experimental results here.)
1	37.5	
2	37.5	
3	12.5	

19. (A) $p(x) = \binom{6}{x}(.05)^x(.95)^{6-x}$

21. .035   **23.** (A) $p(x) = \binom{6}{x}(.6)^x(.4)^{6-x}$

(B)

x	p(x)
0	.735
1	.232
2	.031
3	.002
4	.000
5	.000
6	.000

(C) p(x)

(B)

x	p(x)
0	.004
1	.037
2	.138
3	.276
4	.311
5	.187
6	.047

(C) p(x)

(D) $\mu = .30$, $\sigma = .53$

(D) $\mu = 3.6$, $\sigma = 1.2$

25. (A) .001   (B) .264   (C) .897

27. (A) $p(x) = \binom{5}{x}(.2)^x(.8)^{5-x}$ (B)

x	p(x)
0	.328
1	.410
2	.205
3	.051
4	.006
5	.000

(C) p(x)

(D) $\mu = 1$, $\sigma = .89$

29. .29 (better than one chance out of four!)

Exercise 17-3

1. $f(x) \geqslant 0$ from graph

$\int_0^4 f(x)\, dx = 1$

3. $\int_2^3 \frac{1}{8}x\, dx = \frac{5}{16} = .3125$ **5.** $\int_3^4 \frac{1}{8}x\, dx = \frac{7}{16} = .4375$ **7.** $\int_5^\infty f(x)\, dx = 0$

9. $F(x) = \begin{cases} 0 & x < 0 \\ \frac{1}{16}x^2 & 0 \leqslant x \leqslant 4 \\ 1 & x > 4 \end{cases}$ **11.** $F(2) - F(0) = \frac{1}{4} - 0 = \frac{1}{4}$ **13.** $f(x) \geqslant 0$ from graph

$\int_0^\infty \frac{2}{(1+x)^3}\, dx = 1$

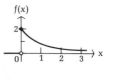

15. $\int_3^\infty \frac{2}{(1+x)^3}\, dx = \frac{1}{16} = .0625$ **17.** $\int_1^1 \frac{2}{(1+x)^3}\, dx = 0$ **19.** $F(x) = \begin{cases} 0 & x < 0 \\ 1 - [1/(1+x)^2] & x \geqslant 0 \end{cases}$

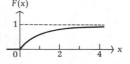

21. $F(x) = \begin{cases} 0 & x < 0 \\ \frac{3}{4}x^2 - \frac{1}{4}x^3 & 0 \leqslant x \leqslant 2 \\ 1 & x > 2 \end{cases}$

23. $F(x) = \begin{cases} 0 & x < 1 \\ x \ln x - x + 1 & 1 \leqslant x \leqslant e \\ 1 & x > e \end{cases}$

$F(2) - F(1) = 2 \ln 2 - 1 \approx .3863$

25. $F(x) = \begin{cases} 1 - xe^{-x} - e^{-x} & x \geqslant 0 \\ 0 & \text{otherwise} \end{cases}$

$1 - F(1) = 2e^{-1} \approx .7358$

27. $f(x) = \begin{cases} 2x & 0 \leqslant x \leqslant 1 \\ 0 & \text{otherwise} \end{cases}$

29. $f(x) = \begin{cases} 12x - 24x^2 + 12x^3 & 0 \leqslant x \leqslant 1 \\ 0 & \text{otherwise} \end{cases}$

31. $F(x = \begin{cases} 0 & x < 0 \\ \frac{1}{2}x^2 & 0 \leqslant x < 1 \\ 2x - \frac{1}{2}x^2 - 1 & 1 \leqslant x \leqslant 2 \\ 1 & x > 2 \end{cases}$

33. $F(x) = \begin{cases} 0 & x < -1 \\ \frac{1}{2} - \frac{1}{2}x^2 & -1 \leqslant x < 0 \\ \frac{1}{2} + \frac{1}{2}x^2 & 0 \leqslant x \leqslant 1 \\ 1 & x > 1 \end{cases}$

35. (A) $\displaystyle\int_0^1 \frac{1}{10}e^{-x/10}\,dx = 1 - e^{-1/10} \approx .0952$ (B) $\displaystyle\int_4^\infty \frac{1}{10}e^{-x/10}\,dx = e^{-2/5} \approx .6703$

37. (A) $\displaystyle\int_4^{10} 0.003x\sqrt{100 - x^2}\,dx = \frac{(84)^{3/2}}{1{,}000} \approx .7699$ (B) $\displaystyle\int_0^8 0.003x\sqrt{100 - x^2}\,dx = .784$ (C) $\sqrt{100 - (100)^{2/3}} \approx 8{,}858$ pounds

39. (A) $\displaystyle\int_7^{10} \frac{1}{5000}(10x^3 - x^4)\,dx = .47178$ (B) $\displaystyle\int_0^5 \frac{1}{5000}(10x^3 - x^4)\,dx = \frac{3}{16} = .1875$

41. (A) $1 - F(30) = 2.5e^{-1.5} \approx .5578$ (B) $1 - F(80) = 5e^{-4} \approx .0916$

Exercise 17-4

1. $\displaystyle\int_0^2 \frac{1}{2}x^2\,dx = \frac{4}{3} \approx 1.333$ **3.** $\sqrt{2}/3 \approx .4714$ **5.** $\displaystyle\int_{(4-\sqrt{2})/3}^{(4+\sqrt{2})/3} f(x)\,dx = 4\sqrt{2}/9 \approx .6285$ **7.** $\sqrt{2} \approx 1.414$

9. $\displaystyle\int_1^\infty (4/x^4)\,dx = \frac{4}{3} \approx 1.333$ **11.** $\sqrt{2}/3 \approx .4714$ **13.** $\displaystyle\int_{(4-2\sqrt{2})/3}^{(4+2\sqrt{2})/3} f(x)\,dx = 1 - \left(\frac{3}{4+2\sqrt{2}}\right)^4 \approx .9627$ **15.** $2^{1/4} \approx 1.189$

17. $\mu = \displaystyle\int_2^5 \frac{1}{3}x\,dx = \frac{7}{2} = 3.5$; $V(X) = \displaystyle\int_2^5 \frac{1}{3}x^2\,dx - (\frac{7}{2})^2 = \frac{3}{4} = .75$; $\sigma = \sqrt{3}/2 \approx .866$

19. $\mu = \displaystyle\int_0^3 \frac{x}{2\sqrt{1+x}}\,dx = \frac{4}{3} \approx 1.333$; $V(X) = \displaystyle\int_0^3 \frac{x^2}{2\sqrt{1+x}}\,dx - (\frac{4}{3})^2 = \frac{34}{45} \approx .7556$; $\sigma = \sqrt{\frac{34}{45}} \approx .8692$

21. $e^{1/2} \approx 1.649$ **23.** 1 **25.** $a\mu + b$ **27.** 0 **29.** $x_1 = 1$; $x_2 = \sqrt{2}$; $x_3 = \sqrt{3}$

31. (A) $E(X) = \frac{1}{8}\displaystyle\int_6^{10}(10x - x^2)\,dx = \frac{22}{3} \approx \$7.333$ thousand (B) $x_m = 10 - 2\sqrt{2} \approx \$7.172$ thousand

33. $E(X) = \displaystyle\int_0^\infty [x/(1+x^2)^{3/2}]\,dx = 1$ million gallons **35.** $\mu = \displaystyle\int_0^{10} \frac{1}{5{,}000}(10x^4 - x^5)\,dx = \frac{20}{3} \approx 6.7$ minutes

37. $E(X) = \displaystyle\int_0^3 (\frac{4}{9}x^3 - \frac{4}{27}x^4)\,dx = \frac{9}{5} = 1.8$ hours

Exercise 17-5

1. $f(x) = \begin{cases} \frac{1}{2} & 0 \le x \le 2 \\ 0 & \text{otherwise} \end{cases}$ **3.** $\mu = 1$; $x_m = 1$; $\sigma = 1/\sqrt{3} \approx .5774$ **5.** $f(x) = \begin{cases} 20x^3(1-x) & 0 \le x \le 1 \\ 0 & \text{otherwise} \end{cases}$

7. $\mu = \frac{2}{3}$; $\sigma = \frac{1}{3}\sqrt{\frac{2}{7}} \approx .1782$ **9.** $f(x) = \begin{cases} 2e^{-2x} & x \ge 0 \\ 0 & \text{otherwise} \end{cases}$ **11.** $\mu = \frac{1}{2}$; $x_m = \frac{1}{2}\ln 2 \approx .3466$; $\sigma = \frac{1}{2}$

13. $f(x) = \begin{cases} \frac{15}{4}x^{1/2}(1-x) & 0 \le x \le 1 \\ 0 & \text{otherwise} \end{cases}$ **15.** $\mu = \frac{3}{7} \approx .4286$; $\sigma = \frac{2}{7}\sqrt{\frac{2}{3}} \approx .2333$ **17.** $\frac{1}{3}$

19. $F(x) = \begin{cases} 0 & x < 0 \\ \frac{7}{3}x^{4/3} - \frac{4}{3}x^{7/3} & 0 \le x \le 1 \\ 1 & x > 1 \end{cases}$ **21.** $\frac{2}{\ln 2} \approx 2.885$ **23.** $F(x) = \begin{cases} 1 - e^{-(x/2)(\ln 2)} & x \ge 0 \\ 0 & \text{otherwise} \end{cases}$

25. (A) $F(\lambda) - F(0) = 1 - e^{-1} \approx .6321$ (B) $F(2\lambda) - F(0) = 1 - e^{-2} \approx .8647$ (C) $F(3\lambda) - F(0) = 1 - e^{-3} \approx .9502$
27. (A) $1/(b-x)$ (B) $1/\lambda$ **29.** $F(40) - F(25) = \frac{3}{8} = .375$ **31.** (A) $\beta = 1$ (B) $F(.75) - F(0) = \frac{27}{32} \approx .8438$
33. $F(2) - F(0) = 1 - e^{-2/3} \approx .4866$ **35.** (A) $E(X) = \mu = \frac{3}{8} = 37.5\%$ (B) $F(1) - F(.5) = 1 - (2.2)(.5)^{1.2} + (1.2)(.5)^{2.2} \approx .3036$
37. (A) $E(X) = \mu = -1/(\ln .7) \approx 2.8$ years (B) $1 - F(\mu) = e^{-1} \approx .3679$
39. (A) $E(X) = \mu = .9 = 90\%$ (B) $1 - F(.95) = 1 - 19(.95)^{18} + 18(.95)^{19} \approx .2453$

Exercise 17-6

1. .3413 **3.** .4987 **5.** .3159 **7.** .4932 **9.** .4332 **11.** .4995 **13.** .1915 **15.** .2881 **17.** .6812 **19.** .3142 **21.** .2266
23. .97 **25.** .7888 **27.** .5328 **29.** .0122 **31.** .1056 **33.** .8262 **35.** .5 **37.** .0116 **39.** 2.27% **41.** .0124 **43.** .0082
45. .0143 **47.** 2.27%
49. A's, 80.2 or higher; B's, 74.2–80.2; C's, 65.8–74.2; D's, 59.8–65.8; F's, 59.8 or lower (The instructor decides borderline cases using additional criteria.)

Exercise 17-7   Chapter Review

1. $S = \{1, 2, 3, 4\}$ **2.** $\frac{1}{2} = .5$ **3.** $E(X) = \frac{11}{4} = 2.75$; $V(X) = \frac{15}{16} = .9375$; $\sigma = \sqrt{15}/4 \approx .9682$ **4.** $-\$0.25$

x_i	1	2	3	4
p_i	$\frac{1}{8}$	$\frac{2}{8}$	$\frac{3}{8}$	$\frac{2}{8}$

5. (A)   p(x)   (B) $\mu = 1.2$, $\sigma = .85$ **6.** $\int_0^1 (1 - \frac{1}{2}x)\,dx = \frac{3}{4} = .75$ f(x)

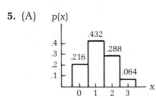

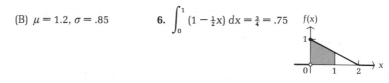

7. $\mu = \int_0^2 (x - \frac{1}{2}x^2)\,dx = \frac{2}{3} \approx .6667$; $V(X) = \int_0^2 (x^2 - \frac{1}{2}x^3)\,dx - (\frac{2}{3})^2 = \frac{2}{9} \approx .2222$; $\sigma = \sqrt{2}/3 \approx .4714$

8. $F(x) = \begin{cases} 0 & x < 0 \\ x - \frac{1}{4}x^2 & 0 \le x \le 2 \\ 1 & x > 2 \end{cases}$ **9.** $2 - \sqrt{2} \approx .5858$ **10.** .4938 **11.** .4641

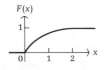

12. (A) p(x)

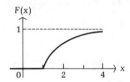

(B) $\mu = 3$, $\sigma = 1.22$ **13.** $\int_1^4 \frac{5}{2}x^{-7/2}\,dx = \frac{31}{32} \approx .9688$ f(x)

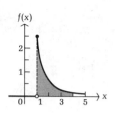

14. $\mu = \int_1^\infty \frac{5}{2}x^{-5/2}\,dx = \frac{5}{3} \approx 1.667$; $V(X) = \int_1^\infty \frac{5}{2}x^{-3/2}\,dx - (\frac{5}{3})^2 = \frac{20}{9} \approx 2.222$; $\sigma = \frac{2}{3}\sqrt{5} \approx 1.491$

15. $F(x) = \begin{cases} 1 - x^{-5/2} & x \geq 1 \\ 0 & \text{otherwise} \end{cases}$ **16.** $2^{2/5} \approx 1.32$ **17.** $f(x) = \begin{cases} 42x^5(1-x) & 0 \leq x \leq 1 \\ 0 & \text{otherwise} \end{cases}$

F(x)

f(x)

18. $F(.75) - F(.25) = 7(.75)^6 - 6(.75)^7 - 7(.25)^6 + 6(.25)^7 \approx .4436$ **19.** $F(x) = \begin{cases} 0 & x < 0 \\ 7x^6 - 6x^7 & 0 \leq x \leq 1 \\ 1 & x > 1 \end{cases}$ F(x)

20. $\mu = \frac{3}{4} = .75$; $\sigma = \sqrt{3}/12 \approx .1443$ **21.** $f(x) = \begin{cases} \frac{1}{2}e^{-x/2} & x \geq 0 \\ 0 & \text{otherwise} \end{cases}$ **22.** $\int_0^2 \frac{1}{2}e^{-x/2}\,dx = 1 - e^{-1} \approx .6321$

23. $F(x) = \begin{cases} 1 - e^{-x/2} & x \geq 0 \\ 0 & \text{otherwise} \end{cases}$ **24.** $\mu = 2$; $\sigma = 2$; $x_m = 2\ln 2 \approx 1.386$ **25.** $\mu = 600$, $\sigma = 15.49$ **26.** .9988

27. (A) .9105 **(B)** .0668   **28.** 7   **29.** $\mu = \int_0^\infty \frac{50x}{(x+5)^3}\,dx = 5$; $x_m = 5\sqrt{2} - 5 \approx 2.071$ **30.** $a\sigma^2 + a\mu^2 + b\mu + c$ **31.** .0188

32. (A) $\frac{1}{50}\int_0^5 (1 - .01x)\,dx = \frac{3}{4} = .75$ **(B)** 80 pounds   **33. (A)** $\int_{.2}^1 6x(1-x)\,dx = .896$ **(B)** 50%

34. (A) $1 - F(4) = e^{-1} \approx .3679$ **(B)** $F(1) - F(0) = 1 - e^{-.25} \approx .2212$ **35.** .0227

36. (A) $1 - F(5) = \frac{2}{3} \approx .6667$ **(B)** 10 months   **37. (A)** $1 - F(2) = e^{-4} \approx .0183$ **(B)** $\frac{1}{2}$ month   **38.** .1251

39. $F(1) - F(.5) = \frac{15}{16} \approx .9375$ **40.** .0122

   Practice Test: Chapter 17

1. $\mu = 3.2$; $V(X) = 1.96$; $\sigma = 1.4$

2. $\mu = \frac{1}{4}\int_0^2 (x + x^2)\,dx = \frac{7}{6} \approx 1.167$; $V(X) = \frac{1}{4}\int_0^2 (x^2 + x^3)\,dx - (\frac{7}{6})^2 = \frac{11}{36} \approx .3056$; $\sigma = \frac{1}{6}\sqrt{11} \approx .5528$

3. $\int_1^5 \dfrac{10}{9x^2}\,dx = \tfrac{8}{9} \approx .8889$ **4.** $F(x) = \begin{cases} 0 & x \leqslant 1 \\ \frac{10}{9}[1 - (1/x)] & 1 \leqslant x \leqslant 10 \\ 1 & x > 10 \end{cases}$

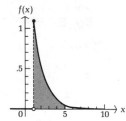

$x_m = \frac{20}{11} \approx 1.8182$

5. (A) $\mu = 60$ (B) $\sigma = 6.48$ **6.** .8764   **7.** (A) .5762 (B) .0668   **8.** $\int_5^\infty \frac{1}{5} e^{-x/5}\,dx = e^{-1} \approx .3679$ **9.** $\frac{1}{3}$

10. $E(X) = \frac{1}{8} \int_2^6 (6x - x^2)\,dx = \frac{10}{3}$, expected demand is 333 dozen; $x_m = 6 - 2\sqrt{2}$, median demand is approx. 317 dozen

Appendix A

Exercise A-1

1. (A) $d = 3$; 14, 17   (B) Not an arithmetic progression   (C) Not an arithmetic progression
(D) $d = -10$; $-22, -32$
3. $a_2 = 11$, $a_3 = 15$ **5.** $a_{21} = 82$, $S_{31} = 1{,}922$ **7.** $S_{20} = 930$ **9.** 2,400   **11.** 1,120
13. Use $a_1 = 1$ and $d = 2$ in $S_n = (n/2)[2a_1 + (n - 1)d]$ **15.** Firm A: \$280,500; firm B: \$278,500
17. $\$48 + \$46 + \cdots + \$4 + \$2 = \$600$

Exercise A-2

1. (A) $r = -2$; $a_4 = -8$, $a_5 = 16$ (B) Not a geometric progression   (C) $r = \frac{1}{2}$, $a_4 = \frac{1}{4}$, $a_5 = \frac{1}{8}$
(D) Not a geometric progression
3. $a_2 = -6$, $a_3 = 12$, $a_4 = -24$ **5.** $S_7 = 547$ **7.** $a_{10} = 199.90$ **9.** $r = 1.09$ **11.** $S_{10} = 1{,}242$, $S_\infty = 1{,}250$
13. (B) $S_\infty = \frac{8}{5} = 1.6$ **15.** 0.999   **17.** About \$11,670,000   **19.** \$31,027; \$251,600

Exercise A-3

1. 720   **3.** 10   **5.** 1,320   **7.** 10   **9.** 6   **11.** 1,140   **13.** 10   **15.** 6   **17.** 1   **19.** 816

21. $\binom{4}{0} a^4 + \binom{4}{1} a^3 b + \binom{4}{2} a^2 b^2 + \binom{4}{3} ab^3 + \binom{4}{4} b^4 = a^4 + 4a^3 b + 6a^2 b^2 + 4ab^3 + b^4$
23. $x^6 - 6x^5 + 15x^4 - 20x^3 + 15x^2 - 6x + 1$ **25.** $32a^5 - 80a^4 b + 80a^3 b^2 - 40a^2 b^3 + 10ab^4 - b^5$ **27.** $3{,}060x^{14}$
29. $5{,}005p^9 q^6$ **31.** $264x^2 y^{10}$ **33.** $\binom{n}{0} = \dfrac{n!}{0!n!} = 1$; $\binom{n}{n} = \dfrac{n!}{n!0!} = 1$ **35.** 1   5   10   10   5   1; 1   6   15   20   15   6   1

Index

Applications Index

Basic Differentiation

For C a constant:

1. $D_x C = 0$

2. $D_x u^n = n u^{n-1} D_x u$

3. $D_x(u \pm v) = D_x u \pm D_x v$

4. $D_x(Cu) = C D_x u$

5. $D_x(uv) = u D_x v + v D_x u$

6. $D_x \left(\dfrac{u}{v} \right) = \dfrac{v D_x u - u D_x v}{v^2}$

7. $D_x \ln u = \dfrac{1}{u} D_x u$

8. $D_x e^u = e^u D_x u$

9. $D_x b^u = b^u \ln b \, D_x u$

Basic Integration

For k and C constants:

1. $\displaystyle \int du = u + C$

2. $\displaystyle \int k f(u) \, du = k \int f(u) \, du$

3. $\displaystyle \int [f(u) \pm g(u)] \, du = \int f(u) \, du \pm \int g(u) \, du$

4. $\displaystyle \int u \, dv = uv - \int v \, du$

5. $\displaystyle \int u^n \, du = \dfrac{u^{n+1}}{n+1} + C, \quad n \neq -1$

6. $\displaystyle \int \dfrac{du}{u} = \ln|u| + C$

7. $\displaystyle \int e^u \, du = e^u + C$

8. $\displaystyle \int b^u \, du = \dfrac{b^u}{\ln b} + C$

Table of Integrals

- Integrals Involving $(au + b)(cu + d)$; $\quad a \neq 0, \quad c \neq 0,$
  and $\quad \Delta = bc - ad \neq 0$

1. $\displaystyle \int \frac{1}{(au + b)(cu + d)} \, du = \frac{1}{\Delta} \ln \left| \frac{cu + d}{au + b} \right|$

2. $\displaystyle \int \frac{u}{(au + b)(cu + d)} \, du = \frac{1}{\Delta} \left(\frac{b}{a} \ln|au + b| - \frac{d}{c} \ln|cu + d| \right)$

3. $\displaystyle \int \frac{u^2}{(au + b)(cu + d)} \, du = \frac{1}{ac} u - \frac{1}{\Delta} \left(\frac{b^2}{a^2} \ln|au + b| - \frac{d^2}{c^2} \ln|cu + d| \right)$

4. $\displaystyle \int \frac{1}{(au + b)^2(cu + d)} \, du = \frac{1}{\Delta} \frac{1}{au + b} + \frac{c}{\Delta^2} \ln \left| \frac{cu + d}{au + b} \right|$

5. $\displaystyle \int \frac{u}{(au + b)^2(cu + d)} \, du = -\frac{b}{a\Delta} \frac{1}{au + b} - \frac{d}{\Delta^2} \ln \left| \frac{cu + d}{au + b} \right|$

6. $\displaystyle \int \frac{1}{(au + b)^2(cu + d)^2} \, du = \frac{2ac}{\Delta^3} \ln \left| \frac{au + b}{cu + d} \right| - \frac{1}{\Delta^2} \left(\frac{a}{au + b} + \frac{c}{cu + d} \right)$

7. $\displaystyle \int \frac{u}{(au + b)^2(cu + d)^2} \, du = \frac{ad + bc}{\Delta^3} \ln \left| \frac{cu + d}{au + b} \right| + \frac{1}{\Delta^2} \left(\frac{b}{au + b} + \frac{d}{cu + d} \right)$

- Integrals Involving $\sqrt{a^2 - u^2}, \quad a > 0$

8. $\displaystyle \int \frac{1}{u \sqrt{a^2 - u^2}} \, du = -\frac{1}{a} \ln \left| \frac{a + \sqrt{a^2 - u^2}}{u} \right|$

9. $\displaystyle \int \frac{1}{u^2 \sqrt{a^2 - u^2}} \, du = -\frac{\sqrt{a^2 - u^2}}{a^2 u}$

10. $\displaystyle \int \frac{\sqrt{a^2 - u^2}}{u} \, du = \sqrt{a^2 - u^2} - a \ln \left| \frac{a + \sqrt{a^2 - u^2}}{u} \right|$